NEURAL NETWORKS

A Comprehensive Foundation

Second Edition

Simon Haykin

McMaster University
Hamilton, Ontario, Canada

Prentice Hall
Upper Saddle River, New Jersey 07458

Library of Congress Cataloging-in-Publication Data

Haykin, Simon, 1931–
  Neural networks : a comprehensive foundation
  Simon Haykin. — 2nd ed.
      p.   cm.
  Includes bibliographical references and index.
  ISBN 0-13-273350-1
  1. Neural networks (Computer science) I. Title.
QA76.87.H39 1999   98-7011
006.3'2—dc21   CIP

Publisher: Tom Robbins
Acquisitions Editor: Alice Dworkin
Editorial / Production/Composition: WestWords, Inc.
Editor-in-Chief: Marcia Horton
Assistant Vice President of Production and Manufacturing: David W. Riccardi
Managing Editor: Bayani Mendoza de Leon
Full Service/Manufacturing Coordinator: Donna M. Sullivan
Creative Director: Jayne Conte
Cover Designer: Bruce Kenselaar
Editorial Assistant: Nancy Garcia
Copy Editor: Julie Hollist

© 1999 by Prentice-Hall, Inc.
Simon & Schuster/A Viacom Company
Upper Saddle River, New Jersey 07458

The author and publisher of this book have used their best efforts in preparing this book. These efforts
include the development, research, and test of the theories and programs to determine their effectiveness. The author and publisher make no warranty of any kind, expressed or implies, with regard to these
programs or the documentation contained in this book. The author and publisher shall not be liable in
any event for incidental or consequential damages in connection with, or arising out of, the furnishing,
performance, or use of these programs.

Printed in the United States of America

10 9 8 7 6 5 4 3 2 1

ISBN 0-13-273350-1

Prentice-Hall International (UK) Limited, *London*
Prentice-Hall of Australia Pty. Limited, *Sydney*
Prentice-Hall Canada Inc., *Toronto*
Prentice-Hall Hispanoamericana, S.A., *Mexico*
Prentice-Hall of India Private Limited, *New Delhi*
Prentice-Hall of Japan, Inc., *Tokyo*
Simon & Schuster Asia Pte. Ltd., *Singapore*
Editora Prentice-Hall do Brasil, Ltda., *Rio de Janeiro*

To the countless researchers in neural networks
for their original contributions,

the many reviewers for their critical inputs,

my many graduate students for their keen interest,

and

my wife, Nancy, for her patience and tolerance.

Contents

3   Single Layer Perceptrons   117

4   Multilayer Perceptrons   156

5   Radial-Basis Function Networks   256

6   Support Vector Machines   318

10   Information-Theoretic Models   484

11   Stochastic Machines And Their Approximates Rooted In Statistical Mechanics   545

Preface

Neural Networks, or artificial neural networks to be more precise, represent a technology that is rooted in many disciplines: neurosciences, mathematics, statistics, physics, computer science, and engineering. Neural networks find applications in such diverse fields as modeling, time series analysis, pattern recognition, signal processing, and control by virtue of an important property: the ability to *learn* from input data with or without a teacher.

This book provides a comprehensive foundation of neural networks, recognizing the multidisciplinary nature of the subject. The material presented in the book is supported with examples, computer-oriented experiments, end-of-chapter problems, and a bibliography.

The book consists of four parts, organized as follows:

1. *Introductory material,* consisting of Chapters 1 and 2. Chapter 1 describes, largely in qualitative terms, what neural networks are, their properties, compositions, and how they relate to artificial intelligence. This chapter ends with some historical notes. Chapter 2 provides an overview of the many facets of the learning process and its statistical properties. This chapter introduces an important concept: the *Vapnik–Chervonenkis (VC) dimension* used as a measure for the capacity of a family of classification functions realized by a learning machine.

2. *Learning machines with a teacher,* consisting of Chapters 3 through 7. Chapter 3 studies the simplest class of neural networks in this part: networks involving one or more output neurons but no hidden ones. The least-mean-square (LMS) algorithm (highly popular in the design of linear adaptive filters) and the perceptron-convergence theorem are described in this chapter. Chapter 4 presents an exhaustive treatment of *multilayer perceptrons* trained with the *back-propagation algorithm.* This algorithm (representing a generalization of the LMS algorithm) has emerged as the workhorse of neural networks. Chapter 5 presents detailed mathematical treatment of another class of layered neural networks: *radial-basis function networks,* whose composition involves a single layer of basis functions. This chapter emphasizes the role of regularization theory in the design of RBF

networks. Chapter 6 describes a relatively new class of learning machines known as *support vector machines,* whose theory builds on the material presented in Chapter 2 on statistical learning theory. The second part of the book finishes in Chapter 7 with a discussion of *committee machines,* whose composition involves several learners as components. In this chapter we describe *ensemble averaging, boosting,* and *hierarchical mixture of experts* as three different methods of building a committee machine.

3. *Learning machines without a teacher,* consisting of Chapters 8 through 12. Chapter 8 applies *Hebbian learning* to *principal components analysis.* Chapter 9 applies another form of self-organized learning, namely *competitive learning,* to the construction of computational maps known as *self-organizing maps.* These two chapters distinguish themselves by emphasizing learning rules that are rooted in neurobiology. Chapter 10 looks to *information theory* for the formulation of unsupervised learning algorithms, and emphasizes their applications to *modeling, image processing,* and *independent components analysis.* Chapter 11 describes self-supervised learning machines rooted in *statistical mechanics,* a subject that is closely allied to information theory. Chapter 12, the last chapter in the third part of the book, introduces *dynamic programming* and its relationship to *reinforcement learning.*

4. *Nonlinear dynamical systems,* consisting of Chapters 13 through 15. Chapter 13 describes a class of dynamical systems consisting of short-term memory and layered feedforward network structures. Chapter 14 emphasizes the issue of stability that arises in nonlinear dynamical systems involving the use of *feedback.* Examples of *associative memory* are discussed in this chapter. Chapter 15 describes another class of nonlinear dynamical systems, namely *recurrent networks,* that rely on the use of feedback for the purpose of input–output mapping.

The book concludes with an epilogue that briefly describes the role of neural networks in the construction of *intelligent machines* for pattern recognition, control, and signal processing.

The organization of the book offers a great deal of flexibility for use in graduate courses on neural networks. The final selection of topics can only be determined by the interests of the instructors using the book. To help in this selection process, a study guide is included in the accompanying manual.

There are a total of 15 computer-oriented experiments distributed throughout the book. Thirteen of these experiments use MATLAB. The files for the MATLAB experiments can be directly downloaded from

ftp://ftp.mathworks.com/pub/books/haykin

or alternatively

http://www.mathworks.com/books/

In this second case, the user will have to click on "Neural/Fuzzy" and then on the title of the book. The latter approach provides a nicer interface.

Each chapter ends with a set of problems. Many of the problems are of a challenging nature, designed not only to test the user of the book for how well the material covered in the book has been understood, but also to extend that material. Solutions to all of the problems are described in an accompanying manual. Copies of this manual are only available to instructors who adopt the book, which can be obtained by writing to the publisher of the book, Prentice Hall.

The book should appeal to engineers, computer scientists, and physicists. It is hoped that researchers in other disciplines such as psychology and neurosciences will also find the book useful.

Simon Haykin
Hamilton, Ontario
February, 1998.

Acknowledgments

I am deeply indebted to the many reviewers who have given freely of their time to read through the book, in part or in full. In particular, I would like to express my deep gratitude to Dr. Kenneth Rose, University of California at Santa Barbara, for his many constructive inputs and invaluable help.

I am grateful to Dr. S. Amari, RIKEN, Japan; Dr. Sue Becker, McMaster University; Dr. Ron Racine, McMaster University; Dr. Sean Holden, University College, London; Dr. Michael Turmon, JPL, Pasadena; Dr. Babak Hassibi, Stanford University; Dr. Paul Yee, formerly of McMaster University; Dr. Edgar Osuna, MIT; Dr. Bernard Schölkopf, Max Planck Institute, Germany; Dr. Michael Jordan, MIT; Dr. Radford Neal, University of Toronto; Dr. Zoubin Gharhamani, University of Toronto; Dr. Marc Van Hulle, Katholieke Universiteit Leuven, Belgium; Dr. John Tsitsiklis, MIT; Dr. Jose Principe, University of Florida, Gainsville; Mr. Gint Puskorius, Ford Research Laboratory, Dearborn, Mich.; Dr. Lee Feldkamp, Ford Research Laboratory, Dearborn, Mich.; Dr. Lee Giles, NEC Research Institute, Princeton, NJ; Dr. Mikel Forcada, Universitat d'Alacant, Spain; Dr. Eric Wan, Oregon Graduate Institute of Science and Technology; Dr. Yann LeCun, AT&T Research, NJ; Dr. Jean-Francois Cardoso, Ecole Nationale, Paris; Dr. Anthony Bell, formerly of Salk Institute, San Diego; and Dr. Stefan Kremer, University of Guelph. They all helped me immeasurably in improving the presentation of material in different parts of the book.

I also wish to thank Dr. Ralph Linsker, IBM, Watson Research Center; Dr. Yaser Abu-Mostafa, Cal Tech.; Dr. Stuart Geman, Brown University; Dr. Alan Gelford, University of Connecticut; Dr. Yoav Freund, AT&T Research; Dr. Bart Kosko, University of Southern California; Dr. Narish Sinha, McMaster University; Dr. Grace Wahba, University of Wisconsin; Dr. Kostas Diamantaras, Aristotelian University of Thessaloniki, Greece; Dr. Robert Jacobs, University of Rochester; Dr. Peter Dayan, MIT; Dr. Dimitris Bertsekas, MIT; Dr. Andrew Barto, University of Massachusetts; Dr. Don Hush, University of New Mexico; Dr. Yoshua Bengio, University of Montreal; Dr. Andrew Cichoki, RIKEN, Japan; Dr. H. Yang, Oregon Graduate Institute of Science and Technology; Dr. Scott Douglas, University of Utah; Dr. Pierre Comon,

Thomson-Sintra Asm., France; Dr. Terrence Sejnowski, Salk Institute; Dr. Harris Drucker, Monmouth College; Dr. Nathan Intrator, Tel Aviv University, Israel; Dr. Vladimir Vapnik, AT&T Research, NJ; Dr. Teuvo Kohonen, Helsinki University of Technology, Finland; Dr. Vladimir Cherkassky, University of Minnesota; Dr. Sebastian Seung, AT&T Research, NJ; Dr. Steve Luttrell, DERA, Great Malvern, United Kingdom; Dr. David Lowe, Aston University, United Kingdom; Dr. N. Ansari, New Jersey Institute of Technology; Dr. Danil Prokhorov, Ford Research Laboratory, Dearborn, Mich.; Dr. Shigeru Katagiri, ATR Human Information Processing Research Lab, Japan; Dr. James Anderson, Brown University; Dr. Irwin Sandberg, University of Texas at Austin; Dr. Thomas Cover, Stanford University; Dr. Walter Freeman, University of California at Berkeley; Dr. Charles Micchelli, IBM Research, Yorktown Heights; Dr. Kari Torkkola, Motorola Phoenix Corp.; Dr. Andreas Andreou, Johns Hopkins University; Dr. Martin Beckerman, Oak Ridge National Laboratory; and Dr. Thomas Anastasio, University of Illinois, Urbana.

I am deeply indebted to my graduate student Hugh Pasika for performing many of the MATLAB experiments in the book, and for preparing the Web site for the book. The help received from my graduate student Himesh Madhuranath, Dr. Sadasivan Puthusserypady, Dr. J. Nie, Dr. Paul Yee, and Mr. Gint Puskorius (Ford Research) in performing five of the experiments is much appreciated.

I am most grateful to Hugh Pasika for proofreading the entire book. In this regard, I also thank Dr. Robert Dony (University of Guelph), Dr. Stefan Kremer (University of Guelph), and Dr. Sadasivan Puthusserypaddy for proofreading selected chapters of the book.

I am grateful to my publisher Tom Robbins and editor Alice Dworkin for their full support and encouragement. The careful copy editing of the manuscript by Julie Hollist is much appreciated. I would like to thank the tireless efforts of Jennifer Maughan and the staff of WestWords Inc. in Logan, Utah in the production of the book.

I wish to record my deep gratitude to Brigitte Maier, Thode Library, McMaster University, for her untiring effort to search for and find very difficult references that have made the bibliography all the more complete. The help of Science and Engineering Librarian Peggy Findlay and Reference Librarian Regina Bendig is also much appreciated.

Last but by no means least, I am most grateful to my secretary Lola Brooks for typing so many different versions of the manuscript. Without her dedicated help, the writing of this book and its production would have taken a great deal longer.

Abbreviations and Symbols

ABBREVIATIONS

AI	artificial intelligence
APEX	adaptive principal components extraction
AR	autoregressive
BBTT	back propagation through time
BM	Boltzmann machine
BP	back propagation
b/s	bits per second
BOSS	bounded, one-sided saturation
BSB	brain-state-in-a-box
BSS	Blind source (signal) separation
CART	classification and regression tree
cmm	correlation matrix memory
CV	cross-validation
DEKF	decoupled extended Kalman filter
DFA	deterministic finite-state automata
DSP	digital signal processor
EKF	extended Kalman filter
EM	expectation-maximization
FIR	finite-duration impulse response
FM	frequency-modulated (signal)
GEKF	global extended Kalman filter
GCV	generalized cross-validation
GHA	generalized Hebbian algorithm
GSLC	generalized sidelobe canceler

HME	hierarchical mixture of experts
HMM	hidden Markov model
Hz	hertz
ICA	independent components analysis
Infomax	maximum mutual information
KR	kernel regression
LMS	least-mean-square
LR	likelihood ratio
LTP	long-term potentiation
LTD	long-term depression
LR	likelihood ratio
LVQ	learning vector quantization
MCA	minor components analysis
MDL	minimum description length
ME	mixture of experts
MFT	mean-field theory
MIMO	multiple input–multiple output
ML	maximum likelihood
MLP	multilayer perceptron
MRAC	model reference adaptive control
NARMA	nonlinear autoregressive moving average
NARX	nonlinear autoregressive with exogenous inputs
NDP	neuron-dynamic programming
NW	Nadaraya–Watson (estimator)
NWKR	Nadaraya–Watson kernal regression
OBD	optimal brain damage
OBS	optimal brain surgeon
OCR	optical character recognition
ODE	ordinary differential equation
PAC	probably approximately correct
PCA	principal components analysis
pdf	probability density function
pmf	probability mass function
RBF	radial basis function
RMLP	recurrent multilayer perceptron
RTRL	real-time recurrent learning
SIMO	single input–multiple output
SISO	single input–single output
SNR	signal-to-noise ratio
SOM	self-organizing map

SRN	simple recurrent network (also referred to as Elman's recurrent network)
SVD	singular value decomposition
SVM	support vector machine
TDNN	time-delay neural network
TLFN	time lagged feedforward network
VC	Vapnik–Chervononkis (dimension)
VLSI	very-large-scale integration
XOR	exclusive OR

IMPORTANT SYMBOLS

a	action
$\mathbf{a}^T\mathbf{b}$	inner product of vectors $\mathbf{a}$ and $\mathbf{b}$
$\mathbf{a}\mathbf{b}^T$	outer product of vectors $\mathbf{a}$ and $\mathbf{b}$
$\binom{l}{m}$	binomial coefficient
$A \cup B$	unions of A and B
B	inverse of temperature
b_k	bias applied to neuron k
$\cos(\mathbf{a},\mathbf{b})$	cosine of the angle between vectors $\mathbf{a}$ and $\mathbf{b}$
D	depth of memory
$D_{f\|g}$	Kullback–Leibler divergence between probability density functions f and g
$\widetilde{\mathbf{D}}$	adjoint of operator $\mathbf{D}$
E	energy function
E_i	energy of state i in statistical mechanics
E	statistical expectation operator
$\langle E \rangle$	average energy
erf	error function
erfc	complimentary error function
exp	exponential
$\mathcal{E}_{av}$	average squared error or sum of squared errors
$\mathcal{E}(n)$	instantaneous value of the sum of squared errors
$\mathcal{E}_{total}$	total sum of error squares
F	free energy
$f_{\mathbf{X}}(\mathbf{x})$	probability density function of random vector $\mathbf{X}$
$\mathcal{F}*$	subset (network) with the smallest minimum empirical risk
$\mathbf{H}$	Hessian matrix
$\mathbf{H}^{-1}$	inverse of matrix $\mathbf{H}$
i	square root of -1, also denoted by j
$\mathbf{I}$	identity matrix
$\mathbf{I}$	Fisher's information matrix
J	mean-square error

$\mathbf{J}$	Jacobian matrix
$\mathbf{K}(n,n-1)$	error covariance matrix in Kalman filter theory
$\mathbf{K}^{1/2}$	square root of matrix $\mathbf{K}$
$\mathbf{K}^{T/2}$	transpose of square root of matrix $\mathbf{K}$
k_B	Boltzmann constant
log	logarithm
$L(\mathbf{w})$	log-likelihood function of weight vector $\mathbf{w}$
$\mathscr{L}(\mathbf{w})$	log-likelihood function of weight vector $\mathbf{w}$ based on a single example
$\mathbf{M}_c$	controllability matrix
$\mathbf{M}_o$	observability matrix
n	discrete time
p_i	probability of state i in statistical mechanics
p_{ij}	transition probability from state i to state j
$\mathbf{P}$	stochastic matrix
P_c	probability of correct classification
P_e	probability of error
$P(e|\mathscr{C})$	conditional probability of error e given that the input is drawn from class $\mathscr{C}$
p_α^+	probability that the visible neurons of a Boltzmann machine are in state α, given that the network is in its clamped condition (i.e., positive phase)
p_α^-	probability that the visible neurons of a Boltzmann machine are in state α, given that the network is in its free-running condition (i.e., negative phase)
$\hat{r}_x(j,k;n)$	estimate of autocorrelation function of $x_j(n)$ and $x_k(n)$
$\hat{r}_{dx}(k;n)$	estimate of cross-correlation function of $d(n)$ and $x_k(n)$
$\mathbf{R}$	correlation matrix of an input vector
t	continuous time
T	temperature
$\mathscr{T}$	training set (sample)
tr	trace of a matrix operator
var	variance operator
$V(\mathbf{x})$	Lyapunov function of state vector $\mathbf{x}$
v_j	induced local field or activation potential of neuron j
$\mathbf{w}_o$	optimum value of synaptic weight vector
w_{kj}	synaptic weight of synapse j belonging to neuron k
$\mathbf{w}^*$	optimum weight vector
$\overline{\mathbf{x}}$	equilibrium value of state vector $\mathbf{x}$
$\langle x_j \rangle$	average of state x_j in a "thermal" sense
$\hat{x}$	estimate of x, signified by the use of a caret (hat)
$|x|$	absolute value (magnitude) of x
x^*	complex conjugate of x, signified by asterisk as superscript
$\|\mathbf{x}\|$	Euclidean norm (length) of vector $\mathbf{x}$
$\mathbf{x}^T$	transpose of vector $\mathbf{x}$, signified by the superscript T
z^{-1}	unit delay operator
Z	partition function

$\delta_j(n)$	local gradient of neuron j at time n
Δw	small change applied to weight w
∇	gradient operator
∇^2	Laplacian operator
$\nabla_w J$	gradient of J with respect to w
$\nabla \cdot \mathbf{F}$	divergence of vector $\mathbf{F}$
η	learning-rate parameter
κ	cumulant
μ	policy
θ_k	threshold applied to neuron k (i.e., negative of bias b_k)
λ	regularization parameter
λ_k	kth eigenvalue of a square matrix
$\varphi_k(\cdot)$	nonlinear activation function of neuron k
$\in$	symbol for "belongs to"
$\cup$	symbol for "union of"
$\cap$	symbol for "intersection of"
$*$	symbol for convolution
$+$	superscript symbol for pseudoinverse of a matrix

Open and closed intervals

- The open interval (a, b) of a variable x signifies that $a < x < b$.
- The closed interval $[a, b]$ of a variable x signifies that $a \leq x < b$.
- The closed-open interval $[a, b)$ of a variable x signifies that $a \leq x < b$; likewise for the open-closed interval $(a, b]$.

Minima and Maxima

- The symbol $\arg \min_{\mathbf{w}} f(\mathbf{w})$ signifies the minimum of the function $f(\mathbf{w})$ with respect to the argument vector $\mathbf{w}$.
- The symbol $\arg \max_{\mathbf{w}} f(\mathbf{w})$ signifies the maximum of the function $f(\mathbf{w})$ with respect to the argument vector $\mathbf{w}$.

CHAPTER 1

Introduction

1.1 WHAT IS A NEURAL NETWORK?

Work on artificial neural networks, commonly referred to as "neural networks," has been motivated right from its inception by the recognition that the human brain computes in an entirely different way from the conventional digital computer. The brain is a highly *complex, nonlinear, and parallel computer* (information-processing system). It has the capability to organize its structural constituents, known as *neurons*, so as to perform certain computations (e.g., pattern recognition, perception, and motor control) many times faster than the fastest digital computer in existence today. Consider, for example, human *vision*, which is an information-processing task (Marr, 1982; Levine, 1985; Churchland and Sejnowski, 1992). It is the function of the visual system to provide a *representation* of the environment around us and, more important, to supply the information we need to *interact* with the environment. To be specific, the brain routinely accomplishes perceptual recognition tasks (e.g., recognizing a familiar face embedded in an unfamiliar scene) in approximately 100–200 ms, whereas tasks of much lesser complexity may take days on a conventional computer.

For another example, consider the *sonar* of a bat. Sonar is an active echo-location system. In addition to providing information about how far away a target (e.g., a flying insect) is, a bat sonar conveys information about the relative velocity of the target, the size of the target, the size of various features of the target, and the azimuth and elevation of the target (Suga, 1990a, b). The complex neural computations needed to extract all this information from the target echo occur within a brain the size of a plum. Indeed, an echo-locating bat can pursue and capture its target with a facility and success rate that would be the envy of a radar or sonar engineer.

How, then, does a human brain or the brain of a bat do it? At birth, a brain has great structure and the ability to build up its own rules through what we usually refer to as "experience." Indeed, experience is built up over time, with the most dramatic development (i.e., hard-wiring) of the human brain taking place during the first two years from birth; but the development continues well beyond that stage.

A "developing" neuron is synonymous with a plastic brain: *Plasticity* permits the developing nervous system to adapt to its surrounding environment. Just as plasticity appears to be essential to the functioning of neurons as information-processing units in the human brain, so it is with neural networks made up of artificial neurons. In its most general form, a *neural network* is a machine that is designed to *model* the way in which the brain performs a particular task or function of interest; the network is usually implemented by using electronic components or is simulated in software on a digital computer. Our interest in this book is confined largely to an important class of neural networks that perform useful computations through a process of *learning*. To achieve good performance, neural networks employ a massive interconnection of simple computing cells referred to as "neurons" or "processing units." We may thus offer the following definition of a neural network viewed as an adaptive machine[1]:

> *A neural network is a massively parallel distributed processor made up of simple processing units, which has a natural propensity for storing experiential knowledge and making it available for use. It resembles the brain in two respects:*
>
> 1. *Knowledge is acquired by the network from its environment through a learning process.*
> 2. *Interneuron connection strengths, known as synaptic weights, are used to store the acquired knowledge.*

The procedure used to perform the learning process is called a *learning algorithm*, the function of which is to modify the synaptic weights of the network in an orderly fashion to attain a desired design objective.

The modification of synaptic weights provides the traditional method for the design of neural networks. Such an approach is the closest to linear adaptive filter theory, which is already well established and successfully applied in many diverse fields (Widrow and Stearns, 1985; Haykin, 1996). However, it is also possible for a neural network to modify its own topology, which is motivated by the fact that neurons in the human brain can die and that new synaptic connections can grow.

Neural networks are also referred to in literature as *neurocomputers*, *connectionist networks*, *parallel distributed processors*, etc. Throughout the book we use the term "neural networks"; occasionally the term "neurocomputer" or "connectionist network" is used.

Benefits of Neural Networks

It is apparent that a neural network derives its computing power through, first, its massively parallel distributed structure and, second, its ability to learn and therefore generalize. *Generalization* refers to the neural network producing reasonable outputs for inputs not encountered during training (learning). These two information-processing capabilities make it possible for neural networks to solve complex (large-scale) problems that are currently intractable. In practice, however, neural networks cannot provide the solution by working individually. Rather, they need to be integrated into a consistent system engineering approach. Specifically, a complex problem of interest is *decomposed* into a number of relatively simple tasks, and neural networks are assigned a subset of the tasks that *match* their inherent capabilities. It is important to recognize, however, that we have a long way to go (if ever) before we can build a computer architecture that mimics a human brain.

The use of neural networks offers the following useful properties and capabilities:

1. *Nonlinearity*. An artificial neuron can be linear or nonlinear. A neural network, made up of an interconnection of nonlinear neurons, is itself nonlinear. Moreover, the nonlinearity is of a special kind in the sense that it is *distributed* throughout the network. Nonlinearity is a highly important property, particularly if the underlying physical mechanism responsible for generation of the input signal (e.g., speech signal) is inherently nonlinear.

2. *Input–Output Mapping.* A popular paradigm of learning called *learning with a teacher or supervised learning* involves modification of the synaptic weights of a neural network by applying a set of labeled *training samples* or *task examples*. Each example consists of a unique *input signal* and a corresponding *desired response*. The network is presented with an example picked at random from the set, and the synaptic weights (free parameters) of the network are modified to minimize the difference between the desired response and the actual response of the network produced by the input signal in accordance with an appropriate statistical criterion. The training of the network is repeated for many examples in the set until the network reaches a steady state where there are no further significant changes in the synaptic weights. The previously applied training examples may be reapplied during the training session but in a different order. Thus the network learns from the examples by constructing an *input–output mapping* for the problem at hand. Such an approach brings to mind the study of *nonparametric statistical inference*, which is a branch of statistics dealing with model-free estimation, or, from a biological viewpoint, *tabula rasa* learning (Geman et. al., 1992); the term "nonparametric" is used here to signify the fact that no prior assumptions are made on a statistical model for the input data. Consider, for example, a *pattern classification* task, where the requirement is to assign an input signal representing a physical object or event to one of several prespecified categories (classes). In a nonparametric approach to this problem, the requirement is to "estimate" arbitrary decision boundaries in the input signal space for the pattern-classification task using a set of examples, and to do so *without* invoking a probabilistic distribution model. A similar point of view is implicit in the supervised learning paradigm, which suggests a close analogy between the input–output mapping performed by a neural network and nonparametric statistical inference.

3. *Adaptivity.* Neural networks have a built-in capability to *adapt* their synaptic weights to changes in the surrounding environment. In particular, a neural network trained to operate in a specific environment can be easily *retrained* to deal with minor changes in the operating environmental conditions. Moreover, when it is operating in a *nonstationary* environment (i.e., one where statistics change with time), a neural network can be designed to change its synaptic weights in real time. The natural architecture of a neural network for pattern classification, signal processing, and control applications, coupled with the adaptive capability of the network, make it a useful tool in adaptive pattern classification, adaptive signal processing, and adaptive control. As a general rule, it may be said that the more adaptive we make a system, all the time ensuring that the system remains stable, the more robust its performance will likely be when the system is required to operate in a nonstationary environment. It should be emphasized, however, that adaptivity does not always lead to robustness; indeed, it may do the very opposite. For example, an adaptive system with short time constants may change rapidly and therefore tend to respond to spurious disturbances, causing a drastic degradation in system performance. To realize the full benefits of adaptivity, the principal time constants of the system should be long enough for the system to ignore spurious disturbances and yet short enough to respond to meaningful changes in the environment; the problem described here is referred to as the *stability–plasticity dilemma* (Grossberg, 1988b).

4. *Evidential Response*. In the context of pattern classification, a neural network can be designed to provide information not only about which particular pattern to *select*, but also about the *confidence* in the decision made. This latter information may be used to reject ambiguous patterns, should they arise, and thereby improve the classification performance of the network.

5. *Contextual Information*. Knowledge is represented by the very structure and activation state of a neural network. Every neuron in the network is potentially affected by the global activity of all other neurons in the network. Consequently, contextual information is dealt with naturally by a neural network.

6. *Fault Tolerance*. A neural network, implemented in hardware form, has the potential to be inherently *fault tolerant*, or capable of robust computation, in the sense that its performance degrades gracefully under adverse operating conditions. For example, if a neuron or its connecting links are damaged, recall of a stored pattern is impaired in quality. However, due to the distributed nature of information stored in the network, the damage has to be extensive before the overall response of the network is degraded seriously. Thus, in principle, a neural network exhibits a graceful degradation in performance rather than catastrophic failure. There is some empirical evidence for robust computation, but usually it is uncontrolled. In order to be assured that the neural network is in fact fault tolerant, it may be necessary to take corrective measures in designing the algorithm used to train the network (Kerlirzin and Vallet, 1993).

7. *VLSI Implementability*. The massively parallel nature of a neural network makes it potentially fast for the computation of certain tasks. This same feature makes a neural network well suited for implementation using *very-large-scale-integrated* (VLSI) technology. One particular beneficial virtue of VLSI is that it provides a means of capturing truly complex behavior in a highly hierarchical fashion (Mead, 1989).

8. *Uniformity of Analysis and Design.* Basically, neural networks enjoy universality as information processors. We say this in the sense that the same notation is used in all domains involving the application of neural networks. This feature manifests itself in different ways:

- Neurons, in one form or another, represent an ingredient *common* to all neural networks.
- This commonality makes it possible to *share* theories and learning algorithms in different applications of neural networks.
- Modular networks can be built through a *seamless integration of modules*.

9. *Neurobiological Analogy*. The design of a neural network is motivated by analogy with the brain, which is a living proof that fault tolerant parallel processing is not only physically possible but also fast and powerful. Neurobiologists look to (artificial) neural networks as a research tool for the interpretation of neurobiological phenomena. On the other hand, engineers look to neurobiology for new ideas to solve problems more complex than those based on conventional hard-wired design techniques. These two viewpoints are illustrated by the following two respective examples:

- In Anastasio (1993), linear system models of the vestibulo-ocular reflex are compared to neural network models based on *recurrent networks* that are described in Section 1.6 and discussed in detail in Chapter 15. The *vestibulo-ocular reflex (VOR)* is part of the oculomotor system. The function of VOR is to maintain visual (i.e., retinal) image stability by making eye rotations that are opposite to head rotations. The VOR is mediated by premotor neurons in the vestibular nuclei that receive and process head rotation signals from vestibular sensory neurons and send the results to the eye muscle motor neurons. The VOR is well suited for modeling because its input (head rotation) and its output (eye rotation) can be precisely specified. It is also a relatively simple reflex and the neurophysiological properties of its constituent neurons have been well described. Among the three neural types, the premotor neurons (reflex interneurons) in the vestibular nuclei are the most complex and therefore most interesting. The VOR has previously been modeled using lumped, linear system descriptors and control theory. These models were useful in explaining some of the overall properties of the VOR, but gave little insight into the properties of its constituent neurons. This situation has been greatly improved through neural network modeling. Recurrent network models of VOR (programmed using an algorithm called real-time recurrent learning that is described in Chapter 15) can reproduce and help explain many of the static, dynamic, nonlinear, and distributed aspects of signal processing by the neurons that mediate the VOR, especially the vestibular nuclei neurons (Anastasio, 1993).

- The *retina,* more than any other part of the brain, is where we begin to put together the relationships between the outside world represented by a visual sense, its *physical image* projected onto an array of receptors, and the first *neural images.* The retina is a thin sheet of neural tissue that lines the posterior hemisphere of the eyeball. The retina's task is to convert an optical image into a neural image for transmission down the optic nerve to a multitude of centers for further analysis. This is a complex task, as evidenced by the synaptic organization of the retina. In all vertebrate retinas the transformation from optical to neural image involves three stages (Sterling, 1990):

   (i) Photo transduction by a layer of receptor neurons.

   (ii) Transmission of the resulting signals (produced in response to light) by chemical synapses to a layer of bipolar cells.

   (iii) Transmission of these signals, also by chemical synapses, to output neurons that are called ganglion cells.

At both synaptic stages (i.e., from receptor to bipolar cells, and from bipolar to ganglion cells), there are specialized laterally connected neurons called *horizontal cells* and *amacrine cells,* respectively. The task of these neurons is to modify the transmission across the synaptic layers. There are also centrifugal elements called *inter-plexiform cells;* their task is to convey signals from the inner synaptic layer back to the outer one. A few researchers have built electronic chips that mimic the structure of the retina (Mahowald and Mead, 1989; Boahen and Ardreou, 1992; Boahen, 1996). These electronic chips are called *neuromorphic* integrated circuits, a term coined by Mead (1989). A neuromorphic imaging sensor consists of an array of photoreceptors combined with analog circuitry at each picture element (pixel). It emulates the retina in that it can adapt locally to changes in brightness, detect edges, and detect motion. The neurobiological analogy, exemplified by neuromorphic integrated circuits is useful in another important way: It provides a hope and belief, and to a certain extent an existence of proof, that physical understanding of neurobiological structures could have a productive influence on the art of electronics and VLSI technology.

With inspiration from neurobiology in mind, it seems appropriate that we take a brief look at the human brain and its structural levels of organization.

1.2 HUMAN BRAIN

The human nervous system may be viewed as a three-stage system, as depicted in the block diagram of Fig. 1.1 (Arbib, 1987). Central to the system is the *brain*, represented by the *neural (nerve) net*, which continually receives information, perceives it, and makes appropriate decisions. Two sets of arrows are shown in the figure. Those pointing from left to right indicate the *forward* transmission of information-bearing signals through the system. The arrows pointing from right to left signify the presence of *feedback* in the system. The *receptors* convert stimuli from the human body or the external environment into electrical impulses that convey information to the neural net (brain). The *effectors* convert electrical impulses generated by the neural net into discernible responses as system outputs.

The struggle to understand the brain has been made easier because of the pioneering work of Ramón y Cajál (1911), who introduced the idea of *neurons* as structural constituents of the brain. Typically, neurons are five to six orders of magnitude slower than silicon logic gates; events in a silicon chip happen in the nanosecond (10^{-9} s) range, whereas neural events happen in the millisecond (10^{-3} s) range. However, the brain makes up for the relatively slow rate of operation of a neuron by having a truly staggering number of neurons (nerve cells) with massive interconnections between them. It is estimated that there are approximately 10 billion neurons in the human cortex, and 60 trillion synapses or connections (Shepherd and Koch, 1990). The net result is that the brain is an enormously efficient structure. Specifically, the *energetic efficiency* of the brain is approximately 10^{-16} joules (J) per operation per second, whereas the corresponding value for the best computers in use today is about 10^{-6} joules per operation per second (Faggin, 1991).

Synapses are elementary structural and functional units that mediate the interactions between neurons. The most common kind of synapse is a *chemical synapse*, which operates as follows. A presynaptic process liberates a *transmitter* substance that diffuses across the synaptic junction between neurons and then acts on a postsynaptic process. Thus a synapse converts a presynaptic electrical signal into a chemical signal and then

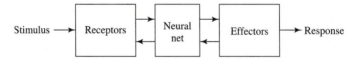

FIGURE 1.1 Block diagram representation of nervous system.

back into a postsynaptic electrical signal (Shepherd and Koch, 1990). In electrical terminology, such an element is said to be a *nonreciprocal two-port device*. In traditional
descriptions of neural organization, it is assumed that a synapse is a simple connection
that can impose *excitation* or *inhibition*, but not both on the receptive neuron.

Earlier we mentioned that plasticity permits the developing nervous system to
adapt to its surrounding environment (Eggermont, 1990; Churchland and Sejnowski,
1992). In an adult brain, plasticity may be accounted for by two mechanisms: the creation of new synaptic connections between neurons, and the modification of existing
synapses. *Axons*, the transmission lines, and *dendrites*, the receptive zones, constitute
two types of cell filaments that are distinguished on morphological grounds; an axon
has a smoother surface, fewer branches, and greater length, whereas a dendrite (so
called because of its resemblance to a tree) has an irregular surface and more branches
(Freeman, 1975). Neurons come in a wide variety of shapes and sizes in different parts
of the brain. Figure 1.2 illustrates the shape of a *pyramidal cell*, which is one of the
most common types of cortical neurons. Like many other types of neurons, it receives
most of its inputs through dendritic spines; see the segment of dendrite in the insert in
Fig. 1.2 for detail. The pyramidal cell can receive 10,000 or more synaptic contacts and
it can project onto thousands of target cells.

The majority of neurons encode their outputs as a series of brief voltage pulses.
These pulses, commonly known as *action potentials* or *spikes,* originate at or close to
the cell body of neurons and then propagate across the individual neurons at constant
velocity and amplitude. The reasons for the use of action potentials for communication
among neurons are based on the physics of axons. The axon of a neuron is very long
and thin and is characterized by high electrical resistance and very large capacitance.
Both of these elements are distributed across the axon. The axon may therefore be
modeled as an RC transmission line, hence the common use of "cable equation" as the
terminology for describing signal propagation along an axon. Analysis of this propagation mechanism reveals that when a voltage is applied at one end of the axon it decays
exponentially with distance, dropping to an insignificant level by the time it reaches
the other end. The action potentials provide a way to circumvent this transmission
problem (Anderson, 1995).

In the brain there are both small-scale and large-scale anatomical organizations,
and different functions take place at lower and higher levels. Figure 1.3 shows a hierarchy of interwoven levels of organization that has emerged from the extensive work
done on the analysis of local regions in the brain (Shepherd and Koch, 1990;
Churchland and Sejnowski, 1992). The *synapses* represent the most fundamental level,
depending on molecules and ions for their action. At the next levels we have neural
microcircuits, dendritic trees, and then neurons. A *neural microcircuit* refers to an
assembly of synapses organized into patterns of connectivity to produce a functional
operation of interest. A neural microcircuit may be likened to a silicon chip made up of
an assembly of transistors. The smallest size of microcircuits is measured in microme-
ters (μm), and their fastest speed of operation is measured in milliseconds. The neural
microcircuits are grouped to form *dendritic subunits* within the *dendritic trees* of
individual neurons. The whole *neuron*, about 100 μm in size, contains several dendritic
subunits. At the next level of complexity we have *local circuits* (about 1 mm in size)
made up of neurons with similar or different properties; these neural assemblies perform

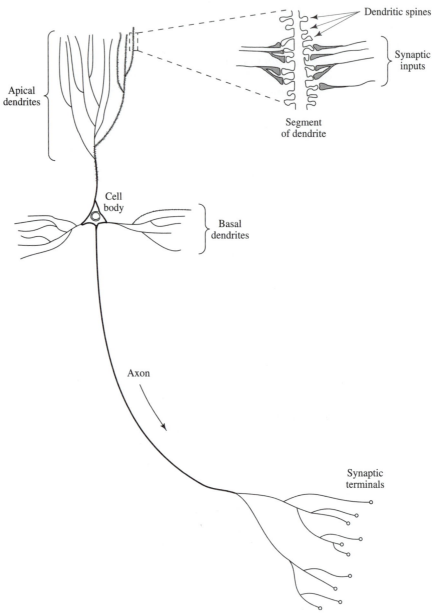

Dendritic spines

Synaptic inputs

Segment of dendrite

Apical dendrites

Cell body

Basal dendrites

Axon

Synaptic terminals

FIGURE 1.2   The pyramidal cell.

operations characteristic of a localized region in the brain. This is followed by *interregional circuits* made up of pathways, columns, and topographic maps, which involve multiple regions located in different parts of the brain.

Topographic maps are organized to respond to incoming sensory information. These maps are often arranged in sheets, as in the *superior colliculus*, where the visual,

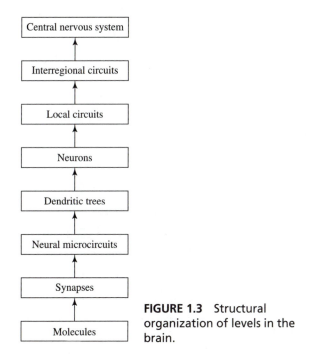

FIGURE 1.3 Structural organization of levels in the brain.

auditory, and somatosensory maps are stacked in adjacent layers in such a way that stimuli from corresponding points in space lie above or below each other. Figure 1.4 presents a cytoarchitectural map of the cerebral cortex as worked out by Brodmann (Brodal, 1981). This figure shows clearly that different sensory inputs (motor, somatosensory, visual, auditory, etc.) are mapped onto corresponding areas of the cerebral cortex in an orderly fashion. At the final level of complexity, the topographic maps and other interregional circuits mediate specific types of behavior in the *central nervous system*.

It is important to recognize that the structural levels of organization described herein are a unique characteristic of the brain. They are nowhere to be found in a digital computer, and we are nowhere close to re-creating them with artificial neural networks. Nevertheless, we are inching our way toward a hierarchy of computational levels similar to that described in Fig. 1.3. The artificial neurons we use to build our neural networks are truly primitive in comparison to those found in the brain. The neural networks we are presently able to design are just as primitive compared to the local circuits and the interregional circuits in the brain. What is really satisfying, however, is the remarkable progress that we have made on so many fronts during the past two decades. With neurobiological analogy as the source of inspiration, and the wealth of theoretical and technological tools that we are bringing together, it is certain that in another decade our understanding of artificial neural networks will be much more sophisticated than it is today.

Our primary interest in this book is confined to the study of artificial neural networks from an engineering perspective.[2] We begin the study by describing the models of (artificial) neurons that form the basis of the neural networks considered in subsequent chapters of the book.

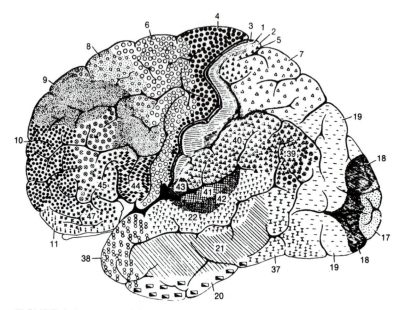

FIGURE 1.4 Cytoarchitectural map of the cerebral cortex. The different areas are identified by the thickness of their layers and types of cells within them. Some of the most important specific areas are as follows. Motor cortex: motor strip, area 4; premotor area, area 6; frontal eye fields, area 8. Somatosensory cortex: areas 3, 1, 2. Visual cortex: areas 17, 18, 19. Auditory cortex: area 41 and 42. (From A. Brodal, 1981; with permission of Oxford University Press.)

1.3 MODELS OF A NEURON

A *neuron* is an information-processing unit that is fundamental to the operation of a neural network. The block diagram of Fig. 1.5 shows the *model* of a neuron, which forms the basis for designing (artificial) neural networks. Here we identify three basic elements of the neuronal model:

1. A set of *synapses* or *connecting links*, each of which is characterized by a *weight* or *strength* of its own. Specifically, a signal x_j at the input of synapse j connected to neuron k is multiplied by the synaptic weight w_{kj}. It is important to make a note of the manner in which the subscripts of the synaptic weight w_{kj} are written. The first subscript refers to the neuron in question and the second subscript refers to the input end of the synapse to which the weight refers. Unlike a synapse in the brain, the synaptic weight of an artificial neuron may lie in a range that includes negative as well as positive values.

2. An *adder* for summing the input signals, weighted by the respective synapses of the neuron; the operations described here constitute a *linear combiner*.

3. An *activation function* for limiting the amplitude of the output of a neuron. The activation function is also referred to as a *squashing function* in that it squashes (limits) the permissible amplitude range of the output signal to some finite value.

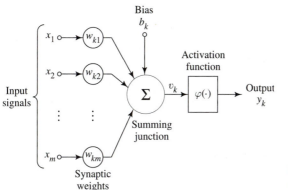

FIGURE 1.5 Nonlinear model of a neuron.

Typically, the normalized amplitude range of the output of a neuron is written as the closed unit interval $[0,1]$ or alternatively $[-1,1]$.

The neuronal model of Fig. 1.5 also includes an externally applied *bias,* denoted by b_k. The bias b_k has the effect of increasing or lowering the net input of the activation function, depending on whether it is positive or negative, respectively.

In mathematical terms, we may describe a neuron k by writing the following pair of equations:

$$u_k = \sum_{j=1}^{m} w_{kj}x_j \tag{1.1}$$

and

$$y_k = \varphi(u_k + b_k) \tag{1.2}$$

where $x_1, x_2, ..., x_m$ are the input signals; $w_{k1}, w_{k2}, ..., w_{km}$ are the synaptic weights of neuron k; u_k is the *linear combiner output* due to the input signals; b_k is the bias; $\varphi(\,\cdot\,)$ is the *activation function*; and y_k is the output signal of the neuron. The use of bias b_k has the effect of applying an *affine transformation* to the output u_k of the linear combiner in the model of Fig. 1.5, as shown by

$$v_k = u_k + b_k \tag{1.3}$$

In particular, depending on whether the bias b_k is positive or negative, the relationship between the *induced local field* or *activation potential* v_k of neuron k and the linear combiner output u_k is modified in the manner illustrated in Fig. 1.6; hereafter the term "induced local field" is used. Note that as a result of this affine transformation, the graph of v_k versus u_k no longer passes through the origin.

The bias b_k is an external parameter of artificial neuron k. We may account for its presence as in Eq. (1.2). Equivalently, we may formulate the combination of Eqs. (1.1) to (1.3) as follows:

$$v_k = \sum_{j=0}^{m} w_{kj}x_j \tag{1.4}$$

and

$$y_k = \varphi(v_k) \tag{1.5}$$

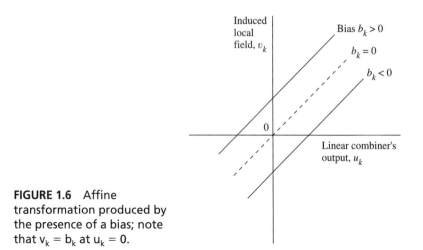

FIGURE 1.6 Affine transformation produced by the presence of a bias; note that $v_k = b_k$ at $u_k = 0$.

In Eq. (1.4) we have added a new synapse. Its input is

$$x_0 = +1 \tag{1.6}$$

and its weight is

$$w_{k0} = b_k \tag{1.7}$$

We may therefore reformulate the model of neruon k as in Fig. 1.7. In this figure, the effect of the bias is accounted for by doing two things: (1) adding a new input signal fixed at $+1$, and (2) adding a new synaptic weight equal to the bias b_k. Although the models of Figs. 1.5 and 1.7 are different in appearance, they are mathematically equivalent.

Types of Activation Function

The activation function, denoted by $\varphi(v)$, defines the output of a neuron in terms of the induced local field v. Here we identify three basic types of activation functions:

1. *Threshold Function.* For this type of activation function, described in Fig. 1.8a, we have

$$\varphi(v) = \begin{cases} 1 & \text{if } v \geq 0 \\ 0 & \text{if } v < 0 \end{cases} \tag{1.8}$$

In engineering literature, this form of a threshold function is commonly referred to as a *Heaviside function.* Correspondingly, the output of neuron k employing such a threshold function is expressed as

$$y_k = \begin{cases} 1 & \text{if } v_k \geq 0 \\ 0 & \text{if } v_k < 0 \end{cases} \tag{1.9}$$

where v_k is the induced local field of the neuron; that is,

$$v_k = \sum_{j=1}^{m} w_{kj} x_j + b_k \tag{1.10}$$

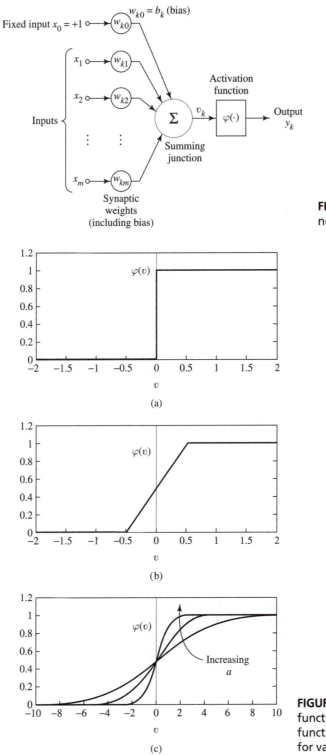

FIGURE 1.7 Another nonlinear model of a neuron.

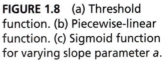

FIGURE 1.8 (a) Threshold function. (b) Piecewise-linear function. (c) Sigmoid function for varying slope parameter a.

Such a neuron is referred to in the literature as the *McCulloch–Pitts model*, in recognition of the pioneering work done by McCulloch and Pitts (1943). In this model, the output of a neuron takes on the value of 1 if the induced local field of that neuron is nonnegative, and 0 otherwise. This statement describes the *all-or-none property* of the McCulloch–Pitts model.

2. *Piecewise-Linear Function.* For the piecewise-linear function described in Fig. 1.8b we have

$$\varphi(v) = \begin{cases} 1, & v \geq +\frac{1}{2} \\ v, & +\frac{1}{2} > v > -\frac{1}{2} \\ 0, & v \leq -\frac{1}{2} \end{cases} \tag{1.11}$$

where the amplification factor inside the linear region of operation is assumed to be unity. This form of an activation function may be viewed as an *approximation* to a non-linear amplifier. The following two situations may be viewed as special forms of the piecewise-linear function:

- A *linear combiner* arises if the linear region of operation is maintained without running into saturation.
- The piecewise-linear function reduces to a *threshold function* if the amplification factor of the linear region is made infinitely large.

3. *Sigmoid Function.* The sigmoid function, whose graph is s-shaped, is by far the most common form of activation function used in the construction of artifical neural networks. It is defined as a strictly increasing function that exhibits a graceful balance between linear and nonlinear behavior.[3] An example of the sigmoid function is the *logistic function,*[4] defined by

$$\varphi(v) = \frac{1}{1 + \exp(-av)} \tag{1.12}$$

where a is the *slope parameter* of the sigmoid function. By varying the parameter a, we obtain sigmoid functions of different slopes, as illustrated in Fig. 1.8c. In fact, the slope at the origin equals $a/4$. In the limit, as the slope parameter approaches infinity, the sigmoid function becomes simply a threshold function. Whereas a threshold function assumes the value of 0 or 1, a sigmoid function assumes a continuous range of values from 0 to 1. Note also that the sigmoid function is differentiable, whereas the threshold function is not. (Differentiability is an important feature of neural network theory, as described in Chapter 4.)

The activation functions defined in Eqs. (1.8), (1.11), and (1.12) range from 0 to $+1$. It is sometimes desirable to have the activation function range from -1 to $+1$, in which case the activation function assumes an antisymmetric form with respect to the origin; that is, the activation function is an odd function of the induced local field. Specifically, the threshold function of Eq. (1.8) is now defined as

$$\varphi(v) = \begin{cases} 1 & \text{if } v > 0 \\ 0 & \text{if } v = 0 \\ -1 & \text{if } v < 0 \end{cases} \tag{1.13}$$

which is commonly referred to as the *signum function*. For the corresponding form of a sigmoid function we may use the *hyperbolic tangent function*, defined by

$$\varphi(v) = \tanh(v) \tag{1.14}$$

Allowing an activation function of the sigmoid type to assume negative values as prescribed by Eq. (1.14) has analytic benefits (as shown in Chapter 4).

Stochastic Model of a Neuron

The neuronal model described in Fig. 1.7 is deterministic in that its input-output behavior is precisely defined for all inputs. For some applications of neural networks, it is desirable to base the analysis on a stochastic neuronal model. In an analytically tractable approach, the activation function of the McCulloch–Pitts model is given a probabilistic interpretation. Specifically, a neuron is permitted to reside in only one of two states: +1 or −1, say. The decision for a neuron to *fire* (i.e., switch its state from "off" to "on") is probabilistic. Let x denote the state of the neuron, and $P(v)$ denote the *probability* of firing, where v is the induced local field of the neuron. We may then write

$$x = \begin{cases} +1 & \text{with probability } P(v) \\ -1 & \text{with probability } 1 - P(v) \end{cases}$$

A standard choice for $P(v)$ is the sigmoid-shaped function (Little, 1974):

$$P(v) = \frac{1}{1 + \exp(-v/T)} \tag{1.15}$$

where T is a *pseudotemperature* that is used to control the noise level and therefore the uncertainty in firing. It is important to realize, however, that T is *not* the physical temperature of a neural network, be it a biological or an artificial neural network. Rather, as already stated, we should think of T merely as a parameter that controls the thermal fluctuations representing the effects of synaptic noise. Note that when $T \rightarrow 0$, the stochastic neuron described by Eq. (1.15) reduces to a noiseless (i.e., deterministic) form, namely the McCulloch–Pitts model.

1.4   NEURAL NETWORKS VIEWED AS DIRECTED GRAPHS

The *block diagram* of Fig. 1.5 or that of Fig. 1.7 provides a functional description of the various elements that constitute the model of an artificial neuron. We may simplify the appearance of the model by using the idea of signal-flow graphs without sacrificing any of the functional details of the model. Signal-flow graphs with a well-defined set of rules were originally developed by Mason (1953, 1956) for linear networks. The presence of nonlinearity in the model of a neuron limits the scope of their application to neural networks. Nevertheless, signal-flow graphs do provide a neat method for the portrayal of the flow of signals in a neural network, which we pursue in this section.

A *signal-flow graph* is a network of directed *links* (*branches*) that are interconnected at certain points called *nodes*. A typical node j has an associated *node signal* x_j. A typical directed link originates at node j and terminates on node k; it has an associated *transfer function* or *transmittance* that specifies the manner in which the signal y_k at node k depends on the signal x_j at node j. The flow of signals in the various parts of the graph is dictated by three basic rules:

Rule 1. A signal flows along a link only in the direction defined by the arrow on the link.

Two different types of links may be distinguished:

- *Synaptic links*, whose behavior is governed by a *linear* input–output relation. Specifically, the node signal x_j is multiplied by the synaptic weight w_{kj} to produce the node signal y_k, as illustrated in Fig. 1.9a.
- *Activation links*, whose behavior is governed in general by a *nonlinear* input–output relation. This form of relationship is illustrated in Fig 1.9b, where $\varphi(\cdot)$ is the nonlinear activation function.

Rule 2. A node signal equals the algebraic sum of all signals entering the pertinent node via the incoming links.

This second rule is illustrated in Fig. 1.9c for the case of *synaptic convergence* or *fan-in*.

Rule 3. The signal at a node is transmitted to each outgoing link originating from that node, with the transmission being entirely independent of the transfer functions of the outgoing links.

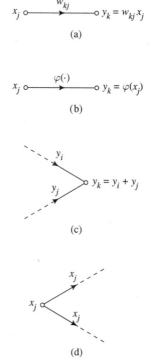

FIGURE 1.9 Illustrating basic rules for the construction of signal-flow graphs.

This third rule is illustrated in Fig. 1.9d for the case of *synaptic divergence* or *fan-out*.

For example, using these rules we may construct the signal-flow graph of Fig 1.10 as the model of a neuron, corresponding to the block diagram of Fig. 1.7. The representation shown in Fig. 1.10 is clearly simpler in appearance than that of Fig. 1.7, yet it contains all the functional details depicted in the latter diagram. Note that in both figures, the input $x_0 = +1$ and the associated synaptic weight $w_{k0} = b_k$, where b_k is the bias applied to neuron k.

Indeed, based on the signal-flow graph of Fig. 1.10 as the model of a neuron, we may now offer the following mathematical definition of a neural network:

> *A neural network is a directed graph consisting of nodes with interconnecting synaptic and activation links, and is characterized by four properties:*
>
> 1. *Each neuron is represented by a set of linear synaptic links, an externally applied bias, and a possibly nonlinear activation link. The bias is represented by a synaptic link connected to an input fixed at $+1$.*
> 2. *The synaptic links of a neuron weight their respective input signals.*
> 3. *The weighted sum of the input signals defines the induced local field of the neuron in question.*
> 4. *The activation link squashes the induced local field of the neuron to produce an output.*

The state of the neuron may be defined in terms of its induced local field or its output signal.

A directed graph so defined is *complete* in the sense that it describes not only the signal flow from neuron to neuron, but also the signal flow inside each neuron. When, however, the focus of attention is restricted to signal flow from neuron to neuron, we may use a reduced form of this graph by omitting the details of signal flow inside the individual neurons. Such a directed graph is said to be *partially complete*. It is characterized as follows:

1. *Source nodes* supply input signals to the graph.
2. Each neuron is represented by a single node called a *computation node*.
3. The *communication links* interconnecting the source and computation nodes of the graph carry no weight; they merely provide directions of signal flow in the graph.

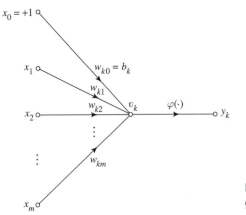

FIGURE 1.10 Signal-flow graph of a neuron.

A partially complete directed graph defined in this way is referred to as an *architectural graph,* describing the layout of the neural network. It is illustrated in Fig. 1.11 for the simple case of a single neuron with *m* source nodes and a single node fixed at +1 for the bias. Note that the computation node representing the neuron is shown shaded, and the source node is shown as a small square. This convention is followed throughout the book. More elaborate examples of architectural layouts are presented in Section 1.6.

To sum up, we have three graphical representations of a neural network:

- Block diagram, providing a functional description of the network.
- Signal-flow graph, providing a complete description of signal flow in the network.
- Architectural graph, describing the network layout.

1.5 FEEDBACK

Feedback is said to exist in a dynamic system whenever the output of an element in the system influences in part the input applied to that particular element, thereby giving rise to one or more closed paths for the transmission of signals around the system. Indeed, feedback occurs in almost every part of the nervous system of every animal (Freeman, 1975). Moreover, it plays a major role in the study of a special class of neural networks known as *recurrent networks*. Figure 1.12 shows the signal-flow graph of a *single-loop feedback system,* where the input signal $x_j(n)$, internal signal $x_j'(n)$, and output signal $y_k(n)$ are functions of the discrete-time variable n. The system is assumed to be *linear*, consisting of a forward path and a feedback path that are characterized by the "operators" A and B, respectively. In particular, the output of the forward channel determines in part its own output through the feedback channel. From Fig 1.12 we readily note the following input–output relationships:

$$y_k(n) = A[x_j'(n)] \tag{1.16}$$

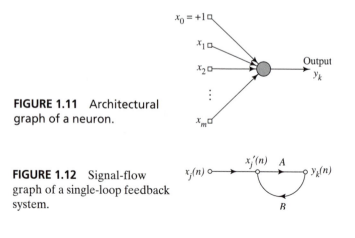

FIGURE 1.11   Architectural graph of a neuron.

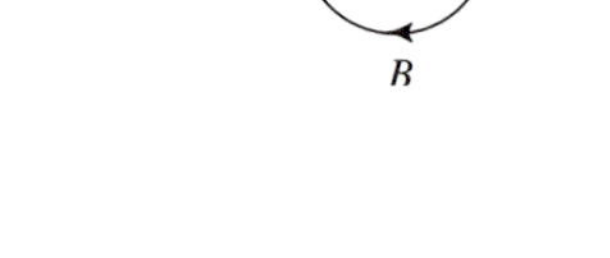

FIGURE 1.12   Signal-flow graph of a single-loop feedback system.

$$x_j'(n) = x_j(n) + B[y_k(n)] \tag{1.17}$$

where the square brackets are included to emphasize that A and B act as operators. Eliminating $x_j'(n)$ between Eqs. (1.16) and (1.17), we get

$$y_k(n) = \frac{A}{1 - AB}[x_j(n)] \tag{1.18}$$

We refer to $A/(1 - AB)$ as the *closed-loop operator* of the system, and to AB as the *open-loop operator*. In general, the open-loop operator is noncommutative in that $BA \neq AB$.

Consider, for example, the single-loop feedback system shown in Fig. 1.13, for which A is a fixed weight, w; and B is a *unit-delay operator*, z^{-1}, whose output is delayed with respect to the input by one time unit. We may then express the closed-loop operator of the system as

$$\frac{A}{1 - AB} = \frac{w}{1 - wz^{-1}}$$

$$= w(1 - wz^{-1})^{-1}$$

Using the binomial expansion for $(1 - wz^{-1})^{-1}$, we may rewrite the closed-loop operator of the system as

$$\frac{A}{1 - AB} = w \sum_{l=0}^{\infty} w^l z^{-l} \tag{1.19}$$

Hence, substituting Eq. (1.19) in (1.18), we get

$$y_k(n) = w \sum_{l=0}^{\infty} w^l z^{-l}[x_j(n)] \tag{1.20}$$

where again we have included square brackets to emphasize the fact that z^{-1} is an operator. In particular, from the definition of z^{-1} we have

$$z^{-l}[x_j(n)] = x_j(n - l) \tag{1.21}$$

where $x_j(n - l)$ is a sample of the input signal delayed by l time units. Accordingly, we may express the output signal $y_k(n)$ as an infinite weighted summation of present and past samples of the input signal $x_j(n)$, as shown by

$$y_k(n) = \sum_{l=0}^{\infty} w^{l+1} x_j(n - l) \tag{1.22}$$

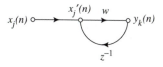

FIGURE 1.13   Signal-flow graph of a first-order, infinite-duration impulse response (IIR) filter.

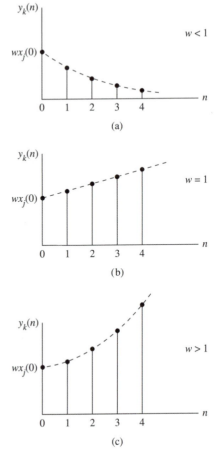

FIGURE 1.14   Time response of Fig. 1.13 for three different values of forward weight w. (a) Stable. (b) Linear divergence. (c) Exponential divergence.

We now see clearly that the dynamic behavior of the system is controlled by the weight w. In particular, we may distinguish two specific cases:

1. $|w| < 1$, for which the output signal $y_k(n)$ is exponentially *convergent*; that is, the system is *stable*. This is illustrated in Fig. 1.14a for a positive w.
2. $|w| \geq 1$, for which the output signal $y_k(n)$ is *divergent*; that is, the system is *unstable*. If $|w| = 1$ the divergence is linear as in Fig. 1.14b, and if $|w| > 1$ the divergence is exponential as in Fig 1.14c.

Stability features prominently in the study of feedback systems.

The case of $|w| < 1$ corresponds to a system with *infinite memory* in the sense that the output of the system depends on samples of the input extending into the infinite past. Moreover, the memory is *fading* in that the influence of a past sample is reduced exponentially with time n.

The analysis of the dynamic behavior of neural networks involving the application of feedback is unfortunately complicated by virute of the fact that the processing units used for the construction of the network are usually *nonlinear*. Further consideration of this issue is deferred to the latter part of the book.

1.6 NETWORK ARCHITECTURES

The manner in which the neurons of a neural network are structured is intimately linked with the learning algorithm used to train the network. We may therefore speak of learning algorithms (rules) used in the design of neural networks as being *structured*. The classification of learning algorithms is considered in the next chapter, and the development of different learning algorithms is taken up in subsequent chapters of the book. In this section we focus our attention on network architectures (structures).

In general, we may identify three fundamentally different classes of network architectures:

1.   Single-Layer Feedforward Networks

In a *layered* neural network the neurons are organized in the form of layers. In the simplest form of a layered network, we have an *input layer* of source nodes that projects onto an *output layer* of neurons (computation nodes), but not vice versa. In other words, this network is strictly a *feedforward* or *acyclic* type. It is illustrated in Fig. 1.15 for the case of four nodes in both the input and output layers. Such a network is called a *single-layer network*, with the designation "single-layer" referring to the output layer of computation nodes (neurons). We do not count the input layer of source nodes because no computation is performed there.

2.   Multilayer Feedforward Networks

The second class of a feedforward neural network distinguishes itself by the presence of one or more *hidden layers*, whose computation nodes are correspondingly called *hidden neurons* or *hidden units*. The function of hidden neurons is to intervene between the external input and the network output in some useful manner. By adding one or

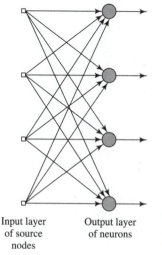

Input layer   Output layer
of source   of neurons
nodes

FIGURE 1.15 Feedforward or acyclic network with a single layer of neurons.

more hidden layers, the network is enabled to extract higher-order statistics. In a rather loose sense the network acquires a *global* perspective despite its local connectivity due to the extra set of synaptic connections and the extra dimension of neural interactions (Churchland and Sejnowski, 1992). The ability of hidden neurons to extract higher-order statistics is particularly valuable when the size of the input layer is large.

The source nodes in the input layer of the network supply respective elements of the activation pattern (input vector), which constitute the input signals applied to the neurons (computation nodes) in the second layer (i.e., the first hidden layer). The output signals of the second layer are used as inputs to the third layer, and so on for the rest of the network. Typically the neurons in each layer of the network have as their inputs the output signals of the preceding layer only. The set of output signals of the neurons in the output (final) layer of the network constitutes the overall response of the network to the activation pattern supplied by the source nodes in the input (first) layer. The architectural graph in Fig. 1.16 illustrates the layout of a multilayer feedforward neural network for the case of a single hidden layer. For brevity the network in Fig. 1.16 is referred to as a 10-4-2 network because it has 10 source nodes, 4 hidden neurons, and 2 output neurons. As another example, a feedforward network with m source nodes, h_1 neurons in the first hidden layer, h_2 neurons in the second hidden layer, and q neurons in the output layer is referred to as an m-h_1-h_2-q network.

The neural network in Fig. 1.16 is said to be *fully connected* in the sense that every node in each layer of the network is connected to every other node in the adjacent forward layer. If, however, some of the communication links (synaptic connections) are missing from the network, we say that the network is *partially connected*.

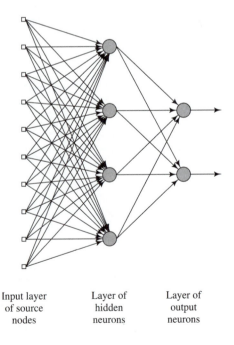

FIGURE 1.16 Fully connected feedforward or acyclic network with one hidden layer and one output layer.

| Input layer of source nodes | Layer of hidden neurons | Layer of output neurons |

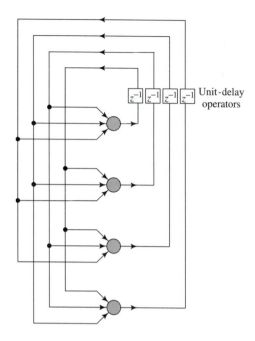

FIGURE 1.17   Recurrent network with no self-feedback loops and no hidden neurons.

3.   Recurrent Networks

A *recurrent neural network* distinguishes itself from a feedforward neural network in that it has at least one *feedback* loop. For example, a recurrent network may consist of a single layer of neurons with each neuron feeding its output signal back to the inputs of all the other neurons, as illustrated in the architectural graph in Fig. 1.17. In the structure depicted in this figure there are *no* self-feedback loops in the network; self-feedback refers to a situation where the output of a neuron is fed back into its own input. The recurrent network illustrated in Fig. 1.17 also has *no* hidden neurons. In Fig. 1.18 we illustrate another class of recurrent networks with hidden neurons. The feedback connections shown in Fig. 1.18 originate from the hidden neurons as well as from the output neurons.

The presence of feedback loops, whether in the recurrent structure of Fig. 1.17 or that of Fig. 1.18, has a profound impact on the learning capability of the network and on its performance. Moreover, the feedback loops involve the use of particular branches composed of *unit-delay elements* (denoted by z^{-1}), which result in a nonlinear dynamical behavior, assuming that the neural network contains nonlinear units.

1.7   KNOWLEDGE REPRESENTATION

In Section 1.1 we used the term "knowledge" in the definition of a neural network without an explicit description of what we mean by it. We now take care of this matter by offering the following generic definition (Fischler and Firschein, 1987):

> *Knowledge refers to stored information or models used by a person or machine to interpret, predict, and appropriately respond to the outside world.*

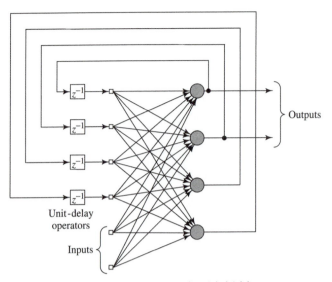

FIGURE 1.18   Recurrent network with hidden neurons.

The primary characteristics of *knowledge representation* are twofold: (1) what information is actually made explicit; and (2) how the information is physically encoded for subsequent use. By the very nature of it, therefore, knowledge representation is goal directed. In real-world applications of "intelligent" machines, it can be said that a good solution depends on a good representation of knowledge (Woods, 1986). So it is with neural networks that represent a special class of intelligent machines. Typically, however, the possible forms of representation from the inputs to internal network parameters are highly diverse, which tends to make the development of a satisfactory solution by means of a neural network a real design challenge.

A major task for a neural network is to learn a model of the world (environment) in which it is embedded and to maintain the model sufficiently consistent with the real world so as to achieve the specified goals of the application of interest. Knowledge of the world consists of two kinds of information:

1. The known world state, represented by facts about what is and what has been known; this form of knowledge is referred to as *prior information*.
2. Observations (measurements) of the world, obtained by means of sensors designed to probe the environment in which the neural network is supposed to operate. Ordinarily these observations are inherently noisy, being subject to errors due to sensor noise and system imperfections. In any event, the observations so obtained provide the pool of information from which the *examples* used to train the neural network are drawn.

The examples can be *labeled* or *unlabeled*. In labeled examples, each example representing an *input signal* is paired with a corresponding *desired response* (i.e., target output). On the other hand, unlabeled examples consist of different realizations of the input signal by itself. In any event, a set of examples, labeled or otherwise, represents knowledge about the environment of interest that a neural network can learn through training.

A set of input–output pairs, with each pair consisting of an input signal and the corresponding desired response, is referred to as a *set of training data* or *training sample*. To illustrate how such a data set can be used, consider, for example, the *handwritten digit recognition problem*. In this problem, the input signal consists of an image with black or white pixels, with each image representing one of 10 digits that are well separated from the background. The desired response is defined by the "identity" of the particular digit whose image is presented to the network as the input signal. Typically, the training sample consists of a large variety of handwritten digits that are representative of a real-world situation. Given such a set of examples, the design of a neural network may proceed as follows:

- First, an appropriate architecture is selected for the neural network, with an input layer consisting of source nodes equal in number to the pixels of an input image, and an output layer consisting of 10 neurons (one for each digit). A subset of examples is then used to train the network by means of a suitable algorithm. This phase of the network design is called *learning*.

- Second, the recognition performance of the trained network is *tested* with data not seen before. Specifically, an input image is presented to the network, but this time it is not told the indentity of the digit to which that particular image belongs. The performance of the network is then assessed by comparing the digit recognition reported by the network with the actual identity of the digit in question. This second phase of the network operation is called *generalization*, a term borrowed from psychology.

Herein lies a fundamental difference between the design of a neural network and that of its classical information-processing counterpart (pattern classifier). In the latter case, we usually proceed by first formulating a mathematical model of environmental observations, validating the model with real data, and then building the design on the basis of the model. In contrast, the design of a neural network is based directly on real-life data, with the *data set being permitted to speak for itself*. Thus, the neural network not only provides the implicit model of the environment in which it is embedded, but also performs the information-processing function of interest.

The examples used to train a neural network may consist of both *positive* and *negative* examples. For instance, in a passive sonar detection problem, positive examples pertain to input training data that contain the target of interest (e.g., a submarine). Now, in a passive sonar environment, the possible presence of marine life in the test data is known to cause occasional false alarms. To alleviate this problem, negative examples (e.g., echos from marine life) are included in the training data to teach the network not to confuse marine life with the target.

In a neural network of specified architecture, knowledge representation of the surrounding environment is defined by the values taken on by the free parameters (i.e., synaptic weights and biases) of the network. The form of this knowledge representation constitutes the very design of the neural network, and therefore holds the key to its performance.

The subject of knowledge representation inside an artificial network is, however, very complicated. Nevertheless, there are four rules for knowledge representation that are of a general commonsense nature (Anderson, 1988).

Rule 1. Similar inputs from similar classes should usually produce similar representations inside the network, and should therefore be classified as belonging to the same category.

There are a plethora of measures for determining the "similarity" between inputs. A commonly used measure of similarity is based on the concept of Euclidian distance. To be specific, let $\mathbf{x}_i$ denote an m-by-1 vector

$$\mathbf{x}_i = [x_{i1}, x_{i2}, ..., x_{im}]^T$$

all of whose elements are real; the superscript T denotes matrix *transposition*. The vector $\mathbf{x}_i$ defines a point in an m-dimensional space called *Euclidean space* and denoted by $\mathbb{R}^m$. The *Euclidean distance* between a pair of m-by-1 vectors $\mathbf{x}_i$ and $\mathbf{x}_j$ is defined by

$$d(\mathbf{x}_i, \mathbf{x}_j) = \|\mathbf{x}_i - \mathbf{x}_j\|$$
$$= \left[\sum_{k=1}^{m} (x_{ik} - x_{jk})^2 \right]^{1/2} \tag{1.23}$$

where x_{ik} and x_{jk} are the kth elements of the input vectors $\mathbf{x}_i$ and $\mathbf{x}_j$, respectively. Correspondingly, the similarity between the inputs represented by the vectors $\mathbf{x}_i$ and $\mathbf{x}_j$ is defined as the *reciprocal* of the Euclidean distance $d(\mathbf{x}_i, \mathbf{x}_j)$. The closer the individual elements of the input vectors $\mathbf{x}_i$ and $\mathbf{x}_j$ are to each other, the smaller the Euclidean distance $d(\mathbf{x}_i, \mathbf{x}_j)$ will be, and therefore the greater the similarity between the vectors $\mathbf{x}_i$ and $\mathbf{x}_j$ will be. Rule 1 states that if the vectors $\mathbf{x}_i$ and $\mathbf{x}_j$ are similar, they should be assigned to the same category (class).

Another measure of similarity is based on the idea of a *dot product* or *inner product* that is also borrowed from matrix algebra. Given a pair of vectors $\mathbf{x}_i$ and $\mathbf{x}_j$ of the same dimension, their inner product is $\mathbf{x}_i^T\mathbf{x}_j$ written in expanded form as follows:

$$(\mathbf{x}_i, \mathbf{x}_j) = \mathbf{x}_i^T\mathbf{x}_j$$
$$= \sum_{k=1}^{m} x_{ik}x_{jk} \tag{1.24}$$

The inner product $(\mathbf{x}_i, \mathbf{x}_j)$ divided by $\|\mathbf{x}_i\| \, \|\mathbf{x}_j\|$ is the cosine of the angle subtended between the vectors $\mathbf{x}_i$ and $\mathbf{x}_j$.

The two measures of similarity defined here are indeed intimately related to each other, as illustrated in Fig. 1.19. The Euclidean distance $\|\mathbf{x}_i - \mathbf{x}_j\|$ between the vectors $\mathbf{x}_i$ and $\mathbf{x}_j$ is related to the "projection" of the vector $\mathbf{x}_i$ onto the vector $\mathbf{x}_j$. Figure 1.19 shows

FIGURE 1.19 Illustrating the relationship between inner product and Euclidean distance as measures of similiarity between patterns.

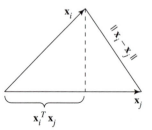

clearly that the smaller the Euclidean distance $\|\mathbf{x}_i - \mathbf{x}_j\|$ and therefore the more similar the vectors $\mathbf{x}_i$ and $\mathbf{x}_j$ are, the larger the inner product $\mathbf{x}_i^T\mathbf{x}_j$ will be.

To put this relationship on a formal basis, we first normalize the vectors $\mathbf{x}_i$ and $\mathbf{x}_j$ to have unit length, that is,

$$\|\mathbf{x}_i\| = \|\mathbf{x}_j\| = 1$$

We may then use Eq. (1.23) to write

$$
\begin{aligned}
d^2(\mathbf{x}_i, \mathbf{x}_j) &= (\mathbf{x}_i - \mathbf{x}_j)^T(\mathbf{x}_i - \mathbf{x}_j) \\
&= 2 - 2\mathbf{x}_i^T\mathbf{x}_j
\end{aligned}
\tag{1.25}
$$

Equation (1.25) shows that minimization of the Euclidean distance $d(\mathbf{x}_i,\mathbf{x}_j)$ corresponds to maximization of the inner product $(\mathbf{x}_i,\mathbf{x}_j)$ and, therefore, the similarity between the vectors $\mathbf{x}_i$ and $\mathbf{x}_j$.

The Euclidean distance and inner product described here are defined in deterministic terms. What if the vectors $\mathbf{x}_i$ and $\mathbf{x}_j$ are drawn from two different populations (pools) of data? To be specific, suppose that the difference between these two populations lies solely in their mean vectors. Let $\boldsymbol{\mu}_i$ and $\boldsymbol{\mu}_j$ denote the mean values of the vectors $\mathbf{x}_i$ and $\mathbf{x}_j$, respectively. That is,

$$\boldsymbol{\mu}_i = E[\mathbf{x}_i] \tag{1.26}$$

where E is the statistical expectation operator. The mean vector $\boldsymbol{\mu}_j$ is similarly defined. For a measure of the distance between these two populations, we may use the *Mahalanobis distance* denoted by d_{ij}. The squared value of this distance from $\mathbf{x}_i$ to $\mathbf{x}_j$ is defined by (Duda and Hart, 1973):

$$d_{ij}^2 = (\mathbf{x}_i - \boldsymbol{\mu}_i)^T \boldsymbol{\Sigma}^{-1}(\mathbf{x}_j - \boldsymbol{\mu}_j) \tag{1.27}$$

where $\boldsymbol{\Sigma}^{-1}$ is the inverse of the covariance matrix $\boldsymbol{\Sigma}$. It is assumed that the covariance matrix is the same for both populations, as shown by

$$
\begin{aligned}
\boldsymbol{\Sigma} &= E[(\mathbf{x}_i - \boldsymbol{\mu}_i)(\mathbf{x}_i - \boldsymbol{\mu}_i)^T] \\
&= E[(\mathbf{x}_j - \boldsymbol{\mu}_j)(\mathbf{x}_j - \boldsymbol{\mu}_j)^T]
\end{aligned}
\tag{1.28}
$$

For the special case when $\mathbf{x}_j = \mathbf{x}_i, \boldsymbol{\mu}_i = \boldsymbol{\mu}_j = \boldsymbol{\mu}$ and $\boldsymbol{\Sigma} = \mathbf{I}$, where $\mathbf{I}$ is the identity matrix, the Mahalanobis distance reduces to the Euclidean distance between the sample vector $\mathbf{x}_i$ and the mean vector $\boldsymbol{\mu}$.

Rule 2. Items to be categorized as separate classes should be given widely different representations in the network.

The second rule is the exact opposite of Rule 1.

Rule 3. If a particular feature is important, then there should be a large number of neurons involved in the representation of that item in the network.

Consider, for example, a radar application involving the detection of a target (e.g., aircraft) in the presence of clutter (i.e., radar reflections from undesirable targets such as buildings, trees, and weather formations). The detection performance of such a radar system is measured in terms of two probabilities:

- *Probability of detection*, defined as the probability that the system decides that a target is present when it is.
- *Probability of false alarm*, defined as the probability that the system decides that a target is present when it is not.

According to the *Neyman-Pearson criterion*, the probability of detection is maximized, subject to the constraint that the probability of false alarm does not exceed a prescribed value (Van Trees, 1968). In such an application, the actual presence of a target in the received signal represents an important feature of the input. Rule 3, in effect, states that there should be a large number of neurons involved in making the decision that a target is present when it actually is. By the same token, there should be a very large number of neurons involved in making the decision that the input consists of clutter only when it actually does. In both situations the large number of neurons assures a high degree of accuracy in decision making and tolerance with respect to faulty neurons.

Rule 4. Prior information and invariances should be built into the design of a neural network, thereby simplifying the network design by not having to learn them.

Rule 4 is particularly important because proper adherence to it results in a neural network with a *specialized (restricted) structure*. This is highly desirable for several reasons (Russo, 1991):

1. Biological visual and auditory networks are known to be very specialized.
2. A neural network with specialized structure usually has a smaller number of free parameters available for adjustment than a fully connected network. Consequently, the specialized network requires a smaller data set for training, learns faster, and often generalizes better.
3. The rate of information transmission through a specialized network (i.e., the network throughput) is accelerated.
4. The cost of building a specialized network is reduced because of its smaller size, compared to its fully connected counterpart.

How to Build Prior Information into Neural Network Design

An important issue that has to be addressed, of course, is how to develop a specialized structure by building prior information into its design. Unfortunately, there are currently no well-defined rules for doing this; rather, we have some *ad-hoc* procedures that are known to yield useful results. In particular, we may use a combination of two techniques (LeCun et al., 1990a):

1. *Restricting the network architecture* through the use of local connections known as *receptive fields*.[5]
2. *Constraining the choice of synaptic weights* through the use of *weight-sharing*.[6]

These two techniques, particularly the latter one, have a profitable side benefit: the number of free parameters in the network is reduced significantly.

To be specific, consider the partially connected feedforward network of Fig. 1.20. This network has a restricted architecture by construction. The top six source nodes

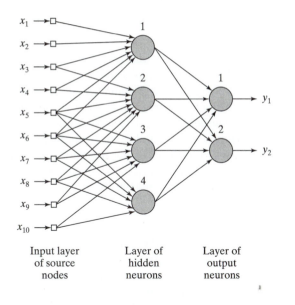

FIGURE 1.20 Illustrating the combined use of a receptive field and weight-sharing. All four hidden neurons share the same set of weights for their synaptic connections.

constitute the receptive field for hidden neuron 1 and so on for the other hidden neurons in the network. To satisfy the weight-sharing constraint, we merely have to use the same set of synaptic weights for each one of the neurons in the hidden layer of the network. Then, for the example shown in Fig. 1.20 with six local connections per hidden neuron and a total of four hidden neurons, we may express the induced local field of hidden neuron j as follows:

$$v_j = \sum_{i=1}^{6} w_i x_{i+j-1}, \quad j = 1, 2, 3, 4 \tag{1.29}$$

where $\{w_i\}_{i=1}^{6}$ constitute the same set of weights shared by all four hidden neurons, and x_k is the signal picked up from source node $k = i + j - 1$. Equation (1.29) is in the form of a *convolution sum*. It is for this reason that a feedforward network using local connections and weight-sharing in the manner described herein is referred to as a *convolutional network*.

The issue of building prior information into the design of a neural network pertains to one part of Rule 4; the remaining part of the rule involves the issue of invariances.

How to Build Invariances into Neural Network Design

Consider the following physical phenomena:

- When an object of interest rotates, the image of the object as perceived by an observer usually changes in a corresponding way.
- In a coherent radar that provides amplitude as well as phase information about its surrounding environment, the echo from a moving target is shifted in frequency due to the Doppler effect that arises due to the radial motion of the target in relation to the radar.

- The utterance from a person may be spoken in a soft or loud voice, and in a slow or quick manner.

In order to build an object recognition system, a radar target recognition system and a speech recognition system for dealing with these phenomena, respectively, the system must be capable of coping with a range of *transformations* of the observed signal (Barnard and Casasent, 1991). Accordingly, a primary requirement of pattern recognition is to design a classifier that is *invariant* to such transformations. In other words, a class estimate represented by an output of the classifier must not be affected by transformations of the observed signal applied to the classifier input.

There are at least three techniques for rendering classifier-type neural networks invariant to transformations (Barnard and Casasent, 1991):

1. *Invariance by Structure.* Invariance may be imposed on a neutral network by structuring its design appropriately. Specifically, synaptic connections between the neurons of the network are created so that transformed versions of the same input are forced to produce the same output. Consider, for example, the classification of an input image by a neural network that is required to be independent of in-plane rotations of the image about its center. We may impose rotational invariance on the network structure as follows. Let w_{ji} be the synaptic weight of neuron j connected to pixel i in the input image. If the condition $w_{ji} = w_{jk}$ is enforced for all pixels i and k that lie at equal distances from the center of the image, then the neural network is invariant to in-plane rotations. However, in order to maintain rotational invariance, the synaptic weight w_{ji} has to be duplicated for every pixel of the input image at the same radial distance from the origin. This points to a shortcoming of invariance by structure: The number of synaptic connections in the neural network becomes prohibitively large even for images of moderate size.

2. *Invariance by Training.* A neural network has a natural ability for pattern classification. This ability may be exploited directly to obtain transformation invariance as follows. The network is trained by presenting it a number of different examples of the same object, with the examples being chosen to correspond to different transformations (i.e., different aspect views) of the object. Provided that the number of examples is sufficiently large, and if the the network is trained to learn to discriminate the different aspect views of the object, we may then expect the network to generalize correctly to transformations other than those shown to it. However, from an engineering perspective, invariance by training has two disadvantages. First, when a neural network has been trained to recognize an object in an invariant fashion with respect to known transformations, it is not obvious that this training will also enable the network to recognize other objects of different classes invariantly. Second, the computational demand imposed on the network may be too severe to cope with, especially if the dimensionality of the feature space is high.

3. *Invariant Feature Space.* The third technique of creating an invariant classifier-type neural network is illustrated in Fig. 1.21. It rests on the premise that it may be pos-

FIGURE 1.21 Block diagram of an invariant feature-space type of system.

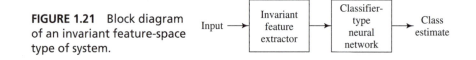

sible to extract *features* that characterize the essential information content of an input data set, and which are invariant to transformations of the input. If such features are used, then the network as a classifier is relieved from the burden of having to delineate the range of transformations of an object with complicated decision boundaries. Indeed, the only differences that may arise between different instances of the same object are due to unavoidable factors such as noise and occlusion. The use of an invariant feature space offers three distinct advantages. First, the number of features applied to the network may be reduced to realistic levels. Second, the requirements imposed on network design are relaxed. Third, invariance for all objects with respect to known transformations is assured (Barnard and Casasent, 1991). However, this approach requires prior knowledge of the problem for it to work.

In conclusion, the use of an invariant-feature space as described may offer a most suitable technique for neural classifiers.

To illustrate the idea of invariant-feature space, consider the example of a coherent radar system used for air surveillance, where the targets of interest include aircraft, weather systems, flocks of migrating birds, and ground objects. The radar echoes from these targets possess different spectral characteristics. Moreover, experimental studies have shown that such radar signals can be modeled fairly closely as an *autoregressive* (AR) *process* of moderate order (Haykin and Deng, 1991). An AR model is a special form of regressive model defined for complex-valued data by

$$x(n) = \sum_{i=1}^{M} a_i^* x(n - i) + e(n) \qquad (1.30)$$

where the $\{a_i\}_{i=1}^{M}$ are the *AR coefficients*, M is the *model order*, $x(n)$ is the *input*, and $e(n)$ is the *error* described as white noise. Basically, the AR model of Eq. (1.30) is represented by a *tapped-delay-line filter* as illustrated in Fig. 1.22a for $M = 2$. Equivalently, it may be represented by a *lattice filter* as shown in Fig. 1.22b, the coefficients of which are called *reflection coefficients*. There is a one-to-one correspondence between the AR coefficients of the model in Fig. 1.22a and the reflection coefficients of the model in Fig. 1.22b. The two models depicted assume that the input $x(n)$ is complex valued, as in the case of a coherent radar, in which case the AR coefficients and the reflection coefficients are all complex valued. The asterisk in Eq. (1.30) and Fig. 1.22 signifies *complex conjugation*. For now, it suffices to say that the coherent radar data may be described by a set of *autoregressive coefficients*, or by a corresponding set of *reflection coefficients*. The latter set of coefficients has a computational advantage in that efficient algorithms exist for their computation directly from the input data. The feature extraction problem, however, is complicated by the fact that moving objects produce varying Doppler frequencies that depend on their radial velocities measured with respect to the radar, and that tend to obscure the spectral content of the reflection coefficients as feature discriminants. To overcome this difficulty, we must build *Doppler invariance* into the computation of the reflection coefficients. The phase angle of the first reflection coefficient turns out to be equal to the Doppler frequency of the radar signal. Accordingly, Doppler frequency *normalization* is applied to all coefficients so as to remove the mean Doppler shift. This is done by defining a new set of reflection

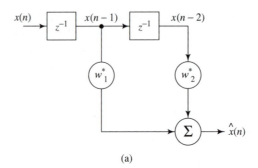

(a)

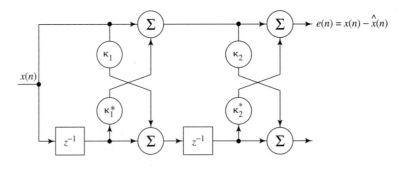

(b)

FIGURE 1.22 Autoregressive model of order 2: (a) tapped-delay-line model; (b) lattice filter model. (The asterisk denotes complex conjugation.)

coefficients $\{\kappa'_m\}$ related to the set of ordinary reflection coefficients $\{\kappa_m\}$ computed from the input data as follows:

$$\kappa'_m = \kappa_m e^{-jm\theta} \quad \text{for } m = 1, 2, ..., M \tag{1.31}$$

where θ is the phase angle of the first reflection coefficient. The operation described in Eq. (1.31) is referred to as *heterodyning*. A set of *Doppler-invariant radar features* is thus represented by the normalized reflection coefficients $\kappa'_1, \kappa'_2, ..., \kappa'_M$, with κ'_1 being the only real-valued coefficient in the set. As mentioned previously, the major categories of radar targets of interest in air surveillance are weather, birds, aircraft, and ground. The first three targets are moving, whereas the last one is not. The heterodyned spectral parameters of radar echoes from ground have echoes similar in characteristic to those from aircraft. A ground echo can be discriminated from an aircraft echo because of its small Doppler shift. Accordingly, the radar classifier includes a postprocessor as shown in Fig. 1.23, which operates on the classified results (encoded labels) for the purpose of identifying the ground class (Haykin and Deng, 1991). Thus, the *preprocessor* in Fig. 1.23 takes care of Doppler shift-invariant feature extraction at the classifier input, whereas the *postprocessor* uses the stored Doppler signature to distinguish between aircraft and ground returns.

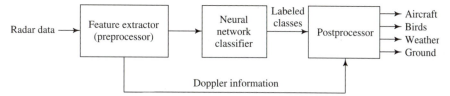

FIGURE 1.23 Doppler shift-invariant classifier of radar signals.

A much more fascinating example of knowledge representation in a neural network is found in the biological sonar system of echo-locating bats. Most bats use *frequency-modulated* (FM or "chirp") signals for the purpose of acoustic imaging; in an FM signal the instantaneous frequency of the signal varies with time. Specifically, the bat uses its mouth to broadcast short-duration FM sonar signals and uses its auditory system as the sonar receiver. Echoes from targets of interest are represented in the auditory system by the activity of neurons that are selective to different combinations of acoustic parameters. There are three principal neural dimensions of the bat's auditory representation (Simmons, 1991; Simmons and Saillant, 1992):

- *Echo frequency*, which is encoded by "place" originating in the frequency map of the cochlea; it is preserved throughout the entire auditory pathway as an orderly arrangement across certain neurons tuned to different frequencies.
- *Echo amplitude*, which is encoded by other neurons with different dynamic ranges; it is manifested both as amplitude tuning and as the number of discharges per stimulus.
- *Echo delay*, which is encoded through neural computations (based on cross-correlation) that produce delay-selective responses; it is manifested as target-range tuning.

The two principal characteristics of a target echo for image-forming purposes are *spectrum* for target shape, and *delay* for target range. The bat perceives "shape" in terms of the arrival time of echoes from different reflecting surfaces (glints) within the target. For this to occur, *frequency* information in the echo spectrum is converted into estimates of the *time* structure of the target. Experiments conducted by Simmons and coworkers on the big brown bat, *Eptesicus fuscus*, critically identify this conversion process as consisting of parallel time-domain and frequency-to-time-domain transforms whose converging outputs create the common delay of range axis of a perceived image of the target. It appears that the unity of the bat's perception is due to certain properties of the transforms themselves, despite the separate ways in which the auditory time representation of the echo delay and frequency representation of the echo spectrum are initially performed. Moreover, feature invariances are built into the sonar image-forming process so as to make it essentially independent of the target's motion and the bat's own motion.

Returning to the main theme of this section, namely, that of knowledge representation in a neural network, this issue is directly related to that of network architecture described in Section 1.6. Unfortunately, there is no well-developed theory for optimizing the architechture of a neural network required to interact with an environment of interest, or for evaluating the way in which changes in the network architecture affect the representation of knowledge inside the network. Indeed, satisfactory answers to these issues are usually found through an exhaustive experimental study, with the designer of the neural network becoming an essential part of the structural learning loop.

No matter how the design is performed, knowledge about the problem domain of interest is acquired by the network in a comparatively straightforward and direct manner through training. The knowledge so acquired is represented in a compactly distributed form as weights across the synaptic connections of the network. While this form of knowledge representation enables the neural network to adapt and generalize, unfortunately the neural network suffers from the inherent inability to explain, in a comprehensive manner, the computational process by which the network makes a decision or reports its output. This can be a serious limitation, particularly in those applications where safety is of prime concern, as in air traffic control or medical diagnosis, for example. In applications of this kind, it is not only highly desirable but also absolutely essential to provide some form of *explanation capability*. One way in which this provision can be made is to integrate a neural network and artificial intelligence into a hybrid system, as discussed in the next section.

1.8   ARTIFICIAL INTELLIGENCE AND NEURAL NETWORKS

The goal of *artificial intelligence* (AI) is the development of paradigms or algorithms that require machines to perform cognitive tasks, at which humans are currently better. This statement on AI is adopted from Sage, 1990. Note that it is not the only accepted definition of AI.

An AI system must be capable of doing three things: (1) store knowledge, (2) apply the knowledge stored to solve problems, and (3) acquire new knowledge through experience. An AI system has three key components: representation, reasoning, and learning (Sage, 1990), as depicted in Fig. 1.24.

1. *Representation*. The most distinctive feature of AI is probably the pervasive use of a language of *symbol* structures to represent both general knowledge about a problem domain of interest and specific knowledge about the solution to the problem. The symbols are usually formulated in familiar terms, which makes the symbolic repre-

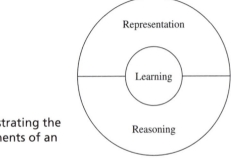

FIGURE 1.24   Illustrating the three key components of an AI system.

sentations of AI relatively easy to understand by a human user. Indeed, the clarity of symbolic AI makes it well suited for human–machine communication.

"Knowledge," as used by AI researchers, is just another term for data. It may be of a declarative or procedural kind. In a *declarative* representation, knowledge is represented as a static collection of facts, with a small set of general procedures used to manipulate the facts. A characteristic feature of declarative representations is that they appear to possess a meaning of their own in the eyes of the human user, independent of their use within the AI system. In a *procedural* representation, on the other hand, knowledge is embodied in an executable code that acts out the meaning of the knowledge. Both kinds of knowledge, declarative and procedural, are needed in most problem domains of interest.

2. *Reasoning.* In its most basic form, *reasoning* is the ability to solve problems. For a system to qualify as a reasoning system it must satisfy certain conditions (Fischler and Firschein,1987):

- The system must be able to express and solve a broad range of problems and problem types.
- The system must be able to make *explicit* and *implicit* information known to it.
- The system must have a *control* mechanism that determines which operations to apply to a particular problem, when a solution to the problem has been obtained, or when further work on the problem should be terminated.

Problem solving may be viewed as a *searching* problem. A common way to deal with "search" is to use *rules*, *data*, and *control* (Nilsson, 1980). The rules operate on the data, and the control operates on the rules. Consider, for example, the "traveling salesman problem," where the requirement is to find the shortest tour that goes from one city to another, with all the cities on the tour being visited only once. In this problem the data are made up of the set of possible tours and their costs in a weighted graph, the rules define the ways to proceed from city to city, and the control decides which rules to apply and when to apply them.

In many situations encountered in practice (e.g., medical diagnosis), the available knowledge is incomplete or inexact. In such situations, *probabilistic reasoning* procedures are used, thereby permitting AI systems to deal with uncertainty (Russell and Norvig, 1995; Pearl, 1988).

3. *Learning.* In the simple model of machine learning depicted in Fig. 1.25, the environment supplies some information to a *learning element*. The learning element then uses this information to make improvements in a *knowledge base*, and finally the

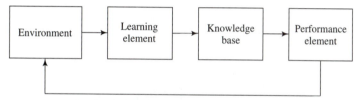

FIGURE 1.25 Simple model of machine learning.

performance element uses the knowledge base to perform its task. The kind of information tion supplied to the machine by the environment is usually imperfect, with the result that the learning element does not know in advance how to fill in missing details or to ignore details that are unimportant. The machine therefore operates by guessing, and then receiving *feedback* from the performance element. The feedback mechanism enables the machine to evaluate its hypotheses and revise them if necessary.

Machine learning may involve two rather different kinds of information processing: inductive and deductive. In *inductive* information processing, general patterns and rules are determined from raw data and experience. In *deductive* information processing, however, general rules are used to determine specific facts. Similarity-based learning uses induction, whereas the proof of a theorem is a deduction from known axioms and other existing theorems. Explanation-based learning uses both induction and deduction.

The importance of knowledge bases and the difficulties experienced in learning have led to the development of various methods for augmenting knowledge bases. Specifically, if there are experts in a given field, it is usually easier to obtain the compiled experience of the experts than to try to duplicate and direct experience that gave rise to the expertise. This, indeed, is the idea behind *expert systems*.

Having familiarized ourselves with symbolic AI machines, how would we compare them to neural networks as cognitive models? For this comparison, we follow three subdivisions: level of explanation, style of processing, and representational structure (Memmi, 1989).

1. *Level of Explanation*. In classical AI, the emphasis is on building *symbolic representations* that are presumably so called because they stand for something. From the viewpoint of cognition, AI assumes the existence of mental representations, and it models cognition as the *sequential processing* of symbolic representations (Newell and Simon, 1972).

The emphasis in neural networks, on the other hand, is on the development of *parallel distributed processing* (PDP) *models*. These models assume that information processing takes place through the interaction of a large number of neurons, each of which sends excitatory and inhibitory signals to other neurons in the network (Rumelhart and McClelland, 1986). Moreover, neural networks place great emphasis on neurobiological explanation of cognitive phenomena.

2. *Processing Style*. In classical AI, the processing is *sequential*, as in typical computer programming. Even when there is no predetermined order (scanning the facts and rules of an expert system, for example), the operations are performed in a step-by-step manner. Most probably, the inspiration for sequential processing comes from the sequential nature of natural language and logical inference, as much as from the structure of the von Neumann machine. We should not forget that classical AI was born shortly after the von Neumann machine, during the same intellectual era.

In contrast, *parallelism* is not only conceptually essential to the processing of information in neural networks, but also the source of their flexibility. Moreover, parallelism may be massive (hundreds of thousands of neurons), which gives neural networks a remarkable form of robustness. With the computation spread over many neurons, it usually does not matter much if the states of some neurons in the network deviate from their expected values. Noisy or incomplete inputs may still be recognized, a damaged network may still be able to function satisfactorily, and learning does not have to be perfect. Performance of the network degrades gracefully within a certain range. The network is made even more robust by virtue of the "coarse coding" (Hinton, 1981), where each feature is spread over several neurons.

3. *Representational Structure.* With a language of thought pursued as a model for classical AI, we find that symbolic representations possess a *quasi-linguistic structure.* Like expressions of natural language, the expressions of classical AI are generally complex, built in a systematic fashion from simple symbols. Given a limited stock of symbols, meaningful new expressions may be composed by virtue of the *compositionality* of symbolic expressions and the analogy between syntactic structure and semantics.

The nature and structure of representations is, however, a crucial problem for neural networks. In the March 1988 Special Issue of the journal *Cognition*, Fodor and Pylyshyn make some potent criticisms about the computational adequacy of neural networks in dealing with cognition and linguistics. They argue that neural networks are on the wrong side of two basic issues in cognition: the nature of *mental representations*, and the nature of *mental processes*. According to Fodor and Pylyshyn, for classical AI theories but *not* neural networks:

- Mental representations characteristically exhibit a combinatorial constituent structure and combinatorial semantics.
- Mental processes are characteristically sensitive to the combinatorial structure of the representations on which they operate.

In summary, we may describe symbolic AI as the formal manipulation of a language of algorithms and data representations in a *top-down* fashion. We may describe neural networks, however, as parallel distributed processors with a natural ability to learn, and which usually operate in a *bottom-up* fashion. For the implementation of cognitive tasks, it therefore appears that rather than seek solutions based on symbolic AI or neural networks alone, a more potentially useful approach would be to build *structured connectionist models* or *hybrid systems* that integrate them together. By so doing, we are able to combine the desirable features of adaptivity, robustness, and uniformity offered by neural networks with the representation, inference, and universality that are inherent features of symbolic AI (Feldman, 1992; Waltz, 1997). Indeed, it is with this objective in mind that several methods have been developed for the extraction of rules from trained neural networks. In addition to the understanding of how symbolic and connectionist approaches can be integrated for building intelligent machines, there are several other reasons for the extraction of rules from neural networks (Andrews and Diederich, 1996):

- To validate neural network components in software systems by making the internal states of the neural network accessible and understandable to users.
- To improve the generalization performance of neural networks by (1) identifying regions of the input space where the training data are not adequately represented, or (2) indicating the circumstances where the neural network may fail to generalize.
- To discover salient features of the input data for data exploration (mining).
- To provide a means for traversing the boundary between the connectionist and symbolic approaches to the development of intelligent machines.
- To satisfy the critical need for safety in a special class of systems where safety is a mandatory condition.

1.9 HISTORICAL NOTES

We conclude this introductory chapter on neural networks with some historical notes.[7]

The modern era of neural networks began with the pioneering work of McCulloch and Pitts (1943). McCulloch was a psychiatrist and neuroanatomist by training; he spent some 20 years thinking about the representation of an event in the nervous system. Pitts was a mathematical prodigy, who joined McCulloch in 1942. According to Rall (1990), the 1943 paper by McCulloch and Pitts arose within a neural modeling community that had been active at the University of Chicago for at least five years prior to 1943, under the leadership of Rashevsky.

In their classic paper, McCulloch and Pitts describe a logical calculus of neural networks that united the studies of neurophysiology and mathematical logic. Their formal model of a neuron was assumed to follow an "all-or-none" law. With a sufficient number of such simple units, and synaptic connections set properly and operating synchronously, McCulloch and Pitts showed that a network so constituted would, in principle, compute any computable function. This was a very significant result and with it, it is generally agreed that the disciplines of neural networks and of artificial intelligence were born.

The 1943 paper by McCulloch and Pitts was widely read at the time and still is. It influenced von Neumann to use idealized switch-delay elements derived from the McCulloch–Pitts neuron in the construction of the EDVAC (Electronic Discrete Variable Automatic Computer) that developed out of the ENIAC (Electronic Numerical Integrator and Computer) (Aspray and Burks, 1986). The ENIAC was the first general purpose electronic computer, which was built at the Moore School of Electrical Engineering of the University of Pennsylvania from 1943 to 1946. The McCulloch–Pitts theory of formal neural networks featured prominently in the second of four lectures delivered by von Neumann at the University of Illinois in 1949.

In 1948, Wiener's famous book *Cybernetics* was published, describing some important concepts for control, communications, and statistical signal processing. The second edition of the book was published in 1961, adding new material on learning and self-organization. In Chapter 2 of both editions of this book, Wiener appears to grasp the physical significance of statistical mechanics in the context of the subject matter, but it was left to Hopfield (more than 30 years later) to bring the linkage between statistical mechanics and learning systems to full fruition.

The next major development in neural networks came in 1949 with the publication of Hebb's book *The Organization of Behavior*, in which an explicit statement of a physiological learning rule for *synaptic modification* was presented for the first time. Specifically, Hebb proposed that the connectivity of the brain is continually changing as an organism learns differing functional tasks, and that *neural assemblies* are created by such changes. Hebb followed up an early suggestion by Ramón y Cajál and introduced his now famous *postulate of learning*, which states that the effectiveness of a variable synapse between two neurons is increased by the repeated activation of one neuron by the other across that synapse. Hebb's book was immensely influential among psychologists, but unfortunately it had little or no impact on the engineering community.

Hebb's book has been a source of inspiration for the development of computational models of *learning and adaptive systems*. The paper by Rochester, Holland,

Haibt, and Duda (1956) is perhaps the first attempt to use computer simulation to test a well-formulated neural theory based on Hebb's postulate of learning; the simulation results reported in that paper clearly show that inhibition must be added for the theory to actually work. In that same year, Uttley (1956) demonstrated that a neural network with modifiable synapses may learn to classify simple sets of binary patterns into corresponding classes. Uttley introduced the so-called *leaky integrate and fire neuron*, which was later formally analyzed by Caianiello (1961). In later work, Uttley (1979) hypothesized that the effectiveness of a variable synapse in the nervous system depends on the statistical relationship between the fluctuating states on either side of that synapse, thereby linking up with Shannon's information theory.

In 1952, Ashby's book, *Design for a Brain: The Origin of Adaptive Behavior*, was published, which is just as fascinating to read today as it must have been then. The book was concerned with the basic notion that adaptive behavior is not inborn but rather learned, and that through learning the behavior of an animal (system) usually changes for the better. The book emphasized the dynamic aspects of the living organism as a machine and the related concept of stability.

In 1954, Minsky wrote a "neural network" doctorate thesis at Princeton University, which was entitled "Theory of Neural-Analog Reinforcement Systems and Its Application to the Brain-Model Problem." In 1961, an excellent early paper by Minsky on AI entitled "Steps Toward Artificial Intelligence," was published; this latter paper contains a large section on what is now termed neural networks. In 1967 Minsky's book, *Computation: Finite and Infinite Machines*, was published. This clearly written book extended the 1943 results of McCulloch and Pitts and put them in the context of automata theory and the theory of computation.

Also in 1954, the idea of a *nonlinear adaptive filter* was proposed by Gabor, one of the early pioneers of communication theory, and the inventor of holography. He went on to build such a machine with the aid of collaborators, the details of which are described in Gabor et al. (1960). Learning was accomplished by feeding samples of a stochastic process into the machine, together with the target function that the machine was expected to produce.

In the 1950s work on *associative memory* was initiated by Taylor (1956). This was followed by the introduction of the *learning matrix* by Steinbuch (1961); this matrix consists of a planar network of switches interposed between arrays of "sensory" receptors and "motor" effectors. In 1969, an elegant paper on nonholographic associative memory by Willshaw, Buneman, and Longuet-Higgins was published. This paper presents two ingenious network models: a simple optical system realizing a correlation memory, and a closely related neural network suggested by the optical memory. Other significant contributions to the early development of associative memory include papers by Anderson (1972), Kohonen (1972), and Nakano (1972), who independently and in the same year introduced the idea of a *correlation matrix memory* based on the *outer product* learning rule.

Von Neumann was one of the great figures in science in the first half of the twentieth century. The *von Neumann architecture*, basic to the design of a digital computer, is named in his honor. In 1955 he was invited by Yale University to give the Silliman Lectures during 1956. He died in 1957, and the unfinished manuscript of the

Silliman Lectures was published later as a book, *The Computer and the Brain* (1958). This book is interesting because it suggests what von Neumann might have done had he lived; he had started to become aware of the profound differences between brains and computers.

An issue of particular concern in the context of neural networks is that of designing a reliable network with neurons that may be viewed as unreliable components. This important problem was solved by von Neumann (1956) using the idea of redundancy, which motivated Winograd and Cowan (1963) to suggest the use of a *distributed* redundant representation for neural networks. Winograd and Cowan showed how a large number of elements could collectively represent an individual concept, with a corresponding increase in robustness and parallelism.

Some 15 years after the publication of McCulloch and Pitt's classic paper, a new approach to the pattern recognition problem was introduced by Rosenblatt (1958) in his work on the *perceptron*, a novel method of supervised learning. The crowning achievement of Rosenblatt's work was the so-called *perceptron convergence theorem*, the first proof for which was outlined by Rosenblatt (1960b); proofs of the theorem also appeared in Novikoff (1963) and others. In 1960, Widrow and Hoff introduced the *least mean-square (LMS) algorithm* and used it to formulate the *Adaline* (adaptive linear element). The difference between the perceptron and the Adaline lies in the training procedure. One of the earliest trainable layered neural networks with multiple adaptive elements was the Madaline (multiple-adaline) structure proposed by Widrow and his students (Widrow, 1962). In 1967, Amari used the stochastic gradient method for adaptive pattern classification. In 1965, Nilsson's book, *Learning Machines*, was published, which is still the best-written exposition of linearly separable patterns in hypersurfaces. During the classical period of the perceptron in the 1960s, it seemed as if neural networks could do anything. But then came the book by Minsky and Papert (1969), who used mathematics to demonstrate that there are fundamental limits on what single-layer perceptrons can compute. In a brief section on multilayer perceptrons, they stated that there was no reason to assume that any of the limitations of single-layer perceptrons could be overcome in the multilayer version.

An important problem encountered in the design of a multilayer perceptron is the *credit assignment problem* (i.e., the problem of assigning credit to hidden neurons in the network). The terminology "credit assignment" was first used by Minsky (1961), under the title "Credit Assignment Problem for Reinforcement Learning Systems." By the late 1960s, most of the ideas and concepts necessary to solve the perceptron credit assignment problem were already formulated, as were many of the ideas underlying the recurrent (attractor neural) networks that are now referred to as Hopfield networks. However, we had to wait until the 1980s for the solutions of these basic problems to emerge. According to Cowan (1990), there were three reasons for this lag of more than 10 years:

- One reason was technological—there were no personal computers or workstations for experimentation. For example, when Gabor developed his nonlinear learning filter, it took his research team an additional six years to build the filter with analog devices (Gabor, 1954; Gabor et al., 1960).

- The other reason was in part psychological, in part financial. The 1969 monograph by Minsky and Papert certainly did not encourage anyone to work on perceptrons, or agencies to support the work on them.
- The analogy between neural networks and lattice spins was premature. The *spin-glass model* by Sherrington and Kirkpatrick was not invented until 1975.

These factors contributed in one way or another to the dampening of continued interest in neural networks in the 1970s. Many of the researchers, except for those in psychology and the neurosciences, deserted the field during that decade. Indeed, only a handful of the early pioneers maintained their commitment to neural networks. From an engineering perspective, we may look back on the 1970s as a decade of dormancy for neural networks.

An important activity that did emerge in the 1970s was *self-organizing maps* using competitive learning. The computer simulation work done by von der Malsburg (1973) was perhaps the first to demonstrate self-organization. In 1976 Willshaw and von der Malsburg published the first paper on the formation of self-organizing maps, motivated by topologically ordered maps in the brain.

In the 1980s major contributions to the theory and design of neural networks were made on several fronts, and with it there was a resurgence of interest in neural networks.

Grossberg (1980), building on his earlier work on competitive learning (Grossberg, 1972, 1976a, b), established a new principle of self-organization known as *adaptive resonance theory* (ART). Basically, the theory involves a bottom-up recognition layer and a top-down generative layer. If the input pattern and learned feedback pattern match, a dynamical state called "adaptive resonance" (i.e., amplification and prolongation of neural activity) takes place. This *principle of forward/backward projections* has been rediscovered by other investigators under different guises.

In 1982, Hopfield used the idea of an energy function to formulate a new way of understanding the computation performed by recurrent networks with symmetric synaptic connections. Moreover, he established the isomorphism between such a recurrent network and an *Ising model* used in statistical physics. This analogy paved the way for a deluge of physical theory (and physicists) to enter neural modeling, thereby transforming the field of neural networks. This particular class of neural networks with feedback attracted a great deal of attention in the 1980s, and in the course of time it has come to be known as *Hopfield networks*. Although Hopfield networks may not be realistic models for neurobiological systems, the principle they embody, namely that of storing information in dynamically stable networks, is profound. The origin of this principle may in fact be traced back to pioneering work of many other investigators:

- Cragg and Tamperley (1954, 1955) made the observation that just as neurons can be "fired" (activated) or "not fired" (quiescent), so can atoms in a lattice have their spins pointing "up" or "down."
- Cowan (1967) introduced the "sigmoid" firing characteristic and the smooth firing condition for a neuron that was based on the logistic function.
- Grossberg (1967, 1968) introduced the *additive model* of a neuron, involving nonlinear difference/differential equations, and explored the use of the model as a basis for short-term memory.

- Amari (1972) independently introduced the additive model of a neuron, and used it to study the dynamic behavior of randomly connected neuron-like elements.
- Wilson and Cowan (1972) derived coupled nonlinear differential equations for the dynamics of spatially localized populations containing both excitatory and inhibitory model neurons.
- Little and Shaw (1975) described a *probabilistic model* of a neuron, either firing or not firing an action potential, and used the model to develop a theory of short-term memory.
- Anderson, Silverstein, Ritz, and Jones (1977) proposed the *brain-state-in-a-box* (*BSB*) *model*, consisting of a simple associative network coupled to nonlinear dynamics.

It is therefore not surprising that the publication of Hopfield's paper in 1982 generated a great deal of controversy. Nevertheless, it is in the same paper that the principle of storing information in dynamically stable networks is first made explicit. Moreover, Hopfield showed that he had the insight from the spin-glass model in statistical mechanics to examine the special case of recurrent networks with symmetric connectivity, thereby guaranteeing their convergence to a stable condition. In 1983, Cohen and Grossberg established a general principle for assessing the stability of a *content-addressable memory* that includes the continuous-time version of the Hopfield network as a special case. A distinctive feature of an attractor neural network is the natural way in which *time*, an essential dimension of learning, manifests itself in the nonlinear dynamics of the network. In this context, the Cohen–Grossberg theorem is of profound importance.

Another important development in 1982 was the publication of Kohonen's paper on self-organizing maps (Kohonen, 1982) using a one- or two-dimensional lattice structure, which was different in some respects from the earlier work by Willshaw and von der Malsburg. Kohonen's model has received far more attention in an analytic context and with respect to applications in the literature, than the Willshaw–von der Malsburg model, and has become the benchmark against which other innovations in this field are evaluated.

In 1983, Kirkpatrick, Gelatt, and Vecchi described a new procedure called *simulated annealing*, for solving combinatorial optimization problems. Simulated annealing is rooted in statistical mechanics. It is based on a simple technique that was first used in computer simulation by Metropolis et al. (1953). The idea of simulated annealing was later used by Ackley, Hinton, and Sejnowski (1985) in the development of a stochastic machine known as the *Boltzmann machine*, which was the *first* successful realization of a multilayer neural network. Although the Boltzmann machine learning algorithm proved not as computationally efficient as the back-propagation algorithm, it broke the psychological logjam by showing that the speculation in Minsky and Papert (1969) was incorrectly founded. The Boltzmann machine also laid the groundwork for the subsequent development of *sigmoid belief networks* by Neal (1992), which accomplished two things: (1) significant improvement in learning, and (2) linking neural networks to belief networks (Pearl, 1988). A further improvement in the learning performance of sigmoid belief networks was made by Saul, Jakkola, and

Jordan (1996) by using mean-field theory, a technique also rooted in statistical mechanics.

A paper by Barto, Sutton, and Anderson on *reinforcement learning* was published in 1983. Although they were not the first to use reinforcement learning (Minsky considered it in his 1954 Ph.D. thesis, for example), this paper has generated a great deal of interest in reinforcement learning and its application in control. Specifically, they demonstrated that a reinforcement learning system could learn to balance a broomstick (i.e., a pole mounted on a cart) in the absence of a helpful teacher. The system required only a failure signal that occurs when the pole falls past a critical angle from the vertical, or when the cart reaches the end of a track. In 1996, the book *Neurodynamic Programming* by Bertsekas and Tsitsiklis was published. This book put reinforcement on a proper mathematical basis by linking it with Bellman's dynamic programming.

In 1984 Braitenberg's book, *Vehicles: Experiments in Synthetic Psychology*, was published. In this book, Braitenberg advocates the *principle of goal-directed, self-organized performance:* the understanding of a complex process is best achieved by a synthesis of putative elementary mechanisms, rather than by a top-down analysis. Under the guise of science fiction, Braitenberg illustrates this important principle by describing various machines with simple internal architecture. The properties of the machines and their behavior are inspired by facts about animal brains, a subject he studied directly or indirectly for more than 20 years.

In 1986 the development of the *back-propagation algorithm* was reported by Rumelhart, Hinton, and Williams (1986). In that same year, the celebrated two-volume book, *Parallel Distributed Processing: Explorations in the Microstructures of Cognition*, edited by Rumelhart and McClelland, was published. This latter book has been a major influence in the use of back-propagation learning, which has emerged as the most popular learning algorithm for the training of multilayer perceptrons. In fact, back-propagation learning was discovered independently in two other places about the same time (Parker, 1985; LeCun, 1985). After the discovery of the back-propagation algorithm in the mid-1980s, it turned out that the algorithm had been described earlier by Werbos in his Ph.D. thesis at Harvard University in August 1974; Werbos's Ph.D. thesis was the first documented description of efficient reverse-mode gradient computation that was applied to general network models with neural networks arising as a special case. The basic idea of back propagation may be traced further back to the book *Applied Optimal Control* by Bryson and Ho (1969). In Section 2.2 entitled "Multistage Systems" of that book, a derivation of back propagation using a Lagrangian formalism is described. In the final analysis, however, much of the credit for the back-propagation algorithm has to be given to Rumelhart, Hinton, and Williams (1986) for proposing its use for machine learning and for demonstrating how it could work.

In 1988 Linsker described a new principle for self-organization in a perceptual network (Linsker, 1988a). The principle is designed to preserve maximum information about input activity patterns, subject to such constraints as synaptic connections and synapse dynamic range. A similar suggestion had been made independently by several vision researchers. However, it was Linsker who used abstract concepts rooted in information theory (originated by Shannon in 1948) to formulate the *maximum mutual information* (Infomax) principle. Linsker's paper reignited interest in the application of information theory to neural networks. In particular, the application of information theory to the *blind source separation problem* by Bell and Sejnowski (1995) has prompted many researchers to explore other information-theoretic models for solving a broad class of problems known collectively as *blind deconvolution*.

Also in 1988, Broomhead and Lowe described a procedure for the design of layered feedforward networks using *radial basis functions* (RBF), which provide an alternative to multilayer perceptrons. The basic idea of radial basis functions goes back at least to the *method of potential functions* that was originally proposed by Bashkirov, Braverman, and Muchnik (1964), and the theoretical properties of which were developed by Aizerman, Braverman, and Rozonoer (1964a, b). A description of the method of potential functions is presented in the classic book, *Pattern Classification and Scene Analysis*, by Duda and Hart (1973). Nevertheless, the paper by Broomhead and Lowe has led to a great deal of research effort linking the design of neural networks to an important area in numerical analysis and also linear adaptive filters. In 1990, Poggio and Girosi (1990a) further enriched the theory of RBF networks by applying Tikhonov's regularization theory.

In 1989, Mead's book, *Analog VLSI and Neural Systems*, was published. This book provides an unusual mix of concepts drawn from neurobiology and VLSI technology. Above all, it includes chapters on silicon retina and silicon cochlea, written by Mead and coworkers, which are vivid examples of Mead's creative mind.

In the early 1990s, Vapnik and coworkers invented a computationally powerful class of supervised learning networks, called *support vector machines*, for solving pattern recognition, regression, and density estimation problems (Boser, Guyon, and Vapnik, 1992; Cortes and Vapnik, 1995; Vapnik, 1995, 1998). This new method is based on results in the theory of learning with finite sample sizes. A novel feature of support vector machines is the natural way in which the *Vapnik-Chervonenkis (VC) dimension* is embodied in their design. The VC dimension provides a measure for the capacity of a neural network to learn from a set of examples (Vapnik and Chervonenkis, 1971; Vapnik, 1982).

It is now well established that *chaos* constitutes a key aspect of physical phenomena. A question raised by many is: Is there a key role for chaos in the study of neural networks? In a biological context, Freeman (1995) believes that the answer to this question is in the affirmative. According to Freeman, patterns of neural activity are not imposed from outside the brain; rather, they are constructed from within. In particular, chaotic dynamics offers a basis for describing the conditions that are required for emergence of self-organized patterns in and among populations of neurons.

Perhaps more than any other publication, the 1982 paper by Hopfield and the 1986 two-volume book by Rumelhart and McLelland were the most influential publications responsible for the resurgence of interest in neural networks in the 1980s. Neural networks have certainly come a long way from the early days of McCulloch and Pitts. Indeed, they have established themselves as an interdisciplinary subject with deep roots in the neurosciences, psychology, mathematics, the physical sciences, and engineering. Needless to say, they are here to stay, and will continue to grow in theory, design, and applications.

NOTES AND REFERENCES

1. This definition of a neural network is adapted from Aleksander and Morton (1990).

2. For a complementary perspective on neural networks with emphasis on neural modeling, cognition, and neurophysiological considerations, see Anderson (1995). For a highly readable account of the computational aspects of the brain, see Churchland and Sejnowski (1992). For more detailed descriptions of neural mechanisms and the human brain, see Kandel and Schwartz (1991), Shepherd (1990a, b), Koch and Segev (1989), Kuffler et al. (1984), and Freeman (1975).

3. For a thorough account of sigmoid functions and related issues, see Menon et al. (1996).

4. The logistic function, or more precisely the *logistic distribution function*, derives its name from a transcendental "law of logistic growth" that resulted in a huge literature. Measured in appropriate units, all growth processes were supposed to be represented by the logistic distribution function

$$F(t) = \frac{1}{1 + e^{\alpha t - \beta}}$$

where t represents time, and α and β are constants. It turned out, however, that not only the logistic distribution but also the Gaussian and other distributions can apply to the same data with the same or better goodness of fit (Feller, 1968).

5. According to Kuffler et al. (1984), the term "receptive field" was coined originally by Sherrington (1906) and reintroduced by Hartline (1940). In the context of a visual system, the receptive field of a neuron refers to the restricted area on the retinal surface, which influences the discharges of that neuron due to light.

6. It appears that the weight-sharing technique was originally described in Rumelhart et al. (1986b).

7. The historical notes presented here are largely (but not exclusively) based on the following sources: (1) the paper by Saarinen et al. (1992); (2) the chapter contribution by Rall (1990); (3) the paper by Widrow and Lehr (1990); (4) the papers by Cowan (1990) and Cowan and Sharp (1988); (5) the paper by Grossberg (1988c); (6) the two-volume book on neurocomputing (Anderson et al., 1990; Anderson and Rosenfeld, 1988); (7) the chapter contribution of Selfridge et al. (1988); (8) the collection of papers by von Neumann on computing and computer theory (Aspray and Burks, 1986); (9) the handbook on brain theory and neural networks edited by Arbib (1995); (10) Chapter 1 of the book by Russell and Norvig (1995); and (11) the article by Taylor (1997).

PROBLEMS

Models of a neuron

1.1 An example of the logistic function is defined by

$$\varphi(v) = \frac{1}{1 + \exp(-av)}$$

whose limiting values are 0 and 1. Show that the derivative of $\varphi(v)$ with respect to v is given by

$$\frac{d\varphi}{dv} = a\varphi(v)[1 - \varphi(v)]$$

What is the value of this derivative at the origin?

1.2 An odd sigmoid function is defined by

$$\varphi(v) = \frac{1 - \exp(-av)}{1 + \exp(-av)}$$

$$= \tanh\left(\frac{av}{2}\right)$$

where tanh denotes a hyperbolic tangent. The limiting values of this second sigmoid function are -1 and $+1$. Show that the derivative of $\varphi(v)$ with respect to v is given by

$$\frac{d\varphi}{dv} = \frac{a}{2}[1 - \varphi^2(v)]$$

What is the value of this derivative at the origin? Suppose that the slope parameter a is made infinitely large. What is the resulting form of $\varphi(v)$?

1.3 Yet another odd sigmoid function is the algebraic sigmoid:

$$\varphi(v) = \frac{v}{\sqrt{1 + v^2}}$$

whose limiting values are -1 and $+1$. Show that the derivative of $\varphi(v)$ with respect to v is given by

$$\frac{d\varphi}{dv} = \frac{\varphi^3(v)}{v^3}$$

What is the value of this derivative at the origin?

1.4 Consider the following two functions:

(i) $\varphi(v) = \dfrac{1}{\sqrt{2\pi}} \displaystyle\int_{-\infty}^{v} \exp\left(-\frac{x^2}{2}\right) dx$

(ii) $\varphi(v) = \dfrac{2}{\pi} \tan^{-1}(v)$

Explain why both of these functions fit the requirements of a sigmoid function. How do these two functions differ from each other?

1.5 Which of the five sigmoid functions described in Problems 1.1 to 1.4 would qualify as a cumulative (probability) distribution function? Justify your answer.

1.6 Consider the pseudolinear activation function $\varphi(v)$ shown in Fig. P1.6.
   (a) Formulate $\varphi(v)$ as a function of v.
   (b) What happens to $\varphi(v)$ if a is allowed to approach zero?

1.7 Repeat Problem 1.6 for the pseudolinear activation function $\varphi(v)$ shown in Fig. P1.7.

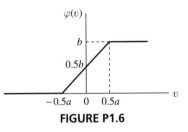

FIGURE P1.6

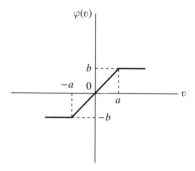

FIGURE P1.7

1.8 A neuron has an activation function $\varphi(v)$ defined by the logistic function of Problem 1.1, where v is the induced local field, and the slope parameter a is available for adjustment. Let $x_1, x_2, ..., x_m$ denote the input signals applied to the source nodes of the neuron, and b denote the bias. For convenience of presentation, we would like to absorb the slope parameter a in the induced local field v by writing

$$\varphi(v) = \frac{1}{1 + \exp(-v)}$$

How would you modify the inputs $x_1, x_2, ..., x_m$ to produce the same output as before? Justify your answer.

1.9 A neuron j receives inputs from four other neurons whose activity levels are $10, -20, 4,$ and -2. The respective synaptic weights of neuron j are $0.8, 0.2, -1.0,$ and -0.9. Calculate the output of neuron j for the following two situations:

(a) The neuron is linear.

(b) The neuron is represented by a McCulloch–Pitts model.
Assume that the bias applied to the neuron is zero.

1.10 Repeat Problem 1.9 for a neuron model based on the logistic function

$$\varphi(v) = \frac{1}{1 + \exp(-v)}$$

1.11 (a) Show that the McCulloch–Pitts formal model of a neuron may be approximated by a sigmoidal neuron (i.e., neuron using a sigmoid activation function with large synaptic weights).

(b) Show that a linear neuron may be approximated by a sigmoidal neuron with small synaptic weights.

Network architectures

1.12 A fully connected feedforward network has 10 source nodes, 2 hidden layers, one with 4 neurons and the other with 3 neurons, and a single output neuron. Construct an architectural graph of this network.

1.13 (a) Figure P1.13 shows the signal-flow graph of a 2-2-2-1 feedforward network. The function $\varphi(\cdot)$ denotes a logistic function. Write the input–output mapping defined by this network.

(b) Suppose that the output neuron in the signal-flow graph of Fig. P1.13 operates in its linear region. Write the input–output mapping defined by this new network.

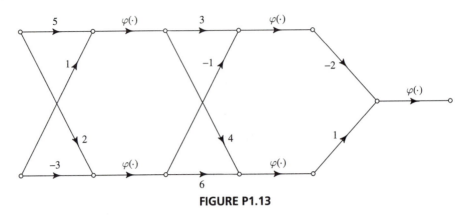

FIGURE P1.13

1.14 The network described in Fig. P1.13 has no biases. Suppose that biases equal to -1 and $+1$ are applied to the top and bottom neurons of the first hidden layer, and biases equal to $+1$ and -2 are applied to the top and bottom neurons of the second hidden layer. Write the new form of the input–output mapping defined by the network.

1.15 Consider a multilayer feedforward network, all the neurons of which operate in their linear regions. Justify the statement that such a network is equivalent to a single-layer feedforward network.

1.16 Construct a fully recurrent network with 5 neurons, but with no self-feedback.

1.17 Figure P1.17 shows the signal-flow graph of a recurrent network made up of two neurons. Write the nonlinear difference equation that defines the evolution of $x_1(n)$ or that of $x_2(n)$. These two variables define the outputs of the top and bottom neurons, respectively. What is the order of this equation?

1.18 Figure P1.18 shows the signal-flow graph of a recurrent network consisting of two neurons with self-feedback. Write the coupled system of two first-order nonlinear difference equations that describe the operation of the system.

1.19 A recurrent network has 3 source nodes, 2 hidden neurons, and 4 output neurons. Construct an architectural graph that describes such a network.

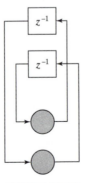

FIGURE P1.17

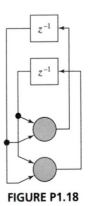

FIGURE P1.18

Knowledge representation

1.20 A useful form of preprocessing is based on the *autoregressive (AR) model* described by the difference equation (for real-valued data)

$$y(n) = w_1 y(n-1) + w_2 y(n-2) + \cdots + w_M y(n-M) + v(n)$$

where $y(n)$ is the model output; $v(n)$ is a sample drawn from a white-noise process of zero mean and some prescribed variance; $w_1, w_2, ..., w_M$ are the *AR* model coefficients; and M is the model order. Show that the use of this model provides two forms of geometric invariance: (a) scale, and (b) time translation. How could these two invariances be used in neural networks?

1.21 Let $\mathbf{x}$ be an input vector, and $\mathbf{s}(\alpha, \mathbf{x})$ be a transformation operator acting on $\mathbf{x}$ and depending on some parameter α. The operator $\mathbf{s}(\alpha, \mathbf{x})$ satisfies two requirements:

- $\mathbf{s}(0, \mathbf{x}) = \mathbf{x}$
- $\mathbf{s}(\alpha, \mathbf{x})$ is differentiable with respect to α.

The *tangent vector* is defined by the partial derivative $\partial \mathbf{s}(\alpha, \mathbf{x})/\partial \alpha$ (Simard et al., 1992).

Suppose that $\mathbf{x}$ represents an image, and α is a rotation parameter. How would you compute the tangent vector for the case when α is small? The tangent vector is locally invariant with respect to rotation of the original image; why?

CHAPTER 2

Learning Processes

2.1 INTRODUCTION

The property that is of primary significance for a neural network is the ability of the network to *learn* from its environment, and to *improve* its performance through learning. The improvement in performance takes place over time in accordance with some prescribed measure. A neural network learns about its environment through an interactive process of adjustments applied to its synaptic weights and bias levels. Ideally, the network becomes more knowledgeable about its environment after each iteration of the learning process.

There are too many activities associated with the notion of "learning" to justify defining it in a precise manner. Moreover, the process of learning is a matter of viewpoint, which makes it all the more difficult to agree on a precise definition of the term. For example, learning as viewed by a psychologist is quite different from learning in a classroom sense. Recognizing that our particular interest is in neural networks, we use a definition of learning that is adapted from Mendel and McClaren (1970).

We define learning in the context of neural networks as:

Learning is a process by which the free parameters of a neural network are adapted through a process of stimulation by the environment in which the network is embedded. The type of learning is determined by the manner in which the parameter changes take place.

This definition of the learning process implies the following sequence of events:

1. The neural network is *stimulated* by an environment.
2. The neural network *undergoes changes* in its free parameters as a result of this stimulation.
3. The neural network *responds in a new way* to the environment because of the changes that have occurred in its internal structure.

A prescribed set of well-defined rules for the solution of a learning problem is called a *learning algorithm*.[1] As one would expect, there is no unique learning algorithm for the design of neural networks. Rather, we have a "kit of tools" represented by a diverse variety of learning algorithms, each of which offers advantages of its own. Basically, learning algorithms differ from each other in the way in which the adjustment to a synaptic weight of a neuron is formulated. Another factor to be considered is the manner in which a neural network (learning machine), made up of a set of interconnected neurons, relates to its environment. In this latter context we speak of a *learning paradigm* that refers to a *model* of the environment in which the neural network operates.

Organization of the Chapter

The chapter is organized in four interrelated parts. In the first part, consisting of Sections 2.2 through 2.6, we discuss five basic learning rules: error-correction learning, memory-based learning, Hebbian learning, competitive learning, and Boltzmann learning. Error-correction learning is rooted in optimum filtering. Memory-based learning operates by memorizing the training data explicitly. Hebbian learning and competitive learning are both inspired by neurobiological considerations. Boltzmann learning is different because it is based on ideas borrowed from statistical mechanics.

The second part of the chapter explores learning paradigms. Section 2.7 discusses the credit-assignment problem, which is basic to the learning process. Sections 2.8 and 2.9 present overviews of the two fundamental learning paradigms: (1) learning *with* a teacher, and (2) learning *without* a teacher.

The third part of the chapter, consisting of Sections 2.10 through 2.12, examines the issues of learning tasks, memory, and adaptation.

The final part of the chapter, consisting of Sections 2.13 through 2.15, deals with probabilistic and statistical aspects of the learning process. Section 2.13 discusses the bias/variance dilemma. Section 2.14 discusses statistical learning theory, based on the notion of VC-dimension that provides a measure of machine capacity. Section 2.14 introduces another important concept: probably approximately correct (PAC) learning, which provides a conservative model for the learning process.

The chapter concludes with some final remarks in Section 2.16.

2.2   ERROR-CORRECTION LEARNING

To illustrate our first learning rule, consider the simple case of a neuron k constituting the only computational node in the output layer of a feedforward neural network, as depicted in Fig. 2.1a. Neuron k is driven by a *signal vector* $\mathbf{x}(n)$ produced by one or more layers of hidden neurons, which are themselves driven by an input vector (stimulus) applied to the source nodes (i.e., input layer) of the neural network. The argument n denotes discrete time, or more precisely, the time step of an iterative process involved in adjusting the synaptic weights of neuron k. The *output signal* of neuron k is denoted by $y_k(n)$. This output signal, representing the only output of the neural network, is compared to a *desired response* or *target output,* denoted by $d_k(n)$. Consequently, an *error signal,* denoted by $e_k(n)$, is produced. By definition, we thus have

$$e_k(n) = d_k(n) - y_k(n) \qquad (2.1)$$

The error signal $e_k(n)$ actuates a *control mechanism,* the purpose of which is to apply a sequence of corrective adjustments to the synaptic weights of neuron k. The corrective adjustments are designed to make the output signal $y_k(n)$ come closer to the desired

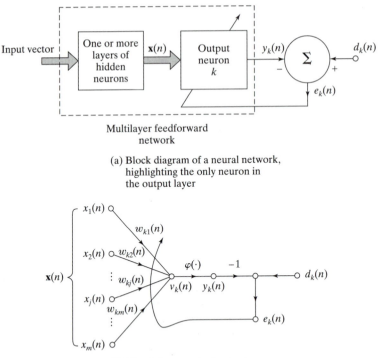

Multilayer feedforward
network

(a) Block diagram of a neural network,
highlighting the only neuron in
the output layer

(b) Signal-flow graph of output neuron

FIGURE 2.1   Illustrating error-correction learning.

response $d_k(n)$ in a step-by-step manner. This objective is achieved by minimizing a *cost function* or *index of performance*, $\mathscr{E}(n)$, defined in terms of the error signal $e_k(n)$ as:

$$\mathscr{E}(n) = \frac{1}{2}e_k^2(n) \qquad (2.2)$$

That is, $\mathscr{E}(n)$ is the *instantaneous value of the error energy*. The step-by-step adjustments to the synaptic weights of neuron k are continued until the system reaches a *steady state* (i.e., the synaptic weights are essentially stabilized). At that point the learning process is terminated.

The learning process described herein is obviously referred to as *error-correction learning*. In particular, minimization of the cost function $\mathscr{E}(n)$ leads to a learning rule commonly referred to as the *delta rule* or *Widrow–Hoff rule,* named in honor of its originators (Widrow and Hoff, 1960). Let $w_{kj}(n)$ denote the value of synaptic weight w_{kj} of neuron k excited by element $x_j(n)$ of the signal vector $\mathbf{x}(n)$ at time step n. According to the delta rule, the adjustment $\Delta w_{kj}(n)$ applied to the synaptic weight w_{kj} at time step n is defined by

$$\Delta w_{kj}(n) = \eta e_k(n)x_j(n) \qquad (2.3)$$

where η is a positive constant that determines the *rate of learning* as we proceed from one step in the learning process to another. It is therefore natural that we refer to η as the *learning-rate parameter.* In other words, the delta rule may be stated as:

The adjustment made to a synaptic weight of a neuron is proportional to the product of the error signal and the input signal of the synapse in question.

Keep in mind that the delta rule, as stated herein, presumes that the error signal is *directly measurable*. For this measurement to be feasible we clearly need a supply of desired response from some external source, which is directly accessible to neuron k. In other words, neuron k is *visible* to the outside world, as depicted in Fig. 2.1a. From this figure we also observe that error-correction learning is in fact *local* in nature. This is merely saying that the synaptic adjustments made by the delta rule are localized around neuron k.

Having computed the synaptic adjustment $\Delta w_{kj}(n)$, the updated value of synaptic weight w_{kj} is determined by

$$w_{kj}(n + 1) = w_{kj}(n) + \Delta w_{kj}(n) \qquad (2.4)$$

In effect, $w_{kj}(n)$ and $w_{kj}(n + 1)$ may be viewed as the *old* and *new* values of synaptic weight w_{kj}, respectively. In computational terms we may also write

$$w_{kj}(n) = z^{-1}[w_{kj}(n + 1)] \qquad (2.5)$$

where z^{-1} is the *unit-delay operator*. That is, z^{-1} represents a *storage element*.

Figure 2.1b shows a signal-flow graph representation of the error-correction learning process, focusing on the activity surrounding neuron k. The input signal x_j and induced local field v_k of neuron k are referred to as the *presynaptic* and *postsynaptic signals* of the jth synapse of neuron k, respectively. From Fig. 2.1b we see that error-correction learning is an example of a *closed-loop feedback system*. From control theory we know that the stability of such a system is determined by those parameters that constitute the feedback loops of the system. In our case we only have a single feedback loop, and one of those parameters of particular interest is the learning-rate parameter η. It is therefore important that η is carefully selected to ensure that the stability or convergence of the iterative learning process is achieved. The choice of η also has a profound influence on the accuracy and other aspects of the learning process. In short, the learning-rate parameter η plays a key role in determining the performance of error-correction learning in practice.

Error-correction learning is discussed in much greater detail in Chapter 3, which discusses single-layer feedforward networks and in Chapter 4, which details multilayer feedforward networks.

2.3 MEMORY-BASED LEARNING

In *memory-based learning,* all (or most) of the past experiences are explicitly stored in a large memory of correctly classified input–output examples: $\{(\mathbf{x}_i, d_i)\}_{i=1}^{N}$, where $\mathbf{x}_i$ denotes an input vector and d_i denotes the corresponding desired response. Without loss of generality, we have restricted the desired response to be a scalar. For example, in a binary pattern classification problem there are two classes/hypotheses, denoted by $\mathscr{C}_1$ and $\mathscr{C}_2$, to be considered. In this example, the desired response d_i takes the value 0 (or -1) for class $\mathscr{C}_1$ and the value 1 for class $\mathscr{C}_2$. When classification of a test vector $\mathbf{x}_{\text{test}}$ (not seen before) is required, the algorithm responds by retrieving and analyzing the training data in a "local neighborhood" of $\mathbf{x}_{\text{test}}$.

All memory-based learning algorithms involve two essential ingredients:

- Criterion used for defining the local neighborhood of the test vector $\mathbf{x}_{\text{test}}$.
- Learning rule applied to the training examples in the local neighborhood of $\mathbf{x}_{\text{test}}$.

The algorithms differ from each other in the way in which these two ingredients are defined.

In a simple yet effective type of memory-based learning known as the *nearest neighbor rule*,[2] the local neighborhood is defined as the training example that lies in the immediate neighborhood of the test vector $\mathbf{x}_{\text{test}}$. In particular, the vector

$$\mathbf{x}_N' \in \{\mathbf{x}_1, \mathbf{x}_2, \ldots, \mathbf{x}_N\} \tag{2.6}$$

is said to be the nearest neighbor of $\mathbf{x}_{\text{test}}$ if

$$\min_i d(\mathbf{x}_i, \mathbf{x}_{\text{test}}) = d(\mathbf{x}_N', \mathbf{x}_{\text{test}}) \tag{2.7}$$

where $d(\mathbf{x}_i, \mathbf{x}_{\text{test}})$ is the Euclidean distance between the vectors $\mathbf{x}_i$ and $\mathbf{x}_{\text{test}}$. The class associated with the minimum distance, that is, vector $\mathbf{x}_N'$, is reported as the classification of $\mathbf{x}_{\text{test}}$. This rule is independent of the underlying distribution responsible for generating the training examples.

Cover and Hart (1967) have formally studied the nearest neighbor rule as a tool for pattern classification. The analysis presented therein is based on two assumptions:

- The classified examples $(\mathbf{x}_i, d_i)$ are *independently and identically distributed (iid)*, according to the joint probability distribution of the example $(\mathbf{x}, d)$.
- The sample size N is infinitely large.

Under these two assumptions, it is shown that the probability of classification error incurred by the nearest neighbor rule is bounded above by twice the *Bayes probability of error*, that is, the minimum probability of error over all decision rules. Bayes probability of error is discussed in Chapter 3. In this sense, it may be said that half the classification information in a training set of infinite size is contained in the nearest neighbor, which is a surprising result.

A variant of the nearest neighbor classifier is the *k-nearest neighbor classifier*, which proceeds as follows:

- Identify the k classified patterns that lie nearest to the test vector $\mathbf{x}_{\text{test}}$ for some integer k.
- Assign $\mathbf{x}_{\text{test}}$ to the class (hypothesis) that is most frequently represented in the k nearest neighbors to $\mathbf{x}_{\text{test}}$ (i.e., use a majority vote to make the classification).

Thus the k-nearest neighbor classifier acts like an averaging device. In particular, it discriminates against a single outlier, as illustrated in Fig. 2.2 for $k = 3$. An *outlier* is an observation that is improbably large for a nominal model of interest.

In Chapter 5 we discuss another important type of memory-based classifier known as the radial-basis function network.

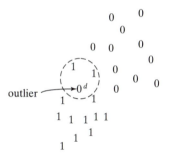

FIGURE 2.2   The area lying inside the dashed circle includes two points pertaining to class 1 and an outlier from class 0. The point d corresponds to the test vector $\mathbf{x}_{\text{test}}$. With $k = 3$, the *k-nearest* neighbor *classifier* assigns class 1 to point d even though it lies closest to the *outlier*.

2.4 HEBBIAN LEARNING

Hebb's postulate of learning is the oldest and most famous of all learning rules; it is named in honor of the neuropsychologist Hebb (1949). Quoting from Hebb's book, *The Organization of Behavior* (1949, p.62):

> When an axon of cell A is near enough to excite a cell B and repeatedly or persistently takes part in firing it, some growth process or metabolic changes take place in one or both cells such that A's efficiency as one of the cells firing B, is increased.

Hebb proposed this change as a basis of associative learning (at the cellular level), which would result in an enduring modification in the activity pattern of a spatially distributed "assembly of nerve cells."

This statement is made in a neurobiological context. We may expand and rephrase it as a two-part rule (Stent, 1973; Changeux and Danchin, 1976):

1. *If two neurons on either side of a synapse (connection) are activated simultaneously (i.e., synchronously), then the strength of that synapse is selectively increased.*
2. *If two neurons on either side of a synapse are activated asynchronously, then that synapse is selectively weakened or eliminated.*

Such a synapse is called a *Hebbian synapse*.[3] (The original Hebb rule did not contain part 2.) More precisely, we define a Hebbian synapse as a synapse that uses a *time-dependent, highly local, and strongly interactive mechanism to increase synaptic efficiency as a function of the correlation between the presynaptic and postsynaptic activities.* From this definition we may deduce the following four key mechanisms (properties) that characterize a Hebbian synapse (Brown et al., 1990):

1. *Time-dependent mechanism.* This mechanism refers to the fact that the modifications in a Hebbian synapse depend on the exact time of occurrence of the presynaptic and postsynaptic signals.
2. *Local mechanism.* By its very nature, a synapse is the transmission site where information-bearing signals (representing ongoing activity in the presynaptic and postsynaptic units) are in *spatiotemporal* contiguity. This locally available information is used by a Hebbian synapse to produce a local synaptic modification that is input specific.

3. *Interactive mechanism.* The occurrence of a change in a Hebbian synapse depends on signals on both sides of the synapse. That is, a Hebbian form of learning depends on a "true interaction" between presynaptic and postsynaptic signals in the sense that we cannot make a prediction from either one of these two activities by itself. Note also that this dependence or interaction may be deterministic or statistical in nature.

4. *Conjunctional* or *correlational mechanism.* One interpretation of Hebb's postulate of learning is that the condition for a change in synaptic efficiency is the conjunction of presynaptic and postsynaptic signals. Thus, according to this interpretation, the co-occurrence of presynaptic and postsynaptic signals (within a short interval of time) is sufficient to produce the synaptic modification. It is for this reason that a Hebbian synapse is sometimes referred to as a *conjunctional synapse.* For another interpretation of Hebb's postulate of learning, we may think of the interactive mechanism characterizing a Hebbian synapse in statistical terms. In particular, the correlation over time between presynaptic and postsynaptic signals is viewed as being responsible for a synaptic change. Accordingly, a Hebbian synapse is also referred to as a *correlational synapse.* Correlation is indeed the basis of learning (Eggermont, 1990).

Synaptic Enhancement and Depression

The definition of a Hebbian synapse presented here does not include additional processes that may result in weakening of a synapse connecting a pair of neurons. Indeed, we may generalize the concept of a Hebbian modification by recognizing that positively correlated activity produces synaptic strengthening, and that either uncorrelated or negatively correlated activity produces synaptic weakening (Stent, 1973). Synaptic depression may also be of a noninteractive type. Specifically, the interactive condition for synaptic weakening may simply be noncoincident presynaptic or postsynaptic activity.

We may go one step further by classifying synaptic modifications as *Hebbian, anti-Hebbian, and non-Hebbian* (Palm, 1982). According to this scheme, a Hebbian synapse increases its strength with positively correlated presynaptic and postsynaptic signals, and decreases its strength when these signals are either uncorrelated or negatively correlated. Conversely, an anti-Hebbian synapse weakens positively correlated presynaptic and postsynaptic signals, and strengthens negatively correlated signals. In both Hebbian and anti-Hebbian synapses, however, the modification of synaptic efficiency relies on a mechanism that is time-dependent, highly local, and strongly interactive in nature. In that sense, an anti-Hebbian synapse is still Hebbian in nature, though not in function. A non-Hebbian synapse, on the other hand, does not involve a Hebbian mechanism of either kind.

Mathematical Models of Hebbian Modifications

To formulate Hebbian learning in mathematical terms, consider a synaptic weight w_{kj} of neuron k with presynaptic and postsynaptic signals denoted by x_j and y_k, respectively. The adjustment applied to the synaptic weight w_{kj} at time step n is expressed in the general form

$$\Delta w_{kj}(n) = F(y_k(n), x_j(n)) \tag{2.8}$$

where $F(\cdot,\cdot)$ is a function of both postsynaptic and presynaptic signals. The signals $x_j(n)$ and $y_k(n)$ are often treated as dimensionless. The formula of Eq. (2.8) admits many forms, all of which qualify as Hebbian. In what follows, we consider two such forms.

Hebb's hypothesis.   The simplest form of Hebbian learning is described by

$$\Delta w_{kj}(n) = \eta y_k(n)x_j(n) \tag{2.9}$$

where η is a positive constant that determines the *rate of learning*. Equation (2.9) clearly emphasizes the correlational nature of a Hebbian synapse. It is sometimes referred to as the *activity product rule*. The top curve of Fig. 2.3 shows a graphical representation of Eq. (2.9) with the change Δw_{kj} plotted versus the output signal (postsynaptic activity) y_k. From this representation we see that the repeated application of the input signal (presynaptic activity) x_j leads to an increase in y_k and therefore *exponential growth* that finally drives the synaptic connection into saturation. At that point no information will be stored in the synapse and selectivity is lost.

Covariance hypothesis.   One way of overcoming the limitation of Hebb's hypothesis is to use the *covariance hypothesis* introduced in Sejnowski (1977a, b). In this hypothesis, the presynaptic and postsynaptic signals in Eq. (2.9) are replaced by the departure of presynaptic and postsynaptic signals from their respective average values over a certain time interval. Let $\bar{x}$ and $\bar{y}$ denote the *time-averaged values* of the presynaptic signal x_j and postsynaptic signal y_k, respectively. According to the covariance hypothesis, the adjustment applied to the synaptic weight w_{kj} is defined by

$$\Delta w_{kj} = \eta(x_j - \bar{x})(y_k - \bar{y}) \tag{2.10}$$

where η is the learning-rate parameter. The average values x and y constitute presynaptic and postsynaptic thresholds, which determine the sign of synaptic modification. In particular, the covariance hypothesis allows for the following:

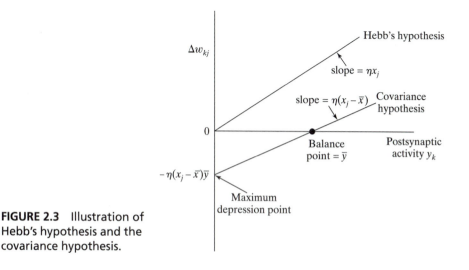

FIGURE 2.3   Illustration of Hebb's hypothesis and the covariance hypothesis.

- Convergence to a nontrivial state, which is reached when $x_k = \bar{x}$ or $y_j = \bar{y}$.
- Prediction of both synaptic *potentiation* (i.e., increase in synaptic strength) and synaptic *depression* (i.e., decrease in synaptic strength).

Figure 2.3 illustrates the difference between Hebb's hypothesis and the covariance hypothesis. In both cases the dependence of Δw_{kj} on y_k is linear; however, the intercept with the y_k-axis in Hebb's hypothesis is at the origin, whereas in the covariance hypothesis it is at $y_k = \bar{y}$.

We make the following important observations from Eq. (2.10):

1. Synaptic weight w_{kj} is enhanced if there are sufficient levels of presynaptic and postsynaptic activities, that is, the conditions $x_j > \bar{x}$ and $y_k > \bar{y}$ are both satisfied.
2. Synaptic weight w_{kj} is depressed if there is either
   - a presynaptic activation (i.e., $x_j > \bar{x}$) in the absence of sufficient postsynaptic activation (i.e., $y_k < \bar{y}$), or
   - a postsynaptic activation (i.e., $y_k > \bar{y}$) in the absence of sufficient presynaptic activation (i.e., $x_j < \bar{x}$).

This behavior may be regarded as a form of temporal competition between the incoming patterns.

There is strong physiological evidence[4] for Hebbian learning in the area of the brain called the *hippocampus*. The hippocampus plays an important role in certain aspects of learning or memory. This physiological evidence makes Hebbian learning all the more appealing.

2.5 COMPETITIVE LEARNING

In *competitive learning*,[5] as the name implies, the output neurons of a neural network compete among themselves to become active (fired). Whereas in a neural network based on Hebbian learning several output neurons may be active simultaneously, in competitive learning only a single output neuron is active at any one time. It is this feature that makes competitive learning highly suited to discover statistically salient features that may be used to classify a set of input patterns.

There are three basic elements to a competitive learning rule (Rumelhart and Zipser, 1985):

- A set of neurons that are all the same except for some randomly distributed synaptic weights, and which therefore *respond differently* to a given set of input patterns.
- A *limit* imposed on the "strength" of each neuron.
- A mechanism that permits the neurons to *compete* for the right to respond to a given subset of inputs, such that only *one* output neuron, or only one neuron per group, is active (i.e., "on") at a time. The neuron that wins the competition is called a *winner-takes-all neuron*.

Accordingly the individual neurons of the network learn to specialize on ensembles of similar patterns; in so doing they become *feature detectors* for different classes of input patterns.

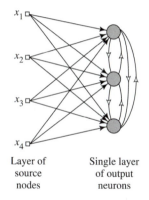

FIGURE 2.4 Architectural graph of a simple competitive learning network with feedforward (excitatory) connections from the source nodes to the neurons, and lateral (inhibitory) connections among the neurons; the lateral connections are signified by open arrows.

Layer of source nodes

Single layer of output neurons

In the simplest form of competitive learning, the neural network has a single layer of output neurons, each of which is fully connected to the input nodes. The network may include feedback connections among the neurons, as indicated in Fig. 2.4. In the network architecture described herein, the feedback connections perform *lateral inhibition*,[6] with each neuron tending to inhibit the neuron to which it is laterally connected. In contrast, the feedforward synaptic connections in the network of Fig. 2.4 are all *excitatory*.

For a neuron k to be the winning neuron, its induced local field v_k for a specified input pattern $\mathbf{x}$ must be the largest among all the neurons in the network. The output signal y_k of winning neuron k is set equal to one; the output signals of all the neurons that lose the competition are set equal to zero. We thus write

$$y_k = \begin{cases} 1 & \text{if } v_k > v_j \text{ for all } j, j \neq k \\ 0 & \text{otherwise} \end{cases} \tag{2.11}$$

where the induced local field v_k represents the combined action of all the forward and feedback inputs to neuron k.

Let w_{kj} denote the synaptic weight connecting input node j to neuron k. Suppose that each neuron is allotted a *fixed* amount of synaptic weight (i.e., all synaptic weights are positive), which is distributed among its input nodes; that is,

$$\sum_j w_{kj} = 1 \qquad \text{for all } k \tag{2.12}$$

A neuron then learns by shifting synaptic weights from its inactive to active input nodes. If a neuron does not respond to a particular input pattern, no learning takes place in that neuron. If a particular neuron wins the competition, each input node of that neuron relinquishes some proportion of its synaptic weight, and the weight relinquished is then distributed equally among the active input nodes. According to the standard *competitive learning rule*, the change Δw_{kj} applied to synaptic weight w_{kj} is defined by

$$\Delta w_{kj} = \begin{cases} \eta(x_j - w_{kj}) & \text{if neuron } k \text{ wins the competition} \\ 0 & \text{if neuron } k \text{ loses the competition} \end{cases} \tag{2.13}$$

where η is the learning-rate parameter. This rule has the overall effect of moving the synaptic weight vector $\mathbf{w}_k$ of winning neuron k toward the input pattern $\mathbf{x}$.

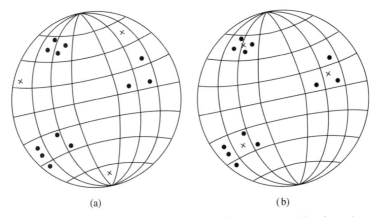

FIGURE 2.5 Geometric interpretation of the competitive learning process. The dots represent the input vectors, and the crosses represent the synaptic weight vectors of three output neurons. (a) Initial state of the network. (b) Final state of the network.

We may use the geometric analogy depicted in Fig. 2.5 to illustrate the essence of competitive learning (Rumelhart and Zipser, 1985). It is assumed that each input pattern (vector) **x** has some constant Euclidean length so that we may view it as a point on an N-dimensional unit sphere where N is the number of input nodes. N also represents the dimension of each synaptic weight vector $\mathbf{w}_k$. It is further assumed that all neurons in the network are constrained to have the same Euclidean length (norm), as shown by

$$\sum_j w_{kj}^2 = 1 \qquad \text{for all } k \qquad (2.14)$$

When the synaptic weights are properly scaled they form a set of vectors that fall on the same N-dimensional unit sphere. In Fig. 2.5a we show three natural groupings (clusters) of the stimulus patterns represented by dots. This figure also includes a possible initial state of the network (represented by crosses) that may exist before learning. Figure 2.5b shows a typical final state of the network that results from the use of competitive learning. In particular, each output neuron has discovered a cluster of input patterns by moving its synaptic weight vector to the center of gravity of the discovered cluster (Rumelhart and Zipser, 1985; Hertz et al., 1991). This figure illustrates the ability of a neural network to perform *clustering* through competitive learning. However, for this function to be performed in a "stable" fashion the input patterns must fall into sufficiently distinct groupings to begin with. Otherwise the network may be unstable because it will no longer respond to a given input pattern with the same output neuron.

2.6 BOLTZMANN LEARNING

The Boltzmann learning rule, named in honor of Ludwig Boltzmann, is a stochastic learning algorithm derived from ideas rooted in statistical mechanics.[7] A neural network designed on the basis of the Boltzmann learning rule is called a *Boltzmann machine* (Ackley et al., 1985; Hinton and Sejnowski, 1986).

In a Boltzmann machine the neurons constitute a recurrent structure, and they operate in a binary manner since, for example, they are either in an "on" state denoted by $+1$ or in an "off" state denoted by -1. The machine is characterized by an *energy function, E,* the value of which is determined by the particular states occupied by the individual neurons of the machine, as shown by

$$E = -\frac{1}{2} \sum_{\substack{j \\ j \neq k}} \sum_k w_{kj} x_k x_j \tag{2.15}$$

where x_j is the state of neuron j, and w_{kj} is the synaptic weight connecting neuron j to neuron k. The fact that $j \neq k$ means simply that none of the neurons in the machine has self-feedback. The machine operates by choosing a neuron at random—for example, neuron k—at some step of the learning process, then flipping the state of neuron k from state x_k to state $-x_k$ at some temperature T with probability

$$P(x_k \rightarrow -x_k) = \frac{1}{1 + \exp(-\Delta E_k / T)} \tag{2.16}$$

where ΔE_k is the *energy change* (i.e., the change in the energy function of the machine) resulting from such a flip. Notice that T is not a physical temperature, but rather a *pseudotemperature,* as explained in Chapter 1. If this rule is applied repeatedly, the machine will reach *thermal equilibrium.*

The neurons of a Boltzmann machine partition into two functional groups: *visible* and *hidden*. The visible neurons provide an interface between the network and the environment in which it operates, whereas the hidden neurons always operate freely. There are two modes of operation to be considered:

- *Clamped condition,* in which the visible neurons are all clamped onto specific states determined by the environment.
- *Free-running condition,* in which all the neurons (visible and hidden) are allowed to operate freely.

Let ρ_{kj}^+ denote the *correlation* between the states of neurons j and k, with the network in its clamped condition. Let ρ_{kj}^- denote the *correlation* between the states of neurons j and k with the network in its free-running condition. Both correlations are averaged over all possible states of the machine when it is in thermal equilibrium. Then, according to the *Boltzmann learning rule,* the change Δw_{kj} applied to the synaptic weight w_{kj} from neuron j to neuron k is defined by (Hinton and Sejnowski, 1986)

$$\Delta w_{kj} = \eta(\rho_{kj}^+ - \rho_{kj}^-), \qquad j \neq k \tag{2.17}$$

where η is a learning-rate parameter. Note that both ρ_{kj}^+ and ρ_{kj}^- range in value from -1 to $+1$.

A brief review of statistical mechanics is presented in Chapter 11; in that chapter we also present a detailed treatment of the Boltzmann machine and other stochastic machines.

2.7   CREDIT-ASSIGNMENT PROBLEM

When studying learning algorithms for distributed systems, it is useful to consider the notion of *credit assignment* (Minsky, 1961). Basically, the credit-assignment problem is the problem of assigning *credit* or *blame* for overall outcomes to each of the internal decisions made by a learning machine and which contributed to those outcomes. (The credit assignment problem is also referred to as the *loading problem,* the problem of "loading" a given set of training data into the free parameters of the network.)

In many cases the dependence of outcomes on internal decisions is mediated by a sequence of actions taken by the learning machine. In other words, internal decisions affect which particular actions are taken, and then the actions, not the internal decisions, directly influence overall outcomes. In these situations, we may decompose the credit-assignment problem into two subproblems (Sutton, 1984):

1. The assignment of credit for outcomes to actions. This is called the *temporal credit-assignment problem* in that it involves the instants of time *when* the actions that deserve credit were actually taken.
2. The assignment of credit for actions to internal decisions. This is called the *structural credit-assignment problem* in that it involves assigning credit to the *internal structures* of actions generated by the system.

The structural credit-assignment problem is relevant in the context of a multicomponent learning machine when we must determine precisely which particular component of the system should have its behavior altered and by how much in order to improve overall system performance. On the other hand, the temporal credit-assignment problem is relevant when there are many actions taken by a learning machine that result in certain outcomes, and we must determine which of these actions were responsible for the outcomes. The combined temporal and structural credit-assignment problem faces any distributed learning machine that attempts to improve its performance in situations involving temporally extended behavior (Williams, 1988).

The credit-assignment problem, for example, arises when error-correction learning is applied to a multilayer feedforward neural network. The operation of each hidden neuron, as well as that of each output neuron in such a network is important to its correct overall operation on a learning task of interest. That is, in order to solve the prescribed task the network must assign certain forms of behavior to all of its neurons through the specification of error-correction learning. With this background in mind, consider the situation described in Fig. 2.1a. Since the output neuron k is visible to the outside world, it is possible to supply a desired response to this neuron. As far as the output neuron is concerned, it is a straightforward matter to adjust the synaptic weights of the output neuron in accordance with error-correction learning, as outlined in Section 2.2. But how do we assign credit or blame for the action of the hidden neurons when the error-correction learning process is used to adjust the respective synaptic weights of these neurons? The answer to this fundamental question requires more detailed attention; it is presented in Chapter 4, where algorithmic details of the design of multilayer feedforward neural networks are described.

2.8 LEARNING WITH A TEACHER

We now turn our attention to learning paradigms. We begin by considering *learning with a teacher,* which is also referred to as *supervised learning.* Figure 2.6 shows a block diagram that illustrates this form of learning. In conceptual terms, we may think of the teacher as having knowledge of the environment, with that knowledge being represented by a set of *input–output examples.* The environment is, however, *unknown* to the neural network of interest. Suppose now that the teacher and the neural network are both exposed to a training vector (i.e., example) drawn from the environment. By virtue of built-in knowledge, the teacher is able to provide the neural network with a desired response for that training vector. Indeed, the desired response represents the optimum action to be performed by the neural network. The network parameters are adjusted under the combined influence of the training vector and the error signal. The *error signal* is defined as the difference between the desired response and the actual response of the network. This adjustment is carried out iteratively in a step-by-step fashion with the aim of eventually making the neural network *emulate* the teacher; the emulation is presumed to be optimum in some statistical sense. In this way knowledge of the environment available to the teacher is transferred to the neural network through training as fully as possible. When this condition is reached, we may then dispense with the teacher and let the neural network deal with the environment completely by itself.

The form of supervised learning we have just described is the error-correction learning discussed previously in Section 2.2. It is a closed-loop feedback system, but the unknown environment is not in the loop. As a performance measure for the system we may think in terms of the mean-square error or the sum of squared errors over the training sample, defined as a function of the free parameters of the system. This function may be visualized as a multidimensional *error-performance surface* or simply *error surface,* with the free parameters as coordinates. The true error surface is *averaged* over all possible input–output examples. Any given operation of the system under the

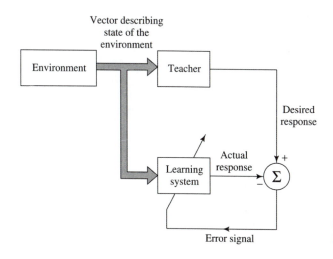

FIGURE 2.6 Block diagram of learning with a teacher.

teacher's supervision is represented as a point on the error surface. For the system to improve performance over time and therefore learn from the teacher, the operating point has to move down successively toward a minimum point of the error surface; the minimum point may be a local minimum or a global minimum. A supervised learning system is able to do this with the useful information it has about the *gradient* of the error surface corresponding to the current behavior of the system. The gradient of an error surface at any point is a vector that points in the direction of *steepest descent*. In fact, in the case of supervised learning from examples, the system may use an *instantaneous estimate* of the gradient vector, with the example indices presumed to be those of time. The use of such an estimate results in a motion of the operating point on the error surface that is typically in the form of a "random walk." Nevertheless, given an algorithm designed to minimize the cost function, an adequate set of input–output examples, and enough time permitted to do the training, a supervised learning system is usually able to perform such tasks as pattern classification and function approximation.

2.9 LEARNING WITHOUT A TEACHER

In supervised learning, the learning process takes place under the tutelage of a teacher. However, in the paradigm known as *learning without a teacher,* as the name implies, there is *no* teacher to oversee the learning process. That is to say, there are no labeled examples of the function to be learned by the network. Under this second paradigm, two subdivisions are identified:

1. Reinforcement learning/Neurodynamic programming

In *reinforcement learning,*[8] the learning of an input–output mapping is performed through continued interaction with the environment in order to minimize a scalar index of performance. Figure 2.7 shows the block diagram of one form of a reinforcement learning system built around a *critic* that converts a *primary reinforcement signal* received from the environment into a higher quality reinforcement signal called the *heuristic reinforcement signal,* both of which are scalar inputs (Barto et al., 1983). The

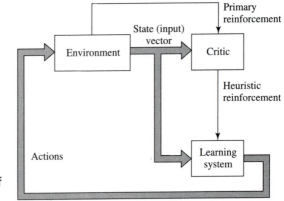

FIGURE 2.7 Block diagram of reinforcement learning.

system is designed to learn under *delayed reinforcement,* which means that the system observes a temporal sequence of stimuli (i.e., state vectors) also received from the environment, which eventually result in the generation of the heuristic reinforcement signal. The goal of learning is to minimize a *cost-to-go function,* defined as the expectation of the cumulative cost of *actions* taken over a sequence of steps instead of simply the immediate cost. It may turn out that certain actions taken earlier in that sequence of time steps are in fact the best determinants of overall system behavior. The function of the *learning machine,* which constitutes the second component of the system, is to *discover* these actions and to feed them back to the environment.

Delayed-reinforcement learning is difficult to perform for two basic reasons:

- There is no teacher to provide a desired response at each step of the learning process.
- The delay incurred in the generation of the primary reinforcement signal implies that the learning machine must solve a *temporal credit assignment problem.* By this we mean that the learning machine must be able to assign credit and blame individually to each action in the sequence of time steps that led to the final outcome, while the primary reinforcement may only evaluate the outcome.

Notwithstanding these difficulties, delayed-reinforcement learning is very appealing. It provides the basis for the system to interact with its environment, thereby developing the ability to learn to perform a prescribed task solely on the basis of the outcomes of its experience that result from the interaction.

Reinforcement learning is closely related to *dynamic programming,* which was developed by Bellman (1957) in the context of optimal control theory. Dynamic programming provides the mathematical formalism for sequential decision making. By casting reinforcement learning within the framework of dynamic programming, the subject matter becomes all the richer for it, as demonstrated in Bertsekas and Tsitsiklis (1996). An introductory treatment of dynamic programming and its relationship to reinforcement learning is presented in Chapter 12.

2. Unsupervised Learning

In *unsupervised* or *self-organized* learning there is no external teacher or critic to oversee the learning process, as indicated in Fig. 2.8. Rather, provision is made for a *task-independent measure* of the quality of representation that the network is required to learn, and the free parameters of the network are optimized with respect to that measure. Once the network has become tuned to the statistical regularities of the input data, it develops the ability to form internal representations for encoding features of the input and thereby to create new classes automatically (Becker, 1991).

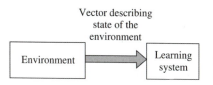

FIGURE 2.8   Block diagram of unsupervised learning.

To perform unsupervised learning we may use a competitive learning rule. For example, we may use a neural network that consists of two layers—an input layer and a competitive layer. The input layer receives the available data. The competitive layer consists of neurons that compete with each other (in accordance with a learning rule) for the "opportunity" to respond to features contained in the input data. In its simplest form, the network operates in accordance with a "winner-takes-all" strategy. As described in Section 2.5, in such a strategy the neuron with the greatest total input "wins" the competition and turns on; all the other neurons then switch off.

Different algorithms for unsupervised learning are described in Chapters 8 through 11.

2.10 LEARNING TASKS

In previous sections of this chapter we have discussed different learning algorithms and learning paradigms. In this section, we describe some basic learning tasks. The choice of a particular learning algorithm is influenced by the learning task that a neural network is required to perform. In this context we identify six learning tasks that apply to the use of neural networks in one form or another.

Pattern Association

An *associative memory* is a brainlike distributed memory that learns by *association*. Association has been known to be a prominent feature of human memory since Aristotle, and all models of cognition use association in one form or another as the basic operation (Anderson, 1995).

Association takes one of two forms: *autoassociation* or *heteroassociation*. In autoassociation, a neural network is required to *store* a set of patterns (vectors) by repeatedly presenting them to the network. The network is subsequently presented a partial description or distorted (noisy) version of an original pattern stored in it, and the task is to *retrieve* (*recall*) that particular pattern. Heteroassociation differs from autoassociation in that an arbitrary set of input patterns is *paired* with another arbitrary set of output patterns. Autoassociation involves the use of unsupervised learning, whereas the type of learning involved in heteroassociation is supervised.

Let $\mathbf{x}_k$ denote a *key pattern* (vector) applied to an associative memory and $\mathbf{y}_k$ denote a *memorized pattern* (vector). The pattern association performed by the network is described by

$$\mathbf{x}_k \rightarrow \mathbf{y}_k, \qquad k = 1, 2, ..., q \qquad (2.18)$$

where q is the number of patterns stored in the network. The key pattern $\mathbf{x}_k$ acts as a stimulus that not only determines the storage location of memorized pattern $\mathbf{y}_k$, but also holds the key for its retrieval.

In an autoassociative memory, $\mathbf{y}_k = \mathbf{x}_k$, so the input and output (data) spaces of the network have the same dimensionality. In a heteroassociative memory, $\mathbf{y}_k \neq \mathbf{x}_k$; hence, the dimensionality of the output space in this second case may or may not equal the dimensionality of the input space.

There are two phases involved in the operation of an associative memory:

FIGURE 2.9 Input–output relation of pattern associator.

- *Storage phase,* which refers to the training of the network in accordance with Eq. (2.18).
- *Recall phase,* which involves the retrieval of a memorized pattern in response to the presentation of a noisy or distorted version of a key pattern to the network.

Let the stimulus (input) $\mathbf{x}$ represent a noisy or distorted version of a key pattern $\mathbf{x}_j$. This stimulus produces a response (output) $\mathbf{y}$, as indicated in Fig. 2.9. For perfect recall, we should find that $\mathbf{y} = \mathbf{y}_j$, where $\mathbf{y}_j$ is the memorized pattern associated with the key pattern $\mathbf{x}_j$. When $\mathbf{y} \neq \mathbf{y}_j$ for $\mathbf{x} = \mathbf{x}_j$, the associative memory is said to have made an *error* in recall.

The number q of patterns stored in an associative memory provides a direct measure of the *storage capacity* of the network. In designing an associative memory, the challenge is to make the storage capacity q (expressed as a percentage of the total number N of neurons used to construct the network) as large as possible and yet insist that a large fraction of the memorized patterns is recalled correctly.

Pattern Recognition

Humans are good at pattern recognition. We receive data from the world around us via our senses and are able to recognize the source of the data. We are often able to do so almost immediately and with practically no effort. For example, we can recognize the familiar face of a person even though that person has aged since our last encounter, identify a familiar person by his or her voice on the telephone despite a bad connection, and distinguish a boiled egg that is good from a bad one by smelling it. Humans perform pattern recognition through a learning process; so it is with neural networks.

Pattern recognition is formally defined as *the process whereby a received pattern/signal is assigned to one of a prescribed number of classes (categories).* A neural network performs pattern recognition by first undergoing a training session, during which the network is repeatedly presented a set of input patterns along with the category to which each particular pattern belongs. Later, a new pattern is presented to the network that has not been seen before, but which belongs to the same population of patterns used to train the network. The network is able to identify the class of that particular pattern because of the information it has extracted from the training data. Pattern recognition performed by a neural network is statistical in nature, with the patterns being represented by points in a multidimensional *decision space*. The decision space is divided into regions, each one of which is associated with a class. The decision boundaries are determined by the training process. The construction of these boundaries is made statistical by the inherent variability that exists within and between classes.

In generic terms, pattern-recognition machines using neural networks may take one of two forms:

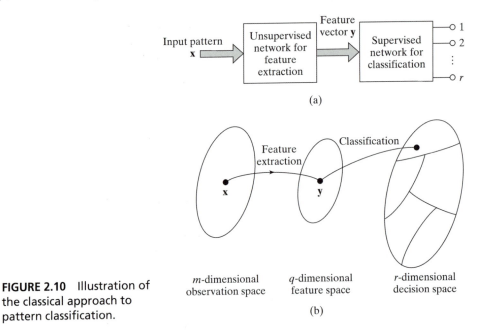

FIGURE 2.10 Illustration of the classical approach to pattern classification.

- The machine is split into two parts, an unsupervised network for *feature extraction* and a supervised network for *classification,* as shown in Fig. 2.10a. Such a method follows the traditional approach to statistical pattern recognition (Duda and Hart, 1973; Fukunaga, 1990). In conceptual terms, a pattern is represented by a set of m observables, which may be viewed as a point $\mathbf{x}$ in an m-dimensional *observation (data) space.* Feature extraction is described by a transformation that maps the point $\mathbf{x}$ into an intermediate point $\mathbf{y}$ in a q-dimensional *feature space* with $q < m$, as indicated in Fig. 2.10b. This transformation may be viewed as one of dimensionality reduction (i.e., data compression), the use of which is justified on the grounds that it simplifies the task of classification. The classification is itself described as a transformation that maps the intermediate point $\mathbf{y}$ into one of the classes in an r-dimensional decision space, where r is the number of classes to be distinguished.
- The machine is designed as a single multilayer feedforward network using a supervised learning algorithm. In this second approach, the task of feature extraction is performed by the computational units in the hidden layer(s) of the network.

Which of these two approaches is adopted in practice depends on the application of interest.

Function Approximation

The third learning task of interest is that of function approximation. Consider a nonlinear input–output mapping described by the functional relationship

$$\mathbf{d} = \mathbf{f}(\mathbf{x}) \tag{2.19}$$

where the vector $\mathbf{x}$ is the input and the vector $\mathbf{d}$ is the output. The vector-valued function $\mathbf{f}(\cdot)$ is assumed to be unknown. To make up for the lack of knowledge about the function $\mathbf{f}(\cdot)$, we are given the set of labeled examples:

$$\mathcal{T} = \{(\mathbf{x}_i, \mathbf{d}_i)\}_{i=1}^N \tag{2.20}$$

The requirement is to design a neural network that approximates the unknown function $\mathbf{f}(\cdot)$ such that the function $\mathbf{F}(\cdot)$ describing the input–output mapping actually realized by the network is close enough to $\mathbf{f}(\cdot)$ in a Euclidean sense over all inputs, as shown by

$$\|\mathbf{F}(\mathbf{x}) - \mathbf{f}(\mathbf{x})\| < \epsilon \qquad \text{for all } \mathbf{x} \tag{2.21}$$

where ϵ is a small positive number. Provided that the size N of the training set is large enough and the network is equipped with an adequate number of free parameters, then the approximation error ϵ can be made small enough for the task.

The approximation problem described here is a perfect candidate for supervised learning with $\mathbf{x}_i$ playing the role of input vector and $\mathbf{d}_i$ serving the role of desired response. We may turn this issue around and view supervised learning as an approximation problem.

The ability of a neural network to approximate an unknown input–output mapping may be exploited in two important ways:

- *System identification.* Let Eq. (2.19) describe the input–output relation of an unknown memoryless *multiple input-multiple output (MIMO) system*; by a "memoryless" system we mean a system that is time invariant. We may then use the set of labeled examples in Eq. (2.20) to train a neural network as a model of the system. Let $\mathbf{y}_i$ denote the output of the neural network produced in response to an input vector $\mathbf{x}_i$. The difference between $\mathbf{d}_i$ (associated with $\mathbf{x}_i$) and the network output $\mathbf{y}_i$ provides the error signal vector $\mathbf{e}_i$, as depicted in Fig. 2.11. This error signal is in turn used to adjust the free parameters of the network to minimize the squared difference between the outputs of the unknown system and the neural network in a statistical sense, and is computed over the entire training set.
- *Inverse system.* Suppose next we are given a known memoryless MIMO system whose input–output relation is described by Eq. (2.19). The requirement in this

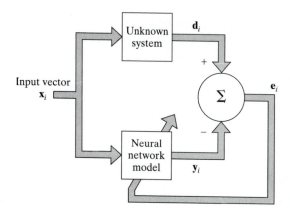

FIGURE 2.11 Block diagram of system identification.

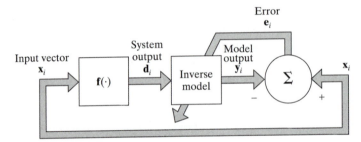

FIGURE 2.12 Block diagram of inverse system modeling.

case is to construct an *inverse system* that produces the vector **x** in response to the vector **d**. The inverse system may thus be described by

$$\mathbf{x} = \mathbf{f}^{-1}(\mathbf{d}) \tag{2.22}$$

where the vector-valued function $\mathbf{f}^{-1}(\cdot)$ denotes the inverse of $\mathbf{f}(\cdot)$. Note, however, that $\mathbf{f}^{-1}(\cdot)$ is not the reciprocal of $\mathbf{f}(\cdot)$; rather, the use of superscript -1 is merely a flag to indicate an inverse. In many situations encountered in practice, the vector-valued function $\mathbf{f}(\cdot)$ is much too complex, and inhibits a straightforward formulation of the inverse function $\mathbf{f}^{-1}(\cdot)$. Given the set of labeled examples in Eq. (2.20), we may construct a neural network approximation of $\mathbf{f}^{-1}(\cdot)$ by using the scheme shown in Fig. 2.12. In the situation described here, the roles of $\mathbf{x}_i$ and $\mathbf{d}_i$ are interchanged: the vector $\mathbf{d}_i$ is used as the input and $\mathbf{x}_i$ is treated as the desired response. Let the error signal vector $\mathbf{e}_i$ denote the difference between $\mathbf{x}_i$ and the actual output $\mathbf{y}_i$ of the neural network produced in response to $\mathbf{d}_i$. As with the system identification problem, this error signal vector is used to adjust the free parameters of the neural network to minimize the squared difference between the outputs of the unknown inverse system and the neural network in a statistical sense, and is computed over the complete training set.

Control

The control of a *plant* is another learning task that can be done by a neural network; by a "plant" we mean a process or critical part of a system that is to be maintained in a controlled condition. The relevance of learning to control should not be surprising because, after all, the human brain is a computer (i.e., information processor), the outputs of which as a whole system are *actions*. In the context of control, the brain is living proof that it is possible to build a generalized controller that takes full advantage of parallel distributed hardware, can control many thousands of actuators (muscle fibers) in parallel, can handle nonlinearity and noise, and can optimize over a long-range planning horizon (Werbos, 1992).

Consider the *feedback control system* shown in Fig. 2.13. The system involves the use of unity feedback around a plant to be controlled; that is, the plant output is fed back directly to the input.[9] Thus, the plant output **y** is subtracted from a *reference signal* **d** supplied from an external source. The error signal **e** so produced is applied to a neural

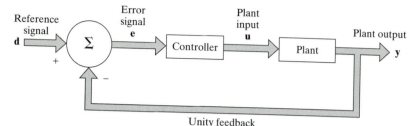

FIGURE 2.13   Block diagram of feedback control system.

controller for the purpose of adjusting its free parameters. The primary objective of the controller is to supply appropriate inputs to the plant to make its output **y** track the reference signal **d**. In other words, the controller has to invert the plant's input–output behavior.

We note that in Fig. 2.13 the error signal **e** has to propagate through the neural controller before reaching the plant. Consequently, to perform adjustments on the free parameters of the plant in accordance with an error-correction learning algorithm we need to know the Jacobian matrix

$$\mathbf{J} = \left\{ \frac{\partial y_k}{\partial u_j} \right\} \tag{2.23}$$

where y_k is an element of the plant output **y** and u_j is an element of the plant input **u**. Unfortunately, the partial derivatives $\partial y_k / \partial u_j$ for various k and j depend on the operating point of the plant and are therefore not known. We may use one of two approaches to account for them:

- *Indirect learning.* Using actual input–output measurements on the plant, a neural network model is first constructed to produce a copy of it. This model is in turn used to provide an estimate of the Jacobian matrix **J**. The partial derivatives constituting this Jacobian matrix are subsequently used in the error-correction learning algorithm for computing the adjustments to the free parameters of the neural controller (Nguyen and Widrow, 1989; Suykens et al., 1996; Widrow and Walach, 1996).
- *Direct learning.* The signs of the partial derivatives $\partial y_k / \partial u_j$ are generally known and usually remain constant over the dynamic range of the plant. This suggests that we may approximate these partial derivatives by their individual signs. Their absolute values are given a distributed representation in the free parameters of the neural controller (Saerens and Soquet, 1991; Schiffman and Geffers, 1993). The neural controller is thereby enabled to learn the adjustments to its free parameters directly from the plant.

Filtering

The term *filter* often refers to a device or algorithm used to extract information about a prescribed quantity of interest from a set of noisy data. The noise may arise from a variety of sources. For example, the data may have been measured by means of noisy sensors or may represent an information-bearing signal that has been corrupted by transmission through a communication channel. Another example is that of a useful signal component corrupted by an interfering signal picked up from the surrounding environment. We may use a filter to perform three basic information processing tasks:

1. *Filtering.* This task refers to the extraction of information about a quantity of interest at discrete time n by using data measured up to and including time n.
2. *Smoothing.* This second task differs from filtering in that information about the quantity of interest need not be available at time n, and data measured later than time n can be used in obtaining this information. This means that in smoothing there is a *delay* in producing the result of interest. Since, in the smoothing process, we are able to use data obtained not only up to time n but also after time n, we expect smoothing to be more accurate than filtering in some statistical sense.
3. *Prediction.* This task is the forecasting side of information processing. The aim here is to derive information about what the quantity of interest will be like at some time $n + n_0$ in the future, for some $n_0 > 0$, by using data measured up to and including time n.

A filtering problem humans are familiar with is the *cocktail party problem.*[10] We have a remarkable ability to focus on a speaker in the noisy environment of a cocktail party, despite the fact that the speech signal originating from that speaker is buried in an undifferentiated noise background due to other interfering conversations in the room. It is thought that some form of preattentive, preconscious analysis must be involved in resolving the cocktail party problem (Velmans, 1995). In the context of (artificial) neural networks, a similar filtering problem arises under the umbrella of *blind signal separation* (Comon, 1994; Bell and Sejnowski, 1995; Amari et al., 1996). To formulate the blind signal separation problem, consider a set of unknown source signals $\{s_i(n)\}_{i=1}^{m}$ that are mutually independent of each other. These signals are linearly mixed by an unknown sensor to produce the m-by-1 observation vector (see Fig. 2.14)

$$\mathbf{x}(n) = \mathbf{A}\mathbf{u}(n) \tag{2.24}$$

where

$$\mathbf{u}(n) = [u_1(n), u_2(n), ..., u_m(n)]^T \tag{2.25}$$

$$\mathbf{x}(n) = [x_1(n), x_2(n), ..., x_m(n)]^T \tag{2.26}$$

and $\mathbf{A}$ is an unknown nonsingular *mixing matrix* of dimensions m-by-m. Given the observation vector $\mathbf{x}(n)$, the requirement is to recover the original signals $u_1(n)$, $u_2(n), ..., u_m(n)$ in an unsupervised manner.

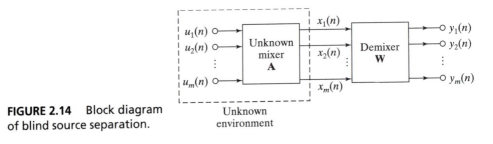

FIGURE 2.14 Block diagram of blind source separation.

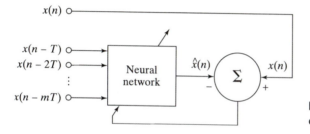

FIGURE 2.15 Block diagram of nonlinear prediction.

Turning now to the prediction problem, the requirement is to predict the present value $x(n)$ of a process, given past values of the process that are uniformly spaced in time as shown by $x(n - T), x(n - 2T), ..., x(n - mT)$, where T is the sampling period and m is the prediction order. Prediction may be solved by using error-correction learning in an unsupervised manner since the training examples are drawn directly from the process itself, as depicted in Fig. 2.15, where $x(n)$ serves the purpose of desired response. Let $\hat{x}(n)$ denote the one-step prediction produced by the neural network at time n. The error signal $e(n)$ is defined as the difference between $x(n)$ and $\hat{x}(n)$, which is used to adjust the free parameters of the neural network. On this basis, prediction may be viewed as a form of *model building* in the sense that the smaller we make the prediction error in a statistical sense, the better the network serves as a model of the underlying physical process responsible for generating the data. When this process is *nonlinear,* the use of a neural network provides a powerful method for solving the prediction problem because of the nonlinear processing units that could be built into its construction. The only possible exception to the use of nonlinear processing units, however, is the output unit of the network: If the dynamic range of the time series $\{x(n)\}$ is unknown, the use of a linear output unit is the most reasonable choice.

Beamforming

Beamforming is a *spatial* form of filtering and is used to distinguish between the spatial properties of a target signal and background noise. The device used to do the beamforming is called a *beamformer.*

The task of beamforming is compatible with the use of a neural network, for which we have relevant cues from psychoacoustic studies of human auditory responses (Bregman, 1990) and studies of feature mapping in the cortical layers of auditory systems of echo-locating bats (Suga, 1990a; Simmons and Sailant, 1992). The echo-locating bat illuminates the surrounding environment by broadcasting short-duration frequency-modulated (FM) sonar signals, and then uses its auditory system (including a pair of ears) to focus attention on its prey (e.g., flying insect). The ears provide the bat with some form of spatial filtering (interferometry to be precise), which is then exploited by the auditory system to produce *attentional selectivity.*

Beamforming is commonly used in radar and sonar systems where the primary task is to detect and track a target of interest in the combined presence of receiver noise and interfering signals (e.g., jammers). This task is complicated by two factors.

- The target signal originates from an unknown direction.
- There is no *a priori* information available on the interfering signals.

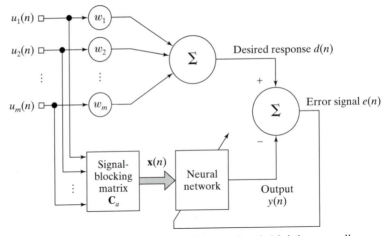

FIGURE 2.16   Block diagram of generalized sidelobe canceller.

One way of coping with situations of this kind is to use a *generalized sidelobe canceller* (GSLC), the block diagram of which is shown in Fig. 2.16. The system consists of the following components (Griffiths and Jim, 1982; Van Veen, 1992; Haykin, 1996):

- An *array of antenna elements,* which provides a means of sampling the observed signal at discrete points in space.
- A *linear combiner* defined by a set of fixed weights $\{w_i\}_{i=1}^m$, the output of which is a desired response. This linear combiner acts like a "spatial filter," characterized by a radiation pattern (i.e., a polar plot of the amplitude of the antenna output versus the incidence angle of an incoming signal). The mainlobe of this radiation pattern is pointed along a prescribed direction, for which the GSLC is *constrained* to produce a distortionless response. The output of the linear combiner, denoted by $d(n)$, provides a desired response for the beamformer.
- A *signal-blocking matrix* $\mathbf{C}_a$, the function of which is to cancel interference that leaks through the sidelobes of the radiation pattern of the spatial filter representing the linear combiner.
- A *neural network* with adjustable parameters, which is designed to accommodate statistical variations in the interfering signals.

The adjustments to the free parameters of the neural network are performed by an error-correcting learning algorithm that operates on the error signal $e(n)$, defined as the difference between the linear combiner output $d(n)$ and the actual output $y(n)$ of the neural network. Thus the GSLC operates under the supervision of the linear combiner that assumes the role of a "teacher." As with ordinary supervised learning, notice that the linear combiner is outside the feedback loop acting on the neural network. A beamformer that uses a neural network for learning is called a *neural beamformer* or *neuro-beamformer*. This class of learning machines comes under the general heading of *attentional neurocomputers* (Hecht-Nielsen, 1990).

   The diversity of the six learning tasks discussed here is testimony to the *universality* of neural networks as information-processing systems. In a fundamental sense, these learning tasks are all problems of learning a *mapping* from (possibly noisy) examples of the mapping. Without the imposition of prior knowledge, each of the tasks is in fact *ill posed* in the sense of nonuniqueness of possible solution mappings. One method of making the solution well posed is to use the theory of regularization as described in Chapter 5.

2.11 MEMORY

Discussion of learning tasks, particularly the task of pattern association, leads us naturally to think about *memory*. In a neurobiological context, memory refers to the relatively enduring neural alterations induced by the interaction of an organism with its environment (Teyler, 1986). Without such a change there can be no memory. Furthermore, for the memory to be useful it must be accessible to the nervous system in order to influence future behavior. However, an activity pattern must initially be stored in memory through a *learning process*. Memory and learning are intricately connected. When a particular activity pattern is learned, it is stored in the brain where it can be recalled later when required. Memory may be divided into "short-term" and "long-term" memory, depending on the retention time (Arbib, 1989). *Short-term memory* refers to a compilation of knowledge representing the "current" state of the environment. Any discrepancies between knowledge stored in short-term memory and a "new" state are used to update the short-term memory. *Long-term memory,* on the other hand, refers to knowledge stored for a long time or permanently.

In this section we study an associative memory that offers the following characteristics:

- The memory is distributed.
- Both the stimulus (key) pattern and the response (stored) pattern of an associative memory consist of data vectors.
- Information is stored in memory by setting up a spatial pattern of neural activities across a large number of neurons.
- Information contained in a stimulus not only determines its storage location in memory but also an address for its retrieval.
- Although neurons do not represent reliable and low-noise computing cells, the memory exhibits a high degree of resistance to noise and damage of a diffusive kind.
- There may be interactions between individual patterns stored in memory. (Otherwise the memory would have to be exceptionally large for it to accommodate the storage of a large number of patterns in perfect isolation from each other.) There is therefore the distinct possibility for the memory to make *errors* during the recall process.

In a *distributed memory,* the basic issue of interest is the simultaneous or near-simultaneous activities of many different neurons, which are the result of external or internal stimuli. The neural activities form a spatial pattern inside the memory that contains information about the stimuli. The memory is therefore said to perform a distributed mapping that transforms an activity pattern in the input space into another activity pattern in the output space. We may illustrate some important properties of a

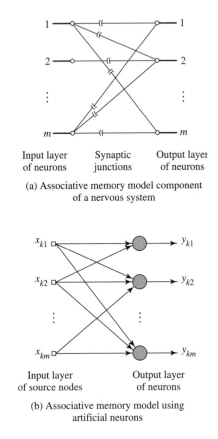

(a) Associative memory model component
of a nervous system

(b) Associative memory model using
artificial neurons

FIGURE 2.17 Associative memory models.

distributed memory mapping by considering an idealized neural network that consists of two layers of neurons. Figure 2.17a illustrates a network that may be regarded as a *model component of a nervous system* (Cooper, 1973; Scofield and Cooper, 1985). Each neuron in the input layer is connected to every one of the neurons in the output layer. The actual synaptic connections between the neurons are complex and redundant. In the model of Fig. 2.17a, a single ideal junction is used to represent the integrated effect of all the synaptic contacts between the dendrites of a neuron in the input layer and the axon branches of a neuron in the output layer. The level of activity of a neuron in the input layer may affect the level of activity of every other neuron in the output layer.

The corresponding situation for an artificial neural network is depicted in Fig. 2.17b. Here we have an input layer of source nodes and an output layer of neurons acting as computation nodes. In this case, the synaptic weights of the network are included as integral parts of the neurons in the output layer. The connecting links between the two layers of the network are simply wires.

In the following mathematical analysis, the neural networks in Figs. 2.17a and 2.17b are both assumed to be *linear*. The implication of this assumption is that each neuron acts as a linear combiner, as depicted in the signal-flow graph of Fig. 2.18. To proceed with the analysis, suppose that an activity pattern $\mathbf{x}_k$ occurs in the input layer of the network and that an activity pattern $\mathbf{y}_k$ occurs simultaneously in the output layer. The issue we wish to consider here is that of learning from the association

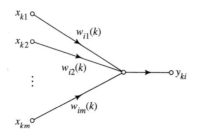

FIGURE 2.18 Signal-flow graph model of a linear neuron labeled *i*.

between the patterns $\mathbf{x}_k$ and $\mathbf{y}_k$. The patterns $\mathbf{x}_k$ and $\mathbf{y}_k$ are represented by vectors, written in their expanded forms as:

$$\mathbf{x}_k = [x_{k1}, x_{k2}, ..., x_{km}]^T$$

and

$$\mathbf{y}_k = [y_{k1}, y_{k2}, ..., y_{km}]^T$$

For convenience of presentation we have assumed that the input space dimensionality (i.e., the dimension of vector $\mathbf{x}_k$) and the output space dimensionality (i.e., the dimension of vector $\mathbf{y}_k$) are the same, equal to m. From here on we refer to m as *network dimensionality* or simply *dimensionality*. Note that m equals the number of source nodes in the input layer or neurons in the output layer. For a neural network with a large number of neurons, which is typically the case, the dimensionality m can be large.

The elements of both $\mathbf{x}_k$ and $\mathbf{y}_k$ can assume positive and negative values. This is a valid proposition in an artificial neural network. It may also occur in a nervous system by considering the relevant physiological variable to be the difference between an actual activity level (e.g., firing rate of a neuron) and a nonzero spontaneous activity level.

With the networks of Fig. 2.17 assumed to be linear, the association of key vector $\mathbf{x}_k$ with memorized vector $\mathbf{y}_k$ may be described in matrix form as:

$$\mathbf{y}_k = \mathbf{W}(k)\mathbf{x}_k, \qquad k = 1, 2, ..., q \tag{2.27}$$

where $\mathbf{W}(k)$ is a weight matrix determined solely by the input–output pair $(\mathbf{x}_k, \mathbf{y}_k)$.

To develop a detailed description of the weight matrix $\mathbf{W}(k)$, consider Fig. 2.18 that shows a detailed arrangement of neuron i in the output layer. The output y_{ki} of neuron i due to the combined action of the elements of the key pattern $\mathbf{x}_k$ applied as stimulus to the input layer is given by

$$y_{ki} = \sum_{j=1}^{m} w_{ij}(k)x_{kj}, \qquad i = 1, 2, ..., m \tag{2.28}$$

where the $w_{ij}(k), j = 1, 2, ..., m$, are the synaptic weights of neuron i corresponding to the kth pair of associated patterns. Using matrix notation, we may express y_{ki} in the equivalent form

$$y_{ki} = [w_{i1}(k), w_{i2}(k), ..., w_{im}(k)] \begin{bmatrix} x_{k1} \\ x_{k2} \\ \vdots \\ x_{km} \end{bmatrix}, \qquad i = 1, 2, ..., m \tag{2.29}$$

The column vector on the right-hand side of Eq. (2.29) is recognized as the key vector $\mathbf{x}_k$. By substituting Eq. (2.29) in the definition of the m-by-1 stored vector $\mathbf{y}_k$, we get

$$
\begin{bmatrix} y_{k1} \\ y_{k2} \\ \vdots \\ y_{km} \end{bmatrix} = \begin{bmatrix} w_{11}(k) & w_{12}(k) & \cdots & w_{1m}(k) \\ w_{21}(k) & w_{22}(k) & \cdots & w_{2m}(k) \\ \vdots & \vdots & \vdots & \vdots \\ w_{m1}(k) & w_{m2}(k) & \cdots & w_{mm}(k) \end{bmatrix} \begin{bmatrix} x_{k1} \\ x_{k2} \\ \vdots \\ x_{km} \end{bmatrix} \tag{2.30}
$$

Equation (2.30) is the expanded form of the matrix transformation or mapping described in Eq. (2.27). In particular, the m-by-m weight matrix $\mathbf{W}(k)$ is defined by

$$
\mathbf{W}(k) = \begin{bmatrix} w_{11}(k) & w_{12}(k) & \cdots & w_{1m}(k) \\ w_{21}(k) & w_{22}(k) & \cdots & w_{2m}(k) \\ \vdots & \vdots & \vdots & \vdots \\ w_{m1}(k) & w_{m2}(k) & \cdots & w_{mm}(k) \end{bmatrix} \tag{2.31}
$$

The individual presentations of the q pairs of associated patterns $\mathbf{x}_k \rightarrow \mathbf{y}_k$, $k = 1, 2, \ldots, q$, produce corresponding values of the individual matrix, namely, $\mathbf{W}(1), \mathbf{W}(2), \ldots, \mathbf{W}(q)$. Recognizing that this pattern association is represented by the weight matrix $\mathbf{W}(k)$, we may define an m–by–m *memory matrix* that describes the summation of the weight matrices for the entire set of pattern associations as follows:

$$
\mathbf{M} = \sum_{k=1}^{q} \mathbf{W}(k) \tag{2.32}
$$

The memory matrix $\mathbf{M}$ defines the overall connectivity between the input and output layers of the associative memory. In effect, it represents the *total experience* gained by the memory as a result of the presentations of q input–output patterns. Stated in another way, the memory matrix $\mathbf{M}$ contains a piece of every input–output pair of activity patterns presented to the memory.

The definition of the memory matrix given in Eq. (2.32) may be restructured in the form of a recursion as shown by

$$
\mathbf{M}_k = \mathbf{M}_{k-1} + \mathbf{W}(k), \qquad k = 1, 2, \ldots, q \tag{2.33}
$$

where the initial value $\mathbf{M}_0$ is zero (i.e., the synaptic weights in the memory are all initially zero), and the final value $\mathbf{M}_q$ is identically equal to $\mathbf{M}$ as defined in Eq. (2.32). According to the recursive formula of Eq. (2.33), the term $\mathbf{M}_{k-1}$ is the old value of the memory matrix resulting from $(k-1)$ pattern associations, and $\mathbf{M}_k$ is the updated value in light of the increment $\mathbf{W}(k)$ produced by the kth association. Note, however, that when $\mathbf{W}(k)$ is added to $\mathbf{M}_{k-1}$, the increment $\mathbf{W}(k)$ loses its distinct identity among the mixture of contributions that form $\mathbf{M}_k$. In spite of the synaptic mixing of different associations, information about the stimuli may not have been lost, as demonstrated in the sequel. Notice also that as the number q of stored patterns increases, the influence of a new pattern on the memory as a whole is progressively reduced.

Correlation Matrix Memory

Suppose that the associative memory of Fig. 2.17b has learned the memory matrix $\mathbf{M}$ through the associations of key and memorized patterns described by $\mathbf{x}_k \rightarrow \mathbf{y}_k$, where $k = 1, 2, ..., q$. We may postulate $\hat{\mathbf{M}}$, denoting an *estimate* of the memory matrix $\mathbf{M}$ in terms of these patterns as (Anderson, 1972, 1983; Cooper, 1973):

$$\hat{\mathbf{M}} = \sum_{k=1}^{q} \mathbf{y}_k \mathbf{x}_k^T \tag{2.34}$$

The term $\mathbf{y}_k \mathbf{x}_k^T$ represents the *outer product* of the key pattern $\mathbf{x}_k$ and the memorized pattern $\mathbf{y}_k$. This outer product is an "estimate" of the weight matrix $\mathbf{W}(k)$ that maps the output pattern $\mathbf{y}_k$ onto the input pattern $\mathbf{x}_k$. Since the pattern $\mathbf{x}_k$ and $\mathbf{y}_k$ are both m-by-1 vectors by assumption, it follows that their output product $\mathbf{y}_k \mathbf{x}_k^T$, and therefore the estimate $\hat{\mathbf{M}}$, is an m-by-m matrix. This dimensionality is in perfect agreement with that of the memory matrix $\mathbf{M}$ defined in Eq. (2.32). The format of the summation of the estimate $\hat{\mathbf{M}}$ bears a direct relation to that of the memory matrix defined in that equation.

A typical term of the outer product $\mathbf{y}_k \mathbf{x}_k^T$ is written as $y_{ki} x_{kj}$, where x_{kj} is the output of source node j in the input layer, and y_{ki} is the output of neuron i in the output layer. In the context of synaptic weight $w_{ij}(k)$ for the kth association, source node j acts as a presynaptic node and neuron i in the output layer acts as a postsynaptic node. Hence, the "local" learning process described in Eq. (2.34) may be viewed as a *generalization of Hebb's postulate of learning*. It is also referred to as the *outer product rule* in recognition of the matrix operation used to construct the memory matrix $\hat{\mathbf{M}}$. Correspondingly, an associative memory so designed is called a *correlation matrix memory*. Correlation, in one form or another, is indeed the basis of learning, association, pattern recognition, and memory recall in the human nervous system (Eggermont, 1990.)

Equation (2.34) may be reformulated in the equivalent form

$$\hat{\mathbf{M}} = [\mathbf{y}_1, \mathbf{y}_2, ..., \mathbf{y}_q] \begin{bmatrix} \mathbf{x}_1^T \\ \mathbf{x}_2^T \\ \vdots \\ \mathbf{x}_q^T \end{bmatrix} \tag{2.35}$$

$$= \mathbf{Y}\mathbf{X}^T$$

where

$$\mathbf{X} = [\mathbf{x}_1, \mathbf{x}_2, ..., \mathbf{x}_q] \tag{2.36}$$

and

$$\mathbf{Y} = [\mathbf{y}_1, \mathbf{y}_2, ..., \mathbf{y}_q] \tag{2.37}$$

The matrix $\mathbf{X}$ is an m-by-q matrix composed of the entire set of key patterns used in the learning process; it is called the *key matrix*. The matrix $\mathbf{Y}$ is an m-by-q matrix composed of the corresponding set of memorized patterns; it is called the *memorized matrix*.

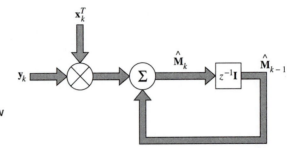

FIGURE 2.19 Signal-flow graph representation of Eq. (2.38).

Equation (2.35) may also be restructured in the form of a recursion as follows:

$$\hat{\mathbf{M}}_k = \hat{\mathbf{M}}_{k-1} + \mathbf{y}_k \mathbf{x}_k^T, \qquad k = 1, 2, \ldots, q \tag{2.38}$$

A signal-flow graph representation of this recursion is depicted in Fig. 2.19. According to this signal-flow graph and the recursive formula of Eq. (2.38), the matrix $\hat{\mathbf{M}}_{k-1}$ represents an old estimate of the memory matrix; and $\hat{\mathbf{M}}_k$ represents its updated value in the light of a new association performed by the memory on the patterns $\mathbf{x}_k$ and $\mathbf{y}_k$. Comparing the recursion of Eq. (2.38) with that of Eq. (2.33), we see that the outer product $\mathbf{y}_k \mathbf{x}_k^T$ represents an estimate of the weight matrix $\mathbf{W}(k)$ corresponding to the kth association of key and memorized patterns, $\mathbf{x}_k$ and $\mathbf{y}_k$.

Recall

The fundamental problem posed by the use of an associative memory is the address and recall of patterns stored in memory. To explain one aspect of this problem, let $\hat{\mathbf{M}}$ denote the memory matrix of an associative memory, which has been completely learned through its exposure to q pattern associations in accordance with Eq. (2.34). Let a key pattern $\mathbf{x}_j$ be picked at random and reapplied as *stimulus* to the memory, yielding the *response*

$$\mathbf{y} = \hat{\mathbf{M}}\mathbf{x}_j \tag{2.39}$$

Substituting Eq. (2.34) in (2.39), we get

$$\begin{aligned}
\mathbf{y} &= \sum_{k=1}^{m} \mathbf{y}_k \mathbf{x}_k^T \mathbf{x}_j \\
&= \sum_{k=1}^{m} (\mathbf{x}_k^T \mathbf{x}_j) \mathbf{y}_k
\end{aligned} \tag{2.40}$$

where, in the second line, it is recognized that $\mathbf{x}_k^T \mathbf{x}_j$ is a scalar equal to the *inner product* of the key vectors $\mathbf{x}_k$ and $\mathbf{x}_j$. We may rewrite Eq. (2.40) as

$$\mathbf{y} = (\mathbf{x}_j^T \mathbf{x}_j) \mathbf{y}_j + \sum_{\substack{k=1 \\ k \neq j}}^{m} (\mathbf{x}_k^T \mathbf{x}_j) \mathbf{y}_k \tag{2.41}$$

Let each of the key patterns $\mathbf{x}_1, \mathbf{x}_2, ..., \mathbf{x}_q$ be normalized to have unit energy; that is,

$$E_k = \sum_{l=1}^{m} x_{kl}^2$$
$$= \mathbf{x}_k^T \mathbf{x}_k \tag{2.42}$$
$$= 1, \qquad k = 1, 2, ..., q$$

Accordingly, we may simplify the response of the memory to the stimulus (key pattern) $\mathbf{x}_j$ as

$$\mathbf{y} = \mathbf{y}_j + \mathbf{v}_j \tag{2.43}$$

where

$$\mathbf{v}_j = \sum_{\substack{k=1 \\ k \neq j}}^{m} (\mathbf{x}_k^T \mathbf{x}_j) \mathbf{y}_k \tag{2.44}$$

The first term on the right-hand side of Eq. (2.43) represents the "desired" response $\mathbf{y}_j$; it may therefore be viewed as the "signal" component of the actual response $\mathbf{y}$. The second term $\mathbf{v}_j$ is a "noise vector" that arises because of the *crosstalk* between the key vector $\mathbf{x}_j$ and all the other key vectors stored in memory. The noise vector $\mathbf{v}_j$ is responsible for making errors on recall.

In the context of a linear signal space, we may define the *cosine of the angle* between a pair of vectors $\mathbf{x}_j$ and $\mathbf{x}_k$ as the inner product of $\mathbf{x}_j$ and $\mathbf{x}_k$ divided by the product of their individual Euclidean *norms* or *lengths* as shown by

$$\cos(\mathbf{x}_k, \mathbf{x}_j) = \frac{\mathbf{x}_k^T \mathbf{x}_j}{\|\mathbf{x}_k\| \|\mathbf{x}_j\|} \tag{2.45}$$

The symbol $\|\mathbf{x}_k\|$ signifies the Euclidean norm of vector $\mathbf{x}_k$, defined as the square root of the energy of $\mathbf{x}_k$:

$$\|\mathbf{x}_k\| = (\mathbf{x}_k^T \mathbf{x}_k)^{1/2}$$
$$= E_k^{1/2} \tag{2.46}$$

Returning to the situation, note that the key vectors are normalized to have unit energy in accordance with Eq. (2.42). We may therefore reduce the definition of Eq. (2.45) to

$$\cos(\mathbf{x}_k, \mathbf{x}_j) = \mathbf{x}_k^T \mathbf{x}_j \tag{2.47}$$

We may then redefine the noise vector of Eq. (2.44) as

$$\mathbf{v}_j = \sum_{\substack{k=1 \\ k \neq j}}^{m} \cos(\mathbf{x}_k, \mathbf{x}_j) \mathbf{y}_k \tag{2.48}$$

We now see that if the key vectors are *orthogonal* (i.e., perpendicular to each other in a Euclidean sense), then

$$\cos(\mathbf{x}_k, \mathbf{x}_j) = 0, \qquad k \neq j \tag{2.49}$$

and therefore the noise vector $\mathbf{v}_j$ is identically zero. In such a case, the response $\mathbf{y}$ equals $\mathbf{y}_j$. The *memory associates perfectly* if the key vectors from an *orthonormal set; that is,* if they satisfy the following pair of conditions:

$$\mathbf{x}_k^T \mathbf{x}_j = \begin{cases} 1, & k = j \\ 0, & k \neq j \end{cases} \tag{2.50}$$

Suppose now that the key vectors do form an orthonormal set, as prescribed in Eq. (2.50). What is then the limit on the *storage capacity* of the associative memory? Stated in another way, what is the largest number of patterns that can be reliably stored? The answer to this fundamental question lies in the rank of the memory matrix $\hat{\mathbf{M}}$. The *rank* of a matrix is defined as the number of independent columns (rows) of the matrix. That is, if r is the rank of such a rectangular matrix of dimensions l-by-m, we then have $r \leq \min(l, m)$. In the case of a correlation memory, the memory matrix $\hat{\mathbf{M}}$ is an m-by-m matrix, where m is the dimensionality of the input space. Hence the rank of the memory matrix $\mathbf{M}$ is limited by the dimensionality m. We may thus formally state that the number of patterns that can be reliably stored in a correlation matrix memory can never exceed the input space dimensionality.

In real-life situations, we often find that the key patterns presented to an associative memory are neither orthogonal nor highly separated from each other. Consequently, a correlation matrix memory characterized by the memory matrix of Eq. (2.34) may sometimes get confused and make *errors*. That is, the memory occasionally recognizes and associates patterns never seen or associated before. To illustrate this property of an associative memory, consider a set of key patterns.

$$\{\mathbf{x}_{\text{key}}\}: \mathbf{x}_1, \mathbf{x}_2, \dots, \mathbf{x}_q$$

and a corresponding set of memorized patterns,

$$\{\mathbf{y}_{\text{mem}}\}: \mathbf{y}_1, \mathbf{y}_2, \dots, \mathbf{y}_q$$

To express the closeness of the key patterns in a linear signal space, we introduce the concept of *community*. We define the community of the set of patterns $\{\mathbf{x}_{\text{key}}\}$ as the lower bound on the inner products $\mathbf{x}_k^T \mathbf{x}_j$ of any two patterns $\mathbf{x}_j$ and $\mathbf{x}_k$ in the set. Let $\hat{\mathbf{M}}$ denote the memory matrix resulting from the training of the associative memory on a set of key patterns represented by $\{\mathbf{x}_{\text{key}}\}$ and a corresponding set of memorized patterns $\{\mathbf{y}_{\text{mem}}\}$ in accordance with Eq. (2.34). The response of the memory, $\mathbf{y}$, to a stimulus $\mathbf{x}_j$ selected from the set $\{\mathbf{x}_{\text{key}}\}$ is given by Eq. (2.39), where it is assumed that each pattern in the set $\{\mathbf{x}_{\text{key}}\}$ is a unit vector (i.e., a vector with unit energy). Let it be further assumed that

$$\mathbf{x}_k^T \mathbf{x}_j \geq \gamma \qquad \text{for } k \neq j \tag{2.51}$$

If the lower bound γ is large enough, the memory may fail to distinguish the response $\mathbf{y}$ from that of any other key pattern contained in the set $\{\mathbf{x}_{\text{key}}\}$. If the key patterns in this set have the form

$$\mathbf{x}_j = \mathbf{x}_0 + \mathbf{v} \tag{2.52}$$

where $\mathbf{v}$ is a stochastic vector, it is likely that the memory will recognize $\mathbf{x}_0$ and associate with it a vector $\mathbf{y}_0$ rather than any of the actual pattern pairs used to train it in the first place; $\mathbf{x}_0$ and $\mathbf{y}_0$ denote a pair of patterns never seen before. This phenomenon may be termed *animal logic,* which is not logic at all (Cooper, 1973).

2.12 ADAPTATION

In performing a task of interest, we often find that *space* is one fundamental dimension of the learning process; *time* is the other. The *spatiotemporal* nature of learning is exemplified by many of the learning tasks (e.g., control, beamforming) discussed in Section 2.10. Species ranging from insects to humans have an inherent capacity to represent the temporal structure of experience. Such a representation makes it possible for an animal to *adapt* its behavior to the temporal structure of an event in its behavioral space (Gallistel, 1990).

When a neural network operates in a *stationary* environment (i.e., an environment whose statistical characteristics do not change with time), the essential statistics of the environment can, in theory, be *learned* by the network under the supervision of a teacher. In particular, the synaptic weights of the network can be computed by having the network undergo a training session with a set of data that is representative of the environment. Once the training process has completed, the synaptic weights of the network should capture the underlying statistical structure of the environment, which would justify "freezing" their values thereafter. Thus a learning system relies on *memory,* in one form or another, to recall and exploit past experiences.

Frequently, however, the environment of interest is *nonstationary,* which means that the statistical parameters of the information-bearing signals generated by the environment vary with time. In situations of this kind, the traditional methods of supervised learning may prove to be inadequate because the network is not equipped with the necessary means to *track* the statistical variations of the environment in which it operates. To overcome this shortcoming, it is desirable for a neural network to continually *adapt* its free parameters to variations in the incoming signals in a *real-time* fashion. Thus an *adaptive system* responds to every distinct input as a novel one. In other words the learning process encountered in an adaptive system never stops, with learning going on while signal processing is being performed by the system. This form of learning is called *continuous learning* or *learning-on-the-fly*.

Linear adaptive filters, built around a linear combiner (i.e., a single neuron operating in its linear mode), are designed to perform continuous learning. Despite their simple structure (and perhaps because of it), they are widely used in such diverse applications as radar, sonar, communications, seismology, and biomedical signal processing. The theory of linear adaptive filters has reached a highly mature stage of development (Haykin, 1996; Widrow and Stearns, 1985). However, the same cannot be said about nonlinear adaptive filters.[11]

With continuous learning as the property of interest and a neural network as the vehicle for its implementation, the question we need to address is: How can a neural network adapt its behavior to the varying temporal structure of the incoming signals in its behavioral space? One way of addressing this fundamental issue is to recognize that statistical characteristics of a nonstationary process usually change slowly enough for the process to be considered *pseudostationary* over a window of short enough duration. Examples include:

- The mechanism responsible for the production of a speech signal may be considered essentially stationary over a period of 10 to 30 milliseconds.
- Radar returns from an ocean surface remain essentially stationary over a period of several seconds.
- With long-range weather forecasting in mind, weather related data may be viewed as essentially stationary over a period of minutes.
- In the context of long-range trends extending into months and years, stock market data may be considered as essentially stationary over a period of days.

We may thus exploit the pseudostationary property of a stochastic process to extend the utility of a neural network by *retraining* it at some regular intervals to account for statistical fluctuations of the incoming data. Such an approach may, for example, be suitable for processing stock market data.

For a more refined *dynamic* approach to learning, we may proceed as follows:

- Select a window short enough for the input data to be considered pseudostationary, and use the data to train the network.
- When a new data sample is received, update the window by dropping the oldest data sample and shifting the remaining data samples back by one time unit to make room for the new sample.
- Use the updated data window to retrain the network.
- Repeat the procedure on a continuing basis.

We may thus build temporal structure into the design of a neural network by having the network undergo *continual training* with *time-ordered examples*. According to this dynamic approach, a neural network is viewed as a *nonlinear adaptive filter* that represents a generalization of linear adaptive filters. However, for this dynamic approach to nonlinear adaptive filters to be feasible, the resources available must be *fast* enough to complete all the described computations in one sampling period. Only then can the filter keep up with changes in the input.

2.13 STATISTICAL NATURE OF THE LEARNING PROCESS

The last part of the chapter deals with statistical aspects of learning. In this context we are not interested in the evolution of the weight vector $\mathbf{w}$ as a neural network is cycled through a learning algorithm. We instead focus on the deviation between a "target" function $f(\mathbf{x})$ and the "actual" function $F(\mathbf{x}, \mathbf{w})$ realized by the neural network where the vector $\mathbf{x}$ denotes the input signal. The deviation is expressed in statistical terms.

A neural network is merely one form in which *empirical knowledge* about a physical phenomenon or environment of interest may be encoded through training. By "empirical" knowledge we mean a set of measurements that characterizes the phenomenon. To be more specific, consider the example of a stochastic phenomenon described by a random vector $\mathbf{X}$ consisting of a set of *independent variables,* and a random scalar D representing a *dependent variable*. The elements of the random vector $\mathbf{X}$ may have different physical meanings of their own. The assumption that the dependent variable D is a scalar has been made merely to simplify the exposition without any loss of generality. Suppose also that we have N realizations of the random vector $\mathbf{X}$ denoted by

$\{\mathbf{x}_i\}_{i=1}^N$, and a corresponding set of realizations of the random scalar D denoted by $\{d_i\}_{i=1}^N$. These realizations (measurements) constitute the training sample denoted by

$$\mathcal{T} = \{(\mathbf{x}_i, d_i)\}_{i=1}^N \tag{2.53}$$

Ordinarily we do *not* have knowledge of the exact functional relationship between $\mathbf{X}$ and D, so we proceed by proposing the model (White, 1989a)

$$D = f(\mathbf{X}) + \epsilon \tag{2.54}$$

where $f(\cdot)$ is a *deterministic* function of its argument vector, and ϵ is a random *expectational error* that represents our "ignorance" about the dependence of D and $\mathbf{X}$. The statistical model described by Eq. (2.54) is called a *regressive model*; it is depicted in Fig. 2.20a. The expectational error ϵ is, in general, a random variable with zero mean and positive probability of occurrence. On this basis, the regressive model of Fig. 2.20a has two useful properties:

1. *The mean value of the expectational error ϵ, given any realization $\mathbf{x}$, is zero*; that is,

$$E[\epsilon|\mathbf{x}] = 0 \tag{2.55}$$

where E is the statistical expectation operator. As a corollary to this property, we may state that the *regression function $f(\mathbf{x})$ is the conditional mean of the model output D, given that the input $\mathbf{X} = \mathbf{x}$*, as shown by

$$f(\mathbf{x}) = E[D|\mathbf{x}] \tag{2.56}$$

This property follows directly from Eq. (2.54) in light of Eq. (2.55).

2. *The expectational error ϵ is uncorrelated with the regression function $f(\mathbf{X})$*; that is

$$E[\epsilon f(\mathbf{X})] = 0 \tag{2.57}$$

This property is the well-known *principle* of *orthogonality*, which states that all the information about D available to us through the input $\mathbf{X}$ has been encoded

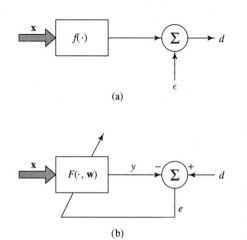

(a)

(b)

FIGURE 2.20   (a) Regressive model (mathematical). (b) Neural network model (physical).

into the regression function $f(\mathbf{X})$. Equation (2.57) is readily demonstrated by writing:

$$E[\epsilon f(\mathbf{X})] = E[E[\epsilon f(\mathbf{X})|\mathbf{x}]]$$
$$= E[f(\mathbf{X})E[\epsilon|\mathbf{x}]]$$
$$= E[f(\mathbf{X}) \cdot 0]$$
$$= 0$$

The regressive model of Fig. 2.20a is a "mathematical" description of a stochastic environment. Its purpose is to use the vector $\mathbf{X}$ to explain or predict the dependent variable D. Figure 2.20b is the corresponding "physical" model of the environment. The purpose of this second model, based on a neural network, is to encode the empirical knowledge represented by the training sample $\mathcal{T}$ into a corresponding set of synaptic weight vectors, $\mathbf{w}$, as shown by

$$\mathcal{T} \rightarrow \mathbf{w} \tag{2.58}$$

In effect, the neural network provides an "approximation" to the regressive model of Fig. 2.20a. Let the actual response of the neural network, produced in response to the input vector $\mathbf{x}$, be denoted by the random variable

$$Y = F(\mathbf{X}, \mathbf{w}) \tag{2.59}$$

where $F(\cdot, \mathbf{w})$ is the input–output function realized by the neural network. Given the training data $\mathcal{T}$ of Eq. (2.53), the weight vector $\mathbf{w}$ is obtained by minimizing the cost function

$$\mathcal{E}(\mathbf{w}) = \frac{1}{2}\sum_{i=1}^{N}(d_i - F(\mathbf{x}_i, \mathbf{w}))^2 \tag{2.60}$$

where the factor $1/2$ has been used to be consistent with earlier notations and those used in subsequent chapters. Except for the scaling factor $1/2$, the cost function $\mathcal{E}(\mathbf{w})$ is the squared difference between the desired response d and the actual response y of the neural network, averaged over the entire training data set $\mathcal{T}$. The use of Eq. (2.60) as the cost function implies the use of "batch" training, by which we mean that the adjustments to the synaptic weights of the network are performed over the entire set of training examples rather than on an example-by-example basis.

Let the symbol $E_{\mathcal{T}}$ denote the *average operator* taken over the entire training sample $\mathcal{T}$. The variables or their functions that come under the average operator $E_{\mathcal{T}}$ are denoted by $\mathbf{x}$ and d; the pair $(\mathbf{x}, d)$ represents an example in the training sample $\mathcal{T}$. In contrast, the statistical expectation operator E acts on the whole ensemble of random variables $\mathbf{X}$ and D, which includes $\mathcal{T}$ as a subset. The difference between the operators E and $E_{\mathcal{T}}$ should be carefully identified in what follows.

In light of the transformation described in Eq. (2.58), we may interchangably use $F(\mathbf{x}, \mathbf{w})$ and $F(\mathbf{x}, \mathcal{T})$ and therefore rewrite Eq. (2.60) in the equivalent form

$$\mathcal{E}(\mathbf{w}) = \frac{1}{2}E_{\mathcal{T}}[(d - F(\mathbf{x}, \mathcal{T}))^2] \tag{2.61}$$

By adding and subtracting $f(\mathbf{x})$ to the argument $(d - F(\mathbf{x}, \mathcal{T}))$ and then using Eq. (2.54), we may write

$$d - F(\mathbf{x}, \mathcal{T}) = (d - f(\mathbf{x})) + (f(\mathbf{x}) - F(\mathbf{x}, \mathcal{T}))$$
$$= \epsilon + (f(\mathbf{x}) - F(\mathbf{x}, \mathcal{T}))$$

By substituting this expression in Eq. (2.61) and then expanding terms, we may recast the cost function $\mathscr{E}(\mathbf{w})$ in the equivalent form

$$\mathscr{E}(\mathbf{w}) = \frac{1}{2}E_{\mathcal{T}}[\epsilon^2] + \frac{1}{2}E_{\mathcal{T}}[f(\mathbf{x}) - F(\mathbf{x}, \mathcal{T}))^2] + E_{\mathcal{T}}[\epsilon(f(\mathbf{x}) - F(\mathbf{x}, \mathcal{T}))] \qquad (2.62)$$

However, the last expectation term on the right-hand side of Eq. (2.62) is zero for two reasons:

- The expectational error ϵ is uncorrelated with the regression function $f(\mathbf{x})$ by virtue of Eq. (2.57), interpreted in terms of the operator $E_{\mathcal{T}}$.
- The expectational error ϵ pertains to the regressive model of Fig. 2.20a, whereas the approximating function $F(\mathbf{x}, \mathbf{w})$ pertains to the neural network model of Fig. 2.20b.

Accordingly, Eq. (2.62) reduces to

$$\mathscr{E}(\mathbf{w}) = \frac{1}{2}E_{\mathcal{T}}[\epsilon^2] + \frac{1}{2}E_{\mathcal{T}}[(f(\mathbf{x}) - F(\mathbf{x}, \mathcal{T}))^2] \qquad (2.63)$$

The first term on the right-hand side of Eq. (2.63) is the variance of the expectational (regressive modeling) error ϵ, evaluated over the training sample $\mathcal{T}$. This term represents the *intrinsic error* because it is independent of the weight vector $\mathbf{w}$. It may be ignored as far as the minimization of the cost function $\mathscr{E}(\mathbf{w})$ with respect to $\mathbf{w}$ is concerned. Hence, the particular value of the weight vector $\mathbf{w}^*$ that minimizes the cost function $\mathscr{E}(\mathbf{w})$ will also minimize the ensemble average of the squared distance between the regression function $f(\mathbf{x})$ and the approximating function $F(\mathbf{x}, \mathbf{w})$. In other words, the *natural measure* of the effectiveness of $F(\mathbf{x}, \mathbf{w})$ as a predictor of the desired response d is defined by

$$L_{av}(f(\mathbf{x}), F(\mathbf{x}, \mathbf{w})) = E_{\mathcal{T}}[f(\mathbf{x}) - F(\mathbf{x}, \mathcal{T}))^2] \qquad (2.64)$$

This result is fundamentally important because it provides the mathematical basis for the tradeoff between the bias and variance resulting from the use of $F(\mathbf{x}, \mathbf{w})$ as the approximation to $f(\mathbf{x})$ (Geman et al., 1992).

Bias/Variance Dilemma

Invoking the use of Eq. (2.56), we may redefine the squared distance between $f(\mathbf{x})$ and $F(\mathbf{x}, \mathbf{w})$ as:

$$L_{av}(f(\mathbf{x}), F(\mathbf{x}, \mathbf{w})) = E_{\mathcal{T}}[(E[D|\mathbf{X} = \mathbf{x}] - F(\mathbf{x}, \mathcal{T}))^2] \qquad (2.65)$$

This expression may also be viewed as the average value of the *estimation error* between the regression function $f(\mathbf{x}) = E[D|\mathbf{X} = \mathbf{x}]$ and the approximating function $F(\mathbf{x}, \mathbf{w})$, evaluated over the entire training sample $\mathcal{T}$. Notice that the conditional mean

$E[D|\mathbf{X} = \mathbf{x}]$ has a constant expectation with respect to the training data sample $\mathcal{T}$. Next we find that

$$E[D|\mathbf{X} = \mathbf{x}] - F(\mathbf{x}, \mathcal{T}) = (E[D|\mathbf{X} = \mathbf{x}] - E_{\mathcal{T}}[F(\mathbf{x}, \mathcal{T})]) + (E_{\mathcal{T}}[F(\mathbf{x}, \mathcal{T})] - F(\mathbf{x}, \mathcal{T}))$$

where we have simply added and subtracted the average $E_{\mathcal{T}}[F(\mathbf{x}, \mathcal{T})]$. By proceeding in a manner similar to that described for deriving Eq. (2.62) from (2.61), we may reformulate Eq. (2.65) as the sum of two terms (see Problem 2.22):

$$L_{\text{av}}(f(\mathbf{x}), F(\mathbf{x}, \mathcal{T})) = B^2(\mathbf{w}) + V(\mathbf{w}) \qquad (2.66)$$

where $B(\mathbf{w})$ and $V(\mathbf{w})$ are themselves defined by

$$B(\mathbf{w}) = E_{\mathcal{T}}[F(\mathbf{x}, \mathcal{T})] - E[D|\mathbf{X} = \mathbf{x}] \qquad (2.67)$$

and

$$V(\mathbf{w}) = E_{\mathcal{T}}[(F(\mathbf{x}, \mathcal{T}) - E_{\mathcal{T}}[F(\mathbf{x}, \mathcal{T})])^2] \qquad (2.68)$$

We now make two important observations:

1. The term $B(\mathbf{w})$ is the *bias* of the average value of the approximating function $F(\mathbf{x}, \mathcal{T})$, measured with respect to the regression function $f(\mathbf{x}) = E[D|\mathbf{X} = \mathbf{x}]$. This term represents the inability of the neural network defined by the function $F(\mathbf{x}, \mathbf{w})$ to accurately approximate the regression function $f(\mathbf{x}) = E[D|\mathbf{X} = \mathbf{x}]$. We may therefore view the bias $B(\mathbf{w})$ as an *approximation error*.
2. The term $V(\mathbf{w})$ is the *variance* of the approximating function $F(\mathbf{x}, \mathbf{w})$, measured over the entire training sample $\mathcal{T}$. This second term represents the inadequacy of the information contained in the training sample $\mathcal{T}$ about the regression function $f(\mathbf{x})$. We may therefore view the variance $V(\mathbf{w})$ as the manifestation of an *estimation error*.

Figure 2.21 illustrates the relations between the target and approximating functions, and shows how the estimation errors, namely bias and variance, accumulate. To

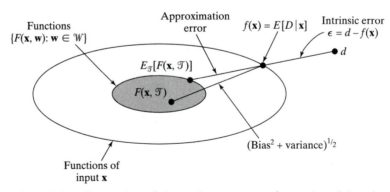

FIGURE 2.21   Illustration of the various sources of error in solving the regression problem.

achieve good overall performance, the bias $B(\mathbf{w})$ and the variance $V(\mathbf{w})$ of the approximating function $F(\mathbf{x}, \mathbf{w}) = F(\mathbf{x}, \mathcal{T})$ would both have to be small.

Unfortunately, we find that in a neural network that learns by example and does so with a training sample of fixed size, the price for achieving a small bias is a large variance. For a single neural network, it is only when the size of the training sample becomes infinitely large that we can hope to eliminate both bias and variance at the same time. We then have a *bias/variance dilemma,* and the consequence is prohibitively slow convergence (Geman et al., 1992). The bias/variance dilemma may be circumvented if we are willing to *purposely* introduce bias, which then makes it possible to eliminate the variance or to reduce it significantly. Needless to say, we must be sure that the bias built into the network design is harmless. In the context of pattern classification, for example, the bias is said to be "harmless" in the sense that it will contribute significantly to mean-square error only if we try to infer regressions that are not in the anticipated class. In general, bias must be *designed* for each specific application of interest. A practical way of achieving such an objective is to use a *constrained* network architecture, which usually performs better than a general-purpose architecture. For example, the constraints and therefore the bias may take the form of prior knowledge built into the network design using (1) *weight-sharing* where several synapses of the network are controlled by a single weight, and/or (2) *local receptive fields* assigned to individual neurons in the network, as demonstrated in the application of a multilayer perceptron to the optical character recognition problem (LeCun et al., 1990a). These network design issues were briefly discussed in Section 1.7.

2.14   STATISTICAL LEARNING THEORY

In this section we continue the statistical characterization of neural networks by describing a *learning theory* that addresses the fundamental issue of how to control the generalization ability of a neural network in mathematical terms. The discussion is presented in the context of supervised learning.

A model of supervised learning consists of three interrelated components, illustrated in Fig. 2.22 and abstracted in mathematical terms as follows (Vapnik, 1992, 1998):

1. *Environment.* The environment is stationary, supplying a vector $\mathbf{x}$ with a fixed but unknown cumulative (probability) distribution function $F_{\mathbf{X}}(\mathbf{x})$.
2. *Teacher.* The teacher provides a desired response d for every input vector $\mathbf{x}$ received from the environment, in accordance with a conditional cumulative distribution function $F_{\mathbf{X}}(\mathbf{x}|d)$ that is also fixed but unknown. The desired response d and input vector $\mathbf{x}$ are related by

$$d = f(\mathbf{x}, v) \tag{2.69}$$

where v is a noise term, permitting the teacher to be "noisy."
3. *Learning machine (algorithm).* The learning machine (neural network) is capable of implementing a set of input–output mapping functions described by

$$y = F(\mathbf{x}, \mathbf{w}) \tag{2.70}$$

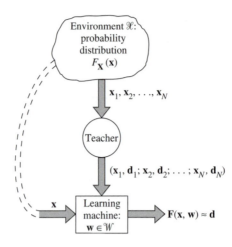

FIGURE 2.22 Model of the supervised learning process.

where y is the actual response produced by the learning machine in response to the input $\mathbf{x}$, and $\mathbf{w}$ is a set of free parameters (synaptic weights) selected from the parameter (weight) space $\mathcal{W}$.

Equations (2.69) and (2.70) are written in terms of the examples used to perform the training.

The supervised learning problem is that of selecting the particular function $F(\mathbf{x}, \mathbf{w})$ that approximates the desired response d in an optimum fashion, with "optimum" being defined in some statistical sense. The selection itself is based on the set of *N independent, identically distributed (iid)* training examples described in Eq. (2.53) and reproduced here for convenience of presentation:

$$\mathcal{T} = \{(\mathbf{x}_i, d_i)\}_{i=1}^{N}$$

Each example pair is drawn by the learning machine from $\mathcal{T}$ with a joint cumulative (probability) distribution function $F_{\mathbf{X},D}(\mathbf{x}, d)$, which, like the other distribution functions, is also fixed but unknown. The feasibility of supervised learning depends on this question: Do the training examples $\{(\mathbf{x}_i, d_i)\}$ contain sufficient information to construct a learning machine capable of good generalization performance? An answer to this fundamental question lies in the use of tools pioneered by Vapnik and Chervonenkis (1971). Specifically, we proceed by viewing the supervised learning problem as an *approximation problem,* which involves finding the function $F(\mathbf{x}, \mathbf{w})$ that is the best possible approximation to the desired function $f(\mathbf{x})$.

Let $L(d, F(\mathbf{x}, \mathbf{w}))$ denote a measure of the *loss* or *discrepancy* between the desired response d corresponding to an input vector $\mathbf{x}$ and the actual response $F(\mathbf{x}, \mathbf{w})$ produced by the learning machine. A popular definition for the loss $L(d, F(\mathbf{x}, \mathbf{w}))$ is the *quadratic loss function* defined as the squared distance between $d = f(\mathbf{x})$ and the approximation $F(\mathbf{x}, \mathbf{w})$ as shown by[12]

$$L(d, F(\mathbf{x}, \mathbf{w})) = (d - F(\mathbf{x}, \mathbf{w}))^2 \qquad (2.71)$$

The squared distance of Eq. (2.64) is the ensemble-averaged extension of $L(d, F(\mathbf{x}, \mathbf{w}))$, with the averaging being performed over all the example pairs $(\mathbf{x}, d)$.

Most of the literature on statistical learning theory deals with a specific loss. The strong point of the statistical learning theory presented here is that it does *not* depend critically on the form of the loss function $L(d, F(\mathbf{x}, \mathbf{w}))$. Later in the section we do restrict the discussion to a specific loss function.

The expected value of the loss is defined by the *risk functional*

$$R(\mathbf{w}) = \int L(d, F(\mathbf{x}, \mathbf{w})) dF_{\mathbf{X},D}(\mathbf{x}, d) \tag{2.72}$$

where the integral is a multi-fold integral taken over all possible values of the example pair $(\mathbf{x}, d)$. The goal of supervised learning is to minimize the risk functional $R(\mathbf{w})$ over the class of approximating functions $\{F(\mathbf{x}, \mathbf{w}), \mathbf{w} \in \mathcal{W}\}$. However, evaluation of the risk functional $R(\mathbf{w})$ is complicated because the joint cumulative distribution function $F_{\mathbf{X},D}(\mathbf{x}, d)$ is usually unknown. In supervised learning, the only information available is contained in the training data set $\mathcal{T}$. To overcome this mathematical difficulty, we use the inductive principle of empirical risk minimization (Vapnik, 1982). This principle relies entirely on availability of the training data set $\mathcal{T}$, which makes it perfectly suited to the design philosophy of neural networks.

Some Basic Definitions

Before proceeding further, we digress briefly to introduce some basic definitions that we use in the material to follow.

Convergence in probability.   Consider a sequence of random variables $a_1, a_2, ..., a_N$. This sequence of random variables is said to *converge in probability* to a random variable a_0 if for any $\delta > 0$, the probabilistic relation

$$P(|a_N - a_0| > \delta) \xrightarrow{P} 0 \qquad \text{as } N \to \infty \tag{2.73}$$

holds.

Supremum and infimum.   The supremum of a nonempty set $\mathcal{A}$ of scalars, denoted by sup $\mathcal{A}$, is defined as the smallest scalar x such that $x \geq y$ for all $y \in \mathcal{A}$. If no such scalar exists, we say that the supremum of the nonempty set $\mathcal{A}$ is ∞. Similarly, the infimum of set $\mathcal{A}$, denoted by inf $\mathcal{A}$, is defined as the largest scalar x such that $x \leq y$ for all $y \in \mathcal{A}$. If no such scalar exists, we say that the infimum of the nonempty set $\mathcal{A}$ is ∞.

Empirical risk functional.   Given the training sample $\mathcal{T} = \{(\mathbf{x}_i, d_i)\}_{i=1}^{N}$, the empirical risk functional is defined in terms of the loss function $L(d_i, F(\mathbf{x}_i, \mathbf{w}))$ as

$$R_{\text{emp}}(\mathbf{w}) = \frac{1}{N} \sum_{i=1}^{N} L(d_i, F(\mathbf{x}_i, \mathbf{w})) \tag{2.74}$$

Strict Consistency. Consider the set $\mathcal{W}$ of functions $L(d, F(\mathbf{x}, \mathbf{w}))$ whose underlying distribution is defined by the joint cumulative distribution function $F_{\mathbf{X},D}(\mathbf{x}, d)$. Let $\mathcal{W}(c)$ be any nonempty subset of this set of functions, such that

$$\mathcal{W}(c) = \left\{ \mathbf{w}: \int L(d, F(\mathbf{x}, \mathbf{w})) \geq c \right\} \tag{2.75}$$

where $c \in (-\infty, \infty)$. The empirical risk functional is said to be *strictly (nontrivially) consistent* if for any subset $\mathcal{W}(c)$ the following convergence in probability

$$\inf_{\mathbf{w} \in \mathcal{W}(c)} R_{\text{emp}}(\mathbf{w}) \xrightarrow{P} \inf_{\mathbf{w} \in \mathcal{W}(c)} R(\mathbf{w}) \qquad \text{as } N \to \infty \tag{2.76}$$

holds.

With these definitions we may resume the discussion of Vapnik's statistical learning theory.

Principle of Empirical Risk Minimization

The basic idea of the principle of *empirical risk minimization* is to work with the empirical risk functional $R_{\text{emp}}(\mathbf{w})$ defined in Eq. (2.74). This new functional differs from the risk functional $R(\mathbf{w})$ of Eq. (2.72) in two desirable ways:

1. It does *not* depend on the unknown distribution function $F_{\mathbf{X},D}(\mathbf{x}, d)$ in an explicit sense.
2. In theory it can be minimized with respect to the weight vector $\mathbf{w}$.

Let $\mathbf{w}_{\text{emp}}$ and $F(\mathbf{x}, \mathbf{w}_{\text{emp}})$ denote the weight vector and the corresponding mapping that minimize the empirical risk functional $R_{\text{emp}}(\mathbf{w})$ in Eq. (2.74). Similarly, let $\mathbf{w}_o$ and $F(\mathbf{x}, \mathbf{w}_o)$ denote the weight vector and the corresponding mapping that minimize the actual risk functional $R(\mathbf{w})$ in Eq. (2.72). Both $\mathbf{w}_{\text{emp}}$ and $\mathbf{w}_o$ belong to the weight space $\mathcal{W}$. The problem we must now consider is the conditions under which the approximate mapping $F(\mathbf{x}, \mathbf{w}_{\text{emp}})$ is "close" to the desired mapping $F(\mathbf{x}, \mathbf{w}_o)$ as measured by the mismatch between $R(\mathbf{w}_{\text{emp}})$ and $R(\mathbf{w}_o)$.

For some fixed $\mathbf{w} = \mathbf{w}^*$, the risk functional $R(\mathbf{w}^*)$ determines the *mathematical expectation* of a random variable defined by

$$Z_{\mathbf{w}^*} = L(d, F(\mathbf{x}, \mathbf{w}^*)) \tag{2.77}$$

In contrast, the empirical risk functional $R_{\text{emp}}(\mathbf{w}^*)$ is the *empirical (arithmetic) mean* of the random variable $Z_{\mathbf{w}^*}$. According to the *law of large numbers,* which constitutes one of the main theorems of probability theory, in general cases we find that as the size N of the training sample $\mathcal{T}$ is made infinitely large, the empirical mean of the random variable $Z_{\mathbf{w}^*}$ converges to its expected value. This observation provides theoretical justification for the use of the empirical risk functional $R_{\text{emp}}(\mathbf{w})$ in place of the risk functional $R(\mathbf{w})$. However, just because the empirical mean of $Z_{\mathbf{w}^*}$ converges to its expected value, there is no reason to expect that the weight vector $\mathbf{w}_{\text{emp}}$ that minimizes the empirical risk functional $R_{\text{emp}}(\mathbf{w})$ will also minimize the risk functional $R(\mathbf{w})$.

We may satisfy this requirement in an approximate fashion by proceeding as follows. If the empirical risk functional $R_{\text{emp}}(\mathbf{w})$ approximates the original risk functional

$R(\mathbf{w})$ *uniformly* in $\mathbf{w}$ with some *precision* ϵ, then the minimum of $R_{emp}(\mathbf{w})$ deviates from the minimum of $R(\mathbf{w})$ by an amount not exceeding 2ϵ. Formally, this means that we must impose a stringent condition, such that for any $\mathbf{w} \in \mathcal{W}$ and $\epsilon > 0$, the probabilistic relation

$$P\left(\sup_{\mathbf{w}}|R(\mathbf{w}) - R_{emp}(\mathbf{w})| > \epsilon\right) \to 0 \qquad \text{as } N \to \infty \qquad (2.78)$$

holds (Vapnik, 1982). When Eq. (2.78) is satisfied, we say that a *uniform convergence in the weight vector* $\mathbf{w}$ *of the empirical mean risk to its expected value occurs.* Equivalently, provided that for any prescribed precision ϵ we can assert the inequality

$$P\left(\sup_{\mathbf{w}}|R(\mathbf{w}) - R_{emp}(\mathbf{w})| > \epsilon\right) < \alpha \qquad (2.79)$$

for some $\alpha > 0$, then as a consequence the following inequality also holds:

$$P(R(\mathbf{w}_{emp}) - R(\mathbf{w}_o) > 2\epsilon) < \alpha \qquad (2.80)$$

In other words, if the condition (2.79) holds, then with probability at least $(1 - \alpha)$, the solution $F(\mathbf{x}, \mathbf{w}_{emp})$ that minimizes the empirical risk functional $R_{emp}(\mathbf{w})$ will give an actual risk $R(\mathbf{w}_{emp})$ that deviates from the true minimum possible actual risk $R(\mathbf{w}_o)$ by an amount not exceeding 2ϵ. Indeed, the condition (2.79) implies that with probability $(1 - \alpha)$ the following two inequalities are satisfied simultaneously (Vapnik, 1982):

$$R(\mathbf{w}_{emp}) - R_{emp}(\mathbf{w}_{emp}) < \epsilon \qquad (2.81)$$

$$R_{emp}(\mathbf{w}_o) - R(\mathbf{w}_o) < \epsilon \qquad (2.82)$$

These two equations define the differences between the true risk and empirical risk functionals at $\mathbf{w} = \mathbf{w}_{emp}$ and $\mathbf{w} = \mathbf{w}_o$, respectively. Furthermore, since $\mathbf{w}_{emp}$ and $\mathbf{w}_o$ are the minimum points of $R_{emp}(\mathbf{w})$ and $R(\mathbf{w})$, respectively, it follows that

$$R_{emp}(\mathbf{w}_{emp}) \leq R_{emp}(\mathbf{w}_o) \qquad (2.83)$$

By adding the inequalities (2.81) and (2.82), and then using (2.83), we may write the following inequality

$$R(\mathbf{w}_{emp}) - R(\mathbf{w}_o) < 2\epsilon \qquad (2.84)$$

Also, since the inequalities (2.81) and (2.82) are both satisfied simultaneously with probability $(1 - \alpha)$, so is the inequality (2.84). We may also state that with probability α the inequality

$$R(\mathbf{w}_{emp}) - R(\mathbf{w}_o) > 2\epsilon$$

holds, which is a restatement of (2.80).

We are now ready to make a formal statement of the *principle of empirical risk minimization* in three interrelated parts (Vapnik, 1982, 1998):

1. In place of the risk functional $R(\mathbf{w})$, construct the empirical risk functional

$$R_{emp}(\mathbf{w}) = \frac{1}{N}\sum_{i=1}^{N} L(d_i, F(\mathbf{x}_i, \mathbf{w}))$$

on the basis of the training set of i.i.d. examples

$$(\mathbf{x}_i, d_i), \qquad i = 1, 2, ..., N$$

2. Let $\mathbf{w}_{emp}$ denote the weight vector that minimizes the empirical risk functional $R_{emp}(\mathbf{w})$ over the weight space $\mathcal{W}$. Then $R(\mathbf{w}_{emp})$ converges in probability to the minimum possible values of the actual risk $R(\mathbf{w})$, $\mathbf{w} \in \mathcal{W}$, as the size N of the training sample is made infinitely large, provided that the empirical risk functional $R_{emp}(\mathbf{w})$ converges uniformly to the actual risk functional $R(\mathbf{w})$.

3. Uniform convergence as defined by

$$P\left(\sup_{\mathbf{w} \in \mathcal{W}} |R(\mathbf{w}) - R_{emp}(\mathbf{w})| > \epsilon \right) \to 0 \qquad \text{as } N \to \infty$$

is a necessary and sufficient condition for the consistency of the principle of empirical risk minimization.

For a physical interpretation of this important principle, we offer the following observation. Prior to the training of a learning machine, all approximating functions are equally likely. As the training of the learning machine progresses, the likelihood of those approximating functions $F(\mathbf{x}_i, \mathbf{w})$ that are consistent with the training data set $\{\mathbf{x}_i, d_i\}_{i=1}^{N}$ is increased. As the size N of the training data set grows, and the input space is thereby "densely" populated, the minimum point of the empirical risk functional $R_{emp}(\mathbf{w})$ converges in probability to the minimum point of the true risk functional $R(\mathbf{w})$.

VC Dimension

The theory of uniform convergence of the empirical risk functional $R_{emp}(\mathbf{w})$ to the actual risk functional $R(\mathbf{w})$ includes bounds on the rate of convergence, which are based on an important parameter called the *Vapnik–Chervonenkis dimension,* or simply the *VC dimension,* named in honor of its originators, Vapnik and Chervonenkis (1971). The VC dimension is a measure of the *capacity* or *expressive power* of the family of classification functions realized by the learning machine.

To describe the concept of VC dimension in a manner suitable for our purposes, consider a binary pattern classification problem, for which the desired response is written as $d \in \{0, 1\}$. We use the term *dichotomy* to refer to a binary classification function or decision rule. Let $\mathcal{F}$ denote the ensemble of dichotomies implemented by a learning machine, that is,

$$\mathcal{F} = \{F(\mathbf{x}, \mathbf{w}): \mathbf{w} \in \mathcal{W}, F: \mathbb{R}^m \mathcal{W} \to \{0, 1\}\} \tag{2.85}$$

Let $\mathcal{L}$ denote the set of N points in the m-dimensional space $\mathcal{X}$ of input vectors, that is,

$$\mathcal{L} = \{\mathbf{x}_i \in \mathcal{X}; i = 1, 2, ..., N\} \tag{2.86}$$

A dichotomy implemented by the learning machine partitions $\mathcal{L}$ into two disjoint subsets $\mathcal{L}_0$ and $\mathcal{L}_1$, such that we may write

$$F(\mathbf{x},\mathbf{w}) = \begin{cases} 0 & \text{for } \mathbf{x} \in \mathcal{L}_0 \\ 1 & \text{for } \mathbf{x} \in \mathcal{L}_1 \end{cases} \tag{2.87}$$

Let $\Delta_{\mathcal{F}}(\mathcal{L})$ denote the number of distinct dichotomies implemented by the learning machine, and $\Delta_{\mathcal{F}}(l)$ denote the maximum of $\Delta_{\mathcal{F}}(\mathcal{L})$ over all $\mathcal{L}$ with $|\mathcal{L}| = l$, where $|\mathcal{L}|$ is the number of elements of $\mathcal{L}$. We say that $\mathcal{L}$ is *shattered* by $\mathcal{F}$ if $\Delta_{\mathcal{F}}(\mathcal{L}) = 2^{|\mathcal{L}|}$, that is, if all possible dichotomies of $\mathcal{L}$ can be induced by functions in $\mathcal{F}$. We refer to $\Delta_{\mathcal{F}}(l)$ as the *growth function.*

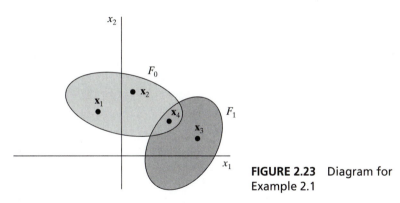

FIGURE 2.23 Diagram for Example 2.1

Example 2.1

Figure 2.23 illustrates a two-dimensional input space $\mathscr{X}$ consisting of four points $\mathbf{x}_1, \mathbf{x}_2, \mathbf{x}_3$, and $\mathbf{x}_4$. The decision boundaries of functions F_0 and F_1, indicated in the figure, correspond to the classes (hypotheses) 0 and 1 being true, respectively. From Fig. 2.23 we see that the function F_0 induces the dichotomy

$$\mathscr{D}_0 = \{\mathscr{S}_0 = \{\mathbf{x}_1, \mathbf{x}_2, \mathbf{x}_4\}, \mathscr{S}_1 = \{\mathbf{x}_3\}\}$$

On the other hand, the function F_1 induces the dichotomy

$$\mathscr{D}_1 = \{\mathscr{S}_0 = \{\mathbf{x}_1, \mathbf{x}_2\}, \mathscr{S}_1 = \{\mathbf{x}_3, \mathbf{x}_4\}\}$$

With the set $\mathscr{S}$ consisting of four points, the cardinality $|\mathscr{S}| = 4$. Hence,

$$\Delta_{\mathscr{F}}(\mathscr{S}) = 2^4 = 16$$

∎

Returning to the general discussion delineated by the ensemble $\mathscr{F}$ of dichotomies in Eq. (2.85) and the corresponding set of points $\mathscr{L}$ in Eq. (2.86), we may now formally define the VC dimension as (Vapnik and Chervonenkis, 1971; Kearns and Vazirani, 1994; Vidyasagar, 1997; Vapnik, 1998):

> The VC dimension of an ensemble of dichotomies $\mathscr{F}$ is the cardinality of the largest set $\mathscr{L}$ that is shattered by $\mathscr{F}$.

In other words, the VC dimension of $\mathscr{F}$, written as VCdim($\mathscr{F}$), is the largest N such that $\Delta_{\mathscr{F}}(N) = 2^N$. Stated in more familiar terms, the VC dimension of the set of classification functions $\{F(\mathbf{x}, \mathbf{w}): \mathbf{w} \in \mathcal{W}\}$ is the maximum number of training examples that can be learned by the machine without error for all possible binary labelings of the classification functions.

Example 2.2

Consider a simple decision rule in an m-dimensional space $\mathscr{X}$ of input vectors, which is described by

$$\mathscr{F}: y = \varphi(\mathbf{w}^T\mathbf{x} + b) \tag{2.88}$$

where $\mathbf{x}$ is an m-dimensional weight vector and b is a bias. The activation function φ is a threshold function; that is,

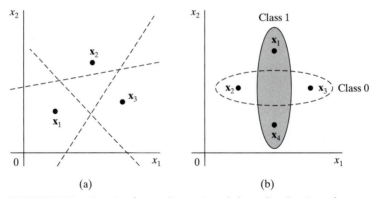

FIGURE 2.24   A pair of two-dimensional data distributions for Example 2.2.

$$\varphi(v) = \begin{cases} 1, & v \geq 0 \\ 0, & v < 0 \end{cases}$$

The VC dimension of the decision rule in Eq. (2.88) is given by

$$\text{VC dim}(\mathcal{F}) = m + 1 \tag{2.89}$$

To demonstrate this result, consider the situations described in Fig. 2.24 pertaining to a two-dimensional input space (i.e., $m = 2$). In Fig. 2.24a, we have three points x_1, x_2, and x_3. Three different possible labelings of these points are included in Fig. 2.24a, from which we readily see that a maximum of three lines can shatter these points. In Fig. 2.24b we have points x_1, x_2, x_3, and x_4, with points x_2 and x_3 labeled as 0 and points x_1 and x_4 labeled as 1. This time, however, we see that points x_1 and x_4 cannot be shattered from x_2 and x_3 by a line. The VC dimension of the decision rule described in Eq. (2.88) with $m = 2$ is therefore 3, which is in accord with the formula of Eq. (2.89).   ∎

Example 2.3

With the VC dimension providing a measure of the capacity of a set of classification (indicator) functions, we may be led to expect that a learning machine with many free parameters would have a high VC dimension, whereas a learning machine with few free parameters would have a low VC dimension. We now present a counterexample[13] to this statement.

Consider the one parameter family of indicator functions, defined by

$$f(x, a) = \text{sgn}(\sin(ax)), \qquad a \in \mathbb{R}$$

where $\text{sgn}(\cdot)$ is the signum function. Suppose we choose any number N and the requirement is to find N points that can be shattered. This requirement is satisfied by the set of functions $f(x, a)$ by choosing

$$x_i = 10^{-i}, \qquad i = 1, 2, ..., N$$

To separate these data points into two classes determined by the sequence

$$d_1, d_2, ..., d_N, \qquad d_i \in \{-1, 1\}$$

it is sufficient that we choose the parameter a according to the formula:

$$a = \pi\left(1 + \sum_{i=1}^{N} \frac{(1 - d_i)10^i}{2}\right)$$

We thus conclude that the VC dimension of the family of indicator functions $f(x, a)$ with a single free paramenter a is infinite.

■

Importance of the VC dimension and its Estimation

The VC dimension is a purely *combinatorial concept* that has no connection with the geometric notion of dimension. It plays a central role in statistical learning theory as will be shown in the material presented in the next two subsections. The VC dimension is also important from a design point of view. Roughly speaking, the number of examples needed to learn a class of interest reliably is proportional to the VC dimension of that class. Therefore, an estimate of the VC dimension is of primary concern.

In some cases the VC dimension is determined by the free parameters of a neural network. In most practical cases, however, it is difficult to evaluate the VC dimension by analytic means. Nevertheless, *bounds* on the VC dimension of neural networks are often *tractable*. In this context, the following two results are of special interest:[14]

1. *Let $\mathcal{N}$ denote an arbitrary feedforward network built up from neurons with a threshold (Heaviside) activation function:*

$$\varphi(v) = \begin{cases} 1 & \text{for } v \geq 0 \\ 0 & \text{for } v < 0 \end{cases}$$

   The VC dimension of $\mathcal{N}$ is $O(W \log W)$ where W is the total number of free parameters in the network.

This first result is due to Cover (1968) and Baum and Haussler (1989).

2. *Let $\mathcal{N}$ denote a multilayer feedforward network whose neurons use a sigmoid activation function*

$$\varphi(v) = \frac{1}{1 + \exp(-v)}$$

   The VC dimension of $\mathcal{N}$ is $O(W^2)$, where W is the total number of free parameters in the network.

This second result is due to Koiran and Sontag (1996). They arrived at this result by first showing that networks consisting of two types of neurons, one linear and the other using a threshold activation function, already have a VC dimension proportional to W^2. This is a rather surprising result, since a purely linear network has a VC dimension proportional to W as shown in Example 2.2, while a purely threshold neural network has a VC dimension proportional to $W \log W$ by virtue of result 1. The desired result pertaining to a sigmoid neural network is then obtained by invoking two approximations. First, neurons with threshold activation functions are approximated by sigmoidal ones with large synaptic weights. Second, linear neurons are approximated by sigmoidal neurons with small synaptic weights.

The important point to note here is that multilayer feedforward networks have a *finite* VC dimension.

Constructive Distribution-free Bounds on the Generalization Ability of Learning Machines

At this point in the discussion we find it instructive to consider the specific case of binary pattern classification, for which the desired response is defined by $d \in \{0, 1\}$. In a corresponding way the loss function has only two possible values as shown by

$$L(d, F(\mathbf{x}, \mathbf{w})) = \begin{cases} 0 & \text{if } F(\mathbf{x}, \mathbf{w}) = d \\ 1 & \text{otherwise} \end{cases} \tag{2.90}$$

Under these conditions the risk functional $R(\mathbf{w})$ and the empirical risk functional $R_{\text{emp}}(\mathbf{w})$ defined in Eqs. (2.72) and (2.74) respectively, assume the following interpretations:

- The risk functional $R(\mathbf{w})$ is the *probability of classification error* (i.e., error rate), denoted by $P(\mathbf{w})$.
- The empirical risk functional $R_{\text{emp}}(\mathbf{w})$ is the *training error* (i.e., frequency of errors made during the training session), denoted by $v(\mathbf{w})$.

Now, according to the *law of large numbers* (Gray and Davisson, 1986), the empirical frequency of occurrence of an event converges almost surely to the actual probability of that event as the number of trials (assumed to be independent and identically distributed) is made infinitely large. In the context of the discussion presented here, this result means that for any weight vector $\mathbf{w}$, which does not depend on the training set, and for any precision $\epsilon > 0$, the following condition holds (Vapnik, 1982):

$$P\big(|P(\mathbf{w}) - v(\mathbf{w})| > \epsilon\big) \to 0 \qquad \text{as } N \to \infty \tag{2.91}$$

where N is the size of the training set. Note, however, that the condition (2.91) does not imply that the classification rule (i.e., a particular weight vector $\mathbf{w}$) that minimizes the training error $v(\mathbf{w})$ will also minimize the probability of classification error $P(\mathbf{w})$. For a training set of sufficiently large size N, the proximity between $v(\mathbf{w})$ and $P(\mathbf{w})$ follows from a stronger condition, which stipulates that the following condition holds for any $\epsilon > 0$ (Vapnik, 1982):

$$P(\sup_{\mathbf{w}} |P(\mathbf{w}) - v(\mathbf{w})| > \epsilon) \to 0 \qquad \text{as } N \to \infty \tag{2.92}$$

In such a case, we speak of *the uniform convergence of the frequency of training errors to the probability* that $v(\mathbf{w}) = P(\mathbf{w})$.

The notion of VC dimension provides a bound on the rate of uniform convergence. Specifically, for the set of classification functions with VC dimension h, the following inequality holds (Vapnik, 1982, 1998):

$$P\big(\sup_{\mathbf{w}} |P(\mathbf{w}) - v(\mathbf{w})| > \epsilon\big) < \left(\frac{2eN}{h}\right)^h \exp(-\epsilon^2 N) \tag{2.93}$$

where N is the size of the training sample and e is the base of the natural logarithm. We want to make the right-hand side of the inequality (2.93) small for large N in order to achieve uniform convergence. The factor $\exp(-\epsilon^2 N)$ is helpful in this regard, since it decays exponentially with increasing N. The remaining factor $(2eN/h)^h$ represents a *bound* on the growth function $\Delta_{\mathcal{F}}(l)$ for the family of functions $\mathcal{F} = \{F(\mathbf{x}, \mathbf{w}); \mathbf{w} \in \mathcal{W}\}$ for $l \geq h \geq 1$ as obtained by *Sauer's lemma*.[15] Provided that this function does *not* grow too fast, the right-hand side will go to zero as N goes to infinity; this requirement is satisfied if the VC dimension h is finite. In other words, a finite VC dimension is a necessary and sufficient condition for uniform convergence of the principle of empirical risk minimization. If the input space $\mathcal{X}$ has finite cardinality, any family of dichotomies $\mathcal{F}$ will have finite VC dimension with respect to $\mathcal{X}$, although the reverse is not necessarily true.

Let α denote the probability of occurrence of the event

$$\sup_{\mathbf{w}} |P(\mathbf{w}) - \nu(\mathbf{w})| \geq \epsilon$$

Then, with probability $1 - \alpha$, we may state that for all weight vectors $\mathbf{w} \in \mathcal{W}$ the following inequality holds:

$$P(\mathbf{w}) < \nu(\mathbf{w}) + \epsilon \tag{2.94}$$

Using the bound described in Eq. (2.93) and the definition for the probability α, we may thus set

$$\alpha = \left(\frac{2eN}{h}\right)^h \exp(-\epsilon^2 N) \tag{2.95}$$

Let $\epsilon_0(N, h, \alpha)$ denote the special value of ϵ that satisfies Eq. (2.95). Hence, we readily obtain the following important result (Vapnik, 1992):

$$\epsilon_0(N, h, \alpha) = \sqrt{\frac{h}{N}\left[\log\left(\frac{2N}{h}\right) + 1\right] - \frac{1}{N}\log\alpha} \tag{2.96}$$

We refer to $\epsilon_0(N, h, \alpha)$ as a *confidence interval*, the value of which depends on the size N of the training sample, the VC dimension h, and the probability α.

The bound described in (2.93) with $\epsilon = \epsilon_0(N, h, \alpha)$ is achieved for the worst case $P(\mathbf{w}) = \frac{1}{2}$, but not, unfortunately, for small $P(\mathbf{w})$, which is the case of interest in practice. For small $P(\mathbf{w})$, a more useful bound is obtained by considering a modification of the inequality (2.93) as follows (Vapnik, 1982, 1998):

$$P\left(\sup_{\mathbf{w}} \frac{|P(\mathbf{w}) - \nu(\mathbf{w})|}{\sqrt{P(\mathbf{w})}} > \epsilon\right) < \left(\frac{2eN}{h}\right)^h \exp\left(-\frac{\epsilon^2 N}{4}\right) \tag{2.97}$$

In the literature, different results are reported for the bound in (2.97), depending on which particular form of inequality is used for its derivation. Nevertheless, they all have a similar form. From (2.97) it follows that with probability $1 - \alpha$, and simultaneously for all $\mathbf{w} \in \mathcal{W}$ (Vapnik, 1992, 1998),

$$P(\mathbf{w}) \leq \nu(\mathbf{w}) + \epsilon_1(N, h, \alpha, \nu) \tag{2.98}$$

where $\epsilon_1(N, h, \alpha, v)$ is a new confidence interval defined in terms of the former confidence interval $\epsilon_0(N, h, \alpha)$ as follows (see Problem 2.25):

$$\epsilon_1(N, h, \alpha, v) = 2\epsilon_0^2(N, h, \alpha)\left(1 + \sqrt{1 + \frac{v(\mathbf{w})}{\epsilon_0^2(N, h, \alpha)}}\right) \tag{2.99}$$

This second confidence interval depends on the training error $v(\mathbf{w})$. For $v(\mathbf{w}) = 0$ it reduces to the special form

$$\epsilon_1(N, h, \alpha, 0) = 4\epsilon_0^2(N, h, \alpha) \tag{2.100}$$

We may now summarize the two bounds we have derived for the rate of uniform convergence:

1. In general, we have the following bound on the rate of uniform convergence:

$$P(\mathbf{w}) \le v(\mathbf{w}) + \epsilon_1(N, h, \alpha, v)$$

where $\epsilon_1(N, h, \alpha, v)$ is as defined in Eq. (2.99).

2. For a small training error $v(\mathbf{w})$ close to zero, we have

$$P(\mathbf{w}) \lesssim v(\mathbf{w}) + 4\epsilon_0^2(N, h, \alpha)$$

which provides a fairly precise bound for real-case learning.

3. For a large training error $v(\mathbf{w})$ close to unity, we have the bound

$$P(\mathbf{w}) \lesssim v(\mathbf{w}) + \epsilon_0(N, h, \alpha)$$

Structural Risk Minimization

The *training error* is the frequency of errors made by a learning machine of some weight vector $\mathbf{w}$ during the training session. Similarly, the *generalization error* is defined as the frequency of errors made by the machine when it is tested with examples not seen before. Here it is assumed that the test data are drawn from the same population as the training data. Let these two errors be denoted by $v_{\text{train}}(\mathbf{w})$ and $v_{\text{gene}}(\mathbf{w})$, respectively. Note that $v_{\text{train}}(\mathbf{w})$ is the *same* as the $v(\mathbf{w})$ used in the previous subsection; we used $v(\mathbf{w})$ there to simplify the notation. Let h be the VC dimension of a family of classification functions $\{F(\mathbf{x}, \mathbf{w}); \mathbf{w} \in \mathcal{W}\}$ with respect to the input space $\mathcal{X}$. Then, in light of the theory on the rate of uniform convergence, we may state that with probability $1 - \alpha$, for a number of training examples $N > h$, and simultaneously for all classification functions $F(\mathbf{x}, \mathbf{w})$, the generalization error $v_{\text{gene}}(\mathbf{w})$ is lower than a *guaranteed risk* defined by the sum of a pair of competing terms (Vapnik, 1992, 1998)

$$v_{\text{guarant}}(\mathbf{w}) = v_{\text{train}}(\mathbf{w}) + \epsilon_1(N, h, \alpha, v_{\text{train}}) \tag{2.101}$$

where the confidence interval $\epsilon_1(N, h, \alpha, v_{\text{train}})$ is itself defined by Eq. (2.99). For a fixed number of training examples N, the training error decreases monotonically as the capacity or VC dimension h is increased, whereas the confidence interval increases monotonically. Accordingly, both the guaranteed risk and the generalization error go through a minimum. These trends are illustrated in a generic way in Fig. 2.25. Before the minimum point is reached, the learning problem is *overdetermined* in the sense that the machine capacity h is too small for the amount of training detail. Beyond the mini-

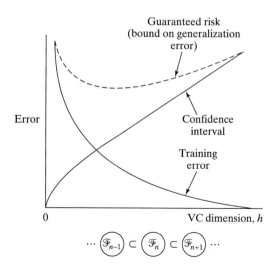

FIGURE 2.25   Illustration of the relationship between training error, confidence interval, and guaranteed risk.

mum point, the learning problem is *underdetermined* because the machine capacity is too large for the amount of training data.

The challenge in solving a supervised learning problem is therefore to realize the best generalization performance by matching the machine capacity to the available amount of training data for the problem at hand. The *method of structural risk minimization* provides an inductive procedure for achieving this goal by making the VC dimension of the learning machine a *control* variable (Vapnik, 1992, 1998). To be specific, consider an ensemble of pattern classifiers $\{F(\mathbf{x}, \mathbf{w}); \mathbf{w} \in \mathcal{W}\}$, and define a nested structure of n such machines

$$\mathcal{F}_k = \{F(\mathbf{x}, \mathbf{w}); \mathbf{w} \in \mathcal{W}_k\}, \qquad k = 1, 2, ..., n \qquad (2.102)$$

such that we have (see Fig. 2.25)

$$\mathcal{F}_1 \subset \mathcal{F}_2 \subset \cdots \subset \mathcal{F}_n \qquad (2.103)$$

where the symbol $\subset$ signifies "is contained in." Correspondingly, the VC dimensions of the individual pattern classifiers satisfy the condition

$$h_1 \leq h_2 \leq \cdots \leq h_n \qquad (2.104)$$

which implies that the VC dimension of each pattern classifier is finite. Then, the method of structural risk minimization may proceed as follows:

- The empirical risk (i.e., training error) for each pattern classifier is minimized.
- The pattern classifier $\mathcal{F}^*$ with the smallest guaranteed risk is identified; this particular machine provides the best compromise between the training error (i.e., quality of approximation of the training data) and the confidence interval, (i.e., complexity of the approximating function) which compete with each other.

Our goal is to find a network structure such that decreasing the VC dimension occurs at the expense of the smallest possible increase in training error.

The principle of structural risk minimization may be implemented in a variety of ways. For example, we may vary the VC dimension h by varying the number of hidden neurons. Specifically, we evaluate an ensemble of fully connected multilayer feedforward networks, in which the number of neurons in one of the hidden layers is increased in a monotonic fashion. The principle of structural risk minimization states that the best network in this ensemble is the one for which the guaranteed risk is the minimum.

The VC dimension is not only central to the principle of structural risk minimization but also to an equally powerful learning model called probably approximately correct (PAC). This latter model, discussed in the next section, completes the last part of the chapter dealing with probabilistic and statistical aspects of learning.

2.15 PROBABLY APPROXIMATELY CORRECT MODEL OF LEARNING

The *probably approximately correct (PAC)* learning model is credited to Valiant (1984). As the name implies, the PAC model is a probabilistic framework for the study of learning and generalization in binary classification systems. It is closely related to supervised learning.

We begin with an environment $\mathcal{X}$. A set of $\mathcal{X}$ is called a *concept,* and a set of subsets of $\mathcal{X}$ is called a *concept class.* An *example* of a concept is an object in the domain of interest, together with a class label. If the example is a member of the concept, we refer to it as a *positive example*; if the object is *not* a member of the concept, we refer to it as a *negative example.* A concept for which examples are provided is called a *target concept.* We may acquire a sequence of training data of length N for a target concept c as shown by

$$\mathcal{T} = \{(\mathbf{x}_i, c(\mathbf{x}_i)\}_{i=1}^N \tag{2.105}$$

which may contain repeated examples. The examples $\mathbf{x}_1, \mathbf{x}_2, ..., \mathbf{x}_N$ are drawn from the environment $\mathcal{X}$ at random, according to some fixed but unknown probability distribution. The following points are also noteworthy in Eq. (2.105):

- The target concept $c(\mathbf{x}_i)$ is treated as a function from $\mathcal{X}$ to $\{0, 1\}$. Moreover, $c(\mathbf{x}_i)$ is assumed to be unknown.
- The examples are usually assumed to be statistically independent, which means that the joint probability density function of any two examples, $\mathbf{x}_i$ and $\mathbf{x}_j$, say, is equal to the product of their individual probability density functions.

In the context of our previous terminology, the environment $\mathcal{X}$ may be identified with the input space of a neural network, and the target concept may be identified with the desired response for the network.

The set of concepts derived from the environment $\mathcal{X}$ is referred to as a *concept space* $\mathcal{C}$. For example, the concept space may contain "the letter A," "the letter B," and so on. Each of these concepts may be coded differently to generate a set of positive examples and a set of negative examples. In the framework of supervised learning, however, we have another set of concepts. A learning machine typically represents a set of functions, with each function corresponding to a specific state. For example, the machine may be designed to recognize "the letter A," "the letter B," and so on. The set of all functions (i.e., concepts) determined by the learning machine is referred to as a *hypothesis space* $\mathcal{G}$. The hypothesis space may or may not be equal to the concept space. In a way the notions of concept space and hypothesis space are analogous to the function $f(\mathbf{x})$ and approximating function $F(\mathbf{x}, \mathbf{w})$, respectively, that were used in the previous section.

Suppose then we are given a target concept $c(\mathbf{x}) \in \mathscr{C}$, which takes only the value 0 or 1. We wish to learn this concept by means of a neural network by training it on the data set $\mathscr{T}$ defined in Eq. (2.105). Let $g(\mathbf{x}) \in \mathscr{G}$ denote the hypothesis corresponding to the input–output mapping that results from this training. One way of assessing the success of the learning process is to measure how close the hypothesis $g(\mathbf{x})$ is to the target concept $c(\mathbf{x})$. There will naturally be errors incurred, making $g(\mathbf{x}) \neq c(\mathbf{x})$. The reason errors are incurred is that we are trying to learn a function on the basis of limited information available about that function. The probability of training error is defined by

$$\nu_{\text{train}} = P(\mathbf{x} \in \mathscr{X} : g(\mathbf{x}) \neq c(\mathbf{x})) \tag{2.106}$$

The probability distribution in this equation must be the same as the one responsible for generating the examples. The goal of PAC learning is to ensure that ν_{train} is *usually small*. The domain that is available to the learning algorithm is controlled by the size N of the training sample $\mathscr{T}$. In addition, the learning algorithm is supplied with two control parameters:

- *Error parameter* $\epsilon \in (0, 1]$. This parameter specifies the error allowed in a good approximation of the target concept $c(\mathbf{x})$ by the hypothesis $g(\mathbf{x})$.
- *Confidence parameter* $\delta \in (0,1]$. This second parameter controls the likelihood of constructing a good approximation.

We may thus visualize the PAC learning model as depicted in Fig. 2.26.

With this background we may now formally state the PAC learning model (Valiant, 1984; Kearns and Vazirani, 1994; Vidyasagar, 1997):

Let $\mathscr{C}$ be a concept class over the environment $\mathscr{X}$. The concept class $\mathscr{C}$ is said to be PAC learnable if there exists an algorithm $\mathscr{L}$ with the following property: For every target concept $c \in \mathscr{C}$, for every probability distribution on $\mathscr{X}$, and for all $0 < \epsilon < 1/2$ and $0 < \delta < 1/2$, if the learning algorithm $\mathscr{L}$ is supplied the set of training examples $\mathscr{T} = \{(\mathbf{x}_i, c(\mathbf{x}_i))\}_{i=1}^{N}$ and the parameters ϵ and δ, then with probability at least $1 - \delta$, the learning algorithm $\mathscr{L}$ outputs a hypothesis g with error $\nu_{\text{train}} \leq \epsilon$. This probability is taken over the random examples drawn from the set $\mathscr{T}$ and any internal randomization that may exist in the learning algorithm $\mathscr{L}$. The sample size N must be greater than a function of δ and ϵ.

In other words, provided that the size N of the training sample $\mathscr{T}$ is large enough, after the neural network has been trained on that data set it is "probably" the case that the

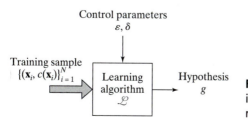

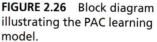

FIGURE 2.26   Block diagram illustrating the PAC learning model.

input–output mapping computed by the network is "approximately correct." Note that although there is a dependence on δ and ε, the number of examples, N, does not have to be dependent on the target concept c or the underlying probability distribution of $\mathscr{X}$.

Sample Complexity

In PAC learning theory, an issue of particular interest with practical implications is the issue of *sample complexity*. The focus in this issue is on how many random examples should be presented to the learning algorithm for it to acquire sufficient information to learn an unknown target concept c chosen from the concept class $\mathscr{C}$. Or, how large should the size N of the training set $\mathscr{T}$ be?

The issue of sample complexity is closely linked with the VC dimension. However, before proceeding further on this issue, we need to define the notion of a consistent concept. Let $\mathscr{T} = \{(\mathbf{x}_i, d_i)\}_{i=1}^N$ be any set of labeled examples, where each $\mathbf{x}_i \in \mathscr{X}$ and each $d_i \in (0,1)$. Let c be a target concept over the environment $\mathscr{X}$. Then, concept c is said to be *consistent* with the training set $\mathscr{T}$ (or, equivalently, $\mathscr{T}$ is consistent with c) if for all $1 \leq i \leq N$ we have $c(\mathbf{x}_i) = d_i$ (Kearns and Vazarini, 1994). Now, as far as PAC learning is concerned, it is *not* the size of the set of input–output functions computable by a neural network that is crucial, but rather it is the VC dimension of the network. More precisely, we have a key result presented in two parts (Blumer et al., 1989; Anthony and Biggs, 1992; Vidyasagar, 1997):

Consider a neural network with a finite VC dimension $h \geq 1$.

1. Any consistent learning algorithm for that neural network is a PAC learning algorithm.
2. There is a constant K such that a sufficient size of training set $\mathscr{T}$ for any such algorithm is

$$N = \frac{K}{\epsilon}\left(h\log\left(\frac{1}{\epsilon}\right) + \log\left(\frac{1}{\delta}\right)\right) \tag{2.107}$$

where ϵ is the error parameter and δ is the confidence parameter.

The generality of this result is impressive: it is applicable to a supervised learning process regardless of the type of learning algorithm used, and the underlying probability distribution for generating the labeled examples. It is the broad generality of this result that has made it a subject of intensive research interest in neural network literature. Comparison of results predicted from bounds on measures based on the VC dimension with experimental results reveals a wide numerical discrepancy.[16] In a sense this should not be surprising because the discrepancy is merely a reflection of the *distribution-free, worst-case* nature of the theoretical measures, and on *average* we can always do better.

Computational Complexity

Another issue of primary concern in PAC learning is that of computational complexity. This issue concerns the computational effectiveness of a learning algorithm. More precisely, *computational complexity* deals with the worst-case "running time" needed to train a neural network (learning machine), given a set of labeled examples of some finite size N.

In a practical situation, the running time of an algorithm naturally depends on the speed with which the underlying computations are performed. From a theoretical perspective, however, the intention is to have a definition of running time that is independent of the device used to do the computations. With this intention in mind, running time, and therefore computational complexity, is usually measured in terms of the number of operations (additions, multiplications, and storage) needed to perform the computation.

In assessing the computational complexity of a learning algorithm, we like to know how it varies with the example size m (i.e., size of the input layer of the neural network being trained). For the algorithm to be computationally *efficient* in this context, the running time should be $O(m^r)$ for some fixed integer $r \geq 1$. In such a case, the running time is said to increase polynomially with m, and the algorithm itself is said to be a *polynomial time algorithm*. Learning tasks performed by a polynomial time algorithm are usually regarded as "easy" (Anthony and Biggs, 1992).

The other parameter that requires attention is the error parameter ϵ. Whereas in the case of sample complexity the parameter ϵ is fixed but arbitrary, in assessing the computational complexity of a learning algorithm we like to know how it varies with ϵ. Intuitively, we expect that as ϵ is decreased, the learning task under study would become more difficult. It follows then that some condition would have to be imposed on the time taken for the algorithm to produce a probably approximately correct output. For efficient computation, the appropriate condition is to have the running time polynomial in $1/\epsilon$.

By putting these considerations together, we may make the following formal statement on computational complexity (Anthony and Biggs, 1992):

> A learning algorithm is computationally efficient with respect to error parameter ϵ, example size m, and size N of the training set if its running time is polynomial in N and if there is a value of $N_0(\delta, \epsilon)$ sufficient for PAC learning that is polynomial in both m and ϵ^{-1}.

2.16 SUMMARY AND DISCUSSION

In this chapter we discussed some important issues relating to the many facets of the learning process in the context of neural networks. In so doing we have laid down the foundations for much of the material in the rest of this book. The five learning rules, *error-correction learning, memory-based learning, Hebbian learning, competitive learning,* and *Boltzmann learning,* are basic to the design of neural networks. Some of these algorithms require the use of a teacher and some do not. The important point is that these rules enable us to go far beyond the reach of linear adaptive filters in both capability and universality.

In the study of supervised learning, a key provision is a "teacher" capable of supplying exact corrections to the network outputs when an error occurs as in error-correction learning; or "clamping" the free-running input and output units of the network to the environment as in Boltzmann learning. Neither of these models is possible in biological organisms, which have neither the exact reciprocal nervous connections needed for the back propagation of error corrections (in a multilayer feedforward network) nor the nervous means for the imposition of behavior from outside. Nevertheless, supervised learning has established itself as a powerful paradigm for the design of artificial neural networks, as is demonstrated in Chapters 3 through 7.

In contrast, self-organized (unsupervised) learning rules such as Hebbian learning and competitive learning are motivated by neurobiological considerations. However, to improve our understanding of self-organized learning, we also need to look at Shannon's *information theory* for relevant ideas. Here we should mention the *maximum mutual information (Infomax) principle* due to Linsker (1988a, b), which provides the mathematical formalism for the processing of information in a self-organized neural network in a manner somewhat analogous to the transmission of information in a communication channel. The Infomax principle and its variants are discussed in Chapter 10.

A discussion of learning methods would be incomplete without mentioning the *Darwinian selective learning model* (Edelman, 1987; Reeke et al., 1990). *Selection* is a powerful biological principle with applications in both evolution and development. It is at the heart of the immune system (Edelman, 1973), the best understood biological recognition system. The Darwinian selective learning model is based on the theory of neural group selection. It presupposes that the nervous system operates by a form of selection akin to natural selection in evolution but takes place within the brain during the lifetime of each animal. According to this theory, the basic operational units of the nervous system are not single neurons but rather local groups of strongly interconnected cells. The membership of neurons in a group is changed by alterations in the neurons' synaptic weights. Local competition and cooperation among cells are clearly necessary to produce local order in the network. A collection of neuronal groups is referred to as a *repertoire*. Groups in a repertoire respond best to overlapping but similar input patterns due to the random nature of neural growth. One or more neuronal groups in a repertoire respond to every input pattern, thereby ensuring some response to unexpected input patterns that may be important. Darwinian selective learning is different from the learning algorithms commonly used in neural network design in that it assumes that there are many subnetworks by design, and that only those with the desired response are selected during the training process.

We complete this discussion with some concluding remarks on statistical and probabilistic aspects of learning. The VC dimension has established itself as a central parameter in statistical learning theory. It is basic to structural risk minimization and the probably approximately correct (PAC) model of learning. The VC dimension is an integral part of the underlying theory of so-called support vector machines, discussed in Chapter 6. In Chapter 7 we discuss a class of committee machines based on boosting, the theory of which is rooted in PAC learning.

As we progress through the rest of the book there will be many occasions and good reasons for revisiting the material presented in this chapter on the fundamentals of learning processes.

NOTES AND REFERENCES

1. The word "algorithm" is derived from the name of the Persian mathematician Mohammed al-Kowârisimi, who lived during the ninth century and who is credited with developing the step-by-step rules for the addition, subtraction, multiplication, and divi-

sion of ordinary decimal numbers. When his name was written in Latin it became Algorismus, from which *algorithm* is derived (Harel, 1987).

2. The nearest neighbor rule embodies a huge literature; see the collection of papers edited by Dasarathy (1991). This book includes the seminal work of Fix and Hodges (1951) and many other important papers on nearest neighbor pattern-classification techniques.

3. For a detailed review of Hebbian synapses, including a historical account, see Brown et al. (1990) and Frégnac and Schulz (1994). For additional review material, see Constantine-Paton et al. (1990).

4. Long-Term Potentiation—Physiological Evidence for the Hebbian Synapse
Hebb (1949) provided us with a way to think about synaptic memory mechanisms, but it was nearly a quarter of a century before experimental evidence was obtained in support of his proposals. In 1973, Bliss and Lomo published a paper describing a form of activation-induced synaptic modification in an area of the brain called the *hippocampus*. They applied pulses of electrical stimulation to the major pathway entering this structure while recording the synaptically evoked responses. When they were confident that they had characterized a stable baseline response morphology, they applied brief, high frequency trains of pulses to the same pathway. When they resumed application of the test pulses, they found the responses to be much larger in amplitude. Of most interest to memory researchers was the finding that this effect was very long lasting. They called the phenomenon *long-term potentiation* (LTP).

There are now hundreds of papers published every year on the LTP phenomenon, and we know much about the underlying mechanisms. We know, for example, that the potentiation effects are restricted to the activated pathways. We also know that LTP shows a number of associative properties. What we mean by associative properties is that there are interaction effects between *co-active* pathways. In particular, if a weak input that would not normally induce an LTP effect is paired with a strong input, the weak input can be potentiated. This is called an *associative property* because it is similar to the associative properties of learning systems. In Pavlov's conditioning experiments, for example, a neutral (weak) auditory stimulus was paired with a strong (food) stimulus. The pairing resulted in the appearance of a *conditioned response,* salivation in response to the auditory stimulus.

Much of the experimental work in this area has focused on the associative properties of LTP. Most of the synapses that have been shown to support LTP utilize glutamate as the neurotransmitter. It turns out, however, that there are a number of different receptors in the postsynaptic neuron that respond to glutamate. These receptors all have different properties, but we will consider just two of them. The main synaptic response is induced by activation of the AMPA receptor (these receptors are named according to the drugs to which they respond most strongly, but they are all glutamate receptors). When a response is recorded in an LTP experiment, it is primarily attributable to the activation of AMPA receptors. After synaptic activation the glutamate is released and binds with the receptors in the postsynaptic membrane. Ion channels that are part of the AMPA receptors open up, leading to the current flow that is the basis of the synaptic response.

The second type of glutamate receptor, the NMDA receptor, has some interesting properties. Glutamate binding with the NMDA receptor is not enough to open the associated ion channel. That channel remains blocked until a sufficiently large voltage change has been produced by synaptic activity (involving the AMPA receptors). Consequently, while AMPA receptors are chemically dependent, the NMDA receptors are both chemically dependent and voltage dependent. We need one other piece of information to see the importance of this difference. The ion channel associated with the AMPA receptor is linked to the movement of sodium ions (which produces the synaptic currents). The ion channel linked to the NMDA receptor allows calcium to move into the cell. While calcium movement also contributes to the membrane currents, its main role is as a signal that triggers a chain of events leading to a long-lasting increase in the strength of the response associated with the AMPA receptor.

We now have our mechanism for the Hebbian synapse. The NMDA receptor requires *both* presynaptic activity (glutamate release) *and* postsynaptic activity. How would that normally come about? By ensuring that there is a sufficiently strong input. Thus, when we pair a weak input with a strong input, the weak input releases its own glutamate, while the strong input ensures that there is a sufficiently strong voltage change to activate the NMDA receptors associated with the weak synapse.

Although Hebb's original proposal was for a unidirectional learning rule, neural networks are considerably more flexible if a *bidirectional* learning rule is used. It is an advantage to have synapses in which the synaptic weight can be decreased as well as increased. It is reassuring to know that there is also experimental evidence for a synaptic depression mechanism. If weak inputs are activated without the combined activation of strong inputs, the synaptic weight is often weakened. This is most typically seen in response to low-frequency activation of synaptic systems, and the phenomenon is called *long-term depression* (LTD). There is also some evidence for what is called a *heterosynaptic* depression effect. While LTD is a depression that is restricted to the activated input, heterosynaptic depression is restricted to the nonactivated input.

5. The idea of competitive learning may be traced back to the early works of von der Malsburg (1973) on the self-organization of orientation-sensitive nerve cells in the striate cortex, Fukushima (1975) on a self-organizing multilayer neural network known as the *neocognitron,* Willshaw and von der Malsburg (1976) on the formation of patterned neural connections by self-organization, and Grossberg (1972, 1976a,b) on adaptive pattern classification. Also, there is substantial evidence for competitive learning playing an important role in the formation of topographic maps in the brain (Durbin et al., 1989), and recent experimental work by Ambros-Ingerson et al. (1990) provides further physiological justification for competitive learning.

6. The use of lateral inhibition, as indicated in Fig. 2.4, is fashioned from neurobiological systems. Most sensory tissues, namely, retina of the eye, cochlea of the ear, and pressure-sensitive nerves of the skin, are organized in such a way that stimulation of any given location produces inhibition in the surrounding nerve cells (Arbib, 1989; Fischler and Firschein, 1987). In human perception, lateral inhibition manifests itself in a phenomenon called *Mach bands,* named after the physicist Ernest Mach (1865). For example, if we look at a sheet of paper half white and half black, we will see parallel to the boundary a "brighter than bright" band on the white side and a "darker than dark" band on the black side, even though in reality both of them have a uniform density. Mach bands are not physically present; rather, they are a visual illusion, representing "overshoots" and "undershoots" caused by the differentiating action of lateral inhibition.

7. The importance of statistical thermodynamics in the study of computing machinery was well recognized by John von Neumann. This is evidenced by the third of his five lectures on *Theory and Organization of Complicated Automata* at the University of Illinois in 1949. In his third lecture, on "Statistical Theories of Information," von Neumann said:

> Thermodynamical concepts will probably enter into this new theory of information. There are strong indications that information is similar to entropy and that degenerative processes of entropy are paralleled by degenerative processes in the processing of information. It is likely that you cannot define the function of an automaton, or its efficiency, without characterizing the milieu in which it works by

means of statistical traits like the ones used to characterize a milieu in thermody-namics. The statistical variables of the automaton's milieu will, of course, be some-what more involved than the standard thermodynamic variable of temperature, but they will be similar in character.

8. It appears that the term "reinforcement learning" was coined by Minsky (1961) in his early studies of artificial intelligence, and then independently in control theory by Waltz and Fu (1965). However, the basic idea of "reinforcement" has its origins in experimental studies of animal learning in psychology (Hampson, 1990). In this context it is particu-larly illuminating to recall Thorndike's classical *law of effect* (Thorndike, 1911, p 244):

> Of several responses made to the same situation, those which are accompanied or closely followed by satisfaction to the animal will, other things being equal, be more firmly connected with the situation, so that, when it recurs, they will be more likely to recur; those which are accompanied or closely followed by discomfort to the animal will, other things being equal, have their connections with that situation weakened, so that, when it recurs, they will be less likely to occur. The greater the satisfaction or discomfort, the greater the strengthening or weakening of the bond.

Although it cannot be claimed that this principle provides a complete model of biological behavior, its simplicity and common sense approach have made it an influential learning rule in the classical approach to *reinforcement learning*.

9. The plant output is typically a physical variable. To control the plant, we clearly need to know the value of this variable; that is, we must measure the plant output. The system used for the measurement of a physical variable is called a *sensor*. To be precise there-fore, the block diagram of Fig. 2.13 should include a sensor in its feedback path. We have omitted the sensor which, by implication, means that the transfer function of the sensor is assumed to be unity.

10. The "cocktail party phenomenon" refers to the remarkable human ability to selectively attend to and follow one source of auditory input in a noisy environment (Cherry, 1953; Cherry and Taylor, 1954). This ability manifests itself in a combination of three processes performed in the auditory system:
- *Segmentation*. The incoming auditory signal is segmented into individual channels, with each channel providing meaningful information about a listener's environment. Among the heuristics used by the listener to do this segmentation, *spatial location* is perhaps the most important (Moray, 1959).
- *Attention*. This pertains to the ability of the listener to focus attention on one channel while blocking attention to irrelevant channels (Cherry, 1953).
- *Switching*. This third process involves the ability to switch attention from one channel to another, which is probably mediated in a top-down manner by "gating" the incom-ing auditory signal (Wood and Cowan, 1995).

The conclusion to be drawn from these points is that the processing performed on the incoming auditory signal is indeed of a *spatiotemporal* kind.

11. The problem of designing an optimum linear filter that provides the theoretical frame-work for linear adaptive filters was first conceived by Kolmogorov (1942) and solved shortly afterward independently by Wiener (1949). On the other hand, a formal solution to the optimum nonlinear filtering problem is mathematically intractable. Nevertheless, in the 1950s a great deal of brilliant work was done by Zadeh (1953), Wiener and his col-laborators (Wiener, 1958), and others that did much to clarify the nature of the problem.

Gabor was the first to conceive the idea of nonlinear adaptive filter in 1954, and went on to build such a filter with the aid of collaborators (Gabor et al., 1960). Basically, Gabor proposed a shortcut through the mathematical difficulties of nonlinear adaptive filtering by constructing a filter that optimizes its response through learning. The output of the filter is expressed in the form

$$y(n) = \sum_{n=0}^{N} w_n x(n) + \sum_{n=0}^{N} \sum_{m=0}^{N} w_{n,m} x(n) x(m) + \cdots$$

where $x(0), x(1), \dots, x(N)$ are samples of the filter input. (This polynomial is now referred to as the *Gabor–Kolmogorov polynomial* or *Volterra series*.) The first term of the polynomial represents a linear filter characterized by a set of coefficients $\{w_n\}$. The second term characterized by a set of dyadic coefficients $\{w_{n,m}\}$ is nonlinear; this term contains the products of two samples of the filter input, and so on for the higher-order terms. The coefficients of the filter are adjusted via gradient descent to minimize the mean-square value of the difference between a target (desired) response $d(N)$ and the actual filter output $y(N)$.

12. The cost function $L(d, F(\mathbf{x}, \mathbf{w}))$ defined in Eq. (2.71) applies to a scalar d. In the case of a vector $\mathbf{d}$ as the desired response, the approximating function takes the vector-valued form $\mathbf{F}(\mathbf{x}, \mathbf{w})$. In this case we use the squared Euclidean distance

$$L(\mathbf{d}, \mathbf{F}(\mathbf{x}, \mathbf{w})) = \|\mathbf{d} - \mathbf{F}(\mathbf{x}, \mathbf{w})\|^2$$

as the loss function. The function $\mathbf{F}(\cdot, \cdot)$ is a vector-valued function of its arguments.

13. According to Burges (1998), Example 2.3 that first appeared in Vapnik (1995) is due to E. Levin and J. S. Denker.

14. The upper bound of order $W \log W$ for the VC dimension of a feedforward neural network constructed from linear threshold units (perceptrons) was obtained by Baum and Haussler (1989). Subsequently, Maass (1993) showed that a lower bound also of order $W \log W$ holds for this class of networks.

    The first upper bound on the VC dimension of a sigmoidal neural network was derived in Macintyre and Sontag (1993). Subsequently, Koiran and Sontag (1996) addressed an open question raised in Maass (1993):

    "Is the VC dimension of analog neural nets with the sigmoidal activation function $\sigma(y) = 1/1 + e^{-y}$ bounded by a polynomial in the number of programmable parameters?"

    Koiran and Sontag answered this question in the affirmative in their 1996 paper, as described in the text.

    This question has also been answered in the affirmative in Karpinski and Macintyre (1997). In this latter paper, a complicated method based on differential topology is used to show that the VC dimension of a sigmoidal neural network used as pattern classifier is bounded above by $O(W^4)$. There is a large gap between this upper bound and the lower bound derived in Koiran and Sontag (1996). In Karpinski and Macintyre (1997) it is conjectured that their upper bound could be lowered.

15. *Sauer's lemma* may be stated as (Sauer, 1972; Anthony and Biggs, 1992; Vidyasagar, 1997):

    Let $\mathcal{F}$ denote the ensemble of dichotomies implemented by a learning machine. If VCdim$(\mathcal{F}) = h$ with h finite, and $l \geq h \geq 1$, then the growth function $\Delta_{\mathcal{F}}(l)$ is bounded above by $(el/h)^h$ where e is the base of the natural logarithm.

16. In this note we present summaries of four important studies reported in the literature on sample complexity and the related issue of generalization.

    First, Cohn and Tesauro (1992) present a detailed experimental study on the practical value of bounds on sample complexity based on the VC dimension as a design tool for pattern classifiers. In particular, the experiments were designed to test the relationship between the generalization performance of a neural network and the *distribution-free, worst-case bound* derived from Vapnik's statistical learning theory. The bound considered therein is defined by Vapnik (1982)

$$v_{\text{gene}} \geq O\left(\frac{h}{N} \log\left(\frac{h}{N}\right)\right) \tag{1}$$

where v_{gene} is the generalization error, h is the VC dimension, and N is the size of the training set. The results presented by Cohn and Tesauro show that the average generalization performance is significantly better than that predicted from Eq. (1).

Second, Holden and Niranjan (1995) extend the earlier study of Cohn and Tesauro by addressing a similar question. However, there are three important differences that should be pointed out:

- All the experiments were performed on neural networks with known exact results or very good bounds on the VC dimension.
- Specific account of the learning algorithm was taken.
- The experiments were based on real-life data.

Although the results reported were found to provide sample complexity predictions of a significantly more practical value than those provided by earlier theories, there are still significant shortcomings in the theory that need to be overcome.

Third, Baum and Haussler (1989) report on the size N of the training sample needed to train a single-layer feedforward network of linear-threshold neurons for good generalization. It is assumed that the training examples are chosen from an arbitrary probability distribution, and that the test examples for evaluating the generalization performance are also drawn from the same distribution. Then, according to Baum and Haussler, the network will almost certainly provide generalization, provided two conditions are satisfied:

(1) The number of errors made on the training set is less than $\epsilon/2$.

(2) The number of examples, N, used in training is

$$N \geq O\left(\frac{W}{\epsilon} \log\left(\frac{W}{\epsilon}\right)\right) \tag{2}$$

where W is the number of synaptic weights in the network. Equation (2) provides a *distribution-free, worst-case bound* on the size N. Here again there can be a huge numerical gap between the actual size of the training sample needed and that calculated from the bound of Eq. (2).

Finally, Bartlett (1997) addresses the issue that in pattern-classification tasks using large neural networks we often find that a network is able to perform successfully with training samples that are considerably smaller in size than the number of weights in the network, as reported in Cohn and Tesauro (1992). In Bartlett's paper it is shown that for such tasks on which neural networks generalize well and if the synaptic weights are not too large, it is the size of the weights rather than the number of weights that determines the generalization performance of the network.

PROBLEMS

Learning Rules

2.1 The delta rule described in Eq. (2.3) and Hebb's rule described in Eq. (2.9) represent two different methods of learning. List the features that distinguish these two rules from each other.

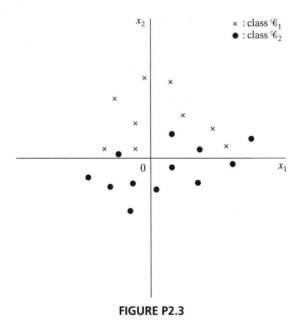

FIGURE P2.3

2.2 The error-correction learning rule may be implemented by using inhibition to subtract the desired response (target value) from the output, and then applying the anti-Hebbian rule (Mitchison, 1989). Discuss this interpretation of error-correction learning.

2.3 Figure P2.3 shows a two-dimensional set of data points. Part of the data points belongs to class $\mathscr{C}_1$ and the other part belongs to class $\mathscr{C}_2$. Construct the decision boundary produced by the nearest neighbor rule applied to this data sample.

2.4 Consider a group of people whose collective opinion on a topic of interest is defined as the weighted average of the individual opinions of its members. Suppose that if, over the course of time, the opinion of a member in the group tends to agree with the collective opinion of the group, the opinion of that member is given more weight. If, on the other hand, the particular member consistently disagrees with the collective opinion of the group, that member's opinion is given less weight. This form of weighting is equivalent to positive-feedback control, which has the effect of producing a consensus of opinion among the group (Linsker, 1988a).

Discuss the analogy between the situation described and Hebb's postulate of learning.

2.5 A generalized form of Hebb's rule is described by the relation

$$\Delta w_{kj}(n) = \alpha F(y_k(n))G(x_j(n)) - \beta w_{kj}(n)F(y_k(n))$$

where $x_j(n)$ and $y_k(n)$ are the presynaptic and postsynaptic signals, respectively; $F(\cdot)$ and $G(\cdot)$ are functions of their respective arguments; and $\Delta w_{kj}(n)$ is the change produced in the synaptic weight w_{kj} at time n in response to the signals $x_j(n)$ and $y_j(n)$. Find (a) the balance point and (b) the maximum depression that are defined by this rule.

2.6 An input signal of unit amplitude is applied repeatedly to a synaptic connection whose initial value is also unity. Calculate the variation in the synaptic weight with time using the following two rules:

(a) The simple form of Hebb's rule described in Eq. (2.9) assuming the learning-rate parameter $\eta = 0.1$.

(b) The covariance rule described in Eq. (2.10), assuming that the presynaptic activity $\bar{x} = 0$ and the postsynaptic activity $\bar{y} = 1.0$.

2.7 The Hebbian synapse described in Eq. (2.9) involves the use of positive feedback. Justify the validity of this statement.

2.8 Consider the covariance hypothesis for self-organized learning described in Eq. (2.10). Assuming ergodicity (i.e., time averages can be substituted for ensemble averages), show that the expected value of Δw_{kj} in Eq. (2.10) may be expressed as

$$E[\Delta w_{kj}] = \eta(E[y_k x_j] - \bar{y}\,\bar{x})$$

How would you interpret this result?

2.9 According to Linsker (1986), Hebb's postulate of learning may be formulated as:

$$\Delta w_{ki} = \eta(y_k - y_o)(x_i - x_o) + a_1$$

where x_j and y_k are the presynaptic and postsynaptic signals, respectively and a_1, η, x_o, and y_o are all constants. Assume that neuron k is linear, as shown by

$$y_k = \sum_j w_{kj} x_j + a_2$$

where a_2 is another constant. Assume the same probability distribution for all the input signals, that is, $E[x_i] = E[x_j] = \mu$. Let the matrix $\mathbf{C}$ denote the covariance matrix of the input signals with its ij-th element defined by

$$c_{ij} = E[(x_i - \mu)(x_j - \mu)]$$

Determine Δw_{ki}.

2.10 Formulate the expression for the output y_j of neuron j in the network of Fig. 2.4. You may use the following:

$x_i = i$th input signal

$w_{ji} =$ synaptic weight from input i to neuron j

$c_{kj} =$ weight of lateral connection from neuron k to neuron j

$v_j =$ induced local field of neuron j

$y_j = \varphi(v_j)$

What is the condition that would have to be satisfied for neuron j to be the winning neuron?

2.11 Repeat Problem 2.10, assuming that each output neuron includes self-feedback.

2.12 The connection pattern for lateral inhibition, namely "excitation close and inhibition further away," may be modeled as the difference between two Gaussian curves. The two curves have the same area, but the positive curve for excitation has a higher and narrower peak than the negative curve for inhibition. That is, we may express the connection pattern as

$$W(x) = \frac{1}{\sqrt{2\pi}\,\sigma_e} e^{-x^2/2\sigma_e^2} - \frac{1}{\sqrt{2\pi}\,\sigma_i} e^{-x^2/2\sigma_i^2}$$

where x is the distance from the neuron responsible for the lateral inhibition. The pattern $W(x)$ is used to scan a page, one half of which is white and the other half is black; the boundary between the two halves is perpendicular to the x-axis.

Plot the output that results from this scanning process with $\sigma_e = 1$ and $\sigma_i = 2$.

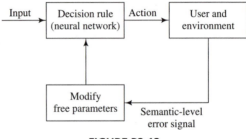

FIGURE P2.13

Learning Paradigms

2.13 Figure P2.13 shows the block diagram of an *adaptive language-acquisition system* (Gorin, 1992). The synaptic connections in the neural network part of the system are strengthened or weakened, depending on feedback as to the appropriateness of the machine's response to input stimuli. This system may be viewed as an example of reinforcement learning. Rationalize the validity of this statement.

2.14 To which of the two paradigms, learning with a teacher and learning without a teacher, do the following algorithms
  (a) nearest neighbor rule
  (b) *k*-nearest neighbor rule
  (c) Hebbian learning
  (d) Boltzmann learning rule
  belong? Justify your answers.

2.15 Unsupervised learning can be implemented in an off-line or on-line fashion. Discuss the physical implications of these two possibilities.

2.16 Consider the difficulties that a learning machine faces in assigning credit for the outcome (win, loss, or draw) of a game of chess. Discuss the notions of temporal credit assignment and structural credit assignment in the context of this game.

2.17 A supervised learning task may be viewed as a reinforcement learning task by using as the reinforcement signal some measure of the closeness of the actual response of the system to the desired response. Discuss this relationship between supervised learning and reinforcement learning.

Memory

2.18 Consider the following orthonormal sets of key patterns, applied to a correlation matrix memory:

$$\mathbf{x}_1 = [1, 0, 0, 0]^T$$
$$\mathbf{x}_2 = [0, 1, 0, 0]^T$$
$$\mathbf{x}_3 = [0, 0, 1, 0]^T$$

The respective stored patterns are

$$\mathbf{y}_1 = [5, 1, 0]^T$$
$$\mathbf{y}_2 = [-2, 1, 6]^T$$
$$\mathbf{y}_3 = [-2, 4, 3]^T$$

  (a) Calculate the memory matrix $\mathbf{M}$.

(b) Show that the memory associates perfectly.

2.19 Consider again the correlation matrix memory of Problem 2.18. The stimulus applied to the memory is a noisy version of the key pattern $\mathbf{x}_1$, as shown by

$$\mathbf{x} = [0.8, -0.15, 0.15, -0.20]^T$$

(a) Calculate the memory response $\mathbf{y}$.

(b) Show that the response $\mathbf{y}$ is closest to the stored pattern $\mathbf{y}_1$ in a Euclidean sense.

2.20 An autoassociative memory is trained on the following key vectors:

$$\mathbf{x}_1 = \tfrac{1}{4}\left[-2, -3, \sqrt{3}\right]^T$$
$$\mathbf{x}_2 = \tfrac{1}{4}\left[2, -2, -\sqrt{8}\right]^T$$
$$\mathbf{x}_3 = \tfrac{1}{4}\left[3, -1, \sqrt{6}\right]^T$$

(a) Calculate the angles between these vectors. How close are they to orthogonality with respect to each other?

(b) Using the generalization of Hebb's rule (i.e., the outer product rule), calculate the memory matrix of the network. Investigate how close to perfect the memory autoassociates.

(c) A masked version of the key vector $\mathbf{x}_1$, namely,

$$\mathbf{x} = \left[0, -3, \sqrt{3}\right]^T$$

is applied to the memory. Calculate the response of the memory, and compare your result with the desired response $\mathbf{x}_1$.

Adaptation

2.21 Figure P2.21 shows the block diagram of an adaptive system. The input signal to the *predictive model* is defined by past values of a process, as shown by

$$\mathbf{x}(n - 1) = [x(n - 1), x(n - 2), ..., x(n - m)]$$

The model output, $\hat{x}(n)$, represents an *estimate* of the present value, $x(n)$, of the process. The *comparator* computes the error signal

$$e(n) = x(n) - \hat{x}(n)$$

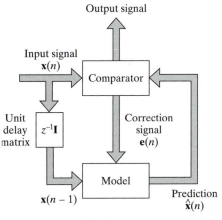

FIGURE P2.21

which in turn applies a correction to the adjustable parameters of the model. It also supplies an output signal for transfer to the next level of neural processing for interpretation. By repeating this operation on a level-by-level basis, the information processed by the system tends to be of progressively higher quality (Mead, 1990).

Fill in the details of the level of signal processing next to that described in Fig. P2.21.

Statistical learning theory

2.22 Following a procedure similar to that described for deriving Eq. (2.62) from (2.61), derive the formula for the ensemble-averaged function $L_{av}(f(\mathbf{x}), F(\mathbf{x}, \mathcal{T}))$ defined in Eq. (2.66).

2.23 In this problem we wish to calculate the VC dimension of a rectangular region aligned with one of the axes in a plane. Show that the VC dimension of this concept is four. You may do this by considering the following:

(a) Four points in a plane and a dichotomy realized by an axis-aligned rectangle.

(b) Four points in a plane, for which there is no realizable dichotomy by an axis-aligned rectangle.

(c) Five points in a plane, for which there is also no realizable dichotomy by an axis-aligned rectangle.

2.24 Consider a linear binary pattern classifier whose input vector $\mathbf{x}$ has dimension m. The first element of the vector $\mathbf{x}$ is constant and set to unity so that the corresponding weight of the classifier introduces a bias. What is the VC dimension of the classifier with respect to the input space?

2.25 The inequality (2.97) defines a bound on the rate of uniform convergence, which is basic to the principle of empirical risk minimization.

(a) Justify the validity of Eq. (2.98), assuming that the inequality (2.97) holds.

(b) Derive Eq. (2.99) that defines the confidence interval ϵ_1.

2.26 Continuing with Example 2.3, show that the four uniformly spaced points of Fig. P2.26 cannot be shattered by the one parameter family of indicator functions $f(x,a)$, $a \in \mathbb{R}$.

2.27 Discuss the relationship between the bias-variance dilemma and structural risk minimization in the context of nonlinear regression.

2.28 (a) An algorithm used to train a multilayer feedforward network whose neurons use a sigmoid function is PAC learnable. Justify the validity of this statement.

(b) Can you make a similar statement for an arbitrary neural network whose neurons use a threshold activation function? Justify your answer.

FIGURE P2.26

CHAPTER 3

Single-Layer Perceptrons

3.1 INTRODUCTION

In the formative years of neural networks (1943–1958), several researchers stand out for their pioneering contributions:

- McCulloch and Pitts (1943) for introducing the idea of neural networks as computing machines.
- Hebb (1949) for postulating the first rule for self-organized learning.
- Rosenblatt (1958) for proposing the perceptron as the first model for learning with a teacher (i.e., supervised learning).

The impact of the McCulloch–Pitts paper on neural networks was highlighted in Chapter 1. The idea of Hebbian learning was discussed at some length in Chapter 2. In this chapter we discuss Rosenblatt's *perceptron*.

The perceptron is the simplest form of a neural network used for the classification of patterns said to be *linearly separable* (i.e., patterns that lie on opposite sides of a hyperplane). Basically, it consists of a single neuron with adjustable synaptic weights and bias. The algorithm used to adjust the free parameters of this neural network first appeared in a learning procedure developed by Rosenblatt (1958, 1962) for his perceptron brain model.[1] Indeed, Rosenblatt proved that if the patterns (vectors) used to train the perceptron are drawn from two linearly separable classes, then the perceptron algorithm converges and positions the decision surface in the form of a hyperplane between the two classes. The proof of convergence of the algorithm is known as the *perceptron convergence theorem*. The perceptron built around a *single neuron* is limited to performing pattern classification with only two classes (hypotheses). By expanding the output (computation) layer of the perceptron to include more than one neuron, we may correspondingly form classification with more than two classes. However, the classes have to be linearly separable for the perceptron to work properly. The important point is that insofar as the basic theory of the perceptron as a pattern classifier is concerned, we need consider only the case of a single neuron. The extension of the theory to the case of more than one neuron is trivial.

The single neuron also forms the basis of an *adaptive filter*, a functional block that is basic to the ever-expanding subject of *signal processing*. The development of

117

adaptive filtering owes much to the classic paper of Widrow and Hoff (1960) for pioneering the so-called *least-mean-square (LMS) algorithm*, also known as the *delta rule*. The LMS algorithm is simple to implement yet highly effective in application. Indeed, it is the workhorse of *linear* adaptive filtering, linear in the sense that the neuron operates in its linear mode. Adaptive filters have been successfully applied in such diverse fields as antennas, communication systems, control systems, radar, sonar, seismology, and biomedical engineering (Widrow and Stearns, 1985; Haykin, 1996).

The LMS algorithm and the perceptron are naturally related. It is therefore proper for us to study them together in one chapter.

Organization of the Chapter

The chapter is organized in two parts. The first part, consisting of Sections 3.2 through 3.7 deals with linear adaptive filters and the LMS algorithm. The second part, consisting of Sections 3.8 through 3.10 deals with Rosenblatt's perceptron. From a presentation point of view, we find it more convenient to discuss linear adaptive filters first and then Rosenblatt's perceptron, reversing the historical order in which they appeared.

In Section 3.2 we address the adaptive filtering problem, followed by Section 3.3, a review of three unconstrained optimization techniques: the method of steepest descent, Newton's method, and Gauss-Newton method, which are particularly relevant to the study of adaptive filters. In Section 3.4 we discuss a linear least-squares filter, which asymptotically approaches the Wiener filter as the data length increases. The Wiener filter provides an ideal framework for the performance of linear adaptive filters operating in a stationary environment. In Section 3.5 we describe the LMS algorithm, including a discussion of its virtues and limitations. In Section 3.6 we explore the idea of learning curves commonly used to assess the performance of adaptive filters. This is followed by a discussion of annealing schedules for the LMS algorithm in Section 3.7.

Then moving on to Rosenblatt's perceptron, Section 3.8 presents some basic considerations involved in its operation. In Section 3.9 we describe the algorithm for adjusting the synaptic weight vector of the perceptron for pattern classification of linearly separable classes, and demonstrate convergence of the algorithm. In Section 3.10 we consider the relationship between the perceptron and the Bayes classifier for a Gaussian environment.

The chapter concludes with summary and discussion in Section 3.11.

3.2   ADAPTIVE FILTERING PROBLEM

Consider a *dynamical system,* the mathematical characterization of which is *unknown*. All that we have available on the system is a set of labeled input–output data generated by the system at discrete instants of time at some uniform rate. Specifically, when an m-dimensional stimulus $\mathbf{x}(i)$ is applied across m input nodes of

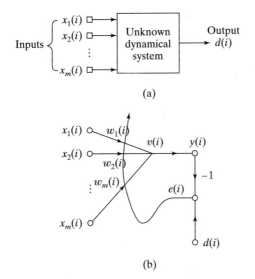

(a)

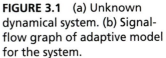

(b)

FIGURE 3.1   (a) Unknown dynamical system. (b) Signal-flow graph of adaptive model for the system.

the system, the system responds by producing a scalar output $d(i)$, where $i = 1, 2, ...,$ $n, ...$ as depicted in Fig. 3.1a. Thus, the external behavior of the system is described by the data set

$$\mathcal{T}: \{\mathbf{x}(i), d(i); i = 1, 2, ..., n, ...\} \tag{3.1}$$

where

$$\mathbf{x}(i) = [x_1(i), x_2(i), ..., x_m(i)]^T$$

The samples comprising $\mathcal{T}$ are identically distributed according to an unknown probability law. The dimension m pertaining to the input vector $\mathbf{x}(i)$ is referred to as the *dimensionality of the input space* or simply as *dimensionality*.

The stimulus $\mathbf{x}(i)$ can arise in one of two fundamentally different ways, one spatial and the other temporal:

- The m elements of $\mathbf{x}(i)$ originate at different points in space; in this case we speak of $\mathbf{x}(i)$ as a *snapshot* of data.
- The m elements of $\mathbf{x}(i)$ represent the set of present and $(m - 1)$ past values of some excitation that are *uniformly spaced in time*.

The problem we address is how to design a multiple input-single output *model* of the unknown dynamical system by building it around a single linear neuron. The neuronal model operates under the influence of an algorithm that *controls* necessary adjustments to the synaptic weights of the neuron, with the following points in mind:

- The algorithm starts from an *arbitrary setting* of the neuron's synaptic weights.
- Adjustments to the synaptic weights, in response to statistical variations in the system's behavior, are made on a *continuous* basis (i.e., time is incorporated into the constitution of the algorithm).

- Computations of adjustments to the synaptic weights are completed inside a time interval that is one sampling period long.

The neuronal model described is referred to as an *adaptive filter*. Although the description is presented in the context of a task clearly recognized as one of *system identification*, the characterization of the adaptive filter is general enough to have wide application.

Figure 3.1b shows a signal-flow graph of the adaptive filter. Its operation consists of two continuous processes:

1. *Filtering process*, which involves the computation of two signals:
   - An output, denoted by $y(i)$, that is produced in response to the m elements of the stimulus vector $\mathbf{x}(i)$, namely, $x_1(i), x_2(i), ..., x_m(i)$.
   - An error signal, denoted by $e(i)$, that is obtained by comparing the output $y(i)$ to the corresponding output $d(i)$ produced by the unknown system. In effect, $d(i)$ acts as a *desired response* or *target signal*.
2. *Adaptive process*, which involves the automatic adjustment of the synaptic weights of the neuron in accordance with the error signal $e(i)$.

Thus, the combination of these two processes working together constitutes a *feedback loop* acting around the neuron.

Since the neuron is linear, the output $y(i)$ is exactly the same as the induced local field $v(i)$; that is,

$$y(i) = v(i) = \sum_{k=1}^{m} w_k(i) x_k(i) \tag{3.2}$$

where $w_1(i), w_2(i), ..., w_m(i)$ are the m synaptic weights of the neuron, measured at time i. In matrix form we may express $y(i)$ as an inner product of the vectors $\mathbf{x}(i)$ and $\mathbf{w}(i)$ as follows:

$$y(i) = \mathbf{x}^T(i)\mathbf{w}(i) \tag{3.3}$$

where

$$\mathbf{w}(i) = [w_1(i), w_2(i), ..., w_m(i)]^T$$

Note that the notation for a synaptic weight has been simplified here by *not* including an additional subscript to identify the neuron, since we only have a single neuron to deal with. This practice is followed throughout the chapter. The neuron's output $y(i)$ is compared to the corresponding output $d(i)$ received from the unknown system at time i. Typically, $y(i)$ is different from $d(i)$; hence, their comparison results in the error signal:

$$e(i) = d(i) - y(i) \tag{3.4}$$

The manner in which the error signal $e(i)$ is used to control the adjustments to the neuron's synaptic weights is determined by the cost function used to derive the adaptive filtering algorithm of interest. This issue is closely related to that of optimization. It is therefore appropriate to present a review of unconstrained optimization methods. The material is applicable not only to linear adaptive filters but also to neural networks in general.

3.3 UNCONSTRAINED OPTIMIZATION TECHNIQUES

Consider a cost function $\mathscr{E}(\mathbf{w})$ that is a *continuously differentiable* function of some unknown weight (parameter) vector $\mathbf{w}$. The function $\mathscr{E}(\mathbf{w})$ maps the elements of $\mathbf{w}$ into real numbers. It is a measure of how to choose the weight (parameter) vector $\mathbf{w}$ of an adaptive filtering algorithm so that it behaves in an optimum manner. We want to find an optimal solution $\mathbf{w}^*$ that satisfies the condition

$$\mathscr{E}(\mathbf{w}^*) \leq \mathscr{E}(\mathbf{w}) \tag{3.5}$$

That is, we need to solve an *unconstrained optimization problem*, stated as follows:

Minimize the cost function $\mathscr{E}(\mathbf{w})$ with respect to the weight vector $\mathbf{w}$ (3.6)

The necessary condition for optimality is

$$\nabla\mathscr{E}(\mathbf{w}^*) = \mathbf{0} \tag{3.7}$$

where ∇ is the *gradient operator*:

$$\nabla = \left[\frac{\partial}{\partial w_1}, \frac{\partial}{\partial w_2}, \dots, \frac{\partial}{\partial w_m} \right]^T \tag{3.8}$$

and $\nabla\mathscr{E}(\mathbf{w})$ is the *gradient vector* of the cost function:

$$\nabla\mathscr{E}(\mathbf{w}) = \left[\frac{\partial\mathscr{E}}{\partial w_1}, \frac{\partial\mathscr{E}}{\partial w_2}, \dots, \frac{\partial\mathscr{E}}{\partial w_m} \right]^T \tag{3.9}$$

A class of unconstrained optimization algorithms that is particularly well suited for the design of adaptive filters is based on the idea of local *iterative descent*:

Starting with an initial guess denoted by $\mathbf{w}(0)$, generate a sequence of weight vectors $\mathbf{w}(1)$, $\mathbf{w}(2)$, …, such that the cost function $\mathscr{E}(\mathbf{w})$ is reduced at each iteration of the algorithm, as shown by

$$\mathscr{E}(\mathbf{w}(n + 1)) < \mathscr{E}(\mathbf{w}(n)) \tag{3.10}$$

where $\mathbf{w}(n)$ is the old value of the weight vector and $\mathbf{w}(n + 1)$ is its updated value.

We hope that the algorithm will eventually converge onto the optimal solution $\mathbf{w}^*$. We say "hope" because there is a distinct possibility that the algorithm will diverge (i.e., become unstable) unless special precautions are taken.

In this section we describe three unconstrained optimization methods that rely on the idea of iterative descent in one form or another (Bertsekas, 1995a).

Method of Steepest Descent

In the method of steepest descent, the successive adjustments applied to the weight vector $\mathbf{w}$ are in the direction of steepest descent, that is, in a direction opposite to the gradient vector $\nabla\mathscr{E}(\mathbf{w})$. For convenience of presentation we write

$$\mathbf{g} = \nabla\mathscr{E}(\mathbf{w}) \tag{3.11}$$

Accordingly, the steepest descent algorithm is formally described by

$$\mathbf{w}(n + 1) = \mathbf{w}(n) - \eta\mathbf{g}(n) \tag{3.12}$$

where η is a positive constant called the *stepsize* or *learning-rate parameter,* and $\mathbf{g}(n)$ is the gradient vector evaluated at the point $\mathbf{w}(n)$. In going from iteration n to $n + 1$ the algorithm applies the *correction*

$$\Delta\mathbf{w}(n) = \mathbf{w}(n + 1) - \mathbf{w}(n)$$
$$= -\eta\mathbf{g}(n) \tag{3.13}$$

Equation (3.13) is in fact a formal statement of the error-correction rule described in Chapter 2.

To show that the formulation of the steepest descent algorithm satisfies the condition of (3.10) for iterative descent, we use a first-order Taylor series expansion around $\mathbf{w}(n)$ to approximate $\mathscr{E}(\mathbf{w}(n + 1))$ as

$$\mathscr{E}(\mathbf{w}(n + 1)) \simeq \mathscr{E}(\mathbf{w}(n)) + \mathbf{g}^T(n)\Delta\mathbf{w}(n)$$

the use of which is justified for small η. Substituting Eq. (3.13) in this approximate relation yields

$$\mathscr{E}(\mathbf{w}(n + 1)) \simeq \mathscr{E}(\mathbf{w}(n)) - \eta\mathbf{g}^T(n)\mathbf{g}(n)$$
$$= \mathscr{E}(\mathbf{w}(n)) - \eta\|\mathbf{g}(n)\|^2$$

which shows that, for a positive learning-rate parameter η, the cost function is decreased as the algorithm progresses from one iteration to the next. The reasoning presented here is approximate in that this end result is only true for small enough learning rates.

The method of steepest descent converges to the optimal solution $\mathbf{w}^*$ slowly. Moreover, the learning-rate parameter η has a profound influence on its convergence behavior:

- When η is small, the transient response of the algorithm is *overdamped,* in that the trajectory traced by $\mathbf{w}(n)$ follows a smooth path in the W-plane, as illustrated in Fig. 3.2a.
- When η is large, the transient response of the algorithm is *underdamped,* in that the trajectory of $\mathbf{w}(n)$ follows a zigzagging (oscillatory) path, as illustrated in Fig. 3.2b.
- When η exceeds a certain critical value, the algorithm becomes unstable (i.e., it diverges).

Newton's Method

The basic idea of *Newton's method* is to minimize the quadratic approximation of the cost function $\mathscr{E}(\mathbf{w})$ around the current point $\mathbf{w}(n)$; this minimization is performed at each iteration of the algorithm. Specifically, using a *second-order* Taylor series expansion of the cost function around the point $\mathbf{w}(n)$, we may write

$$\Delta\mathscr{E}(\mathbf{w}(n)) = \mathscr{E}(\mathbf{w}(n + 1)) - \mathscr{E}(\mathbf{w}(n))$$
$$\simeq \mathbf{g}^T(n)\Delta\mathbf{w}(n) + \frac{1}{2}\Delta\mathbf{w}^T(n)\mathbf{H}(n)\Delta\mathbf{w}(n) \tag{3.14}$$

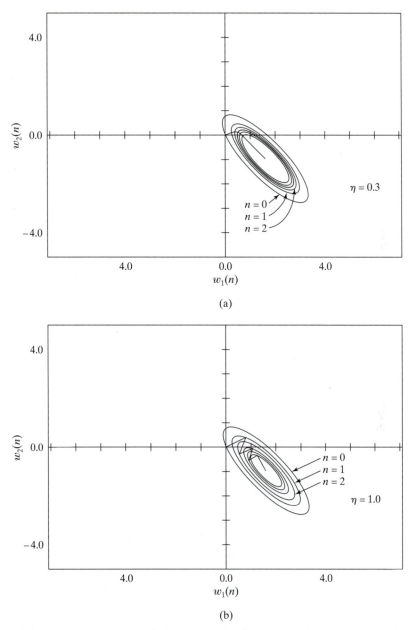

FIGURE 3.2 Trajectory of the method of steepest descent in a two-dimensional space for two different values of learning-rate parameter: (a) $\eta = 0.3$, (b) $\eta = 1.0$. The coordinates w_1 and w_2 are elements of the weight vector **w**.

As before, $\mathbf{g}(n)$ is the m-by-1 gradient vector of the cost function $\mathscr{E}(\mathbf{w})$ evaluated at the point $\mathbf{w}(n)$. The matrix $\mathbf{H}(n)$ is the m-by-m *Hessian matrix* of $\mathscr{E}(\mathbf{w})$, also evaluated at $\mathbf{w}(n)$. The Hessian of $\mathscr{E}(\mathbf{w})$ is defined by

$$\mathbf{H} = \nabla^2 \mathscr{E}(\mathbf{w})$$

$$= \begin{bmatrix} \dfrac{\partial^2 \mathscr{E}}{\partial w_1^2} & \dfrac{\partial^2 \mathscr{E}}{\partial w_1 \partial w_2} & \cdots & \dfrac{\partial^2 \mathscr{E}}{\partial w_1 \partial w_m} \\[2ex] \dfrac{\partial^2 \mathscr{E}}{\partial w_2 \partial w_1} & \dfrac{\partial^2 \mathscr{E}}{\partial w_2^2} & \cdots & \dfrac{\partial^2 \mathscr{E}}{\partial w_2 \partial w_m} \\[2ex] \vdots & \vdots & & \vdots \\[2ex] \dfrac{\partial^2 \mathscr{E}}{\partial w_m \partial w_1} & \dfrac{\partial^2 \mathscr{E}}{\partial w_m \partial w_2} & \cdots & \dfrac{\partial^2 \mathscr{E}}{\partial w_m^2} \end{bmatrix} \tag{3.15}$$

Equation (3.15) requires the cost function $\mathscr{E}(\mathbf{w})$ to be twice continuously differentiable with respect to the elements of $\mathbf{w}$. Differentiating[2] Eq. (3.14) with respect to $\Delta \mathbf{w}$, the change $\Delta \mathscr{E}(\mathbf{w})$ is minimized when

$$\mathbf{g}(n) + \mathbf{H}(n)\Delta\mathbf{w}(n) = \mathbf{0}$$

Solving this equation for $\Delta \mathbf{w}(n)$ yields

$$\Delta\mathbf{w}(n) = -\mathbf{H}^{-1}(n)\mathbf{g}(n)$$

That is,

$$\begin{aligned} \mathbf{w}(n+1) &= \mathbf{w}(n) + \Delta\mathbf{w}(n) \\ &= \mathbf{w}(n) - \mathbf{H}^{-1}(n)\mathbf{g}(n) \end{aligned} \tag{3.16}$$

where $\mathbf{H}^{-1}(n)$ is the inverse of the Hessian of $\mathscr{E}(\mathbf{w})$.

Generally speaking, Newton's method converges quickly asymptotically and does *not* exhibit the zigzagging behavior that sometimes characterizes the method of steepest descent. However, for Newton's method to work, the Hessian $\mathbf{H}(n)$ has to be a *positive definite matrix*[3] for all n. Unfortunately, in general there is no guarantee that $\mathbf{H}(n)$ is positive definite at every iteration of the algorithm. If the Hessian $\mathbf{H}(n)$ is not positive definite, modification of Newton's method is necessary (Powell, 1987; Bertsekas, 1995a).

Gauss–Newton Method

The *Gauss–Newton method* is applicable to a cost function that is expressed as the sum of error squares. Let

$$\mathscr{E}(\mathbf{w}) = \frac{1}{2} \sum_{i=1}^{n} e^2(i) \tag{3.17}$$

where the scaling factor $1/2$ is included to simplify matters in subsequent analysis. All the error terms in this formula are calculated on the basis of a weight vector $\mathbf{w}$ that is fixed over the entire observation interval $1 \leq i \leq n$.

The error signal $e(i)$ is a function of the adjustable weight vector $\mathbf{w}$. Given an operating point $\mathbf{w}(n)$, we linearize the dependence of $e(i)$ on $\mathbf{w}$ by writing

$$e'(i, \mathbf{w}) = e(i) + \left[\frac{\partial e(i)}{\partial \mathbf{w}} \right]^T_{\mathbf{w} = \mathbf{w}(n)} (\mathbf{w} - \mathbf{w}(n)), \quad i = 1, 2, ..., n \qquad (3.18)$$

Equivalently, by using matrix notation we may write

$$\mathbf{e}'(n, \mathbf{w}) = \mathbf{e}(n) + \mathbf{J}(n)(\mathbf{w} - \mathbf{w}(n)) \qquad (3.19)$$

where $\mathbf{e}(n)$ is the error vector

$$\mathbf{e}(n) = [e(1), e(2), ..., e(n)]^T$$

and $\mathbf{J}(n)$ is the n-by-m *Jacobian matrix* of $\mathbf{e}(n)$:

$$\mathbf{J}(n) = \begin{bmatrix} \dfrac{\partial e(1)}{\partial w_1} & \dfrac{\partial e(1)}{\partial w_2} & \cdots & \dfrac{\partial e(1)}{\partial w_m} \\[2mm] \dfrac{\partial e(2)}{\partial w_1} & \dfrac{\partial e(2)}{\partial w_2} & \cdots & \dfrac{\partial e(2)}{\partial w_m} \\[2mm] \vdots & \vdots & & \vdots \\[2mm] \dfrac{\partial e(n)}{\partial w_1} & \dfrac{\partial e(n)}{\partial w_2} & \cdots & \dfrac{\partial e(n)}{\partial w_m} \end{bmatrix}_{\mathbf{w} = \mathbf{w}(n)} \qquad (3.20)$$

The Jacobian $\mathbf{J}(n)$ is the transpose of the m-by-n gradient matrix $\nabla \mathbf{e}(n)$, where

$$\nabla \mathbf{e}(n) = [\nabla e(1), \nabla e(2), ..., \nabla e(n)]$$

The updated weight vector $\mathbf{w}(n+1)$ is then defined by

$$\mathbf{w}(n + 1) = \arg \min_{\mathbf{w}} \left\{ \frac{1}{2} \| \mathbf{e}'(n, \mathbf{w}) \|^2 \right\} \qquad (3.21)$$

Using Eq. (3.19) to evaluate the squared Euclidean norm of $\mathbf{e}'(n, \mathbf{w})$, we get

$$\frac{1}{2} \| \mathbf{e}'(n, \mathbf{w}) \|^2 = \frac{1}{2} \| \mathbf{e}(n) \|^2 + \mathbf{e}^T(n) \mathbf{J}(n)(\mathbf{w} - \mathbf{w}(n))$$

$$+ \frac{1}{2}(\mathbf{w} - \mathbf{w}(n))^T \mathbf{J}^T(n) \mathbf{J}(n)(\mathbf{w} - \mathbf{w}(n))$$

Hence, differentiating this expression with respect to $\mathbf{w}$ and setting the result equal to zero, we obtain

$$\mathbf{J}^T(n) \mathbf{e}(n) + \mathbf{J}^T(n) \mathbf{J}(n)(\mathbf{w} - \mathbf{w}(n)) = \mathbf{0}$$

Solving this equation for $\mathbf{w}$, we may thus write in light of Eq. (3.21):

$$\mathbf{w}(n+1) = \mathbf{w}(n) - (\mathbf{J}^T(n)\mathbf{J}(n))^{-1}\mathbf{J}^T(n)\mathbf{e}(n) \tag{3.22}$$

which describes the pure form of the Gauss–Newton method.

Unlike Newton's method that requires knowledge of the Hessian matrix of the cost function $\mathcal{E}(n)$, the Gauss–Newton method only requires the Jacobian matrix of the error vector $\mathbf{e}(n)$. However, for the Gauss–Newton iteration to be computable, the matrix product $\mathbf{J}^T(n)\mathbf{J}(n)$ must be nonsingular.

With regard to the latter point, we recognize that $\mathbf{J}^T(n)\mathbf{J}(n)$ is always nonnegative definite. To ensure that it is nonsingular, the Jacobian $\mathbf{J}(n)$ must have row *rank n*; that is, the n rows of $\mathbf{J}(n)$ in Eq. (3.20) must be linearly independent. Unfortunately, there is no guarantee that this condition will always hold. To guard against the possibility that $\mathbf{J}(n)$ is rank deficient, the customary practice is to add the diagonal matrix $\delta\mathbf{I}$ to the matrix $\mathbf{J}^T(n)\mathbf{J}(n)$. The parameter δ is a small positive constant chosen to ensure that

$$\mathbf{J}^T(n)\mathbf{J}(n) + \delta\mathbf{I} : \text{positive definite for all } n$$

On this basis, the Gauss–Newton method is implemented in the slightly modified form:

$$\mathbf{w}(n+1) = \mathbf{w}(n) - (\mathbf{J}^T(n)\mathbf{J}(n) + \delta\mathbf{I})^{-1}\mathbf{J}^T(n)\mathbf{e}(n) \tag{3.23}$$

The effect of this modification is progressively reduced as the number of iterations, n, is increased. Note also that the recursive equation (3.23) is the solution of the *modified* cost function:

$$\mathcal{E}(\mathbf{w}) = \frac{1}{2}\left\{\delta\|\mathbf{w} - \mathbf{w}(0)\|^2 + \sum_{i=1}^{n} e^2(i)\right\} \tag{3.24}$$

where $\mathbf{w}(0)$ is the *initial value* of the weight vector $\mathbf{w}(i)$.

We are now equipped with the optimization tools we need to address the specific issues involved in linear adaptive filtering.

3.4   LINEAR LEAST-SQUARES FILTER

As the name implies, a *linear least-squares filter* has two distinctive characteristics. First, the single neuron around which it is built is linear, as shown in the model of Fig. 3.1b. Second, the cost function $\mathcal{E}(\mathbf{w})$ used to design the filter consists of the sum of error squares, as defined in Eq. (3.17). On this basis, using Eqs. (3.3) and (3.4), we may express the error vector $\mathbf{e}(n)$ as follows:

$$\begin{aligned} \mathbf{e}(n) &= \mathbf{d}(n) - [\mathbf{x}(1), \mathbf{x}(2), ..., \mathbf{x}(n)]^T\mathbf{w}(n) \\ &= \mathbf{d}(n) - \mathbf{X}(n)\mathbf{w}(n) \end{aligned} \tag{3.25}$$

where $\mathbf{d}(n)$ is the n-by-1 *desired response vector*:

$$\mathbf{d}(n) = [d(1), d(2), ..., d(n)]^T$$

and $\mathbf{X}(n)$ is the n-by-m *data matrix*:

$$\mathbf{X}(n) = [\mathbf{x}(n), \mathbf{x}(2), ..., \mathbf{x}(n)]^T$$

Differentiating Eq. (3.25) with respect to $\mathbf{w}(n)$ yields the gradient matrix

$$\nabla \mathbf{e}(n) = -\mathbf{X}^T(n)$$

Correspondingly, the Jacobian of $\mathbf{e}(n)$ is

$$\mathbf{J}(n) = -\mathbf{X}(n) \tag{3.26}$$

Since the error equation (3.19) is already linear in the weight vector $\mathbf{w}(n)$, the Gauss–Newton method converges in a single iteration, as shown here. Substituting Eqs. (3.25) and (3.26) in (3.22) yields

$$
\begin{aligned}
\mathbf{w}(n + 1) &= \mathbf{w}(n) + (\mathbf{X}^T(n)\mathbf{X}(n))^{-1}\mathbf{X}^T(n)(\mathbf{d}(n) - \mathbf{X}(n)\mathbf{w}(n)) \\
&= (\mathbf{X}^T(n)\mathbf{X}(n))^{-1}\mathbf{X}^T(n)\mathbf{d}(n)
\end{aligned}
\tag{3.27}
$$

The term $(\mathbf{X}^T(n)\mathbf{X}(n))^{-1}\mathbf{X}^T(n)$ is recognized as the *pseudoinverse* of the data matrix $\mathbf{X}(n)$ as shown in Golub and Van Loan (1996), and Haykin (1996); that is,

$$\mathbf{X}^+(n) = (\mathbf{X}^T(n)\mathbf{X}(n))^{-1}\mathbf{X}^T(n) \tag{3.28}$$

Hence, we may rewrite Eq. (3.27) in the compact form

$$\mathbf{w}(n + 1) = \mathbf{X}^+(n)\mathbf{d}(n) \tag{3.29}$$

This formula represents a convenient way of saying: "The weight vector $\mathbf{w}(n + 1)$ solves the linear least-squares problem defined over an observation interval of duration n."

Wiener Filter: Limiting form of the Linear Least-Squares Filter for an Ergodic Environment

A case of particular interest is when the input vector $\mathbf{x}(i)$ and desired response $d(i)$ are drawn from an *ergodic* environment that is also stationary. We may then substitute long-term sample averages or time-averages for expectations or ensemble averages (Gray and Davisson, 1986). Such an environment is partially described by the second-order statistics:

- *Correlation matrix* of the input vector $\mathbf{x}(i)$; it is denoted by $\mathbf{R}_x$
- *Cross-correlation vector* between the input vector $\mathbf{x}(i)$ and desired response $d(i)$; it is denoted by $\mathbf{r}_{xd}$.

These two quantities are defined as follows, respectively:

$$
\begin{aligned}
\mathbf{R}_x &= E[\mathbf{x}(i)\mathbf{x}^T(i)] \\
&= \lim_{n \to \infty} \frac{1}{n} \sum_{i=1}^{n} \mathbf{x}(n)\mathbf{x}^T(i) \\
&= \lim_{n \to \infty} \frac{1}{n} \mathbf{X}^T(n)\mathbf{X}(n)
\end{aligned}
\tag{3.30}
$$

$$\mathbf{r}_{\mathbf{x}d} = E[\mathbf{x}(i)d(i)]$$

$$= \lim_{n \to \infty} \frac{1}{n} \sum_{i=1}^{n} \mathbf{x}(i)d(i) \tag{3.31}$$

$$= \lim_{n \to \infty} \frac{1}{n} \mathbf{X}^{T}(n)\mathbf{d}(n)$$

where E denotes the statistical expectation operator. Accordingly, we may reformulate the linear least-squares solution of Eq. (3.27) as follows:

$$\begin{aligned}
\mathbf{w}_o &= \lim_{n \to \infty} \mathbf{w}(n + 1) \\
&= \lim_{n \to \infty} (\mathbf{X}^{T}(n)\mathbf{X}(n))^{-1} \mathbf{X}^{T}(n)\mathbf{d}(n) \\
&= \lim_{n \to \infty} \frac{1}{n} (\mathbf{X}^{T}(n)\mathbf{X}(n))^{-1} \lim_{n \to \infty} \frac{1}{n} \mathbf{X}^{T}(n)\mathbf{d}(n) \\
&= \mathbf{R}_{\mathbf{x}}^{-1} \mathbf{r}_{\mathbf{x}d}
\end{aligned} \tag{3.32}$$

where $\mathbf{R}_{\mathbf{x}}^{-1}$ is the inverse of the correlation matrix $\mathbf{R}_{\mathbf{x}}$. The weight vector $\mathbf{w}_o$ is called the *Wiener solution* to the linear optimum filtering problem in recognition of the contributions of Norbert Wiener to this problem (Widrow and Stearns, 1985; Haykin, 1996). Accordingly, we may make the following statement:

> *For an ergodic process, the linear least-squares filter asymptotically approaches the Wiener filter as the number of observations approaches infinity.*

Designing the Wiener filter requires knowledge of the second-order statistics: the correlation matrix $\mathbf{R}_{\mathbf{x}}$ of the input vector $\mathbf{x}(n)$ and the cross-correlation vector $\mathbf{r}_{\mathbf{x}d}$ between $\mathbf{x}(n)$ and the desired response $d(n)$. However, this information is not available in many important situations encountered in practice. We may deal with an unknown environment by using a *linear adaptive filter*, adaptive in the sense that the filter is able to adjust its free parameters in response to statistical variations in the environment. A highly popular algorithm for doing this kind of adjustment on a continuing basis is the least-mean-square algorithm, which is intimately related to the Wiener filter.

3.5 LEAST-MEAN-SQUARE ALGORITHM

The *least-mean-square (LMS) algorithm* is based on the use of *instantaneous values* for the cost function, namely,

$$\mathscr{E}(\mathbf{w}) = \frac{1}{2} e^{2}(n) \tag{3.33}$$

where $e(n)$ is the error signal measured at time n. Differentiating $\mathscr{E}(\mathbf{w})$ with respect to the weight vector $\mathbf{w}$ yields

$$\frac{\partial \mathscr{E}(\mathbf{w})}{\partial \mathbf{w}} = e(n) \frac{\partial e(n)}{\partial \mathbf{w}} \tag{3.34}$$

As with the linear least-squares filter, the LMS algorithm operates with a linear neuron so we may express the error signal as

$$e(n) = d(n) - \mathbf{x}^T(n)\mathbf{w}(n) \tag{3.35}$$

Hence,

$$\frac{\partial e(n)}{\partial \mathbf{w}(n)} = -\mathbf{x}(n)$$

and

$$\frac{\partial \mathscr{E}(\mathbf{w})}{\partial \mathbf{w}(n)} = -\mathbf{x}(n)e(n)$$

Using this latter result as an *estimate* for the gradient vector, we may write

$$\hat{\mathbf{g}}(n) = -\mathbf{x}(n)e(n) \tag{3.36}$$

Finally, using Eq. (3.36) for the gradient vector in Eq. (3.12) for the method of steepest descent, we may formulate the LMS algorithm as follows:

$$\hat{\mathbf{w}}(n+1) = \hat{\mathbf{w}}(n) + \eta\mathbf{x}(n)e(n) \tag{3.37}$$

where η is the learning-rate parameter. The feedback loop around the weight vector $\hat{\mathbf{w}}(n)$ in the LMS algorithm behaves like a *low-pass filter,* passing the low frequency components of the error signal and attenuating its high frequency components (Haykin, 1996). The average time constant of this filtering action is inversely proportional to the learning-rate parameter η. Hence, by assigning a small value to η, the adaptive process will progress slowly. More of the past data are then remembered by the LMS algorithm, resulting in a more accurate filtering action. In other words, the inverse of the learning-rate parameter η is a measure of the *memory* of the LMS algorithm.

In Eq. (3.37) we have used $\hat{\mathbf{w}}(n)$ in place of $\mathbf{w}(n)$ to emphasize the fact that the LMS algorithm produces an *estimate* of the weight vector that would result from the use of the method of steepest descent. As a consequence, in using the LMS algorithm we sacrifice a distinctive feature of the steepest descent algorithm. In the steepest descent algorithm the weight vector $\mathbf{w}(n)$ follows a well-defined trajectory in weight space for a prescribed η. In contrast, in the LMS algorithm the weight vector $\hat{\mathbf{w}}(n)$ traces a random trajectory. For this reason, the LMS algorithm is sometimes referred to as a "stochastic gradient algorithm." As the number of iterations in the LMS algorithm approaches infinity, $\hat{\mathbf{w}}(n)$ performs a random walk (Brownian motion) about the Wiener solution $\mathbf{w}_o$. The important point is the fact that, unlike the method of steepest descent, the LMS algorithm does *not* require knowledge of the statistics of the environment.

A summary of the LMS algorithm is presented in Table 3.1, which clearly illustrates the simplicity of the algorithm. As indicated in this table, for the *initialization* of the algorithm, it is customary to set the initial value of the weight vector in the algorithm equal to zero.

TABLE 3.1 Summary of the LMS Algorithm

Training Sample:	Input signal vector $= \mathbf{x}(n)$
	Desired response $= d(n)$

User-selected parameter: η
Initialization. Set $\hat{\mathbf{w}}(0) = \mathbf{0}$.
Computation. For $n = 1, 2, ...,$ compute

$$e(n) = d(n) - \hat{\mathbf{w}}^T(n)\mathbf{x}(n)$$

$$\hat{\mathbf{w}}(n + 1) = \hat{\mathbf{w}}(n) + \eta\,\mathbf{x}(n)e(n)$$

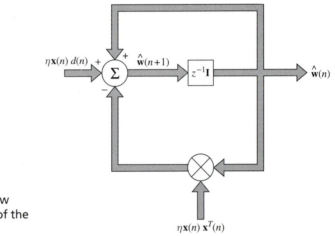

FIGURE 3.3 Signal-flow graph representation of the LMS algorithm.

Signal-Flow Graph Representation of the LMS Algorithm

By combining Eqs. (3.35) and (3.37) we may express the evolution of the weight vector in the LMS algorithm as follows:

$$\hat{\mathbf{w}}(n + 1) = \hat{\mathbf{w}}(n) + \eta\mathbf{x}(n)[d(n) - \mathbf{x}^T(n)\hat{\mathbf{w}}(n)]$$
$$= [\mathbf{I} - \eta\mathbf{x}(n)\mathbf{x}^T(n)]\hat{\mathbf{w}}(n) + \eta\mathbf{x}(n)d(n) \tag{3.38}$$

where $\mathbf{I}$ is the identity matrix. In using the LMS algorithm, we recognize that

$$\hat{\mathbf{w}}(n) = z^{-1}[\hat{\mathbf{w}}(n + 1)] \tag{3.39}$$

where z^{-1} is the *unit-delay operator*, implying storage. Using Eqs. (3.38) and (3.39), we may thus represent the LMS algorithm by the signal-flow graph depicted in Fig. 3.3. This signal-flow graph reveals that the LMS algorithm is an example of a *stochastic feedback system*. The presence of feedback has a profound impact on the convergence behavior of the LMS algorithm.

Convergence Considerations of the LMS Algorithm

From control theory we know that the stability of a feedback system is determined by the parameters that constitute its feedback loop. From Fig. 3.3 we see that it is the lower feedback loop that adds variability to the behavior of the LMS algorithm. In particular, there are two distinct quantities, the learning-rate parameter η and the input vector $\mathbf{x}(n)$, that determine the transmittance of this feedback loop. We therefore deduce that the convergence behavior (i.e., stability) of the LMS algorithm is influenced by the statistical characteristics of the input vector $\mathbf{x}(n)$ and the value assigned to the learning-rate parameter η. Casting this observation in a different way, we may state that for a specified environment that supplies the input vector $\mathbf{x}(n)$, we have to exercise care in the selection of the learning-rate parameter η for the LMS algorithm to be convergent.

The first criterion for convergence of the LMS algorithm is *convergence of the mean*, described by

$$E[\hat{\mathbf{w}}(n)] \to \mathbf{w}_o \qquad \text{as } n \to \infty \tag{3.40}$$

where $\mathbf{w}_o$ is the Wiener solution. Unfortunately, such a convergence criterion is of little practical value, since a sequence of zero-mean, but otherwise arbitrary, random vectors converges in this sense.

From a practical point of view, the convergence issue that really matters is *convergence in the mean square*, described by

$$E[e^2(n)] \to \text{constant as } n \to \infty \tag{3.41}$$

Unfortunately, a detailed convergence analysis of the LMS algorithm in the mean square is rather complicated. To make the analysis mathematically tractable, the following assumptions are usually made:

1. The successive input vectors $\mathbf{x}(1)$, $\mathbf{x}(2)$, ... are statistically independent of each other.
2. At time step n, the input vector $\mathbf{x}(n)$ is statistically independent of all previous samples of the desired response, namely $d(1), d(2), ..., d(n-1)$.
3. At time step n, the desired response $d(n)$ is dependent on $\mathbf{x}(n)$, but statistically independent of all previous values of the desired response.
4. The input vector $\mathbf{x}(n)$ and desired response $d(n)$ are drawn from Gaussian-distributed populations.

A statistical analysis of the LMS algorithm so based is called the *independence theory* (Widrow et al., 1976).

By invoking the elements of independence theory and assuming that the learning-rate parameter η is sufficiently small, it is shown in Haykin (1996) that the LMS is convergent in the mean square provided that η satisfies the condition

$$0 < \eta < \frac{2}{\lambda_{\max}} \tag{3.42}$$

where $\lambda_{\max}$ is the *largest eigenvalue* of the correlation matrix $\mathbf{R}_x$. In typical applications of the LMS algorithm, however, knowledge of $\lambda_{\max}$ is not available. To overcome this difficulty, the *trace* of $\mathbf{R}_x$ may be taken as a conservative estimate for $\lambda_{\max}$, in which case the condition of Eq. (3.42) may be reformulated as

$$0 < \eta < \frac{2}{\text{tr}[\mathbf{R}_x]} \tag{3.43}$$

where $\text{tr}[\mathbf{R}_x]$ denotes the trace of matrix $\mathbf{R}_x$. By definition, the trace of a square matrix is equal to the sum of its diagonal elements. Since each diagonal element of the correlation matrix $\mathbf{R}_x$ equals the mean-square value of the corresponding sensor input, we may restate the condition for convergence of the LMS algorithm in the mean square as follows:

$$0 < \eta < \frac{2}{\text{sum of mean-square values of the sensor inputs}} \qquad (3.44)$$

Provided the learning-rate parameter satisfies this condition, the LMS algorithm is also assured of convergence of the mean. That is, convergence in the mean square implies convergence of the mean, but the converse is not necessarily true.

Virtues and Limitations of the LMS Algorithm

An important virtue of the LMS algorithm is its simplicity, as exemplified by the summary of the algorithm presented in Table 3.1. Moreover, the LMS algorithm is model independent and therefore *robust*, which means that small model uncertainty and small disturbances (i.e., disturbances with small energy) can only result in small estimation errors (error signals). In precise mathematical terms, the LMS algorithm is optimal in accordance with the H^∞ (*or minimax*) *criterion* (Hassibi et al., 1993, 1996). The basic philosophy of optimality in the H^∞ sense is to cater to the worst-case scenario[4]:

> *If you do not know what you are up against, plan for the worst and optimize.*

For a long time the LMS algorithm was regarded as an instantaneous approximation to the gradient-descent algorithm. However, the H^∞ optimality of LMS provides this widely used algorithm with a rigorous footing. In particular, it explains its ability to work satisfactorily in a stationary as well as in a nonstationary environment. By a "nonstationary" environment we mean one where the statistics vary with time. In such an environment, the optimum Wiener solution takes on a time-varying form, and the LMS algorithm now has the additional task of *tracking* variations in the parameters of the Wiener filter.

The primary limitations of the LMS algorithm are its slow rate of convergence and sensitivity to variations in the eigenstructure of the input (Haykin, 1996). The LMS algorithm typically requires a number of iterations equal to about 10 times the dimensionality of the input space for it to reach a steady-state condition. The slow rate of convergence becomes particularly serious when the dimensionality of the input space becomes high. As for sensitivity to changes in environmental conditions, the LMS algorithm is particularly sensitive to variations in the *condition number* or *eigenvalue spread* of the correlation matrix $\mathbf{R}_x$ of the input vector $\mathbf{x}$. The condition number of $\mathbf{R}_x$, denoted by $\chi(\mathbf{R}_x)$, is defined by

$$\chi(\mathbf{R}_x) = \frac{\lambda_{\max}}{\lambda_{\min}} \qquad (3.45)$$

where $\lambda_{\max}$ and $\lambda_{\min}$ are the maximum and minimum eigenvalues of the matrix $\mathbf{R}_x$, respectively. The sensitivity of the LMS algorithm to variations in the condition number $\chi(\mathbf{R}_x)$ becomes particularly acute when the training sample to which the input vector $\mathbf{x}(n)$ belongs is *ill conditioned*, that is, when the condition number $\chi(\mathbf{R}_x)$ is high.[5]

Note that in the LMS algorithm the *Hessian matrix,* defined as the second derivative of the cost function $\mathscr{E}(\mathbf{w})$ with respect to $\mathbf{w}$, is equal to the correlation matrix $\mathbf{R}_x$; see Problem 3.8. Thus, in the discussion presented here, we could have just as well spoken in terms of the Hessian as the correlation matrix $\mathbf{R}_x$.

3.6 LEARNING CURVES

An informative way of examining the convergence behavior of the LMS algorithm, or an adaptive filter in general, is to plot the *learning curve* of the filter under varying environmental conditions. The learning curve is a *plot of the mean-square value of the estimation error,* $\mathscr{E}_{av}(n)$, *versus the number of iterations, n.*

Imagine an experiment involving an *ensemble* of adaptive filters, with each filter operating under the control of a specific algorithm. It is assumed that the details of the algorithm, including initialization, are the same for all the filters. The differences between the filters arise from the *random* manner in which the input vector $\mathbf{x}(n)$ and the desired response $d(n)$ are drawn from the available training sample. For each filter we plot the squared value of the estimation error (i.e., the difference between the desired response and the actual filter output) versus the number of iterations. A *sample* learning curve so obtained consists of *noisy* exponentials, the noise being due to the inherently stochastic nature of the adaptive filter. To compute the *ensemble-averaged learning curve* (i.e., plot of $\mathscr{E}_{av}(n)$ versus n), we take the average of these sample learning curves over the ensemble of adaptive filters used in the experiment, thereby smoothing out the effects of noise.

Assuming that the adaptive filter is stable, we find that the ensemble-averaged learning curve starts from a large value $\mathscr{E}_{av}(0)$ determined by the initial conditions, then decreases at some rate depending on the type of filter used, and finally converges to a steady-state value $\mathscr{E}_{av}(\infty)$, as illustrated in Fig. 3.4. On the basis of this learning curve we may define the *rate of convergence* of the adaptive filter as the number of iterations, n, required for $\mathscr{E}_{av}(n)$ to decrease to some arbitrarily chosen value, such as 10 percent of the initial value $\mathscr{E}_{av}(0)$.

Another useful characteristic of an adaptive filter that is deduced from the ensemble-averaged learning curve is the *misadjustment,* denoted by $\mathcal{M}$. Let $\mathscr{E}_{min}$ denote the minimum mean-square error produced by the Wiener filter, designed on the basis of known values of the correlation matrix $\mathbf{R}_x$ and cross-correlation vector $\mathbf{r}_{xd}$. We may define the *misadjustment* for the adaptive filter as follows (Widrow and Stearns, 1985; Haykin, 1996):

$$\mathcal{M} = \frac{\mathscr{E}(\infty) - \mathscr{E}_{min}}{\mathscr{E}_{min}}$$

$$= \frac{\mathscr{E}(\infty)}{\mathscr{E}_{min}} - 1 \tag{3.46}$$

The misadjustment $\mathcal{M}$ is a dimensionless quantity, providing a measure of how close the adaptive filter is to optimality in the mean-square error sense. The smaller $\mathcal{M}$ is compared to unity, the more *accurate* is the adaptive filtering action of the algorithm. It is customary to express the misadjustment $\mathcal{M}$ as a percentage. Thus, for example, a misadjustment of 10 percent means that the adaptive filter produces a mean-square error

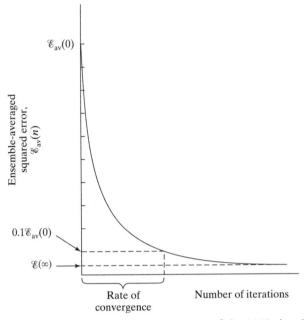

FIGURE 3.4 Idealized learning curve of the LMS algorithm.

(after adaptation is completed) that is 10 percent greater than the minimum mean-square error $\mathscr{E}_{\min}$ produced by the corresponding Wiener filter. Such performance is ordinarily considered to be satisfactory in practice.

Another important characteristic of the LMS algorithm is the *settling time*. However, there is no unique definition for the settling time. We may, for example, approximate the learning curve by a single exponential with *average time constant* τ_{av}, and so use τ_{av} as a rough measure of the settling time. The smaller the value of τ_{av} is, the faster the settling time will be (i.e., the faster the LMS algorithm will converge to a "steady-state" condition).

To a good degree of approximation, the misadjustment $\mathcal{M}$ of the LMS algorithm is directly proportional to the learning-rate parameter η, whereas the average time constant τ_{av} is inversely proportional to the learning-rate parameter η (Widrow and Stearns, 1985; Haykin, 1996). We therefore have conflicting results in the sense that if the learning-rate parameter is reduced so as to reduce the misadjustment, then the settling time of the LMS algorithm is increased. Conversely, if the learning-rate parameter is increased so as to accelerate the learning process, then the misadjustment is increased. Careful attention must be given to the choice of the learning parameter η in the design of the LMS algorithm in order to produce a satisfactory overall performance.

3.7 LEARNING-RATE ANNEALING SCHEDULES

The difficulties encountered with the LMS algorithm may be attributed to the fact that the learning-rate parameter is maintained constant throughout the computation, as shown by

$$\eta(n) = \eta_0 \qquad \text{for all } n \tag{3.47}$$

This is the simplest possible form the learning-rate parameter can assume. In contrast, in *stochastic approximation,* which goes back to the classic paper by Robbins and Monro (1951), the learning-rate parameter is time-varying. The particular time-varying form most commonly used in stochastic approximation literature is described by

$$\eta(n) = \frac{c}{n} \tag{3.48}$$

where c is a constant. Such a choice is indeed sufficient to guarantee convergence of the stochastic approximation algorithm (Ljung, 1977; Kushner and Clark, 1978). However, when the constant c is large, there is a danger of parameter blowup for small n.

As an alternative to Eqs. (3.47) and (3.48), we may use the *search-then-converge schedule,* defined by Darken and Moody (1992)

$$\eta(n) = \frac{\eta_0}{1 + (n/\tau)} \tag{3.49}$$

where η_0 and τ are user-selected constants. In the early stages of adaptation involving a number of iterations n, small compared to the *search time constant* τ, the learning-rate parameter $\eta(n)$ is approximately equal to η_0, and the algorithm operates essentially as the "standard" LMS algorithm, as indicated in Fig. 3.5. Hence, by choosing a high value for η_0 within the permissible range, we hope that the adjustable weights of the filter will find and hover about a "good" set of values. Then, for a number of iterations n large compared to the search time constant τ, the learning-rate parameter $\eta(n)$ approximates as c/n, where $c = \tau\eta_0$, as illustrated in Fig. 3.5. The algorithm now operates as a traditional stochastic approximation algorithm, and the weights converge to their optimum values. Thus the search-then-converge schedule has the potential to combine the desirable features of the standard LMS with traditional stochastic approximation theory.

3.8   PERCEPTRON

We now come to the second part of the chapter that deals with Rosenblatt's perceptron, henceforth referred to simply as the *perceptron.* Whereas the LMS algorithm described in the preceding sections is built around a linear neuron, the perceptron is built around a nonlinear neuron, namely, the *McCulloch–Pitts model* of a neuron. From Chapter 1 we recall that such a neuronal model consists of a linear combiner followed by a hard limiter (performing the signum function), as depicted in Fig. 3.6. The summing node of the neuronal model computes a linear combination of the inputs applied to its synapses, and also incorporates an externally applied bias. The resulting sum, that is, the induced local field, is applied to a hard limiter. Accordingly, the neuron produces an output equal to $+1$ if the hard limiter input is positive, and -1 if it is negative.

In the signal-flow graph model of Fig. 3.6, the synaptic weights of the perceptron are denoted by $w_1, w_2, \ldots, w_m$. Correspondingly, the inputs applied to the perceptron

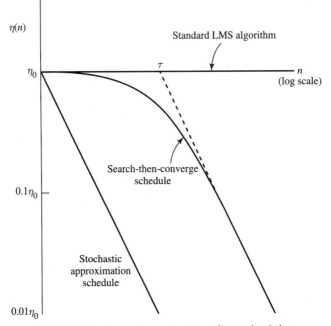

FIGURE 3.5 Learning-rate annealing schedules.

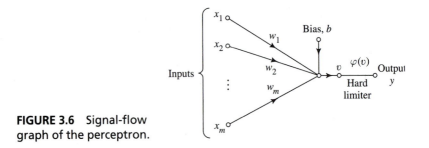

FIGURE 3.6 Signal-flow
graph of the perceptron.

are denoted by $x_1, x_2, ..., x_m$. The externally applied bias is denoted by b. From the model we find that the hard limiter input or induced local field of the neuron is

$$v = \sum_{i=1}^{m} w_i x_i + b \tag{3.50}$$

The goal of the perceptron is to correctly classify the set of externally applied stimuli $x_1, x_2, ..., x_m$ into one of two classes, $\mathscr{C}_1$ or $\mathscr{C}_2$. The decision rule for the classification is to assign the point represented by the inputs $x_1, x_2, ..., x_m$ to class $\mathscr{C}_1$ if the perceptron output y is $+1$ and to class $\mathscr{C}_2$ if it is -1.

To develop insight into the behavior of a pattern classifier, it is customary to plot a map of the decision regions in the m-dimensional signal space spanned by the m

input variables x_1, x_2, x_m. In the simplest form of the perceptron there are two decision regions separated by a *hyperplane* defined by

$$\sum_{i=1}^{m} w_i x_i + b = 0 \qquad (3.51)$$

This is illustrated in Fig. 3.7 for the case of two input variables x_1 and x_2, for which the decision boundary takes the form of a straight line. A point (x_1, x_2) that lies above the boundary line is assigned to class $\mathscr{C}_1$ and a point (x_1, x_2) that lies below the boundary line is assigned to class $\mathscr{C}_2$. Note also that the effect of the bias b is merely to shift the decision boundary away from the origin.

The synaptic weights $w_1, w_2, \ldots, w_m$ of the perceptron can be adapted on an iteration-by-iteration basis. For the adaptation we may use an error-correction rule known as the perceptron convergence algorithm.

3.9 PERCEPTRON CONVERGENCE THEOREM

To derive the error-correction learning algorithm for the perceptron, we find it more convenient to work with the modified signal-flow graph model in Fig. 3.8. In this second model, which is equivalent to that of Fig. 3.6, the bias $b(n)$ is treated as a synaptic weight driven by a fixed input equal to $+1$. We may thus define the $(m + 1)$-by-1 input vector

$$\mathbf{x}(n) = [+1, x_1(n), x_2(n), \ldots, x_m(n)]^T$$

where n denotes the iteration step in applying the algorithm. Correspondingly we define the $(m + 1)$-by-1 weight vector as

$$\mathbf{w}(n) = [b(n), w_1(n), w_2(n), \ldots, w_m(n)]^T$$

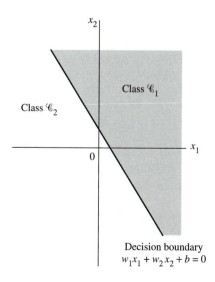

FIGURE 3.7 Illustration of the hyperplane (in this example, a straight line) as decision boundary for a two-dimensional, two-class pattern-classification problem.

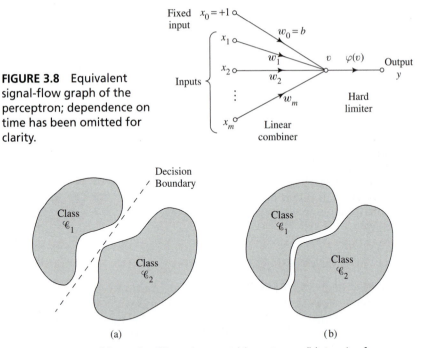

FIGURE 3.8 Equivalent signal-flow graph of the perceptron; dependence on time has been omitted for clarity.

FIGURE 3.9   (a) A pair of linearly separable patterns. (b) A pair of non-linearly separable patterns.

Accordingly, the linear combiner output is written in the compact form

$$v(n) = \sum_{i=0}^{m} w_i(n)x_i(n)$$

$$= \mathbf{w}^T(n)\mathbf{x}(n)$$

(3.52)

where $w_0(n)$ represents the bias $b(n)$. For fixed n, the equation $\mathbf{w}^T\mathbf{x} = 0$, plotted in an m-dimensional space (plotted for some prescribed bias) with coordinates $x_1, x_2, ..., x_m$, defines a hyperplane as the decision surface between two different classes of inputs.

For the perceptron to function properly, the two classes $\mathcal{C}_1$ and $\mathcal{C}_2$ must be *linearly separable*. This, in turn, means that the patterns to be classified must be sufficiently separated from each other to ensure that the decision surface consists of a hyperplane. This requirement is illustrated in Fig. 3.9 for the case of a two-dimensional perceptron. In Fig. 3.9a, the two classes $\mathcal{C}_1$ and $\mathcal{C}_2$ are sufficiently separated from each other for us to draw a hyperplane (in this case a straight line) as the decision boundary. If, however, the two classes $\mathcal{C}_1$ and $\mathcal{C}_2$ are allowed to move too close to each other, as in Fig. 3.9b, they become nonlinearly separable, a situation that is beyond the computing capability of the perceptron.

Suppose then that the input variables of the perceptron originate from two linearly separable classes. Let $\mathcal{X}_1$ be the subset of training vectors $\mathbf{x}_1(1), \mathbf{x}_1(2), ...$ that belong to class $\mathcal{C}_1$, and let $\mathcal{X}_2$ be the subset of training vectors $\mathbf{x}_2(1), \mathbf{x}_2(2), ...$ that belong to class $\mathcal{C}_2$. The union of $\mathcal{X}_1$ and $\mathcal{X}_2$ is the complete training set $\mathcal{X}$. Given the sets of vectors $\mathcal{X}_1$ and $\mathcal{X}_2$ to train the classifier, the training process involves the adjustment of the weight vector $\mathbf{w}$ in such a way that the two classes $\mathcal{C}_1$ and $\mathcal{C}_2$ are linearly separable. That is, there exists a weight vector $\mathbf{w}$ such that we may state

$$\mathbf{w}^T\mathbf{x} > 0 \text{ for every input vector } \mathbf{x} \text{ belonging to class } \mathcal{C}_1$$
$$\mathbf{w}^T\mathbf{x} \leq 0 \text{ for every input vector } \mathbf{x} \text{ belonging to class } \mathcal{C}_2$$
(3.53)

In the second line of Eq. (3.53) we have arbitrarily chosen to say that the input vector $\mathbf{x}$ belongs to class $\mathcal{C}_2$ if $\mathbf{w}^T\mathbf{x} = 0$. Given the subsets of training vectors $\mathcal{X}_1$ and $\mathcal{X}_2$, the training problem for the elementary perceptron is then to find a weight vector $\mathbf{w}$ such that the two inequalities of Eq. (3.53) are satisfied.

The algorithm for adapting the weight vector of the elementary perceptron may now be formulated as follows:

1. If the nth member of the training set, $\mathbf{x}(n)$, is correctly classified by the weight vector $\mathbf{w}(n)$ computed at the nth iteration of the algorithm, no correction is made to the weight vector of the perceptron in accordance with the rule:

$$\mathbf{w}(n+1) = \mathbf{w}(n) \quad \text{if } \mathbf{w}^T\mathbf{x}(n) > 0 \text{ and } \mathbf{x}(n) \text{ belongs to class } \mathcal{C}_1$$
$$\mathbf{w}(n+1) = \mathbf{w}(n) \quad \text{if } \mathbf{w}^T\mathbf{x}(n) \leq 0 \text{ and } \mathbf{x}(n) \text{ belongs to class } \mathcal{C}_2$$
(3.54)

2. Otherwise, the weight vector of the perceptron is updated in accordance with the rule

$$\mathbf{w}(n+1) = \mathbf{w}(n) - \eta(n)\mathbf{x}(n) \quad \text{if } \mathbf{w}^T(n)\mathbf{x}(n) > 0 \text{ and } \mathbf{x}(n) \text{ belongs to class } \mathcal{C}_2$$
$$\mathbf{w}(n+1) = \mathbf{w}(n) + \eta(n)\mathbf{x}(n) \quad \text{if } \mathbf{w}^T(n)\mathbf{x}(n) \leq 0 \text{ and } \mathbf{x}(n) \text{ belongs to class } \mathcal{C}_1$$
(3.55)

where the *learning-rate parameter* $\eta(n)$ controls the adjustment applied to the weight vector at iteration n.

If $\eta(n) = \eta > 0$, where η is a constant independent of the iteration number n, we have a *fixed increment adaptation rule* for the perceptron.

In the sequel we first prove the convergence of a fixed increment adaptation rule for which $\eta = 1$. Clearly the value of η is unimportant, so long as it is positive. A value of $\eta \neq 1$ merely scales the pattern vectors without affecting their separability. The case of a variable $\eta(n)$ is considered later.

The proof is presented for the initial condition $\mathbf{w}(0) = \mathbf{0}$. Suppose that $\mathbf{w}^T(n)\mathbf{x}(n) < 0$ for $n = 1, 2, \ldots$, and the input vector $\mathbf{x}(n)$ belongs to the subset $\mathcal{X}_1$. That is, the perceptron incorrectly classifies the vectors $\mathbf{x}(1), \mathbf{x}(2), \ldots$, since the second condition of Eq. (3.53) is violated. Then, with the constant $\eta(n) = 1$, we may use the second line of Eq. (3.55) to write

$$\mathbf{w}(n+1) = \mathbf{w}(n) + \mathbf{x}(n) \quad \text{for } \mathbf{x}(n) \text{ belonging to class } \mathcal{C}_1 \qquad (3.56)$$

Given the initial condition $\mathbf{w}(0) = \mathbf{0}$, we may iteratively solve this equation for $\mathbf{w}(n+1)$, obtaining the result

$$\mathbf{w}(n+1) = \mathbf{x}(1) + \mathbf{x}(2) + \cdots + \mathbf{x}(n) \qquad (3.57)$$

Since the classes $\mathscr{C}_1$ and $\mathscr{C}_2$ are assumed to be linearly separable, there exists a solution $\mathbf{w}_0$ for which $\mathbf{w}^T\mathbf{x}(n) > 0$ for the vectors $\mathbf{x}(1), \dots, \mathbf{x}(n)$ belonging to the subset $\mathscr{X}_1$. For a fixed solution $\mathbf{w}_0$, we may then define a positive number α as

$$\alpha = \min_{\mathbf{x}(n) \in \mathscr{X}_1} \mathbf{w}_0^T\mathbf{x}(n) \tag{3.58}$$

Hence, multiplying both sides of Eq. (3.57) by the row vector $\mathbf{w}_0^T$, we get

$$\mathbf{w}_0^T\mathbf{w}(n + 1) = \mathbf{w}_0^T\mathbf{x}(1) + \mathbf{w}_0^T\mathbf{x}(2) + \cdots + \mathbf{w}_0^T\mathbf{x}(n)$$

Accordingly, in light of the definition given in Eq. (3.58), we have

$$\mathbf{w}_0^T\mathbf{w}(n + 1) \geq n\alpha \tag{3.59}$$

Next we make use of an inequality known as the Cauchy–Schwarz inequality. Given two vectors $\mathbf{w}_0$ and $\mathbf{w}(n + 1)$, the *Cauchy–Schwarz inequality* states that

$$\|\mathbf{w}_0\|^2 \|\mathbf{w}(n + 1)\|^2 \geq [\mathbf{w}_0^T\mathbf{w}(n + 1)]^2 \tag{3.60}$$

where $\|\cdot\|$ denotes the Euclidean norm of the enclosed argument vector, and the inner product $\mathbf{w}_0^T\mathbf{w}(n + 1)$ is a scalar quantity. We now note from Eq. (3.59) that $[\mathbf{w}_0^T\mathbf{w}(n + 1)]^2$ is equal to or greater than $n^2\alpha^2$. From Eq. (3.60) we note that $\|\mathbf{w}_0\|^2 \|\mathbf{w}(n + 1)\|^2$ is equal to or greater than $[\mathbf{w}_0^T\mathbf{w}(n + 1)]^2$. It follows therefore that

$$\|\mathbf{w}_0\|^2 \|\mathbf{w}(n + 1)\|^2 \geq n^2\alpha^2$$

or equivalently,

$$\|\mathbf{w}(n + 1)\|^2 \geq \frac{n^2\alpha^2}{\|\mathbf{w}_0\|^2} \tag{3.61}$$

We next follow another development route. In particular, we rewrite Eq. (3.56) in the form

$$\mathbf{w}(k + 1) = \mathbf{w}(k) + \mathbf{x}(k) \quad \text{for } k = 1, \dots, n \quad \text{and} \quad \mathbf{x}(k) \in \mathscr{X}_1 \tag{3.62}$$

By taking the squared Euclidean norm of both sides of Eq. (3.62), we obtain

$$\|\mathbf{w}(k + 1)\|^2 = \|\mathbf{w}(k)\|^2 + \|\mathbf{x}(k)\|^2 + 2\mathbf{w}^T(k)\mathbf{x}(k) \tag{3.63}$$

But, under the assumption that the perceptron incorrectly classifies an input vector $\mathbf{x}(k)$ belonging to the subset $\mathscr{X}_1$, we have $\mathbf{w}^T(k)\mathbf{x}(k) < 0$. We therefore deduce from Eq. (3.63) that

$$\|\mathbf{w}(k + 1)\|^2 \leq \|\mathbf{w}(k)\|^2 + \|\mathbf{x}(k)\|^2$$

or equivalently,

$$\|\mathbf{w}(k + 1)\|^2 - \|\mathbf{w}(k)\|^2 \leq \|\mathbf{x}(k)\|^2, \quad k = 1, \dots, n \tag{3.64}$$

Adding these inequalities for $k = 1, \dots, n$, and invoking the assumed initial condition $\mathbf{w}(0) = \mathbf{0}$, we get the following inequality:

$$\|\mathbf{w}(n + 1)\|^2 \leq \sum_{k=1}^{n} \|\mathbf{x}(k)\|^2 \tag{3.65}$$

$$\leq n\beta$$

where β is a positive number defined by

$$\beta = \max_{\mathbf{x}(k) \in \mathcal{X}_1} \|\mathbf{x}(k)\|^2 \tag{3.66}$$

Equation (3.65) states that the squared Euclidean norm of the weight vector $\mathbf{w}(n + 1)$ grows at most linearly with the number of iterations n.

The second result of Eq. (3.65) is clearly in conflict with the earlier result of Eq. (3.61) for sufficiently large values of n. Indeed, we can state that n cannot be larger than some value n_{max} for which Eqs. (3.61) and (3.65) are both satisfied with the equality sign. That is, n_{max} is the solution of the equation

$$\frac{n_{max}^2 \alpha^2}{\|\mathbf{w}_0\|^2} = n_{max}\beta$$

Solving for n_{max}, given a solution vector $\mathbf{w}_0$, we find that

$$n_{max} = \frac{\beta \|\mathbf{w}_0\|^2}{\alpha^2} \tag{3.67}$$

We have thus proved that for $\eta(n) = 1$ for all n, and $\mathbf{w}(0) = \mathbf{0}$, and given that a solution vector $\mathbf{w}_0$ exists, the rule for adapting the synaptic weights of the perceptron must terminate after at most n_{max} iterations. Note also from Eqs. (3.58), (3.66), and (3.67) that there is *no* unique solution for $\mathbf{w}_0$ or n_{max}.

We may now state the *fixed-increment covergence theorem* for the perceptron as follows (Rosenblatt, 1962):

Let the subsets of training vectors $\mathcal{X}_1$ and $\mathcal{X}_2$ be linearly separable. Let the inputs presented to the perceptron originate from these two subsets. The perceptron converges after some n_0 iterations, in the sense that

$$\mathbf{w}(n_0) = \mathbf{w}(n_0 + 1) = \mathbf{w}(n_0 + 2) = \cdots$$

is a solution vector for $n_0 \leq n_{max}$.

Consider next the *absolute error-correction procedure* for the adaptation of a single-layer perceptron, for which $\eta(n)$ is variable. In particular, let $\eta(n)$ be the smallest integer for which

$$\eta(n)\mathbf{x}^T(n)\mathbf{x}(n) > |\mathbf{w}^T(n)\mathbf{x}(n)|$$

With this procedure we find that if the inner product $\mathbf{w}^T(n)\mathbf{x}(n)$ at iteration n has an incorrect sign, then $\mathbf{w}^T(n + 1)\mathbf{x}(n)$ at iteration $n + 1$ would have the correct sign. This suggests that if $\mathbf{w}^T(n)\mathbf{x}(n)$ has an incorrect sign, we may modify the training sequence at iteration $n + 1$ by setting $\mathbf{x}(n + 1) = \mathbf{x}(n)$. In other words, each pattern is presented repeatedly to the perceptron until that pattern is classified correctly.

Note also that the use of an initial value $\mathbf{w}(0)$ different from the null condition merely results in a decrease or increase in the number of iterations required to converge, depending on how $\mathbf{w}(0)$ relates to the solution $\mathbf{w}_0$. Regardless of the value assigned to $\mathbf{w}(0)$, the perceptron is assured of convergence.

In Table 3.2 we present a summary of the *perceptron convergence algorithm* (Lippmann, 1987). The symbol sgn($\cdot$), used in step 3 of the table for computing the actual response of the perceptron, stands for the *signum function:*

$$\text{sgn}(v) = \begin{cases} +1 & \text{if } v > 0 \\ -1 & \text{if } v < 0 \end{cases} \tag{3.68}$$

We may thus express the *quantized response* $y(n)$ of the perceptron in the compact form

$$y(n) = \text{sgn}(\mathbf{w}^T(n)\mathbf{x}(n)) \tag{3.69}$$

TABLE 3.2   Summary of the Perceptron Convergence Algorithm

Variables and Parameters:

$\mathbf{x}(n) = (m+1)$-by-1 input vector

$= [+1, x_1(n), x_2(n), ..., x_m(n)]^T$

$\mathbf{w}(n) = (m + 1)$-by-1 weight vector

$= [b(n), w_1(n), w_2(n), ..., w_m(n)]^T$

$b(n) = $ bias

$y(n) = $ actual response (quantized)

$d(n) = $ desired response

$\eta = $ learning-rate parameter, a positive constant less than unity

1. *Initialization.* Set $\mathbf{w}(0) = \mathbf{0}$. Then perform the following computations for time step $n = 1, 2, ...$

2. *Activation.* At time step n, activate the perceptron by applying continuous-valued input vector $\mathbf{x}(n)$ and desired response $d(n)$.

3. *Computation of Actual Response.* Compute the actual response of the perceptron:

$$y(n) = \text{sgn}[\mathbf{w}^T(n)\mathbf{x}(n)]$$

where sgn($\cdot$) is the signum function.

4. *Adaptation of Weight Vector.* Update the weight vector of the perceptron:

$$\mathbf{w}(n + 1) = \mathbf{w}(n) + \eta[d(n) - y(n)]\mathbf{x}(n)$$

where

$$d(n) = \begin{cases} +1 & \text{if } \mathbf{x}(n) \text{ belongs to class } \mathscr{C}_1 \\ -1 & \text{if } \mathbf{x}(n) \text{ belongs to class } \mathscr{C}_2 \end{cases}$$

5. *Continuation.* Increment time step n by one and go back to step 2.

Notice that the input vector $\mathbf{x}(n)$ is an $(m + 1)$-by-1 vector whose first element is fixed at $+1$ throughout the computation. Correspondingly, the weight vector $\mathbf{w}(n)$ is an $(m + 1)$-by-1 vector whose first element equals the bias $b(n)$. One other important point in Table 3.2: We have introduced a *quantized desired response* $d(n)$, defined by

$$d(n) = \begin{cases} +1 & \text{if } \mathbf{x}(n) \text{ belongs to class } \mathscr{C}_1 \\ -1 & \text{if } \mathbf{x}(n) \text{ belongs to class } \mathscr{C}_2 \end{cases} \tag{3.70}$$

Thus, the adaptation of the weight vector $\mathbf{w}(n)$ is summed up nicely in the form of the *error-correction learning rule:*

$$\mathbf{w}(n + 1) = \mathbf{w}(n) + \eta[d(n) - y(n)]\mathbf{x}(n) \tag{3.71}$$

where η is the *learning-rate parameter*, and the difference $d(n) - y(n)$ plays the role of an *error signal*. The learning-rate parameter is a positive constant limited to the range $0 < \eta \le 1$. When assigning a value to it inside this range, we must keep in mind two conflicting requirements (Lippmann, 1987):

- *Averaging* of past inputs to provide stable weight estimates, which requires a small η
- *Fast adaptation* with respect to real changes in the underlying distributions of the process responsible for the generation of the input vector $\mathbf{x}$, which requires a large η

3.10 RELATION BETWEEN THE PERCEPTRON AND BAYES CLASSIFIER FOR A GAUSSIAN ENVIRONMENT

The perceptron bears a certain relationship to a classical pattern classifier known as the Bayes classifier. When the environment is Gaussian, the Bayes classifier reduces to a linear classifier. This is the same form taken by the perceptron. However, the linear nature of the perceptron is *not* contingent on the assumption of Gaussianity. In this section we study this relationship, and thereby develop further insight into the operation of the perceptron. We begin the discussion with a brief review of the Bayes classifier.

Bayes Classifier

In the *Bayes classifier* or *Bayes hypothesis testing procedure*, we minimize the *average risk*, denoted by $\mathscr{R}$. For a two-class problem, represented by classes $\mathscr{C}_1$ and $\mathscr{C}_2$, the average risk is defined by Van Trees (1968):

$$\begin{aligned} \mathscr{R} = c_{11}p_1 \int_{\mathscr{X}_1} f_{\mathbf{X}}(\mathbf{x}|\mathscr{C}_1)d\mathbf{x} + c_{22}p_2 \int_{\mathscr{X}_2} f_{\mathbf{X}}(\mathbf{x}|\mathscr{C}_2)d\mathbf{x} \\ + c_{21}p_1 \int_{\mathscr{X}_2} f_{\mathbf{X}}(\mathbf{x}|\mathscr{C}_1)d\mathbf{x} + c_{12}p_2 \int_{\mathscr{X}_1} f_{\mathbf{X}}(\mathbf{x}|\mathscr{C}_2)d\mathbf{x} \end{aligned} \tag{3.72}$$

where the various terms are defined as follows:

p_i = *a priori probability* that the observation vector $\mathbf{x}$ (representing a realization of the random vector $\mathbf{X}$) is drawn from subspace $\mathscr{X}_i$, with $i = 1, 2$, and $p_1 + p_2 = 1$.

c_{ij} = cost of deciding in favor of class $\mathscr{C}_i$ represented by subspace $\mathscr{X}_i$ when class $\mathscr{C}_j$ is true (i.e., observation vector $\mathbf{x}$ is drawn from subspace $\mathscr{X}_j$), with $(i, j) = 1, 2$.

$f_{\mathbf{X}}(\mathbf{x}|\mathscr{C}_i)$ = conditional probability density function of the random vector $\mathbf{X}$, given that the observation vector $\mathbf{x}$ is drawn from subspace $\mathscr{X}_i$, with $i = 1, 2$.

The first two terms on the right-hand side of Eq. (3.72) represent *correct* decisions (i.e., correct classifications), whereas the last two terms represent *incorrect* decisions (i.e., misclassifications). Each decision is weighted by the product of two factors: the cost involved in making the decision, and the relative frequency (i.e., *a priori* probability) with which it occurs.

The intention is to determine a strategy for the *minimum average risk*. Because we require that a decision be made, each observation vector $\mathbf{x}$ must be assigned in the overall observation space $\mathscr{X}$ to either $\mathscr{X}_1$ or $\mathscr{X}_2$. Thus,

$$\mathscr{X} = \mathscr{X}_1 + \mathscr{X}_2 \tag{3.73}$$

Accordingly, we may rewrite Eq. (3.72) in the equivalent form

$$\mathscr{R} = c_{11}p_1 \int_{\mathscr{X}_1} f_{\mathbf{X}}(\mathbf{x}|\mathscr{C}_1)d\mathbf{x} + c_{22}p_2 \int_{\mathscr{X}-\mathscr{X}_1} f_{\mathbf{X}}(\mathbf{x}|\mathscr{C}_2)d\mathbf{x}$$
$$+ c_{21}p_1 \int_{\mathscr{X}-\mathscr{X}_1} f_{\mathbf{X}}(\mathbf{x}|\mathscr{C}_1)\mathbf{x} + c_{12}p_2 \int_{\mathscr{X}_1} f_{\mathbf{X}}(\mathbf{x}|\mathscr{C}_2)d\mathbf{x} \tag{3.74}$$

where $c_{11} < c_{21}$ and $c_{22} < c_{12}$. We now observe the fact that

$$\int_{\mathscr{X}} f_{\mathbf{X}}(\mathbf{x}|\mathscr{C}_1)d\mathbf{x} = \int_{\mathscr{X}} f_{\mathbf{X}}(\mathbf{x}|\mathscr{C}_2)d\mathbf{x} = 1 \tag{3.75}$$

Hence, Eq. (3.74) reduces to

$$\mathscr{R} = c_{21}p_1 + c_{22}p_2$$
$$+ \int_{\mathscr{X}_1} [p_2(c_{12} - c_{22}) f_{\mathbf{X}}(\mathbf{x}|\mathscr{C}_2) - p_1(c_{21} - c_{11}) f_{\mathbf{X}}(\mathbf{x}|\mathscr{C}_1)]d\mathbf{x} \tag{3.76}$$

The first two terms on the right-hand side of Eq. (3.76) represent a fixed cost. Since the requirement is to minimize the average risk $\mathscr{R}$, we may therefore deduce the following strategy from Eq. (3.76) for optimum classification:

1. All values of the observation vector $\mathbf{x}$ for which the integrand (i.e., the expression inside the square brackets) is negative should be assigned to subspace $\mathscr{X}_1$ (i.e., class $\mathscr{C}_1$) for the integral would then make a negative contribution to the risk $\mathscr{R}$.
2. All values of the observation vector $\mathbf{x}$ for which the integrand is positive should be excluded from subspace $\mathscr{X}_1$ (i.e., assigned to class $\mathscr{C}_2$) for the integral would then make a positive contribution to the risk $\mathscr{R}$.

3. Values of **x** for which the integrand is zero have no effect on the average risk $\mathcal{R}$ and may be assigned arbitrarily. We shall assume that these points are assigned to subspace $\mathcal{X}_2$ (i.e., class $\mathcal{C}_2$).

On this basis, we may formulate the Bayes classifier as follows:

If the condition

$$p_1(c_{21} - c_{11})f_{\mathbf{X}}(\mathbf{x}|\mathcal{C}_1) > p_2(c_{12} - c_{22})f_{\mathbf{X}}(\mathbf{x}|\mathcal{C}_2)$$

holds, assign the observation vector **x** *to subspace* $\mathcal{X}_1$ *(i.e., class* $\mathcal{C}_1$*). Otherwise assign* **x** *to* $\mathcal{X}_2$ *(i.e., class* $\mathcal{C}_2$*).*

To simplify matters, define

$$\Lambda(\mathbf{x}) = \frac{f_{\mathbf{X}}(\mathbf{x}|\mathcal{C}_1)}{f_{\mathbf{X}}(\mathbf{x}|\mathcal{C}_2)} \tag{3.77}$$

and

$$\xi = \frac{p_2(c_{12} - c_{22})}{p_1(c_{21} - c_{11})} \tag{3.78}$$

The quantity $\Lambda(\mathbf{x})$, the ratio of two conditional probability density functions, is called the *likelihood ratio*. The quantity ξ is called the *threshold* of the test. Note that both $\Lambda(\mathbf{x})$ and ξ are always positive. In terms of these two quantities, we may now reformulate the Bayes classifier by stating:

If, for an observation vector **x**, *the likelihood ratio* $\Lambda(\mathbf{x})$ *is greater than the threshold* ξ, *assign* **x** *to class* $\mathcal{C}_1$. *Otherwise, assign it to class* $\mathcal{C}_2$.

Figure 3.10a depicts a block-diagram representation of the Bayes classifier. The important points in this block diagram are twofold:

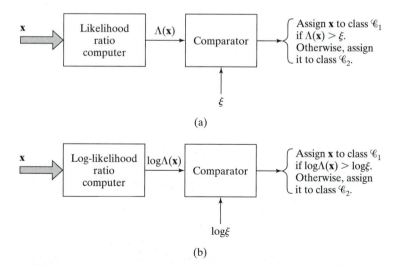

(a)

(b)

FIGURE 3.10 Two equivalent implementations of the Bayes classifier: (a) Likelihood ratio test, (b) Log-likelihood ratio test.

1. The data processing involved in designing the Bayes classifier is confined entirely to the computation of the likelihood ratio $\Lambda(\mathbf{x})$.

2. This computation is completely invariant to the values assigned to the *a priori* probabilities and costs involved in the decision-making process. These quantities merely affect the value of the threshold ξ.

From a computational point of view, we find it more convenient to work with the logarithm of the likelihood ratio rather than the likelihood ratio itself. We are permitted to do this for two reasons. First, the logarithm is a monotonic function. Second, the likelihood ratio $\Lambda(\mathbf{x})$ and threshold ξ are both positive. Therefore, the Bayes classifier may be implemented in the equivalent form shown in Fig. 3.10b. For obvious reasons, the test embodied in this latter figure is called the *log-likelihood ratio test*.

Bayes Classifier for a Gaussian Distribution

Consider now the special case of a two-class problem, for which the underlying distribution is Gaussian. The random vector $\mathbf{X}$ has a mean value that depends on whether it belongs to class $\mathscr{C}_1$ or class $\mathscr{C}_2$, but the covariance matrix of $\mathbf{X}$ is the same for both classes. That is to say:

$$\text{Class } \mathscr{C}_1: \quad E[\mathbf{X}] = \boldsymbol{\mu}_1$$
$$E[(\mathbf{X} - \boldsymbol{\mu}_1)(\mathbf{X} - \boldsymbol{\mu}_1)^T] = \mathbf{C}$$
$$\text{Class } \mathscr{C}_2: \quad E[\mathbf{X}] = \boldsymbol{\mu}_2$$
$$E[(\mathbf{X} - \boldsymbol{\mu}_2)(\mathbf{X} - \boldsymbol{\mu}_2)^T] = \mathbf{C}$$

The covariance matrix $\mathbf{C}$ is nondiagonal, which means that the samples drawn from classes $\mathscr{C}_1$ and $\mathscr{C}_2$ are *correlated*. It is assumed that $\mathbf{C}$ is nonsingular, so that its inverse matrix $\mathbf{C}^{-1}$ exists.

With this background we may express the conditional probability density function of $\mathbf{X}$ as follows:

$$f_{\mathbf{X}}(\mathbf{x}|\mathscr{C}_i) = \frac{1}{(2\pi)^{m/2}(\det(\mathbf{C}))^{1/2}} \exp\left(-\frac{1}{2}(\mathbf{x} - \boldsymbol{\mu}_i)^T \mathbf{C}^{-1}(\mathbf{x} - \boldsymbol{\mu}_i)\right), \quad i = 1, 2 \quad (3.79)$$

where m is the dimensionality of the observation vector $\mathbf{x}$.

It is further assumed that

1. The two classes $\mathscr{C}_1$ and $\mathscr{C}_2$ are equiprobable:

$$p_1 = p_2 = \frac{1}{2} \quad (3.80)$$

2. Misclassifications carry the same cost, and no cost is incurred on correct classifications:

$$c_{21} = c_{12} \quad \text{and} \quad c_{11} = c_{22} = 0 \quad (3.81)$$

We now have the information we need to design the Bayes classifier for the two-class problem. Specifically, by substituting Eq. (3.79) in (3.77) and taking the natural logarithm, we get (after simplifications):

$$\log \Lambda(\mathbf{x}) = -\frac{1}{2}(\mathbf{x} - \boldsymbol{\mu}_1)^T \mathbf{C}^{-1}(\mathbf{x} - \boldsymbol{\mu}_1) + \frac{1}{2}(\mathbf{x} - \boldsymbol{\mu}_2)^T \mathbf{C}^{-1}(\mathbf{x} - \boldsymbol{\mu}_2)$$
(3.82)
$$= (\boldsymbol{\mu}_1 - \boldsymbol{\mu}_2)^T \mathbf{C}^{-1}\mathbf{x} + \frac{1}{2}(\boldsymbol{\mu}_2^T \mathbf{C}^{-1}\boldsymbol{\mu}_2 - \boldsymbol{\mu}_1^T \mathbf{C}^{-1}\boldsymbol{\mu}_1)$$

By substituting Eqs. (3.80) and (3.81) in Eq. (3.78) and taking the natural logarithm, we get

$$\log \xi = 0 \qquad (3.83)$$

Equations (3.82) and (3.83) state that the Bayes classifier for the problem at hand is a *linear classifier*, as described by the relation

$$y = \mathbf{w}^T\mathbf{x} + b \qquad (3.84)$$

where

$$y = \log \Lambda(\mathbf{x}) \qquad (3.85)$$

$$\mathbf{w} = \mathbf{C}^{-1}(\boldsymbol{\mu}_1 - \boldsymbol{\mu}_2) \qquad (3.86)$$

$$b = \frac{1}{2}(\boldsymbol{\mu}_2^T \mathbf{C}^{-1}\boldsymbol{\mu}_2 - \boldsymbol{\mu}_1^T \mathbf{C}^{-1}\boldsymbol{\mu}_1) \qquad (3.87)$$

More specifically, the classifier consists of a linear combiner with weight vector $\mathbf{w}$ and bias b, as shown in Fig. 3.11.

On the basis of Eq. (3.84), we may now describe the log-likelihood ratio test for our two-class problem as follows:

If the output y of the linear combiner (including the bias b) is positive, assign the observation vector $\mathbf{x}$ *to class* $\mathscr{C}_1$. *Otherwise, assign it to class* $\mathscr{C}_2$.

The operation of the Bayes classifier for the Gaussian environment described herein is analogous to that of the perceptron in that they are both linear classifiers; see Eqs. (3.71) and (3.84). There are, however, some subtle and important differences between them, which should be carefully examined (Lippmann, 1987):

- The perceptron operates on the premise that the patterns to be classified are *linearly separable*. The Gaussian distribution of the two patterns assumed in the derivation of the Bayes classifier certainly do *overlap* each other and are therefore

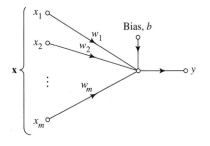

FIGURE 3.11   Signal-flow graph of Gaussian classifier.

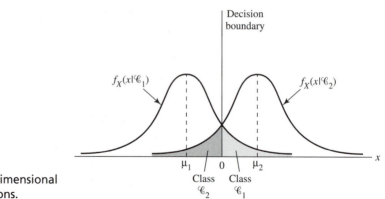

FIGURE 3.12 Two overlapping, one-dimensional Gaussian distributions.

not separable. The extent of the overlap is determined by the mean vectors $\boldsymbol{\mu}_1$ and $\boldsymbol{\mu}_2$, and the covariance matrix **C**. The nature of this overlap is illustrated in Fig. 3.12 for the special case of a scalar random variable (i.e., dimensionality $m = 1$). When the inputs are nonseparable and their distributions overlap as illustrated, the perceptron convergence algorithm develops a problem because decision boundaries between the different classes may oscillate continuously.

- The Bayes classifier minimizes the probability of classification error. This minimization is independent of the overlap between the underlying Gaussian distributions of the two classes. For example, in the special case illustrated in Fig. 3.12, the Bayes classifier always positions the decision boundary at the point where the Gaussian distributions for the two classes $\mathscr{C}_1$ and $\mathscr{C}_2$ cross each other.

- The perceptron convergence algorithm is *nonparametric* in the sense that it makes no assumptions concerning the form of the underlying distributions. It operates by concentrating on errors that occur where the distributions overlap. It may therefore work well when the inputs are generated by nonlinear physical mechanisms and when their distributions are heavily skewed and non-Gaussian. In contrast, the Bayes classifier is *parametric*; its derivation is contingent on the assumption that the underlying distributions are Gaussian, which may limit its area of application.

- The perceptron convergence algorithm is both adaptive and simple to implement; its storage requirement is confined to the set of synaptic weights and bias. On the other hand, the design of the Bayes classifier is fixed; it can be made adaptive, but at the expense of increased storage requirements and more complex computations.

3.11 SUMMARY AND DISCUSSION

The perceptron and an adaptive filter using the LMS algorithm are naturally related, as evidenced by their weight updates. Indeed, they represent different implementations of a *single-layer perceptron based on error-correction-learning*. The term "single-layer" is used here to signify that in both cases the computation layer consists of a single neuron—hence the title of the chapter. However, the perceptron and LMS algorithm differ from each other in some fundamental respects:

- The LMS algorithm uses a linear neuron, whereas the perceptron uses the McCulloch–Pitts formal model of a neuron.
- The learning process in the perceptron is performed for a finite number of iterations and then stops. In contrast, *continuous learning* takes place in the LMS algorithm in the sense that learning happens while signal processing is being performed in a manner that never stops.

A hard limiter constitutes the nonlinear element of the McCulloch–Pitts neuron. It is tempting to raise the question: Would the perceptron perform better if it used a sigmoidal nonlinearity in place of the hard limiter? It turns out that the steady-state, decision-making characteristics of the perceptron are basically the same, regardless of whether we use hard-limiting or soft-limiting as the source of nonlinearity in the neuronal model (Shynk, 1990; Shynk and Bershad, 1991). We may therefore state formally that so long as we limit ourselves to the model of a neuron that consists of a linear combiner followed by a nonlinear element, then regardless of the form of nonlinearity used, a single-layer perceptron can perform pattern classification only on linearly separable patterns.

We close this discussion of single-layer perceptrons on a historical note. The perceptron and the LMS algorithm emerged roughly about the same time, during the late 1950s. The LMS algorithm has truly survived the test of time. Indeed, it has established itself as the workhorse of adaptive signal processing because of its simplicity in implementation and its effectiveness in application. The importance of Rosenblatt's perceptron is largely historical.

The first real critique of Rosenblatt's perceptron was presented by Minsky and Selfridge (1961). Minsky and Selfridge pointed out that the perceptron as defined by Rosenblatt could not even generalize toward the notion of binary parity, let alone make general abstractions. The computational limitations of Rosenblatt's perceptron were subsequently put on a solid mathematical foundation in the famous book, *Perceptrons,* by Minsky and Papert (1969, 1988). After the presentation of some brilliant and highly detailed mathematical analyses of the perceptron, Minsky and Papert proved that the perceptron as defined by Rosenblatt is inherently incapable of making some global generalizations on the basis of locally learned examples. In the last chapter of their book, Minsky and Papert make the conjecture that the limitations they had discovered for Rosenblatt's perceptron would also hold true for its variants, more specifically, multilayer neural networks. Quoting from Section 13.2 of their book (1969):

> The perceptron has shown itself worthy of study despite (and even because of !) its severe limitations. It has many features to attract attention: its linearity; its intriguing learning theorem its clear paradigmatic simplicity as a kind of parallel computation. There is no reason to suppose that any of these virtues carry over to the many-layered version. Nevertheless, we consider it to be an important research problem to elucidate (or reject) our intuitive judgement that the extension to multilayer systems is sterile.

This conclusion was largely responsible for casting serious doubt on the computational capabilities of not only the perceptron but neural networks in general up to the mid-1980s.

History has shown, however, that the conjecture made by Minsky and Papert seems to be unjustified in that we now have several advanced forms of neural networks that are computationally more powerful than Rosenblatt's perceptron. For example, multilayer perceptrons trained with the back-propagation algorithm discussed in Chapter 4, the radial basis-function networks discussed in Chapter 5, and support vector machines discussed in Chapter 6, overcome the computational limitations of the single-layer perceptron in their own individual ways.

NOTES AND REFERENCES

1. The network organization of the original version of the perceptron as envisioned by Rosenblatt (1962) has three types of units: sensory units, association units, and response units. The connections from the sensory units to the association units have fixed weights, and the connections from the association units to the response units have variable weights. The association units act as preprocessors designed to extract a pattern from the environmental input. Insofar as the variable weights are concerned, the operation of Rosenblatt's original perceptron is essentially the same as that for the case of a single response unit (i.e., single neuron).

2. **Differentiation with respect to a vector**

   Let $f(\mathbf{w})$ denote a real-valued function of parameter vector $\mathbf{w}$. The derivative of $f(\mathbf{w})$ with respect to $\mathbf{w}$ is defined by the vector:

   $$\frac{\partial f}{\partial \mathbf{w}} = \left[\frac{\partial f}{\partial w_1}, \frac{\partial f}{\partial w_2}, \ldots, \frac{\partial f}{\partial w_m} \right]^T$$

   where m is the dimension of vector $\mathbf{w}$. The following two cases are of special interest:

   CASE 1   The function $f(\mathbf{w})$ is defined by the inner product:

   $$f(\mathbf{w}) = \mathbf{x}^T \mathbf{w}$$

   $$= \sum_{i=1}^{m} x_i w_i$$

   Hence,

   $$\frac{\partial f}{\partial w_i} = x_i, \qquad i = 1, 2, \ldots, m$$

   or equivalently, in matrix form:

   $$\frac{\partial f}{\partial \mathbf{w}} = \mathbf{x} \tag{1}$$

   CASE 2   The function $f(\mathbf{w})$ is defined by the quadratic form:

   $$f(\mathbf{w}) = \mathbf{w}^T \mathbf{R} \mathbf{w}$$

   $$= \sum_{i=1}^{m} \sum_{j=1}^{m} w_i r_{ij} w_j$$

   where r_{ij} is the ij-th element of the m-by-m matrix $\mathbf{R}$. Hence,

   $$\frac{\partial f}{\partial w_i} = \sum_{j=1}^{m} r_{ij} w_j, \qquad i = 1, 2, \ldots, m$$

   or equivalently, in matrix form:

   $$\frac{\partial f}{\partial \mathbf{w}} = \mathbf{R} \mathbf{w} \tag{2}$$

Equations (1) and (2) provide two useful rules for the differentiation of a real-valued function with respect to a vector.

3. **Positive definite matrix**

An m-by-m matrix $\mathbf{R}$ is said to be nonnegative definite if it satisfies the condition

$$\mathbf{a}^T \mathbf{R} \mathbf{a} \geq 0 \quad \text{for any vector } \mathbf{a} \in \mathbb{R}^m$$

If this condition is satisfied with the inequality sign, the matrix $\mathbf{R}$ is said to be positive definite.

An important property of a positive definite matrix $\mathbf{R}$ is that it is *nonsingular,* that is, the inverse matrix $\mathbf{R}^{-1}$ exists.

Another important property of a positive definite matrix $\mathbf{R}$ is that its eigenvalues, or roots or the characteristic equation

$$\det(\mathbf{R}) = 0$$

are all positive.

4. **Robustness**

The H^∞ criterion is due to Zames (1981), and it is developed in Zames and Francis (1983). The criterion is discussed in Doyle et al. (1989), Green and Limebeer (1995), and Hassibi et al. (1998).

5. To overcome the limitations of the LMS algorithm, namely, slow rate of convergence and sensitivity to variations in the condition number of the correlation matrix $\mathbf{R_x}$, we may use the *recursive least-squares (RLS) algorithm*, which follows from a recursive implementation of the linear least-squares filter described in Section 3.4. The RLS algorithm is a special case of the Kalman filter, which is known to be the optimum linear filter for a nonstationary environment. Most importantly, the Kalman filter exploits all past data extending up to and including the time instant at which the computations are made. For more details about the RLS algorithm and its relationship to the Kalman filter, see Haykin (1996). The Kalman filter is discussed in Chapter 15.

PROBLEMS

Unconstrained optimization

3.1 Explore the method of steepest descent involving a single weight w by considering the following cost function:

$$\mathcal{E}(w) = \frac{1}{2}\sigma^2 - r_{xd}w + \frac{1}{2}r_x w^2$$

where σ^2, r_{xd}, and r_x are constants.

3.2 Consider the cost function

$$\mathcal{E}(\mathbf{w}) = \frac{1}{2}\sigma^2 - \mathbf{r}_{xd}^T \mathbf{w} + \frac{1}{2}\mathbf{w}^T \mathbf{R_x} \mathbf{w}$$

where σ^2 is some constant, and

$$\mathbf{r}_{xd} = \begin{bmatrix} 0.8182 \\ 0.354 \end{bmatrix}$$

$$\mathbf{R_x} = \begin{bmatrix} 1 & 0.8182 \\ 0.8182 & 1 \end{bmatrix}$$

(a) Find the optimum value $\mathbf{w}^*$ for which $\mathcal{E}(\mathbf{w})$ reaches its minimum value.

(b) Use the method of steepest descent to compute $\mathbf{w}^*$ for the following two values of learning-rate parameter:

(i) $\eta = 0.3$

(ii) $\eta = 1.0$

For each case, plot the trajectory traced by the evolution of the weight vector $\mathbf{w}(n)$ in the W-plane.

Note: The trajectories obtained for cases (i) and (ii) of part (b) should correspond to the pictures displayed in Fig. 3.2.

3.3 Consider the cost function of Eq. (3.24) that represents a modified form of the sum of error squares defined in Eq. (3.17). Show that the application of the Gauss–Newton method to Eq. (3.24) yields the weight-update described in Eq. (3.23).

LMS Algorithm

3.4 The correlation matrix $\mathbf{R}_x$ of the input vector $\mathbf{x}(n)$ in the LMS algorithm is defined by

$$\mathbf{R}_x = \begin{bmatrix} 1 & 0.5 \\ 0.5 & 1 \end{bmatrix}$$

Define the range of values for the learning-rate parameter η of the LMS algorithm for it to be convergent in the mean square.

3.5 The *normalized LMS algorithm* is described by the following recursion for the weight vector:

$$\hat{\mathbf{w}}(n + 1) = \hat{\mathbf{w}}(n) + \frac{\eta}{\|\mathbf{x}(n)\|^2} e(n)\mathbf{x}(n)$$

where η is a positive constant and $\|\mathbf{x}(n)\|$ is the Euclidean norm of the input vector $\mathbf{x}(n)$. The error signal $e(n)$ is defined by

$$e(n) = d(n) - \hat{\mathbf{w}}^T(n)\mathbf{x}(n)$$

where $d(n)$ is the desired response. For the normalized LMS algorithm to be convergent in the mean square, show that

$$0 < \eta < 2$$

3.6 The LMS algorithm is used to implement the generalized sidelobe canceler shown in Fig. 2.16. Set up the equations that define the operation of this system, assuming the use of a single neuron for the neural network.

3.7 Consider a linear predictor with its input vector made up of the samples $x(n - 1)$, $x(n - 2)$, ..., $x(n - m)$, where m is the prediction order. The requirement is to use the LMS algorithm to make a prediction $\hat{x}(n)$ of the input sample $x(n)$. Set up the recursions that may be used to compute the tap weight $w_1, w_2, \ldots, w_m$ of the predictor.

3.8 The ensemble-averaged counterpart to the sum of error squares viewed as a cost function is the mean-square value of the error signal:

$$J(\mathbf{w}) = \frac{1}{2} E[e^2(n)]$$

$$= \frac{1}{2} E[(d(n) - \mathbf{x}^T(n)\mathbf{w})^2]$$

(a) Assuming that the input vector $\mathbf{x}(n)$ and desired response $d(n)$ are drawn from a stationary environment, show that

$$J(\mathbf{w}) = \frac{1}{2}\sigma_d^2 - \mathbf{r}_{xd}^T \mathbf{w} + \frac{1}{2}\mathbf{w}^T \mathbf{R}_x \mathbf{w}$$

where

$$\sigma_d^2 = E[d^2(n)]$$
$$\mathbf{r}_{xd} = E[\mathbf{x}(n)\, d(n)]$$
$$\mathbf{R}_x = E[\mathbf{x}(n)\mathbf{x}^T(n)]$$

(b) For this cost function, show that the gradient vector and Hessian matrix of $J(\mathbf{w})$ are as follows, respectively:

$$\mathbf{g} = -\mathbf{r}_{xd} + \mathbf{R}_x \mathbf{w}$$
$$\mathbf{H} = \mathbf{R}_x$$

(c) In the *LMS/Newton algorithm*, the gradient vector $\mathbf{g}$ is replaced by its instantaneous value (Widrow and Stearns, 1985). Show that this algorithm, incorporating a learning-rate parameter η, is described by

$$\hat{\mathbf{w}}(n+1) = \hat{\mathbf{w}}(n) + \eta\, \mathbf{R}_x^{-1} \mathbf{x}(n)\,(d(n) - \mathbf{x}^T(n)\mathbf{w}(n))$$

The inverse of the correlation matrix $\mathbf{R}_x$, assumed to be positive definite, is calculated ahead of time.

3.9 In this problem we revisit the correlation matrix memory discussed in Section 2.11. A shortcoming of this memory is that when a key pattern $\mathbf{x}_j$ is presented to it, the actual response $\mathbf{y}$ produced by the memory may not be close enough (in a Euclidean sense) to the desired response (memorized pattern) $\mathbf{y}_j$ for the memory to associate perfectly. This shortcoming is inherited from the use of Hebbian learning that has no provision for feedback from the output to the input. As a remedy for this shortcoming, we may incorporate an error-correction mechanism into the design of the memory, forcing it to associate properly (Anderson, 1983).

Let $\hat{\mathbf{M}}(n)$ denote the memory matrix learned at iteration n of the error-correction learning process. The memory matrix $\hat{\mathbf{M}}(n)$ learns the information represented by the associations:

$$\mathbf{x}_k \rightarrow \mathbf{y}_k, \quad k = 1, 2, \dots, q$$

(a) Adapting the LMS algorithm for the problem at hand, show that the updated value of the memory matrix is defined by

$$\hat{\mathbf{M}}(n+1) = \hat{\mathbf{M}}(n) + \eta[\mathbf{y}_k - \hat{\mathbf{M}}(n)\mathbf{x}_k]\mathbf{x}_k^T$$

where η is the learning-rate parameter.

(b) For autoassociation, $\mathbf{y}_k = \mathbf{x}_k$. For this special case, show that as the number of iterations, n, approaches infinity, the memory autoassociates perfectly, as shown by

$$\mathbf{M}(\infty)\mathbf{x}_k = \mathbf{x}_k, \quad k = 1, 2, \dots, q$$

(c) The result described in part (b) may be viewed as an *eigenvalue problem*. In that context, $\mathbf{x}_k$ represents an eigenvector of $\mathbf{M}(\infty)$. What are the eigenvalues of $\mathbf{M}(\infty)$?

3.10 In this problem we investigate the effect of bias on the condition number of a correlation matrix, and therefore the performance of the LMS algorithm.

Consider a random vector $\mathbf{X}$ with covariance matrix

$$\mathbf{C} = \begin{bmatrix} c_{11} & c_{12} \\ c_{21} & c_{22} \end{bmatrix}$$

and mean vector

$$\boldsymbol{\mu} = \begin{bmatrix} \mu_1 \\ \mu_2 \end{bmatrix}$$

(a) Calculate the condition number of the covariance matrix $\mathbf{C}$.
(b) Calculate the condition number of the correlation matrix $\mathbf{R}$.
Comment on the effect of the bias $\boldsymbol{\mu}$ on the performance of the LMS algorithm.

Rosenblatt's Perceptron

3.11 In this problem, we consider another method for deriving the update equation for Rosenblatt's perceptron. Define the *perceptron criterion function* (Duda and Hart, 1973):

$$J_p(\mathbf{w}) = \sum_{\mathbf{x} \in \mathcal{X}(\mathbf{w})} (-\mathbf{w}^T \mathbf{x})$$

where $\mathcal{X}(\mathbf{w})$ is the set of samples misclassified by the choice of weight vector $\mathbf{w}$. Note that $J_p(\mathbf{w})$ is defined as zero if there are no misclassified samples, and the output is misclassified if $\mathbf{w}_{\mathbf{x}}^T \leq 0$.

(a) Demonstrate geometrically that $J_p(\mathbf{w})$ is proportional to the sum of Euclidean distances from the misclassified samples to the decision boundary.
(b) Determine the gradient of $J_p(\mathbf{w})$ with respect to the weight vector $\mathbf{w}$.
(c) Using the result obtained in part (b), show that the weight-update for the perceptron is:

$$\mathbf{w}(n + 1) = \mathbf{w}(n) + \eta(n) \sum_{\mathbf{x} \in \mathcal{X}(\mathbf{w}(n))} \mathbf{x}$$

where $\mathcal{X}(\mathbf{w}(n))$ is the set of samples misclassified by the use of weight vector $\mathbf{w}(n)$, and $\eta(n)$ is the learning-rate parameter. Show that this result, for the case of a single-sample correction, is basically the same as that described by Eqs. (3.54) and (3.55).

3.12 Verify that Eqs. (3.68)–(3.71), summarizing the perceptron convergence algorithm, are consistent with Eqs. (3.54) and (3.55).

3.13 Consider two one-dimensional, Gaussian-distributed classes $\mathcal{C}_1$ and $\mathcal{C}_2$ that have a common variance equal to 1. Their mean values are

$$\mu_1 = -10$$

$$\mu_2 = +10$$

These two classes are essentially linearly separable. Design a classifier that separates these two classes.

3.14 Suppose that in the signal-flow graph of the perceptron shown in Fig. 3.6 the hard limiter is replaced by the sigmoidal nonlinearity:

$$\varphi(v) = \tanh\left(\frac{v}{2}\right)$$

where v is the induced local field. The classification decisions made by the perceptron are defined as follows:

> *Observation vector* $\mathbf{x}$ *belongs to class* $\mathscr{C}_1$ *if the output* $y > \theta$ *where* θ *is a threshold; otherwise,* $\mathbf{x}$ *belongs to class* $\mathscr{C}_2$.

Show that the decision boundary so constructed is a hyperplane.

3.15 (a) The perceptron may be used to perform numerous logic functions. Demonstrate the implementation of the binary logic functions AND, OR, and COMPLEMENT.

(b) A basic limitation of the perceptron is that it cannot implement the EXCLUSIVE OR function. Explain the reason for this limitation.

3.16 Equations (3.86) and (3.87) define the weight vector and bias of the Bayes classifier for a Gaussian environment. Determine the composition of this classifier for the case when the covariance matrix $\mathbf{C}$ is defined by

$$\mathbf{C} = \sigma^2 \mathbf{I}$$

where σ^2 is a constant.

CHAPTER 4

Multilayer Perceptrons

4.1 INTRODUCTION

In this chapter we study multilayer feedforward networks, an important class of neural networks. Typically, the network consists of a set of sensory units (source nodes) that constitute the *input layer,* one or more *hidden layers* of computation nodes, and an *output layer* of computation nodes. The input signal propagates through the network in a forward direction, on a layer-by-layer basis. These neural networks are commonly referred to as *multilayer perceptrons* (MLPs), which represent a generalization of the single-layer perceptron considered in Chapter 3.

Multilayer perceptrons have been applied successfully to solve some difficult and diverse problems by training them in a supervised manner with a highly popular algorithm known as the *error back-propagation algorithm.* This algorithm is based on the *error-correction learning rule.* As such, it may be viewed as a generalization of an equally popular adaptive filtering algorithm: the ubiquitous least-mean-square (LMS) algorithm described in Chapter 3 for the special case of a single linear neuron.

Basically, error back-propagation learning consists of two passes through the different layers of the network: a forward pass and a backward pass. In the *forward pass,* an activity pattern (input vector) is applied to the sensory nodes of the network, and its effect propagates through the network layer by layer. Finally, a set of outputs is produced as the actual response of the network. During the forward pass the synaptic weights of the networks are all *fixed.* During the *backward pass,* on the other hand, the synaptic weights are all *adjusted* in accordance with an error-correction rule. Specifically, the actual response of the network is subtracted from a desired (target) response to produce an *error signal.* This error signal is then propagated backward through the network, against the direction of synaptic connections—hence the name "error back-propagation." The synaptic weights are adjusted to make the actual response of the network move closer to the desired response in a statistical sense. The error back-propagation algorithm is also referred to in the literature as the *back-propagation algorithm,* or simply *back-prop.* Henceforth we will refer to it as the back-propagation algorithm. The learning process performed with the algorithm is called *back-propagation learning.*

A multilayer perceptron has three distinctive characteristics:

1. The model of each neuron in the network includes a *nonlinear activation function.* The important point to emphasize here is that the nonlinearity is *smooth* (i.e., differentiable everywhere), as opposed to the hard-limiting used in Rosenblatt's perceptron. A commonly used form of nonlinearity that satisfies this requirement is a *sigmoidal nonlinearity*[1] defined by the *logistic function:*

$$y_j = \frac{1}{1 + \exp(-v_j)}$$

where v_j is the induced local field (i.e., the weighted sum of all synaptic inputs plus the bias) of neuron j, and y_j is the output of the neuron. The presence of non-linearities is important because otherwise the input–output relation of the network could be reduced to that of a single-layer perceptron. Moreover, the use of the logistic function is biologically motivated, since it attempts to account for the refractory phase of real neurons.

2. The network contains one or more layers of *hidden neurons* that are not part of the input or output of the network. These hidden neurons enable the network to learn complex tasks by extracting progressively more meaningful features from the input patterns (vectors).

3. The network exhibits a high degrees of *connectivity,* determined by the synapses of the network. A change in the connectivity of the network requires a change in the population of synaptic connections or their weights.

It is through the combination of these characteristics together with the ability to learn from experience through training that the multilayer perceptron derives it computing power. These same characteristics, however, are also responsible for the deficiencies in our present state of knowledge on the behavior of the network. First, the presence of a distributed form of nonlinearity and the high connectivity of the network make the theoretical analysis of a multilayer perceptron difficult to undertake. Second, the use of hidden neurons makes the learning process harder to visualize. In an implicit sense, the learning process must decide which features of the input pattern should be represented by the hidden neurons. The learning process is therefore made more difficult because the search has to be conducted in a much larger space of possible functions, and a choice has to be made between alternative representations of the input pattern (Hinton, 1989).

The usage of the term "back-propagation" appears to have evolved after 1985, when its use was popularized through the publication of the seminal book entitled *Parallel Distributed Processing* (Rumelhart and McClelland, 1986). For historical notes on the back-propagation algorithm, see Section 1.9.

The development of the back-propagation algorithm represents a landmark in neural networks in that it provides a *computationally efficient* method for the training of multilayer perceptrons. Although we cannot claim that the back-propagation algorithm provides an optimal solution for all solvable problems, it has put to rest the pessimism about learning in multilayer machines that may have been inferred from the book by Minsky and Papert (1969).

Organization of the Chapter

In this chapter, we study basic aspects of the multilayer perceptron as well as back-propagation learning. The chapter is organized in seven parts. In the first part encompassing Sections 4.2 through 4.6, we discuss matters relating to back-propagation learning. We begin with some preliminaries in Section 4.2 to pave the way for the derivation of the back-propagation algorithm. In Section 4.3 we present a detailed derivation of the algorithm, using the chain rule of calculus; we take a traditional approach in the derivation presented here. A summary of the back-propagation algorithm is presented in Section 4.4. In Section 4.5 we illustrate the use of the back-propagation algorithm by solving the XOR problem, an interesting problem that cannot be solved by the single-layer perceptron. In Section 4.6 we present some heuristics or practical guidelines for making the back-propagation algorithm perform better.

The second part, encompassing Sections 4.7 through 4.9, explores the use of multilayer perceptrons for pattern recognition. In Section 4.7 we address the development of a rule for the use of a multilayer perceptron to solve the statistical pattern-recognition problem. In Section 4.8 we use a computer experiment to illustrate the application of back-propagation learning to distinguish between two classes of overlapping two-dimensional Gaussian distributions. The important role of hidden neurons as feature detectors is discussed in Section 4.9.

The third part of the chapter, encompassing Sections 4.10 through 4.12 deals with the error surface. In Section 4.10 we discuss the fundamental role of back-propagation learning in computing partial derivatives of an approximate function. We then discuss computational issues relating to the Hessian matrix of the error surface in Section 4.11.

The fourth part of the chapter deals with various matters relating to the performance of a multilayer perceptron trained with the back-propagation algorithm. In Section 4.12 we discuss the issue of generalization, the very essence of learning. Section 4.13 discusses the approximation of continuous functions by means of multilayer perceptrons. The use of cross-validation as a statistical design tool is discussed in Section 4.14. In Section 4.15 we describe procedures to orderly "prune" a multilayer perceptron while maintaining (and frequently improving) overall performance. Network pruning is desirable when computational complexity is of primary concern.

The fifth part of the chapter completes the study of back-propagation learning. In Section 4.16 we summarize the important advantages and limitations of back-propagation learning. In Section 4.17 we investigate heuristics that provide guidelines for how to accelerate the rate of convergence of back-propagation learning.

In the sixth part of the chapter we take a different viewpoint on learning. With improved learning as the objective, we discuss the issue of supervised learning as a problem in numerical optimization in Section 4.18. In particular, we describe the conjugate-gradient algorithm and quasi-Newton methods for supervised learning.

The last part of the chapter, Section 4.19, deals with the multilayer perceptron itself. There we describe an interesting neural network structure, the *convolutional multilayer perceptron*. This network has been successfully used in the solution of difficult pattern-recognition problems.

The chapter concludes with some general discussion in Section 4.20.

4.2   SOME PRELIMINARIES

Figure 4.1 shows the architectural graph of a multilayer perceptron with two hidden layers and an output layer. To set the stage for a description of the multilayer perceptron in its general form, the network shown here is *fully connected*. This means that a neuron in any layer of the network is connected to all the nodes/neurons in the previous layer. Signal flow through the network progresses in a forward direction, from left to right and on a layer-by-layer basis.

Figure 4.2 depicts a portion of the multilayer perceptron. Two kinds of signals are identified in this network (Parker, 1987):

1. *Function Signals.* A function signal is an input signal (stimulus) that comes in at the input end of the network, propagates forward (neuron by neuron) through the network, and emerges at the output end of the network as an output signal. We refer to such a signal as a "function signal" for two reasons. First, it is presumed to perform a useful function at the output of the network. Second, at each neuron of the network through which a function signal passes, the signal is

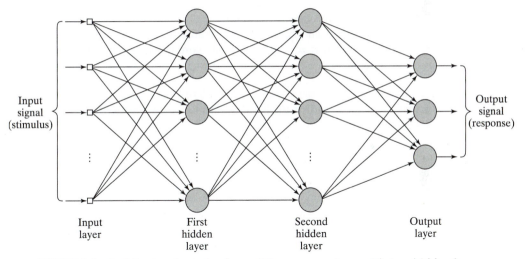

FIGURE 4.1   Architectural graph of a multilayer perceptron with two hidden layers.

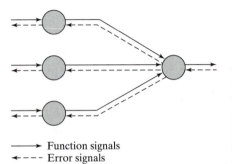

→ Function signals
← - - - Error signals

FIGURE 4.2   Illustration of the directions of two basic signal flows in a multilayer perceptron: forward propagation of function signals and back-propagation of error signals.

calculated as a function of the inputs and associated weights applied to that neuron. The function signal is also referred to as the input signal.

2. *Error Signals.* An error signal originates at an output neuron of the network, and propagates backward (layer by layer) through the network. We refer to it as an "error signal" because its computation by every neuron of the network involves an error-dependent function in one form or another.

The output neurons (computational nodes) constitute the output layers of the network. The remaining neurons (computational nodes) constitute hidden layers of the network. Thus the hidden units are not part of the output or input of the network—hence their designation as "hidden." The first hidden layer is fed from the input layer made up of sensory units (source nodes); the resulting outputs of the first hidden layer are in turn applied to the next hidden layer; and so on for the rest of the network.

Each hidden or output neuron of a multilayer perceptron is designed to perform two computations:

1. The computation of the function signal appearing at the output of a neuron, which is expressed as a continuous nonlinear function of the input signal and synaptic weights associated with that neuron.

2. The computation of an estimate of the gradient vector (i.e., the gradients of the error surface with respect to the weights connected to the inputs of a neuron), which is needed for the backward pass through the network.

The derivation of the back-propagation algorithm is rather involved. To ease the mathematical burden involved in this derivation, we first present a summary of the notations used in the derivation.

Notation

- The indices $i, j,$ and k refer to different neurons in the network; with signals propagating through the network from left to right, neuron j lies in a layer to the right of neuron $i,$ and neuron k lies in a layer to the right of neuron j when neuron j is a hidden unit.

- In iteration (time step) $n,$ the nth training pattern (example) is presented to the network.

- The symbol $\mathcal{E}(n)$ refers to the instantaneous sum of error squares or error energy at iteration $n.$ The average of $\mathcal{E}(n)$ over all values of n (i.e., the entire training set) yields the average error energy $\mathcal{E}_{av}.$

- The symbol $e_j(n)$ refers to the error signal at the output of neuron j for iteration $n.$

- The symbol $d_j(n)$ refers to the desired response for neuron j and is used to compute $e_j(n).$

- The symbol $y_j(n)$ refers to the function signal appearing at the output of neuron j at iteration $n.$

- The symbol $w_{ji}(n)$ denotes the synaptic weight connecting the output of neuron i to the input of neuron j at iteration $n.$ The correction applied to this weight at iteration n is denoted by $\Delta w_{ji}(n).$

- The induced local field (i.e., weighted sum of all synaptic inputs plus bias) of neuron j at iteration n is denoted by $v_j(n)$; it constitutes the signal applied to the activation function associated with neuron j.
- The activation function describing the input–output functional relationship of the nonlinearity associated with neuron j is denoted by $\varphi_j(\cdot)$.
- The bias applied to neuron j is denoted by b_j; its effect is represented by a synapse of weight $w_{j0} = b_j$ connected to a fixed input equal to $+1$.
- The ith element of the input vector (pattern) is denoted by $x_i(n)$.
- The kth element of the overall output vector (pattern) is denoted by $o_k(n)$.
- The learning-rate parameter is denoted by η.
- The symbol m_l denotes the size (i.e., number of nodes) in layer l of the multilayer perceptron; $l = 0, 1, \ldots, L$, where L is the "depth" of the network. Thus m_0 denotes the size of the input layer, m_1 denotes the size of the first hidden layer, and m_L denotes the size of the output layer. The notation $m_L = M$ is also used.

4.3   BACK-PROPAGATION ALGORITHM

The error signal at the output of neuron j at iteration n (i.e., presentation of the nth training example) is defined by

$$e_j(n) = d_j(n) - y_j(n), \qquad \text{neuron } j \text{ is an output node} \tag{4.1}$$

We define the instantaneous value of the error energy for neuron j as $\frac{1}{2}e_j^2(n)$. Correspondingly, the instantaneous value $\mathcal{E}(n)$ of the total error energy is obtained by summing $\frac{1}{2}e_j^2(n)$ over *all neurons in the output layer;* these are the only "visible" neurons for which error signals can be calculated directly. We may thus write

$$\mathcal{E}(n) = \frac{1}{2} \sum_{j \in C} e_j^2(n) \tag{4.2}$$

where the set C includes all the neurons in the output layer of the network. Let N denote the total number of patterns (examples) contained in the training set. The *average squared error energy* is obtained by summing $\mathcal{E}(n)$ over all n and then normalizing with respect to the set size N, as shown by

$$\mathcal{E}_{av} = \frac{1}{N} \sum_{n=1}^{N} \mathcal{E}(n) \tag{4.3}$$

The instantaneous error energy $\mathcal{E}(n)$, and therefore the average error energy $\mathcal{E}_{av}$, is a function of all the free parameters (i.e., synaptic weights and bias levels) of the network. For a given training set, $\mathcal{E}_{av}$ represents the *cost function* as a measure of learning performance. The objective of the learning process is to adjust the free parameters of the network to minimize $\mathcal{E}_{av}$. To do this minimization, we use an approximation similar in rationale to that used for the derivation of the LMS algorithm in Chapter 3. Specifically, we consider a simple method of training in which the weights are updated on a *pattern-by-pattern* basis until one *epoch,* that is, one complete presentation of the entire training set has been dealt with. The adjustments to the weights are made in accordance with the respective errors computed for *each* pattern presented to the network.

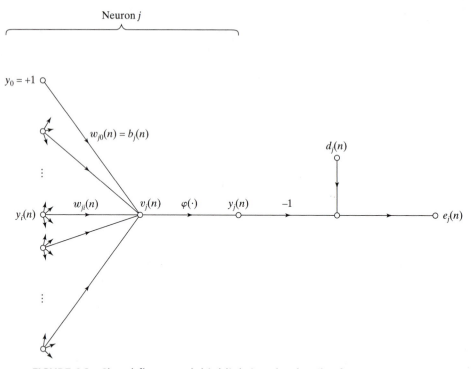

FIGURE 4.3   Signal-flow graph highlighting the details of output neuron j.

The arithmetic average of these individual weight changes over the training set is there-fore an *estimate* of the true change that would result from modifying the weights based on minimizing the cost function $\mathcal{E}_{\text{av}}$ over the entire training set. We will address the quality of the estimate later in this section.

Consider then Fig. 4.3., which depicts neuron j being fed by a set of function sig-nals produced by a layer of neurons to its left. The induced local field $v_j(n)$ produced at the input of the activation function associated with neuron j is therefore

$$v_j(n) = \sum_{i=0}^{m} w_{ji}(n) y_i(n) \tag{4.4}$$

where m is the total number of inputs (excluding the bias) applied to neuron j. The synap-tic weight w_{jo} (corresponding to the fixed input $y_0 = +1$) equals the bias b_j applied to neu-ron j. Hence the function signal $y_j(n)$ appearing at the output of neuron j at iteration n is

$$y_j(n) = \varphi_j(v_j(n)) \tag{4.5}$$

In a manner similar to the LMS algorithm, the back-propagation algorithm applies a correction $\Delta w_{ji}(n)$ to the synaptic weight $w_{ji}(n)$, which is proportional to the partial derivative $\partial \mathcal{E}(n)/\partial w_{ji}(n)$. According to the *chain rule* of calculus, we may express this gradient as:

$$\frac{\partial \mathcal{E}(n)}{\partial w_{ji}(n)} = \frac{\partial \mathcal{E}(n)}{\partial e_j(n)} \frac{\partial e_j(n)}{\partial y_j(n)} \frac{\partial y_j(n)}{\partial v_j(n)} \frac{\partial v_j(n)}{\partial w_{ji}(n)} \tag{4.6}$$

The partial derivative $\partial \mathcal{E}(n)/\partial w_{ji}(n)$ represents a *sensitivity factor*, determining the direction of search in weight space for the synaptic weight w_{ji}.

Differentiating both sides of Eq. (4.2) with respect to $e_j(n)$, we get

$$\frac{\partial \mathcal{E}(n)}{\partial e_j(n)} = e_j(n) \qquad (4.7)$$

Differentiating both sides of Eq. (4.1) with respect to $y_j(n)$, we get

$$\frac{\partial e_j(n)}{\partial y_j(n)} = -1 \qquad (4.8)$$

Next, differentiating Eq. (4.5) with respect to $v_j(n)$, we get

$$\frac{\partial y_j(n)}{\partial v_j(n)} = \varphi_j'(v_j(n)) \qquad (4.9)$$

where the use of prime (on the right-hand side) signifies differentiation with respect to the argument. Finally, differentiating Eq. (4.4) with respect to $w_{ji}(n)$ yields

$$\frac{\partial v_j(n)}{\partial w_{ji}(n)} = y_i(n) \qquad (4.10)$$

The use of Eqs. (4.7) to (4.10) in (4.6) yields

$$\frac{\partial \mathcal{E}(n)}{\partial w_{ji}(n)} = -e_j(n)\varphi_j'(v_j(n))y_i(n) \qquad (4.11)$$

The correction $\Delta w_{ji}(n)$ applied to $w_{ji}(n)$ is defined by the *delta rule*:

$$\Delta w_{ji}(n) = -\eta \frac{\partial \mathcal{E}(n)}{\partial w_{ji}(n)} \qquad (4.12)$$

where η is the *learning-rate parameter* of the back-propagation algorithm. The use of the minus sign in Eq. (4.12) accounts for *gradient descent* in weight space (i.e., seeking a direction for weight change that reduces the value of $\mathcal{E}(n)$). Accordingly, the use of Eq. (4.11) in (4.12) yields

$$\Delta w_{ji}(n) = \eta \delta_j(n)y_i(n) \qquad (4.13)$$

where the *local gradient* $\delta_j(n)$ is defined by

$$\delta_j(n) = -\frac{\partial \mathcal{E}(n)}{\partial v_j(n)}$$

$$= -\frac{\partial \mathcal{E}(n)}{\partial e_j(n)} \frac{\partial e_j(n)}{\partial y_j(n)} \frac{\partial y_j(n)}{\partial v_j(n)} \qquad (4.14)$$

$$= e_j(n)\varphi_j'(v_j(n))$$

The local gradient points to required changes in synaptic weights. According to Eq. (4.14), the local gradient $\delta_j(n)$ for output neuron j is equal to the product of the corresponding error signal $e_j(n)$ for that neuron and the derivative $\varphi_j'(v_j(n))$ of the associated activation function.

From Eqs. (4.13) and (4.14) we note that a key factor involved in the calculation of the weight adjustment $\Delta w_{ji}(n)$ is the error signal $e_j(n)$ at the output of neuron j. In this context we may identify two distinct cases, depending on where in the network neuron j is located. In case 1, neuron j is an output node. This case is simple to handle because each output node of the network is supplied with a desired response of its own, making it a straightforward matter to calculate the associated error signal. In case 2, neuron j is a hidden node. Even though hidden neurons are not directly accessible, they share responsibility for any error made at the output of the network. The question, however, is to know how to penalize or reward hidden neurons for their share of the responsibility. This problem is the *credit-assignment problem* considered in Section 2.7. It is solved in an elegant fashion by back-propagating the error signals through the network.

Case 1   Neuron *j* Is an Output Node

When neuron j is located in the output layer of the network, it is supplied with a desired response of its own. We may use Eq. (4.1) to compute the error signal $e_j(n)$ associated with this neuron; see Fig. 4.3. Having determined $e_j(n)$, it is a straightforward matter to compute the local gradient $\delta_j(n)$ using Eq. (4.14).

Case 2   Neuron *j* Is a Hidden Node

When neuron j is located in a hidden layer of the network, there is no specified desired response for that neuron. Accordingly, the error signal for a hidden neuron would have to be determined recursively in terms of the error signals of all the neurons to which that hidden neuron is directly connected; this is where the development of the back-propagation algorithm gets complicated. Consider the situation depicted in Fig. 4.4, which depicts neuron j as a hidden node of the network. According to Eq. (4.14), we may redefine the local gradient $\delta_j(n)$ for hidden neuron j as

$$
\begin{aligned}
\delta_j(n) &= -\frac{\partial \mathscr{E}(n)}{\partial y_j(n)} \frac{\partial y_j(n)}{\partial v_j(n)} \\[2mm]
&= -\frac{\partial \mathscr{E}(n)}{\partial y_j(n)} \varphi_j'(v_j(n)), \qquad \text{neuron } j \text{ is hidden}
\end{aligned}
\tag{4.15}
$$

where in the second line we have used Eq. (4.9). To calculate the partial derivative $\partial \mathscr{E}(n)/\partial y_j(n)$, we may proceed as follows. From Fig. 4.4 we see that

$$
\mathscr{E}(n) = \frac{1}{2} \sum_{k \in C} e_k^2(n), \qquad \text{neuron } k \text{ is an output node}
\tag{4.16}
$$

which is Eq. (4.2) with index k used in place of index j. We have done so in order to avoid confusion with the use of index j that refers to a hidden neuron under case 2. Differentiating Eq. (4.16) with respect to the function signal $y_j(n)$, we get

$$
\frac{\partial \mathscr{E}(n)}{\partial y_j(n)} = \sum_k e_k \frac{\partial e_k(n)}{\partial y_j(n)}
\tag{4.17}
$$

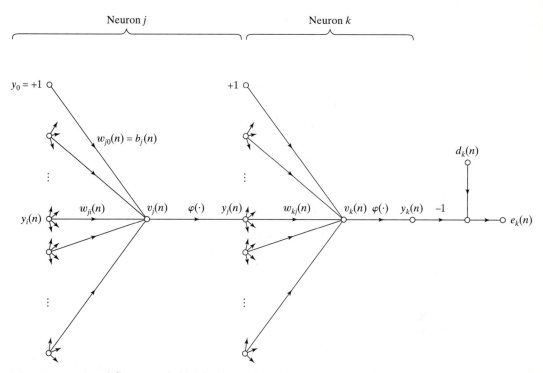

FIGURE 4.4   Signal-flow graph highlighting the details of output neuron k connected to hidden neuron j.

Next we use the chain rule for the partial derivative $\partial e_k(n)/\partial y_j(n)$, and rewrite Eq. (4.17) in the equivalent form

$$\frac{\partial \mathcal{E}(n)}{\partial y_j(n)} = \sum_k e_k(n) \frac{\partial e_k(n)}{\partial v_k(n)} \frac{\partial v_k(n)}{\partial y_j(n)} \tag{4.18}$$

However, from Fig. 4.4, we note that

$$e_k(n) = d_k(n) - y_k(n)$$
$$= d_k(n) - \varphi_k(v_k(n)), \qquad \text{neuron } k \text{ is an output node} \tag{4.19}$$

Hence

$$\frac{\partial e_k(n)}{\partial v_k(n)} = -\varphi_k'(v_k(n)) \tag{4.20}$$

We also note from Fig. 4.4 that for neuron k the induced local field is

$$v_k(n) = \sum_{j=0}^{m} w_{kj}(n) y_j(n) \tag{4.21}$$

where m is the total number of inputs (excluding the bias) applied to neuron k. Here again, the synaptic weight $w_{k0}(n)$ is equal to the bias $b_k(n)$ applied to neuron k, and the corresponding input is fixed at the value $+1$. Differentiating Eq. (4.21) with respect to $y_j(n)$ yields

$$\frac{\partial v_k(n)}{\partial y_j(n)} = w_{kj}(n) \tag{4.22}$$

By using Eqs. (4.20) and (4.22) in (4.18) we get the desired partial derivative:

$$
\begin{aligned}
\frac{\partial \mathcal{E}(n)}{\partial y_j(n)} &= -\sum_k e_k(n) \varphi_k'(v_k(n)) w_{kj}(n) \\
&= -\sum_k \delta_k(n) w_{kj}(n)
\end{aligned}
\tag{4.23}
$$

where in the second line we have used the definition of the local gradient $\delta_k(n)$ given in Eq. (4.14) with the index k substituted for j.

Finally, using Eq. (4.23) in (4.15), we get the *back-propagation formula* for the local gradient $\delta_j(n)$ as described:

$$\delta_j(n) = \varphi_j'(v_j(n)) \sum_k \delta_k(n) w_{kj}(n), \qquad \text{neuron } j \text{ is hidden} \tag{4.24}$$

Figure 4.5 shows the signal-flow graph representation of Eq. (4.24), assuming that the output layer consists of m_L neurons.

The factor $\varphi_j'(v_j(n))$ involved in the computation of the local gradient $\delta_j(n)$ in Eq. (4.24) depends solely on the activation function associated with hidden neuron j. The remaining factor involved in this computation, namely the summation over k, depends on two sets of terms. The first set of terms, the $\delta_k(n)$, requires knowledge of the error signals $e_k(n)$, for all neurons that lie in the layer to the immediate right of hidden neuron j, and that are directly connected to neuron j: see Fig. 4.4. The second set of terms, the $w_{kj}(n)$, consists of the synaptic weights associated with these connections.

We now summarize the relations that we have derived for the back-propagation algorithm. First, the correction $\Delta w_{ji}(n)$ applied to the synaptic weight connecting neuron i to neuron j is defined by the delta rule:

$$
\begin{pmatrix} \text{Weight} \\ \text{correction} \\ \Delta w_{ji}(n) \end{pmatrix}
=
\begin{pmatrix} \text{learning-} \\ \text{rate parameter} \\ \eta \end{pmatrix}
\cdot
\begin{pmatrix} \text{local} \\ \text{gradient} \\ \delta_j(n) \end{pmatrix}
\cdot
\begin{pmatrix} \text{input signal} \\ \text{of neuron } j \\ y_i(n) \end{pmatrix}
\tag{4.25}
$$

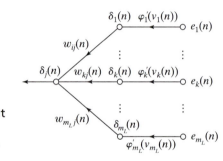

FIGURE 4.5 Signal-flow graph of a part of the adjoint system pertaining to back-propagation of error signals.

Second, the local gradient $\delta_j(n)$ depends on whether neuron j is an output node or a hidden node:

1. If neuron j is an output node, $\delta_j(n)$ equals the product of the derivative $\varphi_j'(v_j(n))$ and the error signal $e_j(n)$, both of which are associated with neuron j; see Eq.(4.14).
2. If neuron j is a hidden node, $\delta_j(n)$ equals the product of the associated derivative $\varphi_j'(v_j(n))$ and the weighted sum of the δs computed for the neurons in the next hidden or output layer that are connected to neuron j; see Eq. (4.24).

The Two Passes of Computation

In the application of the back-propagation algorithm, two distinct passes of computation are distinguished. The first pass is referred to as the forward pass, and the second is referred to as the backward pass.

In the *forward pass* the synaptic weights remain unaltered throughout the network, and the function signals of the network are computed on a neuron-by-neuron basis. The function signal appearing at the output of neuron j is computed as

$$y_j(n) = \varphi(v_j(n)) \tag{4.26}$$

where $v_j(n)$ is the induced local field of neuron j, defined by

$$v_j(n) = \sum_{i=0}^{m} w_{ji}(n) y_i(n) \tag{4.27}$$

where m is the total number of inputs (excluding the bias) applied to neuron j, and $w_{ji}(n)$ is the synaptic weight connecting neuron i to neuron j, and $y_i(n)$ is the input signal of neuron j or equivalently, the function signal appearing at the output of neuron i. If neuron j is in the first hidden layer of the network, $m = m_0$ and the index i refers to the ith input terminal of the network, for which we write

$$y_i(n) = x_i(n) \tag{4.28}$$

where $x_i(n)$ is the ith element of the input vector (pattern). If, on the other hand, neuron j is in the output layer of the network, $m = m_L$ and the index j refers to the jth output terminal of the network, for which we write

$$y_j(n) = o_j(n) \tag{4.29}$$

where $o_j(n)$ is the jth element of the output vector (pattern). This output is compared with the desired response $d_j(n)$, obtaining the error signal $e_j(n)$ for the jth output neuron. Thus the forward phase of computation begins at the first hidden layer by presenting it with the input vector, and terminates at the output layer by computing the error signal for each neuron of this layer.

The backward pass, on the other hand, starts at the output layer by passing the error signals leftward through the network, layer by layer, and recursively computing the δ (i.e., the local gradient) for each neuron. This recursive process permits the synaptic weights of the network to undergo changes in accordance with the delta rule of Eq.(4.25). For a neuron located in the output layer, the δ is simply equal to the error signal of that neuron multiplied by the first derivative of its nonlinearity. Hence we use

Eq. (4.25) to compute the changes to the weights of all the connections feeding into the output layer. Given the δs for the neurons of the output layer, we next use Eq. (4.24) to compute the δs for all the neurons in the penultimate layer and therefore the changes to the weights of all connections feeding into it. The recursive computation is continued, layer by layer, by propagating the changes to all synaptic weights in the network.

Note that for the presentation of each training example, the input pattern is fixed ("clamped") throughout the round-trip process, encompassing the forward pass followed by the backward pass.

Activation Function

The computation of the δ for each neuron of the multilayer perceptron requires knowledge of the derivative of the activation function $\varphi(\cdot)$ associated with that neuron. For this derivative to exist, we require the function $\varphi(\cdot)$ to be continuous. In basic terms, *differentiability* is the only requirement that an activation function has to satisfy. An example of a continuously differentiable nonlinear activation function commonly used in multilayer perceptrons is *sigmoidal nonlinearity;* two forms are described:

1. *Logistic Function.* This form of sigmoidal nonlinearity in its general form is defined by

$$\varphi_j(v_j(n)) = \frac{1}{1 + \exp(-av_j(n))} \qquad a > 0 \text{ and } -\infty < v_j(n) < \infty \qquad (4.30)$$

where $v_j(n)$ is the induced local field of neuron j. According to this nonlinearity, the amplitude of the output lies inside the range $0 \leq y_j \leq 1$. Differentiating Eq. (4.30) with respect to $v_j(n)$, we get

$$\varphi_j'(v_j(n)) = \frac{a \exp(-av_j(n))}{[1 + \exp(-av_j(n))]^2} \qquad (4.31)$$

With $y_j(n) = \varphi_j(v_j(n))$, we may eliminate the exponential term $\exp(-av_j(n))$ from Eq. (4.31), and so express the derivative $\varphi_j'(v_j(n))$ as

$$\varphi_j'(v_j(n)) = ay_j(n)[1 - y_j(n)] \qquad (4.32)$$

For a neuron j located in the output layer, $y_j(n) = o_j(n)$. Hence, we may express the local gradient for neuron j as

$$\delta_j(n) = e_j(n)\varphi_j'(v_j(n))$$
$$= a[d_j(n) - o_j(n)]o_j(n)[1 - o_j(n)], \qquad \text{neuron } j \text{ is an output node} \qquad (4.33)$$

where $o_j(n)$ is the function signal at the output of neuron j, and $d_j(n)$ is the desired response for it. On the other hand, for an arbitrary hidden neuron j, we may express the local gradient as

$$\delta_j(n) = \varphi_j'(v_j(n)) \sum_k \delta_k(n)w_{kj}(n)$$
$$\qquad (4.34)$$
$$= ay_j(n)[1 - y_j(n)] \sum_k \delta_k(n)w_{kj}(n), \qquad \text{neuron } j \text{ is hidden}$$

Note from Eq. (4.32) that the derivative $\varphi'_j(v_j(n))$ attains its maximum value at $y_j(n) = 0.5$, and its minimum value (zero) at $y_j(n) = 0$, or $y_j(n) = 1.0$. Since the amount of change in a synaptic weight of the network is proportional to the derivative $\varphi'_j(v_j(n))$, it follows that for a sigmoid activation function the synaptic weights are changed the most for those neurons in the network where the function signals are in their midrange. According to Rumelhart et al. (1986a), it is this feature of back-propagation learning that contributes to its stability as a learning algorithm.

   2. *Hyperbolic tangent function.* Another commonly used form of sigmoidal non-linearity is the hyperbolic tangent function, which in its most general form is defined by

$$\varphi_j(v_j(n)) = a \tanh(bv_j(n)), \qquad (a,b) > 0 \tag{4.35}$$

where a and b are constants. In reality, the hyperbolic tangent function is just the logistic function rescaled and biased. Its derivative with respect to $v_j(n)$ is given by

$$\begin{aligned} \varphi'_j(v_j(n)) &= ab\,\mathrm{sech}^2(bv_j(n)) \\ &= ab\big(1 - \tanh^2(bv_j(n))\big) \\ &= \frac{b}{a}[a - y_j(n)][a + y_j(n)] \end{aligned} \tag{4.36}$$

For a neuron j located in the output layer, the local gradient is

$$\begin{aligned} \delta_j(n) &= e_j(n)\varphi'_j(v_j(n)) \\ &= \frac{b}{a}[d_j(n) - o_j(n)][a - o_j(n)][a + o_j(n)] \end{aligned} \tag{4.37}$$

For a neuron j in a hidden layer, we have

$$\begin{aligned} \delta_j(n) &= \varphi'_j(v_j(n))\sum_k \delta_k(n)w_{kj}(n) \\ &= \frac{b}{a}[a - y_j(n)][a + y_j(n)]\sum_k \delta_k(n)w_{kj}(n), \qquad \text{neuron } j \text{ is hidden} \end{aligned} \tag{4.38}$$

By using Eqs. (4.33) and (4.34) for the logistic function and Eqs. (4.37) and (4.38) for the hyperbolic tangent function, we may calculate the local gradient δ_j without requiring explicit knowledge of the activation function.

Rate of Learning

The back-propagation algorithm provides an "approximation" to the trajectory in weight space computed by the method of steepest descent. The smaller we make the learning-rate parameter η, the smaller the changes to the synaptic weights in the network will be from one iteration to the next, and the smoother will be the trajectory in weight space. This improvement, however, is attained at the cost of a slower rate of learning. If, on the other hand, we make the learning-rate parameter η too large in order to speed up the rate of learning, the resulting large changes in the synaptic weights assume such a form that the network may become unstable (i.e., oscillatory). A simple method of increasing the rate of learning yet avoiding the danger of instability

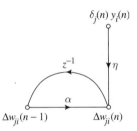

FIGURE 4.6 Signal-flow graph illustrating the effect of momentum constant α.

is to modify the delta rule of Eq.(4.13) by including a momentum term,[2] as shown by (Rumelhart et al, 1986a)

$$\Delta w_{ji}(n) = \alpha \Delta w_{ji}(n-1) + \eta \delta_j(n) y_i(n) \tag{4.39}$$

where α is usually a positive number called the *momentum constant*. It controls the feedback loop acting around $\Delta w_{ji}(n)$, as illustrated in Fig. 4.6 where z^{-1} is the unit-delay operator. Equation (4.39) is called the *generalized delta rule*[3]; it includes the delta rule of Eq. (4.13) as a special case (i.e., $\alpha = 0$).

In order to see the effect of the sequence of pattern presentations on the synaptic weights due to the momentum constant α, we rewrite Eq. (4.39) as a time series with index t. The index t goes from the initial time 0 to the current time n. Equation (4.39) may be viewed as a first-order difference equation in the weight correction $\Delta w_{ji}(n)$. Solving this equation for $\Delta w_{ji}(n)$ we have

$$\Delta w_{ji}(n) = \eta \sum_{t=0}^{n} \alpha^{n-t} \delta_j(t) y_i(t) \tag{4.40}$$

which represents a time series of length $n + 1$. From Eqs. (4.11) and (4.14) we note the product $\delta_j(n) y_i(n)$ is equal to $-\partial \mathcal{E}(n)/\partial w_{ji}(n)$. Accordingly, we may rewrite Eq. (4.40) in the equivalent form

$$\Delta w_{ji}(n) = -\eta \sum_{t=0}^{n} \alpha^{n-t} \frac{\partial \mathcal{E}(t)}{\partial w_{ji}(t)} \tag{4.41}$$

Based on this relation, we may make the following insightful observations (Watrous, 1987; Jacobs, 1988):

1. The current adjustment $\Delta w_{ji}(n)$ represents the sum of an exponentially weighted time series. For the time series to be *convergent*, the momentum constant must be restricted to the range $0 \leq |\alpha| < 1$. When α is zero, the back-propagation algorithm operates without momentum. Also the momentum constant α can be positive or negative, although it is unlikely that a negative α would be used in practice.

2. When the partial derivative $\partial \mathcal{E}(t)/\partial w_{ji}(t)$ has the same algebraic sign on consecutive iterations, the exponentially weighted sum $\Delta w_{ji}(n)$ grows in magnitude, and so the weight $w_{ji}(n)$ is adjusted by a large amount. The inclusion of momentum in the back-propagation algorithm tends to *accelerate descent* in steady downhill directions.

3. When the partial derivative $\partial \mathcal{E}(t)/\partial w_{ji}(t)$ has opposite signs on consecutive iterations, the exponentially weighted sum $\Delta w_{ji}(n)$ shrinks in magnitude, so the weight $w_{ji}(n)$ is adjusted by a small amount. The inclusion of momentum in the back-propagation algorithm has a *stabilizing effect* in directions that oscillate in sign.

The incorporation of momentum in the back-propagation algorithm represents a minor modification to the weight update, yet it may have some beneficial effects on the learning behavior of the algorithm. The momentum term may also have the benefit of preventing the learning process from terminating in a shallow local minimum on the error surface.

In deriving the back-propagation algorithm, it was assumed that the learning-rate parameter is a constant denoted by η. In reality, however, it should be defined as η_{ji}; that is, the learning-rate parameter should be *connection-dependent*. Indeed, many interesting things can be done by making the learning-rate parameter different for different parts of the network. We provide more detail on this issue in subsequent sections.

It is also noteworthy that in the application of the back-propagation algorithm we may choose all the synaptic weights in the network to be adjustable, or we may constrain any number of weights in the network to remain fixed during the adaptation process. In the latter case, the error signals are back-propagated through the network in the usual manner; however, the fixed synaptic weights are left unaltered. This can be done simply by making the learning-rate parameter η_{ji} for synaptic weight w_{ji} equal to zero.

Sequential and Batch Modes of Training

In a practical application of the back-propagation algorithm, learning results from the many presentations of a prescribed set of training examples to the multilayer perceptron. As mentioned earlier, one complete presentation of the entire training set during the learning process is called an *epoch*. The learning process is maintained on an epoch-by-epoch basis until the synaptic weights and bias levels of the network stabilize and the average squared error over the entire training set converges to some minimum value. It is good practice to *randomize the order of presentation of training examples* from one epoch to the next. This randomization tends to make the search in weight space stochastic over the learning cycles, thus avoiding the possibility of limit cycles in the evolution of the synaptic weight vectors; limit cycles are discussed in Chapter 14.

For a given training set, back-propagation learning may thus proceed in one of two basic ways:

1. *Sequential Mode.* The *sequential mode* of back-propagation learning is also referred to as *on-line, pattern, or stochastic mode*. In this mode of operation weight updating is performed after the presentation of each training example; this is the very mode of operation for which the derivation of the back-propagation algorithm presented applies. To be specific, consider an epoch consisting of N training examples (patterns) arranged in the order $(\mathbf{x}(1), \mathbf{d}(1)), \ldots, (\mathbf{x}(N), \mathbf{d}(N))$. The first example pair $(\mathbf{x}(1), \mathbf{d}(1))$ in the epoch is presented to the network, and the sequence of forward and backward computations described previously is performed, resulting in certain adjustments to the synaptic weights and bias levels of the network. Then the second example pair $(\mathbf{x}(2), \mathbf{d}(2))$ in the epoch is presented, and the sequence of forward and backward computations is repeated, resulting in further adjustments to the synaptic weights and bias levels. This process is continued until the last example pair $(\mathbf{x}(N), \mathbf{d}(N))$ in the epoch is accounted for.

2. *Batch Mode.* In the *batch mode* of back-propagation learning, weight updating is performed *after* the presentation of *all* the training examples that constitute an epoch. For a particular epoch, we define the cost function as the average squared error of Eqs. (4.2) and (4.3), reproduced here in the composite form:

$$\mathcal{E}_{av} = \frac{1}{2N} \sum_{n=1}^{N} \sum_{j \in C} e_j^2(n) \tag{4.42}$$

where the error signal $e_j(n)$ pertains to output neuron j for training example n and which is defined by Eq. (4.1). The error $e_j(n)$ equals the difference between $d_j(n)$ and $y_j(n)$, which represents the jth element of the desired response vector $\mathbf{d}(n)$ and the corresponding value of the network output, respectively. In Eq. (4.42) the inner summation with respect to j is performed over all the neurons in the output layer of the network, whereas the outer summation with respect to n is performed over the entire training set in the epoch at hand. For a learning-rate parameter η, the adjustment applied to synaptic weight w_{ji}, connecting neuron i to neuron j, is defined by the delta rule

$$\Delta w_{ji} = -\eta \frac{\partial \mathcal{E}_{av}}{\partial w_{ji}}$$

$$= -\frac{\eta}{N} \sum_{n=1}^{N} e_j(n) \frac{\partial e_j(n)}{\partial w_{ji}} \tag{4.43}$$

To calculate the partial derivative $\partial e_j(n)/\partial w_{ji}$ we proceed in the same way as before. According to Eq. (4.43), in the batch mode the weight adjustment Δw_{ji} is made only after the entire training set has been presented to the network.

From an "on-line" operational point of view, the sequential mode of training is preferred over the batch mode because it requires *less* local storage for each synaptic connection. Moreover, given that the patterns are presented to the network in a random manner, the use of pattern-by-pattern updating of weights makes the search in weight space *stochastic* in nature. This in turn makes it less likely for the back-propagation algorithm to be trapped in a local minimum.

In the same way, the stochastic nature of the sequential mode makes it difficult to establish theoretical conditions for convergence of the algorithm. In contrast, the use of batch mode of training provides an accurate estimate of the gradient vector; convergence to a local minimum is thereby guaranteed under simple conditions. Also, the composition of the batch mode makes it easier to parallelize than the sequential mode.

When the training data are *redundant* (i.e., the data set contains several copies of exactly the same pattern), we find that unlike the batch mode, the sequential mode is able to take advantage of this redundancy because the examples are presented one at a time. This is particularly so when the data set is large and highly redundant.

In summary, despite the fact that the sequential mode of back-propagation learning has several disadvantages, it is highly popular (particularly for solving pattern-classification problems) for two important practical reasons:

- The algorithm is simple to implement.
- It provides effective solutions to large and difficult problems.

Stopping Criteria

In general, the back-propagation algorithm cannot be shown to converge, and there are no well-defined criteria for stopping its operation. Rather, there are some reasonable criteria, each with its own practical merit, which may be used to terminate the weight adjustments. To formulate such a criterion, it is logical to think in terms of the unique properties of a *local* or *global minimum* of the error surface[4]. Let the weight vector $\mathbf{w}^*$ denote a minimum, be it local or global. A necessary condition for $\mathbf{w}^*$ to be a minimum is that the gradient vector $\mathbf{g}(\mathbf{w})$ (i.e., first-order partial derivative) of the error surface with respect to the weight vector $\mathbf{w}$ be zero at $\mathbf{w} = \mathbf{w}^*$. Accordingly, we may formulate a sensible convergence criterion for back-propagation learning as follows (Kramer and Sangiovanni-Vincentelli, 1989):

> *The back-propagation algorithm is considered to have converged when the Euclidean norm of the gradient vector reaches a sufficiently small gradient threshold.*

The drawback of this convergence criterion is that, for successful trials, learning times may be long. Also, it requires the computation of the gradient vector $\mathbf{g}(\mathbf{w})$.

Another unique property of a minimum that we can use is the fact that the cost function or error measure $\mathscr{E}_{av}(\mathbf{w})$ is stationary at the point $\mathbf{w} = \mathbf{w}^*$. We may therefore suggest a different criterion of convergence:

> *The back-propagation algorithm is considered to have converged when the absolute rate of change in the average squared error per epoch is sufficiently small.*

The rate of change in the average squared error is typically considered to be small enough if it lies in the range of 0.1 to 1 percent per epoch. Sometimes a value as small as 0.01 percent per epoch is used. Unfortunately, this criterion may result in a premature termination of the learning process.

There is another useful and theoretically supported criterion for convergence. After each learning iteration, the network is tested for its generalization performance. The learning process is stopped when the generalization performance is adequate, or when it is apparent that the generalization performance has peaked; see Section 4.14 for more details.

4.4 SUMMARY OF THE BACK-PROPAGATION ALGORITHM

Figure 4.1 presents the architectural layout of a multilayer perceptron. The corresponding signal-flow graph for back-propagation learning, incorporating both the forward and backward phases of the computations involved in the learning process, is presented in Fig. 4.7 for the case of $L = 2$ and $m_0 = m_1 = m_2 = 3$. The top part of the signal-flow graph accounts for the forward pass. The lower part of the signal-flow graph accounts for the backward pass, which is referred to as a *sensitivity graph* for computing the local gradients in the back-propagation algorithm (Narendra and Parthasarathy, 1990).

Earlier we mentioned that the sequential updating of weights is the preferred method for on-line implementation of the back-propagation algorithm. For this mode of operation, the algorithm cycles through the training sample $\{(\mathbf{x}(n), \mathbf{d}(n))\}_{n=1}^{N}$ as follows:

1. *Initialization.* Assuming that no prior information is available, pick the synaptic weights and thresholds from a uniform distribution whose mean is zero and whose

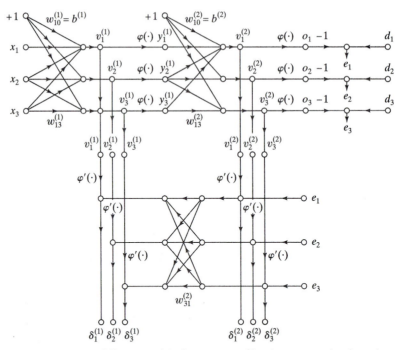

FIGURE 4.7 Signal-flow graphical summary of back-propagation learning. Top part of the graph: forward pass. Bottom part of the graph: backward pass.

variance is chosen to make the standard deviation of the induced local fields of the neurons lie at the transition between the linear and saturated parts of the sigmoid activation function.

2. *Presentations of Training Examples.* Present the network with an epoch of training examples. For each example in the set, ordered in some fashion, perform the sequence of forward and backward computations described under points 3 and 4, respectively.

3. *Forward Computation.* Let a training example in the epoch be denoted by $(\mathbf{x}(n), \mathbf{d}(n))$, with the input vector $\mathbf{x}(n)$ applied to the input layer of sensory nodes and the desired response vector $\mathbf{d}(n)$ presented to the output layer of computation nodes. Compute the induced local fields and function signals of the network by proceeding forward through the network, layer by layer. The induced local field $v_j^{(l)}(n)$ for neuron j in layer l is

$$v_j^{(l)}(n) = \sum_{i=0}^{m_0} w_{ji}^{(l)}(n) y_i^{(l-1)}(n) \tag{4.44}$$

where $y_i^{(l-1)}(n)$ is the output (function) signal of neuron i in the previous layer $l-1$ at iteration n and $w_{ji}^{(l)}(n)$ is the synaptic weight of neuron j in layer l that is fed from neuron i in layer $l-1$. For $i = 0$, we have $y_0^{(l-1)}(n) = +1$ and $w_{j0}^{(l)}(n) = b_j^{(l)}(n)$ is the bias applied to neuron j in layer l. Assuming the use of a sigmoid function, the output signal of neuron j in layer l is

$$y_j^{(l)} = \varphi_j(v_j(n))$$

If neuron j is in the first hidden layer (i.e., $l = 1$), set

$$y_j^{(0)}(n) = x_j(n)$$

where $x_j(n)$ is the jth element of the input vector $\mathbf{x}(n)$. If neuron j is in the output layer (i.e., $l = L$, where L is referred to as the *depth* of the network), set

$$y_j^{(L)} = o_j(n)$$

Compute the error signal

$$e_j(n) = d_j(n) - o_j(n) \tag{4.45}$$

where $d_j(n)$ is the jth element of the desired response vector $\mathbf{d}(n)$.

4. *Backward Computation.* Compute the δs (i.e., local gradients) of the network, defined by

$$\delta_j^{(l)}(n) = \begin{bmatrix} e_j^{(L)}(n)\,\varphi_j'(v_j^{(L)}(n)) & \text{for neuron } j \text{ in output layer } L \\[2ex] \varphi_j'(v_j^{(l)}(n)) \sum_k \delta_k^{(l+1)}(n)\,w_{kj}^{(l+1)}(n) & \text{for neuron } j \text{ in hidden layer } l \end{bmatrix} \tag{4.46}$$

where the prime in $\varphi_j'(\cdot)$ denotes differentiation with respect to the argument. Adjust the synaptic weights of the network in layer l according to the generalized delta rule:

$$w_{ji}^{(l)}(n+1) = w_{ji}^{(l)}(n) + \alpha[w_{ji}^{(l)}(n-1)] + \eta\delta_j^{(l)}(n)y_i^{(l-1)}(n) \tag{4.47}$$

where η is the learning-rate parameter and α is the momentum constant.

5. *Iteration.* Iterate the forward and backward computations under points 3 and 4 by presenting new epochs of training examples to the network until the stopping criterion is met.

Notes: The order of presentation of training examples should be randomized from epoch to epoch. The momentum and learning-rate parameter are typically adjusted (and usually decreased) as the number of training iterations increases. Justification for these points will be presented later.

4.5   XOR PROBLEM

In the elementary (single-layer) perceptron there are no hidden neurons. Consequently, it cannot classify input patterns that are not linearly separable. However, nonlinearly separable patterns are of common occurrence. For example, this situation arises in the *Exclusive OR (XOR) problem*, which may be viewed as a special case of a more general problem, namely that of classifying points in the *unit hypercube*. Each point in the hypercube is either in class 0 or class 1. However, in the special case of the XOR problem, we need consider only the four corners of the *unit square* that correspond to the input patterns (0,0), (0,1), (1,1), and (1,0). The first and third input patterns are in class 0, as shown by

$$0 \oplus 0 = 0$$

and

$$1 \oplus 1 = 0$$

where $\oplus$ denotes the Exclusive OR Boolean function operator. The input patterns (0,0) and (1,1) are at opposite corners of the unit square, yet they produce the identical output 0. On the other hand, the input patterns (0,1) and (1,0) are also at opposite corners of the square, but they are in class 1, as shown by

$$0 \oplus 1 = 1$$

and

$$1 \oplus 0 = 1$$

We first recognize that the use of a single neuron with two inputs results in a straight line for a decision boundary in the input space. For all points on one side of this line, the neuron outputs 1; for all points on the other side of the line, it outputs 0. The position and orientation of the line in the input space are determined by the synaptic weights of the neuron connected to the input nodes, and the bias applied to the neuron. With the input patterns (0,0) and (1,1) located on opposite corners of the unit square, and likewise for the other two input patterns (0,1) and (1,0), it is clear that we cannot construct a straight line for a decision boundary so that (0,0) and (0,1) lie in one decision region, and (0,1) and (1,0) lie in the other decision region. In other words, an elementary perceptron cannot solve the XOR problem.

We may solve the XOR problem by using a single hidden layer with two neurons, as in Fig. 4.8a. (Touretzky and Pomerleau, 1989). The signal-flow graph of the network is shown in Fig. 4.8b. The following assumptions are made here:

- Each neuron is represented by a McCulloch–Pitts model, which uses a threshold function for its activation function.
- Bits 0 and 1 are represented by the levels 0 and +1, respectively.

The top neuron, labeled 1 in the hidden layer, is characterized as:

$$w_{11} = w_{12} = +1$$

$$b_1 = -\frac{3}{2}$$

The slope of the decision boundary constructed by this hidden neuron is equal to -1, and positioned as in Fig. 4.9a. The bottom neuron, labeled 2 in the hidden layer, is characterized as:

$$w_{21} = w_{22} = +1$$

$$b_2 = -\frac{1}{2}$$

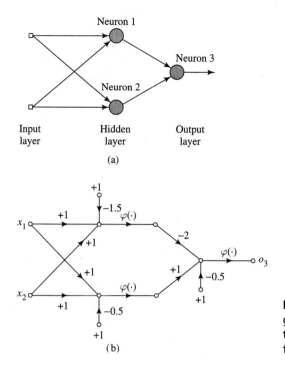

Neuron 1

Neuron 3

Neuron 2

Input
layer

Hidden
layer

Output
layer

(a)

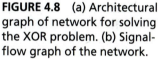

FIGURE 4.8   (a) Architectural
graph of network for solving
the XOR problem. (b) Signal-
flow graph of the network.

(b)

The orientation and position of the decision boundary constructed by this second hidden neuron are as shown in Fig. 4.9b.

The output neuron, labeled 3 in Fig. 4.8a, is characterized as:

$$w_{31} = -2$$

$$w_{32} = +1$$

$$b_3 = -\frac{1}{2}$$

The function of the output neuron is to construct a linear combination of the decision boundaries formed by the two hidden neurons. The result of this computation is shown in Fig. 4.9c. The bottom hidden neuron has an excitatory (positive) connection to the output neuron, whereas the top hidden neuron has a stronger inhibitory (negative) connection to the output neuron. When both hidden neurons are off, which occurs when the input pattern is (0,0), the output neuron remains off. When both hidden neurons are on, which occurs when the input pattern is (1,1), the output neuron is switched off again because the inhibitory effect of the larger negative weight connected to the top hidden neuron overpowers the excitatory effect of the positive weight connected to the bottom hidden neuron. When the top hidden neuron is off and the bottom hidden neuron is on, which occurs when the input pattern is (0,1) or (1,0), the output neuron is switched on due to the excitatory effect of the positive weight connected to the bottom hidden neuron. Thus the network of Fig. 4.8a does indeed solve the XOR problem.

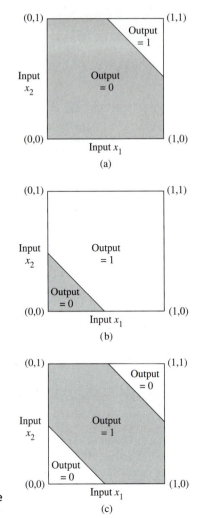

FIGURE 4.9 (a) Decision boundary constructed by hidden neuron 1 of the network in Fig. 4.8. (b) Decision boundary constructed by hidden neuron 2 of the network. (c) Decision boundaries constructed by the complete network.

4.6 HEURISTICS FOR MAKING THE BACK-PROPAGATION ALGORITHM PERFORM BETTER

It is often said that the design of a neural network using the back-propagation algorithm is more of an art than a science in the sense that many of the numerous factors involved in the design are the results of one's own personal experience. There is some truth in this statement. Nevertheless, there are methods that will significantly improve the back-propagation algorithm's performance, as described here.

1. *Sequential versus batch update.* As mentioned previously, the sequential mode of back-propagation learning (involving pattern-by-pattern updating) is computationally faster than the batch mode. This is especially true when the training data set is large and highly redundant. (Highly redundant data pose computational problems for the estimation of the Jacobian required for the batch update.)

2. *Maximizing information content.* As a general rule, every training example presented to the back-propagation algorithm should be chosen on the basis that its information content is the largest possible for the task at hand (LeCun, 1993). Two ways of achieving this aim are:

- The use of an example that results in the largest training error.
- The use of an example that is radically different from all those previously used.

These two heuristics are motivated by a desire to search more of the weight space.

In pattern-classification tasks using sequential back-propagation learning, a simple technique that is commonly used is to randomize (i.e., shuffle) the order in which the examples are presented to the multilayer perceptron from one epoch to the next. Ideally, the randomization ensures that the successive examples in an epoch presented to the network rarely belong to the same class.

For a more refined technique, we may use an *emphasizing scheme*, which involves more difficult patterns than easy ones being presented to the network (LeCun, 1993). Whether a particular pattern is easy or difficult can be identified by examining the error it produces, compared to previous iterations of the algorithm. However, there are two problems with the use of an emphasizing scheme that should be carefully examined:

- The distribution of examples within an epoch presented to the network is distorted.
- The presence of an outlier or a mislabeled example can have a catastrophic consequence on the performance of the algorithm; learning such outliers compromises the generalization ability of the network over more probable regions of the input space.

3. *Activation function.* A multilayer perceptron trained with the back-propagation algorithm may, in general, learn faster (in terms of the number of training iterations required) when the sigmoid activation function built into the neuron model of the network is antisymmetric than when it is nonsymmetric; see Section 4.11 for details. We say that an activation function $\varphi(v)$ is *antisymmetric* (i.e., odd function of its argument) if

$$\varphi(-v) = -\varphi(v)$$

as depicted in Fig. 4.10a. This condition is not satisfied by the standard logistic function depicted in Fig. 4.10b.

A popular example of an antisymmetric activation function is a sigmoidal nonlinearity in the form of a *hyperbolic tangent*, defined by

$$\varphi(v) = a \tanh(bv)$$

where a and b are constants. Suitable values for the constants a and b are (LeCun, 1989, 1993)

$$a = 1.7159$$

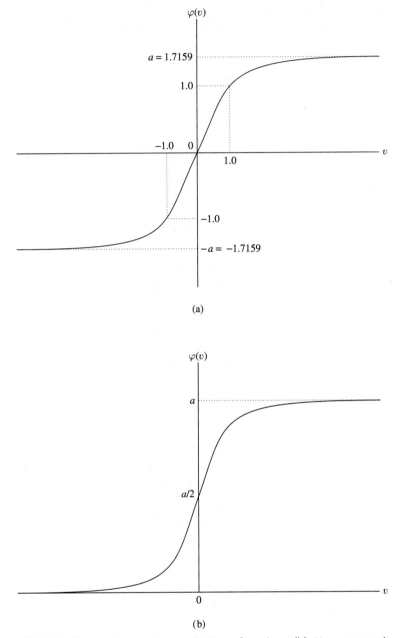

(a)

(b)

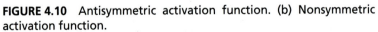

FIGURE 4.10 Antisymmetric activation function. (b) Nonsymmetric activation function.

and

$$b = \frac{2}{3}$$

The hyperbolic tangent function so defined has the following useful properties:

- $\varphi(1) = 1$ and $\varphi(-1) = -1$
- At the origin the slope (i.e., effective gain) of the activation function is close to unity, as shown by

$$\varphi(0) = ab$$

$$= 1.7159 \times 2/3$$

$$= 1.1424$$

- The second derivative of $\varphi(v)$ attains its maximum value at $v = 1$.

4. *Target values*. It is important that the target values (desired response) be chosen within the range of the sigmoid activation function. More specifically, the desired response d_j for neuron j in the output layer of the multilayer perceptron should be *offset* by some amount ϵ away from the limiting value of the sigmoid activation function, depending on whether the limiting value is positive or negative. Otherwise the back-propagation algorithm tends to drive the free parameters of the network to infinity, and thereby slow down the learning process by driving the hidden neurons into saturation. To be specific, consider the antisymmetric activation function of Fig. 4.10a. For the limiting value $+a$, we set

$$d_j = a - \epsilon$$

and for the limiting value of $-a$, we set

$$d_j = -a + \epsilon$$

where ϵ is an appropriate positive constant. For the choice of $a = 1.7159$ referred to earlier, we may set $\epsilon = 0.7159$, in which case the target value (desired response) d_j can be conveniently chosen as ± 1, as indicated in Fig. 4.10a.

5. *Normalizing the inputs*. Each input variable should be *preprocessed* so that its mean value, averaged over the entire training set, is close to zero, or else it is small compared to its standard deviation (LeCun, 1993). To appreciate the practical significance of this rule, consider the extreme case where the input variables are consistently positive. In this situation, the synaptic weights of a neuron in the first hidden layer can only increase together or decrease together. Accordingly, if the weight vector of that neuron is to change direction, it can only do so by zigzagging its way through the error surface, which is typically slow and should therefore be avoided.

In order to accelerate the back-propagation learning process, the normalization of the inputs should also include two other measures (LeCun, 1993):

- The input variables contained in the training set should be *uncorrelated*; this can be done by using principal components analysis, as detailed in Chapter 8.
- The decorrelated input variables should be scaled so that their *covariances are approximately equal*, thereby ensuring that the different synaptic weights in the network learn at approximately the same speed.

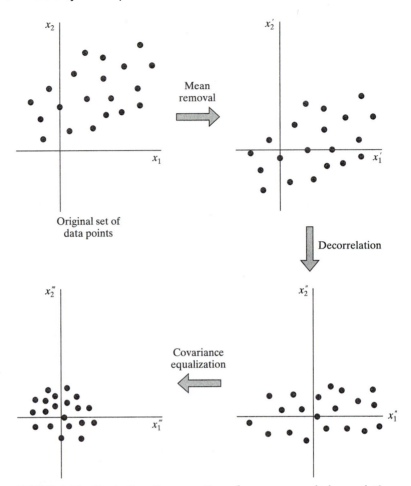

FIGURE 4.11   Illustrating the operation of mean removal, decorrelation, and covariance equalization for a two-dimensional input space.

Figure 4.11 illustrates the results of three normalization steps: mean removal, decorrelation, and covariance equalization, applied in that order.

6. *Initialization.* A good choice for the initial values of the synaptic weights and thresholds of the network can be of tremendous help in a successful network design. The key question is: What is a good choice?

When the synaptic weights are assigned large initial values, it is highly likely that the neurons in the network will be driven into saturation. If this happens, the local gradients in the back-propagation algorithm assume small values, which in turn will cause the learning process to slow down. However, if the synaptic weights are assigned small initial values, the back-propagation algorithm may operate on a very flat area around the origin of the error surface; this is particularly true in the case of antisymmetric activation functions such as the hyperbolic tangent function. Unfortunately, the origin is a *saddle point*, which refers to a stationary point where the curvature of the error surface across the saddle is negative and the curvature along the saddle is positive. For these reasons the use of both large and small values for initializing the synaptic weights should be avoided. The proper choice of initialization lies somewhere between these two extreme cases.

To be specific, consider a multilayer perceptron using the hyperbolic tangent function for its activation functions. Let the bias applied to each neuron in the network be set to zero. We may then express the induced local field of neuron j as

$$v_j = \sum_{i=1}^{m} w_{ji} y_i$$

Let it be assumed that the inputs applied to each neuron in the network have zero mean and unit variance, as shown by

$$\mu_y = E[y_i] = 0 \qquad \text{for all } i$$

and

$$\sigma_y^2 = E[(y_i - \mu_i)^2] = E[y_i^2] = 1 \qquad \text{for all } i$$

Let it be further assumed that the inputs are uncorrelated, as shown by

$$E[y_i y_k] = \begin{cases} 1 & \text{for } k = i \\ 0 & \text{for } k \neq i \end{cases}$$

and that the synaptic weights are drawn from a uniformly distributed set of numbers with zero mean

$$\mu_w = E[w_{ji}] = 0 \qquad \text{for all } (j,i) \text{ pairs}$$

and variance

$$\sigma_w^2 = E[(w_{ji} - \mu_w)^2] = E[w_{ji}^2] \qquad \text{for all } (j,i) \text{ pairs}$$

Accordingly, we may express the mean and variance of the induced local field v_j as

$$\mu_v = E[v_j] = E\left[\sum_{i=1}^{m} w_{ji} y_i\right] = \sum_{i=1}^{m} E[w_{ji}] E[y_i] = 0$$

and

$$\sigma_v^2 = E[(v_j - \mu_v)^2] = E[v_j^2]$$

$$= E\left[\sum_{i=1}^{m} \sum_{k=1}^{m} w_{ji} w_{jk} y_i y_k\right]$$

$$= \sum_{i=1}^{m} \sum_{k=1}^{m} E[w_{ji} w_{jk}] E[y_i y_k] \qquad (4.48)$$

$$= \sum_{i=1}^{m} E[w_{ji}^2]$$

$$= m\sigma_w^2$$

where m is the number of synaptic connections of a neuron.

In light of this result, we may now describe a good strategy for initializing the synaptic weights so that the standard deviation of the induced local field of a neuron lies in the transition area between the linear and saturated parts of its sigmoid activation function. For example, for the case of a hyperbolic tangent function with its parameters a and b as specified previously, this objective is satisfied by setting $\sigma_v = 1$ in Eq. (4.48), in which case we obtain (LeCun, 1993)

$$\sigma_w = m^{-1/2} \tag{4.49}$$

Thus it is desirable for the uniform distribution, from which the synaptic weights are selected, to have a mean of zero and a variance equal to the reciprocal of the number of synaptic connections of a neuron.

7. *Learning from hints.* Learning from a set of training examples deals with an unknown input–output mapping function $f(\cdot)$. In effect, the learning process exploits the information contained in the examples about the function $f(\cdot)$ to *infer* an approximate implementation of it. The process of learning from examples may be generalized to *include learning from hints,* which is achieved by allowing prior information that we may have about the function $f(\cdot)$ to be included in the learning process (Abu-Mostafa, 1995). Such information may include invariance properties, symmetries, or any other knowledge about the function $f(\cdot)$ that may be used to accelerate the search for its approximate realization, and more importantly, to improve the quality of the final estimate. The use of Eq. (4.49) is an example of how this is achieved.

8. *Learning rates.* All neurons in the multilayer perceptron should ideally learn at the same rate. The last layers usually have larger local gradients than the layers at the front end of the network. Hence, the learning-rate parameter η should be assigned a smaller value in the last layers than in the front layers. Neurons with many inputs should have a smaller learning-rate parameter than neurons with few inputs so as to maintain a similar learning time for all neurons in the network. In LeCun (1993), it is suggested that for a given neuron, the learning rate should be inversely proportional to the square root of synaptic connections made to that neuron. We discuss learning rates more fully in Section 4.17.

4.7   OUTPUT REPRESENTATION AND DECISION RULE

In theory, for an M-class *classification problem* in which the union of the M distinct classes forms the entire input space, we need a total of M outputs to represent all possible classification decisions, as depicted in Fig. 4.12. In this figure the vector $\mathbf{x}_j$ denotes the jth *prototype* (i.e., unique sample) of an m-dimensional random vector $\mathbf{x}$ to be classified by a multilayer perceptron. The kth of M possible classes to which $\mathbf{x}$ can belong is denoted by $\mathcal{C}_k$. Let y_{kj} be the kth output of the network produced in response to the prototype $\mathbf{x}_j$, as shown by

$$y_{k,j} = F_k(\mathbf{x}_j), \quad k = 1, 2, \ldots, M \tag{4.50}$$

FIGURE 4.12   Block diagram of a pattern classifier.

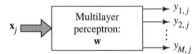

where the function $F_k(\cdot)$ defines the mapping learned by the network from the input to the kth output. For convenience of presentation, let

$$\begin{aligned}
\mathbf{y}_j &= [y_{1,j}, y_{2,j}, ..., y_{M,j}]^T \\
&= [F_1(\mathbf{x}_j), F_2(\mathbf{x}_j), ..., F_M(\mathbf{x}_j)]^T \qquad (4.51) \\
&= \mathbf{F}(\mathbf{x}_j)
\end{aligned}$$

where $\mathbf{F}(\cdot)$ is a vector-valued function. A basic question we wish to address in this section is:

> *After a multilayer perceptron is trained, what should the optimum decision rule be for classifying the M outputs of the network?*

Clearly, any reasonable output decision rule should be based on knowledge of the vector-valued function:

$$\mathbf{F}: \mathbb{R}^m \ni \mathbf{x} \rightarrow \mathbf{y} \in \mathbb{R}^M \qquad (4.52)$$

In general, all that is certain about the vector-valued function $\mathbf{F}(\cdot)$ is that it is a continuous function that minimizes the *empirical risk functional:*

$$R = \frac{1}{2N} \sum_{j=1}^{N} \|\mathbf{d}_j - \mathbf{F}(\mathbf{x}_j)\|^2 \qquad (4.53)$$

where $\mathbf{d}_j$ is the desired (target) output pattern for the prototype $\mathbf{x}_j$, $\|\cdot\|$ is the Euclidean norm of the enclosed vector, and N is the total number of examples presented to the network in training. The essence of the criterion of Eq.(4.53) is the same as the cost function of Eq. (4.3). The vector-valued function $\mathbf{F}(\cdot)$ is strongly dependent on the choice of examples $(\mathbf{x}_j, \mathbf{d}_j)$ used to train the network, so that different values of $(\mathbf{x}_j, \mathbf{d}_j)$ will indeed lead to different vector-valued function $\mathbf{F}(\cdot)$. Note that the terminology $(\mathbf{x}_j, \mathbf{d}_j)$ used here is the same as that of $(\mathbf{x}(j), \mathbf{d}(j))$ used previously.

Suppose now that the network is trained with *binary* target values (that incidently correspond to the upper and lower bounds on the network outputs when using the logistic function), written as:

$$d_{kj} = \begin{cases} 1 & \text{when the prototype } \mathbf{x}_j \text{ belongs to class } \mathscr{C}_k \\ 0 & \text{when the prototype } \mathbf{x}_j \text{ does not belong to class } \mathscr{C}_k \end{cases} \qquad (4.54)$$

Based on this notation, class $\mathscr{C}_k$ is represented by the M-dimensional target vector

$$\begin{bmatrix} 0 \\ \vdots \\ 1 \\ \vdots \\ 0 \end{bmatrix} \leftarrow k\text{th element}$$

It is tempting to suppose that a multilayer perceptron classifier trained with the backpropagation algorithm on a finite set of independently and identically distributed (i.i.d.) examples may lead to an asymptotic approximation of the underlying *a posteriori* class probabilities. This property may be justified on the following grounds (White, 1989a; Richard and Lippmann, 1991):

- The *law of large numbers* is invoked to show that as the size of the training set, N, approaches infinity, the weight vector $\mathbf{w}$ that minimizes cost functional R of Eq. (4.53) approaches the optimum weight vector $\mathbf{w}^*$ that minimizes the expectation of the random quantity $\frac{1}{2}\|\mathbf{d} - \mathbf{F}(\mathbf{w},\mathbf{x})\|^2$, where $\mathbf{d}$ is the desired response vector and $\mathbf{F}(\mathbf{w}, \mathbf{x})$ is the approximation realized by a multilayer perceptron with weight vector $\mathbf{w}$ and vector $\mathbf{x}$ as input (White, 1989a). The function $\mathbf{F}(\mathbf{w}, \mathbf{x})$, showing explicit dependence on the weight vector $\mathbf{w}$, is the same as $\mathbf{F}(\mathbf{x})$ used previously.
- The optimum weight vector $\mathbf{w}^*$ has the property that the corresponding vector of actual network outputs, $\mathbf{F}(\mathbf{w}^*, \mathbf{x})$, is a mean-squared, error-minimizing approximation to the conditional expectation of the desired response vector, given the input vector $\mathbf{x}$ (White, 1989a). This issue is discussed in Chapter 2.
- For a 1 of M *pattern classification problem,* the kth element of the desired response vector equals one if the input vector $\mathbf{x}$ belongs to class $\mathcal{C}_k$ and zero otherwise. Hence, the conditional expectation of the desired response vector, given $\mathbf{x}$, equals the *a posteriori* class probability $P(\mathcal{C}_k|\mathbf{x})$, $k = 1, 2, ..., M$ (Richard and Lippmann, 1991).

It follows therefore that a multilayer perceptron classifier (using the logistic function for nonlinearity) does indeed approximate the *a posteriori* class probabilities, provided that the size of the training set is large enough and that the back-propagation learning process does not get stuck at a local minimum. We may now answer the question we posed earlier. Specifically, we may say that an appropriate output decision rule is the (approximate) Bayes rule generated by the *a posteriori* probability estimates:

Classify the random vector $\mathbf{x}$ as belonging to class $\mathcal{C}_k$ if

$$F_k(\mathbf{x}) > F_j(\mathbf{x}) \qquad \text{for all } j \neq k \tag{4.55}$$

where $F_k(\mathbf{x})$ and $F_j(\mathbf{x})$ are elements of the vector-valued mapping function

$$\mathbf{F}(\mathbf{x}) = \begin{bmatrix} F_1(\mathbf{x}) \\ F_2(\mathbf{x}) \\ \vdots \\ F_M(\mathbf{x}) \end{bmatrix}$$

A unique largest output value exists with probability 1 when the underlying posterior class distributions are distinct. (Here it is assumed that infinite-precision arithmetic is used; ties are possible with finite precision.) This decision rule has the advantage of rendering *unambiguous* decisions over the common *ad hoc* rule of selecting class membership based on the concept of output "firing." That is, the vector $\mathbf{x}$ is assigned membership in a particular class if the corresponding output value is greater than some fixed threshold (usually 0.5 for the logistic form of activation function), which can lead to multiple class assignments.

In Section 4.6 we pointed out that the binary target values [0,1], corresponding to the logistic function of Eq. (4.30), are perturbed by a small amount ϵ as a practical measure, to avoid the saturation of synaptic weights (due to finite numerical precision) during training of the network. As a result of this perturbation, the target values are now nonbinary, and the asymptotic approximations $F_k(\mathbf{x})$ are no longer exactly the *a posteriori* probabilities $P(\mathcal{C}_k|\mathbf{x})$ of the M classes of interest (Hampsire and

Pearlmutter, 1990). Instead, $P(\mathscr{C}_k|\mathbf{x})$ is linearly mapped to the closed interval $[\epsilon, 1 - \epsilon]$, such that $P(\mathscr{C}_k|\mathbf{x}) = 0$ is mapped to an output of ϵ and $P(\mathscr{C}_k|\mathbf{x}) = 1$ is mapped to an output of $1 - \epsilon$. Because this linear mapping preserves relative ordering, it does *not* affect the result of applying the output decision rule of Eq. (4.55).

It is also interesting that when a decision boundary is formed by thresholding the outputs of a multilayer perceptron against some fixed values, the overall shape and orientation of the decision boundary may be explained heuristically (for the case of a single hidden layer) in terms of the number of hidden neurons and the ratios of synaptic weights connected to them (Lui, 1990). Such an analysis, however, is not applicable to a decision boundary formed in accordance with the output decision rule of Eq. (4.55). A more appropriate approach is to consider the hidden neurons as *nonlinear feature detectors* that attempt to map classes from the original input space $\mathbb{R}^{m_0}$, where the classes may not be linearly separable, into the space of hidden-layer activations, where it is more likely for them to be linearly separable.

4.8 COMPUTER EXPERIMENT

In this section we use a computer experiment to illustrate the learning behavior of a multilayer perceptron used as a pattern classifier. The objective of the experiment is to distinguish between two classes of "overlapping," two-dimensional, Gaussian-distributed patterns labeled 1 and 2. Let $\mathscr{C}_1$ and $\mathscr{C}_2$ denote the set of events for which a random vector $\mathbf{x}$ belongs to patterns 1 and 2, respectively. We may then express the conditional probability density functions for the two classes:

Class $\mathscr{C}_1$:
$$f_{\mathbf{x}}(\mathbf{x}|\mathscr{C}_1) = \frac{1}{2\pi\sigma_1^2} \exp\left(-\frac{1}{2\sigma_1^2} \|\mathbf{x} - \boldsymbol{\mu}_1\|^2\right) \tag{4.56}$$

where

$$\boldsymbol{\mu}_1 = \text{mean vector} = [0,0]^T$$
$$\sigma_1^2 = \text{variance} = 1$$

Class $\mathscr{C}_2$:
$$f_{\mathbf{x}}(\mathbf{x}|\mathscr{C}_2) = \frac{1}{2\pi\sigma_2^2} \exp\left(-\frac{1}{2\sigma_2^2} \|\mathbf{x} - \boldsymbol{\mu}_2\|^2\right) \tag{4.57}$$

where

$$\boldsymbol{\mu}_2 = [2,0]^T$$
$$\sigma_2^2 = 4$$

The two classes are assumed to be equiprobable; that is,

$$p_1 = p_2 = \frac{1}{2}$$

Figure 4.13a shows three-dimensional plots of the two Gaussian distributions defined by Eqs. (4.56) and (4.57). The input vector is $\mathbf{x} = [x_1, x_2]^T$, and the dimensionality of the

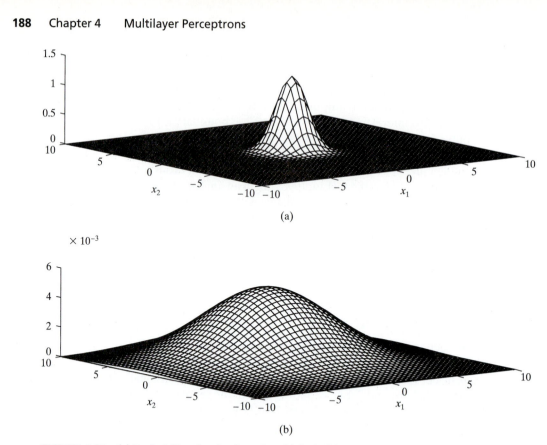

FIGURE 4.13 (a) Probability density function $f_x(\mathbf{x}|\mathscr{C}_1)$; (b) Probability density function $f_x(\mathbf{x}|\mathscr{C}_2)$.

input space is $m_0 = 2$. Figure 4.14 shows individual scatter diagrams for classes $\mathscr{C}_1$ and $\mathscr{C}_2$ and the joint scatter diagram representing the superposition of scatter plots of 500 points taken from each of the two processes. This latter diagram clearly shows that the two distributions overlap each other significantly, indicating that there is inevitably a significant probability of misclassification.

Bayesian Decision Boundary

The Bayes criterion for optimum classification is discussed in Chapter 3. Assuming that for a two-class problem, (1) classes $\mathscr{C}_1$ and $\mathscr{C}_2$ are equiprobable, (2) the costs for correct classifications are zero, and (3) the costs for misclassifications are equal, we find that the optimum decision boundary is found by applying the likelihood ratio test:

$$\Lambda(\mathbf{x}) \underset{\mathscr{C}_1}{\overset{\mathscr{C}_2}{\lessgtr}} \xi \tag{4.58}$$

where $\Lambda(\mathbf{x})$ is the *likelihood ratio*, defined by

$$\Lambda(\mathbf{x}) = \frac{f_{\mathbf{X}}(\mathbf{x}|\mathscr{C}_1)}{f_{\mathbf{X}}(\mathbf{x}|\mathscr{C}_2)} \tag{4.59}$$

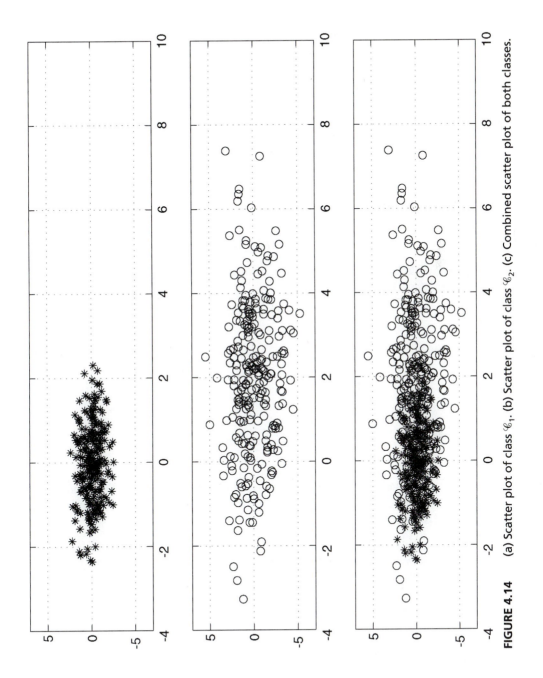

FIGURE 4.14   (a) Scatter plot of class $\mathscr{C}_1$. (b) Scatter plot of class $\mathscr{C}_2$. (c) Combined scatter plot of both classes.

and ξ is the *threshold of the test*, defined by

$$\xi = \frac{p_2}{p_1} = 1 \tag{4.60}$$

For the example being considered, we have

$$\Lambda(\mathbf{x}) = \frac{\sigma_2^2}{\sigma_1^2} \exp\left(-\frac{1}{2\sigma_1^2}\|\mathbf{x} - \boldsymbol{\mu}_1\|^2 + \frac{1}{2\sigma_2^2}\|\mathbf{x} - \boldsymbol{\mu}_2\|^2\right)$$

The optimum (Bayesian) decision boundary is therefore defined by

$$\frac{\sigma_2^2}{\sigma_1^2} \exp\left(-\frac{1}{2\sigma_1^2}\|\mathbf{x} - \boldsymbol{\mu}_1\|^2 + \frac{1}{2\sigma_2^2}\|\mathbf{x} - \boldsymbol{\mu}_2\|^2\right) = 1$$

or equivalently,

$$\frac{1}{\sigma_2^2}\|\mathbf{x} - \boldsymbol{\mu}_2\|^2 - \frac{1}{\sigma_1^2}\|\mathbf{x} - \boldsymbol{\mu}_1\|^2 = 4 \log\left(\frac{\sigma_1}{\sigma_2}\right) \tag{4.61}$$

Using straightforward manipulations, we may redefine the optimum decision boundary of Eq. (4.61) simply as

$$\|\mathbf{x} - \mathbf{x}_c\|^2 = r^2 \tag{4.62}$$

where

$$\mathbf{x}_c = \frac{\sigma_2^2\boldsymbol{\mu}_1 - \sigma_1^2\boldsymbol{\mu}_2}{\sigma_2^2 - \sigma_1^2} \tag{4.63}$$

and

$$r^2 = \frac{\sigma_1^2\sigma_2^2}{\sigma_2^2 - \sigma_1^2}\left[\frac{\|\boldsymbol{\mu}_1 - \boldsymbol{\mu}_2\|^2}{\sigma_2^2 - \sigma_1^2} + 4\log\left(\frac{\sigma_2}{\sigma_1}\right)\right] \tag{4.64}$$

Equation (4.62) represents a circle with center $\mathbf{x}_c$ and radius r. Let Ω_1 define the region lying inside this circle. The Bayesian classification rule for the problem at hand may be stated as follows:

> Classify the observation vector **x** *as belonging to class* $\mathscr{C}_1$ *if the likelihood ratio* $\Lambda(\mathbf{x})$ *is greater than the threshold* ξ, *and to class* $\mathscr{C}_2$ *otherwise.*

For the particular parameters of this experiment, we have a circular decision boundary whose center is located at

$$\mathbf{x}_c = \begin{bmatrix} -2/3 \\ 0 \end{bmatrix}$$

and whose radius is

$$r \simeq 2.34$$

Let c denote the set of correct classification outcomes, and e the set of erroneous classification outcomes. The *probability of error* (misclassification), P_e, of a classifier operating according to the Bayesian decision rule is

$$P_e = p_1 P(e|\mathscr{C}_1) + p_2 P(e|\mathscr{C}_2) \tag{4.65}$$

where $P(e|\mathscr{C}_1)$ is the conditional probability of error given that the classifier input vector was drawn from the distribution of class $\mathscr{C}_1$, and similarly for $P(e|\mathscr{C}_2)$; p_1 and p_2 are the *a priori* probabilities of classes $\mathscr{C}_1$ and $\mathscr{C}_2$, respectively. For our problem we may numerically evaluate the probability integrals to obtain

$$P(e|\mathscr{C}_1) \simeq 0.1056$$

and

$$P(e|\mathscr{C}_2) \simeq 0.2642$$

With $p_1 = p_2 = 1/2$, the probability of misclassification is therefore

$$P_e \simeq 0.1849$$

Equivalently, the *probability of correct classification* is

$$P_c = 1 - P_e$$
$$\simeq 0.8151$$

Experimental Determination of Optimal Multilayer Perceptron

Table 4.1 lists the variable parameters of a multilayer perceptron (MLP) that involves a single layer of hidden neurons, and that is trained with the back-propagation algorithm operating in the sequential mode. Since the ultimate objective of a pattern classifier is to achieve an acceptable rate of correct classification, this criterion is used to judge when the variable parameters of the MLP (used as a pattern classifier) are optimal.

Optimal Number of Hidden Neurons.   Reflecting practical approaches to the problem of determining the optimal number of hidden neurons, m_1, the criterion used is the smallest number of hidden neurons that yields a performance "close" to the Bayesian classifier—usually within 1 percent. Thus the experimental study begins with two hidden neurons as the starting point for the simulation results summarized in

TABLE 4.1   Variable Parameters of Multilayer Perceptron

Parameter	Symbol	Typical Range
Number of hidden neurons	m_1	$(2, \infty)$
Learning-rate parameter	η	$(0, 1)$
Momentum constant	α	$(0, 1)$

TABLE 4.2   Simulation Results for Two Hidden Neurons[a]

Run Number	Training Set Size	Number of Epochs	Mean-Square Error	Probability of Correct Classification, P_c
1	500	320	0.2375	80.36%
2	2000	80	0.2341	80.33%
3	8000	20	0.2244	80.47%

[a]Learning rate parameter $\eta = 0.1$ and momentum $\alpha = 0$.

Table 4.2. Since the purpose of the first set of simulations is merely to ascertain the sufficiency of two hidden neurons or otherwise, the learning-rate parameter η and momentum constant α are arbitrarily set to some nominal values. For each simulation run, a training set of examples, randomly generated from the Gaussian distributions for classes $\mathscr{C}_1$ and $\mathscr{C}_2$ with equal probability, is repeatedly cycled through the network, with each training cycle representing an *epoch*. The number of epochs is chosen so that the total number of training examples used for each run is constant. By so doing, any potential effects arising from variations of the training set sizes are averaged out.

In Table 4.2 and subsequent tables, the *mean-square error* is computed precisely as the error functional defined in Eq. (4.53). We emphasize that the mean-square error is included in these tables only as a matter of record, since a *small mean-square error does not necessarily imply good generalization* (i.e., good performance with data not seen before).

After convergence of a network trained with a total number of N patterns, the probability of correct classification can in theory be calculated as follows:

$$P(c,N) = p_1 P(c,N|\mathscr{C}_1) + p_2 P(c,N|\mathscr{C}_2) \tag{4.66}$$

where $p_1 = p_2 = 1/2$, and

$$P(c,N|\mathscr{C}_1) = \int_{\Omega_1(N)} f_{\mathbf{X}}(\mathbf{x}|\mathscr{C}_1)\,d\mathbf{x} \tag{4.67}$$

$$P(c,N|\mathscr{C}_2) = 1 - \int_{\Omega_1(N)} f_{\mathbf{X}}(\mathbf{x}|\mathscr{C}_2)\,d\mathbf{x} \tag{4.68}$$

and $\Omega_1(N)$ is the region in decision space over which the multilayer perceptron (trained with N patterns) classifies the vector $\mathbf{x}$ (representing a realization of the random vector $\mathbf{X}$) as belonging to class $\mathscr{C}_1$. This region is usually found experimentally by evaluating the mapping function learned by the network and then applying the output decision rule of Eq. (4.55). Unfortunately, the numerical evaluation of $P(c,N|\mathscr{C}_1)$ and $P(c,N|\mathscr{C}_2)$ is problematic because closed-form expressions describing the decision boundary $\Omega_1(N)$ cannot easily be found.

Accordingly, we resort to the use of an experimental approach that involves testing the trained multilayer perceptron against another independent set of examples that are again drawn randomly from the distributions for classes $\mathscr{C}_1$ and $\mathscr{C}_2$ with equal probability. Let A be a random variable that counts the number of patterns out of the N test patterns that are classified correctly. Then the ratio

$$p_N = \frac{A}{N}$$

is a random variable that provides the maximum-likelihood unbiased estimate of the actual classification performance p of the network. Assuming that p is constant over the N input–output pairs, we may apply the *Chernoff bound* (Devroye, 1991) to the estimator p_N of p, obtaining

$$P(|p_N - p| > \epsilon) < 2 \exp(-2\epsilon^2 N) = \delta$$

Application of the Chernoff bound yields $N \approx 26,500$ for $\epsilon = 0.01$, and $\delta = 0.01$ (i.e., 99 percent certainty that the estimate p has the given tolerance). We thus picked a test set of size $N = 32,000$. The last column of Table 4.2 presents the probability of correct classification estimated for this test set size, with each result being the average of 10 independent trials of the experiment.

The classification performance presented in Table 4.2 for a multilayer perceptron using two hidden neurons is already reasonably close to the Bayesian performance $P_c = 81.51$ percent. On this basis we may conclude that for the pattern-classification problem described here the use of two hidden neurons is adequate. To emphasize this conclusion, in Table 4.3 we present the results of simulations repeated for the case of four hidden neurons, with all other parameters held constant. Although the mean-square error in Table 4.3 for four hidden neurons is slightly lower than that in Table 4.2 for two hidden neurons, the average rate of correct classification does not show improvement; in fact, it is slightly worse. For the remainder of the computer experiment described here, the number of hidden neurons is held at two.

Optimal Learning and Momentum Constants. For the "optimal" values of the learning-rate parameter η and momentum constant α, we may use any one of three definitions:

1. The η and α that on average yield convergence to a local minimum in the error surface of the network with the least number of epochs.
2. The η and α that, for either the worst-case or on average, yield convergence to the global minimum in the error surface with the least number of epochs.

TABLE 4.3   Simulation Results for Multilayer Perceptron Using Four Hidden Neurons[a]

Run Number	Training Set Size	Number of Epochs	Mean-Square Error	Probability of Correct Classification
1	500	320	0.2199	80.80%
2	2000	80	0.2108	80.81%
3	8000	20	0.2142	80.19%

[a]Learning-rate parameter $\eta = 0.1$ and momentum constant $\alpha = 0$.

3. The η and α that on average yield convergence to the network configuration that has the best generalization over the entire input space, with the least number of epochs.

The terms "average" and "worst-case" used here refer to the distribution of the training input–output pairs. Definition 3 is the ideal in practice; however, it is difficult to apply since minimizing the mean-square error is usually the mathematical criterion for optimality during network training and, as stated previously, a lower mean-square error over a training set does not necessarily imply good generalization. From a research point of view, definition 2 is more interesting than definition 1. For example, in Luo (1991), rigorous results are presented for the optimal adaptation of the learning-rate paramater η such that the smallest number of epochs is needed for the multilayer perceptron to approximate the globally optimum synaptic weight matrix to a desired accuracy, albeit for the special case of linear neurons. In general, however, heuristic and experimental procedures dominate the optimal selection of η and α when using definition 1. For the experiment described here, we therefore consider optimality in the sense of definition 1.

Using a multilayer perceptron with two hidden neurons, combinations of learning-rate parameter $\eta \in \{0.01, 0.1, 0.5, 0.9\}$ and momentum constant $\alpha \in \{0.0, 0.1, 0.5, 0.9\}$ are simulated to observe their effect on network convergence. Each combination is trained with the same set of initial random weights and the same set of 500 examples so that the results of the experiment may be compared directly. The learning process was continued for 700 epochs, after which it was terminated; this length of training was considered to be adequate for the back-propagation algorithm to reach a local minimum on the error surface. The ensemble-averaged learning curves so computed are plotted in Figs. 4.15a–4.15d, which are individually grouped by η.

The experimental learning curves shown here suggest the following trends:

- While, in general, a small learning-rate parameter η results in slower convergence, it can locate "deeper" local minima in the error surface than a larger η. This finding is intuitively satisfying, since a smaller η implies that the search for a minimum should cover more of the error surface than would be the case for a larger η.
- For $\eta \to 0$, the use of $\alpha \to 1$ produces increasing speed of convergence. On the other hand, for $\eta \to 1$, the use of $\alpha \to 0$ is required to ensure learning stability.
- The use of the constants $\eta = \{0.5, 0.9\}$ and $\alpha = 0.9$ causes oscillations in the mean-squared error during learning and a higher value for the final mean-square error at convergence, both of which are undesirable effects.

In Fig. 4.16 we show plots of the "best" learning curves selected from each group of the learning curves plotted in Fig. 4.16, so as to determine an "overall" best learning curve, "best" being defined in the sense of point 1 described previously. From Fig. 4.16, it appears that the optimal learning-rate parameter η_{opt} is about 0.1 and the optimal momentum constant α_{opt} is about 0.5. Thus, Table 4.4 summarizes the "optimal" values of network parameters used in the remainder of the experiment. The fact that the final mean-square error of each curve in Fig. 4.16 does not vary significantly over the range of η and α suggests a "well-behaved" (i.e., relatively smooth) error surface for the problem.

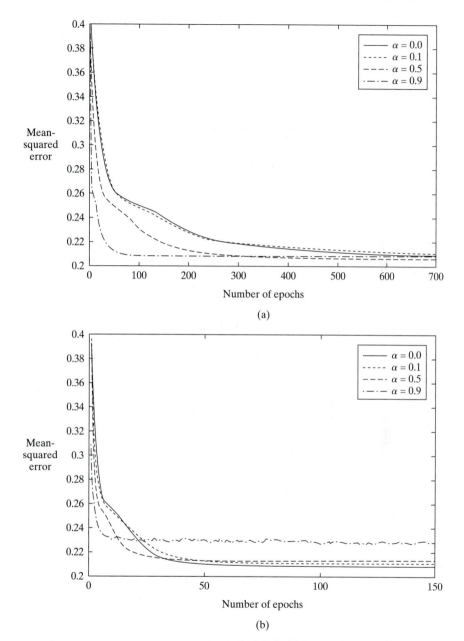

(a)

(b)

FIGURE 4.15 Ensemble-averaged learning curves for varying momentum α, and the following values of learning-rate parameters: (a) $\eta = 0.01$, (b) $\eta = 0.1$ (c) $\eta = 0.5$, and (d) $\eta = 0.9$.

(continued on p. 196)

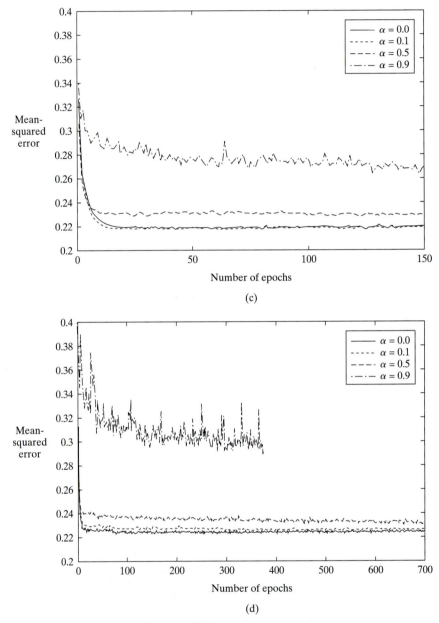

FIGURE 4.15   (continued from p. 195)

Evaluation of Optimal Network Design.   Given the "optimized" multilayer perceptron having the parameters summarized in Table 4.4, the final network is evaluated to determine its decision boundary, ensemble-averaged learning curve, and probability of correct classification. With finite-size training sets, the network function learned with the optimal parameters is "stochastic" in nature. Accordingly, these performance measures are ensemble-averaged over 20 independently trained networks. Each training set consists

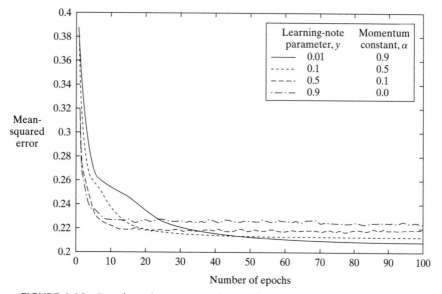

FIGURE 4.16   Best learning curves selected from the four parts of Fig. 4.15.

TABLE 4.4   Configuration of Optimized Multilayer Perceptron

Parameter	Symbol	Value
Optimum number of hidden neurons	m_{opt}	2
Optimum learning-rate parameter	η_{opt}	0.1
Optimum momentum constant	α_{opt}	0.5

of 1000 examples, drawn from the distributions for classes $\mathscr{C}_1$ and $\mathscr{C}_2$ with equal probability, and which are presented to the network in random order. As before, the training was continued for 700 epochs. For the experimental determination of the probabilities of correct classification, the same test set of 32,000 examples used previously is used again.

Figure 4.17a shows three of the "best" decision boundaries for three networks in the ensemble of 20. Figure 4.17b shows three of the "worst" decision boundaries for three other networks in the same ensemble. The shaded (circular) Bayesian decision boundary is included in both figures for reference. From these figures we observe that the decision boundaries constructed by the back-propagation algorithm are convex with respect to the region where they classify the observation vector $\mathbf{x}$ as belonging to class $\mathscr{C}_1$ or class $\mathscr{C}_2$.

The ensemble statistics of the performance measures, probability of correct classification and final mean-squared error, computed over the training sample are listed in Table 4.5. The probability of correct classification for the optimum Bayes classifier is 81.51%.

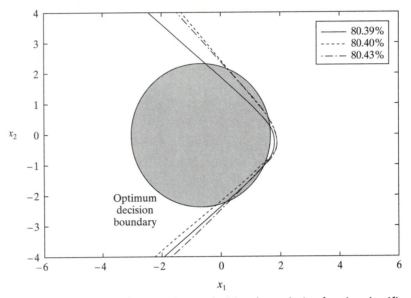

FIGURE 4.17A Plot of three "best" decision boundaries for the classification accuracies: 80.39, 80.40, and 80.43%.

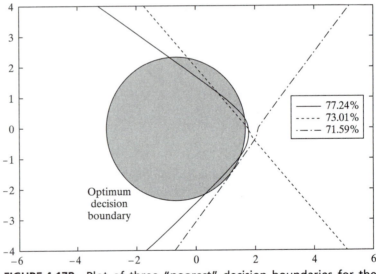

FIGURE 4.17B Plot of three "poorest" decision boundaries for the following classification accuracies: 77.24, 73.01, and 71.59%.

TABLE 4.5 Ensemble Statistics of Performance Measures (Sample Size = 20)

Performance Measure	Mean	Standard Deviation
Probability of correct classification	79.70%	0.44%
Final mean-square error	0.2277	0.0118

4.9 FEATURE DETECTION

Hidden neurons play a critical role in the operation of a multilayer perceptron with back-propagation learning because they act as *feature detectors*. As the learning process progresses, the hidden neurons begin to gradually "discover" the salient features that characterize the training data. They do so by performing a nonlinear transformation on the input data into a new space called the *hidden space*, or *feature space*; these two terminologies are used interchangeably throughout the book. In this new space the classes of interest in a pattern-classification task, for example, may be more easily separated from each other than in the original input space. This statement is well illustrated by the XOR problem considered in Section 4.5.

To put matters into a mathematical context, consider a multilayer perceptron with a single nonlinear layer of m_1 hidden neurons, and a linear layer of $m_2 = M$ output neurons. The choice of linear neurons in the output layer is motivated by the desire to focus attention on the role of hidden neurons on the operation of the multilayer perceptron. Let the synaptic weights of the network be adjusted to minimize the mean-square error between the target output (desired response) and the actual output of the network produced in response to a m_0-dimensional input vector (pattern), with the ensemble averaging performed over a total of N patterns. Let $z_j(n)$ denote the output of hidden neuron j due to the presentation of input pattern n. The $z_j(n)$ is a nonlinear function of the pattern (vector) applied to the input layer of the network by virtue of the sigmoid activation function built into each hidden neuron.

The output of neuron k in the output layer is

$$y_k(n) = \sum_{j=0}^{m_1} w_{kj} z_j(n), \qquad \begin{matrix} k = 1, 2, ..., M \\ n = 1, 2, ..., N \end{matrix} \tag{4.69}$$

where w_{k0} represents the bias applied to neuron k. The cost function to be minimized is

$$\mathcal{E}_{av} = \frac{1}{2N} \sum_{n=1}^{N} \sum_{k=1}^{M} (d_k(n) - y_k(n))^2 \tag{4.70}$$

Note that the use of a batch mode of operation is assumed here. Using Eqs. (4.69) and (4.70), it is easy to reformulate the cost function $\mathcal{E}_{av}$ in the compact matrix form:

$$\mathcal{E}_{av} = \frac{1}{2N} \|\tilde{\mathbf{D}} - \mathbf{W}\tilde{\mathbf{Z}}\|^2 \tag{4.71}$$

where $\mathbf{W}$ is the M-by-m_1 matrix of synaptic weights pertaining to the output layer of the network. The matrix $\tilde{\mathbf{Z}}$ is the m_1-by-N matrix of hidden neuronal outputs (with their mean values subtracted off), which are produced by the individual N input patterns applied to the input layer of the network; that is,

$$\tilde{\mathbf{Z}} = \{(z_j(n) - \mu_{z_j}); \quad j = 1, 2, ..., m_1: \quad n = 1, 2, ..., N\}$$

where μ_{z_j} is the mean value of $z_j(n)$. Correspondingly, the matrix $\tilde{\mathbf{D}}$ is the M-by-N matrix of target patterns (desired responses) presented to the output layer of the network; that is,

$$\tilde{\mathbf{D}} = \{(d_k(n) - \mu_{d_k}); \quad k = 1, 2, ..., M; \quad n = 1, 2, ..., N\}$$

where μ_{d_k} is the mean value of $d_k(n)$. The minimization of $\mathcal{E}_{av}$ defined by Eq. (4.70) is recognized as a linear least-squares problem, the solution of which is given by

$$\mathbf{W} = \tilde{\mathbf{D}}\,\tilde{\mathbf{Z}}^{+} \tag{4.72}$$

where $\tilde{\mathbf{Z}}^{+}$ is the pseudo-inverse of matrix $\tilde{\mathbf{Z}}$. The minimum value of $\mathcal{E}_{av}$ is given by (see Problem 4.7)

$$\mathcal{E}_{av,\,min} = \frac{1}{2N}\,\text{tr}\left[\tilde{\mathbf{D}}\,\tilde{\mathbf{D}}^T\,\tilde{\mathbf{D}}\tilde{\mathbf{Z}}^T(\tilde{\mathbf{Z}}\,\tilde{\mathbf{Z}}^T)^{+}\,\tilde{\mathbf{Z}}\,\tilde{\mathbf{D}}^T\right] \tag{4.73}$$

where $\text{tr}[\cdot]$ denotes the trace operator. Since the target patterns represented by matrix $\tilde{\mathbf{D}}$ are all fixed, minimization of the cost function $\mathcal{E}_{av}$ with respect to the synaptic weights of the multilayer perceptron is equivalent to maximizing the *discriminant function* (Webb and Lowe, 1990)

$$\mathcal{D} = \text{tr}[\mathbf{C}_b\mathbf{C}_t^{+}] \tag{4.74}$$

where the matrices $\mathbf{C}_b$ and $\mathbf{C}_t$ are themselves defined as:

- The m_1-by-m_1 matrix $\mathbf{C}_t$ is the *total covariance matrix* of the hidden neuronal outputs due to the presentations of N input patterns:

$$\mathbf{C}_t = \tilde{\mathbf{Z}}\tilde{\mathbf{Z}}^T \tag{4.75}$$

  The matrix $\mathbf{C}_t^{+}$ is the pseudo-inverse of matrix $\mathbf{C}_t$.
- The m_1-by-m_1 matrix $\mathbf{C}_b$ is defined by

$$\mathbf{C}_b = \tilde{\mathbf{Z}}\tilde{\mathbf{D}}^T\,\tilde{\mathbf{D}}\tilde{\mathbf{Z}}^T \tag{4.76}$$

Note that the discriminant function $\mathcal{D}$ defined in Eq. (4.74) is determined entirely by the hidden neurons of the multilayer perceptron. There is also no restriction on the number of hidden layers constituting the nonlinear transformation responsible for generating the discriminant function $\mathcal{D}$. In a multilayer perceptron with more than one hidden layer, the matrix $\tilde{\mathbf{Z}}$ refers to the entire set of patterns in the space defined by the final layer of hidden neurons.

For an interpretation of matrix $\mathbf{C}_b$, consider the specific choice of a *one-from-M coding scheme* (Webb and Lowe, 1990). That is, the target value (desired response) in a particular pattern is unity if the chosen pattern belongs to that class, and zero otherwise, as shown in (see page 185)

$$d(n) = \begin{bmatrix} 0 \\ \vdots \\ 0 \\ 1 \\ 0 \\ \vdots \\ 0 \end{bmatrix} \leftarrow k\text{th element}, \qquad d(n) \in \mathcal{C}_k$$

Thus, if there are M classes, $\mathscr{C}_k, k = 1, 2, \ldots, M$ with N_k patterns in class $\mathscr{C}_k$ and

$$\sum_{k=1}^{M} N_k = N$$

we may then expand the matrix $\mathbf{C}_b$ for this particular coding scheme into the form

$$\mathbf{C}_b = \sum_{k=1}^{M} N_k^2 \left(\boldsymbol{\mu}_{z,k} - \boldsymbol{\mu}_z \right) \left(\boldsymbol{\mu}_{z,k} - \boldsymbol{\mu}_z \right)^T \tag{4.77}$$

where the m_1-by-1 vector $\boldsymbol{\mu}_{z,k}$ is the mean value of the vector of the hidden neuronal outputs over all N_k patterns in class $\mathscr{C}_k$, and the vector $\boldsymbol{\mu}_z$ is the mean value of the vector of hidden neuronal outputs over all N input presentations. According to Eq. (4.77), we may interpret $\mathbf{C}_b$ as the *weighted between-class covariance matrix* at the outputs of the hidden layer.

Thus, for a one-from-M coding scheme, the multilayer perceptron maximizes a discriminant function that is the trace of the product of two matrices: the weighted between-class covariance matrix and the pseudo-inverse of the total covariance matrix. This result is interesting because it illustrates how a multilayer perceptron with back-propagation learning incorporates the proportions of samples within individual classes as *priors*.

Relation to Fisher's Linear Discriminent

The discriminant function $\mathscr{D}$ defined in Eq. (4.74) is unique to multilayer perceptrons. It bears a close resemblance to *Fisher's linear discriminant*, which describes a linear transformation from a multidimensional problem to a one-dimensional problem. Consider a variable y formed as a linear combination of the elements of an input vector $\mathbf{x}$; that is, it is defined as the inner product of $\mathbf{x}$ and a vector of adjustable parameters, $\mathbf{w}$ (that includes a bias as its first element), as shown by

$$y = \mathbf{w}^T \mathbf{x}$$

The vector $\mathbf{x}$ is drawn from one of two populations, $\mathscr{C}_1$ and $\mathscr{C}_2$, which differ from each other by virtue of their mean vectors $\boldsymbol{\mu}_1$ and $\boldsymbol{\mu}_2$, respectively. The *Fisher criterion* for discriminating between these two classes is defined by

$$J(\mathbf{w}) = \frac{\mathbf{w}^T \mathbf{C}_b \mathbf{w}}{\mathbf{w}^T \mathbf{C}_t \mathbf{w}}$$

where $\mathbf{C}_b$ is the *between-class covariance matrix* defined by

$$\mathbf{C}_b = (\boldsymbol{\mu}_2 - \boldsymbol{\mu}_1)(\boldsymbol{\mu}_2 - \boldsymbol{\mu}_1)^T$$

and $\mathbf{C}_t$ is the total *within-class covariance matrix* defined by

$$\mathbf{C}_t = \sum_{n \in \mathscr{C}_1} (\mathbf{x}_n - \boldsymbol{\mu}_1)(\mathbf{x}_n - \boldsymbol{\mu}_1)^T + \sum_{n \in \mathscr{C}_2} (\mathbf{x}_n - \boldsymbol{\mu}_2)(\mathbf{x}_n - \boldsymbol{\mu}_2)^T$$

The within-class covariance matrix $\mathbf{C}_t$ is proportional to the sample covariance matrix of the training set. It is symmetric and nonnegative definite, and is usually nonsingular if the size of the training set is large. The between-class covariance matrix $\mathbf{C}_b$ is also symmetric and nonnegative definite, but singular. A property of particular interest is that the matrix product $\mathbf{C}_b\mathbf{w}$ is always in the direction of the difference mean vector $\boldsymbol{\mu}_1 - \boldsymbol{\mu}_2$. This property follows directly from the definition of $\mathbf{C}_b$.

The expression defining $J(\mathbf{w})$ is known as the *generalized Rayleigh quotient*. The vector $\mathbf{w}$ that maximizes $J(\mathbf{w})$ must satisfy the condition

$$\mathbf{C}_b\mathbf{w} = \lambda\mathbf{C}_t\mathbf{w} \tag{4.76}$$

Equation (4.76) is a generalized eigenvalue problem. Recognizing that in our case the matrix product $\mathbf{C}_b\mathbf{w}$ is always in the direction of the difference vector $\boldsymbol{\mu}_1 - \boldsymbol{\mu}_2$, we find that the solution for Eq. (4.76) is simply

$$\mathbf{w} = \mathbf{C}_t^{-1}(\boldsymbol{\mu}_1 - \boldsymbol{\mu}_2) \tag{4.77}$$

which is referred to as *Fisher's linear discriminant* (Duda and Hart, 1973).

Returning to the issue of feature detection, recall that the discriminant function $\mathcal{D}$ of Eq. (4.74) relates the between-class and total covariance matrices of patterns transformed into the hidden space of the network. The discriminant function $\mathcal{D}$ plays a role similar to that of Fisher's linear discriminant. This is precisely the reason why these neural networks are able to perform the task of pattern classification so well.

4.10   BACK-PROPAGATION AND DIFFERENTIATION

Back-propagation is a specific technique for implementing *gradient descent* in weight space for a multilayer feedforward network. The basic idea is to efficiently compute *partial derivatives* of an approximating function $F(\mathbf{w},\mathbf{x})$ realized by the network with respect to all the elements of the adjustable weight vector $\mathbf{w}$ for a given value of input vector $\mathbf{x}$. Herein lies the computational power of the back-propagation algorithm.[5]

To be specific, consider a multilayer perceptron with an input layer of m_0 nodes, two hidden layers, and a single output neuron, as depicted in Fig. 4.18. The elements of the weight vector $\mathbf{w}$ are ordered by layer (starting from the first hidden layer), then by neurons in a layer, and then by the number of a synapse within a neuron. Let $w_{ji}^{(l)}$ denote the synaptic weight from neuron i to neuron j in layer $l = 0, 1, 2, \dots$. For $l = 1$, corresponding to the first hidden layer, the index i refers to a source node rather than to a neuron. For $l = 3$, corresponding to the output layer in Fig. 4.18, we have $j = 1$. We wish to evaluate the derivatives of the function $F(\mathbf{w},\mathbf{x})$ with respect to all the elements of the weight vector $\mathbf{w}$, for a specified input vector $\mathbf{x} = [x_1, x_2, \cdots, x_{m_0}]^T$. Note that for $l = 2$ (i.e., a single hidden layer), the function $F(\mathbf{w},\mathbf{x})$ has a form similar to that on the right-hand side of Eq. (4.69). We have included the weight vector $\mathbf{w}$ as an argument of the function F to focus attention on it.

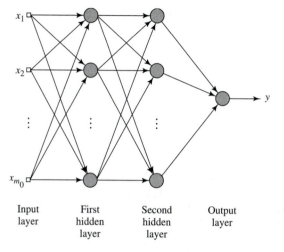

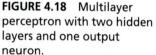

FIGURE 4.18 Multilayer perceptron with two hidden layers and one output neuron.

Input layer First hidden layer Second hidden layer Output layer

The multilayer perceptron of Fig. 4.18 is parameterized by an *architecture* $\mathcal{A}$ (representing a discrete parameter) and a *weight vector* $\mathbf{w}$ (made up of continuous elements). Let $\mathcal{A}_j^{(l)}$ denote that part of the architecture extending from the input layer ($l = 0$) to node j in layer $l = 1, 2, 3$. Accordingly, we may write

$$F(\mathbf{w}, \mathbf{x}) = \varphi(\mathcal{A}_1^{(3)}) \tag{4.80}$$

where φ is the activation function. However, $\mathcal{A}_1^{(3)}$ is to be interpreted merely as an architectural symbol rather than a variable. Thus, adapting Eqs. (4.1), (4.2), (4.11), and (4.23) for use in this situation, we obtain the following results:

$$\frac{\partial F(\mathbf{w}, \mathbf{x})}{\partial w_{1k}^{(3)}} = \varphi'(\mathcal{A}_1^{(3)}) \varphi(\mathcal{A}_k^{(2)}) \tag{4.81}$$

$$\frac{\partial F(\mathbf{w}, \mathbf{x})}{\partial w_{kj}^{(2)}} = \varphi'(\mathcal{A}_1^{(3)}) \, \varphi'(\mathcal{A}_k^{(2)}) \, \varphi(\mathcal{A}_j^{(1)}) w_{1k}^{(3)} \tag{4.82}$$

$$\frac{\partial F(\mathbf{w}, \mathbf{x})}{\partial w_{ji}^{(1)}} = \varphi'(\mathcal{A}_1^{(3)}) \varphi'(\mathcal{A}_j^{(1)}) x_i \left[\sum_k w_{1k}^{(3)} \varphi'(\mathcal{A}_k^{(2)}) w_{kj}^{(2)} \right] \tag{4.83}$$

where φ' is the partial derivative of the nonlinearity φ with respect to its input, and x_i is the ith element of the input vector $\mathbf{x}$. In a similar way we may derive the equations for the partial derivatives of a general network with more hidden layers and more neurons in the output layer.

Equations (4.81) through (4.83) provide the basis for calculating the sensitivity of the network function $F(\mathbf{w}, \mathbf{x})$ with respect to variations in the elements of the weight vector $\mathbf{w}$. Let ω denote an element of the weight vector $\mathbf{w}$. The *sensitivity* of $F(\mathbf{w}, \mathbf{x})$ with respect to ω is formally defined by

$$S_\omega^F = \frac{\partial F/F}{\partial \omega/\omega}, \qquad \omega \in \mathbf{w}$$

It is for this reason that we referred to the lower part of the signal-flow graph in Fig. 4.7 as a "sensitivity graph."

Jacobian Matrix

Let W denote the total number of free parameters (i.e., synaptic weights and biases) of a multilayer perceptron, which are ordered in the manner described to form the weight vector $\mathbf{w}$. Let N denote the total number of examples used to train the network. Using back-propagation we may compute a set of W partial derivatives of the approximating function $F[\mathbf{w}, \mathbf{x}(n)]$ with respect to the elements of the weight vector $\mathbf{w}$ for a specific example $\mathbf{x}(n)$ in the training set. Repeating these computations for $n = 1, 2, ..., N$, we end up with an N-by-W matrix of partial derivatives. This matrix is called the *Jacobian* $\mathbf{J}$ of the multilayer perceptron evaluated at $\mathbf{x}(n)$. Each row of the Jacobian corresponds to a particular example in the training set.

There is experimental evidence to suggest that many neural network training problems are intrinsically *ill-conditioned,* leading to a Jacobian $\mathbf{J}$ that is almost rank deficient (Saarinen et al., 1991). The *rank* of a matrix is equal to the number of linearly independent columns or rows in the matrix, whichever one is smallest. The Jacobian $\mathbf{J}$ is said to be *rank-deficient* if its rank is less than $\min(N, W)$. Any rank deficiency in the Jacobian causes the back-propagation algorithm to obtain only partial information of the possible search directions, and also causes training times to be long.

4.11 HESSIAN MATRIX

The *Hessian matrix* of the cost function $\mathscr{E}_{av}(\mathbf{w})$, denoted by $\mathbf{H}$, is defined as the second derivative of $\mathscr{E}_{av}(\mathbf{w})$ with respect to the weight vector $\mathbf{w}$, as shown by

$$\mathbf{H} = \frac{\partial^2 \mathscr{E}_{av}(\mathbf{w})}{\partial \mathbf{w}^2} \tag{4.84}$$

The Hessian matrix plays an important role in the study of neural networks; specifically, we may mention the following:[6]

1. The eigenvalues of the Hessian matrix have a profound influence on the dynamics of back-propagation learning.
2. The inverse of the Hessian matrix provides a basis for pruning (i.e., deleting) insignificant synaptic weights from a multilayer perceptron, as discussed in Section 4.15.
3. The Hessian matrix is basic to the formulation of second-order optimization methods as an alternative to back-propagation learning, as discussed in Section 4.18.

An iterative procedure for the computation[7] of the Hessian matrix is presented in Section 4.15. In this section we confine our attention to point 1.

In Chapter 3 we indicated that the eigenstructure of the Hessian matrix has a profound influence on the convergence properties of the LMS algorithm. So it is also with the back-propagation algorithm, but in a much more complicated way. Typically the Hessian matrix of the error surface pertaining to a multilayer perceptron trained with the back-propagation algorithm has the following composition of eigenvalues (LeCun, et al., 1991; LeCun, 1993):

- A small number of small eigenvalues.
- A large number of medium-sized eigenvalues.
- A small number of large eigenvalues.

The factors affecting this composition may be grouped as follows:

- Nonzero-mean input signals or nonzero-mean induced neuronal output signals.
- Correlations between the elements of the input signal vector and correlations between induced neuronal output signals.
- Wide variations in the second derivatives of the cost function with respect to synaptic weights of neurons in the network, as we proceed from one layer to the next. The second derivatives are often smaller in the lower layers, with the synaptic weights in the first hidden layer learning slowly and those in the last layers learning quickly.

From Chapter 3 we recall that the *learning time* of the LMS algorithm is sensitive to variations in the condition number $\lambda_{max}/\lambda_{min}$, where λ_{max} is the largest eigenvalue of the Hessian and λ_{min} is its smallest nonzero eigenvalue. Experimental results show that a similar result holds for the back-propagation algorithm, which is a generalization of the LMS algorithm. For inputs with nonzero mean, the ratio $\lambda_{max}/\lambda_{min}$ is larger than its corresponding value for zero-mean inputs: the larger the mean of the inputs, the larger the ratio $\lambda_{max}/\lambda_{min}$ (see Problem 3.10). This observation has a serious implication for the dynamics of back-propagation learning.

For the learning time to be minimized, the use of nonzero-mean inputs should be avoided. Now, insofar as the signal vector **x** applied to a neuron in the first hidden layer of a multilayer perceptron (i.e., the signal vector applied to the input layer) is concerned, it is easy to remove the mean from each element of **x** before its application to the network. But what about the signals applied to the neurons in the remaining hidden and output layers of the network? The answer to this question lies in the type of activation function used in the network. If the activation function is nonsymmetric, as in the case of the logistic function, the output of each neuron is restricted to the interval [0,1]. Such a choice introduces a source of *systematic bias* for those neurons located beyond the first hidden layer of the network. To overcome this problem we need to use an antisymmetric activation function such as the hyperbolic tangent function. With this latter choice, the output of each neuron is permitted to assume both positive and negative values in the interval [−1,1], in which case it is likely for its mean to be zero. If the network connectivity is large, back-propagation learning with antisymmetric activation functions can yield faster convergence than a similar process with nonsymmetric activation functions, for which there is also empirical evidence (LeCun et al., 1991). This provides justification for heuristic 3 described in Section 4.6.

4.12   GENERALIZATION

In back-propagation learning, we typically start with a training sample and use the back-propagation algorithm to compute the synaptic weights of a multilayer perceptron by loading (encoding) as many of the training examples as possible into the network. The hope is that the neural network so designed will generalize. A network is said to *generalize* well when the input–output mapping computed by the network is correct (or nearly so) for test data never used in creating or training the network; the term "generalization" is borrowed from psychology. Here it is assumed that the test data are drawn from the same population used to generate the training data.

The learning process (i.e., training of a neural network) may be viewed as a "curve-fitting" problem. The network itself may be considered simply as a nonlinear input–output mapping. Such a viewpoint then permits us to look on generalization not as a mystical property of neural networks but rather simply as the effect of a good non-linear interpolation of the input data (Wieland and Leighton, 1987). The network performs useful interpolation primarily because multilayer perceptrons with continuous activation functions lead to output functions that are also continuous.

Figure 4.19a illustrates how generalization may occur in a hypothetical network. The nonlinear input–output mapping represented by the curve depicted in this figure is computed by the network as a result of learning the points labeled as "training data." The point marked on the curve as "generalization" is thus seen as the result of interpolation performed by the network.

A neural network that is designed to generalize well will produce a correct input–output mapping even when the input is slightly different from the examples used to train the network, as illustrated in the figure. When, however, a neural network learns too many input–output examples, the network may end up memorizing the training data. It may do so by finding a feature (due to noise, for example) that is present in the training data but not true of the underlying function that is to be modeled. Such a phenomenon is referred to as *overfitting* or *overtraining*. When the network is overtrained, it loses the ability to generalize between similar input–output patterns.

Ordinarily, loading data into a multilayer perceptron in this way requires the use of more hidden neurons than is actually necessary, with the result that undesired contributions in the input space due to noise are stored in synaptic weights of the network. An example of how poor generalization due to memorization in a neural network may occur is illustrated in Fig. 4.19b for the same data depicted in Fig. 4.19a. "Memorization" is essentially a "look-up table," which implies that the input–output mapping computed by the neural network is not smooth. As pointed out in Poggio and Girosi (1990a), smoothness of input–output mapping is closely related to such model-selection criteria as the *Occam's razor,* the essence of which is to select the "simplest" function in the absence of any prior knowledge to the contrary. In the context of our present discussion, the simplest function means the smoothest function that approximates the mapping for a given error criterion, because such a choice generally demands the fewest computational resources. Smoothness is also natural in many applications, depending on the scale of the phenomenon being studied. It is therefore important to seek a smooth nonlinear mapping for ill-posed input–output relationships, so that the network is able to classify novel patterns correctly with respect to the training patterns (Wieland and Leighton, 1987).

Sufficient Training Set Size for a Valid Generalization

Generalization is influenced by three factors: (1) the size of the training set, and how representative it is of the environment of interest, (2) the architecture of the neural network, and (3) the physical complexity of the problem at hand. Clearly, we have no

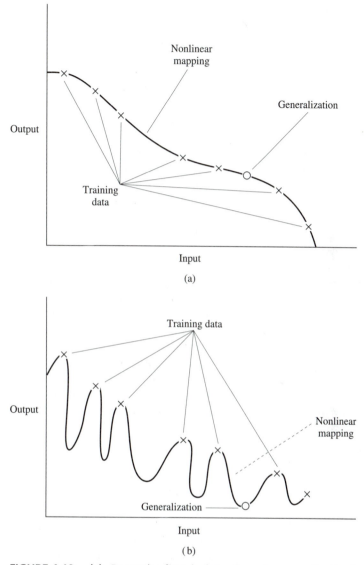

FIGURE 4.19 (a) Properly fitted data (good generalization) (b) Overfitted data (poor generalization).

control over the latter. In the context of the other two factors, we may view the issue of generalization from two different perspectives (Hush and Horne, 1993):

- The architecture of the network is fixed (hopefully in accordance with the physical complexity of the underlying problem), and the issue to be resolved is that of determining the size of the training set needed for a good generalization to occur.
- The size of the training set is fixed, and the issue of interest is that of determining the best architecture of network for achieving good generalization.

Both of these viewpoints are valid in their own individual ways. In the present discussion we focus on the first viewpoint.

The adequacy of the training sample size or the sample complexity problem is discussed in Chapter 2. As pointed out in that chapter, the VC dimension provides the theoretical basis for a principled solution to this important design problem. In particular, we have *distribution-free, worst-case* formulas for estimating the size of the training sample that is sufficient for a good generalization performance; see Section 2.14. Unfortunately, we often find that there is a huge numerical gap between the size of the training sample actually needed and that predicted by these formulas. It is this gap that has made the sample complexity problem a continuing open research area.

In practice, it seems that all we really need for a good generalization is to have the size of the training set, N, satisfy the condition

$$N = O\left(\frac{W}{\epsilon}\right) \tag{4.85}$$

where W is the total number of free parameters (i.e., synaptic weights and biases) in the network, and ϵ denotes the fraction of classification errors permitted on test data (as in pattern classification), and $O(\cdot)$ denotes the order of quantity enclosed within. For example, with an error of 10 percent the number of training examples needed should be about 10 times the number of free parameters in the network.

Equation (4.85) is in accordance with *Widrow's rule of thumb* for the LMS algorithm, which states that the settling time for adaptation in linear adaptive temporal filtering is approximately equal to the memory span of an adaptive tapped-delay-line filter divided by the misadjustment (Widrow and Stearns, 1985). The misadjustment in the LMS algorithm plays a role somewhat analogous to the error ϵ in Eq. (4.85). Further justification for this empirical rule is presented in the next section.

4.13 APPROXIMATIONS OF FUNCTIONS

A multilayer perceptron trained with the back-propagation algorithm may be viewed as a practical vehicle for performing a *nonlinear input–output mapping* of a general nature. To be specific, let m_0 denote the number of input (source) nodes of a multilayer perceptron, and let $M = m_L$ denote the number of neurons in the output layer of the network. The input–output relationship of the network defines a mapping from an m_0-dimensional Euclidean input space to an M-dimensional Euclidean output space, which is infinitely continuously differentiable when the activation function is likewise. In assessing the capability of the multilayer perceptron from this viewpoint of input–output mapping, the following fundamental question arises:

> *What is the minimum number of hidden layers in a multilayer perceptron with an input–output mapping that provides an approximate realization of any continuous mapping?*

Universal Approximation Theorem

The answer to this question is embodied in the *universal approximation theorem*[8] for a nonlinear input–output mapping, which may be stated as:

> *Let $\varphi(\cdot)$ be a nonconstant, bounded, and monotone-increasing continuous function. Let I_{m_0} denote the m_0-dimensional unit hypercube $[0,1]^{m_0}$. The space of continuous functions on I_{m_0}*

is denoted by $C(I_{m_0})$. Then, given any function $f \ni C(I_{m_0})$ and $\epsilon > 0$, there exist an integer M and sets of real constants α_i, b_i, and w_{ij}, where $i = 1, \ldots, m_1$ and $j = 1, \ldots, m_0$ such that we may define

$$F(x_1, \ldots, x_{m_0}) = \sum_{i=1}^{m_1} \alpha_i \varphi \left(\sum_{j=1}^{m_0} w_{ij} x_j + b_i \right) \tag{4.86}$$

as an approximate realization of the function $f(\,\cdot\,)$; that is,

$$|F(x_1, \ldots, x_{m_0}) - f(x_1, \ldots, x_{m_0})| < \epsilon$$

for all $x_1, x_2, \ldots, x_{m_0}$ that lie in the input space.

The universal approximation theorem is directly applicable to multilayer perceptrons. We first note that the logistic function $1/[1 + \exp(-v)]$ used as the nonlinearity in a neuronal model for the construction of a multilayer perceptron is indeed a nonconstant, bounded, and monotone-increasing function; it therefore satisfies the conditions imposed on the function $\varphi(\cdot)$. Next, we note that Eq. (4.86) represents the output of a multilayer perceptron described as follows:

1. The network has m_0 input nodes and a single hidden layer consisting of m_1 neurons; the inputs are denoted by $x_1, \ldots, x_{m_0}$.
2. Hidden neuron i has synaptic weights $w_{i_1}, \ldots, w_{m_0}$, and bias b_i.
3. The network output is a linear combination of the outputs of the hidden neurons, with $\alpha_1, \ldots, \alpha_{m_1}$ defining the synaptic weights of the output layer.

The universal approximation theorem is an *existence theorem* in the sense that it provides the mathematical justification for the approximation of an arbitrary continuous function as opposed to exact representation. Equation (4.86), which is the backbone of the theorem, merely generalizes approximations by finite Fourier series. In effect, the theorem states that a *single hidden layer is sufficient for a multilayer perceptron to compute a uniform ϵ approximation to a given training set represented by the set of inputs $x_1, \ldots, x_{m_0}$ and a desired (target) output $f(x_1, \ldots, x_{m_0})$*. However, the theorem does not say that a single hidden layer is optimum in the sense of learning time, ease of implementation, or (more importantly) generalization.

Bounds on Approximation Errors

Barron (1993) has established the approximation properties of a multilayer perceptron, assuming that the network has a single layer of hidden neurons using sigmoid functions and a linear output neuron. The network is trained using the back-propagation algorithm and then tested with new data. During training, the network learns specific points of a target function f in accordance with the training data, and thereby produces the approximating function F defined in Eq. (4.86). When the network is exposed to test data that have not been seen before, the network function F acts as an "estimator" of new points of the target function; that is, $F = \hat{f}$.

A smoothness property of the target function f is expressed in terms of its Fourier representation. In particular, the average of the norm of the frequency vector weighted by the Fourier magnitude distribution is used as a measure for the extent to which the function f oscillates. Let $\tilde{f}(\omega)$ denote the multidimensional Fourier transform of the function $f(\mathbf{x})$, $\mathbf{x} \in \mathbb{R}^{m_0}$; the m_0-by-1 vector $\boldsymbol{\omega}$ is the frequency vector. The function $f(x)$ is defined in terms of its Fourier transform $\tilde{f}(\boldsymbol{\omega})$ by the inverse formula:

$$f(x) = \int_{\mathbb{R}^{m_0}} \tilde{f}(\boldsymbol{\omega})\exp(j\boldsymbol{\omega}^T\mathbf{x})\, d\boldsymbol{\omega} \tag{4.87}$$

where $j = \sqrt{-1}$. For the complex-valued function $\tilde{f}(\boldsymbol{\omega})$ for which $\boldsymbol{\omega}\tilde{f}(\boldsymbol{\omega})$ is integrable, we define the *first absolute moment* of the Fourier magnitude distribution of the function f as:

$$C_f = \int_{\mathbb{R}^{m_0}} |\tilde{f}(\boldsymbol{\omega})| \times \|\boldsymbol{\omega}\|^{1/2}\, d\boldsymbol{\omega} \tag{4.88}$$

where $\|\boldsymbol{\omega}\|$ is the Euclidean norm of $\boldsymbol{\omega}$ and $|\tilde{f}(\boldsymbol{\omega})|$ is the absolute value of $\tilde{f}(\boldsymbol{\omega})$. The first absolute moment C_f quantifies the *smoothness* or *regularity* of the function f.

The first absolute moment C_f provides the basis for a *bound* on the error that results from the use of a multilayer perceptron represented by the input–output mapping function $F(\mathbf{x})$ of Eq. (4.86) to approximate $f(\mathbf{x})$. The approximation error is measured by the *integrated squared error* with respect to an arbitrary probability measure μ on the ball $B_r = \{\mathbf{x}: \|\mathbf{x}\| \le r\}$ of radius $r > 0$. On this basis we may state the following proposition for a bound on the approximation error due to Barron (1993):

For every continuous function $f(\mathbf{x})$ with first moment C_f finite, and every $m_1 \ge 1$, there exists a linear combination of sigmoid functions $F(\mathbf{x})$ of the form defined in Eq. (4.86), such that

$$\int_{B_r} (f(\mathbf{x}) - F(\mathbf{x}))^2 \mu(dx) \le \frac{C_f'}{m_1}$$

where $C_f' = (2 r C_f)^2$.

When the function $f(\mathbf{x})$ is observed at a set of values of the input vector $\mathbf{x}$ denoted by $\{\mathbf{x}_i\}_{i=1}^N$ that are restricted to lie inside the ball B_r, the result provides the following bound on the *empirical risk:*

$$R = \frac{1}{N} \sum_{i=1}^N (f(\mathbf{x}_i) - F(\mathbf{x}_i))^2 \le \frac{C_f'}{m_1} \tag{4.89}$$

In Barron (1992), the approximation result of Eq. (4.89) is used to express the bound on the risk R resulting from the use of a multilayer perceptron with m_0 input nodes and m_1 hidden neurons as follows:

$$R \le O\left(\frac{C_f^2}{m_1}\right) + O\left(\frac{m_0 m_1}{N}\log N\right) \tag{4.90}$$

The two terms in the bound on the risk R express the tradeoff between two conflicting requirements on the size of the hidden layer:

1. *Accuracy of best approximation.* For this requirement to be satisfied, m_1, the size of the hidden layer, must be large in accordance with the universal approximation theorem.
2. *Accuracy of empirical fit to the approximation.* To satisfy this second requirement, we must use a small ratio m_1/N. For a fixed size of training sample, N, the

size of the hidden layer, m_1, should be kept small, which is in conflict with the first requirement.

The bound on the risk R described in Eq. (4.90) has other interesting implications. Specifically, we see that an exponentially large sample size, large in the dimensionality m_0 of the input space, is *not* required to get an accurate estimate of the target function, provided that the first absolute moment C_f remains finite. This result makes multilayer perceptrons as universal approximators even more important in practical terms.

The error between the empirical fit and the best approximation may be viewed as an *estimation error* along the lines described in Chapter 2. Let ϵ_0 denote the mean-square value of this estimation error. Then ignoring the logarithmic factor $\log N$ in the second term of the bound in Eq. (4.90), we may infer that the size N of the training sample needed for a good generalization is about $m_0 m_1/\epsilon_0$. This result has a mathematical structure similar to the empirical rule of Eq. (4.85), bearing in mind that $m_0 m_1$ is equal to the total number of free parameters W in the network. In other words, we may generally say that for good generalization, the number of training examples N should be larger than the ratio of the total number of free parameters in the network to the mean-square value of the estimation error.

Curse of Dimensionality

Another interesting result that emerges from the bounds described in (4.90) is that when the size of the hidden layer is optimized (i.e., the risk R is minimized with respect to N) by setting

$$m_1 \simeq C_f \left(\frac{N}{m_0 \log N} \right)^{1/2}$$

then the risk R is bounded by $O(C_f \sqrt{m_0(\log N/N)})$. A surprising aspect of this result is that in terms of the first-order behavior of the risk R, the rate of convergence expressed as a function of the training sample size N is of order $(1/N)^{1/2}$ (times a logarithmic factor). In contrast, for traditional smooth functions (e.g., polynomials and trigonometric functions) we have a different behavior. Let s denote a measure of smoothness, defined as the number of continuous derivatives of a function of interest. Then, for traditional smooth functions we find that the minimax rate of convergence of the total risk R is of order $(1/N)^{2s/(2s+m_0)}$. The dependence of this rate on the dimensionality of the input space, m_0, is a curse of dimensionality, which severely restricts the practical application of these functions. The use of a multilayer perceptron for function approximation appears to offer an advantage over traditional smooth functions; this advantage is, however, subject to the condition that the first absolute moment C_f remains finite; this is a smoothness constraint.

The *curse of dimensionality* was introduced by Richard Bellman in his studies of adaptive control processes (Bellman, 1961). For a geometric interpretation of this notion, let $\mathbf{x}$ denote an m_0-dimensional input vector, and $\{(\mathbf{x}_i, d_i)\}, i = 1, 2, ..., N$, denote the training sample. The *sampling density* is proportional to N^{1/m_0}. Let a function $f(\mathbf{x})$ represent a surface lying in the m_0-dimensional input space, which passes near the data points $\{(\mathbf{x}_i, d_i)\}_{i=1}^N$. Now, if the function $f(\mathbf{x})$ is arbitrarily complex and (for the most part) completely unknown, we need *dense* sample (data) points to learn it well. Unfortunately, dense samples are hard to find in "high dimensions," hence the curse of dimensionality. In particular, there is an *exponential* growth in complexity as a result of an increase in dimensionality, which in turn leads to the deterioration of the space-filling properties for uniformly randomly distributed points in higher-dimension spaces. The basic reason for the curse of dimensionality is (Friedman, 1995):

> A function defined in high-dimensional space is likely to be much more complex than a function defined in a lower-dimensional space, and those complications are harder to discern.

The only practical way to beat the curse of dimensionality is to incorporate *prior knowledge* about the function over and above the training data, which is known to be *correct*.

In practice, it may also be argued that to have any hope of good estimation in a high-dimensional space we must provide increasing smoothness of the unknown underlying function with increasing input dimensionality (Niyogi and Girosi, 1996). This viewpoint is pursued further in Chapter 5.

Practical Considerations

The universal approximation theorem is important from a theoretical viewpoint, because it provides the *necessary mathematical tool* for the viability of feedforward networks with a single hidden layer as a class of approximate solutions. Without such a theorem, we could conceivably be searching for a solution that cannot exist. However, the theorem is not constructive, that is, it does not actually specify how to determine a multilayer perceptron with the stated approximation properties.

The universal approximation theorem assumes that the continuous function to be approximated is given and that a hidden layer of unlimited size is available for the approximation. Both of these assumptions are violated in most practical applications of multilayer perceptrons.

The problem with multilayer perceptrons using a single hidden layer is that the neurons therein tend to interact with each other globally. In complex situations this interaction makes it difficult to improve the approximation at one point without worsening it at some other point. On the other hand, with two hidden layers the approximation (curve-fitting) process becomes more manageable. In particular, we may proceed as follows (Funahashi, 1989; Chester, 1990):

1. *Local features* are extracted in the first hidden layer. Specifically, some neurons in the first hidden layer are used to partition the input space into regions, and other neurons in that layer learn the local features characterizing those regions.
2. *Global features* are extracted in the second hidden layer. Specifically, a neuron in the second hidden layer combines the outputs of neurons in the first hidden layer operating on a particular region of the input space, and thereby learns the global features for that region and outputs zero elsewhere.

This two-stage approximation process is similar in philosophy to the spline technique for curve fitting, in the sense that the effects of neurons are isolated and the approximations in different regions of the input space may be individually adjusted. A *spline* is an example of a piecewise polynomial approximation.

Sontag (1992) provides further justification for the use of two hidden layers in the context of *inverse problems*. Specifically, the following inverse problem is considered:

> Given a continuous vector-valued function $\mathbf{f}: \mathbb{R}^m \to \mathbb{R}^M$, a compact subset $\mathscr{C} \subseteq \mathbb{R}^M$ that is included in the image of $\mathbf{f}$, and an $\epsilon > 0$, find a vector-valued function $\varphi: \mathbb{R}^M \to \mathbb{R}^m$ such that the following condition is satisfied:
>
> $$\|\varphi(\mathbf{f}(\mathbf{u})) - \mathbf{u}\| < \epsilon \qquad \text{for } \mathbf{u} \in \mathscr{C}$$

This problem arises in *inverse kinematics* (dynamics), where the observed state $\mathbf{x}(n)$ of a system is a function of current actions $\mathbf{u}(n)$ and the previous state $\mathbf{x}(n-1)$ of the system, as shown by

$$\mathbf{x}(n) = \mathbf{f}(\mathbf{x}(n-1), \mathbf{u}(n))$$

It is assumed that $\mathbf{f}$ is *invertible*, so that we may solve for $\mathbf{u}(n)$ as a function of $\mathbf{x}(n)$ for any $\mathbf{x}(n-1)$. The function $\mathbf{f}$ represents the direct kinematics, whereas the function φ represents the inverse kinematics. In practical terms, the motivation is to find a function φ that is computable by a multilayer perceptron. In general, discontinuous functions φ are needed to solve the inverse kinematics problem. It is interesting that even if the use of neuronal models with discontinuous activation functions is permitted, one hidden layer is *not* enough to guarantee the solution of all such inverse problems, whereas multilayer perceptrons with two hidden layers are sufficient for every possible $\mathbf{f}$, $\mathscr{C}$, and ϵ (Sontag, 1992).

4.14   CROSS-VALIDATION

The essence of back-propagation learning is to encode an input–output mapping (represented by a set of labeled examples) into the synaptic weights and thresholds of a multilayer perceptron. The hope is that the network becomes well trained so that it learns enough about the past to generalize to the future. From such a perspective the learning process amounts to a choice of network parameterization for this data set. More specifically, we may view the network selection problem as choosing, within a set of candidate model structures (parameterizations), the "best" one according to a certain criterion.

In this context, a standard tool in statistics known as *cross-validation* provides an appealing guiding principle[9] (Stone, 1974, 1978). First the available data set is randomly partitioned into a training set and a test set. The training set is further partitioned into two disjoint subsets:

- *Estimation subset,* used to select the model.
- *Validation subset,* used to test or validate the model.

The motivation here is to validate the model on a data set different from the one used for parameter estimation. In this way we may use the training set to assess the performance of various candidate models, and thereby choose the "best" one. There is, however, a distinct possibility that the model with the best-performing parameter values so selected may end up overfitting the validation subset. To guard against this possibility, the generalization performance of the selected model is measured on the test set, which is different from the validation subset.

The use of cross-validation is appealing particularly when we have to design a large neural network with good generalization as the goal. For example, we may use cross-validation to determine the multilayer perceptron with the best number of hidden neurons, and when it is best to stop training, as described in the next two subsections.

Model Selection

The idea of selecting a model in accordance with cross-validation follows a philosophy similar to that of structural risk minimization described in Chapter 2. Consider then a nested *structure* of Boolean function classes denoted by

$$\mathcal{F}_1 \subset \mathcal{F}_2 \subset \cdots \subset \mathcal{F}_n$$
$$\mathcal{F}_k = \{F_k\} \tag{4.91}$$
$$= \{F(\mathbf{x}, \mathbf{w}); \mathbf{w} \in \mathcal{W}_k\}, \qquad k = 1, 2, \ldots, n$$

In words, the kth function class $\mathcal{F}_k$ encompasses a family of multilayer perceptrons with similar architecture and weight vectors $\mathbf{w}$ drawn from a multidimensional weight space $\mathcal{W}_k$. A member of this class, characterized by the function or hypothesis $F_k = F(\mathbf{x}, \mathbf{w})$, $\mathbf{w} \in \mathcal{W}_k$, maps the input vector $\mathbf{x}$ into $\{0, 1\}$, where $\mathbf{x}$ is drawn from an input space $\mathcal{X}$ with some unknown probability P. Each multilayer perceptron in the structure described is trained with the back-propagation algorithm, which takes care of training the parameters of the multilayer perceptron. The model selection problem is essentially that of choosing the multilayer perceptron with the best value of W, the number of free parameters (i.e., synaptic weights and biases). More precisely, given that the scalar desired response for an input vector $\mathbf{x}$ is $d = \{0, 1\}$, we define the generalization error as

$$\epsilon_g(F) = P(F(\mathbf{x}) \neq d) \qquad \text{for } \mathbf{x} \in \mathcal{X}$$

We are given a training set of labeled examples

$$\mathcal{T} = \{(\mathbf{x}_i, d_i)\}_{i=1}^N$$

The objective is to select the particular hypothesis $F(\mathbf{x}, \mathbf{w})$, which minimizes the generalization error $\epsilon_g(F)$ that results when it is given inputs from the test set.

In what follows we assume that the structure described by Eq. (4.91) has the property that for any sample size N we can always find a multilayer perceptron with a large enough number of free parameters $W_{max}(N)$, such that the training data set $\mathcal{T}$ can be fitted adequately. This is merely restating the universal approximation theorem of Section 4.13. We refer to $W_{max}(N)$ as the *fitting number*. The significance of $W_{max}(N)$ is that a reasonable model selection procedure would choose a hypothesis $F(\mathbf{x}, \mathbf{w})$ that requires $W \leq W_{max}(N)$; otherwise the network complexity would be increased.

Let a parameter r, lying in the range between 0 and 1, determine the split of the training data set $\mathcal{T}$ between the estimation subset and validation subset. With $\mathcal{T}$ consisting of N examples, $(1 - r)N$ examples are allotted to the estimation subset and the remaining rN examples are allotted to the validation subset. The estimation subset, denoted by $\mathcal{T}'$, is used to train a nested sequence of multilayer perceptrons, resulting in the hypotheses $\mathcal{F}_1, \mathcal{F}_2, \ldots, \mathcal{F}_n$ of increasing complexity. With $\mathcal{T}'$ made up of $(1 - r)N$ examples, we consider values of W smaller than or equal to the corresponding fitting number $W_{max}((1 - r)N)$.

The use of cross-validation results in the choice

$$\mathscr{F}_{cv} = \min_{k=1, 2, ..., \nu} \{e_t''(\mathscr{F}_k)\} \tag{4.92}$$

where ν corresponds to $W_\nu \leq W_{max}((1-r)N)$, and $e_t''(\mathscr{F}_k)$ is the classification error produced by hypothesis $\mathscr{F}_k$ when it is tested on the validation subset $\mathscr{T}''$, consisting of rN examples.

The key issue is how to specify the parameter r that determines the split of the training set $\mathscr{T}$ between the estimation subset $\mathscr{T}'$ and validation subset $\mathscr{T}''$. In a study described in Kearns (1996) involving an analytic treatment of this issue using the VC dimension and supported with detailed computer simulations, several qualitative properties of the optimum r are identified:

- When the complexity of the target function, defining the desired response d in terms of the input vector $\mathbf{x}$, is small compared to the sample size N, the performance of cross-validation is relatively insensitive to the choice of r.
- As the target function becomes more complex relative to the sample size N, the choice of optimum r has a more pronounced effect on cross-validation performance, and its own value decreases.
- A single *fixed* value of r works *nearly* optimally for a wide range of target function complexity.

On the basis of the results reported in Kearns (1996), a fixed value of r equal to 0.2 appears to be a sensible choice, which means that 80 percent of the training set $\mathscr{T}$ is assigned to the estimation subset and the remaining 20 percent is assigned to the validation subset.

Earlier we spoke of a nested sequence of multilayer perceptrons of increasing complexity. For prescribed input and output layers, such a sequence can be created, for example, by having $\nu = p + q$ fully-connected multilayer perceptrons structured as follows:

- p multilayer perceptrons with a single hidden layer of increasing size $h_1' < h_2' < ... < h_p'$.
- q multilayer perceptrons with two hidden layers; the first hidden layer is of size h_p' and the second hidden layer is of increasing size $h_1'' < h_2'' < ... < h_q''$.

As we go from one multilayer perceptron to the next, there is a corresponding increase in the number of free parameters W. The model selection procedure based on cross-validation as described provides us with a principled approach to determine the number of hidden neurons in a multilayer perceptron. Although the procedure was described in the context of binary classification, it applies equally well to other applications of the multilayer perceptron.

Early Stopping Method of Training

Ordinarily, a multilayer perceptron trained with the back-propagation algorithm learns in stages, moving from the realization of fairly simple to more complex mapping functions as the training session progresses. This is exemplified by the fact that in a typical situation the mean-square error decreases with an increasing number of epochs during training: it starts off at a large value, decreases rapidly, and then continues to decrease slowly as the network makes its way to a local minimum on the error surface. With good generalization as the goal, it is very difficult to figure out when it is best to stop training if we were to look at the learning curve for training all by itself. In particular, in light of what was said in Section 4.12 on generalization, it is possible for the network to end up overfitting the training data if the training session is not stopped at the right point.

We may identify the onset of overfitting through the use of cross-validation, for which the training data are split into an estimation subset and a validation subset. The estimation subset of examples is used to train the network in the usual way, except for a minor modification: the training session is stopped periodically (i.e., every so many epochs), and the network is tested on the validation subset after each period of training. More specifically, the periodic estimation-followed-by-validation process proceeds as follows:

- After a period of estimation (training), the synaptic weights and bias levels of the multilayer perceptron are all fixed, and the network is operated in its forward mode. The validation error is thus measured for each example in the validation subset.
- When the validation phase is completed, the estimation (training) is resumed for another period, and the process is repeated.

This procedure is referred to as the *early stopping method of training*.[10]

Figure 4.20 shows conceptualized forms of two learning curves, one pertaining to measurements on the estimation subset and the other pertaining to the validation subset. Typically, the model does not do as well on the validation subset as it does on the estimation subset, on which its design was based. The *estimation learning curve* decreases monotonically for an increasing number of epochs in the usual manner. In contrast, the *validation learning curve* decreases monotonically to a minimum, it then starts to increase as the training continues. When we look at the estimation learning curve it may appear that we could do better by going beyond the minimum point on

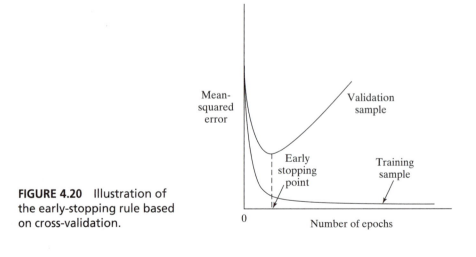

FIGURE 4.20 Illustration of the early-stopping rule based on cross-validation.

the validation learning curve. In reality, however, what the network is learning beyond this point is essentially noise contained in the training data. This heuristic suggests that the minimum point on the validation learning curve be used as a sensible criterion for stopping the training session.

What if the training data are noise free? How then would we justify early stopping for a deterministic scenario? Part of the answer in this case is that if both the estimation and validation errors cannot be simultaneously driven to zero, it implies that the network does not have the capacity to model the function exactly. The best that we can do in that situation is to try to minimize, for example, the integrated-squared error that is (roughly) equivalent to minimizing the usual global mean-square error with a uniform input density.

A statistical theory of the overfitting phenomenon presented in Amari et al. (1996) provides a word of caution for using the early stopping method of training. The theory is based on batch learning, and supported with detailed computer simulations involving a multilayer perceptron classifier with a single hidden layer. Two modes of behavior are identified depending on the size of the training set:

Nonasymptotic mode, for which $N < W$, where N is the size of the training set and W is the number of free parameters in the network. For this mode of behavior, the early stopping method of training does improve the generalization performance of the network over exhaustive training (i.e., when the complete set of examples is used for training and the training session is not stopped). This result suggests that overfitting may occur when $N < 30W$, and there is practical merit in the use of cross-validation to stop training. The optimum value of parameter r that determines the split of the training data between estimation and validation subsets is defined by

$$r_{\text{opt}} = 1 - \frac{\sqrt{2W - 1} - 1}{2(W - 1)}$$

For large W, this formula approximates to

$$r_{\text{opt}} \simeq 1 - \frac{1}{\sqrt{2W}}, \qquad \text{large } W \tag{4.93}$$

For example, for $W = 100$, $r_{\text{opt}} = 0.07$, which means that 93 percent of the training data are allotted to the estimation subset and the remaining 7 percent are allotted to the validation subset.

Asymptotic mode, for which $N > 30W$. For this mode of behavior, the improvement in generalization performance produced by the use of the early stopping method of training over exhaustive training is small. In other words, exhaustive learning is satisfactory when the size of the training sample is large compared to the number of network parameters.

Variants of Cross-Validation

The approach to cross-validation described is referred to as the *hold out method*. There are other variants of cross-validation that find their own uses in practice, particularly when there is a scarcity of labeled examples. In such a situation we may use *multifold*

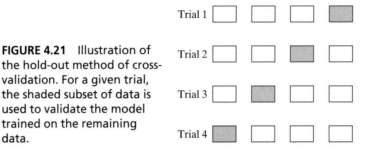

FIGURE 4.21 Illustration of the hold-out method of cross-validation. For a given trial, the shaded subset of data is used to validate the model trained on the remaining data.

cross-validation by dividing the available set of N examples into K subsets, $K > 1$; this assumes that K is divisible into N. The model is trained on all the subsets except for one, and the validation error is measured by testing it on the subset left out. This procedure is repeated for a total of K trials, each time using a different subset for validation, as illustrated in Fig. 4.21 for $K = 4$. The performance of the model is assessed by averaging the squared error under validation over all the trials of the experiment. There is a disadvantage to multifold cross-validation: it may require an excessive amount of computation since the model has to be trained K times, where $1 < K \leq N$.

When the available number of labeled examples, N, is severely limited, we may use the extreme form of multifold cross-validation known as the *leave-one-out method*. In this case, $N - 1$ examples are used to train the model, and the model is validated by testing it on the example left out. The experiment is repeated for a total of N times, each time leaving out a different example for validation. The squared error under validation is then averaged over the N trials of the experiment.

4.15 NETWORK PRUNING TECHNIQUES

To solve real-world problems with neural networks usually requires the use of highly structured networks of a rather large size. A practical issue that arises in this context is that of minimizing the size of the network while maintaining good performance. A neural network with minimum size is less likely to learn the idiosyncrasies or noise in the training data, and may thus generalize better to new data. We may achieve this design objective in one of two ways:

- *Network growing,* in which case we start with a small multilayer perceptron, small for accomplishing the task at hand, and then add a new neuron or a new layer of hidden neurons only when we are unable to meet the design specification.[11]
- *Network pruning,* in which case we start with a large multilayer perceptron with an adequate performance for the problem at hand, and then prune it by weakening or eliminating certain synaptic weights in a selective and orderly fashion.

In this section we focus on network pruning. In particular, we describe two approaches, one based on a form of "regularization," and the other based on the "deletion" of certain synaptic connections from the network.

Complexity-Regularization

In designing a multilayer perceptron by whatever method, we are in effect building a nonlinear *model* of the physical phenomenon responsible for the generation of the input–output examples used to train the network. Insofar as the network design is statistical in nature, we need an appropriate tradeoff between reliability of the training data and goodness of the model (i.e., a method for solving the bias-variance dilemma). In the context of back-propagation learning, or any other supervised learning procedure for that matter, we may realize this tradeoff by minimizing the total risk expressed as:

$$R(\mathbf{w}) = \mathcal{E}_s(\mathbf{W}) + \lambda \mathcal{E}_c(\mathbf{w}) \tag{4.94}$$

The first term, $\mathcal{E}_s(\mathbf{w})$, is the standard *performance measure,* which depends on both the network (model) and the input data. In back-propagation learning it is typically defined as a mean-square error whose evaluation extends over the output neurons of the network and which is carried out for all the training examples on an epoch-by-epoch basis. The second term, $\mathcal{E}_c(\mathbf{w})$, is the *complexity penalty,* which depends on the network (model) alone; its inclusion imposes on the solution prior knowledge that we may have on the models being considered. In fact, the form of the total risk defined in Eq. (4.94) is simply a statement of Tikhonov's *regularization theory;* this subject is detailed in Chapter 5. For the present discussion, it suffices to think of λ as a *regularization parameter,* which represents the relative importance of the complexity-penalty term with respect to the performance-measure term. When λ is zero, the back-propagation learning process is unconstrained, with the network being completely determined from the training examples. When λ is made infinitely large, on the other hand, the implication is that the constraint imposed by the complexity penalty is by itself sufficient to specify the network, which is another way of saying that the training examples are unreliable. In practical applications of the weight-decay procedure, the regularization parameter λ is assigned a value somewhere between these two limiting cases. The viewpoint described here for the use of complexity regularization for improved generalization is entirely consistent with the structural risk minimization procedure discussed in Chapter 2.

In a general setting, one choice of complexity-penalty term $\mathcal{E}_c(\mathbf{w})$ is the kth order smoothing integral

$$\mathcal{E}_c(\mathbf{w}, k) = \frac{1}{2} \int \left\| \frac{\partial^k}{\partial \mathbf{x}^k} F(\mathbf{x}, \mathbf{w}) \right\|^2 \mu(\mathbf{x}) d\mathbf{x} \tag{4.95}$$

where $F(\mathbf{x}, \mathbf{w})$ is the input–output mapping performed by the model, and $\mu(\mathbf{x})$ is some weighting function that determines the region of the input space over which the function $F(\mathbf{x}, \mathbf{w})$ is required to be smooth. The motivation is to make the kth derivative of $F(\mathbf{x}, \mathbf{w})$ with respect to the input vector $\mathbf{x}$ small. The larger we choose k, the smoother (i.e., less complex) the function $F(\mathbf{x}, \mathbf{w})$ will become.

In the sequel, we describe three different complexity regularizations (of increasing sophistication) for multilayer perceptrons.

Weight Decay.   In the *weight-decay procedure* (Hinton, 1989), the complexity penalty term is defined as the squared norm of the weight vector $\mathbf{w}$ (i.e., all the free parameters) in the network, as shown by

$$\mathcal{E}_c(\mathbf{w}) = \|\mathbf{w}\|^2$$

$$= \sum_{i \in \mathcal{C}_{\text{total}}} w_i^2 \tag{4.96}$$

where the set $\mathcal{C}_{\text{total}}$ refers to all the synaptic weights in the network. This procedure operates by forcing some of the synaptic weights in the network to take values close to zero, while permitting other weights to retain their relatively large values. Accordingly, the weights of the network are grouped roughly into two categories: those that have a large influence on the network (model), and those that have little or no influence on it. The weights in the latter category are referred to as *excess weights*. In the absence of complexity regularization, these weights result in poor generalization by virtue of their high likelihood of taking on completely arbitrary values or causing the network to overfit the data in order to produce a slight reduction in the training error (Hush and Horne, 1993). The use of complexity regularization encourages the excess weights to assume values close to zero, and thereby improve generalization.

In the weight-decay procedure, all the weights in the multilayer perceptron are treated equally. That is, the prior distribution in weight space is assumed to be centered at the origin. Strictly speaking, weight decay is not the correct form of complexity regularization for a multilayer perceptron since it does not fit into the rationale described in Eq. (4.95). Nevertheless, it is simple and appears to work well in some applications.

Weight Elimination.   In this second complexity-regularization procedure, the complexity penalty is defined by (Weigend et al., 1991)

$$\mathcal{E}_c(\mathbf{w}) = \sum_{i \in \mathcal{C}_{\text{total}}} \frac{(w_i/w_0)^2}{1 + (w_i/w_0)^2} \tag{4.97}$$

where w_0 is a preassigned parameter, and w_i refers to the weight of some synapse i in the network. The set $\mathcal{C}_{\text{total}}$ refers to all the synaptic connections in the network. An individual penalty term varies with w_i/w_0 in a symmetric fashion, as shown in Fig. 4.22. When $|w_i| \ll w_0$, the complexity penalty (cost) for that weight approaches zero. The implication of this condition is that insofar as learning from examples is concerned, the ith synaptic weight is unreliable and should therefore be eliminated from the network. On the other hand, when $|w_i| \gg w_0$, the complexity penalty (cost) for that weight approaches the maximum value of unity, which means that w_i is important to the back-propagation learning process. We thus see that the complexity penalty term of Eq. (4.97) does serve the desired purpose of identifying the synaptic weights of the network that are of significant influence. Note also that the weight-elimination procedure includes the weight-decay procedure as a special case; specifically, for large w_0, Eq. (4.97) reduces to the form shown in Eq. (4.96) except for a scaling factor.

Strictly speaking, the weight-elimination procedure is also not the correct form of complexity regularization for multilayer perceptrons because it does not fit the

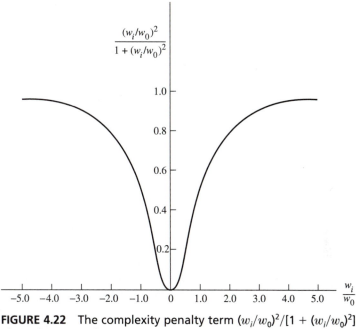

FIGURE 4.22 The complexity penalty term $(w_i/w_0)^2/[1 + (w_i/w_0)^2]$ plotted versus w_i/w_0.

description specified in Eq. (4.95). Nevertheless, with the proper choice of parameter w_0, it permits some weights in the network to assume values that are larger than with weight decay (Hush, 1997).

Approximate Smoother.   In Moody and Rögnvaldsson (1997), the following complexity-penalty term is proposed for a multilayer perceptron with a single hidden layer and a single neuron in the output layer:

$$\mathscr{E}_c(\mathbf{w}) = \sum_{j=1}^{M} w_{oj}^2 \|\mathbf{w}_j\|^p \tag{4.98}$$

where the w_{oj} are the weights in the output layer, and $\mathbf{w}_j$ is the weight vector for the jth neuron in the hidden layer; the power p is defined by

$$p = \begin{cases} 2k - 1 & \text{for a global smoother} \\ 2k & \text{for a local smoother} \end{cases} \tag{4.99}$$

where k is the order of differentiation of $F(\mathbf{x}, \mathbf{w})$ with respect to $\mathbf{x}$.

The approximate smoother appears to be more accurate than weight decay or weight elimination for the complexity regularization of a multilayer perceptron. Unlike those earlier methods, it does two things:

1. It distinguishes between the roles of synaptic weights in the hidden layer and those in the output layer.
2. It captures the interactions between these two sets of weights.

However, it has a much more complicated form than weight decay or weight elimina-tion, and is therefore more demanding in computational complexity.

Hessian-based Network Pruning

The basic idea of this second approach to network pruning is to use information on second-order derivatives of the error surface in order to make a trade-off between net-work complexity and training-error performance. In particular, a local model of the error surface is constructed for analytically predicting the effect of perturbations in synaptic weights. The starting point in the construction of such a model is the local approximation of the cost function $\mathscr{E}_{av}$ using a *Taylor series* about the operating point, described as follows:

$$\mathscr{E}_{av}(\mathbf{w} + \Delta\mathbf{w}) = \mathscr{E}_{av}(\mathbf{w}) + \mathbf{g}^T(\mathbf{w})\Delta\mathbf{w} + \frac{1}{2}\Delta\mathbf{w}^T\mathbf{H}\Delta\mathbf{w} + O(\|\Delta\mathbf{w}\|^3) \qquad (4.100)$$

where $\Delta\mathbf{w}$ is a perturbation applied to the operating point $\mathbf{w}$, and $\mathbf{g}(\mathbf{w})$ is the gradient vector evaluated at $\mathbf{w}$. The Hessian is also evaluated at the point $\mathbf{w}$, and therefore, to be correct we should denote it by $\mathbf{H}(\mathbf{w})$. We have not done so in Eq. (4.100) merely to sim-plify the notation.

The requirement is to identify a set of parameters whose deletion from the multi-layer perceptron will cause the least increase in the value of the cost function $\mathscr{E}_{av}$. To solve this problem in practical terms, we make the following approximations:

1. *Extremal Approximation.* We assume that parameters are deleted from the network only after the training process has converged (i.e., the network is fully trained). The implication of this assumption is that the parameters have a set of values corresponding to a local minimum or global minimum of the error surface. In such a case, the gradient vector $\mathbf{g}$ may be set equal to zero and the term $\mathbf{g}^T\Delta\mathbf{w}$ on the right-hand side of Eq. (4.100) may therefore be ignored. Otherwise the saliency measures (defined later) will be invalid for the problem at hand.

2. *Quadratic Approximation.* We assume that the error surface around a local minimum or global minimum is nearly "quadratic." Hence the higher-order terms in Eq. (4.100) may also be neglected.

Under these two assumptions, Eq. (4.100) is approximated simply as

$$\Delta\mathscr{E}_{av} = \mathscr{E}(\mathbf{w} + \Delta\mathbf{w}) - \mathscr{E}(\mathbf{w})$$

$$\simeq \frac{1}{2}\Delta\mathbf{w}^T\mathbf{H}\Delta\mathbf{w} \qquad (4.101)$$

The *optimal brain damage* (OBD) *procedure* (LeCun et al., 1990b) simplifies the computations by making a further assumption: The Hessian matrix $\mathbf{H}$ is a diagonal matrix. However, no such assumption is made in the *optimal brain surgeon* (*OBS*) *pro-cedure* (Hassibi et al., 1992); accordingly, it contains the OBD procedure as a special case. From here on, we follow the OBS strategy.

The goal of OBS is to set one of the synaptic weights to zero to minimize the incremental increase in $\mathscr{E}_{av}$ given in Eq. (4.101). Let $w_i(n)$ denote this particular synaptic weight. The elimination of this weight is equivalent to the condition

$$\Delta w_i + w_i = 0$$

or

$$\mathbf{1}_i^T \Delta \mathbf{w} + w_i = 0 \tag{4.102}$$

where $\mathbf{1}_i$ is the *unit vector* whose elements are all zero, except for the ith element, which is equal to unity. We may now restate the goal of OBS as (Hassibi et al., 1992):

Minimize the quadratic form $\frac{1}{2}\Delta \mathbf{w}^T \mathbf{H} \Delta \mathbf{w}$ with respect to the incremental change in the weight vector, $\Delta \mathbf{w}$, subject to the constraint that $\mathbf{1}_i^T \Delta \mathbf{w} + w_i$ is zero, and then minimize the result with respect to the index i.

There are two levels of minimization going on here. One minimization is over the synaptic weight vectors that remain after the ith weight vector is set equal to zero. The second minimization is over which particular vector is pruned.

To solve this constrained optimization problem, we first construct the *Lagrangian*

$$S = \frac{1}{2}\Delta \mathbf{w}^T \mathbf{H} \Delta \mathbf{w} - \lambda(\mathbf{1}_i^T \Delta \mathbf{w} + w_i) \tag{4.103}$$

where λ is the *Lagrange multiplier*. Then, taking the derivative of the Lagrangian S with respect to $\Delta \mathbf{w}$, applying the constraint of Eq. (4.102), and using matrix inversion, we find that the optimum change in the weight vector $\mathbf{w}$ is

$$\Delta \mathbf{w} = -\frac{w_i}{[\mathbf{H}^{-1}]_{i,i}} \mathbf{H}^{-1} \mathbf{1}_i \tag{4.104}$$

and the corresponding optimum value of the Lagrangian S for element w_i is

$$S_i = \frac{w_i^2}{2[\mathbf{H}^{-1}]_{i,i}} \tag{4.105}$$

where $\mathbf{H}^{-1}$ is the inverse of the Hessian matrix $\mathbf{H}$, and $[\mathbf{H}^{-1}]_{i,i}$ is the i i-th element of this inverse matrix. The Lagrangian S_i optimized with respect to $\Delta \mathbf{w}$, subject to the constraint that the ith synaptic weight w_i be eliminated, is called the *saliency* of w_i. In effect, the saliency S_i represents the increase in the mean-square error (performance measure) that results from the deletion of w_i. Note that the saliency S_i is proportional to w_i^2. Thus small weights have a small effect on the mean-square error. However, from Eq. (4.105) we see that the saliency S_i is also inversely proportional to the diagonal elements of the inverse Hessian. Thus if $[\mathbf{H}^{-1}]_{i,i}$ is small, then even small weights may have a substantial effect on the mean-square error.

In the OBS procedure, the weight corresponding to the smallest saliency is the one selected for deletion. Moreover, the corresponding optimal changes in the remainder of the weights are given in Eq. (4.104), which show that they should be updated along the direction of the ith column of the inverse of the Hessian.

In their paper, Hassibi et al. report that on some benchmark problems the OBS procedure resulted in smaller networks than those obtained using the weight-decay procedure. It is also reported that as a result of applying the OBS procedure to the NETtalk multilayer perceptron involving a single hidden layer and 18,000 weights , the network was pruned to a mere 1560 weights, a dramatic reduction in the size of the network. NETtalk, due to Sejnowski and Rosenberg (1987), is described in Chapter 13.

Computing the inverse Hessian matrix. The inverse Hessian matrix $\mathbf{H}^{-1}$ is fundamental to the formulation of the OBS procedure. When the number of free parameters, W, in the network is large, the problem of computing $\mathbf{H}^{-1}$ may be intractable. In what follows we describe a manageable procedure for computing $\mathbf{H}^{-1}$, assuming that the multilayer perceptron is fully trained to a local minimum on the error surface (Hassibi et al., 1992).

To simplify the presentation, suppose that the multilayer perceptron has a single output neuron. Then, for a given training set we may express the cost function as

$$\mathcal{E}_{av}(\mathbf{w}) = \frac{1}{2N} \sum_{n=1}^{N} (d(n) - o(n))^2$$

where $o(n)$ is the actual output of the network on the presentation of the nth example, $d(n)$ is the corresponding desired response, and N is the total number of examples in the training set. The output $o(n)$ may itself be expressed as

$$o(n) = F(\mathbf{w}, \mathbf{x})$$

where F is the input–output mapping function realized by the multilayer perceptron, $\mathbf{x}$ is the input vector, and $\mathbf{w}$ is the synaptic weight vector of the network. The first derivative of $\mathcal{E}_{av}$ with respect to $\mathbf{w}$ is therefore

$$\frac{\partial \mathcal{E}_{av}}{\partial \mathbf{w}} = -\frac{1}{N} \sum_{n=1}^{N} \frac{\partial F(\mathbf{w}, \mathbf{x}(n))}{\partial \mathbf{w}} (d(n) - o(n)) \tag{4.106}$$

and the second derivative of $\mathcal{E}_{av}$ with respect to $\mathbf{w}$ or the Hessian matrix is

$$\begin{aligned}
\mathbf{H}(N) &= \frac{\partial^2 \mathcal{E}_{av}}{\partial \mathbf{w}^2} \\
&= \frac{1}{N} \sum_{n=1}^{N} \left\{ \left(\frac{\partial F(\mathbf{w}, \mathbf{x}(n))}{\partial \mathbf{w}} \right) \left(\frac{\partial F(\mathbf{w}, \mathbf{x}(n))}{\partial \mathbf{w}} \right)^T \right. \\
&\quad \left. - \frac{\partial^2 F(\mathbf{w}, \mathbf{x}(n))}{\partial \mathbf{w}^2} (d(n) - o(n)) \right\}
\end{aligned} \tag{4.107}$$

where we have emphasized the dependence of the Hessian matrix on the size of the training sample, N.

Under the assumption that the network is fully trained, that is, the cost function $\mathcal{E}_{av}$ has been adjusted to a local minimum on the error surface, it is reasonable to say that $o(n)$ is close to $d(n)$. Under this condition we may ignore the second term and approximate Eq. (4.107) as

$$\mathbf{H}(N) \simeq \frac{1}{N} \sum_{n=1}^{N} \left(\frac{\partial F(\mathbf{w}, \mathbf{x}(n))}{\partial \mathbf{w}} \right) \left(\frac{\partial F(\mathbf{w}, \mathbf{x}(n))}{\partial \mathbf{w}} \right)^T \tag{4.108}$$

To simplify the notation, define the W-by-1 vector

$$\boldsymbol{\xi}(n) = \frac{1}{\sqrt{N}} \frac{\partial F(\mathbf{w}, \mathbf{x}(n))}{\partial \mathbf{w}} \tag{4.109}$$

which may be computed using the procedure described in Section 4.10. We may then rewrite Eq. (4.108) in the form of a recursion as:

$$\mathbf{H}(n) = \sum_{k=1}^{n} \boldsymbol{\xi}(k)\boldsymbol{\xi}^T(k)$$

$$= \mathbf{H}(n - 1) + \boldsymbol{\xi}(n)\boldsymbol{\xi}^T(n), \qquad n = 1, 2, ..., N \tag{4.110}$$

This recursion is in the right form for application of the so-called *matrix inversion lemma*, also known as *Woodbury's equality.*

Let $\mathbf{A}$ and $\mathbf{B}$ denote two positive definite matrices related by

$$\mathbf{A} = \mathbf{B}^{-1} + \mathbf{C}\mathbf{D}\mathbf{C}^T$$

where $\mathbf{C}$ and $\mathbf{D}$ are two other matrices. According to the matrix inversion lemma, the inverse of matrix $\mathbf{A}$ is defined by

$$\mathbf{A}^{-1} = \mathbf{B} - \mathbf{B}\mathbf{C}(\mathbf{D} + \mathbf{C}^T\mathbf{B}\mathbf{C})^{-1}\mathbf{C}^T\mathbf{B}$$

For the problem described in Eq. (4.110) we have

$$\mathbf{A} = \mathbf{H}(n)$$

$$\mathbf{B}^{-1} = \mathbf{H}(n - 1)$$

$$\mathbf{C} = \boldsymbol{\xi}(n)$$

$$\mathbf{D} = 1$$

Application of the matrix inversion lemma therefore yields the desired formula for recursive computation of the inverse Hessian:

$$\mathbf{H}^{-1}(n) = \mathbf{H}^{-1}(n - 1) - \frac{\mathbf{H}^{-1}(n - 1)\boldsymbol{\xi}(n)\boldsymbol{\xi}^T(n)\mathbf{H}^{-1}(n - 1)}{1 + \boldsymbol{\xi}^T(n)\mathbf{H}^{-1}(n - 1)\boldsymbol{\xi}(n)} \tag{4.111}$$

Note that the denominator in Eq. (4.111) is a scalar; it is therefore straightforward to calculate its reciprocal. Thus, given the past value of the inverse Hessian, $\mathbf{H}^{-1}(n - 1)$, we may compute its updated value $\mathbf{H}^{-1}(n)$ on the presentation of the nth example represented by the vector $\boldsymbol{\xi}(n)$. This recursive computation is continued until the entire set of N examples has been accounted for. To initialize the algorithm we need to make $\mathbf{H}^{-1}(0)$ large, since it is being constantly reduced according to Eq. (4.111). This requirement is satisfied by setting

$$\mathbf{H}^{-1}(0) = \delta^{-1}\mathbf{I} \tag{4.112}$$

where δ is a small positive number and $\mathbf{I}$ is the identity matrix. This form of initialization assures that $\mathbf{H}^{-1}(n)$ is always positive definite. The effect of δ becomes progressively smaller as more and more examples are presented to the network.

A summary of the brain surgeon algorithm is presented in Table 4.6 (Hassibi and Stork, 1992).

TABLE 4.6 Summary of the Optimal Brain Surgeon Algorithm

---

1. Train the given multilayer perceptron to minimum mean-square error.
2. Use the procedure described in Section 4.10 to compute the vector

$$\xi(n) = \frac{1}{\sqrt{N}} \frac{\partial F(\mathbf{w}, \mathbf{x}(n))}{\partial \mathbf{w}}$$

where $F(\mathbf{w}, \mathbf{x}(n))$ is the input–output mapping realized by the multilayer perceptron with an overall weight vector $\mathbf{w}$, and $\mathbf{x}(n)$ is the input vector.
3. Use the recursion (4.111) to compute the inverse Hessian $\mathbf{H}^{-1}$.
4. Find the i that corresponds to the smallest saliency:

$$S_i = \frac{w_i^2}{2[\mathbf{H}^{-1}]i,i}$$

where $[\mathbf{H}^{-1}]_{i,i}$ is the (i, i)th element of $\mathbf{H}^{-1}$. If the saliency S_i is much smaller than the mean-square $\mathcal{E}_{av}$, then delete synaptic weight w_i, and proceed to step 4. Otherwise, go to step 5.
5. Update all the synaptic weights in the network by applying the adjustment:

$$\Delta \mathbf{w} = -\frac{w_i}{[\mathbf{H}^{-1}]_{i,i}} \mathbf{H}^{-1} \mathbf{1}_i$$

Go to step 2.
6. Stop the computation when no more weights can be deleted from the network without a large increase in the mean-square error. (It may be desirable to retrain the network at this point).

4.16 VIRTUES AND LIMITATIONS OF BACK-PROPAGATION LEARNING

The back-propagation algorithm has emerged as the most popular algorithm for the supervised training of multilayer perceptrons. Basically, it is a gradient (derivative) technique and *not* an optimization technique. Back-propagation has two distinct properties:

- It is *simple* to compute locally.
- It performs *stochastic* gradient descent in weight space (for pattern-by-pattern updating of synaptic weights).

These two properties of back-propagation learning in the context of a multilayer perceptron are responsible for its advantages and disadvantages.

Connectionism

The back-propagation algorithm is an example of a *connectionist paradigm* that relies on local computations to discover the information-processing capabilities of neural networks. This form of computational restriction is referred to as the *locality constraint,* in the sense that the computation performed by the neuron is influenced solely by those neurons that are in physical contact with it. The use of local computations in the design of artificial neural networks is usually advocated for three principal reasons:

1. Artificial neural networks that perform local computations are often held up as metaphors for biological neural networks.

2. The use of local computations permits a graceful degradation in performance due to hardware errors, and therefore provides the basis for a fault-tolerant network design.
3. Local computations favor the use of parallel architectures as an efficient method for the implementation of artificial neural networks.

Taking these three points in reverse order, point 3 is entirely justified in the case of back-propagation learning. In particular, the back-propagation algorithm has been implemented successfully on parallel computers by many investigators, and VLSI architectures have been developed for the hardware realization of multilayer perceptrons (Hammerstrom, 1992a, 1992b). Point 2 is justified so long as certain precautions are taken in the application of the back-propagation algorithm, as described in Kerlirzin and Vallet (1993). As for point 1, relating to the biological plausibility of back-propagation learning, it has been seriously questioned on the following grounds (Shepherd, 1990b; Crick, 1989; Stork, 1989):

1. The reciprocal synaptic connections between the neurons of a multilayer perceptron may assume weights that are excitatory or inhibitory. In the real nervous system, however, neurons usually appear to be the one or the other. This is one of the most serious of the unrealistic assumptions made in neural network models.
2. In a multilayer perceptron, hormonal and other types of global communications are ignored. In real nervous systems, these types of global communication are critical for state-setting functions such as arousal, attention, and learning.
3. In back-propagation learning, a synaptic weight is modified by a presynaptic activity and an error (learning) signal independent of postsynaptic activity. There is evidence from neurobiology to suggest otherwise.
4. In a neurobiological sense, the implementation of back-propagation learning requires the rapid transmission of information backward along an axon. It appears highly unlikely that such an operation actually takes place in the brain.
5. Back-propagation learning implies the existence of a "teacher," which in the context of the brain would presumably be another set of neurons with novel properties. The existence of such neurons is biologically implausible.

However, these neurobiological misgivings do not belittle the engineering importance of back-propagation learning as a tool for information processing, as evidenced by its successful application in numerous highly diverse fields, including the simulation of neurobiological phenomena (see, for example, Robinson (1992)).

Feature Detection

As discussed in Section 4.9, the hidden neurons of a multilayer perceptron trained with the back-propagation algorithm play a critical role as feature detectors. A novel way in which this important property of the multilayer perceptron can be exploited is in its use as a *replicator* or *identity map* (Rumelhart et al., 1986b; Cottrel et al., 1987). Figure 4.23 illustrates how this can be accomplished for the case of a multilayer perceptron using a single hidden layer. The network layout satisfies the following structural requirements, as illustrated in Fig. 4.23a:

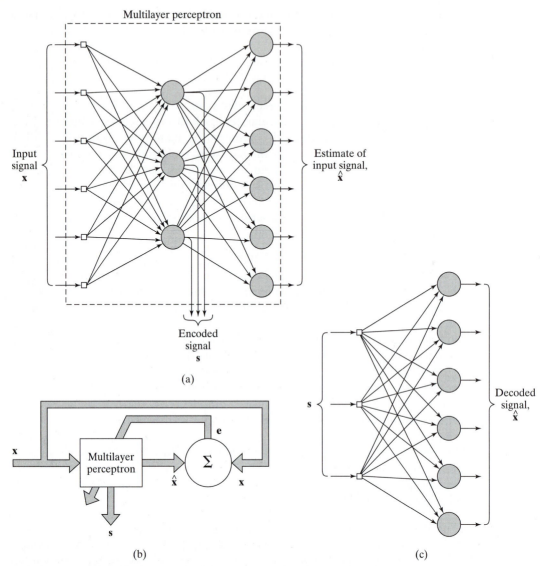

FIGURE 4.23 (a) Replicator network (identity map) with a single hidden layer used as an encoder. (b) Block diagram for the supervised training of the replicator network. (c) Part of the replicator network used as a decoder.

- The input and output layers have the same size, m.
- The size of the hidden layer, M, is smaller than m.
- The network is fully connected.

A given pattern, $\mathbf{x}$, is simultaneously applied to the input layer as the stimulus and to the output layer as the desired response. The actual response of the output layer, $\hat{\mathbf{x}}$, is intended to be an "estimate" of $\mathbf{x}$. The network is trained using the back-propagation algorithm in the usual way, with the estimation error vector $(\mathbf{x} - \hat{\mathbf{x}})$ treated as the error signal, as illustrated in Fig. 4.23b. The training is performed in an *unsupervised* manner (i.e., without the need for a teacher). By virtue of the special structure built into the design of the multilayer perceptron, the network is *constrained* to perform identity mapping through its hidden layer. An *encoded* version of the input pattern, denoted by **s**, is produced at the output of the hidden layer, as depicted in Fig. 4.23a. In effect, the fully trained multilayer perceptron performs the role of an "encoder." To reconstruct an estimate $\hat{\mathbf{x}}$ of the original input pattern **x** (i.e., to perform *decoding*), we apply the encoded signal to the hidden layer of the replicator network, as illustrated in Fig. 4.23c. In effect, this latter network performs the role of a "decoder." The smaller we make the size M of the hidden layer compared to the size m of the input/output layer, the more effective the configuration of Fig. 4.23a will be as a *data compression system.*[12]

Function Approximation

A multilayer perceptron trained with the back-propagation algorithm manifests itself as a *nested sigmoidal scheme*, written in the following compact form for the case of a single output:

$$F(\mathbf{x},\mathbf{w}) = \varphi\left(\sum_k w_{ok} \varphi\left(\sum_j w_{kj} \varphi\left(\cdots \varphi\left(\sum_i w_{li} x_i \right) \right) \right) \right) \qquad (4.113)$$

where $\varphi(\cdot)$ is a common sigmoid activation function, w_{ok} is the synaptic weight from neuron k in the last hidden layer to the single output neuron o, and so on for the other synaptic weights, and x_i is the ith element of the input vector **x**. The weight vector **w** denotes the entire set of synaptic weights ordered by layer, then neurons in a layer, and then synapses in a neuron. The scheme of nested nonlinear functions described in Eq. (4.113) is unusual in classical approximation theory. It is a *universal approximator* as discussed in Section 4.13.

In the context of approximation, the use of back-propagation learning offers another useful property. Intuition suggests that a multilayer perceptron with smooth activation functions should have output function derivatives that can also approximate the derivatives of an unknown input–output mapping. A proof of this result is presented in Hornik et al. (1990). In fact, it is shown that multilayer perceptrons can approximate functions that are not differentiable in the classical sense, but possess a generalized derivative as in the case of piecewise differentiable functions. The approximation results reported by Hornik et al. provide a previously missing theoretical justification for the use of multilayer perceptrons in applications that require the approximation of a function and its derivatives.

Computational Efficiency

The *computational complexity* of an algorithm is usually measured in terms of the number of multiplications, additions, and storage involved in its implementation, as discussed in Chapter 2. A learning algorithm is said to be *computationally efficient* when its computational complexity is *polynomial* in the number of adjustable parameters that are to be updated from one iteration to the next. On this basis it can be said that the back-propagation algorithm is computationally efficient. Specifically, in using it to train a multilayer perceptron containing a total of W synaptic weights (including biases), its computational complexity is linear in W. This important property of the back-propagation algorithm can be readily verified by examining the computations involved in performing the forward and backward passes summarized in Section 4.5. In the forward pass, the only computations involving the synaptic weights are those that pertain to the induced local fields of the various neurons in the network. Here we see from Eq. (4.44) that these computations are all linear in the synaptic weights of the network. In the backward pass, the only computations involving the synaptic weights are those that pertain to (1) the local gradients of the hidden neurons, and (2) the updating of the synaptic weights themselves, as shown in Eqs. (4.46) and (4.47), respectively. Here we also see that these computations are all linear in the synaptic weights of the network. The conclusion is therefore that the computational complexity of the back-propagation algorithm is linear in W, that is, it is $O(W)$.

Sensitivity Analysis

Another computational benefit gained from the use of back-propagation learning is the efficient manner in which we can carry out a sensitivity analysis of the input–output mapping realized by the algorithm. The *sensitivity* of an input–output mapping function F with respect to a parameter of the function, denoted by ω, is defined by

$$S_\omega^F = \frac{\partial F/F}{\partial \omega/\omega} \tag{4.114}$$

Consider then a multilayer perceptron trained with the back-propagation algorithm. Let the function $F(\mathbf{w})$ be the input–output mapping realized by this network; $\mathbf{w}$ denotes the vector of all synaptic weights (including biases) contained in the network. In Section 4.10 we showed that the partial derivatives of the function $F(\mathbf{w})$ with respect to all the elements of the weight vector $\mathbf{w}$ can be computed efficiently. In particular, examining Eqs. (4.81) to (4.83) together with Eq. (4.114), we see that the complexity involved in computing each of these partial derivatives is linear in W, the total number of weights contained in the network. This linearity holds irrespective of where the synaptic weight in question appears in the chain of computations.

Robustness

In Chapter 3 we pointed out that the LMS algorithm is robust in the sense that disturbances with small energy can only give rise to small estimation errors. If the underlying observation model is linear, the LMS algorithm is an H^∞-optimal filter (Hassibi et al., 1993, 1996). What this means is that the LMS algorithm minimizes the *maximum energy gain* from the disturbances to the estimation errors.

If, on the other hand, the underlying observation model is nonlinear, Hassibi and Kailath. (1995) have shown that the back-propagation algorithm is a *locally* H^∞-optimal filter. The term "local" used here means that the initial value of the weight vector used in the back-propagation algorithm is sufficiently close to the optimum value $\mathbf{w}^*$ of the weight vector to ensure that the algorithm does not get trapped in a poor local minimum. In conceptual terms, it is satisfying to see that the LMS and back-propagation algorithms belong to the same class of H^∞-optimal filters.

Convergence

The back-propagation algorithm uses an "instantaneous estimate" for the gradient of the error surface in weight space. The algorithm is therefore *stochastic* in nature; that is, it has a tendency to zigzag its way about the true direction to a minimum on the error surface. Indeed, back-propagation learning is an application of a statistical method known as *stochastic approximation* that was originally proposed by Robbins and Monro (1951). Consequently, it tends to converge slowly. We may identify two fundamental causes for this property (Jacobs, 1988):

1. The error surface is fairly flat along a weight dimension, which means that the derivative of the error surface with respect to that weight is small in magnitude. In such a situation, the adjustment applied to the weight is small, and consequently many iterations of the algorithm may be required to produce a significant reduction in the error performance of the network. Alternatively, the error surface is highly curved along a weight dimension, in which case the derivative of the error surface with respect to that weight is large in magnitude. In this second situation, the adjustment applied to the weight is large, which may cause the algorithm to overshoot the minimum of the error surface.

2. The direction of the negative gradient vector (i.e., the negative derivative of the cost function with respect to the vector of weights) may point away from the minimum of the error surface: hence the adjustments applied to the weights may induce the algorithm to move in the wrong direction.

Consequently, the rate of convergence in back-propagation learning tends to be relatively slow, which in turn may make it computationally excruciating. According to the empirical study of Saarinen et al. (1992), the local convergence rates of the back-propagation algorithm are *linear,* which is justified on the grounds that the Jacobian matrix is almost rank deficient, and so is the Hessian matrix. These are consequences of the intrinsically ill-conditioned nature of neural-network training problems. Saarinen et al. interpret the linear local convergence rates of back-propagation learning in one of two ways:

- It is vindication of back-propagation (gradient descent) in the sense that higher-order methods may not converge much faster while requiring more computational effort; or
- Large-scale neural-network training problems are so inherently difficult to perform that no supervised learning strategy is feasible, and other approaches such as the use of preprocessing may be necessary.

We explore the issue of convergence more fully in Section 4.17, and explore the issue of preprocessing the input in Chapter 8.

Local Minima

Another peculiarity of the error surface that impacts the performance of the back-propagation algorithm is the presence of *local minima* (i.e., isolated valleys) in addition to global minima. Since back-propagation learning is basically a hill climbing technique, it runs the risk of being trapped in a local minimum where every small change in synaptic weights increases the cost function. But somewhere else in the weight space there exists another set of synaptic weights for which the cost function is smaller than the local minimum in which the network is stuck. It is clearly undesirable to have the learning process terminate at a local minimum, especially if it is located far above a global minimum.

The issue of local minima in back-propagation learning has been raised in the epilogue of the enlarged edition on the classic book by Minsky and Papert (1988), where most of the attention is focused on a discussion of the two-volume book, *Parallel Distributed Processing,* by Rumelhart and McClelland (1986). In Chapter 8 of the latter book it is claimed that getting trapped in a local minimum is rarely a practical problem for back-propagation learning. Minsky and Papert counter by pointing out that the entire history of pattern recognition shows otherwise. Gori and Tesi (1992) describe a simple example where, although a nonlinearly separable set of patterns could be learned by the chosen network with a single hidden layer, back-propagation learning can get stuck in a local minimum.[13]

Scaling

In principle, neural networks such as multilayer perceptrons trained with the back-propagation algorithm offer the potential of universal computing machines. However, for that potential to be fully realized, we have to overcome the *scaling problem,* which addresses the issue of how well the network behaves (e.g., as measured by the time required for training or the best generalization performance attainable) as the computational task increases in size and complexity. Among the many possible ways of measuring the size or complexity of a computational task, the predicate order defined by Minsky and Papert (1969, 1988) provides the most useful and important measure.

To explain what we mean by a predicate, let $\psi(X)$ denote a function that can have only two values. Ordinarily we think of the two values of $\psi(X)$ as 0 and 1. But by taking the values to be FALSE or TRUE, we may think of $\psi(X)$ as a *predicate,* that is, a variable statement whose falsity or truth depends on the choice of argument X. For example, we may write

$$\psi_{\text{CIRCLE}}(X) = \begin{cases} 1 & \text{if the figure } X \text{ is a circle} \\ 0 & \text{if the figure } X \text{ is not a circle} \end{cases} \tag{4.115}$$

Using the idea of a predicate, Tesauro and Janssens (1988) performed an empirical study involving the use of a multilayer perceptron trained with the back-propagation algorithm to learn to compute the parity function. The *parity function* is a Boolean predicate defined by

$$\psi_{\text{PARITY}}(X) = \begin{cases} 1 & \text{if } |X| \text{ is an odd number} \\ 0 & \text{otherwise} \end{cases} \tag{4.116}$$

and whose order is equal to the number of inputs. The experiments performed by Tesauro and Janssens appear to show that the time required for the network to learn to compute the parity function scales exponentially with the number of inputs (i.e., the predicate order of the computation), and that projections of the use of the back-propagation algorithm to learn arbitrarily complicated functions may be overly optimistic.

It is generally agreed that it is inadvisable for a multilayer perceptron to be fully connected. In this context, we may therefore raise the following question: Given that a multilayer perceptron should not be fully connected, how should the synaptic connections of the network be allocated? This question is of no major concern in the case of small-scale applications, but it is certainly crucial to the successful application of back-propagation learning for solving large-scale, real-world problems.

One effective method of alleviating the scaling problem is to develop insight into the problem at hand (possibly through neurobiological analogy) and use it to put ingenuity into the architectural design of the multilayer perceptron. Specifically, the network architecture and the constraints imposed on synaptic weights of the network should be designed so as to incorporate prior information about the task into the makeup of the network. This design strategy is illustrated in Section 4.19 for the optical character recognition problem.

4.17 ACCELERATED CONVERGENCE OF BACK-PROPAGATION LEARNING

In the previous section we identified the main causes for the possible slow rate of convergence of the back-propagation algorithm. In this section we describe some *heuristics* that provide useful guidelines for thinking about how to accelerate the convergence of back-propagation learning through learning rate adaptation. Details of the heuristics are as follows (Jacobs, 1988):

HEURISTIC 1. Every adjustable network parameter of the cost function should have its own individual learning-rate parameter.

Here we note that the back-propagation algorithm may be slow to converge because the use of a fixed learning-rate parameter may not suit all portions of the error surface. In other words, a learning-rate parameter appropriate for the adjustment of one synaptic weight is not necessarily appropriate for the adjustment of other synaptic weights in the network. Heuristic 1 recognizes this fact by assigning a different learning-rate parameter to each adjustable synaptic weight (parameter) in the network.

HEURISTIC 2. Every learning-rate parameter should be allowed to vary from one iteration to the next.

The error surface typically behaves differently along different regions of a single weight dimension. In order to match this variation, heuristic 2 states that the learning-rate parameter needs to vary from iteration to iteration. It is interesting that this heuristic is well founded in the case of linear units (Luo, 1991).

HEURISTIC 3. When the derivative of the cost function with respect to a synaptic weight has the same algebraic sign for several consecutive iterations of the algorithm, the learning-rate parameter for that particular weight should be increased.

The current operating point in weight space may lie on a relatively flat portion of the error surface along a particular weight dimension. This may in turn account for the derivative of the cost function (i.e., the gradient of the error surface) with respect to that weight maintaining the same algebraic sign, and therefore pointing in the same direction, for several consecutive iterations of the algorithms. Heuristic 3 states that in such a situation the number of iterations required to move across the flat portion of the error surface may be reduced by appropriately increasing the learning-rate parameter.

HEURISTIC 4. When the algebraic sign of the derivative of the cost function with respect to a particular synaptic weight alternates for several consecutive interations of the algorithm, the learning-rate parameter for that weight should be decreased.

When the current operating point in weight space lies on a portion of the error surface along a weight dimension of interest that exhibits peaks and valleys (i.e., the surface is highly curved), then it is possible for the derivative of the cost function with respect to that weight to change its algebraic sign from one iteration to the next. In order to prevent the weight adjustment from oscillating, heuristic 4 states that the learning-rate parameter for that particular weight should be decreased appropriately.

It is noteworthy that the use of a different and time-varying learning-rate parameter for each synaptic weight in accordance with these heuristics modifies the back-propagation algorithm in a fundamental way. Specifically, the modified algorithm no longer performs a steepest-descent search. Rather, the adjustments applied to the synaptic weights are based on (1) the partial derivatives of the error surface with respect to the weights, and (2) estimates of the curvatures of the error surface at the current operating point in weight space along the various weight dimensions.

Furthermore, all four heuristics satisfy the locality constraint, which is an inherent characteristic of back-propagation learning. Unfortunately, adherence to the locality constraint limits the domain of usefulness of these heuristics because error surfaces exist for which they do not work. Nevertheless, modifications of the back-propagation algorithm in accordance with these heuristics do have practical value.[14]

4.18   SUPERVISED LEARNING VIEWED AS AN OPTIMIZATION PROBLEM

In this section we take a viewpoint on supervised learning that is quite different from that pursued in previous sections of the chapter. Specifically, we view the supervised training of a multilayer perceptron as a problem in *numerical optimization*. In this context we first point out that the error surface of a multilayer perceptron with supervised learning is a highly nonlinear function of the synaptic weight vector $\mathbf{w}$. Let $\mathscr{E}_{av}(\mathbf{w})$ denote the cost function, averaged over the training sample. Using the Taylor series we may expand $\mathscr{E}_{av}(\mathbf{w})$ about the current point on the error surface $\mathbf{w}(n)$ for example, as described in Eq. (4.100), reproduced here with dependences on n included:

$$\mathscr{E}_{av}(\mathbf{w}(n) + \Delta\mathbf{w}(n)) = \mathscr{E}_{av}(\mathbf{w}(n)) + \mathbf{g}^T(n)\Delta\mathbf{w}(n) + \frac{1}{2}\Delta\mathbf{w}^T(n)\mathbf{H}(n)\Delta\mathbf{w}(n)$$

$$+ \text{(third- and higher-order terms)}$$

(4.117)

where $\mathbf{g}(n)$ is the local gradient vector defined by

$$\mathbf{g}(n) = \frac{\partial\mathscr{E}_{av}(\mathbf{w})}{\partial\mathbf{w}}\bigg|_{\mathbf{w}=\mathbf{w}(n)}$$

(4.118)

and $\mathbf{H}(n)$ is the local Hessian matrix defined by

$$\mathbf{H}(n) = \left.\frac{\partial^2 \mathscr{E}_{av}(\mathbf{w})}{\partial \mathbf{w}^2}\right|_{\mathbf{w}=\mathbf{w}(n)} \qquad (4.119)$$

The use of an ensemble-averaged cost function $\mathscr{E}_{av}(\mathbf{w})$ presumes a batch mode of learning.

In the steepest-descent method, exemplified by the back-propagation algorithm, the adjustment $\Delta\mathbf{w}(n)$ applied to the synaptic weight vector $\mathbf{w}(n)$ is defined by

$$\Delta\mathbf{w}(n) = -\eta\mathbf{g}(n) \qquad (4.120)$$

where η is the learning-rate parameter. In effect, the steepest-descent method operates on the basis of a *linear approximation* of the cost function in the local neighborhood of the operating point $\mathbf{w}(n)$. In so doing, it relies on the gradient vector $\mathbf{g}(n)$ as the only source of local information about the error surface. This restriction has a beneficial effect: simplicity of implementation. Unfortunately, it also has a detrimental effect: a slow rate of convergence, which can be excruciating, particularly in the case of large-scale problems. The inclusion of the momentum term in the update equation for the synaptic weight vector is a crude attempt at using second-order information about the error surface, which is of some help. However, its use makes the training process more delicate to manage by adding one more item to the list of parameters that have to be "tuned" by the designer.

In order to produce a significant improvement in the convergence performance of a multilayer perceptron (compared to back-propagation learning), we have to use *higher-order information* in the training process. We may do so by invoking a *quadratic approximation* of the error surface around the current point $\mathbf{w}(n)$. We then find from Eq. (4.117) that the optimum value of the adjustment $\Delta\mathbf{w}(n)$ applied to the synaptic weight vector $\mathbf{w}(n)$ is given by

$$\Delta\mathbf{w}^*(n) = \mathbf{H}^{-1}(n)\mathbf{g}(n) \qquad (4.121)$$

where $\mathbf{H}^{-1}(n)$ is the inverse of the Hessian matrix $\mathbf{H}(n)$, assuming that it exists. Equation (4.121) is the essence of *Newton's method*. If the cost function $\mathscr{E}_{av}(\mathbf{w})$ is quadratic (i.e., the third- and higher-order terms in Eq. (4.117) are zero), Newton's method converges to the optimum solution in one iteration. However, the practical application of Newton's method to the supervised training of a multilayer perceptron is handicapped by the following factors:

- It requires calculation of the inverse Hessian matrix $\mathbf{H}^{-1}(n)$, which can be computationally expensive.
- For $\mathbf{H}^{-1}(n)$ to be computable, $\mathbf{H}(n)$ has to be nonsingular. In the case when $\mathbf{H}(n)$ is positive definite, the error surface around the current point $\mathbf{w}(n)$ is describable by a "convex bowl." Unfortunately, there is no guarantee that the Hessian matrix of the error surface of a multilayer perceptron will always fit this description. Moreover, there is the potential problem of the Hessian matrix being rank deficient (i.e., not all the columns of $\mathbf{H}$ being linearly independent), which results from the intrinsically ill-conditioned nature of neural network training problems (Saarinen et al., 1992); this only makes the computational task more difficult.

- When the cost function $\mathscr{E}_{av}(\mathbf{w})$ is nonquadratic, there is no guarantee for convergence of Newton's method, which makes it unsuitable for the training of a multilayer perceptron.

To overcome some of these difficulties, we may use a *quasi-Newton method,* which only requires an estimate of the gradient vector $\mathbf{g}$. This modification of Newton's method maintains a positive definite estimate of the inverse matrix $\mathbf{H}^{-1}$ directly without matrix inversion. By using such an estimate, a quasi-Newton method is assured of going downhill on the error surface. However, we still have a computational complexity that is $O(W^2)$, where W is the size of weight vector $\mathbf{w}$. Quasi-Newton methods are therefore computationally impractical, except for the training of very small-scale neural networks. A description of quasi-Newton methods is presented later in the section.

Another class of second-order optimization methods includes the conjugate-gradient method, which may be regarded as being somewhat intermediate between the method of steepest descent and Newton's method. Use of the conjugate-gradient method is motivated by the desire to accelerate the typically slow rate of convergence experienced with the method of steepest descent, while avoiding the computational requirements associated with the evaluation, storage, and inversion of the Hessian matrix in Newton's method. Among second-order optimization methods, it is widely acknowledged that the conjugate-gradient method is perhaps the only method that is applicable to large-scale problems, that is, problems with hundreds or thousands of adjustable parameters (Fletcher, 1987). It is therefore well suited for the training of multilayer perceptrons, with typical applications including function approximation, control, and time series analysis (i.e., regression).

Conjugate-Gradient Method

The conjugate-gradient method belongs to a class of second-order optimization methods known collectively as *conjugate-direction methods.* We begin the discussion of these methods by considering the minimization of the *quadratic function*

$$f(\mathbf{x}) = \frac{1}{2}\mathbf{x}^T\mathbf{A}\mathbf{x} - \mathbf{b}^T\mathbf{x} + c \qquad (4.122)$$

where $\mathbf{x}$ is a W-by-1 parameter vector, $\mathbf{A}$ is a W-by-W symmetric, positive definite matrix, $\mathbf{b}$ is a W-by-1 vector, and c is a scalar. Minimization of the quadratic function $f(\mathbf{x})$ is achieved by assigning to $\mathbf{x}$ the unique value

$$\mathbf{x}^* = \mathbf{A}^{-1}\mathbf{b} \qquad (4.123)$$

Thus minimizing $f(\mathbf{x})$ and solving the linear system of equations $\mathbf{A}\mathbf{x}^* = \mathbf{b}$ are equivalent problems.

Given the matrix $\mathbf{A}$, we say that a set of nonzero vectors $\mathbf{s}(0), \mathbf{s}(1), \ldots, \mathbf{s}(W-1)$ is $\mathbf{A}$-*conjugate* (i.e., noninterfering with each other in the context of matrix $\mathbf{A}$) if the following condition is satisfied:

$$\mathbf{s}^T(n)\mathbf{A}\mathbf{s}(j) = 0 \qquad \text{for all } n \text{ and } j \text{ such that } n \neq j \qquad (4.124)$$

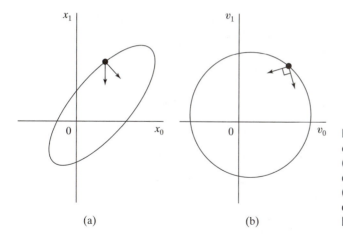

FIGURE 4.24   Interpretation of A-conjugate vectors. (a) Elliptic locus in two-dimesional weight space. (b) Transformation of the elliptic locus into a circular locus.

(a)   (b)

If $\mathbf{A}$ is equal to the identity matrix, conjugacy is equivalent to the usual notion of orthogonality.

Example 4.1

For an interpretation of A-conjugate vectors, consider the situation described in Fig. 4.24a, pertaining to a two-dimensional problem. The elliptic locus shown in this figure corresponds to a plot of Eq. (4.122) for

$$\mathbf{x} = [x_0, x_1]^T$$

at some constant value assigned to the quadratic function $f(\mathbf{x})$. Figure 4.24a also includes a pair of direction vectors that are conjugate with respect to the matrix $\mathbf{A}$. Suppose that we define a new parameter vector $\mathbf{v}$ related to $\mathbf{x}$ by the transformation

$$\mathbf{v} = \mathbf{A}^{1/2}\mathbf{x}$$

where $\mathbf{A}^{1/2}$ is the square root of $\mathbf{A}$. Then the elliptic locus of Fig. 4.24a is transformed into a circular locus, as shown in Fig. 4.24b. Correspondingly, the pair of A-conjugate direction vectors in Fig. 4.24a is transformed into a pair of orthogonal direction vectors in Fig. 4.24b.

■

An important property of $\mathbf{A}$-conjugate vectors is that they are *linearly independent*. We prove this property by contradiction. Let one of these vectors, say $\mathbf{s}(0)$, be expressed as a linear combination of the remaining $W - 1$ vectors as follows:

$$\mathbf{s}(0) = \sum_{j=1}^{W-1} \alpha_j \mathbf{s}(j)$$

Multiplying by $\mathbf{A}$ and then taking the inner product of $\mathbf{As}(0)$ with $\mathbf{s}(0)$ yields

$$\mathbf{s}^T(0)\mathbf{As}(0) = \sum_{j=1}^{W-1} \alpha_j \mathbf{s}^T(0)\mathbf{As}(j) = 0$$

However, it is impossible for the quadratic form $\mathbf{s}^T(0)\mathbf{As}(0)$ to be zero for two reasons: the matrix $\mathbf{A}$ is positive definite by assumption, and the vector $\mathbf{s}(0)$ is nonzero by definition. It follows therefore that the **A**-conjugate vectors **s**(0),**s**(1), ..., **s**($W - 1$) cannot be linearly dependent; that is, they must be linearly independent.

For a given set of **A**-conjugate vectors **s**(0), **s**(1), ..., **s**($W - 1$), the corresponding *conjugate direction method* for unconstrained minimization of the quadratic error function $f(\mathbf{x})$ is defined by (Luenberger, 1973; Fletcher, 1987; Bertsekas, 1995)

$$\mathbf{x}(n + 1) = \mathbf{x}(n) + \eta(n)\mathbf{s}(n), \qquad n = 0, 1, ..., W - 1 \tag{4.125}$$

where **x**(0) is an arbitrary starting vector, and $\eta(n)$ is a scalar defined by

$$f(\mathbf{x}(n) + \eta(n)\mathbf{s}(n)) = \min_{\eta} f(\mathbf{x}(n) + \eta\mathbf{s}(n)) \tag{4.126}$$

The procedure of choosing η so as to minimize the function $f(\mathbf{x}(n) + \eta\mathbf{s}(n))$ for some fixed n is referred to as a line search, which represents a one-dimensional minimization problem.

In light of Eqs. (4.124), (4.125) and (4.126), we now offer some observations:

1. Since the **A**-conjugate vectors **s**(0), **s**(1), ..., **s**($W - 1$) are linearly independent, they form a basis that spans the vector space of **w**.
2. The update equation (4.125) and the line minimization of Eq. (4.126) lead to the same formula for the learning-rate parameter, namely,

$$\eta(n) = -\frac{\mathbf{s}^T(n)\,\mathbf{A}\boldsymbol{\epsilon}(n)}{\mathbf{s}^T(n)\mathbf{A}\mathbf{s}(n)}, \qquad n = 0, 1, ..., W - 1 \tag{4.127}$$

where $\boldsymbol{\epsilon}(n)$ is the *error vector* defined by

$$\boldsymbol{\epsilon}(n) = \mathbf{x}(n) - \mathbf{x}^* \tag{4.128}$$

3. Starting from an arbitrary point **x**(0), the conjugate direction method is guaranteed to find the optimum solution **x*** of the quadratic equation $f(\mathbf{x}) = 0$ in at most W iterations.

The principal property of the conjugate-direction method is described as (Luenberger, 1984; Fletcher, 1987; Bertsekas, 1995):

> At successive iterations, the conjugate-direction method minimizes the quadratic function $f(\mathbf{x})$ over a progressively expanding linear vector space that eventually includes the global minimum of $f(\mathbf{x})$.

In particular, for each iteration n, the iterate **x**($n + 1$) minimizes the function $f(\mathbf{x})$ over a linear vector space $\mathcal{D}_n$ that passes through some arbitrary point **x**(0) and is spanned by the **A**-conjugate vectors **s**(0), **s**(1), ..., **s**(n), as shown by

$$\mathbf{x}(n + 1) = \arg\min_{\mathbf{x} \in \mathcal{D}_n} f(\mathbf{x}) \tag{4.129}$$

where the space $\mathcal{D}_n$ is defined by

$$\mathcal{D}_n = \left\{ \mathbf{x}(n) \,\middle|\, \mathbf{x}(n) = \mathbf{x}(0) + \sum_{j=0}^{n} \eta(j)\mathbf{s}(j) \right\} \tag{4.130}$$

For the conjugate-direction method to work, we require the availability of a set of **A**-conjugate vectors **s**(0), **s**(1), ..., **s**($W - 1$). In a special form of this method known as the *conjugate-gradient method*,[15] the successive direction vectors are generated as

A-conjugate versions of the successive gradient vectors of the quadratic function $f(\mathbf{x})$ as the method progresses, hence the name of the method. Thus, except for $n = 0$, the set of direction vectors $\{\mathbf{s}(n)\}$ is not specified beforehand, but rather it is determined in a sequential manner at successive steps of the method.

Define the *residual* as the steepest descent direction:

$$\mathbf{r}(n) = \mathbf{b} - \mathbf{A}\mathbf{x}(n) \tag{4.131}$$

Then to proceed, we use a linear combination of $\mathbf{r}(n)$ and $\mathbf{s}(n - 1)$, as shown by

$$\mathbf{s}(n) = \mathbf{r}(n) + \beta(n)\mathbf{s}(n - 1), \quad n = 1, 2, ..., W - 1 \tag{4.132}$$

where $\beta(n)$ is a scaling factor to be determined. Multiplying this equation by $\mathbf{A}$, taking the inner product of the resulting expression with $\mathbf{s}(n - 1)$, invoking the $\mathbf{A}$-conjugate property of the direction vectors, and then solving the resulting expression for $\beta(n)$, we get

$$\beta(n) = -\frac{\mathbf{s}^T(n - 1)\mathbf{A}\mathbf{r}(n)}{\mathbf{s}^T(n - 1)\mathbf{A}\mathbf{s}(n - 1)} \tag{4.133}$$

Using Eqs. (4.132) and (4.133), we find that the vectors $\mathbf{s}(0), \mathbf{s}(1), ..., \mathbf{s}(W - 1)$ so generated are indeed $\mathbf{A}$-conjugate.

Generation of the direction vectors in accordance with the recursive equation (4.132) depends on the coefficient $\beta(n)$. The formula of Eq. (4.133) for evaluating $\beta(n)$, as it presently stands, requires knowledge of matrix $\mathbf{A}$. For computational reasons, it would be desirable to evaluate $\beta(n)$ without explicit knowledge of $\mathbf{A}$. This evaluation can be achieved by using one of two formulas (Fletcher, 1987):

1. *Polak–Ribiére formula*, for which $\beta(n)$ is defined by

$$\beta(n) = \frac{\mathbf{r}^T(n)(\mathbf{r}(n) - \mathbf{r}(n - 1))}{\mathbf{r}^T(n - 1)\mathbf{r}(n - 1)} \tag{4.134}$$

2. *Fletcher–Reeves formula*, for which $\beta(n)$ is defined by

$$\beta(n) = \frac{\mathbf{r}^T(n)\mathbf{r}(n)}{\mathbf{r}^T(n - 1)\mathbf{r}(n - 1)} \tag{4.135}$$

To use the conjugate-gradient method to attack the unconstrained minimization of the cost function $\mathscr{E}_{av}(\mathbf{w})$ pertaining to the unsupervised training of multilayer perceptrons, we do two things:

- Approximate the cost function $\mathscr{E}_{av}(\mathbf{w})$ by a quadratic function. That is, the third- and higher-order terms in Eq. (4.117) are ignored, which means that we are operating close to a local minimum on the error surface. On this basis, comparing Eqs. (4.117) and (4.122), we can make the associations indicated in Table 4.7.
- Formulate the computation of coefficients $\beta(n)$ and $\eta(n)$ in the conjugate-gradient algorithm so as to only require gradient information.

The latter point is particularly important in the context of multilayer perceptrons because it avoids using the Hessian matrix $\mathbf{H}(n)$, the evaluation of which is plagued with computational difficulties.

TABLE 4.7 Correspondence Between $f(\mathbf{x})$ and $\mathcal{E}_{av}(\mathbf{w})$

Quadratic function $f(\mathbf{x})$	Cost function $\mathcal{E}_{av}(\mathbf{w})$
Parameter vector $\mathbf{x}(n)$	Synaptic weight vector $\mathbf{w}(n)$
Gradient vector $\partial f(\mathbf{x})/\partial \mathbf{x}$	Gradient vector $\mathbf{g} = \partial \mathcal{E}_{av}/\partial \mathbf{w}$
Matrix $\mathbf{A}$	Hessian matrix $\mathbf{H}$

To compute the coefficient $\beta(n)$ that determines the search direction $\mathbf{s}(n)$ without explicit knowledge of the Hessian matrix $\mathbf{H}(n)$, we can use the Polak–Ribiére formula of Eq. (4.134) or the Fletcher–Reeves formula of Eq. (4.135). Both of these formulas involve the use of residuals only. In the linear form of the conjugate-gradient method, assuming a quadratic function, the Polak–Ribiére and Fletcher–Reeves formulas are equivalent. On the other hand, in the case of a nonquadratic cost function, they are no longer equivalent.

For nonquadratic optimization problems, the Polak–Ribiére form of the conjugate-gradient algorithm is typically superior to the Fletcher–Reeves form of the algorithm, for which we offer the following heuristic explanation (Bertsekas, 1995). Due to the presence of third- and higher-order terms in the cost function $\mathcal{E}_{av}(\mathbf{w})$ and possible inaccuracies in the line search, conjugacy of the generated search directions is progressively lost. This may in turn cause the algorithm to "jam" in the sense that the generated direction vector $\mathbf{s}(n)$ is nearly orthogonal to the residual $\mathbf{r}(n)$. When this phenomenon occurs, we have $\mathbf{r}(n) = \mathbf{r}(n - 1)$, in which case the scalar $\beta(n)$ will be nearly zero. Correspondingly, the direction vector $\mathbf{s}(n)$ will be close to $\mathbf{r}(n)$, thereby breaking the jam. In contrast, when the Fletcher–Reeves formula is used, the conjugate gradient algorithm typically continues to jam under similar conditions.

In rare cases, however, the Polak–Ribiére method can cycle indefinitely without converging. Fortunately, convergence of the Polak–Ribiére method can be guaranteed by choosing (Shewchuk, 1994)

$$\beta = \max\{\beta_{PR}, 0\} \tag{4.136}$$

where β_{PR} is the value defined by the Polak–Ribiére formula of Eq. (4.134). Using the value of β defined in Eq. (4.136) is equivalent to restarting the conjugate gradient algorithm if $\beta_{PR} < 0$. To restart the algorithm is equivalent to forgetting the last search direction and starting it anew in the direction of steepest descent (Shewchuk, 1994).

Consider next the issue of computing the parameter $\eta(n)$, which determines the learning rate of the conjugate-gradient algorithm. As with $\beta(n)$, the preferred method for computing $\eta(n)$ is one that avoids having to use the Hessian matrix $\mathbf{H}(n)$. We recall that the line minimization based on Eq. (4.126) leads to the same formula for $\eta(n)$ as that derived from the update equation (4.125). We therefore need a *line search*,[16] the purpose of which is to minimize the function $\mathcal{E}_{av}(\mathbf{w} + \eta\mathbf{s})$ with respect to η. That is, given fixed values of the vectors $\mathbf{w}$ and $\mathbf{s}$, the problem is to vary η such that this function is minimized. As η varies, the argument $\mathbf{w} + \eta\mathbf{s}$ traces a line in the W-dimensional vector space of $\mathbf{w}$, hence the name "line search." A *line search algorithm* is an iterative procedure that generates a sequence of estimates $\{\eta(n)\}$ for each iteration of the conjugate-gradient algorithm. The line search is terminated when a satisfactory solution is found. A line search must be performed along each search direction.

Several line search algorithms have been proposed in the literature, and a good choice is important because it has a profound impact on the performance of the conjugate-gradient algorithm in which it is embedded. There are two phases to any line search algorithm (Fletcher, 1987):

- *Bracketing phase,* which searches for a *bracket,* that is, a nontrivial interval that is known to contain a minimum.
- *Sectioning phase,* in which the bracket is *sectioned* (i.e., divided), thereby generating a sequence of brackets whose length is progressively reduced.

We now describe a *curve-fitting procedure* that takes care of these two phases in a straightforward manner.

Let $\mathscr{E}_{av}(\eta)$ denote the cost function of the multilayer perceptron, expressed as a function of η. It is assumed that $\mathscr{E}_{av}(\eta)$ is strictly *unimodal* (i.e., it has a single minimum in the neighborhood of the current point $\mathbf{w}(n)$) and is twice continuously differentiable. We initiate the search procedure by searching along the line until we find three points η_1, η_2, and η_3 such that the following condition is satisfied:

$$\mathscr{E}_{av}(\eta_1) \geq \mathscr{E}_{av}(\eta_3) \geq \mathscr{E}_{av}(\eta_2) \qquad \text{for } \eta_1 < \eta_2 < \eta_3 \tag{4.137}$$

as illustrated in Fig. 4.25. Since $\mathscr{E}_{av}(\eta)$ is a continuous function of η, the choice described in Eq. (4.137) ensures that the bracket $[\eta_1, \eta_3]$ contains a minimum of the function $\mathscr{E}_{av}(\eta)$. Provided that the function $\mathscr{E}_{av}(\eta)$ is sufficiently smooth, we may consider this function to be parabolic in the immediate neighborhood of the minimum. Accordingly, we may use *inverse parabolic interpolation* to do the sectioning (Press et al., 1988). Specifically, a parabolic function is fitted through the three original points η_1, η_2, and η_3, as illustrated in Fig. 4.26, where the solid line corresponds to $\mathscr{E}_{av}(\eta)$ and the dashed line corresponds to the first iteration of the sectioning procedure. Let the minimum of the parabola passing through the three points η_1, η_2, and η_3 be denoted by η_4. In the example illustrated in Fig. 4.26, we have $\mathscr{E}_{av}(\eta_4) < \mathscr{E}_{av}(\eta_2)$ and $\mathscr{E}_{av}(\eta_4) < \mathscr{E}_{av}(\eta_1)$. Point η_3 is replaced in favor of η_4, making $[\eta_1, \eta_4]$ the new bracket. The process is repeated by constructing a new parabola through the points η_1, η_2, and η_4. The bracketing-followed-by-sectioning procedure, as illustrated, is repeated several times until a point close enough to the minimum of $\mathscr{E}_{av}(\eta)$ is located, at which time the line search is terminated.

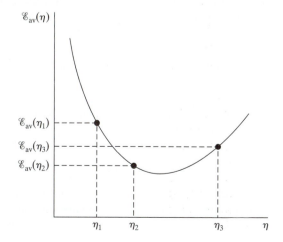

FIGURE 4.25 Illustration of the line search.

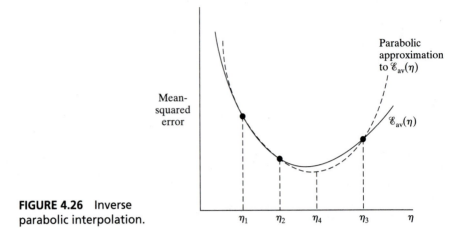

FIGURE 4.26 Inverse parabolic interpolation.

Brent's method constitutes a highly refined version of the three-point curve-fitting procedure just described (Press et al., 1988). At any particular stage of the computation, Brent's method keeps track of six points on the function $\mathscr{E}_{av}(\eta)$, which may not all be necessarily distinct. As before, parabolic interpolation is attempted through three of these points. For the interpolation to be acceptable, certain criteria involving the remaining three points must be satisfied. The net result is a robust line search algorithm.

Summary of the Nonlinear Conjugate Gradient Algorithm

All the ingredients we need to formally describe the nonlinear (nonquadratic) form of the conjugate-gradient algorithm for the supervised training of a multilayer perceptron are now in place. A summary of the algorithm is presented in Table 4.8.

Quasi-Newton Methods

Resuming the discussion on quasi-Newton methods, we find that these are basically gradient methods described by the update equation:

$$\mathbf{w}(n+1) = \mathbf{w}(n) + \eta(n)\mathbf{s}(n) \tag{4.138}$$

where the direction vector $\mathbf{s}(n)$ is defined in terms of the gradient vector $\mathbf{g}(n)$ by

$$\mathbf{s}(n) = -\mathbf{S}(n)\mathbf{g}(n) \tag{4.139}$$

The matrix $\mathbf{S}(n)$ is a positive definite matrix that is adjusted from one iteration to the next. This is done in order to make the direction vector $\mathbf{s}(n)$ approximate the *Newton direction*, namely

$$-(\partial^2 \mathscr{E}_{av}/\partial \mathbf{w}^2)^{-1}(\partial \mathscr{E}_{av}/\partial \mathbf{w})$$

Quasi-Newton methods use second-order (curvature) information about the error surface without actually requiring knowledge of the Hessian matrix $\mathbf{H}$. They do

TABLE 4.8 Summary of the Nonlinear Conjugate Gradient Algorithm for the Supervised Training of a Multilayer Perceptron

Initialization
Unless prior knowledge on the weight vector **w** is available, choose the initial value **w**(0) using a procedure similar to that described for the back-propagation algorithm.

Computation
1. For **w**(0), use back-propagation to compute the gradient vector **g**(0).
2. Set **s**(0) = **r**(0) = −**g**(0).
3. At time step n, use a line search to find $\eta(n)$ that minimizes $\mathscr{E}_{av}(\eta)$ sufficiently, representing the cost function $\mathscr{E}_{av}$ expressed as a function of η for fixed values of **w** and **s**.
4. Test to determine if the Euclidean norm of the residual **r**(n) has fallen below a specified value, that is, a small fraction of the initial value $\|\mathbf{r}(0)\|$.
5. Update the weight vector:

$$\mathbf{w}(n+1) = \mathbf{w}(n) + \eta(n)\mathbf{s}(n)$$

6. For **w**(n + 1), use back-propagation to compute the updated gradient vector **g**(n + 1).
7. Set **r**(n + 1) = −**g**(n + 1).
8. Use the Polak–Ribiére method to calculate $\beta(n+1)$:

$$\beta(n+1) = \max\left\{\frac{\mathbf{r}^T(n+1)(\mathbf{r}(n+1) - \mathbf{r}(n))}{\mathbf{r}^T(n)\mathbf{r}(n)}, 0\right\}$$

9. Update the direction vector:

$$\mathbf{s}(n+1) = \mathbf{r}(n+1) + \beta(n+1)\mathbf{s}(n)$$

10. Set $n = n + 1$, and go back to step 3.

Stopping criterion. Terminate the algorithm when the following condition is satisfied:

$$\|\mathbf{r}(n)\| \leq \epsilon\|\mathbf{r}(0)\|$$

where ϵ is a prescribed small number.

---

so by using two successive iterates **w**(n) and **w**(n + 1), together with the respective gradient vectors **g**(n) and **g**(n + 1). Let

$$\mathbf{q}(n) = \mathbf{g}(n+1) - \mathbf{g}(n) \tag{4.140}$$

and

$$\Delta\mathbf{w}(n) = \mathbf{w}(n+1) - \mathbf{w}(n) \tag{4.141}$$

We may then derive curvature information by using the approximate formula:

$$\mathbf{q}(n) \simeq \left(\frac{\partial}{\partial\mathbf{w}}\mathbf{g}(n)\right)\Delta\mathbf{w}(n) \tag{4.142}$$

In particular, given W linearly independent weight increments $\Delta\mathbf{w}(0), \Delta\mathbf{w}(1), ..., \Delta\mathbf{w}(W-1)$ and the respective gradient increments $\mathbf{q}(0), \mathbf{q}(1), ..., \mathbf{q}(W-1)$, we may approximate the Hessian matrix **H** as:

$$\mathbf{H} \simeq [\mathbf{q}(0), \mathbf{q}(1), ..., \mathbf{q}(W-1)][\Delta\mathbf{w}(0), \Delta\mathbf{w}(1), ..., \Delta\mathbf{w}(W-1)]^{-1} \tag{4.143}$$

We may also approximate the inverse Hessian matrix as:

$$\mathbf{H}^{-1} \simeq [\Delta\mathbf{w}(0), \Delta\mathbf{w}(1), ..., \Delta\mathbf{w}(W - 1)] \, [\mathbf{q}(0), \mathbf{q}(1), ..., \mathbf{q}(W - 1)]^{-1} \quad (4.144)$$

When the cost function $\mathcal{E}_{av}(\mathbf{w})$ is quadratic, Eqs. (4.143) and (4.144) are exact.

In the most popular class of quasi-Newton methods, the matrix $\mathbf{S}(n + 1)$ is obtained from its previous value $\mathbf{S}(n)$, the vectors $\Delta\mathbf{w}(n)$ and $\mathbf{q}(n)$, by using the recursion (Fletcher, 1987; Bertsekas, 1995):

$$\mathbf{S}(n + 1) = \mathbf{S}(n) + \frac{\Delta\mathbf{w}(n)\Delta\mathbf{w}^T(n)}{\mathbf{q}^T(n)\mathbf{q}(n)} - \frac{\mathbf{S}(n)\mathbf{q}(n)\mathbf{q}^T(n)\mathbf{S}(n)}{\mathbf{q}^T(n)\mathbf{S}(n)\mathbf{q}(n)}$$

$$+ \, \xi(n)[\mathbf{q}^T(n)\mathbf{S}(n)\mathbf{q}(n)] \, [\mathbf{v}(n)\mathbf{v}^T(n)] \quad (4.145)$$

where

$$\mathbf{v}(n) = \frac{\Delta\mathbf{w}(n)}{\Delta\mathbf{w}^T(n)\Delta\mathbf{w}(n)} - \frac{\mathbf{S}(n)\mathbf{q}(n)}{\mathbf{q}^T(n)\mathbf{S}(n)\mathbf{q}(n)} \quad (4.146)$$

and

$$0 \le \xi(n) \le 1 \quad \text{for all } n \quad (4.147)$$

The algorithm is initiated with some arbitrary positive definite matrix $\mathbf{S}(0)$. The particular form of the quasi-Newton method is parameterized by how the scalar $\eta(n)$ is defined, as indicated here (Fletcher, 1987):

- For $\xi(n) = 0$ for all n, we obtain the *Davidon–Fletcher–Powell (DFP) algorithm,* which is historically the first quasi-Newton method.
- For $\xi(n) = 1$ for all n, we obtain the *Broyden–Fletcher–Goldfarb–Shanno algorithm,* which is considered to be the best form of quasi-Newton methods currently known.

Comparison of Quasi-Newton Methods with Conjugate-Gradient Methods

We conclude this brief discussion of quasi-Newton methods by comparing them with conjugate-gradient methods in the context of nonquadratic optimization problems (Bertsekas, 1995):

- Both quasi-Newton and conjugate-gradient methods avoid the need for using the Hessian matrix. However, quasi-Newton methods go one step further by generating an approximation to the inverse Hessian matrix. Accordingly, when the line search is accurate and we are in close proximity to a local minimum with a positive definite Hessian, a quasi-Newton method tends to approximate Newton's method, thereby attaining faster convergence than would be possible with the conjugate-gradient method.
- Quasi-Newton methods are not as sensitive to accuracy in the line search stage of the optimization as the conjugate-gradient method.
- Quasi-Newton methods require storage of the matrix $\mathbf{S}(n)$, in addition to the matrix-vector multiplication overhead associated with the computation of the

director vector $\mathbf{s}(n)$. The net result is that the computational complexity of quasi-Newton's methods is $O(W^2)$. In contrast, the computational complexity of the conjugate-gradient method is $O(W)$. Thus, when the dimension W (i.e., size of the weight vector $\mathbf{w}$) is large, conjugate-gradient methods are preferable to quasi-Newton methods in computational terms.

It is because of this latter point that the use of quasi-Newton methods is restricted, in practice, to the design of small-scale neural networks.

4.19 CONVOLUTIONAL NETWORKS

Up to this point, we have been concerned with the algorithmic design of multilayer perceptrons and related issues. In this section we focus on the structural layout of the multilayer perceptron itself. In particular, we describe a special class of multilayer perceptrons known collectively as *convolutional networks;* the idea behind these networks was briefly highlighted in Chapter 1.

A *convolutional network* is a multilayer perceptron designed specifically to recognize two-dimensional shapes with a high degree of invariance to translation, scaling, skewing, and other forms of distortion. This difficult task is learned in a supervised manner by means of a network whose structure includes the following forms of *constraints* (LeCun and Bengio, 1995):

1. *Feature extraction.* Each neuron takes its synaptic inputs from a local *receptive field* in the previous layer, thereby forcing it to extract local features. Once a feature has been extracted, its exact location becomes less important so long as its position relative to other features is approximately preserved.

2. *Feature mapping.* Each computational layer of the network is composed of multiple *feature maps,* with each feature map being in the form of a plane within which the individual neurons are *constrained* to share the same set of synaptic weights. This second form of structural constraint has the following beneficial effects:

- *Shift invariance,* forced into the operation of a feature map through the use of *convolution* with a kernel of small size, followed by a sigmoid (squashing) function.
- *Reduction in the number of free parameters,* accomplished through the use of *weight sharing.*

3. *Subsampling.* Each convolutional layer is followed by a computational layer that performs *local averaging* and *subsampling,* whereby the resolution of the feature map is reduced. This operation has the effect of reducing the sensitivity of the feature map's output to shifts and other forms of distortion.

The development of convolutional networks, as described, is neurobiologically motivated, which goes back to the pioneering work of Hubel and Wiesel (1962, 1977) on locally sensitive and orientation-selective neurons in the visual cortex of a cat.

We emphasize that all weights in all layers of a convolutional network are learned through training. Moreover, the network learns to extract its own features automatically.

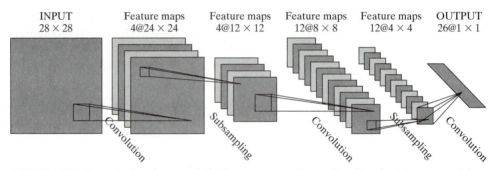

FIGURE 4.27 Convolutional network for image processing such as handwriting recognition. (Reproduced with permission of MIT Press).

Figure 4.27 shows the architectural layout of a convolutional network made up of an input layer, four hidden layers, and an output layer. This network is designed to perform *image processing* (e.g., recognition of handwritten characters). The input layer, made up of 28×28 sensory nodes, receives the images of different characters that have been approximately centered and normalized in size. Thereafter, the computational layouts alternate between convolution and subsampling, as described here:

- The first hidden layer performs convolution. It consists of four feature maps, with each feature map consisting of 24×24 neurons. Each neuron is assigned a receptive field of size 5×5.
- The second hidden layer performs subsampling and local averaging. It also consists of four feature maps, but each feature map is now made up of 12×12 neurons. Each neuron has a receptive field of size 2×2, a trainable coefficient, a trainable bias, and a sigmoid activation function. The trainable coefficient and bias control the operating point of the neuron; for example, if the coefficient is small, the neuron operates in a quasi-linear mode.
- The third hidden layer performs a second convolution. It consists of 12 feature maps, with each feature map consisting of 8×8 neurons. Each neuron in this hidden layer may have synaptic connections from several feature maps in the previous hidden layer. Otherwise it operates in a manner similar to the first convolutional layer.
- The fourth hidden layer performs a second subsampling and local averaging. It consists of 12 feature maps, but with each feature map consisting of 4×4 neurons. Otherwise it operates in a manner similar to the first subsampling layer.
- The output layer performs one final stage of convolution. It consists of 26 neurons, with each neuron assigned to one of 26 possible characters. As before, each neuron is assigned a receptive field of size 4×4.

With the successive computational layers alternating between convolution and subsampling, we get a "bipyramidal" effect. That is, at each convolutional or subsampling layer, the number of feature maps is increased while the spatial resolution is reduced, compared to the corresponding previous layer. The idea of convolution followed by subsampling is inspired by the notion of "simple" cells followed by "complex" cells[17] that was first described in Hubel and Wiesel (1962).

The multilayer perceptron described in Fig. 4.27 contains approximately 100,000 synaptic connections but only about 2600 free parameters. This dramatic reduction in the number of free parameters is achieved through the use of weight sharing. The capacity of the learning machine (measured in terms of the VC dimension) is thereby reduced, which in turn improves its generalization ability (LeCun, 1989). What is even more remarkable is that the adjustments to the free parameters are made by using the stochastic (sequential) form of back-propagation learning.

Another noteworthy point is that the use of weight sharing makes it possible to implement the convolutional network in parallel form. Here is another advantage of the convolutional network over a fully connected multilayer perceptron.

The lesson to be learned from the convolutional network of Fig. 4.27 is two-fold. First, a multilayer perceptron of manageable size is able to learn a complex, high-dimensional, nonlinear mapping by *constraining* its design through the incorporation of prior knowledge about the task at hand. Second, the synaptic weights and bias levels can be learned by cycling the simple back-propagation algorithm through the training set.

4.20 SUMMARY AND DISCUSSION

Back-propagation learning has emerged as the *standard* algorithm for the training of multilayer perceptrons, against which other learning algorithms are often benchmarked. The back-propagation algorithm derives its name from the fact that the partial derivatives of the cost function (performance measure) with respect to the free parameters (synaptic weights and biases) of the network are determined by back-propagating the error signals (computed by the output neurons) through the network, layer by layer. In so doing, it solves the credit-assignment problem in a most elegant fashion. The computing power of the algorithm lies in its two main attributes:

- *Local* method for updating the synaptic weights and biases of the multilayer perceptron.
- *Efficient* method for computing *all* the partial derivatives of the cost function with respect to these free parameters.

For a given epoch of training data, the back-propagation algorithm operates in one of two modes: sequential or batch. In the sequential mode the synaptic weights of all neurons in the network are adjusted on a pattern-by-pattern basis. Consequently, estimation of the gradient vector of the error surface used in the computation is stochastic (random) in nature, hence the name "stochastic back-propagation" that is also used to refer to the sequential mode of back-propagation learning. On the other hand, in the batch mode the adjustments to all synaptic weights and biases are made on an epoch-by-epoch basis, with the result that a more accurate estimate of the gradient vector is used in the computation. Despite its disadvantages, the sequential (stochastic) form of back-propagation learning is most frequently used for the design of neural networks, particularly on large problems. To achieve best results, careful tuning of the algorithm is required.

The specific details involved in the design of a multilayer perceptron naturally depend on the application of interest. We may, however, make two distinctions:

1. In pattern classification involving nonlinearly separable patterns, *all* the neurons in the network are *nonlinear*. The nonlinearity is achieved by using a sigmoid function, two commonly used forms of which are (a) the nonsymmetric logistic function, and (b) the antisymmetric hyperbolic tangent function. Each neuron is responsible for producing a hyperplane of its own in decision space. Through a supervised learning process, the combination of hyperplanes formed by all the neurons in the network is iteratively adjusted in order to separate patterns drawn from the different classes and not seen before, with the fewest classification errors on average. For pattern classification, the stochastic back-propagation algorithm is the most widely used algorithm to perform the training, particularly on large problems (e.g., optical character recognition).

2. In nonlinear regression, the output *range* of the multilayer perceptron should be sufficiently large to contain the process values; if this information is not available, then use of linear output neurons is the most sensible choice. As for learning algorithms, we offer the following observations:

- The sequential (stochastic) mode of back-propagation learning is *much* slower than the batch mode.
- The batch mode of back-propagation learning is slower than the conjugate-gradient method. Note, however, that the latter method can only be used in the batch mode.

We conclude this discussion with some final remarks on *performance measures.* The derivation of the back-propagation algorithm presented in this chapter is based on minimizing the cost function $\mathscr{E}_{av}$, defined as the sum of error squares that is averaged over the entire training set, one way or another. The important virtue of this criterion is its generality and mathematical tractability. However, in many situations encountered in practice, minimizing the cost function $\mathscr{E}_{av}$ corresponds to optimizing an intermediate quantity that is not the ultimate objective of the system, and may therefore lead to a suboptimal performance. For example, in capital market trading systems, an investor's ultimate goal or that of a trader is to maximize the expected return at minimum risk (Choey and Weigend, 1996; Moody and Wu, 1996). The *Sharpe ratio* or *reward-to-volatility ratio* as a performance measure of risk-adjusted return is intuitively more appealing than $\mathscr{E}_{av}$.

NOTES AND REFERENCES

1. Sigmoid functions are so called because their graphs are "s-shaped." Menon et al. (1996) present a detailed study of two classes of sigmoids:
   - *Simple sigmoids,* defined to be odd, asymptotically bounded, and completely monotone functions of one variable.
   - *Hyperbolic sigmoids,* representing a proper subset of simple sigmoids and a natural generalization of the hyperbolic tangent function.

2. For the special case of the LMS algorithm, it has been shown that use of the momentum constant α reduces the stable range of the learning-rate parameter η, and could thus lead to instability if η is not adjusted appropriately. Moreover, the misadjustment increases with increasing α; for details, see Roy and Shynk (1990).

3. For a derivation of the back-propagation algorithm including the momentum constant from first principles, see Hagiwara (1992).

4. A vector $\mathbf{w}^*$ is said to be a *local minimum* of an input–output function F if it is no worse than its *neighbors,* that is, if there exists an ϵ such that (Bertsekas, 1995)

$$F(\mathbf{w}^*) \le F(\mathbf{w}) \qquad \text{for all } \mathbf{w} \text{ with } \|\mathbf{w} - \mathbf{w}^*\| < \epsilon$$

The vector $\mathbf{w}^*$ is said to be a *global minimum* of the function F if it is no worse than *all* other vectors; that is,

$$F(\mathbf{w}^*) \le F(\mathbf{w}) \qquad \text{for all } \mathbf{w} \in \mathbb{R}^n$$

where n is the dimension of $\mathbf{w}$.

5. The first documented description of the use of back-propagation for efficient gradient evaluation is due to Werbos (1974). The material presented in Section 4.10 follows the treatment given in Saarinen et al. (1992); a more general discussion of the topic is presented by Werbos (1990).

6. Other aspects of neural network design that benefit from knowledge of the Hessian matrix include (Bishop, 1995):
   (1) The Hessian forms the basis of a procedure for the retraining of a multilayer perceptron after a small change has been made in the training data.
   (2) In the context of Bayesian learning,
       • the inverse of the Hessian matrix may be used to assign error bars to the nonlinear prediction made by a trained neural network, and
       • the eigenvalues of the Hessian matrix may be used to determine suitable values for regularization parameters.

7. Buntine and Weigend (1994) present a review of exact and approximate algorithms for computing the Hessian matrix, with particular reference to neural networks; see also the paper by Battiti (1992).

8. The universal approximation theorem may be viewed as a natural extension of the *Weierstrass theorem* (Weierstrass, 1885). This theorem states that *any continuous function over a closed interval on the real axis can be expressed in that interval as an absolutely and uniformly convergent series of polynomials.*

   Research interest in the virtues of multilayer perceptrons as devices for the representation of arbitrary continuous functions was perhaps first put into focus by Hecht-Nielsen (1987), who invoked an improved version of Kolomogorov's superposition theorem due to Sprecher (1965). Then Gallant and White (1988) showed that a single-hidden-layer multilayer perceptron with monotone "cosine" squashing at the hidden layer and no squashing at the output embeds as a special case of a "Fourier network" that yields a Fourier series approximation to a given function as its output. However, in the context of traditional multilayer perceptrons, it was Cybenko who demonstrated rigorously for the first time that a single hidden layer is sufficient to uniformly approximate any continuous function with support in a unit hypercube; this work was published as a University of Illinois Technical Report in 1988, and republished as a paper one year later (Cybenko, 1988, 1989). In 1989, two other papers were published independently on multilayer perceptrons as universal approximators, one by Funahashi and the other by Hornik, Stinchcombe, and White. For subsequent contributions to the approximation problem, see Light (1992b).

9. The history of the development of cross-validation is documented in Stone (1974). The idea of cross-validation had been around at least since the 1930s, but refinement of the technique was accomplished in the 1960s and 1970s. Two important papers from that era are Stone (1974) and Geisser (1975), who independently and almost simultaneously propounded it. The technique was termed the "cross-validating method" by Stone and the "predictive sample reuse method" by Geisser.

10. The earliest references on the early-stopping method of training include Morgan and Bourlard (1990) and Weigend et al. (1990). Perhaps the most detailed statistical analysis of the early-stopping method of training for multilayer perceptrons is presented in Amari et al. (1996a). The study is supported with computer simulations on an 8-8-4 classifier with 108 adjustable parameters and a very large data set (50,000 examples).

11. The *cascade-correlation learning architecture* (Fahlman and Lebiere, 1990) is an example of the network-growing approach. The procedure begins with a minimal network that has some inputs and one or more output nodes as indicated by input/output considerations, but no hidden nodes. The LMS algorithm, for example, may be used to train the network. The hidden neurons are added to the network one by one, thereby obtaining a multilayer structure. Each new hidden neuron receives a synaptic connection from each of the input nodes and also from each preexisting hidden neuron. When a new hidden neuron is added, the synaptic weights on the input side of that neuron are frozen; only the synaptic weights on the output side are trained repeatedly. The added hidden neuron then becomes a permanent feature detector in the network. The procedure of adding new hidden neurons is continued in the manner described until satisfactory performance is attained.

    In yet another network-growing approach described in Lee et al. (1990), a third level of computation termed the *structure-level adaptation* is added to the forward pass (function-level adaptation) and backward pass (parameter-level adaptation). In this third level of computation the structure of the network is adapted by changing the number of neurons and the structural relationship among neurons in the network. The criterion used here is that when the estimation error (after convergence) is larger than a desired value, a new neuron is added to the network in a position where it is most needed. The desirable position for the new neuron is determined by monitoring the learning behavior of the network. In particular, if after a long period of parameter adaptation (training), the synaptic weight vector pertaining to the inputs of a neuron continues to fluctuate significantly, it may be inferred that the neuron in question does not have enough representation power to learn its proper share of the task. The structure-level adaptation also includes a provision for the possible annihilation of neurons. A neuron is annihilated when it is not a functioning element of the network or it is a redundant element of the network. This method of network growing appears to be computationally intensive.

12. Hecht-Nielsen (1995) describes a replicator neural network in the form of a multilayer perceptron with three hidden layers and an output layer:
    - The activation functions of neurons in the second and fourth (hidden) layers are defined by the hyperbolic tangent function:

    $$\varphi^{(2)}(v) = \varphi^{(4)}(v) = \tanh(v)$$

    where v is the induced local field of a neuron in those layers.
    - The activation function for each neuron in the middle (hidden) layer is given by

    $$\varphi^{(3)}(v) = \frac{1}{2} + \frac{1}{2(N-1)} \sum_{j=1}^{N-1} \tanh\left(a\left(v - \frac{j}{N}\right)\right)$$

    where a is a gain parameter, and v is the induced local field of a neuron in that layer. The function $\varphi^{(3)}(v)$ describes a smooth staircase activation function with N treadles, thereby essentially quantizing the vector of the respective neuronal outputs into $K = N^n$, where n is the number of neurons in the middle hidden layer.

- The neurons in the output layer are linear, with their activation functions defined by

$$\varphi^{(5)}(v) = v$$

Based on this neural network structure, Hecht-Nielsen describes a theorem, showing that optimal data compression for arbitrary input data vectors can be carried out.

13. What we basically need is a theoretical framework of back-propagation learning that explains the local-minima problem. This is a difficult task to accomplish. Nevertheless, some progress has been reported in the literature on this issue. Baldi and Hornik (1989) have considered the problem of learning in layered feedforward neural networks with linear activation functions using back-propagation learning. The main result of their paper is that the error surface has only one minimum, corresponding to an orthogonal projection onto the subspace spanned by the first principal eigenvectors of a covariance matrix associated with the training patterns; all other critical points of the error surface are saddle points. Gori and Tesi (1992) have considered the more general case of back-propagation learning that involves the use of nonlinear neurons. The main result of their paper is that for linearly separable patterns, convergence to an optimal solution (i.e., global minimum) is ensured by using the batch mode of back-propagation learning, and the network exceeds Rosenblatt's perceptron in generalization to new examples.

14. A modification of the back-propagation algorithm that builds on heuristics 1 through 4 is known as the *delta-bar-delta-learning rule* (Jacobs, 1988), the derivation of which follows a procedure similar to that described in Section 4.3 for the conventional form of the back-propagation algorithm. Implementation of the delta-bar-delta learning rule may be simplified by exploiting an idea similar to the gradient reuse method (Hush and Sales, 1988; Haykin and Deng, 1991).

    Salomon and van Hemmen (1996) describe a dynamic self-adaptation procedure for accelerating the back-propagation learning process. The underlying idea is to take the learning rate of the previous time step, increase and decrease it slightly, evaluate the cost function for both new values of the learning-rate parameter and then choose the particular one that gives the lower value of the cost function.

15. The classic reference for the conjugate-gradient method is Hestenes and Stiefel (1952). For a discussion of the convergence behavior of the conjugate-gradient algorithm, see Luenberger (1984) and Bertsekas (1995). For a tutorial treatment of the many facets of the conjugate-gradient algorithm, see Shewchuk (1994). For a readable account of the algorithm in the context of neural networks, see Johansson et al. (1990).

16. The conventional form of the conjugate-gradient algorithm requires the use of a line search, which can be time consuming because of its trial and error nature. Møller (1993) describes a modified version of the conjugate-gradient algorithm, called the scaled conjugate-gradient algorithm, which avoids the use of a line search. Essentially, the line search is replaced by a one-dimensional Levenberg–Marquardt form of algorithm. The motivation for using such methods is to circumvent the difficulty caused by nonpositive definite Hessian matrices (Fletcher, 1987).

17. Hubel and Wiesel's notion of "simple" and "complex" cells were first exploited in the neural network literature by Fukushima (1980, 1995) in the design of a learning machine called the *neocognitron*. This learning machine, however, operates in a self-organized manner, whereas the convolutional network described in Fig. 4.27 operates in a supervised manner using labeled examples.

PROBLEMS

XOR Problem

4.1 Figure P4.1 shows a neural network involving a single hidden neuron, for solving the XOR problem; this network may be viewed as an alternative to that considered in Section 4.5. Show that the network of Fig. P4.1 solves the XOR problem by constucting (a) decision regions, and (b) a truth table for the network.

4.2 Use the back-propagation algorithm for computing a set of synaptic weights and bias levels for a neural network structured as in Fig. 4.8 to solve the XOR problem. Assume the use of a logistic function for the nonlinearity.

Back-propagation learning

4.3 The inclusion of a momentum term in the weight update may be viewed as a mechanism for satisfying heuristics 3 and 4 that provide guidelines for accelerating the convergence of the back-propagation algorithm, which was discussed in Section 4.17. Demonstrate the validity of this statement.

4.4 The momentum constant α is normally assigned a positive value in the range $0 \leq \alpha < 1$. Investigate the difference that would be made in the behavior of Eq. (4.41) with respect to time t if α was assigned a negative value in the range $-1 < \alpha \leq 0$.

4.5 Consider the simple example of a network involving a single weight, for which the cost function is

$$\mathcal{E}(w) = k_1(w - w_0)^2 + k_2$$

where w_0, k_1, and k_2 are constants. A back-propagation algorithm with momentum is used to minimize $\mathcal{E}(w)$.

Explore the way in which the inclusion of the momentum constant α influences the learning process, with particular reference to the number of epochs required for convergence versus α.

4.6 In Section 4.7 we presented qualitative arguments for the property of a multilayer perceptron classifier (using a logistic function for nonlinearity) that its outputs provide estimates of the *a posteriori* class probabilities. This property assumes that the size of the training set is large enough, and that the back-propagation algorithm used to train the network does not get stuck at a local minimum. Fill in the mathematical details of this property.

4.7 Starting from the cost function defined in Eq. (4.70), derive the minimizing solution of Eq. (4.72) and the minimum value of the cost function defined in Eq. (4.73).

4.8 Equations (4.81) through (4.83) define the partial derivatives of the approximating function $F(\mathbf{w}, \mathbf{x})$ realized by the multilayer perceptron in Fig. 4.18. Derive these equations from the following scenario:

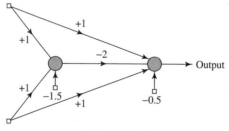

FIGURE P4.1

(a) *Cost function:*

$$\mathscr{E}(n) = \frac{1}{2}[d - F(\mathbf{w},\mathbf{x})]^2$$

(b) *Output of neuron j:*

$$y_j = \varphi\left(\sum_i w_{ji}\, y_i\right)$$

where w_{ji} is synaptic weight from neuron i to neuron j, and y_i is output of neuron i;

(c) *Nonlinearity:*

$$\varphi(v) = \frac{1}{1 + \exp(-v)}$$

Cross-validation

4.9 It may be argued that cross-validation is a case study in structural risk minimization that is discussed in Chapter 2. Describe a neural network example using cross-validation that supports this argument.

4.10 In multifold cross-validation there is no clear separation between the training data and test (validation) data as there is in the hold-out method. Is it possible for the use of multifold cross-validation to produce a biased estimate? Justify your answer.

Network-pruning techniques

4.11 Statistical criterion for model selection, such as Rissanen's *minimum description length (MDL) criterion* and *an information-theoretic criterion (AIC)* due to Akaike, share a common form of composition:

$$\begin{pmatrix} \text{Model-complexity} \\ \text{criterion} \end{pmatrix} = \begin{pmatrix} \text{log-likelihood} \\ \text{function} \end{pmatrix} + \begin{pmatrix} \text{model-complexity} \\ \text{penalty} \end{pmatrix}$$

Discuss how the weight-decay and weight-elimination methods used for network pruning fit into this formalism.

4.12 **(a)** Derive the formula for the saliency S_i given in Eq. (4.105).

(b) Assume that the Hessian matrix of the average squared error of a multilayer perceptron with respect to its weights may be approximated by a diagonal matrix as follows:

$$\mathbf{H} = \text{diag}[h_{11}, h_{22}, ..., h_{WW}]$$

where W is the total number of weights in the network. Determine the saliency S_i of weight w_i in the network.

Accelerated convergence of back-propagation learning

4.13 The *delta-bar-delta learning rule* (Jacobs, 1988) represents a modified form of the back-propagation algorithm that builds on the heuristics described in Section 4.17. In this rule, each synaptic weight in the network is assigned a learning-rate parameter of its own. The cost function, $E(n)$, is therefore modified in a corresponding fashion. In other words, although $E(n)$ is mathematically similar to the cost function $\mathscr{E}(n)$ in Eq. (4.2), the parameter space pertaining to the new cost function $E(n)$ involves different learning rates.

(a) Derive an expression for the partial derivative $\partial E(n)/\partial \eta_{ji}(n)$, where $\eta_{ji}(n)$ is the learning-rate parameter associated with synaptic weight $w_{ji}(n)$.

(b) Hence, demonstrate that the adjustments made to the learning-rate parameters based on the result of part (a) are in perfect accord with heuristics 3 and 4 of Section 4.17.

Second-order optimization methods

4.14 The use of a momentum term in the weight update described in Eq. (4.39) may be considered as an approximation to the conjugate-gradient method (Battiti, 1992). Discuss the validity of this statement.

4.15 Starting with the formula for $\beta(n)$ in Eq. (4.133), derive the *Hesteness–Stiefel formula*:

$$\beta(n) = \frac{\mathbf{r}^T(n)(\mathbf{r}(n) - \mathbf{r}(n-1))}{\mathbf{x}^T(n-1)\mathbf{r}(n-1)}$$

where $s(n)$ is the direction vector and $\mathbf{r}(n)$ is the residual in the conjugate-gradient method. Use this result to derive the Polak–Ribiére formula of Eq. (4.134) and the Fletcher–Reeves formula of Eq. (4.135).

Computer experiments

4.16 Investigate the use of back-propagation learning using a sigmoidal nonlinearity to achieve one-to-one mappings, as described here:

1. $f(x) = \dfrac{1}{x}$, $1 \leq x \leq 100$

2. $f(x) = \log_{10} x$, $1 \leq x \leq 10$

3. $f(x) = \exp(-x)$, $1 \leq x \leq 10$

4. $f(x) = \sin x$, $0 \leq x \leq \dfrac{\pi}{2}$

For each mapping, do the following:

(a) Set up two sets of data, one for network training, and the other for testing.

(b) Use the training data set to compute the synaptic weights of the network, assumed to have a single hidden layer.

(c) Evaluate the computation accuracy of the network by using the test data.

Use a single hidden layer but with a variable number of hidden neurons. Investigate how the network performance is affected by varying the size of the hidden layer.

4.17 The data presented in Table P4.17 show the weights of eye lenses of wild Australian rabbits as a function of age. No simple analytical function can exactly interpolate these data, because we do not have a single-valued function. Instead, we have a nonlinear least-squares model of this data set, using a negative exponential, as described by

$$y = 233.846\,(1 - \exp(-0.006042x)) + \epsilon$$

where ϵ is an error term.

Using the back-propagation algorithm, design a multilayer perceptron that provides a nonlinear least-squares approximation to this data set. Compare your result against the least-squares model described.

TABLE P4.17   Weights of Eye Lenses of Wild Australian Rabbits

Ages (days)	Weights (mg)	Ages (days)	Weights (mg)	Ages (days)	Weights (mg)	Ages (days)	Weights (mg)
15	21.66	75	94.6	218	174.18	338	203.23
15	22.75	82	92.5	218	173.03	347	188.38
15	22.3	85	105	219	173.54	354	189.7
18	31.25	91	101.7	224	178.86	357	195.31
28	44.79	91	102.9	225	177.68	375	202.63
29	40.55	97	110	227	173.73	394	224.82
37	50.25	98	104.3	232	159.98	513	203.3
37	46.88	125	134.9	232	161.29	535	209.7
44	52.03	142	130.68	237	187.07	554	233.9
50	63.47	142	140.58	246	176.13	591	234.7
50	61.13	147	155.3	258	183.4	648	244.3
60	81	147	152.2	276	186.26	660	231
61	73.09	150	144.5	285	189.66	705	242.4
64	79.09	159	142.15	300	186.09	723	230.77
65	79.51	165	139.81	301	186.7	756	242.57
65	65.31	183	153.22	305	186.8	768	232.12
72	71.9	192	145.72	312	195.1	860	246.7
75	86.1	195	161.1	317	216.41		

Radial-Basis Function Networks

5.1 INTRODUCTION

The design of a supervised neural network may be pursued in a variety of ways. The back-propagation algorithm for the design of a multilayer perceptron (under supervision) as described in the previous chapter may be viewed as the application of a recursive technique known in statistics as *stochastic approximation*. In this chapter we take a completely different approach by viewing the design of a neural network as a *curve-fitting (approximation) problem* in a high-dimensional space. According to this viewpoint, learning is equivalent to finding a surface in a multidimensional space that provides a best fit to the training data, with the criterion for "best fit" being measured in some statistical sense. Correspondingly, generalization is equivalent to the use of this multidimensional surface to interpolate the test data. Such a viewpoint is the motivation behind the method of radial-basis functions in the sense that it draws upon research work on traditional strict interpolation in a multidimensional space. In the context of a neural network, the hidden units provide a set of "functions" that constitute an arbitrary "basis" for the input patterns (vectors) when they are expanded into the hidden space; these functions are called *radial-basis functions*.[1] Radial-basis functions were first introduced in the solution of the real multivariate interpolation problem. The early work on this subject is surveyed in Powell (1985), and more recent work is surveyed in Light (1992b). It is now one of the main fields of research in numerical analysis.

The construction of a *radial-basis function (RBF) network*, in its most basic form, involves three layers with entirely different roles. The input layer is made up of source nodes (sensory units) that connect the network to its environment. The second layer, the *only* hidden layer in the network, applies a nonlinear transformation from the input space to the hidden space; in most applications the hidden space is of high dimensionality. The output layer is linear, supplying the response of the network to the activation pattern (signal) applied to the input layer. A mathematical justification for the rationale of a nonlinear transformation followed by a linear transformation may be traced back to an early paper by Cover (1965). According to this paper, a pattern-classification problem cast in a high-dimensional space is more likely to be linearly separable than in a low-dimensional space—hence the reason for frequently making the dimension of the hidden space in an RBF network high. Another important point is the fact that the dimension of the hidden space is directly related to the capacity of the network to approximate a smooth input–output mapping (Mhaskar, 1996; Niyogi and Girosi, 1996); the higher the dimension of the hidden space, the more accurate the approximation will be.

Organization of the Chapter

The main body of the chapter is organized as follows. We lay the foundations for the construction of an RBF network in Sections 5.2 and 5.4. We do this in two stages. First, we describe Cover's theorem on the separability of patterns; the XOR problem is used here to illustrate the application of this theorem. In Section 5.3 we consider the interpolation problem and its relationship to RBF networks.

     After developing an understanding of how the RBF network functions, we move on to the second part of the chapter that consists of Sections 5.4 through 5.9. In Section 5.4 we discuss the viewpoint that supervised learning is an ill-posed hypersurface reconstruction problem. In Section 5.5 we present a detailed treatment of Tikhonov's regularization theory and its application to RBF networks. This theory naturally leads to the formulation of regularization networks in Section 5.6. This class of RBF networks is computationally demanding. To reduce computational complexity, in Section 5.7 we discuss a modified form of regularization networks referred to as generalized RBF networks. In Section 5.8 we revisit the XOR problem and show how it can be solved using an RBF network. In Section 5.9 we complete the study of regularization theory by describing the method of generalized cross-validation for selecting a suitable value for the regularization parameter.

     Section 5.10 discusses the approximation properties of RBF networks. Section 5.11 presents a comparison between RBF networks and multilayer preceptrons, both of which are important examples of layered feedforward networks.

     In Section 5.12 we discuss kernel regression estimation as the basis of another viewpoint of RBF networks. We relate RBF networks to a large body of the statistics literature dealing with density estimation and kernel regression theory.

     The last part of the chapter consists of Sections 5.13 and 5.14. In Section 5.13 we describe four different learning strategies for the design of RBF networks. In Section 5.14 we describe a computer experiment on pattern classification using RBF networks.

     The chapter concludes with some final thoughts on RBF networks in Section 5.15.

5.2   COVER'S THEOREM ON THE SEPARABILITY OF PATTERNS

When a radial-basis function (RBF) network is used to perform a *complex* pattern-classification task, the problem is basically solved by transforming it into a high-dimensional space in a nonlinear manner. The underlying justification is found in *Cover's theorem* on the *separability of patterns*, which, in qualitative terms, may be stated as follows (Cover, 1965):

A complex pattern-classification problem cast in a high-dimensional space nonlinearly is more likely to be linearly separable than in a low-dimensional space.

From the work we did on single-layer perceptrons in Chapter 3, we know that once we have linearly separable patterns, the classification problem is relatively easy to solve. Accordingly, we may develop a great deal of insight into the operation of an RBF network as a pattern classifier by studying the separability of patterns.

Consider a family of surfaces where each naturally divides an input space into two regions. Let $\mathcal{X}$ denote a set of N patterns (vectors) $\mathbf{x}_1, \mathbf{x}_2, ..., \mathbf{x}_N$, each of which is assigned to one of two classes $\mathcal{X}_1$ and $\mathcal{X}_2$. This *dichotomy* (binary partition) of the points is said to be separable with respect to the family of surfaces if a surface exists in the family that separates the points in the class $\mathcal{X}_1$ from those in the class $\mathcal{X}_2$. For each pattern $\mathbf{x} \in \mathcal{X}$, define a vector made up of a set of real-valued functions $\{\varphi_i(\mathbf{x})|i = 1, 2, ..., m_1\}$, as shown by

$$\boldsymbol{\varphi}(\mathbf{x}) = [\varphi_1(\mathbf{x}), \varphi_2(\mathbf{x}), ..., \varphi_{m_1}(\mathbf{x})]^T \tag{5.1}$$

Suppose that the pattern $\mathbf{x}$ is a vector in an m_0-dimensional input space. The vector $\boldsymbol{\varphi}(\mathbf{x})$ then maps points in m_0-dimensional input space into corresponding points in a new space of dimension m_1. We refer to $\varphi_i(\mathbf{x})$ as a *hidden function*, because it plays a role similar to that of a hidden unit in a feedforward neural network. Correspondingly, the space spanned by the set of hidden functions $\{\varphi_i(\mathbf{x})\}_{i=1}^{m_1}$ is referred to as the *hidden space* or *feature space*.

A dichotomy $\{\mathcal{X}_1, \mathcal{X}_2\}$ of $\mathcal{X}$ is said to be φ-*separable* if there exists an m_1-dimensional vector $\mathbf{w}$ such that we may write (Cover, 1965)

$$\begin{aligned}
\mathbf{w}^T\boldsymbol{\varphi}(\mathbf{x}) > 0, \qquad \mathbf{x} \in \mathcal{X}_1 \\
\mathbf{w}^T\boldsymbol{\varphi}(\mathbf{x}) < 0, \qquad \mathbf{x} \in \mathcal{X}_2
\end{aligned} \tag{5.2}$$

The hyperplane defined by the equation

$$\mathbf{w}^T\boldsymbol{\varphi}(\mathbf{x}) = 0$$

describes the separating surface in the φ-space (i.e., hidden space). The inverse image of this hyperplane, that is,

$$\mathbf{x}: \mathbf{w}^T\boldsymbol{\varphi}(\mathbf{x}) = 0 \tag{5.3}$$

defines the *separating surface* in the input space.

Consider a natural class of mappings obtained by using a linear combination of r-wise products of the pattern vector coordinates. The separating surfaces corresponding to such mappings are referred to as *rth-order rational varieties*. A rational variety of order r in a space of dimension m_0 is described by an rth degree homogeneous equation in the coordinates of the input vector $\mathbf{x}$, as shown by

$$\sum_{0 \le i_1 \le i_2 \le \cdots \le i_r \le m_0} a_{i_1 i_2 \ldots i_r} x_{i_1} x_{i_2} \ldots x_{i_r} = 0 \tag{5.4}$$

where x_i is the ith component of input vector $\mathbf{x}$, and x_0 is set equal to unity in order to express the equation in a homogeneous form. An rth order product of entries x_i of

$\mathbf{x}$, that is, $x_{i_1} x_{i_2} \ldots x_{i_r}$, is called a *monomial*. For an input space of dimensionality m_0, there are

$$\frac{(m_0 - r)!}{m_0! r!}$$

monomials in Eq. (5.4). Examples of the type of separating surfaces described by Eq. (5.4) are *hyperplanes* (first-order rational varieties), *quadrices* (second-order rational varieties), and *hyperspheres* (quadrics with certain linear constraints on the coefficients). These examples are illustrated in Fig. 5.1 for a configuration of five points in a two-dimensional input space. In general, linear separability implies spherical separability which implies quadric separability; however, the converses are not necessarily true.

In a probabilistic experiment, the separability of a set of patterns becomes a random event that depends on the dichotomy chosen and the distribution of the patterns in the input space. Suppose that the activation patterns $\mathbf{x}_1, \mathbf{x}_2, \ldots, \mathbf{x}_N$ are chosen independently, according to a probability measure imposed on the input space. Suppose also that all the possible dichotomies of $\mathcal{X} = \{\mathbf{x}_i\}_{i=1}^N$ are equiprobable. Let $P(N, m_1)$ denote the probability that a particular dichotomy picked at random is φ-separable, where the class of separating surfaces chosen has m_1 degrees of freedom. Following Cover (1965), we may then state that

$$P(N, m_1) = \left(\frac{1}{2}\right)^{N-1} \sum_{m=0}^{m_1-1} \binom{N-1}{m} \tag{5.5}$$

where the binomial coefficients comprising $N - 1$ and m are themselves defined for all integer l and m by

$$\binom{l}{m} = \frac{l(l-1)\cdots(l-m+1)}{m!}$$

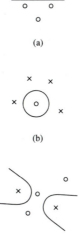

(a)

(b)

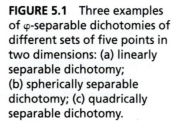

FIGURE 5.1   Three examples of φ-separable dichotomies of different sets of five points in two dimensions: (a) linearly separable dichotomy; (b) spherically separable dichotomy; (c) quadrically separable dichotomy.

(c)

Equation (5.5) embodies the essence of *Cover's separability theorem* for random patterns.[2] It is a statement of the fact that the cumulative binomial distribution corresponding to the probability that $(N - 1)$ flips of a fair coin will result in $(m_1 - 1)$ or fewer heads.

Although the hidden-unit surfaces envisioned in the derivation of Eq. (5.5) are of a polynomial form and therefore different from those commonly used in radial-basis function networks, the essential content of the equation has general applicability. Specifically, the higher we make the dimension m_1 of the hidden space, the closer the probability $P(N, m_1)$ will be to unity. To sum up, Cover's theorem on the separability of patterns encompasses two basic ingredients:

1. Nonlinear formulation of the hidden function defined by $\varphi_i(\mathbf{x})$, where $\mathbf{x}$ is the input vector and $i = 1, 2, ..., m_1$.
2. High dimensionality of the hidden space compared to the input space; that dimensionality is determined by the value assigned to m_1 (i.e., the number of hidden units).

In general, as stated previously, a complex pattern-classification problem cast in high-dimensional space nonlinearly is more likely to be linearly separable than in a low-dimensional space. We emphasize, however, that in some cases the use of nonlinear mapping (i.e., point 1) may be sufficient to produce linear separability without having to increase the dimensionality of the hidden-unit space, as illustrated in the following example.

Example 5.1. The XOR Problem

To illustrate the significance of the idea of φ-separability of patterns, consider the simple yet important XOR problem. In the XOR problem there are four points (patterns): $(1, 1)$, $(0, 1)$, $(0, 0)$, and $(1, 0)$, in a two-dimensional input space, as depicted in Fig. 5.2a. The requirement is to construct a pattern classifier that produces the binary output 0 in response to the input pattern $(1, 1)$, or $(0, 0)$, and the binary output 1 in response to the input pattern $(0, 1)$ or $(1, 0)$. Thus points that are closest in the input space, in terms of the Hamming distance, map to regions that are maximally apart in the output space.

Define a pair of Gaussian hidden functions as follows:

$$\varphi_1(\mathbf{x}) = e^{-\|\mathbf{x} - \mathbf{t}_1\|^2}, \qquad \mathbf{t}_1 = [1, 1]^T$$

$$\varphi_2(\mathbf{x}) = e^{-\|\mathbf{x} - \mathbf{t}_2\|^2}, \qquad \mathbf{t}_2 = [0, 0]^T$$

We may then construct the results summarized in Table 5.1 for the four different input patterns of interest. The input patterns are mapped onto the $\varphi_1 - \varphi_2$ plane as shown in Fig. 5.2b. Here we now see that the input patterns $(0, 1)$ and $(1, 0)$ are linearly separable from the remaining input patterns $(1, 1)$ and $(0, 0)$. Thereafter, the XOR problem may be readily solved by using the functions $\varphi_1(\mathbf{x})$ and $\varphi_2(\mathbf{x})$ as the inputs to a linear classifier such as the perceptron.

■

In this example there is no increase in the dimensionality of the hidden space compared to the input space. In other words, nonlinearity exemplified by the use of Gaussian hidden functions is sufficient to transform the XOR problem into a linearly separable one.

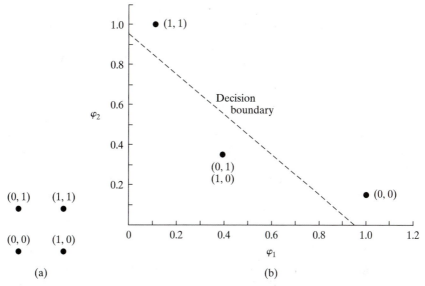

FIGURE 5.2 (a) The four patterns of the XOR problem; (b) Decision-making diagram.

TABLE 5.1 Specification of the Hidden Functions for the XOR Problem of Example 5.1

Input Pattern, $\mathbf{x}$	First Hidden Function, $\varphi_1(\mathbf{x})$	Second Hidden Function, $\varphi_2(\mathbf{x})$
(1,1)	1	0.1353
(0,1)	0.3678	0.3678
(0,0)	0.1353	1
(1,0)	0.3678	0.3678

Separating Capacity of a Surface

Equation (5.5) has an important bearing on the expected maximum number of randomly assigned patterns that are linearly separable in a multidimensional space. To explore this issue, let $\mathbf{x}_1, \mathbf{x}_2, ..., \mathbf{x}_N$ be a sequence of random patterns (vectors) as previously described. Let N be a random variable defined as the largest integer such that this sequence is φ-separable, where φ has m_1 degrees of freedom. Then, from Eq. (5.5) we deduce that the probability that $N = n$ is given by

$$\text{Prob}(N = n) = P(n, m_1) - P(n + 1, m_1)$$

$$= \left(\frac{1}{2}\right)^n \binom{n-1}{m_1 - 1}, \quad n = 0, 1, 2, ..., \tag{5.6}$$

For an interpretation of this result, we recall the definition of a *negative binomial distribution*. This distribution equals the probability that k failures precede the rth success in a long, repeated sequence of *Bernoulli trials*. In such a probabilistic experiment there are only two possible outcomes for each trial, success or failure, and their probabilities remain the same throughout the experiment. Let p and q denote the probabilities of success and failure, respectively, with $p + q = 1$. The negative binomial distribution is defined by (Feller, 1968)

$$f(k;r,p) = p^r q^k \binom{r+k-1}{k}$$

For the special case of $p = q = 1/2$ (i.e., success and failure are equiprobable) and $k + r = n$, the negative binomial distribution reduces to

$$f\left(k;n - k, \frac{1}{2}\right) = \left(\frac{1}{2}\right)^n \binom{n-1}{k}, \quad n = 0, 1, 2, \dots$$

With this definition, we now see that the result described in Eq. (5.6) is just the negative binomial distribution, shifted by m_1 units to the right, and with parameters m_1 and $1/2$. Thus N corresponds to the "waiting time" for the m_1-th failure in a sequence of tosses of a fair coin. The expectation of the random variable N and its median are respectively:

$$E[N] = 2m_1 \tag{5.7}$$

and

$$\text{Median}[N] = 2m_1 \tag{5.8}$$

We therefore have a corollary to Cover's theorem in the form of a celebrated asymptotic result that may be stated as (Cover, 1965):

> The expected maximum number of randomly assigned patterns (vectors) that are linearly separable in a space of dimensionality m_1 is equal to $2m_1$.

This result suggests that $2m_1$ is a natural definition of the *separating capacity* of a family of decision surfaces having m_1 degrees of freedom. In a way, the separating capacity of a surface is closely related to the notion of VC dimension that is discussed in Chapter 2.

5.3   INTERPOLATION PROBLEM

The important point that emerges from Cover's theorem on the separability of patterns is that in solving a nonlinearly separable pattern-classification problem, there is usually practical benefit to be gained by mapping the input space into a new space of high enough dimension. Basically, a nonlinear mapping is used to transform a nonlinearly separable classification problem into a linearly separable one. In a similiar way, we may use a nonlinear mapping to transform a difficult nonlinear filtering problem into an easier one that involves linear filtering.

Consider then a feedforward network with an input layer, a single hidden layer, and an output layer consisting of a single unit. We have purposely chosen a single output unit to simplify the exposition without loss of generality. The network is designed to perform a *nonlinear mapping* from the input space to the hidden space, followed by a *linear mapping* from the hidden space to the output space. Let m_o denote the dimension of the input space. Then, in an overall fashion, the network represents a map from the m_o-dimensional input space to the single-dimensional output space, written as

$$s: \mathbb{R}^{m_0} \rightarrow \mathbb{R}^1 \qquad (5.9)$$

We may think of the map s as a *hypersurface* (graph) $\Gamma \subset \mathbb{R}^{m_o+1}$, just as we think of the elementary map $s: \mathbb{R}^1 \rightarrow \mathbb{R}^1$, where $s(x) = x^2$, as a parabola drawn in $\mathbb{R}^2$ space. The surface Γ is a multidimensional plot of the output as a function of the input. In a practical situation, the surface Γ is unknown and the training data are usually contaminated with noise. The training phase and generalization phase of the learning process may be respectively viewed as follows (Broomhead and Lowe, 1988):

- The training phase constitutes the optimization of a fitting procedure for the surface Γ, based on known data points presented to the network in the form of input–output examples (patterns).
- The generalization phase is synonymous with interpolation between the data points, with the interpolation being performed along the constrained surface generated by the fitting procedure as the optimum approximation to the true surface Γ.

Thus we are led to the theory of *multivariable interpolation* in high-dimensional space, which has a long history (Davis, 1963). The interpolation problem, in its *strict* sense, may be stated:

> *Given a set of N different points $\{x_i \in \mathbb{R}^{m_0} | i = 1, 2, ..., N\}$ and a corresponding set of N real numbers $\{d_i \in \mathbb{R}^1 | i = 1, 2, ..., N\}$, find a function $F: \mathbb{R}^N \rightarrow \mathbb{R}^1$ that satisfies the interpolation condition:*

$$F(x_i) = d_i, \qquad i = 1, 2, ..., N \qquad (5.10)$$

For strict interpolation as specified here, the interpolating surface (i.e., function F) is constrained to pass through *all* the training data points.

The *radial-basis functions* (RBF) technique consists of choosing a function F that has the following form (Powell, 1988):

$$F(\mathbf{x}) = \sum_{i=1}^{N} w_i \varphi(\|\mathbf{x} - \mathbf{x}_i\|) \qquad (5.11)$$

where $\{\varphi(\|\mathbf{x} - \mathbf{x}_i\|) | i = 1, 2, ..., N\}$ is a set of N arbitrary (generally nonlinear) functions, known as *radial-basis functions,* and $\|\cdot\|$ denotes a *norm* that is usually Euclidean. The known data points $\mathbf{x}_i \in \mathbb{R}^{m_0}, i = 1, 2, ..., N$ are taken to be the *centers* of the radial-basis functions.

Inserting the interpolation conditions of Eq. (5.10) in (5.11), we obtain the following set of simultaneous linear equations for the unknown coefficients (weights) of the expansion $\{w_i\}$:

$$\begin{bmatrix} \varphi_{11} & \varphi_{12} & \cdots & \varphi_{1N} \\ \varphi_{21} & \varphi_{22} & \cdots & \varphi_{2N} \\ \vdots & \vdots & \vdots & \vdots \\ \varphi_{N1} & \varphi_{N2} & \cdots & \varphi_{NN} \end{bmatrix} \begin{bmatrix} w_1 \\ w_2 \\ \vdots \\ w_N \end{bmatrix} = \begin{bmatrix} d_1 \\ d_2 \\ \vdots \\ d_N \end{bmatrix} \qquad (5.12)$$

where

$$\varphi_{ji} = \varphi(\|\mathbf{x}_j - \mathbf{x}_i\|), \quad (j, i) = 1, 2, ..., N \tag{5.13}$$

Let

$$\mathbf{d} = [d_1, d_2, ..., d_N]^T$$

$$\mathbf{w} = [w_1, w_2, ..., w_N]^T$$

The N-by-1 vectors $\mathbf{d}$ and $\mathbf{w}$ represent the *desired response vector* and *linear weight vector,* respectively, where N is the *size of the training sample.* Let $\mathbf{\Phi}$ denote an N-by-N matrix with elements φ_{ji}:

$$\mathbf{\Phi} = \{\varphi_{ji} | (j, i) = 1, 2, ..., N\} \tag{5.14}$$

We call this matrix the *interpolation matrix.* We may then rewrite Eq. (5.12) in the compact form

$$\mathbf{\Phi}\mathbf{w} = \mathbf{x} \tag{5.15}$$

Assuming that $\mathbf{\Phi}$ is nonsingular and therefore that the inverse matrix $\mathbf{\Phi}^{-1}$ exists, we may go on to solve Eq. (5.15) for the weight vector $\mathbf{w}$ as shown by

$$\mathbf{w} = \mathbf{\Phi}^{-1}\mathbf{x} \tag{5.16}$$

The vital question is: How can we be sure that the interpolation matrix $\mathbf{\Phi}$ is nonsingular? It turns out that for a large class of radial-basis functions and under certain conditions, the answer to this question is given in the following important theorem.

Micchelli's Theorem

In Micchelli (1986), the following theorem is proved:

> Let $\{\mathbf{x}_i\}_{i=1}^N$ be a set of distinct points in $\mathbb{R}^{m_0}$. Then the N-by-N interpolation matrix $\mathbf{\Phi}$, whose ji-th element is $\varphi_{ji} = \varphi(\|\mathbf{x}_j - \mathbf{x}_i\|)$, is nonsingular.

There is a large class of radial-basis functions that is covered by Micchelli's theorem; it includes the following functions that are of particular interest in the study of RBF networks:

1. *Multiquadrics:*

$$\varphi(r) = (r^2 + c^2)^{1/2} \quad \text{for some } c > 0 \text{ and } r \in \mathbb{R} \tag{5.17}$$

2. *Inverse multiquadrics:*

$$\varphi(r) = \frac{1}{(r^2 + c^2)^{1/2}} \quad \text{for some } c > 0 \text{ and } r \in \mathbb{R} \tag{5.18}$$

3. *Gaussian functions:*

$$\varphi(r) = \exp\left(-\frac{r^2}{2\sigma^2}\right) \quad \text{for some } \sigma > 0 \text{ and } r \in \mathbb{R} \tag{5.19}$$

The multiquadrics and inverse multiquadrics are both due to Hardy (1971).

For the radial-basis functions listed in Eqs. (5.17) to (5.19) to be nonsingular, the points $\{x_i\}_{i=1}^{N}$ must all be different (i.e., distinct). This is all that is required for nonsingularity of the interpolation matrix Φ, whatever the values of size N of the data points or dimensionality m_0 of the vectors (points) x_i.

The inverse multiquadrics of Eq. (5.18) and the Gaussian functions of (5.19) share a common property: They are both *localized* functions, in the sense that $\varphi(r) \to 0$ as $r \to \infty$. In both of these cases the interpolation matrix Φ is positive definite. By contrast, the multiquadrics of Eq. (5.17) are *nonlocal* in that $\varphi(r)$ becomes unbounded as $r \to \infty$; and the corresponding interpolation matrix Φ has N-1 *negative* eigenvalues and only one positive eigenvalue, with the result that it is *not* positive definite (Micchelli, 1986). What is remarkable, however, is that an interpolation matrix Φ based on Hardy's multiquadrics is nonsingular, and therefore suitable for use in the design of RBF networks.

What is even more remarkable is that radial-basis functions that *grow* at infinity, such as multiquadrics, can be used to approximate a smooth input–output mapping with greater accuracy than those that yield a positive-definite interpolation matrix. This surprising result is discussed in Powell (1988).

5.4   SUPERVISED LEARNING AS AN ILL-POSED HYPERSURFACE RECONSTRUCTION PROBLEM

The strict interpolation procedure described may not be a good strategy for the training of RBF networks for certain classes of tasks because of poor generalization to new data for the following reason: When the number of data points in the training sample is much larger than the number of degrees of freedom of the underlying physical process, and we are constrained to have as many radial-basis functions as data points, the problem is overdetermined. Consequently, the network may end up fitting misleading variations due to idiosyncrasies or noise in the input data, thereby resulting in degraded generalization performance (Broomhead and Lowe, 1988).

To develop a deep understanding of the overfitting problem and how to cure it, we first go back to the viewpoint that the design of a neural network trained to retrieve an output pattern when presented with an input pattern is equivalent to learning a hypersurface (i.e., multidimensional mapping) that defines the output in terms of the input. In other words, *learning is viewed as a problem of hypersurface reconstruction, given a set of data points that may be sparse.*

According to Keller (1976) and Kirsch (1996), two related problems are said to be the *inverse* of each other if the formulation of each of them requires partial or full knowledge of the other. Ordinarily we find that one of the problems has been studied earlier, and perhaps in greater detail than the other. This particular problem is called the *direct problem,* in which case the other one is called the *inverse problem.* However, from a mathematical perspective, there is another more important difference between a direct problem and an inverse problem. Specifically, a problem of interest may be well-posed or ill-posed. The term "well-posed" has been used in applied mathematics since the time of Hadamard in the early 1900s. To explain this terminology, assume that we have a domain X and a range Y taken to be metric spaces, and that are related by a fixed but unknown mapping f. The problem of reconstructing the mapping f is said to be *well-posed* if three conditions are satisfied (Tikhonov and Arsenin, 1977: Morozov, 1993; Kirsch, 1996):

1. *Existence.* For every input vector $\mathbf{x} \in \mathcal{X}$, there does exist an output $y = f(\mathbf{x})$, where $y \in \mathcal{Y}$.
2. *Uniqueness.* For any pair of input vectors $\mathbf{x}, \mathbf{t} \in \mathcal{X}$, we have $f(\mathbf{x}) = f(\mathbf{t})$ if, and only if, $\mathbf{x} = \mathbf{t}$.
3. *Continuity.* The mapping is continuous, that is, for any $\varepsilon > 0$ there exists $\delta = \delta(\varepsilon)$ such that the condition $\rho_x(\mathbf{x}, \mathbf{t}) < \delta$ implies that $\rho_y(f(\mathbf{x}), f(\mathbf{t})) < \varepsilon$, where $\rho(\cdot, \cdot)$ is the symbol for distance between the two arguments in their respective spaces. This criterion is illustrated in Fig 5.3. The property of continuity is also referred to as *stability*.

If any of these conditions is not satisfied, the problem is said to be *ill-posed*. Basically, an ill-posed problem means that large data sets may contain a surprisingly small amount of information about the desired solution.

In the context of our present situation, the physical phenomenon responsible for generating the training data (e.g., speech, pictures, radar signals, sonar signals, seismic data) is a well-posed direct problem. However, learning from such physical forms of data, viewed as a hypersurface reconstruction problem, is an ill-posed inverse problem for the following reasons. First, the existence criterion may be violated in that a distinct output may not exist for every input. Second, there may not be as much information in the training sample as we really need to reconstruct the input–output mapping uniquely, hence the uniqueness criterion is likely to be violated. Third, the unavoidable presence of noise or imprecision in real-life training data adds uncertainty to the reconstructed input–output mapping. In particular, if the noise level in the input is too high, it is possible for the neural network to produce an output outside of the range $\mathcal{Y}$ for a specified input $\mathbf{x}$ in the domain $\mathcal{X}$; in other words, there is likelihood for the continuity criterion to be violated. If a learning problem lacks the property of continuity, then the computed input–output mapping has nothing to do with the true solution to the learning problem. There is no way to overcome this difficulty unless some prior information about the input–output mapping is available. In this context, it is rather appropriate that we remind ourselves of a statement made by Lanczos on linear differential operators (Lanczos, 1964): "A lack of information cannot be remedied by any mathematical trickery."

The important issue of how to make an ill-posed problem into a well-posed one via regularization is discussed in the next section.[3]

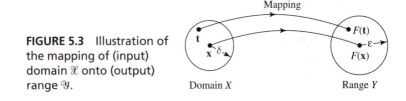

FIGURE 5.3 Illustration of the mapping of (input) domain $\mathcal{X}$ onto (output) range $\mathcal{Y}$.

5.5 REGULARIZATION THEORY

In 1963, Tikhonov proposed a new method called *regularization* for solving ill-posed problems.[4] In the context of a hypersurface reconstruction problem, the basic idea of regularization is to *stabilize* the solution by means of some auxiliary nonnegative functional that embeds prior information about the solution. The most common form of prior information involves the assumption that the input–output mapping function (i.e., solution to the reconstruction problem) is *smooth*, in the sense that similar inputs correspond to similar outputs.

To be specific, let the set of input–output data (i.e., training sample) available for approximation be described by

$$\text{Input signal:} \qquad \mathbf{x}_i \in \mathbb{R}^{m_0}, \quad i = 1, 2, ..., N$$
$$\text{Desired response:} \qquad d_i \in \mathbb{R}^1, \quad i = 1, 2, ..., N \tag{5.20}$$

Note that the output is assumed to be one-dimensional. This assumption does not in any way limit the general applicability of the regularization theory being developed here. Let the approximating function be denoted by $F(\mathbf{x})$, where (for convenience of presentation) we have omitted the weight vector $\mathbf{w}$ of the network from the argument of the function F. Basically, Tikhonov's regularization theory involves two terms:

1. *Standard Error Term.* This first term, denoted by $\mathscr{E}_s(F)$, measures the standard error (distance) between the desired (target) response d_i and the actual response y_i for training example $i = 1, 2, ..., N$. Specifically, we define

$$\mathscr{E}_s(F) = \frac{1}{2} \sum_{i=1}^{N} (d_i - y_i)^2$$
$$= \frac{1}{2} \sum_{i=1}^{N} [d_i - F(\mathbf{x}_i)]^2 \tag{5.21}$$

    where we have introduced the scaling factor $\frac{1}{2}$ for the sake of consistency with material presented in previous chapters.

2. *Regularizing Term.* This second term, denoted for $\mathscr{E}_c(F)$, depends on the "geometric" properties of the approximating function $F(\mathbf{x})$. Specifically, we write

$$\mathscr{E}_c(F) = \frac{1}{2} \|\mathbf{D}F\|^2 \tag{5.22}$$

    where $\mathbf{D}$ is a *linear differential operator*. Prior information about the form of the solution [i.e., the input–output mapping function $F(\mathbf{x})$] is embedded in the operator $\mathbf{D}$, which naturally makes the selection of $\mathbf{D}$ problem-dependent. We also refer to $\mathbf{D}$ as a *stabilizer* because it stabilizes the solution to the regularization problem, making it smooth and thereby satisfying the property of continuity. However, smoothness implies continuity, but the reverse is not necessarily true.

The analytic approach used to handle the situation described in Eq. (5.22) builds on the concept of a *function space*,[5] which refers to a *normed space*[6] of functions. In such a space of many (strictly speaking, infinitely many) dimensions, a continuous function is represented by a *vector*. By using this geometrical image, an insightful link is established between matrices and linear differential operators. The analysis of linear systems thereby becomes translatable to the analysis of linear differential equations (Lanczos, 1964).

Thus, the symbol $\|\cdot\|$ in Eq. (5.22) denotes a norm imposed on the function space to which $\mathbf{D}F(\mathbf{x})$ belongs. Ordinarily the function space used here is the L_2 *space* that consists of all real-valued functions $f(\mathbf{x})$, $\mathbf{x} \in \mathbb{R}^{m_0}$, for which $\|f(\mathbf{x})\|^2$ is Lebesgue integrable. The function $f(\mathbf{x})$ used here denotes the actual function that defines the underlying physical process responsible for generating the set of input–output data pairs $\{(\mathbf{x}_i, d_i)\}_{i=1}^{N}$; see note 7 for more details.

The quantity to be minimized in regularization theory is

$$\mathscr{E}(F) = \mathscr{E}_s(F) + \lambda\mathscr{E}_c(F)$$

$$= \frac{1}{2} \sum_{i=1}^{N} [d_i - F(\mathbf{x}_i)]^2 + \frac{1}{2} \lambda \|\mathbf{D}F\|^2 \tag{5.23}$$

where λ is a positive real number called the *regularization parameter* and $\mathscr{E}(F)$ is called the *Tikhonov functional*. A functional maps functions (defined in some suitable function space) onto the real line. The minimizer of the Tikhonov functional $\mathscr{E}(F)$ (i.e., the solution to the regularization problem) is denoted by $F_\lambda(\mathbf{x})$.

In a sense, we may view the regularization parameter λ as an indicator of the sufficiency of the given data set as examples that specify the solution $F_\lambda(\mathbf{x})$. In particular, the limiting case $\lambda \to 0$ implies that the problem is unconstrained, with the solution $F_\lambda(\mathbf{x})$ being completely determined from the examples. The other limiting case, $\lambda \to \infty$, on the other hand, implies that the prior smoothness constraint imposed by the differential operator $\mathbf{D}$ is by itself sufficient to specify the solution $F_\lambda(\mathbf{x})$, which is another way of saying that the examples are unreliable. In practical applications, the regularization parameter λ is assigned a value somewhere between these two limiting conditions, so that both the sample data and the prior information contribute to the solution $F_\lambda(\mathbf{x})$. Thus the regularizing term $\mathscr{E}_c(F)$ represents a *model complexity-penalty function,* the influence of which on the final solution is controlled by the regularization parameter λ.

Another way of viewing regularization is that it provides a practical solution to the bias-variance dilemma that is discussed in Chapter 2. Specifically, the optimum choice of the regularization parameter λ is designed to steer the solution to the learning problem toward a satisfactory balance between model bias and model variance by incorporating the right amount of prior information into it.

Fréchet Differential of the Tikhonov Functional

The *principle of regularization* may now be stated as:

Find the function $F_\lambda(\mathbf{x})$ that minimizes the Tikhonov functional $\mathscr{E}(F)$, defined by

$$\mathscr{E}(F) = \mathscr{E}_s(F) + \lambda\mathscr{E}_c(F)$$

where $\mathscr{E}_s(F)$ is the standard error term, $\mathscr{E}_c(F)$ is the regularizing term, and λ is the regularization parameter.

To proceed with the minimization of the cost functional $\mathscr{E}(F)$, we need a rule for evaluating the differential of $\mathscr{E}(F)$. We can take care of this matter by using the *Fréchet differential*. In elementary calculus, the tangent to a curve is a straight line that gives the best approximation of the curve in the neighborhood of the point of tangency. Similarly, the Fréchet differential of a functional may be interpreted as the best local linear approximation. Thus the Fréchet differential of the functional $\mathscr{E}(F)$ is formally defined by (Dorny, 1975; Debnath and Mikusiński, 1990; de Figueiredo and Chen, 1993):

$$d\mathscr{E}(F, h) = \left[\frac{d}{d\beta} \mathscr{E}(F + \beta h)\right]_{\beta=0} \tag{5.24}$$

where $h(\mathbf{x})$ is a fixed function of the vector $\mathbf{x}$. In Eq. (5.24), the ordinary rules of differentiation are used. A necessary condition for the function $F(\mathbf{x})$ to be a relative extremum of the functional $\mathscr{E}(F)$ is that the Fréchet differential $d\mathscr{E}(F, h)$ must be zero at $F(\mathbf{x})$ for all $h \in \mathscr{H}$, as shown by

$$d\mathscr{E}(F, h) = d\mathscr{E}_s(F, h) + \lambda d\mathscr{E}_c(F, h) = 0 \tag{5.25}$$

where $d\mathscr{E}_s(F, h)$ and $d\mathscr{E}_c(F, h)$ are the Fréchet differentials of the functionals $\mathscr{E}_s(F)$ and $\mathscr{E}_c(F)$, respectively.

Evaluating the Fréchet differential of the standard error term $\mathscr{E}_s(F, h)$ of Eq. (5.21), we have

$$\begin{aligned}
d\mathscr{E}_s(F, h) &= \left[\frac{d}{d\beta} \mathscr{E}_s(F + \beta h)\right]_{\beta=0} \\
&= \left[\frac{1}{2}\frac{d}{d\beta} \sum_{i=1}^{N} [d_i - F(\mathbf{x}_i) - \beta h(\mathbf{x}_i)]^2\right]_{\beta=0} \\
&= -\sum_{i=1}^{N} [d_i - F(\mathbf{x}_i) - \beta h(\mathbf{x}_i)]h(\mathbf{x}_i)|_{\beta=0} \\
&= -\sum_{i=1}^{N} [d_i - F(\mathbf{x}_i)]h(\mathbf{x}_i)
\end{aligned} \tag{5.26}$$

At this point in the discussion, we find it instructive to invoke the *Riesz representation theorem,* which may be stated as follows (Debnath and Mikusiński, 1990; Kirsch, 1996):

Let f be a bounded linear functional in a Hilbert space (i.e., an inner product space that is complete)[8] denoted by $\mathscr{H}$. There exists one $h_0 \in \mathscr{H}$ such that

$$f = (h, h_0)_{\mathscr{H}} \quad \text{for all } h \in \mathscr{H}$$

Moreover, we have

$$\|f\|_{\widetilde{\mathscr{H}}} = \|h_0\|_{\mathscr{H}}$$

where $\widetilde{\mathscr{H}}$ is the dual or conjugate of the Hilbert space $\mathscr{H}$.

The symbol $(\cdot, \cdot)_{\mathscr{H}}$ used here stands for the *inner (scalar) product* of two functions in $\mathscr{H}$ space. Hence, in light of the Riesz representation theorem, we may rewrite the Fréchet differential $d\mathscr{E}_s(F, h)$ of Eq. (5.26) in the equivalent form

$$d\mathscr{E}_s(F, h) = -\left(h, \sum_{i=1}^{N} (d_i - F)\delta_{\mathbf{x}_i}\right)_{\mathscr{H}} \tag{5.27}$$

where $\delta_{\mathbf{x}_i}$ denotes the *Dirac delta distribution* of $\mathbf{x}$, centered at $\mathbf{x}_i$; that is,

$$\delta_{\mathbf{x}_i}(\mathbf{x}) = \delta(\mathbf{x} - \mathbf{x}_i) \tag{5.28}$$

Consider next the evaluation of the Fréchet differential of the regularizing term $\mathscr{E}_c(F)$ of Eq. (5.22). Proceeding in a manner similar to that just described, we have

$$d\mathscr{E}_c(F, h) = \frac{d}{d\beta} \mathscr{E}_c(F + \beta h)|_{\beta=0}$$

$$= \frac{1}{2} \frac{d}{d\beta} \int_{\mathbb{R}^{m_0}} (\mathbf{D}[F + \beta h])^2 d\mathbf{x}|_{\beta=0}$$

$$= \int_{\mathbb{R}^{m_0}} \mathbf{D}[F + \beta h] \mathbf{D}h \, d\mathbf{x}|_{\beta=0} \qquad (5.29)$$

$$= \int_{\mathbb{R}^{m_0}} \mathbf{D}F \, \mathbf{D}h \, d\mathbf{x}$$

$$= (\mathbf{D}h, \mathbf{D}F)_{\mathscr{H}}$$

where $(\mathbf{D}h, \mathbf{D}F)_{\mathscr{H}}$ is the inner product of the two functions $\mathbf{D}h(\mathbf{x})$ and $\mathbf{D}F(\mathbf{x})$ that result from the action of the differential operator $\mathbf{D}$ on $h(\mathbf{x})$ and $F(\mathbf{x})$, respectively.

Euler–Lagrange Equation

Given a linear differential operator $\mathbf{D}$, we can find a uniquely determined *adjoint operator*, denoted by $\tilde{\mathbf{D}}$, such that for any pair of functions $u(\mathbf{x})$ and $v(\mathbf{x})$ which are sufficiently differentiable and which satisfy proper boundary conditions, we can write (Lanczos, 1964)

$$\int_{\mathbb{R}^m} u(\mathbf{x})\mathbf{D}v(\mathbf{x})d\mathbf{x} = \int_{\mathbb{R}^m} v(\mathbf{x})\tilde{\mathbf{D}}u(\mathbf{x})d\mathbf{x} \qquad (5.30)$$

Equation (5.30) is called *Green's identity*; it provides a mathematical basis for defining the adjoint operator $\tilde{\mathbf{D}}$ in terms of the given differential $\mathbf{D}$. Viewing $\mathbf{D}$ as a matrix, the adjoint operator $\tilde{\mathbf{D}}$ plays a role similar to that of a matrix transpose.

Comparing the left-hand side of Eq. (5.30) with the fourth line of Eq. (5.29), we may make the following identifications:

$$u(\mathbf{x}) = \mathbf{D}F(\mathbf{x})$$

$$\mathbf{D}v(\mathbf{x}) = \mathbf{D}h(\mathbf{x})$$

Using Green's identity, we may rewrite Eq. (5.29) in the equivalent form

$$d\mathscr{E}_c(F, h) = \int_{\mathbb{R}^{m_0}} h(\mathbf{x})\tilde{\mathbf{D}}\mathbf{D}F(\mathbf{x})d\mathbf{x}$$

$$= (h, \tilde{\mathbf{D}}\mathbf{D}F)_{\mathscr{H}} \qquad (5.31)$$

where $\tilde{\mathbf{D}}$ is the adjoint of $\mathbf{D}$.

Returning to the extremum condition described in Eq. (5.25) and substituting the Fréchet differentials of Eqs. (5.27) and (5.31) in that equation, we may now express the Fréchet differential $d\mathscr{E}(F, h)$ as

$$d\mathscr{E}(F, h) = \left(h, \left[\tilde{\mathbf{D}}\mathbf{D}F - \frac{1}{\lambda}\sum_{i=1}^{N}(d_i - F)\delta_{\mathbf{x}_i}\right]\right)_{\mathscr{H}} \qquad (5.32)$$

Since the regularization parameter λ is ordinarily assigned a value somewhere in the open interval $(0, \infty)$, the Fréchet differential $d\mathcal{E}(F, h)$ is zero for every $h(\mathbf{x})$ in $\mathcal{H}$ space if and only if the following condition is satisfied in the distributional sense:

$$\tilde{\mathbf{D}}\mathbf{D}F_\lambda - \frac{1}{\lambda} \sum_{i=1}^{N} (d_i - F)\delta_{\mathbf{x}_i} = 0$$

or equivalently,

$$\tilde{\mathbf{D}}\mathbf{D}F_\lambda(\mathbf{x}) = \frac{1}{\lambda} \sum_{i=1}^{N} [d_i - F(\mathbf{x}_i)]\delta(\mathbf{x} - \mathbf{x}_i) \tag{5.33}$$

Equation (5.33) is the *Euler–Lagrange equation* for the Tikhonov functional $\mathcal{E}(F)$; it defines a necessary condition for the Tikhonov functional $\mathcal{E}(F)$ to have an extremum at $F_\lambda(\mathbf{x})$ (Debnath and Mikusiński, 1990).

Green's Function

Equation (5.33) represents a partial differential equation in the approximating function F. The solution of this equation is known to consist of the integral transformation of the right-hand side of the equation.

Let $G(\mathbf{x}, \boldsymbol{\xi})$ denote a function in which both vectors $\mathbf{x}$ and $\boldsymbol{\xi}$ appear on equal footing but for different purposes: $\mathbf{x}$ as a parameter and $\boldsymbol{\xi}$ as an argument. For a given linear differential operator $\mathbf{L}$, we stipulate that the function $G(\mathbf{x}, \boldsymbol{\xi})$ satisfies the following conditions (Courant and Hilbert, 1970):

1. For a fixed $\boldsymbol{\xi}$, $G(\mathbf{x}, \boldsymbol{\xi})$ is a function of $\mathbf{x}$ and satisfies the prescribed boundary conditions.
2. Except at the point $\mathbf{x} = \boldsymbol{\xi}$, the derivatives of $G(\mathbf{x}, \boldsymbol{\xi})$ with respect to $\mathbf{x}$ are all continuous; the number of derivatives is determined by the order of the operator $\mathbf{L}$.
3. With $G(\mathbf{x}, \boldsymbol{\xi})$ considered as a function of $\mathbf{x}$, it satisfies the partial differential equation

$$\mathbf{L}G(\mathbf{x}, \boldsymbol{\xi}) = 0 \tag{5.34}$$

everywhere except at the point $\mathbf{x} = \boldsymbol{\xi}$, where it has a singularity. That is, the function $G(\mathbf{x}, \boldsymbol{\xi})$ satisfies the following partial differential equation (taken in the sense of distributions)

$$\mathbf{L}G(\mathbf{x}, \boldsymbol{\xi}) = \delta(\mathbf{x} - \boldsymbol{\xi}) \tag{5.35}$$

where, as defined previously, $\delta(\mathbf{x} - \boldsymbol{\xi})$ is the Dirac delta function positioned at the point $\mathbf{x} = \boldsymbol{\xi}$.

The function $G(\mathbf{x}, \boldsymbol{\xi})$ thus described is called the *Green's function* for the differential operator $\mathbf{L}$. The Green's function plays a role for a linear differential operator that is similar to that for the inverse matrix for a matrix equation.

Let $\varphi(\mathbf{x})$ denote a continuous or piecewise continuous function of $\mathbf{x} \in \mathbb{R}^{m_0}$. Then the function

$$F(\mathbf{x}) = \int_{\mathbb{R}^{m_0}} G(\mathbf{x}, \boldsymbol{\xi})\varphi(\boldsymbol{\xi})d\boldsymbol{\xi} \tag{5.36}$$

is a solution of the differential equation

$$\mathbf{L}F(\mathbf{x}) = \varphi(\mathbf{x}) \tag{5.37}$$

where $G(\mathbf{x}, \boldsymbol{\xi})$ is the Green's function for the linear differential operator $\mathbf{L}$ (Courant and Hilbert, 1970).

To prove the validity of $F(\mathbf{x})$ as a solution of Eq. (5.37), apply the differential operator $\mathbf{L}$ to Eq. (5.36), obtaining

$$\begin{aligned} \mathbf{L}F(\mathbf{x}) &= \mathbf{L} \int_{\mathbb{R}^{m_0}} G(\mathbf{x}, \boldsymbol{\xi})\varphi(\boldsymbol{\xi})d(\boldsymbol{\xi}) \\ &= \int_{\mathbb{R}^{m_0}} \mathbf{L}G(\mathbf{x}, \boldsymbol{\xi})\varphi(\boldsymbol{\xi})d\boldsymbol{\xi} \end{aligned} \tag{5.38}$$

The differential operator $\mathbf{L}$ treats $\boldsymbol{\xi}$ as a constant, acting on the kernel $G(\mathbf{x}; \boldsymbol{\xi})$ only as a function of $\mathbf{x}$. By using Eq. (5.35) in (5.38) we get

$$\mathbf{L}F(\mathbf{x}) = \int_{\mathbb{R}^{m_0}} \delta(\mathbf{x} - \boldsymbol{\xi})\varphi(\boldsymbol{\xi})d\boldsymbol{\xi}$$

Finally, using the *sifting property* of the Dirac delta function, namely

$$\int_{\mathbb{R}^{m_0}} \varphi(\boldsymbol{\xi})\delta(\mathbf{x} - \boldsymbol{\xi})d(\boldsymbol{\xi}) = \varphi(\mathbf{x})$$

we obtain $\mathbf{L}F(\mathbf{x}) = \varphi(\mathbf{x})$, as described in Eq. (5.37).

Solution to the Regularization Problem

Returning to the issue at hand, namely, that of solving the Euler–Lagrange equation (5.33), set

$$\mathbf{L} = \tilde{\mathbf{D}}\mathbf{D} \tag{5.39}$$

and

$$\varphi(\boldsymbol{\xi}) = \frac{1}{\lambda} \sum_{i=1}^{N} [d_i - F(\mathbf{x}_i)]\delta(\boldsymbol{\xi} - \mathbf{x}_i) \tag{5.40}$$

Then we may use Eq. (5.36) to write

$$\begin{aligned} F_\lambda(\mathbf{x}) &= \int_{\mathbb{R}^{m_0}} G(\mathbf{x}, \boldsymbol{\xi})\left\{ \frac{1}{\lambda} \sum_{i=1}^{N} [d_i - F(\mathbf{x}_i)]\delta(\boldsymbol{\xi} - \mathbf{x}_i) \right\}d\boldsymbol{\xi} \\ &= \frac{1}{\lambda} \sum_{i=1}^{N} [d_i - F(\mathbf{x}_i)] \int_{\mathbb{R}^{m_0}} G(\mathbf{x}, \boldsymbol{\xi})\delta(\boldsymbol{\xi} - \mathbf{x}_i)d\boldsymbol{\xi} \end{aligned}$$

where in the last line we have interchanged the order of integration and summation. Finally, using the sifting property of the Dirac delta function, we get the desired solution to the Euler–Lagrange equation (5.33) as follows:

$$F_\lambda(\mathbf{x}) = \frac{1}{\lambda} \sum_{i=1}^{N} [d_i - F(\mathbf{x}_i)]G(\mathbf{x}, \mathbf{x}_i) \tag{5.41}$$

Equation (5.41) states that the minimizing solution $F_\lambda(\mathbf{x})$ to the regularization problem is a linear superposition of N Green's functions. The $\mathbf{x}_i$ represent the *centers of the expansion,* and the weights $[d_i - F(\mathbf{x}_i)]/\lambda$ represent the *coefficients of the expansion.* In other words, the solution to the regularization problem lies in an N-dimensional subspace of the space of smooth functions, and the set of Green's functions $\{G(\mathbf{x}, \mathbf{x}_i)\}$ centered at $\mathbf{x}_i$, $i = 1, 2, ..., N$, constitutes a basis for this subspace. (Poggio and Girosi, 1990a). Note that the coefficients of expansion in Eq. (5.41) are first, *linear* in the estimation error defined as the difference between the desired response d_i and the corresponding output $F(\mathbf{x}_i)$ computed by the network, and second, inversely proportional to the regularization parameter λ.

Determination of the Expansion Coefficients

The next issue to be resolved is the determination of the unknown coefficients in the expansion of Eq. (5.41). Let

$$w_i = \frac{1}{\lambda} [d_i - F(\mathbf{x}_i)], \quad i = 1, 2, ..., N \tag{5.42}$$

We may recast the minimizing solution of Eq. (5.41) simply as follows:

$$F_\lambda(\mathbf{x}) = \sum_{i=1}^{N} w_i G(\mathbf{x}, \mathbf{x}_i) \tag{5.43}$$

Evaluating Eq. (5.43) at $\mathbf{x}_j, j = 1, 2, ..., N$, we get

$$F_\lambda(\mathbf{x}_j) = \sum_{i=1}^{N} w_i G(\mathbf{x}_j, \mathbf{x}_i), \quad j = 1, 2, ..., N \tag{5.44}$$

We now introduce the following definitions:

$$\mathbf{F}_\lambda = [F_\lambda(\mathbf{x}_1), F_\lambda(\mathbf{x}_2), ..., F_\lambda(\mathbf{x}_N)]^T \tag{5.45}$$

$$\mathbf{d} = [d_1, d_2, ..., d_N]^T \tag{5.46}$$

$$\mathbf{G} = \begin{bmatrix} G(\mathbf{x}_1, \mathbf{x}_1) & G(\mathbf{x}_1, \mathbf{x}_2) & \cdots & G(\mathbf{x}_1, \mathbf{x}_N) \\ G(\mathbf{x}_2, \mathbf{x}_1) & G(\mathbf{x}_2, \mathbf{x}_2) & \cdots & G(\mathbf{x}_2, \mathbf{x}_N) \\ \vdots & \vdots & & \vdots \\ G(\mathbf{x}_N, \mathbf{x}_1) & G(\mathbf{x}_N, \mathbf{x}_2) & \cdots & G(\mathbf{x}_N, \mathbf{x}_N) \end{bmatrix} \tag{5.47}$$

$$\mathbf{w} = [w_1, w_2, ..., w_N]^T \tag{5.48}$$

Then we may rewrite Eqs. (5.42) and (5.44) in matrix form as follows, respectively:

$$\mathbf{w} = \frac{1}{\lambda} (\mathbf{d} - \mathbf{F}_\lambda) \tag{5.49}$$

and

$$\mathbf{F}_\lambda = \mathbf{G}\mathbf{w} \tag{5.50}$$

Eliminating $\mathbf{F}_\lambda$ between Eqs. (5.49) and (5.50) and rearranging terms, we get

$$(\mathbf{G} + \lambda\mathbf{I})\mathbf{w} = \mathbf{d} \tag{5.51}$$

where $\mathbf{I}$ is the N-by-N identity matrix. We call matrix $\mathbf{G}$ the *Green's matrix*.

The differential operator $\mathbf{L}$ defined in Eq. (5.39) is *self-adjoint*, in the sense that its adjoint is equal to operator $\mathbf{L}$ itself. It follows therefore that the associated Green's function $\mathbf{G}(\mathbf{x}, \mathbf{x}_i)$ is a *symmetric function*, as shown by

$$G(\mathbf{x}_i, \mathbf{x}_j) = G(\mathbf{x}_j, \mathbf{x}_i) \quad \text{for all } i \text{ and } j \tag{5.52}$$

Equation (5.52) states that the positions of the two points $\mathbf{x}$ and $\boldsymbol{\xi}$ can be interchanged without affecting the value of the Green's function $G(\mathbf{x}, \boldsymbol{\xi})$. Equivalently, the Green's matrix $\mathbf{G}$ defined in Eq. (5.47) is a *symmetric matrix;* that is,

$$\mathbf{G}^T = \mathbf{G} \tag{5.53}$$

We now invoke the interpolation theorem, which is described in Section 5.3 in the context of the interpolation matrix $\mathbf{\Phi}$. We first note that Green's matrix $\mathbf{G}$ plays a role in regularization theory similar to that of $\mathbf{\Phi}$ in RBF interpolation theory. Both $\mathbf{G}$ and $\mathbf{\Phi}$ are N-by-N symmetric matrices. Accordingly, we may state that the matrix $\mathbf{G}$, for certain classes of Green's functions, is positive definite provided that the data points $\mathbf{x}_1, \mathbf{x}_2, ..., \mathbf{x}_N$ are distinct. The classes of Green's functions covered by Micchelli's theorem include inverse multiquadrics and Gaussian functions, but not multiquadrics. In practice, we may always choose λ sufficiently large to ensure that $\mathbf{G} + \lambda\mathbf{I}$ is positive definite and therefore invertible. This, in turn, means that the linear system of equations (5.51) will have a unique solution given by (Poggio and Girosi, 1990a)

$$\mathbf{w} = (\mathbf{G} + \lambda\mathbf{I})^{-1}\mathbf{d} \tag{5.54}$$

Thus, having selected the differential operator $\mathbf{D}$ and therefore having identified the associated Green's function $G(\mathbf{x}_j, \mathbf{x}_i)$, where $i = 1, 2, ..., N$, we may use Eq. (5.54) to obtain the weight vector $\mathbf{w}$ for a specified desired response vector $\mathbf{d}$ and an appropriate value of regularization parameter λ.

In conclusion, we may state that the solution to the regularization problem is given by the expansion [9]

$$F_\lambda(\mathbf{x}) = \sum_{i=1}^{N} w_i G(\mathbf{x}, \mathbf{x}_i) \tag{5.55}$$

where $G(\mathbf{x}, \mathbf{x}_i)$ is the Green's function for the self-adjoint differential operator $\mathbf{L} = \tilde{\mathbf{D}}\mathbf{D}$, and w_i is the ith element of the weight vector $\mathbf{w}$; these two quantities are themselves defined by Eq. (5.35) and (5.54), respectively. Equation (5.55) states the following (Poggio and Girosi, 1990a):

- The regularization approach is equivalent to the expansion of the solution in terms of a set of Green's functions, whose characterization depends only on the form adopted for the stabilizer $\mathbf{D}$ and the associated boundary conditions.
- The number of Green's functions used in the expansion is equal to the number of examples used in the training process.

It should be noted, however, that the solution of the regularization problem given in Eq. (5.55) is incomplete, as it represents a solution *modulo* term $g(\mathbf{x})$ that lies in the null space of the operator $\mathbf{D}$ (Poggio and Girosi, 1990a). We say this because all the functions that lie in the null space of $\mathbf{D}$ are "invisible" to the smoothing term $\|\mathbf{D}F\|^2$ in the cost functional $\mathscr{E}(F)$ of Eq. (5.23); by the *null space* of $\mathbf{D}$, we mean the set of all functions $g(\mathbf{x})$ for which $\mathbf{D}g$ is zero. The exact form of the additional term $g(\mathbf{x})$ is problem-dependent in the sense that it depends on the stabilizer chosen and the boundary conditions of the problem at hand. For example, it is not needed in the case of a stabilizer $\mathbf{D}$ corresponding to a bell-shaped Green's function such as a Gaussian or inverse multiquadric. For this reason, and since its inclusion does not modify the main conclusions, we will disregard it in the sequel.

The characterization of the Green's function $G(\mathbf{x}, \mathbf{x}_i)$, for a specified center $\mathbf{x}_i$, depends only on the form of the stabilizer $\mathbf{D}$, that is, on the *a priori* assumption made concerning the input–output mapping. If the stabilizer $\mathbf{D}$ is *translationally invariant*, the Green's function $G(\mathbf{x}, \mathbf{x}_i)$ centered at $\mathbf{x}_i$ will depend only on the difference between the arguments $\mathbf{x}$ and $\mathbf{x}_i$; that is,

$$G(\mathbf{x}, \mathbf{x}_i) = G(\mathbf{x} - \mathbf{x}_i) \tag{5.56}$$

If the stabilizer $\mathbf{D}$ is both *translationally and rotationally invariant*, the Green's function $G(\mathbf{x}, \mathbf{x}_i)$ will depend only on the *Euclidean norm* of the difference vector $\mathbf{x} - \mathbf{x}_i$, as shown by

$$G(\mathbf{x}, \mathbf{x}_i) = G(\|\mathbf{x} - \mathbf{x}_i\|) \tag{5.57}$$

Under these conditions, the Green's function must be a *radial-basis function*. In such a case, the regularized solution of Eq. (5.55) takes on the following special form (Poggio and Girosi, 1990a):

$$F_\lambda(\mathbf{x}) = \sum_{i=1}^{N} w_i G(\|\mathbf{x} - \mathbf{x}_i\|) \tag{5.58}$$

The solution described in Eq. (5.58) constructs a linear function space that depends on the known data points according to the Euclidean distance measure.

The solution described by Eq. (5.58) is termed *strict interpolation*, since all the N data points available for training are used to generate the interpolating function $F(\mathbf{x})$. It is important, however, to realize that this solution differs from that of Eq. (5.11) in a fundamental respect: The solution of Eq. (5.58) is *regularized* by virtue of the definition given in Eq. (5.54) for the weight vector $\mathbf{w}$. It is only when we set the regularization parameter λ equal to zero that the two solutions may become one and the same.

Multivariate Gaussian Functions

The Green's function $G(\mathbf{x}, \mathbf{x}_i)$, whose linear differential operator $\mathbf{D}$ is both translationally and rotationally invariant and which satisfies the condition of Eq. (5.57), is of particular interest in practice. An example of such a Green's function is the *multivariate Gaussian function* defined by

$$G(\mathbf{x}, \mathbf{x}_i) = \exp\left(-\frac{1}{2\sigma_i^2} \|\mathbf{x} - \mathbf{x}_i\|^2\right) \tag{5.59}$$

where $\mathbf{x}_i$ denotes the center of the function and σ_i denotes its width. The self-adjoint operator $\mathbf{L} = \tilde{\mathbf{D}}\mathbf{D}$ defining the Green's function of Eq. (5.59) is given by (Poggio and Girosi, 1990a)

$$\mathbf{L} = \sum_{n=0}^{\infty} (-1)^n \alpha_n \nabla^{2n} \tag{5.60}$$

where

$$\alpha_n = \frac{\sigma_i^{2n}}{n!2^n} \tag{5.61}$$

and ∇^{2n} is the *iterated Laplace operator* in m_0 dimensions, with

$$\nabla^2 = \frac{\partial^2}{\partial x_1^2} + \frac{\partial^2}{\partial x_2^2} + \cdots + \frac{\partial^2}{\partial x_{m_0}^2} \tag{5.62}$$

With the number of terms permitted to go to infinity in Eq. (5.60), $\mathbf{L}$ ceases to be a differential operator in the standard sense. For this reason, the operator $\mathbf{L}$ in Eq. (5.60) is referred to as a *pseudo-differential operator*.

Since by definition, $\mathbf{L} = \tilde{\mathbf{D}}\mathbf{D}$, we deduce from Eq. (5.60) that the operator $\mathbf{D}$ and its adjoint $\tilde{\mathbf{D}}$ are as follows, respectively (see note 10):

$$\mathbf{D} = \sum_n \alpha_n^{1/2} \left(\frac{\partial}{\partial x_1} + \frac{\partial}{\partial x_2} + \cdots + \frac{\partial}{\partial x_{m_0}} \right)^n$$
$$= \sum_{a+b+\cdots+k=n} \alpha_n^{1/2} \frac{\partial^n}{\partial x_1^a \partial x_2^b \ldots \partial x_{m_0}^k}, \tag{5.63}$$

and

$$\tilde{\mathbf{D}} = \sum_n (-1)^n \alpha_n^{1/2} \left(\frac{\partial}{\partial x_1} + \frac{\partial}{\partial x_2} + \cdots + \frac{\partial}{\partial x_{m_0}} \right)^n$$
$$= \sum_{a+b+\cdots k=n} (-1) \alpha_n^{1/2} \frac{\partial^n}{\partial x_1^a \partial x_2^b \ldots \partial x_{m_0}^k} \tag{5.64}$$

Thus the regularized solution described in Eq. (5.58) is attained by using a stabilizer that includes all of its possible partial derivatives.

Using Eqs. (5.59) to (5.61) in (5.35) with ξ set equal to $\mathbf{x}_i$, we may write

$$\sum_n (-1)^n \frac{\sigma_i^{2n}}{n!2^n} \nabla^{2n} \exp\left(-\frac{1}{2\sigma_i^2} \|\mathbf{x} - \mathbf{x}_i\|^2 \right) = \delta(\mathbf{x} - \mathbf{x}_i) \tag{5.65}$$

With the Green's function $G(\mathbf{x}, \mathbf{x}_i)$ defined by the special form of Eq. (5.59), the regularized solution given in Eq. (5.55) takes the form of a linear superposition of multivariate Gaussian functions as follows:

$$F_\lambda(\mathbf{x}) = \sum_{i=1}^{N} w_i \exp\left(-\frac{1}{2\sigma_i^2} \|\mathbf{x} - \mathbf{x}_i\|^2 \right) \tag{5.66}$$

where the linear weights, w_i, are themselves defined by (5.42).

In Eq. (5.66), the individual Gaussian members of the sum defining the approximating function $F(\mathbf{x})$ are assigned different variances. To simplify matters, the condition $\sigma_i = \sigma$ for all i is often imposed on $F(\mathbf{x})$. Even though RBF networks thus designed are of a somewhat restricted kind, they are still universal approximators (Park and Sandberg, 1991).

5.6   REGULARIZATION NETWORKS

The expansion of the regularized approximating function $F_\lambda(\mathbf{x})$ given in Eq. (5.55) in terms of the Green's function $G(\mathbf{x}, \mathbf{x}_i)$ centered at $\mathbf{x}_i$ suggests the network structure shown in Fig. 5.4 as a method for its implementation. For obvious reasons, this network is called a *regularization network* (Poggio and Girosi, 1990a). As with the network described in section 5.1, it consists of three layers. The first layer is composed of input (source) nodes whose number is equal to the dimension m_0 of the input vector $\mathbf{x}$ (i.e., the number of independent variables of the problem). The second layer is a hidden layer, composed of nonlinear units that are connected *directly* to all of the nodes in the input layer. There is one hidden unit for each data point $\mathbf{x}_i$, $i = 1, 2, ..., N$, where N is the size of the training sample. The activation functions of the individual hidden units are defined by the Green's functions. Accordingly, the output of the ith hidden unit is $G(\mathbf{x}, \mathbf{x}_i)$. The output layer consists of a single linear unit, being fully connected to the hidden layer. By "linearity" we mean that the output of the network is a linearly weighted sum of the outputs of the hidden units. The weights of the output layer are the unknown coefficients of the expansion, defined in terms of the Green's functions $G(\mathbf{x}, \mathbf{x}_i)$ and the regularization parameter λ as seen in Eq. (5.54). Figure 5.4 depicts the architecture of the regularization network for a single output. Clearly, such an architecture can be readily extended to accommodate any number of network outputs desired.

The regularization network shown in Fig. 5.5 assumes that the Green's function $G(\mathbf{x}, \mathbf{x}_i)$ is *positive definite* for all i. Provided that this condition is satisfied, which it is in the case of the $G(\mathbf{x}, \mathbf{x}_i)$ having the Gaussian form given in Eq. (5.59), for example, then

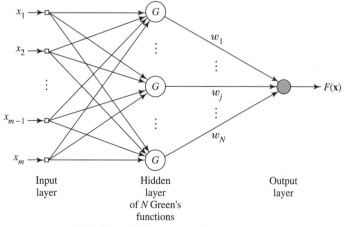

FIGURE 5.4   Regularization network.

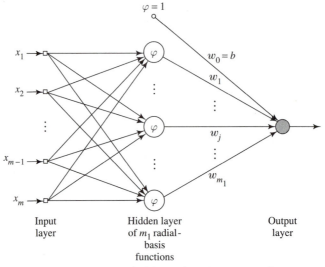

FIGURE 5.5   Radial-basis function network.

the solution produced by this network will be an "optimal" interpolant in the sense that it minimizes the functional $\mathscr{E}(F)$. Moreover, from the viewpoint of approximation theory, the regularization network has three desirable properties (Poggio and Girosi, 1990a):

1. The regularization network is a *universal approximator* in that it can approximate arbitrarily well any multivariate continuous function on a compact subset of $\mathbb{R}^{m_0}$, given a sufficiently large number of hidden units.
2. Since the approximation scheme derived from regularization theory is linear in the unknown coefficients, it follows that the regularization network has the *best-approximation property*. This means that given an unknown nonlinear function f, there always exists a choice of coefficients that approximates f better than all other possible choices.
3. The solution computed by the regularization network is *optimal*. Optimality here means that the regularization network minimizes a functional that measures how much the solution deviates from its true value as represented by the training data.

5.7   GENERALIZED RADIAL-BASIS FUNCTION NETWORKS

The one-to-one correspondence between the training input data $\mathbf{x}_i$ and the Green's function $G(\mathbf{x}, \mathbf{x}_i)$ for $i = 1, 2, \ldots, N$ produces a regularization network that may sometimes be considered prohibitively expensive to implement in computational terms for large N. Specifically, the computation of the linear weights of the network (i.e., the coefficients of the expansion in Eq. (5.55) requires the inversion of an N-by-N matrix, which therefore grows polynomially with N (roughly as N^3). Furthermore, the likelihood of *ill conditioning* is higher for larger matrices; the *condition number* of a matrix is defined as the ratio of the largest eigenvalue to the smallest eigenvalue of the matrix. To overcome these computational difficulties, the complexity of the network would have to be reduced, which requires an approximation to the regularized solution.

The approach taken involves searching for a suboptimal solution in a lower-dimensional space that approximates the regularized solution of Eq. (5.55). This is done by using a standard technique known in variational problems as *Galerkin's method*. According to this technique, the approximated solution $F^*(\mathbf{x})$ is expanded on a finite basis, as shown by (Poggio and Girosi, 1990a)

$$F^*(\mathbf{x}) = \sum_{i=1}^{m_1} w_i \varphi_i(\mathbf{x}) \qquad (5.67)$$

where $\{\varphi_i(\mathbf{x}|i = 1, 2, ..., m_1\}$ is a new set of basis functions that we assume to be linearly independent without loss of generality. Typically, the number of basis functions is less than the number of data points (i.e., $m_1 \leq N$, and the w_i constitute a new set of weights. With radial-basis functions in mind, we set

$$\varphi_i(\mathbf{x}) = G(\|\mathbf{x} - \mathbf{t}_i\|), \quad i = 1, 2, ..., m_1 \qquad (5.68)$$

where the set of centers $\{\mathbf{t}_i|i = 1, 2, ..., m_1\}$ is to be determined. This particular choice of basis functions is the only one that guarantees that in the case of $m_1 = N$, and

$$\mathbf{t}_i = \mathbf{x}_i, \qquad i = 1, 2, ..., N$$

the correct solution of Eq. (5.58) is consistently recovered. Thus, using Eq. (5.68) in (5.67), we may redefine $F^*(\mathbf{x})$ as

$$\begin{aligned} F^*(\mathbf{x}) &= \sum_{i=1}^{m_1} w_i G(\mathbf{x}, \mathbf{t}_i) \\ &= \sum_{i=1}^{m_1} w_i G(\|\mathbf{x} - \mathbf{t}_i\|) \end{aligned} \qquad (5.69)$$

Given the expansion of Eq. (5.69) for the approximating function $F^*(\mathbf{x})$, the problem we now address is the determination of the new set of weights $\{w_i|i = 1, 2, ..., m_1\}$ so as to minimize the new cost functional $\mathscr{E}(F^*)$ defined by

$$\mathscr{E}(F^*) = \sum_{i=1}^{N} \left(d_i - \sum_{j=1}^{m_1} w_j G(\|\mathbf{x}_i - \mathbf{t}_j\|) \right)^2 + \lambda \|\mathbf{D}F^*\|^2 \qquad (5.70)$$

The first term on the right-hand side of Eq. (5.70) may be expressed as the squared Euclidean norm $\|\mathbf{d} - \mathbf{Gw}\|^2$, where

$$\mathbf{d} = [d_1, d_2, ..., d_N]^T \qquad (5.71)$$

$$\mathbf{G} = \begin{bmatrix} G(\mathbf{x}_1, \mathbf{t}_1) & G(\mathbf{x}_1, \mathbf{t}_2) & \cdots & G(\mathbf{x}_1, \mathbf{t}_{m_1}) \\ G(\mathbf{x}_2, \mathbf{t}_1) & G(\mathbf{x}_2, \mathbf{t}_2) & \cdots & G(\mathbf{x}_2, \mathbf{t}_{m_1}) \\ \vdots & \vdots & & \vdots \\ G(\mathbf{x}_N, \mathbf{t}_1) & G(\mathbf{x}_N, \mathbf{t}_2) & \cdots & G(\mathbf{x}_N, \mathbf{t}_{m_1}) \end{bmatrix} \qquad (5.72)$$

$$\mathbf{w} = [w_1, w_2, ..., w_{m_1}]^T \qquad (5.73)$$

The desired response vector $\mathbf{d}$ is N-dimensional as before. However, the matrix $\mathbf{G}$ of Green's functions and the weight vector $\mathbf{w}$ have different dimensions; the matrix $\mathbf{G}$ is now N-by-m_1 and therefore no longer symmetric, and the vector $\mathbf{w}$ is m_1-by-1. From Eq. (5.69) we note that the approximating function F^* is a linear combination of the Green's functions for the stabilizer $\mathbf{D}$. Accordingly, we may express the second term on the right-hand side of Eq. (5.70) as

$$\|\mathbf{D}F^*\|^2 = (\mathbf{D}F^*, \mathbf{D}F^*)_{\mathcal{H}}$$

$$= \left[\sum_{i=1}^{m_1} w_i G(\mathbf{x}, \mathbf{t}_i), \tilde{\mathbf{D}}\mathbf{D} \sum_{i=1}^{m_1} w_i G(\mathbf{x}; \mathbf{t}_i) \right]_{\mathcal{H}}$$

$$= \left[\sum_{i=1}^{m_1} w_i G(\mathbf{x}, \mathbf{t}_i), \sum_{i=1}^{m_1} w_i \delta_{\mathbf{t}_i} \right]_{\mathcal{H}} \qquad (5.74)$$

$$= \sum_{j=1}^{m_1} \sum_{i=1}^{m_1} w_j w_i G(\mathbf{t}_j, \mathbf{t}_i)$$

$$= \mathbf{w}^T \mathbf{G}_0 \mathbf{w}$$

where in the second and third lines we made use of the definition of an adjoint operator and Eq. (5.35), respectively. The matrix $\mathbf{G}_0$ is a symmetric m_1-by-m_1 matrix, defined by

$$\mathbf{G}_0 = \begin{bmatrix} G(\mathbf{t}_1, \mathbf{t}_1) & G(\mathbf{t}_1, \mathbf{t}_2) & \cdots & G(\mathbf{t}_1, \mathbf{t}_{m_1}) \\ G(\mathbf{t}_2, \mathbf{t}_1) & G(\mathbf{t}_2, \mathbf{t}_2) & \cdots & G(\mathbf{t}_2, \mathbf{t}_{m_1}) \\ \vdots & \vdots & & \vdots \\ G(\mathbf{t}_{m_1}, \mathbf{t}_1) & G(\mathbf{t}_{m_1}, \mathbf{t}_2) & \cdots & G(\mathbf{t}_{m_1}, \mathbf{t}_{m_1}) \end{bmatrix} \qquad (5.75)$$

Thus the minimization of Eq. (5.70) with respect to the weight vector $\mathbf{w}$ yields the result (see Problem 5.5)

$$(\mathbf{G}^T\mathbf{G} + \lambda \mathbf{G}_0)\mathbf{w} = \mathbf{G}^T\mathbf{d} \qquad (5.76)$$

As the regularization parameter λ approaches zero, the weight vector $\mathbf{w}$ converges to the pseudoinverse (minimum-norm) solution to the overdetermined least-squares data-fitting problem for $m_1 < N$, as shown by (Broomhead and Lowe, 1988)

$$\mathbf{w} = \mathbf{G}^+\mathbf{d}, \quad \lambda = 0 \qquad (5.77)$$

where $\mathbf{G}^+$ is the pseudoinverse of matrix $\mathbf{G}$; that is,

$$\mathbf{G}^+ = (\mathbf{G}^T\mathbf{G})^{-1}\mathbf{G}^T \qquad (5.78)$$

Weighted Norm

The norm in the approximate solution of Eq. (5.69) is ordinarily intended to be a Euclidean norm. When, however, the individual elements of the input vector $\mathbf{x}$ belong to different classes, it is more appropriate to consider a general *weighted norm*, the squared form of which is defined by (Poggio and Girosi, 1990a)

$$\|\mathbf{x}\|_C^2 = (\mathbf{C}\mathbf{x})^T(\mathbf{C}\mathbf{x})$$

$$= \mathbf{x}^T\mathbf{C}^T\mathbf{C}\mathbf{x} \qquad (5.79)$$

where $\mathbf{C}$ is an m_0-by-m_0 *norm weighting matrix*, and m_0 is the dimension of the input vector $\mathbf{x}$.

Using the definition of weighted norm, we may now rewrite the approximation to the regularized solution given in Eq. (5.69) in the more generalized form (Lowe, 1989; Poggio and Girosi, 1990a)

$$F^*(\mathbf{x}) = \sum_{i=1}^{m_1} w_i G(\|\mathbf{x} - \mathbf{t}_i\|_C) \tag{5.80}$$

The use of a weighted norm may be interpreted in two ways. We may simply view it as applying an *affine transformation* to the original input space. In principle, allowing for such a transformation cannot degrade results from the default case, since it actually corresponds to an identity norm-weighting matrix. On the other hand, the weighted norm follows directly from a slight generalization of the m_0-dimensional Laplacian in the definition of the pseudo-differential operator $\mathbf{D}$ in Eq. (5.63); see Problem 5.6. The use of a weighted norm may also be justified in the context of Gaussian radial-basis functions on the following grounds. A Gaussian radial-based function $G(\|\mathbf{x} - \mathbf{t}_i\|_C)$ centered at $\mathbf{t}_i$ and with norm weighting matrix $\mathbf{C}$ may be expressed as

$$\begin{aligned} G(\|\mathbf{x} - \mathbf{t}_i\|_C &= \exp[-(\mathbf{x} - \mathbf{t}_i)^T \mathbf{C}^T \mathbf{C}(\mathbf{x} - \mathbf{t}_i)] \\ &= \exp\left[-\frac{1}{2}(\mathbf{x} - \mathbf{t}_i)^T \mathbf{\Sigma}^{-1}(\mathbf{x} - \mathbf{t}_i) \right] \end{aligned} \tag{5.81}$$

where the inverse matrix $\mathbf{\Sigma}^{-1}$ is defined by

$$\frac{1}{2}\mathbf{\Sigma}^{-1} = \mathbf{C}^T\mathbf{C} \tag{5.82}$$

Equation (5.81) represents a multivariate Gaussian distribution with mean vector $\mathbf{t}_i$ and covariance matrix $\mathbf{\Sigma}$. As such, it represents a generalization of the distribution described in Eq. (5.59).

The solution to the approximation problem given in Eq. (5.70) provides the framework for the *generalized radial-basis function (RBF) network* having the structure shown in Fig. 5.5. In this network, provision is made for a bias (i.e., data-independent variable) applied to the output unit. This is done simply by setting one of the linear weights in the output layer of the network equal to the bias and treating the associated radial-basis function as a constant equal to $+1$.

In structural terms, the generalized RBF network of Fig. 5.5 is similar to the regularization RBF network of Fig. 5.4. However, they differ from each other in two important ways:

1. The number of nodes in the hidden layer of the generalized RBF network of Fig. 5.5 is m_1, where m_1 is ordinarily smaller than the number N of examples available for training. On the other hand, the number of hidden nodes in the regularization RBF network of Fig. 5.4 is exactly N.
2. In the generalized RBF network of Fig. 5.5, the linear weights associated with the output layer, and the positions of the centers of the radial-basis functions and the

norm weighting matrix associated with the hidden layer, are all unknown parameters that have to be learned. However, the activation functions of the hidden layer in the regularization RBF network of Fig. 5.4 are known, being defined by a set of Green's functions centered at the training data points; the linear weights of the output layer are the only unknown parameters of the network.

Receptive Field

The covariance matrix Σ determines the receptive field of the Gaussian radial-basis function $G(\|\mathbf{x} - \mathbf{t}_i\|_C)$ given in Eq. (5.81). For a prescribed center $\mathbf{t}_i$, the *receptive field* of $G(\|\mathbf{x} - \mathbf{t}_i\|_C)$ is formally defined as the support of the function

$$\Psi(\mathbf{x}) = G(\|\mathbf{x} - \mathbf{t}_i\|_C) - a \tag{5.83}$$

where a is some positive constant (Xu et al., 1994). In other words, the receptive field of $G(\|\mathbf{x} - \mathbf{t}_i\|_C)$ is that particular subset of the domain of the input vector $\mathbf{x}$ for which $G(\|\mathbf{x} - \mathbf{t}_i\|_C)$ takes sufficiently large values, greater than the prescribed level a.

In a manner corresponding to the way in which the norm-weighting matrix $\mathbf{C}$ was defined, we may identify three different scenarios pertaining to the covariance matrix Σ and its influence on the shape, size, and orientation of the receptive field:

1. $\Sigma = \sigma^2 \mathbf{I}$, where $\mathbf{I}$ is the identity matrix and σ^2 is a common variance. In this case, the receptive field of $G(\|\mathbf{x} - \mathbf{t}_i\|_C)$ consists of a hypersphere centered at $\mathbf{t}_i$ and with a radius determined by σ.
2. $\Sigma = \text{diag}(\sigma_1^2, \sigma_2^2, ..., \sigma_{m_0}^2)$, where σ_j^2 is the variance of the jth element of the input vector $\mathbf{x}$ and $j = 1, 2, ..., m_0$. In this second case, the receptive field of $G(\|\mathbf{x} - \mathbf{t}_i\|_C)$ consists of a hyperellipse whose individual axes coincide with those of the input space, and with its extension along the jth axis being determined by σ_j.
3. Σ is a nondiagonal matrix. By definition, Σ is a positive definite matrix. We may therefore use the similarity transformation of matrix algebra to decompose Σ as follows:

$$\Sigma = \mathbf{Q}^T \Lambda \mathbf{Q} \tag{5.84}$$

where Λ is a diagonal matrix and $\mathbf{Q}$ is an orthonormal rotation matrix. The matrix Λ determines the shape and size of the receptive field, while the matrix $\mathbf{Q}$ determines its orientation.

5.8   XOR PROBLEM (REVISITED)

Consider again the XOR (Exclusive OR) problem, which we solved in Chapter 4 using a multilayer perceptron with a single hidden layer. Here we are going to present a solution to this same problem by using an RBF network.

The RBF network to be investigated consists of a pair of Gaussian functions, defined as:

$$G(\|\mathbf{x} - \mathbf{t}_i\|) = \exp(-\|\mathbf{x} - \mathbf{t}_i\|^2), \quad i = 1, 2 \tag{5.85}$$

where the centers $\mathbf{t}_1$ and $\mathbf{t}_2$ are

$$\mathbf{t}_1 = [1, 1]^T$$
$$\mathbf{t}_2 = [0, 0]^T$$

For the characterization of the output unit, we assume the following:

1. The output unit uses *weight-sharing*, which is justified by virtue of the symmetry of the problem; this is a form of prior information being built into the design of the network. With only two hidden units, we therefore only have a single weight w to be determined.
2. The output unit includes a bias b (i.e., data-independent variable). The significance of this bias is that the desired output values of the XOR function have nonzero mean.

Thus the structure of the RBF network proposed for solving the XOR problem is as shown in Fig. 5.6. The input–output relation of the network is defined by

$$y(\mathbf{x}) = \sum_{i=1}^{2} wG(\|\mathbf{x} - \mathbf{t}_i\|) + b \tag{5.86}$$

To fit the training data of Table 5.2, we require that

$$y(\mathbf{x}_j) = d_j, \quad j = 1, 2, 3, 4 \tag{5.87}$$

where $\mathbf{x}_j$ is an input vector and d_j is the corresponding value of the desired output. Let

$$g_{ji} = G(\|\mathbf{x}_j - \mathbf{t}_i\|), \quad j = 1, 2, 3, 4; \ i = 1, 2 \tag{5.88}$$

Then, using the values of Table 5.2 in Eq. (5.88), we get the following set of equations written in matrix form:

$$\mathbf{Gw} = \mathbf{d} \tag{5.89}$$

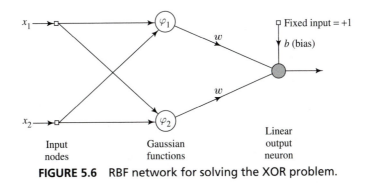

FIGURE 5.6   RBF network for solving the XOR problem.

TABLE 5.2 Input–Output Transformation Computed for XOR Problem

Data Point, j	Input Pattern, $\mathbf{x}_j$	Desired Output, d_j
1	$(1,1)$	0
2	$(0,1)$	1
3	$(0,0)$	0
4	$(1,0)$	1

where

$$
\mathbf{G} = \begin{bmatrix}
1 & 0.1353 & 1 \\
0.3678 & 0.3678 & 1 \\
0.1353 & 1 & 1 \\
0.3678 & 0.3678 & 1
\end{bmatrix}
\tag{5.90}
$$

$$
\mathbf{d} = \begin{bmatrix} 0 & 1 & 0 & 1 \end{bmatrix}^T
\tag{5.91}
$$

$$
\mathbf{w} = \begin{bmatrix} w & w & b \end{bmatrix}^T
\tag{5.92}
$$

The problem described here is *overdetermined in the sense that we have more data points than free parameters*. This explains why the matrix $\mathbf{G}$ is not square. Consequently, no unique inverse exists for the matrix $\mathbf{G}$. To overcome this difficulty, we use the *minimum-norm* solution of Eq. (5.78), and so write

$$
\begin{aligned}
\mathbf{w} &= \mathbf{G}^+\mathbf{d} \\
&= (\mathbf{G}^T\mathbf{G})^{-1}\mathbf{G}^T\mathbf{d}
\end{aligned}
\tag{5.93}
$$

Note that $\mathbf{G}^T\mathbf{G}$ is a square matrix with a unique inverse of its own. Substituting Eq. (5.90) in (5.93), we get

$$
\mathbf{G}^+ = \begin{bmatrix}
1.8292 & -1.2509 & 0.6727 & -1.2509 \\
0.6727 & -1.2509 & 1.8292 & -1.2509 \\
-0.9202 & 1.4202 & -0.9202 & 1.4202
\end{bmatrix}
\tag{5.94}
$$

Finally, substituting Eqs. (5.91) and (5.94) in (5.93), we get

$$
\mathbf{w} = \begin{bmatrix}
-2.5018 \\
-2.5018 \\
+2.8404
\end{bmatrix}
$$

which completes the specification of the RBF network.

5.9 ESTIMATION OF THE REGULARIZATION PARAMETER

The regularization parameter λ plays a central role in the regularization theory of radial-basis function networks presented in Sections 5.5 through 5.7. To derive the full benefit of this theory we need an equally principled approach to the estimation of λ.

To fix ideas, consider a *nonlinear regression problem*, described by a model whose observable output y_i at time step i in response to an input vector $\mathbf{x}_i$ is defined by

$$y_i = f(\mathbf{x}_i) + \epsilon_i, \qquad i = 1, 2, ..., N \tag{5.95}$$

where $f(\mathbf{x}_i)$ is a "smooth curve," and ϵ_i is a sample drawn from a white noise process of zero mean and variance σ^2. That is

$$E[\epsilon_i] = 0 \quad \text{for all } i \tag{5.96}$$

and

$$E[\epsilon_i \epsilon_k] = \begin{cases} \sigma^2 & \text{for } k = i \\ 0 & \text{otherwise} \end{cases} \tag{5.97}$$

The problem is to reconstruct the underlying function of the model, $f(\mathbf{x}_i)$, given the training sample $\{(\mathbf{x}_i, y_i)\}_{i=1}^{N}$.

Let $F_\lambda(\mathbf{x})$ be the regularized estimate of $f(\mathbf{x})$ for some value of the regularization parameter λ. That is, $F_\lambda(\mathbf{x})$ is the minimizer of the Tikhonov functional formulated for the nonlinear regression problem as:

$$\mathcal{E}(F) = \frac{1}{2} \sum_{i=1}^{N} [y_i - F(\mathbf{x}_i)]^2 + \frac{\lambda}{2} \| \mathbf{D}F(\mathbf{x}) \|^2 \tag{5.98}$$

It is a nontrivial matter to choose a suitable value for λ, which controls the tradeoff between two conflicting issues:

- "Roughness" of the solution as measured by the term $\| \mathbf{D}F(\mathbf{x}) \|^2$
- "Infidelity" of the data, as measured by the term $\sum_{i=1}^{N} [y_i - F(\mathbf{x}_i)]^2$

A good choice for the regularization parameter λ is the subject matter of this section.

Average Squared Error

Let $R(\lambda)$ denote the "average squared error over a given data set" between two functions: the regression function $f(\mathbf{x})$ pertaining to the model and the approximating function $F_\lambda(\mathbf{x})$ representing the solution for some λ, evaluated over the entire data set. That is,

$$R(\lambda) = \frac{1}{N} \sum_{i=1}^{N} [f(\mathbf{x}_i) - F_\lambda(\mathbf{x}_i)]^2 \tag{5.99}$$

The *optimum* λ is the particular value of λ that minimizes $R(\lambda)$.

Let $F_\lambda(\mathbf{x}_k)$ be expressed as a linear combination of the given set of observables as follows:

$$F_\lambda(\mathbf{x}_k) = \sum_{i=1}^{N} a_{ki}(\lambda) y_i \tag{5.100}$$

In matrix form, we may equivalently write

$$\mathbf{F}_\lambda = \mathbf{A}(\lambda) \mathbf{y} \tag{5.101}$$

where

$$\mathbf{F}_\lambda = [F_\lambda(\mathbf{x}_1), F_\lambda(\mathbf{x}_2), \dots, F_\lambda(\mathbf{x}_N)]^T$$

$$\mathbf{y} = [y_1, y_2, \dots, y_N]^T$$

and

$$\mathbf{A}(\lambda) = \begin{bmatrix} a_{11} & a_{12} & \cdots & a_{1N} \\ a_{21} & a_{22} & \cdots & a_{2N} \\ \vdots & \vdots & & \vdots \\ a_{N1} & a_{N2} & \cdots & a_{NN} \end{bmatrix} \tag{5.102}$$

The N-by-N matrix $\mathbf{A}(\lambda)$ is called the *influence matrix*.

Using this matrix notation, we may rewrite Eq. (5.99) in the form

$$R(\lambda) = \frac{1}{N}\|\mathbf{f} - \mathbf{F}_\lambda\|^2$$

$$= \frac{1}{N}\|\mathbf{f} - \mathbf{A}(\lambda)\mathbf{y}\|^2 \tag{5.103}$$

where the N-by-1 vector $\mathbf{f}$ is

$$\mathbf{f} = [f(\mathbf{x}_1), f(\mathbf{x}_2), \dots, f(\mathbf{x}_N)]^T$$

We can go one step further in our matrix formulation by rewriting Eq. (5.95) in the form

$$\mathbf{y} = \mathbf{f} + \boldsymbol{\epsilon} \tag{5.104}$$

where

$$\boldsymbol{\epsilon} = [\epsilon_1, \epsilon_2, \dots, \epsilon_N]^T$$

Hence, using Eq. (5.104) in (5.103) and then expanding terms, we obtain

$$R(\lambda) = \frac{1}{N}\|(\mathbf{I} - \mathbf{A}(\lambda))\mathbf{f} - \mathbf{A}(\lambda)\boldsymbol{\epsilon}\|^2$$

$$= \frac{1}{N}\|(\mathbf{I} - \mathbf{A}(\lambda))\mathbf{f}\|^2 - \frac{2}{N}\boldsymbol{\epsilon}^T\mathbf{A}(\lambda)(\mathbf{I} - \mathbf{A}(\lambda))\mathbf{f} \tag{5.105}$$

$$+ \frac{1}{N}\|\mathbf{A}(\lambda)\boldsymbol{\epsilon}\|^2$$

where $\mathbf{I}$ is the N-by-N identity matrix. To find the expected value of $R(\lambda)$, note the following points:

- The first term on the right-hand side of Eq. (5.105) is a constant and is therefore unaffected by the expectation operator.
- The expectation of the second term is zero by virtue of Eq. (5.96)

- The expectation of the scalar $\|\mathbf{A}(\lambda)\boldsymbol{\epsilon}\|^2$ is

$$
\begin{aligned}
E[\|\mathbf{A}(\lambda)\boldsymbol{\epsilon}\|^2 &= E[\boldsymbol{\epsilon}^T\mathbf{A}^T(\lambda)\mathbf{A}(\lambda)\boldsymbol{\epsilon}] \\
&= \mathrm{tr}\{E[\boldsymbol{\epsilon}^T\mathbf{A}^T(\lambda)\mathbf{A}(\lambda)\boldsymbol{\epsilon}]\} \\
&= E\{\mathrm{tr}[\boldsymbol{\epsilon}^T\mathbf{A}^T(\lambda)\mathbf{A}(\lambda)\boldsymbol{\epsilon}]\}
\end{aligned}
\tag{5.106}
$$

where we first used the fact that the trace of a scalar is the same as the scalar itself, and then interchanged the order of expectation and trace operation.

Next we may use this rule in matrix algebra: Given two matrices $\mathbf{B}$ and $\mathbf{C}$ of compatible dimensions, the trace of $\mathbf{BC}$ is equal to the trace of $\mathbf{CB}$. Thus, setting $\mathbf{B} = \boldsymbol{\epsilon}^T$ and $\mathbf{C} = \mathbf{A}^T(\lambda)\mathbf{A}(\lambda)\boldsymbol{\epsilon}$, we may rewrite Eq. (5.106) in the equivalent form

$$
\begin{aligned}
E[\|\mathbf{A}(\lambda)\mathbf{f}\|^2] &= E\{\mathrm{tr}[\mathbf{A}^T(\lambda)\mathbf{A}(\lambda)\boldsymbol{\epsilon}\boldsymbol{\epsilon}^T]\} \\
&= \sigma^2\mathrm{tr}[\mathbf{A}^T(\lambda)\mathbf{A}(\lambda)]
\end{aligned}
\tag{5.107}
$$

where in the last line we have made use of Eq. (5.97). Finally, noting that the trace of $\mathbf{A}^T(\lambda)\mathbf{A}(\lambda)$ is the same as the trace of $\mathbf{A}^2(\lambda)$, we may write

$$
E[\|\mathbf{A}(\lambda)\mathbf{f}\|^2 = \sigma^2\mathrm{tr}[\mathbf{A}^2(\lambda)]
\tag{5.108}
$$

Putting these three results together, we may express the expected value of $R(\lambda)$ as

$$
E[R(\lambda)] = \frac{1}{N}\|\mathbf{I} - \mathbf{A}(\lambda)\mathbf{f}\|^2 + \frac{\sigma^2}{N}\,\mathrm{tr}[\mathbf{A}^2(\lambda)]
\tag{5.109}
$$

The average squared error over a given data set, $R(\lambda)$, however, is not a practical measure because it requires knowledge of the regression function $f(\mathbf{x})$, the function that is to be reconstructed. As an estimate of $E[R(\lambda)]$, we introduce the following definition (Craven and Wahba, 1979)

$$
\hat{R}(\lambda) = \frac{1}{N}\|(\mathbf{I} - \mathbf{A}(\lambda))\mathbf{y}\|^2 + \frac{\sigma^2}{N}\,\mathrm{tr}[\mathbf{A}^2(\lambda)] - \frac{\sigma^2}{N}\,\mathrm{tr}[(\mathbf{I} - \mathbf{A}(\lambda))^2]
\tag{5.110}
$$

This estimate is *unbiased*, in that (following a procedure similar to that described for deriving Eq. (5.109)), we may show that

$$
E[\hat{R}(\lambda)] = E[R(\lambda)]
\tag{5.111}
$$

Accordingly, the minimizer of the estimate $\hat{R}(\lambda)$ can be taken as a good choice for the regularization parameter λ.

Generalized Cross-Validation

A drawback of the estimate $\hat{R}(\lambda)$ is that it requires knowledge of the noise variance σ^2. In situations encountered in practice, σ^2 is usually not known. To deal with situations of this kind, we may use the concept of generalized cross-validation that was originated by Craven and Wahba (1979).

We begin by adapting the ordinary leave-one-out form of cross-validation (described in Chapter 4) to the problem at hand. Specifically, let $F_\lambda^{[k]}(\mathbf{x})$ be the minimizer of the functional

$$\mathscr{E}(F) = \frac{1}{2} \sum_{\substack{i=1 \\ i \neq k}}^{N} [y_i - F_\lambda(\mathbf{x}_i)]^2 + \frac{\lambda}{2} \|\mathbf{D}F(\mathbf{x})\|^2 \tag{5.112}$$

where the kth term $[y_k - F_\lambda(\mathbf{x}_k)]$ has been left out of the standard error term. By leaving out this term, we may take the ability of $F_\lambda^{[k]}(\mathbf{x})$ to "predict" the missing data point y_k as a measure of the goodness of λ. Accordingly, we may introduce the following measure of goodness

$$V_0(\lambda) = \frac{1}{N} \sum_{k=1}^{N} [y_k - F_\lambda^{[k]}(\mathbf{x}_k)]^2 \tag{5.113}$$

which depends on the data alone. The *ordinary cross-validation estimate* of λ is thus defined to be the minimizer of $V_0(\lambda)$ (Wahba, 1990).

A useful property of $F_\lambda^{[k]}(\mathbf{x}_k)$ is that if the data point y_k is replaced by the prediction $F_\lambda^{[k]}(\mathbf{x}_k)$, and the original Tikhonov functional $\mathscr{E}(F)$ of Eq. (5.98) is minimized using the data points $y_1, y_2, \ldots, y_{k-1}, y_k, y_{k+1}, \ldots, y_N$, we get $F_\lambda^{[k]}(\mathbf{x}_k)$ for the solution. This property, together with the fact that for each input vector $\mathbf{x}$ the minimizer $F_\lambda(\mathbf{x})$ of $\mathscr{E}(F)$ depends linearly on y_k, allows us to write:

$$F_\lambda^{[k]}(\mathbf{x}_k) = F_\lambda(\mathbf{x}_k) + (F_\lambda^{[k]}(\mathbf{x}_k) - y_k)\frac{\partial F_\lambda(\mathbf{x}_k)}{\partial y_k} \tag{5.114}$$

From Eq. (5.100), defining the entries of the influence matrix $\mathbf{A}(\lambda)$, we readily see that

$$\frac{\partial F_\lambda(\mathbf{x}_k)}{\partial y_k} = a_{kk}(\lambda) \tag{5.115}$$

where $a_{kk}(\lambda)$ is the kth diagonal element of $\mathbf{A}(\lambda)$. Hence, using Eq. (5.115) in (5.114), and solving the resulting equation for $F_\lambda^{[k]}(\mathbf{x}_k)$, we obtain

$$\begin{aligned} F_\lambda^{[k]}(\mathbf{x}_k) &= \frac{F_\lambda(\mathbf{x}_k) - a_{kk}(\lambda)y_k}{1 - a_{kk}(\lambda)} \\ &= \frac{F_\lambda(\mathbf{x}_k) - y_k}{1 - a_{kk}(\lambda)} + y_k \end{aligned} \tag{5.116}$$

Substituting Eq. (5.116) in (5.113), we may redefine $V_0(\lambda)$ as

$$V_0(\lambda) = \frac{1}{N} \sum_{k=1}^{N} \left[\frac{y_k - F_\lambda(\mathbf{x}_k)}{1 - a_{kk}(\lambda)} \right]^2 \tag{5.117}$$

Typically, $a_{kk}(\lambda)$ is different for different k, which means that the data points in $V_0(\lambda)$ are not treated equally. To circumvent this undesirable feature of ordinary cross-validation, Craven and Wahba (1979) introduced the *generalized cross-validation* (GCV), using a rotation of coordinates.[11] Specifically, the ordinary cross-validation function $V_0(\lambda)$ of Eq. (5.117) is modified as:

$$V(\lambda) = \frac{1}{N} \sum_{k=1}^{N} \omega_k \left[\frac{y_k - F_\lambda(\mathbf{x}_k)}{1 - a_{kk}(\lambda)} \right]^2 \tag{5.118}$$

where the weights, ω_k, are themselves defined by

$$\omega_k = \left[\frac{1 - a_{kk}(\lambda)}{\frac{1}{N} \text{tr}[\mathbf{I} - \mathbf{A}(\lambda)]} \right]^2 \tag{5.119}$$

Then, the generalized cross-validation function $V(\lambda)$ becomes

$$V(\lambda) = \frac{\frac{1}{N} \sum_{k=1}^{N} [y_k - F_\lambda(\mathbf{x}_k)]^2}{\left[\frac{1}{N} \text{tr}[\mathbf{I} - \mathbf{A}(\lambda)] \right]^2} \tag{5.120}$$

Finally, using Eq. (5.100) in (5.120) yields

$$V(\lambda) = \frac{\frac{1}{N} \|(\mathbf{I} - \mathbf{A}(\lambda))\mathbf{y}\|^2}{\left[\frac{1}{N} \text{tr}[\mathbf{I} - \mathbf{A}(\lambda)] \right]^2} \tag{5.121}$$

which relies solely on quantities related to the data for its computation.

An Optimal Property of the Generalized Cross-Validation Function V(λ)

Let λ denote the minimizer of the expected value of the generalized cross-validation function $V(\lambda)$. The *expectation inefficiency* of the method of generalized cross-validation is defined by

$$I^* = \frac{E[R(\lambda)]}{\min_\lambda E[R(\lambda)]} \tag{5.122}$$

where $R(\lambda)$ is the average squared error over the data set given in Eq. (5.99). Naturally, the asymptotic value of I^* satisfies the condition

$$\lim_{N \to \infty} I^* = 1 \tag{5.123}$$

In other words, for large N, the average squared error $R(\lambda)$ with λ estimated by minimizing the generalized cross-validation function $V(\lambda)$ should be close to the minimum possible value of $R(\lambda)$, which makes $V(\lambda)$ a good method for estimating λ.

Summarizing Comments

The general idea is to choose the regularization parameter λ so as to minimize the average squared error over the data set, $R(\lambda)$. Unfortunately, this cannot be accomplished directly, since $R(\lambda)$ involves the unknown regression function $f(\mathbf{x})$. With this being so, there are two possibilities that may be pursued in practice:

- If the noise variance σ^2 is known, we may use the minimizer of the estimate $\hat{R}(\lambda)$ of Eq. (5.110) as the optimum choice of λ, optimum in the sense that it also minimizes $R(\lambda)$.
- If σ^2 is not known, we may use the minimizer of the generalized cross-validation function $V(\lambda)$ of Eq. (5.121) as a good choice of λ, which produces an expected mean square error that approaches the minimum possible expected mean square error as $N \to \infty$.

The important point to note here is that the theory justifying the use of generalized cross-validation for estimating λ is an asymptotic one. Good results can therefore be expected only when the available data set is long enough for the signal to be distinguishable from noise.

Practical experience with generalized cross-validation appears to show that it is robust against nonhomogeneity of variances and non-Gaussian noise (Wahba, 1990). However, the method is quite likely to produce unsatisfactory estimates of the regularization parameter λ if the noise process is highly correlated.

Finally, some comments pertaining to the computation of the generalized cross-validation function $V(\lambda)$ are in order. For given trial values of the regularization parameter λ, finding the denominator term $[\mathrm{tr}[\mathbf{I}-\mathbf{A}(\lambda)]/N]^2$ in the formula of Eq. (5.121) is the most expensive part of the work involved in computing $V(\lambda)$. The "randomized trace method" described in Wahba et al. (1995) may be used to compute $\mathrm{tr}[\mathbf{A}(\lambda)]$; it is feasible to apply this method to very large systems.

5.10 APPROXIMATION PROPERTIES OF RBF NETWORKS

In Chapter 4 we discuss the approximation properties of multilayer perceptrons. Radial-basis function networks exhibit good approximation properties of their own, paralleling those of multilayer perceptrons. The family of RBF networks is broad enough to uniformly approximate any continuous function on a compact set.[12]

Universal Approximation Theorem

Let $G: \mathbb{R}^{m_0} \to \mathbb{R}$ be an integrable bounded function such that G is continuous and

$$\int_{\mathbb{R}^{m_0}} G(\mathbf{x})d\mathbf{x} \neq 0$$

Let $\mathcal{S}_G$ denote the family of RBF networks consisting of functions $F: \mathbb{R}^{m_0} \to \mathbb{R}$ represented by

$$F(\mathbf{x}) = \sum_{i=1}^{m_1} w_i G\left(\frac{\mathbf{x} - \mathbf{t}_i}{\sigma}\right)$$

where $\sigma > 0$, $w_i \in \mathbb{R}$ and $\mathbf{t}_i \in \mathbb{R}^{m_0}$ for $i = 1, 2, ..., m_1$. We may then state the *universal approximation theorem* for RBF networks (Park and Sandberg, 1991):

> *For any continuous input–output mapping function $f(\mathbf{x})$ there is an RBF network with a set of centers $\{\mathbf{t}_i\}_{i=1}^{m_1}$ and a common width $\sigma > 0$ such that the input–output mapping function $F(\mathbf{x})$ realized by the RBF network is close to $f(\mathbf{x})$ in the L_p norm, $p \in [1, \infty]$.*

Note that in the universal approximation theorem as stated, the kernel $G : \mathbb{R}^{m_0} \to \mathbb{R}$ is not required to satisfy the property of radial symmetry. The theorem is therefore stronger than necessary for RBF networks. Most importantly, it provides the theoretical basis for the design of neural networks using radial basis functions for practical applications.

Curse of Dimensionality (Revisited)

In addition to the universal approximation property of RBF networks, there is the issue of the rate of approximation attainable by these networks that must be considered. From the discussion presented in Chapter 4, we recall that the intrinsic *complexity* of a class of approximating functions increases exponentially in the ratio m_0/s, where m_0 is the *input dimensionality* (i.e., dimension of the input space) and s is a *smoothness index* measuring the number of constraints imposed on an approximating function in that particular class. Bellman's *curse of dimensionality* tells us that, irrespective of the approximation technique employed, if the smoothness index s is maintained constant, the number of parameters needed for the approximating function to attain a prescribed degree of accuracy increases exponentially with the input dimensionality m_0. The only way that we can achieve a rate of convergence independent of the input dimensionality m_0, and therefore be immune to the curse of dimensionality, is for the smoothness index s to increase with the number of parameters in the approximating function so as to compensate for the increase in complexity. This point is illustrated in Table 5.3, adapted from Girosi and Anzellotti (1992). Table 5.3 summarizes the constraints on function

TABLE 5.3   Two Approximation Techniques and Corresponding Function Spaces with the Same Rate of Convergence $O(1/\sqrt{m_1})$, Where m_1 is the Size of the Hidden Space.

Function Space	Norm	Approximation Technique
$\int_{\mathbb{R}^{m_0}} \|\mathbf{s}\| \tilde{F}(\mathbf{s}) d\mathbf{s} < \infty$ where $\tilde{F}(\mathbf{s})$ is the multidimensional Fourier transform of the approximating function $F(\mathbf{x})$	$L_2(\Omega)$	(a) multilayer perceptrons $F(\mathbf{x}) = \sum_{i=1}^{m_1} a_i \varphi(\mathbf{w}_i^T \mathbf{x} + b_i)$ where $\varphi(.)$ is the sigmoid activation function
Sobolev space of functions whose derivatives up to order $2m > m_0$ are integrable	$L_2(\mathbb{R}^2)$	(b) RBF networks: $F(\mathbf{x}) = \sum_{i=1}^{m_1} a_i \exp\left(-\frac{\|\mathbf{x} - \mathbf{t}_i\|^2}{2\sigma^2}\right)$

space that have to be satisfied by two approximating techniques, multilayer perceptrons and RBF networks, for the rate of convergence to be independent of the input dimensionality m_0. Naturally, the constraints imposed on these two approximating techniques are different, reflecting the different paths followed in their formulations. In the case of RBF networks, the result holds in the *Sobolev space*[13] of functions whose derivations up to order $2m > m_0$ are integrable. In other words, the number of derivatives of the approximating function that are integrable is required to increase with the input dimensionality m_0 in order to make the rate of convergence independent of m_0. As explained in Chapter 4, a similar constraint applies to multilayer perceptrons, but in a rather deceptive way. The conclusion to be drawn from Table 5.3 may therefore be stated as:

> *The space of approximating functions attainable with multilayer perceptrons and RBF networks becomes increasingly constrained as the input dimensionality m_0 is increased.*

The net result is that the curse of dimensionality can be broken neither by neural networks whether they are multilayer perceptrons or RBF networks, nor by any other nonlinear technique of a similar nature.

Relationship between Sample Complexity, Computational Complexity, and Generalization Performance

A discussion of the approximation problem would be incomplete without some consideration being given to the fact that, in practice, we do not have an infinite amount of data, but rather a training sample of some finite size. By the same token, we do not have a neural network with infinite computational complexity, but rather a finite one. Accordingly, there are two components to the generalization error of a neural network trained on a data set of finite size and tested on data not seen before, as discussed in Chapter 2. One component, called the *approximation error*, results from the limited capacity of the network to represent a target function of interest. The other component, called the *estimation error*, results from the limited amount of information contained in the training sample about the target function. Using this form of decomposition, Niyogi and Girosi (1996) have derived a *bound* on the generalization error produced by a Gaussian RBF network, expressed in terms of the size of the hidden layer and the size of the training sample. The derivation is for the case of learning a regression function in a model of the kind described in Eq. (5.95); the regression function belongs to a certain Sobolev space.

This bound, formulated in the terminology of PAC learning described in Chapter 2, may be stated as follows (Niyogi and Girosi, 1996):

> Let G denote the class of Gaussian RBF networks with m_0 input (source) nodes and m_1 hidden units. Let $f(\mathbf{x})$ denote a regression function that belongs to a certain Sobolev space. Assume that the training sample $\mathcal{T} = \{(\mathbf{x}_i, d_i)\}_{i=1}^N$ is obtained by random sampling of the regressive model based on $f(\mathbf{x})$. Then, for any confidence parameter $\delta \in (0,1]$, the generalization error produced by the network is bounded from above by

$$O\left(\frac{1}{m_1}\right) + O\left(\frac{m_0 m_1}{N} \log(m_1 N) + \frac{1}{N} \log\left(\frac{1}{\delta}\right)^{1/2}\right) \tag{5.124}$$

> with probability greater than $1 - \delta$.

From the bound of Eq. (5.124), we may make the following deductions:

- The generalization error converges to zero only if the number of hidden units, m_1, increases more slowly than the size N of the training sample.
- For a given size N of training sample, the optimum number of hidden units, m_1^*, behaves as (see Problem 5.11)

$$m_1^* \propto N^{1/3} \tag{5.125}$$

- The RBF network exhibits a rate of approximation $O(1/m_1)$ that is similar to that derived by Barron (1993) for the case of a multilayer perceptron with sigmoid activation functions; see the discussion in Section 4.12.

5.11   COMPARISON OF RBF NETWORKS AND MULTILAYER PERCEPTRONS

Radial-basis function (RBF) networks and multilayer perceptrons are examples of nonlinear layered feedforward networks. They are both universal approximators. It is therefore not surprising to find that there always exists an RBF network capable of accurately mimicking a specified MLP, or vice versa. However, these two networks differ from each other in several important respects.

1. An RBF network (in its most basic form) has a single hidden layer, whereas an MLP may have one or more hidden layers.
2. Typically the computation nodes of an MLP, located in a hidden or an output layer, share a common neuronal model. On the other hand, the computation nodes in the hidden layer of an RBF network are quite different and serve a different purpose from those in the output layer of the network.
3. The hidden layer of an RBF network is nonlinear, whereas the output layer is linear. However, the hidden and output layers of an MLP used as a pattern classifier are usually all nonlinear. When the MLP is used to solve nonlinear regression problems, a linear layer for the output is usually the preferred choice.
4. The argument of the activation function of each hidden unit in an RBF network computes the *Euclidean norm* (*distance*) between the input vector and the center of that unit. Meanwhile, the activation function of each hidden unit in an MLP computes the *inner product* of the input vector and the synaptic weight vector of that unit.
5. MLPs construct *global* approximations to nonlinear input–output mapping. On the other hand, RBF networks using exponentially decaying localized nonlinearities (e.g., Gaussian functions) construct *local* approximations to nonlinear input–output mappings.

This in turn means that for the approximation of a nonlinear input–output mapping, the MLP may require a smaller number of parameters than the RBF network for the same degree of accuracy.

The linear characteristics of the output layer of the RBF network mean that such a network is more closely related to Rosenblatt's perceptron than to the multilayer perceptron. However, the RBF network differs from the perceptron in that it is capable of implementing arbitrary nonlinear transformations of the input space. This is well illustrated by the XOR problem, which cannot be solved by any linear perceptron but can be solved by an RBF network.

5.12 KERNEL REGRESSION AND ITS RELATION TO RBF NETWORKS

The theory of RBF networks presented so far has built on the notion of interpolation. In this section we take another viewpoint, namely, *kernel regression* building on the notion of density estimation.

To be specific, consider again the nonlinear regression model of Eq. (5.95), reproduced here for convenience of presentation:

$$y_i = f(\mathbf{x}_i) + \epsilon_i, \quad i = 1, 2, \ldots, N$$

As a reasonable estimate of the unknown regression function $f(\mathbf{x})$, we may take the mean of observables (i.e., values of the model output y) near a point $\mathbf{x}$. For this approach to be successful, however, the local average should be confined to observations in a small neighborhood (i.e., receptive field) around the point $\mathbf{x}$, because in general, observations corresponding to points away from $\mathbf{x}$ will have different mean values. More precisely, we recall from the discussion presented in Chapter 2 that $f(\mathbf{x})$ is equal to the conditional mean of y given $\mathbf{x}$ (i.e., the regression of y on $\mathbf{x}$), as shown by

$$f(\mathbf{x}) = E[y|\mathbf{x}]$$

Using the formula for the expectation of a random variable, we may write

$$f(\mathbf{x}) = \int_{-\infty}^{\infty} y f_Y(y|\mathbf{x}) dy \tag{5.126}$$

where $f_Y(y|\mathbf{x})$ is the conditional probability density function (pdf) of Y, given $\mathbf{x}$. From probability theory, we have

$$f_Y(y|\mathbf{x}) = \frac{f_{X,Y}(\mathbf{x}, y)}{f_X(\mathbf{x})} \tag{5.127}$$

where $f_X(\mathbf{x})$ is the pdf of $\mathbf{x}$ and $f_{X,Y}(\mathbf{x}, y)$ is the joint pdf of $\mathbf{x}$ and y. Hence, using Eq. (5.127) in (5.126), we obtain the following formula for the regression function

$$f(\mathbf{x}) = \frac{\int_{-\infty}^{\infty} y f_{X,Y}(\mathbf{x}, y) dy}{f_X(\mathbf{x})} \tag{5.128}$$

Our particular interest is in a situation where the joint probability density function $f_{X,Y}(\mathbf{x}, y)$ is unknown. All that we have available is the training sample, $\{(\mathbf{x}_i, y_i)\}_{i=1}^{N}$. To estimate $f_{X,Y}(\mathbf{x}, y)$ and therefore $f_X(\mathbf{x})$, we may use a nonparametric estimator known as the *Parzen–Rosenblatt density estimator* (Rosenblatt, 1956, 1970; Parzen, 1962). Basic to the formulation of this estimator is a *kernel*, denoted by $K(\mathbf{x})$, which has properties similar to those associated with a probability density function:

- *The kernel $K(\mathbf{x})$ is a continuous, bounded, and real function of $\mathbf{x}$, and symmetric about the origin where it attains its maximum value.*
- *The total volume under the surface of the kernel $K(\mathbf{x})$ is unity; that is, for an m-dimensional vector $\mathbf{x}$,*

$$\int_{\mathbb{R}^m} K(\mathbf{x})d\mathbf{x} = 1 \tag{5.129}$$

Assuming that $\mathbf{x}_1, \mathbf{x}_2, \ldots, \mathbf{x}_N$ are independent random vectors and identically distributed, we may formally define the Parzen-Rosenblatt density estimate of $f_\mathbf{X}(\mathbf{x})$ as:

$$\hat{f}_\mathbf{X}(\mathbf{x}) = \frac{1}{Nh^{m_0}} \sum_{i=1}^{N} K\left(\frac{\mathbf{x} - \mathbf{x}_i}{h}\right) \quad \text{for } \mathbf{x} \in \mathbb{R}^{m_0} \tag{5.130}$$

where the smoothing parameter h is a positive number called *bandwidth* or simply *width*; h controls the size of the kernel. (The parameter h used here should *not* be confused with the h used to define the Fréchet derivative in Section 5.5.) An important property of the Parzen–Rosenblatt density estimator is that it is a *consistent estimator*[14] (i.e., asymptotically unbiased) in the sense that if $h = h(N)$ is chosen as a function of N such that

$$\lim_{N \to \infty} h(N) = 0,$$

then

$$\lim_{N \to \infty} E[\hat{f}_\mathbf{X}(\mathbf{x})] = f_\mathbf{X}(\mathbf{x})$$

For this latter equation to hold, $\mathbf{x}$ should be a point of continuity for $\hat{f}_\mathbf{X}(\mathbf{x})$.

In a manner similar to that described in Eq. (5.130), we may formulate the Parzen–Rosenblatt density estimate of the joint probability density function $f_{\mathbf{X},Y}(\mathbf{x},y)$ as follows:

$$\hat{f}_{\mathbf{X},Y}(\mathbf{x},y) = \frac{1}{Nh^{m_0+1}} \sum_{i=1}^{N} K\left(\frac{\mathbf{x} - \mathbf{x}_i}{h}\right) K\left(\frac{y - y_i}{h}\right) \quad \text{for } \mathbf{x} \in \mathbb{R}^{m_0} \text{ and } y \in \mathbb{R} \tag{5.131}$$

Integrating $\hat{f}_{\mathbf{X},Y}(\mathbf{x},y)$ with respect to y, we get the $f_\mathbf{X}(\mathbf{x})$ of Eq. (5.130), and so we should. Moreover,

$$\int_{-\infty}^{\infty} y\hat{f}_{\mathbf{X},Y}(\mathbf{x},y)dy = \frac{1}{Nh^{m_0+1}} \sum_{i=1}^{N} K\left(\frac{\mathbf{x} - \mathbf{x}_i}{h}\right) \int_{-\infty}^{\infty} yK\left(\frac{y - y_i}{h}\right)dy$$

Changing the variable of integration by putting $z = (y - y_i)/h$, and using the symmetric property of the kernel $K(\cdot)$, we obtain the result

$$\int_{-\infty}^{\infty} y\hat{f}_{\mathbf{X},Y}(\mathbf{x}, y)dy = \frac{1}{Nh^{m_0}} \sum_{i=1}^{N} y_i K\left(\frac{\mathbf{x} - \mathbf{x}_i}{h}\right) \tag{5.132}$$

Thus, using Eqs. (5.132) and (5.130) as estimates of the quantities in the numerator and denominator of Eq. (5.128), respectively, we obtain the following estimate of the regression function $f(\mathbf{x})$, after canceling common terms:

$$\mathbf{F}(\mathbf{x}) = \hat{f}(\mathbf{x})$$

$$= \frac{\sum_{i=1}^{N} y_i K\left(\dfrac{\mathbf{x} - \mathbf{x}_i}{h}\right)}{\sum_{j=1}^{N} K\left(\dfrac{\mathbf{x} - \mathbf{x}_j}{h}\right)} \tag{5.133}$$

where in the denominator, for clarity of presentation, we have used j instead of i as the index of summation. As with an ordinary RBF network, the kernel regression estimator $F(\mathbf{x})$ defined in Eq. (5.133) is a universal approximator.

There are two ways in which the approximating function $F(\mathbf{x})$ may be viewed:

1. *Nadaraya–Watson regression estimator.* Define the *normalized weighting function*

$$W_{N,i}(\mathbf{x}) = \frac{K\left(\dfrac{\mathbf{x} - \mathbf{x}_i}{h}\right)}{\sum_{j=1}^{N} K\left(\dfrac{\mathbf{x} - \mathbf{x}_j}{h}\right)}, \qquad i = 1, 2, ..., N \tag{5.134}$$

with

$$\sum_{i=1}^{N} W_{N,i}(\mathbf{x}) = 1 \quad \text{for all } \mathbf{x} \tag{5.135}$$

We may then rewrite the kernel regression estimator of Eq. (5.133) in the simplified form

$$F(\mathbf{x}) = \sum_{i=1}^{N} W_{N,i}(\mathbf{x}) y_i \tag{5.136}$$

which describes $F(\mathbf{x})$ as a *weighted average* of the y-observables. The particular form of weighting function $W_{N,i}(\mathbf{x})$ given in Eq. (5.136) was originally proposed by Nadaraya (1964) and Watson (1964). Accordingly, the approximating function of Eq. (5.136) is often called the *Nadaraya–Watson regression estimator (NWRE)*.[15]

2. *Normalized RBF network.* For the second viewpoint, we assume *spherical symmetry* of the kernel $K(\mathbf{x})$, in which case we may set (Krżyzak et al., 1996)

$$K\left(\frac{\mathbf{x} - \mathbf{x}_i}{h}\right) = K\left(\frac{\|\mathbf{x} - \mathbf{x}_i\|}{h}\right) \quad \text{for all } i \tag{5.137}$$

where $\|\cdot\|$ denotes the Euclidean norm of the enclosed vector. Correspondingly, we define the *normalized radial basis function*

$$\psi_N(\mathbf{x}, \mathbf{x}_i) = \frac{K\left(\dfrac{\|\mathbf{x} - \mathbf{x}_i\|}{h}\right)}{\sum_{j=1}^{N} K\left(\dfrac{\|\mathbf{x} - \mathbf{x}_j\|}{h}\right)}, \qquad i = 1, 2, ..., N \tag{5.138}$$

with

$$\sum_{i=1}^{N} \psi_N(\mathbf{x}, \mathbf{x}_i) = 1 \quad \text{for all } \mathbf{x} \tag{5.139}$$

The subscript N in $\psi_N(\mathbf{x}, \mathbf{x}_i)$ signifies the use of *normalization*.

For the regression problem considered here, we recognize that the "linear weights," w_i, applied to the basic functions $\psi_N(\mathbf{x}, \mathbf{x}_i)$ are simply the observables, y_i, of the regression model for the input data $\mathbf{x}_i$. Thus, letting

$$y_i = w_i, \quad i = 1, 2, \ldots, N$$

we may reformulate the approximating function of Eq. (5.133) in the general form

$$F(\mathbf{x}) = \sum_{i=1}^{N} w_i \psi_N(\mathbf{x}, \mathbf{x}_i) \tag{5.140}$$

Equation (5.140) represents the input–output mapping of a *normalized radial-basis function (RBF) network* (Moody and Darken, 1989; Xu et al., 1994). Note that

$$0 \le \psi_N(\mathbf{x}, \mathbf{x}_i) \le 1 \quad \text{for all } \mathbf{x} \text{ and } \mathbf{x}_i \tag{5.141}$$

Accordingly, $\psi_N(\mathbf{x}, \mathbf{x}_i)$ may be interpreted as the probability of an event described by the input vector $\mathbf{x}$, conditional on $\mathbf{x}_i$.

The basic difference between the normalized radial-basis function $\psi_N(\mathbf{x}, \mathbf{x}_i)$ of Eq. (5.138) and an ordinary radial-basis function is a denominator term that constitutes the *normalization factor*. This normalization factor is an estimate of the underlying pdf of the input vector $\mathbf{x}$. Consequently, the basis functions $\psi_N(\mathbf{x}, \mathbf{x}_i)$ for $i = 1, 2, \ldots, N$ sum to unity for all $\mathbf{x}$, as described in Eq. (5.139). In contrast, there is no guarantee that this condition is satisfied by the basis (Green's) functions of the ordinary RBF network of Eq. (5.57).

The derivation of the input–output mapping $F(\mathbf{x})$ described in Eq. (5.138) was presented here by using the notion of density estimation. Like the hypersurface reconstruction problem, density estimation is an ill-posed problem. To make it well-posed, some form of regularization has to be used. The Parzen–Rosenblatt density estimator, and therefore the Nadaraya–Watson regression estimator, can be derived within the framework of regularization theory (Vapnik, 1982). Naturally, the cost functional to be minimized for density estimation is different from the deterministic Tikhonov functional of Eq. (5.23). The cost functional for density estimation consists of the sum of two terms: a square error term involving the unknown probability density function, and an appropriate form of stabilizing functional.

Multivariate Gaussian Distribution

A variety of kernel functions is possible in general. However, both theoretical and practical considerations limit the choice. As with the Green's function, a widely used kernel is the multivariate Gaussian distribution:

$$K(\mathbf{x}) = \frac{1}{(2\pi)^{m_0/2}} \exp\left(-\frac{\|\mathbf{x}\|^2}{2}\right) \tag{5.142}$$

where m_0 is the dimension of the input vector $\mathbf{x}$. The spherical symmetry of the kernel $K(\mathbf{x})$ is clearly apparent in Eq. (5.142). Assuming the use of a common width (spread) σ that plays the role of smoothing parameter h for a Gaussian distribution, and centering the kernel on a data point $\mathbf{x}_i$, we may write

$$K\left(\frac{\mathbf{x} - \mathbf{x}_i}{h}\right) = \frac{1}{(2\pi\sigma^2)^{m_0/2}} \exp\left(\frac{\|\mathbf{x} - \mathbf{x}_i\|^2}{2\sigma^2}\right), \qquad i = 1, 2, \ldots, N \qquad (5.143)$$

Thus, using Eq. (5.143), the Nadaraya–Watson regression estimator takes the following form (Specht, 1991):

$$F(\mathbf{x}) = \frac{\displaystyle\sum_{i=1}^{N} y_i \exp\left(-\frac{\|\mathbf{x} - \mathbf{x}_i\|^2}{2\sigma^2}\right)}{\displaystyle\sum_{j=1}^{N} \exp\left(-\frac{\|\mathbf{x} - \mathbf{x}_j\|^2}{2\sigma^2}\right)} \qquad (5.144)$$

where the denominator term representing the Parzen–Rosenblatt density estimator, consists of the sum of N multivariate Gaussian distributions centered on the data points $\mathbf{x}_1, \mathbf{x}_2, \ldots, \mathbf{x}_N$.

Correspondingly, using Eq. (5.143) in (5.138) and then (5.140), the input–output mapping function of the normalized RBF network takes the following form:

$$F(\mathbf{x}) = \frac{\displaystyle\sum_{i=1}^{N} w_i \exp\left(-\frac{\|\mathbf{x} - \mathbf{x}_i\|^2}{2\sigma^2}\right)}{\displaystyle\sum_{j=1}^{N} \exp\left(-\frac{\|\mathbf{x} - \mathbf{x}_j\|^2}{2\sigma^2}\right)} \qquad (5.145)$$

In Eqs. (5.144) and (5.145) the centers of the normalized radial-basis functions coincide with the data points $\{\mathbf{x}_i\}_{i=1}^{N}$. As with ordinary radial-basis functions, a smaller number of normalized radial-basis functions can be used, with their centers treated as free parameters to be chosen according to some heuristic (Moody and Darken, 1989) or determined in a principled manner (Poggio and Girosi, 1990a).

5.13   LEARNING STRATEGIES

The learning process undertaken by a radial-basis function (RBF) network irrespective of its theoretical background, may be visualized as follows. The linear weights associated with the output unit(s) of the network tend to evolve on a different "time scale" compared to the nonlinear activation functions of the hidden units. Thus, as the hidden layer's activation functions evolve slowly in accordance with some *nonlinear* optimization strategy, the output layer's weights adjust themselves rapidly through a *linear* optimization strategy. The important point is that the different layers of an RBF network perform different tasks, and so it is reasonable to separate the optimization of the hidden and output layers of the network by using different techniques, and perhaps by operating on different time scales (Lowe, 1991a).

There are different learning strategies that we can follow in the design of an RBF network, depending on how the centers of the radial-basis functions of the network are specified. Here we identify four approaches. The first three design strategies pertain to an RBF network whose formulation is based on interpolation theory. The last design strategy combines elements of regularization theory and kernel regression estimation theory.

1. Fixed Centers Selected at Random

The simplest approach is to assume *fixed* radial-basis functions defining the activation functions of the hidden units. The locations of the centers may be chosen *randomly* from the training data set. This is considered to be a "sensible" approach, provided that the training data are distributed in a representative manner for the problem at hand (Lowe, 1989). For the radial-basis functions themselves, we may employ an *isotropic* Gaussian function whose standard deviation is fixed according to the spread of the centers. Specifically, a (normalized) radial-basis function centered at t_i is defined as

$$G(\|\mathbf{x} - \mathbf{t}_i\|^2) = \exp\left(-\frac{m_1}{d_{max}^2}\|\mathbf{x} - \mathbf{t}_i\|^2\right), \qquad i = 1, 2, \ldots, m_1 \qquad (5.146)$$

where m_1 is the number of centers and d_{max} is the maximum distance between the chosen centers. In effect, the standard deviation (i.e., width) of all the Gaussian radial-basis functions is fixed at

$$\sigma = \frac{d_{max}}{\sqrt{2m_1}} \qquad (5.147)$$

This formula ensures that the individual radial-basis functions are not too peaked or too flat; both of these two extreme conditions should be avoided. As an alternative to Eq. (5.147) we may use individually scaled centers with broader widths in areas of lower data density, which requires experimentation with the training data.

The only parameters that would need to be learned in this approach are the linear weights in the output layer of the network. A straightforward procedure for doing this is to use the *pseudoinverse method* (Broomhead and Lowe, 1988). Specifically, we have (see also Eqs. (5.77) and (5.78))

$$\mathbf{w} = \mathbf{G}^+ \mathbf{d} \qquad (5.148)$$

where $\mathbf{d}$ is the desired response vector in the training set. The matrix $\mathbf{G}^+$ is the pseudoinverse of the matrix $\mathbf{G}$, which is itself defined as

$$\mathbf{G} = \{g_{ji}\} \qquad (5.149)$$

where

$$g_{ji} = \exp\left(-\frac{m_1}{d^2}\|\mathbf{x}_j - \mathbf{t}_i\|^2\right), \qquad j = 1, 2, \ldots, N; \quad i = 1, 2, \ldots, m_1 \qquad (5.150)$$

where $\mathbf{x}_j$ is the jth input vector of the training sample.

Basic to all algorithms for the computation of a pseudoinverse of a matrix is the *singular-value decomposition* (*SVD*) (Golub and Van Loan, 1996):

*If **G** is a real N-by-M matrix, there exist orthogonal matrices*

$$\mathbf{U} = \{\mathbf{u}_1, \mathbf{u}_2, ..., \mathbf{u}_N\}$$

and

$$\mathbf{V} = \{\mathbf{v}_1, \mathbf{v}_2, ..., \mathbf{v}_M\}$$

such that

$$\mathbf{U}^T\mathbf{G}\mathbf{V} = \text{diag}(\sigma_1, \sigma_2, ..., \sigma_K), \qquad K = \min(M, N) \tag{5.151}$$

where

$$\sigma_1 \geq \sigma_2 \geq \cdots \geq \sigma_K > 0$$

The column vectors of the matrix **U** are called the *left singular vectors* of **G**, and the column vectors of the matrix **V** are called its *right singular vectors*. The $\sigma_1, \sigma_2, ..., \sigma_K$ are called the *singular values* of the matrix **G**. According to the singular value decomposition theorem, the *M-by-N* pseudoinverse of matrix **G** is defined by

$$\mathbf{G}^+ = \mathbf{V}\mathbf{\Sigma}^+\mathbf{U}^T \tag{5.152}$$

where $\mathbf{\Sigma}^+$ is itself an *N-by-N* matrix defined in terms of the singular values of **G** by

$$\mathbf{\Sigma}^+ = \text{diag}\left(\frac{1}{\sigma_1}, \frac{1}{\sigma_2}, ..., \frac{1}{\sigma_K}, 0, ..., 0\right) \tag{5.153}$$

Efficient algorithms for the computation of a pseudoinverse matrix are discussed in Golub and Van Loan (1996).

It is interesting that experience with the random selection of centers shows that this method is relatively insensitive to the use of regularization; see Problem 5.14 for a computer experiment on pattern classification using this method. This kind of performance suggests that the random selection of centers as a method for the design of RBF networks from a large training set of fixed size is perhaps a regularization method in its own right.

2. Self-Organized Selection of Centers

The main problem with the method of fixed centers just described is the fact that it may require a large training set for a satisfactory level of performance. One way of overcoming this limitation is to use a *hybrid learning process*, consisting of two different stages (Moody and Darken, 1989; Lippmann, 1989b; Chen et al., 1992):

- *Self-organized learning* stage, the purpose of which is to estimate appropriate locations for the centers of the radial basis functions in the hidden layer.
- *Supervised learning* stage, which completes the design of the network by estimating the linear weights of the output layer.

Although batch processing can be used to implement these two stages of learning, it is preferable to take an adaptive (iterative) approach.

For the self-organized learning process we need a *clustering* algorithm that partitions the given set of data points into subgroups, each of which should be as homogeneous as possible. One such algorithm is the *k-means clustering algorithm* (Duda and Hart, 1973), which places the centers of the radial-basis functions in only those regions of the input space $\mathcal{X}$ where significant data are present. Let m_1 denote the number of radial-basis functions; the determination of a suitable value for m_1 may require experimentation. Let $\{\mathbf{t}_k(n)\}_{k=1}^{m_1}$ denote the centers of the radial-basis functions at iteration n of the algorithm. Then, the *k*-means clustering algorithm proceeds as follows:

1. *Initialization.* Choose random values for the initial centers $\mathbf{t}_k(0)$; the only restriction is that these initial values be different. It may also be desirable to keep the Euclidean norm of the centers small.
2. *Sampling.* Draw a sample vector $\mathbf{x}$ from the input space $\mathcal{X}$ with a certain probability. The vector $\mathbf{x}$ is input into the algorithm at iteration n.
3. *Similarity matching.* Let $k(\mathbf{x})$ denote the index of the best-matching (winning) center for input vector $\mathbf{x}$. Find $k(\mathbf{x})$ at iteration n by using the minimum-distance Euclidean criterion:

$$k(\mathbf{x}) = \arg\min_k \|\mathbf{x}(n) - \mathbf{t}_k(n)\|, \quad k = 1, 2, ..., m_1 \tag{5.154}$$

   where $\mathbf{t}_k(n)$ is the center of the kth radial-basis function at iteration n.
4. *Updating.* Adjust the centers of the radial-basis functions, using the update rule:

$$\mathbf{t}_k(n+1) = \begin{cases} \mathbf{t}_k(n) + \eta[\mathbf{x}(n) - \mathbf{t}_k(n)], & k = k(\mathbf{x}) \\ \mathbf{t}_k(n), & \text{otherwise} \end{cases} \tag{5.155}$$

   where η is a *learning-rate parameter* that lies in the range $0 < \eta < 1$.
5. Continuation. Increment n by 1, go back to step 2, and continue the procedure until no noticeable changes are observed in the centers $\mathbf{t}_k$.

The *k*-means clustering algorithm just described is, in fact, a special case of a competitive (winner-takes-all) learning process known as the *self-organizing map*, which is discussed in Chapter 9. This latter algorithm also lends itself for implementing the self-organized learning stage.

A limitation of the *k*-means clustering algorithm is that it can only achieve a local optimum solution that depends on the initial choice of cluster centers. Consequently, computing resources may be wasted in that some initial centers get stuck in regions of the input space $\mathcal{X}$ with a scarcity of data points and may therefore never have the chance to move to new locations where they are needed. The net result is possibly an unnecessarily large network. To overcome this limitation of the conventional *k*-means clustering algorithm, Chen (1995) proposes the use of an *enhanced k-means clustering algorithm* due to Chinunrueng and Séquin (1994), which is based on a cluster variation-weighted measure that enables the algorithm to converge to an optimum or near-optimum configuration, independent of the initial center locations.

Having identified the individual centers of the Gaussian radial-basis functions and their common width using the *k*-means clustering algorithm or its enhanced version, the next and final stage of the hybrid learning process is to estimate the weights of the output layer. A simple method for this estimation is the least-mean-square (LMS) algorithm described in Chapter 3. The vector of output signals produced by the hidden units constitutes the input vector to the LMS algorithm. Note also that the *k*-means clustering algorithm for the hidden units and the LMS algorithm for the output unit(s) may proceed with their own individual computations in a concurrent fashion, thereby accelerating the training process.

3. Supervised Selection of Centers

In the third approach, the centers of the radial-basis functions and all other free parameters of the network undergo a supervised learning process; in other words, the RBF network takes on its most generalized form. A natural candidate for such a process is error-correction learning, which is most conveniently implemented using a gradient-descent procedure that represents a generalization of the LMS algorithm.

The first step in the development of such a learning procedure is to define the instantaneous value of the cost function.

$$\mathscr{E} = \frac{1}{2} \sum_{j=1}^{N} e_j^2 \tag{5.156}$$

where *N* is the size of the training sample used to do the learning, and e_j is the error signal defined by

$$e_j = d_j - F^*(\mathbf{x}_j)$$

$$= d_j - \sum_{i=1}^{M} w_i G(\|\mathbf{x}_j - \mathbf{t}_i\|_{C_i}) \tag{5.157}$$

The requirement is to find the free parameters w_i, $\mathbf{t}_i$, and $\mathbf{\Sigma}_i^{-1}$ (the latter being related to the norm-weighting matrix $\mathbf{C}_i$) so as to minimize $\mathscr{E}$. The results of this minimization are summarized in Table 5.4; the derivations of these results are presented as an exercise to the reader in Problem 5.13. The following points are noteworthy in Table 5.4.

- The cost function $\mathscr{E}$ is convex with respect to the linear parameters w_i, but nonconvex with respect to the centers $\mathbf{t}_i$ and matrix $\mathbf{\Sigma}_i^{-1}$; in the latter case, the search for the optimum values of $\mathbf{t}_i$ and $\mathbf{\Sigma}_i^{-1}$ may get stuck at a local minimum in parameter space.
- The update equations for w_i, $\mathbf{t}_i$, and $\mathbf{\Sigma}_i^{-1}$ are (in general) assigned different learning-rate parameters η_1, η_2, and η_3, respectively.
- Unlike the back-propagation algorithm, the gradient-descent procedure described in Table 5.4 for an RBF network does *not* involve error back-propagation.
- The gradient vector $\partial\mathscr{E}/\partial\mathbf{t}_i$ has an effect similar to a *clustering effect* that is task-dependent (Poggio and Girosi, 1990a).

For the *initialization* of the gradient-descent procedure, it is often desirable to begin the search in parameter space from a *structured* initial condition that limits the region of parameter space to be searched to an already known useful area, which may be achieved by implementing a standard pattern-classification method (Lowe, 1991a). In so doing, the likelihood of converging to an undesirable local minimum in weight space is reduced. For example, we may begin with a *Gaussian classifier*, which assumes

TABLE 5.4   Adaptation Formulas for the Linear Weights and the Positions and Spreads of Centers for RBF Network[a]

1.   *Linear weights* (output layer)

$$\frac{\partial \mathscr{E}(n)}{\partial w_i(n)} = \sum_{j=1}^{N} e_j(n) G(\|\mathbf{x}_j - \mathbf{t}_i(n)\|_{C_i})$$

$$w_i(n + 1) = w_i(n) - \eta_1 \frac{\partial \mathscr{E}(n)}{\partial w_i(n)}, \qquad i = 1, 2, \ldots, \mathbf{m}_1$$

2.   *Positions of centers* (hidden layer)

$$\frac{\partial \mathscr{E}(n)}{\partial \mathbf{t}_i(n)} = 2w_i(n) \sum_{j=1}^{N} e_j(n) G'(\|\mathbf{x}_j - \mathbf{t}_i(n)\|_{C_i}) \boldsymbol{\Sigma}_i^{-1} [\mathbf{x}_j - \mathbf{t}_i(n)]$$

$$\mathbf{t}_i(n + 1) = \mathbf{t}_i(n) - \eta_2 \frac{\partial \mathscr{E}(n)}{\partial \mathbf{t}_i(n)}, \qquad i = 1, 2, \ldots, \mathbf{m}_1$$

3.   *Spreads of centers* (hidden layer)

$$\frac{\partial \mathscr{E}(n)}{\partial \boldsymbol{\Sigma}_i^{-1}(n)} = - w_i(n) \sum_{j=1}^{N} e_j(n) G'(\|\mathbf{x}_j - \mathbf{t}_i(n)\|_{C_i}) \mathbf{Q}_{ji}(n)$$

$$\mathbf{Q}_{ji}(n) = [\mathbf{x}_j - \mathbf{t}_i(n)][\mathbf{x}_j - \mathbf{t}_i(n)]^T$$

$$\boldsymbol{\Sigma}_i^{-1}(n + 1) = \boldsymbol{\Sigma}_i^{-1}(n) - \eta_3 \frac{\partial \mathscr{E}(n)}{\partial \boldsymbol{\Sigma}_i^{-1}(n)}$$

[a]The term $e_j(n)$ is the error signal of output unit j at time n. The term $G'(\cdot)$ is the first derivative of the Green's function $G(\cdot)$ with respect to its argument.

that each pattern in each class is drawn from a Gaussian distribution; this special form of pattern classifier based on the Bayes hypothesis testing procedure is discussed in Chapter 3.

A question that arises at this stage of the discussion is: What can be gained by adapting the positions of the centers of the radial-basis functions? The answer to this question naturally depends on the application of interest. Nevertheless, on the basis of some results reported in the literature, there is practical merit to the idea of allowing the centers to move. Work done by Lowe (1989) on speech recognition using RBF networks indicates that nonlinear optimization of the parameters that define the activation functions of the hidden layer is beneficial when a minimal network configuration is required. However, according to Lowe, the same performance on generalization may be achieved by using a larger RBF network; that is, a network with a larger number of fixed centers in the hidden layer, and only adapting the output layer of the network by linear optimization.

Wettschereck and Dietterich (1992) have compared the performance of (Gaussian) radial-basis function networks with fixed centers to that of generalized radial-basis function networks with adjustable centers; in the latter case, the positions of the centers are determined by supervised learning. The performance comparison was made for the NETtalk task. The original NETtalk experiment was carried out by Sejnowski and Rosenberg (1987) using a multilayer perceptron trained with the back-propagation algorithm; it is described in Chapter 13. The purpose of the experiment performed by Wettschereck and Dietterich was to understand how a neural network could learn to map English spelling into its phonetic pronunciation. The experimental study by Wettschereck and Dietterich in the NETtalk domain may be summarized as follows:

- RBF networks (with unsupervised learning of the centers' locations and supervised learning of the output-layer weights) did *not* generalize nearly as well as multilayer perceptrons trained with the back-propagation algorithm.
- Generalized RBF networks (with supervised learning of the centers' locations as well as the output-layer weights) were able to exceed substantially the generalization performance of multilayer perceptrons.

4. Strict Interpolation with Regularization

A method for designing RBF networks that combines elements of the regularization theory of Section 5.5 and the kernel regression estimation theory described in Section 5.12 in a principled way is described in Yee (1998). The method involves the combined use of the following four ingredients:

1. Radial-basis function, G, admissable (possibly with some scaling) as the kernel of a (mean-square) consistent Nadaraya–Watson regression estimate (NWRE).
2. Diagonal input norm-weighting matrix, Σ^{-1}, common to all centers with entries

$$\Sigma = \text{diag}(h_1, h_2, \dots, h_{m_0}) \tag{5.156}$$

where $h_1, h_2, \dots, h_{m_0}$ are the per dimension *bandwidths* of a consistent NWRE with (scaled) kernel G, as previously set forth, and m_0 is the dimensionality of the input space. For example, we may set $h_i = \alpha_i \sigma_i^2, i = 1, 2, \dots, m_0$, where σ_i^2 is the sample variance of the ith input variable estimated from the available training input data. The positive input *scale factors* $\alpha_1, \alpha_2, \dots, \alpha_{m_0}$ can then be determined by using a suitable cross-validation (CV) procedure, as explained in Section 5.9.
3. Regularized strict interpolation, which involves training for the linear weights according to Eq. (5.54).
4. Selection of the regularization parameter λ and the input scale factors $\alpha_1, \alpha_2, \dots, \alpha_{m_0}$, which is accomplished via an *asymptotically optimal* method such as the leave-out-one CV method defined in Eq. (5.117) or the GCV method defined in Eq. (5.121). The selected parameters may be interpreted as follows:
   - The larger the value of λ selected, the larger is the noise corrupting the measurement of parameters.
   - When the radial-basis function G is a unimodal kernel (e.g., the Gaussian kernel), the smaller the value of a particular α_i, the more "sensitive" the overall network output is to the associated input dimension. Conversely, the larger the value of a particular α_i, the less "relevant" the associated input dimension is to explaining the variation of the overall network output with respect to changes in the input. Hence, we can use the selected α_i to rank the relative *significance* of the input variables and thereby indicate which input variables are suitable candidates for dimensionality reduction, if necessary.

The justification for this design procedure is discussed in detail in Yee (1998). For our purposes here, we can motivate these design choices as follows. It can be shown that the NWRE corresponds to a special class of regularized RBF networks, in the sense that any NWRE may be approximated with vanishing mean-square and absolute error by a suitably constructed sequence of regularized RBF networks for which the regularization parameter sequence $\{\lambda_N\}$ is allowed to grow (at an appropriate rate) to infinity with N, the size of training sample. On the other hand, as $N \to \infty$, we have (under some mild conditions) the convergence of the risk defined in Eq. (5.99) to the (global) mean-squared error. If we use an asymptotically optimal parameter selection procedure for the regularization parameter sequence, then, by construction, the resulting sequence of RBF networks must have (asymptotically) minimum mean-squared error over all possible choices of regularization parameter sequence, including the one corresponding to the NWRE. If conditions then hold such that the NWRE is known to be mean-squared error consistent, the same must also be true for the regularized RBF network designed according to the same procedure. In other words, regularized RBF networks designed according to this procedure can *inherit* the consistency properties of the NWRE. This consequence allows us to leverage the known consistency results of the NWRE in areas such as *time series regression*, where *dependent* and *nonstationary* processes are often encountered and where the usual neural network assumptions of i.i.d. training data and stationary processes are invalid. In summary, by *synthesizing* elements of both regularization theory and kernel regression estimation theory, the design procedure outlined here offers a practical prescription for theoretically supported regularized RBF network design and application.

5.14 COMPUTER EXPERIMENT: PATTERN CLASSIFICATION

In this section we use a computer experiment to illustrate the design of a regularized RBF network based on the use of strict interpolation. The computer experiment involves a binary classification problem based on data drawn from two equiprobable overlapping two-dimensional Gaussian distributions corresponding to classes $\mathscr{C}_1$ and $\mathscr{C}_2$. Details of the Gaussian distributions are the same as those described in Section 4.8. Class $\mathscr{C}_1$ is characterized by mean vetor $[0, 0]^T$ and common variance 1, whereas class $\mathscr{C}_2$ is characterized by mean vector $[0, 2]^T$ and common variance 4. The experiment described in this section may thus be viewed as the regularization RBF counterpart to the back-propagation learning experiment of Section 4.8.

With two classes $\mathscr{C}_1$ and $\mathscr{C}_2$, the regularized RBF network is constructed to have two output functions, one for each class. Also, binary-valued class indicator outputs are used as the desired output values, as shown by

$$d_k^{(p)} = \begin{cases} 1 & \text{if pattern } p \text{ belongs to class } \mathscr{C}_k \\ 0 & \text{otherwise} \end{cases}$$

where $k = 1, 2$.

Before we proceed with the experiment, however, we must resolve the issue of an output decision rule for performing the pattern classification. In Yee (1998) it is shown that the outputs of a regularized RBF network classifier provide estimates of the posterior class probabilities. This is true only under the condition that the network is trained with the binary-valued class indicator vector type of desired outputs. We may now proceed to apply the decision rule of Eq. (4.55) for this class of networks:

Select the class corresponding to the maximum output function.

The method of strict interpolation for the selection of centers is tested with different values of regularization parameter λ. For a prescribed λ, Eq. (5.54) is used to compute the weight vector of the output layer in the RBF network, as shown by

$$\mathbf{w} = (\mathbf{G} + \lambda\mathbf{I})^{-1}\mathbf{d}$$

where $\mathbf{G}$ is an N-by-N Green's matrix whose ji-th element is equal to the radially symmetric Green's function $G(\mathbf{x}_j, \mathbf{x}_i)$, N is the sample size, and $\mathbf{d}$ is the desired response vector.

For each regularization parameter λ, the ensemble comprises 50 independent networks, each of which is tested against the same reference set of 1000 patterns.

Table 5.5 presents the ensemble statistics for the probability of correct classification P_c, computed for the case of $m_1 = 20$ centers. The ensemble statistics are computed for different values of the regularization parameter λ. Table 5.6 presents the corresponding results computed for the case of a larger regularized RBF network with $m_1 = 100$ centers.

Figure 5.7 displays the decision boundaries formed by the network outputs for a regularization parameter $\lambda = 10$, for which we have the best statistics. The two parts of Fig. 5.7 correspond to the best- and worst-performing network within the ensemble under test; both parts of the figure are for the case of 100 units.

TABLE 5.5 Size of Hidden Layer $m_1 = 20$ Centers: Details of Probability of Correct Classification, P_c(%) for varying Regularization Parameter

	Regularization Parameter, λ					
Ensemble Statistic	0	0.1	1	10	100	1000
Mean	57.49	72.42	74.42	73.80	72.46	72.14
Std. dev.[a]	7.47	4.11	3.51	4.17	4.98	5.09
Minimum	44.20	61.60	65.80	63.10	60.90	60.50
Maximum	72.70	78.30	78.90	79.20	79.40	79.40

[a]Std. dev.: Standard deviation.

TABLE 5.6 Size of Hidden Layer $m_1 = 100$ Centers: Details of Probability of Correct Classification, P_c(%) for Varying Regularization Parameter

	Regularization Parameter, λ					
Ensemble Statistic	0	0.1	1	10	100	1000
Mean	50.58	77.03	77.72	77.87	76.47	75.33
Std. dev.[a]	4.70	1.45	0.94	0.91	1.62	2.25
Minimum	41.00	70.60	75.10	75.10	72.10	70.10
Maximum	61.30	79.20	79.80	79.40	78.70	78.20

[a]Std. dev.: Standard deviation.

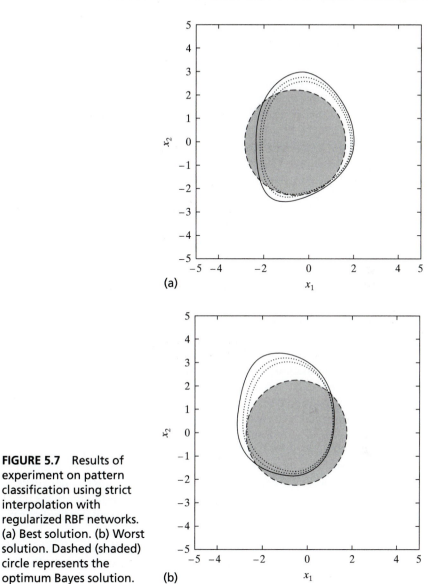

FIGURE 5.7 Results of experiment on pattern classification using strict interpolation with regularized RBF networks. (a) Best solution. (b) Worst solution. Dashed (shaded) circle represents the optimum Bayes solution.

Comparing Tables 5.5 and 5.6, we make the following observations:

1. For both $m_1 = 20$ and $m_1 = 100$ centers, the classification performance of the network for $\lambda = 0$ is relatively poor.
2. The use of regularization has a dramatic influence on the classification performance of the RBF network.
3. For $\lambda \geq 0.1$, the classification performance of the network is somewhat insensitive to an increase in the regularization parameter λ. For $m_1 = 20$ centers, best performance is obtained at $\lambda = 1$, and for $m_1 = 100$ centers, it is obtained at $\lambda = 10$.
4. Increasing the number of centers from $m_1 = 20$ to $m_1 = 100$ improves the classification performance by about 4.5 percent.

5.15 SUMMARY AND DISCUSSION

The structure of an RBF network is unusual in that the constitution of its hidden units is entirely different from that of its output units. With radial-basis functions providing the foundation for the design of the hidden units, the theory of RBF networks is linked closely with that of radial-basis functions, which is one of the main fields of study in numerical analysis (Singh, 1992). Another interesting point is that with linear weights of the output layer providing a set of adjustable parameters, much can be gained by ploughing through the extensive literature on linear adaptive filters (Haykin, 1996).

Unlike multilayer perceptrons trained with the back-propagation algorithm, the design of RBF networks follows a principled approach. In particular, Tikhonov's regularization theory presented in Section 5.5 provides a sound mathematical basis for the formulation of RBF networks. The Green's function $G(\mathbf{x}, \boldsymbol{\xi})$ plays a central role in this theory. The form of the Green's function as the basis function of the network is determined by the form of the *smoothing constraint* imposed in the application of regularization theory. The smoothing constraint specified by the differential operator $\mathbf{D}$ of Eq. (5.63) leads to the formulation of a multivariate Gaussian function for the Green's function. By formulating a different composition for the differential operator $\mathbf{D}$, we naturally end up with a different form of Green's function. Keep in mind that once the requirement of fewer basis functions than data points is relaxed, the reduction in computational complexity becomes an important factor in determining the smoothing regularizer. Here is another possible reason for the use of some other function (e.g., the thin-plate-spline function described in Problem 5.1) as the basis function in the design of the regularized RBF network in Fig. 5.5. Whatever the choice of basis functions, to derive the full benefits of regularization theory applied to the design of RBF networks, we need a principled approach for estimating the regularization parameter λ. The generalized cross-validation described in Section 5.9 fills this need. The theory justifying the use of generalized cross-validation is an asymptotic one, which stipulates that the training set must be large enough for a good estimate of λ to be attainable.

Another principled approach for the design of RBF networks is via kernel regression. This approach involves the use of density estimation, for which the radial-basis functions sum to unity exactly. Multivariate Gaussian distributions provide a convenient method for satisfying this requirement.

In conclusion, the input–output mapping function of a Gaussian RBF network bears a close resemblance to that realized by a mixture of experts. This latter model is discussed in Chapter 7.

NOTES AND REFERENCES

1. Radial-basis functions were first introduced in the solution of the real multivariate interpolation problem. The early work on this subject is surveyed in Powell (1985). It is now one of the main fields of research in numerical analysis.

   Broomhead and Lowe (1988) were the first to exploit the use of radial-basis functions in the design of neural networks. Another major contribution to the theory and design of radial-basis function networks is due to Poggio and Girosi (1990a). This latter paper emphasizes the use of regularization theory applied to this class of networks as a method for improved generalization to new data.

2. The proof of Cover's theorem follows from these considerations (Cover, 1965):

- *Schlafli's theorem* or *function-counting theorem*, which states that the number of homogeneously linearly separable dichotomies of N vectors in general position in Euclidean-m_1 space is equal to

$$C(N, m_1) = 2 \sum_{m=0}^{m_1-1} \binom{N-1}{m}$$

A set of vectors $\mathscr{X} = \{\mathbf{x}_i\}_{i=1}^N$ is said to be in *general position* in Euclidean-m_1 space if every subset of m_1 or fewer vectors is linearly independent.

- *Reflection invariance* of the joint probability distribution of $\mathscr{X}$, which implies that the probability (conditional on $\mathscr{X}$) that a random dichotomy be separable is equal to the unconditional probability that a particular dichotomy of $\mathscr{X}$ (all N vectors in one category) be separable.

The function-counting theorem has been independently proved in different forms and applied to specific configurations of perceptrons (i.e., linear threshold units) by Cameron (1960), Joseph (1960), and Winder (1961). In Cover (1968), this theorem is applied to evaluate the capacity of a network of perceptrons in terms of the total number of adjustable parameters, which is shown to be lower-bounded by $N/(1 + \log_2 N)$ where N is the number of input patterns.

3. Another approach for regularization by incorporating prior information into an input–output mapping is through the use of Bayesian interpolation; for a detailed exposition of this approach see MacKay (1992a,b) and Neal (1995).

4. Regularization theory is usually credited to Tikhonov (1963). A similar approach was described in Phillips (1962). For this reason, the theory is sometimes referred to as *Tikhonov–Phillips regularization*.

A form of regularization in the actuarial literature was considered in Whittaker (1923); the smoothing process considered therein was referred to as *graduation* or *adjustment* of the observations.

For a discussion of regularization theory in book form, see Tikhonov and Arsenin (1977), Mozorov (1993), and Kirch (1996).

5. The concept "function space" evolved in consequence of Hilbert's fundamental investigation of a certain class of integral equations. While Fredholm, the originator of Fredholm integrals, formulated the problem in essentially algebraic language, Hilbert recognized the close relation of the problem with the analytic geometry of second-order surfaces in a Euclidean space of many dimensions (Lanczos, 1964).

6. A *normed space* is a linear vector space in which a real-valued function $\|\mathbf{x}\|$, called the *norm* of $\mathbf{x}$, is defined. The norm $\|\mathbf{x}\|$ has the following properties:

$$\|\mathbf{x}\| > 0 \text{ for } \mathbf{x} \neq \mathbf{0}$$

$$\|\mathbf{0}\| = 0$$

$$\|a\mathbf{x}\| = |a| \cdot \|\mathbf{x}\|, \qquad a = \text{constant}$$

$$\|\mathbf{x} + \mathbf{y}\| \leq \|\mathbf{x}\| + \|\mathbf{y}\|$$

The norm $\|\mathbf{x}\|$ plays the role of "length" of $\mathbf{x}$.

7. Strictly speaking, we require the function $f(\mathbf{x})$ responsible for the data generation to be a member of a *reproducing kernel Hilbert space* (RKHS) with a reproducing kernel in the form of the Dirac delta distribution δ (Tapia and Thompson, 1978). We do this because we require that the Dirac delta distribution δ be in the *decreasing, infinitely continuously differentiable functions*, that is, the classical space $\mathcal{S}$ of test functions for the *Schwarz theory of distributions*, with finite **D**-induced norm, as shown by

$$H_p = \{f \in \mathcal{S} : \|\mathbf{D}f\| < \infty\}$$

Generally speaking, engineers usually think of only the L_2 space whenever Hilbert space is mentioned, perhaps on the grounds that L_2 space is isomorphic to any Hilbert space. But the *norm* is the most important feature of a Hilbert space, and *isometrics* (i.e., norm-preserving isomorphism) are more important than simply additive isomorphism (Kailath, 1974). The theory of RKHS shows that there are many other different and quite useful Hilbert spaces besides the L_2 space. For a tutorial review of RKHS, see Kailath (1971).

8. An *inner product space* is a linear vector space in which the inner product of **u** and **v**, denoted by $(\mathbf{u},\mathbf{v})$, is induced with the following properties:

$$(\mathbf{u}, \mathbf{v}) = (\mathbf{v}, \mathbf{u})$$

$$(a\mathbf{u}, \mathbf{v}) = a(\mathbf{u}, \mathbf{v}), \quad a = \text{constant}$$

$$(\mathbf{u} + \mathbf{v}, \mathbf{w}) = (\mathbf{u}, \mathbf{w}) + (\mathbf{v}, \mathbf{w})$$

$$(\mathbf{u}, \mathbf{u}) > 0 \quad \text{for } \mathbf{u} \neq \mathbf{0}$$

An inner product space $\mathcal{H}$ is said to be *complete*, and referred to as a Hilbert space, if every Cauchy sequence picked from $\mathcal{H}$ converges in norm to a limit in $\mathcal{H}$. A sequence of vectors $\{\mathbf{x}_n\}$ is called a *Cauchy sequence* if for every $\epsilon > 0$ there exists a number M such that (Debnath and Mikusiński 1990)

$$\|\mathbf{x}_m - \mathbf{x}_n\| < \epsilon \quad \text{for all } (m, n) > M$$

9. In Girosi et al. (1995), a different method for deriving Eq. (5.55) is presented by relating the regularizing term $\mathcal{E}_c(F)$ directly to the smoothness of the approximating function $F(\mathbf{x})$.

Smoothness is viewed as a measure of the oscillatory nature of a function. In particular, a function is said to be smoother than another function if it is less oscillatory. In other words, the smoother a function is, the smaller its high-frequency content will be. With this measure of smoothness in mind, let $F(\mathbf{s})$ be the multidimensional Fourier transform of $F(\mathbf{x})$, with $\mathbf{s}$ denoting a multidimensional transform variable. Let $H(\mathbf{s})$ denote a positive function that tends to zero as $\|\mathbf{s}\|$ approaches infinity, that is $1/H(\mathbf{s})$ represents the action of a "high-pass filter." Then, according to Girosi et al. (1995), we may define a smoothness functional representing the regularizing term as:

$$\mathcal{E}_C(F) = \frac{1}{2} \int_{\mathbb{R}^{m_0}} \frac{|F(\mathbf{s})|^2}{H(\mathbf{s})} d\mathbf{s}$$

where m_0 is the dimension of $\mathbf{x}$. By virtue of Parseval's theorem of Fourier theory, this functional is a measure of the power contained in the output of the high-pass filter $1/H(\mathbf{s})$. Thus, by casting the regularization problem in the Fourier domain and using properties of the Fourier transform, the solution of Eq. (5.55) is derived.

10. The most general form of a linear differential operator is

$$\mathbf{D} = p(x_1, x_2, \ldots, x_{m_0}) \frac{\partial^n}{\partial x_1^a \partial x_2^b \cdots \partial x_{m_0}^k}, \quad a + b + \cdots + k = n$$

where $x_1, x_2, \ldots, x_{m_0}$ are the elements of vector $\mathbf{x}$, and $p(x_1, x_2, \ldots, x_{m_0})$ is some function of these elements. The adjoint operator of $\mathbf{D}$ is (Morse and Feshback, 1953)

$$\tilde{\mathbf{D}} = (-1)^n \frac{\partial^n}{\partial x_1^a \partial x_2^b \cdots \partial x_{m_0}^k} [p(x_1, x_2, \ldots, x_{m_0})], \quad a + b + \cdots + k = n$$

11. To obtain generalized cross-validation from ordinary cross-validation, we may consider a *ridge regression problem* described in Wahba (1990):

$$\mathbf{y} = \mathbf{X}\boldsymbol{\alpha} + \boldsymbol{\epsilon} \tag{1}$$

where $\mathbf{X}$ is an N-by-N matrix of inputs, and the noise vector $\boldsymbol{\epsilon}$ has a mean vector of zero and a covariance matrix equal to $\sigma^2\mathbf{I}$. Using the singular value decomposition of $\mathbf{X}$, we may write

$$\mathbf{X} = \mathbf{U}\mathbf{D}\mathbf{V}^T$$

where $\mathbf{U}$ and $\mathbf{V}$ are orthogonal matrices and $\mathbf{D}$ is a diagonal matrix. Let

$$\tilde{\mathbf{y}} = \mathbf{U}^T\mathbf{y}$$

$$\boldsymbol{\beta} = \mathbf{V}^T\boldsymbol{\alpha}$$

and

$$\tilde{\boldsymbol{\epsilon}} = \mathbf{U}^T\boldsymbol{\epsilon}$$

We may then use $\mathbf{U}$ and $\mathbf{V}$ to transform Eq. (1) into

$$\tilde{\mathbf{y}} = \mathbf{D}\boldsymbol{\beta} + \tilde{\boldsymbol{\epsilon}} \tag{2}$$

The diagonal matrix $\mathbf{D}$ (not to be confused with a differential operator) is chosen to have its singular values come in pairs. Then there is an orthogonal matrix $\mathbf{W}$ for which $\mathbf{W}\mathbf{D}\mathbf{W}^T$ is a *circulant matrix*; that is,

$$\mathbf{A} = \mathbf{W}\mathbf{D}\mathbf{W}^T$$

$$= \begin{bmatrix} a_0 & a_1 & \cdots & a_{N-1} \\ a_{N-1} & a_0 & \cdots & a_{N-2} \\ a_{N-2} & a_{N-1} & \cdots & a_{N-3} \\ \vdots & \vdots & & \vdots \\ a_1 & a_2 & \cdots & a_0 \end{bmatrix}$$

which is constant down the diagonal. Let

$$\mathbf{z} = \mathbf{W}\tilde{\mathbf{y}}$$

$$\boldsymbol{\gamma} = \mathbf{W}\boldsymbol{\beta}$$

and

$$\boldsymbol{\xi} = \mathbf{W}\tilde{\boldsymbol{\epsilon}}$$

We may then use $\mathbf{W}$ to transform Eq. (2) into

$$\mathbf{z} = \mathbf{A}\boldsymbol{\gamma} + \boldsymbol{\xi} \tag{3}$$

The diagonal matrix $\mathbf{D}$ has "maximally uncoupled" rows, while the circulant matrix $\mathbf{A}$ has "maximally coupled" rows.

With these transformations at hand, we may now state that generalized cross-validation is equivalent to transforming the ridge regression problem of Eq. (1) into the maximally coupled form Eq. (3), then doing ordinary cross-validation on $\mathbf{z}$, and finally transforming back to the original coordinate system (Wahba, 1990).

12. In an appendix to a chapter contribution in Powell (1992) that is based on a lecture presented in 1990, credit is given to a result due to A.C. Brown. The result, apparently obtained in 1981, states that an RBF network can map an arbitrary function from a closed domain in $\mathbb{R}^{m_0}$ to $\mathbb{R}$.

   Hartman et al. (1990) consider Gaussian functions and approximations on compact subsets of $\mathbb{R}^{m_0}$ that are convex; therein it is shown that RBF networks with a single hidden layer of Gaussian units are universal approximators. However, the most rigorous proof of the universal approximation property of RBF networks is presented in Park and Sandberg (1991); this latter work was completed before the publication of the paper by Hartman et al.

13. Let Ω be a bounded domain in $\mathbb{R}^n$ with boundary Γ. Consider the set $\mathcal{S}$ of real-valued functions that are continuous and have a continuous gradient on $\overline{\Omega} = \Omega + \Gamma$. The bilinear form

$$\int_\Omega (\mathrm{grad}\, u : \mathrm{grad}\, v + uv) dx$$

   is clearly an admissible inner product on $\mathcal{S}$. The completion of $\mathcal{S}$ in the norm generated by this inner product is known as the *Sobolev space* (Debnath and Mikusiński, 1990). Sobolev spaces play an important role in the theory of partial differential equations and are therefore important examples of Hilbert spaces.

14. For a proof of the asymptotically unbiased property of the Parzen–Rosenblatt density estimator, see Parzen (1962) and Cacoullos (1966).

15. The Nadaraya–Watson regression estimator has been the subject of extensive study in statistics literature. In a broader context, nonparametric functional estimation occupies a central place in statistics; see Härdle (1990), and the collection of papers in Roussas (1991).

PROBLEMS

Radial-basis functions

5.1 The *thin-plate-spline* function is described by

$$\varphi(r) = \left(\frac{r}{\sigma}\right)^2 \log\left(\frac{r}{\sigma}\right) \text{ for some } \sigma > 0 \text{ and } r \in \mathbb{R}$$

Justify the use of this function as a translationally and rotationally invariant Green's function.

5.2 The set of values given in Section 5.8 for the weight vector $\mathbf{w}$ of the RBF network of Fig. 5.6 presents one possible solution for the XOR problem. Investigate another set of values for the weight vector $\mathbf{w}$ for solving this problem.

5.3 In Section 5.8 we presented a solution of the XOR problem using an RBF network with two hidden units. In this problem we consider an exact solution of the XOR problem using an RBF network with four hidden units, with each radial-basis function center being determined by each piece of input data. The four possible input patterns are defined by $(0, 0), (0, 1), (1, 1), (1, 0)$, which represent the cyclically ordered corners of a square.

(a) Construct the interpolation matrix $\mathbf{\Phi}$ for the resulting RBF network. Hence, compute the inverse matrix $\mathbf{\Phi}^{-1}$.

(b) Calculate the linear weights of the output layer of the network.

5.4 The Gaussian function is the only radial-basis function that is factorizable.

Using this property of the Gaussian function, show that a Green's function $G(\mathbf{x}, \mathbf{t})$ defined as a multivariate Gaussian distribution may be factorized as follows:

$$G(\mathbf{x}, \mathbf{t}) = \prod_{i=1}^{m} G(x_i, t_i)$$

where x_i and t_i are the ith elements of the m-by-1 vectors $\mathbf{x}$ and $\mathbf{t}$.

Regularized networks

5.5 Consider the cost functional

$$\mathscr{E}(F^*) = \sum_{i=1}^{N} \left[d_i - \sum_{j=1}^{m_1} w_j G(\|\mathbf{x}_j - \mathbf{t}_i\|) \right]^2 + \lambda \|\mathbf{D}F^*\|^2$$

which refers to the approximating function

$$F^*(\mathbf{x}) = \sum_{i=1}^{m_1} w_i G(\|\mathbf{x} - \mathbf{t}_i\|)$$

Using the Frèchet differential, show that the cost functional $\mathscr{E}(F^*)$ is minimized when

$$(\mathbf{G}^T\mathbf{G} + \lambda\mathbf{G}_0)\mathbf{w} = \mathbf{G}^T\mathbf{d}$$

where the N-by-m_1 matrix $\mathbf{G}$, the m_1-by-m_1 matrix $\mathbf{G}_0$, the m_1-by-1 vector $\mathbf{w}$, and the N-by-1 vector $\mathbf{d}$ are defined by Eqs. (5.72), (5.75), (5.73), and (5.46), respectively.

5.6 Suppose that we define

$$(\tilde{\mathbf{D}}\mathbf{D})_{\mathbf{U}} = \sum_{k=0}^{\infty} (-1)^k \frac{\nabla_{\mathbf{U}}^{2k}}{k!2^k}$$

where

$$\nabla_{\mathbf{U}}^2 = \sum_{j=1}^{m_0} \sum_{i=1}^{m_0} u_{ji} \frac{\partial^2}{\partial x_j \partial x_i}$$

The m_0-by-m_0 matrix $\mathbf{U}$, with its ji-th element denoted by u_{ji}, is symmetric and positive definite. Hence the inverse matrix $\mathbf{U}^{-1}$ exists, and so it permits the following decomposition via the similarity transformation:

$$\mathbf{U}^{-1} = \mathbf{V}^T\mathbf{\Sigma}\mathbf{V}$$

$$= \mathbf{V}^T\mathbf{\Sigma}^{1/2}\mathbf{\Sigma}_1^{1/2}\mathbf{V}$$

$$= \mathbf{C}^T\mathbf{C}$$

where $\mathbf{V}$ is an orthogonal matrix, $\mathbf{\Sigma}$ is a diagonal matix, $\mathbf{\Sigma}^{1/2}$ is the square root of $\mathbf{\Sigma}$, and the matrix $\mathbf{C}$ is defined by

$$\mathbf{C} = \mathbf{\Sigma}^{1/2}\mathbf{V}$$

The problem is to solve for the Green's function $G(\mathbf{x}, \mathbf{t})$ that satisfies the following condition (in the distributional sense):

$$(\tilde{\mathbf{D}}\mathbf{D})_U G(\mathbf{x}, \mathbf{t}) = \delta(\mathbf{x} - \mathbf{t})$$

Using the mutidimensional Fourier transform to solve this equation for $G(\mathbf{x}, \mathbf{t})$, show that

$$G(\mathbf{x}, \mathbf{t}) = \exp\left(-\frac{1}{2}\|\mathbf{x} - \mathbf{t}\|_C^2\right)$$

where

$$\|\mathbf{x}\|_C^2 = \mathbf{x}^T\mathbf{C}^T\mathbf{C}\mathbf{x}$$

5.7 Consider a regularizing term defined by

$$\int_{\mathbb{R}^{m_0}} \|\mathbf{D}F(\mathbf{x})\|^2 d\mathbf{x} = \sum_{k=0}^{\infty} a_k \int_{\mathbb{R}^{m_0}} \|D^k F(\mathbf{x})\|^2 d\mathbf{x}$$

where

$$a_k = \frac{\sigma^{2k}}{k!2^k}$$

and the linear differential operator D is defined in terms of the gradient operator ∇ and the Laplacian operator ∇^2 as follows:

$$D^{2k} = (\nabla^2)^k$$

and

$$D^{2k+1} = \nabla(\nabla^2)^k$$

Show that

$$\mathbf{D}F(\mathbf{x}) = \sum_{k=0}^{\infty} \frac{\sigma^{2k}}{k!2^k} \nabla^{2k} F(\mathbf{x})$$

5.8 In Section 5.5 we derived the approximating function $F_\lambda(\mathbf{x})$ of Eq. (5.66) by using the relationship of Eq. (5.65). In this problem we wish to start with the relationship of Eq. (5.65) and use the multidimensional Fourier transformation to derive Eq. (5.66). Perform this derivation by using the following definition of the multidimensional Fourier transform of the Green's function $G(\mathbf{x})$:

$$G(\mathbf{s}) = \int_{\mathbb{R}^{m_0}} G(\mathbf{x}) \exp(-i\,\mathbf{s}^T\mathbf{x}) d\mathbf{x}$$

where $i = \sqrt{-1}$ and $\mathbf{s}$ is the m_0-dimensional transform variable.

5.9 Consider the nonlinear regression problem described in Eq. (5.95). Let a_{ik} denote the ik-th element of the inverse matrix $(\mathbf{G} + \lambda\mathbf{I})^{-1}$. Hence, starting with Eq. (5.58), show that the estimate of the regression function $f(\mathbf{x})$ may be expressed as

$$\hat{f}(\mathbf{x}) = \sum_{k=1}^{N} \psi(\mathbf{x}, \mathbf{x}_k) y_k$$

where y_k is the model output for the input $\mathbf{x}_k$, and

$$\psi(\mathbf{x}, \mathbf{x}_k) = \sum_{i=1}^{N} G(\|\mathbf{x} - \mathbf{x}_i\|)a_{ik}, \qquad k = 1, 2, \ldots, N$$

where $G(\|\cdot\|)$ is the Green's function.

5.10 *Spline functions* are examples of piecewise polynomial approximators (Schumaker, 1981). The basic idea behind the method of splines is as follows. An approximation region of interest is broken up into a finite number of subregions via the use of *knots*; the knots can be fixed, in which case the approximators are *linearly* parameterized, or they can be variable, in which case the approximators are *nonlinearly* parameterized. In both cases, in each region of the approximation a polynomial of degree at most n is used, with the additional requirement that the overall function be $n - 1$ times differentiable. Polynomial splines are relatively smooth functions that are easy to store, manipulate, and evaluate on a computer.

Among spline functions used in practice, *cubic splines* are perhaps the most popular. The cost functional for a cubic spline, pertaining to a one-dimensional input, is defined by

$$\mathcal{E}(F) = \frac{1}{2}\sum_{i=1}^{N}[y_i - f(x_i)]^2 + \frac{\lambda}{2}\int_{x_1}^{x_N}\left[\frac{d^2f(x)}{dx^2}\right]^2 dx$$

where, in the language of splines, λ denotes a *smoothing* parameter.

(a) Justify the following properties of the solution $f_\lambda(x)$ to this problem:

    (1) $f_\lambda(x)$ is a cubic polynomial between two successive values of x.

    (2) $f_\lambda(x)$ and its first two derivatives are all continuous, except at the boundary points where the second derivative of $f_\lambda(x)$ is zero.

(b) Since $\mathcal{E}(f)$ has a unique minimum, we must have

$$\mathcal{E}(f_\lambda + \alpha g) \geq \mathcal{E}(f_\lambda)$$

for any g drawn from the same class of twice-differentiable functions as f_λ and for any real-valued constant α. This means that $\mathcal{E}(f_\lambda + \alpha g)$, interpreted as a function of α, must have a local minimum at $\alpha = 0$. Hence, show that

$$\int_{x_1}^{x_N}\left(\frac{d^2f_\lambda(x)}{dx^2}\right)\left(\frac{d^2g(x)}{dx^2}\right)dx = \frac{1}{2}\sum_{i=1}^{N}[y - f_\lambda(x_i)]g(x_i)$$

which is the Euler–Lagrange equation for the cubic spline problem.

Rate of approximation

5.11 Equation (5.124) defines the upper bound on the generalization error of a Gaussian RBF network designed to learn a regression function that belongs to a certain Sobolev space. Using this bound, derive the formula of Eq. (5.125) for the optimum size of this network for a specified size of training sample.

Kernel estimation

5.12 Suppose that you are given a "noiseless" training sample $\{f(\mathbf{x}_i)\}_{i=1}^{N}$, and that the requirement is to design a network that generalizes to data samples that are corrupted by additive noise and therefore not included in the training set. Let $F(\mathbf{x})$ denote the approximating function realized by such a network, which is chosen so that the expected squared error

$$J(F) = \frac{1}{2}\sum_{i=1}^{N}\int_{\mathbb{R}^{m_0}}[f(\mathbf{x}_i) - F(\mathbf{x}_i, \boldsymbol{\xi})]^2 f_{\boldsymbol{\xi}}(\boldsymbol{\xi})d\boldsymbol{\xi}$$

is minimum, where $f_\xi(\xi)$ is the probability density function of a noise distribution in the input space $\mathbb{R}^{m_0}$. Show that the solution of this least-squares problem is given by (Webb, 1994)

$$F(\mathbf{x}) = \frac{\sum\limits_{i=1}^{N} f(\mathbf{x}_i) f_\xi(\mathbf{x} - \mathbf{x}_i)}{\sum\limits_{i=1}^{N} f_\xi(\mathbf{x} - \mathbf{x}_i)}$$

Compare this estimator to the Nadaraya–Watson regression estimator.

Supervised selection of centers

5.13 Consider the cost functional

$$\mathcal{E} = \frac{1}{2} \sum_{j=1}^{N} e_j^2$$

where

$$e_j = d_j - F^*(x_j)$$

$$= d_j - \sum_{i=1}^{m_1} w_i G(\|\mathbf{x}_j - \mathbf{t}_i\|_{C_i})$$

The free parameters are the linear weights w_i, the centers $\mathbf{t}_i$ of the Green's functions, and the inverse covariance matrix $\mathbf{\Sigma}_i^{-1} = \mathbf{C}_i^T \mathbf{C}_i$, where $\mathbf{C}_i$ is the norm weighting matrix. The problem is to find the values of these free parmeters that minimize the cost functional $\mathcal{E}$. Derive the following partial derivatives:

(a) $\dfrac{\partial \mathcal{E}}{\partial w_i} = \sum\limits_{j=1}^{N} e_j G(\|\mathbf{x}_j - \mathbf{t}_i\|_{C_i})$

(b) $\dfrac{\partial \mathcal{E}}{\partial \mathbf{t}_i} = 2w_i \sum\limits_{j=1}^{N} e_j G'(\|\mathbf{x}_j - \mathbf{t}_i\|_{C_i}) \mathbf{\Sigma}_i^{-1} (\mathbf{x}_j - \mathbf{t}_i)$

(c) $\dfrac{\partial \mathcal{E}}{\partial \mathbf{\Sigma}_i^{-1}} = -w_i \sum\limits_{j=1}^{N} e_j G'(\|\mathbf{x}_j - \mathbf{t}_i\|_{C_i}) \mathbf{Q}_{ji}$

where $G'(\cdot)$ is the derivative of $G(\cdot)$ with respect to its argument, and

$$\mathbf{Q}_{ji} = (\mathbf{x}_j - \mathbf{t}_i)(\mathbf{x}_j - \mathbf{t}_i)^T$$

For the rules for differentiating a scalar with respect to a vector, see note 2 of Chapter 3.

Computer Experiments

5.14 In this problem we continue with the computer experiment in Section 5.13 to study the random selection of centers for the design of an RBF network used as a binary pattern classifier. The purpose of the experiment is to demonstrate that the generalization performance of the network so trained is relatively good.

The network is intended to solve the binary pattern-classification problem described in Section 5.13, where the requirement is to classify data drawn from a mixture model consisting of two equiprobable overlapping two-dimensional Gaussian distributions. One distribution has a mean vector $[0,0]^T$ and common variance 1, whereas the other distribution has a mean vector $[0,2]^T$ and common variance 4. The "select class with maximum function output" decision rule is used for the classification.

(a) Consider a random selection of centers using $m_1 = 20$ centers. Compute the mean, standard deviation, and minimum and maximum values of the probability of correct

classification P_c for different values of regularization parameter $\lambda = 0, 0.1, 1, 10, 100,$
1000. For the computation of ensemble statistics, use 50 independent network trials
per ensemble, with each one tested against a fixed reference set of 1000 patterns.

(b) Construct the decision boundary computed for the configuration described in part (a)
for regularization parameter $\lambda = 1$.

(c) Repeat the computations described in part (a) for $m_1 = 10$ centers (selected at
random).

(d) In light of your results, discuss the merit of random selection of centers as a method
for the design of RBF networks, and the role of regularization in the performance of
the network as a pattern classifier.

(e) Compare your results with those presented in Section 5.13 that were computed using
the method of strict interpolation. In particular, confirm that the random selection of
centers is relatively insensitive to the regularization parameter.

5.15 It may be argued that in the case of the experiment described in Section 5.13 involving
the classification of a pair of Gaussian-distributed classes, the RBF network considered
there performed well since it uses Gaussian radial-basis functions to approximate the
underlying Gaussian class conditional distributions. In this problem we use a computer
experiment to explore the design of a strict-interpolation Gaussian RBF network for dis-
tinctly discontinuous class conditional distributions. Specifically, consider two equiprobable classes $\mathscr{C}_1$ and $\mathscr{C}_2$ whose distributions

- $U(\mathscr{C}_1)$, where $\mathscr{C}_1 \triangleq \Omega_1$ is a circle of radius $r = 2.34$ centered at $\mathbf{x}_c = [-2, 30]^T$
- $U(\mathscr{C}_2)$, where $\mathscr{C}_2 \subset \mathbb{R}^2$ is a square region centered at $\mathbf{x}_c$ with side length $r = \sqrt{2\pi}$

Here $U(\Omega)$ denotes a uniform distribution over $\Omega \subset \mathbb{R}^2$. These parameters are chosen
so that the decision region for class $\mathscr{C}_1$ is the same as in the Gaussian-distributed case
considered in Section 5.13. Investigate the use of regularization as a means of improving
the classification performance of a Gaussian RBF network using strict interpolation.

CHAPTER 6

Support Vector Machines

6.1 INTRODUCTION

In Chapter 4 we studied multilayer perceptrons trained with the back-propagation algorithm. In Chapter 5 we studied another class of layered feedforward networks, radial-basis function networks. Both of these neural networks are universal approximators in their own ways. In this chapter we discuss another category of universal feedforward networks, known as *support vector machines (SVM),* pioneered by Vapnik (Boser, Guyon, and Vapnik, 1992; Cortes and Vapnik, 1995; Vapnik, 1995, 1998). Like multilayer perceptrons and radial-basis function networks, support vector machines can be used for pattern classification and nonlinear regression.

Basically, the support vector machine is a *linear machine* with some very nice properties. To explain how it works, it is perhaps easiest to start with the case of separable patterns that could arise in the context of pattern classification. In this context, the main idea of a support vector machine is to construct a hyperplane as the decision surface in such a way that the margin of separation between positive and negative examples is maximized. The machine achieves this desirable property by following a principled approach rooted in the statistical learning theory that is discussed in Chapter 2. More precisely, the support vector machine is an approximate implementation of the *method of structural risk minimization.* This induction principle is based on the fact that the error rate of a learning machine on test data (i.e., the generalization error rate) is bounded by the sum of the training-error rate and a term that depends on the *Vapnik–Chervonenkis (VC) dimension;* in the case of separable patterns, a support vector machine produces a value of zero for the first term and minimizes the second term. Accordingly, the support vector machine can provide a good generalization performance on pattern classification problems despite the fact that it *does not* incorporate problem-domain knowledge. This attribute is unique to support vector machines.

A notion that is central to the construction of the support vector learning algorithm is the inner-product kernel between a "support vector" x_i and the vector x drawn from the input space. The support vectors consist of a small subset of the training data extracted by the algorithm. Depending on how this inner-product kernel is generated, we may construct different learning machines characterized by nonlinear decision surfaces of their own. In particular, we may use the support vector learning algorithm to construct the following three types of learning machines (among others):

- Polynomial learning machines
- Radial-basis function networks
- Two-layer perceptrons (i.e., with a single hidden layer)

That is, for each of these feedforward networks we may use the support vector learning algorithm to implement the learning process using a given set of training data, automatically determining the required number of hidden units. Stated in another way: Whereas the back-propagation algorithm is devised specifically to train a multilayer perceptron, the support vector learning algorithm is of a more generic nature because it has wider applicability.

Organization of the Chapter

The main body of the chapter is organized in three parts. In the first part we describe the basic ideas behind a support vector machine. Specifically, in Section 6.2 we discuss the construction of optimal hyperplanes for the simple case of linearly separable patterns. This is followed by considering the more difficult case of nonseparable patterns in Section 6.3.

In so doing, we pave the way for the second part of the chapter, which presents a detailed discussion of the support vector machine for solving pattern-recognition tasks. This is done in Section 6.4. In Section 6.5 we revisit the XOR problem to illustrate the construction of a support vector machine. In Section 6.6 we revisit the computer experiment on pattern classification that was studied in Chapters 4 and 5, thereby providing a comparative evaluation of support vector machines with multilayer perceptrons trained on the back-propagation algorithm and standard radial-basis function networks.

The last part of the chapter deals with the nonlinear regression problem. In Section 6.7 we describe a loss function that is well suited for such a problem. Then in Section 6.8 we discuss the construction of a support vector machine for nonlinear regression.

The chapter concludes with some final remarks in Section 6.9.

6.2 OPTIMAL HYPERPLANE FOR LINEARLY SEPARABLE PATTERNS

Consider the training sample $\{(\mathbf{x}_i, d_i)\}_{i=1}^{N}$, where $\mathbf{x}_i$ is the input pattern for the ith example and d_i is the corresponding desired response (target output). To begin with, we assume that the pattern (class) represented by the subset $d_i = +1$ and the pattern represented by the subset $d_i = -1$ are "linearly separable." The equation of a decision surface in the form of a hyperplane that does the separation is

$$\mathbf{w}^T\mathbf{x} + b = 0 \tag{6.1}$$

where $\mathbf{x}$ is an input vector, $\mathbf{w}$ is an adjustable weight vector, and b is a bias. We may thus write

$$
\begin{align}
\mathbf{w}^T\mathbf{x}_i + b \geq 0 & \qquad \text{for } d_i = +1 \\
\mathbf{w}^T\mathbf{x}_i + b < 0 & \qquad \text{for } d_i = -1
\end{align}
\tag{6.2}
$$

The assumption of linearly separable patterns is made here to explain the basic idea behind a support vector machine in a rather simple setting; this assumption will be relaxed in Section 6.3.

For a given weight vector $\mathbf{w}$ and bias b, the separation between the hyperplane defined in Eq. (6.1) and the closest data point is called the *margin of separation*, denoted by ρ. The goal of a support vector machine is to find the particular hyperplane for which the margin of separation ρ is maximized. Under this condition, the decision surface is referred to as the *optimal hyperplane*. Figure 6.1 illustrates the geometric construction of an optimal hyperplane for a two-dimensional input space.

Let $\mathbf{w}_o$ and b_o denote the optimum values of the weight vector and bias, respectively. Correspondingly, the *optimal hyperplane*, representing a multidimensional linear decision surface in the input space, is defined by

$$
\mathbf{w}_o^T\mathbf{x} + b_o = 0
\tag{6.3}
$$

which is a rewrite of Eq. (6.1). The discriminant function

$$
g(\mathbf{x}) = \mathbf{w}_o^T\mathbf{x} + b_o
\tag{6.4}
$$

gives an algebraic measure of the *distance* from $\mathbf{x}$ to the optimal hyperplane (Duda and Hart, 1973). Perhaps the easiest way to see this is to express $\mathbf{x}$ as

$$
\mathbf{x} = \mathbf{x}_p + r\frac{\mathbf{w}_o}{\|\mathbf{w}_o\|}
$$

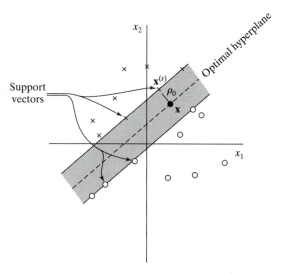

FIGURE 6.1 Illustration of the idea of an optimal hyperplane for linearly separable patterns.

where $\mathbf{x}_p$ is the normal projection of $\mathbf{x}$ onto the optimal hyperplane, and r is the desired algebraic distance; r is positive if $\mathbf{x}$ is on the positive side of the optimal hyperplane and negative if $\mathbf{x}$ is on the negative side. Since, by definition, $g(\mathbf{x}_p) = 0$, it follows that

$$g(\mathbf{x}) = \mathbf{w}_o^T\mathbf{x} + b_o = r\|\mathbf{w}_o\|$$

or

$$r = \frac{g(\mathbf{x})}{\|\mathbf{w}_o\|} \tag{6.5}$$

In particular, the distance from the origin (i.e., $\mathbf{x} = \mathbf{0}$) to the optimal hyperplane is given by $b_o/\|\mathbf{w}_o\|$. If $b_o > 0$, the origin is on the positive side of the optimal hyperplane; if $b_o < 0$, it is on the negative side. If $b_o = 0$, the optimal hyperplane passes through the origin. A geometric interpretation of these algebraic results is given in Fig. 6.2.

The issue at hand is to find the parameters $\mathbf{w}_o$ and b_o for the optimal hyperplane, given the training set $\mathcal{T} = \{(\mathbf{x}_i, d_i)\}$. In light of the results portrayed in Fig. 6.2, we see that the pair $(\mathbf{w}_o, b_o)$ must satisfy the constraint:

$$\begin{aligned} \mathbf{w}_o^T\mathbf{x}_i + b_o &\geq 1 \qquad \text{for } d_i = +1 \\ \mathbf{w}_o^T\mathbf{x}_i + b_o &\leq -1 \qquad \text{for } d_i = -1 \end{aligned} \tag{6.6}$$

Note that if Eq. (6.2) holds, that is, the patterns are linearly separable, we can always rescale $\mathbf{w}_o$ and b_o such that Eq. (6.6) holds; this scaling operation leaves Eq. (6.3) unaffected.

The particular data points $(\mathbf{x}_i, d_i)$ for which the first or second line of Eq. (6.6) is satisfied with the equality sign are called *support vectors*, hence the name "support vector machine." These vectors play a prominent role in the operation of this class of learning machines. In conceptual terms, the support vectors are those data points that lie closest to the decision surface and are therefore the most difficult to classify. As such, they have a direct bearing on the optimum location of the decision surface.

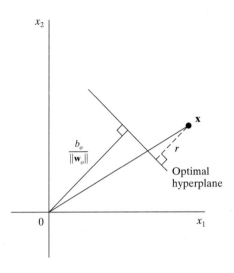

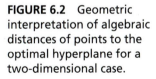

FIGURE 6.2 Geometric interpretation of algebraic distances of points to the optimal hyperplane for a two-dimensional case.

Consider a support vector $\mathbf{x}^{(s)}$ for which $d^{(s)} = +1$. Then by definition, we have

$$g(\mathbf{x}^{(s)}) = \mathbf{w}_o^T \mathbf{x}^{(s)} \mp b_o = \mp 1 \qquad \text{for } d^{(s)} = \mp 1 \tag{6.7}$$

From Eq. (6.5) the *algebraic distance* from the support vector $\mathbf{x}^{(s)}$ to the optimal hyperplane is

$$r = \frac{g(\mathbf{x}^{(s)})}{\|\mathbf{w}_o\|}$$

$$= \begin{cases} \dfrac{1}{\|\mathbf{w}_o\|} & \text{if } d^{(s)} = +1 \\[2ex] -\dfrac{1}{\|\mathbf{w}_o\|} & \text{if } d^{(s)} = -1 \end{cases} \tag{6.8}$$

where the plus sign indicates that $\mathbf{x}^{(s)}$ lies on the positive side of the optimal hyperplane and the minus sign indicates that $\mathbf{x}^{(s)}$ lies on the negative side of the optimal hyperplane. Let ρ denote the optimum value of the *margin of separation* between the two classes that constitute the training set $\mathcal{T}$. Then, from Eq. (6.8) it follows that

$$\rho = 2r$$

$$= \frac{2}{\|\mathbf{w}_o\|} \tag{6.9}$$

Equation (6.9) states that maximizing the margin of separation between classes is equivalent to minimizing the Euclidean norm of the weight vector $\mathbf{w}$.

In summary, the optimal hyperplane defined by Eq. (6.3) is *unique* in the sense that the optimum weight vector $\mathbf{w}_o$ provides the maximum possible separation between positive and negative examples. This optimum condition is attained by minimizing the Euclidean norm of the weight vector $\mathbf{w}$.

Quadratic Optimization for Finding the Optimal Hyperplane

Our goal is to develop a computationally efficient procedure for using the training sample $\mathcal{T} = \{(\mathbf{x}_i, d_i)\}_{i=1}^N$ to find the optimal hyperplane, subject to the constraint

$$d_i(\mathbf{w}^T \mathbf{x}_i + b) \geq 1 \qquad \text{for } i = 1, 2, ..., N \tag{6.10}$$

This constraint combines the two lines of Eq. (6.6) with $\mathbf{w}$ used in place of $\mathbf{w}_o$.

The constrained optimization problem that we have to solve may now be stated as:

Given the training sample $\{(\mathbf{x}_i, d_i)\}_{i=1}^N$, find the optimum values of the weight vector $\mathbf{w}$ and bias b such that they satisfy the constraints

$$d_i(\mathbf{w}^T \mathbf{x}_i + b) \geq 1 \qquad \text{for } i = 1, 2, ..., N$$

and the weight vector $\mathbf{w}$ minimizes the cost function:

$$\Phi(\mathbf{w}) = \frac{1}{2} \mathbf{w}^T \mathbf{w}$$

The scaling factor $1/2$ is included here for convenience of presentation. This constrained optimization problem is called the *primal problem*. It is characterized as follows:

- The cost function $\Phi(\mathbf{w})$ is a *convex* function[1] of $\mathbf{w}$.
- The constraints are *linear* in $\mathbf{w}$.

Accordingly, we may solve the constrained optimization problem using the *method of Lagrange multipliers* (Bertsekas, 1995).

First, we construct the *Lagrangian function*:

$$J(\mathbf{w}, b, \alpha) = \frac{1}{2} \mathbf{w}^T \mathbf{w} - \sum_{i=1}^{N} \alpha_i \big[d_i(\mathbf{w}^T \mathbf{x}_i + b) - 1 \big] \tag{6.11}$$

where the auxiliary nonnegative variables α_i are called *Lagrange multipliers*. The solution to the constrained optimization problem is determined by the *saddle point* of the Lagrangian function $J(\mathbf{w}, b, \alpha)$, which has to be *minimized* with respect to $\mathbf{w}$ and b; it also has to be *maximized* with respect to α. Thus, differentiating $J(\mathbf{w}, b, \alpha)$ with respect to $\mathbf{w}$ and b and setting the results equal to zero, we get the following two *conditions of optimality*:

Condition 1: $\dfrac{\partial J(\mathbf{w}, b, \alpha)}{\partial \mathbf{w}} = \mathbf{0}$

Condition 2: $\dfrac{\partial J(\mathbf{w}, b, \alpha)}{\partial b} = 0$

Application of optimality condition 1 to the Lagrangian function of Eq. (6.11) yields (after rearrangement of terms)

$$\mathbf{w} = \sum_{i=1}^{N} \alpha_i d_i \mathbf{x}_i \tag{6.12}$$

Application of optimality condition 2 to the Lagrangian function of Eq. (6.11) yields

$$\sum_{i=1}^{N} \alpha_i d_i = 0 \tag{6.13}$$

The solution vector $\mathbf{w}$ is defined in terms of an expansion that involves the N training examples. Note, however, that although this solution is unique by virtue of the convexity of the Lagrangian, the same cannot be said about the Lagrange coefficients, α_i.

It is also important to note that at the saddle point, for each Lagrange multiplier α_i, the product of that multiplier with its corresponding constraint vanishes, as shown by

$$\alpha_i \big[d_i(\mathbf{w}^T \mathbf{x}_i + b) - 1 \big] = 0 \qquad \text{for } i = 1, 2, ..., N \tag{6.14}$$

Therefore, only those multipliers exactly meeting Eq. (6.14) can assume *nonzero* values. This property follows from the *Kuhn–Tucker conditions* of optimization theory (Fletcher, 1987; Bertsekas, 1995).

As noted earlier, the primal problem deals with a convex cost function and linear constraints. Given such a constrained optimization problem, it is possible to construct another problem called the *dual problem*. This second problem has the same optimal value as the primal problem, but with the Lagrange multipliers providing the optimal solution. In particular, we may state the following *duality theorem* (Bertsekas, 1995):

(a) If the primal problem has an optimal solution, the dual problem also has an optimal solution, and the corresponding optimal values are equal.
(b) In order for $\mathbf{w}_o$ to be an optimal primal solution and α_o to be an optimal dual solution, it is necessary and sufficient that $\mathbf{w}_o$ is feasible for the primal problem, and

$$\Phi(\mathbf{w}_o) = J(\mathbf{w}_o, b_o, \alpha_o) = \min_{\mathbf{w}} J(\mathbf{w}, b_o, \alpha_o)$$

To postulate the dual problem for our primal problem, we first expand Eq. (6.11), term by term, as follows:

$$J(\mathbf{w}, b, \alpha) = \frac{1}{2}\mathbf{w}^T\mathbf{w} - \sum_{i=1}^{N} \alpha_i d_i \mathbf{w}^T\mathbf{x}_i - b\sum_{i=1}^{N} \alpha_i d_i + \sum_{i=1}^{N} \alpha_i \tag{6.15}$$

The third term on the right-hand side of Eq. (6.15) is zero by virtue of the optimality condition of Eq. (6.13). Furthermore, from Eq. (6.12) we have

$$\mathbf{w}^T\mathbf{w} = \sum_{i=1}^{N} \alpha_i d_i \mathbf{w}^T\mathbf{x}_i = \sum_{i=1}^{N} \sum_{j=1}^{N} \alpha_i \alpha_j d_i d_j \mathbf{x}_i^T\mathbf{x}_j$$

Accordingly, setting the objective function $J(\mathbf{w}, b, \alpha) = Q(\alpha)$, we may reformulate Eq. (6.15) as

$$Q(\alpha) = \sum_{i=1}^{N} \alpha_i - \frac{1}{2}\sum_{i=1}^{N}\sum_{j=1}^{N} \alpha_i \alpha_j d_i d_j \mathbf{x}_i^T\mathbf{x}_j \tag{6.16}$$

where the α_i are nonnegative.

We may now state the dual problem:

Given the training sample $\{(\mathbf{x}_i, d_i)\}_{i=1}^{N}$, *find the Lagrange multipliers* $\{\alpha_i\}_{i=1}^{N}$ *that maximize the objective function*

$$Q(\alpha) = \sum_{i=1}^{N} \alpha_i - \frac{1}{2}\sum_{i=1}^{N}\sum_{j=1}^{N} \alpha_i \alpha_j d_i d_j \mathbf{x}_i^T\mathbf{x}_j$$

subject to the constraints

(1) $\sum_{i=1}^{N} \alpha_i d_i = 0$

(2) $\alpha_i \geq 0$ for $i = 1, 2, ..., N$

Note that the dual problem is cast entirely in terms of the training data. Moreover, the function $Q(\alpha)$ to be maximized depends *only* on the input patterns in the form of a set of dot products, $\{\mathbf{x}_i^T\mathbf{x}_j\}_{(i,j)=1}^{N}$.

Having determined the optimum Lagrange multipliers, denoted by $\alpha_{o,i}$, we may compute the optimum weight vector $\mathbf{w}_o$ using Eq. (6.12) and so write

$$\mathbf{w}_o = \sum_{i=1}^{N} \alpha_{o,i} d_i \mathbf{x}_i \tag{6.17}$$

To compute the optimum bias b_o, we may use the $\mathbf{w}_o$ thus obtained and take advantage of Eq. (6.7) pertaining to a positive support vector, and thus write

$$b_o = 1 - \mathbf{w}_o^T\mathbf{x}^{(s)} \qquad \text{for } d^{(s)} = 1 \tag{6.18}$$

Statistical Properties of the Optimal Hyperplane

From the statistical learning theory presented in Chapter 2, we recall that the VC dimension of a learning machine determines the way in which a nested structure of approximating functions should be used. We also recall that the VC dimension of a set of separating hyperplanes in a space of dimensionality m is equal to $m + 1$. However, in order to apply the method of structural risk minimization described in Chapter 2 we need to construct a set of separating hyperplanes of varying VC dimension such that the empirical risk (i.e., the training classification error) and the VC dimension are both minimized at the same time. In a support vector machine a structure is imposed on the set of separating hyperplanes by constraining the Euclidean norm of the weight vector $\mathbf{w}$. Specifically, we may state the following theorem (Vapnik, 1995, 1998):

> Let D denote the diameter of the smallest ball containing all the input vectors $\mathbf{x}_1, \mathbf{x}_2, ..., \mathbf{x}_N$. The set of optimal hyperplanes described by the equation
>
> $$\mathbf{w}_o^T \mathbf{x} + b_o = 0$$
>
> has a VC dimension h bounded from above as
>
> $$h \leq \min \left\{ \left\lceil \frac{D^2}{\rho^2} \right\rceil, m_0 \right\} + 1 \tag{6.19}$$
>
> where the ceiling sign $\lceil \cdot \rceil$ means the smallest integer greater than or equal to the number enclosed within, ρ is the margin of separation equal to $2/\|\mathbf{w}_o\|$, and m_0 is the dimensionality of the input space.

This theorem tells us that we may exercise control over the VC dimension (i.e., complexity) of the optimal hyperplane, independently of the dimensionality m_0 of the input space, by properly choosing the margin of separation ρ.

Suppose then we have a nested structure described in terms of the separating hyperplanes as follows:

$$S_k = \{ \mathbf{w}^T \mathbf{x} + b : \|\mathbf{w}\|^2 \leq c_k \}, \qquad k = 1, 2, ... \tag{6.20}$$

By virtue of the upper bound on the VC dimension h defined in Eq. (6.19), the nested structure described in Eq. (6.20) may be reformulated in terms of the margin of separation in the equivalent form

$$S_k = \left\{ \left\lceil \frac{r^2}{\rho^2} \right\rceil + 1 : \rho^2 \geq a_k \right\}, \qquad k = 1, 2, ... \tag{6.21}$$

The a_k and c_k are constants.

From Chapter 2 we also recall that in order to achieve a good generalization capability, we should select the particular structure with the smallest VC dimension and training error, in accordance with the principle of structural risk minimization. From Eqs. (6.19) and (6.21) we see that this requirement can be satisfied by using the optimal hyperplane (i.e., the separating hyperplane with the largest margin of separation ρ). Equivalently, in light of Eq. (6.9), we should use the optimum weight vector $\mathbf{w}_o$ having the minimum Euclidean norm. Thus, the choice of the optimal hyperplane as the decision surface for a set of linearly separable patterns is not only intuitively satisfying but also in complete fulfillment of the principle of structural risk minimization of a support vector machine.

6.3 OPTIMAL HYPERPLANE FOR NONSEPARABLE PATTERNS

The discussion thus far has focused on linearly separable patterns. In this section we consider the more difficult case of nonseparable patterns. Given such a set of training data, it is not possible to construct a separating hyperplane without encountering classification errors. Nevertheless, we would like to find an optimal hyperplane that minimizes the probability of classification error, averaged over the training set.

The margin of separation between classes is said to be *soft* if a data point $(\mathbf{x}_i, d_i)$ violates the following condition (see Eq. (6.10)):

$$d_i(\mathbf{w}^T\mathbf{x}_i + b) \geq +1, \qquad i = 1, 2, \ldots, N$$

This violation can arise in one of two ways:

- The data point $(\mathbf{x}_i, d_i)$ falls inside the region of separation but on the right side of the decision surface, as illustrated in Fig. 6.3a.
- The data point $(\mathbf{x}_i, d_i)$ falls on the wrong side of the decision surface, as illustrated in Fig. 6.3b.

Note that we have correct classification in case 1, but misclassification in case 2.

To set the stage for a formal treatment of nonseparable data points, we introduce a new set of nonnegative scalar variables, $\{\xi_i\}_{i=1}^N$, into the definition of the separating hyperplane (i.e., decision surface) as shown here:

$$d_i(\mathbf{w}^T\mathbf{x}_i + b) \geq 1 - \xi_i, \qquad i = 1, 2, \ldots, N \tag{6.22}$$

The ξ_i are called *slack variables*; they measure the deviation of a data point from the ideal condition of pattern separability. For $0 \leq \xi_i \leq 1$, the data point falls inside the region of separation but on the right side of the decision surface, as illustrated in Fig. 6.3a. For $\xi_i > 1$, it falls on the wrong side of the separating hyperplane, as illustrated in Fig. 6.3b. The support vectors are those particular data points that satisfy Eq. (6.22)

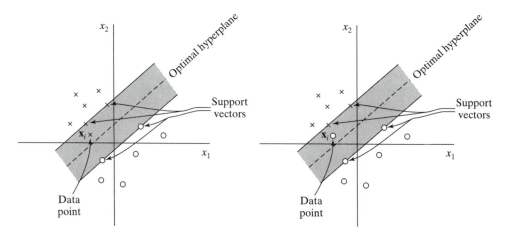

FIGURE 6.3 (a) Data point $\mathbf{x}_i$ (belonging to class $\mathscr{C}_1$) falls inside the region of separation, but on the right side of the decision surface. (b) Data point $\mathbf{x}_i$ (belonging to class $\mathscr{C}_2$) falls on the wrong side of the decision surface.

precisely even if $\xi_i > 0$. Note that if an example with $\xi_i > 0$ is left out of the training set, the decision surface would change. The support vectors are thus defined in exactly the same way for both linearly separable and nonseparable cases.

Our goal is to find a separating hyperplane for which the misclassification error, averaged on the training set, is minimized. We may do this by minimizing the functional

$$\Phi(\xi) = \sum_{i=1}^{N} I(\xi_i - 1)$$

with respect to the weight vector $\mathbf{w}$, subject to the constraint described in Eq. (6.22) and the constraint on $\|\mathbf{w}\|^2$. The function $I(\xi)$ is an *indicator function*, defined by

$$I(\xi) = \begin{cases} 0 & \text{if } \xi \le 0 \\ 1 & \text{if } \xi > 0 \end{cases}$$

Unfortunately, minimization of $\Phi(\xi)$ with respect to $\mathbf{w}$ is a nonconvex optimization problem that is NP-complete.[2]

To make the optimization problem mathematically tractable, we approximate the functional $\Phi(\xi)$ by writing

$$\Phi(\xi) = \sum_{i=1}^{N} \xi_i$$

Moreover, we simplify the computation by formulating the functional to be minimized with respect to the weight vector $\mathbf{w}$ as follows:

$$\Phi(\mathbf{w}, \xi) = \frac{1}{2} \mathbf{w}^T \mathbf{w} + C \sum_{i=1}^{N} \xi_i \qquad (6.23)$$

As before, minimizing the first term in Eq. (6.23) is related to minimizing the VC dimension of the support vector machine. As for the second term $\sum_i \xi_i$, it is an upper bound on the number of test errors. Formulation of the cost function $\Phi(\mathbf{w}, \xi)$ in Eq. (6.23) is therefore in perfect accord with the principle of structural risk minimization.

The parameter C controls the tradeoff between complexity of the machine and the number of nonseparable points; it may therefore be viewed as a form of a "regularization" parameter. The parameter C has to be selected by the user. This can be done in one of two ways:

- The parameter C is determined *experimentally* via the standard use of a training/ (validation) test set, which is a crude form of resampling.
- It is determined *analytically* by estimating the VC dimension via Eq. (6.19) and then by using bounds on the generalization performance of the machine based on the VC dimension.

In any event, the functional $\Phi(\mathbf{w}, \xi)$ is optimized with respect to $\mathbf{w}$ and $\{\xi_i\}_{i=1}^{N}$, subject to the constraint described in Eq. (6.22), and $\xi_i \ge 0$. In so doing, the squared norm of $\mathbf{w}$ is treated as a quantity to be jointly minimized with respect to the nonseparable points rather than as a constraint imposed on the minimization of the number of nonseparable points.

The optimization problem for nonseparable patterns just stated, includes the optimization problem for linearly separable patterns as a special case. Specifically, setting $\xi_i = 0$ for all i in both Eqs. (6.22) and (6.23) reduces them to the corresponding forms for the linearly separable case.

We may now formally state the primal problem for the nonseparable case as:

Given the training sample $\{(\mathbf{x}_i, d_i)\}_{i=1}^N$, find the optimum values of the weight vector $\mathbf{w}$ and bias b such that they satisfy the constraint

$$d_i(\mathbf{w}^T\mathbf{x}_i + b) \geq 1 - \xi_i \quad \text{for } i = 1, 2, ..., N$$

$$\xi_i \geq 0 \quad \text{for all } i$$

and such that the weight vector $\mathbf{w}$ and the slack variables ξ_i minimize the cost functional

$$\Phi(\mathbf{w}, \xi) = \frac{1}{2}\mathbf{w}^T\mathbf{w} + C\sum_{i=1}^N \xi_i$$

where C is a user-specified positive parameter.

Using the method of Lagrange multipliers and proceeding in a manner similar to that described in Section 6.2, we may formulate the dual problem for nonseparable patterns as (see Problem 6.3):

Given the training sample $\{(\mathbf{x}_i, d_i)\}_{i=1}^N$, find the Lagrange multipliers $\{\alpha_i\}_{i=1}^N$ that maximize the objective function

$$Q(\alpha) = \sum_{i=1}^N \alpha_i - \frac{1}{2}\sum_{i=1}^N\sum_{j=1}^N \alpha_i\alpha_j d_i d_j \mathbf{x}_i^T\mathbf{x}_j$$

subject to the constraints

(1) $\sum_{i=1}^N \alpha_i d_i = 0$

(2) $0 \leq \alpha_i \leq C \quad \text{for } i = 1, 2, ..., N$

where C is a user-specified positive parameter.

Note that neither the slack variables ξ_i nor their Lagrange multipliers appear in the dual problem. The dual problem for the case of nonseparable patterns is thus similar to that for the simple case of linearly separable patterns except for a minor but important difference. The objective function $Q(\alpha)$ to be maximized is the same in both cases. The nonseparable case differs from the separable case in that the constraint $\alpha_i \geq 0$ is replaced with the more stringent constraint $0 \leq \alpha_i \leq C$. Except for this modification, the constrained optimization for the nonseparable case and computations of the optimum values of the weight vector $\mathbf{w}$ and bias b proceed in the same way as in the linearly separable case. Note also that the support vectors are defined in exactly the same way as before.

The optimum solution for the weight vector $\mathbf{w}$ is given by

$$\mathbf{w}_o = \sum_{i=1}^{N_s} \alpha_{o,i} d_i \mathbf{x}_i \tag{6.24}$$

where N_S is the number of support vectors. The determination of the optimum values of the bias also follows a procedure similar to that described before. Specifically, the Kuhn–Tucker conditions are now defined by

$$\alpha_i\big[d_i(\mathbf{w}^T\mathbf{x}_i + b) - 1 + \xi_i\big] = 0, \qquad i = 1, 2, \ldots, N \tag{6.25}$$

and

$$\mu_i\xi_i = 0, \qquad i = 1, 2, \ldots, N \tag{6.26}$$

Equation (6.25) is a rewrite of Eq. (6.14) except for the replacement of the unity term by $(1 - \xi_i)$. As for Eq. (6.26), the μ_i are Lagrange multipliers that have been introduced to enforce the nonnegativity of the slack variables ξ_i for all i. At the saddle point the derivative of the Lagrangian function for the primal problem with respect to the slack variable ξ_i is zero, the evaluation of which yields

$$\alpha_i + \mu_i = C \tag{6.27}$$

By combining Eqs. (6.26) and (6.27), we see that

$$\xi_i = 0 \quad \text{if} \quad \alpha_i < C \tag{6.28}$$

We may determine the optimum bias b_o by taking any data point $(\mathbf{x}_i, d_i)$ in the training set for which we have $0 < \alpha_{o,i} < C$ and therefore $\xi_i = 0$, and using that data point in Eq. (6.25). However, from a numerical perspective it is better to take the mean value of b_o resulting from all such data points in the training sample (Burges, 1998).

6.4 HOW TO BUILD A SUPPORT VECTOR MACHINE FOR PATTERN RECOGNITION

With the material on how to find the optimal hyperplane for nonseparable patterns at hand, we are now in a position to formally describe the construction of a support vector machine for a pattern-recognition task.

Basically, the idea of a support vector machine[3] hinges on two mathematical operations summarized here and illustrated in Fig. 6.4:

1. Nonlinear mapping of an input vector into a high-dimensional *feature space* that is hidden from both the input and output.
2. Construction of an optimal hyperplane for separating the features discovered in step 1.

The rationale for each of these two operations is explained in what follows.

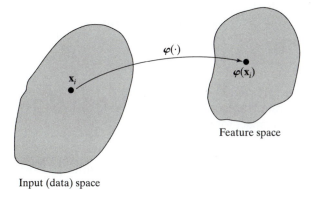

$\varphi(\cdot)$

$\mathbf{x}_i$

$\varphi(\mathbf{x}_i)$

Feature space

Input (data) space

FIGURE 6.4   Nonlinear map $\varphi(\cdot)$ from the input space to the feature space.

Operation 1 is performed in accordance with Cover's theorem on the separability of patterns, which is discussed in Chapter 5. Consider an input space made up of *non-linearly separable patterns*. Cover's theorem states that such a multidimensional space may be transformed into a new feature space where the patterns are linearly separable with high probability, provided two conditions are satisfied. First, the transformation is nonlinear. Second, the dimensionality of the feature space is high enough. These two conditions are embodied in operation 1. Note, however, Cover's theorem does *not* discuss the optimality of the separating hyperplane. It is only by using an optimal separating hyperplane that the VC dimension is minimized and generalization is achieved.

This latter matter is where the second operation comes in. Specifically, operation 2 exploits the idea of building an optimal separating hyperplane in accordance with the theory described in Section 6.3, but with a fundamental difference: The separating hyperplane is now defined as a linear function of vectors drawn from the feature space rather than the original input space. Most importantly, construction of this hyperplane is performed in accordance with the principle of structural risk minimization that is rooted in VC dimension theory. The construction hinges on the evaluation of an inner-product kernel.

Inner-Product Kernel

Let $\mathbf{x}$ denote a vector drawn from the input space, assumed to be of dimension m_0. Let $\{\varphi_j(\mathbf{x})\}_{j=1}^{m_1}$ denote a set of nonlinear transformations from the input space to the feature space: m_1 is the dimension of the feature space. It is assumed that $\varphi_j(\mathbf{x})$ is defined *a priori* for all j. Given such a set of nonlinear transformations, we may define a hyperplane acting as the decision surface as follows:

$$\sum_{j=1}^{m_1} w_j \varphi_j(\mathbf{x}) + b = 0 \tag{6.29}$$

where $\{w_j\}_{j=1}^{m_1}$ denotes a set of linear weights connecting the feature space to the output space, and b is the bias. We may simplify matters by writing

$$\sum_{j=0}^{m_1} w_j \varphi_j(\mathbf{x}) = 0 \tag{6.30}$$

where it is assumed that $\varphi_0(\mathbf{x}) = 1$ for all $\mathbf{x}$, so that w_0 denotes the bias b. Equation (6.30) defines the decision surface computed in the feature space in terms of the linear weights of the machine. The quantity $\varphi_j(\mathbf{x})$ represents the input supplied to the weight w_j *via* the feature space. Define the vector

$$\boldsymbol{\varphi}(\mathbf{x}) = [\varphi_0(\mathbf{x}), \varphi_1(\mathbf{x}), ..., \varphi_{m_1}(\mathbf{x})]^T \tag{6.31}$$

where, by definition, we have

$$\varphi_0(\mathbf{x}) = 1 \quad \text{for all } \mathbf{x} \tag{6.32}$$

In effect, the vector $\boldsymbol{\varphi}(\mathbf{x})$ represents the "image" induced in the feature space due to the input vector $\mathbf{x}$, as illustrated in Fig. 6.4. Thus, in terms of this image we may define the decision surface in the compact form:

$$\mathbf{w}^T \boldsymbol{\varphi}(\mathbf{x}) = 0 \tag{6.33}$$

Adapting Eq. (6.12) to our present situation involving a feature space where we now seek "linear" separability of features, we may write

$$\mathbf{w} = \sum_{i=1}^{N} \alpha_i d_i \varphi(\mathbf{x}_i) \tag{6.34}$$

where the *feature vector* $\varphi(\mathbf{x}_i)$ corresponds to the input pattern $\mathbf{x}_i$ in the *i*th example. Therefore, substituting Eq. (6.34) in (6.33), we may define the decision surface computed in the feature space as:

$$\sum_{i=1}^{N} \alpha_i d_i \varphi^T(\mathbf{x}_i)\varphi(\mathbf{x}) = 0 \tag{6.35}$$

The term $\varphi^T(\mathbf{x}_i)\varphi(\mathbf{x})$ represents the inner product of two vectors induced in the feature space by the input vector $\mathbf{x}$ and the input pattern $\mathbf{x}_i$ pertaining to the *i*th example. We may therefore introduce the *inner-product kernel* denoted by $K(\mathbf{x}, \mathbf{x}_i)$ and defined by

$$
\begin{aligned}
K(\mathbf{x}, \mathbf{x}_i) &= \varphi^T(\mathbf{x})\varphi(\mathbf{x}_i) \\
&= \sum_{j=0}^{m_1} \varphi_j(\mathbf{x})\varphi_j(\mathbf{x}_i) \qquad \text{for } i = 1, 2, ..., N
\end{aligned}
\tag{6.36}
$$

From this definition we immediately see that the inner-product kernel is a *symmetric function* of its arguments, as shown by

$$K(\mathbf{x}, \mathbf{x}_i) = K(\mathbf{x}_i, \mathbf{x}) \qquad \text{for all } i \tag{6.37}$$

Most importantly, we may use the inner-product kernel $K(\mathbf{x}, \mathbf{x}_i)$ to construct the optimal hyperplane in the feature space without having to consider the feature space itself in explicit form. This is readily seen by using Eq. (6.36) in (6.35), whereby the optimal hyperplane is now defined by

$$\sum_{i=1}^{N} \alpha_i d_i K(\mathbf{x}, \mathbf{x}_i) = 0 \tag{6.38}$$

Mercer's Theorem

The expansion of Eq. (6.36) for the inner-product kernel $K(\mathbf{x}, \mathbf{x}_i)$ is an important special case of *Mercer's theorem* that arises in functional analysis. This theorem may be formally stated as (Mercer, 1908; Courant and Hilbert, 1970):

Let $K(\mathbf{x}, \mathbf{x}')$ be a continuous symmetric kernel that is defined in the closed interval $a \leq \mathbf{x} \leq b$ and likewise for $\mathbf{x}'$. The kernel $K(\mathbf{x}, \mathbf{x}')$ can be expanded in the series

$$K(\mathbf{x}, \mathbf{x}') = \sum_{i=1}^{\infty} \lambda_i \varphi_i(\mathbf{x})\varphi_i(\mathbf{x}') \tag{6.39}$$

with positive coefficients, $\lambda_i > 0$ for all *i*. For this expansion to be valid and for it to converge absolutely and uniformly, it is necessary and sufficient that the condition

$$\int_b^a \int_b^a K(\mathbf{x}, \mathbf{x}')\psi(\mathbf{x})\psi(\mathbf{x}')d\mathbf{x}d\mathbf{x}' \geq 0$$

holds for all $\psi(\cdot)$ for which

$$\int_b^a \psi^2(\mathbf{x})d\mathbf{x} < \infty$$

The functions $\varphi_i(\mathbf{x})$ are called *eigenfunctions* of the expansion and the numbers λ_i are called *eigenvalues.* The fact that all of the eigenvalues are positive means that the kernal $K(\mathbf{x}, \mathbf{x}')$ is *positive definite.*

In light of Mercer's theorem, we may now make the following observations:

- For $\lambda_i \neq 1$, the ith image $\sqrt{\lambda_i}\, \varphi_i(\mathbf{x})$ induced in the feature space by the input vector $\mathbf{x}$ is an eigenfunction of the expansion.
- In theory, the dimensionality of the feature space (i.e., the number of eigenvalues/eigenfunctions) can be infinitely large.

Mercer's theorem only tells us whether or not a candidate kernel is actually an inner-product kernel in some space and therefore admissible for use in a support vector machine. However, it says nothing about how to construct the functions $\varphi_i(\mathbf{x})$; we have to do that ourselves.

From the defining equation (6.23), we see that the support vector machine includes a form of regularization in an *implicit* sense. In particular, the use of a kernel $K(\mathbf{x}, \mathbf{x}')$ defined in accordance with Mercer's theorem corresponds to regularization with an operator $\mathbf{D}$ such that the kernel $K(\mathbf{x}, \mathbf{x}')$ is the Green's function of $\widetilde{\mathbf{D}}\mathbf{D}$, where $\widetilde{\mathbf{D}}$ is the adjoint of $\mathbf{D}$ (Smola and Schölkopf, 1998). Regularization theory is discussed in Chapter 5.

Optimum Design of a Support Vector Machine

The expansion of the inner-product kernel $K(\mathbf{x}, \mathbf{x}_i)$ in Eq. (6.36) permits us to construct a decision surface that is nonlinear in the input space, but *its image in the feature space is linear.* With this expansion at hand, we may now state the dual form for the constrained optimization of a support vector machine as follows:

Given the training sample $\{(\mathbf{x}_i, d_i)\}_{i=1}^N$, find the Lagrange multipliers $\{\alpha_i\}_{i=1}^N$ that maximize the objective function

$$Q(\alpha) = \sum_{i=1}^N \alpha_i - \frac{1}{2}\sum_{i=1}^N \sum_{j=1}^N \alpha_i\alpha_j d_i d_j K(\mathbf{x}_i, \mathbf{x}_j) \tag{6.40}$$

subject to the constraints:

(1) $\displaystyle\sum_{i=1}^N \alpha_i d_i = 0$

(2) $0 \leq \alpha_i \leq C$ for $i = 1, 2, ..., N$

where C is a user-specified positive parameter.

Note that constraint (1) arises from optimization of the Lagrangian $Q(\alpha)$ with respect to the bias $b = w_0$ for $\varphi_0(\mathbf{x}) = 1$. The dual problem just stated is of the same form as that for the case of nonseparable patterns considered in Section 6.3, except for the fact that the inner product $\mathbf{x}_i^T\mathbf{x}_j$ used therein has been replaced by the inner-product kernel $K(\mathbf{x}_i, \mathbf{x}_j)$. We may view $K(\mathbf{x}_i, \mathbf{x}_j)$ as the ij-th element of a symmetric N-by-N matrix $\mathbf{K}$, as shown by

$$\mathbf{K} = \{K(\mathbf{x}_i, \mathbf{x}_j)\}_{(i,j)=1}^N \tag{6.41}$$

Having found the optimum values of the Lagrange multipliers, denoted by $\alpha_{o,i}$, we may determine the corresponding optimum value of the linear weight vector, $\mathbf{w}_o$, connecting the feature space to the output space by adapting the formula of Eq. (6.17) to the new situation. Specifically, recognizing that the image $\varphi(\mathbf{x}_i)$ plays the role of input to the weight vector $\mathbf{w}$, we may define $\mathbf{w}_o$ as

$$\mathbf{w}_o = \sum_{i=1}^{N} \alpha_{o,i} d_i \varphi(\mathbf{x}_i) \tag{6.42}$$

where $\varphi(\mathbf{x}_i)$ is the image induced in the feature space due to $\mathbf{x}_i$. Note the first component of $\mathbf{w}_o$ represents the optimum bias b_o.

Examples of Support Vector Machine

The requirement on the kernel $K(\mathbf{x}, \mathbf{x}_i)$ is to satisfy Mercer's theorem. Within this requirement there is some freedom in how it is chosen. In Table 6.1 we summarize the inner-product kernels for three common types of support vector machines: polynomial learning machine, radial-basis function network, and two-layer perceptron. The following points are noteworthy:

1. The inner-product kernels for polynomial and radial-basis function types of support vector machines always satisfy Mercer's theorem. In contrast, the inner-product kernel for a two-layer perceptron type of support vector machine is somewhat restricted, as indicated in the last row of Table 6.1. This latter entry is a testament to the fact that the determination of whether or not a given kernel satisfies Mercer's theorem can indeed be a difficult matter; see Problem 6.8.
2. For all three machine types, the dimensionality of the feature space is determined by the number of support vectors extracted from the training data by the solution to the constrained optimization problem.
3. The underlying theory of a support vector machine avoids the need for heuristics often used in the design of conventional radial-basis function networks and multilayer perceptrons:

TABLE 6.1   Summary of Inner-Product Kernels

Type of support vector machine	Inner product kernel $K(\mathbf{x}, \mathbf{x}_i)$, $i = 1, 2, \ldots, N$	Comments
Polynomial learning machine	$(\mathbf{x}^T\mathbf{x}_i + 1)^p$	Power p is specified *a priori* by the user
Radial-basis function network	$\exp\left(-\dfrac{1}{2\sigma^2}\|\mathbf{x} - \mathbf{x}_i\|^2\right)$	The width σ^2, common to all the kernels, is specified *a priori* by the user
Two-layer perceptron	$\tanh(\beta_0 \mathbf{x}^T\mathbf{x}_i + \beta_1)$	Mercer's theorem is satisfied only for some values of β_0 and β_1

- In the radial-basis function type of a support vector machine, the number of radial-basis functions and their centers are determined automatically by the number of support vectors and their values, respectively.
- In the two-layer perceptron type of a support vector machine, the number of hidden neurons and their weight vectors are determined automatically by the number of support vectors and their values, respectively.

Figure 6.5 displays the architecture of a support vector machine.

Irrespective of how a support vector machine is implemented, it differs from the conventional approach to the design of a multilayer perceptron in a fundamental way. In the conventional approach, model complexity is controlled by keeping the number of features (i.e., hidden neurons) small. On the other hand, the support vector machine offers a solution to the design of a learning machine by controlling model complexity independently of dimensionality, as summarized here (Vapnik, 1995, 1998):

- *Conceptual problem.* Dimensionality of the feature (hidden) space is purposely made very large to enable the construction of a decision surface in the form of a hyperplane in that space. For good generalization performance, the model complexity is controlled by imposing certain constraints on the construction of the separating hyperplane, which results in the extraction of a fraction of the training data as support vectors.
- *Computational problem.* Numerical optimization in a high-dimensional space suffers from the curse of dimensionality. This computational problem is avoided by using the notion of an inner-product kernel (defined in accordance with Mercer's theorem) and solving the dual form of the constrained optimization problem formulated in the input (data) space.

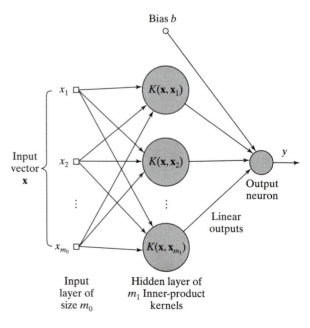

FIGURE 6.5 Architecture of support vector machine.

6.5 EXAMPLE: XOR PROBLEM (REVISITED)

To illustrate the procedure for the design of a support vector machine, we revisit the XOR (Exclusive OR) problem discussed in Chapters 4 and 5. Table 6.2 presents a summary of the input vectors and desired responses for the four possible states.

To proceed, let (Cherkassky and Mulier, 1998)

$$K(\mathbf{x}, \mathbf{x}_i) = (1 + \mathbf{x}^T\mathbf{x}_i)^2 \tag{6.43}$$

With $\mathbf{x} = [x_1, x_2]^T$ and $\mathbf{x}_i = [x_{i1}, x_{i2}]^T$, we may thus express the inner-product kernel $K(\mathbf{x}, \mathbf{x}_i)$ in terms of *monomials* of various orders as follows:

$$K(\mathbf{x}, \mathbf{x}_i) = 1 + x_1^2 x_{i1}^2 + 2x_1 x_2 x_{i1} x_{i2} + x_2^2 x_{i2}^2 + 2x_1 x_{i1} + 2x_2 x_{i2}$$

The image of the input vector $\mathbf{x}$ induced in the feature space is therefore deduced to be

$$\varphi(\mathbf{x}) = \left[1, x_1^2, \sqrt{2}x_1 x_2, x_2^2, \sqrt{2}x_1, \sqrt{2}x_2\right]^T$$

Similarly,

$$\varphi(\mathbf{x}_i) = \left[1, x_{i1}^2, \sqrt{2}x_{i1}x_{i2}, x_{i2}^2, \sqrt{2}x_{i1}, \sqrt{2}x_{i2}\right]^T, \qquad i = 1, 2, 3, 4$$

From Eq. (6.41), we also find that

$$\mathbf{K} = \begin{bmatrix} 9 & 1 & 1 & 1 \\ 1 & 9 & 1 & 1 \\ 1 & 1 & 9 & 1 \\ 1 & 1 & 1 & 9 \end{bmatrix}$$

The objective function for the dual form is therefore (see Eq. (6.40))

$$Q(\alpha) = \alpha_1 + \alpha_2 + \alpha_3 + \alpha_4 - \frac{1}{2}\left(9\alpha_1^2 - 2\alpha_1\alpha_2 - 2\alpha_1\alpha_3 + 2\alpha_1\alpha_4\right.$$

$$\left. + 9\alpha_2^2 + 2\alpha_2\alpha_3 - 2\alpha_2\alpha_4 + 9\alpha_3^2 - 2\alpha_3\alpha_4 + 9\alpha_4^2\right)$$

Optimizing $Q(\alpha)$ with respect to the Lagrange multipliers yields the following set of simultaneous equations:

$$9\alpha_1 - \alpha_2 - \alpha_3 + \alpha_4 = 1$$
$$-\alpha_1 + 9\alpha_2 + \alpha_3 - \alpha_4 = 1$$
$$-\alpha_1 + \alpha_2 + 9\alpha_3 - \alpha_4 = 1$$
$$\alpha_1 - \alpha_2 - \alpha_3 + 9\alpha_4 = 1$$

TABLE 6.2 XOR Problem

Input vector, $\mathbf{x}$	Desired response, d
$(-1, -1)$	-1
$(-1, +1)$	$+1$
$(+1, -1)$	$+1$
$(+1, +1)$	-1

Hence, the optimum values of the Lagrange multipliers are

$$\alpha_{o,1} = \alpha_{o,2} = \alpha_{o,3} = \alpha_{o,4} = \frac{1}{8}$$

This result indicates that in this example all four input vectors $\{\mathbf{x}_i\}_{i=1}^4$ are support vectors. The optimum value of $Q(\alpha)$ is

$$Q_o(\alpha) = \frac{1}{4}$$

Correspondingly, we may write

$$\frac{1}{2}\|\mathbf{w}_o\|^2 = \frac{1}{4}$$

or

$$\|\mathbf{w}_o\| = \frac{1}{\sqrt{2}}$$

From Eq. (6.42), we find that the optimum weight vector is

$$\mathbf{w}_o = \frac{1}{8}[-\varphi(\mathbf{x}_1) + \varphi(\mathbf{x}_2) + \varphi(\mathbf{x}_3) - \varphi(\mathbf{x}_4)]$$

$$= \frac{1}{8}\left[-\begin{bmatrix} 1 \\ 1 \\ \sqrt{2} \\ 1 \\ -\sqrt{2} \\ -\sqrt{2} \end{bmatrix} + \begin{bmatrix} 1 \\ 1 \\ -\sqrt{2} \\ 1 \\ -\sqrt{2} \\ \sqrt{2} \end{bmatrix} + \begin{bmatrix} 1 \\ 1 \\ -\sqrt{2} \\ 1 \\ \sqrt{2} \\ -\sqrt{2} \end{bmatrix} - \begin{bmatrix} 1 \\ 1 \\ \sqrt{2} \\ 1 \\ \sqrt{2} \\ \sqrt{2} \end{bmatrix}\right]$$

$$= \begin{bmatrix} 0 \\ 0 \\ -1/\sqrt{2} \\ 0 \\ 0 \\ 0 \end{bmatrix}$$

The first element of $\mathbf{w}_o$ indicates that the bias b is zero. The optimal hyperplane is defined by (see Eq. (6.33))

$$\mathbf{w}_o^T \varphi(\mathbf{x}) = 0$$

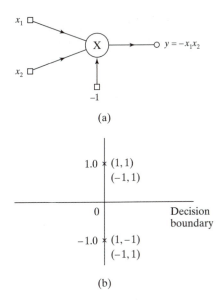

(a)

(b)

<figure>FIGURE 6.6   (a) Polynomial machine for solving the XOR problem. (b) Induced images in the feature space due to the four data points of the XOR problem.</figure>

That is,

$$\left[0, 0, \frac{-1}{\sqrt{2}}, 0, 0, 0\right] \begin{bmatrix} 1 \\ x_1^2 \\ \sqrt{2}x_1x_2 \\ x_2^2 \\ \sqrt{2}x_1 \\ \sqrt{2}x_2 \end{bmatrix} = 0$$

which reduces to

$$-x_1x_2 = 0$$

The polynomial form of support vector machine for the XOR problem is as shown in Fig. 6.6a. For both $x_1 = x_2 = -1$ and $x_1 = x_2 = +1$, the output $y = -1$; and for both $x_1 = -1, x_2 = +1$ and $x_1 = +1$ and $x_2 = -1$, we have $y = +1$. Thus the XOR problem is solved as indicated in Fig. 6.6b.

6.6 COMPUTER EXPERIMENT

In this computer experiment we revisit the pattern-classification problem that we studied in Chapters 4 and 5. The experiment involved the classification of two overlapping two-dimensional Gaussian distributions labeled 1 (class $\mathcal{C}_1$) and 2 (Class $\mathcal{C}_2$). The scatter plots for these two sets of data are shown in Fig. 4.14. The probability of correct classification produced by the Bayesian (optimum) classifier is calculated to be

$$p_c = 81.51 \text{ percent}$$

TABLE 6.3   Summary of the Results of Two-Class Pattern-Classification Experiment Using the Support Vector Machine

	Common width, $\sigma^2 = 4$ Regularization parameter, $C = 0.1$				
Probability of correct classification, p_c	81.22	81.28	81.55	81.49	81.45
Number of support vectors, N_S	298	287	283	287	286

Table 6.3 presents the summary of the results obtained from a computer experiment performed on this data set using the support vector machine. For the inner-product kernel, we used the radial-basis function:

$$K(\mathbf{x}, \mathbf{x}_i) = \exp\left(-\frac{\|\mathbf{x} - \mathbf{x}_i\|^2}{2\sigma^2}\right), \qquad i = 1, 2, ..., N$$

where the same width $\sigma^2 = 4$ was used for all points in the data set. The machine was trained on a total of $N = 500$ data points drawn at random from the population of data representing the two classes. The value used for the regularization parameter was $C = 0.1$.

The results presented in Table 6.3 pertain to five different trials of the experiment, with each trial involving the use of 500 data points for training and 32,000 data points for testing. The probability of correct classification, averaged over these five trials, is 81.40 percent. This average is almost equal to that realized by the Bayesian classifier. The fact that this optimum result was exceeded by 0.05 percent on one of the trials is attributed to experimental errors.

The near-perfect classification performance attained by the support vector machine is further confirmed by the decision boundary shown in Fig. 6.7, which was achieved by one of the five realizations of the machine picked at random. In this figure we have also included the decision boundary for the Bayesian classifier, which consists of a circle of center $\mathbf{x}_c = [-2/3, 0]^T$ and radius $r = 2.34$. Figure 6.6 clearly demonstrates that the support vector machine is capable of constructing a decision boundary between the two classes $\mathcal{C}_1$ and $\mathcal{C}_2$ that is almost as good as the optimum decision boundary.

Returning to the summary of results presented in Table 6.3, the second row displays the sizes of the five different realizations of the support vector machine. These results indicate that for this experiment, the support vector machine learning algorithm selected close to 60 percent of the data points as support vectors.

In the case of nonseparable patterns, all the training errors give rise to support vectors of their own; this follows from the Kuhn–Tucker conditions. For the present experiment, the classification error is about 20 percent. With a sample size of 500, we therefore find that about one third of the support vectors were in fact due to classification errors.

Summarizing Remarks

Comparing the results of this simple computer experiment on the support vector machine with the corresponding results reported in Section 4.8 on the multilayer perceptron trained on the same data sample using the back-propagation algorithm, we can make the following observations:

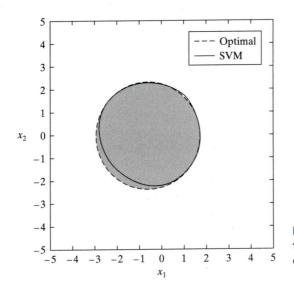

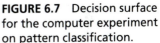

FIGURE 6.7   Decision surface for the computer experiment on pattern classification.

1. The support vector machine has the inherent ability to solve a pattern-classification problem in a manner *close to the optimum* for the problem of interest. Moreover, it is able to achieve such a remarkable performance with *no* problem domain knowledge built into the design of the machine.

2. The multilayer perceptron trained using the back-propagation algorithm, on the other hand, provides a *computationally efficient* solution to the pattern-classification problem of interest. For the two-class experiment described here we were able to realize a probability of correct classification of about 79.70 percent using a multilayer perceptron with only two hidden neurons.

In making this summary, we highlighted the individual virtues of these two approaches to pattern classification. However, for a balanced summary we must also identify their individual shortcomings. In the case of a support vector machine, the near-to-perfect classification performance is achieved at the cost of a significant demand on computational complexity. On the other hand, for a multilayer perceptron trained on the back-propagation algorithm to attain a classification performance comparable to that of the support vector machine for the same pattern-classification tasks, we need to do two things: build problem-domain knowledge into the design of the multilayer perceptron, and tune a multitude of design parameters, a practice that can be excruciating for difficult learning tasks.

6.7 ϵ-INSENSITIVE LOSS FUNCTION

Up to this point in the chapter we have focused on the use of support vector machines for solving pattern-recognition tasks. We now consider the use of support vector machines to solve nonlinear regression problems. To prepare for this discussion we will first address the issue of a suitable optimization criterion for this latter class of learning tasks.

In Chapter 4 on multilayer perceptrons and Chapter 5 on radial-basis function networks, we used a quadratic loss function as the criterion for optimizing these networks.

The main reason for using this criterion is mathematical, that is, for computational convenience. However, a least-squares estimator is sensitive to the presence of outliers (i.e., observations that are improbably large for a nominal model), and it performs poorly when the underlying distribution of the additive noise has a long tail. To overcome these limitations, we need a *robust* estimator that *is insensitive to small changes in the model.*

With robustness as a design goal, any quantitative measure of robustness must be concerned with the maximum degradation of performance that is possible for an ϵ-deviation from the nominal noise model. According to this viewpoint, an *optimal robust estimation procedure* minimizes the maximum degradation, and will therefore be a *minimax* procedure of some kind (Huber, 1981). For the case when the additive noise has a probability density function that is symmetric about the origin, the minimax procedure[4] for solving the nonlinear regression problem uses the absolute error as the quantity to be minimized (Huber, 1964). That is, the loss function has the form

$$L(d, y) = |d - y| \tag{6.44}$$

where d is the desired response and y is the estimator output.

To construct a support vector machine for approximating a desired response d, we may use an extension of the loss function in Eq. (6.44), originally proposed in Vapnik (1995, 1998), as described here

$$L_\epsilon(d, y) = \begin{cases} |d - y| - \epsilon, & \text{for } |d - y| \geq \epsilon \\ 0 & \text{otherwise} \end{cases} \tag{6.45}$$

where ϵ is a prescribed parameter. The loss function $L_\epsilon(d, y)$ is called the *ϵ-insensitive loss function.* It is equal to zero if the absolute value of the deviation of the estimator output y from the desired response d is less than or equal to zero, and it is equal to the absolute value of the deviation minus ϵ otherwise. The loss function of Eq. (6.44) is a special case of the ϵ-insensitive loss function for $\epsilon = 0$. Figure 6.8 illustrates the dependence of $L_\epsilon(d, y)$ on the error $d - y$.

6.8 SUPPORT VECTOR MACHINES FOR NONLINEAR REGRESSION

Consider a *nonlinear regressive model* in which the dependence of a scalar d on a vector $\mathbf{x}$ is described by

$$d = f(\mathbf{x}) + v \tag{6.46}$$

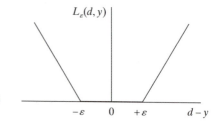

FIGURE 6.8 ϵ-insensitive loss function.

The scalar-valued nonlinear function $f(\mathbf{x})$ is defined by the conditional expectation $E[D|\mathbf{x}]$, as discussed in Chapter 2; D is a random variable with a realization denoted by d. The additive noise term v is statistically independent of the input vector $\mathbf{x}$. The function $f(\cdot)$ and the statistics of noise v are unknown. All that we have available is a set of training data $\{(\mathbf{x}_i, d_i)\}_{i=1}^N$, where $\mathbf{x}_i$ is a sample value of the input vector $\mathbf{x}$ and d_i is the corresponding value of the model output d. The problem is to provide an estimate of the dependence of d on $\mathbf{x}$.

To proceed, we postulate an estimate of d, denoted by y, which is expanded in terms of a set of nonlinear basis functions $\{\varphi_j(\mathbf{x})\}_{j=0}^{m_1}$ as follows:

$$y = \sum_{j=0}^{m_1} w_j \varphi_j(\mathbf{x}) \tag{6.47}$$

$$= \mathbf{w}^T \boldsymbol{\varphi}(\mathbf{x})$$

where

$$\boldsymbol{\varphi}(\mathbf{x}) = [\varphi_0(\mathbf{x}), \varphi_1(\mathbf{x}), \ldots, \varphi_{m_1}(\mathbf{x})]^T$$

and

$$\mathbf{w} = [w_0, w_1, \ldots, w_{m_1}]^T$$

As before, it is assumed that $\varphi_0(\mathbf{x}) = 1$, so that the weight w_0 represents the bias b. The issue to be resolved is to minimize the empirical risk

$$R_{\text{emp}} = \frac{1}{N} \sum_{i=1}^N L_\epsilon(d_i, y_i) \tag{6.48}$$

subject to the inequality

$$\|\mathbf{w}\|^2 \le c_0 \tag{6.49}$$

where c_0 is a constant. The ϵ-insensitive loss function $L_\epsilon(d_i, y_i)$ is as previously defined in Eq. (6.45). We may reformulate this constrained optimization problem by introducing two sets of nonnegative *slack variables* $\{\xi_i\}_{i=1}^N$ and $\{\xi_i'\}_{i=1}^N$ that are defined as follows:

$$d_i - \mathbf{w}^T \boldsymbol{\varphi}(\mathbf{x}_i) \le \epsilon + \xi_i, \qquad i = 1, 2, \ldots, N \tag{6.50}$$

$$\mathbf{w}^T \boldsymbol{\varphi}(\mathbf{x}_i) - d_i \le \epsilon + \xi_i', \qquad i = 1, 2, \ldots, N \tag{6.51}$$

$$\xi_i \ge 0, \qquad i = 1, 2, \ldots, N \tag{6.52}$$

$$\xi_i' \ge 0, \qquad i = 1, 2, \ldots, N \tag{6.53}$$

The slack variables ξ_i and ξ_i' describe the ϵ-insensitive loss function defined in Eq. (6.45). This constrained optimization problem may therefore be viewed as equivalent to that of minimizing the cost functional

$$\Phi(\mathbf{w}, \xi, \xi') = C\left(\sum_{i=1}^N (\xi_i + \xi_i') \right) + \frac{1}{2} \mathbf{w}^T \mathbf{w} \tag{6.54}$$

subject to the constraints of Eqs. (6.50) to (6.53). By incorporating the term $\mathbf{w}^T \mathbf{w}/2$ in the functional $\Phi(\mathbf{w}, \xi, \xi')$ of Eq. (6.54), we dispense with the need for the inequality constraint of Eq. (6.49). The constant C in Eq. (6.54) is a user-specified parameter. Accordingly, we may define the Lagrangian function:

$$J(\mathbf{w}, \xi, \xi', \alpha, \alpha', \gamma, \gamma') = C \sum_{i=1}^{N} (\xi_i + \xi_i') + \frac{1}{2} \mathbf{w}^T \mathbf{w} - \sum_{i=1}^{N} \alpha_i \left[\mathbf{w}^T \varphi(\mathbf{x}_i) - d_i + \epsilon + \xi_i \right]$$

$$- \sum_{i=1}^{N} \alpha_i' \left[d_i - \mathbf{w}^T \varphi(\mathbf{x}_i) + \epsilon + \xi_i' \right] \tag{6.55}$$

$$- \sum_{i=1}^{N} (\gamma_i \xi_i + \gamma_i' \xi_i')$$

where the α_i and the α_i' are the Lagrange multipliers. The last term on the right-hand side of Eq. (6.55), involving γ_i and γ_i', is included to ensure that the optimality constraints on the Lagrange multipliers α_i and α_i' assume variable forms. The requirement is to minimize $J(\mathbf{w}, \xi, \xi'\ \alpha, \alpha', \gamma, \gamma')$ with respect to the weight vector $\mathbf{w}$ and slack variables ξ and ξ'; it must also be maximized with respect to α and α' and also with respect to γ and γ'. By carrying out this optimization we have in respective ways:

$$\mathbf{w} = \sum_{i=1}^{N} (\alpha_i - \alpha_i') \varphi(\mathbf{x}_i) \tag{6.56}$$

$$\gamma_i = C - \alpha_i \tag{6.57}$$

and

$$\gamma_i' = C - \alpha_i' \tag{6.58}$$

The optimization of $J(\mathbf{w}, \xi, \xi', \alpha, \alpha', \gamma, \gamma')$ just described is the primal problem for regression. To formulate the corresponding dual problem, we substitute Eqs. (6.56) through (6.58) in Eq. (6.55), and thus get the convex functional (after simplifying terms):

$$Q(\alpha_i, \alpha_i') = \sum_{i=1}^{N} d_i (\alpha_i - \alpha_i') - \epsilon \sum_{i=1}^{N} (\alpha_i + \alpha_i')$$

$$- \frac{1}{2} \sum_{i=1}^{N} \sum_{j=1}^{N} (\alpha_i - \alpha_i')(\alpha_j - \alpha_j') K(\mathbf{x}_i, \mathbf{x}_j) \tag{6.59}$$

where $K(\mathbf{x}_i, \mathbf{x}_j)$ is the inner-product kernel defined in accordance with Mercer's theorem:

$$K(\mathbf{x}_i, \mathbf{x}_j) = \varphi^T(\mathbf{x}_i) \varphi(\mathbf{x}_j)$$

The solution to our constrained optimization problem is thus obtained by maximizing $Q(\alpha, \alpha')$ with respect to the Lagrange multipliers α and α', subject to a new set of constraints that incorporates the constant C included in the definition of the function $\Phi(\mathbf{w}, \xi, \xi')$ of Eq. (6.54).

We may now state the dual problem for nonlinear regression using a support vector machine as follows:

Given the training sample $\{(x_i, d_i)\}_{i=1}^{N}$ *find the Lagrange multipliers* $\{\alpha_i\}_{i=1}^{N}$ *and* $\{\alpha_i'\}_{i=1}^{N}$ *that maximize the objective function*

$$Q(\alpha_i, \alpha_i') = \sum_{i=1}^{N} d_i(\alpha_i - \alpha_i') - \epsilon \sum_{i=1}^{N} (\alpha_i + \alpha_i')$$

$$- \frac{1}{2} \sum_{i=1}^{N} \sum_{j=1}^{N} (\alpha_i - \alpha_i')(\alpha_j - \alpha_j') K(\mathbf{x}_i, \mathbf{x}_j)$$

subject to the following constraints:

(1) $\sum_{i=1}^{N} (\alpha_i - \alpha_i') = 0$

(2) $0 \leq \alpha_i \leq C, \quad i = 1, 2, ..., N$
    $0 \leq \alpha_i' \leq C, \quad i = 1, 2, ..., N$

where C is a user-specified constant.

Constraint (1) arises from optimization of the Lagrangian with respect to the bias $b = w_0$ for $\varphi_0(\mathbf{x}) = 1$. Thus, having obtained the optimum values of α_i, and α_i', we may then use Eq. (6.56) to determine the optimum value of the weight vector $\mathbf{w}$ for a pre-scribed map $\varphi(\mathbf{x})$. Note that, as in the solution to the pattern-recognition problem, only some of the coefficients in the expansion of Eq. (6.56) have values different from zero; in particular, the data points for which $\alpha_i \neq \alpha_i'$ define the support vectors for the machine.

The two parameters ϵ and C are free parameters that control the VC dimension of the approximating function

$$F(\mathbf{x}, \mathbf{w}) = \mathbf{w}^T \mathbf{x}$$

$$= \sum_{i=1}^{N} (\alpha_i - \alpha_i') K(\mathbf{x}, \mathbf{x}_i) \tag{6.60}$$

Both ϵ and C must be selected by the user. In a conceptual sense, the choice of ϵ and C raises the same issues of complexity control as the choice of parameter C for pattern classification. In practice, however, complexity control for regression is a more difficult problem for the following reasons:

- The parameters ϵ and C must be *tuned simultaneously.*
- Regression is intrinsically more difficult than pattern classification.

A principled approach for the selection of ϵ and C is still an open research area.

Finally, as with a support vector machine for pattern recognition, a support vec-tor machine for nonlinear regression may be implemented in the form of a polynomial learning machine, radial-basis function network, or two-layer perceptron. The inner product kernels for these three methods of implementation are presented in Table 6.1.

6.9 SUMMARY AND DISCUSSION

The support vector machine is an elegant and highly principled learning method for the design of a feedforward network with a single hidden layer of nonlinear units. Its derivation follows the principle of structural risk minimization that is rooted in VC dimension theory, which makes its derivation even more profound. As the name implies, the design of the machine hinges on the extraction of a subset of the training data that serves as support vectors and therefore represents a stable characteristic of the data. The support vector machine includes the polynomial learning machine, radial-basis function network, and two-layer perceptron as special cases. Thus, although these methods provide different representations of intrinsic statistical regularities contained in the training data, they all stem from a common root in a support vector machine setting.

Unlike the popular back-propagation algorithm, the support vector learning algorithm operates only in a batch mode. There is another important difference between these two algorithms. The back-propagation algorithm minimizes a quadratic loss function, regardless of what the learning task is. In contrast, the support vector learning algorithm for pattern recognition is quite different from that for nonlinear regression, as indicated here:

- When performing a pattern-recognition task, the support vector learning algorithm minimizes the number of training samples that fall inside the margin of separation between positive and negative examples; this is only approximately true, since the slack variables ξ_i are used instead of the indicator function $I(\xi_i - 1)$. Although such a criterion is not exactly the same as that of minimizing the probability of classification error, it is considered to be more appropriate than the mean-square error criterion that is behind the back-propagation algorithm.
- When performing a nonlinear regression task, the support vector learning algorithm minimizes an ϵ-insensitive loss function that is an extension of the mean absolute error criterion of minimax theory. The algorithm is thereby robustified.

Whatever the learning task, the support vector machine provides a method for controlling model complexity independently of dimensionality. In particular, the model complexity problem is solved in a high-dimensional space by using a penalized hyperplane defined in the feature (hidden) space as the decision surface; the result is good generalization performance. The curse of dimensionality is bypassed by focusing on the dual problem for performing the constrained optimization problem. An important reason for using the dual setting is to avoid having to define and compute the parameters of the optimal hyperplane in a data space of possibly high dimensionality.

Ordinarily, the training of a support vector machine consists of a quadratic programming problem[5] that is attractive for two reasons:

- It is guaranteed to find a global extremum of the error surface, where the error refers to the difference between the desired response and the output of the support vector machine.
- The computation can be performed efficiently.

Most importantly, by using a suitable inner-product kernel, the support vector machine automatically computes all the important network parameters pertaining to that choice of a kernel. For example, in the case of a radial-basis function network, the kernel is a Gaussian function. For this method of implementation, the number of radial-basis functions and their centers, and the linear weights and bias levels are all computed automatically. The centers of the radial-basis functions are defined by the support vectors picked by the quadratic optimization strategy. The support vectors are typically a fraction of the total number of examples constituting the training sample. We may thus view the design of an RBF network obtained using the support-vector machine learning procedure as a *sparse* version of the corresponding design resulting from the use of a strict interpolation strategy described in the previous chapter.

Several commercial optimization libraries[6] can be used to solve the quadratic programming problem. However, these libraries are of limited use. The memory requirements of the quadratic programming problem grow with the square of the size of the training sample. Consequently, in real-life applications that may involve several thousand data points, the quadratic programming problem cannot be solved by a straightforward use of a commercial optimization library. Osuna et al. (1997) have developed a novel decomposition algorithm that attains optimality by solving a sequence of much smaller subproblems. In particular, the decomposition algorithm takes advantage of the support vector coefficients that are active on either side of their bounds defined by $\alpha_i = 0$ or $\alpha_i = C$. It is reported therein that the decomposition algorithm performs satisfactorily in applications with as many as 100,000 data points.

In terms of running time, support vector machines are currently slower than other neural networks (e.g., multilayer perceptrons trained with the back-propagation algorithm) for a similar generalization performance. There are two reasons for this slower behavior:

1. There is no control over the number of data points selected by the learning algorithm for use as support vectors.
2. There is no provision for incorporating prior knowledge about the task at hand into the design of the learning machine.

Modifications of the support vector machine intended to deal with these shortcomings are now briefly discussed.

The issue of how to control the selection of support vectors is a difficult one, particularly when the patterns to be classified are nonseparable and the training data are noisy. In general, attempts to remove known errors from the data before training or to remove them from the expansion after training will not give the same optimal hyperplane because the errors are needed for penalizing nonseparability. In Osuna and Girosi (1998), the problem of reducing the running time of a support vector machine for pattern classification has been investigated. Two novel approaches for dealing with this problem are described:

- The support vector machine is itself used as a nonlinear regression tool to approximate the decision surface (separating the classes) with a user-specified accuracy.
- The procedure for training the support vector machine is reformulated to yield the same exact decision surface while using a smaller number of basis functions.

In the first approach, the solution is simplified by approximating it with a linear combination of a *subset* of the basis functions. The resulting machine is a natural extension of the support vector machine for function approximation. This extension is designed to find the minimum of a cost functional of the following form:

$$\mathscr{E}(F) = \sum_{i=1}^{N} |d_i - F(\mathbf{x}_i)|_\epsilon + \frac{1}{2C} \psi(F)$$

where $F(\cdot)$ is the approximating function, $\psi(\cdot)$ is a smoothness functional, and $|x|_\epsilon$ is the ϵ-insensitive cost function defined by

$$|x|_\epsilon = \begin{cases} 0 & \text{if } |x| < \epsilon \\ |x| - \epsilon & \text{otherwise} \end{cases}$$

The ϵ-insensitive cost function has the effect of making the solution robust to outliers and insensitive to errors below a certain threshold ϵ. The minimum of the cost functional $\mathscr{E}(F)$ has the form

$$F(\mathbf{x}) = \sum_{i=1}^{N} c_i G(\mathbf{x}, \mathbf{x}_i)$$

where $G(\cdot, \cdot)$ is a kernel that depends on the particular choice of the smoothness function $\psi(\cdot)$, and the coefficients c_i are computed by solving a quadratic programming problem. The solution is typically *sparse*; that is, only a small number of the c_i will be different from zero, and their number is controlled by the parameter ϵ. In the second approach, the primal problem is reformulated in such a way that it has the same initial structure as the original primal problem, but with one difference: The inner-product kernel $K(\mathbf{x}, \mathbf{x}')$ is now incorporated in the formulation. Both approaches also lend themselves to use for reducing the complexity of support vector machines for nonlinear regression.

Finally, turning to the issue of prior knowledge, it is widely recognized that the performance of a learning machine can be improved by incorporating prior knowledge about the task to be learned in the design of the machine (Abu-Mostafa, 1995). In general, two different ways of exploiting prior knowledge have been pursued in the literature:

- An additional term is included in the cost function, thereby forcing the learning machine to construct a function that incorporates the prior knowledge. This is precisely what is done in the use of regularization.
- Virtual examples are generated from the given training sample. The motivation here is that the learning machine can extract the prior knowledge more readily from the artificially enlarged training data.

In the second approach, the learning process may be slowed down due to correlations in the artificial data and the increased size of the training data set. However, the second approach has an advantage over the first approach because it may be readily implemented for all kinds of prior knowledge and learning machines. One way of implementing the second approach is to proceed as follows (Schölkopf et al., 1996):

1. A support vector machine is trained on the given data to extract a set of support vectors in the usual way.

2. Artificial examples, called *virtual support vectors*, are generated by applying prior knowledge in the form of desired invariance transformations to the support vectors obtained under step 1.

3. Another support vector machine is trained on the artificially enlarged set of examples.

This method has the potential of yielding a significant gain in classification accuracy at a moderate cost in time: It requires two training runs instead of a single one, but it constructs classification rules using more support vectors.

NOTES AND REFERENCES

1. Let $\mathscr{C}$ be a subset of $\mathbb{R}^m$. The subset $\mathscr{C}$ is said to be *convex* if

$$\alpha x + (1 - \alpha)y \in \mathscr{C} \quad \text{for all} \quad (x, y) \in \mathscr{C} \quad \text{and} \quad \alpha \in [0, 1]$$

A function $f:\mathscr{C} \to \mathbb{R}$ is said to be a *convex function* if

$$f(\alpha x + (1 - \alpha)y) \le \alpha f(x) + (1 - \alpha)f(y) \quad \text{for all} \quad (x, y) \in \mathscr{C} \quad \text{and} \quad \alpha \in [0, 1]$$

2. With computational complexity as the issue of interest, we may identify two classes of algorithms:
- *Polynomial time algorithms*, which require a running time that is a polynomial function of the problem size. For example, the fast Fourier transform (FFT) algorithm, commonly used for spectrum analysis, is a polynomial time algorithm as it requires a running time of order $n \log n$, where n is a measure of the problem size.
- *Exponential time algorithms*, which require a running time that is an exponential function of the problem size. For example, an exponential time algorithm may take time 2^n, where n is a measure of the problem size.

On this basis we may view polynomial time algorithms as "efficient" algorithms and exponential time algorithms as "inefficient" algorithms.

There are many computational problems that arise in practice, for which no efficient algorithms have been devised. Many if not all of these seemingly intractable problems are said to belong to a class of problems referred to as *NP-complete problems*. The term "NP" stands for "nondeterministic polynomial."

For more detailed discussion of NP-complete problems, see Cook (1971), Garey and Johnson (1979), and Cormen et al. (1990).

3. The idea of an inner-product kernel was first used in Aizerman et al. (1964a, 1964b) in the formulation of the method of potential functions, which represents the forerunner to radial-basis function networks. About the same time, Vapnik and Chervonenkis (1965) developed the idea of an optimal hyperplane. The combined use of these two powerful concepts in formulating the support vector machine was pioneered by Vapnik and coworkers in 1992; see Boser, Guyon and Vapnik (1992) and Cortes and Vapnik (1995). A full mathematical account of the support vector machine was first described in Vapnik (1995), and subsequently in a more expanded form in Vapnik (1998).

4. Huber's minimax theory is based on neighborhoods that are not global by virtue of excluding asymmetric distributions. Nevertheless, this theory deals successfully with a large part of traditional statistics, particularly, regression.

5. In Schurmars (1997), the use of linear programming is explored by adopting the L_1 norm, $\|\mathbf{w}\|_1$, in place of the L_2 norm, $\|\mathbf{w}\|_2$, that is used in support vector machines. The L_1 norm of the weight vector $\mathbf{w}$ is defined by

$$\|\mathbf{w}\|_1 = \sum_i |w_i|$$

where w_i is the ith element of $\mathbf{w}$. Maximum margin classification using the L_1 norm appears to be biased toward axis-oriented hyperplanes, that is, toward weight vectors with few nonzero elements.

6. Commercial libraries for quadratic programming include the following:
   - MINOS5.4: (Murtagh and Saunders, 1978)
   - LSSOL (Gill et al., 1986)
   - LOQO (Vanderbei, 1994)
   - QPOPT and SQOPT (Gill and Murray, 1991).

PROBLEMS

Optimal separating hyperplane

6.1 Consider the case of a hyperplane for linearly separable patterns, which is defined by the equation

$$\mathbf{w}^T \mathbf{x} + b = 0$$

where $\mathbf{w}$ denotes the weight vector, b denotes the bias, and $\mathbf{x}$ denotes the input vector. The hyperplane is said to correspond to a *canonical pair* $(\mathbf{w}, b)$ if, for the set of input patterns $\{\mathbf{x}_i\}_{i=1}^N$, the additional requirement

$$\min_{i=1,2,\dots,N} |\mathbf{w}^T \mathbf{x}_i + b| = 1$$

is satisfied. Show that this requirement leads to a margin of separation between the two classes equal to $2/\|\mathbf{w}\|$.

6.2 Justify the following statement in the context of nonseparable patterns: Misclassification implies nonseparability of patterns, but the converse is *not* necessarily true.

6.3 Starting with the primal problem for the optimization of the separating hyperplane for nonseparable patterns, formulate the dual problem as described in Section 6.3.

6.4 In this problem we explore the "leave-one-out method," discussed in Chapter 4, for estimating the expected test error produced by an optimal hyperplane for the case of non-separable patterns. Discuss the various possibilities that can arise in the use of this method by eliminating any one pattern from the training sample and constructing a solution based on the remaining patterns.

6.5 The location of the optimal hyperplane in the data space is determined by the data points selected as support vectors. If the data are noisy, one's first reaction might be to question the robustness of the margin of separation to the presence of noise. Yet careful study of the optimal hyperplane reveals that the margin of separation is actually robust to noise. Discuss the rationale for this robust behavior.

Inner-product kernel

6.6 The inner-product kernel $K(\mathbf{x}_i, \mathbf{x}_j)$ is evaluated over a training sample $\mathcal{T}$ of size N, yielding the N-by-N matrix:

$$\mathbf{K} = \{K_{ij}\}_{(i,j)=1}^N$$

where $K_{ij} = K(\mathbf{x}_i, \mathbf{x}_j)$. The matrix $\mathbf{K}$ is positive in that all of its elements have positive values. Using the similarity transformation:

$$\mathbf{K} = \mathbf{Q}\mathbf{\Lambda}\mathbf{Q}^T$$

where $\mathbf{\Lambda}$ is a diagonal matrix of eigenvalues and $\mathbf{Q}$ is a matrix made up of the corresponding eigenvectors, formulate an expression for the inner-product kernel $K(\mathbf{x}_i, \mathbf{x}_j)$ in terms of the eigenvalues and eigenvectors of matrix $\mathbf{K}$. What conclusions can you draw from this representation?

6.7 (a) Prove the *unitary invariance property* of the inner-product kernel $K(\mathbf{x}, \mathbf{x}_i)$; that is,

$$K(\mathbf{x}, \mathbf{x}_i) = K(\mathbf{Q}\mathbf{x}, \mathbf{Q}\mathbf{x}_i)$$

where $\mathbf{Q}$ is a unitary matrix defined by

$$\mathbf{Q}^{-1} = \mathbf{Q}^T$$

(b) Demonstrate that all three inner-product kernels described in Table 6.1 satisfy this property.

6.8 The inner-product kernel for a two-layer perceptron is defined by

$$K(\mathbf{x}, \mathbf{x}_i) = \tanh(\beta_0 \mathbf{x}^T \mathbf{x}_i + \beta_1)$$

Explore some values for the constants β_0 and β_1 for which Mercer's theorem is not satisfied.

Pattern classification

6.9 The inner-product kernel for a polynomial learning machine used to solve the XOR problem is defined by

$$K(\mathbf{x}, \mathbf{x}_i) = (1 + \mathbf{x}^T \mathbf{x}_i)^p$$

What is the minimum value of power p for which the XOR problem is solved? Assume that p is a positive integer. What is the result of using a value for p larger than the minimum?

6.10 Figure P6.10 shows the XOR function operating on a three-dimensional pattern $\mathbf{x}$, as described here

$$\text{XOR}(x_1, x_2, x_3) = x_1 \oplus x_2 \oplus x_3$$

where the symbol $\oplus$ denotes the Exclusive OR Boolean function operator. Design a polynomial learning machine to separate the two classes of points represented by the output of this operator.

6.11 Throughout the chapter we discussed the use of a support vector machine for binary classification. Discuss how a support vector machine can be used to solve an M-ary pattern-classification problem, where $M > 2$.

Nonlinear regression

6.12 The dual problem described in Section 6.8 for using a support vector machine to solve the nonlinear regression problem includes the following constraint:

$$\sum_{i=1}^{N} (\alpha_i - \alpha_i') = 0$$

where the α_i and α_i' are the Lagrange multipliers. Show that this constraint arises from minimization of the Lagrangian with respect to the bias b, that is, the first element w_0 of the weight vector $\mathbf{w}$ corresponding to $\varphi_0(\mathbf{x}) = 1$.

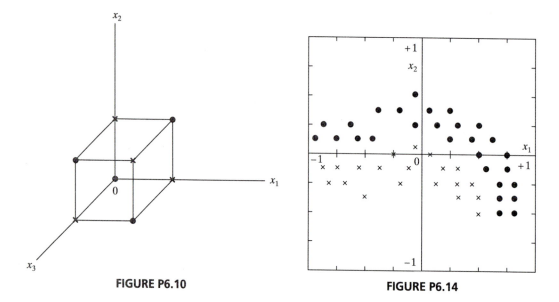

FIGURE P6.10   **FIGURE P6.14**

Virtues and limitations

6.13 **(a)** Compare the virtues and limitations of support vector machines with those of radial-basis function (RBF) networks with respect to the following tasks:

(1) Pattern classification

(2) Nonlinear regression

(b) Do the same for support vector machines versus multilayer perceptrons trained using the back-propagation algorithm.

Computer experiments

6.14 Figure P6.14 shows a set of data points corresponding to two classes, $\mathscr{C}_1$ and $\mathscr{C}_2$. Both coordinates, x_1 and x_2, range from -1 to $+1$. Using a single radial-basis function kernel

$$K(\mathbf{x}, \mathbf{t}) = \exp(-\|\mathbf{x} - \mathbf{t}\|^2)$$

construct the optimal hyperplane and identify the support vectors for this data set.

6.15 The computer experiment described in Section 6.6 was for the classification of two over-lapping Gaussian distributions. The following "regularization" parameter was used in that experiment: $C = 0.1$. The common width of the radial-basis functions used for con-structing the inner-product kernels was $\sigma^2 = 4$. Repeat the computer experiment described in that section for the following two values of the regularization parameter:

(a) $C = 0.05$

(b) $C = 0.2$

Comment on your results in light of the findings reported in Section 6.6.

6.16 In applying radial-basis function networks to nonlinear regression problems, we often find that the use of an unlocalized basis function such as the multiquadric results in a more accurate solution than the use of a localized basis function such as the Gaussian function. It may be conjectured that a similar situation arises in the case of support vec-tor machines, because the use of an (unbounded) polynomial learning machine may prove more accurate than a (bounded) radial-basis function machine. Using a computer experiment on a nonlinear regression problem, explore the validity of this conjecture.

CHAPTER 7

Committee Machines

7.1 INTRODUCTION

In the previous three chapters we describe three different approaches to supervised learning. The MLP trained with the back-propagation algorithm, discussed in Chapter 4, relies on a form of global optimization for its design. The RBF network, discussed in Chapter 5, relies on local optimization for its design. The support vector machine, discussed in Chapter 6, exploits VC-dimension theory for its design. In this chapter we discuss another class of methods for solving supervised learning tasks. The approach used here is based on a commonly used engineering principle: divide and conquer.

According to the *principle of divide and conquer,* a complex computational task is solved by dividing it into a number of computationally simple tasks and then combining the solutions to those tasks. In supervised learning, computational simplicity is achieved by distributing the learning task among a number of *experts,* which in turn divides the input space into a set of subspaces. The combination of experts is said to constitute a *committee machine.* Basically, it fuses knowledge acquired by experts to arrive at an overall decision that is supposedly superior to that attainable by any one of them acting alone. The idea of a committee machine may be traced back to Nilsson (1965); the network structure considered therein consisted of a layer of elementary perceptrons followed by a vote-taking perceptron in the second layer.

Committee machines are universal approximators. They may be classified into two major categories:

1. *Static structures*. In this class of committee machines, the responses of several predictors (experts) are combined by means of a mechanism that does *not* involve the input signal, hence the designation "static." This category includes the following methods:
   - *Ensemble averaging,* where the outputs of different predictors are linearly combined to produce an overall output.
   - *Boosting,* where a weak learning algorithm is converted into one that achieves arbitrarily high accuracy.

2. *Dynamic structures*. In this second class of committee machines, the input signal is directly involved in actuating the mechanism that integrates the outputs of the

individual experts into an overall output, hence the designation "dynamic." Here we mention two kinds of dynamic structures:

- *Mixture of experts*, in which the individual responses of the experts are nonlinearly combined by means of a single gating network.
- *Hierarchical mixture of experts*, in which the individual responses of the experts are nonlinearly combined by means of several gating networks arranged in a hierarchical fashion.

In the mixture of experts, the principle of divide and conquer is applied just once, whereas in the hierarchical mixture of experts it is applied several times, resulting in a corresponding number of levels of hierarchy.

The mixture of experts and hierarchical mixture of experts may also be viewed as examples of modular networks. A formal definition of the notion of *modularity* is (Osherson et al., 1990):

> A neural network is said to be modular if the computation performed by the network can be decomposed into two or more modules (subsystems) that operate on distinct inputs without communicating with each other. The outputs of the modules are mediated by an integrating unit that is not permitted to feed information back to the modules. In particular, the integrating unit both (1) decides how the outputs of the modules should be combined to form the final output of the system, and (2) decides which modules should learn which training patterns.

This definition of modularity rules out the static class of committee machines since there is no integrating unit at the output that has a decision-making role.

Organization of the Chapter

This chapter is organized in two parts. The class of static structures is covered in the first part, encompassing Sections 7.2 through 7.5. Section 7.2 discusses the method of ensemble averaging, followed by a computer experiment in Section 7.3. Section 7.4 discusses the boosting technique, followed by a computer experiment in Section 7.5.

The class of dynamic structures is covered in the second part of the chapter, encompassing Sections 7.6 through 7.13. Specifically, Section 7.6 discusses the mixture of experts (ME) as an associative Gaussian mixture model. Section 7.7 discusses the more general case, namely the hierarchical mixture of experts (HME). This latter model is closely related to standard decision trees. Then Section 7.8 describes how a standard decision tree may be used to solve the model selection problem (i.e., number of gating and expert networks) for HME. In Section 7.9, we define some *a posteriori* probabilities that assist us in the formulation of learning algorithms for the HME model. In Section 7.10 we lay the foundation for solving the parameter estimation problem by formulating the likelihood function for the HME model. Section 7.11 presents an overview of learning strategies. This is followed by a detailed discussion of the so-called EM algorithm in Section 7.12 and its application to the HME model in Section 7.13.

The chapter concludes with some final remarks in Section 7.14.

7.2   ENSEMBLE AVERAGING

Figure 7.1 shows a number of differently trained neural networks (i.e., experts), which share a common input and whose individual outputs are somehow combined to produce an overall output y. To simplify the presentation, the outputs of the experts are assumed to be scalar-valued. Such a technique is referred to as an *ensemble averaging method*[1]. The motivation for its use is two-fold:

- If the combination of experts in Fig. 7.1 were replaced by a single neural network, we would have a network with a correspondingly large number of adjustable parameters. The training time for such a large network is likely to be longer than for the case of a set of experts trained in parallel.
- The risk of overfitting the data increases when the number of adjustable parameters is large compared to cardinality (i.e., size of the set) of the training data.

In any event, in using a committee machine as depicted in Fig. 7.1, the expectation is that the differently trained experts converge to different local minima on the error surface, and overall performance is improved by combining the outputs in some way.

Consider first the case of a single neural network that has been trained on a given data set. Let $\mathbf{x}$ denote an input vector not seen before, and let d denote the corresponding desired response (representing a class label or numerical response); $\mathbf{x}$ and d represent realizations of the random vector $\mathbf{X}$ and random variable D, respectively. Let $F(\mathbf{x})$ denote the input–output function realized by the network. Then, in light of the material on the bias/variance dilemma presented in Chapter 2, we may decompose the mean-square error between $F(\mathbf{x})$ and the conditional expectation $E[D|\mathbf{X} = \mathbf{x}]$ into its bias and variance components as follows:

$$E_{\mathcal{D}}[(F(\mathbf{x}) - E[D|\mathbf{X} = \mathbf{x}])^2] = B_{\mathcal{D}}(F(\mathbf{x})) + V_{\mathcal{D}}(F(\mathbf{x})) \qquad (7.1)$$

where $B_{\mathcal{D}}(F(\mathbf{x}))$ is the bias squared:

$$B_{\mathcal{D}}(F(\mathbf{x})) = (E_{\mathcal{D}}[F(\mathbf{x})] - E[D|\mathbf{X} = \mathbf{x}])^2 \qquad (7.2)$$

and $V_{\mathcal{D}}(F(\mathbf{x}))$ is the variance:

$$V_{\mathcal{D}}(F(\mathbf{x})) = E_{\mathcal{D}}\left[(F(\mathbf{x}) - E_{\mathcal{D}}[F(\mathbf{x})])^2\right] \qquad (7.3)$$

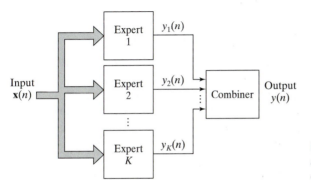

FIGURE 7.1 Block diagram of a committee machine based on ensemble-averaging.

The expectation $E_{\mathcal{D}}$ is taken over the space $\mathcal{D}$, defined as the *space encompassing the distribution of all training sets (i.e., inputs and target outputs) and the distribution of all initial conditions*.

There are different ways of individually training the expert networks in Fig. 7.1 and also different ways of combining their outputs. For the discussion presented here, we will consider the situation where the expert networks have an identical configuration, but they are trained starting from different initial conditions. For the combiner at the output of the committee machine in Fig. 7.1, we will use a simple *ensemble averager*.[2] Let $\mathcal{I}$ denote the *space of all initial conditions*. Let $F_I(\mathbf{x})$ denote the average of the input–output functions of the expert networks in Fig. 7.1 over a "representative" number of initial conditions. By analogy with Eq. (7.1), we may write

$$E_{\mathcal{I}}[(F_I(\mathbf{X}) - E[D \mid \mathbf{X} = \mathbf{x}])^2] = B_{\mathcal{I}}(F(\mathbf{x})) + V_{\mathcal{I}}(F(\mathbf{x})) \tag{7.4}$$

where $B_{\mathcal{I}}(F(\mathbf{x}))$ is the squared bias defined over the space $\mathcal{I}$:

$$B_{\mathcal{I}}(F(\mathbf{x})) = (E_{\mathcal{I}}[F_I(\mathbf{x})] - E[D \mid \mathbf{X} = \mathbf{x}])^2 \tag{7.5}$$

and $V_{\mathcal{I}}(F(\mathbf{x}))$ is the corresponding variance:

$$V_{\mathcal{I}}(F(\mathbf{x})) = E_{\mathcal{I}}\big[(F_I(\mathbf{x}) - E_{\mathcal{I}}[F(\mathbf{x})])^2\big] \tag{7.6}$$

The expectation $E_{\mathcal{I}}$ is taken over the space $\mathcal{I}$.

From the definition of space $\mathcal{D}$, we may view it as the product of the space of initial conditions, $\mathcal{I}$, and the *remnant space* denoted by $\mathcal{D}'$. Accordingly, we may write, again by analogy with Eq. (7.1):

$$E_{\mathcal{D}'}\big[(F_I(\mathbf{x}) - E[D \mid \mathbf{X} = \mathbf{x}])^2\big] = B_{\mathcal{D}'}(E_I(\mathbf{x})) + V_{\mathcal{D}'}(F_I(\mathbf{x})) \tag{7.7}$$

where $B_{\mathcal{D}'}(F(\mathbf{x}))$ is the squared bias defined over the remnant space $\mathcal{D}'$:

$$B_{\mathcal{D}'}(F_I(\mathbf{x})) = (E_{\mathcal{D}'}[F_I(\mathbf{x})] - E[D \mid \mathbf{X} = \mathbf{x}])^2 \tag{7.8}$$

and $V_{\mathcal{D}'}(F_I(\mathbf{x}))$ is the corresponding variance:

$$V_{\mathcal{D}'}(F_I(\mathbf{x})) = E_{\mathcal{D}'}\big[(F_I(\mathbf{x}) - E_{\mathcal{D}'}[F_I(\mathbf{x})])^2\big] \tag{7.9}$$

From the definitions of spaces $\mathcal{D}$, $\mathcal{I}$, and $\mathcal{D}'$, we readily see that

$$E_{\mathcal{D}'}[F_I(\mathbf{x})] = E_{\mathcal{D}}[F(\mathbf{x})] \tag{7.10}$$

It follows therefore that Eq. (7.8) may be rewritten in the equivalent form:

$$B_{\mathcal{D}'}(F_I(\mathbf{x})) = (E_{\mathcal{D}}[F(\mathbf{x})] - E[D \mid \mathbf{X} = \mathbf{x}])^2$$
$$= B_{\mathcal{D}}(F(\mathbf{x})) \tag{7.11}$$

Consider next the variance $V_{\mathcal{D}'}(F_I(\mathbf{x}))$ of Eq. (7.9). Since the variance of a random variable is equal to the mean-square value of that random variable minus its bias squared, we may equivalently write

$$V_{\mathcal{D}'}(F_I(\mathbf{x})) = E_{\mathcal{D}'}[(F_I(\mathbf{x}))^2] - (E_{\mathcal{D}'}[F_I(\mathbf{x})])^2$$
$$= E_{\mathcal{D}'}[(F_I(\mathbf{x}))^2] - (E_{\mathcal{D}}[F(\mathbf{x})])^2 \tag{7.12}$$

where in the last line we have made use of Eq. (7.10). Similarly, we may redefine Eq. (7.3) in the equivalent form

$$V_{\mathcal{D}}(F_I(\mathbf{x})) = E_{\mathcal{D}}[(F(\mathbf{x}))^2] - (E_{\mathcal{D}}[F(\mathbf{x})])^2 \tag{7.13}$$

Note that the mean-square value of the function $F(\mathbf{x})$ over the entire space $\mathcal{D}$ is destined to be equal to or greater than the mean-square value of the ensemble-averaged function $F_I(\mathbf{x})$ over the remnant space $\mathcal{D}'$. That is,

$$E_{\mathcal{D}}[F(\mathbf{x})^2] \geq E_{\mathcal{D}'}[(F_I(\mathbf{x}))^2]$$

In light of this inequality, comparison of Eqs. (7.12) and (7.13) immediately reveals that

$$V_{\mathcal{D}'}(F_I(\mathbf{x})) \leq V_{\mathcal{D}}(F(\mathbf{x})) \tag{7.14}$$

Thus, from Eqs. (7.11) and (7.14) we draw two conclusions (Naftaly et al., 1997):

1. The bias of the ensemble-averaged function $F_I(\mathbf{x})$, pertaining to the committee machine of Fig. 7.1, is exactly the same as that of the function $F(\mathbf{x})$ pertaining to a single neural network.
2. The variance of the ensemble-averaged function $F_I(\mathbf{x})$ is less than that of the function $F(\mathbf{x})$.

These theoretical findings point to a training strategy for reducing the overall error produced by a committee machine due to *varying initial conditions* (Naftaly et al., 1997): The constituent experts of the machine are purposely *overtrained*, the use of which is justified on the following grounds. Insofar as the individual experts are concerned, the bias is reduced at the cost of variance. Subsequently, however, the variance is reduced by ensemble averaging the experts over the initial conditions, leaving the bias unchanged.

7.3 COMPUTER EXPERIMENT I

In this computer experiment on the ensemble-averaging method, we revisit the pattern classification problem considered in the previous three chapters. The problem pertains to the classification of two overlapping two-dimensional Gaussian distributions. The two distributions have different mean vectors and different variances. The statistics of distribution 1 (class $\mathcal{C}_1$) are

$$\boldsymbol{\mu}_1 = [0, 0]^T$$
$$\sigma_1^2 = 1$$

The statistics of distribution 2 (class $\mathcal{C}_2$) are

$$\boldsymbol{\mu}_2 = [2, 0]^T$$
$$\sigma_2^2 = 4$$

Scatter plots for these two distributions are shown in Fig. 4.13.

The two classes are assumed to be equiprobable. The costs for misclassifications are assumed to be equal, and the costs for correct classifications are assumed to be zero. On this basis, the (optimum) Bayes classifier achieves a probability of correct classification $p_c = 81.51$ percent. Details of this calculation are also presented in Chapter 4.

In the computer experiment described in Chapter 4 we were able to achieve a probability of correct classification close to 80 percent using a multilayer perceptron with two hidden neurons and trained using the back-propagation algorithm. In this experiment we study a committee machine composed as follows:

- Ten experts.
- Each expert made up of a multilayer perceptron with two hidden neurons.

All the experts were individually trained using the back-propagation algorithm. The parameters used in the algorithm were

Learning-rate parameter, $\eta = 0.1$
Momentum constant, $\alpha = 0.5$

The size of the training sample was 500 patterns. All the experts were trained on the same data set, but were initialized differently. In particular, the initial values of the synaptic weights and thresholds were picked at random from a uniform distribution inside the range $[-1, 1]$.

Table 7.1 presents a summary of the classification performances of the 10 experts trained on 500 patterns using the test set. The probability of correct classification obtained by simply taking the arithmetic average of the 10 results presented in Table 7.1 is $p_{c,av} = 79.37$ percent. On the other hand, by using the ensemble-averaging method, that is, by simply summing the individual outputs of the 10 experts and then computing the probability of correct classification, we obtained the result: $p_{c,ens} = 80.27$ percent. This result represents an improvement of 0.9 percent over $p_{c,av}$. The improvement of $p_{c,ens}$ over $p_{c,av}$ was maintained in all the trials of the experiment. The classification results were all computed using 32,000 test patterns.

TABLE 7.1   Classification Performances of Individual Experts Used in a Committee Machine

Expert	Correct classification percentage
Net 1	80.65
Net 2	76.91
Net 3	80.06
Net 4	80.47
Net 5	80.44
Net 6	76.89
Net 7	80.55
Net 8	80.47
Net 9	76.91
Net 10	80.38

Summarizing the results of this experiment, we may say: The classification performance is improved by overtraining the individual multilayer perceptrons (experts), summing their individual numerical outputs to produce the overall output of the committee machine, and then making a decision.

7.4 BOOSTING

As mentioned in the introduction, boosting is another method that belongs to the "static" class of committee machines. Boosting is quite different from ensemble averaging. In a committee machine based on ensemble averaging, all the experts in the machine are trained on the same data set; they may differ from each other in the choice of initial conditions used in network training. By contrast, in a boosting machine the experts are trained on data sets with entirely different distributions; it is a general method that can be used to improve the performance of *any* learning algorithm.

Boosting[3] can be implemented in three fundamentally different ways:

1. *Boosting by filtering.* This approach involves filtering the training examples by different versions of a weak learning algorithm. It assumes the availability of a large (in theory, infinite) source of examples, with the examples being either discarded or kept during training. An advantage of this approach is that it allows for a small memory requirement compared to the other two approaches.
2. *Boosting by subsampling.* This second approach works with a training sample of *fixed* size. The examples are "resampled" according to a given probability distribution during training. The error is calculated with respect to the fixed training sample.
3. *Boosting by reweighting.* This third approach also works with a fixed training sample, but it assumes that the weak learning algorithm can receive "weighted" examples. The error is calculated with respect to the weighted examples.

In this section we describe two different boosting algorithms. One algorithm, due to Schapire (1990), belongs to approach 1. The other algorithm, known as AdaBoost due to Freund and Schapire (1996a, 1996b), belongs to approach 2.

Boosting by Filtering

The original idea of boosting described in Schapire (1990) is rooted in a *distribution-free* or *probably approximately correct (PAC) model of learning*. From the discussion of PAC learning presented in Chapter 2, we recall that a *concept* is a Boolean function in some domain of instances that contains encodings of all objects of interest. In PAC learning, a learning machine tries to identify an unknown binary concept on the basis of randomly chosen examples of the concept. To be more specific, the goal of the learning machine is to find a hypothesis or prediction rule with an error rate of at most ϵ, for arbitrarily small positive values of ϵ, and this should hold uniformly for all input distributions. It is for this reason that the PAC learning model is also referred to as a *strong learning model*. Since the examples are of a random nature, it is likely that the learning machine will be unable to learn anything about the unknown concept due to the presentation of a highly unrepresentative example. We therefore require the learning model to succeed only in finding a good approximation to the unknown concept with a probability of $1 - \delta$, where δ is a small positive number.

In a variant of the PAC learning model, called a *weak learning model*, the requirement to learn an unknown concept is relaxed dramatically. The learning machine is now required to find a hypothesis with an error rate only slightly less than $1/2$. When a hypothesis guesses a binary label in an entirely random manner on every example, it can be right or wrong with equal probability. That is, it achieves an error rate of exactly $1/2$. It follows therefore that a weak learning model has to perform only slightly better than random guessing. The notion of weak learnability was introduced by Kearns and Valiant (1989), who posed the *hypothesis boosting problem* that is embodied in the following question:

Are the notions of strong and weak learning equivalent?

In other words, is any concept class that is weakly learnable also strongly learnable? This question, which is perhaps surprising, was answered in the affirmative by Schapire (1990). The proof presented therein was constructive. Specifically, an algorithm was devised for directly converting a weak learning model into a strong learning model. This is achieved by modifying the distribution of examples in such a way that a strong learning model is built around a weak one.

In boosting by filtering, the committee machine consists of three experts or sub-hypotheses. The algorithm used to train them is called a *boosting algorithm*. The three experts are arbitrarily labeled "first," "second," and "third." The three experts are individually trained as follows:

1. The first expert is trained on a set consisting of N_1 examples.
2. The trained first expert is used to *filter* another set of examples by proceeding in the following manner:
   - Flip a fair coin; this in effect simulates a random guess.
   - If the result is *heads*, pass new patterns through the first expert and discard correctly classified patterns until a pattern is misclassified. That misclassified pattern is added to the training set for the second expert.
   - If the result is *tails*, do the opposite. Specifically, pass new patterns through the first expert and discard incorrectly classified patterns until a pattern is classified correctly. That correctly classified pattern is added to the training set for the second expert.
   - Continue this process until a total of N_1 examples has been filtered by the first expert. This set of filtered examples constitutes the training set for the second expert.

   By following this coin flipping procedure, it is ensured that if the first expert is tested on the second set of examples, it would have an error rate of $1/2$. In other words, the second set of N_1 examples available for training the second expert has a distribution entirely different from the first set of N_1 examples used to train the first expert. In this way, the second expert is forced to learn a distribution different from that learned by the first expert.

3. Once the second expert has been trained in the usual way, a third training set is formed for the third expert by proceeding in the following manner:
   - Pass a new pattern through both the first and second experts. If the two experts agree in their decisions, discard that pattern. If, on the other hand, they disagree, the pattern is added to the training set for the third expert.
   - Continue with this process until a total of N_1 examples has been filtered jointly by the first and second experts. This set of jointly filtered examples constitutes the training set for the third expert.

   The third expert is then trained in the usual way, and the training of the entire committee machine is thereby completed.

This three-point filtering procedure is illustrated in Fig. 7.2.

Let N_2 denote the number of examples that must be filtered by the first expert to obtain the training set of N_1 examples for the second expert. Note that N_1 is fixed, and N_2 depends on the generalization error rate of the first expert. Let N_3 denote the number of examples that must be jointly filtered by the first and second experts to obtain the training set of N_1 examples for the third expert. With N_1 examples also needed to train the first expert, the total size of data set needed to train the entire committee machine is $N_4 = N_1 + N_2 + N_3$. However, the computational cost is based on $3N_1$ examples because N_1 is the number of examples actually used to train each of the three experts. We may therefore say that the boosting algorithm described herein is indeed "smart" in the sense that the committee machine requires a large set of examples for its operation, but only a subset of that data set is used to perform the actual training.

Another noteworthy point is that the filtering operation performed by the first expert and the joint filtering operation performed by the first and second experts make the second and third experts, respectively, focus on "hard-to-learn" parts of the distribution.

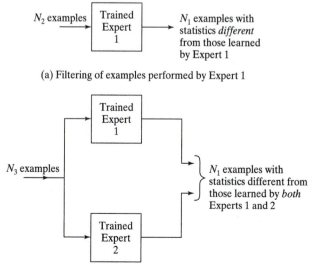

(a) Filtering of examples performed by Expert 1

(b) Filtering of examples performed by Experts 2 and 3

FIGURE 7.2 Illustration of boosting by filtering.

In the theoretical derivation of the boosting algorithm originally presented in Schapire (1990), simple *voting* was used to evaluate the performance of the committee machine on test patterns not seen before. Specifically, a test pattern is presented to the committee machine. If the first and second experts in the committee machine agree in their respective decisions, that class label is used. Otherwise, the class label discovered by the third expert is used. However, in experimental work presented in Drucker et al. (1993, 1994), it has been determined that *addition* of the respective outputs of the three experts yields a better performance than voting. For example, in the optical character recognition (OCR) problem, the addition operation is performed simply by adding the "digit 0" outputs of the three experts, and likewise for the other nine digit outputs.

Suppose that the three experts (i.e., subhypotheses) have an error rate of $\epsilon < 1/2$ with respect to the distributions on which they are individually trained; that is, all three of them are weak learning models. In Schapire (1990), it is proved that the overall error rate of the committee machine is *bounded* by

$$g(\epsilon) = 3\epsilon^2 - 2\epsilon^3 \tag{7.15}$$

The bound $g(\epsilon)$ is plotted versus ϵ in Fig. 7.3. From this figure we see that the bound is significantly smaller than the original error rate ϵ. By applying the boosting algorithm recursively, the error rate can be made arbitrarily small. In other words, a weak learning model, which performs only slightly better than random guessing, is converted into a strong learning model. It is in that sense that we may say that strong and weak learnability are indeed equivalent.

AdaBoost

A practical limitation of boosting by filtering is that it often requires a large training sample. This limitation can be overcome by using another boosting algorithm called *AdaBoost* (Freund and Schapire, 1996a, 1996b), which belongs to boosting by resampling. The sampling framework of AdaBoost is the natural framework of batch learning; most importantly, it permits the training data to be reused.

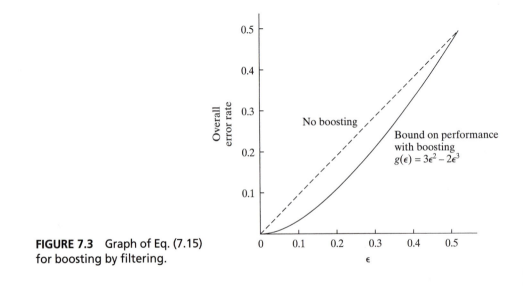

FIGURE 7.3 Graph of Eq. (7.15) for boosting by filtering.

As with the boosting by filtering algorithm, AdaBoost has access to a weak learning model. The goal of the new algorithm is to find a final mapping function or hypothesis with low error rate relative to a given distribution $\mathcal{D}$ over the labeled training examples. It differs from other boosting algorithms in two respects:

- AdaBoost adjusts *adaptively* to the errors of the weak hypothesis returned by the weak learning model, hence the name of the algorithm.
- The bound on performance of AdaBoost depends only on the performance of the weak learning model on those distributions that are actually generated during the learning process.

AdaBoost operates as follows. On iteration n, the boosting algorithm provides the weak learning model with a distribution $\mathcal{D}_n$ over the training sample $\mathcal{T}$. In response, the weak learning model computes a hypothesis $\mathcal{F}_n : \mathbf{X} \rightarrow Y$ that correctly classifies a fraction of the training examples. The error is measured with respect to the distribution $\mathcal{D}_n$. The process continues for T iterations, and finally the boosting machine combines the hypotheses $\mathcal{F}_1, \mathcal{F}_2, \ldots, \mathcal{F}_T$ into a single final hypothesis $\mathcal{F}_{\text{fin}}$.

To calculate (1) the distribution $\mathcal{D}_n$ on iteration n, and (2) the final hypothesis $\mathcal{F}_{\text{fin}}$, the simple procedure summarized in Table 7.2 is used. The initial distribution $\mathcal{D}_1$ is uniform over the training sample $\mathcal{T}$, as shown by

$$\mathcal{D}_1(i) = \frac{1}{n} \qquad \text{for all } i$$

Given the distribution $\mathcal{D}_n$ and weak hypothesis $\mathcal{F}_n$ on iteration n of the algorithm, the next distribution $\mathcal{D}_{n+1}$ is computed by multiplying the weight of example i by some number $\beta_n \in [0, 1)$ if $\mathcal{F}_n$ classifies the input vector $\mathbf{x}_i$ correctly; otherwise, the weight is left unchanged. The weights are then renormalized by dividing by the normalizing constant Z_n. In effect, the "easy" examples in the training set $\mathcal{T}$ that are correctly classified by many of the previous weak hypotheses are given lower weights, whereas the "hard" examples that are often misclassified are given higher weights. Thus the AdaBoost algorithm focuses the most weight on those examples that appear to be hardest for it to classify.

As for the final hypothesis $\mathcal{F}_{\text{fin}}$, it is computed as a weighted vote (i.e., weighted linear threshold) of the weak hypotheses $\mathcal{F}_1, \mathcal{F}_2, \ldots, \mathcal{F}_T$. That is, for a given input vector $\mathbf{x}$, the final hypothesis $\mathcal{F}_{\text{fin}}$ outputs the label d that maximizes the sum of the weights of the weak hypotheses predicting that label. The weight of hypothesis $\mathcal{F}_n$ is defined to be $\log(1/\beta_n)$, so that greater weight is given to hypotheses with lower error.

The important theoretical property of AdaBoost is stated in the following theorem (Freund and Schapire, 1996a):

Suppose the weak learning model, when called by AdaBoost, generates hypotheses with errors $\epsilon_1, \epsilon_2, \ldots, \epsilon_T$, where the error ϵ_n on iteration n of the AdaBoost algorithm is defined by

$$\epsilon_n = \sum_{i : \mathcal{F}_n(\mathbf{x}_i) \neq d_i} \mathcal{D}_n(i)$$

Assume that $\epsilon_n \leq 1/2$, and let $\gamma_n = 1/2 - \epsilon_n$. Then the following upper bound holds on the error of the final hypothesis:

$$\frac{1}{N} |\{i : \mathcal{F}_{\text{fin}}(\mathbf{x}_i) \neq d_i\}| \leq \prod_{n=1}^{T} \sqrt{1 - 4\gamma_n^2} \leq \exp\left(-2 \sum_{n=1}^{T} \gamma_n^2\right) \tag{7.16}$$

TABLE 7.2 Summary of AdaBoost

Input: Training sample $\{(\mathbf{x}_i, d_i)\}_{i=1}^{N}$

                                  Distribution $\mathscr{D}$ over the N labeled examples

                                  Weak learning model

                                  Integer T specifying the number of iterations of the algorithm

Initialization: Set $\mathscr{D}_1(i)=1/N$ for all i

Computation: Do the following for $n = 1, 2, \ldots, T$:

      1.   Call the weak learning model, providing it with the distribution $\mathscr{D}_n$.

      2.   Get back hypothesis $\mathscr{F}_n : \mathbf{X} \to Y$

      3.   Calculate the error of hypothesis $\mathscr{F}_n$:

$$\epsilon_n = \sum_{i:\mathscr{F}_n(\mathbf{x}_i) \neq d_i} \mathscr{D}_n(i)$$

      4.   Set $\beta_n = \dfrac{\epsilon_n}{1 - \epsilon_n}$

      5.   Update the distribution $\mathscr{D}_n$:

$$\mathscr{D}_{n+1}(i) = \frac{\mathscr{D}_n(i)}{Z_n} \times \begin{cases} \beta_n & \text{if } \mathscr{F}_n(\mathbf{x}_i) = d_i \\ 1 & \text{otherwise} \end{cases}$$

               where Z_n is a normalization constant (chosen so that $\mathscr{D}_{n+1}(i)$ is a probability distribution).

Output: The final hypothesis is

$$\mathscr{F}_n(\mathbf{x}) = \arg\max_{d \in \mathscr{D}} \sum_{n:\mathscr{F}_n(\mathbf{x})=d} \log \frac{1}{\beta_n}$$

This theorem shows that if the weak hypotheses constructed by the weak learning model consistently have error only slightly better than $1/2$, then the training error of the final hypothesis $\mathscr{F}_{\text{fin}}$ drops to zero exponentially fast. However, this does not mean that the generalization error on test data is necessarily small. Experiments presented in Freund and Schapire (1996a) indicate two things. First, the theoretical bound on the training error is often weak. Second, the generalization error tends to be much better than what the theory would suggest.

Table 7.2 presents a summary of AdaBoost for a binary classification problem.

When the number of possible classes (labels) is $M > 2$, the boosting problem becomes more intricate because the probability that random guessing gives the correct label is $1/M$, which is now smaller than $1/2$. For boosting to be able to use any hypothesis that is slightly better than random guessing in such a situation, we need to somehow change the algorithm and the definition of what is meant to be a "weakly learning" algorithm. Ways of invoking that change are described in Freund and Schapire (1997) and Schapire (1997).

Error Performance

Experiments with AdaBoost reported in Breiman (1996b) show that when the training error and test error are plotted as a function of the number of boosting iterations, we often find that the test error continues to decrease after the training error has reduced

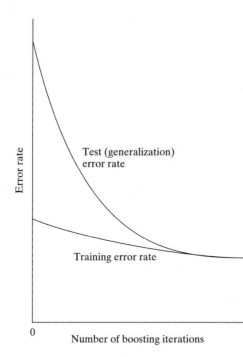

Test (generalization)
error rate

Training error rate

Error rate

0   Number of boosting iterations

FIGURE 7.4 Conceptualized error performance of the AdaBoost algorithm.

essentially to zero. This phenomenon is illustrated in Fig. 7.4. A similar result was reported earlier in Drucker et al. (1994) for boosting by filtering.

The phenomenon displayed in Fig. 7.4 is very surprising in light of what we know about the generalization performance of a single neural network. From Chapter 4 we recall that in the case of a multilayer perceptron trained with the back-propagation algorithm, the error on test (validation) data decreases, reaches a minimum, and then increases due to overfitting; see Fig. 4.20. The behavior displayed in Fig. 7.4 is remarkably different in that as the networks become more and more complex through increased training, the generalization error continues to decrease. Such a phenomenon appears to contradict *Occam's razor*, which states that a learning machine should be as simple as possible in order to achieve a good generalization performance.

In Schapire et al. (1997), an explanation is given for this phenomenon as it applies to AdaBoost. The key idea of the analysis presented therein is that in evaluating the generalization error produced by a boosting machine, not only the training error should be considered but also the *confidence* of classifications. The analysis presented therein reveals a relation between boosting and support vector machines; support vector machines are considered in the previous chapter. In particular, the classification *margin*, for example, is defined as the difference between the weight assigned to the correct label pertaining to that example and the maximal weight assigned to any single incorrect label. From this definition, it is easy to see that the margin is a number in the range $[-1, 1]$ and that an example is correctly classified if and only if its margin is positive. Thus Schapire et al. show that the phenomenon observed in Fig. 7.4 is indeed related to the distribution of margins of the training examples with respect to the generated voting classification error. It should be emphasized again that the margin analysis presented in Schapire et al. (1997) is specific to AdaBoost and does not apply to other boosting algorithms.

7.5 COMPUTER EXPERIMENT II

In this experiment we explore the boosting by filtering algorithm to solve a fairly difficult pattern classification task. The classification problem is two-dimensional, involving nonconvex decision regions, as shown in Fig. 7.5. One class of patterns consists of data points lying inside the region labeled $\mathscr{C}_1$, and the other class of patterns consists of data points inside the region labeled $\mathscr{C}_2$. The requirement is to design a committee machine that decides whether a test pattern belongs to class $\mathscr{C}_1$ or to class $\mathscr{C}_2$.

The committee machine used to solve this problem consists of three experts. Each expert consists of a 2-5-2 multilayer perceptron that has two input nodes, five hidden neurons, and two output neurons. The back-propagation algorithm was used to perform the training. Figure 7.6 displays the scatter plots of the data used to train the three experts. The data shown in Fig. 7.6a were used to train expert 1. The data shown in Fig. 7.6b were filtered by expert 1 after its training was completed; this data set was used to train expert 2. The data shown in Fig. 7.6c were filtered by the combined action of trained experts 1 and 2; this data set was used to train expert 3. The size of training sample for each expert was $N_1 = 1000$ patterns. Examining these three figures we observe:

- The training data for expert 1 in Fig. 7.6a are uniformly distributed.
- The training data for expert 2 in Fig. 7.6b exhibit concentrations of data points in areas labeled A and B that are seemingly difficult for the first expert to classify. The number of data points in these two regions is equal to the number of the correctly classified points.

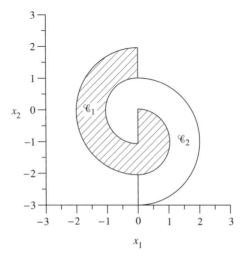

FIGURE 7.5 Pattern configurations for experiment on boosting.

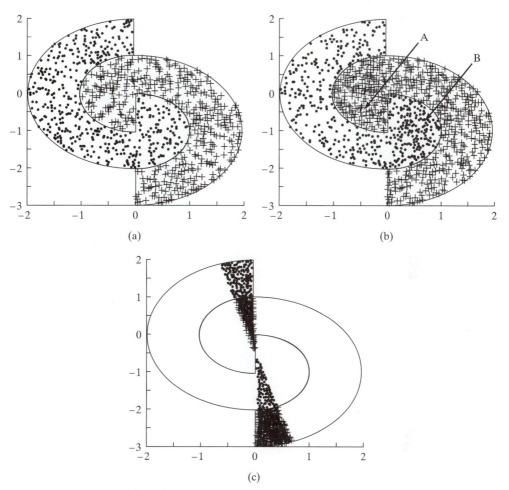

FIGURE 7.6 Scatter plots for expert training in computer experiment on boosting: (a) Expert 1. (b) Expert 2. (c) Expert 3.

- The training data for expert 3 in Fig. 7.6c exhibit an even greater concentration of data points seemingly difficult for both experts 1 and 2 to classify.

Figures 7.7a, 7.7b, and 7.7c display the decision boundaries formed by experts 1, 2, and 3, respectively. Figure 7.7d displays the overall decision boundary formed by the combined action of all three experts, which is obtained by simply summing their individual outputs. Note that the difference between the decision regions of Figs. 7.7a and 7.7b pertaining to experts 1 and 2 defines the distribution of data points in Fig. 7.6c used to train expert 3.

The probabilities of correct classification for the three experts on test data were:

Expert 1 :   75.15 percent
Expert 2 :   71.44 percent
Expert 3 :   68.90 percent

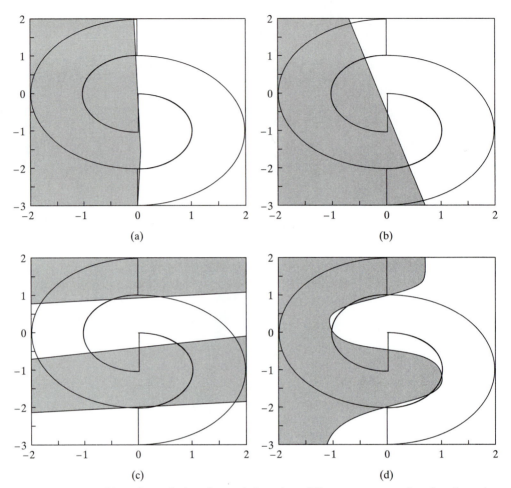

FIGURE 7.7 Decision boundaries formed by the different experts in the boosting experiment. (a) Expert 1. (b) Expert 2. (c) Expert 3. (d) Entire committee machine.

The overall probability of correct classification for the entire committee machine was 91.79 percent, which was computed using 32,000 patterns for test data. The overall decision boundary constructed by the boosting algorithm for the three experts shown in Fig. 7.7d is further evidence for this good classification performance.

7.6 ASSOCIATIVE GAUSSIAN MIXTURE MODEL

In the second part of the chapter, beginning with this section, we study the second class of committee machines, namely dynamic structures. The term "dynamic" is used here in the sense that integration of knowledge acquired by the experts is accomplished under the action of the input signal.

To begin the discussion, consider a modular network in which the learning process proceeds by fusing self-organized and supervised forms of learning in a seamless fashion. The experts are technically performing supervised learning in that their individual outputs are combined to model the desired response. There is, however, a sense in which the experts are also performing self-organized learning; that is, they self-organize to find a good partitioning of the input space so that each expert does well at modeling its own subspace, and as a whole group they model the input space well.

In the learning scheme just described, there is a point of departure from the schemes considered in the previous three chapters in that a specific model is assumed for generating the training data.

Probabilistic Generative Model

To fix ideas, consider a *regression* problem in which a regressor $\mathbf{x}$ produces a response denoted by random variable D; a realization of this random variable is denoted by d. Without loss of generality, we have adopted a scalar form of regression, merely to simplify the presentation. Specifically, we assume that the generation of response d is governed by the following probabilistic model (Jordan and Jacobs, 1995):

1. An input vector $\mathbf{x}$ is picked at random from some prior distribution.
2. A particular rule, say the kth rule, is selected in accordance with the conditional probability $P(k|\mathbf{x}, \mathbf{a}^{(0)})$, given $\mathbf{x}$ and some parameter vector $\mathbf{a}^{(0)}$.
3. For rule k, $k = 1, 2, ..., K$, the model response d is linear in $\mathbf{x}$, with an additive error ϵ_k modeled as a Gaussian distributed random variable with zero mean and unit variance:

$$E[\epsilon_k] = 0 \qquad \text{for all } k \tag{7.17}$$

and

$$\text{var}[\epsilon_k] = 1 \qquad \text{for all } k \tag{7.18}$$

Under point 3, the unity variance assumption is made just for didactic simplicity. In general, each expert can have a different output variance that can be learned from the training data.

The probabilistic generation of D is determined by the conditional probability $P(D = d|\mathbf{x}, \mathbf{w}_k^{(0)})$, given $\mathbf{x}$ and some parameter vector $\mathbf{w}_k^{(0)}$, for $k = 1, 2, ..., K$. We do not require that the probabilistic generative model just described must have a direct correspondence to a physical reality. Rather, we simply require that the probabilistic decisions embodied therein represent an *abstract* model, which with increasing precision specifies the location of the *conditional mean of response d on a nonlinear manifold* that relates the input vector to mean output (Jordan, 1994).

According to this model, the response D can be generated in K different ways, corresponding to the K choices of label k. Thus, the conditional probability of generating response $D = d$, given input vector $\mathbf{x}$, is equal to

$$P(D = d|\mathbf{x}, \boldsymbol{\theta}^{(0)}) = \sum_{k=1}^{K} P(D = d|\mathbf{x}, \mathbf{w}_k^{(0)}) P(k|\mathbf{x}, \mathbf{a}^{(0)}) \tag{7.19}$$

where $\boldsymbol{\theta}^{(0)}$ is the *generative model parameter vector* denoting the combination of $\mathbf{a}^{(0)}$ and $\{\mathbf{w}_k^{(0)}\}_{k=1}^{K}$ The superscript 0 in $\mathbf{a}^{(0)}$ and $\mathbf{w}_k^{(0)}$ is intended to distinguish the generative model parameters from those of the mixture of experts model considered next.

Mixture of Experts Model

Consider the network configuration of Fig. 7.8, referred to as a *mixture of experts (ME) model*.[4] Specifically, it consists of K supervised modules called *expert networks* or simply *experts,* and an integrating unit called a *gating network* that performs the function of a mediator among the expert networks. It is assumed here that the different experts work best in different regions of the input space in accordance with the probabilistic generative model described, hence the need for the gating network.

With the regression problem assumed to be scalar, each expert network consists of a linear filter. Figure 7.9 shows the signal-flow graph of a single neuron constituting expert k. Thus, the output produced by expert k is the inner product of the input vector $\mathbf{x}$ and synaptic weight vector $\mathbf{w}_k$ of this neuron, as shown by

$$y_k = \mathbf{w}_k^T \mathbf{x}, \qquad k = 1, 2, \ldots, K \tag{7.20}$$

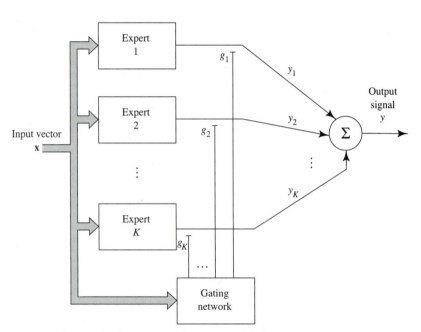

FIGURE 7.8 Block diagram of the ME model; the scalar outputs of the experts are mediated by a gating network.

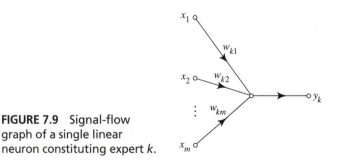

FIGURE 7.9 Signal-flow graph of a single linear neuron constituting expert k.

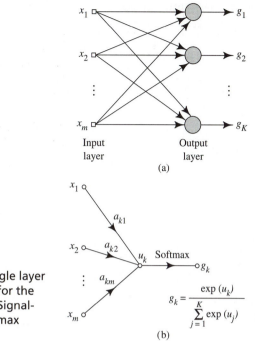

FIGURE 7.10 (a) Single layer of softmax neurons for the gating network. (b) Signal-flow graph of a softmax neuron.

The gating network consists of a single layer of K neurons, with each neuron assigned to a specific expert. Figure 7.10a shows the architectural graph of the gating network, and Fig. 7.10b shows the signal-flow graph of neuron k in that network. Unlike the experts, the neurons of the gating network are nonlinear, with their activation functions defined by

$$g_k = \frac{\exp(u_k)}{\displaystyle\sum_{j=1}^{K} \exp(u_j)}, \qquad k = 1, 2, ..., K \tag{7.21}$$

where u_k is the inner product of the input vector $\mathbf{x}$ and synaptic weight vector $\mathbf{a}_k$; that is,

$$u_k = \mathbf{a}_k^T \mathbf{x}, \qquad k = 1, 2, ..., K \tag{7.22}$$

The "normalized" exponential transformation of Eq. (7.21) may be viewed as a multi-input generalization of the logistic function. It preserves the rank order of its input values, and is a differentiable generalization of the "winner-takes-all" operation of picking the maximum value. For this reason, the activation function of Eq. (7.21) is referred to as *softmax* (Bridle, 1990a). Note that the linear dependence of u_k on the input $\mathbf{x}$ makes the outputs of the gating network nonlinear functions of $\mathbf{x}$.

For a probabilistic interpretation of the gating network's role, we may view it as a "classifier" that maps the input vector $\mathbf{x}$ into *multinomial probabilities* so that the different experts will be able to match the desired response (Jordan and Jacobs, 1995).

Most importantly, the use of softmax as the activation function for the gating network ensures that these probabilities satisfy the following requirements:

$$0 \leq g_k \leq 1 \qquad \text{for all } k \tag{7.23}$$

and

$$\sum_{k=1}^{K} g_k = 1 \tag{7.24}$$

Let y_k denote the output of the kth expert in response to the input vector $\mathbf{x}$. The overall output of the ME model is

$$y = \sum_{k=1}^{K} g_k y_k \tag{7.25}$$

where, as pointed out earlier, g_k is a nonlinear function of $\mathbf{x}$. Given that rule k of the probabilistic model is selected and the input is $\mathbf{x}$, an individual output y_k is treated as the conditional mean of the random variable D, as shown by

$$\begin{aligned} E[D \mid \mathbf{x}, k] &= y_k \\ &= \mathbf{w}_k^T \mathbf{x}, \qquad k = 1, 2, \ldots, K \end{aligned} \tag{7.26}$$

With μ_k denoting the conditional mean of D, we may write

$$\mu_k = y_k, \qquad k = 1, 2, \ldots, K \tag{7.27}$$

The variance of D is the same as that of the error ϵ_k. Thus, invoking the use of Eq. (7.18), we may write

$$\text{var}[D \mid \mathbf{x}, k] = 1, \qquad k = 1, 2, \ldots, K \tag{7.28}$$

The probability density function of D, given the input vector $\mathbf{x}$ and given that the kth rule of the probabilistic generative model (i.e., expert k) is selected, may therefore be described as:

$$f_D(d \mid \mathbf{x}, k, \boldsymbol{\theta}) = \frac{1}{\sqrt{2\pi}} \exp\left(-\frac{1}{2}(d - y_k)^2\right), \qquad k = 1, 2, \ldots, K \tag{7.29}$$

where $\boldsymbol{\theta}$ is a parameter vector denoting the parameters of the gating network and those of the experts in the ME model. The probability density function of D, given $\mathbf{x}$, is the *mixture* of the probability density functions $\{f_D(d \mid \mathbf{x}, k, \boldsymbol{\theta})\}_{k=1}^{K}$, with the mixing parameters being the multinominal probabilities determined by the gating network. We may thus write

$$\begin{aligned} f_D(d \mid \mathbf{x}, \boldsymbol{\theta}) &= \sum_{k=1}^{K} g_k f_D(d \mid \mathbf{x}, k, \boldsymbol{\theta}) \\ &= \frac{1}{\sqrt{2\pi}} \sum_{k=1}^{K} g_k \exp\left(-\frac{1}{2}(d - y_k)^2\right) \end{aligned} \tag{7.30}$$

The probability distribution of Eq. (7.30) is called an *associative Gaussian mixture model*. Its nonassociative counterpart is the traditional Gaussian mixture model (Titterington et al., 1985; McLachlan and Basford, 1988), which is briefly described in

Chapter 5. An associative model differs from a nonassociative model in that the conditional means μ_k and mixing parameters g_k are *not* fixed; rather, they are all functions of the input vector **x**. The associative Gaussian mixture model of Eq. (7.30) may therefore be viewed as a generalization of the traditional Gaussian mixture model.

The important aspects of the ME model shown in Fig. 7.8, assuming that it is properly tuned through training, are :

1. The output y_k of the *k*th expert provides an estimate of the conditional mean of the random variable representing the desired response *D*, given **x** and that rule *k* of the probabilistic generative model holds.
2. The output g_k of the gating network defines the multinomial probability that the output of expert *k* matches the value $D = d$, on the basis of knowledge gained from **x** alone.

Working with the probability distribution of Eq. (7.30) and given the training sample $\{(\mathbf{x}_i, d_i)\}_{i=1}^{N}$, the problem is to *learn* the conditional means $\mu_k = y_k$ and the mixing parameters g_k, $k = 1, 2, \ldots, K$, in an optimum manner, so that $f_D(d|\mathbf{x}, \boldsymbol{\theta})$ provides a good estimate of the underlying probability density function of the environment responsible for generating the training data.

Example 7.1   Regression Surface

Consider an ME model with two experts, and a gating network with two outputs denoted by g_1 and g_2. The output g_1 is defined by (see Eq. (7.21))

$$g_1 = \frac{\exp(u_1)}{\exp(u_1) + \exp(u_2)}$$

$$= \frac{1}{1 + \exp(-(u_1 - u_2))} \tag{7.31}$$

Let $\mathbf{a}_1$ and $\mathbf{a}_2$ denote the two weight vectors of the gating network. We may then write

$$u_k = \mathbf{x}^T\mathbf{a}_k, \qquad k = 1, 2$$

and therefore rewrite Eq. (7.31) as:

$$g_1 = \frac{1}{1 + \exp(-\mathbf{x}^T(\mathbf{a}_1 - \mathbf{a}_2))} \tag{7.32}$$

The other output g_2 of the gating network is

$$g_2 = 1 - g_1$$

$$= \frac{1}{1 + \exp(-\mathbf{x}^T(\mathbf{a}_2 - \mathbf{a}_1))}$$

Thus, both g_1 and g_2 are in the form of a logistic function, but with a difference. The orientation of g_1 is determined by the direction of the difference vector $(\mathbf{a}_1 - \mathbf{a}_2)$, whereas the orientation of g_2 is determined by the direction of the difference vector $(\mathbf{a}_2 - \mathbf{a}_1)$, that is the negative of that for gate g_1. Along the *ridge* defined by $\mathbf{a}_1 = \mathbf{a}_2$, we have $g_1 = g_2 = 1/2$, and the two experts contribute equally to the output of the ME model. Away from the ridge, one or the other of the two experts assumes the dominant role.

■

7.7 HIERARCHICAL MIXTURE OF EXPERTS MODEL

The ME model of Fig. 7.8 works by dividing the input space into different subspaces, with a single gating network responsible for distributing information (gathered from the training data) to the various experts. The *hierarchical mixtures of experts (HME) model,* illustrated in Fig. 7.11, is a natural extension of the ME model. The illustration is for an HME model of four experts. The architecture of the HME model is like a *tree,* in which the gating networks sit at the various nonterminals of the tree and the experts sit at the leaves of the tree. The HME model differs from the ME model in that the input space is divided into a *nested* set of subspaces, with the information combined and redistributed among the experts under the control of several gating networks arranged in a *hierarchical* manner.

The HME model of Fig. 7.11 has two *levels of hierarchy* or two *layers of gating networks.* By continuing with the application of the principle of divide and conquer in a manner similar to that illustrated, we may construct an HME model with any number of levels of hierarchy. Note that according to the convention described in Fig. 7.11, the numbering of gating *levels* starts from the output node of the tree.

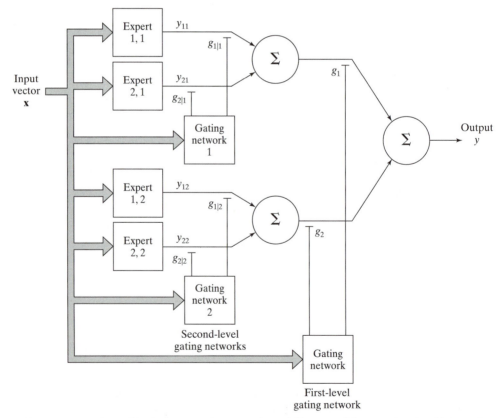

FIGURE 7.11 Hierarchical mixture of experts (HME) illustrated for two levels of hierarchy.

The formulation of the HME model of Fig. 7.11 can be viewed in two ways (Jordan, 1994):

1. *The HME model is a product of the divide and conquer strategy.* If we believe that it is a good strategy to divide the input space into regions, then it is an equally good strategy to divide each of those regions into subregions. We can continue recursively in this manner until we reach a stage where the complexity of the approximating surfaces is a good fit to the "local" complexity of the training data. The HME model should therefore perform at least as well as, and often better than, the ME model for this reason: A higher level gating network in the HME model effectively combines information and redistributes it to the experts in the particular subtree controlled by that gating network. Consequently, each parameter in the subtree in question shares strength with other parameters contained in the same subtree, thereby contributing to possible improvement in the overall performance of the HME model.

2. *The HME model is a soft-decision tree.* According to this second viewpoint, the mixture of experts is just a one-level decision tree, sometimes referred to as a *decision stump*. In a more general setting, the HME model is viewed as the probabilistic framework for a decision tree, with the output node of the HME model referred to as the *root* of the tree. The methodology of a *standard* decision tree constructs a tree that leads to *hard* (e.g., yes-no) decisions in different regions of the input space. This is to be contrasted with the soft decisions performed by an HME model. Consequently, the HME model may outperform the standard decision tree for two reasons:
   - A hard decision inevitably results in loss of information, whereas a soft decision tries to preserve information. For example, a soft binary decision conveys information about distance from the decision boundry (i.e., the point at which the decision is 0.5), whereas a hard decision cannot. We may therefore say that unlike the standard decision tree, the HME model adheres to the *information preservation rule*. This empirical rule states that the information content of an input signal should be preserved in a computationally efficient manner until the system is ready for final decision-making or parameter estimation (Haykin, 1996).
   - Standard decision trees suffer from a *greediness* problem. Once a decision is made in such a tree, it is frozen and never changes thereafter. The HME model lessens the greediness problem because the decisions made throughout the tree are continually altered. Unlike the standard decision tree, it *is* possible in the HME model to recover from a poor decision somewhere further along the tree.

The second viewpoint, that is, a soft decision tree is the preferred way to think about an HME model. With the HME viewed as the probabilistic basis for a decision tree, it allows us to calculate a likelihood for any given data set, and maximize that likelihood with respect to the parameters that determine the splits between various regions in the input space. Thus, by building on what we already know about standard decision trees, we may have a practical solution to the model selection problem as discussed in the next section.

7.8 MODEL SELECTION USING A STANDARD DECISION TREE

As with every other neural network, a satisfactory solution to the parameter estimation problem hinges on the selection of a suitable model for the problem at hand. In the case of an HME model, the model selection involves the choice of the number and arrangement of the decision nodes in the tree. One practical solution to this particular model selection problem is to run a standard decision tree algorithm on the training data, and to adopt the tree so obtained as the initializing step for the learning algorithm used to determine the parameters of the HME model (Jordan, 1994).

The HME model has clear similarities with standard decision trees, such as the *classification and regression tree* (CART) due to Breiman et al., (1984). Figure 7.12 shows an example of CART, where the space of input data, $\mathcal{X}$ is repeatedly partitioned by a sequence of binary *splits* into *terminal nodes*. Comparing Fig. 7.12 with Fig. 7.11, we readily see the following similarities between CART and HME:

- The rules for selecting splits at intermediate (i.e., nonterminal) nodes of CART play an analogous role to gating networks in the HME model.
- The terminal nodes in CART play an analogous role to expert networks in the HME model.

By starting with CART for a classification or regression problem of interest, we take advantage of the *discrete* nature of CART to provide an efficient search among alternative trees. By using a tree so chosen as the initializing step in the learning algorithm for parameter estimation, we take advantage of the *continuous* probabilistic basis of the HME model to yield an improved "soft" estimate of the desired response.

The CART Algorithm

In light of what has just been said, a brief description of the CART algorithm is in order. The description is presented in the context of regression. Starting with the training data $\{(\mathbf{x}_i, d_i)\}_{i=1}^N$, we may use CART to construct a binary tree T for least-squares regression, by proceeding as follows (Breiman et al., 1984):

1. *Selection of splits.* Let a node t denote a subset of the current tree T. Let $\bar{d}(t)$ denote the average of d_i for all cases $(\mathbf{x}_i, d_i)$ falling into t, that is,

$$\bar{d}(t) = \frac{1}{N(t)} \sum_{\mathbf{x}_i \in t} d_i \tag{7.33}$$

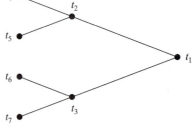

FIGURE 7.12 Binary decision tree, described as follows:
- Nodes t_2 and t_3 are *descendents* of node t_1.
- Nodes t_4 and t_5 are descendents of node t_2; and likewise for t_6 and t_7 in relation to t_3.

where the sum is over all d_i such that $\mathbf{x}_i \in t$ and $N(t)$ is the total number of cases in t. Define

$$\mathcal{E}(t) = \frac{1}{N} \sum_{\mathbf{x}_i \in t} (d_i - \bar{d}(t))^2 \tag{7.34}$$

and

$$\mathcal{E}(T) = \sum_{t \in T} \mathcal{E}(t) \tag{7.35}$$

For node t, the sum $\sum_{\mathbf{x}_i \in t} (d_i - \bar{d}(t))^2$ represents the "within node sum of squares"; that is, it is the total squared deviations of all the d_i in t from their average $\bar{d}(t)$. Summing these deviations over $t \in T$ gives the total node sum of squares and dividing it by N gives the average.

Given any set of splits S of a current node t in T, the best split $s*$ is that split in S that most decreases $\mathcal{E}(T)$. To be more precise, suppose that for any split s of node t into t_L (a new node to the left of t) and t_R (another new node to the right of t), we let

$$\Delta\mathcal{E}(s,t) = \mathcal{E}(T) - \mathcal{E}(t_L) - \mathcal{E}(t_R) \tag{7.36}$$

The best split $s*$ is then taken to be the particular split for which we have

$$\Delta\mathcal{E}(s*,t) = \max_{s \in S} \Delta\mathcal{E}(t,s) \tag{7.37}$$

A regression tree so constructed is designed to maximize the decrease in $\mathcal{E}(T)$.

2. *Determination of a terminal node.* A node t is declared a terminal node if this condition is satisfied:

$$\max_{s \in S} \Delta\mathcal{E}(s,t) < \beta \tag{7.38}$$

where β is a prescribed *threshold*.

3. *Least-squares estimation of a terminal node's parameters.* Let t denote a terminal node in the final binary tree T, and let $\mathbf{X}(t)$ denote the matrix composed of $\mathbf{x}_i \in t$. Let $\mathbf{d}(t)$ denote the corresponding vector composed of all the d_i in t. Define

$$\mathbf{w}(t) = \mathbf{X}^+(t)\mathbf{d}(t) \tag{7.39}$$

where $\mathbf{X}^+(t)$ is the pseudoinverse of matrix $\mathbf{X}(t)$. The use of $\mathbf{w}(t)$ yields a least-squares estimate of $d(t)$ at the output of terminal node t. Using the weights calculated from Eq. (7.39), the split selection problem is solved by looking for the least sum of squared residuals (errors) with respect to the regression surfaces, rather than with respect to the means.

Using CART to Initialize the HME Model

Suppose that the CART algorithm has been applied to a set of training data, resulting in a binary decision tree for this problem. We may describe a split produced by CART as a multidimensional surface defined by

$$\mathbf{a}^T\mathbf{x} + b = 0$$

where $\mathbf{x}$ is the input vector, $\mathbf{a}$ denotes a parameter vector, and b denotes a bias.

Consider next the corresponding situation in an HME model. From Example 7.1 we note that the regression surface produced by a gating network in a binary tree may be expressed as:

$$g = \frac{1}{1 + \exp\left(-(\mathbf{a}^T \mathbf{x} + b)\right)} \tag{7.40}$$

which defines a split, particularly when $g = 1/2$. Let the weight vector (difference) $\mathbf{a}$ for this particular gating network be written as

$$\mathbf{a} = \|\mathbf{a}\| \cdot \frac{\mathbf{a}}{\|\mathbf{a}\|} \tag{7.41}$$

where $\|\mathbf{a}\|$ denotes the length (i.e., Euclidean norm) of $\mathbf{a}$, and $\mathbf{a}/\|\mathbf{a}\|$ is a normalized unit-length vector. Using Eq. (7.41) in (7.40), we may thus rewrite a parameterized split at a gating network as:

$$g = \frac{1}{1 + \exp\left(-\|\mathbf{a}\|\left(\left(\dfrac{\mathbf{a}}{\|\mathbf{a}\|}\right)^T \mathbf{x} + \dfrac{b}{\|\mathbf{a}\|}\right)\right)} \tag{7.42}$$

where we see that $\mathbf{a}/\|\mathbf{a}\|$ determines the *direction* of the split and $\|\mathbf{a}\|$ determines the sharpness of the split. From the discussion presented in Chapter 2, we observe that the length of vector $\mathbf{a}$ acts effectively as the *reciprocal of temperature*. The important point to note from Eq. (7.42) is that a gating network made up of a linear filter followed by a softmax form of nonlinearity is able to mimic a split in the style of CART. Moreover, we have an additional degree of freedom, namely, the length of parameter vector $\mathbf{a}$. In a standard decision tree, this additional parameter is irrelevant because a threshold (i.e., hard decision) is used to create a split. In contrast, the length of $\mathbf{a}$ has a profound influence on the split sharpness produced by a gating network in the HME model. Specifically, for a synaptic weight vector $\mathbf{a}$ of fixed direction, we may state that:

- when $\mathbf{a}$ is long (i.e., the temperature is low), the split is sharp, and
- when $\mathbf{a}$ is short (i.e., the temperature is high), the split is soft.

If, in the limit we have $\|\mathbf{a}\| = 0$, the split vanishes and $g = 1/2$ on both sides of the vanished (fictitious) split. The effect of setting $\|\mathbf{a}\| = 0$ is equivalent to pruning the nonterminal node from the tree, because the gating network in question is no longer splitting. In the very extreme case when $\|\mathbf{a}\|$ is small (i.e., the temperature is high) at every nonterminal node, the entire HME model acts like a single node; that is, the HME is reduced to a linear regression model (assuming linear experts). As the synaptic weight vectors of the gating networks start to grow in length, the HME starts to make (soft) splits, thereby enlarging the number of degrees of freedom available to the model.

We may thus initialize the HME by proceeding as follows:

1. Apply CART to the training data.
2. Set the synaptic weight vectors of the experts in the HME model equal to the least-squares estimates of the parameter vectors at the corresponding terminal nodes of the binary tree resulting from the application of CART.

3. For the gating networks:
   (a) set the synaptic weight vectors to point in directions that are orthogonal to the corresponding splits in the binary tree obtained from CART, and
   (b) set the lengths (i.e., Euclidean norms) of the synaptic weight vectors equal to *small* random vectors.

7.9 A PRIORI AND A POSTERIORI PROBABILITIES

The multinomial probabilities g_k and $g_{j|k}$ pertaining to the first-level and second-level gating networks, respectively, may be viewed as *a priori* probabilities, in the sense that their values are solely dependent on the input vector (stimulus) $\mathbf{x}$. In a corresponding way, we may define *a posteriori* probabilities $h_{j|k}$ and h_k whose values depend on both the input vector $\mathbf{x}$ and the responses of the experts to $\mathbf{x}$. This latter set of probabilities is useful in the development of learning algorithms for HME models.

Referring to the HME model of Fig. 7.11, we define the *a posteriori* probabilities at the nonterminal nodes of the tree as (Jordan and Jacobs, 1994):

$$h_k = \frac{g_k \sum_{j=1}^{2} g_{j|k} \exp\left(-\frac{1}{2}(d - y_{jk})^2\right)}{\sum_{k=1}^{2} g_k \sum_{j=1}^{2} g_{j|k} \exp\left(-\frac{1}{2}(d - y_{jk})^2\right)} \tag{7.43}$$

and

$$h_{j|k} = \frac{g_{j|k} \exp\left(-\frac{1}{2}(d - y_{jk})^2\right)}{\sum_{j=1}^{2} g_{j|k} \exp\left(-\frac{1}{2}(d - y_{jk})^2\right)} \tag{7.44}$$

The product of h_k and $h_{j|k}$ defines the *joint a posteriori* probability that expert (j, k) produces an output y_{jk} that matches the desired response d, as given by

$$h_{jk} = h_k h_{j|k}$$

$$= \frac{g_k g_{j|k} \exp\left(-\frac{1}{2}(d - y_{jk})^2\right)}{\sum_{k=1}^{2} g_k \sum_{j=1}^{2} g_{j|k} \exp\left(-\frac{1}{2}(d - y_{jk})^2\right)} \tag{7.45}$$

The probability h_{jk} satisfies the following two conditions

$$0 \leq h_{jk} \leq 1 \qquad \text{for all } (j, k) \tag{7.46}$$

and

$$\sum_{j=1}^{2} \sum_{k=1}^{2} h_{jk} = 1 \tag{7.47}$$

The implication of Eq. (7.47) is that credit is distributed across the experts on a competitive basis. Moreover, we note from Eq. (7.45) that the closer y_{jk} is to d, the more likely it is that expert (j, k) is given credit for its output having matched d, which is intuitively satisfying.

An important feature of the HME model that deserves special mention is the *recursiveness* in the computations involved in calculating the *a posteriori* probabilities. In examining Eqs. (7.43) and (7.44), we see that the denominator of $h_{j|k}$ in Eq. (7.44) appears in the numerator of h_k in Eq. (7.43). In an HME model we want to calculate the *a posteriori* probability for every nonterminal node in the tree. This is where recursiveness is of particular value. Specifically, the calculation of the *a posteriori* probabilities of all the nonterminal nodes in the tree is achieved in a single pass as described here:

- In moving through the tree toward the root node, level by level, the *a posteriori* probability at any nonterminal node of the tree is obtained simply by combining the *a posteriori* probabilities of its "children".

7.10 MAXIMUM LIKELIHOOD ESTIMATION

Turning next to the issue of parameter estimation for the HME model, we first note that its probabilistic interpretation is somewhat different from that of the ME model. With the HME model formulated as a binary tree, it is assumed that the environment responsible for generating the data involves a *nested sequence of soft (binary) decisions, terminating on the regression of the input vector* $\mathbf{x}$ *onto the output d*. In particular, we assume that in the *probabilistic generative model* for the HME, the decisions are modeled as multinomial random variables (Jordan and Jacobs, 1994). That is, for each input $\mathbf{x}$ we interpret $g_i(\mathbf{x}, \boldsymbol{\theta}_i^0)$ as the multinomial probabilities associated with the first decision, and $g_{j|i}(\mathbf{x}, \boldsymbol{\theta}_{ji}^0)$ as the conditional multinomial distributions associated with the second decision. As before, the superscript 0 signifies true values of the generative model parameters. The decisions form a decision tree. As with the ME model, softmax is used as the activation function of the gating networks throughout the HME model. In particular, the activation g_k of the kth output neuron in the *top-level gating network* is defined by

$$g_k = \frac{\exp(u_k)}{\exp(u_1) + \exp(u_2)}, \qquad k = 1, 2 \qquad (7.48)$$

where u_k is the weighted sum of the inputs applied to that neuron. Similarly, the activation of the jth output neuron in the kth gating network in the second level of hierarchy is defined by

$$g_{j|k} = \frac{\exp(u_{jk})}{\exp(u_{1k}) + \exp(u_{2k})}, \qquad (j, k) = 1, 2 \qquad (7.49)$$

where u_{jk} is the weighted sum of the inputs applied to this particular neuron.

For the sake of presentation we will work with an HME model with only two levels of hierarchy (i.e., two layers of gating networks), as indicated in Fig. 7.11. As with the ME model, each of the experts in the HME model is assumed to consist of a single layer of linear neurons. Let y_{jk} denote the output of expert (j, k). We may then express the overall output of the HME model as

$$y = \sum_{k=1}^{2} g_k \sum_{j=1}^{2} g_{j|k} \, y_{jk} \tag{7.50}$$

Following a procedure similar to that described for the ME model in Section 7.6, we may formulate the probability density function of the random variable D representing the desired response for the HME model of Fig. 7.11, given the input $\mathbf{x}$, as follows:

$$f_D(d \,|\, \mathbf{x}, \boldsymbol{\theta}) = \frac{1}{\sqrt{2\pi}} \sum_{k=1}^{2} g_k \sum_{j=1}^{2} g_{j|k} \exp\left(-\frac{1}{2}(d - y_{jk})^2\right) \tag{7.51}$$

Thus, for a given set of training data, Eq. (7.51) defines a model for the underlying distribution of the data. The vector $\boldsymbol{\theta}$ encompasses all the synaptic weights involved in the characterization of the gating and expert networks of the HME model.

The designation *likelihood function*, denoted by $l(\boldsymbol{\theta})$, is given to the probability density function $f_D(d|\mathbf{x}, \boldsymbol{\theta})$, *viewed as a function of the parameter vector* $\boldsymbol{\theta}$. We thus write

$$l(\boldsymbol{\theta}) = f_D(d \,|\, \mathbf{x}, \boldsymbol{\theta}) \tag{7.52}$$

Although the conditional joint probability density function and the likelihood function have exactly the same formula, it is vital that we appreciate the physical difference between them. In the case of $f_D(d|\mathbf{x}, \boldsymbol{\theta})$, the input vector $\mathbf{x}$ and parameter vector $\boldsymbol{\theta}$ are fixed but the desired response d is variable. However, in the case of the likelihood function $l(\boldsymbol{\theta})$, both $\mathbf{x}$ and d are fixed but $\boldsymbol{\theta}$ is variable.

In practice, we find it more convenient to work with the natural logarithm of the likelihood function rather than the likelihood itself. Using $L(\boldsymbol{\theta})$ to denote the *log-likelihood function*, we write

$$\begin{aligned} L(\boldsymbol{\theta}) &= \log[l(\boldsymbol{\theta})] \\ &= \log[f_D(d \,|\, \mathbf{x}, \boldsymbol{\theta})] \end{aligned} \tag{7.53}$$

The logarithm of $l(\boldsymbol{\theta})$ is a monotonic transformation of $l(\boldsymbol{\theta})$. This means that whenever $l(\boldsymbol{\theta})$ increases, its logarithm $L(\boldsymbol{\theta})$ also increases. Since $l(\boldsymbol{\theta})$ is a formula for a conditional probability density function, it can never become negative. It follows therefore that there is no problem in evaluating $L(\boldsymbol{\theta})$. Hence, an estimate $\hat{\boldsymbol{\theta}}$ of parameter vector $\boldsymbol{\theta}$ can be obtained as a solution of the *likelihood equation*

$$\frac{\partial}{\partial \boldsymbol{\theta}} \, l(\boldsymbol{\theta}) = \mathbf{0}$$

or equivalently the *log-likelihood equation*

$$\frac{\partial}{\partial \boldsymbol{\theta}} \, L(\boldsymbol{\theta}) = \mathbf{0} \tag{7.54}$$

The term "maximum likelihood estimate" with the desired asymptotic properties[5] usually refers to a root of the likelihood equation that globally maximizes the likelihood function $l(\theta)$. The estimate $\hat{\theta}$ used in practice, however, may in actual fact be a local maximum and not a global maximum. In any event, maximum likelihood estimation, due to Fisher (1925), is based on a relatively simple idea:

> *Different populations generate different data samples and any given data sample is more likely to have come from some population than from some others.*

More specifically, the unknown parameter vector θ is estimated by its *most plausible value*, given the input vector $\mathbf{x}$. In other words, the maximum likelihood estimate $\hat{\theta}$ is that value of the parameter vector θ for which the conditional probability density function $f_D(d|\mathbf{x}, \theta)$ is maximum.

7.11 LEARNING STRATEGIES FOR THE HME MODEL

The probabilistic description of the HME model in Section 7.10 has led us to the log-likelihood function $L(\theta)$ as the objective function to be maximized. The key question is how to perform this maximization. As with every optimization problem, there is no unique approach to the maximization of $L(\theta)$. Rather, we have several approaches at our disposal, two of which are summarized here (Jacobs and Jordan, 1991; Jordan and Jacobs, 1994,):

1. *Stochastic gradient approach.* This approach yields an on-line algorithm for the maximization of $L(\theta)$. Its formulation for a two-level HME model, as depicted in Fig. 7.11, hinges on formulas for the following ingredients:

- The gradient vector $\partial L/\partial \mathbf{w}_{jk}$ for the vector of synaptic weights in expert (j, k).
- The gradient vector $\partial L/\partial \mathbf{a}_k$ for the vector of synaptic weights in output neuron k of the top-level gating network.
- The gradient vector $\partial L/\partial \mathbf{a}_{jk}$ for the vector of synaptic weights in the output neuron of the second-level gating network associated with expert (j, k).

It is a straightforward matter to show that (see Problem 7.9):

$$\frac{\partial L}{\partial \mathbf{w}_{jk}} = h_{j|k}(n)\, h_k(n)(d(n) - y_{jk}(n))\mathbf{x}(n) \tag{7.55}$$

$$\frac{\partial L}{\partial \mathbf{a}_k} = (h_k(n) - g_k(n))\mathbf{x}(n) \tag{7.56}$$

$$\frac{\partial L}{\partial \mathbf{a}_{jk}} = h_k(n)(h_{j|k}(n) - g_{j|k}(n))\mathbf{x}(n) \tag{7.57}$$

Equation (7.55) states that during the training process, the synaptic weights of expert (j, k) are adjusted to correct for the error between the output y_{jk} and the desired response d, in proportion to the joint *a posteriori* probability h_{jk} that expert (j, k) produces an output that matches d. Equation (7.56) states that the synaptic weights of output neuron k in the top-level gating network are adjusted so as to force the *a priori* probabilities $g_k(n)$ to move toward the corresponding *a posteriori* probabilities $h_k(n)$. Equation (7.57) states that the synaptic weights of the output neuron of the second-level gating network associated with expert (j, k) are adjusted to correct for the error between the *a priori* probability $g_{j|k}$ and the corresponding *a posteriori probability $h_{j|k}$* in proportion to the *a posteriori* probability $h_k(n)$.

According to Eqs. (7.55) to (7.57), the synaptic weights of the HME model are updated after the presentation of each pattern (stimulus). By summing the gradient vectors shown here over n, we may formulate the batch version of the gradient ascent algorithm for maximizing the log-likelihood function $L(\theta)$.

2. *Expectation-maximization approach.* The *expectation-maximization (EM) algorithm,* due to Dempster et al. (1977), provides an iterative procedure for computing maximum likelihood estimates in situations where, except for some missing data, the issue of maximum likelihood estimation would be a straightforward matter. The EM algorithm derives its name from the fact that on each iteration of the algorithm there are two steps:

- *Expectation step* or *E-step,* which uses the observed data set of an *incomplete data problem* and the current value of the parameter vector to manufacture data so as to postulate an augmented or so-called *complete data* set.
- *Maximization step* or *M-step,* which consists of deriving a new estimate of the parameter vector by maximizing the log-likelihood function of the complete data manufactured in the E-step.

Thus, starting from a suitable value for the parameter vector, the E-step and M-step are repeated on an alternate basis until convergence.

The situations where the EM algorithm is applicable include not only those that involve naturally incomplete data, but also a variety of other situations where the incompleteness of data is not at all evident in or natural to the problem of interest. Indeed, computation of the maximum likelihood estimate is often greatly facilitated by artificially formulating it as an incomplete data problem. This is done because the EM algorithm is able to exploit the reduced complexity of the maximum likelihood estimation, given the complete data (McLachlan and Krishnan, 1997). The HME model is one such example application. In this case, missing data in the form of certain indicator variables are artificially introduced into the HME model to facilitate the maximum likelihood estimation of the unknown parameter vector, as described in Section 7.12.

An important feature of the HME model, whether designed using the stochastic gradient approach or the EM algorithm, is two-fold:

- Each gating network in the model is continually computing the *a posteriori* probability for every data point in the training set.
- The adjustments applied to the synaptic weights of the expert and gating networks in the model, from one iteration to the next, are functions of the *a posteriori* probability thus computed and the corresponding *a priori* probability.

Accordingly, if an expert network lower down in the tree fails to do a good job in fitting the training data in its local neighborhood, the regression (discriminant) surface of a gating network higher up in the tree will be moved around. This movement can, in turn, help the experts on the next iteration of the learning algorithm to fit the data better by shifting the subspaces in which they are supposed to do their data fitting. This is the process by which the HME model is able to ameliorate the greediness problem inherent to a standard decision tree like CART.

7.12 EM ALGORITHM

The EM algorithm is remarkable in part because of the simplicity and generality of the underlying theory, and in part because of the wide range of applications that fall under its umbrella.[6] In this section we present a description of the EM algorithm in a generic sense. We go on to consider its application to the parameter estimation problem in the HME model in the next section.

Let the vector z denote the missing or unobservable data. Let r denote the complete-data vector, made up of some observable data d and the missing data vector z. There are therefore two data spaces, $\mathcal{R}$ and $\mathcal{D}$, to be considered, with the mapping from $\mathcal{R}$ to $\mathcal{D}$ being many-to-one. However, instead of observing the complete data vector r, we are actually only able to observe the incomplete data $d = d(r)$ in $\mathcal{D}$.

Let $f_c(r|\theta)$ denote the conditional pdf of r, given a parameter vector θ. It follows therefore that the conditional pdf of random variable D, given θ, is defined by

$$f_D(d|\theta) = \int_{\mathcal{R}(d)} f_c(r|\theta)dr \tag{7.58}$$

where $\mathcal{R}(d)$ is the subspace of $\mathcal{R}$ that is determined by $d = d(r)$. The EM algorithm is directed at finding a value of θ that maximizes the *incomplete-data log-likelihood function*

$$L(\theta) = \log f_D(d|\theta)$$

This problem, however, is solved indirectly by working *iteratively* with the *complete-data log-likelihood function*

$$L_c(\theta) = \log f_c(r|\theta) \tag{7.59}$$

which is a random variable, because the missing data vector z is unknown.

To be more specific, let $\hat{\theta}(n)$ denote the value of the parameter vector θ on iteration n of the EM algorithm. In the E-step of this iteration, we calculate the expectation

$$Q(\theta,\hat{\theta}(n)) = E[L_c(\theta)] \tag{7.60}$$

where the expectation is performed with respect to $\hat{\theta}(n)$. In the M-step of this same iteration, we maximize $Q(\theta,\hat{\theta}(n))$ with respect to θ over the parameter (weight) space $\mathcal{W}$, and so find the updated parameter estimate $\hat{\theta}(n + 1)$, as shown by

$$\hat{\theta}(n + 1) = \arg\max_{\theta} Q(\theta,\hat{\theta}(n)) \tag{7.61}$$

The algorithm is started with some initial value $\hat{\theta}(0)$ of the parameter vector θ. The E-step and M-step are then alternately repeated in accordance with Eqs. (7.60) and (7.61), respectively, until the difference between $L(\hat{\theta}(n + 1))$ and $L(\hat{\theta}(n))$ drops to some arbitrary small value; at that point the computation is terminated.

Note that after an iteration of the EM algorithm, the incomplete-data log-likelihood function is *not* decreased, as shown by (see Problem 7.10)

$$L(\hat{\theta}(n + 1) \geq L\hat{\theta}(n)) \qquad \text{for } n = 0, 1, 2, ..., \tag{7.62}$$

Equality usually means that we are at a stationary point of the log-likelihood function.[7]

7.13 APPLICATION OF THE EM ALGORITHM TO THE HME MODEL

Having familiarized ourselves with the EM algorithm, we are now ready to solve the parameter estimation problem in the HME model using the EM algorithm.[8]

Let $g_k^{(i)}$ and $g_{j|k}^{(i)}$ denote the (conditional) multinomial probabilities associated with the decisions taken by the first-level gating network k and second-level gating network (j, k) of the HME model in Fig. 7.11, respectively, when it operates under example i of the training set. Then, from Eq. (7.51) we readily see that the corresponding value of the conditional pdf of the random variable D, given example $\mathbf{x}_i$ and parameter vector $\boldsymbol{\theta}$, is given by

$$f_D(d_i|\mathbf{x}_i,\boldsymbol{\theta}) = \frac{1}{\sqrt{2\pi}} \sum_{k=1}^{2} g_k^{(i)} \sum_{j=1}^{2} g_{j|k}^{(i)} \exp\left(-\frac{1}{2}(d_i - y_{jk}^{(i)})^2\right) \tag{7.63}$$

where $y_{jk}^{(i)}$ is the output produced by expert (j, k) in response to the ith example of the training set. Assuming that all the N examples contained in the training set are statistically independent, we may formulate the log-likelihood function for the incomplete data problem as follows:

$$L(\boldsymbol{\theta}) = \log\left[\prod_{i=1}^{N} f_D(d_i|\mathbf{x}_i,\boldsymbol{\theta})\right] \tag{7.64}$$

Using Eq. (7.63) in (7.64) and ignoring the constant $-(1/2)\log(2\pi)$, we may write

$$L(\boldsymbol{\theta}) = \sum_{i=1}^{N} \log\left[\sum_{k=1}^{2} g_k^{(i)} \sum_{j=1}^{2} g_{j|k}^{(i)} \exp\left(-\frac{1}{2}(d_i - y_{jk}^{(i)})^2\right)\right] \tag{7.65}$$

To compute the maximum likelihood estimate of $\boldsymbol{\theta}$, we have to find a stationary point (i.e., local or global maximum) of $L(\boldsymbol{\theta})$. Unfortunately, the log-likelihood function $L(\boldsymbol{\theta})$, as defined in Eq. (7.65), does not lend itself readily to this kind of computation.

To overcome this computational difficulty, we *artificially* augment the observable data $\{d_i\}_{i=1}^{N}$ by including a corresponding set of missing data in accordance with the EM algorithm. We do so by introducing *indicator variables* that pertain to the probability model of the HME architecture as follows (Jordan and Jacobs, 1994):

- $z_k^{(i)}$ and $z_{j|k}^{(i)}$ are interpreted as the labels that correspond to decisions made in the probability model for the ith example in the training set. These variables are defined in such a way that only a single one of the $z_k^{(i)}$ is equal to one and only a single one of the $z_{j|k}^{(i)}$ is equal to one for all i. Both $z_k^{(i)}$ and $z_{j|k}^{(i)}$ are treated as statistically independent discrete random variables with their respective expectations defined by

$$\begin{aligned} E[z_k^{(i)}] &= P\left[z_k^{(i)} = 1|\mathbf{x}_i, d_i, \hat{\boldsymbol{\theta}}(n)\right] \\ &= h_k^{(i)} \end{aligned} \tag{7.66}$$

and

$$\begin{aligned} E[z_{j|k}^{(i)}] &= P\left[z_{j|k}^{(i)} = 1|\mathbf{x}_i, d_i, \hat{\boldsymbol{\theta}}(n)\right] \\ &= h_{j|k}^{(i)} \end{aligned} \tag{7.67}$$

where $\hat{\boldsymbol{\theta}}(n)$ is the estimate of parameter vector $\boldsymbol{\theta}$ at iteration n of the EM algorithm.

- $z_{jk}^{(i)} = z_{j|k}^{(i)} z_k^{(i)}$ is interpreted as the label that specifies the expert (j, k) in the probability model for the ith example in the training sample. It is also treated as a discrete random variable with its expectation, defined by

$$E\left[z_{jk}^{(i)}\right] = E\left[z_{j|k}^{(i)} z_k^{(i)}\right]$$

$$= E\left[z_{j|k}^{(i)}\right] E\left[z_k^{(i)}\right] \tag{7.68}$$

$$= h_{j|k}^{(i)} h_k^{(i)} = h_{jk}^{(i)}$$

The $h_k^{(i)}$, $h_{j|k}^{(i)}$ and $h_{jk}^{(i)}$ in Eqs. (7.66) to (7.68) are the *a posteriori* probabilities introduced in Section 7.9; the superscript i has been added to them to signify the training example in question. See Problem 7.13 for a justification of these three equations.

By adding the missing data thus defined to the observable data, the maximum likelihood estimation problem is considerably simplified. More specifically, let $f_c(d_i, z_{jk}^{(i)} | \mathbf{x}_i, \boldsymbol{\theta})$ denote the conditional pdf of the complete data made up of d_i and $z_{jk}^{(i)}$, given $\mathbf{x}_i$ and parameter vector $\boldsymbol{\theta}$. We then write

$$f_c(d_i, z_{jk}^{(i)} | \mathbf{x}_i, \boldsymbol{\theta}) = \prod_{j=1}^{2} \prod_{k=1}^{2} (g_k^{(i)} g_{j|k}^{(i)}(i) f_{jk}(d_i)) \tag{7.69}$$

where $f_{jk}(d_i)$ is the conditional pdf of d_i, given that expert (j, k) in the HME model is chosen; that is, $f_{jk}(i)$ is given by the Gaussian distribution:

$$f_{jk}(d_i) = \frac{1}{\sqrt{2\pi}} \exp\left(-\frac{1}{2}(d_i - y_{jk}^{(i)})^2\right) \tag{7.70}$$

Note that the formula of Eq. (7.69) corresponds to a hypothetical experiment, containing indicator variables represented by $z_{jk}^{(i)}$ that are nonobservable in a physical data sense. In any event, the log-likelihood function for the complete-data problem, accounting for the entire training set, is given by

$$L_c(\boldsymbol{\theta}) = \log\left[\prod_{i=1}^{N} f_c(d_i, z_{jk}^{(i)} | \mathbf{x}_i, \boldsymbol{\theta})\right]$$

$$= \log\left[\prod_{i=1}^{N} \prod_{j=1}^{2} \prod_{k=1}^{2} (g_k^{(i)} g_{j|k}^{(i)} f_{jk}(d_i))^{z_{jk}^{(i)}}\right] \tag{7.71}$$

$$= \sum_{i=1}^{N} \sum_{j=1}^{2} \sum_{k=1}^{2} z_{jk}^{(i)}\left[\log g_k^{(i)} + \log g_{j|k}^{(i)} + \log f_{jk}(d_i)\right]$$

Using Eq. (7.70) in (7.71) and ignoring the constant $-(1/2)\log(2\pi)$, we may therefore write

$$L_c(\boldsymbol{\theta}) = \sum_{i=1}^{N} \sum_{j=1}^{2} \sum_{k=1}^{2} z_{jk}^{(i)}\left[\log g_k^{(i)} + \log g_{j|k}^{(i)} - \frac{1}{2}(d_i - y_{jk}^{(i)})^2\right] \tag{7.72}$$

Comparing Eq. (7.72) with (7.65), we immediately see the computational benefit gained by adding the indicator variables as missing data to the set of observables: The maximum likelihood estimation problem has been decoupled into a set of regression problems for the individual experts and a separate set of multinomial classification problems for the gating networks.

To proceed with the application of the EM algorithm, we first invoke the E-step of the algorithm by taking the expectation of the complete data log-likelihood function $L_c(\theta)$, as shown by

$$
\begin{aligned}
Q(\theta, \hat{\theta}(n)) &= E[L_c(\theta)] \\
&= \sum_{i=1}^{N} \sum_{j=1}^{2} \sum_{k=1}^{2} E\left[z_{jk}^{(i)}\right] \cdot \left(\log g_k^{(i)} + \log g_{j|k}^{(i)} - \frac{1}{2}(d_i - y_{jk}^{(i)})^2\right)
\end{aligned}
\tag{7.73}
$$

where the expectation operator is shown acting on the indicator variable $z_{jk}^{(i)}$ as this is the only unobservable variable. Hence, using Eq. (7.68) in (7.73), we obtain (Jordan and Jacobs, 1994):

$$
Q(\theta, \hat{\theta}(n)) = \sum_{i=1}^{N} \sum_{j=1}^{2} \sum_{k=1}^{2} h_{jk}^{(i)}\left(\log g_k^{(i)} + \log g_{j|k}^{(i)} - \frac{1}{2}(d_i - y_{jk}^{(i)})^2\right)
\tag{7.74}
$$

The M-step of the algorithm requires maximizing $Q(\theta, \hat{\theta}(n))$ with respect to θ. The parameter vector θ is made up of two sets of synaptic weights: one belonging to the gating networks and the other belonging to the experts. From our earlier discussion we note the following:

- The synaptic weights of the experts determine $y_{jk}^{(i)}$, which also enters the definition of $h_{jk}^{(i)}$. The expression $Q(\theta, \hat{\theta}(n))$ is therefore influenced by the experts only through the term $h_{jk}^{(i)}(d_i - y_{jk}^{(i)})^2$.
- The synaptic weights of the gating networks determine the probabilities $g_{jk}^{(i)}$, $g_{j|k}^{(i)}$ and $h_{jk}^{(i)}$. The expression $Q(\theta, \hat{\theta}(n))$ is therefore influenced by the gating networks only through the term $h_{jk}^{(i)}(\log g_k^{(i)} + \log g_{j|k}^{(i)})$.

Accordingly, the M-step of the algorithm reduces to the following three optimization problems for an HME with two levels of hierarchy:

$$
\mathbf{w}_{jk}(n+1) = \arg\min_{\mathbf{w}_{jk}} \sum_{i=1}^{N} h_{jk}^{(i)}(d_i - y_{jk}^{(i)})^2
\tag{7.75}
$$

$$
\mathbf{a}_j(n+1) = \arg\max_{\mathbf{a}_j} \sum_{i=1}^{N} \sum_{k=1}^{2} h_k^{(i)} \log g_k^{(i)}
\tag{7.76}
$$

and

$$
\mathbf{a}_{jk}(n+1) = \arg\max_{\mathbf{a}_{jk}} \sum_{i=1}^{N} \sum_{l=1}^{2} h_l^{(i)} \sum_{m=1}^{2} h_{m|l}^{(i)} \log g_{m|l}^{(i)}
\tag{7.77}
$$

The optimizations in Eqs. (7.75) to (7.77) are done with h fixed; h is a function of the parameters but the derivatives are not taken through h. Note also that all the quantities in the right-hand sides of these equations refer to measurements made at step n.

The optimization in Eq. (7.75), pertaining to the experts, is a weighted least-squares estimation problem. The remaining two optimization in Eqs. (7.76) and (7.77), pertaining to the gating networks, are maximum likelihood estimation problems.[9] Note also that although the equations are formulated for two levels of hierarchy, they may be readily extended to an arbitrary number of levels of heirarchy.

7.14 SUMMARY AND DISCUSSION

In the study of modeling, pattern classification, and regression problems, we have two extreme cases to consider:

1. *Simple models,* which provide insight into the problem of interest but lack accuracy.
2. *Complex models,* which provide accurate results but lack insight.

It is perhaps impossible to combine simplicity and accuracy in a single model. In the context of the discussion presented in the second part of this chapter, CART is an example of a simple model that uses hard decisions to partition the input space into a piecewise set of subspaces, with each subspace having its own expert. Unfortunately, the use of hard decisions results in some information loss and therefore loss in performance. The multilayer perceptron (MLP), on the other hand, is a complex model with a nested form of nonlinearity designed to preserve the information content of the training data. However, it uses a black box approach to globally fit a single function into the data, thereby losing insight into the problem. The HME, representing a dynamic type of committee machine, is a compromise model between these two extremes, sharing common features with both CART and MLP:

- The architecture of HME is similar to that of CART, but differs from it in soft partitioning rather than hard partitioning the input space.
- The HME uses a nested form of nonlinearity similar to that of MLP, not for the purpose of input–output mapping, but rather for partitioning the input space.

In this chapter we emphasize the use of two tools for the design of an HME model:

- CART as the architectural basis for dealing with the model selection problem.
- EM algorithm for solving the parameter estimation problem by iteratively computing maximum likelihood estimates of the model parameters.

The EM algorithm is usually guaranteed to move *uphill* in likelihood. Therefore, by using CART to initialize the EM algorithm in the manner described in Section 7.8, we may expect the EM algorithm to yield a better generalization performance than would be possible with the initial condition established by CART.

The EM algorithm is important and fundamental if the application of interest is that of maximum likelihood estimation, such as in *modeling.* An interesting modeling application is described in Jacobs, Jordan, and Barto (1991b), where an ME model is trained to do the what/where task. In this task, the model is required to determine what an object is, and where it is in the visual field. Two experts were used in the study, each of which is specialized for one aspect of the task. For a specific input, both experts generate outputs. But, the gating network decides the appropriate mixture for that input. The successful results reported by Jacobs et al. demonstrate that it is possible for a task assignment to be innately determined, not on the basis of the task *per se,* but rather by the match between a task's requirements and the computational properties of the model (Elman et al., 1996).

We finish this discussion by returning to the other class of committee machines studied in the first part of the chapter. Whereas the ME model or HME model relies on the use of gating network(s) activated by the input signal for fusing the knowledge acquired by the various experts in the model, a committee machine, based on the use of ensemble averaging or, alternatively, boosting, relies on the learning algorithm itself to do the integration, as summarized here:

1. Ensemble averaging improves error performance in a clever way by the combined use of two effects:
   - Reduction of error due to bias by purposely overfitting the individual experts in the committee machine.
   - Reduction of error due to variance by using different initial conditions in the training of the individual experts, and then ensemble-averaging their respective outputs.
2. Boosting improves error performance in an ingenious way of its own. In this case, individual experts are required to perform only slightly better than random guessing. The weak learnability of the experts is converted into strong learnability whereby the committee machine's error is made arbitrarily small. This remarkable conversion is achieved by *filtering* the distribution of the input data in a manner causing the weak learning models (i.e., experts) to eventually learn the entire distribution, or by *resampling* the training examples according to a certain probability distribution as in AdaBoost. The advantage of AdaBoost over boosting by filtering is that it works with a training sample of fixed size.

NOTES AND REFERENCES

1. *Ensemble-averaging methods* are discussed in Perrone (1993), where a large bibliography on the subject is included. Other references on this subject include Wolpert (1992), and Hashem (1997).
2. The use of ensemble-averaging for the design of a committee machine over a set of different initial conditions has been suggested by several neural network practitioners. However, the statistical analysis presented in Naftaly et al. (1997) and the procedure described therein for training a committee machine designed by ensemble-averaging over the space of initial conditions appear to be the first of their kind. In that paper, experimental results are presented based on the sunspot data and energy-prediction competition data. In both cases, significant reduction in variance is demonstrated by averaging over the space of initial conditions.

   According to Naftaly et al. (1997), the use of popular training constraints such as weight decay and early stopping is *not* recommended in the design of a committee machine by ensemble averaging over the space of initial conditions.
3. The main references on boosting theory and related experimental studies, more or less in chronological order, are as follows: Schapire (1990), Drucker et al. (1993, 1994), Freund (1995), Breiman (1996b), Freund and Schapire (1996a, 1996b, 1997), Schapire (1997), and Schapire et al. (1997). The first references on the three basic approaches to boosting are as follows:
   - Filtering: Schapire (1990)
   - Resampling: Freund and Schapire (1996a)
   - Reweighting: Freund (1995)

4. The idea of using a mixture of experts for realizing a complex mapping function was first discussed by Jacobs, Jordan, Nowlan, and Hinton in their 1991a paper. The development of this model was motivated by (1) a proposal described in Nowlan (1990), viewing competitive adaptation in unsupervised learning as an attempt to fit a mixture of simple probability distributions (such as Gaussians) into a set of data points, and (2) ideas developed in the Ph.D. thesis of Jacobs (1990) using a similar modular architecture but a different cost function.

5. Maximum likelihood estimators have some desirable properties. Under quite general conditions, the following *asymptotic* properties may be proved (Kmenta, 1971):

   (i) *Maximum likelihood estimators are consistent.* Let $L(\boldsymbol{\theta})$ denote the log-likelihood function and θ_i denote an element of the parameter vector $\boldsymbol{\theta}$. The partial derivative $\partial L / \partial \theta_i$ is called a *score*. We say that a maximum likelihood estimator is consistent in the sense that the value of θ_i, for which the score $\partial L / \partial \theta_i$ is identically zero, *converges in probability* to the true value of θ_i as the sample size used in the estimation approaches infinity.

   (ii) *Maximum likelihood estimators are asymptotically efficient.* That is,

   $$\lim_{N \to \infty} \left\{ \frac{\text{var}\,[\theta_i - \hat{\theta}_i]}{I_{ii}} \right\} = 1 \qquad \text{for all } i$$

   where N is the sample size, $\hat{\theta}_i$ is the maximum likelihood estimate of θ_i, and I_{ii} is the ith diagonal element of the inverse of *Fisher's information matrix*. Fisher's information matrix is defined by

   $$\mathbf{J} = - \begin{bmatrix} E\left[\dfrac{\partial^2 L}{\partial \theta_1^2}\right] & E\left[\dfrac{\partial^2 L}{\partial \theta_1 \partial \theta_2}\right] & \cdots & E\left[\dfrac{\partial^2 L}{\partial \theta_1 \partial \theta_M}\right] \\[2ex] E\left[\dfrac{\partial^2 L}{\partial \theta_2 \partial \theta_1}\right] & E\left[\dfrac{\partial^2 L}{\partial \theta_2^2}\right] & \cdots & E\left[\dfrac{\partial^2 L}{\partial \theta_2 \partial \theta_M}\right] \\[2ex] \vdots & \vdots & & \vdots \\[2ex] E\left[\dfrac{\partial^2 L}{\partial \theta_M \partial \theta_1}\right] & E\left[\dfrac{\partial^2 L}{\partial \theta_M \partial \theta_2}\right] & \cdots & E\left[\dfrac{\partial^2 L}{\partial \theta_M^2}\right] \end{bmatrix}$$

   where M is the dimension of parameter vector $\boldsymbol{\theta}$.

   (iii) *Maximum likelihood estimators are asymptotically Gaussian.* That is, as the sample size approaches infinity, each element of the maximum likelihood estimate $\boldsymbol{\theta}$ assumes a Gaussian distribution.

   In practice, we find the large sample (i.e., asymptotic) properties of maximum likelihood estimators hold rather well for sample size $N \geq 50$.

6. The paper by Newcomb (1886), considering the estimation of parameters of a mixture of two univariate Gaussian distributions, appears to be the earliest reference to an EM-type of process reported in the literature.

   The name "EM algorithm" was coined by Dempster, Laird, and Rubin in their 1977 fundamental paper. In that paper, formulation of the EM algorithm for computing maximum likelihood estimates from incomplete data at various levels of generality was presented for the first time.

   The first unified account of the theory, methodology, and applications of the EM algorithm, its history, and extensions was presented in book form by McLachlan and Krishnan (1997).

7. Under fairly general conditions, the likelihood values computed by the EM algorithm converge to stationary values. Wu (1983) presents a detailed account of the convergence

properties of the EM algorithm. However, the EM algorithm will *not* always lead to a local or a global maximum of the likelihood function. In Chapter 3 of the book by McLachlan and Krishnan (1997), two examples are presented where this is not the case. In one example the algorithm converges to a saddle point, and in the other example the algorithm converges to a local *minimum* of the likelihood function.

8. The EM algorithm can also handle Bayesian *maximum a posterior (MAP) estimation* by incorporating *prior* information on the parameter vector; see Problem 7.11. Using Bayes' rule, we may express the conditional probability density function for parameter vector $\boldsymbol{\theta}$, given a set of observations $\mathbf{x}$, as

$$f_{\boldsymbol{\Theta}}(\boldsymbol{\theta}|\mathbf{x}) = \frac{f_{\mathbf{X}}(\mathbf{x}|\boldsymbol{\theta})f_{\boldsymbol{\Theta}}(\boldsymbol{\theta})}{f_{\mathbf{X}}(\mathbf{x})}$$

From this relation, we readily see that maximizing the *a posteriori* density $f_{\boldsymbol{\Theta}}(\boldsymbol{\theta}|\mathbf{x})$ is equivalent to maximizing the product $f_{\mathbf{X}}(\mathbf{x}|\boldsymbol{\theta})f_{\boldsymbol{\Theta}}(\boldsymbol{\theta})$, since $f_{\mathbf{X}}(\mathbf{x})$ is independent of $\boldsymbol{\theta}$. The probability density function $f_{\boldsymbol{\Theta}}(\boldsymbol{\theta})$ represents the prior information available on $\boldsymbol{\theta}$. Maximizing $f_{\boldsymbol{\Theta}}(\boldsymbol{\theta}|\mathbf{x})$ provides the *most probable* estimate of the parameter vector $\boldsymbol{\theta}$, given $\mathbf{x}$. Two points are noteworthy in the context of this estimate:

- Maximum likelihood estimation, represented by maximizing $f_{\mathbf{X}}(\mathbf{x}|\boldsymbol{\theta})$ with respect to $\boldsymbol{\theta}$, is a reduced form of maximum *a posteriori* estimation, reduced in the sense that it is void of prior information.
- The use of prior information is synonymous with regularization, which (we recall from Chapter 5) corresponds to a smooth input–output mapping.

In Waterhouse et al. (1996), a Bayesian framework for estimating parameters of a mixture of experts model is presented. The Bayesian approach described therein overcomes a phenomenon known as "overfitting", which leads to an estimate with high variance when using maximum likelihood inference.

9. An efficient algorithm, known as the *iteratively reweighted least-squares (IRLS) algorithm*, is available to solve the maximum likelihood estimation problems described in Eqs. (7.76) and (7.77); for a description of the IRLS algorithm, see McCullagh and Nelder (1989).

PROBLEMS

Ensemble Averaging

7.1 Consider a committee machine consisting of K experts. The input–output function of the kth expert is denoted by $F_k(\mathbf{x})$, where $\mathbf{x}$ is the input vector and $k = 1, 2, ..., K$. The individual outputs of the experts are linearly combined to form the overall output y, defined by

$$y = \sum_{k=1}^{K} w_k F_k(\mathbf{x})$$

where w_k is a linear weight assigned to $F_k(\mathbf{x})$. The requirement is to evaluate w_k so that y provides a least-squares estimate of the desired response d corresponding to $\mathbf{x}$. Given a set of training data $\{(\mathbf{x}_i, d_i)\}_{i=1}^{N}$, determine the required values of the w_k's to solve this parameter estimation problem.

Boosting

7.2 Compare the computational advantages and disadvantages of boosting by filtering and AdaBoost.

7.3 Ordinarily, boosting performs best on weak learning models, that is, learning models with relatively low generalization error rates. Suppose, however, you are given a strong learning model, that is, a learning model with high generalization error rate. Assume that you are dealing with a training sample of fixed size. How do boosting by filtering and AdaBoost cope with this situation?

Mixture of Experts

7.4 Consider a piecewise-linear task described by

$$F(x_1, x_2, ..., x_{10}) = \begin{cases} 3x_2 + 2x_3 + x_4 + 3 + \epsilon & \text{if } x_1 = 1 \\ 3x_5 + 2x_6 + x_7 - 3 + \epsilon & \text{if } x_1 = -1 \end{cases}$$

For comparison, the following network configurations are used:

1. Multilayer perceptron: "10→10→1" network
2. Mixture of experts: Gating networks: 10→2;
Expert networks: 10→1

Compare the computational complexities of these two networks.

7.5 The ME model described by the conditional probability density function of Eq. (7.30) is based on a scalar regression model, in which the error is Gaussian distributed with zero mean and unit variance.

(a) Reformulate this equation for the more general case of an ME model corresponding to a multiple regression model, in which the desired response is a vector with dimension q and the error is a multivariate Gaussian distribution with zero mean and covariance matrix Σ.

(b) How is the ME model for this reformulation different from the ME model shown in Fig. 7.8?

7.6 Derive the stochastic gradient algorithm for the training of the mixure of experts model.

Hierarchical Mixture of Experts

7.7 (a) Construct the block diagram of an HME model with three levels of hierarchy. Assume the use of a binary decision tree for the model.

(b) Write the *a posteriori* probabilities for the nonterminal nodes of the HME described in part (a). Demonstrate the recursiveness of the computations involved in evaluating these probabilities.

(c) Formulate the conditional probability density function for the HME model described in part (a).

7.8 Discuss the similarities and differences between HME models and radial-basis function (RBF) networks.

7.9 Derive the equations that describe the stochastic gradient algorithm for the training of an HME model with two levels of hierarchy. Assume a binary decision tree for the model.

EM Algorithm and its Application to the HME Model

7.10 Prove the monotonic-increasing property of the EM algorithm described in Eq. (7.62). For this derivation, do the following:

(a) Let

$$k(\mathbf{r}|d, \boldsymbol{\theta}) = \frac{f_c(\mathbf{r}|\boldsymbol{\theta})}{f_D(d|\boldsymbol{\theta})}$$

denote the conditional probability density function of the augmented data vector $\mathbf{r}$, given the observation d and parameter vector $\boldsymbol{\theta}$. Hence, the incomplete data log-likelihood function may be expressed as

$$L(\boldsymbol{\theta}) = L_c(\boldsymbol{\theta}) - \log k(\mathbf{r}|d, \boldsymbol{\theta})$$

where $L_c(\theta) = \log f_c(r|\theta)$ is the complete data log-likelihood function. By taking the expectation of $L(\theta)$ with respect to the conditional distribution of r, given d, show that

$$L(\theta) = Q(\theta,\hat{\theta}(n)) - K(\theta,\hat{\theta}(n))$$

where

$$K(\theta,\hat{\theta}(n)) = E[\log k(r|d,\hat{\theta})]$$

Hence, show that

$$L(\hat{\theta}(n + 1)) - L(\hat{\theta}(n)) = \left[Q(\hat{\theta}(n + 1),\hat{\theta}(n)) - Q(\hat{\theta}(n),\hat{\theta}(n))\right]$$
$$- \left[K(\hat{\theta}(n + 1),\hat{\theta}(n)) - K(\hat{\theta}(n),\hat{\theta}(n))\right]$$

(b) *Jensen's inequality* states that if $f(\cdot)$ is a convex function and u is a random variable, then

$$E[g(u)] \geq g(E[u])$$

where E is the expectation operator; moreover, if $g(\cdot)$ is strictly convex, then equality in this relation implies that $u = E[u]$ with probability 1 (Cover and Thomas, 1991). Using Jensen's inequality, show that

$$K(\hat{\theta}(n + 1), \hat{\theta}(n)) - K(\hat{\theta}(n),\hat{\theta}(n)) \leq 0$$

Hence, show that Eq. (7.62) holds for $n = 0, 1, 2, \dots$.

7.11 The EM algorithm is easily modified to accommodate the maximum *a posteriori* (MAP) estimate of a parameter vector θ. Using Bayes' rule, modify the E-step and M-step of the EM algorithm to provide for this estimation.

7.12 For an HME trained with the EM algorithm and an MLP trained with the back-propagation algorithm to provide a similar level of performance for a given task, we would intuitively expect the computational complexity of the HME to exceed that of the MLP. Argue in favor of or against the plausibility of this statement.

7.13 Justify the relations between the indicator variables and corresponding *a posteriori* probabilities described in Eqs. (7.66) to (7.68).

7.14 Equation (7.75) describes the weighted least-squares for the optimization of the expert networks in the HME model of Fig. 7.11, assuming that the desired response d is scalar. How is this relation modified for the case of a multidimensional desired response?

CHAPTER 8

Principal Components Analysis

8.1 INTRODUCTION

An important feature of neural networks is the ability they have to *learn* from their environment, and through learning to *improve* performance in some sense. In the previous four chapters, the focus was on algorithms for supervised learning, for which a set of targets of interest is provided by an external teacher. The targets take the form of a desired input–output mapping, which the network is required to approximate. In this chapter and the next three chapters, we study algorithms for *self-organized learning* or *unsupervised learning*. The purpose of an algorithm for self-organized learning is to *discover* significant patterns or features in the input data, and to do the discovery *without* a teacher. To do so, the algorithm is provided with a set of rules of a *local* nature, which enables it to learn to compute an input–output mapping with specific desirable properties; the term "local" means that the change applied to the synaptic weight of a neuron is confined to the immediate neighborhood of that neuron. The modeling of network stuctures used for self-organized learning tends to follow neurobiological structures to a much greater extent than for supervised learning. This may not be surprising, because the process of network organization is fundamental to the organization of the brain.

The structure of a self-organizing system may take on a variety of different forms. It may, for example, consist of an *input* (*source*) *layer* and an *output* (*representation*) *layer,* with feedforward connections from input to output and lateral connections between neurons in the output layer. Another example is a feedforward network with multiple layers, in which the self-organization proceeds on a layer-by-layer basis. In both examples, the learning process consists of repeatedly modifying the synaptic weights of all the connections in the system in response to input (activation) patterns and in accordance with prescribed rules, until a final configuration develops.

This chapter on self-organizing systems is restricted to Hebbian learning. The primary focus of the chapter is *principal components analysis,* which is a standard technique commonly used for data reduction in statistical pattern recognition and signal processing.

Organization of the Chapter

The material in this chapter is organized as follows. In Section 8.2 we use qualitative arguments to describe the basic principles of self-organization. This is followed by introductory material on principal components analysis in Section 8.3, which is also basic to the self-organizing systems discussed in the rest of the chapter.

With this basic background material on hand, we then proceed to study some specific self-organizing systems. In Section 8.4 we describe a simple model consisting of a single neuron, which extracts the first principal component in a self-organized manner. In Section 8.5 we describe a more elaborate self-organizing system in the form of a feedforward network with a single layer of neurons, which extracts all the principal components by building on the previous simple model. This procedure is illustrated by a computer experiment on image coding presented in Section 8.6. In Section 8.7 we describe another self-organizing system for a similar function; this system is even more elaborate because it also includes lateral connections.

In Section 8.8 we present a classification of algorithms for principal components analysis using neural networks. This is followed by Section 8.9 on the classification of data reduction algorithms into adaptive and batch methods.

In Section 8.10 we describe a nonlinear form of principal components analysis that builds on the idea of an inner-product kernel defined in accordance with Mercer's theorem, which is discussed in Chapter 6 on support vector machines.

The chapter concludes in Section 8.11 with some final thoughts on principal components analysis.

8.2   SOME INTUITIVE PRINCIPLES OF SELF-ORGANIZATION

As mentioned previously, self-organized (unsupervised) learning consists of repeatedly modifying the synaptic weights of a neural network in response to activation patterns and in accordance with prescribed rules, until a final configuration develops. The key question, of course, is how a useful configuration can finally develop from self-organization. The answer lies essentially in the following observation (Turing, 1952):

> Global order can arise from local interactions.

This observation is of fundamental importance; it applies to the brain and to artificial neural networks. In particular, many originally random local interactions between neighboring neurons of a network can coalesce into states of global order and ultimately lead to coherent behavior in the form of spatial patterns or temporal rhythms; these are the essence of self-organization.

Network organization takes place at two different levels that interact with each other in the form of a *feedback* loop. The two levels are:

- *Activity.* Certain activity patterns are produced by a given network in response to input signals.
- *Connectivity.* Connection strengths (synaptic weights) of the network are modified in response to neuronal signals in the activity patterns, due to synaptic plasticity.

The feedback between changes in synaptic weights and changes in activity patterns must be *positive* in order to achieve self-organization (instead of stabilization) of the network. Accordingly, we may abstract the first principle of self-organization (von der Malsburg, 1990a):

PRINCIPLE 1. Modifications in synaptic weights tend to self-amplify.

The process of self-amplification is constrained by the requirement that modifications in synaptic weights must be based on locally available signals, namely presynaptic signals and postsynaptic signals. The requirements of self-reinforcement and locality specify the mechanism whereby a strong synapse leads to coincidence of presynaptic and postsynaptic signals. In turn, the synapse is increased in strength by such a coincidence. The mechanism described here is in fact a restatement of Hebb's postulate of learning!

In order to stabilize the system there must be some form of competition for "limited" resources (e.g., number of inputs, energy resources). Specifically, an increase in the strength of some synapses in the network must be compensated for by a decrease in others. Accordingly, only the "successful" synapses can grow, while the less successful ones tend to weaken and may eventually disappear. This observation leads us to abstract the second principle of self-organization (von der Malsburg, 1990a):

PRINCIPLE 2. Limitation of resources leads to competition among synapses and therefore the selection of the most vigorously growing synapses (i.e., the fittest) at the expense of the others.

This principle is also made possible by synaptic plasticity.

For our next observation we note that a single synapse on its own cannot efficiently produce favorable events. To do so, we need cooperation among a set of synapses converging onto a particular neuron and carrying coincident signals strong enough to activate that neuron. We may therefore abstract the third principle of self-organization (von der Malsburg, 1990a):

PRINCIPLE 3. Modifications in synaptic weights tend to cooperate.

The presence of a vigorous synapse can enhance the fitness of other synapses, in spite of the overall competition in the network. This form of cooperation may arise due to synaptic plasticity, or due to simultaneous stimulation of presynaptic neurons brought on by the existence of the right conditions in the external environment.

All three principles of self-organization described thus far relate only to the neural network itself. However, for self-organized learning to perform a useful information-processing function, there must be *redundancy* in the activation patterns supplied to the network by the environment. The issue of redundancy is discussed in Shannon's framework of information theory in Chapter 10. For now it suffices to postulate the last principle of self-organized learning as follows (Barlow, 1989):

PRINCIPLE 4. Order and structure in the activation patterns represent redundant information that is acquired by the neural network in the form of knowledge, which is a necessary prerequisite to self-organized learning.

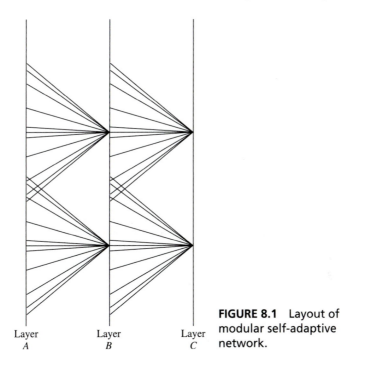

FIGURE 8.1   Layout of modular self-adaptive network.

Layer
A

Layer
B

Layer
C

Some of this knowledge may be obtained by observations of statistical parameters such as the mean, variance, and correlation matrix of the input data.

Principles 1 through 4 on self-organized learning provide the neurobiological basis for the adaptive algorithms for principal components analysis described in this chapter and for Kohonen's self-organizing map presented in the next chapter. These principles are also incorporated in many other self-organized models that are motivated by neurobiological considerations. One such model that deserves to be mentioned is *Linsker's model* of the mammalian visual system (Linsker, 1986).

Self-Organized Feature Analysis

Processing information in the visual system is performed in stages. In particular, simple features such as contrast and edge orientation are analyzed in the early stages of the system, whereas more elaborate complex features are analyzed in later stages. Figure 8.1 shows the gross structure of a modular network that resembles the visual system. In Linsker's model, the neurons of the network in Fig. 8.1 are organized into two-dimensional layers, with local forward connections from one layer to the next. Each neuron receives information from a limited number of neurons located in an overlying region of the previous layer, which constitutes the *receptive field* of that neuron. The receptive fields of the network play a crucial role in the synaptic development process because they make it possible for neurons in one layer to respond to *spatial correlations* of the neuronal activities in the previous layer. Two assumptions of a structural nature are made:

1. The positions of the synaptic connections are fixed for the entire neuronal development process once they have been chosen.
2. Each neuron acts as a linear combiner.

The model combines aspects of Hebb-like synaptic modification with cooperative and competitive learning in such a way that the network's outputs optimally discriminate among an ensemble of inputs, with the self-organized learning proceeding on a *layer-by-layer basis*. That is, the learning process permits the self-organized feature-analyzing properties of each layer to develop fully before proceeding to the next layer. In Linsker (1986), simulation results are presented that are qualitatively similar to properties found in the early stages of visual processing in cats and monkeys. Recognizing the highly complex nature of the visual system, it is indeed remarkable that the simple model considered by Linsker is capable of developing similar feature-analyzing neurons. The point is not to imply that feature-analyzing neurons in the mammalian visual system develop in exactly the manner described in Linsker's model. Rather, such structures may be produced by a relatively simple layered network whose synaptic connections develop in accordance with a Hebbian form of learning.

Our primary interest in this chapter, however, is in principal components analysis and how it can be performed using self-organizing systems based on Hebbian learning.

8.3 PRINCIPAL COMPONENTS ANALYSIS

A common problem in statistical pattern recognition is that of feature selection or feature extraction. *Feature selection* refers to a process whereby a *data space* is transformed into a *feature space* that, in theory, has exactly the same dimension as the original data space. However, the transformation is designed in such a way that the data set may be represented by a reduced number of "effective" features and yet retain most of the intrinsic information content of the data; in other words, the data set undergoes a *dimensionality reduction*. To be specific, suppose we have an m-dimensional vector $\mathbf{x}$ and wish to transmit it using l numbers, where $l < m$. If we simply truncate the vector $\mathbf{x}$, we will cause a mean-square error equal to the sum of the variances of the elements eliminated from $\mathbf{x}$. So we ask the following question: Does there exist an invertible *linear* transformation $\mathbf{T}$ such that the truncation of $\mathbf{Tx}$ is optimum in the mean-squared error sense? Clearly, the transformation $\mathbf{T}$ should have the property that some of its components have low variance. *Principal components analysis* (also known as the *Karhunen–Loève transformation* in communication theory) maximizes the rate of decrease of variance and is therefore the right choice. In this chapter we derive Hebbian-based learning algorithms that can perform principal components analysis[1] on a data vector of interest.

Let $\mathbf{X}$ denote an m-dimensional *random vector* representing the environment of interest. We assume that the random vector $\mathbf{X}$ has zero mean:

$$E[\mathbf{X}] = \mathbf{0}$$

where E is the statistical expectation operator. If $\mathbf{X}$ has a nonzero mean, we subtract the mean from it before proceeding with the analysis. Let $\mathbf{q}$ denote a *unit vector,* also of dimension m, onto which the vector $\mathbf{X}$ is to be *projected*. This projection is defined by the inner product of the vectors $\mathbf{X}$ and $\mathbf{q}$, as shown by

$$A = \mathbf{X}^T\mathbf{q} = \mathbf{q}^T\mathbf{X} \tag{8.1}$$

subject to the constraint

$$\|\mathbf{q}\| = (\mathbf{q}^T\mathbf{q})^{1/2} = 1 \tag{8.2}$$

The projection A is a *random variable* with a mean and variance related to the statistics of the random vector $\mathbf{X}$. Under the assumption that the random vector $\mathbf{X}$ has zero mean, it follows that the mean value of the projection A is zero too:

$$E[A] = \mathbf{q}^T E[\mathbf{X}] = 0$$

The variance of A is therefore the same as its mean-square value, and so we may write

$$\begin{aligned} \sigma^2 &= E[A^2] \\ &= E[(\mathbf{q}^T\mathbf{X})(\mathbf{X}^T\mathbf{q})] \\ &= \mathbf{q}^T E[\mathbf{X}\mathbf{X}^T]\mathbf{q} \\ &= \mathbf{q}^T\mathbf{R}\mathbf{q} \end{aligned} \tag{8.3}$$

The m-by-m matrix $\mathbf{R}$ is the *correlation matrix* of the random vector $\mathbf{X}$, formally defined as the expectation of the outer product of the vector $\mathbf{X}$ with itself, as shown by

$$\mathbf{R} = E[\mathbf{X}\mathbf{X}^T] \tag{8.4}$$

We observe that the correlation matrix $\mathbf{R}$ is *symmetric,* which means that

$$\mathbf{R}^T = \mathbf{R} \tag{8.5}$$

From this property it follows that if $\mathbf{a}$ and $\mathbf{b}$ are any m-by-1 vectors, then

$$\mathbf{a}^T\mathbf{R}\mathbf{b} = \mathbf{b}^T\mathbf{R}\mathbf{a} \tag{8.6}$$

From Eq. (8.3) we see that the variance σ^2 of the projection A is a function of the unit vector $\mathbf{q}$; we may thus write

$$\begin{aligned} \psi(\mathbf{q}) &= \sigma^2 \\ &= \mathbf{q}^T\mathbf{R}\mathbf{q} \end{aligned} \tag{8.7}$$

on the basis of which we may think of $\psi(\mathbf{q})$ as a *variance probe.*

Eigenstructure of Principal Components Analysis

The next issue to be considered is that of finding those unit vectors $\mathbf{q}$ along which $\psi(\mathbf{q})$ has *extremal* or *stationary* values (local maxima or minima), subject to a constraint on the Euclidean norm of $\mathbf{q}$. The solution to this problem lies in the eigenstructure of the correlation matrix $\mathbf{R}$. If $\mathbf{q}$ is a unit vector such that the variance probe $\psi(\mathbf{q})$ has an extremal value, then for any small perturbation $\delta\mathbf{q}$ of the unit vector $\mathbf{q}$, we find that, to a first order in $\delta\mathbf{q}$,

$$\psi(\mathbf{q} + \delta\mathbf{q}) = \psi(\mathbf{q}) \tag{8.8}$$

Now, from the definition of the variance probe given in Eq. (8.7), we have

$$\psi(\mathbf{q} + \delta\mathbf{q}) = (\mathbf{q} + \delta\mathbf{q})^T \mathbf{R}(\mathbf{q} + \delta\mathbf{q})$$
$$= \mathbf{q}^T \mathbf{R}\mathbf{q} + 2(\delta\mathbf{q})^T \mathbf{R}\mathbf{q} + (\delta\mathbf{q})^T \mathbf{R}\,\delta\mathbf{q}$$

where in the second line, we have made use of Eq. (8.6). Ignoring the second-order term $(\delta\mathbf{q})^T \mathbf{R}\,\delta\mathbf{q}$ and invoking the definition of Eq. (8.7), we may therefore write

$$\psi(\mathbf{q} + \delta\mathbf{q}) = \mathbf{q}^T \mathbf{R}\mathbf{q} + 2(\delta\mathbf{q})^T \mathbf{R}\mathbf{q}$$
$$= \psi(\mathbf{q}) + 2(\delta\mathbf{q})^T \mathbf{R}\mathbf{q} \tag{8.9}$$

Hence, the use of Eq. (8.8) in (8.9) implies that

$$(\delta\mathbf{q})^T \mathbf{R}\mathbf{q} = 0 \tag{8.10}$$

Just any perturbations $\delta\mathbf{q}$ of $\mathbf{q}$ are not admissible; rather, we are restricted to use only those perturbations for which the Euclidean norm of the perturbed vector $\mathbf{q} + \delta\mathbf{q}$ remains equal to unity; that is

$$\|\mathbf{q} + \delta\mathbf{q}\| = 1$$

or equivalently,

$$(\mathbf{q} + \delta\mathbf{q})^T(\mathbf{q} + \delta\mathbf{q}) = 1$$

Hence, in light of Eq. (8.2), we require that to a first order in $\delta\mathbf{q}$,

$$(\delta\mathbf{q})^T \mathbf{q} = 0 \tag{8.11}$$

This means that the perturbations $\delta\mathbf{q}$ must be orthogonal to $\mathbf{q}$, and therefore only a change in the direction of $\mathbf{q}$ is permitted.

By convention, the elements of the unit vector $\mathbf{q}$ are dimensionless in a physical sense. If, therefore, we are to combine Eqs. (8.10) and (8.11), we must introduce a scaling factor λ into the latter equation with the same dimensions as the entries in the correlation matrix $\mathbf{R}$. We may then write

$$(\delta\mathbf{q})^T \mathbf{R}\mathbf{q} - \lambda(\delta\mathbf{q})^T \mathbf{q} = 0$$

or equivalently,

$$(\delta\mathbf{q})^T(\mathbf{R}\mathbf{q} - \lambda\mathbf{q}) = 0 \tag{8.12}$$

For the condition of Eq. (8.12) to hold, it is necessary and sufficient to have

$$\mathbf{R}\mathbf{q} = \lambda\mathbf{q} \tag{8.13}$$

This is the equation that governs the unit vectors $\mathbf{q}$ for which the variance probe $\psi(\mathbf{q})$ has extremal values.

Equation (8.13) is recognized as the *eigenvalue problem,* commonly encountered in linear algebra (Strang, 1980). The problem has nontrivial solutions (i.e., $\mathbf{q} \neq \mathbf{0}$) only for special values of λ that are called the *eigenvalues* of the correlation matrix $\mathbf{R}$. The associated values of $\mathbf{q}$ are called *eigenvectors.* A correlation matrix is characterized by real, nonnegative eigenvalues. The associated eigenvectors are unique, assuming that the eigenvalues are distinct. Let the eigenvalues of the m-by-m matrix $\mathbf{R}$ be denoted by

$\lambda_1, \lambda_2, ..., \lambda_m$, and the associated eigenvectors be denoted by $\mathbf{q}_1, \mathbf{q}_2, ..., \mathbf{q}_m$, respectively. We may then write

$$\mathbf{Rq}_j = \lambda_j \mathbf{q}_j, \qquad j = 1, 2, ..., m \tag{8.14}$$

Let the corresponding eigenvalues be arranged in decreasing order:

$$\lambda_1 > \lambda_2 > \cdots > \lambda_j > \cdots > \lambda_m \tag{8.15}$$

so that $\lambda_1 = \lambda_{max}$. Let the associated eigenvectors be used to construct an m-by-m matrix:

$$\mathbf{Q} = [\mathbf{q}_1, \mathbf{q}_2, ..., \mathbf{q}_j, ..., \mathbf{q}_m] \tag{8.16}$$

We may then combine the set of m equations represented in (8.14) into a single equation:

$$\mathbf{RQ} = \mathbf{Q\Lambda} \tag{8.17}$$

where $\mathbf{\Lambda}$ is a diagonal matrix defined by the eigenvalues of matrix $\mathbf{R}$:

$$\mathbf{\Lambda} = \text{diag}[\lambda_1, \lambda_2, ..., \lambda_j, ..., \lambda_m] \tag{8.18}$$

The matrix $\mathbf{Q}$ is an *orthogonal (unitary) matrix* in the sense that its column vectors (i.e., the eigenvectors of $\mathbf{R}$) satisfy the *conditions of orthonormality*:

$$\mathbf{q}_i^T \mathbf{q}_j = \begin{cases} 1, & j = i \\ 0, & j \neq i \end{cases} \tag{8.19}$$

Equation (8.19) requires distinct eigenvalues. Equivalently, we may write

$$\mathbf{Q}^T \mathbf{Q} = \mathbf{I}$$

from which we deduce that the inverse of matrix $\mathbf{Q}$ is the same as its transpose, as shown by

$$\mathbf{Q}^T = \mathbf{Q}^{-1} \tag{8.20}$$

This means that we may rewrite Eq. (8.17) in a form known as the *orthogonal similarity transformation*:

$$\mathbf{Q}^T \mathbf{RQ} = \mathbf{\Lambda} \tag{8.21}$$

or in expanded form,

$$\mathbf{q}_j^T \mathbf{Rq}_k = \begin{cases} \lambda_j, & k = j \\ 0, & k \neq j \end{cases} \tag{8.22}$$

The orthogonal similarity (unitary) transformation of Eq. (8.21) transforms the correlation matrix $\mathbf{R}$ into a diagonal matrix of eigenvalues. The correlation matrix $\mathbf{R}$ may itself be expressed in terms of its eigenvalues and eigenvectors as:

$$\mathbf{R} = \sum_{i=1}^{m} \lambda_i \mathbf{q}_i \mathbf{q}_i^T \tag{8.23}$$

which is referred to as the *spectral theorem*. The outer product $\mathbf{q}_i \mathbf{q}_i^T$ is of *rank* 1 for all i.

Equations (8.21) and (8.23) are two equivalent representations of the *eigende-composition* of the correlation matrix **R**.

Principal components analysis and eigendecomposition of matrix **R** are basically one and the same, just viewing the problem in different ways. This equivalence follows from Eqs. (8.7) and (8.23) where we see that the variance probes and eigenvalues are indeed equal, as shown by

$$\psi(\mathbf{q}_j) = \lambda_j, \qquad j = 1, 2, ..., m \tag{8.24}$$

We may now summarize the two important findings we have made from the eigenstructure of principal components analysis:

- The eigenvectors of the correlation matrix **R** pertaining to the zero-mean random vector **X** define the unit vectors $\mathbf{q}_j$, representing the principal directions along which the variance probes $\psi(\mathbf{q}_j)$ have their extremal values.
- The associated eigenvalues define the extremal values of the variance probes $\psi(\mathbf{u}_j)$.

Basic Data Representations

Let the *data vector* **x** denote a realization of the random vector **X**.

With m possible solutions for the unit vector **q**, we find that there are m possible projections of the data vector **x** to be considered. Specifically, from Eq. (8.1) we note that

$$a_j = \mathbf{q}_j^T \mathbf{x} = \mathbf{x}^T \mathbf{q}_j, \qquad j = 1, 2, ..., m \tag{8.25}$$

where the a_j are the projections of **x** onto the principal directions represented by the unit vectors $\mathbf{u}_j$. The a_j are called the *principal components;* they have the same physical dimensions as the data vector **x**. The formula in Eq. (8.25) may be viewed as one of *analysis.*

To reconstruct the original data vector **x** exactly from the projections a_j, we proceed as follows. First, we combine the set of projections $\{a_j | j = 1, 2, ..., m\}$ into a single vector, as shown by

$$\begin{aligned} \mathbf{a} &= [a_1, a_2, ..., a_m]^T \\ &= [\mathbf{x}^T \mathbf{q}_1, \mathbf{x}^T \mathbf{q}_2, ..., \mathbf{x}^T \mathbf{q}_m]^T \\ &= \mathbf{Q}^T \mathbf{x} \end{aligned} \tag{8.26}$$

Next, we premultiply both sides of Eq. (8.26) by the matrix **Q**, and then use the relation of Eq. (8.20). Accordingly, the original data vector **x** may be reconstructed as follows:

$$\begin{aligned} \mathbf{x} &= \mathbf{Q}\mathbf{a} \\ &= \sum_{j=1}^{m} a_j \mathbf{q}_j \end{aligned} \tag{8.27}$$

which may be viewed as the formula for *synthesis.* In this sense, the unit vectors $\mathbf{q}_j$ represent a *basis* of the data space. Indeed, Eq. (8.27) is nothing but a coordinate transformation, according to which a point **x** in the data space is transformed into a corresponding point **a** in the feature space.

Dimensionality Reduction

From the perspective of statistical pattern recognition, the practical value of principal components analysis is that it provides an effective technique for *dimensionality reduction*. In particular, we may reduce the number of features needed for effective data representation by discarding those linear combinations in Eq. (8.27) that have small variances and retain only those terms that have large variances. Let $\lambda_1, \lambda_2, ..., \lambda_l$ denote the largest l eigenvalues of the correlation matrix $\mathbf{R}$. We may then approximate the data vector $\mathbf{x}$ by *truncating* the expansion of Eq. (8.27) after l terms as follows:

$$\hat{\mathbf{x}} = \sum_{j=1}^{l} a_j \mathbf{q}_j$$

$$= [\mathbf{q}_1, \mathbf{q}_1, ..., \mathbf{q}_l] \begin{bmatrix} a_1 \\ a_2 \\ \vdots \\ a_l \end{bmatrix}, \qquad l \leq m \qquad (8.28)$$

Given the original data vector $\mathbf{x}$, we may use Eq. (8.25) to compute the set of principal components retained in Eq. (8.28) as follows:

$$\begin{bmatrix} a_1 \\ a_2 \\ \vdots \\ a_l \end{bmatrix} = \begin{bmatrix} \mathbf{q}_1^T \\ \mathbf{q}_2^T \\ \vdots \\ \mathbf{q}_l^T \end{bmatrix} \mathbf{x}, \qquad l \leq m \qquad (8.29)$$

The linear projection of Eq. (8.29) from $\mathbb{R}^m$ to $\mathbb{R}^l$ (i.e., the mapping from the data space to the feature space) represents an *encoder* for the approximate representation of the data vector $\mathbf{x}$ as illustrated in Fig. 8.2a. Correspondingly, the linear projection of Eq. (8.28)

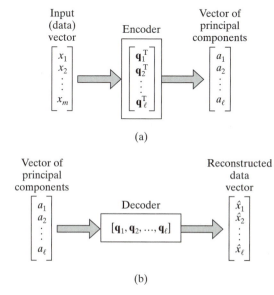

(a)

(b)

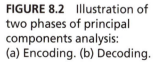

FIGURE 8.2   Illustration of two phases of principal components analysis: (a) Encoding. (b) Decoding.

from $\mathbb{R}^l$ to $\mathbb{R}^m$ (i.e., the mapping from the feature space back to the data space) represents a *decoder* for the approximate reconstruction of the original data vector $\mathbf{x}$, as illustrated in Fig. 8.2b. Note that the *dominant* (i.e., largest) eigenvalues $\lambda_1, \lambda_2, ..., \lambda_l$ do not enter the computations described in Eqs. (8.28) and (8.29); they merely determine the number of principal components used for encoding and decoding, respectively.

The *approximation error vector* $\mathbf{e}$ equals the difference between the original data vector $\mathbf{x}$ and the approximating data vector $\hat{\mathbf{x}}$, as shown by

$$\mathbf{e} = \mathbf{x} - \hat{\mathbf{x}} \tag{8.30}$$

Substituting Eqs. (8.27) and (8.28) in (8.30) yields

$$\mathbf{e} = \sum_{j=l+1}^{m} a_j \mathbf{q}_j \tag{8.31}$$

The error vector $\mathbf{e}$ *is orthogonal to the approximating data vector* $\hat{\mathbf{x}}$, as illustrated in Fig. 8.3. In other words, the inner product of the vectors $\hat{\mathbf{x}}$ and $\mathbf{e}$ is zero. This property is shown by using Eqs. (8.28) and (8.31) as follows:

$$\begin{aligned}
\mathbf{e}^T \hat{\mathbf{x}} &= \sum_{i=l+1}^{m} a_i \mathbf{q}_i^T \sum_{j=1}^{l} a_j \mathbf{q}_j \\
&= \sum_{i=l+1}^{m} \sum_{j=1}^{l} a_i a_j \mathbf{q}_i^T \mathbf{q}_j \\
&= 0
\end{aligned} \tag{8.32}$$

where we have made use of the second condition in Eq. (8.19). Equation (8.32) is known as the *principle of orthogonality*.

The total variance of the m components of the data vector $\mathbf{x}$ is, via Eq. (8.7) and the first line of Eq. (8.22),

$$\sum_{j=1}^{m} \sigma_j^2 = \sum_{j=1}^{m} \lambda_j \tag{8.33}$$

where σ_j^2 is the variance of the jth principal component a_j. The total variance of the l elements of the approximating vector $\hat{\mathbf{x}}$ is

$$\sum_{j=1}^{l} \sigma_j^2 = \sum_{j=1}^{l} \lambda_j \tag{8.34}$$

The total variance of the $(l - m)$ elements in the approximation error vector $\mathbf{x} - \hat{\mathbf{x}}$ is therefore

$$\sum_{j=l+1}^{m} \sigma_j^2 = \sum_{j=l+1}^{m} \lambda_j \tag{8.35}$$

FIGURE 8.3 Illustration of the relationship between vector $\mathbf{x}$, its reconstructed version $\hat{\mathbf{x}}$, and error vector $\mathbf{e}$.

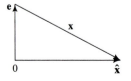

The eigenvalues $\lambda_{l+1}, \ldots, \lambda_m$ are the *smallest* $(m - l)$ eigenvalues of the correlation matrix $\mathbf{R}$; they correspond to the terms discarded from the expansion of Eq. (8.28) used to construct the approximating vector $\hat{\mathbf{x}}$. The closer all these eigenvalues are to zero, the more effective the dimensionality reduction (resulting from the application of the principal components analysis to the data vector $\mathbf{x}$) will be in preserving the information content of the original input data. Thus, to perform dimensionality reduction on some input data, we *compute the eigenvalues and eigenvectors of the correlation matrix of the input data vector, and then project the data orthogonally onto the subspace spanned by the eigenvectors belonging to the dominant eigenvalues.* This method of data representation is commonly referred to as *subspace decomposition* (Oja, 1983).

Example 8.1   Bivariate Data Set

To illustrate the application of principal components analysis, consider the example of a bivariate (two-dimensional) data set depicted in Fig. 8.4, where it is assumed that both feature axes are approximately of the same scale. The horizontal and vertical axes of the diagram represent the natural coordinates of the data set. The rotated axes labeled 1 and 2 result from the application of principal components analysis to this data set. From Fig. 8.4 we see that projecting the data set onto axis 1 captures the salient feature of the data, namely the fact that the data set is bimodal (i.e., there are two clusters in its structure). Indeed, the variance of the projections of the data points onto axis 1 is greater than that for any other projection axis in the figure. By contrast, the

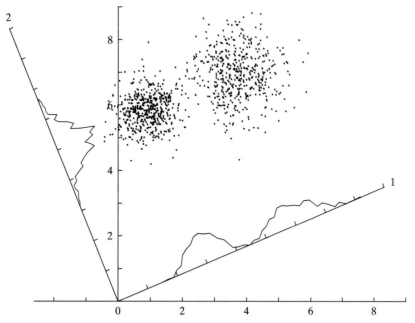

FIGURE 8.4   A cloud of data points is shown in two dimensions, and the density plots formed by projecting this cloud onto each of two axes, 1 and 2, are indicated. The projection onto axis 1 has maximum variance, and clearly shows the bimodal, or clustered character of the data.

inherent bimodal nature of the data set is completely obscured when it is projected onto the orthogonal axis 2.

The important point to note from this simple example is that although the cluster structure of the data set is evident from the two-dimensional plot of the raw data displayed in the framework of the horizontal and vertical axes, this is not always the case in practice. In the more general case of high-dimensional data sets, it is quite conceivable to have the intrinsic cluster structure of the data concealed, and to see it we must perform a statistical analysis similar to principal components analysis (Linsker, 1988a).

■

8.4 HEBBIAN-BASED MAXIMUM EIGENFILTER

There is a close correspondence between the behavior of self-organized neural networks and the statistical method of principal components analysis. In this section we demonstrate this correspondence by establishing a remarkable result: A single linear neuron with a Hebbian-type adaptation rule for its synaptic weights can evolve into a filter for the first principal component of the input distribution (Oja, 1982).

To proceed with the demonstration, consider the simple neuronal model depicted in Fig. 8.5a. The model is *linear* in the sense that the model output is a linear combination of its inputs. The neuron receives a set of m input signals $x_1, x_2, ..., x_m$ through a corresponding set of m synapses with weights $w_1, w_2, ..., w_m$, respectively. The resulting model output y is thus defined by

$$y = \sum_{i=1}^{m} w_i x_i \tag{8.36}$$

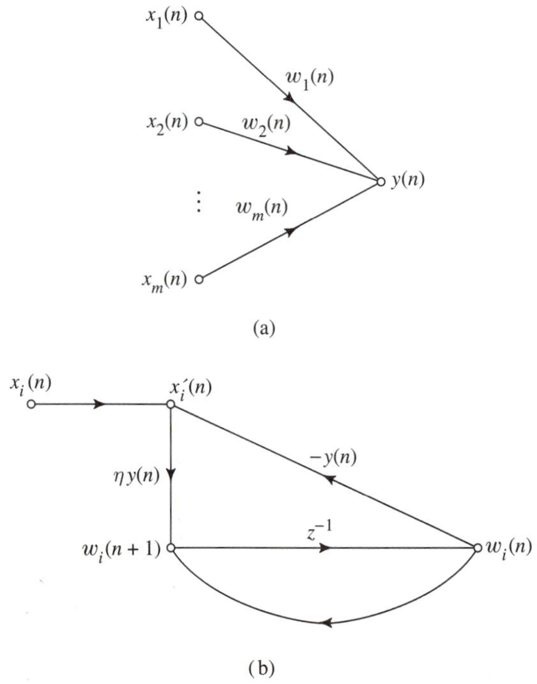

FIGURE 8.5 Signal-flow graph representation of maximum eigenfilter.
(a) Graph of Eq. (8.36).
(b) Graph of Eqs. (8.41) and (8.42).

Note that in the situation described here we have a single neuron to deal with, so there is no need to use double subscripts to identify the synaptic weights of the network.

In accordance with Hebb's postulate of learning, a synaptic weight w_i varies with time, growing strong when the presynaptic signal x_i and postsynaptic signal y coincide with each other. Specifically, we write

$$w_i(n + 1) = w_i(n) + \eta y(n)x_i(n), \qquad i = 1, 2, ..., m \tag{8.37}$$

where n denotes discrete time and η is the *learning-rate parameter*. However, this learning rule in its basic form leads to unlimited growth of the synaptic weight w_i, which is unacceptable on physical grounds. We may overcome this problem by incorporating some form of *saturation* or *normalization* in the learning rule for the adaptation of synaptic weights. The use of normalization has the effect of introducing competition among the synapses of the neuron over limited resources, which, from Principle 2 of self-organization, is essential for stabilization. From a mathematical point of view, a convenient form of normalization is described by (Oja, 1982):

$$w_i(n + 1) = \frac{w_i(n) + \eta y(n)x_i(n)}{(\sum_{i=1}^{m} [w_i(n) + \eta y(n)x_i(n)]^2)^{1/2}} \tag{8.38}$$

where the summation in the denominator extends over the complete set of synapses associated with the neuron. Assuming that the learning-rate parameter η is small, we may expand Eq. (8.38) as a power series in η, and so write

$$w_i(n + 1) = w_i(n) + \eta y(n)[x_i(n) - y(n)w_i(n)] + O(\eta^2) \tag{8.39}$$

where the term $O(\eta^2)$ represents second- and higher-order effects in η. For small η, we may justifiably ignore this term, and therefore approximate Eq. (8.38) to first order in η as follows:

$$w_i(n + 1) = w_i(n) + \eta y(n)[x_i(n) - y(n)w_i(n)] \tag{8.40}$$

The term $y(n)x_i(n)$ on the right-hand side of Eq. (8.40) represents the usual Hebbian modifications to synaptic weight w_i, and therefore accounts for the self-amplification effect dictated by Principle 1 of self-organization. The inclusion of the negative term $-y(n)w_i(n)$ is responsible for stabilization in accordance with Principle 2; it modifies the input $x_i(n)$ into a form that is dependent on the associated synaptic weight $w_i(n)$ and the output $y(n)$, as shown by

$$x_i'(n) = x_i(n) - y(n)w_i(n) \tag{8.41}$$

which may be viewed as the *effective input* of the ith synapse. We may now use the definition given in Eq. (8.41) to rewrite the learning rule of Eq. (8.40) as follows:

$$w_i(n + 1) = w_i(n) + \eta y(n)x_i'(n) \tag{8.42}$$

The overall operation of the neuron is represented by a combination of two signal-flow graphs, as shown in Fig. 8.5. The signal-flow graph of Fig. 8.5a shows the dependence of the output $y(n)$ on the weights $w_1(n), w_2(n), ..., w_m(n)$, in accordance with Eq. (8.36). The signal-flow graph of Fig. 8.5b provides a portrayal of Eqs. (8.41) and (8.42); the transmittance z^{-1} in the middle portion of the graph represents a unit-delay operator. The output signal $y(n)$ produced in Fig. 8.5a acts as a transmittance in Fig. 8.5b. The graph of

Fig. 8.5b clearly exhibits the following two forms of internal feedback acting on the neuron:

- Positive feedback for self-amplification and therefore growth of the synaptic weight $w_i(n)$, according to its external input $x_i(n)$.
- Negative feedback due to $-y(n)$ for controlling the growth, thereby resulting in stabilization of the synaptic weight $w_i(n)$.

The product term $-y(n)w_i(n)$ is related to a *forgetting* or *leakage factor* that is frequently used in learning rules, but with a difference: The forgetting factor becomes more pronounced with a stronger response $y(n)$. This kind of control appears to have neurobiological support (Stent, 1973).

Matrix Formulation of the Algorithm

For convenience of presentation, let

$$\mathbf{x}(n) = [x_1(n), x_2(n), ..., x_m(n)]^T \tag{8.43}$$

and

$$\mathbf{w}(n) = [w_1(n), w_2(n), ..., w_m(n)]^T \tag{8.44}$$

The input vector $\mathbf{x}(n)$ and the synaptic weight vector $\mathbf{w}(n)$ are typically both realizations of random vectors. Using this vector notation we may rewrite Eq. (8.36) in the form of an inner product as follows:

$$y(n) = \mathbf{x}^T(n)\mathbf{w}(n) = \mathbf{w}^T(n)\mathbf{x}(n) \tag{8.45}$$

Similarly, we may rewrite Eq. (8.40) as

$$\mathbf{w}(n + 1) = \mathbf{w}(n) + \eta y(n)[\mathbf{x}(n) - y(n)\mathbf{w}(n)] \tag{8.46}$$

Hence, substituting Eq. (8.45) in (8.46) yields

$$\mathbf{w}(n + 1) = \mathbf{w}(n) + \eta[\mathbf{x}(n)\mathbf{x}^T(n)\mathbf{w}(n) - \mathbf{w}^T(n)\mathbf{x}(n)\mathbf{x}^T(n)\mathbf{w}(n)\mathbf{w}(n)] \tag{8.47}$$

The learning algorithm of Eq. (8.47) represents a *nonlinear stochastic difference equation,* which makes convergence analysis of the algorithm mathematically difficult. To pave the way for this convergence analysis, we will digress briefly to introduce a general tool for convergence analysis of stochastic approximation algorithms.

Asymptotic Stability Theorem

The self-organized learning algorithm of Eq. (8.47) is a special case of the generic stochastic approximation algorithm

$$\mathbf{w}(n + 1) = \mathbf{w}(n) + \eta(n)h(\mathbf{w}(n),\mathbf{x}(n)), \qquad n = 0, 1, 2, ..., \tag{8.48}$$

The sequence $\eta(\cdot)$ is assumed to be a sequence of positive scalars.

The *update function* $h(\cdot,\cdot)$ is a deterministic function with some regularity conditions imposed on it. This function, together with the scalar sequence $\eta(\cdot)$, specify the complete structure of the algorithm.

The goal of the procedure described here is to associate a *deterministic ordinary differential equation (ODE)* with the stochastic nonlinear difference equation (8.48). The stability properties of the differential equation are then tied to the convergence properties of the algorithm. This procedure is a fairly general tool and has wide applicability. It was developed independently by Ljung (1977) and by Kushner and Clark (1978), who used different approaches.[2]

To begin with, the procedure assumes that the stochastic approximation algorithm described by Eq. (8.48) satisfies the following set of conditions, using our terminology:

1. The $\eta(n)$ is a decreasing sequence of positive real numbers, such that we have:

   (a)
   $$\sum_{n=1}^{\infty} \eta(n) = \infty \tag{8.49}$$

   (b)
   $$\sum_{n=1}^{\infty} \eta^p(n) < \infty \qquad \text{for } p > 1 \tag{8.50}$$

   (c)
   $$\eta(n) \to 0 \qquad \text{as } n \to \infty \tag{8.51}$$

2. The sequence of parameter (synaptic weight) vectors $\mathbf{w}(\cdot)$ is bounded with probability 1.
3. The update function $h(\mathbf{w}, \mathbf{x})$ is continuously differentiable with respect to $\mathbf{w}$ and $\mathbf{x}$, and its derivatives are bounded in time.
4. The limit

$$\bar{h}(\mathbf{w}) = \lim_{n \to \infty} E[h(\mathbf{w}, \mathbf{X})] \tag{8.52}$$

   exists for each $\mathbf{w}$; the statistical expectation operator E is over the random vector $\mathbf{X}$ with a realization denoted by $\mathbf{x}$.
5. There is a locally asymptotically stable (in the sense of Lyapunov) solution to the ordinary differential equation

$$\frac{d}{dt} \mathbf{w}(t) = \bar{h}(\mathbf{w}(t)) \tag{8.53}$$

   where t denotes continuous time; stability in the sense of Lyapunov is discussed in Chapter 14.
6. Let $\mathbf{q}_1$ denote the solution to Eq. (8.53) with a basin of attraction $\mathscr{B}(\mathbf{q})$; basin of attraction is defined in Chapter 14. Then the parameter vector $\mathbf{w}(n)$ enters a compact subset $\mathcal{A}$ of the basin of attraction $\mathscr{B}(\mathbf{q})$ infinitely often, with probability 1.

The six conditions described here are all reasonable. In particular, condition 1(a) is a necessary condition that makes it possible for the algorithm to move the estimate to a desired limit, regardless of the initial conditions. Condition 1(b) gives a condition on how fast $\eta(n)$ must tend to zero; it is considerably less restrictive than the usual condition

$$\sum_{n=1}^{\infty} \eta^2(n) < \infty$$

Condition 4 is the basic assumption that makes it possible to associate a differential equation with the algorithm of Eq. (8.48).

Consider then, a stochastic approximation algorithm described by the recursive equation (8.48), subject to assumptions 1 through 6. We may then state the *asymptotic stability theorem* for this class of stochastic approximation algorithms as follows (Ljung, 1977; Kushner and Clark, 1978):

$$\lim_{n \to \infty} \mathbf{w}(n) = \mathbf{q}_1, \quad \text{infinitely often with probability 1} \tag{8.54}$$

We emphasize, however, that although the procedure described here can provide us with information about asymptotic properties of the algorithm (8.48), it usually does not tell us how large the number of iterations n has to be for the results of the analysis to be applicable. Moreover, in tracking problems where a time-varying parameter vector is to be tracked using algorithm (8.48), it is not feasible to require

$$\eta(n) \to 0 \quad \text{as } n \to \infty$$

as stipulated by condition 1(c). We may overcome this latter difficulty by assigning some small, positive value to η, the size of which usually depends on the application of interest. This is usually done in the practical use of stochastic approximation algorithms in neural networks.

Stability Analysis of the Maximum Eigenfilter

In the ODE approach to stability, we have the tool we need to investigate the convergence behavior of the recursive algorithm of Eq. (8.46) pertaining to a maximum eigenfilter, as described here.

To satisfy condition 1 of the asymptotic stability theorem, we let

$$\eta(n) = \frac{1}{n}$$

Next, we note from Eq. (8.47) that the update function $h(\mathbf{w}, \mathbf{x})$ is defined by

$$
\begin{aligned}
h(\mathbf{w}, \mathbf{x}) &= \mathbf{x}(n)y(n) - y^2(n)\mathbf{w}(n) \\
&= \mathbf{x}(n)\mathbf{x}^T(n)\mathbf{w}(n) - [\mathbf{w}^T(n)\mathbf{x}(n)\mathbf{x}^T(n)\mathbf{w}(n)]\mathbf{w}(n)
\end{aligned}
\tag{8.55}
$$

which clearly satisfies condition 3 of the theorem. Equation (8.55) results from the use of a realization $\mathbf{x}$ of the random vector $\mathbf{X}$ in the update function $h(\mathbf{w}, \mathbf{X})$. For condition 4 we take the expectation of $h(\mathbf{w}, \mathbf{X})$ over $\mathbf{X}$, and thus write

$$
\begin{aligned}
\overline{h} &= \lim_{n \to \infty} E[\mathbf{X}(n)\mathbf{X}^T(n)\mathbf{w}(n) - (\mathbf{w}^T(n)\mathbf{X}(n)\mathbf{X}^T(n)\mathbf{w}(n))\mathbf{w}(n)] \\
&= \mathbf{R}\mathbf{w}(\infty) - [\mathbf{w}^T(\infty)\mathbf{R}\mathbf{w}(\infty)]\mathbf{w}(\infty)
\end{aligned}
\tag{8.56}
$$

where $\mathbf{R}$ is the correlation matrix of the stochastic process represented by the random vector $\mathbf{X}(n)$, and $\mathbf{w}(\infty)$ is the limiting value of the synaptic weight vector.

In accordance with condition 5 and in light of Eqs. (8.53) and (8.56), we seek stable points of the nonlinear differential equation

$$
\begin{aligned}
\frac{d}{dt}\mathbf{w}(t) &= \overline{h}(\mathbf{w}(t)) \\
&= \mathbf{R}\mathbf{w}(t) - [\mathbf{w}^T(t)\mathbf{R}\mathbf{w}(t)]\mathbf{w}(t)
\end{aligned}
\tag{8.57}
$$

Let $\mathbf{w}(t)$ be expanded in terms of the complete orthonormal set of eigenvectors of the correlation matrix $\mathbf{R}$ as follows:

$$\mathbf{w}(t) = \sum_{k=1}^{m} \theta_k(t)\mathbf{q}_k \tag{8.58}$$

where $\mathbf{q}_k$ is the kth normalized eigenvector of the matrix $\mathbf{R}$, and the coefficient $\theta_k(t)$ is the time-varying projection of the vector $\mathbf{w}(t)$ onto $\mathbf{q}_k$. Substituting Eq. (8.58) in (8.57), and using the basic definitions

$$\mathbf{R}\mathbf{q}_k = \lambda_k\mathbf{q}_k$$

and

$$\mathbf{q}_k^T\mathbf{R}\mathbf{q}_k = \lambda_k$$

where λ_k is the eigenvalue associated with $\mathbf{q}_k$, we finally get

$$\sum_{k=1}^{m} \frac{d\theta_k(t)}{dt}\mathbf{q}_k = \sum_{k=1}^{m} \lambda_k\theta_k(t)\mathbf{q}_k - \left[\sum_{l=1}^{m} \lambda_l\theta_l^2(t)\right]\sum_{k=1}^{m} \theta_k(t)\mathbf{q}_k \tag{8.59}$$

Equivalently, we may write

$$\frac{d\theta_k(t)}{dt} = \lambda_k\theta_k(t) - \theta_k(t)\sum_{l=1}^{m} \lambda_l\theta_l^2(t), \quad k = 1, 2, \ldots, m \tag{8.60}$$

We have thus reduced the convergence analysis of the stochastic approximation algorithm of (8.48) to the stability analysis of a system of ordinary differential equations (8.60) involving the *principal modes* $\theta_k(t)$.

There are two cases to be considered here, depending on the value assigned to the index k. Case I corresponds to $1 < k \leq m$, and case II corresponds to $k = 1$; m is the dimension of both $\mathbf{x}(n)$ and $\mathbf{w}(n)$. These two cases are considered in turn.

Case I. $1 < k \leq m$. For the treatment of this case we define

$$\alpha_k(t) = \frac{\theta_k(t)}{\theta_1(t)}, \quad 1 < k \leq m \tag{8.61}$$

Hence it is assumed that $\theta_1(t) \neq 0$, which is true with probability 1 provided that *the initial values $\mathbf{w}(0)$ are chosen at random.* Then, differentiating both sides of Eq. (8.61) with respect to time t, we get

$$\frac{d\alpha_k(t)}{dt} = \frac{1}{\theta_1(t)}\frac{d\theta_k(t)}{dt} - \frac{\theta_k(t)}{\theta_1^2(t)}\frac{d\theta_1(t)}{dt}$$

$$= \frac{1}{\theta_1(t)}\frac{d\theta_k(t)}{dt} - \frac{\alpha_k(t)}{\theta_1(t)}\frac{d\theta_1(t)}{dt}, \quad 1 < k \leq m \tag{8.62}$$

Next, using Eq. (8.60) in (8.62), applying the definition of Eq. (8.61), and then simplifying the result, we get

$$\frac{d\alpha_k(t)}{dt} = -(\lambda_1 - \lambda_k)\alpha_k(t), \quad 1 < k \leq m \tag{8.63}$$

With the eigenvalues of the correlation matrix $\mathbf{R}$ assumed to be distinct and arranged in decreasing order, we have

$$\lambda_1 > \lambda_2 > \cdots > \lambda_k > \cdots > \lambda_m > 0 \tag{8.64}$$

It follows therefore that the eigenvalue difference $\lambda_1 - \lambda_k$, representing the reciprocal of a time constant in Eq. (8.63), is positive, so we find that for case I:

$$\alpha_k(t) \to 0 \quad \text{as } t \to \infty \quad \text{for } 1 < k \le m \tag{8.65}$$

Case II. $k = 1$. From Eq. (8.60), this second case is described by the differential equation

$$
\begin{aligned}
\frac{d\theta_1(t)}{dt} &= \lambda_1\theta_1(t) - \theta_1(t) \sum_{l=1}^{m} \lambda_l\theta_l^2(t) \\
&= \lambda_1\theta_1(t) - \lambda_1\theta_1^3(t) - \theta_1(t) \sum_{l=2}^{m} \lambda_l\theta_l^2(t) \tag{8.66} \\
&= \lambda_1\theta_1(t) - \lambda_1\theta_1^3(t) - \theta_1^3(t) \sum_{l=2}^{m} \lambda_l\alpha_l^2(t)
\end{aligned}
$$

However, from case I we know that $\alpha_l \to 0$ for $l \neq 1$ as $t \to \infty$. Hence the last term on the right-hand side of Eq. (8.66) approaches zero as time t approaches infinity. Ignoring this term, Eq. (8.66) simplifies to

$$\frac{d\theta_1(t)}{dt} = \lambda_1\theta_1(t)[1 - \theta_1^2(t)] \quad \text{for } t \to \infty \tag{8.67}$$

It must be emphasized, however, that Eq. (8.67) holds only in an asymptotic sense.

Equation (8.67) represents an *autonomous system* (i.e., a system with no explicit time dependence). The stability of such a system is best handled using a positive-definite function called the *Lyapunov function,* a detailed treatment of which is deferred to Chapter 14. Let $\mathbf{s}$ denote the state vector of an autonomous system, and $V(t)$ denote a Lyapunov function of the system. An equilibrium state $\bar{\mathbf{s}}$ of the system is asymptotically stable if

$$\frac{d}{dt} V(t) < 0 \quad \text{for } \mathbf{s} \in \mathcal{U} - \bar{\mathbf{s}}$$

where $\mathcal{U}$ is a small neighborhood around $\bar{\mathbf{s}}$.

For the problem at hand, we assert that the differential equation (8.67) has a Lyapunov function defined by

$$V(t) = [\theta_1^2(t) - 1]^2 \tag{8.68}$$

To validate this assertion, we must show that $V(t)$ satisfies two conditions:

1. $\dfrac{dV(t)}{dt} < 0 \quad \text{for all } t$ $\hspace{3cm}$ (8.69)

2. $V(t)$ has a minimum $\hspace{5.5cm}$ (8.70)

Differentiating Eq. (8.68) with respect to time, we get

$$\frac{dV(t)}{dt} = 4\theta_1(t)[\theta_1(t) - 1]\frac{d\theta_1}{dt} \tag{8.71}$$

$$= -4\lambda_1\theta_1^2(t)\,[\theta_1^2(t) - 1]^2 \qquad \text{for } t \to \infty$$

where in the second line we have made use of Eq. (8.67). Since the eigenvalue λ_1 is positive, we find from Eq. (8.71) that the condition of Eq. (8.69) is true for t approaching infinity. Furthermore, from Eq. (8.71) we note that $V(t)$ has a minimum [i.e., $dV(t)/dt$ is zero] at $\theta_1(t) = \pm 1$, and so the condition of Eq. (8.70) is also satisfied. We may therefore conclude the analysis of case II by stating that

$$\theta_1(t) \to \pm 1 \qquad \text{as } t \to \infty \tag{8.72}$$

In light of the result described in Eq. (8.72) and the definition of Eq. (8.71), we may restate the result of case I given in Eq. (8.65) in its final form:

$$\theta_k(t) \to 0 \qquad \text{as } t \to \infty \qquad \text{for } 1 < k \le m \tag{8.73}$$

The overall conclusion drawn from the analysis of cases I and II is twofold:

- The only principal mode of the stochastic approximation algorithm described in Eq. (8.47) that will converge is $\theta_1(t)$; all the other modes of the algorithm will decay to zero.
- The mode $\theta_1(t)$ will converge to ± 1.

Hence, condition 5 of the asymptotic stability theorem is satisfied. Specifically, in light of the expansion described in Eq. (8.58), we may formally state that

$$\mathbf{w}(t) \to \mathbf{q}_1 \qquad \text{as } t \to \infty$$

where $\mathbf{q}_1$ is the normalized eigenvector associated with the largest eigenvalue λ_1 of the correlation matrix $\mathbf{R}$.

We must next show that, in accordance with condition 6 of the asymptotic stability theorem, there exists a subset $\mathscr{A}$ of the set of all vectors, such that

$$\lim_{n \to \infty} \mathbf{w}(n) = \mathbf{q}_1 \qquad \text{infinitely often with probability 1}$$

To do so, we must first satisfy condition 2, which we do by *hard-limiting* the entries of $\mathbf{w}(n)$ so that their magnitudes remain below some threshold a. We may then define the norm of $\mathbf{w}(n)$ by writing

$$\|\mathbf{w}(n)\| = \max_j |w_j(n)| \le a \tag{8.74}$$

Let $\mathscr{A}$ be the compact subset of $\mathbb{R}^m$ defined by the set of vectors with norm less than or equal to a. It is straightforward to show that (Sanger, 1989b)

If $\|\mathbf{w}(n)\| \le a$, and the constant a is sufficiently large, then $\|\mathbf{w}(n + 1)\| < \|\mathbf{w}(n)\|$ with probability 1.

Thus, as the number of iterations n increases, $\mathbf{w}(n)$ will eventually be within $\mathscr{A}$, and it will remain inside $\mathscr{A}$ (infinitely often) with probability 1. Since the basin of attraction

$\mathscr{B}(\mathbf{q}_1)$ includes all vectors with bounded norm, we have $\mathscr{A} \in \mathscr{B}(\mathbf{q}_1)$. In other words, condition 6 is satisfied.

We have now satisfied all six conditions of the asymptotic stability theorem, and thereby shown that (subject to the aforementioned assumptions) the stochastic approximation algorithm of (8.47) will cause $\mathbf{w}(n)$ to coverge with probability 1 to the eigenvector $\mathbf{q}_1$ associated with the largest eigenvalue λ_1 of the correlation matrix $\mathbf{R}$. This is not the only fixed point of the algorithm, but it is the only one that is asymptotically stable.

Summarizing Properties of the Hebbian-Based Maximum Eigenfilter

The convergence analysis just presented shows that a single linear neuron governed by the self-organized learning rule of Eq. (8.39), or equivalently that of Eq. (8.46), adaptively extracts the first principal component of a stationary input. This first principal component corresponds to the largest eigenvalue λ_1 of the correlation matrix of the random vector $\mathbf{X}(n)$; in fact, λ_1 is related to the variance of the model output $y(n)$, as shown here.

Let $\sigma^2(n)$ denote the variance of random variable $Y(n)$ with a realization of it denoted by $y(n)$, that is,

$$\sigma^2(n) = E[Y^2(n)] \tag{8.75}$$

where $Y(n)$ has zero mean for a zero-mean input. Letting $n \to \infty$ in Eq. (8.46) and using the fact that, in a corresponding way, $\mathbf{w}(n)$ approaches $\mathbf{q}_1$, we obtain

$$\mathbf{x}(n) = y(n)\mathbf{q}_1 \qquad \text{for } n \to \infty$$

Using this relation, we can show that the variance $\sigma^2(n)$ approaches λ_1 as the number of iterations n approaches infinity; see Problem 8.2.

In summary, a Hebbian-based linear neuron whose operation is described by Eq. (8.46) converges with probability 1 to a fixed point, which is characterized as follows (Oja, 1982):

1. The variance of the model output approaches the largest eigenvalue of the correlation matrix $\mathbf{R}$, as shown by

$$\lim_{n \to \infty} \sigma^2(n) = \lambda_1 \tag{8.76}$$

2. The synaptic weight vector of the model approaches the associated eigenvector, as shown by

$$\lim_{n \to \infty} \mathbf{w}(n) = \mathbf{q}_1 \tag{8.77}$$

with

$$\lim_{n \to \infty} \|\mathbf{w}(n)\| = 1 \tag{8.78}$$

These results assume that the correlation matrix $\mathbf{R}$ is positive definite with the largest eigenvalue λ_1 having multiplicity 1. They also hold for a nonnegative definite correlation matrix $\mathbf{R}$ provided that $\lambda_1 > 0$ with multiplicity 1.

Example 8.2   Matched Filter

Consider a random vector $\mathbf{X}(n)$ composed as follows:

$$\mathbf{X}(n) = \mathbf{s} + \mathbf{V}(n)$$

where $\mathbf{s}$ is a fixed unit vector representing the *signal component,* and $\mathbf{V}(n)$ is a zero-mean *white-noise component.* The correlation matrix of the input vector is

$$\mathbf{R} = E[\mathbf{X}(n)\mathbf{X}^T(n)]$$

$$= \mathbf{s}\mathbf{s}^T + \sigma^2\mathbf{I}$$

where σ^2 is the variance of the elements of the noise vector $\mathbf{V}(n)$, and $\mathbf{I}$ is the identity matrix. The largest eigenvalue of the correlation matrix $\mathbf{R}$ is therefore

$$\lambda_1 = 1 + \sigma^2$$

The associated eigenvector $\mathbf{q}_1$ is

$$\mathbf{q}_1 = \mathbf{s}$$

It is readily shown that this solution satisfies the eigenvalue problem

$$\mathbf{R}\mathbf{q}_1 = \lambda_1\mathbf{q}_1$$

Hence, for the situation described in this example, the self-organized linear neuron (upon convergence to its stable condition) acts as a *matched filter* in the sense that its impulse response (represented by the synaptic weights) is matched to the signal component $\mathbf{s}$ of the input vector $\mathbf{X}(n)$. ∎

8.5   HEBBIAN-BASED PRINCIPAL COMPONENTS ANALYSIS

The Hebbian-based maximum eigenfilter of the previous section extracts the first principal component of the input. This single linear neuronal model may be expanded into a feedforward network with a single layer of linear neurons for the purpose of principal components analysis of arbitrary size on the input (Sanger, 1989b).

To be specific, consider the feedforward network shown in Fig. 8.6. The following two assumptions of a structural nature are made:

1. Each neuron in the output layer of the network is *linear.*
2. The network has m inputs and l outputs, both of which are specified. Moreover, the network has fewer outputs than inputs (i.e., $l < m$).

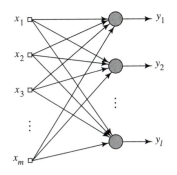

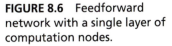

FIGURE 8.6   Feedforward network with a single layer of computation nodes.

The only aspect of the network that is subject to training is the set of synaptic weights $\{w_{ji}\}$ connecting source nodes i in the input layer to computation nodes j in the output layer, where $i = 1, 2, ..., m$, and $j = 1, 2, ..., l$.

The output $y_j(n)$ of neuron j at time n, produced in response to the set of inputs $\{x_i(n)|i = 1, 2, ..., m\}$ is given by (see Fig 8.7a)

$$y_j(n) = \sum_{i=1}^{m} w_{ji}(n)x_i(n), \qquad j = 1, 2, ..., l \tag{8.79}$$

The synaptic weight $w_{ji}(n)$ is adapted in accordance with a generalized form of Hebbian learning, as shown by (Sanger, 1989b):

$$\Delta w_{ji}(n) = \eta \left[y_j(n)x_i(n) - y_j(n) \sum_{k=1}^{j} w_{ki}(n)y_k(n) \right], \qquad \begin{matrix} i = 1, 2, ..., m \\ j = 1, 2, ..., l \end{matrix} \tag{8.80}$$

where $\Delta w_{ji}(n)$ is the change applied to the synaptic weight $w_{ji}(n)$ at time n, and η is the learning-rate parameter. The *generalized Hebbian algorithm* (GHA) of Eq. (8.80) for a layer of l neurons includes the algorithm of Eq. (8.39) for a single neuron as a special case, that is, $j = 1$.

To develop insight into the behavior of the generalized Hebbian algorithm, we rewrite Eq. (8.80) in the form

$$\Delta w_{ji}(n) = \eta y_j(n)[x_i'(n) - w_{ji}(n)y_j(n)], \qquad \begin{matrix} i = 1, 2, ..., m \\ j = 1, 2, ..., l \end{matrix} \tag{8.81}$$

where $x_i'(n)$ is a modified version of the ith element of the input vector $\mathbf{x}(n)$; it is a function of the index j, as shown by

$$x_i'(n) = x_i(n) - \sum_{k=1}^{j-1} w_{ki}(n)y_k(n) \tag{8.82}$$

For a specified neuron j, the algorithm described in Eq. (8.81) has exactly the same mathematical form as that of Eq. (8.39), except for the fact that the input signal $x_i(n)$ is replaced by its modified value $x_i'(n)$ in Eq. (8.82). We may go one step further and rewrite Eq. (8.81) in a form that corresponds to Hebb's postulate of learning, as shown by

$$\Delta w_{ji}(n) = \eta y_j(n)x_i''(n) \tag{8.83}$$

where

$$x_i''(n) = x_i' - w_{ji}(n)y_j(n) \tag{8.84}$$

Thus noting that

$$w_{ji}(n + 1) = w_{ji}(n) + \Delta w_{ji}(n) \tag{8.85}$$

and

$$w_{ji}(n) = z^{-1}[w_{ji}(n + 1)] \tag{8.86}$$

where z^{-1} is the unit-delay operator, we may construct the signal-flow graph of Fig. 8.7b for the generalized Hebbian algorithm. From this graph we see that the algorithm lends itself to a *local* form of implementation, provided that it is formulated as in

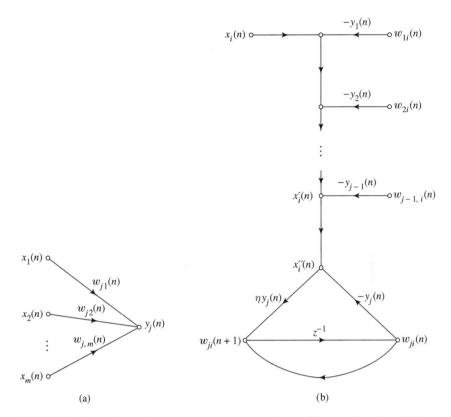

FIGURE 8.7 The signal-flow graph representation of generalized Hebbian algorithm. (a) Graph of Eq. (8.79). (b) Graph of Eqs. (8.80) through (8.81).

Eq. (8.85). Note also that $y_j(n)$, responsible for feedback in the signal-flow graph of Fig. 8.7b, is itself determined by Eq. (8.79); signal-flow graph representation of this latter equation is shown in Fig. 8.7a.

For a heuristic understanding of how the generalized Hebbian algorithm actually operates, we first use matrix notation to rewrite the version of the algorithm defined in Eq. (8.81) as follows:

$$\Delta\mathbf{w}_j(n) = \eta y_j(n)\mathbf{x}'(n) - \eta y_j^2(n)\mathbf{w}_j(n), \qquad j = 1, 2, ..., l \tag{8.87}$$

where

$$\mathbf{x}'(n) = \mathbf{x}(n) - \sum_{k=1}^{j-1} \mathbf{w}_k(n)y_k(n) \tag{8.88}$$

The vector $\mathbf{x}'(n)$ represents a modified form of the input vector. Based on the representation given in Eq. (8.87), we make the following observations (Sanger, 1989b):

1. For the first neuron of the feedforward network shown in Fig. 9.6, we have

$$j = 1: \qquad \mathbf{x}'(n) = \mathbf{x}(n)$$

In this case, the generalized Hebbian algorithm reduces to that of Eq. (8.46) for a single neuron. From the material presented in Section 8.5 we already know that this neuron will discover the first principal component of the input vector $\mathbf{x}(n)$.

2. For the second neuron of the network in Fig. 8.6, we write

$$j = 2: \quad \mathbf{x}'(n) = \mathbf{x}(n) - \mathbf{w}_1(n)y_1(n)$$

Provided that the first neuron has already converged to the first principal component, the second neuron sees an input vector $\mathbf{x}'(n)$ from which the first eigenvector of the correlation matrix $\mathbf{R}$ has been removed. The second neuron therefore extracts the first principal component of $\mathbf{x}'(n)$, which is equivalent to the second principal component of the original input vector $\mathbf{x}(n)$.

3. For the third neuron we write

$$j = 3: \quad \mathbf{x}'(n) = \mathbf{x}(n) - \mathbf{w}_1(n)y_1(n) - \mathbf{w}_2(n)y_2(n)$$

Suppose that the first two neurons have already converged to the first and second principal components, as explained in steps 1 and 2. The third neuron now sees an input vector $\mathbf{x}'(n)$ from which the first two eigenvectors have been removed. Therefore, it extracts the first principal component of the vector $\mathbf{x}'(n)$, which is equivalent to the third principal component of the original input vector $\mathbf{x}(n)$.

4. Proceeding in this fashion for the remaining neurons of the feedforward network in Fig. 8.6, it is now apparent that each output of the network trained in accordance with the generalized Hebbian algorithm of Eq. (8.81) represents the response to a particular eigenvector of the correlation matrix of the input vector, and that the individual outputs are ordered by decreasing eigenvalue.

This method of computing eigenvectors is similar to a technique known as *Hotelling's deflation technique* (Kreyszig, 1988); it follows a procedure similar to Gram–Schmidt orthogonalization (Strang, 1980).

The neuron-by-neuron description presented here is intended merely to simplify the explanation. In practice, all the neurons in the generalized Hebbian algorithm tend to converge together.

Convergence Considerations

Let $\mathbf{W}(n) = \{w_{ji}(n)\}$ denote the l-by-m synaptic weight matrix of the feedforward network shown in Fig. 8.6; that is,

$$\mathbf{W}(n) = [\mathbf{w}_1(n), \mathbf{w}_2(n), ..., \mathbf{w}_l(n)]^T \tag{8.89}$$

Let the learning-rate parameter of the generalized Hebbian algorithm of Eq. (8.81) take a time-varying form $\eta(n)$, such that in the limit we have

$$\lim_{n \to \infty} \eta(n) = 0 \quad \text{and} \quad \sum_{n=0}^{\infty} \eta(n) = \infty \tag{8.90}$$

We may then rewrite this algorithm in the matrix form

$$\Delta \mathbf{W}(n) = \eta(n)\{\mathbf{y}(n)\mathbf{x}^T(n) - \text{LT}[\mathbf{y}(n)\mathbf{y}^T(n)]\mathbf{W}(n)\} \tag{8.91}$$

where the operator LT[·] sets all the elements above the diagonal of its matrix argument to zero, thereby making that matrix *lower triangular.* Under these conditions, and invoking the assumptions made in Section 8.4, convergence of the GHA algorithm is proved by following a procedure similar to that presented in the previous section for the maximum eigenfilter. Thus we may state the following theorem (Sanger, 1989b):

> If the synaptic weight matrix $\mathbf{W}(n)$ is assigned random values at time step $n = 0$, then with probability 1, the generalized Hebbian algorithm of Eq. (8.91) will converge to a fixed point with $\mathbf{W}^T(n)$ approaching a matrix whose columns are the first l eigenvectors of the m-by-m correlation matrix $\mathbf{R}$ of the m-by-l input vector, ordered by decreasing eigenvalue.

The practical significance of this theorem is that it guarantees the generalized Hebbian algorithm to find the first l eigenvectors of the correlation matrix $\mathbf{R}$, assuming that the associated eigenvalues are distinct. Equally important is the fact that we do not need to compute the correlation matrix $\mathbf{R}$. Rather, the first l eigenvectors of $\mathbf{R}$ are computed by the algorithm directly from the input data. The resulting computational savings can be enormous especially if the dimensionality m of the input space is very large, and the required number of the eigenvectors associated with the l largest eigenvalues of the correlation matrix $\mathbf{R}$ is a small fraction of m.

The convergence theorem is formulated in terms of a time-varying learning-rate parameter $\eta(n)$. In practice, the learning-rate parameter is chosen to be a small constant η, in which case convergence is guaranteed with mean-squared error in synaptic weights of order η.

In Chatterjee et al. (1998), the convergence properties of the GHA algorithm described in Eq. (8.91) are investigated. The analysis presented therein shows that increasing η leads to faster convergence and larger asymptotic mean-square error, which is intuitively satisfying. In that paper, the tradeoff between the accuracy of computation and speed of learning is made explicit, among other things.

Optimality of the Generalized Hebbian Algorithm

Suppose that in the limit we write

$$\Delta \mathbf{w}_j(n) \to \mathbf{0} \quad \text{and} \quad \mathbf{w}_j(n) \to \mathbf{q}_j \quad \text{as } n \to \infty \quad \text{for } j = 1, 2, ..., l \quad\quad (8.92)$$

and that we have

$$\|\mathbf{w}_j(n)\| = 1 \quad\quad \text{for all } j \quad\quad (8.93)$$

Then the limiting values $\mathbf{q}_1, \mathbf{q}_2, ..., \mathbf{q}_l$ of the synaptic weight vectors of the neurons in the feedforward network of Fig. 8.5 represent the *normalized eigenvectors* associated with l dominant eigenvalues of the correlation matrix $\mathbf{R}$, and which are ordered in descending eigenvalue. At equilibrium we may therefore write

$$\mathbf{q}_j^T \mathbf{R} \mathbf{q}_k = \begin{cases} \lambda_j, & k = j \\ 0, & k \neq j \end{cases} \quad\quad (8.94)$$

where $\lambda_1 > \lambda_2 > \cdots > \lambda_l$.

For the output of neuron j, we have the limiting value

$$\lim_{n \to \infty} y_j(n) = \mathbf{x}^T(n)\mathbf{q}_j = \mathbf{q}_j^T\mathbf{x}(n) \tag{8.95}$$

Let $Y_j(n)$ denote a random variable with a realization denoted by the output $y_j(n)$. The cross-correlation between the random variables $Y_j(n)$ and $Y_k(n)$ at equilibrium is given by

$$
\begin{aligned}
\lim_{n \to \infty} E[Y_j(n)Y_k(n)] &= E[\mathbf{q}_j^T\mathbf{X}(n)\mathbf{X}^T(n)\mathbf{q}_k] \\
&= \mathbf{q}_j^T\mathbf{R}\mathbf{q}_k \\
&= \begin{cases} \lambda_j, & k = j \\ 0, & k \neq j \end{cases}
\end{aligned} \tag{8.96}
$$

Hence, we may state that at equilibrium the generalized Hebbian algorithm of Eq. (8.91) acts as an *eigen-analyzer* of the input data.

Let $\hat{\mathbf{x}}(n)$ denote the particular value of the input vector $\mathbf{x}(n)$ for which the limiting conditions of Eq. (8.92) are satisfied for $j = l - 1$. Hence, from the matrix form of Eq. (8.80), we find that in the limit

$$\hat{\mathbf{x}}(n) = \sum_{k=1}^{l} y_k(n)\mathbf{q}_k \tag{8.97}$$

This means that given two sets of quantities, the limiting values $\mathbf{q}_1, \mathbf{q}_2, \ldots, \mathbf{q}_l$ of the synaptic weight vectors of the neurons in the feedforward network of Fig. 8.5 and the corresponding outputs $y_1(n), y_2(n), \ldots, y_l(n)$, we may then construct a *linear least-squares estimate* $\hat{\mathbf{x}}(n)$ of the input vector $\mathbf{x}(n)$. In effect, the formula of Eq. (8.97) may be viewed as one of *data reconstruction,* as depicted in Fig. 8.8. Note that in light of the discussion presented in Section 8.3, this method of data reconstruction is subject to an approximation error vector that is orthogonal to the estimate $\hat{\mathbf{x}}(n)$.

Summary of the GHA

The computations involved in the generalized Hebbian algorithm (GHA) are simple; they may be summarized as follows:

1. Initialize the synaptic weights of the network, w_{ji}, to small random values at time $n = 1$. Assign a small positive value to the learning-rate parameter η.

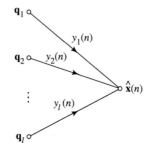

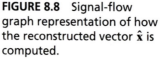

FIGURE 8.8 Signal-flow graph representation of how the reconstructed vector $\hat{\mathbf{x}}$ is computed.

2. For $n = 1, j = 1, 2, ..., l,$ and $i = 1, 2, ..., m,$ compute

$$y_j(n) = \sum_{i=1}^{m} w_{ji}(n)x_i(n)$$

$$\Delta w_{ji}(n) = \eta \left[y_j(n)x_i(n) - y_j(n) \sum_{k=1}^{j} w_{ki}(n)y_k(n) \right]$$

where $x_i(n)$ is the ith component of the m-by-1 input vector $\mathbf{x}(n)$ and l is the desired number of principal components.
3. Increment n by 1, go to step 2, and continue until the synaptic weights w_{ji} reach their steady-state values. For large n, the synaptic weight w_{ji} of neuron j converges to the ith component of the eigenvector associated with the jth eigenvalue of the correlation matrix of the input vector $\mathbf{x}(n)$.

8.6   COMPUTER EXPERIMENT: IMAGE CODING

We complete discussion of the generalized Hebbian learning algorithm by examining its use for solving an *image coding* problem.

Figure 8.9a shows an image of parents used for training; this image emphasizes *edge* information. It was digitized to form a 256 × 256 image with 256 gray levels. The image was coded using a linear feedforward network with a single layer of 8 neurons, each with 64 inputs. To train the network, 8 × 8 nonoverlapping blocks of the image were used. The experiment was performed with 2000 scans of the picture and a small learning rate $\eta = 10^{-4}$.

Figure 8.9b shows the 8 × 8 masks representing the synaptic weights learned by the network. Each of the eight masks displays the set of synaptic weights associated with a particular neuron of the network. Specifically, excitatory synapses (positive weights) are shown white, whereas inhibitory synapses (negative weights) are shown black; gray indicates zero weights. In our notation, the masks represent the columns of the 64 × 8 synaptic weight matrix $\mathbf{W}^T$ after the generalized Hebbian algorithm has converged.

To code the image, the following procedure was used:

- Each 8 × 8 block of the image was multiplied by each of the 8 masks shown in Fig. 8.9b, thereby generating 8 coefficients for image coding; Fig. 8.9c shows the reconstructed image based on the dominant 8 principal components without quantization.
- Each coefficient was uniformly quantized with a number of bits approximately proportional to the logarithm of the variance of that coefficient over the image. Thus, the first three masks were assigned 6 bits each, the next two masks 4 bits each, the next two masks 3 bits each, and the last mask 2 bits. Based on this representation, a total of 34 bits were needed to code each 8 × 8 block of pixels, resulting in a data rate of 0.53 bits per pixel.

To reconstruct the image from the quantized coefficients, all the masks were weighted by their quantized coefficients, then added to reconstitute each block of the image. The reconstructed parents' image with 15 to 1 compression ratio is shown in Fig. 8.9d.

Original image

Weights

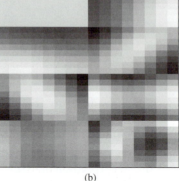

(a)

(b)

Using first 8 components

15 to 1 compression

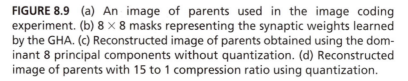

(c)

(d)

FIGURE 8.9 (a) An image of parents used in the image coding experiment. (b) 8 × 8 masks representing the synaptic weights learned by the GHA. (c) Reconstructed image of parents obtained using the dominant 8 principal components without quantization. (d) Reconstructed image of parents with 15 to 1 compression ratio using quantization.

For a variation on the first image, we next applied the generalized Hebbian algorithm to the image of an ocean scene shown in Fig. 8.10a. This second image emphasizes *textural* information. Figure 8.10b shows the 8 × 8 masks of synaptic weights learned by the network by proceeding in the same manner described; note the difference between these masks and those of Fig. 8.9b. Figure 8.10c shows the reconstructed image of the ocean scene based on the dominant 8 principal components without quantization. To study the effect of quantization, the outputs of the first 2 masks were quantized using 5 bits each, the third with 3 bits, and the remaining 5 masks with 2 bits each. Thus a total of 23 bits were needed to code each 8 × 8 block of pixels, resulting in a bit rate of 0.36 bits per pixel. Figure 8.10d shows the reconstructed image of the ocean scene, using its own masks quantized in the manner just described. The compression ratio of this image was 22 to 1.

(a)

(c)

(d)

(e)

FIGURE 8.10 (a) Image of ocean scene. (b) 8 × 8 masks representing the synpatic weights learned by the GHA algorithm applied to the ocean scene. (c) Reconstructed image of ocean scene using 8 dominant principal components. (d) Reconstructed image of ocean scene with 22 to 1 compression ratio, using masks of part (b) with quantization. (e) Reconstructed image of ocean scene using the masks of Fig. 8.9(b) for encoding, with quantization for a compression of 22 to 1, same as that in part (d).

To test the "generalization" performance of the generalized Hebbian algorithm, we finally used the masks of Fig. 8.9b to decompose the ocean scene of Fig. 8.10a and then applied the same quantization procedure that was used to generate the reconstructed image of Fig. 8.10d. The result of this image reconstruction is shown in Fig. 8.10e with a compression ratio of 22 to 1, the same as that in Fig. 8.10d. While the reconstructed images in Figs. 8.10d and 8.10e do bear a striking accord with each other, it can be seen that Fig. 8.10d possesses a greater amount of "true" textural information and thus looks less "blocky" than Fig. 8.10e. The reason for this behavior lies in the network weights. For the training performed on the images of the parents and the ocean scene, the first four weights are very similar. However, for the parents image the final four weights encode edge information, but in the case of the ocean scene these weights encode the textural information. Thus when encoding of the ocean scene occurs with the edge-type weights, the reconstruction of textural data is crude, thereby resulting in a blocky appearance.

8.7 ADAPTIVE PRINCIPAL COMPONENTS ANALYSIS USING LATERAL INHIBITION

The generalized Hebbian algorithm described in the previous section relies on the exclusive use of feedforward connections for principal components analysis. In this section we describe another algorithm called the *adaptive principal components extraction* (APEX) algorithm (Kung and Diamantaras, 1990; Diamantaras and Kung, 1996). The APEX algorithm uses both feedforward and feedback connections.[3] The algorithm is iterative in nature in that if we are given the first $(j - 1)$ principal components, the jth principal component is readily computed.

Figure 8.11 shows the network model used for the derivation of the APEX algorithm. As before, the input vector $\mathbf{x}$ has dimension m, with its components denoted by $x_1, x_2, ..., x_m$. Each neuron in the network is assumed to be linear. As depicted in Fig. 8.11, there are two kinds of synaptic connections in the network:

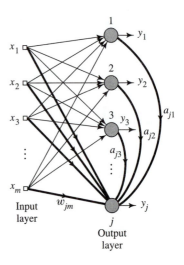

FIGURE 8.11 Network with feedforward and lateral connections for deriving the APEX algorithm.

- *Feedforward connections* from the input nodes to each of the neurons $1, 2, ..., j$, with $j < m$. Of particular interest here are the feedforward connections to neuron j; these connections are represented by the feedforward weight vector

$$\mathbf{w}_j = [w_{j1}(n), w_{j2}(n), ..., w_{jm}(n)]^T$$

  The feedforward connections operate in accordance with a *Hebbian learning rule*; they are *excitatory* and therefore provide for self-amplification.
- *Lateral connections* from the individual outputs of neurons $1, 2, ..., j - 1$ to neuron j, thereby applying *feedback* to the network. These connections are represented by the feedback weight vector

$$\mathbf{a}_j(n) = [a_{j1}(n), a_{j2}(n), ..., a_{j,j-1}(n)]^T$$

  The lateral connections operate in accordance with an *anti-Hebbian learning rule,* which has the effect of making them *inhibitory.*

In Fig. 8.11 the feedforward and feedback connections of neuron j are boldfaced merely to emphasize that neuron j is the subject of study.

The output $y_j(n)$ of neuron j is given by

$$y_j(n) = \mathbf{w}_j^T(n)\mathbf{x}(n) + \mathbf{a}_j^T(n)\mathbf{y}_{j-1}(n) \tag{8.98}$$

where the contribution $\mathbf{w}_j^T(n)\mathbf{x}(n)$ is due to the feedforward connections, and the remaining contribution $\mathbf{a}_j^T(n)\mathbf{y}_{j-1}(n)$ is due to the lateral connections. The feedback signal vector $\mathbf{y}_{j-1}(n)$ is defined by the outputs of neurons $1, 2, ..., j - 1$:

$$\mathbf{y}_{j-1}(n) = [y_1(n), y_2(n), ..., y_{j-1}(n)]^T \tag{8.99}$$

It is assumed that the input vector $\mathbf{x}(n)$ is drawn from a stationary process whose correlation matrix $\mathbf{R}$ has *distinct eigenvalues arranged in decreasing order* as follows:

$$\lambda_1 > \lambda_2 > \cdots > \lambda_{j-1} > \lambda_j > \cdots > \lambda_m \tag{8.100}$$

It is further assumed that neurons $1, 2, ..., j - 1$ of the network in Fig. 8.11 have *already converged to their respective stable conditions,* as shown by

$$\mathbf{w}_k(0) = \mathbf{q}_k, \qquad k = 1, 2, ..., j - 1 \tag{8.101}$$

$$\mathbf{a}_k(0) = \mathbf{0}, \qquad k = 1, 2, ..., j - 1 \tag{8.102}$$

where $\mathbf{q}_k$ is the eigenvector associated with the kth eigenvalue of the correlation matrix $\mathbf{R}$, and time step $n = 0$ refers to the start of computations by neuron j of the network. We may then use Eqs. (8.98), (8.99), (8.101), and (8.102) to write

$$\mathbf{y}_{j-1}(n) = [\mathbf{q}_1^T\mathbf{x}(n), \mathbf{q}_2^T\mathbf{x}(n), ..., \mathbf{q}_{j-1}^T\mathbf{x}(n)]$$
$$= \mathbf{Q}\mathbf{x}(n) \tag{8.103}$$

where $\mathbf{Q}$ is a $(j - 1)$-by-m matrix defined in terms of the eigenvectors $\mathbf{q}_1, \mathbf{q}_2, ..., \mathbf{q}_{j-1}$ associated with the $(j - 1)$ largest eigenvalues $\lambda_1, \lambda_2, ..., \lambda_{j-1}$ of the correlation matrix $\mathbf{R}$; that is,

$$\mathbf{Q} = [\mathbf{q}_1, \mathbf{q}_2, ..., \mathbf{q}_{j-1}]^T \tag{8.104}$$

The requirement is to use neuron j in the network of Fig. 8.11 to compute the next largest eigenvalue λ_j of the correlation matrix $\mathbf{R}$ of the input vector $\mathbf{x}(n)$ and the associated eigenvector $\mathbf{q}_j$.

The update equations for the feedforward weight vector $\mathbf{w}_j(n)$ and the feedback weight vector $\mathbf{a}_j(n)$ for neuron j are defined as, respectively,

$$\mathbf{w}_j(n + 1) = \mathbf{w}_j(n) + \eta[y_j(n)\mathbf{x}(n) - y_j^2(n)\mathbf{w}_j(n)] \tag{8.105}$$

and

$$\mathbf{a}_j(n + 1) = \mathbf{a}_j(n) - \eta[y_j(n)\mathbf{y}_{j-1}(n) + y_j^2(n)\mathbf{a}_j(n)] \tag{8.106}$$

where η is the *learning-rate parameter,* assumed to be the same for both update equations. The term $y_j(n)\mathbf{x}(n)$ on the right-hand side of Eq. (8.106) represents Hebbian learning, whereas the term $-y_j(n)\mathbf{y}_{j-1}(n)$ on the right-hand side of Eq. (8.106) represents anti-Hebbian learning. The remaining terms, $-y_j^2(n)\mathbf{w}_j(n)$ and $y_j^2(n)\mathbf{a}_j(n)$, are included in these two equations to assure the stability of the algorithm. Basically, Eq. (8.105) is the vector form of Oja's leaning rule described in Eq. (8.40), whereas Eq. (8.106) is *new,* accounting for the use of lateral inhibition (Kung and Diamantaras, 1990; Diamantaras and Kung, 1996).

We prove absolute stability of the neural network of Fig. 8.11 by *induction,* as follows:

- First, we prove that if neurons $1, 2, ..., j - 1$ have converged to their stable conditions, then neuron j converges to its own stable condition by extracting the next largest eigenvalue λ_j of the correlation matrix $\mathbf{R}$ of the input vector $\mathbf{x}(n)$ and the associated eigenvector $\mathbf{q}_j$.
- Next, we complete the proof by induction by recognizing that neuron 1 has no feedback and therefore the feedback weight vector $\mathbf{a}_1$ is zero. Hence this particular neuron operates in exactly the same way as Oja's neuron, and from Section 8.4 we know that this neuron is absolutely stable under certain conditions.

The only matter that requires attention is therefore the first point.

To proceed then, we invoke the fundamental assumptions made in Section 8.4, and so state the following theorem in the context of neuron j in the neural network of Fig. 8.11 operating under the conditions described by Eqs. (8.105) and (8.106) (Kung and Diamantaras, 1990; Diamantaras and Kung, 1996):

> Given that the learning-rate parameter η is assigned a sufficiently small value to ensure that the adjustments to the weight vectors proceed slowly, then in the limit, the feedforward weight vector and the average output power (variance) of neuron j approaches the normalized eigenvector $\mathbf{q}_j$ and corresponding eigenvalue λ_j of the correlation matrix $\mathbf{R}$, as shown by, respectively,
>
> $$\lim_{n \to \infty} \mathbf{w}_j(n) = \mathbf{q}_j$$
>
> and
>
> $$\lim_{n \to \infty} \sigma_j^2(n) = \lambda_j$$
>
> where $\sigma_j^2(n) = E[y_j^2(N)]$, and $\lambda_1 > \lambda_2 > \cdots > \lambda_j > \cdots > \lambda_m > 0$. In other words, given the eigenvectors $\mathbf{q}_1, ..., \mathbf{q}_{j-1}$, neuron j in the network of Fig. 8.11 computes the next largest eigenvalue λ_j and associated eigenvector $\mathbf{q}_j$.

To prove this theorem, consider first Eq. (8.105). Using Eqs. (8.98) and (8.99), and recognizing that

$$\mathbf{a}_j^T(n)\mathbf{y}_{j-1}(n) = \mathbf{y}_{j-1}^T(n)\mathbf{a}_j(n)$$

we may recast Eq. (8.105) as follows:

$$\mathbf{w}_j(n+1) = \mathbf{w}_j(n) + \eta[\mathbf{x}(n)\mathbf{x}^T(n)\mathbf{w}_j(n) + \mathbf{x}(n)\mathbf{x}^T(n)\mathbf{Q}^T\mathbf{a}_j(n) - y_j^2(n)\mathbf{w}_j(n)] \quad (8.107)$$

where the matrix $\mathbf{Q}$ is defined by Eq. (8.104). The term $y_j^2(n)$ in Eq. (8.107) has not been touched for a reason that will become apparent. Invoking the fundamental assumptions described in Section 8.4, we find that applying the statistical expectation operator to both sides of Eq. (8.107) yields

$$\mathbf{w}_j(n+1) = \mathbf{w}_j(n) + \eta[\mathbf{R}\mathbf{w}_j(n) + \mathbf{R}\mathbf{Q}^T\mathbf{a}_j(n) - \sigma_j^2(n)\mathbf{w}_j(n)] \quad (8.108)$$

where $\mathbf{R}$ is the correlation matrix of the input vector $\mathbf{x}(n)$, and $\sigma_j^2(n)$ is the average output power of neuron j. Let the synaptic weight vector $\mathbf{w}_j(n)$ be expanded in terms of the entire orthonormal set of eigenvectors of the correlation matrix $\mathbf{R}$ as follows:

$$\mathbf{w}_j(n) = \sum_{k=1}^{m} \theta_{jk}(n)\mathbf{q}_k \quad (8.109)$$

where $\mathbf{q}_k$ is the eigenvector associated with the eigenvalue λ_k of matrix $\mathbf{R}$, and $\theta_{jk}(n)$ is a time-varying coefficient of the expansion. We may then use the basic relation (see Eq. (8.14))

$$\mathbf{R}\mathbf{q}_k = \lambda_k\mathbf{q}_k$$

to express the matrix product $\mathbf{R}\mathbf{w}_j(n)$ as follows:

$$\begin{aligned}
\mathbf{R}\mathbf{w}_j(n) &= \sum_{k=1}^{m} \theta_{jk}(n)\mathbf{R}\mathbf{q}_k \\
&= \sum_{k=1}^{m} \lambda_k\theta_{jk}(n)\mathbf{q}_k
\end{aligned} \quad (8.110)$$

Similarly, using Eq. (8.104), we may express the matrix product $\mathbf{R}\mathbf{Q}^T\mathbf{a}_j(n)$ as

$$\begin{aligned}
\mathbf{R}\mathbf{Q}^T\mathbf{a}_j(n) &= \mathbf{R}[\mathbf{q}_1, \mathbf{q}_2, ..., \mathbf{q}_{j-1}]\mathbf{a}_j(n) \\
&= [\lambda_1\mathbf{q}_1, \lambda_2\mathbf{q}_2, ..., \lambda_{j-1}\mathbf{q}_{j-1}] \begin{bmatrix} a_{j1}(n) \\ a_{j2}(n) \\ \vdots \\ a_{j,j-1}(n) \end{bmatrix} \\
&= \sum_{k=1}^{j-1} \lambda_k a_{jk}(n)\mathbf{q}_k
\end{aligned} \quad (8.111)$$

Hence, substituting Eqs. (8.109), (8.110), and (8.111) in (8.108), and simplifying, we get (Kung and Diamantaras, 1990)

$$\sum_{k=1}^{m} \theta_{jk}(n+1)\mathbf{q}_k = \sum_{k=1}^{m} \{1 + \eta[\lambda_k - \sigma_j^2(n)]\} \theta_{jk}(n)\mathbf{q}_k$$

$$+ \eta \sum_{k=1}^{j-1} \lambda_k a_{jk}(n)\mathbf{q}_k \tag{8.112}$$

Following a procedure similar to that described, it is possible to show that the update equation (8.106) for the feedback weight vector $\mathbf{a}_j(n)$ may be transformed as follows (see Problem 8.7):

$$\mathbf{a}_j(n+1) = -\eta\lambda_k\theta_{jk}(n)\mathbf{1}_k + \{1 - \eta[\lambda_k + \sigma_j^2(n)]\}\mathbf{a}_j(n) \tag{8.113}$$

where $\mathbf{1}_k$ is a vector all of whose j elements are zero, except for the kth element, which is equal to 1. The index k is restricted to lie in the range $1 \le k \le j - 1$.

There are two cases to be considered, depending on the value assigned to index k in relation to $j - 1$. Case I refers to $1 \le k \le j - 1$, which pertains to the analysis of the "old" principal modes of the network. Case II refers to $j \le k \le m$, which pertains to the analysis of the remaining "new" principal modes. The total number of principal modes is m, the dimension of the input vector $\mathbf{x}(n)$.

CASE I: $1 \le k \le j - 1$

In this case we deduce the following update equations for the coefficient $\theta_{jk}(n)$ associated with eigenvector $\mathbf{q}_k$ and the feedback weight $a_{jk}(n)$ from Eqs. (8.112) and (8.113), respectively:

$$\theta_{jk}(n+1) = \eta\lambda_k a_{jk}(n) + \{1 + \eta[\lambda_k - \sigma_j^2(n)]\}\theta_{jk}(n) \tag{8.114}$$

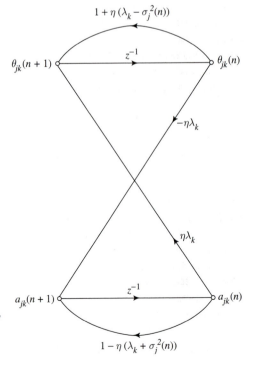

FIGURE 8.12 Signal-flow graph representation of Eqs. (8.114) and (8.115).

and

$$a_{jk}(n + 1) = -\eta\lambda_k\theta_{jk}(n) + \{1 - \eta[\lambda_k + \sigma_j^2(n)]\}a_{jk}(n) \qquad (8.115)$$

Figure 8.12 presents a signal-flow graph representation of Eqs. (8.114) and (8.115). In matrix form we may rewrite Eqs. (8.114) and (8.115) as

$$\begin{bmatrix} \theta_{jk}(n + 1) \\ a_{jk}(n + 1) \end{bmatrix} = \begin{bmatrix} 1 + \eta[\lambda_k - \sigma_j^2(n)] & \eta\lambda_k \\ -\eta\lambda_k & 1 - \eta[\lambda_k + \sigma_j^2(n)] \end{bmatrix}\begin{bmatrix} \theta_{jk}(n) \\ a_{jk}(n) \end{bmatrix} \qquad (8.116)$$

The system matrix described in Eq. (8.116) has a double eigenvalue at

$$\rho_{jk} = [1 - \eta\sigma_j^2(n)]^2 \qquad (8.117)$$

From Eq. (8.117) we can make two important observations:

1. The double eigenvalue ρ_{jk} of the system matrix in Eq. (8.116) is independent of all the eigenvalues λ_k of the correlation matrix **R**, corresponding to $k = 1, 2, ..., j - 1$.
2. For all k, the double eigenvalue ρ_{jk} depends solely on the learning-rate parameter η and the average output power σ_j^2 of neuron j. It is therefore less than unity, provided that η is a sufficiently small, positive number.

Given that $\rho_{jk} < 1$, the coefficients $\theta_{jk}(n)$ of the expansion in Eq. (8.109) and the feedback weights $a_{jk}(n)$ will, for all k, approach zero asymptotically with the same speed, since all the principal modes of the network have the same eigenvalue (Kung and Diamantaras, 1990; Diamantaras and Kung, 1996). This result is a consequence of the property that the orthogonality of eigenvectors of a correlation matrix does *not* depend on the eigenvalues. In other words, the expansion of $w_j(n)$ in terms of the orthonormal set of eigenvectors of the correlation matrix **R** given in Eq. (8.109), which is basic to the result described in Eq. (8.117), is *invariant* to the choice of eigenvalues $\lambda_1 \lambda_2, ..., \lambda_{j-1}$.

CASE II: $j \le k \le m$

In this second case, the feedback weights $a_{jk}(n)$ have *no* influence on the modes of the network, as shown by

$$a_{jk}(n) = 0 \quad \text{for } j \le k \le m \qquad (8.118)$$

Hence, for every principal mode $k \ge j$ we have a very simple equation:

$$\theta_{jk}(n + 1) = \{1 + \eta[\lambda_k - \sigma_j^2(n)]\}\theta_{jk}(n) \qquad (8.119)$$

which follows directly from Eqs. (8.112) and (8.118). According to case I, both $\theta_{jk}(n)$ and $a_{jk}(n)$ will eventually converge to zero for $k = 1, 2, ..., j - 1$. With the random variable $Y_j(n)$ representing the output of neuron j, we may express its average output power as follows:

$$\sigma_j^2(n) = E[Y_j^2(n)]$$
$$= \sum_{k=j}^{m} \lambda_k \theta_{jk}^2(n) \qquad (8.120)$$

where in the last line we have made use of the following relation:

$$q_k^T R q_l = \begin{cases} \lambda_k, & l = k \\ 0, & \text{otherwise} \end{cases}$$

It follows therefore that Eq. (8.119) cannot diverge, because whenever $\theta_{jk}(n)$ becomes large such that $\sigma_j^2(n) > \lambda_k$, then $1 + \eta[\lambda_k - \sigma_j^2(n)]$ becomes smaller than unity, in which case $\theta_{jk}(n)$ will decrease in magnitude. Let the algorithm be initialized, with $\theta_{jj}(0) \neq 0$. Also define

$$r_{jk}(n) = \frac{\theta_{jk}(n)}{\theta_{jj}(n)}, \quad k = j + 1, \dots, m \tag{8.121}$$

We may then use Eq. (8.119) to write

$$r_{jk}(n + 1) = \frac{1 + \eta[\lambda_k - \sigma_j^2(n)]}{1 + \eta[\lambda_j - \sigma_j^2(n)]} \, r_{jk}(n) \tag{8.122}$$

With the eigenvalues of the correlation matrix arranged in the descending order

$$\lambda_1 > \lambda_2 > \dots > \lambda_k > \dots > \lambda_j \dots > \lambda_m$$

it follows that

$$\frac{\theta_{jk}(n)}{\theta_{jj}(n)} < 1 \quad \text{for all } n, \text{ and for } k = j + 1, \dots, m \tag{8.123}$$

Moreover, we note from Eqs. (8.119) and (8.120) that $\theta_{jj}(n + 1)$ remains bounded; therefore,

$$r_{jk}(n) \to 0 \quad \text{as } n \to \infty \text{ for } k = j + 1, \dots, m \tag{8.124}$$

Equivalently, in light of the definition given in Eq. (8.121) we may state that

$$\theta_{jk}(n) \to 0 \quad \text{as } n \to \infty \text{ for } k = j + 1, \dots, m \tag{8.125}$$

Under this condition, Eq. (8.120) simplifies to

$$\sigma_j^2(n) = \lambda_j \theta_{jj}^2(n) \tag{8.126}$$

and so Eq. (8.119) for $k = j$ becomes

$$\theta_{jj}(n + 1) = \{1 + \eta\lambda_j [1 - \theta_{jj}(n)]\} \, \theta_{jj}(n) \tag{8.127}$$

From this equation we immediately deduce that

$$\theta_{jj}(n) \to 1 \quad \text{as } n \to \infty \tag{8.128}$$

The implications of this limiting condition and that of Eq. (8.125) are twofold:

1. From Eq. (8.126) we have

$$\sigma_j^2(n) \to \lambda_j \quad \text{as } n \to \infty \tag{8.129}$$

2. From Eq. (8.109) we have

$$\mathbf{w}_j(n) \to \mathbf{q}_j \quad \text{as } n \to \infty \tag{8.130}$$

In other words, the neural network model of Fig. 8.11 extracts the jth eigenvalue and associated eigenvector of the correlation matrix $\mathbf{R}$ of the input vector $\mathbf{x}(n)$ as the number of iterations n approaches infinity. This of course assumes that neurons $1, 2, \dots, j - 1$ of the network have already converged to the respective eigenvalues and associated eigenvectors of the correlation matrix $\mathbf{R}$.

The treatment of the APEX algorithm presented here rests on the premise that neurons 1, 2, ..., $j - 1$ have converged before neuron j begins to act. This was done merely to explain the operation of the algorithm in a simple way. In practice, however, the neurons in the APEX algorithm tend to converge together.[4]

Learning Rate

In the APEX algorithm described in Eqs. (8.105) and (8.106), the same learning-rate parameter η is used for updating both the feedforward weight vector $\mathbf{w}_j(n)$ and feedback weight vector $\mathbf{a}_j(n)$. The relationship of Eq. (8.117) may be exploited to define an optimum value for the learning-rate parameter for each neuron j by setting the double eigenvalue ρ_{jk} equal to zero. In such a case, we have

$$\eta_{j,\text{opt}}(n) = \frac{1}{\sigma_j^2(n)} \tag{8.131}$$

where $\sigma_j^2(n)$ is the average output power of neuron j. A more practical proposition, however, is to set (Kung and Diamantaras, 1990; Diamantaras and Kung, 1996):

$$\eta_j = \frac{1}{\lambda_{j-1}} \tag{8.132}$$

which yields an underestimated value for the learning-rate parameter, since $\lambda_{j-1} > \lambda_j$ and $\sigma_j^2(n) \to \lambda_j$ as $n \to \infty$. Note that the eigenvalue λ_{j-1} is computed by neuron $j - 1$ and therefore available for use in updating the feedforward and feedback weights of neuron j.

Summary of the APEX Algorithm

1. Initialize the feedforward weight vector $\mathbf{w}_j$ and the feedback weight vector $\mathbf{a}_j$ to small random values at time $n = 1$, where $j = 1, 2, ..., m$. Assign a small positive value to the learning-rate parameter η.
2. Set $j = 1$, and for $n = 1, 2, ...$, compute

$$y_1(n) = \mathbf{w}_1^T(n)\mathbf{x}(n)$$
$$\mathbf{w}_1(n + 1) = \mathbf{w}_1(n) + \eta[y_1(n)\mathbf{x}(n) - y_1^2(n)\mathbf{w}_1(n)]$$

where $\mathbf{x}(n)$ is the input vector. For large n, we have $\mathbf{w}_1(n) \to \mathbf{q}_1$, where $\mathbf{q}_1$ is the eigenvector associated with the largest eigenvalue λ_1 of the correlation matrix of $\mathbf{x}(n)$.
3. Set $j = 2$, and for $n = 1, 2, ...$, compute

$$\mathbf{y}_{j-1}(n) = [y_1(n), y_2(n), ..., y_{j-1}(n)]^T$$
$$y_j(n) = \mathbf{w}_j^T(n)\mathbf{x}(n) + \mathbf{a}_j^T(n)\mathbf{y}_{j-1}(n)$$
$$\mathbf{w}_j(n + 1) = \mathbf{w}_j(n) + \eta[y_j(n)\mathbf{x}(n) - y_j^2(n)\mathbf{w}_j(n)]$$
$$\mathbf{a}_j(n + 1) = \mathbf{a}_j(n) - \eta[y_j(n)\mathbf{y}_{j-1}(n) + y_j^2(n)\mathbf{a}_j(n)]$$

4. Increment j by 1, go to step 3, and continue until $j = m$, where m is the desired number of principal components. (Note that $j = 1$ corresponds to the eigenvector associated with the largest eigenvalue, which is taken care of in step 2.) For large

n we have $\mathbf{w}_j(n) \to \mathbf{q}_j$ and $\mathbf{a}_j(n) \to \mathbf{0}$, where $\mathbf{q}_j$ is the eigenvector associated with the jth eigenvalue of the correlation matrix of $\mathbf{x}(n)$.

8.8 TWO CLASSES OF PCA ALGORITHMS

In addition to the generalized Hebbian algorithm (GHA), discussed in Section 8.5, and the APEX algorithm, discussed in Section 8.7, several other algorithms for principal components analysis (PCA) have been reported in literature.[5] The various PCA algorithms using neural networks may be categorized into two classes: *reestimation algorithms* and *decorrelating algorithms*.

According to this classification, the GHA is a reestimation algorithm in that Eqs. (8.87) and (8.88) may be recast in the equivalent form

$$\mathbf{w}_j(n + 1) = \mathbf{w}_j(n) + \eta y_j(n)[\mathbf{x}(n) - \hat{\mathbf{x}}_j(n)] \tag{8.133}$$

where $\hat{\mathbf{x}}_j(n)$ is the *reestimator* defined by

$$\hat{\mathbf{x}}_j(n) = \sum_{k=1}^{j} \mathbf{w}_k(n) y_k(n) \tag{8.134}$$

In a reestimation algorithm the neural network has only forward connections, whose strengths (weights) are modified in a Hebbian manner. The successive outputs of the network are forced to learn different principal components by subtracting estimates of the earlier components from the input before the data set is involved in the learning process.

In contrast, the APEX algorithm is a decorrelating algorithm. In such an algorithm the neural network has both forward and feedback connections. The strengths of the forward connections follow a Hebbian law, whereas the strengths of the feedback connections follow an anti-Hebbian law. The successive outputs of the network are decorrelated, forcing the network to respond to different principal components.

Principal Subspace

In situations where only the *principal subspace* (i.e., the space of the principal components) is required, we may use a *symmetric model* in which the reestimator $\hat{\mathbf{x}}_j(n)$ in the GHA algorithm is replaced by

$$\hat{\mathbf{x}}(n) = \sum_{k=1}^{l} \mathbf{w}_k(n) y_k(n) \quad \text{for all } l \tag{8.135}$$

In the symmetric model defined by Eqs. (8.133) and (8.135), the network converges to a set of outputs that span the principal subspace, rather than the principal components themselves. At convergence, the weight vectors of the network are orthogonal to each other, as in the GHA. The principal subspace, as described here, may be viewed as a generalization of the classical Oja rule defined in Eq. (8.46).

8.9 BATCH AND ADAPTIVE METHODS OF COMPUTATION

A discussion of principal components analysis would be incomplete without some consideration of the computational aspects of the problem. In this context there are two basic approaches to the computation of principal components: batch and adaptive methods.

The method of eigendecomposition described in Section 8.3 and the related method of singular value decomposition belong to the *batch* category. On the other hand, the GHA and APEX algorithms discussed in Sections 8.5 and 8.7 belong to the *adaptive* category.

In theory, eigendecomposition is based on the ensemble-averaged correlation matrix $\mathbf{R}$ of a random vector $\mathbf{X}(n)$ as described in Section 8.3. In practice, we use an estimate of the correlation matrix $\mathbf{R}$. Let $\{\mathbf{x}(n)\}_{n=1}^{N}$ denote a set of N realizations of the random vector $\mathbf{X}(n)$ at uniformly spaced discrete instants of time. Given such a set of observations, we may then use the *sample mean* as an estimate of the correlation matrix as follows:

$$\hat{\mathbf{R}}(N) = \frac{1}{N} \sum_{n=1}^{N} \mathbf{x}(n)\mathbf{x}^{T}(n) \tag{8.136}$$

Provided that the environment represented by the random vector $\mathbf{X}(n)$ is ergodic, the sample mean $\hat{\mathbf{R}}(N)$ approaches $\mathbf{R}$ as the sample size N approaches infinity. On this basis we may apply the eigendecomposition procedure to the sample mean $\hat{\mathbf{R}}(N)$ and thereby compute its eigenvalues and associated eigenvectors by invoking the use of Eq. (8.22) with $\hat{\mathbf{R}}(N)$ used in place of $\mathbf{R}$.

From a numerical perspective, however, a better method is to use *singular value decomposition* (SVD) by applying it directly to the *data matrix*. For the set of observations $\{\mathbf{x}(n)\}_{n=1}^{N}$, the data matrix is defined by

$$\mathbf{A} = [\mathbf{x}(1), \mathbf{x}(2), ..., \mathbf{x}(N)] \tag{8.137}$$

Then, except for the scaling factor $1/N$, we readily see that the estimate $\hat{\mathbf{R}}(N)$ of the correlation matrix $\mathbf{R}$ is equal to the matrix product $\mathbf{AA}^{T}$. According to the *singular value decomposition theorem* described in Chapter 5, the data matrix $\mathbf{A}(n)$ may be decomposed as follows (Golub and Van Loan, 1996):

$$\mathbf{A} = \mathbf{U\Sigma V}^{T} \tag{8.138}$$

where $\mathbf{U}$ and $\mathbf{V}$ are orthogonal matrices, which means that

$$\mathbf{U}^{-1} = \mathbf{U}^{T} \tag{8.139}$$

and

$$\mathbf{V}^{-1} = \mathbf{V}^{T} \tag{8.140}$$

As for matrix $\mathbf{\Sigma}$, it has a structure of the form

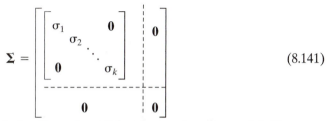

$$\tag{8.141}$$

where $k \leq m$, and where m is the dimension of the observation vector $\mathbf{x}(n)$. The numbers $\sigma_1, \sigma_2, ..., \sigma_k$ are called the *singular values* of the data matrix $\mathbf{A}$. Correspondingly, the columns of the orthogonal matrix $\mathbf{U}$ are called the *left singular vectors,* and the columns of the matrix $\mathbf{V}$ are called the *right singular vectors.* The singular value decomposition of data matrix $\mathbf{A}$ is related to the eigendecomposition of the estimate $\hat{\mathbf{R}}(N)$ of the correlation matrix in the following ways:

- Except for the scaling factor $1/\sqrt{N}$, the singular values of the data matrix $\mathbf{A}$ are the square roots of the eigenvalues of the estimate $\hat{\mathbf{R}}(N)$.
- The left singular vectors of $\mathbf{A}$ are the eigenvectors of $\hat{\mathbf{R}}(N)$.

Now we can see the numerical advantage of singular value decomposition over eigendecomposition. For a prescribed accuracy of computation, the singular value decomposition procedure requires half the numerical precision of the eigendecomposition procedure. Moreover, several algorithms and highly accurate canned routines are available for implementing the singular value decomposition procedure on a computer (Golub and Van Loan, 1996; Haykin, 1996). However, in practice, storage requirements may restrict the use of these routines to values of sample size that are not too excessive.

Turning next to the category of *adaptive methods,* these methods work with an arbitrarily large sample size N. For all practical purposes there is no restriction on N. Adaptive methods are exemplified by Hebbian-based neural networks, whose operation is inspired by ideas from neurobiology. The storage requirement of such methods is relatively modest, since intermediate values of the eigenvalues and associated eigenvectors do not have to be stored. Another attractive feature of adaptive algorithms is that, in a *nonstationary environment,* they have an inherent ability to *track* gradual changes in the optimal solution in an inexpensive way compared to batch methods. However, the main drawback of adaptive algorithms of the stochastic approximation type is their relatively slow rate of convergence, which puts them at a disadvantage compared to classical batch techniques; this is especially so on large stationary problems, even when the adaptive methods are implemented on parallel neural hardware (Kotilainen, 1993).

8.10   KERNEL PRINCIPAL COMPONENTS ANALYSIS

The form of PCA discussed up to this point in the chapter involves computations in the input (data) space. We now consider another form of PCA where the computations are performed in a feature space that is *nonlinearly* related to the input space. The feature space we have in mind is that defined by an *inner-product kernel* in accordance with Mercer's theorem; the notion of inner-product kernels is discussed in Chapter 6 on support vector machines. The idea of *kernel-based principal components analysis* is due to Schölkopf et al. (1998).

Due to the nonlinear relationship between the input space and feature space, kernel PCA in nonlinear. However, unlike other forms of nonlinear PCA,[6] the implementation of kernel PCA relies on linear algebra. We may therefore think of kernel PCA as a natural extension of ordinary PCA.

Let vector $\boldsymbol{\varphi}(\mathbf{x}_j)$ denote the image of an input vector $\mathbf{x}_j$ induced in a feature space defined by the nonlinear map: $\boldsymbol{\varphi}:\mathbb{R}^{m_0}\to\mathbb{R}^{m_1}$, where m_0 is the dimensionality of the input space and m_1 is the dimensionality of the feature space. Given the set of examples $\{\mathbf{x}_i\}_{i=1}^N$, we have a corresponding set of feature vectors $\{\boldsymbol{\varphi}(\mathbf{x}_i)\}_{i=1}^N$. Accordingly, we may define an m_1-by-m_1 correlation matrix in the feature space, denoted by $\widetilde{\mathbf{R}}$, as follows:

$$\widetilde{\mathbf{R}} = \frac{1}{N}\sum_{i=1}^{N}\boldsymbol{\varphi}(\mathbf{x}_i)\boldsymbol{\varphi}^T(\mathbf{x}_i) \tag{8.142}$$

As with ordinary PCA, the first thing we have to do is to ensure that the set of feature vectors $\{\varphi(\mathbf{x}_i)\}_{i=1}^{N}$ has zero mean:

$$\frac{1}{N}\sum_{i=1}^{N}\varphi(\mathbf{x}_i) = \mathbf{0}$$

To satisfy this condition in the feature space is a more difficult proposition than it is in the input space; in Problem 8.10 we describe a procedure for catering to this requirement. Proceeding then on the assumption that the feature vectors have been centered, we may adapt the use of Eq. (8.14) to our present situation by writing

$$\tilde{\mathbf{R}}\tilde{\mathbf{q}} = \tilde{\lambda}\tilde{\mathbf{q}} \tag{8.143}$$

where $\tilde{\lambda}$ is an eigenvalue of the correlation matrix $\tilde{\mathbf{R}}$ and $\tilde{\mathbf{q}}$ is the associated eigenvector. Now we note that all eigenvectors that satisfy Eq. (8.143) for $\tilde{\lambda} \neq 0$ lie in the span of the set of feature vectors $\{\varphi(\mathbf{x}_j)\}_{j}^{N}$. Consequently, there does exist a corresponding set of coefficients $\{\alpha_j\}_{j=1}^{N}$ for which we can write

$$\tilde{\mathbf{q}} = \sum_{j=1}^{N}\alpha_j\varphi(\mathbf{x}_j) \tag{8.144}$$

Thus substituting Eqs. (8.142) and (8.144) into (8.143), we obtain

$$\sum_{i=1}^{N}\sum_{j=1}^{N}\alpha_j\varphi(\mathbf{x}_i)K(\mathbf{x}_i,\mathbf{x}_j) = N\tilde{\lambda}\sum_{j=1}^{N}\alpha_j\varphi(\mathbf{x}_j) \tag{8.145}$$

where $K(\mathbf{x}_i,\mathbf{x}_j)$ is an *inner-product kernel* defined in terms of the feature vectors by

$$K(\mathbf{x}_i,\mathbf{x}_j) = \varphi^T(\mathbf{x}_i)\varphi(\mathbf{x}_j) \tag{8.146}$$

We need to go one step further with Eq. (8.145) so that the relationship is expressed entirely in terms of the inner-product kernel. To do so, we premultiply both sides of Eq. (8.145) by the transposed vector $\varphi^T(\mathbf{x}_k)$, thereby obtaining

$$\sum_{i=1}^{N}\sum_{j=1}^{N}\alpha_j K(\mathbf{x}_k,\mathbf{x}_i)K(\mathbf{x}_i,\mathbf{x}_j) = N\tilde{\lambda}\sum_{j=1}^{N}\alpha_j K(\mathbf{x}_k,\mathbf{x}_j), \quad k = 1, 2, ..., N \tag{8.147}$$

where the definitions of $K(\mathbf{x}_k,\mathbf{x}_i)$ and $K(\mathbf{x}_k,\mathbf{x}_j)$ follow Eq. (8.146).

We now introduce two matrix definitions:

- The N-by-N matrix $\mathbf{K}$, called the *kernel matrix*, whose ij-th element is the inner-product kernel $K(\mathbf{x}_i,\mathbf{x}_j)$.
- The N-by-1 vector $\boldsymbol{\alpha}$, whose jth element is the coefficient α_j.

Accordingly, we may recast Eq. (8.147) in the compact matrix form:

$$\mathbf{K}^2\boldsymbol{\alpha} = N\tilde{\lambda}\,\mathbf{K}\boldsymbol{\alpha} \tag{8.148}$$

where the squared matrix $\mathbf{K}^2$ denotes the product of $\mathbf{K}$ with itself. Since $\mathbf{K}$ is common to both sides of Eq. (8.148), all the solutions of this eigenvalue problem that are of interest are equally well represented in the simpler eigenvalue problem:

$$\mathbf{K}\boldsymbol{\alpha} = N\tilde{\lambda}\boldsymbol{\alpha} \tag{8.149}$$

Let $\lambda_1 \geq \lambda_2 \geq \cdots \geq \lambda_N$ denote the eigenvalues of the kernel matrix $\mathbf{K}$; that is

$$\lambda_j = N\tilde{\lambda}_j, \quad j = 1, 2, ..., N \tag{8.150}$$

where $\tilde{\lambda}_j$ is the jth eigenvalue of the correlation matrix $\tilde{\mathbf{R}}$. Then Eq. (8.149) takes the standard form

$$\mathbf{K}\boldsymbol{\alpha} = \lambda\boldsymbol{\alpha} \tag{8.151}$$

where the coefficient vector $\boldsymbol{\alpha}$ plays the role of the eigenvector associated with the eigenvalue λ of the kernel matrix $\mathbf{K}$. The vector $\boldsymbol{\alpha}$ is normalized by requiring that the eigenvector $\tilde{\mathbf{q}}$ of the correlation matrix $\tilde{\mathbf{R}}$ is normalized to unit length; that is

$$\tilde{\mathbf{q}}_k^T \tilde{\mathbf{q}}_k = 1 \quad \text{for } k = 1, 2, ..., p \tag{8.152}$$

where it is assumed that the eigenvalues are arranged in decreasing order, with λ_p being the smallest nonzero eigenvalue of the kernel matrix $\mathbf{K}$. Using Eq. (8.144) and then invoking Eq. (8.151), we may show that the normalization condition of Eq. (8.152) is equivalent to

$$\boldsymbol{\alpha}_k^T \boldsymbol{\alpha}_k = \frac{1}{\lambda_k}, \quad k = 1, 2, ..., p \tag{8.153}$$

For the extraction of principal components, we need to compute the projections onto the eigenvectors $\tilde{\mathbf{q}}_k$ in feature space, as shown by

$$\tilde{\mathbf{q}}_k^T \boldsymbol{\varphi}(\mathbf{x}) = \sum_{j=1}^{N} \alpha_{k,j} \boldsymbol{\varphi}^T(\mathbf{x}_j)\boldsymbol{\varphi}(\mathbf{x})$$

$$= \sum_{j=1}^{N} \alpha_{k,j} K(\mathbf{x}_j, \mathbf{x}), \quad k = 1, 2, ..., p \tag{8.154}$$

where the vector $\mathbf{x}$ is a "test" point, and $\alpha_{k,j}$ is the jth coefficient of eigenvector $\boldsymbol{\alpha}_k$ associated with the kth eigenvalue of matrix $\mathbf{K}$. The projections of Eq. (8.154) define the *nonlinear principal components* in the m_1-dimensional feature space.

Figure 8.13 illustrates the basic idea of kernel PCA, where the feature space is nonlinearly related to the input space via the transformation $\varphi(\mathbf{x})$. Parts a and b of the figure refer to the input space and feature space, respectively. The contour lines shown in Fig. 8.13b represent constant projections onto a principal eigenvector, which is shown as a dashed arrow. In this figure it is assumed that the transformation $\varphi(\mathbf{x})$ has been chosen in such a way that the images of the data points induced in the future space congregate themselves essentially along the eigenvector. Figure 8.13a shows the *nonlinear* contour lines in the input space that correspond to those in the feature space. Note that we purposely have not shown a pre-image of the eigenvector in the input space, as it may not even exist (Schölkopf et al., 1998).

For inner-product kernels defined in accordance with Mercer's theorem, we are basically performing ordinary PCA in an m_1-dimensional feature space, where the dimension m_1 is a design parameter. All the properties of ordinary PCA that are described in Section 8.3 carry over to kernel PCA. In particular, kernel PCA is linear in

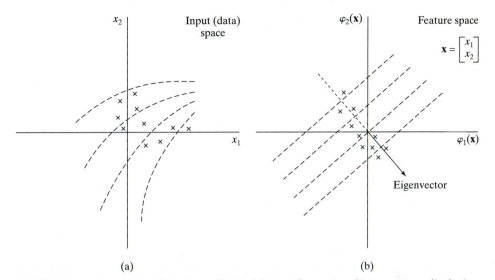

(a)   (b)

FIGURE 8.13   Illustration of the kernel PCA. (a) Two-dimensional input space, displaying a set of data points. (b) Two-dimensional feature space, displaying the induced images of the data points congregating around a principal eigenvector. The uniformly spaced dashed lines in part (b) represent contours of constant projections onto the eigenvector; the corresponding contours are nonlinear in the input space.

the feature space, but nonlinear in the input space. As such, it can be applied to all those domains where ordinary PCA has been used for feature extraction or data reduction, for which nonlinear extension would make sense.

In Chapter 6, we presented three methods for constructing inner-product kernels that were based on the use of polynomials, radial-basis functions, and hyperbolic functions; see Table 6.1. The question of how to select the kernel best suited for a given task (i.e., the appropriate feature space) is an open problem (Schölkopf, 1997).

Summary of the Kernel PCA

1. Given the training examples $\{\mathbf{x}_i\}_{i=1}^{N},$ compute the N-by-N kernel matrix $\mathbf{K} = \{K(\mathbf{x}_i, \mathbf{x}_j)\}$, where

$$K(\mathbf{x}_i, \mathbf{x}_j) = \boldsymbol{\varphi}^T(\mathbf{x}_i)\boldsymbol{\varphi}(\mathbf{x}_j)$$

2. Solve the eigenvalue problem:

$$\mathbf{K}\boldsymbol{\alpha} = \lambda\boldsymbol{\alpha}$$

where λ is an eigenvalue of the kernel matrix $\mathbf{K}$ and $\boldsymbol{\alpha}$ is the associated eigenvector.

3. Normalize the eigenvectors so computed by requiring that

$$\boldsymbol{\alpha}_k^T \boldsymbol{\alpha}_k = \frac{1}{\lambda_k}, \quad k = 1, 2, ..., p$$

where λ_p is the smallest nonzero eigenvalue of matrix $\mathbf{K}$, assuming that the eigenvalues are arranged in decreasing order.

4. For the extraction of principal components of a test point $\mathbf{x}$, compute the projections

$$a_k = \tilde{\mathbf{q}}_k^T \, \boldsymbol{\varphi}(\mathbf{x})$$

$$= \sum_{j=1}^{N} \alpha_{k,j} \, K(\mathbf{x}_j, \mathbf{x}), \qquad k = 1, 2, \ldots, p$$

where $\alpha_{k,j}$ is the jth element of eigenvector $\boldsymbol{\alpha}_k$.

Example 8.3

To provide some intuitive understanding for the operation of kernel PCA, we show in Fig. 8.14 the results of a simple experiment described in Schölkopf et al. (1998). The two-dimensional data, consisting of components x_1 and x_2, used in this experiment were generated as follows: The x_1-values have a uniform distribution in the interval $[-1, 1]$. The x_2-values are nonlinearly related to the x_1-values by the formula:

$$x_2 = x_1^2 + v$$

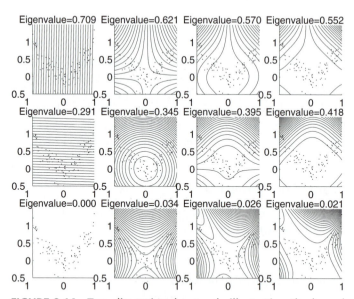

FIGURE 8.14 Two-dimensional example illustrating the kernel PCA. From left to right, the polynomial degree of the kernel is $d = 1, 2, 3, 4$. From top to bottom, the first three eigenvectors in the feature space are shown. The first column corresponds to ordinary PCA, and the other three columns correspond to kernel PCA with polynomial degree $d = 2, 3, 4$. (Reproduced with permission from Dr. Klaus-Robert Müller.)

where v is an additive Gaussian noise of zero mean and variance 0.04.

The results of PCA shown in Fig. 8.14 were obtained using kernel polynomials:

$$K(\mathbf{x}, \mathbf{x}_i) = (\mathbf{x}^T\mathbf{x}_i)^d, \qquad d = 1, 2, 3, 4$$

where $d = 1$ corresponds to linear PCA, and $d = 2, 3, 4$ corresponds to kernel PCA. Linear PCA, shown on the left of Fig. 8.14, results in only two eigenvalues since the dimensionality of the input space is two. In contrast, kernel PCA permits the extraction of higher-order components, as shown by the results depicted in columns 2, 3, and 4 of Fig. 8.14, corresponding to polynomial degree $d = 2, 3, 4$, respectively. The contour lines shown in each part of the figure (except for the zero eigenvalue in the case of linear PCA) represent constant principal values (i.e., constant projections onto the eigenvector associated with the eigenvalue in question).

Based on the results shown in Fig. 8.14, we make the following observations:

- As expected, linear PCA fails to provide an adequate representation of the nonlinear input data.
- In all cases, the first principal component varies monotonically along a parabola that underlies the input data.
- In the kernel PCA, the second and third principal components exhibit a behavior that appears somewhat similar for different values of polynomial degree d.
- In the case of polynomial degree $d = 2$, the third principal component of kernel PCA appears to pick up the variance due to the additive Gaussian noise v. By removing the contribution due to this component, we would in effect be performing some form of noise reduction.

8.11 SUMMARY AND DISCUSSION

In this chapter we present material dealing with the theory of principal components analysis and the use of neural networks for its implementation. It is appropriate that we now reflect over that material and ask: How useful is principal components analysis? The answer to this question, of course, depends on the application of interest.

If the main objective is to achieve good *data compression* while preserving as much information about the inputs as possible, the use of principal components analysis offers a useful self-organized learning procedure. Here we note from the material presented in Section 8.3 that the use of a subspace decomposition method based on the "first l principal components" of the input data provides a linear mapping, which is optimum in the sense that it permits reconstruction of the original input data in the mean-square error sense. Moreover, a representation based on the first l principal components is preferable to an arbitrary subspace representation, because the principal components of the input data are naturally ordered in decreasing eigenvalue or, equivalently, decreasing variance. Accordingly, we may optimize the use of principal components analysis for data compression by employing the greatest numerical precision to encode the first principal component of the input, and progressively less precision to encode the remaining $l - 1$ components.

A related issue is the representation of a data set made up of an aggregate of several *clusters*. For the clusters to be individually visible, the separation between them has to be larger than the internal scatter of the clusters. If it so happens that there are only a few clusters in the data set, then the leading principal axes found by using the principal components analysis will tend to pick projections of clusters with good separations, thereby providing an effective basis for *feature extraction*.

In this latter context we mention a useful application of a principal components analyzer as the *preprocessor* for a supervised neural network (e.g., multilayer perceptron trained with the back-propagation algorithm). The motivation here is to speed up convergence of the learning process by decorrelating the input data. A supervised learning procedure such as back-propagation relies on steepest descent. The convergence process in this form of learning is typically slow due to interacting effects of a multilayer perceptron's synaptic weights on the error signal, even with the use of simple local accelerating procedures such as momentum and adaptive learning rates for individual weights. If, however, the inputs to the multilayer perceptron consist of uncorrelated components, then from the discussion presented in Chapter 4 we note that the Hessian matrix of the cost function $\mathscr{E}(n)$ with respect to the free parameters of the network is more nearly diagonal than would be the case otherwise. With this form of diagonalization in place, the use of simple local accelerating procedures permits a considerable acceleration in the convergence process, which is made possible by appropriate scaling of the learning rates along each weight axis independently (Becker, 1991).

With the Hebbian-based algorithms of this chapter motivated by ideas taken from neurobiology, it is fitting to conclude our discussion by commenting on the role of principal components analysis in biological perceptual systems. Linsker (1990a) questions the "sufficiency" of principal components analysis as a principle for determining the response property developed by a neuron to analyze an ensemble of input "scenes." In particular, the optimality of principal components analysis with respect to the accurate reconstruction of an input signal from a neuron's response is considered to be of questionable relevance. In general, it appears that a brain does much more than simply try to reproduce the input scenes received by its sensory units. Rather, some underlying "meaningful cues" or features are extracted so as to permit high-level interpretations of the inputs. We may therefore sharpen the question we raised at the beginning of this discussion and ask: How useful is principal components analysis for perceptual processing?

Ambros–Ingerson et al. (1990) point out the significance of the algorithms set forth by Oja (1982) and Sanger (1989a) for principal components analysis (i.e., the Hebbian-inspired algorithms discussed in Sections 8.4 and 8.5) in a *hierarchical clustering algorithm*. They put forward the hypothesis that hierarchical clustering may emerge as a fundamental property (at least in part) of memories based on long-term potentiation (LTP)—like synaptic modifications of the kind found in cortico-bulbar networks and circuitry of similar design in other regions of the brain, and which property may be used for recognizing environmental cues. The point is that self-organized principal components analysis may be significant in hierarchical clustering of learned cues in the cerebral cortex not because of its optimal reconstruction property, but rather by virtue of its intrinsic property of picking projections of clusters with good separations.

Another interesting role for principal components analysis in perceptual processing appears in an approach to the *shape-from-shading* problem, proposed in Atick et al. (1996). The problem may be stated as follows: How is the brain able to perceive a three-dimensional shape from the shading patterns cast in a two-dimensional image? Atick et al. propose a hierarchical solution to the shape-from-shading problem, consisting of two notions:

1. The brain, through evolution or prior experience, has discovered that objects can be classified into lower-dimensional object classes with regard to their shape. This notion in effect builds on the fact that the *cues* the brain uses to extract a three-dimensional interpretation are well understood.
2. In light of notion 1, the extraction of a shape from shading patterns is reduced to the much simpler problem of *parameter estimation* in a low-dimensional space.

For example, the gross structure of a human head shape is invariably the same, in the sense that all people have noses representing protrusions, eye sockets representing depressions, and foreheads and cheeks representing flatter regions. This invariance suggests that any given face, expressed as $r(\theta,l)$ in cylindrical coordinates, may be described as the sum of two components:

$$r(\theta,l) = r_0(\theta,l) + \rho(\theta,l)$$

where $r_0(\theta,l)$ denotes a *mean-head* for a particular category of people (e.g., male adults or female adults), and $\rho(\theta,l)$ denotes *perturbations* that capture the identity of a particular person. Typically, $\rho(\theta,l)$ is small compared to $r_0(\theta,l)$. To represent $\rho(\theta,l)$, Atick et al. use principal components analysis, whereby the fluctuations are represented in terms of a set of eigenfunctions (i.e., the two-dimensional counterparts of eigenvectors). Results are presented in Atick et al. (1996), demonstrating the ability of the two-stage hierarchical approach to recover the three-dimensional surface for a given person from a single two-dimensional image of that person.

NOTES AND REFERENCES

1. Principal components analysis (PCA) is perhaps the oldest and best known technique in multivariate analysis (Jolliffe, 1986; Preisendorfer, 1988). It was first introduced by Pearson (1901), who used it in a biological context to recast linear regression analysis into a new form. It was then developed by Hotelling (1933) in work done on psychometry. It appeared once again and quite independently in the setting of probability theory, as considered by Karhunen (1947); and was subsequently generalized by Loéve (1963).

2. The approaches taken by Ljung (1977) and Kushner and Clark (1978) for studying the dynamical behavior of a stochastic approximation algorithm reduce the problem to that of studying the dynamics of an associated differential equation. However, these two approaches are fundamentally different. Ljung's approach involves the use of a Lyapunov function, whereas the approach taken by Kushner and Clark involves a linear interpolation process and invokes the so-called Arzelà–Ascoli theorem (Dunford and Schwartz, 1966). The approach by Kushner and Clark is followed in Diamantaras and Kung (1996) for studying the convergence of the Hebbian-based maximum eigenfilter. The conclusions reached therein are the same as those obtained using Ljung's approach.

3. Földiak (1989) expanded the neural network configuration for principal components analysis by including anti-Hebbian feedback connections. The motivation for this modification was derived from some earlier work by Barlow and Földiak (1989) on adaptation and decorrelation in the visual cortex; there it was argued that if the neurons interact according

to an anti-Hebbian rule, then the outputs of the neurons define a coordinate system in which there are no correlations even when the incoming signals have strong correlations.

The use of lateral inhibitions among the output neurons was also proposed by Rubner and Tavan (1989) and Rubner and Schulten (1990). However, unlike the model proposed by Földiak, the lateral network considered by Rubner et al. is not symmetrically connected. Rather, the lateral network is hierarchical, in that neuron i (say) inhibits all other neurons in the model except for $1, 2, ..., i - 1$, where $i = 1, 2, ...$.

The APEX model studied in Kung and Diamantaras (1990) has the same network topology as that of the model due to Rubner et al., but it uses Oja's single-neuron learning rule (described in Section 8.4) for adjusting the synaptic weights in both the feedforward and lateral connections in the model.

4. A rigorous proof of convergence of the APEX algorithm, with all the neurons tending to converge together, is given in Chen and Liu (1992).

5. For a discussion of several neural models for principal components analysis and their comparison, see the book by Diamantaras and Kung (1996).

6. Nonlinear PCA methods, excluding kernel PCA, may be grouped into three classes (Diamantaras and Kung, 1996):

   • *Hebbian networks,* which are obtained by replacing the linear neurons in Hebbian-based PCA algorithms with nonlinear neurons (Karhunen and Joutsensalo, 1995).
   • *Replicator networks* or *autoencoders,* which are built around multilayer perceptrons: replicator networks are discussed in Chapter 4.
   • *Principal curves,* which are based on an iterative estimation of a curve or surface capturing the structure of the data (Hastie and Stuetzle, 1989). In Ritter et al. (1992) and Cherkassky and Mulier (1995), it is pointed out that Kohonen's self-organizing map can be viewed as a computational procedure for finding a discrete approximation of principal curves; self-organizing maps are discussed in the next chapter.

PROBLEMS

Hebbian-based maximum eigenfilter

8.1 For the matched filter considered in Example 8.2, the eigenvalue λ_1 and associated eigenvector $\mathbf{q}_1$ are defined by

$$\lambda_1 = 1 + \sigma^2$$

$$\mathbf{q}_1 = \mathbf{s}$$

Show that these parameters satisfy the basic relation

$$\mathbf{R}\mathbf{q}_1 = \lambda_1\mathbf{q}_1$$

where $\mathbf{R}$ is the correlation matrix of the input vector $\mathbf{X}$.

8.2 Consider the maximum eigenfilter where the weight vector $\mathbf{w}(n)$ evolves in accordance with Eq. (8.46). Show that the variance of the filter output approaches $\lambda_{\max}$ as n approaches infinity, where $\lambda_{\max}$ is the largest eigenvalue of the correlation matrix of the input vector.

8.3 *Minor components analysis* (MCA) is the opposite of principal components analysis. In MCA, we seek to find those directions that *minimize* the projection variance. The directions so found are the eigenvectors corresponding to the smallest (minimum) eigenvalues of the correlation matrix $\mathbf{R}$ of the input vector $\mathbf{X}(n)$.

In this problem we explore how to modify the single neuron of Section 8.4 so as to find the minor component of $\mathbf{R}$. In particular, we make a change of sign in the learning rule of Eq. (8.40), obtaining (Xu et al., 1992)

$$w_i(n + 1) = w_i(n) - \eta y(n)[x_i(n) - y(n) w_i(n)]$$

Show that if the smallest eigenvalue of the correlation matrix $\mathbf{R}$ is λ_m with multiplicity 1, then

$$\lim_{n \to \infty} \mathbf{w}(n) = \eta \, \mathbf{q}_m$$

where $\mathbf{q}_m$ is the eigenvector associated with λ_m.

Hebbian-based principal components analysis

8.4 Construct a signal-flow graph to represent the vector-valued Eqs. (8.87) and (8.88).

8.5 The ordinary differential equation approach to convergence analysis described in Section 8.4 does not apply directly to the generalized Hebbian-learning algorithm (GHA). However, by expressing the synaptic weight matrix $\mathbf{W}(n)$ in Eq. (8.91) as a vector made up of the individual columns of $\mathbf{W}(n)$, we may interpret the update function $h(\cdot,\cdot)$ in the usual way, then proceed to apply the asymptotic stability theorem. Hence, in light of what has been said here, explore the convergence theorem for the generalized Hebbian-learning algorithm.

8.6 In this problem we explore the use of the generalized Hebbian algorithm to study two-dimensional receptive fields produced by a random input (Sanger, 1990). The random input consists of a two-dimensional field of independent Gaussian noise with zero mean and unit variance, which is convolved with a Gaussian mask (filter) and then multiplied by a Gaussian window. The Gaussian mask has a standard deviation of 2 pixels, and the Gaussian window has a standard deviation of 8 pixels. The resulting random input $x(r, s)$ at position (r, s) may thus be written as follows:

$$x(r, s) = m(r, s)[g(r,s) * w(r, s)]$$

where $w(r, s)$ is the field of independent and identically distributed Gaussian noise, $g(r, s)$ is the Gaussian mask, and $m(r, s)$ is the Gaussian window function. The circular convolution of $g(r, s)$ and $w(r, s)$ is defined by

$$g(r, s) * w(r, s) = \sum_{p=0}^{N-1} \sum_{q=0}^{N-1} g(p, q)w(r - p, s - q)$$

where $g(r, s)$ and $w(r, s)$ are both assumed to be periodic.

Use 2000 samples of the random input $x(r, s)$ to train a single-layer feedforward network by means of the generalized Hebbian algorithm. The network has 4096 inputs arranged as a 64×64 grid of pixels, and 16 outputs. The resulting synaptic weights of the trained network are represented as 64×64 arrays of numbers. Perform the computations described herein and display the 16 arrays of synaptic weights as two-dimensional masks. Comment on your results.

8.7 Equation (8.113) defines the transformed version of the update equation (8.106) for computing the feedback weight vector $\mathbf{a}_j(n)$. The transformation is based on the definition of the synaptic weight vector $\mathbf{w}_j(n)$ in terms of the m principal modes of the network given in Eq. (8.109). Derive Eq. (8.113).

8.8 Consider the system matrix of Eq. (8.116), represented by the signal-flow graph of Fig. 8.12 which corresponds to $1 \leq k \leq j - 1$.

(a) Formulate the characteristic equation of this 2×2 matrix.

(b) Show that the matrix has a double eigenvalue.

(c) Justify the statement that all the principal modes of the network have the same eigenvalue.

8.9 The GHA uses forward connections only, whereas the APEX algorithm uses both forward and lateral connections. Yet despite these differences, the long-term convergence behavior of the APEX algorithm is in theory exactly the same as that of the GHA. Justify the validity of this statement.

Kernel PCA

8.10 Let $\overline{K}_{ij}$ denote the centered counterpart of the ij-th element K_{ij} of kernel matrix $\mathbf{K}$. Show that (Schölkopf, 1997)

$$\overline{K}_{ij} = K_{ij} - \frac{1}{N}\sum_{m=1}^{N} \varphi^T(\mathbf{x}_m)\varphi(\mathbf{x}_j) - \frac{1}{N}\sum_{n=1}^{N} \varphi^T(\mathbf{x}_i)\varphi(\mathbf{x}_n)$$

$$+ \frac{1}{N^2}\sum_{m=1}^{N}\sum_{n=1}^{N} \varphi^T(\mathbf{x}_m)\varphi(\mathbf{x}_n)$$

Suggest a compact representation of this relation in matrix form.

8.11 Show that the normalization of eigenvector $\boldsymbol{\alpha}$ of the kernel matrix $\mathbf{K}$ is equivalent to the requirement that Eq. (8.153) be satisfied.

8.12 Summarize the properties of kernel PCA.

CHAPTER 9

Self-Organizing Maps

9.1 INTRODUCTION

In this chapter we continue our study of self-organizing systems by considering a special class of artificial neural networks known as self-organizing maps. These networks are based on *competitive learning*; the output neurons of the network compete among themselves to be activated or fired, with the result that only *one* output neuron, or one neuron per group, is on at any one time. An output neuron that wins the competition is called a *winner-takes-all neuron* or simply a *winning neuron*. One way of inducing a winner-takes-all competition among the output neurons is to use lateral inhibitory connections (i.e., negative feedback paths) between them; such an idea was originally proposed by Rosenblatt (1958).

In a *self-organizing map*, the neurons are placed at the nodes of a *lattice* that is usually one- or two-dimensional. Higher-dimensional maps are also possible but not as common. The neurons become *selectively tuned* to various input patterns (stimuli) or classes of input patterns in the course of a competitive learning process. The locations of the neurons so tuned (i.e., the winning neurons) become ordered with respect to each other in such a way that a meaningful coordinate system for different input *features* is created over the lattice (Kohonen, 1990a). A self-organizing map is therefore characterized by the formation of a *topographic map* of the input patterns in which the *spatial locations (i.e., coordinates) of the neurons in the lattice are indicative of intrinsic statistical features contained in the input patterns*, hence the name "self-organizing map."

As a neural model, the self-organizing map provides a bridge between two levels of adaptation:

- Adaptation rules formulated at the microscopic level of a single neuron.
- Formation of experimentally better and physically accessible patterns of feature selectivity at the microscopic level of neural layers.

Because a self-organizing map is inherently nonlinear, it may thus be viewed as a nonlinear generalization of principal components analysis (Ritter, 1995).

The development of self-organizing maps as a neural model is motivated by a distinct feature of the human brain: The brain is organized in many places in such a way that different sensory inputs are represented by *topologically ordered computational maps.* In particular, sensory inputs such as tactile (Kaas et al., 1983), visual (Hubel and Wiesel, 1962, 1977), and acoustic (Suga, 1985) are mapped onto different areas of the cerebral cortex in a topologically ordered manner. Thus the computational map constitutes a basic building block in the information-processing infrastructure of the nervous system. A computational map is defined by an array of neurons representing slightly differently tuned processors or filters, which operate on the sensory information-bearing signals in parallel. Consequently, the neurons transform input signals into a *place-coded probability distribution* that represents the computed values of parameters by sites of maximum relative activity within the map (Knudsen et al., 1987). The information so derived is of such a form that it can be readily accessed by higher-order processors using relatively simple connection schemes.

Organization of the Chapter

The material presented in this chapter on computational maps is organized as follows. In Section 9.2 we describe two feature-mapping models, which in their own individual ways are able to explain or capture the essential features of computational maps in the brain. The two models differ from each other in the form of the inputs used.

The rest of the chapter is devoted to detailed considerations of one of these models, commonly referred to as a "self-organizing map" due to Kohonen (1982). In Section 9.3 we use neurobiological considerations to develop a mathematical formalism of Kohonen's model. A summary of the model is presented in Section 9.4. Important properties of the model are described in Section 9.5, which is followed by computer simulations in Section 9.6. The performance of the feature map may finally be fine-tuned through a supervised technique known as learning vector quantization; this technique is described in Section 9.7. Section 9.8 describes a computer experiment on adaptive pattern classification that combines the use of learning vector quantization and the self-organizing map. In Section 9.9 we describe hierarchical vector quantization built around the self-organizing map for data compression. Section 9.10 describes another application of the self-organizing map for building contextual maps that find applications in unsupervised categorization of phonemic classes from text, remote sensing, and data exploration. The chapter concludes with some final remarks in Section 9.12.

9.2 TWO BASIC FEATURE-MAPPING MODELS

Anyone who examines a human brain cannot help but be impressed by the extent to which the brain is dominated by the cerebral cortex. The brain is almost completely enveloped by the cerebral cortex, which obscures the other parts. For sheer complexity, the cerebral cortex probably exceeds any other known structure in the universe (Hubel and Wiesel, 1977). What is equally impressive is the way in which different sensory inputs (motor, somatosensory, visual, auditory, etc.) are *mapped* onto corresponding areas of the cerebral cortex in an *orderly* fashion; to appreciate this point, see the cytoarchitectural maps of the cerebral cortex in Fig. 2.4. The use of computational maps offers the following properties (Knudsen et al., 1987):

- At each stage of representation, each incoming piece of information is kept in its proper context.
- Neurons, dealing with closely related pieces of information, are close together so that they can interact via short synaptic connections.

Our interest lies in building artificial topographic maps that learn through self-organization in a neurobiologically inspired manner. In this context, the one important point that emerges from the very brief discussion of computational maps in the brain is the *principle of topographic map formation*, which may be stated as (Kohonen, 1990a):

> The spatial location of an output neuron in a topographic map corresponds to a particular domain or feature of data drawn from the input space.

This principle has provided the neurobiological motivation for two basically different *feature-mapping models*[1] described herein.

Figure 9.1 displays the layout of the two models. In both cases the output neurons are arranged in a two-dimensional lattice. This kind of topology ensures that each neuron has a set of neighbors. The models differ from each other in the manner in which the input patterns are specified.

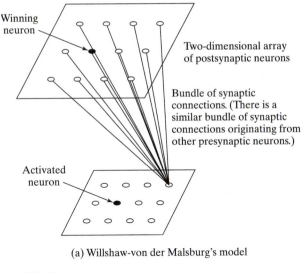

(a) Willshaw-von der Malsburg's model

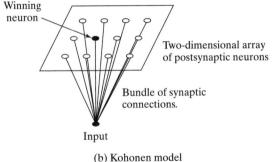

(b) Kohonen model

FIGURE 9.1 Two self-organized feature maps.

The model of Fig. 9.1a was originally proposed by Willshaw and von der Malsburg (1976) on biological grounds to explain the problem of retinotopic mapping from the retina to the visual cortex (in higher vertebrates). Specifically, there are two separate two-dimensional lattices of neurons connected together, with one projecting onto the other. One lattice represents presynaptic (input) neurons, and the other lattice represents postsynaptic (output) neurons. The postsynaptic lattice uses a *short-range excitatory mechanism* as well as a *long-range inhibitory mechanism*. These two mechanisms are local in nature and critically important for self-organization. The two lattices are interconnected by modifiable synapses of a Hebbian type. Strictly speaking, therefore, the postsynaptic neurons are not winner takes all; rather, a threshold is used to ensure that only a few postsynaptic neurons will fire at any one time. Moreover, to prevent a steady buildup in the synaptic weights that may lead to network instability, the total weight associated with each postsynaptic neuron is limited by an upper boundary condition.[2] Thus, for each neuron some synaptic weights increase while others decrease. The basic idea of the Willshaw–von der Malsburg model is for the geometric proximity of presynaptic neurons to be coded in the form of correlations in their electrical activity, and to use these correlations in the postsynaptic lattice so as to connect neighboring presynaptic neurons to neighboring postsynaptic neurons. A topologically ordered mapping is thereby produced by self-organization. Note, however, that the Willshaw–von der Malsburg model is specialized to mapping where the input dimension is the same as the output dimension.

The second model of Fig. 9.1b, introduced by Kohonen (1982), is not meant to explain neurobiological details. The model captures the essential features of computational maps in the brain and yet remains computationally tractable.[3] It appears that the Kohonen model is more general than the Willshaw–von der Malsburg model in the sense that it is capable of performing data compression (i.e., dimensionality reduction on the input).

In reality, the Kohonen model belongs to the class of *vector-coding* algorithms. The model provides a topological mapping that optimally places a fixed number of vectors (i.e., code words) into a higher-dimensional input space, and thereby facilitates data compression. The Kohonen model may therefore be derived in two ways. We may use basic ideas of self-organization, motivated by neurobiological considerations, to derive the model, which is the traditional approach (Kohonen, 1982, 1990a, 1997a). Alternatively, we may use a vector quantization approach that uses a model involving an encoder and a decoder, which is motivated by communication–theoretic considerations (Luttrell, 1989b, 1991a). In this chapter we consider both approaches.

The Kohonen model has received much more attention in the literature than the Willshaw–von der Malsburg model. It possesses certain properties discussed later in the chapter, which make it particularly interesting for understanding and modeling cortical maps in the brain. The remainder of the chapter is devoted to the derivation of the *self-organizing map,* its basic properties and ramifications.

9.3   SELF-ORGANIZING MAP

The principal goal of the self-organizing map (SOM) is to transform an incoming signal pattern of arbitrary dimension into a one- or two-dimensional discrete map, and to perform this transformation adaptively in a topologically ordered fashion. Figure 9.2 shows the schematic diagram of a two-dimensional lattice of neurons commonly used

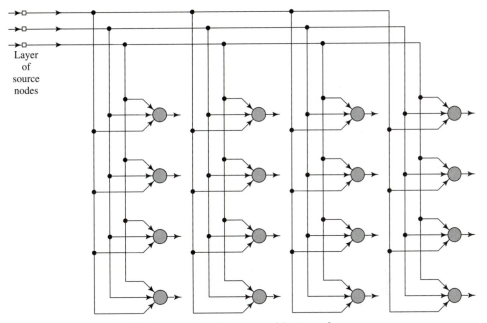

FIGURE 9.2   Two-dimensional lattice of neurons.

as the discrete map. Each neuron in the lattice is fully connected to all the source nodes in the input layer. This network represents a feedforward structure with a single computational layer consisting of neurons arranged in rows and columns. A one-dimensional lattice is a special case of the configuration depicted in Fig. 9.2: in this special case the computational layer consists simply of a single column or row of neurons.

Each input pattern presented to the network typically consists of a localized region or "spot" of activity against a quiet background. The location and nature of such a spot usually varies from one realization of the input pattern to another. All the neurons in the network should therefore be exposed to a sufficient number of different realizations of the input pattern to ensure that the self-organization process has a chance to mature properly.

The algorithm responsible for the formation of the self-organizing map proceeds first by *initializing* the synaptic weights in the network. This can be done by assigning them *small values picked from a random number generator*; in so doing, no prior order is imposed on the feature map. Once the network has been properly initialized, there are three essential processes involved in the formation of the self-organizing map, as summarized here:

1. *Competition.* For each input pattern, the neurons in the network compute their respective values of a discriminant function. This discriminant function provides the basis for competition among the neurons. The particular neuron with the largest value of discriminant function is declared winner of the competition.
2. *Cooperation.* The winning neuron determines the spatial location of a topological neighborhood of excited neurons, thereby providing the basis for cooperation among such neighboring neurons.

3. *Synaptic Adaptation.* This last mechanism enables the excited neurons to increase their individual values of the discriminant function in relation to the input pattern through suitable adjustments applied to their synaptic weights. The adjustments made are such that the response of the winning neuron to the subsequent application of a similar input pattern is enhanced.

The processes of competition and cooperation are in accordance with two of the four principles of self-organization described in Chapter 8. As for the principle of self-amplification, it comes in a modified form of Hebbian learning in the adaptive process. As explained in Chapter 8, the presence of redundancy in the input data (though not mentioned explicitly in describing the SOM algorithm) is needed for learning since it provides knowledge. Detailed descriptions of the processes of competition, cooperation, and synaptic adaptation are now presented.

Competitive Process

Let m denote the dimension of the input (data) space. Let an input pattern (vector) selected at random from the input space be denoted by

$$\mathbf{x} = [x_1, x_2, ..., x_m]^T \tag{9.1}$$

The synaptic weight vector of each neuron in the network has the same dimension as the input space. Let the synaptic weight vector of neuron j be denoted by

$$\mathbf{w}_j = [w_{j1}, w_{j2}, ..., w_{jm}]^T, \qquad j = 1, 2, ..., l \tag{9.2}$$

where l is the total number of neurons in the network. To find the best match of the input vector $\mathbf{x}$ with the synaptic weight vectors $\mathbf{w}_j$, compare the inner products $\mathbf{w}_j^T \mathbf{x}$ for $j = 1, 2, ..., l$ and select the largest. This assumes that the same threshold is applied to all the neurons; the threshold is the negative of bias. Thus, by selecting the neuron with the largest inner product $\mathbf{w}_j^T \mathbf{x}$, we will have in effect determined the location where the topological neighborhood of excited neurons is to be centered.

From Chapter 1 we recall that the best matching criterion, based on maximizing the inner product $\mathbf{w}_j^T \mathbf{x}$, is mathematically equivalent to minimizing the Euclidean distance between the vectors $\mathbf{x}$ and $\mathbf{w}_j$. If we use the index $i(\mathbf{x})$ to identify the neuron that best matches the input vector $\mathbf{x}$, we may then determine $i(\mathbf{x})$ by applying the condition[4]

$$i(\mathbf{x}) = \arg \min_j \|\mathbf{x} - \mathbf{w}_j\|, \quad j = 1, 2, ..., l \tag{9.3}$$

which sums up the essence of the competition process among the neurons. According to Eq. (9.3), $i(\mathbf{x})$ is the subject of attention because we want the identity of neuron i. The particular neuron i that satisfies this condition is called the *best-matching* or *winning neuron* for the input vector $\mathbf{x}$. Equation (9.3) leads to this observation:

> *A continuous input space of activation patterns is mapped onto a discrete output space of neurons by a process of competition among the neurons in the network.*

Depending on the application of interest, the response of the network could be either the index of the winning neuron (i.e., its position in the lattice), or the synaptic weight vector that is closest to the input vector in a Euclidean sense.

Cooperative Process

The winning neuron locates the center of a topological neighborhood of cooperating neurons. The key question is: How do we define a topological neighborhood that is neurobiologically correct? To answer this question, remember that there is neurobiological evidence for *lateral interaction* among a set of excited neurons. In particular, a neuron that is firing tends to excite the neurons in its immediate neighborhood *more* than those farther away from it, which is intuitively satisfying. This observation leads us to make the topological neighborhood around the winning neuron i decay smoothly with lateral distance[5] (Lo et al., 1991, 1993; Ritter et al., 1992). To be specific, let $h_{j,i}$ denote the *topological neighborhood* centered on winning neuron i, and encompassing a set of excited (cooperating) neurons, a typical one of which is denoted by j. Let $d_{i,j}$ denote the *lateral* distance between winning neuron i and excited neuron j. Then we may assume that the topological neighborhood $h_{j,i}$ is a unimodal function of the lateral distance $d_{j,i}$, such that it satisfies two distinct requirements:

- The topological neighborhood $h_{j,i}$ is symmetric about the maximum point defined by $d_{i,j} = 0$; in other words, it attains its maximum value at the winning neuron i for which the distance $d_{j,i}$ is zero.
- The amplitude of the topological neighborhood $h_{j,i}$ decreases monotonically with increasing lateral distance $d_{j,i}$, decaying to zero for $d_{j,i} \to \infty$; this is a necessary condition for convergence.

A typical choice of $h_{j,i}$ that satisfies these requirements is the *Gaussian* function[6]

$$h_{j,i(x)} = \exp\left(-\frac{d_{j,i}^2}{2\sigma^2}\right) \tag{9.4}$$

which is *translation invariant* (i.e., independent of the location of the winning neuron). The parameter σ is the "effective width" of the topological neighborhood as illustrated in Fig. 9.3; it measures the degree to which excited neurons in the vicinity of the winning neuron participate in the learning process. In a qualitative sense, the Gaussian topological neighborhood of Eq. (9.4) is more biologically appropriate than a rectangular one.

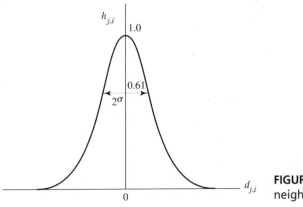

FIGURE 9.3   Gaussian neighborhood function.

Its use also makes the SOM algorithm converge more quickly than a rectangular topological neighborhood would (Lo et al., 1991, 1993; Erwin et al., 1992a).

For cooperation among neighboring neurons to hold, it is necessary that topological neighborhood $h_{j,i}$ be dependent on lateral distance $d_{j,i}$ between winning neuron i and excited neuron j in the output space rather than on some distance measure in the original input space. This is precisely what we have in Eq. (9.4). In the case of a one-dimensional lattice, $d_{j,i}$ is an integer equal to $|j\text{-}i|$. On the other hand, in the case of a two-dimensional lattice it is defined by

$$d_{j,i}^2 = \|\mathbf{r}_j - \mathbf{r}_i\|^2 \tag{9.5}$$

where the discrete vector $\mathbf{r}_j$ defines the position of excited neuron j and $\mathbf{r}_i$ defines the discrete position of winning neuron i, both of which are measured in the discrete output space.

Another unique feature of the SOM algorithm is that the size of the topological neighborhood shrinks with time. This requirement is satisfied by making the width σ of the topological neighborhood function $h_{j,i}$ decrease with time. A popular choice for the dependence of σ on discrete time n is the exponential decay described by (Ritter et al., 1992; Obermayer et al., 1991)

$$\sigma(n) = \sigma_0 \exp\left(-\frac{n}{\tau_1}\right) \qquad n = 0, 1, 2, ..., \tag{9.6}$$

where σ_0 is the value of σ at the initiation of the SOM algorithm, and τ_1 is a *time constant*. Correspondingly, the topological neighborhood assumes a time-varying form of its own, as shown by

$$h_{j,i(\mathbf{x})}(n) = \exp\left(-\frac{d_{j,i}^2}{2\sigma^2(n)}\right), \qquad n = 0, 1, 2, ..., \tag{9.7}$$

where $\sigma(n)$ is defined by Eq. (9.6). Thus, as time n (i.e., the number of iterations) increases, the width $\sigma(n)$ decreases at an exponential rate, and the topological neighborhood shrinks in a corresponding manner. Henceforth we will refer to $h_{j,i(\mathbf{x})}(n)$ as the *neighborhood function*.

Another useful way of viewing the variation of the neighborhood function $h_{j,i(\mathbf{x})}(n)$ around a winning neuron $i(\mathbf{x})$ is as follows (Luttrell, 1989a). The purpose of a wide $h_{j,i(\mathbf{x})}(n)$ is essentially to *correlate* the directions of the weight updates of a large number of excited neurons in the lattice. As the width of $h_{j,i(\mathbf{x})}(n)$ is decreased, so is the number of neurons whose update directions are correlated. This phenomenon becomes particularly obvious when the training of a self-organizing map is played on a computer graphics screen. It is rather wasteful of computer resources to move a large number of degrees of freedom around a winning neuron in a correlated fashion, as in the standard SOM algorithm. Instead it is much better to use a *renormalized* SOM form of training, according to which we work with a much smaller number of *normalized degrees of freedom*. This operation is easily performed in discrete form by having a neighborhood function $h_{j,i(\mathbf{x})}(n)$ of *constant* width, but gradually *increasing* the total number of neurons. The new neurons are inserted halfway between the old ones, and the smoothness properties of the SOM algorithm guarantee that the new ones join the synaptic adaptation in a graceful manner (Luttrell, 1989a). A summary of the renormalized SOM algorithm is presented in Problem 9.13.

Adaptive Process

Now we come to the last process, the synaptic adaptive process, in the self-organized formation of a feature map. For the network to be self-organizing, the synaptic weight vector $\mathbf{w}_j$ of neuron j in the network is required to change in relation to the input vector $\mathbf{x}$. The question is how to make the change. In Hebb's postulate of learning, a synaptic weight is increased with a simultaneous occurrence of presynaptic and postsynaptic activities. The use of such a rule is well suited for associative learning. For the type of unsupervised learning being considered here, however, the Hebbian hypothesis in its basic form is unsatisfactory for the following reason: Changes in connectivities occur in one direction only, which finally drive all the synaptic weights into saturation. To overcome this problem we modify the Hebbian hypothesis by including a *forgetting term*— $g(y_j)\mathbf{w}_j$, where $\mathbf{w}_j$ is the synaptic weight vector of neuron j and $g(y_j)$ is some positive scalar function of the response y_j. The only requirement imposed on the function $g(y_j)$ is that the constant term in the Taylor series expansion of $g(y_j)$ be zero, so that we may write

$$g(y_j) = 0 \qquad \text{for } y_j = 0 \tag{9.8}$$

The significance of this requirement will become apparent momentarily. Given such a function, we may then express the change to the weight vector of neuron j in the lattice as follows:

$$\Delta\mathbf{w}_j = \eta y_j \mathbf{x} - g(y_j)\mathbf{w}_j \tag{9.9}$$

where η is the *learning-rate parameter* of the algorithm. The first term on the right-hand side of Eq. (9.9) is the Hebbian term and the second term is the forgetting term. To satisfy the requirement of Eq. (9.8), we choose a linear function for $g(y_j)$, as shown by

$$g(y_j) = \eta y_j \tag{9.10}$$

We may further simplify Eq. (9.9) by setting

$$y_j = h_{j,i(\mathbf{x})} \tag{9.11}$$

Using Eqs. (9.10) and (9.11) in (9.9), we obtain

$$\Delta\mathbf{w}_j = \eta h_{j,i(\mathbf{x})}(\mathbf{x} - \mathbf{w}_j) \tag{9.12}$$

Finally, using discrete-time formalism, given the synaptic weight vector $\mathbf{w}_j(n)$ of neuron j at time n, the updated weight vector $\mathbf{w}_j(n + 1)$ at time $n + 1$ is defined by (Kohonen, 1982; Ritter et al., 1992; Kohonen, 1997a):

$$\mathbf{w}_j(n + 1) = \mathbf{w}_j(n) + \eta(n)h_{j,i(\mathbf{x})}(n)(\mathbf{x} - \mathbf{w}_j(n)) \tag{9.13}$$

which is applied to all the neurons in the lattice that lie inside the topological neighborhood of winning neuron i. Equation (9.13) has the effect of moving the synaptic weight vector $\mathbf{w}_i$ of winning neuron i toward the input vector $\mathbf{x}$. Upon repeated presentations of the training data, the synaptic weight vectors tend to follow the distribution of the input vectors due to the neighborhood updating. The algorithm therefore leads to a *topological ordering* of the feature map in the input space in the sense that neurons that are adjacent in the lattice will tend to have similar synaptic weight vectors. We have more to say on this issue in Section 9.5.

Equation (9.13) is the desired formula for computing the synaptic weights of the feature map. In addition to this equation, however, we need the heuristic of Eq. (9.7) for selecting the neighborhood function $h_{j,i(\mathbf{x})}(n)$ and another heuristic for selecting the learning-rate parameter $\eta(n)$.

The learning-rate parameter $\eta(n)$ should be time varying as indicated in Eq. (9.13), which is how it should be for stochastic approximation. In particular, it should start at an initial value η_0, and then decrease gradually with increasing time n. This requirement can be satisfied by choosing an exponential decay for $\eta(n)$, as shown by

$$\eta(n) = \eta_0 \exp\left(-\frac{n}{\tau_2}\right), \qquad n = 0, 1, 2, \ldots, \tag{9.14}$$

where τ_2 is another time constant of the SOM algorithm. Even though the exponential decay formulas described in Eqs. (9.6) and (9.14) for the width of the neighborhood function and the learning-rate parameter, respectively, may not be optimal, they are usually adequate for the formation of the feature map in a self-organized manner.

Two Phases of the Adaptive Process: Ordering and Convergence

Starting from an initial state of complete disorder, it is amazing how the SOM algorithm gradually leads to an organized representation of activation patterns drawn from the input space, provided that the parameters of the algorithm are selected properly. We may decompose the adaptation of the synaptic weights in the network, computed in accordance with Eq. (9.13), into two phases: an ordering or self-organizing phase followed by a convergence phase. These two phases of the adaptive process are described as follows (Kohonen, 1982, 1997a):

1. *Self-organizing* or *ordering phase*. It is during this first phase of the adaptive process that the topological ordering of the weight vectors takes place. The ordering phase may take as many as 1000 iterations of the SOM algorithm, and possibly more. Careful considerations must be given to the choice of the learning-rate parameter and neighborhood function:
   - The learning-rate parameter $\eta(n)$ should begin with a value close to 0.1; thereafter it should decrease gradually, but remain above 0.01. These desirable values are satisfied by the following choices in the formula of Eq. (9.14):

$$\eta_0 = 0.1$$

$$\tau_2 = 1000$$

   - The neighborhood function $h_{j,i}(n)$ should initially include almost all neurons in the network centered on the winning neuron i, and then shrink slowly with time.

Specifically, during the ordering phase that may occupy 1000 iterations or more, $h_{j,i}(n)$ is permitted to reduce to a small value of only a couple of neighboring neurons around a winning neuron or to the winning neuron by itself. Assuming the use of a two-dimensional lattice of neurons for the discrete map, we may thus set the initial size σ_0 of the neighborhood function equal to the "radius" of the lattice. Correspondingly, we may set the time constant τ_1 in the formula of Eq. (9.6) as follows:

$$\tau_1 = \frac{1000}{\log \sigma_0}$$

2. *Convergence phase.* This second phase of the adaptive process is needed to fine tune the feature map and therefore provide an accurate statistical quantification of the input space. As a general rule, the number of iterations constituting the convergence phase must be at least 500 times the number of neurons in the network. Thus, the convergence phase may have to go on for thousands and possibly tens of thousands of iterations:

- For good statistical accuracy, the learning parameter $\eta(n)$ should be maintained during the convergence phase at a small value, on the order of 0.01. In any event, it must not be allowed to decrease to zero; otherwise, it is possible for the network to get stuck in a metastable state. A *metastable state* belongs to a configuration of the feature map with a topological defect. The exponential decay of Eq. (9.14) guarantees against the possibility of metastable states.
- The neighborhood function $h_{j,i(\mathbf{x})}$ should contain only the nearest neighbors of a winning neuron, which may eventually reduce to one or zero neighboring neurons.

9.4   SUMMARY OF THE SOM ALGORITHM

The essence of Kohonen's SOM algorithm is that it substitutes a simple geometric computation for the more detailed properties of the Hebb-like rule and lateral interactions. The essential ingredients/parameters of the algorithm are:

- A continuous input space of activation patterns that are generated in accordance with a certain probability distribution.
- A topology of the network in the form of a lattice of neurons, which defines a discrete output space.
- A time-varying neighborhood function $h_{j,i(\mathbf{x})}(n)$ that is defined around a winning neuron $i(\mathbf{x})$.
- A learning-rate parameter $\eta(n)$ that starts at an initial value η_0 and then decreases gradually with time, n, but never goes to zero.

For the neighborhood function and learning-rate parameter, we may use Eqs. (9.7) and (9.14), respectively, for the ordering phase (i.e., the first thousand iterations or so). For good statistical accuracy, $\eta(n)$ should be maintained at a small value (0.01 or less) during the convergence for a fairly long period of time, which is typically thousands of iterations. As for the neighborhood function, it should contain only the nearest neighbors of the winning neuron at the start of the convergence phase, and may eventually shrink to one or zero neighboring neurons.

There are three basic steps involved in the application of the algorithm after initialization: sampling, similarity matching, and updating. These three steps are repeated until formation of the feature map has completed. The algorithm is summarized as follows:

1. *Initialization*. Choose random values for the initial weight vectors $\mathbf{w}_j(0)$. The only restriction here is that the $\mathbf{w}_j(0)$ be different for $j = 1, 2, ..., l$, where l is the number of neurons in the lattice. It may be desirable to keep the magnitude of the weights small.

   Another way of initalizing the algorithm is to select the weight vectors $\{\mathbf{w}_j(0)\}_{j=1}^{l}$ from the available set of input vectors $\{\mathbf{x}_i\}_{i=1}^{N}$ in a random manner.

2. *Sampling*. Draw a sample $\mathbf{x}$ from the input space with a certain probability; the vector $\mathbf{x}$ represents the activation pattern that is applied to the lattice. The dimension of vector $\mathbf{x}$ is equal to m.

3. *Similarity Matching*. Find the best-matching (winning) neuron $i(\mathbf{x})$ at time step n by using the minimum-distance Euclidean criterion:

$$i(\mathbf{x}) = \arg \min_j \|\mathbf{x}(n) - \mathbf{w}_j\|, \quad j = 1, 2, ..., l$$

4. *Updating*. Adjust the synaptic weight vectors of all neurons by using the update formula

$$\mathbf{w}_j(n + 1) = \mathbf{w}_j(n) + \eta(n)h_{j, i(\mathbf{x})}(n)(\mathbf{x}(n) - \mathbf{w}_j(n))$$

   where $\eta(n)$ is the learning-rate parameter, and $h_{j,i(\mathbf{x})}(n)$ is the neighborhood function centered around the winning neuron $i(\mathbf{x})$; both $\eta(n)$ and $h_{j,i(\mathbf{x})}(n)$ are varied dynamically during learning for best results.

5. *Continuation*. Continue with step 2 until no noticeable changes in the feature map are observed.

9.5 PROPERTIES OF THE FEATURE MAP

Once the SOM algorithm has converged, the *feature map* computed by the algorithm displays important statistical characteristics of the input space.

To begin with, let $\mathcal{X}$ denote a *spatially continuous input (data) space*, the topology of which is defined by the metric relationship of the vectors $\mathbf{x} \in \mathcal{X}$. Let $\mathcal{A}$ denote a *spatially discrete output space*, the topology of which is endowed by arranging a set of neurons as the computation nodes of a lattice. Let Φ denote a nonlinear transformation called a *feature map*, which maps the input space $\mathcal{X}$ onto the output space $\mathcal{A}$, as shown by

$$\Phi: \mathcal{X} \rightarrow \mathcal{A} \tag{9.15}$$

Equation (9.15) may be viewed as an abstraction of Eq. (9.3) that defines the location of a winning neuron $i(\mathbf{x})$ developed in response to an input vector $\mathbf{x}$. For example, in a neurobiological context, the input space $\mathcal{X}$ may represent the coordinate set of somatosensory receptors distributed densely over the entire body surface. Correspondingly, the output space $\mathcal{A}$ represents the set of neurons located in that layer of the cerebral cortex to which the somatosensory receptors are confined.

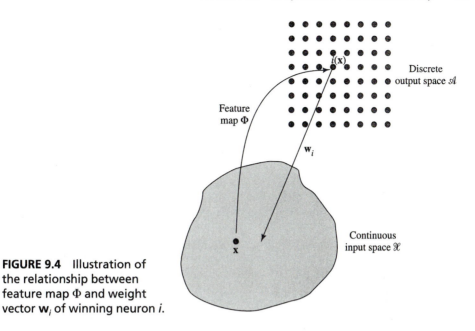

FIGURE 9.4 Illustration of the relationship between feature map Φ and weight vector $\mathbf{w}_i$ of winning neuron i.

Given an input vector $\mathbf{x}$, the SOM algorithm proceeds by first identifying a best-matching or winning neuron $i(\mathbf{x})$ in the output space $\mathcal{A}$, in accordance with the feature map Φ. The synaptic weight vector $\mathbf{w}_i$ of neuron $i(\mathbf{x})$ may then be viewed as a *pointer* for that neuron into the input space $\mathcal{X}$; that is, the synaptic elements of vector $\mathbf{w}_i$ may be viewed as the coordinates of the *image* of neuron i projected in the input space. These two operations are depicted in Fig. 9.4.

The feature map Φ has some important properties:

Property 1. Approximation of the Input Space.   *The feature map Φ, represented by the set of synaptic weight vectors $\{\mathbf{w}_j\}$ in the output space $\mathcal{A}$, provides a good approximation to the input space $\mathcal{X}$.*

The basic aim of the SOM algorithm is to store a large set of input vectors $\mathbf{x} \in \mathcal{X}$ by finding a smaller set of prototypes $\mathbf{w}_j \in \mathcal{A}$, so as to provide a good approximation to the original input space $\mathcal{X}$. The theoretical basis of the idea just described is rooted in *vector quantization theory*, the motivation for which is dimensionality reduction or data compression (Gersho and Gray, 1992). It is therefore appropriate to present a brief discussion of this theory.

Consider Fig. 9.5, where $\mathbf{c}(\mathbf{x})$ acts as an *encoder* of the input vector $\mathbf{x}$ and $\mathbf{x}'(\mathbf{c})$ acts as a *decoder* of $\mathbf{c}(\mathbf{x})$. The vector $\mathbf{x}$ is selected at random from a training sample (i.e., input space $\mathcal{X}$), subject to an underlying probability density function $f_{\mathbf{X}}(\mathbf{x})$. The optimum encoding-decoding scheme is determined by varying the functions $\mathbf{c}(\mathbf{x})$ and $\mathbf{x}'(\mathbf{c})$, so as to minimize the *expected distortion* defined by

$$D = \frac{1}{2} \int_{-\infty}^{\infty} d\mathbf{x} f_{\mathbf{X}}(\mathbf{x}) d(\mathbf{x}, \mathbf{x}') \tag{9.16}$$

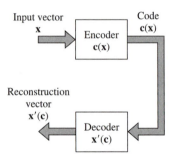

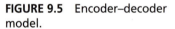

FIGURE 9.5 Encoder–decoder model.

where the factor $1/2$ has been introduced for convenience of presentation, and $d(\mathbf{x},\mathbf{x}')$ is a *distortion* measure. The integration is performed over the entire input space $\mathscr{X}$ assumed to be of dimensionality m. A popular choice for the distortion measure $d(\mathbf{x},\mathbf{x}')$ is the square of the Euclidean distance between the input vector $\mathbf{x}$ and the reconstruction vector $\mathbf{x}'$; that is,

$$d(\mathbf{x}, \mathbf{x}') = \|\mathbf{x} - \mathbf{x}'\|^2$$
$$= (\mathbf{x} - \mathbf{x}')^T(\mathbf{x} - \mathbf{x}') \tag{9.17}$$

Thus we may rewrite Eq. (9.16) as

$$D = \frac{1}{2}\int_{-\infty}^{\infty} d\mathbf{x}f_{\mathbf{X}}(\mathbf{x})\|\mathbf{x} - \mathbf{x}'\|^2 \tag{9.18}$$

The necessary conditions for the minimization of the expected distortion D are embodied in the *generalized Lloyd algorithm*[7] (Gersho and Gray, 1992). The conditions are twofold:

Condition 1. Given the input vector $\mathbf{x}$, choose the code $\mathbf{c} = \mathbf{c}(\mathbf{x})$ to minimize the squared error distortion $\|\mathbf{x} - \mathbf{x}'(\mathbf{c})\|^2$.

Condition 2. Given the code $\mathbf{c}$, compute the reconstruction vector $\mathbf{x}' = \mathbf{x}'(\mathbf{c})$ as the centroid of those input vectors $\mathbf{x}$ that satisfy condition 1.

Condition 1 is recognized as a *nearest-neighbor* encoding rule. Conditions 1 and 2 imply that the average distortion D is stationary (i.e., at a local minimum) with respect to variations in the encoder $\mathbf{c}(\mathbf{x})$ and decoder $\mathbf{x}'(\mathbf{c})$, respectively. To implement vector quantization, the generalized Lloyd algorithm operates in a *batch* training mode. Basically, the algorithm consists of alternately optimizing the encoder $\mathbf{c}(\mathbf{x})$ in accordance with condition 1, and then optimizing the decoder $\mathbf{x}'(\mathbf{c})$ in accordance with condition 2 until the expected distortion D reaches a minimum. In order to overcome the local-minimum problem, it may be necessary to run the generalized Lloyd algorithm several times with different initial code vectors.

The generalized Lloyd algorithm is closely related to the SOM algorithm, as shown in Luttrell (1989b). We may delineate the form of this relationship by considering the scheme shown in Fig. 9.6, where we have introduced a signal-independent *noise*

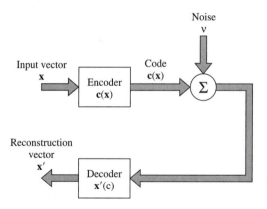

Noise
ν

Input vector
x

Encoder
c(x)

Code
c(x)

Σ

Reconstruction
vector
x'

Decoder
x'(c)

FIGURE 9.6   Noisy encoder–
decoder model.

process ν following the encoder $\mathbf{c(x)}$. The noise ν is associated with a fictitious "communication channel" between the encoder and the decoder, the purpose of which is to account for the possibility that the output code $\mathbf{c(x)}$ may be distorted. On the basis of the model shown in Fig. 9.6, we may consider a *modified* form of expected distortion as follows:

$$D_1 = \frac{1}{2}\int_{-\infty}^{\infty} d\mathbf{x} f_{\mathbf{X}}(\mathbf{x}) \int_{-\infty}^{\infty} d\boldsymbol{\nu}\,\pi(\boldsymbol{\nu})\|\mathbf{x} - \mathbf{x}'(\mathbf{c(x)} + \boldsymbol{\nu})\|^2 \tag{9.19}$$

where $\pi(\boldsymbol{\nu})$ is the probability density function (pdf) of the additive noise $\boldsymbol{\nu}$, and the second integration is over all possible realizations of this noise.

In accordance with the strategy described for the generalized Lloyd algorithm, there are two separate optimizations to be considered for the model of Fig. 9.6, one pertaining to the encoder and the other pertaining to the decoder. To find the optimum encoder for a given $\mathbf{x}$, we need the partial derivative of the expected distortion measure D_1 with respect to the encoded vector $\mathbf{c}$. Using Eq. (9.19), we thus obtain

$$\frac{\partial D_1}{\partial \mathbf{c}} = \frac{1}{2} f_{\mathbf{X}}(\mathbf{x}) \int_{-\infty}^{\infty} d\boldsymbol{\nu}\,\pi(\boldsymbol{\nu}) \frac{\partial}{\partial \mathbf{c}} \|\mathbf{x} - \mathbf{x}'(\mathbf{c})\|^2 \,|_{\mathbf{c}=\mathbf{c(x)}+\boldsymbol{\nu}} \tag{9.20}$$

To find the optimum decoder for a given $\mathbf{c}$, we need the partial derivative of the expected distortion measure D_1 with respect to the decoded vector $\mathbf{x}'(\mathbf{c})$. Using Eq. (9.19), we thus obtain

$$\frac{\partial D_1}{\partial \mathbf{x}'(\mathbf{c})} = -\int_{-\infty}^{\infty} d\mathbf{x} f_{\mathbf{X}}(\mathbf{x}) \pi(\mathbf{c} - \mathbf{c(x)}) (\mathbf{x} - \mathbf{x}'(\mathbf{c})) \tag{9.21}$$

Hence, in light of Eqs. (9.20) and (9.21), conditions 1 and 2 stated earlier for the generalized Lloyd algorithm must be modified as follows (Luttrell, 1989b):

Condition I. Given the input vector $\mathbf{x}$, choose the code $\mathbf{c} = \mathbf{c(x)}$ to minimize the distortion measure

$$D_2 = \int_{-\infty}^{\infty} d\boldsymbol{\nu}\,\pi(\boldsymbol{\nu})\|\mathbf{x} - \mathbf{x}'(\mathbf{c(x)} + \boldsymbol{\nu})\|^2 \tag{9.22}$$

Condition II. Given the code **c**, compute the reconstruction vector $\mathbf{x}'(\mathbf{c})$ to satisfy the condition

$$\mathbf{x}'(\mathbf{c}) = \frac{\displaystyle\int_{-\infty}^{\infty} d\mathbf{x} f_{\mathbf{X}}(\mathbf{x})\pi(\mathbf{c} - \mathbf{c}(\mathbf{x}))\mathbf{x}}{\displaystyle\int_{-\infty}^{\infty} d\mathbf{x} f_{\mathbf{X}}(\mathbf{x})\pi(\mathbf{c} - \mathbf{c}(\mathbf{x}))} \tag{9.23}$$

Equation (9.23) is obtained by setting the partial derivative $\partial D_1/\partial\mathbf{x}'(\mathbf{c})$ in Eq. (9.21) equal to zero and then solving for $\mathbf{x}'(\mathbf{c})$.

The model described in Fig. 9.5 may be viewed as a special case of that shown in Fig. 9.6. In particular, if we set the probability density function $\pi(\boldsymbol{\nu})$ of the noise $\boldsymbol{\nu}$ equal to a Dirac delta function $\delta(\boldsymbol{\nu})$, conditions I and II reduce to conditions 1 and 2 for the generalized Lloyd algorithm, respectively.

To simplify condition I, we assume that $\pi(\boldsymbol{\nu})$ is a smooth function of $\boldsymbol{\nu}$. It may then be shown that, to a second-order of approximation, the distortion measure D_2 defined in Eq. (9.22) consists of two components (Luttrell, 1989b):

- The *conventional* distortion term, defined by the squared error distortion $\|\mathbf{x} - \mathbf{x}'(\mathbf{c})\|^2$
- A *curvature* term that arises from the noise model $\pi(\boldsymbol{\nu})$

Assuming that the curvature term is small, condition I for the model of Fig. 9.6 may be approximated by condition 1 for the noiseless model of Fig. 9.5. This in turn reduces condition I to a nearest-neighbor encoding rule as before.

As for condition II, we may realize it by using stochastic descent learning. In particular, we choose input vectors $\mathbf{x}$ at random from the input space $\mathscr{X}$ using the factor $\int d\mathbf{x} f_{\mathbf{X}}(\mathbf{x})$, and update the reconstruction vector $\mathbf{x}'(\mathbf{c})$ as follows (Luttrell, 1989b):

$$\mathbf{x}'_{\text{new}}(\mathbf{c}) \leftarrow \mathbf{x}'_{\text{old}}(\mathbf{c}) + \eta\pi(\mathbf{c} - \mathbf{c}(\mathbf{x}))[\mathbf{x} - \mathbf{x}'_{\text{old}}(\mathbf{c})] \tag{9.24}$$

where η is the learning-rate parameter, and $\mathbf{c}(\mathbf{x})$ is the nearest-neighbor encoding approximation to condition 1. The update equation (9.24) is obtained by inspection of the partial derivative in Eq. (9.21). This update is applied to all $\mathbf{c}$, for which we have

$$\pi(\mathbf{c} - \mathbf{c}(\mathbf{x})) > 0 \tag{9.25}$$

We may think of the gradient descent procedure described in Eq. (9.24) as a way of minimizing the distortion measure D_1 of Eq. (9.19). That is, Eqs. (9.23) and (9.24) are essentially of the same type, except for the fact that (9.23) is batch and (9.24) is continuous (i.e., in flowthrough form).

The update equation (9.24) is identical to the (continuous) SOM algorithm of Eq. (9.13), bearing in mind the correspondences listed in Table 9.1. Accordingly, we may state that the generalized Lloyd algorithm for vector quantization is the batch training version of the SOM algorithm with zero neighborhood size; for zero neighborhood, $\pi(0) = 1$. Note that in order to obtain the generalized Lloyd algorithm from the batch version of the SOM algorithm we do *not* need to make any approximations because the curvature terms (and all higher-order terms) make no contribution when the neighborhood has *zero* width.

TABLE 9.1   Correspondence between the SOM Algorithm and the
Model of Fig. 9.6

Encoding-Decoding Model of Fig. 9.6	SOM Algorithm
Encoder $c(\mathbf{x})$	Best-matching neuron $i(\mathbf{x})$
Reconstruction vector $\mathbf{x}'(c)$	Synaptic weight vector $\mathbf{w}_j$
Probability density function $\pi(c - c(\mathbf{x}))$	Neighborhood function $h_{j,i(\mathbf{x})}$

The important points to note from the discussion presented here are:

- The SOM algorithm is a vector quantization algorithm, which provides a good approximation to the input space $\mathcal{X}$. This viewpoint provides another approach for deriving the SOM algorithm, as exemplified by Eq. (9.24).
- According to this viewpoint, the neighborhood function $h_{j,i(\mathbf{x})}$ in the SOM algorithm has the form of a probability density function. In Luttrell (1991a), a zero-mean Gaussian model is considered appropriate for the noise $\boldsymbol{v}$ in the model of Fig. 9.6. We thus also have theoretical justification for adopting the Gaussian neighborhood function of Eq. (9.4).

The *batch SOM*[8] is merely a rewrite of Eq. (9.23), with summations used to approximate the integrals in the numerator and denominator of the right-hand side of the equation. Note that in this version of the SOM algorithm the order in which the input patterns are presented to the network has no effect on the final form of the feature map, and there is no need for a learning-rate schedule. But the algorithm still requires the use of a neighborhood function.

Property 2. Topological Ordering.   *The feature map Φ computed by the SOM algorithm is topologically ordered in the sense that the spatial location of a neuron in the lattice corresponds to a particular domain or feature of input patterns.*

The topological ordering property[9] is a direct consequence of the update equation (9.13) that forces the synaptic weight vector $\mathbf{w}_i$ of the winning neuron $i(\mathbf{x})$ to move toward the input vector $\mathbf{x}$. It also has the effect of moving the synaptic weight vectors $\mathbf{w}_j$ of the closest neurons j along with the winning neuron $i(\mathbf{x})$. We may therefore visualize the feature map Φ as an *elastic* or *virtual net* with the topology of a one- or two-dimensional lattice as prescribed in the output space $\mathcal{A}$, and whose nodes have weights as coordinates in the input space $\mathcal{X}$ (Ritter, 1995). The overall aim of the algorithm may thus be stated as:

> Approximate the input space $\mathcal{X}$ by pointers or prototypes in the form of synaptic weight vectors w_j, in such a way that the feature map Φ provides a faithful representation of the important features that characterize the input vectors $\mathbf{x} \in \mathcal{X}$ in terms of a certain criterion.

The feature map Φ is usually displayed in the input space $\mathcal{X}$. Specifically, all the pointers (i.e., synaptic weight vectors) are shown as dots, and the pointers of neighboring neurons are connected with lines in accordance with the topology of the lattice. Thus, by using a line to connect two pointers $\mathbf{w}_i$ and $\mathbf{w}_j$, we are indicating that the corresponding neurons i and j are neighboring neurons in the lattice.

Property 3. Density Matching.   *The feature map* Φ *reflects variations in the statistics of the input distribution: regions in the input space* $\mathcal{X}$ *from which sample vectors* $\mathbf{x}$ *are drawn with a high probability of occurrence are mapped onto larger domains of the output space* $\mathcal{A}$*, and therefore with better resolution than regions in* $\mathcal{X}$ *from which sample vectors* $\mathbf{x}$ *are drawn with a low probability of occurrence.*

Let $f_{\mathbf{X}}(\mathbf{x})$ denote the multidimensional pdf of the random input vector $\mathbf{X}$. This pdf, integrated over the entire input space $\mathcal{X}$, must equal unity, by definition:

$$\int_{-\infty}^{\infty} f_{\mathbf{X}}(\mathbf{x})d\mathbf{x} = 1$$

Let $m(\mathbf{x})$ denote the map *magnification factor*, defined as the number of neurons in a small volume $d\mathbf{x}$ of the input space $\mathcal{X}$. The magnification factor, integrated over the input space $\mathcal{X}$, must contain the total number l of neurons in the network, as shown by

$$\int_{-\infty}^{\infty} m(\mathbf{x})d\mathbf{x} = l \tag{9.26}$$

For the SOM algorithm to *match the input density* exactly, we require that (Amari, 1980)

$$m(\mathbf{x}) \propto f_{\mathbf{X}}(\mathbf{x}) \tag{9.27}$$

This property implies that if a particular region of the input space contains frequently occurring stimuli, it will be represented by a larger area in the feature map than a region of the input space where the stimuli occur less frequently.

Generally in two-dimensional feature maps the magnification factor $m(\mathbf{x})$ is not expressible as a simple function of the probability density function $f_{\mathbf{X}}(\mathbf{x})$ of the input vector $\mathbf{x}$. It is only in the case of a one-dimensional feature map that it is possible to derive such a relationship. For this special case we find that, contrary to earlier supposition (Kohonen, 1982), the magnification factor $m(\mathbf{x})$ is *not* proportional to $f_{\mathbf{X}}(\mathbf{x})$. Two different results are reported in the literature, depending on the encoding method advocated:

1. *Minimum-distortion encoding*, according to which the curvature terms and all higher-order terms in the distortion measure of Eq. (9.22) due to the noise model $\pi(\mathbf{v})$ are retained. This encoding method yields the result

$$m(\mathbf{x}) \propto f_{\mathbf{X}}^{1/3}(\mathbf{x}) \tag{9.28}$$

which is the same as the result obtained for the standard vector quantizer (Luttrell, 1991a).

2. *Nearest-neighbor encoding*, which emerges if the curvature terms are ignored, as in the standard form of the SOM algorithm. This encoding method yields the result (Ritter, 1991)

$$m(\mathbf{x}) \propto f_{\mathbf{X}}^{2/3}(\mathbf{x}) \tag{9.29}$$

Our earlier statement that a cluster of frequently occurring input stimuli is represented by a larger area in the feature map still holds, albeit in a distorted version of the ideal condition described in Eq. (9.27).

As a general rule (confirmed by computer simulations), the feature map computed by the SOM algorithm tends to overrepresent regions of low input density and to underrepresent regions of high input density. In other words, the SOM algorithm fails to provide a faithful representation of the probability distribution that underlies the input data.[10]

Property 4. Feature selection.   *Given data from an input space with a nonlinear distribution, the self-organizing map is able to select a set of best features for approximating the underlying distribution.*

This property is a natural culmination of Properties 1 through 3. It brings to mind the idea of principal components analysis that is discussed in the previous chapter, but with an important difference as illustrated in Fig. 9.7. In Fig. 9.7a we show a two-dimensional distribution of zero-mean data points resulting from a linear input–output mapping corrupted by additive noise. In such a situation, principal components analysis works perfectly fine: It tells us that the best description of the "linear" distribution in Fig. 9.7a is defined by a straight line (i.e., one-dimensional "hyperplane") that passes through the origin and runs parallel to the eigenvector associated with the largest eigenvalue of the correlation matrix of the data. Consider next the situation described in Fig. 9.7b, which is the result of a nonlinear input–output mapping corrupted by additive noise of zero mean. In this second situation, it is impossible for a straight-line approximation computed from principal components analysis to provide an acceptable description of the data. On the other hand, the use of a self-organizing map built on a one-dimensional lattice of neurons is able to overcome this approximation problem by virtue of its topological-ordering property. This latter approximation is illustrated in Fig. 9.7b.

In precise terms we may state that self-organizing feature maps provide a *discrete* approximation of the so-called *principal curves*[11] or *principal surfaces* (Hastie and Stuetzle, 1989), and may therefore be viewed as a nonlinear generalization of principal components analysis.

9.6   COMPUTER SIMULATIONS

Two-Dimensional Lattice Driven by a Two-Dimensional Distribution

We illustrate the behavior of the SOM algorithm by using computer simulations to study a network with 100 neurons, arranged in the form of a two-dimensional lattice with 10 rows and 10 columns. The network is trained with a two-dimensional input vector $\mathbf{x}$, whose elements x_1 and x_2 are uniformly distributed in the region $\{(-1 < x_1 < +1); (-1 < x_2 < +1)\}$. To initialize the network the synaptic weights are chosen from a random set.

Figure 9.8 shows three stages of training as the network learns to represent the input distribution. Figure 9.8a shows the uniform distribution of data used to train the

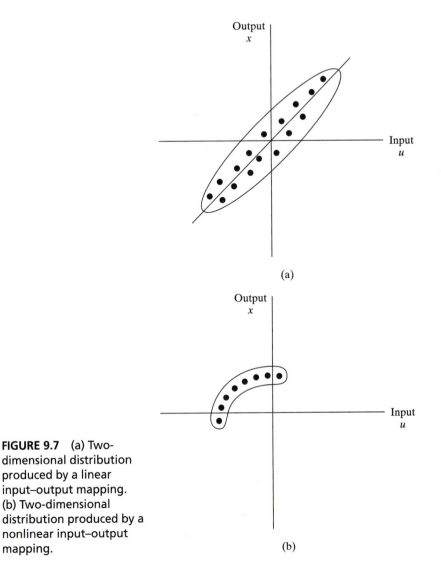

FIGURE 9.7 (a) Two-dimensional distribution produced by a linear input–output mapping. (b) Two-dimensional distribution produced by a nonlinear input–output mapping.

feature map. Figure 9.8b shows the initial values of the synaptic weights, randomly chosen. Figures 9.8c and 9.8d present the values of the synaptic weight vectors, plotted as dots in the input space, after completion of the ordering and convergence phases, respectively. The lines drawn in Fig. 9.8 connect neighboring neurons (across rows and columns) in the network.

The results shown in Fig. 9.8 demonstrate the ordering phase and the convergence phase that characterize the learning process of the SOM algorithm. During the ordering phase the map *unfolds* to form a mesh, as shown in Fig. 9.8c. The neurons are mapped in the correct order at the end of this phase. During the convergence phase the map spreads out to fill the input space. At the end of this second phase, shown in Fig. 9.8d,

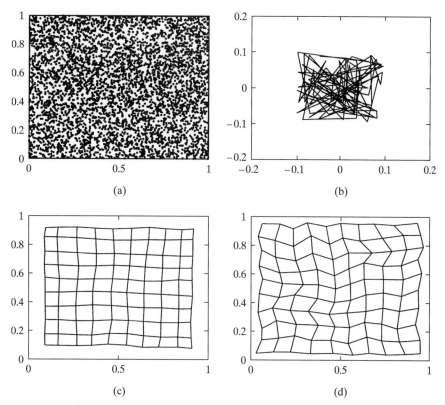

FIGURE 9.8   (a) Input data distribution. (b) Initial condition of the two-dimensional lattice. (c) Condition of the lattice at the end of the ordering phase. (d) Condition of the lattice at the end of the convergence phase.

the statistical distribution of the neurons in the map approaches that of the input vectors, except for some edge effects. Comparing the final state of the feature map in Fig. 9.8d with the uniform distribution of the input in Fig. 9.8a, we see that the tuning of the map during the convergence phase has captured the local irregularities that can be seen in the input distribution.

The topological ordering property of the SOM algorithm is well illustrated in Fig. 9.8d. In particular we observe that the algorithm (after convergence) captures the underlying topology of the uniform distribution at the input. In the computer simulations presented in Fig. 9.8, the input space $\mathcal{X}$ and output space $\mathcal{A}$ are both two-dimensional.

One-Dimensional Lattice Driven by a Two-Dimensional Distribution

We now examine the case when the dimension of the input space $\mathcal{X}$ is greater than the dimension of the output space $\mathcal{A}$. In spite of this mismatch, the feature map Φ is often able to form a topological representation of the input distribution. Figure 9.9 shows

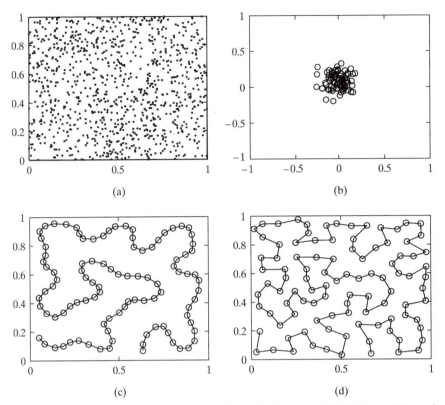

FIGURE 9.9 (a) Two-dimensional input data distribution. (b) Initial condition of the one-dimensional lattice. (c) Condition of the lattice at the end of the ordering phase. (d) Condition of the lattice at the end of the convergence phase.

three different stages in the evolution of a feature map initialized as in Fig. 9.9b and trained with input data drawn from a uniform distribution inside a square as in Fig. 9.9a, but this time the computation is performed with a one-dimensional lattice of 100 neurons. Figures 9.9c and 9.9d show the feature map after the completion of the ordering and convergence phases, respectively. Here we see that the feature map computed by the algorithm is very distorted in order to fill the square as densely as possible and thereby provide a good approximation to the underlying topology of the two-dimensional input space $\mathcal{X}$. The approximating curve shown in Fig. 9.9d resembles a *Peano curve* (Kohonen, 1990a). An operation of the kind exemplified by the feature map of Fig. 9.9, where an input space $\mathcal{X}$ is represented by projecting it onto a lower-dimensional output space $\mathcal{A}$, is referred to as *dimensionality reduction*.

Parameter Specifications for the Simulations

Figure 9.10 presents details of the variations of the neighborhood function $h_{j,i}(n)$ and learning-rate parameter $\eta(n)$ with time (i.e., number of epochs) for the experiments involving a one-dimensional lattice. The neighborhood-function parameter $\sigma(n)$, shown

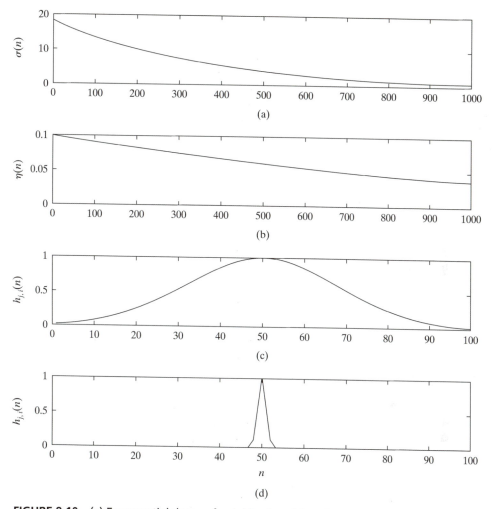

FIGURE 9.10 (a) Exponential decay of neighborhood function parameter $\sigma(n)$. (b) Exponential decay of learning-rate parameter $\eta(n)$. (c) Initial shape of the Gaussian neighborhood function. (d) Shape of the neighborhood function at the end of the ordering phase (i.e., beginning of the convergence phase).

in Fig. 9.10a, starts with an initial value $\sigma_0 = 18$ and then shrinks to about 1 in 1000 iterations during the ordering phase. During that same phase, the learning-rate parameter $\eta(n)$ starts with an initial value $\eta_0 = 0.1$ and then decreases to 0.037. Figure 9.10c shows the initial Gaussian distribution of neurons around a winning neuron located at the midpoint of the one-dimensional lattice. Figure 9.10d shows the shape of the neighborhood function at the end of the ordering phase. During the convergence phase the learning-rate parameter decreases linearly from 0.037 to 0.001 in 5000 iterations. During the same phase the neighborhood function decreases essentially to zero.

The specifications of the ordering phase and convergence phase for the computer simulations in Fig. 9.8 involving the two-dimensional lattice are similar to those used

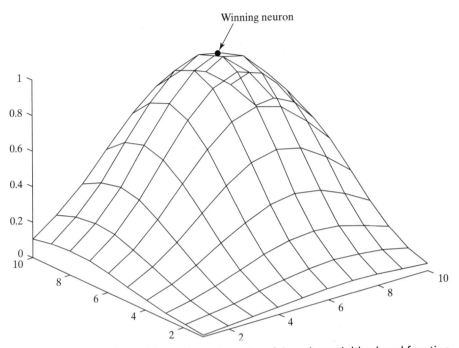

FIGURE 9.11   Initial condition of two-dimensional Gaussian neighborhood function centered on a winning neuron located at the point (7, 8) in a two-dimesional lattice of 10 × 10 neurons.

for the one-dimensional lattice, except for the fact that the neighborhood function is now two-dimensional. The parameter $\sigma(n)$ starts at the initial value $\sigma_0 = 3$ and then decreases to 0.75 in 1000 iterations. Figure 9.11 displays the initial value of the two-dimensional Gaussian neighborhood function $h_{j,i}$ for $\sigma_0 = 3$ and a winning neuron centered on the point (7,8) inside the two-dimensional lattice of 10 × 10 neurons.

9.7   LEARNING VECTOR QUANTIZATION

Vector quantization, discussed previously in Section 9.6, is a technique that exploits the underlying structure of input vectors for the purpose of data compression (Gersho and Gray, 1992). Specifically, an input space is divided into a number of distinct regions, and for each region a reconstruction vector is defined. When the quantizer is presented a new input vector, the region in which the vector lies is first determined, and is then represented by the reproduction vector for that region. Thus, by using an encoded version of this reproduction vector for storage or transmission in place of the original input vector, considerable savings in storage or transmission bandwidth can be realized, at the expense of some distortion. The collection of possible reproduction vectors is called the *code book* of the quantizer, and its members are called *code words*.

A vector quantizer with minimum encoding distortion is called a *Voronoi* or *nearest-neighbor quantizer*, since the *Voronoi cells* about a set of points in an input space correspond to a partition of that space according to the *nearest-neighbor rule* based on the Euclidean metric (Gersho and Gray, 1992). Figure 9.12 shows an example of an input space divided into four Voronoi cells with their associated Voronoi vectors (i.e., reconstruction vectors). Each Voronoi cell contains those points of the input space that are the closest to the Voronoi vector among the totality of such points.

The SOM algorithm provides an approximate method for computing the Voronoi vectors in an unsupervised manner, with the approximation being specified by the synaptic weight vectors of the neurons in the feature map; this is merely restating property 1 of the SOM algorithm discussed in Section 9.6. Computation of the feature map may therefore be viewed as the first of two stages for adaptively solving a pattern classification problem, as depicted in Fig. 9.13. The second stage is provided by learning vector quantization, which provides a mechanism for the final fine tuning of a feature map.

Learning vector quantization[12] (LVQ) is a supervised learning technique that uses class information to move the Voronoi vectors slightly, so as to improve the quality of the classifier decision regions. An input vector $\mathbf{x}$ is picked at random from the input space. If the class labels of the input vector $\mathbf{x}$ and a Voronoi vector $\mathbf{w}$ agree, the Voronoi vector $\mathbf{w}$ is moved in the direction of the input vector $\mathbf{x}$. If, on the other hand, the class labels of the input vector $\mathbf{x}$ and the Voronoi vector $\mathbf{w}$ disagree, the Voronoi vector $\mathbf{w}$ is moved away from the input vector $\mathbf{x}$.

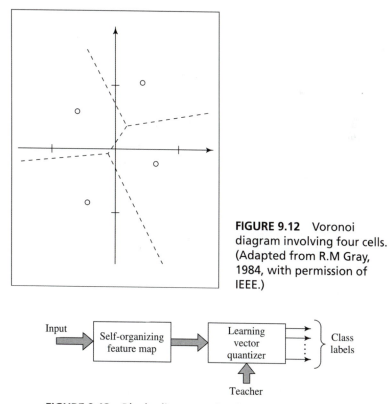

FIGURE 9.12 Voronoi diagram involving four cells. (Adapted from R.M Gray, 1984, with permission of IEEE.)

FIGURE 9.13 Block diagram of adaptive pattern classification, using a self-organizing feature map and learning vector quantizer.

Let $\{\mathbf{w}_j\}_{j=1}^l$ denote the set of Voronoi vectors, and let $\{\mathbf{x}_i\}_{i=1}^N$ denote the set of input (observation) vectors. We assume that there are many more input vectors than Voronoi vectors, which is typically the case in practice. The learning vector quantization (LVQ) algorithm proceeds as follows:

(i) Suppose that the Voronoi vector $\mathbf{w}_c$ is the closest to the input vector $\mathbf{x}_i$. Let $\mathcal{C}_{\mathbf{w}_c}$ denote the class associated with the Voronoi vector $\mathbf{w}_c$ and $\mathcal{C}_{\mathbf{x}_i}$ denote the class label of the input vector $\mathbf{x}_i$. The Voronoi vector $\mathbf{w}_c$ is adjusted as follows:

- If $\mathcal{C}_{\mathbf{w}_c} = \mathcal{C}_{\mathbf{x}_i}$, then

$$\mathbf{w}_c(n + 1) = \mathbf{w}_c(n) + \alpha_n[\mathbf{x}_i - \mathbf{w}_c(n)] \tag{9.30}$$

where $0 < \alpha_n < 1$.
- If, on the other hand, $\mathcal{C}_{\mathbf{w}_c} \neq \mathcal{C}_{\mathbf{x}_i}$, then

$$\mathbf{w}_c(n + 1) = \mathbf{w}_c(n) - \alpha_n[\mathbf{x}_i - \mathbf{w}_c(n)] \tag{9.31}$$

(ii) The other Voronoi vectors are not modified.

It is desirable for the learning constant α_n to decrease monotonically with the number of iterations n. For example, α_n may initially be about 0.1 or smaller, and then decrease linearly with n. After several passes through the input data, the Voronoi vectors typically converge, and the training is complete. However, difficulties may be experienced if the method is applied without proper care.

9.8   COMPUTER EXPERIMENT: ADAPTIVE PATTERN CLASSIFICATION

In pattern classification, the first and most important step is *feature selection* (extraction), which is ordinarily performed in an unsupervised manner. The objective of this first step is to select a reasonably small set of features, in which the essential information content of the input data (to be classified) is concentrated. The self-organizing map, by virtue of property 4 discussed in Section 9.5, is well suited for the task of feature selection, particularly if the input data are generated by a nonlinear process.

The second step in pattern classification is the actual *classification*, where the features selected from the input data are assigned to individual classes. Although a self-organizing map is equipped to perform the role of classification too, the recommended procedure for best performance is to accompany it with a supervised learning scheme for the second stage of classification. The combination of a self-organizing map and a supervised learning scheme forms the basis of an *adaptive pattern classification* that is *hybrid* in nature.

Such a hybrid approach to pattern classification may take different forms, depending on how the supervised learning scheme is implemented. One simple scheme is to use a learning vector quantizer, which is described in the previous section. We thus have the two-stage adaptive pattern classifier shown in Fig. 9.13.

In this experiment we revisit the classification of overlapping two-dimensional, Gaussian-distributed patterns labeled 1 (class $\mathcal{C}_1$) and labeled 2 (class $\mathcal{C}_2$), which was first described in Chapter 4 involving the use of a multilayer perceptron trained with the back-propagation algorithm. The scatter plots for the data used in the experiment are shown in Fig. 4.13.

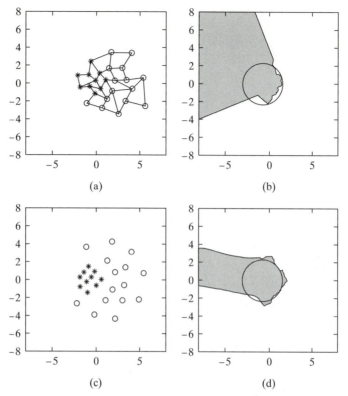

FIGURE 9.14   (a) Self-organizing map after labeling. (b) Decision boundary constructed by the feature map of part a. (c) Labeled map after learning-vector quantization. (d) Decision boundary constructed by the feature map of part c.

Figure 9.14a shows the two-dimensional feature map of 5×5 neurons after training with the SOM algorithm is complete. The feature map has been labeled, with each neuron assigned to one class or the other depending on how it responds to test data drawn from the input distribution. Figure 9.14b shows the decision boundary realized by the feature map operating on its own.

Figure 9.14c shows the modified feature map after it was tuned in a supervised manner using LVQ. Figure 9.14d shows the decision boundary produced by the combined action of the SOM and LVQ algorithms. Comparing these two figures with their counterparts shown in Figs. 9.14a and 9.14b, we see, in a qualitative manner, the beneficial effect obtained by using the LVQ.

Table 9.2 presents a summary of the classification performances of the feature map on its own and the feature map working together with the learning vector quantizer. The results presented here were obtained on 10 independent trials of the experiment, with each experiment involving the use of 30,000 patterns as test data. On each trial of the experiment there was an improvement in classification performance due to the use of LVQ. The average classification performance for the feature map on its own

TABLE 9.2 Summary of Classification Performances (Percentage) for the Computer Experiment on Overlapping Two-Dimensional Gaussian Distribution Using 5×5 Lattice

Trial	Feature map on its own	Cascade combination of feature map and learning vector quantizer
1	79.05	80.18
2	79.79	80.56
3	79.41	81.17
4	79.38	79.84
5	80.30	80.43
6	79.55	80.36
7	79.79	80.86
8	78.48	80.21
9	80.00	80.51
10	80.32	81.06
Average	79.61%	80.52%

is 79.61 percent, and for the combination of the feature map and the learning vector quantizer is 80.52 percent, which represents an improvement of 0.91 percent over the feature map on its own. For a frame of reference, we recall that the performance of the optimum Bayes classifier for this experiment is 81.51 percent.

9.9 HIERARCHICAL VECTOR QUANTIZATION

In discussing property 1 of the self-organizing feature map in Section 9.6, we pointed out that it is closely related to the generalized Lloyd algorithm for vector quantization. Vector quantization is a form of *lossy* data compression, lossy in the sense that some information contained in the input data is lost as a result of the compression. Data compression is rooted in a branch of Shannon's information theory known as *rate distortion theory* (Cover and Thomas, 1991). For the purpose of our present discussion dealing with hierarchical vector quantization, it is appropriate to begin by stating the following fundamental result of rate distortion theory (Gray, 1984):

> Better data compression performance can always be achieved by coding vectors instead of scalars, even if the source of data is memoryless (e.g., it provides a sequence of independent random variables), or if the data compression system has memory (i.e., the action of an encoder depends on past encoder inputs or outputs).

This fundamental result underlies the extensive research effort that has been devoted to vector quantization (Gersho and Gray, 1992).

However, conventional vector quantization algorithms require a prohibitive amount of computation, which has hindered their practical use. The most time consuming part of vector quantization is the encoding operation. For encoding, the input vector must be compared with each code vector in the code book in order to determine which particular code yields the minimum distortion. For a code book containing N code vectors, for example, the time taken for encoding is on the order of N, which can therefore be large for large N. In Luttrell (1989a), a *multistage hierarchical vector quantizer* is described that trades off accuracy for speed of encoding. This scheme is not simply the standard tree search of a code book; it is genuinely new. The multistage hierarchical vector quantizer attempts to factorize the overall vector quantization into a number of suboperations, each of which requires very little computation. Desirably, the factorization is reduced to a single table look-up per suboperation. By clever use of the SOM algorithm to train each stage of the quantizer, the loss in accuracy can be small (as low as a fraction of a decibel), while the gain in speed of computation is large.

Consider two vector quantizers VQ_1 and VQ_2, with VQ_1 feeding its output into VQ_2. The output from VQ_2 is the final encoded version of the original input signal applied to VQ_1. In performing its quantization, it is inevitable for VQ_2 to discard some information. As far as VQ_1 is concerned, the sole effect of VQ_2 is therefore to distort the information output by VQ_1. It thus appears that the appropriate training method for VQ_1 is the SOM algorithm, which accounts for the signal distortion induced by VQ_2 (Luttrell, 1989a). In order to use the generalized Lloyd algorithm to train VQ_2 we need only assume that the output of VQ_2 is not corrupted before we do the reconstruction. Then we do not need to introduce any noise model (at the output of VQ_2) with its associated finite width neighborhood function.

We can generalize this heuristic argument to a multistage vector quantizer. Each stage must be designed to account for the distortion induced by all *subsequent* stages, and model it as noise. To do so, the SOM algorithm is used to train all the stages of the quantizer, except for the last stage for which the generalized Lloyd algorithm is adequate.

Hierarchical vector quantization is a special case of multistage vector quantization (Luttrell, 1989a). As an illustration, consider the quantization of 4×1 input vector

$$\mathbf{x} = [x_1, x_2, x_3, x_4]^T$$

In Fig. 9.15a we show a single-stage vector quantizer for $\mathbf{x}$. Alternatively, we may use a two-stage hierarchical vector quantizer as depicted in Fig. 9.15b. The significant difference between these two schemes is that the input dimension of the quantizer in Fig. 9.15a is four, whereas for the quantizer in Fig. 9.15b it is two. Accordingly, the quantizer of Fig. 9.15b requires a look-up table of smaller size, and is therefore simpler to implement than that of Fig. 9.15a. This is the advantage of a hierarchical quantizer over a conventional quantizer.

Luttrell (1989a) has demonstrated the performance of a multistage hierarchical vector quantizer applied to various stochastic time series, with little loss in encoding accuracy. In Fig. 9.16 we have reproduced Luttrell's results for the case of a correlated Gaussian noise process generated using a *first-order autoregressive (AR) model*:

$$x(n + 1) = \rho x(n) + v(n) \tag{9.32}$$

where ρ is the AR coefficient and the $v(n)$ are independent and identically distributed (iid) Gaussian random variables of zero mean and unit variance. Hence we may show that $x(n)$ is characterized as follows:

$$E[x(n)] = 0 \tag{9.33}$$

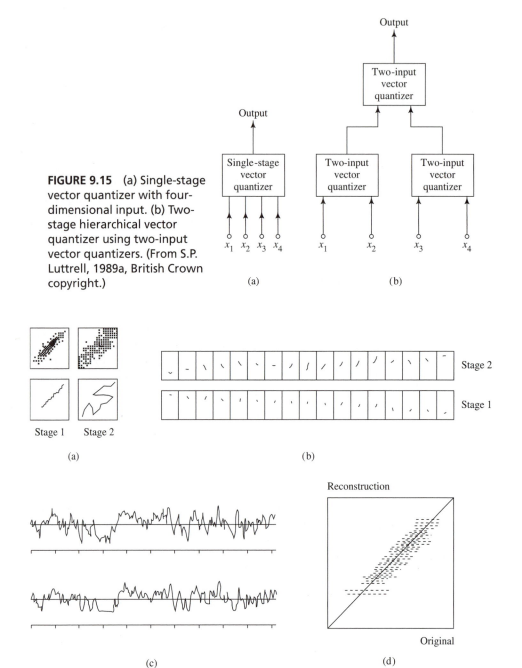

FIGURE 9.15 (a) Single-stage vector quantizer with four-dimensional input. (b) Two-stage hierarchical vector quantizer using two-input vector quantizers. (From S.P. Luttrell, 1989a, British Crown copyright.)

FIGURE 9.16 Two-stage encoding/decoding results for correlated Gaussian noise input. Correlation coefficient $\rho = 0.85$. (From S.P.Luttrell, 1989a, British Crown copyright.)

$$E[x^2(n)] = \frac{1}{1 - \rho^2} \tag{9.34}$$

$$\frac{E[x(n + 1)x(n)]}{E[x^2(n)]} = \rho \tag{9.35}$$

Thus ρ may also be viewed as the *correlation coefficient* of the time series $\{x(n)\}$. To initiate the generation of the time series according to Eq. (9.32), a Gaussian random variable of zero mean and variance $1/(1 - \rho^2)$ was used for $x(0)$, and the value $\rho = 0.85$ was used for the correlation coefficient.

For the vector quantization a hierarchical encoder with a four-dimensional input space, like the binary tree of Fig. 9.15b, was used. For the AR time series $\{x(n)\}$, translational symmetry implies that only *two* distinct look-up tables are needed. The size of each table depends exponentially on the number of input bits, and linearly on the number of output bits. During training, a large number of bits is needed to represent numbers for a correct computation of the updates described in Eq. (9.24); so the look-up tables are not used during training. Once training is complete, however, the number of bits may be reduced to their normal level, and the table entries filled in as required. For the encoder shown in Fig. 9.15b, the input samples were approximated by using four bits per sample. For all stages of the encoder, N $(= 17)$ code vectors were used, so the number of output bits from each lookup table was approximately four, too. Thus the address space size of both the first stage and second stage look-up tables is 256 $(= 2^{4+4})$, which means that the overall memory requirements for representing the tables are modest.

Figure 9.16 shows the encoding-decoding results obtained with $x(n)$ as the input. The lower half of Fig. 9.16a shows the code vectors for each of the two stages as a curve embedded in a two-dimensional input space; the upper half of Fig. 9.16a presents estimates of the corresponding co-occurrence matrices using 16×16 bins. Figure 9.16b presents, as fragments of the time series, the following:

- The code vector computed by the first encoder stage
- The reconstruction vector computed by the second stage that minimizes the mean-squares distortion, while keeping all other variables fixed

Figure 9.16c presents 512 samples of both the original time series (top curve) and its reconstruction (bottom curve) from the output of the last encoder stage; the horizontal scale in Fig. 9.16c is half that in Fig. 9.16b. Finally, Fig. 9.16d presents a co-occurrence matrix created from a pair of samples: an original time series sample and its corresponding reconstruction. The width of the band in Fig. 9.16d indicates the extent of the distortion produced by the hierarchical vector quantization.

Examining the waveforms in Fig. 9.16c, we see that the reconstruction is a good representation of the original time series, except for some positive and negative peaks that were clipped. According to Luttrell (1989a), the normalized mean-squared distortion was computed as 0.15, which is almost as good (0.05 dB loss) as the 8.8 dB obtained with a single-stage four-sample block encoder using one bit per sample (Jayant and Noll, 1984).

9.10 CONTEXTUAL MAPS

There are two fundamentally different ways of visualizing a self-organizing feature map. In one method of visualization the feature map is viewed as an elastic net with the synaptic weight vectors treated as pointers for the respective neurons, which are directed into the input space. This method of visualization is particularly useful for displaying the topological ordering property of the SOM algorithm, as illustrated by the results of computer simulation experiments presented in Section 9.6.

In the second method of visualization, class labels are assigned to neurons in a two-dimensional lattice (representing the output layer of the network), depending on how each *test pattern* (not seen before) excites a particular neuron in the self-organized network. As a result of this second stage of stimulation, the neurons in the two-dimensional lattice are partitioned into a number of *coherent regions*, coherent in the sense that each grouping of neurons represents a distinct set of contiguous symbols or labels (Ritter and Kohonen, 1989). This assumes that the right conditions have been followed for the development of a well-ordered feature map in the first place.

Consider, for example, the set of data given in Table 9.3, which pertains to a number of different animals. Each column of the table is a schematic description of an animal, based on the presence ($= 1$) or absence ($= 0$) of some of the 13 different attributes given on the left. Some attributes such as "feathers" and "two legs" are correlated, while many of the other attributes are uncorrelated. For each animal given at the top of the table we have an *attribute code* $\mathbf{x}_a$ made up of 13 elements. The animal is itself specified by a *symbol code* $\mathbf{x}_s$, the composition of which must *not* convey any information or known similarities between the animals. For the example at hand, $\mathbf{x}_s$ consists of a column vector whose kth element, representing animal $k = 1, 2, \ldots, 16$, is given a fixed value of a; the remaining elements are all set equal to zero. The parameter a deter-

TABLE 9.3 Animal Names and Their Attributes

Animal		Dove	Hen	Duck	Goose	Owl	Hawk	Eagle	Fox	Dog	Wolf	Cat	Tiger	Lion	Horse	Zebra	Cow
is	small	1	1	1	1	1	1	0	0	0	0	1	0	0	0	0	0
	medium	0	0	0	0	0	0	1	1	1	1	0	0	0	0	0	0
	big	0	0	0	0	0	0	0	0	0	0	0	1	1	1	1	1
has	2 legs	1	1	1	1	1	1	1	0	0	0	0	0	0	0	0	0
	4 legs	0	0	0	0	0	0	0	1	1	1	1	1	1	1	1	1
	hair	0	0	0	0	0	0	0	1	1	1	1	1	1	1	1	1
	hooves	0	0	0	0	0	0	0	0	0	0	0	0	0	1	1	1
	mane	0	0	0	0	0	0	0	0	0	1	0	0	1	1	1	0
	feathers	1	1	1	1	1	1	1	0	0	0	0	0	0	0	0	0
likes to	hunt	0	0	0	0	1	1	1	1	0	1	1	1	1	0	0	0
	run	0	0	0	0	0	0	0	0	1	1	0	1	1	1	1	0
	fly	1	0	0	1	1	1	1	0	0	0	0	0	0	0	0	0
	swim	0	0	1	1	0	0	0	0	0	0	0	0	0	0	0	0

mines the relative influence of the symbol code compared to the attribute code. To make sure that the attribute code is the dominant one, a is chosen equal to 0.2. The input vector $\mathbf{x}$ for each animal is a vector of 29 elements, representing a concatenation of the attribute code $\mathbf{x}_a$ and the symbol code $\mathbf{x}_s$, as shown by

$$
\mathbf{x} = \begin{bmatrix} \mathbf{x}_s \\ \mathbf{x}_a \end{bmatrix} = \begin{bmatrix} \mathbf{x}_s \\ \mathbf{0} \end{bmatrix} + \begin{bmatrix} \mathbf{0} \\ \mathbf{x}_a \end{bmatrix}
$$

Finally, each data vector is normalized to unit length. The patterns of the data set thus generated are presented to a two-dimensional lattice of 10×10 neurons, and the synaptic weights of the neurons are adjusted in accordance with the SOM algorithm summarized in Section 9.4. The training is continued for 2000 iterations, whereafter the feature map should have reached a steady state. Next, a test pattern defined by $\mathbf{x} = [\mathbf{x}_s, \mathbf{0}]^T$ containing the symbol code of only one of the animals, is presented to the self-organized network and the neuron with the strongest response is identified. This is repeated for all 16 animals.

Proceeding in the manner just described, we obtain the map shown in Fig. 9.17, where the labeled neurons represent those with the strongest responses to their respective test patterns; the dots represent neurons with weaker responses.

Figure 9.18 shows the result of "simulated electrode penetration mapping" for the same self-organized network. This time, however, each neuron in the network has been marked by the particular animal for which it produces the best response. Figure 9.18 clearly shows that the feature map has essentially captured the "family relationships" among the 16 different animals. There are three distinct clusters, one representing "birds," a second representing "peaceful species," and the third representing animals that are "hunters."

A feature map of the type illustrated in Fig. 9.18 is referred to as a *contextual map* or *semantic map* (Ritter and Kohonen, 1989; Kohonen, 1997a). Such a map resembles cortical maps (i.e., the computational maps formed in the cerebral cortex) that are discussed briefly in Section 9.2. Contextual maps, resulting from the use of the SOM algorithm, find applications in such diverse fields as unsupervised categorization of phonemic classes from text, remote sensing (Kohonen, 1997a), and data exploration or data mining (Kohonen, 1997b).

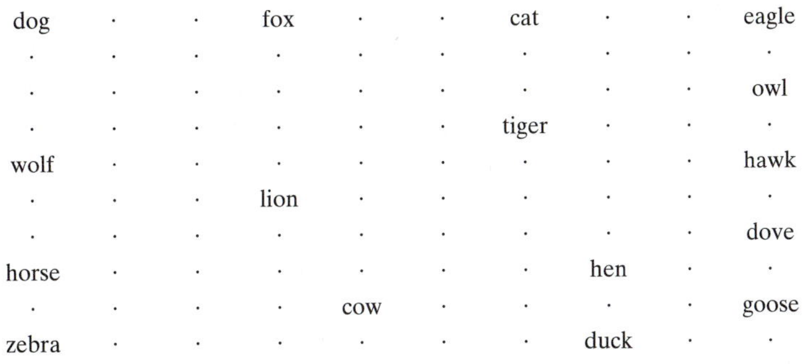

dog	.	.	fox	.	.	cat	.	.	eagle
.	.	.	.	.	.	.	.	.	owl
.	.	.	.	.	.	tiger	.	.	.
wolf	.	.	.	.	.	.	.	.	hawk
.	.	.	lion	.	.	.	.	.	.
.	.	.	.	.	.	.	.	.	dove
horse	.	.	.	.	.	.	hen	.	.
.	.	.	.	cow	.	.	.	.	goose
zebra	.	.	.	.	.	.	duck	.	.

FIGURE 9.17 Feature map containing labeled neurons with strongest responses to their respective inputs.

dog	dog	fox	fox	fox	cat	cat	cat	eagle	eagle
dog	dog	fox	fox	fox	cat	cat	cat	eagle	eagle
wolf	wolf	wolf	fox	cat	tiger	tiger	tiger	owl	owl
wolf	wolf	lion	lion	lion	tiger	tiger	tiger	hawk	hawk
wolf	wolf	lion	lion	lion	tiger	tiger	tiger	hawk	hawk
wolf	wolf	lion	lion	lion	owl	dove	hawk	dove	dove
horse	horse	lion	lion	lion	dove	hen	hen	dove	dove
horse	horse	zebra	cow	cow	cow	hen	hen	dove	dove
zebra	zebra	zebra	cow	cow	cow	hen	hen	duck	goose
zebra	zebra	zebra	cow	cow	cow	duck	duck	duck	goose

FIGURE 9.18 Semantic map obtained through the use of simulated electrode penetration mapping. The map is divided into three regions representing: birds, peaceful species, and hunters.

9.11 SUMMARY AND DISCUSSION

The self-organizing map due to Kohonen (1982) is an ingenious neural network built around a one- or two-dimensional lattice of neurons for capturing the important features contained in an input (data) space of interest. In so doing, it provides a structural representation of the input data by the neurons' weight vectors as prototypes. The SOM algorithm is neurobiologically inspired, incorporating all the mechanisms that are basic to self-organization: competition, cooperation, and self-amplification that are discussed in Chapter 8. It may therefore serve as a generic though degenerate model for describing the emergence of collective ordering phenomena in complex systems after starting from total disorder.

The self-organizing map may also be viewed as a vector quantizer, thereby providing a principled approach for deriving the update rule used to adjust the weight vectors (Luttrell, 1989b). This latter approach clearly emphasizes the role of the neighborhood function as a probability density function.

It should, however, be emphasized that this latter approach, based on the use of average distribution D_1 in Eq. (9.19) as the cost function to be minimized, can be justified only when the feature map is already well ordered. In Erwin et al. (1992b), it is shown that the learning dynamics of a self-organizing map during the ordering phase of the adaptive process (i.e., during the topological ordering of a feature map that is initially highly disordered) *cannot* be described by a stochastic gradient descent on a *single* cost function. But in the case of a one-dimensional lattice, it may be described using a set of cost functions, one for each neuron in the network, which are independently minimized following a stochastic gradient descent.

What is astonishing about Kohonen's SOM algorithm is that it is so simple to implement, yet mathematically so difficult to analyze its properties in a general setting. Some fairly powerful methods have been used to analyze it by several investigators, but they have only produced results of limited applicability. In Cottrell et al. (1997), a survey of results on theoretical aspects of the SOM algorithm is given. In particular, a recent result due to Forte and Pagés (1995, 1997) is highlighted, and states that in the case of a one-dimensional lattice we have a rigorous proof of the "almost sure" convergence of the SOM algorithm to a unique state after completion of the self-organization phase. This important result has been shown to hold for a general class of neighborhood functions. However, the same cannot be said in a multidimensional setting.

One final point of enquiry is in order. With the self-organizing feature map being inspired by ideas derived from cortical maps in the brain, it seems natural to enquire whether such a model could actually explain the formation of cortical maps. Erwin et al. (1995) have performed such an investigation. They have shown that the self-organizing feature map is able to explain the formation of computational maps in the primary visual cortex of the macaque monkey. The input space used in this study has five dimensions: two dimensions for representing the position of a receptive field in retinotopic space, and the remaining three dimensions for representing orientation preference, orientation selectivity, and ocular dominance. The cortical surface is divided into small patches that are considered as computational units (i.e., artificial neurons) of a two-dimensional square lattice. Under certain assumptions, it is shown that Hebbian learning leads to spatial patterns of orientation and ocular dominance that are remarkably similar to those found in the macaque monkey.

NOTES AND REFERENCES

1. The two feature-mapping models of Fig. 9.1 were inspired by the pioneering self-organizing studies of von der Malsburg (1973), who noted that a model of the visual cortex could not be entirely genetically predetermined; rather, a self-organizing process involving synaptic learning may be responsible for the *local* ordering of feature-sensitive cortical cells. However, global topographic ordering was *not* achieved in von der Malsburg's model because the model used a fixed (small) neighborhood. The computer simulation by von der Malsburg was perhaps the first to demonstrate self-organization.

2. Amari (1980) relaxes this restriction on the synaptic weights of the postsynaptic neurons somewhat. The mathematical analysis presented by Amari elucidates the dynamical stability of a cortical map formed by self-organization.

3. Neurobiological feasibility of the self-organizing map (SOM) is discussed in Kohonen (1993, 1997a).

4. The competitive learning rule described in Eq. (9.3) was first introduced into the neural network literature in Grossberg (1969b).

5. In the original form of the SOM algorithm derived by Kohonen (1982), the topological neighborhood is assumed to have a constant amplitude. Let $d_{j,i}$ denote the *lateral distance* between winning neuron i and excited neuron j inside the neighborhood function. The topological neighborhood for the case of a one-dimensional lattice is thus defined by

$$h_{j,i} = \begin{cases} 1, & -K \leq d_{j,i} \leq K \\ 0, & \text{otherwise} \end{cases} \tag{1}$$

where $2K$ is the overall size of the one-dimensional neighborhood of excited neurons. Contrary to neurobiological considerations, the implication of the model described in Eq. (1) is that all the neurons located inside the topological neighborhood fire at the same rate, and the interaction among those neurons is independent of their lateral distance from the winning neuron i.

6. In Erwin et al. (1992b), it is shown that metastable states, representing topological defects in the configuration of a feature map, arise when the SOM algorithm uses a

neighborhood function that is not convex. A Gaussian function is convex, whereas a rectangular function is not. A broad, convex neighborhood function such as a broad Gaussian, leads to relatively shorter topological ordering times than a nonconvex one (e.g., rectangular) due to the absence of metastable states.

7. In the communications and information theory literature, an early method known as the *Lloyd algorithm* was proposed for scalar quantization. The algorithm was first described by Lloyd in an unpublished 1957 report at Bell Laboratories (Lloyd, 1957), then much later appeared in published form (Lloyd, 1982). The Lloyd algorithm is also sometimes referred to as the "Max quantizer." The *generalized Lloyd algorithm* (GLA) for vector quantization is a direct generalization of Lloyd's original algorithm. The generalized Lloyd algorithm is sometimes referred to as the *k-means algorithm* after McQueen (1967) who used it as a tool for statistical clustering. It is also sometimes referred to in data compression literature as the *LBG algorithm* after Linde et al. (1980). For a historical account of the Lloyd algorithm and generalized Lloyd algorithm, see Gersho and Gray (1992).

8. In Kohonen (1993), experimental results are presented showing that the batch version of the SOM algorithm is faster than its on-line version. However, the adaptive capability of the SOM algorithm is lost in using the batch version.

9. The topological property of a self-organizing map may be assessed quantatively in different ways. One such quantitative measure, called the *topographic product*, is described in Bauer and Pawelzik (1992), which may be used to compare the faithful behavior of different feature maps pertaining to different dimensionalities. However, the measure is quantitative only when the dimension of the lattice matches that of the input space.

10. The inability of the SOM algorithm to provide a faithful representation of the distribution that underlies the input data has prompted modifications to the algorithm and the development of new self-organizing algorithms that are faithful to the input.

   Two types of modifications to the SOM algorithm have been reported in the literature:

   (i) *Modification to the competitive process*. In DeSieno (1988), a form of memory is used to track the cumulative activities of individual neurons in the lattice. Specifically, a "conscience" mechanism is added to bias the competitive learning process of the SOM algorithm. This is done in such a way that each neuron, regardless of its location in the lattice, has the chance to win competition with a probability close to the ideal of $1/l$, where l is the total number of neurons. A description of the SOM algorithm with conscience is presented in Problem 9.8.

   (ii) *Modification to the adaptive process*. In this second approach, the update rule for adjusting the weight vector of each neuron under the neighborhood function is modified to control the magnification properties of the feature map. In Bauer et al. (1996), it is shown that through the addition of an adjustable step-size parameter to the update rule, it is possible for the feature map to provide a faithful representation of the input distribution. Lin et al. (1997) follow a similar path by introducing two modifications to the SOM algorithm:

   • The update rule is modified to extract direct dependence on the input vector $\mathbf{x}$ and weight vector $\mathbf{w}_j$ of neuron j in question.

   • The Voronoi partition is replaced with an equivariant partition designed specially for separable input distributions.

   This second modification enables the SOM algorithm to perform blind source separation. (Blind source separation is briefly discussed in Chapter 1, and is discussed in greater detail in Chapter 10.)

The modifications mentioned build on the standard SOM algorithm in one form or another. In Linsker (1989b), a completely different approach is taken. Specifically, a global learning rule for topographic map formation is derived by maximizing the mutual information between the output signal and the signal part of the input corrupted by additive noise. (The notion of mutual information, rooted in Shannon's information theory, is discussed in Chapter 10.) Linsker's model yields a distribution of neurons that matches the input distribution exactly. The use of an information-theoretic approach to topographic map formation in a self-organized manner is also pursued in Van Hulle (1996, 1997).

11. The relationship between the SOM algorithm and principal curves is discussed in Ritter et al. (1992) and Cherkassky and Mulier (1995). The algorithm for finding a principal curve consists of two steps (Hastie and Stuetzl, 1989):

   1. *Projection.* For each data point, find its nearest projection or closest point on the curve.
   2. *Conditional exceptation.* Apply scatter plot smoothing to the projected values along the length of the curve. The recommended procedure is to start the smoothing with a large span and then decrease it gradually.

   These two steps are similar to the vector quantization and neighborhood annealing performed in the SOM algorithm.

12. The idea of learning vector quantization was originated by Kohonen in 1986; three versions of this algorithm are described in Kohonen (1990b; 1997a). The version of the algorithm discussed in Section 9.7 is the first version of learning vector quantization, referred to as LVQ1 by Kohonen.

   The learning vector quantization algorithm is a stochastic approximation algorithm. Baras and LaVigna (1990) discuss the convergence properties of the algorithm using the ordinary differential equation (ODE) approach that is described in Chapter 8.

PROBLEMS

SOM algorithm

9.1 The function $g(y_j)$ denotes a nonlinear function of the response y_j, which is used in the SOM algorithm as described in Eq. (9.9). Discuss the implication of what could happen if the constant term in the Taylor series of $g(y_j)$ is nonzero.

9.2 Assume that $\pi(\boldsymbol{v})$ is a smooth function of the noise $\boldsymbol{v}$ in the model of Fig. 9.6. Using a Taylor expansion of the distortion measure of Eq. (9.19), determine the curvature term that arises from the noise model $\pi(\boldsymbol{v})$.

9.3 It is sometimes said that the SOM algorithm *preserves* the topological relationships that exist in the input space. Strictly speaking, this property can be guaranteed only for an input space of equal or lower dimensionality than that of the neural lattice. Discuss the validity of this statement.

9.4 It is said that the SOM algorithm based on competitive learning lacks any tolerance against hardware failure, yet the algorithm is error tolerant in that a small perturbation applied to the input vector causes the output to jump from the winning neuron to a neighboring one. Discuss the implications of these two statements.

9.5 Consider the batch version of the SOM algorithm obtained by expressing Eq. (9.23) in its discrete form, as shown by

$$\mathbf{w}_j = \frac{\sum_i \pi_{j,i} \mathbf{x}_i}{\sum_i \pi_{j,i}}, \quad j = 1, 2, \ldots, l$$

Show that this version of the SOM algorithm can be expressed in a form similar to the Nadaraya–Watson regression estimator (Cherkassky and Mulier, 1995); this estimator is discussed in Chapter 5.

Learning vector quantization

9.6 In this problem we consider the optimized form of the learning vector quantization algorithm of Section 9.7 (Kohonen, 1997a). We wish to arrange for the effects of the corrections to the Voronoi vectors, made at different times, to have equal influence when referring to the end of the learning period.

(a) First, show that Eqs. (9.30) and (9.31) may be integrated into a single equation, as follows:

$$\mathbf{w}_c(n + 1) = (1 - s_n \alpha_n)\mathbf{w}_c(n) + s_n \alpha_n \mathbf{x}(n)$$

where

$$s_n = \begin{cases} +1 & \text{if the classification is correct} \\ -1 & \text{if the classification is wrong} \end{cases}$$

(b) Hence, show that the optimization criterion described at the beginning of the problem is satisfied if

$$\alpha_n = (1 - s_n \alpha_n)\alpha_{n-1}$$

which yields the optimized value of the learning constant α_n as

$$\alpha_n^{\text{opt}} = \frac{\alpha_{n-1}^{\text{opt}}}{1 + s_n \alpha_{n-1}^{\text{opt}}}$$

9.7 The update rules for both the maximum eigenfilter discussed in Chapter 8 and the self-organizing map employ modifications of Hebb's postulate of learning. Compare these two modifications, highlighting the differences and similarities between them.

9.8 The *conscience algorithm* is a modification of the SOM algorithm, which forces the density matching to be exact (DeSieno, 1988). In the conscience algorithm, summarized in Table P9.8, each neuron keeps track of how many times it has won the competition (i.e., how many times its synaptic weight vector has been the neuron closest to the input vector in Euclidean distance). The notion used here is that if a neuron wins too often, it "feels guilty" and therefore pulls itself out of the competition.

To investigate the improvement produced in density matching by the use of the conscience algorithm, consider a one-dimensional lattice (i.e., linear array) made up of 20 neurons, which is trained with the linear input density plotted in Fig. P9.8.

(a) Using computer simulations, compare the density matching produced by the conscience algorithm with that produced by the SOM algorithm. For the SOM algorithm use $\eta = 0.05$ and for the conscience algorithm use $B = 0.0001$, $C = 1.0$, and $\eta = 0.05$.

(b) As frames of reference for this comparison, include the "exact" match to the input density.

Discuss the results of your computer simulations.

Computer experiments

9.9 In this experiment we use computer simulations to investigate the SOM algorithm applied to a one-dimensional lattice with a two-dimensional input. The lattice consists of 65 neurons. The inputs consist of random points uniformly distributed inside the triangular area shown in Fig. P9.9. Compute the map produced by the SOM algorithm after 0, 20, 100, 1000, 10,000, and 25,000 iterations.

9.10 Consider a two-dimensional lattice of neurons trained with a three-dimensional input distribution. The lattice consists of 10×10 neurons.

(a) The input is uniformly distributed in a thin volume defined by

$$\{(0 < x_1 < 1), (0 < x_2 < 1), (0 < x_3 < 0.2)\}$$

Use the SOM algorithm to compute a two-dimensional projection of the input space after 50, 1000, and 10,000 iterations of the algorithm.

(b) Repeat your computations for the case when the input is uniformly distributed inside a wider parallelpiped volume defined by

$$\{(0 < x_1 < 1), (0 < x_2 < 1), (0 < x_3 < 0.4)\}$$

TABLE P9.8   Summary of the Conscience Algorithm

---

1. Find the synaptic weight vector $\mathbf{w}_i$ closest to the input vector $\mathbf{x}$:

$$\|\mathbf{x} - \mathbf{w}_i\| = \min_j \|\mathbf{x} - \mathbf{w}_j\|, \quad j = 1, 2, \dots, N$$

2. Keep a running total of the fraction of time, p_j, that neuron j wins the competition:

$$p_j^{\text{new}} = p_j^{\text{old}} + B\left(y_j - p_j^{\text{old}}\right)$$

where $0 < B \ll 1$ and

$$y_j = \begin{cases} 1 & \text{if neuron } j \text{ is the winning neuron} \\ 0 & \text{otherwise} \end{cases}$$

The p_j are initialized to zero at the beginning of the algorithm.

3. Find the new winning neuron using the conscience mechanism

$$\|\mathbf{x} - \mathbf{w}_i\| = \min_j(\|\mathbf{x} - \mathbf{w}_j\| - b_j)$$

where b_j is a *bias* term introduced to modify the competition; it is defined by

$$b_j = C\left(\frac{1}{N} - p_j\right)$$

where C is a bias factor and N is the total number of neurons in the network.

4. Update the synaptic weight vector of the winning neuron:

$$\mathbf{w}_i^{\text{new}} = \mathbf{w}_i^{\text{old}} + \eta(\mathbf{x} - \mathbf{w}_i^{\text{old}})$$

where η is the usual learning-rate parameter used in the SOM algorithm.

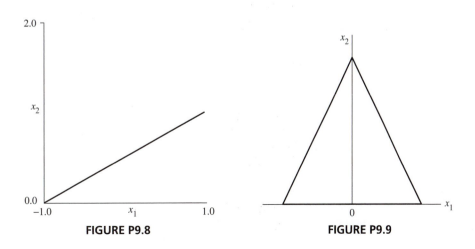

FIGURE P9.8   **FIGURE P9.9**

(c) Repeat your computations one more time for the case when the input is uniformly distributed inside a cube defined by

$$\{(0 < x_1 < 1), (0 < x_2 < 1), (0 < x_3 < 1)\}$$

Discuss the implications of the results of your computer simulations.

9.11 A problem that occasionally arises in the application of the SOM algorithm is the failure of topological ordering by creating a "folded" map. This problem arises when the neighborhood size is permitted to decay too rapidly. The creation of a folded map may be viewed as some form of a "local minimum" of the topological ordering process.

To investigate this phenomenon, consider a two-dimensional lattice of 10×20 neurons trained on a two-dimensional input uniformly distributed inside the square $\{(-1 < x_1 < +1), (-1 < x_2 < +1)\}$. Compute the map produced by the SOM algorithm, permitting the neighborhood function around the winning neuron to decay much faster than that normally used. You may have to repeat the experiment several times in order to see a failure of the ordering process.

9.12 The topological ordering property of the SOM algorithm may be used to form an abstract two-dimensional representation of a high-dimensional input space. To investigate this form of a representation, consider a two-dimensional lattice consisting of 10×10 neurons that is trained with an input consisting of four Gaussian clouds, $\mathcal{C}_1$, $\mathcal{C}_2$, $\mathcal{C}_3$, and $\mathcal{C}_4$, in an eight-dimensional input space. All the clouds have unit variance but different centers. The centers are located at the points $(0, 0, 0, ..., 0)$, $(4, 0, 0, ..., 0)$, $(4, 4, 0, ..., 0)$, and $(0, 4, 0, ..., 0)$. Compute the map produced by the SOM algorithm, with each neuron in the map being labeled with the particular class most represented by the input points around it.

9.13 Table P9.13 presents a summary of the *renormalized SOM algorithm*; a brief description of the algorithm is given in Section 9.3. Compare the conventional and renormalized SOM algorithms, keeping in mind the following two issues:

1. The coding complexity involved in algorithmic implementation.

2. The computer time taken to do the training.

Illustrate the comparison between these two algorithms using data drawn from a uniform distribution inside a square and the following two network configurations:

(a) One-dimensional lattice of 257 neurons.

(b) One-dimensional lattice of 2049 neurons.

In both cases, start with an initial number of code vectors equal to two.

9.14 Consider the signal-space diagram shown in Fig. P9.14 corresponding to *M-ary pulse-amplitude modulation* (PAM) with $M = 8$. The signal points correspond to Gray-encoded data blocks. Each signal point is represented by a rectangular pulse signal with appropriate amplitude scaling:

$$p(t) = \pm \frac{7}{2}, \pm \frac{5}{2}, \pm \frac{3}{2}, \pm \frac{1}{2}, \quad 0 \le t \le T$$

where T is the signaling interval. At the receiver input, white Gaussian noise of zero mean is added to the transmitted signal with varying signal-to-noise ratio (SNR). The SNR is defined as the ratio of the "average" transmitted signal power to the average noise power.

(a) Using a random binary sequence as the transmitter input, generate data representing the received signal for SNR $= 10, 20$, and 30 dB.

(b) For each of these SNRs, set up a self-organizing feature map. For typical values you may use:

• Input vector made up of eight elements obtained by sampling the received signal at a rate equal to eight times the signaling rate (i.e., 8 samples per signaling interval). Do not assume knowledge of timing information.
• One-dimensional lattice of 64 neurons (i.e., eight times the size of the input vector).
(c) Display the feature maps for each of the three SNRs, and thereby demonstrate the topological ordering property of the SOM algorithm.

TABLE P9.13 Summary of Renormalized Training Algorithm (One-Dimensional Version)

---

1. *Initialization.* Set the number of code vectors to be some small number (e.g., use two for simplicity or some other value more representative of the problem at hand). Initialize their positions to be those of a corresponding number of training vectors chosen randomly from the training set.
2. *Selection of an input vector.* Choose an input vector randomly from the training set.
3. *Encoding of the input vector.* Determine the "winning" code vector (i.e., the synaptic weight vector of the winning neuron). To do this, use either the "nearest neighbor" or the "minimum distortion" encoding prescription as required.
4. *Updating of the code book.* Do the usual "winner and its topological neighbors" update. You may find it sufficient to keep the learning-rate parameter η fixed (0.125, say) and to update the winning neuron using η, and its nearest neighbors using $\eta/2$, for example.
5. *Splitting of the code book.*[a] Continue with the code book update (step 4), each time using a new input vector chosen randomly from the training set, until the number of code book updates is about 10–30 times the number of code vectors. When this number is reached, the code book has probably settled down, and it is time to split the code book. You may do so by taking the Peano string of code vectors that you have and interpolate their positions to generate a finer grained approximation to the Peano string; you may simply put an extra code vector halfway between each two existing code vectors.
6. *Completion of training.* The code book update and the code book splitting are continued until the total number of code vectors has reached some predetermined value (e.g., 100), at which time the training is all over.

---

[a]The splitting of the code book approximately doubles the number of code vectors after each epoch, so it does not take too many epochs to get to any prescribed number of code vectors.

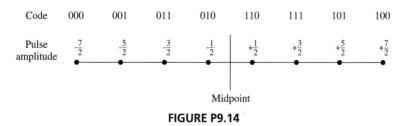

Code	000	001	011	010	110	111	101	100
Pulse amplitude	$-\frac{7}{2}$	$-\frac{5}{2}$	$-\frac{3}{2}$	$-\frac{1}{2}$	$+\frac{1}{2}$	$+\frac{3}{2}$	$+\frac{5}{2}$	$+\frac{7}{2}$

Midpoint

FIGURE P9.14

CHAPTER 10

Information-Theoretic Models

10.1 INTRODUCTION

In a classic paper published in 1948, Claude Shannon laid down the foundations of *information theory*. Shannon's original work on information theory,[1] and its refinement by other researchers, was in direct response to the need for electrical engineers to design communication systems that are both *efficient* and *reliable*. In spite of its practical origins, information theory as we know it today is a deep mathematical theory concerned with the very essence of the *communication process*. The theory provides a framework for the study of fundamental issues such as the efficiency of information representation and the limitations involved in the reliable transmission of information over a communication channel. Moreover, the theory encompasses a multitude of powerful theorems for computing ideal *bounds* on the optimum representation and transmission of information-bearing signals. These bounds are important because they provide benchmarks for the improved design of information processing systems.

The main purpose of this chapter is to discuss *information-theoretic models* that lead to self-organization in a principled manner. In this context, a model that deserves special mention is the *maximum mutual information principle*[2] due to Linsker (1988). This principle states that the synaptic connections of a multilayered neural network develop in such a way as to *maximize the amount of information that is preserved when signals are transformed at each processing stage of the network, subject to certain constraints*. The idea that information theory may offer an explanation for perceptual processing is not new.[3] For instance, we may mention an early paper by Attneave (1954), in which the following information-theoretic function is proposed for the perceptual system:

> A major function of the perceptual machinery is to strip away some of the redundancy of stimulation, to describe or encode information in a form more economical than that in which it impinges on the receptors.

The main idea behind Attneave's paper is the recognition that encoding of data from a scene for the purpose of redundancy reduction is related to the identification of specific features in the scene. This important insight is related to a view of the brain described in Craik (1943), where a model of the external world is constructed so as to incorporate the regularities and constraints of the world.

Organization of the Chapter

The main body of the chapter is organized in two parts. The first part, consisting of Sections 10.2 through 10.5, provides a review of the fundamentals of information theory. In Section 10.2 we discuss the concept of entropy as a quantitative measure of information, which leads naturally to the maximum entropy principle discussed in Section 10.3. Next, in Section 10.4 we discuss the concept of mutual information and its properties, followed by a discussion of the Kullback–Leibler divergence in Section 10.5.

The second part of the chapter, consisting of Sections 10.6 through 10.14, deals with information-theoretic models for self-organizing systems. Section 10.6 highlights mutual information as an objective function to be optimized. The principle of maximum mutual information is discussed in Section 10.7, which is followed by a discussion of the relationship between this principle and that of redundancy reduction in Section 10.8. Sections 10.9 and 10.10 deal with two variants of the principle of maximum mutual information that are suitable for different applications in image processing. Sections 10.11 to 10.14 present three different methods for solving the blind source separation problem.

The chapter concludes with some final remarks in Section 10.15.

10.2 ENTROPY

Following the terminology commonly used in probability theory, we use an uppercase letter to denote a *random variable*, and the corresponding lowercase letter to denote the *value* of the random variable.

Consider then a random variable X, each realization (presentation) of which may be regarded as a *message*. Strictly speaking, if the random variable X is continuous in its amplitude range, then it carries an infinite amount of information. However, on physical and biological grounds we recognize that it is meaningless to think in terms of amplitude measurements with infinite precision, which suggests that the value of X may be uniformly *quantized* into a finite number of discrete levels. Accordingly, we may view X as a *discrete* random variable, modeled as follows:

$$X = \{x_k | k = 0, \pm 1, \ldots, \pm K\} \tag{10.1}$$

where x_k is a discrete number and $(2K + 1)$ is the total number of discrete levels. The separation δx between the discrete levels is assumed to be small enough for the model of Eq. (10.1) to provide an adequate representation for the variable of interest. We may, of course, pass to the continuum limit by letting δx approach zero and K approach infinity, in which case we have a continuous random variable and (as we will see later in the section) sums become integrals.

To complete the model, let the event $X = x_k$ occur with *probability*

$$p_k = P(X = x_k) \tag{10.2}$$

with the requirement that

$$0 \le p_k \le 1 \quad \text{and} \quad \sum_{k=-K}^{K} p_k = 1 \tag{10.3}$$

Suppose that the event $X = x_k$ occurs with probability $p_k = 1$, which therefore requires that $p_i = 0$ for all $i \ne k$. In such a situation there is no "surprise" and therefore no "information" conveyed by the occurrence of the event $X = x_k$, since we know what the message must be. If, on the other hand, the various discrete levels were to occur with different probabilities and, in particular, the probability p_k is low, then there is more "surprise" and therefore "information" when X takes the value x_k rather than another value x_i with higher probability p_i, $i \ne k$. Thus the words "uncertainty," "surprise," and "information" are all related. Before the occurrence of the event $X = x_k$, there is an amount of uncertainty. When the event $X = x_k$ occurs, there is an amount of surprise. After the occurrence of the event $X = x_k$, there is an increase in the amount of information. These three amounts are obviously the same. Moreover, the amount of information is related to the *inverse* of the probability of occurrence.

We define the amount of information gained after observing the event $X = x_k$ with probability p_k as the logarithmic function

$$I(x_k) = \log\left(\frac{1}{p_k}\right) = -\log p_k \tag{10.4}$$

where the base of the logarithm is arbitrary. When the natural logarithm is used the units for information are *nats*, and when the base 2 logarithm is used the units are *bits*. In any case, the definition of information given in Eq. (10.4) exhibits the following properties:

1.
$$I(x_k) = 0 \quad \text{for } p_k = 1 \tag{10.5}$$

Obviously, if we are absolutely certain of the outcome of an event, there is *no* information gained by its occurrence.

2.
$$I(x_k) \ge 0 \quad \text{for } 0 \le p_k \le 1 \tag{10.6}$$

That is, the occurrence of an event $X = x_k$ either provides some or no information, but it never results in a loss of information.

3.
$$I(x_k) > I(x_i) \quad \text{for } p_k < p_i \tag{10.7}$$

That is, the less probable an event is, the more information we gain through its occurrence.

The amount of information $I(x_k)$ is a discrete random variable with probability p_k. The mean value of $I(x_k)$ over the complete range of $2K + 1$ discrete values is given by

$$
\begin{aligned}
H(X) &= E[I(x_k)] \\
&= \sum_{k=-K}^{K} p_k I(x_k) \\
&= -\sum_{k=-K}^{K} p_k \log p_k
\end{aligned}
\tag{10.8}
$$

The quantity $H(X)$ is called the *entropy* of a random variable X permitted to take a finite set of discrete values; it is so called in recognition of the analogy between the definition given in Eq. (10.8) and that of entropy in statistical thermodynamics.[4] The entropy $H(X)$ is a measure of the *average amount of information conveyed per message*. Note, however, that the X in $H(X)$ is not an argument of a function but rather a label for a random variable. Note also that in the definition of Eq. (10.8) we take $0 \log 0$ to be 0.

The entropy $H(X)$ is bounded as follows:

$$0 \le H(X) \le \log(2K + 1) \qquad (10.9)$$

where $(2K + 1)$ is the total number of discrete levels. Furthermore, we may make the following statements:

1. $H(X) = 0$ if and only if the probability $p_k = 1$ for some k, and the remaining probabilities in the set are all zero; this lower bound on entropy corresponds to no *uncertainty*.
2. $H(X) = \log_2(2K + 1)$, if and only if $p_k = 1/(2K + 1)$ for all k (i.e., all the discrete levels are equiprobable); this upper bound on entropy corresponds to *maximum uncertainty*.

The proof of property 2 follows from the following lemma (Gray, 1990):

Given any two probability distributions $\{p_k\}$ and $\{q_k\}$ for a discrete random variable X, then

$$\sum_k p_k \log\left(\frac{p_k}{q_k}\right) \ge 0 \qquad (10.10)$$

which is satisfied with equality if and only if $q_k = p_k$ for all k.

The quantity used in this lemma is of such fundamental importance that we pause to recast it in a form suitable for use in the study of stochastic systems. Let $p_X(x)$ and $q_X(x)$ denote the probabilities that random variable X is in state x under two different operating conditions. The *relative entropy* or *Kullback–Leibler divergence (distance)* between the two *probability mass functions* $p_X(x)$ and $q_X(x)$ is defined by (Kullback, 1968; Gray, 1990; Cover and Thomas, 1991)

$$D_{p\|q} = \sum_{x \in \mathcal{X}} p_X(x) \log\left(\frac{p_X(x)}{q_X(x)}\right) \qquad (10.11)$$

where the sum is over all possible states of the system (i.e., the alphabet $\mathcal{X}$ of the discrete random variable X). The probability mass function $q_X(x)$ plays the role of a *reference measure*.

Differential Entropy of Continuous Random Variables

The discussion of information-theoretic concepts has thus far involved ensembles of random variables that are discrete in their amplitude values. We now extend some of these concepts to continuous random variables.

Consider a continuous random variable X with the *probability density function* $f_X(x)$. By analogy with the entropy of a discrete random variable, we introduce the following definition:

$$h(X) = -\int_{-\infty}^{\infty} f_X(x) \log f_X(x) dx \tag{10.12}$$

$$= -E[\log f_X(x)]$$

We refer to $h(X)$ as the *differential entropy* of X to distinguish it from the ordinary or *absolute entropy*. We do so in recognition of the fact that although $h(X)$ is a useful mathematical quantity to know, it is *not* in any sense a measure of the randomness of X.

We justify the use of Eq. (10.12) as follows. We begin by viewing the continuous random variable X as the limiting form of a discrete random variable that assumes the value $x_k = k \delta x$, where $k = 0, \pm 1, \pm 2, \ldots$, and δx approaches zero. By definition, the continuous random variable X assumes a value in the interval $[x_k, x_k + \delta x]$ with probability $f_X(x_k) \delta x$. Hence, permitting δx to approach zero, the ordinary entropy of the continuous random variable X may be written in the limit as

$$H(X) = -\lim_{\delta x \to 0} \sum_{k=-\infty}^{\infty} f_X(x_k) \delta x \log (f_X(x_k) \delta x)$$

$$= -\lim_{\delta x \to 0} \left[\sum_{k=-\infty}^{\infty} f_X(x_k)(\log f_X(x_k)) \delta x + \log \delta x \sum_{k=-\infty}^{\infty} f_X(x_k) \delta x \right] \tag{10.13}$$

$$= -\int_{-\infty}^{\infty} f_X(x) \log f_X(x) dx - \lim_{\delta x \to 0} \log \delta x \int_{-\infty}^{\infty} f_X(x) dx$$

$$= h(X) - \lim_{\delta x \to 0} \log \delta x$$

where in the last line we have made use of Eq. (10.12) and the fact that the total area under the curve of the probability density function $f_X(x)$ is unity. In the limit as δx approaches zero, $-\log \delta x$ approaches infinity. This means that the entropy of a continuous random variable is infinitely large. Intuitively, we would expect this to be true because a continuous random variable may assume a value anywhere in the open interval $(-\infty, \infty)$ and the uncertainty associated with the variable is on the order of infinity. We avoid the problem associated with the term $\log \delta x$ by adopting $h(X)$ as a differential entropy, with the term $-\log \delta x$ serving as a reference. Moreover, since the information processed by a stochastic system as an entity of interest is actually the difference between two entropy terms that have a common reference, the information will be the same as the difference between the corresponding differential entropy terms. We are therefore perfectly justified in using the term $h(X)$, defined in Eq. (10.13), as the differential entropy of the continuous random variable X.

When we have a continuous random vector $\mathbf{X}$ consisting of n random variables $X_1, X_2, \ldots, X_n$, we define the differential entropy of $\mathbf{X}$ as the *n-fold integral*

$$h(\mathbf{X}) = -\int_{-\infty}^{\infty} f_{\mathbf{X}}(\mathbf{x}) \log f_{\mathbf{X}}(\mathbf{x}) d\mathbf{x} \tag{10.14}$$

$$= -E[\log f_{\mathbf{X}}(\mathbf{x})]$$

where $f_{\mathbf{X}}(\mathbf{x})$ is the joint probability density function of $\mathbf{X}$.

Example 10.1   Uniform Distribution

Consider a random variable X uniformly distributed inside the interval $[0, 1]$, as shown by

$$f_X(x) = \begin{cases} 1 & \text{for } 0 \le x \le 1 \\ 0 & \text{otherwise} \end{cases}$$

By applying Eq. (10.12), we find that the differential entropy of X is

$$h(X) = -\int_{-\infty}^{\infty} 1 \cdot \log 1 \, dx$$

$$= -\int_{-\infty}^{\infty} 1 \cdot 0 \, dx$$

$$= 0$$

The entropy of X is therefore zero.

∎

Properties of Differential Entropy

From the definition of differential entropy $h(X)$ given in Eq. (10.12), we readily see that translation does not change its value; that is,

$$h(X + c) = h(X) \tag{10.15}$$

where c is constant.

Another useful property of $h(X)$ is described by

$$h(aX) = h(X) + \log|a| \tag{10.16}$$

where a is a scaling factor. To prove this property, we first recognize that since the area under the curve of a probability density function is unity, then

$$f_Y(y) = \frac{1}{|a|} f_Y\left(\frac{y}{a}\right) \tag{10.17}$$

Next, using the formula of Eq. (10.12), we may write

$$h(Y) = -E[\log f_Y(y)]$$

$$= -E\left[\log\left(\frac{1}{|a|} f_Y\left(\frac{y}{a}\right)\right)\right]$$

$$= -E\left[\log f_Y\left(\frac{y}{a}\right)\right] + \log|a|$$

By putting $Y = aX$ in this relation we obtain

$$h(aX) = -\int_{-\infty}^{\infty} f_X(x) \log f_X(x) dx + \log|a|$$

from which Eq. (10.16) follows immediately.

Equation (10.16) applies to a scalar random variable. It may be generalized to the case of a random vector $\mathbf{X}$ premultiplied by matrix $\mathbf{A}$ as follows:

$$h(\mathbf{AX}) = h(\mathbf{X}) + \log|\det(\mathbf{A})| \tag{10.18}$$

where $\det(\mathbf{A})$ is the determinant of matrix $\mathbf{A}$.

10.3 MAXIMUM ENTROPY PRINCIPLE

Suppose that we are given a stochastic system with a set of known states but unknown probabilities, and that somehow we learn some *constraints* on the probability distribution of the states. The constraints can be certain ensemble average values or bounds on these values. The problem is to choose a probability model that is optimum in some sense, given this *prior knowledge* about the model. We usually find that there is an infinite number of possible models that satisfy the constraints. Which model should we choose?

The answer to this fundamental question lies in the *maximum entropy (Max Ent) principle*[5] due to Jaynes (1957). The Max Ent principle may be stated as follows (Jaynes, 1957, 1982):

> When an inference is made on the basis of incomplete information, it should be drawn from the probability distribution that maximizes the entropy, subject to constraints on the distribution.

In effect, the notion of entropy defines a kind of measure on the space of probability distributions, such that those distributions of high entropy are favored over others.

From this statement, it is apparent that the Max Ent problem is a constrained optimization problem. To illustrate the procedure for solving such a problem, consider the maximization of the differential entropy

$$h(X) = -\int_{-\infty}^{\infty} f_X(x) \log f_X(x)dx$$

over all probability density functions $f_X(x)$ of a random variable X, subject to the following constraints:

1. $f_X(x) \geq 0$, with equality outside the support of x.

2. $\displaystyle\int_{-\infty}^{\infty} f_X(x)dx = 1$.

3. $\displaystyle\int_{-\infty}^{\infty} f_X(x)g_i(x)dx = \alpha_i$ for $i = 1, 2, ..., m$

where $g_i(x)$ is some function of x. Constraints 1 and 2 simply describe two fundamental properties of a probability density function. Constraint 3 defines the moments of X depending on how the function $g_i(x)$ is formulated. In effect, constraint 3 sums up the prior knowledge available about the random variable X. To solve this constrained optimization problem, we use the *method of Lagrange multipliers*[6] by first formulating the objective function

$$J(f) = \int_{-\infty}^{\infty} \left[-f_X(x) \log f_X(x) + \lambda_0 f_X(x) + \sum_{i=1}^{m} \lambda_i g_i(x) f_X(x) \right] dx \quad (10.19)$$

where $\lambda_0, \lambda_1, \ldots, \lambda_m$ are the *Lagrange multipliers*. Differentiating the integrand with respect to $f_X(x)$ and then setting the result equal to zero, we get

$$-1 - \log f_X(x) + \lambda_0 + \sum_{i=1}^{m} \lambda_i g_i(x) = 0$$

Solving this equation for the unknown $f_X(x)$, we get

$$f_X(x) = \exp\left(-1 + \lambda_0 + \sum_{i=1}^{m} \lambda_i g_i(x)\right) \quad (10.20)$$

The Lagrange multipliers in Eq. (10.20) are chosen in accordance with constraints 2 and 3. Equation (10.20) defines the maximum entropy distribution for this problem.

Example 10.2   One-dimensional Gaussian Distribution

Suppose the prior knowledge available to us is made up of the mean μ and variance σ^2 of a random variable X. By definition, we have

$$\int_{-\infty}^{\infty} (x - \mu)^2 f_X(x) dx = \sigma^2 = \text{constant}$$

Comparing this equation with constraint 3, we readily see that

$$g_1(x) = (x - \mu)^2$$

$$\alpha_1 = \sigma^2$$

Hence, the use of Eq. (10.20) yields

$$f_X(x) = \exp[-1 + \lambda_0 + \lambda_1(x - \mu)^2]$$

Note that λ_1 has to be negative if the integrals of $f_X(x)$ and $(x - \sigma)^2 f_X(x)$ with respect to x are to converge. Substituting this equation in equality constraints 2 and 3, and then solving for λ_0 and λ_1, we get

$$\lambda_0 = 1 - \log(2\pi\sigma^2)$$

and

$$\lambda_1 = -\frac{1}{2\sigma^2}$$

The desired form for $f_X(x)$ is therefore described by

$$f_X(x) = \frac{1}{\sqrt{2\pi}\sigma} \exp\left(-\frac{(x - \mu)^2}{2\sigma^2}\right) \quad (10.21)$$

which is recognized as the probability density of a *Gaussian random variable X of mean μ and variance σ^2*. The maximum value of the differential entropy of such a random variable is given by

$$h(X) = \frac{1}{2}[1 + \log(2\pi\sigma^2)] \quad (10.22)$$

We may summarize the results of this example as follows:

1. *For a given variance σ^2, the Gaussian random variable has the largest differential entropy attainable by any random variable.* That is, if X is a Gaussian random variable and Y is any other random variable with the same mean and variance, then for all Y

$$h(X) \geq h(Y)$$

with the equality holding only if X and Y are the same.

2. *The entropy of a Gaussian random variable X is uniquely determined by the variance of X* (i.e., it is independent of the mean of X).

■

Example 10.3   Multidimensional Gaussian Distribution

In this second example, we want to build on the results of Example 10.2 to evaluate the differential entropy of a *multidimensional Gaussian distribution*. Since the entropy of a Gaussian random variable X is independent of the mean of X, we may justifiably simplify the discussion in this example by considering an m-dimensional vector $\mathbf{X}$ of zero mean. Let the second-order statistics of $\mathbf{X}$ be described by the covariance matrix $\mathbf{\Sigma}$ defined as the expectation of the outer product of $\mathbf{X}$ with itself. The joint probability density function of the random vector $\mathbf{X}$ is given by (Wilks, 1962)

$$f_{\mathbf{X}}(\mathbf{x}) = \frac{1}{(2\pi)^{m/2}(\det(\mathbf{\Sigma}))^{1/2}} \exp\left(-\frac{1}{2}\mathbf{x}^T\mathbf{\Sigma}^{-1}\mathbf{x}\right) \tag{10.23}$$

where $\det(\mathbf{\Sigma})$ is the determinant of $\mathbf{\Sigma}$. Equation (10.14) defines the differential entropy of $\mathbf{X}$. Therefore, substituting Eq. (10.23) in (10.14), we obtain the result

$$h(\mathbf{X}) = \frac{1}{2}[m + m\log(2\pi) + \log|\det(\mathbf{\Sigma})|] \tag{10.24}$$

which includes Eq. (10.22) as a special case. In light of the Max Ent principle, we may thus state that for a given covariance matrix $\mathbf{\Sigma}$, the multivariate Gaussian distribution of Eq. (10.23) has the largest differential entropy attainable by any random vector of zero mean, and that maximum differential entropy is defined by Eq. (10.24).

■

10.4   MUTUAL INFORMATION

In the design of a self-organizing system, the primary objective is to develop an algorithm that is able to learn an input–output relationship of interest on the basis of input patterns alone. In this context, the notion of mutual information is of profound importance because of some highly desirable properties. To set the stage for the discussion, consider a stochastic system with input X and output Y. Both X and Y are permitted to take *discrete* values only, denoted by x and y, respectively. The entropy $H(X)$ is a measure of the prior uncertainty about X. How can we measure the uncertainty about X after observing Y? In order to answer this question, we define the *conditional entropy* of X given Y as follows (Gray, 1990; Cover and Thomas, 1991):

$$H(X|Y) = H(X, Y) - H(Y) \tag{10.25}$$

with the property that

$$0 \le H(X|Y) \le H(X) \tag{10.26}$$

The conditional entropy $H(X|Y)$ represents the *amount of uncertainty remaining about the system input X after the system output Y has been observed.* The other quantity $H(X, Y)$ in Eq. (10.25) is the *joint entropy* of X and Y, which is defined by

$$H(X, Y) = -\sum_{x \in \mathcal{X}} \sum_{y \in \mathcal{Y}} p(x, y) \, \log p(x, y)$$

where $p(x, y)$ is the *joint probability mass function* of discrete random variables X and Y, and $\mathcal{X}$ and $\mathcal{Y}$ are their respective alphabets.

Since the entropy $H(X)$ represents our uncertainty about the system input *before* observing the system output, and the conditional entropy $H(X|Y)$ represents our uncertainty about the system input *after* observing the system output, the difference $H(X) - H(X|Y)$ must represent our uncertainty about the system input that is resolved by observing the system output. This quantity is called the *mutual information* between the random variables X and Y. Denoting it by $I(X; Y)$, we may thus write[7]

$$I(X; Y) = H(X) - H(X|Y)$$

$$= \sum_{x \in \mathcal{X}} \sum_{y \in \mathcal{Y}} p(x, y) \log \left(\frac{p(x, y)}{p(x) \, p(y)} \right) \tag{10.27}$$

Entropy is a special case of mutual information, since we have

$$H(X) = I(X; X)$$

The mutual information $I(X; Y)$ between two discrete random variables X and Y has the following properties (Cover and Thomas, 1991; Gray 1990).

1. *The mutual information between X and Y is symmetric;* that is,

$$I(Y; X) = I(X; Y)$$

   where the mutual information $I(Y; X)$ is a measure of the uncertainty about the system output Y that is resolved by observing the system input X, and the mutual information $I(X; Y)$ is a measure of the uncertainty about the system input that is resolved by observing the system output.

2. *The mutual information between X and Y is always nonnegative;* that is,

$$I(X; Y) \ge 0$$

   In effect, this property states that we cannot lose information, on the average, by observing the system output Y. Moreover, the mutual information is zero if and only if the input and output of the system are statistically independent.

3. *The mutual information between X and Y may be expressed in terms of the entropy of Y as*

$$I(X; Y) = H(Y) - H(Y|X) \tag{10.28}$$

   where $H(Y|X)$ is a *conditional entropy.* The right-hand side of Eq. (10.28) is the ensemble average of the information conveyed by the system output Y, minus the ensemble average of the information conveyed by Y given what we already know the system input X. This latter quantity, $H(Y|X)$, conveys information about the processing noise, rather than about the system input X.

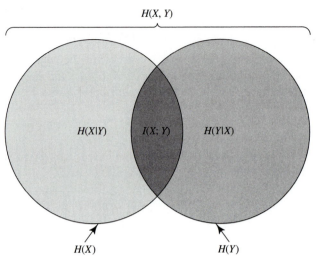

$H(X, Y)$

$H(X|Y)$ $I(X; Y)$ $H(Y|X)$

$H(X)$ $H(Y)$

FIGURE 10.1 Illustration of the relations among the mutual information $I(X; Y)$ and the entropies $H(X)$ and $H(Y)$.

Figure 10.1 provides a visual interpretation of Eqs. (10.27) and (10.28). The entropy of the system input X is represented by the circle on the left. The entropy of the system output Y is represented by the circle on the right. The mutual information between X and Y is represented by the overlap between these two circles.

Mutual Information for Continuous Random Variables

Consider next a pair of continuous random variables X and Y. By analogy with Eq. (10.27), we define the *mutual information* between the random variables X and Y as

$$I(X; Y) = \int_{-\infty}^{\infty} \int_{-\infty}^{\infty} f_{X,Y}(x, y) \log\left(\frac{f_X(x|y)}{f_X(x)}\right) dx\, dy \qquad (10.29)$$

where $f_{X,Y}(x, y)$ is the joint probability density function of X and Y, and $f_X(x|y)$ is the conditional probability density function of X, given $Y = y$. Note that

$$f_{X,Y}(x, y) = f_X(x|y)\, f_Y(y)$$

and so we may also write

$$I(X; Y) = \int_{-\infty}^{\infty} \int_{-\infty}^{\infty} f_{X,Y}(x, y) \log\left(\frac{f_{X,Y}(x, y)}{f_X(x)\, f_Y(y)}\right) dx\, dy$$

Also, by analogy with our previous discussion for discrete random variables, the mutual information $I(X; Y)$ between the continuous random variables X and Y has the following properties:

$$\begin{aligned} I(X; Y) &= h(X) - h(X|Y) \\ &= h(Y) - h(Y|X) \qquad (10.30) \\ &= h(X) + h(Y) - h(X, Y) \end{aligned}$$

$$I(Y; X) = I(X; Y) \tag{10.31}$$

$$I(X; Y) \geq 0 \tag{10.32}$$

The parameter $h(X)$ is the differential entropy of X; likewise for $h(Y)$. The parameter $h(X|Y)$ is the *conditional differential entropy* of X given Y; it is defined by the double integral

$$h(X|Y) = -\int_{-\infty}^{\infty} \int_{-\infty}^{\infty} f_{X,Y}(x, y) \log f_X(x|y) \, dx \, dy \tag{10.33}$$

The parameter $h(Y|X)$ is the conditional differential entropy of Y given X. It is defined in a manner similar to $h(X|Y)$. The parameter $h(X, Y)$ is the joint differential entropy of X and Y.

Note that Eq. (10.32) is satisfied with the equality sign only when the random variables X and Y are *statistically independent*. When this condition is satisfied, the joint probability density function of X and Y can be factored as

$$f_{X,Y}(x, y) = f_X(x) f_Y(y) \tag{10.34}$$

where $f_X(x)$ and $f_Y(y)$ are the marginal probability density functions of X and Y, respectively. Equivalently, we may write

$$f_X(x|y) = f_X(x)$$

which states that knowledge of the outcome of Y can in no way affect the distribution of X. Applying this condition to Eq. (10.29) reduces the mutual information $I(X; Y)$ between X and Y to zero.

The definition of mutual information $I(X; Y)$ given in Eq. (10.29) applies to scalar random variables X and Y. This definition may be readily extended to random vectors $\mathbf{X}$ and $\mathbf{Y}$, and we may thus write $I(\mathbf{X}; \mathbf{Y})$. Specifically, we define $I(\mathbf{X}; \mathbf{Y})$ as the multifold integral:

$$I(\mathbf{X}; \mathbf{Y}) = \int_{-\infty}^{\infty} \int_{-\infty}^{\infty} f_{\mathbf{X},\mathbf{Y}}(\mathbf{x}, \mathbf{y}) \log \left(\frac{f_{\mathbf{X}}(\mathbf{x}|\mathbf{y})}{f_{\mathbf{X}}(\mathbf{x})} \right) d\mathbf{x} \, d\mathbf{y} \tag{10.35}$$

The mutual information $I(\mathbf{X}; \mathbf{Y})$ has properties that parallel those given in Eqs. (10.30) to (10.32) for scalar random variables.

10.5 KULLBACK–LEIBLER DIVERGENCE

In Eq. (10.11) we define the Kullback–Leibler divergence for discrete random variables. This definition may be extended to the general case of continuous random vectors. Let $f_{\mathbf{X}}(\mathbf{x})$ and $g_{\mathbf{X}}(\mathbf{x})$ denote two different probability density functions of an m-by-1 random vector $\mathbf{X}$. In light of Eq. (10.11), we may then define the *Kullback–Leibler divergence* between $f_{\mathbf{X}}(\mathbf{x})$ and $g_{\mathbf{X}}(\mathbf{x})$ as follows (Kullback, 1968; Shore and Johnson, 1980):

$$D_{f_{\mathbf{X}}\|g_{\mathbf{X}}} = \int_{-\infty}^{\infty} f_{\mathbf{X}}(\mathbf{x}) \log \left(\frac{f_{\mathbf{X}}(\mathbf{x})}{g_{\mathbf{X}}(\mathbf{x})} \right) d\mathbf{x} \tag{10.36}$$

The Kullback–Leibler divergence has some unique properties:

1. It always has a positive value or is zero. For the special case when $f_{\mathbf{X}}(\mathbf{x}) = g_{\mathbf{X}}(\mathbf{x})$, we have a perfect match between the two distributions, and $D_{f\|g}$ is exactly zero.
2. It is invariant with respect to the following changes in the components of the vector $\mathbf{x}$:
   - Permutation of the order in which the components are arranged.
   - Amplitude scaling.
   - Monotonic nonlinear transformation.

The mutual information $I(\mathbf{X}; \mathbf{Y})$ between a pair of vectors $\mathbf{X}$ and $\mathbf{Y}$ has an interesting interpretation in terms of the Kullback–Leibler divergence. First, we note that

$$f_{\mathbf{X},\mathbf{Y}}(\mathbf{x}, \mathbf{y}) = f_{\mathbf{Y}}(\mathbf{y}|\mathbf{x})f_{\mathbf{X}}(\mathbf{x}) \tag{10.37}$$

Hence, we may rewrite Eq. (10.35) in the equivalent form

$$I(\mathbf{X}; \mathbf{Y}) = \int_{-\infty}^{\infty} \int_{-\infty}^{\infty} f_{\mathbf{X},\mathbf{Y}}(\mathbf{x}, \mathbf{y}) \log\left(\frac{f_{\mathbf{X},\mathbf{Y}}(\mathbf{x}, \mathbf{y})}{f_{\mathbf{X}}(\mathbf{x})f_{\mathbf{Y}}(\mathbf{y})}\right) d\mathbf{x}\, d\mathbf{y}$$

Comparing this formula with that of Eq. (10.36), we immediately deduce the following result:

$$I(\mathbf{X}; \mathbf{Y}) = D_{f_{\mathbf{X},\mathbf{Y}}\|f_{\mathbf{X}}f_{\mathbf{Y}}} \tag{10.38}$$

In words, the mutual information $I(\mathbf{X}; \mathbf{Y})$ between $\mathbf{X}$ and $\mathbf{Y}$ is equal to the Kullback–Leibler divergence between the joint probability density function $f_{\mathbf{X},\mathbf{Y}}(\mathbf{x}, \mathbf{y})$ and the product of the probability density functions $f_{\mathbf{X}}(\mathbf{x})$ and $f_{\mathbf{Y}}(\mathbf{y})$.

A special case of this latter result is the Kullback–Leibler divergence between the probability density function $f_{\mathbf{X}}(\mathbf{x})$ of an m-by-1 random vector $\mathbf{X}$ and the product of its m marginal probability density functions. Let $\tilde{f}_{X_i}(x_i)$ denote the ith marginal probability density function of element X_i, which is defined by

$$\tilde{f}_{X_i}(x_i) = \int_{-\infty}^{\infty} f_{\mathbf{X}}(\mathbf{x})d\mathbf{x}^{(i)}, \qquad i = 1, 2, ..., m \tag{10.39}$$

where $\mathbf{x}^{(i)}$ is the $(m-1)$-by-1 vector left after removing the ith element from vector $\mathbf{x}$. The Kullback–Leibler divergence between $f_{\mathbf{X}}(\mathbf{x})$ and the *factorial distribution* $\prod_i \tilde{f}_{X_i}(x_i)$ is given by

$$D_{f_{\mathbf{X}}\|\tilde{f}_{\mathbf{x}}} = \int_{-\infty}^{\infty} f_{\mathbf{X}}(\mathbf{x}) \log\left(\frac{f_{\mathbf{X}}(\mathbf{x})}{\displaystyle\prod_{i=1}^{m} \tilde{f}_{X_i}(x_i)}\right) d\mathbf{x} \tag{10.40}$$

which may also be written in the expanded form

$$D_{f_{\mathbf{X}}\|\tilde{f}_{\mathbf{x}}} = \int_{-\infty}^{\infty} f_{\mathbf{X}}(\mathbf{x}) \log f_{\mathbf{X}}(\mathbf{x})d\mathbf{x} - \sum_{i=1}^{m} \int_{-\infty}^{\infty} f_{\mathbf{X}}(\mathbf{x}) \log \tilde{f}_{X_i}(x_i)d\mathbf{x} \tag{10.41}$$

The first integral on the right-hand side of Eq. (10.41) is, by definition, equal to $-h(\mathbf{X})$, where $h(\mathbf{X})$ is the differential entropy of $\mathbf{X}$. To deal with the second term we first note that

$$dx = dx^{(i)}dx_i$$

Hence we may write

$$\int_{-\infty}^{\infty} f_{\mathbf{X}}(\mathbf{x})\log\widetilde{f}_{X_i}(x_i)dx = \int_{-\infty}^{\infty}\log\widetilde{f}_{X_i}(x_i)\int_{-\infty}^{\infty} f_{\mathbf{X}}(\mathbf{x})dx^{(i)}dx_i \tag{10.42}$$

where the inner integral on the right-hand side is with respect to the $(m-1)$-by-1 vector $\mathbf{x}^{(i)}$ and the outer integral is with respect to the scalar x_i. But from Eq. (10.39) we see that the inner integral is in fact equal to the marginal probability density function $\widetilde{f}_{X_i}(x_i)$. Accordingly, we may rewrite Eq. (10.42) in the equivalent form

$$\int_{-\infty}^{\infty} f_{\mathbf{X}}(\mathbf{x})\log\widetilde{f}_{X_i}(x_i)dx = \int_{-\infty}^{\infty}\widetilde{f}_{X_i}(x_i)\log\widetilde{f}_{X_i}(x_i)dx_i$$

$$= -\widetilde{h}(X_i), \qquad i = 1, 2, ..., m \tag{10.43}$$

where $\widetilde{h}(X_i)$ is the ith *marginal entropy* (i.e., the differential entropy based on the marginal probability density function $\widetilde{f}_{X_i}(x_i)$). Finally, using Eq. (10.43) in (10.41) and noting that the first integral in Eq. (10.41) is equal to $-h(\mathbf{X})$, we may simplify the Kullback–Leibler divergence of Eq. (10.41) to

$$D_{f_{\mathbf{X}}\|\widetilde{f}_{\mathbf{X}}} = -h(\mathbf{X}) + \sum_{i=1}^{m}\widetilde{h}(X_i) \tag{10.44}$$

This formula will be of particular use in our study of the blind source separation problem later in the chapter.

Pythagorean Decomposition

Next we consider the Kullback–Leibler divergence between the probability density functions $f_{\mathbf{X}}(\mathbf{x})$ and $f_{\mathbf{U}}(\mathbf{x})$. The m-by-1 random vector $\mathbf{U}$ consists of independent variables, as shown by

$$f_{\mathbf{U}}(\mathbf{x}) = \prod_{i=1}^{m} f_{U_i}(x_i)$$

and the m-by-1 random vector $\mathbf{X}$ is defined in terms of $\mathbf{U}$ by

$$\mathbf{X} = \mathbf{A}\mathbf{U}$$

where $\mathbf{A}$ is a nondiagonal matrix. Let $\widetilde{f}_{X_i}(x_i)$ denote the marginal probability density function of each X_i that is derived from $f_{\mathbf{X}}(\mathbf{x})$. Then the Kullback–Leibler divergence between $f_{\mathbf{X}}(\mathbf{x})$ and $f_{\mathbf{U}}(\mathbf{x})$ admits the following *Pythagorean decomposition:*

$$D_{f_{\mathbf{X}}\|f_{\mathbf{U}}} = D_{f_{\mathbf{X}}\|\widetilde{f}_{\mathbf{X}}} + D_{\widetilde{f}_{\mathbf{X}}\|f_{\mathbf{U}}} \tag{10.45}$$

We refer to this classic relation as a Pythagorean decomposition because it has an information-geometric interpretation (Amari, 1985). A proof of the decomposition is given in note 8.

10.6 MUTUAL INFORMATION AS AN OBJECTIVE FUNCTION TO BE OPTIMIZED

Now that we have developed an adequate understanding of Shannon's information theory, we are ready to discuss its role in the study of self-organizing systems.

To proceed with the discussion, consider a neural system with multiple inputs and outputs. The primary objective here is for the system to be self-organizing, designed for a specific task (e.g., modeling, extraction of statistically salient features, or signal separation). This requirement can be satisfied by choosing the mutual information between certain variables of the system as the *objective function* to be optimized. This particular choice is justified by the following considerations:

- The mutual information has some unique properties as discussed in Section 10.4.
- It can be determined without the need for a teacher so the provision for self-organization is naturally met.

The problem thus becomes one of adjusting the free parameters (i.e., synaptic weights) of the system so as to optimize the mutual information.

Depending on the application of interest, we may identify four different scenarios as illustrated in Fig. 10.2, which can arise in practice. These scenarios are described as follows:

- In scenario 1 depicted in Fig. 10.2a, the input vector $\mathbf{X}$ is composed of the elements $X_1, X_2, ..., X_m$, and the output vector $\mathbf{Y}$ is composed of the elements $Y_1, Y_2, ..., Y_l$. The requirement is to *maximize the information conveyed to the system output $\mathbf{Y}$ about the system input $\mathbf{X}$.*

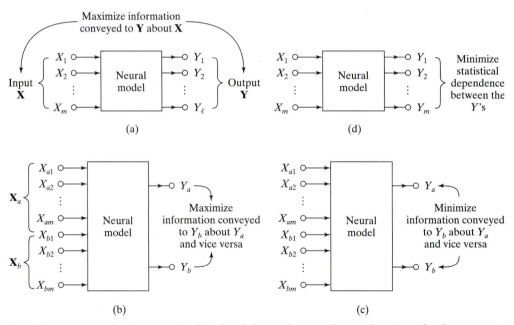

FIGURE 10.2 Four basic scenarios that lend themselves to the application of Infomax and its three variants.

- In scenario 2 depicted in Fig. 10.2b, a pair of input vectors $\mathbf{X}_a$ and $\mathbf{X}_b$ is derived from adjacent but nonoverlapping regions of an image. The inputs $\mathbf{X}_a$ and $\mathbf{X}_b$ produce scalar outputs Y_a and Y_b, respectively. The requirement is to *maximize the information conveyed to Y_a about Y_b* and vice versa.
- In scenario 3 depicted in Fig. 10.2c, the input vectors $\mathbf{X}_a$ and $\mathbf{X}_b$ are derived from a corresponding pair of regions belonging to two separate images. The outputs produced by these two input vectors are denoted by Y_a and Y_b, respectively. The objective is to *minimize the information conveyed to Y_a about Y_b*.
- In scenario 4 depicted in Fig. 10.2d, the input vector $\mathbf{X}$ and output vector $\mathbf{Y}$ are defined in a manner similar to those in Fig. 10.2a but with equal dimensionality (i.e., $l = m$). The objective here is for the *statistical dependence between the components of the output vector $\mathbf{Y}$ to be minimized*.

In all of these situations, mutual information plays a central role. However, the way in which it is formulated depends on the particular situation being considered. In the rest of the chapter, the issues involved in these scenarios and their practical implications are discussed in the same order just presented.

10.7 MAXIMUM MUTUAL INFORMATION PRINCIPLE

The idea of designing a neural processor to maximize the mutual information $I(\mathbf{Y}; \mathbf{X})$ is appealing as the basis for statistical signal processing. This method of optimization is embodied in the *maximum mutual information (Infomax) principle* due to Linsker (1987, 1988a, 1989a), which may be stated formally as follows:

> The transformation of a random vector $\mathbf{X}$ observed in the input layer of a neural system to a random vector $\mathbf{Y}$ produced in the output layer of the system should be so chosen that the activities of the neurons in the output layer jointly maximize information about the activities in the input layer. The objective function to be maximized is the mutual information $I(\mathbf{Y}; \mathbf{X})$ between the vectors $\mathbf{X}$ and $\mathbf{Y}$.

The Infomax principle provides a mathematical framework for self-organization of the signal transmission system described in Fig. 10.2a that is independent of the rule used for its implementation. Also, this principle may be viewed as the neural network counterpart of the concept of *channel capacity,* which defines the Shannon limit on the rate of information transmission through a communication channel.

In the sequel, we illustrate application of the Infomax principle with two examples involving a single noisy neuron. In one example the noise appears at the output, and in the other example it appears at the input.

Example 10.4   A Single Neuron Corrupted by Processing Noise

Consider the simple case of a linear neuron that receives its inputs from a set of m source nodes. Let the output of this neuron in the presence of *processing noise* be expressed as

$$Y = \left(\sum_{i=1}^{m} w_i X_i \right) + N \qquad (10.46)$$

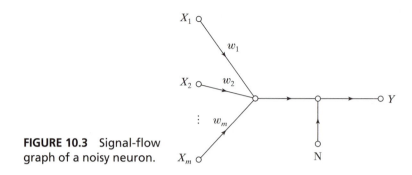

FIGURE 10.3 Signal-flow graph of a noisy neuron.

where w_i is the ith synaptic weight and N is the processing noise, as modeled in Fig. 10.3. It is assumed that:

- The output Y of the neuron is a Gaussian random variable with variance σ_Y^2.
- The processing noise N is also a Gaussian random variable with zero mean and variance σ_N^2.
- The processing noise is uncorrelated with any of the input components. That is,

$$E[NX_i] = 0 \qquad \text{for all } i$$

The Gaussianity of the output Y can be satisfied in one of two ways. The inputs $X_1, X_2, ..., X_m$ are all Gaussian-distributed. Then with the additive noise N assumed to be Gaussian too, the Gaussianity of Y is assured by virtue of the fact that it is the weighted sum of a number of Gaussian-distributed random variables. Alternatively, the inputs $X_1, X_2, ..., X_m$ are identically and independently distributed, in which case the distribution of their weighted sum approaches a Gaussian distribution for large m by the central limit theorem.

To proceed with the analysis, we first note from the second line of Eq. (10.30) that the mutual information $I(Y; \mathbf{X})$ between the output Y of the neuron and the input vector $\mathbf{X}$ is

$$I(Y; \mathbf{X}) = h(Y) - h(Y|\mathbf{X}) \tag{10.47}$$

In view of Eq. (10.46), we note that the probability density function of Y, given the input vector $\mathbf{X}$, is the same as the probability density function of a constant plus a Gaussian-distributed random variable. Accordingly, the conditional entropy $h(Y|\mathbf{X})$ is the "information" that the output neuron conveys about the processing noise N rather than about the signal vector $\mathbf{X}$. We may thus set

$$h(Y|\mathbf{X}) = h(N)$$

and therefore rewrite Eq. (10.47) simply as

$$I(Y; \mathbf{X}) = h(Y) - h(N) \tag{10.48}$$

By applying Eq. (10.22) for the differential entropy of a Gaussian random variable to the problem at hand, we obtain

$$h(Y) = \frac{1}{2}[1 + \log(2\pi\sigma_Y^2)] \tag{10.49}$$

and

$$h(N) = \frac{1}{2}[1 + \log(2\pi\sigma_N^2)] \tag{10.50}$$

After simplification, the use of Eqs. (10.49) and (10.50) in (10.48) yields

$$I(Y; \mathbf{X}) = \frac{1}{2} \log\left(\frac{\sigma_Y^2}{\sigma_N^2}\right) \tag{10.51}$$

where σ_Y^2 depends on σ_N^2.

The ratio σ_Y^2/σ_N^2 may be viewed as a *signal-to-noise ratio*. Imposing the constraint that the noise variance σ_N^2 is fixed, we see from Eq. (10.51) that the mutual information $I(Y; \mathbf{X})$ is maximized by maximizing the variance σ_Y^2 of the neuron output Y. We may therefore state that under certain conditions, maximizing the output variance of a neuron maximizes the mutual information between the output signal of that neuron and its inputs (Linsker, 1988a).

∎

Example 10.5   A Single Neuron Corrupted by Additive Input Noise

Suppose that the noise corrupting the behavior of a linear neuron originates at the input ends of the synapses as shown in the model of Fig. 10.4. According to this second noise model, we have

$$Y = \sum_{i=1}^{m} w_i(X_i + N_i) \tag{10.52}$$

where each noise N_i is assumed to be an independent Gaussian random variable with zero mean and variance σ_N^2. We may rewrite Eq. (10.52) in a form similar to that of Eq. (10.46), as shown by

$$Y = \left(\sum_{i=1}^{m} w_i X_i\right) + N'$$

where N' is a composite noise component, defined by

$$N' = \sum_{i=1}^{m} w_i N_i$$

The noise N' has a Gaussian distribution with zero mean and a variance equal to the sum of the variances of its independent noise components; that is,

$$\sigma_{N'}^2 = \sum_{i=1}^{m} w_i^2 \sigma_N^2$$

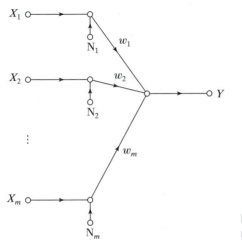

FIGURE 10.4   Another noise model.

As before, we assume that the output Y of the neuron has a Gaussian distribution with variance σ_Y^2. The mutual information $I(Y; \mathbf{X})$ between Y and $\mathbf{X}$ is still given by Eq. (10.47). This time, however, the conditional entropy $h(Y|\mathbf{X})$ is defined by

$$h(Y|\mathbf{X}) = h(N')$$

$$= \frac{1}{2}(1 + 2\pi\sigma_{N'}^2) \tag{10.53}$$

$$= \frac{1}{2}\left[1 + 2\pi\sigma_N^2 \sum_{i=1}^{m} w_i^2\right]$$

Thus, using Eqs. (10.49) and (10.53) in (10.47), and then simplifying terms, we get (Linsker, 1988a)

$$I(Y; \mathbf{X}) = \frac{1}{2}\log\left(\frac{\sigma_Y^2}{\sigma_N^2 \sum_{i=1}^{m} w_i^2}\right) \tag{10.54}$$

Under the constraint that the noise variance σ_N^2 is maintained constant, the mutual information $I(Y; \mathbf{X})$ is now maximized by maximizing the ratio $\sigma_Y^2/\sum_{i=1}^{m} w_i^2$, where σ_Y^2 is a function of w_i. ∎

What can we deduce from Examples 10.4 and 10.5?   First, we see from the material presented in these two examples that the result of applying the Infomax principle is problem-dependent. The equivalence between maximizing the mutual information $I(Y; \mathbf{X})$ and output variance that applies to the model of Fig. 10.3, for a prescribed noise variance σ_N^2, does not carry over to the model of Fig. 10.4. It is only when we impose the constraint $\sum_i w_i^2 = 1$ on the model of Fig. 10.4 that both models behave in a similar manner.

In general, the determination of the mutual information $I(\mathbf{Y}; \mathbf{X})$ between input vector $\mathbf{X}$ and output vector $\mathbf{Y}$ is a difficult task. In Examples 10.4 and 10.5, we made the analysis mathematically tractable by assuming that the noise distributions in a system with one or more sources of noise are *multivariate Gaussian*. This assumption needs to be justified.

In adopting a Gaussian noise model, we are in essence invoking a "surrogate" mutual information computed on the premise that the output vector $\mathbf{Y}$ of a neuron has a multivariate Gaussian distribution with the same mean vector and covariance matrix as the actual distribution. In Linsker (1993), the Kullback–Leibler divergence is used to provide a principled justification for the use of such surrogate mutual information, under the condition that the network has stored information about the mean vector and covariance matrix of the output vector $\mathbf{Y}$, but not about higher-order statistics.

Finally, the analysis presented in both Examples 10.4 and 10.5 was carried out in the context of a single neuron. This was done purposely with a specific point in mind: For the Infomax principle to be mathematically tractable, the optimization should be performed at the local neuronal level. Such optimization is consistent with the essence of self-organization.

Example 10.6

In Examples 10.4 and 10.5 we considered noisy neurons. In this example we consider a noiseless network that transforms a random vector $\mathbf{X}$ of arbitrary distribution to a new random vector $\mathbf{Y}$ of different distribution. Recognizing that $I(\mathbf{X}; \mathbf{Y}) = I(\mathbf{Y}; \mathbf{X})$ and extending Eq. (10.28) to the situation described here, we may express the mutual information between the input vector $\mathbf{X}$ and output vector $\mathbf{Y}$ as follows:

$$I(\mathbf{Y}; \mathbf{X}) = H(\mathbf{Y}) - H(\mathbf{Y}|\mathbf{X})$$

where $H(\mathbf{Y})$ is the entropy of $\mathbf{Y}$ and $H(\mathbf{Y}|\mathbf{X})$ is the conditional entropy of $\mathbf{Y}$ given $\mathbf{X}$. With the mapping from $\mathbf{X}$ to $\mathbf{Y}$ assumed to be noiseless, the conditional entropy $H(\mathbf{Y}|\mathbf{X})$ attains its lowest possible value: it diverges to $-\infty$. This result is due to the differential nature of the entropy of a continuous random variable that was discussed in Section 10.2. However, this difficulty is of no consequence when we consider the *gradient* of the mutual information $I(\mathbf{Y}; \mathbf{X})$ with respect to a weight matrix $\mathbf{W}$ that parameterizes the mapping network. Specifically, we may write:

$$\frac{\partial I(\mathbf{Y}; \mathbf{X})}{\partial \mathbf{W}} = \frac{\partial H(\mathbf{Y})}{\partial \mathbf{W}} \qquad (10.55)$$

because the conditional entropy $H(\mathbf{Y}|\mathbf{X})$ is independent of $\mathbf{W}$. Equation (10.55) shows that for a noiseless mapping network, maximizing the entropy of the network output $\mathbf{Y}$ is equivalent to maximizing the mutual information between $\mathbf{Y}$ and the network input $\mathbf{X}$, with both maximizations being performed with respect to the weight matrix $\mathbf{W}$ of the mapping network (Bell and Sejnowski, 1995).

∎

10.8   INFOMAX AND REDUNDANCY REDUCTION

In Shannon's framework of information theory, order and structure represent *redundancy,* which diminishes uncertainty that is resolved by the receipt of information. The more order and structure we have in the underlying process, the less information we receive by observing that process. Consider for example, the highly structured and redundant sequence of examples *aaaaaa*. On receiving the first example *a*, we can immediately say that the remaining five examples are all the same. The information conveyed by such a sequence of examples is limited to that contained in a single example. In other words, the more redundant a sequence of examples is, the less information content is received from the environment.

From the definition of mutual information $I(\mathbf{Y}; \mathbf{X})$, we know it is a measure of the uncertainty about the output $\mathbf{Y}$ of a system that is resolved by observing the system input $\mathbf{X}$. The Infomax principle operates by maximizing the mutual information $I(\mathbf{Y}; \mathbf{X})$, as a result of which we are more certain about the system output $\mathbf{Y}$ by observing the system input $\mathbf{X}$. In light of the previously mentioned relationship between information and redundancy, we may therefore say that the Infomax principle leads to a reduction in redundancy in the output $\mathbf{Y}$ compared to that in the input $\mathbf{X}$.

The presence of noise is a factor that prompts the use of redundancy and the related method of diversity (Linsker, 1988a). When the additive noise in the input signal is high, we may use redundancy to combat the degrading effects of noise. In such an environment more of the (correlated) components of the input signal are combined by the processor to provide an accurate representation of the input. Also when the output noise (i.e., processor noise) is high, more of the output components are directed by the processor to provide redundant information. The number of independent properties observed at the output of the processor is thereby reduced, but the representation accuracy of each property is increased. We may thus state that *a high level of noise favors redundancy of representation.* When, however, the *noise level is low, diversity of representation is favored* over redundancy. By diversity we mean two or more outputs with different properties being produced by the processor. Problem 10.6 discusses the redundancy/diversity tradeoff from the perspective of Infomax. It is noteworthy that the redundancy/diversity tradeoff is somewhat analogous (though different from) the bias/variance tradeoff discussed in Chapter 2.

Modeling of a Perceptual System

Since the early days of information theory, it has been suggested that the redundancy of sensory messages (stimuli) is important for understanding perception (Attneave, 1954; Barlow, 1959). Indeed, the redundancy of sensory messages provides the *knowledge* that enables the brain to build up its "cognitive maps" or "working models" of the environment around it (Barlow, 1989). Regularities in the sensory messages must somehow be recoded by the brain for it to know what usually happens. However, *redundancy reduction* is the more specific form of *Barlow's hypothesis*. This hypothesis says that the purpose of early processing is to transform the highly redundant sensory input into a more efficient *factorial code*. In other words, the neuronal outputs become *statistically independent* when conditioned on the input.

Inspired by Barlow's hypothesis, Atick and Redlich (1990) postulated the *principle of minimum redundancy* as the basis for an information-theoretic model of the perceptual system shown in Fig. 10.5. The model consists of three components: *input channel, recoding system,* and *output channel.* The output of the input channel is described by

$$\mathbf{X} = \mathbf{S} + \mathbf{N}_1$$

where $\mathbf{S}$ is an ideal signal received by the input channel and $\mathbf{N}_1$ is assumed to be the source of all noise in the input. The signal $\mathbf{X}$ is subsequently transformed (recoded) by a linear matrix operator $\mathbf{A}$. It is then transmitted through the optic nerve, or output channel, producing the output $\mathbf{Y}$, as shown by

$$\mathbf{Y} = \mathbf{AX} + \mathbf{N}_2$$

where $\mathbf{N}_2$ denotes the postencoding intrinsic noise. In the approach taken by Atick and Redlich, it is observed that light signals arriving at the retina contain useful sensory information in a highly redundant form. Moreover, it is hypothesized that the purpose of retinal signal processing is to reduce or eliminate the redundant bits of data due to

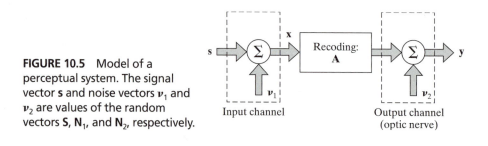

FIGURE 10.5 Model of a perceptual system. The signal vector **s** and noise vectors $\boldsymbol{v}_1$ and $\boldsymbol{v}_2$ are values of the random vectors **S**, $\mathbf{N}_1$, and $\mathbf{N}_2$, respectively.

both correlations and noise, before sending the signal along the optic nerve. To quantify this notion, a *redundancy measure* is defined by

$$R = 1 - \frac{I(\mathbf{Y}; \mathbf{S})}{C(\mathbf{Y})} \tag{10.56}$$

where $I(\mathbf{Y}; \mathbf{S})$ is the mutual information between $\mathbf{Y}$ and $\mathbf{S}$, and $C(\mathbf{Y})$ is the channel capacity of the optic nerve (output channel). Equation (10.56) is justified on the grounds that the information the brain is interested in is the ideal signal $\mathbf{S}$, while the physical channel through which this information must pass is in reality the optic nerve. It is assumed that there is no dimensionality reduction in the input–output mapping performed by the perceptual system, which means that $C(\mathbf{Y}) > I(\mathbf{Y}; \mathbf{S})$. The requirement is to find an input–output mapping (i.e., matrix $\mathbf{A}$) that minimizes the redundancy measure R, subject to the constraint of no information loss, as shown by

$$I(\mathbf{Y}; \mathbf{X}) = I(\mathbf{X}; \mathbf{X}) - \epsilon$$

where ϵ is some small positive parameter. The *channel capacity* $C(\mathbf{Y})$ is defined as the maximum rate of information flow possible through the optic nerve, ranging over all probability distributions of inputs applied to it, and keeping the average input power fixed.

When the signal vector $\mathbf{S}$ and the output vector $\mathbf{Y}$ have the same dimensionality and there is noise in the system, the principle of minimum redundancy and the Infomax principle are mathematically equivalent, provided that a similar constraint is imposed on the computational capability of the output neurons in both cases. To be specific, suppose that the channel capacity is measured in terms of the dynamic range of the output of each neuron in the model of Fig. 10.5. Then, according to the principle of minimum redundancy, the quantity to be minimized is

$$1 - \frac{I(\mathbf{Y}; \mathbf{S})}{C(\mathbf{Y})}$$

for a given permissible information loss, and therefore for a given $I(\mathbf{Y}; \mathbf{S})$. Thus the quantity to be minimized is essentially

$$F_1(\mathbf{Y}; \mathbf{S}) = C(\mathbf{Y}) - \lambda I(\mathbf{Y}; \mathbf{S}) \tag{10.57}$$

On the other hand, according to the Infomax principle the quantity to be maximized in the model of Fig. 10.5 is

$$F_2(\mathbf{Y}; \mathbf{S}) = I(\mathbf{Y}; \mathbf{S}) + \lambda C(\mathbf{Y}) \tag{10.58}$$

Although the functions $F_1(\mathbf{Y}; \mathbf{S})$ and $F_2(\mathbf{Y}; \mathbf{S})$ are different, their optimizations yield identical results: They are both formulations of the method of Lagrange multipliers, with the roles of $I(\mathbf{Y}; \mathbf{S})$ and $C(\mathbf{Y})$ being simply interchanged.

The important point to note from this discussion is that despite the difference in formulations, these two information-theoretic principles lead to similar results. In summary, maximization of the mutual information between the output and input of a neural system does indeed lead to redundancy reduction.[9]

10.9 SPATIALLY COHERENT FEATURES

The Infomax principle, as postulated in Section 10.6, applies to a situation where the mutual information $I(\mathbf{Y}; \mathbf{X})$ between the output vector $\mathbf{Y}$ of a neural system and the input vector $\mathbf{X}$ is the objective function to be maximized, as illustrated in Fig. 10.2a. With appropriate changes in terminology, we may extend this principle to deal with the unsupervised processing of the image of a natural scene (Becker and Hinton, 1992). An unprocessed pixel of such an image contains a wealth of information about the scene of interest, albeit in complex form. In particular, the intensity of each pixel is affected by such intrinsic parameters as depth, reflectance, and surface orientation, as well as background noise and illumination. The goal is to design a self-organizing system that is capable of learning to encode this complex information in a simpler form. To be more specific, the objective is to extract higher-order features that exhibit *simple coherence across space* in such a way that the representation of information in one spatially localized region of the image makes it easy to produce the representation of information in neighboring regions; a region refers to a collection of pixels in the image. The situation described herein pertains to the scenario illustrated in Fig. 10.2b.

We may thus formulate the *first* variant of the Infomax principle[10] as follows (Becker, 1996; Becker and Hinton, 1992):

> The transformation of a pair of vectors $\mathbf{X}_a$ and $\mathbf{X}_b$ (representing adjacent, nonoverlapping regions of an image by a neural system) should be so chosen that the scalar output Y_a of the system due to the input $\mathbf{X}_a$ maximizes information about the second scalar output Y_b due to $\mathbf{X}_b$. The objective function to be maximized is the mutual information $I(Y_a; Y_b)$ between the outputs Y_a and Y_b.

We refer to this principle as a variant of the Infomax principle in the sense that it is not equivalent to Infomax or derived from it, but certainly functions in a similar spirit.

To be specific, consider Fig. 10.6 that shows two neural networks (modules) a and b receiving inputs $\mathbf{X}_a$ and $\mathbf{X}_b$ from adjacent, nonoverlapping regions of an image. The scalars Y_a and Y_b denote the outputs of these two modules due to the respective input vectors $\mathbf{X}_a$ and $\mathbf{X}_b$. Let S denote a signal component common to both Y_a and Y_b, which

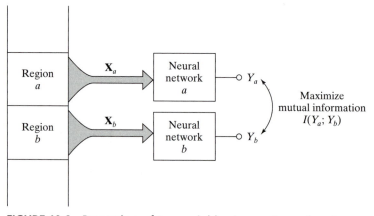

FIGURE 10.6 Processing of two neighboring regions of an image in accordance with the first variant of Infomax.

is representative of the spatial coherence across the two pertinent regions of the original image. We may express Y_a and Y_b as noisy versions of the common signal S, as shown by

$$Y_a = S + N_a \tag{10.59}$$

and

$$Y_b = S + N_b \tag{10.60}$$

The N_a and N_b are additive noise components, assumed to be statistically independent, zero-mean Gaussian-distributed random variables. The signal component S is also assumed to be Gaussian with a distribution of its own. According to Eqs. (10.59) and (10.60), the two modules a and b in Fig. 10.6 make consistent assumptions about each other.

Using the last line of Eq. (10.30), the mutual information between Y_a and Y_b is defined by

$$I(Y_a; Y_b) = h(Y_a) + h(Y_b) - h(Y_a, Y_b) \tag{10.61}$$

According to the formula of Eq. (10.22) for the differential entropy of a Gaussian random variable, the differential entropy $h(Y_a)$ of Y_a is given by

$$h(Y_a) = \frac{1}{2}\left[1 + \log\left(2\pi\sigma_a^2\right)\right] \tag{10.62}$$

where σ_a^2 is the variance of Y_a. Similarly, the differential entropy of Y_b is given by

$$h(Y_b) = \frac{1}{2}\left[1 + \log\left(2\pi\sigma_b^2\right)\right] \tag{10.63}$$

where σ_b^2 is the variance of Y_b. As for the joint differential entropy $h(Y_a, Y_b)$, we use the formula of Eq. (10.24) to write

$$h(Y_a, Y_b) = 1 + \log(2\pi) + \frac{1}{2}\log|\det(\Sigma)| \tag{10.64}$$

The 2-by-2 matrix Σ is the covariance matrix of Y_a and Y_b; it is defined by

$$\Sigma = \begin{bmatrix} \sigma_a^2 & \rho_{ab}\sigma_a\sigma_b \\ \rho_{ab}\sigma_a\sigma_b & \sigma_b^2 \end{bmatrix} \tag{10.65}$$

where ρ_{ab} is the *correlation coefficient* of Y_a and Y_b; that is,

$$\rho_{ab} = \frac{E[(Y_a - E[Y_a])(Y_b - E[Y_b])]}{\sigma_a\sigma_b} \tag{10.66}$$

Hence the determinant of Σ is

$$\det(\Sigma) = \sigma_a^2\sigma_b^2(1 - \rho_{ab}^2) \tag{10.67}$$

and so we may rewrite Eq. (10.64) as

$$h(Y_a, Y_b) = 1 + \log(2\pi) + \frac{1}{2}\log[\sigma_a^2\sigma_b^2(1 - \rho_{ab}^2)] \tag{10.68}$$

By substituting Eqs. (10.62), (10.63), and (10.68) in (10.61), and then simplifying terms, we get

$$I(Y_a; Y_b) = -\frac{1}{2} \log(1 - \rho_{ab}^2) \tag{10.69}$$

From Eq. (10.69) we immediately deduce that maximizing the mutual information $I(Y_a; Y_b)$ is equivalent to maximizing the correlation coefficient ρ_{ab}, which is intuitively satisfying. Note that, by definition, $|\rho_{ab}| \le 1$.

Maximizing the mutual information $I(Y_a; Y_b)$ may be viewed as the nonlinear generalization of canonical correlation in statistics (Becker and Hinton, 1992). Given two input vectors (stimuli) $\mathbf{X}_a$ and $\mathbf{X}_b$ (not necessarily of the same dimensionality), and two corresponding weight vectors, $\mathbf{w}_a$ and $\mathbf{w}_b$, the objective of *canonical correlation analysis* is to find linear combinations $Y_a = \mathbf{w}_a^T \mathbf{X}_a$ and $Y_b = \mathbf{w}_b^T \mathbf{X}_b$ that have maximum correlation between them (Anderson, 1984). Maximizing $I(Y_a; Y_b)$ is a nonlinear generalization of canonical correlation by virtue of the nonlinearity built into the design of the neural modules in Fig. 10.6.

In Becker and Hinton (1992), it is demonstrated that by maximizing the mutual information $I(Y_a; Y_b)$ it is possible to extract stereo disparity (depth) from random dot stereograms. This is a difficult feature extraction problem that could not be solved by a one-layer or linear neural network.

10.10   SPATIALLY INCOHERENT FEATURES

The unsupervised processing of an image considered in the previous section deals with the extraction of spatially coherent features from an image. We now consider the opposite of the situation described therein. To be specific, consider Fig. 10.2c, where the objective is to enhance the *spatial differences* between a pair of corresponding regions derived from two separate images. Whereas the mutual information between the outputs of the modules is maximized in Fig. 10.2b, we do the exact opposite in Fig. 10.2c.

We may thus state the *second* variant of the Infomax principle as follows (Ukrainec and Haykin, 1992, 1996):

> The transformation of a pair of input vectors, $\mathbf{X}_a$ and $\mathbf{X}_b$, representing data derived from corresponding regions in a pair of separate images, by a neural system should be so chosen that the scalar output Y_a of the system due to the input $\mathbf{X}_a$ minimizes information about the second scalar output Y_b due to $\mathbf{X}_b$. The objective function to be minimized is the mutual information $I(Y_a; Y_b)$ between the outputs Y_a and Y_b.

Here again we refer to this principle as a variant of the Infomax principle in that it is not equivalent to Infomax or derived from it, but certainly functions in the spirit of it.[11]

The second variant of the Infomax principle finds application in *radar polarimetry,* for example, where a surveillance radar system produces a pair (or more) of images of an environment of interest by transmitting on one polarization and receiving the backscatter from the environment on the same or different polarization. The polarization can be vertical or horizontal. For example, we may have a pair of radar images, one image representing like polarization (horizontal–horizontal, say), and the other image representing cross-polarization (horizontal on transmit and vertical on receive). Such an application is described in Ukrainec and Haykin (1992, 1996), which pertains to the *enhancement of a polarization target* in a dual-polarized radar system. The sample radar scene used in the study is described as follows. An incoherent radar transmits in a horizontally polarized fashion, and receives radar returns on both horizontal and vertical polarization channels. The target of interest is a *cooperative, polarization-twisting reflector* designed to rotate the incident polarization through 90 degrees. In the normal operation of a radar system, the detection of such a target is made difficult by imperfections in the system as well as reflections from unwanted polarimetric targets on the ground (i.e., radar clutter). We perceive that a nonlinear mapping is needed to account for the non-Gaussian distribution common to radar returns. The target enhancement problem is cast as a variational problem involving the minimization of a quadratic cost functional with constraints. The net result is a processed cross-polarized image that exhibits a significant improvement in target visibility, far more pronounced than that attainable through the use of a linear technique such as principal components analysis. The model used by Ukrainec and Haykin assumes Gaussian statistics for the transformed data, since a model-free estimate of the probability density function is a computationally challenging task. The mutual information between two Gaussian variables Y_a and Y_b is defined by Eq. (10.61). To learn the synaptic weights of the two modules, a variational approach is taken. The requirement is to suppress the radar clutter that is common to the horizontally polarized and vertically polarized radar images. To satisfy this requirement, the mutual information $I(Y_a; Y_b)$ is minimized, subject to a constraint imposed on the synaptic weights as shown by

$$P = (\text{tr}[\mathbf{W}^T\mathbf{W}] - 1)^2 \tag{10.70}$$

where $\mathbf{W}$ is the overall weight matrix of the network, and $\text{tr}[\cdot]$ is the trace of the enclosed matrix. A stationary point is reached when we have

$$\nabla_{\mathbf{W}}I(Y_a; Y_b) + \lambda\nabla_{\mathbf{W}}P = 0 \tag{10.71}$$

where λ is the Lagrange multiplier. A quasi-Newton optimization routine was used to find the minimum; quasi-Newton's methods are discussed in Chapter 4.

Figure 10.7 shows the architecture of the neural network used in Ukrainec and Haykin (1992, 1996). A Gaussian radial-basis function (RBF) network was chosen for each of the two modules because it has the advantage of providing a set of fixed basis-functions (i.e., a nonadaptive hidden layer). The input data are expanded onto the basis functions and then combined using layers of *linear* weights; the dashed lines shown in Fig. 10.7 represent the cross-coupling connections between the two modules. The centers of the Gaussian functions were chosen at evenly spaced intervals to cover the entire input domain, and their widths were chosen using a heuristic. Figure 10.8a shows the raw horizontally polarized and vertically polarized (both on receive) radar images of a parklike setting on the shore of Lake Ontario. The range coordinate is along the horizontal axis of each image, increasing from left to right; the azimuth coordinate is along the vertical axis, increasing down the image. Figure 10.8b shows the combined image obtained by minimizing the mutual information between the horizontally and vertically polarized radar images, as just described. The bright spot clearly visible in this image corresponds to the radar return from a cooperative, polarization-twisting reflector placed along the lake shore. The clutter-suppression performance of

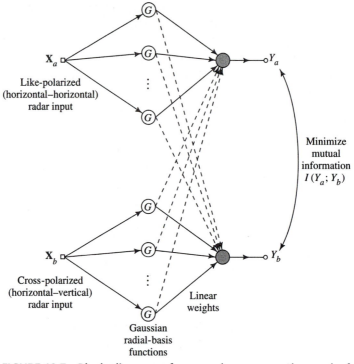

FIGURE 10.7 Block diagram of a neural processor, the goal of which is to suppress background clutter using a pair of polarimetric, noncoherent radar inputs; clutter suppression is attained by minimizing the mutual information between the outputs of the two modules.

the information-theoretic model described here exceeds that of commonly used projections using principal components analysis (Ukrainec and Haykin, 1992, 1996).[12]

10.11 INDEPENDENT COMPONENTS ANALYSIS

We now turn our attention to the last scenario described in Fig. 10.2d. To add more specificity to the signal processing problem stated therein, consider the block diagram in Fig. 10.9. The operation starts with a random source vector $\mathbf{U}(n)$ defined by

$$\mathbf{U} = [U_1, U_2, \ldots, U_m]^T$$

where the m components are supplied by a set of *independent sources.* Temporal sequences are considered here; henceforth the argument n denotes discrete time. The vector $\mathbf{U}$ is applied to a linear system whose input–output characterization is defined by a nonsingular m-by-m matrix $\mathbf{A}$, called the *mixing matrix.* The result is an m-by-1 observation vector $\mathbf{X}(n)$ related to $\mathbf{U}(n)$ as follows (see Fig. 10.10a)

$$\mathbf{X} = \mathbf{AU} \qquad (10.72)$$

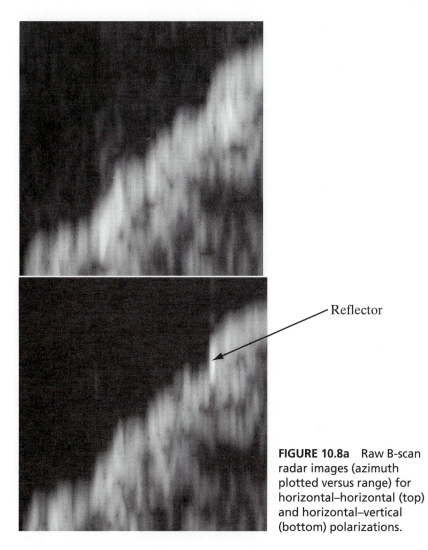

Reflector

FIGURE 10.8a Raw B-scan radar images (azimuth plotted versus range) for horizontal–horizontal (top) and horizontal–vertical (bottom) polarizations.

where

$$\mathbf{X} = [X_1, X_2, ..., X_m]^T$$

The source vector $\mathbf{U}$ and the mixing matrix $\mathbf{A}$ are both unknown: The only thing available to us is the observation vector $\mathbf{X}$. Given $\mathbf{X}$, the problem is to find a *demixing matrix* $\mathbf{W}$ such that the original source vector $\mathbf{U}$ can be recovered from the output vector $\mathbf{Y}$ defined by (see Fig. 10.10b)

$$\mathbf{Y} = \mathbf{WX} \tag{10.73}$$

where

$$\mathbf{Y} = [Y_1, Y_2, ..., Y_m]^T$$

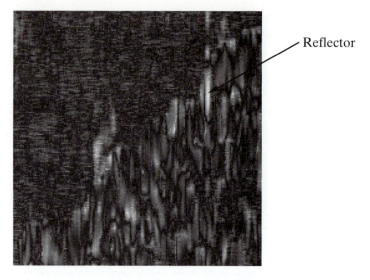

FIGURE 10.8b Composite image computed by minimizing the mutual information between the two polarized radar images of Fig. 10.8a.

Reflector

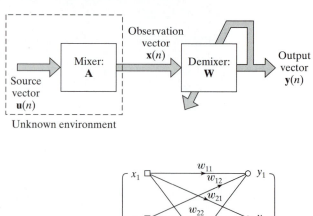

FIGURE 10.9 Block diagram of processor for the blind source separation problem. The vectors **u**, **x**, and **y** are values of the respective random vectors **U**, **X**, and **Y**.

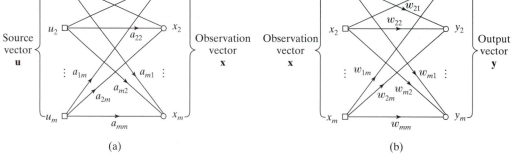

(a)   (b)

FIGURE 10.10 Detailed description of (a) mixing matrix and (b) demixing matrix.

It is ordinarily assumed that the source signals $U_1, U_2, ..., U_m$ are zero-mean signals, which in turn means that the observables $X_1, X_2, ..., X_m$ are also zero-mean signals. The same is true for the demixer outputs $Y_1, Y_2, ..., Y_m$.

We may thus state the *blind source separation problem* as follows:

*Given N independent realizations of the observation vector **X**, find an estimate of the inverse of the mixing matrix **A**.*

Source separation exploits primarily *spatial diversity* in that the different sensors providing realizations of the vector **X** carry different mixtures of the sources. Spectral diversity, if it exists, can also be exploited, but the fundamental approach to source separation is essentially *spatial*: looking for structure across the sensors, not across time (Cardoso, 1998a).

The solution to the blind source separation problem is feasible, except for an arbitrary scaling of each signal component and permutation of indices. In other words, it is possible to find a demixing matrix **W** whose individual rows are a rescaling and permutation of those of the mixing matrix **A**. That is, the solution may be expressed in the form

$$\mathbf{Y} = \mathbf{WX} = \mathbf{WAU} \rightarrow \mathbf{DPU}$$

where **D** is a nonsingular diagonal matrix and **P** is a permutation matrix.

The problem described herein is commonly referred to as the blind (signal) source separation problem,[13] where the term "blind" is used to signify the fact that the only information used to recover the original signal sources is contained in a realization of the observation vector **X**, denoted by **x**. The underlying principle involved in its solution is called *independent components analysis* (ICA) (Comon, 1994), which may be viewed as an extension of principal components analysis (PCA). Whereas PCA can only impose independence up to the second order while constraining the direction vectors to be orthogonal, ICA imposes *statistical independence* on the individual components of the output vector **Y** and has no orthogonality constraint. Note also that in practice, an algorithmic implementation of independent components analysis can only go for "as statistically independent as possible."

The need for blind source separation arises in many diverse applications that include the following:

- *Speech separation.* In this application, the vector **x** consists of several speech signals that have been linearly mixed together, and the requirement is to separate them (Bell and Sejnowski, 1995). A difficult form of this situation, for example, arises in a teleconferencing environment.

- *Array antenna processing.* In this second application, the vector **x** represents the output of a radar array antenna produced by several incident narrowband signals originating from sources of unknown directions (Cardoso and Souloumia, 1993; Swindlehurst et al., 1997). Here again the requirement is to separate the source signals. (By a narrowband signal we mean a band-pass signal whose bandwidth is small compared to the carrier frequency.)

- *Multisensor biomedical records.* In this third application, the vector **x** consists of recordings made by a multitude of sensors used to monitor biological signals of interest. For example, the requirement may be that of separating the heartbeat of a fetus from that of the mother (Cardoso, 1998b).

- *Financial market data analysis.* In this application the vector **x** consists of a set of different stock market data and the requirement is to extract the underlying set of dominant independent components (Back and Weigend, 1998).

In these applications, the blind source separation problem may be compounded by the possible presence of unknown propagation delays, extensive filtering imposed on the sources by their environments, and unavoidable contamination of the observation vector **x** by noise. These impairments mean that (unfortunately) the idealized form of instantaneous mixing of signals described in Eq. (10.72) is very rarely encountered in real-world situations. In what follows, however, we will ignore these impairments in order to develop insight into the fundamental aspects of the blind source separation problem.

Criterion for Statistical Independence

With statistical independence as the property desired from the components of the output vector **Y** for blind source separation, what practical measure can we use for it? One obvious possibility is to choose the mutual information $I(Y_i; Y_j)$ between the random variables Y_i and Y_j constituting any two components of the output vector **Y**. When, in the ideal case, $I(Y_i; Y_j)$ is zero, the components Y_i and Y_j are statistically independent. This would therefore suggest minimizing the mutual information between every pair of the random variables constituting the output vector **Y**. This objective is equivalent to minimizing the Kullback–Leibler divergence between the following two distributions: (1) the probability density function $f_\mathbf{Y}(\mathbf{y}, \mathbf{W})$ parameterized by **W**, and (2) the corresponding factorial distribution defined by

$$\tilde{f}_\mathbf{Y}(\mathbf{y}, \mathbf{W}) = \prod_{i=1}^{m} \tilde{f}_{Y_i}(y_i, \mathbf{W}) \tag{10.74}$$

where $\tilde{f}_{Y_i}(y_i, \mathbf{W})$ is the marginal probability density function of Y_i. In effect, Eq. (10.74) may be viewed as a *constraint* imposed on the learning algorithm, forcing it to contrast $f_\mathbf{Y}(\mathbf{y}, \mathbf{W})$ against the factorial distribution $\tilde{f}_\mathbf{Y}(\mathbf{y}, \mathbf{W})$. We may thus state the third variant to the Infomax principle for independent components analysis as (Comon, 1994):

> Given an m-by-1 vector **X** representing a linear combination of m independent source signals, the transformation of the observation vector **X** by a neural system into a new vector **Y** should be carried out in such a way that the Kullback–Leibler divergence between the parameterized probability denoting function $f_\mathbf{Y}(\mathbf{y}, \mathbf{W})$ and the corresponding factorial distribution $\tilde{f}_\mathbf{Y}(\mathbf{y}, \mathbf{W})$ is minimized with respect to the unknown parameter matrix **W**.

The Kullback–Leibler divergence for the problem described herein is considered in Section 10.5. The formula we are seeking is given in Eq. (10.44). Adapting that formula to our present situation, we may express the Kullback–Leibler divergence between the probability density functions $f_\mathbf{Y}(\mathbf{y}, \mathbf{W})$ and $\tilde{f}_\mathbf{Y}(\mathbf{y}, \mathbf{W})$ as follows:

$$D_{f\|\tilde{f}}(\mathbf{W}) = -h(\mathbf{Y}) + \sum_{i=1}^{m} \tilde{h}(Y_i) \tag{10.75}$$

where $h(\mathbf{Y})$ is the entropy of random vector **Y** at the output of the demixer and $\tilde{h}(Y_i)$ is the marginal entropy of the ith element of **Y**. The Kullback–Leibler divergence $D_{f\|\tilde{f}}$ is the objective (contrast) function that we focus on henceforth for solving the blind source separation problem.

Determination of the Differential Entropy *h*(Y)

The output vector **Y** is related to the input vector **X** by Eq. (10.73), where **W** is the demixing matrix. In light of Eq. (10.18) we may express the differential entropy of **Y** as:

$$h(\mathbf{Y}) = h(\mathbf{WX})$$
$$= h(\mathbf{X}) + \log|\det(\mathbf{W})| \qquad (10.76)$$

where $\det(\mathbf{W})$ is the determinant of $\mathbf{W}$.

Determination of the Marginal Entropy $\tilde{h}(Y_i)$

To determine the Kullback–Leibler divergence $D_{f\|\tilde{f}}$, we also need to know the marginal entropy $\tilde{h}(Y_i)$. To determine $\tilde{h}(Y_i)$ we require knowledge of the marginal distribution of Y_i, which in turn requires integrating out the effects of all the components of the random vector $\mathbf{Y}$ except for the ith component. For a vector $\mathbf{Y}$ of high dimensionality it is usually more difficult to calculate $\tilde{h}(Y_i)$ than $h(\mathbf{Y})$. We may overcome this difficulty by deriving an approximate formula for $\tilde{h}(Y_i)$ in terms of the higher-order moments of the random variable Y_i. This is accomplished by properly truncating one of two expansions:

- Edgeworth series (Comon, 1991)
- Gram–Charlier series (Amari et al., 1996)

In this chapter we follow the latter approach. An exposition of the Gram–Charlier series is presented in note 14. A brief description of the Edgeworth series is also presented in that note.

To be specific, the Gram–Charlier expansion of the parameterized marginal probability density function $\tilde{f}_{Y_i}(y_i, \mathbf{W})$ is described by

$$\tilde{f}_{Y_i}(y_i, \mathbf{W}) = \alpha(y_i)\left[1 + \sum_{k=3}^{\infty} c_k H_k(y_i)\right] \qquad (10.77)$$

where the various terms are defined as follows:

1. The multiplying factor $\alpha(y_i)$ is the probability density function of a normalized Gaussian random variable with zero mean and unit variance; that is,

$$\alpha(y_i) = \frac{1}{\sqrt{2\pi}} e^{-y_i^2/2}$$

2. The $H_k(y_i)$ are *Hermite polynomials*.
3. The coefficients of the expansion, $\{c_k : k = 3, 4, \ldots\}$, are defined in terms of the cumulants of the random variable Y_i.

The natural order of the terms in Eq. (10.77) is *not* the best for the Gram–Charlier series. Rather, the terms listed here in the parentheses should be grouped together (Helstrom, 1968):

$$k = (0), (3), (4, 6), (5, 7, 9), \ldots$$

For the blind-source separation problem, the approximation of the marginal probability density function $\tilde{f}_{Y_i}(y_i)$ by truncating the Gram–Charlier series at $k = (4, 6)$ is considered to be adequate. We may thus write

$$\tilde{f}_{Y_i}(y_i) \simeq \alpha(y_i)\left(1 + \frac{\kappa_{i,3}}{3!} H_3(y_i) + \frac{\kappa_{i,2}^2}{4!} H_4(y_i) + \frac{(\kappa_{i,6} + 10\kappa_{i,3}^2)}{6!} H_6(y_i)\right) \qquad (10.78)$$

where $\kappa_{i,k}$ is the kth order *cumulant* of Y_i. Let $m_{i,k}$ denote the kth-order *moment* of Y_i defined by

$$m_{i,k} = E[Y_i^k]$$

$$= E\left[\left(\sum_{k=1}^{m} w_{ik} X_i\right)^k\right] \tag{10.79}$$

where X_i is the ith element of observation vector $\mathbf{X}$ and w_{ik} is the ik-th element of weight matrix $\mathbf{W}$. Earlier we justified the zero-mean assumption of Y_i for all i. Accordingly, we have $\sigma_i^2 = m_{i,2}$ (i.e., the variance and the mean-square value are equal) and so relate the cumulants of Y_i to its moments as follows:

$$\kappa_{i,3} = m_{i,3} \tag{10.80}$$

$$\kappa_{i,4} = m_{i,4} - 3m_{i,2}^2 \tag{10.81}$$

$$\kappa_{i,6} = m_{i,6} - 10m_{i,3}^2 - 15m_{i,2}m_{i,4} + 30m_{i,2}^3 \tag{10.82}$$

The algorithm of $\tilde{f}_{Y_i}(y_i)$, using the approximation of Eq. (10.78), is given by

$$\log \tilde{f}_{Y_i}(y_i) \simeq \log \alpha(y_i) + \log\left(1 + \frac{\kappa_{i,3}}{3!} H_3(y_i) + \frac{\kappa_{i,2}^2}{4!} H_4(y_i) + \frac{(\kappa_{i,6} + 10\kappa_{i,3}^2)}{6!} H_6(y_i)\right) \tag{10.83}$$

To proceed further, we use the expansion of a logarithm:

$$\log(1 + y) \simeq y - \frac{y^2}{2} \tag{10.84}$$

where all the terms of order three and higher are ignored.

From our previous discussion we recall that the formula for the marginal entropy of Y_i is (see Eq. (10.43))

$$\tilde{h}(Y_i) = -\int_{-\infty}^{\infty} f_{Y_i}(y_i) \log f_{Y_i}(y_i) dy_i, \quad i = 1, 2, \dots, m$$

where m is the number of sources. By making use of the approximations described in Eqs. (10.78), (10.83) and (10.84), and invoking certain integrals that involve the normalized Gaussian density $\alpha(y_i)$ and various Hermite polynomials $H_k(y_i)$, we obtain the following approximate formula for the marginal entropy (Madhuaranth and Haykin, 1998):

$$\tilde{h}(Y_i) \simeq \frac{1}{2}\log(2\pi e) - \frac{\kappa_{i,3}^2}{12} - \frac{\kappa_{i,4}^2}{48} - \frac{(\kappa_{i,6} + 10\kappa_{i,3}^2)^2}{1440}$$

$$+ \frac{3}{8}\kappa_{i,3}^2 \kappa_{i,4} + \frac{\kappa_{i,3}^2(\kappa_{i,6} + 10\kappa_{i,3}^2)}{24} + \frac{\kappa_{i,4}^2(\kappa_{i,6} + 10\kappa_{i,3}^2)}{24} \tag{10.85}$$

$$+ \frac{\kappa_{i,4}(\kappa_{i,6} + 10\kappa_{i,3}^2)^2}{64} + \frac{\kappa_{i,4}^3}{16} + \frac{(\kappa_{i,6} + 10\kappa_{i,3}^2)^3}{432}$$

By substituting Eqs. (10.76) and (10.85) in (10.75), we get the Kullback–Leibler divergence for the problem at hand:

$$
D_{f\|\tilde{f}}(\mathbf{W}) \simeq -h(\mathbf{X}) - \log|\det(\mathbf{W})| + \frac{m}{2}\log(2\pi e)
$$

(10.86)

$$
- \sum_{i=1}^{m}\left(\frac{\kappa_{i,3}^2}{12} + \frac{\kappa_{i,4}^2}{48} + \frac{(\kappa_{i,6} + 10\kappa_{i,3}^2)^2}{1440} - \frac{3}{8}\kappa_{i,3}^2\kappa_{i,4}\right.
$$

$$
- \frac{\kappa_{i,3}^2(\kappa_{i,6} + 10\kappa_{i,3}^2)}{24} - \frac{\kappa_{i,4}^2(\kappa_{i,6} + 10\kappa_{i,3}^2)}{24}
$$

$$
\left. - \frac{\kappa_{i,4}(\kappa_{i,6} + 10\kappa_{i,3}^2)^2}{64} - \frac{\kappa_{i,4}^3}{16} - \frac{(\kappa_{i,6} + 10\kappa_{i,3}^2)^3}{432}\right)
$$

where the cumulants are all functions of the weight matrix $\mathbf{W}$.

Activation Function

To evaluate the Kullback–Leibler divergence described in Eq. (10.86), we need an adaptive procedure for the computation of the higher-order cumulants for the observation vector $\mathbf{x}$. The question is: How do we proceed with this computation, bearing in mind the way in which the approximate formula of Eq. (10.86) is derived? Recall that the derivation is based on the Gram–Charlier expansion, assuming that the random variable Y_i has zero mean and unit variance. We previously justified the zero-mean assumption on the grounds that, to begin with, the source signals typically have zero mean. As for the unit-variance assumption, we may deal with it by taking one of two approaches:

   1. *Constrained approach.* In this approach, the unit-variance assumption is *imposed* on the computation of the higher-order cumulants $\kappa_{i,3}$, $\kappa_{i,4}$, and $\kappa_{i,6}$ for all i (Amari et al., 1996). Unfortunately, there is no guarantee that the variance of Y_i, namely σ_i^2, remains constant, let alone equal to 1, during the computation. From the defining equations (10.81) and (10.82), we note that both $\kappa_{i,4}$ and $\kappa_{i,6}$ depend on $\sigma_i^2 = m_{i,2}$. The result of assuming $\sigma_i^2 = 1$ is that the estimates derived for $\kappa_{i,4}$ and $\kappa_{i,6}$ are highly *biased* and therefore erroneous relative to the estimate of $\kappa_{i,3}$.

   2. *Unconstrained approach.* In this alternative approach, the variance σ_i^2 is treated as an *unknown time-varying parameter,* which is how it actually is in practice (Madhuranath and Haykin, 1998). The effect of deviation in the value of σ_i^2 from 1 is viewed as a scaling variation in the value of random variable Y_i. Most importantly, the estimates derived for $\kappa_{i,4}$ and $\kappa_{i,6}$ account for the variation of σ_i^2 with time. A proper relationship between the estimates of all three higher-order cumulants in Eq. (10.86) is thereby maintained.

An experimental study of blind source separation reported in Madhuranath and Haykin (1998) shows that the unconstrained approach yields a superior performance compared to the constrained approach. In what follows, we follow the unconstrained approach.

To develop a learning algorithm for computing $\mathbf{W}$, we need to differentiate Eq. (10.86) with respect to $\mathbf{W}$ and thereby formulate an activation function for the algorithm.

Let A_{ik} denote the ik-th *cofactor* of matrix $\mathbf{W}$. Using *Laplace's expansion* of $\det(\mathbf{W})$ by the ith row, we may then write (Wylie and Barrett, 1982)

$$\det(\mathbf{W}) = \sum_{k=1}^{m} w_{ik} A_{ik}, \qquad i = 1, 2, \ldots, m \tag{10.87}$$

where w_{ik} is the ik-th element of matrix $\mathbf{W}$. Hence, differentiating the logarithm of $\det(\mathbf{W})$ with respect to w_{ik}, we get

$$\frac{\partial}{\partial w_{ik}} \log(\det(\mathbf{W})) = \frac{1}{\det(\mathbf{W})} \frac{\partial}{\partial w_{ik}} \det(\mathbf{W})$$

$$= \frac{A_{ik}}{\det(\mathbf{W})} \tag{10.88}$$

$$= (\mathbf{W}^{-T})_{ik}$$

where $\mathbf{W}^{-T}$ is the inverse of the transposed matrix $\mathbf{W}^{T}$. The partial derivatives of the other terms (that depend on $\mathbf{W}$) in Eq. (10.86) with respect to w_{ik} are (see Eqs. (10.80) through (10.82))

$$\frac{\partial \kappa_{i,3}}{\partial w_{ik}} = 3 \, E[Y_i^2 X_k]$$

$$\frac{\partial \kappa_{i,4}}{\partial w_{ik}} = 4 \, E[Y_i^3 X_k] - 12 m_{i,2} E[Y_i X_k]$$

$$\frac{\partial}{\partial w_{ik}}(\kappa_{i,6} + 10\kappa_{i,3}^2) = 6 \, E[Y_i^5 X_k] - 30 \, m_{i,4} \, E[Y_i X_k]$$

$$- 60 m_{i,2} E[Y_i^3 X_k] + 180 \, m_{i,2}^2 \, E[Y_i X_k]$$

In deriving an adaptive algorithm the usual approach is to replace expectations with their instantaneous values. Hence, by doing this replacement in these three equations, we get the following approximate results:

$$\frac{\partial \kappa_{i,3}}{\partial w_{ik}} \simeq 3 \, y_i^2 x_k \tag{10.89}$$

$$\frac{\partial \kappa_{i,4}}{\partial w_{ik}} \simeq -8 \, y_i^3 \, x_k \tag{10.90}$$

$$\frac{\partial}{\partial w_{ik}}(\kappa_{i,6} + 10\kappa_{i,3}^2) \simeq 96 y_i^5 \, x_k \tag{10.91}$$

Substituting Eqs. (10.88) through (10.91) in the expression for the derivative of Eq. (10.86) with respect to w_{ik} yields

$$\frac{\partial}{\partial w_{ik}} D_{f\|\tilde{f}}(\mathbf{W}) \simeq -(\mathbf{W}^{-T})_{ik} + \varphi(y_i)x_k \tag{10.92}$$

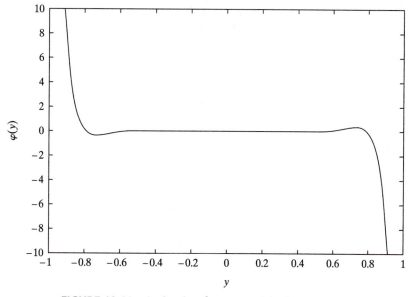

FIGURE 10.11   Activation function $\varphi(y)$ of Eq. (10.93).

where $\varphi(y_i)$ is the nonmonotonic *activation function* of the learning algorithm, defined by (Madhuranath and Haykin, 1998)

$$\varphi(y_i) = \frac{1}{2} y_i^5 + \frac{2}{3} y_i^7 + \frac{15}{2} y_i^9 + \frac{2}{15} y_i^{11} - \frac{112}{3} y_i^{13} + 128 y_i^{15} - \frac{512}{3} y_i^{17} \quad (10.93)$$

Figure 10.11 plots the activation function $\varphi(y_i)$ versus y_i for $-1 < y_i < 1$. This covers the range of values of the demixer output y_i to which the operation of the learning algorithm is ordinarily confined. It is noteworthy that the slope of the activation function is positive in the interval $(-0.734, 0.734)$; this is a requirement for stability of the algorithm as discussed later in the section.

Learning Algorithm for ICA

The objective of the learning algorithm is to minimize the Kullback–Leibler divergence between the probability density function of $\mathbf{Y}$ and the factorial distribution of Y_i for $i = 1, 2, ..., m$. This minimization may be implemented using the method of gradient descent whereby the adjustment applied to weight w_{ik} is defined by

$$\begin{aligned}
\Delta w_{ik} &= -\eta \frac{\partial}{\partial w_{ik}} D_{f \| \tilde{f}} \\
&= \eta \left((\mathbf{W}^{-T})_{ik} - \varphi(y_i) x_k \right)
\end{aligned} \quad (10.94)$$

where η is a learning-rate parameter.

Extending the formula of Eq. (10.94) to the entire weight matrix $\mathbf{W}$ of the demixer, we may express the adjustment $\Delta \mathbf{W}$ applied to $\mathbf{W}$ as follows:

$$\Delta \mathbf{W} = \eta (\mathbf{W}^{-T} - \varphi(\mathbf{y}) \mathbf{x}^T) \quad (10.95)$$

where $\mathbf{x}^T$ is the transpose of the m-by-1 observation vector $\mathbf{x}$, and

$$\boldsymbol{\varphi}(\mathbf{y}) = [\varphi(y_1, \varphi(y_2), ..., \varphi(y_m)]^T \qquad (10.96)$$

The formula for $\Delta \mathbf{W}$ given in Eq. (10.95) can be rewritten by noting that

$$\mathbf{y}^T = \mathbf{x}^T \mathbf{W}^T$$

and thereby reformulating it in the equivalent form

$$\begin{aligned}\Delta \mathbf{W} &= \eta[\mathbf{I} - \boldsymbol{\varphi}(y)\mathbf{x}^T \mathbf{W}^T]\mathbf{W}^{-T} \\ &= \eta[\mathbf{I} - \boldsymbol{\varphi}(\mathbf{y})\mathbf{y}^T]\mathbf{W}^{-T}\end{aligned} \qquad (10.97)$$

where $\mathbf{I}$ is the identity matrix. The update rule for adapting the demixing matrix takes the form

$$\mathbf{W}(n+1) = \mathbf{W}(n) + \eta(n)[\mathbf{I} - \boldsymbol{\varphi}(\mathbf{y}(n))\mathbf{y}^T(n)]\mathbf{W}^{-T}(n) \qquad (10.98)$$

where the parameters are all shown in their time varying forms.

Equivariant Property

The purpose of a blind source separation algorithm is to update the demixing matrix $\mathbf{W}(n)$ such that the output vector

$$\mathbf{y}(n) = \mathbf{W}(n)\mathbf{x}(n) = \mathbf{W}(n)\mathbf{A}\mathbf{u}(n)$$

is as close as possible to the original source vector $\mathbf{u}(n)$ in some statistical sense. To be more specific, consider a global system characterized by a matrix $\mathbf{C}(n)$ that is obtained by multiplying the mixing matrix $\mathbf{A}$ and demixing matrix $\mathbf{W}(n)$ as shown by

$$\mathbf{C}(n) = \mathbf{W}(n)\mathbf{A} \qquad (10.99)$$

Ideally, this global system would satisfy two conditions:

1. The algorithm responsible for adjusting $\mathbf{C}(n)$ converges to an optimum value equal to the permutation matrix.
2. The algorithm is itself described by

$$\mathbf{C}(n+1) = \mathbf{C}(n) + \eta(n)\mathbf{G}(\mathbf{C}(n)\mathbf{u}(n))\mathbf{C}(n) \qquad (10.100)$$

where $\mathbf{G}(\mathbf{C}(n)\mathbf{u}(n))$ is a vector-valued function of $\mathbf{C}(n)\mathbf{u}(n)$. The performance of the algorithm is completely characterized by the system matrix $\mathbf{C}(n)$ and *not* by the individual values of the mixing matrix $\mathbf{A}$ and demixing matrix $\mathbf{W}(n)$. Such an adaptive system is said to be *equivariant* (Cardoso and Laheld, 1996).

The adaptive algorithm of Eq. (10.98) is certainly capable of approximately satisfying the first condition. However, as it stands, it cannot satisfy the second condition. To appreciate this point, we may rewrite Eq. (10.98) in the equivalent form

$$\mathbf{C}(n+1) = \mathbf{C}(n) + \eta(n)\mathbf{G}(\mathbf{C}(n)\mathbf{u}(n))\mathbf{W}^{-T}(n)\mathbf{A} \qquad (10.101)$$

where

$$\mathbf{G}(\mathbf{C}(n)\mathbf{u}(n)) = \mathbf{I} - \boldsymbol{\varphi}(\mathbf{C}(n)\mathbf{u}(n))(\mathbf{C}(n)\mathbf{u}(n))^T \qquad (10.102)$$

The algorithm of Eq. (10.98) falls short of the equivariant condition described in Eq. (10.100) in that the vector-valued function $\mathbf{G}(\mathbf{C}(n)\mathbf{u}(n))$ is postmultiplied by

$\mathbf{W}^{-T}(n)\mathbf{A}$ which, in general, is different from $\mathbf{C}(n)$. We may rectify this situation by interposing the matrix product $\mathbf{W}^T(n)\mathbf{W}(n)$ between them. The term $\mathbf{W}^T\mathbf{W}$, being made up of the product of matrix $\mathbf{W}$ and its transpose, is always positive definite. This is the reason why multiplication by $\mathbf{W}^T\mathbf{W}$ does not change the sign of the minima of the learning algorithm.

The important question is: What is the implication of this modification that is made in order to achieve the equivariant condition? The answer lies in how the gradient descent in parameter space is formulated. Ideally, we should use the *natural gradient*[15] of the objective function $D_{f\|g}(\mathbf{W})$, defined in terms of the usual gradient $\nabla D_{f\|\tilde{f}}$ as:

$$\nabla^* D_{f\|\tilde{f}}(\mathbf{W}) = (\nabla D_{f\|\tilde{f}}(\mathbf{W}))\mathbf{W}^T\mathbf{W} \tag{10.103}$$

The usual gradient matrix $\nabla D_{f\|\tilde{f}}$ is itself defined by Eq. (10.92). In an implicit sense, the gradient $\nabla D_{f\|\tilde{f}}(\mathbf{W})$ is the optimum direction for descent only when the parameter space $\mathcal{W} = \{\mathbf{W}\}$ is Euclidean with an orthonormal coordinate system. In a typical situation involving neural networks, however, the parameter space $\mathcal{W}$ has a coordinate system that is nonorthonormal. The natural gradient $\nabla^* D_{f\|\tilde{f}}(\mathbf{W})$ will provide the *steepest descent* in this latter situation, hence the preference for using it instead of the usual gradient in formulating the stochastic algorithm for blind source separation. For the natural gradient space to be definable, two conditions must be satisfied:

1. The parameter space $\mathcal{W}$ is *Riemannian*.[16] The Riemannian structure is a differentiable manifold with a positive definite metric $\mathbf{W}$.
2. The matrix $\mathbf{W}$ is nonsingular (i.e., invertible).

Both of these conditions are satisfied for the problem at hand.

By modifying the algorithm of Eq. (10.98) in this manner we may write

$$\begin{aligned} \mathbf{W}(n+1) &= \mathbf{W}(n) + \eta(n)[\mathbf{I} - \varphi(\mathbf{y}(n))\mathbf{y}^T](\mathbf{W}(n)\mathbf{W}^T(n))\mathbf{W}^{-T}(n) \\ &= \mathbf{W}(n) + \eta(n)[\mathbf{I} - \varphi(\mathbf{y}(n))\mathbf{y}^T(n)]\mathbf{W}(n) \end{aligned} \tag{10.104}$$

which leads to blind source separation with the equivariance property. Figure 10.12 shows a signal-flow graph representation of Eq. (10.104).

For the adaptive algorithm described in Eq. (10.104) to produce a correct solution to the blind source separation problem described in Fig. 10.9, the following two requirements must be satisfied for all components of the output vector $\mathbf{Y}$:

- The Gram–Charlier expansion used to compute the nonlinearity $\varphi(\cdot)$ includes a sufficient number of terms to produce a good approximation to the marginal entropy $h(Y_i)$; for example, this requirement is satisfied by the activation function of Eq. (10.93).
- The learning rate η is small enough for the estimates of the cumulants of Y_i to be reliable.

Stability Considerations

A discussion of the blind source separation problem is incomplete without consideration of the stability of the adaptive learning algorithm described in Eq. (10.104). In Amari et al. (1997), a general stability analysis of this algorithm is presented for an arbitrary activation function $\varphi(\cdot)$. The analysis is performed in the sense of asymptotic

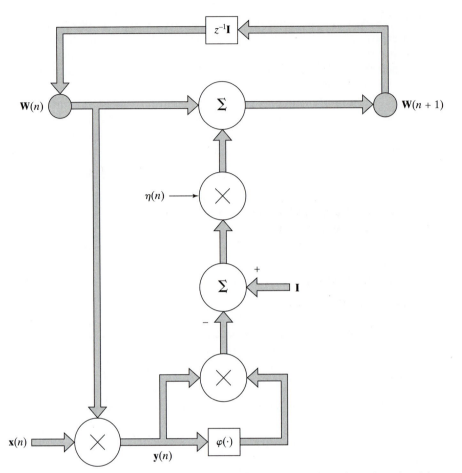

FIGURE 10.12   Signal-flow graph of the blind source separation learning algorithm described in Eq. (10.104).

convergence of the algorithm to the desired equilibrium point where a successful separation of sources is guaranteed.

Equation (10.104) is a discrete-time description of the blind source separation algorithm based on the natural gradient. For the purpose of stability analysis, the algorithm is reformulated in continuous time as follows:

$$\dot{\mathbf{W}}(t) = \eta(t)[\mathbf{I} - \boldsymbol{\varphi}(\mathbf{y}(t))\mathbf{y}^T(t)]\mathbf{W}(t) \tag{10.105}$$

where t denotes continuous time, and $\dot{\mathbf{W}}(t) = \partial\mathbf{W}(t)/\partial t$. The learning-rate parameter $\eta(t)$ is positive for all time t. Let

$$\sigma_i^2 = E\,[y_i^2] \tag{10.106}$$

$$k_i = E\left[\frac{\partial\varphi(y_i)}{\partial y_i}\right] \tag{10.107}$$

$$q_i = E\left[y_i^2 \frac{\partial \varphi(y_i)}{\partial y_i} \right] \tag{10.108}$$

Then, according to Amari et al. (1997), the separating solution is a stable equilibrium point of the adaptive algorithm of Eq. (10.104) for an arbitrary activation function $\varphi(\cdot)$ if and only if the following conditions are satisfied

$$q_i + 1 > 0 \tag{10.109}$$

$$k_i > 0 \tag{10.110}$$

$$\sigma_i^2 \sigma_j^2 k_i k_j > 1 \tag{10.111}$$

for all (i, j) with $i \neq j$. Equations (10.109) to (10.111) are the necessary and sufficient conditions for stability of the adaptive algorithm of Eq. (10.104).

Convergence Considerations

Given that we have satisfied the stability requirements of Eqs. (10.109) through (10.111), what can we say about the convergence behavior of the learning algorithm of Eq. (10.104) based on the activation function of Eq. (10.93)? In light of an experimental study reported in Madhuranath and Haykin (1998), roughly speaking, we may say that there are two phases to the convergence process:

- In phase I, the variance $\sigma_i^2(n)$ of random variable Y_i at the demixer output goes through a period of adjustment, whereafter it reaches a fairly stable value. During this phase, the cumulants $\kappa_{i,3}$, $\kappa_{i,4}$, and $\kappa_{i,6}$ remain essentially unchanged.
- In phase II, the cumulants $\kappa_{i,3}$, $\kappa_{i,4}$, and $\kappa_{i,6}$ go through a period of adjustments of their own, whereafter they attain fairly stable values. At that point, we may say that the algorithm has converged.

It thus appears that an evaluation of the variance and higher-order cumulants of the demixer outputs (i.e., separated source signals) provides the basis of a sensible procedure for studying the convergence behavior of the learning algorithm of Eq. (10.104). It is also of interest to note that it is only during phase II that the algorithm is conforming to the Gram–Charlier expansion.

10.12 COMPUTER EXPERIMENT

Consider the system described in Fig. 10.9 involving the following three independent sources:

$$u_1(n) = 0.1 \sin(400n) \cos(30n)$$
$$u_2(n) = 0.01 \, \text{sgn}(\sin(500n + 9\cos(40n)))$$
$$u_3(n) = \text{noise uniformly distributed in the range } [-1, 1]$$

The mixing matrix $\mathbf{A}$ is

$$\mathbf{A} = \begin{bmatrix} 0.56 & 0.79 & -0.37 \\ -0.75 & 0.65 & 0.86 \\ 0.17 & 0.32 & -0.48 \end{bmatrix}$$

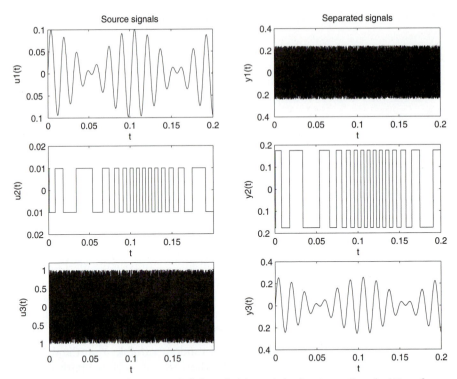

FIGURE 10.13 Waveforms on left-hand side: original source signals. Waveforms on right-hand side: separated source signals.

The waveforms of the source signals are displayed in the left-hand side of Fig. 10.13.

For the demixing, we used the batch version of the update rule described in Eq. (10.104); see Problem 10.14. Batch processing was chosen for the primary reason of improved convergence. The algorithm was implemented using the following conditions:

- *Initialization.* To initialize the algorithm, the weights in the demixing matrix **W** were picked from a random number generator with a uniform distribution inside the range $[0.0, 0.05]$.
- *Learning rate.* The learning-rate parameter was fixed at $\eta = 0.1$.
- *Signal duration.* The time series produced at the mixer output had a sampling period of 10^{-4}s and contained $N = 65{,}000$ samples.

The right-hand side of Fig. 10.13 displays the waveforms of the signals produced at the output of the demixer in Fig. 10.9 after 300 iterations. Except for rescaling and permutation of the unknown source outputs, there are no discernible differences between the two sets of waveforms shown in the left-hand and right-hand sides of Fig. 10.13. For the results presented here, the actual weight matrix used for initialization of the algorithm was

$$\mathbf{W}(0) = \begin{bmatrix} 0.0109 & 0.0340 & 0.0260 \\ 0.0024 & 0.0467 & 0.0415 \\ 0.0339 & 0.0192 & 0.0017 \end{bmatrix}$$

The algorithm converged to the final weight matrix

$$\mathbf{W} = \begin{bmatrix} 0.2222 & 0.0294 & -0.6213 \\ -10.1932 & -9.8141 & -9.7259 \\ 4.1191 & -1.7879 & -6.3765 \end{bmatrix}$$

The corresponding value of the matrix product $\mathbf{WA}$ is

$$\mathbf{WA} = \begin{bmatrix} -0.0032 & -0.0041 & 0.2413 \\ -0.0010 & -17.5441 & -0.0002 \\ 2.5636 & 0.0515 & -0.0009 \end{bmatrix}$$

Rearranging terms in this matrix product so that the output signals appear in the same order as the original source signals, we may write

$$\mathbf{WA} = \begin{bmatrix} 2.5636 & 0.0515 & -0.0009 \\ -0.0010 & -17.5441 & -0.0002 \\ -0.0032 & -0.0041 & 0.2413 \end{bmatrix}$$

The first, second, and third rows of the matrix product correspond to the amplitude-modulated signal, clipped frequency-modulated signal, and noise, respectively. The diagonal elements of $\mathbf{WA}$ define the factors by which the output waveforms on the right-hand side of Fig. 10.13 have been scaled with respect to the original source waveforms on the left-hand side of the figure.

For a quantitative evaluation of the demixer's performance, we may use a global rejection index defined by (Amari et al., 1996)

$$\mathcal{I} = \sum_{i=1}^{m} \left(\sum_{j=1}^{m} \frac{|p_{ij}|}{\max_k |p_{ik}|} - 1 \right) + \sum_{j=1}^{m} \left(\sum_{i=1}^{m} \frac{|p_{ij}|}{\max_k |p_{ki}|} - 1 \right)$$

where $\mathbf{P} = \{p_{ij}\} = \mathbf{WA}$. The performance index $\mathcal{I}$ is a measure of the *diagonality* of matrix $\mathbf{P}$. If the matrix $\mathbf{P}$ is perfectly diagonal, $\mathcal{I} = 0$. For a matrix $\mathbf{P}$ whose elements are not concentrated on the principal diagonal, the performance index $\mathcal{I}$ will be high.

For the waveforms displayed in Fig. 10.13, $\mathcal{I} = 0.0606$.

10.13 MAXIMUM LIKELIHOOD ESTIMATION

The method of independent components analysis (i.e., the third variant of the Infomax principle) described in the previous section is just one of many methods that have been proposed in the literature for blind source separation. In an information-theoretic context, however, there are only two other methods for performing the task of source separation in an unsupervised manner: maximum likelihood and maximum entropy. In this section we discuss maximum likelihood.

Maximum likelihood is a well-established procedure for statistical estimation with some nice properties; see note 5 of Chapter 7. In this procedure we first formulate a log-likelihood function and then optimize it with respect to the parameter vector of the probabilistic model under consideration. From the discussion presented in Chapter 7, we recall that the likelihood function is the probability density function of a data set in a given model, but viewed as a function of the unknown parameters of the model. Referring to Fig. 10.9, let $f_{\mathbf{U}}(\cdot)$ denote the probability density function of the random source vector $\mathbf{U}$. Then the probability density function of the observation vector $\mathbf{X} = \mathbf{A}\mathbf{U}$ at the output of the mixer is defined by (Papoulis, 1984)

$$f_{\mathbf{X}}(\mathbf{x}, \mathbf{A}) = |\det(\mathbf{A})|^{-1} f_{\mathbf{U}}(\mathbf{A}^{-1}\mathbf{x}) \tag{10.112}$$

where $\det(\mathbf{A})$ is the determinant of mixing matrix $\mathbf{A}$. Let $\mathcal{T} = \{\mathbf{x}_k\}_{k=1}^{N}$ denote a set of N independent realizations of the random vector $\mathbf{X}$. We may then write

$$f_{\mathbf{X}}(\mathcal{T}, \mathbf{A}) = \prod_{k=1}^{N} f_{\mathbf{X}}(\mathbf{x}_k, \mathbf{A}) \tag{10.113}$$

We find it more convenient to work with the *normalized* (divided by the sample size N) version of the log-likelihood function, as shown by

$$\frac{1}{N} \log f_{\mathbf{X}}(\mathcal{T}, \mathbf{A}) = \frac{1}{N} \sum_{k=1}^{N} \log f_{\mathbf{X}}(\mathbf{x}_k, \mathbf{A})$$

$$= \frac{1}{N} \sum_{k=1}^{N} \log f_{\mathbf{U}}(\mathbf{A}^{-1}\mathbf{x}_k) - \log|\det(\mathbf{A})|$$

Let $\mathbf{y} = \mathbf{A}^{-1}\mathbf{x}$ be a realization of the random vector $\mathbf{Y}$ at the demixer output, and thus write

$$\frac{1}{N} \log f_{\mathbf{X}}(\mathcal{T}, \mathbf{A}) = \frac{1}{N} \sum_{k=1}^{N} \log f_{\mathbf{U}}(\mathbf{y}_k) - \log|\det(\mathbf{A})| \tag{10.114}$$

Let $\mathbf{A}^{-1} = \mathbf{W}$ and let $f_{\mathbf{Y}}(\mathbf{y}, \mathbf{W})$ denote the probability density function of $\mathbf{Y}$ parameterized by $\mathbf{W}$. Then recognizing that the summation in Eq. (10.114) is the sample average of $\log f_{\mathbf{U}}(\mathbf{y}_k)$, we find from the law of large numbers that, with probability 1, as the sample size N approaches infinity:

$$L(\mathbf{W}) = \lim_{N \to \infty} \frac{1}{N} \sum_{k=1}^{N} \log f_{\mathbf{U}}(\mathbf{y}_k) + \log|\det(\mathbf{W})|$$

$$= E[\log f_{\mathbf{U}}(\mathbf{y}_k)] + \log|\det(\mathbf{W})| \tag{10.115}$$

$$= \int_{-\infty}^{\infty} f_{\mathbf{Y}}(\mathbf{y}, \mathbf{W}) \log f_{\mathbf{U}}(\mathbf{y}) d\mathbf{y} + \log|\det(\mathbf{W})|$$

where the expectation in the second line is with respect to $\mathbf{Y}$. The quantity $L(\mathbf{W})$ is the desired log-likelihood function. By writing

$$f_{\mathbf{U}}(\mathbf{y}) = \left(\frac{f_{\mathbf{U}}(\mathbf{y})}{f_{\mathbf{Y}}(\mathbf{y}, \mathbf{W})} \right) f_{\mathbf{Y}}(\mathbf{y}, \mathbf{W})$$

we may express $L(\mathbf{W})$ in the equivalent form

$$L(\mathbf{W}) = \int_{-\infty}^{\infty} f_{\mathbf{Y}}(\mathbf{y}, \mathbf{W}) \log \left(\frac{f_{\mathbf{U}}(\mathbf{y})}{f_{\mathbf{Y}}(\mathbf{y}, \mathbf{W})} \right) d\mathbf{y} + \int_{-\infty}^{\infty} f_{\mathbf{Y}}(\mathbf{y}, \mathbf{W}) \log f_{\mathbf{Y}}(\mathbf{y}, \mathbf{W}) d\mathbf{y} + \log|\det(\mathbf{W})|$$

$$= -D_{f_{\mathbf{Y}} \| f_{\mathbf{U}}} - h(\mathbf{Y}, \mathbf{W}) + \log|\det(\mathbf{W})| \tag{10.116}$$

where $h(\mathbf{Y}, \mathbf{W})$ is the differential entropy of random vector $\mathbf{Y}$ parameterized by $\mathbf{W}$, and $D_{f_{\mathbf{Y}} \| f_{\mathbf{U}}}$ is the Kullback–Leibler divergence between $f_{\mathbf{Y}}(\mathbf{y}, \mathbf{W})$ and $f_{\mathbf{U}}(\mathbf{y})$. Using Eq. (10.76) in (10.116), we may simplify the expression for the log-likelihood function $L(\mathbf{W})$ as follows (Cardoso, 1998a):

$$L(\mathbf{W}) = -D_{f_{\mathbf{Y}} \| f_{\mathbf{U}}} - h(\mathbf{X}) \tag{10.117}$$

where $h(\mathbf{X})$ is the differential entropy of the random vector $\mathbf{X}$ at the demixer input. The only quantity in Eq. (10.117) that depends on the weight vector $\mathbf{W}$ of the demixer is the Kullback–Leibler divergence $D_{f_{\mathbf{Y}} \| f_{\mathbf{U}}}$. We therefore conclude from Eq. (10.117) that maximizing the log-likelihood function $L(\mathbf{W})$ is equivalent to minimizing the Kullback–Leibler divergence $D_{f_{\mathbf{Y}} \| f_{\mathbf{U}}}$, that is, matching the probability distribution of the demixer output $\mathbf{Y}$ to that of the original source vector $\mathbf{U}$, which is intuitively satisfying.

Relationship between Maximum Likelihood and Independent Components Analysis

Applying the Pythagorean decomposition described in Eq. (10.45) to the problem at hand, we may express the Kullback–Leibler divergence $D_{f_{\mathbf{Y}} \| f_{\mathbf{U}}}$ for maximum likelihood as follows:

$$D_{f_{\mathbf{Y}} \| f_{\mathbf{U}}} = D_{f_{\mathbf{Y}} \| \tilde{f}_{\mathbf{Y}}} + D_{\tilde{f}_{\mathbf{Y}} \| f_{\mathbf{U}}} \tag{10.118}$$

The first Kullback–Leibler divergence $D_{f_{\mathbf{Y}} \| \tilde{f}_{\mathbf{Y}}}$ on the right-hand side of Eq. (10.118) is a measure of *structural mismatch* that characterizes the method of independent components analysis. The second Kullback–Leibler divergence $D_{\tilde{f}_{\mathbf{Y}} \| f_{\mathbf{U}}}$ is a measure of *marginal mismatch* between the marginal distribution of the demixer output $\mathbf{Y}$ and the distribution of the original source vector $\mathbf{U}$. We may thus express the "global" distribution matching criterion for maximum likelihood as follows (Amari, 1997; Cardoso, 1998a):

$$\begin{pmatrix} \text{Total} \\ \text{mismatch} \end{pmatrix} = \begin{pmatrix} \text{Structural} \\ \text{mismatch} \end{pmatrix} + \begin{pmatrix} \text{Marginal} \\ \text{mismatch} \end{pmatrix} \tag{10.119}$$

"Structural mismatch" refers to the structure of a distribution pertaining to a set of independent variables, whereas "marginal mismatch" refers to the mismatch between the individual marginal distributions.

Under the ideal condition $\mathbf{W} = \mathbf{A}^{-1}$ (i.e., perfect blind source separation), both the structural mismatch and marginal mismatch vanish. At that point, maximum likelihood and independent components analysis yield exactly the same solution. The idealized relationship between maximum likelihood and independent components analysis is depicted in Fig. 10.14 (Cardoso, 1996; Amari, 1997). In this figure, $\mathscr{S}$ is the set of *all* probability density functions $f_{\mathbf{Y}}(\mathbf{y})$ of the random vector $\mathbf{Y}$ at the demixer output; $\mathscr{I}$ is the set of all independent probability distributions, that is, those of the product form. Both $\mathscr{S}$ and $\mathscr{I}$ are of infinite dimension. The set $\mathscr{D} = \{f_{\mathbf{Y}}(\mathbf{y}, \mathbf{W})\}$ is the finite set of probability distributions measured at the demixer output. The set $\mathscr{D}$ is m^2-dimensional, where m is the dimension of $\mathbf{Y}$, and the weight matrix $\mathbf{W}$ is a coordinate system in it. From Fig. 10.14 we clearly see that both $D_{f_{\mathbf{Y}} \| \tilde{f}_{\mathbf{Y}}}$ and $D_{\tilde{f}_{\mathbf{Y}} \| f_{\mathbf{U}}}$ are minimized at $\mathbf{W} = \mathbf{A}^{-1}$. It is

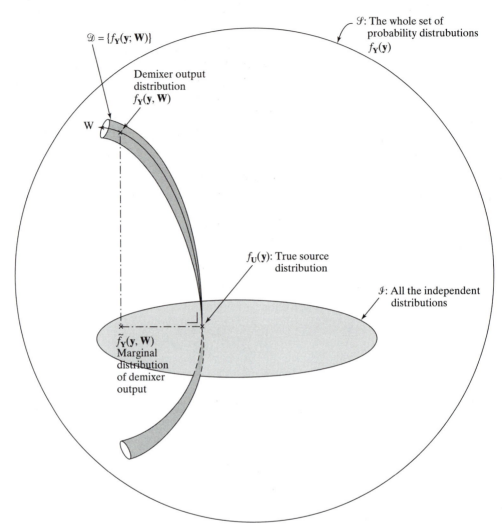

FIGURE 10.14 Illustration of the relationship between maximum likelihood and independent components analysis for blind source separation. Maximum likelihood minimizes $D_{f_Y \| f_U}$, whereas independent components analysis minimizes $D_{f_Y \| \tilde{f}_Y}$.

interesting to prove that the sets $\mathcal{D}$ and $\mathcal{I}$ are indeed orthogonal at their intersection point defined by the true probability density function $f_U(\mathbf{y})$.

A blind source separation algorithm based on maximum likelihood must include a provision for estimating the underlying source distributions when they are unknown, which is typically the case. The parameters for this estimation can be adapted just as we adapt the demixing weight matrix $\mathbf{W}$. In other words, we should perform a *joint estimation* of the mixing matrix and (some characteristics) of the source distributions (Cardoso, 1997, 1998a); an elegant and well developed approach for this joint estimation is presented in Pham et al. (1992, 1997).

10.14 MAXIMUM ENTROPY METHOD

The *maximum entropy method* for blind source separation is due to Bell and Sejnowski (1995). Figure 10.15 shows the block diagram of the system based on this method. As before, the demixer operates on the observation vector $\mathbf{X}$ to produce an output $\mathbf{Y} = \mathbf{WX}$ that is an estimate of the original source vector $\mathbf{U}$. The vector $\mathbf{Y}$ is transformed into a vector $\mathbf{Z}$ by passing it through a component-wise nonlinearity denoted by $\mathbf{G}(\cdot)$, which is monotonic and invertible. Thus, unlike $\mathbf{Y}$, the vector $\mathbf{Z}$ is assured of a bounded differential entropy $h(\mathbf{Z})$ for an arbitrarily large demixer. For a prescribed nonlinearity $\mathbf{G}(\cdot)$, the maximum entropy method produces an estimate of the original source vector $\mathbf{U}$ by maximizing the entropy $h(\mathbf{Z})$ with respect to $\mathbf{W}$. In light of Eq. (10.55) derived in Example 10.6, we see that the maximum entropy method is closely related to the Infomax principle.[17]

The nonlinearity $\mathbf{G}$ is a *diagonal map* described by

$$\mathbf{G}: \begin{bmatrix} y_1 \\ y_2 \\ \vdots \\ y_m \end{bmatrix} \rightarrow \begin{bmatrix} g_1(y_1) \\ g_2(y_2) \\ \vdots \\ g_m(y_m) \end{bmatrix} = \begin{bmatrix} z_1 \\ z_2 \\ \vdots \\ z_m \end{bmatrix} \tag{10.120}$$

We may thus write

$$\mathbf{Z} = \mathbf{G}(\mathbf{Y}) \tag{10.121}$$
$$= \mathbf{G}(\mathbf{WAU})$$

Since the nonlinearity $\mathbf{G}(\cdot)$ is invertible, we may express the original source vector $\mathbf{U}$ in terms of the demixer output vector $\mathbf{Z}$ as

$$\mathbf{U} = \mathbf{A}^{-1}\mathbf{W}^{-1}\mathbf{G}^{-1}(\mathbf{Z}) \tag{10.122}$$
$$= \mathbf{\Psi}(\mathbf{Z})$$

where $\mathbf{G}^{-1}$ is the *inverse nonlinearity*:

$$\mathbf{G}^{-1}: \begin{bmatrix} z_1 \\ z_2 \\ \vdots \\ z_m \end{bmatrix} \rightarrow \begin{bmatrix} g_1^{-1}(z_1) \\ g_2^{-1}(z_2) \\ \vdots \\ g_m^{-1}(z_m) \end{bmatrix} = \begin{bmatrix} y_1 \\ y_2 \\ \vdots \\ y_m \end{bmatrix} \tag{10.123}$$

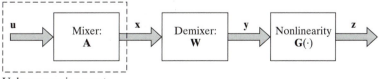

FIGURE 10.15 Block diagram of the maximum entropy method for blind source separation. The vectors **u, x, y** and **z** are values of the random vectors **U, X, Y** and **Z**, respectively.

The probability density function of the output vector $\mathbf{Z}$ is defined in terms of that of the source vector $\mathbf{U}$ by (Papoulis, 1984)

$$f_{\mathbf{Z}}(\mathbf{z}) = \frac{f_{\mathbf{U}}(\mathbf{u})}{|\det(\mathbf{J}(\mathbf{u}))|}\bigg|_{\mathbf{u}=\mathbf{\Psi}(\mathbf{z})} \tag{10.124}$$

where $\det(\mathbf{J}(\mathbf{u}))$ is the determinant of the Jacobian matrix $\mathbf{J}(\mathbf{u})$. The ij-th element of this latter matrix is defined by

$$J_{ij} = \frac{\partial z_i}{\partial u_j} \tag{10.125}$$

Hence the entropy of the random vector $\mathbf{Z}$ at the output of the nonlinearity $\mathbf{G}$ is

$$\begin{aligned} h(\mathbf{Z}) &= -E[\log f_{\mathbf{Z}}(\mathbf{z})] \\ &= -E\left[\log\left(\frac{f_{\mathbf{U}}(\mathbf{u})}{|\det(\mathbf{J}(\mathbf{u}))|}\right)\right]_{\mathbf{u}=\mathbf{\Psi}(\mathbf{z})} \\ &= -D_{f_{\mathbf{U}}\||\det \mathbf{J}|} \quad \text{evaluated at } \mathbf{u} = \mathbf{\Psi}(\mathbf{z}) \end{aligned} \tag{10.126}$$

We thus see that maximizing the entropy $h(\mathbf{Z})$ is equivalent to minimizing the Kullback–Leibler divergence between $f_{\mathbf{U}}(\mathbf{u})$ and a probability density function of $\mathbf{U}$ defined by $|\det(\mathbf{J}(\mathbf{u}))|$.

Suppose now the random variable Z_i (i.e., the ith element of $\mathbf{Z}$) is *uniformly distributed* inside the interval $[0, 1]$ for all i. According to Example 10.1, the entropy $h(\mathbf{Z})$ is then equal to zero. Correspondingly, we find from Eq. (10.126) that

$$f_{\mathbf{U}}(\mathbf{u}) = |\det(\mathbf{J}(\mathbf{u}))| \tag{10.127}$$

Under the ideal condition $\mathbf{W} = \mathbf{A}^{-1}$, this relationship reduces to

$$f_{U_i}(u_i) = \frac{\partial z_i}{\partial y_i}\bigg|_{z_i=g(u_i)} \qquad \text{for all } i \tag{10.128}$$

Conversely, we can say that if Eq. (10.128) is satisfied, then maximizing $h(\mathbf{Z})$ yields $\mathbf{W} = \mathbf{A}^{-1}$ and blind source separation is thereby achieved.

We may now summarize the results obtained on the maximum entropy method for blind source separation as follows (Bell and Sejnowski, 1995):

Let the nonlinearity at the demixer output in Fig. 10.15 be defined in terms of the original source distribution as

$$\begin{aligned} z_i &= g_i(y_i) \\ &= \int_{-\infty}^{z_i} f_{U_i}(u_i)du_i \qquad \text{for } i = 1, 2, ..., m \end{aligned} \tag{10.129}$$

Maximizing the entropy of the random vector $\mathbf{Z}$ at the output of the nonlinearity $\mathbf{G}$ is then equivalent to $\mathbf{W} = \mathbf{A}^{-1}$, which yields perfect blind source separation.

The maximum entropy and maximum likelihood methods for blind source separation are indeed equivalent under the condition that the random variable Z_i is uniformly distributed inside the interval $[0, 1]$ for all i (Cardoso, 1997). To prove this relationship we first use the chain rule of calculus to rewrite Eq. (10.125) in the equivalent form

$$J_{ij} = \sum_{k=1}^{m} \frac{\partial z_i}{\partial y_i} \frac{\partial y_i}{\partial x_k} \frac{\partial x_k}{\partial u_j}$$

$$= \sum_{k=1}^{m} \frac{\partial z_i}{\partial y_i} w_{ik} a_{kj} \tag{10.130}$$

The Jacobian matrix $\mathbf{J}$ may therefore be expressed as

$$\mathbf{J} = \mathbf{DWA}$$

where $\mathbf{D}$ is the diagonal matrix

$$\mathbf{D} = \text{diag}\left(\frac{\partial z_1}{\partial y_1}, \frac{\partial z_2}{\partial y_2}, \dots, \frac{\partial z_m}{\partial y_m}\right)$$

Hence,

$$|\det(\mathbf{J})| = |\det(\mathbf{WA})| \prod_{i=1}^{m} \frac{\partial z_i}{\partial y_i} \tag{10.131}$$

An estimate of the probability density function $f_{\mathbf{U}}(\mathbf{u})$ parameterized by the weight matrix $\mathbf{W}$ and the nonlinearity $\mathbf{G}$, in light of Eq. (10.131), may be written formally as (Roth and Baram, 1996)

$$f_{\mathbf{U}}(\mathbf{u}|\mathbf{W}, \mathbf{G}) = |\det(\mathbf{WA})| \prod_{i=1}^{m} \frac{\partial g_i(y_i)}{\partial y_i} \tag{10.132}$$

We thus see that under this condition, maximizing the log-likelihood function $\log f_{\mathbf{U}}(\mathbf{u}|\mathbf{W}, \mathbf{G})$ is equivalent to maximizing the entropy $h(\mathbf{Z})$ for blind source separation. That is, the methods of maximum entropy and maximum likelihood are equivalent.

Learning Algorithm for Blind Source Separation

Referring to the second line of Eq. (10.126), we note that since the source distribution is typically fixed, maximizing the entropy $h(\mathbf{z})$ requires maximizing the expectation of the denominator term $\log|\det(\mathbf{J}(\mathbf{u}))|$ with respect to the weight matrix $\mathbf{W}$. With an adaptive algorithm for doing this computation as our goal, we may thus consider the instantaneous objective function

$$\Phi = \log|\det(\mathbf{J})| \tag{10.133}$$

Substituting Eq. (10.131) into (10.133) yields

$$\Phi = \log|\det(\mathbf{A})| + \log|\det(\mathbf{W})| + \sum_{i=1}^{m} \log\left(\frac{\partial z_i}{\partial y_i}\right) \tag{10.134}$$

Hence, differentiating Φ with respect to the weight matrix $\mathbf{W}$ of the demixer we get (see Problem 10.16)

$$\frac{\partial \Phi}{\partial \mathbf{W}} = \mathbf{W}^{-T} + \sum_{i=1}^{m} \frac{\partial}{\partial \mathbf{W}} \log\left(\frac{\partial z_i}{\partial y_i}\right) \tag{10.135}$$

To proceed further with this formula we need to specify the nonlinearity fed by the demixer output. A simple form of nonlinearity that may be used here is the logistic function

$$z_i = g(y_i)$$
$$= \frac{1}{1 + e^{-y_i}}, \quad i = 1, 2, \dots, m \tag{10.136}$$

Figure 10.16 presents plots of this nonlinearity and its inverse. This figure shows that the logistic function satisfies the basic requirements of monotonicity and invertibility for blind source separation. Substituting Eq. (10.136) into (10.135) yields

$$\frac{\partial \Phi}{\partial \mathbf{W}} = \mathbf{W}^{-T} + (\mathbf{1} - 2\mathbf{z})\mathbf{x}^{T}$$

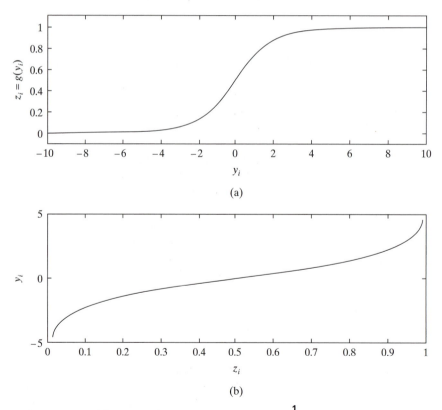

FIGURE 10.16 (a) Logistic function: $z_i = g(y_i) = \dfrac{1}{1 + e^{-y_i}}$. (b) Inverse of logistic function: $y_i = g^{-1}(z_i)$.

where $\mathbf{x}$ is the received signal vector, $\mathbf{z}$ is the nonlinear transformed output vector of the demixer, and $\mathbf{1}$ is a corresponding vector of ones.

The objective of the learning algorithm is to maximize the entropy $h(\mathbf{Z})$. Accordingly, invoking the method of steepest ascent, the change applied to the weight matrix $\mathbf{W}$ is (Bell and Sejnowski, 1995)

$$\Delta\mathbf{W} = \eta \, \frac{\partial\Phi}{\partial\mathbf{W}}$$

$$= \eta \, (\mathbf{W}^{-T} + (\mathbf{1} - 2\mathbf{z})\mathbf{x}^{T}) \qquad (10.137)$$

where η is the learning-rate parameter. As with the independent components analysis we may eliminate the need for inverting the transposed weight matrix $\mathbf{W}^{T}$ by using the natural gradient, which is equivalent to multiplying Eq. (10.137) by the matrix product $\mathbf{W}^{T}\mathbf{W}$. This optimal rescaling yields the desired formula for weight change as

$$\Delta\mathbf{W} = \eta(\mathbf{W}^{-T} + (\mathbf{1} - 2\mathbf{z})\mathbf{x}^{T})\mathbf{W}^{T}\mathbf{W}$$

$$= \eta(\mathbf{I} + (\mathbf{1} - 2\mathbf{z})(\mathbf{Wx})^{T})\mathbf{W} \qquad (10.138)$$

$$= \eta(\mathbf{I} + (\mathbf{1} - 2\mathbf{z})\mathbf{y}^{T})\mathbf{W}$$

where the vector $\mathbf{y}$ is the demixer output. The learning algorithm for computing the weight matrix $\mathbf{W}$ is therefore

$$\mathbf{W}(n + 1) = \mathbf{W}(n) + \eta(\mathbf{I} + (\mathbf{1} - 2\mathbf{z}(n))\mathbf{y}^{T}(n))\mathbf{W}(n) \qquad (10.139)$$

The algorithm is initiated with $\mathbf{W}(0)$ selected from a uniformly distributed set of small numbers.

Theoretical considerations and experimental investigations have shown that the learning algorithm of Eq. (10.139) is limited to separating sources with super-Gaussian distributions (Bell and Sejnowski, 1995); for the definition of super-Gaussian distributions, see note 18. This limitation is a direct consequence of using the logistic function for the nonlinearity at the back end of the system in Fig. 10.15. In particular, the logistic function imposes prior knowledge, namely a super-Gaussian shape, on the source distribution. There is nothing in the maximum entropy method, however, that restricts its use to the logistic function any more than the maximum likelihood method is restricted to some fixed prior. The application of the maximum entropy method may be broadened to a wider spectrum of source distributions by modifying the learning algorithm of Eq. (10.138) so as to provide for joint estimation of the underlying source distribution and the mixing matrix. This requirement is of a similar nature to that discussed for maximum likelihood in the previous section.

10.15 SUMMARY AND DISCUSSION

In this chapter we establish mutual information, rooted in Shannon's information theory, as a basic statistical tool for self-organization. The mutual information between an input process and an output process has some unique properties that suggest its adoption as the objective function to be optimized for self-organized learning. Indeed, some important principles for self organization have emerged from the discussion presented in this chapter:

- *Maximum mutual information (Infomax) principle* (Linsker, 1988). This principle, in its basic form, is well suited for the development of self-organized models and feature maps.
- The *first variant* of Infomax, due to Becker and Hinton (1992), is well suited for image processing where the objective is the discovery of properties of a noisy sensory input exhibiting coherence across both space and time.
- The *second variant* of Infomax, due to Ukrainec and Haykin (1992), finds applications in dual image processing where the objective is to maximize the spatial differentiation between the corresponding regions of two separate images (views) of an environment of interest.
- The third variant of Infomax for *independent components analysis* is due to Comon (1994), although its roots go back to Barlow's hypothesis (Barlow, 1985, 1989). Nevertheless, in Comon (1994) a rigorous formulation of independent components analysis was presented for the first time.
- Maximum entropy method due to Bell and Sejnowski (1995), which is also related to the Infomax principle. Maximum entropy is equivalent to maximum likelihood (Cardoso, 1997).

Independent components analysis and the maximum entropy method provide two alternative methods for blind source separation, with each one of them offering attributes of its own. A blind source separation algorithm based on the maximum entropy method is simple to implement, whereas a corresponding algorithm based on independent components analysis is more elaborate in derivation but may have broader applicability.

A neurobiological motivation that is often cited for blind source separation is the cocktail party phenomenon. This phenomenon refers to the remarkable human ability to selectively tune to and follow an auditory input of interest in a noisy environment. As explained in Chapter 2, the underlying neurobiological model involved in the solution to this very difficult signal processing problem is much more complicated than what is entailed in the idealized model described in Fig. 10.9. The neurobiological model involves both temporal and spatial forms of processing, which are needed in order to cope with unknown delays, reverberation, and noise. Now that we have a reasonably firm understanding of the basic issues involved in the neural solution to the standard blind source separation problem, it is perhaps time that we move on and tackle real-life problems on a scale comparable to the cocktail party phenomenon.

Another open research area worthy of detailed attention is that of *blind deconvolution*. Deconvolution is a signal processing operation that ideally unravels the effects of convolution performed by a linear time-invariant system operating on the input signal. More specifically, in ordinary deconvolution the output signal and the system are both known and the requirement is to reconstruct what the input signal must have been. In blind deconvolution, or in more precise terms, unsupervised deconvolution, *only* the output signal is known and there may also be information on the source statistics; the requirement is to find the input signal, the system, or both. Clearly, blind deconvolution is a more difficult signal processing task than ordinary deconvolution. Although blind deconvolution has indeed received a great deal of attention in the literature (Haykin, 1994a), our understanding of an information-theoretic approach to blind deconvolution that parallels the blind source separation problem is at an early stage of development (Douglas and Haykin, 1997). Moreover, a cost effective solution to the blind equalization of a hostile channel such as the mobile communications channel is just as challenging in its own right as the cocktail party problem.

In summary, *blind adaptation,* be it in the context of source separation or deconvolution, has a long way to go before it can reach a mature state of development comparable to that of supervised learning.

NOTES AND REFERENCES

1. For detailed treatment of information theory, see the book by Cover and Thomas (1991); see also Gray (1990). For a collection of papers on the development of information theory (including the 1948 classic paper by Shannon), see Slepian (1973). Shannon's paper is also reproduced, with minor revisions, in the books by Shannon and Weaver (1949) and Sloane and Wyner (1993).

For a brief review of the important principles of information theory with neural processing in mind, see Atick (1992). For a treatment of information theory from a biology perspective, see Yockey (1992).

2. Linsker's maximum mutual information principle for self-organization is not to be confused with the information-content preservation rule for decision making, a rule of thumb that is briefly discussed in Chapter 7.

3. For a review of the literature on the relation between information theory and perception, see Linsker (1990c) and Atick (1992).

4. The term "entropy," in an information-theoretic context, derives its name from analogy with entropy in thermodynamics; the latter quantity is defined by (see Chapter 11)

$$H = -k_B \sum_\alpha p_\alpha \log p_\alpha$$

where k_B is Boltzmann's constant, and p_α is the probability that the system is in state α. Except for the factor k_B, the formula for entropy H in thermodynamics has exactly the same mathematical form as the definition of entropy given in Eq. (10.8).

5. In Shore and Johnson (1980), it is proved that the maximum entropy principle is correct in the following sense:

> Given prior knowledge in the form of constraints, there is only one distribution satisfying these constraints that can be chosen by a procedure that satisfies the "consistency axioms;" this unique distribution is defined by maximizing entropy.

The consistency axioms are fourfold:
  I. Uniqueness: The result should be unique.
  II. Invariance: The choice of coordinates should not affect the result.
  III. System independence: It should not matter whether independent information about independent systems is accounted for separately in terms of different densities or together in terms of a joint density.
  IV. Subset independence: It should not matter whether an independent subset of system states is treated in terms of a separate conditional density or in terms of the full system density.

In Shore and Johnson (1980), it is shown that the relative entropy or the Kullback–Leibler divergence also satisfies the consistency axioms.

6. For a discussion of the method of Lagrange multipliers, see the book by Dorny (1975).

7. The term $I(X; Y)$ was originally referred to as the *rate of information transmission* by Shannon (1948). Today, however, this term is commonly referred to as the mutual information between the random variables X and Y.

8. To prove the decomposition of Eq. (10.45), we may proceed as follows. By definition we have

$$
\begin{aligned}
D_{f_{\mathbf{X}}\|f_{\mathbf{U}}} &= \int_{-\infty}^{\infty} f_{\mathbf{X}}(\mathbf{x}) \log\left(\frac{f_{\mathbf{X}}(\mathbf{x})}{f_{\mathbf{U}}(\mathbf{x})}\right) d\mathbf{x} \\
&= \int_{-\infty}^{\infty} f_{\mathbf{X}}(\mathbf{x}) \log\left(\frac{f_{\mathbf{X}}(\mathbf{x})}{\tilde{f}_{\mathbf{X}}(\mathbf{x})} \cdot \frac{\tilde{f}_{\mathbf{X}}(\mathbf{x})}{f_{\mathbf{U}}(\mathbf{x})}\right) d\mathbf{x} \\
&= \int_{-\infty}^{\infty} f_{\mathbf{X}}(\mathbf{x}) \log\left(\frac{f_{\mathbf{X}}(\mathbf{x})}{\tilde{f}_{\mathbf{X}}(\mathbf{x})}\right) d\mathbf{x} + \int_{-\infty}^{\infty} f_{\mathbf{X}}(\mathbf{x}) \log\left(\frac{\tilde{f}_{\mathbf{X}}(\mathbf{x})}{f_{\mathbf{U}}(\mathbf{x})}\right) d\mathbf{x} \\
&= D_{f_{\mathbf{X}}\|\tilde{f}_{\mathbf{X}}} + \int_{-\infty}^{\infty} f_{\mathbf{X}}(\mathbf{x}) \log\left(\frac{\tilde{f}_{\mathbf{X}}(\mathbf{x})}{f_{\mathbf{U}}(\mathbf{x})}\right) d\mathbf{x}
\end{aligned}
\tag{1}
$$

From the definitions of $\tilde{f}_{\mathbf{X}}(\mathbf{x})$ and $f_{\mathbf{U}}(\mathbf{u})$, we see that

$$
\begin{aligned}
\log\left(\frac{\tilde{f}_{\mathbf{X}}(\mathbf{x})}{f_{\mathbf{U}}(\mathbf{x})}\right) &= \log\left(\frac{\prod_{i=1}^{m} \tilde{f}_{X_i}(x_i)}{\prod_{i=1}^{m} f_{U_i}(x_i)}\right) \\
&= \sum_{i=1}^{m} \log\left(\frac{\tilde{f}_{X_i}(x_i)}{f_{U_i}(x_i)}\right)
\end{aligned}
$$

Let B denote the integral in the last line of Eq. (1). We may then write

$$
\begin{aligned}
B &= \int_{-\infty}^{\infty} f_{\mathbf{X}}(\mathbf{x}) \log\left(\frac{\tilde{f}_{\mathbf{X}}(\mathbf{x})}{f_{\mathbf{U}}(\mathbf{x})}\right) d\mathbf{x} \\
&= \int_{-\infty}^{\infty} f_{\mathbf{X}}(\mathbf{x}) \log\left(\frac{\prod_{i=1}^{m} \tilde{f}_{X_i}(x_i)}{\prod_{i=1}^{m} f_{U_i}(x_i)}\right) d\mathbf{x} \\
&= \sum_{i=1}^{m} \int_{-\infty}^{\infty} \left(\log\left(\frac{\tilde{f}_{X_i}(x_i)}{f_{U_i}(x_i)}\right) \int_{-\infty}^{\infty} f_{\mathbf{X}}(\mathbf{x}) d\mathbf{x}^{(i)}\right) dx_i \\
&= \sum_{i=1}^{m} \int_{-\infty}^{\infty} \log\left(\frac{\tilde{f}_{X_i}(x_i)}{f_{U_i}(x_i)}\right) \tilde{f}_{X_i}(x_i) dx_i
\end{aligned}
\tag{2}
$$

where, in the last line, we have made use of the defining equation (10.39). The integral in Eq. (2) is the Kullback–Leibler divergence $D_{\tilde{f}_{X_i}\|f_{U_i}}$ for $i = 1, 2, \ldots, m$. To put the expression for B in its final form, we note that the area under $\tilde{f}_X(x_j)$ is unity, and therefore write

$$
\begin{aligned}
B &= \sum_{i=1}^{m} \int_{-\infty}^{\infty} \prod_{j=1}^{m} \tilde{f}_{X}(x_j)\left(\log\left(\frac{\tilde{f}_{X_i}(x_i)}{f_{U_i}(x_i)}\right) dx_i\right) d\mathbf{x}^{(i)} \\
&= \int_{-\infty}^{\infty} \tilde{f}_{\mathbf{X}}(\mathbf{x}) \log\left(\frac{\prod_{i=1}^{m} \tilde{f}_{X_i}(x_i)}{\prod_{i=1}^{m} f_{U_i}(x_i)}\right) d\mathbf{x} \\
&= D_{\tilde{f}_{\mathbf{X}}\|f_{\mathbf{U}}}
\end{aligned}
\tag{3}
$$

where in the first line we have used the definition $d\mathbf{x} = dx_i d\mathbf{x}^{(i)}$ as described in Section 10.5. Thus, substituting Eq. (3) in (1), we obtain the desired decomposition:

$$D_{f_{\mathbf{x}}\|f_U} = D_{f_{\mathbf{x}}\|\tilde{f}_{\mathbf{x}}} + D_{\tilde{f}_{\mathbf{x}}\|f_U}$$

9. Nadal and Parga (1994, 1997) also discuss the relationship between Infomax and redundancy reduction, reaching a similar conclusion that maximization of the mutual information between the input vector and output vector of a neural system leads to data reduction. Haft and van Hemmen (1998) discuss the implementation of Infomax filters for the retina. It is shown that redundancy is essential to the attainment of noise robustness of an internal representation of the environment as it is produced by a sensory system such as the retina.

10. Becker and Hinton (1992) use the acronym I_{max} to refer to the first variant of the Infomax principle.

11. In Uttley (1970) a *negative information pathway* is considered by optimizing the negative of the mutual information between the signals at the input and the output of the pathway. It is shown that such a system adapts to become a discriminator of the more frequent pattern occurring in the set of input signals during adaptation. The model is called "informon," which is loosely related to the second variant of the Infomax principle.

12. The system described in Ukrainec and Haykin (1996) includes a postdetection processor that uses *a priori* information about the reflector location along the water-land boundary of the waterway. A *fuzzy processor* combines primary detection performance with the output from a vision-based edge detector to effectively remove false alarms, thereby resulting in a further improvement in system performance.

13. Blind source-separation may be traced back to the seminal paper by Hérault, Jutten, and Ans (1985). For a historical account of the blind source separation problem, see Nadal and Parga (1997); this paper also emphasizes the neurobiological aspects of the problem. For an insightful overview of blind source separation, with emphasis on the underlying signal processing principles, see Cardoso (1998a).

14. **Approximation of Probability Density Function**
    (a) *Gram–Charlier Expansion*
    Let $\varphi_Y(\omega)$ denote the *characteristic function* of a random variable Y having the probability density function $f_Y(y)$. By definition we have

$$\varphi_Y(\omega) = \int_{-\infty}^{\infty} f_Y(y)e^{j\omega y}\,dy \tag{1}$$

where $j = \sqrt{-1}$ and ω is real. In words, the characteristic function $\varphi_Y(\omega)$ is the Fourier transform of the probability density function $f_Y(y)$, except for a sign change in the exponent. In general, the characteristic function $\varphi_Y(\omega)$ is a complex number whose real and imaginary parts are finite for all ω. If the kth moment of the random variable Y exists, then $\varphi_Y(\omega)$ can be expanded in a power series in a neighborhood of $\omega = 0$ as follows:

$$\varphi_Y(\omega) = 1 + \sum_{k=1}^{\infty} \frac{(j\omega)^k}{k!} m_k \tag{2}$$

where m_k is the kth order *moment* of the random variable Y; it is defined by

$$m_k = E[Y^k]$$
$$= \int_{-\infty}^{\infty} y^k f_Y(y)\,dy \tag{3}$$

Equation (2) is derived simply by substituting the expansion of the exponential function $e^{j\omega y}$ in Eq. (1), interchanging the order of summation and integration, and then invoking the definition of Eq. (3). If the characteristic function $\varphi_Y(\omega)$ can be expanded as in Eq. (2), then we may also expand the logarithm of $\varphi_Y(\omega)$ as follows (Wilks, 1962):

$$\log \varphi_Y(\omega) = \sum_{n=1}^{\infty} \frac{\kappa_n}{n!} (j\omega)^n \tag{4}$$

where κ_n is called the nth order *cumulant* or *semi-invariant* of the random variable Y. Equation (4) is derived by expanding the logarithm of $\varphi_Y(\omega)$ in a Taylor series in $j\omega$ about $\omega = 0$.

To simplify matters, henceforth we make two assumptions:
1. The random variable Y has zero mean, that is, $\mu = 0$.
2. The variance of Y is normalized to unity, that is, $\sigma^2 = 1$.

Correspondingly, we have $\kappa_1 = 0, \kappa_2 = 1$, and the expansion in Eq. (4) becomes

$$\log \varphi_Y(\omega) = \frac{1}{2} (j\omega)^2 + \sum_{n=3}^{\infty} \frac{\kappa_n}{n!} (j\omega)^n \tag{5}$$

Now, let

$$r(\omega) = \sum_{n=3}^{\infty} \frac{\kappa_n}{n!} (j\omega)^n$$

We may then rewrite Eq. (5) as

$$\log \varphi_Y(\omega) = \frac{1}{2} (j\omega)^2 + r(\omega)$$

That is, the characteristic function $\varphi_Y(\omega)$ may be expressed as the product of two exponential terms:

$$\varphi_Y(\omega) = \exp\left(-\frac{\omega^2}{2}\right) \cdot \exp(r(\omega)) \tag{6}$$

By using the power series expansion for the exponential term $\exp(r(\omega))$, we have

$$\exp(r(\omega)) = 1 + \sum_{l=1}^{\infty} \frac{r^l(\omega)}{l!} \tag{7}$$

By substituting Eq. (7) in (6) and collecting terms with like powers of $(j\omega)$ in the resulting double summation, we get new coefficients of the expansion of $\varphi_Y(\omega)$ such as those shown here:

$$c_1 = 0$$

$$c_2 = 0$$

$$c_3 = \frac{\kappa_3}{6}$$

$$c_4 = \frac{\kappa_4}{24}$$

$$c_5 = \frac{\kappa_5}{120}$$

$$c_6 = \frac{1}{720} (\kappa_6 + 10 \kappa_3^2)$$

$$c_7 = \frac{1}{5040} \left(\kappa_7 + 35\,\kappa_4\kappa_3 \right)$$

$$c_8 = \frac{1}{40320} \left(\kappa_8 + 56\kappa_5\kappa_3 + 35\,\kappa_4^2 \right)$$

and so on. We are now in a position to take the inverse Fourier transform of $\varphi_Y(\omega)$ to obtain an expansion for the probability density function $f_Y(y)$. In particular, we may write

$$f_Y(y) = \alpha(y) \left(1 + \sum_{k=3}^{\infty} c_k H_k(y) \right) \tag{8}$$

where $\alpha(y)$ is the probability density function of a *normalized Gaussian random variable of zero mean and unit variance*:

$$\alpha(y) = \frac{1}{\sqrt{2\pi}}\, e^{-y^2/2} \tag{9}$$

The expansion of Eq. (8) is known as the *Gram–Charlier series* of a probability density function in terms of the Gaussian function and its derivatives (Stuart and Ord, 1994). An expansion of this type has intuitive appeal. In particular, if the random variable Y consists of the sum of a number of independently and identically distributed random variables, then as the number of those variables increases, the central limit theorem tells us that the random variable Y is asymptotically Gaussian. The first term of the Gram–Charlier series is indeed Gaussian, which means that for such a sum the remainder of the series approaches zero as the number of variables in the sum increases.

The *Hermite polynomial* $H_k(y)$ appearing in Eq. (8) is defined in terms of the kth derivative of $\alpha(y)$ by

$$\alpha^{(k)}(y) = (-1)^k \alpha(y) H_k(y) \tag{10}$$

Some typical Hermite polynomials are

$$H_0(y) = 1$$
$$H_1(y) = y$$
$$H_2(y) = y^2 - 1$$
$$H_3(y) = y^3 - 3y$$
$$H_4(y) = y^4 - 6y^2 + 3$$
$$H_5(y) = y^5 - 10y^3 + 15y$$
$$H_6(y) = y^6 - 15y^4 + 45y^2 - 15$$

A recursive relation for these polynomials is

$$H_{k+1}(y) = yH_k(y) - kH_{k-1}(y) \tag{11}$$

A particularly useful property of the Hermite polynomials is that $H_k(y)$ and the mth derivative of the Gaussian function $\alpha(y)$ are *biorthogonal*, as shown by

$$\int_{-\infty}^{\infty} H_k(y)\alpha^{(m)}(y)\,dy = (-1)^m m!\delta_{km}, \qquad (k, m) = 0, 1, \ldots \tag{12}$$

The δ_{km} is the Kronecker delta, which is equal to unity if $k = m$ and zero otherwise.

It is important to note that the natural order of the terms is *not* the best for the Gram–Charlier series. Rather, the terms listed here in parentheses should be grouped together (Helstrom, 1968)

$$k = (0), (3), (4, 6), (5, 7, 9) \tag{13}$$

The elements of these groups are usually of the same order of magnitude. If we retain terms through $k = 4$, for example, we should also include the term $k = 6$.

(b)   *Edgeworth Expansion*

As before, let $\alpha(y)$ denote the probability density function of a random variable normalized to zero mean and unit variance. The Edgeworth expansion of the probability density function of a random variable Y about the Gaussian approximate $\alpha(y)$ is given by (Comon, 1994; Stuart and Ord, 1994)

$$
\begin{aligned}
\frac{f_Y(y)}{\alpha(y)} &= 1 + \frac{\kappa_3}{3!} H_3(y) + \frac{\kappa_4}{4!} H_4(y) + \frac{10\kappa_3^2}{6!} H_6(y) + \frac{\kappa_5}{5!} H_5(y) \\
&+ \frac{35\kappa_3\kappa_4}{7!} H_7(y) + \frac{280\kappa_3^3}{9!} H_9(y) + \frac{\kappa_6}{6!} H_6(y) + \frac{56\kappa_3\kappa_5}{8!} H_8(y) \\
&+ \frac{35\kappa_4^2}{8!} H_8(y) + \frac{2100\kappa_3^2\kappa_4}{10!} H_{10}(y) + \frac{15400\kappa_3^4}{12!} H_{12}(y) + \cdots
\end{aligned} \tag{14}
$$

where κ_i denotes the cumulant of order i of the standardized scalar random variable Y, and H_i denotes the Hermite polynomial of order i. Equation (14) is called the *Edgeworth series*.

The key feature of the Edgeworth expansion is that its coefficients decrease uniformly. On the other hand, the terms in the Gram–Charlier expansion of Eq. (8) do not tend uniformly to zero from the viewpoint of numerical errors; that is, in general, no term is negligible compared to a preceding term. It is for this reason that the recommended procedure for truncating the Gram–Charlier expansion is to follow the grouping of terms described in Eq. (13).

15. The idea of using $\nabla^*D = (\nabla D)\mathbf{W}^T\mathbf{W}$ instead of the usual gradient ∇D for solving the source separation problem is described in Cardoso and Laheld (1996). Therein, ∇^*D is referred to as the *relative gradient*. This gradient is exactly the same as the *natural gradient,* the definition of which follows from an information-geometric perspective (Amari, 1998; Amari et al. 1996). A similar algorithm was described earlier in Cichocki and Moszczyński (1992) and Cichocki (et al., 1994).

16. In the Riemannian space of dimension n, for example, the squared norm of a vector $\mathbf{a}$ is defined by

$$\|\mathbf{a}\|^2 = \sum_{i=1}^{n} \sum_{j=1}^{n} g_{ij} a_i a_j$$

where the g_{ij} are functions of the coordinates $x_1, x_2, \ldots, x_n$ of the Riemannian space, $g_{ij} = g_{ji}$, and the right-hand side of this expression is always positive. This expression is a generalization of the Euclidean formula for a squared norm:

$$\|\mathbf{a}\|^2 = \sum_{i=1}^{n} a_i^2$$

For a discussion of the Riemannian structure, see Amari (1987), and Murray and Rice (1993).

17. Bell and Sejnowski (1995) refer to their method of blind source separation as Infomax in light of Eq. (10.55) defining the relationship between the entropy $H(\mathbf{Y})$ and mutual information $I(\mathbf{Y}; \mathbf{X})$. However, the preferred terminology is "maximum entropy method" as it involves maximization of the entropy $h(\mathbf{Z})$, where $\mathbf{Z} = \mathbf{G}(\mathbf{Y})$. A note of caution: The maximum entropy method for blind source separation due to Bell and Sejnowski is *not* to be confused with the maximum entropy method (MEM) due to Burg (1975) for spectrum analysis.

18. A random variable X is said to be *sub-Gaussian* (Benveniste et al., 1987) if:
    - it is uniformly distributed, or
    - its probability density function $f_X(x)$ is expressible in the form $\exp(-g(x))$ where $g(x)$ is an even function that is differentiable, except possibly at the origin and $g(x)$ and $g'(x)/x$ are strictly increasing for $0 < x < \infty$.

    For example, we may have $g(x) = |x|^\beta$ with $\beta > 2$.

    If, however, $g'(x)/x$ is strictly decreasing for $0 < x < \infty$, and the remaining properties mentioned hold, the random variable X is said to be *super-Gaussian* (Benveniste et al., 1987). For example, we may have $g(x) = |x|^\beta$ with $\beta < 2$.

    Sometimes (perhaps in an abusive way) the sign of the kurtosis of a random variable is used as an indicator of its sub-Gaussianity or super-Gaussianity. The *kurtosis* of random variable X is defined by

    $$K_4(x) = \frac{E[X^4]}{(E[X^2])^2} - 3$$

    On this basis, the random variable X is said to be sub-Gaussian or super-Gaussian if the kurtosis $K_4(x)$ is negative or positive, respectively.

PROBLEMS

MaxEnt Principle

10.1 The support of a random variable X (i.e., the range of values for which it is nonzero) is defined by $[a, b]$; there is no other constraint imposed on this random variable. What is the maximum entropy distribution for this random variable? Justify your answer.

Mutual Information

10.2 Derive the properties of the mutual information $I(X; Y)$ between two continuous-valued random values X and Y as described in Section 10.4.

10.3 Consider a random input vector $\mathbf{X}$ made up of a primary component $\mathbf{X}_1$ and a contextual component $\mathbf{X}_2$. Define

$$Y_i = \mathbf{a}_i^T \mathbf{X}_1$$
$$Z_i = \mathbf{b}_i^T \mathbf{X}_2$$

How is the mutual information between $\mathbf{X}_1$ and $\mathbf{X}_2$ related to the mutual information between Y_i and Z_i? Assume that the probability model of $\mathbf{X}$ is defined by the multivariate Gaussian distribution

$$f_\mathbf{X}(\mathbf{x}) = \frac{1}{(2\pi)^{m/2}(\det \Sigma)^{1/2}} \exp((\mathbf{x} - \mu)^T \Sigma^{-1} (\mathbf{x} - \mu))$$

where μ is the mean vector of $\mathbf{X}$ and Σ is its covariance matrix.

10.4 In this problem we explore the use of relative entropy or the Kullback–Leibler divergence to derive a supervised learning algorithm for multilayer perceptrons (Hopfield, 1987b; Baum and Wilczek, 1988). To be specific, consider a multilayer perceptron consisting of an input layer, a hidden layer, and an output layer. Given a case or example α presented to the input, the output of neuron k in the output layer is assigned a probabilistic interpretation:

$$y_{k|\alpha} = p_{k|\alpha}$$

Correspondingly, let $q_{k|\alpha}$ denote the actual (true) value of the conditional probability that the proposition k is true, given the input case α. The relative entropy for the multilayer perceptron is defined by

$$D_{p\|q} = \sum_{\alpha} p_\alpha \sum_k \left(q_{k|\alpha} \log\left(\frac{q_{k|\alpha}}{p_{k|\alpha}}\right) + (1 - q_{k|\alpha})\log\left(\frac{1 - q_{k|\alpha}}{1 - p_{k|\alpha}}\right) \right)$$

where p_α is the *a priori* probability of occurrence of case α.

   Using $D_{p\|q}$ as the cost function to be optimized, derive a learning rule for training the multilayer perceptron.

Infomax Principle

10.5 Consider two channels whose outputs are represented by the random variables X and Y. The requirement is to maximize the mutual information between X and Y. Show that this requirement is achieved by satisfying two conditions:
   (a) The probability of occurrence of X or that of Y is 0.5.
   (b) The joint probability distribution of X and Y is concentrated in a small region of the probability space.

10.6 Consider the noise model of Fig. P10.6, which shows m source nodes in the input layer of a two-neuron network. Both neurons are linear. The inputs are denoted by $X_1, X_2, ..., X_m$, and the resulting outputs are denoted by Y_1 and Y_2. You may make the following assumptions:
   - The additive noise components N_1 and N_2 at the outputs of the network are Gaussian distributed, with zero mean and common variance σ_N^2. They are also uncorrelated with each other.
   - Each noise source is uncorrelated with the input signals.
   - The output signals Y_1 and Y_2 are both Gaussian random variables with zero mean.

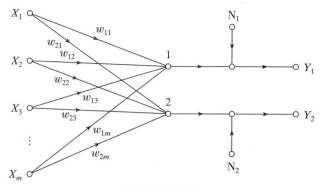

FIGURE P10.6

(a) Determine the mutual information $I(\mathbf{Y}; \mathbf{X})$ between the output vector $\mathbf{Y} = [Y_1,$ $Y_2]^T$ and the input vector $\mathbf{X} = [X_1, X_2, ..., X_m]^T$.

(b) Using the result derived in part (a), investigate the redundancy/diversity tradeoff under the following conditions (Linsker, 1988a):

    (i) Large noise variance, represented by σ_N^2 being large compared to the variances of Y_1 and Y_2.

    (ii) Low noise variance, represented by σ_N^2 being small compared to the variances of Y_1 and Y_2.

10.7 In the variant of the Infomax principle described in Section 10.9, due to Becker and Hinton (1992), the objective is to maximize the mutual information $I(Y_a; Y_b)$ between the outputs Y_a and Y_b of a noisy neural system due to the input vectors $\mathbf{X}_a$ and $\mathbf{X}_b$. In another approach discussed in Becker and Hinton (1992), a different objective is set: maximize the mutual information $I\left(\frac{Y_a + Y_b}{2}; S\right)$ between the average of the outputs Y_a and Y_b and the underlying signal component S common to these two outputs.

Using the noisy model described in Eqs. (10.59) and (10.60), do the following:

(a) Show that

$$I\left(\frac{Y_a + Y_b}{2}; S\right) = \frac{\operatorname{var}[Y_a + Y_b]}{\operatorname{var}[N_a + N_b]}$$

where N_a and N_b are the noise components in Y_a and Y_b, respectively.

(b) Demonstrate the interpretation of this mutual information as a signal-plus-noise to noise ratio.

Independent Components Analysis

10.8 Make a detailed comparison between principal components analysis (discussed in Chapter 8) and independent components analysis (discussed in this chapter).

10.9 Independent components analysis may be used as a preprocessing step for approximate data analysis before detection and classification (Comon, 1994). Discuss the property of independent components analysis that can be exploited for this application.

10.10 *Darmois' theorem* states that the sum of independent variables can be Gaussian distributed only if these variables are themselves Gaussian distributed (Darmois, 1953). Use independent components analysis to prove this theorem.

10.11 In practice, an algorithmic implementation of independent components analysis can only go for "as statistically independent as possible." Contrast the solution to the blind source separation problem using such an algorithm to the solution obtained using a decorrelation method. Assume that the covariance matrix of the observation vector is nonsingular.

10.12 Referring to the scheme described in Fig. 10.9, show that minimizing the mutual information between any two components of the demixer output $\mathbf{Y}$ is equivalent to minimizing the Kullback–Leibler divergence between the paramaterized probability density function $f_{\mathbf{Y}}(\mathbf{y}, \mathbf{W})$ and the corresponding factorial distribution $\tilde{f}_{\mathbf{Y}}(\mathbf{y}, \mathbf{W})$.

10.13 The adaptive algorithm for blind source separation described in Eq. (10.104) has two important properties: (1) the equivariant property, and (2) the property that the weight matrix $\mathbf{W}$ is maintained nonsingular. Property (1) is discussed in some detail in the latter part of Section 10.11. In this problem we consider the second property.

Provided that the initial value $\mathbf{W}(0)$ used in starting the algorithm of Eq. (10.104) satisfies the condition $|\det(\mathbf{W}(0))| \neq 0$, show that

$$|\det(\mathbf{W}(n))| \neq 0 \qquad \text{for all } n$$

This is the necessary and sufficient condition for ensuring that $\mathbf{W}(n)$ is nonsingular for all n.

10.14 In this problem we formulate the batch version of the blind source-separation algorithm described in Eq. (10.104). Specifically we write

$$\Delta\mathbf{W} = \eta\left(\mathbf{I} - \frac{1}{N}\Phi(\mathbf{Y})\mathbf{Y}^T\right)\mathbf{W}$$

where

$$\mathbf{Y} = \begin{bmatrix} y_1(1) & y_1(2) & \cdots & y_1(N) \\ y_2(1) & y_2(2) & \cdots & y_2(N) \\ \vdots & \vdots & & \vdots \\ y_m(1) & y_m(2) & \cdots & y_m(N) \end{bmatrix}$$

and

$$\Phi(\mathbf{Y}) = \begin{bmatrix} \varphi(y_1(1)) & \varphi(y_1(2)) & \cdots & \varphi(y_1(N)) \\ \varphi(y_2(1)) & \varphi(y_2(2)) & \cdots & \varphi(y_2(N)) \\ \vdots & \vdots & & \vdots \\ \varphi(y_m(1)) & \varphi(y_m(2)) & \cdots & \varphi(y_m(N)) \end{bmatrix}$$

where N is the number of available data points. Justify the formulation of the adjustment $\Delta\mathbf{W}$ applied to the weight matrix $\mathbf{W}$ as described.

Maximum Entropy Method

10.15 Consider Fig. 10.15 in which we have

$$\mathbf{Y} = \mathbf{W}\mathbf{X}$$

where

$$\mathbf{Y} = [Y_1, Y_2, ..., Y_m]^T$$
$$\mathbf{X} = [X_1, X_2, ..., X_m]^T$$

and $\mathbf{W}$ is an m-by-m weight matrix. Let

$$\mathbf{Z} = [Z_1, Z_2, ..., Z_m]^T$$

where

$$Z_k = \varphi(Y_k), \qquad k = 1, 2, ..., m$$

(a) Show that the joint entropy of $\mathbf{Z}$ is related to the Kullback–Leibler divergence $D_{f\|\tilde{f}}$ as follows:

$$h(\mathbf{Z}) = -D_{f\|\tilde{f}} + D_{f\|q}$$

where $D_{f\|q}$ is the Kullback–Leibler divergence between (a) the probability density function of the statistically independent (i.e., factorized) version of the output vector $\mathbf{Y}$ and (b) a "probability density function" defined by $\Pi_{i=1}^{m} q(y_i)$.

(b) How is the formula for $h(\mathbf{Z})$ modified for the case when $q(Y_i)$ is equal to the probability density function of the original source output U_i for all i?

10.16 **(a)** Starting with Eq. (10.134), derive the result given in Eq. (10.135).

(b) For the logistic function described in Eq. (10.136), show that the use of Eq. (10.135) yields the formula given in Eq. (10.137).

CHAPTER 11

Stochastic Machines and Their Approximates Rooted in Statistical Mechanics

11.1 INTRODUCTION

For our last class of unsupervised (self-organized) learning systems, we turn to statistical mechanics as the source of ideas. The subject of *statistical mechanics* encompasses the formal study of macroscopic equilibrium properties of large systems of elements that are subject to the microscopic laws of mechanics. The main aim of statistical mechanics is to derive the thermodynamic properties of macroscopic bodies starting from the motion of microscopic elements such as atoms and electrons (Landau and Lifshitz, 1980; Parisi, 1988). The number of degrees of freedom encountered here is enormous, making the use of probabilistic methods mandatory. As with Shannon's information theory, the concept of entropy plays a vital role in the study of statistical mechanics: The more ordered the system, or the more concentrated the underlying probability distribution, the smaller the entropy. By the same token, we can say that the more disordered the system, or the more uniform the underlying probability distribution, the larger the entropy. In 1957, Jaynes showed that entropy can be used not only as the starting point of formulating statistical inference as described in the previous chapter, but also for generating the Gibbs distribution that is basic to the study of statistical mechanics.

Interest in the use of statistical mechanics as a basis for the study of neural networks goes back to the early works of Cragg and Temperley (1954) and Cowan (1968). The *Boltzmann machine* (Hinton and Sejnowski, 1983, 1986; Ackley et al., 1985) is perhaps the first multilayer learning machine inspired by statistical mechanics. The machine is named in recognition of the formal equivalence between Boltzmann's original work on statistical thermodynamics and the network's own dynamic behavior. Basically, the Boltzmann machine is a device for modeling the underlying probability distribution of a given data set, from which conditional distributions for use in tasks such as pattern completion and pattern classification can be derived. Unfortunately, the learning process in the Boltzmann machine is painfully slow. This shortcoming has motivated modifications to the Boltzmann machine and inspired the formulation of new stochastic machines. These issues constitute the bulk of the material presented in this chapter.

Organization of the Chapter

The chapter is organized in three parts. The first part consists of Sections 11.2 to 11.6. Section 11.2 presents a brief review of statistical mechanics. This is followed by Section 11.3, a brief review of a special kind of stochastic process known as the Markov chain that is commonly encountered in the study of statistical mechanics. Sections 11.4, 11.5, and 11.6 describe three stochastic simulation techniques: the Metropolis algorithm, simulated annealing, and Gibbs sampling.

The second part of the chapter, made up of Sections 11.7 to 11.9, discusses three types of stochastic machines. Section 11.7 describes the Boltzmann machine. Section 11.8 describes sigmoid belief networks. Section 11.9 describes another novel stochastic machine known as the Helmholtz machine.

The last part of the chapter, made up of Sections 11.10 to 11.13, discusses approximations to stochastic machines. The approximations are based on the idea of mean-field theory in statistical mechanics. Section 11.10 discusses mean-field theory in general terms. Section 11.11 discusses a naive mean-field theory of the Boltzmann machine, followed by a more principled approach to the mean-field theory of sigmoid belief networks in Section 11.12. Section 11.13 describes deterministic annealing, which is an approximation to simulated annealing.

The chapter concludes with some final remarks in Section 11.14.

11.2   STATISTICAL MECHANICS

Consider a physical system with many degrees of freedom, that can reside in any one of a large number of possible states. Let p_i denote the probability of occurrence of state i, for example, with the following properties:

$$p_i \geq 0 \quad \text{for all } i \tag{11.1}$$

and

$$\sum_i p_i = 1 \tag{11.2}$$

Let E_i denote the *energy* of the system when it is in state i. A fundamental result from statistical mechanics tells us that when the system is in thermal equilibrium with its surrounding environment, state i occurs with a probability defined by

$$p_i = \frac{1}{Z} \exp\left(-\frac{E_i}{k_B T}\right) \tag{11.3}$$

where T is the *absolute temperature* in kelvins, k_B is *Boltzmann's constant*, and Z is a constant that is independent of all states. One degree kelvin corresponds to $-273°$ on the Celsius scale, and $k_B = 1.38 \times 10^{-23}$ joules/kelvin.

Equation (11.2) defines the condition for the normalization of probabilities. Imposing this condition on Eq. (11.3), we get

$$Z = \sum_i \exp\left(-\frac{E_i}{k_B T}\right) \tag{11.4}$$

The normalizing quantity Z is called the *sum over states* or the *partition function*. (The symbol Z is commonly used because the German name for this term is *Zustadsumme*.) The probability distribution of Eq. (11.3) is called the *canonical distribution* or *Gibbs distribution*;[1] the exponential factor $\exp(-E_i/k_B T)$ is called the *Boltzmann factor*.

The following two points are noteworthy from the Gibbs distribution:

1. States of low energy have a higher probability of occurrence than states of high energy.
2. As the temperature T is reduced, the probability is concentrated on a smaller subset of low-energy states.

In the context of neural networks, which is our primary concern, the parameter T may be viewed as a *pseudotemperature* that controls thermal fluctuations representing the effect of "synaptic noise" in a neuron. Its precise scale is therefore irrelevant. Accordingly, we may choose to measure it by setting the constant k_B equal to unity, and thereby redefine the probability p_i and partition function Z as follows, respectively:

$$p_i = \frac{1}{Z}\exp\left(-\frac{E_i}{T}\right) \tag{11.5}$$

and

$$Z = \sum_i \exp\left(-\frac{E_i}{T}\right) \tag{11.6}$$

Henceforth our treatment of statistical mechanics is based on these two definitions, where T is referred to simply as the *temperature of the system*. From Eq. (11.5) we note that $-\log p_i$ may be viewed as a form of "energy" measured at unit temperature.

Free Energy and Entropy

The Helmholtz *free energy* of a physical system, denoted by F, is defined in terms of the partition function Z as follows:

$$F = -T \log Z \tag{11.7}$$

The *average energy* of the system is defined by

$$<E> = \sum_i p_i E_i \tag{11.8}$$

where $<\cdot>$ denotes the ensemble averaging operation. Thus, using Eqs. (11.5) to (11.8), we see that the difference between the average energy and free energy is given by

$$<E> - F = -T \sum_i p_i \log p_i \tag{11.9}$$

The quantity on the right-hand side of Eq. (11.9), except for the temperature T, is recognized as the *entropy* of the system, as shown by

$$H = -\sum_i p_i \log p_i \tag{11.10}$$

We may therefore rewrite Eq. (11.9) in the form

$$<E> - F = TH$$

or, equivalently,

$$F = <E> - TH \tag{11.11}$$

Consider two systems, A and A', placed in thermal contact with each other. Suppose that system A is small compared to system A', so that A' acts as a heat reservoir at some constant temperature, T. The total entropy of the two systems tends to increase in accordance with the relation (Reif, 1967)

$$\Delta H + \Delta H' \geq 0$$

where ΔH and $\Delta H'$ denote the entropy changes of systems A and A', respectively. The implication of this relation, in light of Eq. (11.11), is that the free energy of the system, F, tends to decrease and become a minimum in equilibrium situation. From statistical mechanics we find that the resulting probability distribution is defined by the Gibbs distribution. We thus have an important principle called the *principle of minimal free energy*, which may be stated as follows (Landau and Lifshitz, 1980; Parisi, 1988):

> The minimum of the free energy of a stochastic system with respect to variables of the system is achieved at thermal equilibrium, at which point the system is governed by the Gibbs distribution.

Nature likes to find a physical system with minimum free energy.

11.3 MARKOV CHAINS

Consider a system whose evolution is described by a stochastic process $\{X_n, n = 1, 2, ...\}$, consisting of a family of random variables. The value x_n assumed by the random variable X_n at discrete time n is called the *state* of the system at that time instant. The space of all possible values that the random variables can assume is called the *state space* of the system. If the structure of the stochastic process $\{X_n, n = 1, 2, ...\}$ is such that the conditional probability distribution of X_{n+1} depends only on the value of X_n and is independent of all previous values, we say that the process is a *Markov chain* (Feller, 1950; Ash, 1965). More precisely, we have

$$P(X_{n+1} = x_{n+1}|X_n = x_n, ..., X_1 = x_1) = P(X_{n+1} = x_{n+1}|X_n = x_n) \tag{11.12}$$

which is called the *Markov property*. In words, *a sequence of random variables* $X_1, X_2, ...,$ X_n, X_{n+1} *forms a Markov chain if the probability that the system is in state* x_{n+1} *at time* $n + 1$ *depends exclusively on the probability that the system is in state* x_n *at time n.*

We may therefore think of the Markov chain as a *generative model*, consisting of a number of states linked together (on a pair-wise basis) by possible transitions.

Each time a particular state is visited, the model outputs the symbol associated with that state.

Transition Probabilities

In a Markov chain, the transition from one state to another is *probabilistic*, but the production of an output symbol is deterministic. Let

$$p_{ij} = P(X_{n+1} = j | X_n = i) \tag{11.13}$$

denote the *transition probability* from state i at time n to state j at time $n + 1$. Since the p_{ij} are conditional probabilities, all transition probabilities must satisfy two conditions:

$$p_{ij} \geq 0 \quad \text{for all } (i, j) \tag{11.14}$$

$$\sum_j p_{ij} = 1 \quad \text{for all } i \tag{11.15}$$

We will assume that the transition probabilities are fixed and do not change with time; that is, Eq. (11.13) is satisfied for all time n. In such a case the Markov chain is said to be *homogeneous* in time.

In the case of a system with a finite number of possible states K, for example, the transition probabilities constitute a K-by-K matrix:

$$\mathbf{P} = \begin{bmatrix} p_{11} & p_{12} & \cdots & p_{1K} \\ p_{21} & p_{22} & \cdots & p_{2K} \\ \vdots & \vdots & & \vdots \\ p_{K1} & p_{K2} & \cdots & p_{KK} \end{bmatrix} \tag{11.16}$$

whose individual elements satisfy the conditions described in Eqs. (11.14) and (11.15); the latter condition says that each row of $\mathbf{P}$ must add to one. A matrix of this type is called a *stochastic matrix*. Any stochastic matrix can serve as a matrix of transition probabilities.

The definition of one-step transition probability given in Eq. (11.13) may be generalized to cases where the transition from one state to another takes place in some fixed number of steps. Let $p_{ij}^{(m)}$ denote the *m-step transition probability* from state i to state j:

$$p_{ij}^{(m)} = P(X_{n+m} = x_j | X_n = x_i), \quad m = 1, 2, \ldots \tag{11.17}$$

We may view $p_{ij}^{(m)}$ as the sum over all intermediate states k through which the system passes in its transition from state i to state j. Specifically, $p_{ij}^{(m+1)}$ is related to $p_{ij}^{(m)}$ by the recursive relation:

$$p_{ij}^{(m+1)} = \sum_k p_{ik}^{(m)} p_{kj}, \quad m = 1, 2, \ldots \tag{11.18}$$

with

$$p_{ik}^{(1)} = p_{ik}$$

Equation (11.18) may be generalized as follows:

$$p_{ij}^{(m+n)} = \sum_k p_{ik}^{(m)} p_{kj}^{(n)}, \quad (m, n) = 1, 2, \ldots \tag{11.19}$$

which is a special case of the *Chapman–Kolmogorov identity* (Feller, 1950).

When a state of the chain can only reoccur at time intervals that are multiples of d, where d is the largest such integer, we say that the state has *period d*. A Markov chain is called *aperiodic* if all of its states have period 1.

Recurrent Properties

Suppose a Markov chain starts in state i. The state i is said to be a *recurrent* state if the Markov chain returns to state i with probability 1; that is,

$$f_i = P(\text{ever returning to state } i) = 1$$

If the probability f_i is less than 1, state i is said to be a *transient* state (Leon–Garcia, 1994).

If the Markov chain starts in a recurrent state, that state reoccurs an infinite number of times. If it starts in a transient state, that state reoccurs only a finite number of times, which may be explained as follows. We may view the reoccurrence of state i as a Bernoulli trial with a probability of success equal to f_i. The number of returns is thus a geometric random variable with a mean of $(1 - f_i^{-1})$. If $f_i < 1$, it follows that the number of an infinite number of successes is zero. Therefore, a transient state does not reoccur after some finite number of returns (Leon–Garcia, 1994).

If a Markov chain has some transient states and some recurrent states, then the process will eventually move only among the recurrent states.

Irreducible Markov Chains

The state j of a Markov chain is said to be *accessible* from state i if there is a finite sequence of transitions from i to j with positive probability. If the states i and j are accessible to each other, the states i and j of the Markov chain are said to *communicate* with each other. This communication is described by writing $i \leftrightarrow j$. Clearly, if state i communicates with state j and state j communicates with state k, that is, $i \leftrightarrow j$ and $j \leftrightarrow k$, then state i communicates with state k, that is, $i \leftrightarrow k$.

If two states of a Markov chain communicate with each other, they are said to belong to the same *class*. In general, the states of a Markov chain consist of one or more disjoint classes. If, however, all the states consist of a single class, the Markov chain is said to be *indecomposible* or *irreducible*. In other words, by starting at any state of an irreducible Markov chain we can reach any other state with positive probability. Reducible chains are of little practical interest in most areas of application. Accordingly, we restrict our attention to irreducible chains.

Consider an irreducible Markov chain that starts in a recurrent state i at time $n = 0$. Let $T_i(k)$ denote the time that elapses between the $(k-1)$th and kth returns to state i. The *mean recurrence time* of state i is defined as the expectation of $T_i(k)$ over the returns k. The *steady-state probability* of state i, denoted by π_i, is equal to the reciprocal of the mean recurrence time $E[T_i(k)]$, as shown by

$$\pi_i = \frac{1}{E[T_i(k)]}$$

If $E[T_i(k)] < \infty$, that is, $\pi_i > 0$, the state i is said to be a *positive recurrent* (*persistent*) *state*. If $E[T_i(k)] = \infty$, that is, $\pi_i = 0$, the state i is said to be a *null recurrent* (*persistent*) *state*. The implication of $\pi_i = 0$ is that the Markov chain will eventually reach a point where a return to state i is impossible. Positive recurrence and null recurrence are *different* class properties, which means that a Markov chain with positive recurrent and null recurrent states is reducible.

Ergodic Markov Chains

In principle, *ergodicity* means that we may substitute time averages for ensemble averages. In the context of a Markov chain, ergodicity means that the long-term proportion of time spent by the chain in state i corresponds to the steady-state probability π_i, which may be justified as follows. The *proportion of time spent in state i after k returns*, denoted by $v_i(k)$, is defined by

$$v_i(k) = \frac{k}{\sum_{l=1}^{k} T_i(l)}$$

The return times $T_i(l)$ form a sequence of independently and identically distributed random variables since, by definition, each return time is statistically independent of all previous return times. Moreover, in the case of a recurrent state i the chain returns to state i an infinite number of times. Hence, as the number of returns k approaches infinity, the *law of large numbers* states that the proportion of time spent in state i approaches the steady-state probability, as shown by

$$\lim_{k \to \infty} v_i(k) = \pi_i \quad \text{for } i = 1, 2, ..., K \tag{11.20}$$

A sufficient, but not necessary condition, for a Markov chain to be *ergodic* is for it to be both irreducible and aperiodic.

Convergence to Stationary Distributions

Consider an ergodic Markov chain characterized by a stochastic matrix $\mathbf{P}$. Let the row vector $\boldsymbol{\pi}^{(n-1)}$ denote the *state distribution vector* of the chain at time $n - 1$; the jth element of $\boldsymbol{\pi}^{(n-1)}$ is the probability that the chain is in state x_j at time $n - 1$. The state distribution vector at time n is defined by

$$\boldsymbol{\pi}^{(n)} = \boldsymbol{\pi}^{(n-1)}\mathbf{P} \tag{11.21}$$

By iteration of Eq. (11.21), we obtain

$$\boldsymbol{\pi}^{(n)} = \boldsymbol{\pi}^{(n-1)}\mathbf{P} = \boldsymbol{\pi}^{(n-2)}\mathbf{P}^2 = \boldsymbol{\pi}^{(n-3)}\mathbf{P}^3 = \cdots$$

and finally we may write

$$\boldsymbol{\pi}^{(n)} = \boldsymbol{\pi}^{(0)}\mathbf{P}^n \tag{11.22}$$

where $\boldsymbol{\pi}^{(0)}$ is the *initial value* of the state distribution vector. In words, the state distribution vector of the Markov chain at time n is the product of the initial state distribution vector $\boldsymbol{\pi}^{(0)}$ and the nth power of the stochastic matrix $\mathbf{P}$.

Let $p_{ij}^{(n)}$ denote the ij-th element of $\mathbf{P}^n$. Suppose that as time n approaches infinity, $p_{ij}^{(n)}$ tends to π_j independent of i, where π_j is the steady-state probability of state j. Correspondingly, for large n, the matrix $\mathbf{P}^n$ approaches the limiting form of a square matrix with identical rows as shown by

$$\lim_{n \to \infty} \mathbf{P}^n = \begin{bmatrix} \pi_1 & \pi_2 & \cdots & \pi_K \\ \pi_1 & \pi_2 & \cdots & \pi_K \\ \vdots & \vdots & & \vdots \\ \pi_1 & \pi_2 & \cdots & \pi_K \end{bmatrix}$$

$$= \begin{bmatrix} \boldsymbol{\pi} \\ \boldsymbol{\pi} \\ \vdots \\ \boldsymbol{\pi} \end{bmatrix} \tag{11.23}$$

where $\boldsymbol{\pi}$ is a row vector consisting of $\pi_1, \pi_2, \ldots, \pi_K$. We then find from Eq. (11.22) that (after rearranging terms)

$$\left[\sum_{j=1}^{K} \pi_j^{(0)} - 1 \right] \boldsymbol{\pi} = \mathbf{0}$$

Since, by definition $\sum_{j=1}^{K} \pi_j^{(0)} = 1$, this condition is satisfied by the vector $\boldsymbol{\pi}$ independent of the initial distribution.

We may now state the *ergodicity theorem* for Markov chains as follows (Feller, 1950; Ash, 1965):

Let an ergodic Markov chain with states $x_1, x_2, \ldots, x_K$ and stochastic matrix $\mathbf{P} = \{p_{ij}\}$ be irreducible. The chain then has a unique stationary distribution to which it converges from any initial state; that is, there is a unique set of numbers $\{\pi_j\}_{j=1}^{K}$ such that

1. $\lim_{n \to \infty} p_{ij}^{(n)} = \pi_j$ for all i (11.24)

2. $\pi_j > 0$ for all j (11.25)

3. $\sum_{j=1}^{K} \pi_j = 1$ (11.26)

4. $\pi_j = \sum_{i=1}^{K} \pi_i p_{ij}$ for $j = 1, 2, \ldots, K$ (11.27)

Conversely, suppose that the Markov chain is irreducible and aperiodic, and that there exist numbers $\{\pi_j\}_{j=1}^K$ satisfying Eqs. (11.25) through (11.27). Then the chain is ergodic, the π_j are given by Eq. (11.24), and the mean recurrence time of state j is $1/\pi_j$.

The probability distribution $\{\pi_j\}_{j=1}^K$ is called an *invariant* or *stationary distribution*. It is so called because it persists forever once it is established. In light of the ergodicity theorem, we may thus say the following:

- Starting from an arbitrary initial distribution, the transition probabilities of a Markov chain will converge to a stationary distribution provided that such a distribution exists.
- The stationary distribution of the Markov chain is completely independent of the initial distribution if the chain is ergodic.

Example 11.1

Consider a Markov chain whose *state-transition diagram* is depicted in Fig. 11.1. The chain has two states x_1 and x_2. The stochastic matrix of the chain is

$$\mathbf{P} = \begin{bmatrix} \dfrac{1}{4} & \dfrac{3}{4} \\ \dfrac{1}{2} & \dfrac{1}{2} \end{bmatrix}$$

which satisfies the conditions of Eqs. (11.14) and (11.15).

Suppose the initial condition is

$$\boldsymbol{\pi}^{(0)} = \begin{bmatrix} \dfrac{1}{6} & \dfrac{5}{6} \end{bmatrix}$$

From Eq. (11.21) we find that the state distribution vector at time $n = 1$ is

$$\boldsymbol{\pi}^{(1)} = \boldsymbol{\pi}^{(0)}\mathbf{P}$$

$$= \begin{bmatrix} \dfrac{1}{6} & \dfrac{5}{6} \end{bmatrix} \begin{bmatrix} \dfrac{1}{4} & \dfrac{3}{4} \\ \dfrac{1}{2} & \dfrac{1}{2} \end{bmatrix}$$

$$= \begin{bmatrix} \dfrac{11}{24} & \dfrac{13}{24} \end{bmatrix}$$

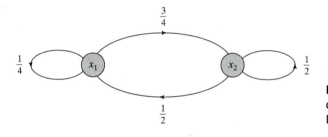

FIGURE 11.1   State-transition diagram of Markov chain for Example 11.1.

Raising the stochastic matrix $\mathbf{P}$ to power $n = 2, 3, 4$, we have

$$\mathbf{P}^2 = \begin{bmatrix} 0.4375 & 0.5625 \\ 0.3750 & 0.6250 \end{bmatrix}$$

$$\mathbf{P}^3 = \begin{bmatrix} 0.4001 & 0.5999 \\ 0.3999 & 0.6001 \end{bmatrix}$$

$$\mathbf{P}^4 = \begin{bmatrix} 0.4000 & 0.6000 \\ 0.4000 & 0.6000 \end{bmatrix}$$

Thus, $\pi_1 = 0.4000$ and $\pi_2 = 0.6000$. In this example, convergence to the stationary distribution is accomplished essentially in $n = 4$ iterations. With both π_1 and π_2 being greater than zero, both states are positive recurrent, and the chain is therefore irreducible. Note also that the chain is aperiodic since the greatest common divisor of all integers $n \geq 1$, such that $(\mathbf{P}^n)_{jj} > 0$, is equal to 1. We therefore conclude that the Markov chain of Fig. 11.1 is ergodic.

∎

Example 11.2

Consider a Markov chain with a stochastic matrix some of whose elements are zero:

$$\mathbf{P} = \begin{bmatrix} 0 & 0 & 1 \\ \dfrac{1}{3} & \dfrac{1}{6} & \dfrac{1}{2} \\ \dfrac{3}{4} & \dfrac{1}{4} & 0 \end{bmatrix}$$

The state transition diagram of the chain is depicted in Fig. 11.2.
    By applying Eq. (11.27) we obtain the following set of simultaneous equations:

$$\pi_1 = \frac{1}{3}\pi_2 + \frac{3}{4}\pi_3$$

$$\pi_2 = \frac{1}{6}\pi_2 + \frac{1}{4}\pi_3$$

$$\pi_3 = \pi_1 + \frac{1}{2}\pi_2$$

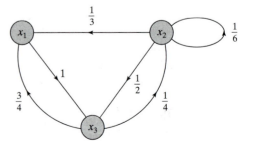

FIGURE 11.2   State-transition diagram of Markov chain for Example 11.2.

By solving these equations for $\pi_1, \pi_2,$ and π_3 we get

$$\pi_1 = 0.3953$$

$$\pi_2 = 0.1395$$

$$\pi_3 = 0.4652$$

The given Markov chain is ergodic with its stationary distribution defined by π_1, π_2 and π_3. ∎

Classification of States

On the basis of the material presented here, we may develop a summary of the classes to which a state can belong as shown in Fig. 11.3 (Feller, 1950; Leon–Garcia, 1994). This figure also includes the associated long-term behavior of the state.

Principle of Detailed Balance

Equations (11.25) and (11.26) merely emphasize the fact that the numbers π_j are probabilities. Equation (11.27) is the critical one because it also has to be satisfied for the Markov chain to be irreducible and therefore, for a stationary distribution to exist. This latter equation is a restatement of the principle of detailed balance that arises in first-order reaction kinetics. The *principle of detailed balance* states that at thermal equilibrium, the rate of occurrence of any transition equals the corresponding rate of occurrence of the inverse transition, as shown by (Reif, 1965):

$$\pi_i p_{ij} = \pi_j p_{ji} \tag{11.28}$$

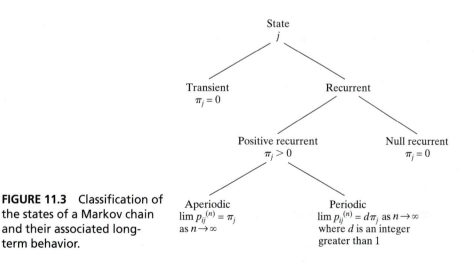

FIGURE 11.3 Classification of the states of a Markov chain and their associated long-term behavior.

To derive the relation of Eq. (11.27), we may manipulate the summation on the right-hand side of this equation as follows:

$$\sum_{i=1}^{K} \pi_i p_{ij} = \sum_{i=1}^{K} \left(\frac{\pi_i}{\pi_j} p_{ij} \right) \pi_j$$

$$= \sum_{i=1}^{K} (p_{ji}) \pi_j$$

$$= \pi_j$$

In the second line of this expression we made use of the principle of detailed balance, and in the last line we made use of the fact that the transition probabilities of a Markov chain satisfy the condition (see Eq. (11.15) with the roles of i and j interchanged):

$$\sum_{i=1}^{K} p_{ji} = 1 \quad \text{for all } j$$

Note that the principle of detailed balance implies that the distribution $\{\pi_j\}$ is a stationary distribution.

11.4 METROPOLIS ALGORITHM

Now that we understand the composition of a Markov chain, we will use it to formulate a stochastic algorithm for simulating the evolution of a physical system to thermal equilibrium. The algorithm is called the *Metropolis algorithm* (Metropolis et al., 1953). It is a modified Monte Carlo method, introduced in the early days of scientific computation for the stochastic simulation of a collection of atoms in equilibrium at a given temperature.

Suppose that the random variable X_n representing an arbitrary Markov chain is in state x_i at time n. We randomly generate a new state x_j, representing a realization of another random variable Y_n. It is assumed that the generation of this new state satisfies the symmetry condition:

$$P(Y_n = x_j | X_n = x_i) = P(Y_n = x_i | X_n = x_j)$$

Let ΔE denote the energy difference resulting from the transition of the system from state $X_n = x_i$ to state $Y_n = x_j$. If the energy difference ΔE is negative, the transition leads to a state with lower energy and the transition is accepted. The new state is then accepted as the starting point for the next step of the algorithm, that is, we put $X_{n+1} = Y_n$. If, on the other hand, the energy difference ΔE is positive, the algorithm proceeds in a probabilistic manner at that point. First, we select a random number ξ uniformly distributed in the range $[0, 1]$. If $\xi < \exp(-\Delta E/T)$, where T is the operating temperature, the transition is accepted and we put $X_{n+1} = Y_n$. Otherwise, the transition is rejected and we put $X_{n+1} = X_n$; that is, the old configuration is reused for the next step of the algorithm.

Choice of Transition Probabilities

Let the arbitrary Markov chain have *a priori* transition probabilities denoted by τ_{ij}, which satisfy three conditions:

1. *Nonnegativity*:

$$\tau_{ij} \geq 0 \quad \text{for all } (i,j)$$

2. *Normalization*:

$$\sum_j \tau_{ij} = 1 \quad \text{for all } i$$

3. *Symmetry*:

$$\tau_{ij} = \tau_{ji} \quad \text{for all } (i,j)$$

Let π_i denote the steady state probability that the Markov chain is in state $x_i, i = 1, 2, \ldots,$ K. We may then use the symmetric $\tau_{ij}s$ and the probability distribution ratio π_j/π_i, to be defined, to formulate the desired set of transition probabilities as (Beckerman, 1997):

$$p_{ij} = \begin{cases} \tau_{ij}\left(\dfrac{\pi_j}{\pi_i}\right) & \text{for } \dfrac{\pi_j}{\pi_i} < 1 \\[2em] \tau_{ij} & \text{for } \dfrac{\pi_j}{\pi_i} \geq 1 \end{cases} \tag{11.29}$$

To ensure that the transition probabilities are normalized to unity, we introduce this additional definition for the probability of no transition:

$$p_{ii} = \tau_{ii} + \sum_{j \neq i} \tau_{ij}\left(1 - \frac{\pi_j}{\pi_i}\right)$$

$$= 1 - \sum_{j \neq i} \alpha_{ij}\tau_{ij} \tag{11.30}$$

where α_{ij} is the moving probability defined by

$$\alpha_{ij} = \min\left(1, \frac{\pi_j}{\pi_i}\right) \tag{11.31}$$

The only outstanding requirement is how to choose the ratio π_j/π_i. To cater to this requirement, we choose the probability distribution that we want the Markov chain to converge to be a Gibbs distribution, as shown by

$$\pi_j = \frac{1}{Z}\exp\left(-\frac{E_j}{T}\right)$$

in which case the probability distribution ratio π_j/π_i takes the simple form

$$\frac{\pi_j}{\pi_i} = \exp\left(-\frac{\Delta E}{T}\right) \tag{11.32}$$

where

$$\Delta E = E_j - E_i \tag{11.33}$$

By using the ratio of probability distributions we have eliminated dependence on the partition function Z.

By construction, the transition probabilities are all nonnegative and normalized to unity, as required by Eqs. (11.14) and (11.15). Moreover, they satisfy the principle of detailed balance defined by Eq. (11.28). This principle is a sufficient condition for thermal equilibrium. To demonstrate that the principle of detailed balance is satisfied, we offer the following considerations:

Case 1: $\Delta E < 0$. Suppose that in going from state x_i to state x_j, the energy change ΔE is negative. From Eq. (11.32) we find that $(\pi_j/\pi_i) > 1$, so the use of Eq. (11.29) yields

$$\pi_i p_{ij} = \pi_i \tau_{ij} = \pi_i \tau_{ji}$$

and

$$\pi_j p_{ji} = \pi_j \left(\frac{\pi_i}{\pi_j} \tau_{ji}\right) = \pi_i \tau_{ji}$$

Hence, the principle of detailed balance is satisfied for $\Delta E < 0$.

Case 2: $\Delta E > 0$. Suppose next that the energy change ΔE in going from state x_i to state x_j is positive. In this case we find that $(\pi_j/\pi_i) < 1$, and the use of Eq. (11.29) yields

$$\pi_i p_{ij} = \pi_i \left(\frac{\pi_j}{\pi_i} \tau_{ij}\right) = \pi_j \tau_{ij} = \pi_j \tau_{ji}$$

and

$$\pi_j p_{ji} = \pi_j \tau_{ji}$$

Here again we see that the principle of detailed balance is satisfied.

To complete the picture, we need to clarify the use of the *a priori* transition probabilities denoted by τ_{ij}. These transition probabilities are in fact the probabilistic model of the random step in the Metropolis algorithm. From the description of the algorithm presented earlier, we recall that the random step is followed by a random decision. We may therefore conclude that the transition probabilities p_{ij} defined in Eqs. (11.29) and (11.30) in terms of the *a priori* transition probabilities, τ_{ij}, and the steady state probabilities, π_j, are indeed the correct choice for the Metropolis algorithm.

It is noteworthy that the stationary distribution generated by the Metropolis algorithm does not uniquely determine the Markov chain. The Gibbs distribution at equilibrium may be generated by using an update rule other than the Monte Carlo rule applied in the Metropolis algorithm. For example, it may be generated using the Boltzmann learning rule due to Ackley et al. (1986); this latter rule is discussed in Section 11.7.

11.5 SIMULATED ANNEALING

Consider the problem of finding a low energy system whose states are ordered in a Markov chain. From Eq. (11.11) we observe that as the temperature T approaches zero, the free energy F of the system approaches the average energy $<E>$. With $F \rightarrow <E>$, we next observe from the principle of minimal free energy that the Gibbs distribution, which is the stationary distribution of the Markov chain, collapses on the global minima of the average energy $<E>$ as $T \rightarrow 0$. In other words, low energy ordered states are strongly favored at low temperatures. These observations prompt us to raise the question: Why not simply apply the Metropolis algorithm for generating a population of configurations representative of the stochastic system at very low temperatures? We do not advocate the use of such a strategy because the rate of convergence of the Markov chain to thermal equilibrium is extremely slow at very low temperatures. Rather, the preferred method for improved computational efficiency is to operate the stochastic system at a high temperature where convergence to equilibrium is fast, and then maintain the system at equilibrium as the temperature is carefully lowered. That is, we use a combination of two related ingredients:

- A schedule that determines the rate at which the temperature is lowered.
- An algorithm—exemplified by the Metropolis algorithm—that iteratively finds the equilibrium distribution at each new temperature in the schedule by using the final state of the system at the previous temperature as the starting point for the new temperature.

The twofold scheme that we have just described is the essence of a widely used stochastic relaxation technique known as *simulated annealing*[2] (Kirkpatrick et al., 1983). The technique derives its name from analogy with an annealing process in physics/chemistry where we start the process at high temperature and then lower the temperature slowly while maintaining thermal equilibrium.

The primary objective of simulated annealing is to find the global minimum of a cost function that characterizes large and complex systems.[3] As such, it provides a powerful tool for solving nonconvex optimization problems, motivated by the following simple idea:

> *When optimizing a very large and complex system (i.e., a system with many degrees of freedom), instead of always going downhill, try to go downhill most of the time.*

Simulated annealing differs from conventional iterative optimization algorithms in two important respects:

- The algorithm need not get stuck, since transition out of a local minimum is always possible when the system operates at a nonzero temperature.
- Simulated annealing is *adaptive* in that gross features of the final state of the system are seen at higher temperatures, while fine details of the state appear at lower temperatures.

Annealing Schedule

As already mentioned, the Metropolis algorithm is the basis for the simulated annealing process, in the course of which the temperature T is decreased slowly. That is, the temperature T plays the role of a *control* parameter. The simulated annealing process will converge to a configuration of minimal energy provided that the temperature is decreased no faster than logarithmically. Unfortunately, such an annealing schedule is extremely slow—too slow to be of practical use. In practice, we must resort to a *finite-time approximation* of the asymptotic convergence of the algorithm. The price paid for the approximation is that the algorithm is no longer guaranteed to find a global mini-mum with probability 1. Nevertheless, the resulting approximate form of the algorithm is capable of producing near optimum solutions for many practical applications.

To implement a finite-time approximation of the simulated annealing algorithm, we must specify a set of parameters governing the convergence of the algorithm. These parameters are combined in a so-called *annealing schedule* or *cooling schedule*. The annealing schedule specifies a finite sequence of values of the temperature and a finite number of transitions attempted at each value of the temperature. The anneal-ing schedule due to Kirkpatrick et al. (1983) specifies the parameters of interest as follows:[4]

- *Initial Value of the Temperature.* The initial value T_0 of the temperature is chosen high enough to ensure that virtually all proposed transitions are accepted by the simulated annealing algorithm
- *Decrement of the Temperature.* Ordinarily, the cooling is performed *exponen-tially*, and the changes made in the value of the temperature are small. In particu-lar, the *decrement function* is defined by

$$T_k = \alpha T_{k-1}, \quad k = 1, 2, \ldots \tag{11.34}$$

  where α is a constant smaller than, but close to, unity. Typical values of α lie between 0.8 and 0.99. At each temperature, enough transitions are attempted so that there are 10 *accepted* transitions per experiment on the average.
- *Final Value of the Temperature.* The system is frozen and annealing stops if the desired number of acceptances is not achieved at three successive temperatures.

The latter criterion may be refined by requiring that the *acceptance ratio*, defined as the number of accepted transitions divided by the number of proposed transitions, is smaller than a prescribed value (Johnson et al., 1989).

Simulated Annealing for Combinatorial Optimization

Simulated annealing is particularly well suited for solving combinatorial optimization problems. The objective of *combinatorial optimization* is to minimize the cost function of a finite, discrete system characterized by a large number of possible solutions. Essentially, simulated annealing uses the Metropolis algorithm to generate a sequence of solutions by invoking an analogy between a physical many-particle system and a combinatorial optimization problem.

In simulated annealing, we interpret the energy E_i in the Gibbs distribution of Eq. (11.5) as a numerical cost and the temperature T as a control parameter. The numerical cost assigns to each configuration in the combinatorial optimization prob-lem a scalar value that describes how desirable that particular configuration is to the solution. The next issue in the simulated annealing procedure to be considered is how to identify configurations and generate new configurations from previous ones in a local manner. This is where the Metropolis algorithm performs its role. We may thus summarize the correspondence between the terminology of statistical physics and that of combinatorial optimization as shown in Table 11.1 (Beckerman, 1997).

11.6   GIBBS SAMPLING

Like the Metropolis algorithm, the *Gibbs sampler*[5] generates a Markov chain with the Gibbs distribution as the equilibrium distribution. However, the transition probabilities associated with the Gibbs sampler are nonstationary (Geman and Geman, 1984). In the final analysis, the choice between the Gibbs sampler and the Metropolis algorithm is based on technical details of the problem at hand.

To proceed with a description of this sampling scheme, consider a K-dimensional random vector $\mathbf{X}$ made up of the components $X_1, X_2, ..., X_K$. Suppose that we have knowledge of the conditional distribution of X_k, given values of all the other components of $\mathbf{X}$ for $k = 1, 2, ..., K$. The problem we wish to address is how to obtain a numerical estimate of the marginal density of the random variable X_k for each k. The Gibbs sampler proceeds by generating a value for the conditional distribution for each component of the random vector $\mathbf{X}$, given the values of all other components of $\mathbf{X}$. Specifically, starting from an arbitrary configuration $\{x_1(0), x_2(0), ..., x_K(0)\}$, we make the following drawings on the first iteration of Gibbs sampling:

$x_1(1)$ is drawn from the distribution of X_1, given $x_2(0), x_3(0), ..., x_K(0)$.

$x_2(1)$ is drawn from the distribution of X_2, given $x_1(1), x_3(0), ..., x_K(0)$.

$\vdots$

$x_k(1)$ is drawn from the distribution of X_k, given $x_1(1), ..., x_{k-1}(1), x_{k+1}(0), ..., x_K(0)$.

$\vdots$

$x_K(1)$ is drawn from the distribution of X_K, given $x_1(1), x_2(1), ..., x_{K-1}(1)$.

We proceed in this same manner on the second iteration and every other iteration of the sampling scheme. The following two points should be carefully noted:

1. Each component of the random vector $\mathbf{X}$ is "visited" in the natural order, with the result that a total of K new variates are generated on each iteration.
2. The new value of component X_{k-1} is used immediately when a new value of X_k is drawn for $k = 2, 3, ..., K$.

From this discussion we see that the Gibbs sampler is an *iterative adaptive* scheme. After n iterations of its use, we arrive at the K variates: $X_1(n), X_2(n), ..., X_K(n)$.

TABLE 11.1   Correspondence between Statistical Physics and Combinatorial Optimization

Statistical physics	Combinatorial optimization
Sample	Problem instance
State (configuration)	Configuration
Energy	Cost function
Temperature	Control parameter
Ground-state energy	Minimum cost
Ground-state configuration	Optimal configuration

Under mild conditions, the following three theorems hold for Gibbs sampling (Geman and Geman, 1984; Gelfand and Smith, 1990):

1. *Convergence theorem. The random variable $X_k(n)$ converges in distribution to the true probability distributions of X_k for $k = 1, 2, ..., K$ as n approaches infinity; that is,*

$$\lim_{n \to \infty} P(X_k^{(n)} \le x | x_k(0)) = F_{X_k}(x) \quad \text{for } k = 1, 2, ..., K \quad (11.35)$$

where $F_{X_k}(x)$ is the marginal probability distribution function of X_k.

In fact, a stronger result is proven in Geman and Geman (1984). Specifically, rather than requiring that each component of the random vector $\mathbf{X}$ be visited in repetitions of the natural order, convergence of Gibbs sampling still holds under an arbitrary visiting scheme provided that this scheme does not depend on the values of the variables and that each component of $\mathbf{X}$ is visited on an "infinitely often" basis.

2. *Rate of convergence theorem. The joint probability distribution of the random variables $X_1(n), X_2(n), ..., X_K(n)$ converges to the true joint probability distribution of $X_1, X_2, ..., X_K$ at a geometric rate in n.*

This theorem assumes that the components of $\mathbf{X}$ are visited in the natural order. When, however, an arbitrary but infinitely often visiting approach is used, then a minor adjustment to the rate of convergence is required.

3. *Ergodic theorem. For any measurable function g, for example, of the random variables $X_1, X_2, ..., X_K$ whose expectation exists, we have*

$$\lim_{n \to \infty} \frac{1}{n} \sum_{i=1}^{n} g(X_1(i), X_2(i), ..., X_K(i)) \to E[g(X_1, X_2, ..., X_K)] \quad (11.36)$$

with probability 1 (i.e., almost surely).

The ergodic theorem tells us how to use the output of the Gibbs sampler to obtain numerical estimations of the desired marginal densities.

Gibbs sampling is used in the Boltzmann machine to sample from distributions over hidden neurons; this stochastic machine is discussed in the next section. In the context of a stochastic machine using binary units (e.g., Boltzmann machine), it is noteworthy that the Gibbs sampler is exactly the same as a variant of the Metropolis algorithm. In the standard form of the Metropolis algorithm, we go downhill with probability 1. In contrast, in the alternative form of the Metropolis algorithm, we go downhill with a probability equal to 1 minus the exponential of the energy gap (i.e., the complement of the uphill rule). In other words, if a change lowers the energy E or leaves it unchanged, that change is accepted; if the change increases the energy, it is accepted with probability $\exp(-\Delta E)$ and is rejected otherwise, with the old state then being repeated (Neal, 1993).

11.7 BOLTZMANN MACHINE

The *Boltzmann machine* is a stochastic machine whose composition consists of stochastic neurons. A *stochastic neuron* resides in one of two possible states in a probabilistic manner, as discussed in Chapter 1. These two states may be designated as $+1$

for the "on" state and -1 for the "off" state, or 1 and 0, respectively. We will adopt the former designation. Another distinguishing feature of the Boltzmann machine is the use of *symmetric synaptic connections* between its neurons. The use of this form of synaptic connections is also motivated by statistical physics considerations.

The stochastic neurons of the Boltzmann machine partition into two functional groups: *visible* and *hidden*, as depicted in Fig. 11.4. The visible neurons[6] provide an interface between the network and the environment in which it operates. During the training phase of the network, the visible neurons are all *clamped* onto specific states determined by the environment. The hidden neurons, on the other hand, always operate freely; they are used to explain underlying constraints contained in the environmental input vectors. The hidden neurons accomplish this task by capturing higher-order statistical correlations in the clamping vectors. The network described here represents a special case of the Boltzmann machine. It may be viewed as an unsupervised learning procedure for modeling a probability distribution that is specified by clamping patterns onto the visible neurons with appropriate probabilities. By so doing, the network can perform *pattern completion*. Specifically, when a partial information-bearing vector is clamped onto a subset of the visible neurons, the network performs completion on the remaining visible neurons, provided that it has learned the training distribution properly (Hinton, 1989).

The primary goal of Boltzmann learning is to produce a neural network that correctly models input patterns according to a Boltzmann distribution. In applying this form of learning, two assumptions are made:

- Each environmental input vector (pattern) persists long enough to permit the network to reach *thermal equilibrium*.
- There is *no* structure in the sequential order in which the environmental vectors are clamped onto the visible units of the network.

A particular set of synaptic weights is said to constitute a perfect model of the environmental structure if it leads to exactly the same probability distribution of the states of

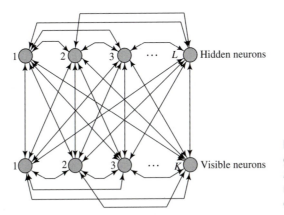

FIGURE 11.4 Architectural graph of Botzmann machine; K is the number of visible neurons and L is the number of hidden neurons.

the visible units (when the network is running freely) as when these units are clamped by the environmental input vectors. In general, unless the number of hidden units is exponentially large compared to the number of visible units, it is impossible to achieve such a perfect model. If, however, the environment has a regular structure, and the network uses its hidden units to capture these regularities, it may achieve a good match to the environment with a manageable number of hidden units.

Gibbs Sampling and Simulated Annealing for the Boltzmann Machine

Let $\mathbf{x}$ denote the state vector of the Boltzmann machine, with its component x_i denoting the state of neuron i. The state $\mathbf{x}$ represents a realization of the random vector $\mathbf{X}$. The synaptic connection from neuron i to neuron j is denoted by w_{ji}, with

$$w_{ji} = w_{ij} \quad \text{for all } (i, j) \tag{11.37}$$

and

$$w_{ii} = 0 \quad \text{for all } i \tag{11.38}$$

Equation (11.37) describes symmetry and Eq. (11.38) emphasizes the absence of self-feedback. The use of a bias is permitted by using the weight w_{j0} from a fictitious node maintained at $+1$ and by connecting it to neuron j for all j.

From an analogy with thermodynamics, the energy of the Boltzmann machine is defined by[7]

$$E(\mathbf{x}) = -\frac{1}{2} \sum_i \sum_{\substack{j \\ i \neq j}} w_{ji} x_i x_j \tag{11.39}$$

Invoking the Gibbs distribution of Eq. (11.5), we may define the probability that the network (assumed to be in equilibrium at temperature T) is in state $\mathbf{x}$ as follows:

$$P(\mathbf{X} = \mathbf{x}) = \frac{1}{Z} \exp\left(-\frac{E(\mathbf{x})}{T}\right) \tag{11.40}$$

where Z is the partition function.

To simplify the presentation, define the single event A and joint events B and C as follows:

$A: X_j = x_j$
$B: \{X_i = x_i\}_{i=1, i \neq j}^K$
$C: \{X_i = x_i\}_{i=1}^K$

In effect, the joint event B excludes A, and the joint event C includes both A and B. The probability of B is the marginal probability of C with respect to A. Hence, using Eqs. (11.39) and (11.40), we may write

$$P(C) = (A, B)$$

$$= \frac{1}{Z} \exp\left(\frac{1}{2T} \sum_i \sum_{\substack{j \\ i \neq j}} w_{ji} x_i x_j\right) \tag{11.41}$$

and

$$P(B) = \sum_A P(A, B)$$

$$= \frac{1}{Z} \sum_{x_j} \exp\left(\frac{1}{2T} \sum_i \sum_{\substack{j \\ i \neq j}} w_{ji} x_i x_j\right)$$

(11.42)

The exponent in Eqs. (11.41) and (11.42) may be expressed as the sum of two components, one involving x_j and the other being independent of x_j. The component involving x_j is given by

$$\frac{x_j}{2T} \sum_{\substack{i \\ i \neq j}} w_{ji} x_i$$

Accordingly, by setting $x_j = x = \pm 1$, we may express the conditional probability of A, given B, as follows:

$$P(A|B) = \frac{P(A, B)}{P(B)}$$

$$= \frac{1}{1 + \exp\left(-\dfrac{x_j}{T} \sum_{\substack{i \\ i \neq j}} w_{ji} x_i\right)}$$

That is, we may write

$$P(X_j = x | \{X_i = x_i\}_{i=1, i \neq j}^K) = \varphi\left(\frac{x}{T} \sum_{\substack{i \\ i \neq j}} w_{ji} x_i\right)$$

(11.43)

where $\varphi(\cdot)$ is a sigmoid function of its argument, as shown by

$$\varphi(v) = \frac{1}{1 + \exp(-v)}$$

(11.44)

Note that although x varies between -1 and $+1$, the whole argument $v = \frac{x}{T}\sum_{i \neq j} w_{ji} x_i$ for large N may vary between $-\infty$ and $+\infty$, as depicted in Fig. 11.5. Note also, in deriving

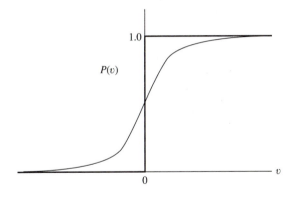

FIGURE 11.5   Sigmoid-shaped function $\varphi(v)$.

Eq. (11.43), the need for the partition function Z has been eliminated. This is highly desirable since a direct computation of Z is infeasible for a network of large complexity.

The use of Gibbs sampling exhibits the joint distribution $P(A, B)$. Basically, as explained in Section 11.6, this stochastic simulation starts with the network assigned an arbitrary state, and the neurons are all repeatedly visited in their natural order. On each visit, a new value for the state of each neuron is chosen in accordance with the probability distribution for that neuron, conditional on the values for the states of all other neurons in the network. Provided that the stochastic simulation is performed long enough, the network will reach thermal equilibrium at temperature T.

Unfortunately, the time taken to reach thermal equilibrium can be much too long. To overcome this difficulty, simulated annealing for a finite sequence of temperatures T_0, $T_1, \ldots, T_{\text{final}}$ is used, as explained in Section 11.5. Specifically, the temperature is initially set to the high value T_0, thereby permitting thermal equilibrium to be reached fast. Thereafter, the temperature T is gradually reduced to the final value T_{final}, at which point the neuronal states will have (hopefully) reached their desired marginal distributions.

Boltzmann Learning Rule

Because the Boltzmann machine is a stochastic machine, it is natural to look to probability theory for an appropriate index of performance. One such criterion is the *likelihood function*.[8] On this basis, the goal of Boltzmann learning is to maximize the likelihood function or, equivalently, the log-likelihood function, in accordance with the *maximum-likelihood principle*.

Let $\mathcal{T}$ denote the set of training examples drawn from the probability distribution of interest. It is assumed that the examples are all two-valued. Repetition of training examples is permitted in proportion to how common certain cases are known to occur. Let a subset of the state vector $\mathbf{x}$, say $\mathbf{x}_\alpha$, denote the state of the visible neurons. The remaining part of the state vector $\mathbf{x}$, say $\mathbf{x}_\beta$, represents the state of the hidden neurons. The state vectors $\mathbf{x}, \mathbf{x}_\alpha$ and $\mathbf{x}_\beta$ are realizations of the random vectors $\mathbf{X}, \mathbf{X}_\alpha$, and $\mathbf{X}_\beta$, respectively. There are two phases to the operation of the Boltzmann machine:

- *Positive phase.* In this phase the network operates in its clamped condition (i.e., under the direct influence of the training set $\mathcal{T}$).
- *Negative phase.* In this second phase, the network is allowed to run freely, and therefore with no environmental input.

Given the synaptic weight vector $\mathbf{w}$ for the whole network, the probability that the visible neurons are in state $\mathbf{x}_\alpha$ is $P(\mathbf{X}_\alpha = \mathbf{x}_\alpha)$. With the many possible values of $\mathbf{x}_\alpha$ contained in the training set $\mathcal{T}$, assumed to be statistically independent, the overall probability distribution is the factorial distribution $\prod_{\mathbf{x}_\alpha \in \mathcal{T}} P(\mathbf{X}_\alpha = \mathbf{x}_\alpha)$. To formulate the log-likelihood function $L(\mathbf{w})$, take the logarithm of this factorial distribution and treat $\mathbf{w}$ as the unknown parameter vector. We may thus write

$$L(\mathbf{w}) = \log \prod_{\mathbf{x}_\alpha \in \mathcal{T}} P(\mathbf{X}_\alpha = \mathbf{x}_\alpha)$$

$$= \sum_{\mathbf{x}_\alpha \in \mathcal{T}} \log P(\mathbf{X}_\alpha = \mathbf{x}_\alpha)$$

$$(11.45)$$

To formulate the expression for the marginal probability $P(\mathbf{X}_\alpha = \mathbf{x}_\alpha)$ in terms of the energy function $E(\mathbf{x})$, we use the following:

- The probability $P(\mathbf{X} = \mathbf{x})$ is equal to $\frac{1}{Z}\exp(-E(\mathbf{x})/T)$ from Eq. (11.40).
- By definition, the state vector $\mathbf{x}$ is the joint combination of $\mathbf{x}_\alpha$ pertaining to the visible neurons and $\mathbf{x}_\beta$ pertaining to the hidden neurons. Hence, the probability of finding the visible neurons in state $\mathbf{x}_\alpha$ with any $\mathbf{x}_\beta$ is given by

$$P(\mathbf{X}_\alpha = \mathbf{x}_\alpha) = \frac{1}{Z}\sum_{\mathbf{x}_\beta}\exp\left(-\frac{E(\mathbf{x})}{T}\right) \tag{11.46}$$

where the random vector $\mathbf{X}_\alpha$ is a subset of $\mathbf{X}$. The partition function Z is itself defined by (see Eq. (11.6)):

$$Z = \sum_{\mathbf{x}}\exp\left(-\frac{E(\mathbf{x})}{T}\right) \tag{11.47}$$

Thus, substituting Eqs. (11.46) and (11.47) in (11.45), we obtain the desired expression for the log-likelihood function:

$$L(\mathbf{w}) = \sum_{\mathbf{x}_\alpha \in \mathcal{T}}\left(\log\sum_{\mathbf{x}_\beta}\exp\left(-\frac{E(\mathbf{x})}{T}\right) - \log\sum_{\mathbf{x}}\exp\left(-\frac{E(\mathbf{x})}{T}\right)\right) \tag{11.48}$$

The dependence on $\mathbf{w}$ is contained in the energy function $E(\mathbf{x})$, as shown in Eq. (11.39).

Differentiating $L(\mathbf{w})$ with respect to w_{ji} in light of Eq. (11.39), we obtain the following result after some manipulation of terms (see Problem 11.8):

$$\frac{\partial L(\mathbf{w})}{\partial w_{ji}} = \frac{1}{T}\sum_{\mathbf{x}_\alpha \in \mathcal{T}}\left(\sum_{\mathbf{x}_\beta}P(\mathbf{X}_\beta = \mathbf{x}_\beta | \mathbf{X}_\alpha = \mathbf{x}_\alpha)x_j x_i - \sum_{\mathbf{x}}P(\mathbf{X} = \mathbf{x})x_j x_i\right) \tag{11.49}$$

To simplify matters, we introduce two definitions:

$$\rho_{ji}^+ = <x_j x_i>^+$$
$$= \sum_{\mathbf{x}_\alpha \in \mathcal{T}}\sum_{\mathbf{x}_\beta}P(\mathbf{X}_\beta = \mathbf{x}_\beta | \mathbf{X}_\alpha = \mathbf{x}_\alpha)x_j x_i \tag{11.50}$$

and

$$\rho_{ji}^- = <x_j x_i>^-$$
$$= \sum_{\mathbf{x}_\alpha \in \mathcal{T}}\sum_{\mathbf{x}}P(\mathbf{X} = \mathbf{x})x_j x_i \tag{11.51}$$

In a loose sense, we may view the first average, ρ_{ji}^+, as the mean firing rate or *correlation* between the states of neurons i and j with the network operating in its clamped or positive phase, and similarly view the second average, ρ_{ji}^-, as the *correlation* between the states of neurons i and j with the network operating in its free-running or negative phase. With these definitions we may simplify Eq. (11.49) to

$$\frac{\partial L(\mathbf{w})}{\partial w_{ji}} = \frac{1}{T}\left(\rho_{ji}^{+} - \rho_{ji}^{-}\right) \tag{11.52}$$

The goal of Boltzmann learning is to maximize the log-likelihood function $L(\mathbf{w})$. We may use *gradient ascent* to achieve that goal by writing

$$\begin{aligned} \Delta w_{ji} &= \epsilon \frac{\partial L(\mathbf{w})}{\partial w_{ji}} \\ &= \eta\left(\rho_{ji}^{+} - \rho_{ji}^{-}\right) \end{aligned} \tag{11.53}$$

where η is a *learning-rate parameter*; it is defined in terms of ϵ and the operating temperature T as:

$$\eta = \frac{\epsilon}{T} \tag{11.54}$$

The gradient ascent rule of Eq. (11.53) is called the *Boltzmann learning rule*. The learning described here is performed in batch; that is, changes to the synaptic weights are made on the presentation of the entire set of training examples.

According to this learning rule, the synaptic weights of a Boltzmann machine are adjusted by using only locally available observations under two different conditions: (1) clamped, and (2) free running. This important feature of Boltzmann learning greatly simplifies the network architecture, particularly when dealing with large networks. Another useful feature of Boltzmann learning, which may come as a surprise, is that the rule for adjusting the synaptic weight from neuron i to neuron j is independent of whether these two neurons are both visible, both hidden, or one of each. All of these nice features of Boltzmann learning result from a key insight by Hinton and Sejnowski (1983, 1986), which ties the abstract mathematical model of the Boltzmann machine to neural networks by using a combination of two things:

- The Gibbs distribution for describing the stochasticity of a neuron.
- The statistical physics-based energy function of Eq. (11.39) for defining the Gibbs distribution.

From a learning point of view, the two terms that constitute the Boltzmann learning rule of Eq. (11.53) have opposite meaning. We may view the first term, corresponding to the clamped condition of the network, as essentially a Hebbian *learning* rule; and view the second term, corresponding to the free-running condition of the network, as an *unlearning* or *forgetting* term. Indeed, the Boltzmann learning rule represents a *generalization* of the *repeated forgetting and relearning rule* described by Pöppel and Krey (1987) for the case of symmetric networks with no hidden neurons.

It is also of interest that since the Boltzmann machine learning algorithm requires that hidden neurons know the difference between stimulated and free-running activities, and given that there is a (hidden) external network that signals to hidden neurons that the machine is being stimulated, we have a primitive form of an *attention* mechanism (Cowan and Sharp, 1988).

Need for the Negative Phase and its Implications

The combined use of a positive and a negative phase stabilizes the distribution of synaptic weights in the Boltzmann machine. This need may be justified in another way. Intuitively, we may say that the need for a negative as well as a positive phase in Boltzmann learning arises due to the presence of the partition function, Z, in the expression for the probability of a neuron's state vector. The implication of this statement is that the direction of steepest descent in energy space is *not* the same as the direction of steepest ascent in probability space. In effect, the negative phase in the learning procedure is needed to account for this discrepancy (Neal, 1992).

The use of a negative phase in Boltzmann learning has two major disadvantages:

1. *Increased computation time.* In the positive phase some of the neurons are clamped to the external environment, whereas in the negative phase all the neurons are free running. Accordingly, the time taken for stochastic simulation of a Boltzmann machine is increased.
2. *Sensitivity to statistical errors.* The Boltzmann learning rule involves the *difference* between two average correlations, one computed for the positive phase and the other computed for the negative phase. When these two correlations are similar, the presence of sampling noise makes the difference between them even more noisy.

We may eliminate these shortcomings of the Boltzmann machine by using a sigmoid belief network. In this new class of stochastic machines, control over the learning procedure is exercised by means other than a negative phase.

11.8   SIGMOID BELIEF NETWORKS

Sigmoid belief networks or *logistic belief nets* were developed by Neal in 1992 in an effort to find a stochastic machine that would share with the Boltzmann machine the capacity to learn arbitrary probability distributions over binary vectors, but would not need the negative phase of the Boltzmann machine learning procedure. This objective was achieved by replacing the symmetric connections of the Boltzmann machine with *directed connections that form an acyclic graph*. Specifically, a sigmoid belief network consists of a multilayer architecture with binary stochastic neurons, as illustrated in Fig. 11.6. The acyclic nature of the machine makes it easy to perform probabilistic calculations. In particular, the network uses the sigmoid function of Eq. (11.43), in analogy with the Boltzmann machine, to calculate the conditional probability of a neuron being activated in response to its own induced local field.

Fundamental Properties of Sigmoid Belief Networks

Let the vector $\mathbf{X}$, consisting of the two-valued random variables $X_1, X_2, ..., X_N$, define a sigmoid belief network composed of N stochastic neurons. The *parents* of element X_j in $\mathbf{X}$ are denoted by

$$\text{pa}(X_j) \subseteq \{X_1, X_2, ..., X_{j-1}\} \tag{11.55}$$

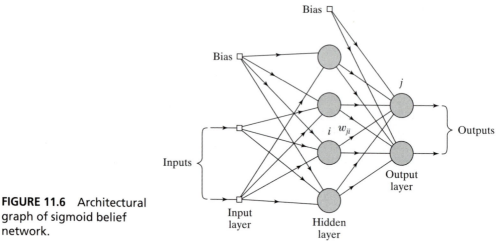

FIGURE 11.6 Architectural graph of sigmoid belief network.

In words, $\text{pa}(X_j)$ is the smallest subset of random vector $\mathbf{X}$ for which we have

$$P(X_j = x_j | X_1 = x_1, \ldots, X_{j-1} = x_{j-1}) = P(X_j = x_j | \text{pa}(X_j)) \qquad (11.56)$$

An important virtue of sigmoid belief networks is their ability to clearly exhibit the conditional dependencies of the underlying probabilistic model of the input data. In particular, the probability that the ith neuron is activated is defined by the sigmoid function (see Eq. (11.43))

$$P(X_j = x_j | \text{pa}(X_j)) = \varphi\left(\frac{x_j}{T} \sum_{i<j} w_{ji} x_i\right) \qquad (11.57)$$

where w_{ji} is the synaptic weight from neuron i to neuron j, as shown in Fig. 11.6. That is, the conditional probability $P(X_j = x_j | \text{pa}(X_j))$ depends on $\text{pa}(X_j)$ solely through a sum of weighted inputs. Thus, Eq. (11.57) provides the basis for the propagation of beliefs through the network.

In performing probability calculations on sigmoid belief networks, the following two points are noteworthy:

1. $w_{ji} = 0$ for all X_i not belonging to $\text{pa}(X_j)$
2. $w_{ji} = 0$ for all $i \geq j$

The first point follows from the definition of parents. The second point follows from the fact that a sigmoid belief network is a directed acyclic graph.

As the name implies, sigmoid belief networks belong to the general class of *belief networks*[9] studied extensively in the literature (Pearl, 1988). The stochastic operation of sigmoid belief networks is somewhat more complex than the Boltzmann machine. Nevertheless, they do lend themselves to the use of gradient-ascent learning in probability space, based on locally available information.

Learning in Sigmoid Belief Networks

Let $\mathcal{T}$ denote a set of training examples drawn from the probability distribution of interest. It is assumed that each example is two-valued, representing certain attributes. Repetition of training examples is permitted, in proportion to how commonly a particular combination of attributes is known to occur. To model the distribution from which $\mathcal{T}$ is drawn, we proceed as follows:

1. Some size for a state vector, $\mathbf{x}$, is decided for the network.
2. A subset of the state vector, say $\mathbf{x}_\alpha$, is selected to represent the attributes in the training cases; that is, $\mathbf{x}_\alpha$ represents the state vector of the visible neurons (i.e., evidence nodes).
3. The remaining part of the state vector $\mathbf{x}$, denoted by $\mathbf{x}_\beta$, defines the state vector of the hidden neurons (i.e., those computational nodes for which we do not have instantiated values).

The design of a sigmoid belief network is highly dependent on the way in which, for a given state vector $\mathbf{x}$, the visible and hidden units are arranged. Therefore, different arrangements of visible and hidden neurons may result in different configurations.

As with the Boltzmann machine, we derive the desired learning rule for a sigmoid belief network by maximizing the log-likelihood function, computed from the training set $\mathcal{T}$. The log-likelihood function, $L(\mathbf{w})$, is defined by Eq. (11.45), reproduced here for convenience of presentation:

$$L(\mathbf{w}) = \sum_{\mathbf{x}_\alpha \in \mathcal{T}} \log P(\mathbf{X}_\alpha = \mathbf{x}_\alpha)$$

where $\mathbf{w}$ is the synaptic weight vector of the network, treated as unknown. The state vector $\mathbf{x}_\alpha$, pertaining to the visible neurons, is a realization of the random vector $\mathbf{X}_\alpha$. Let w_{ji} denote the ji-th element of $\mathbf{w}$ (i.e., synaptic weight from neuron i to neuron j). Differentiating $L(\mathbf{w})$ with respect to w_{ji}, we obtain

$$\frac{\partial L(\mathbf{w})}{\partial w_{ji}} = \sum_{\mathbf{x}_\alpha \in \mathcal{T}} \frac{1}{P(\mathbf{X}_\alpha = \mathbf{x}_\alpha)} \frac{\partial P(\mathbf{X}_\alpha = \mathbf{x}_\alpha)}{\partial w_{ji}}$$

Next we note the following two probabilistic relations:

$$P(\mathbf{X}_\alpha = \mathbf{x}_\alpha) = \sum_{\mathbf{x}_\beta} P(\mathbf{X} = (\mathbf{x}_\alpha, \mathbf{x}_\beta))$$

$$= \sum_{\mathbf{x}_\beta} P(\mathbf{X} = \mathbf{x}) \tag{11.58}$$

where the random vector $\mathbf{X}$ pertains to the whole network and the state vector $\mathbf{x} = (\mathbf{x}_\alpha, \mathbf{x}_\beta)$ is a realization of it, and

$$P(\mathbf{X} = \mathbf{x}) = P(\mathbf{X} = \mathbf{x} | \mathbf{X}_\alpha = \mathbf{x}_\alpha) P(\mathbf{X}_\alpha = \mathbf{x}_\alpha) \tag{11.59}$$

which defines the probability of the joint event $\mathbf{X} = \mathbf{x} = (\mathbf{x}_\alpha, \mathbf{x}_\beta)$.

In light of these two relations, we may redefine the partial derivative $\partial L(\mathbf{w})/\partial w_{ji}$ in the equivalent form:

$$\frac{\partial L(\mathbf{w})}{\partial w_{ji}} = \sum_{\mathbf{x}_\alpha \in \mathcal{T}} \sum_{\mathbf{x}_\beta} \frac{P(\mathbf{X} = \mathbf{x} | \mathbf{X}_\alpha = \mathbf{x}_\alpha)}{P(\mathbf{X} = \mathbf{x})} \frac{\partial P(\mathbf{X} = \mathbf{x})}{\partial w_{ji}} \tag{11.60}$$

In light of Eq. (11.43), we may write

$$P(\mathbf{X} = \mathbf{x}) = \prod_j \varphi\left(\frac{x_j}{T} \sum_{i<j} w_{ji} x_i\right) \tag{11.61}$$

where $\varphi(\cdot)$ is a sigmoid function. We may therefore write

$$\frac{1}{P(\mathbf{X} = \mathbf{x})} \frac{\partial P(\mathbf{X} = \mathbf{x})}{\partial w_{ji}} = \frac{\partial}{\partial w_{ji}} \log P(\mathbf{X} = \mathbf{x})$$

$$= \frac{\partial}{\partial w_{ji}} \sum_j \log\varphi\left(\frac{x_j}{T} \sum_{i<j} w_{ji} x_i\right)$$

$$= \frac{1}{T} \sum_j \frac{1}{\varphi\left(\dfrac{x_j}{T} \sum_{i<j} w_{ji} x_i\right)} \varphi'\left(\frac{x_j}{T} \sum_{i<j} w_{ji} x_i\right) x_j x_i$$

where $\varphi'(\cdot)$ is the first derivative of the sigmoid function $\varphi(\cdot)$ with respect to its argument. But, from the definition of $\varphi(\cdot)$ given in Eq. (11.44) we readily find that

$$\varphi'(v) = \varphi(v)\varphi(-v) \tag{11.62}$$

where $\varphi(-v)$ is obtained from $\varphi(v)$ by replacing v with $-v$. Hence, we may write

$$\frac{1}{P(\mathbf{X} = \mathbf{x})} \frac{\partial P(\mathbf{X} = \mathbf{x})}{\partial w_{ji}} = \frac{1}{T} \sum_j \varphi\left(-\frac{x_j}{T} \sum_{i<j} w_{ji} x_i\right) x_j x_i \tag{11.63}$$

Accordingly, substituting Eq. (11.63) in (11.60), we obtain

$$\frac{\partial L(\mathbf{w})}{\partial w_{ji}} = \frac{1}{T} \sum_{\mathbf{x}_\alpha \in \mathcal{T}} \sum_{\mathbf{x}_\beta} P(\mathbf{X} = \mathbf{x} | \mathbf{X}_\alpha = \mathbf{x}_\alpha)\varphi\left(-\frac{x_j}{T} \sum_{i<j} w_{ji} x_i\right) x_j x_i \tag{11.64}$$

To simplify matters, we define the ensemble average

$$\rho_{ji} = \left\langle \varphi\left(-x_j \sum_{i<j} w_{ji} x_i\right) x_j x_i \right\rangle$$

$$= \sum_{\mathbf{x}_\alpha \in \mathcal{T}} \sum_{\mathbf{x}_\beta} P(\mathbf{X} = \mathbf{x} | \mathbf{X}_\alpha = \mathbf{x}_\alpha)\varphi\left(-\frac{x_j}{T} \sum_{i<j} w_{ji} x_i\right) x_j x_i \tag{11.65}$$

which represents an *average correlation* between the states of neurons i and j, weighted by the factor $\varphi(-\frac{x_j}{T}\sum_{i<j} w_{ji}x_i)$. This average is taken over all possible values of $\mathbf{x}_\alpha$ (drawn from the training set $\mathcal{T}$) as well as all possible values of $\mathbf{x}_\beta$, with $\mathbf{x}_\alpha$ referring to the visible neurons and $\mathbf{x}_\beta$ referring to the hidden neurons.

Gradient ascent in probability space is accomplished by defining the incremental change in synaptic weight w_{ji} as

$$\Delta w_{ji} = \epsilon \frac{\partial L(\mathbf{w})}{\partial w_{ji}}$$

(11.66)

$$= \eta \rho_{ji}$$

where $\eta = \epsilon/T$ is a learning-rate parameter and ρ_{ji} is itself defined by Eq. (11.65). Equation (11.66) is the *learning rule for a sigmoid belief network*.

A summary of the sigmoid belief network learning procedure is presented in Table 11.2, where learning is performed in the batch mode; that is, the changes to the synaptic weights of the network are made on the basis of the entire set of training cases. The summary presented in Table 11.2 does not include the use of simulated annealing, which is why we have set the temperature T to 1 therein. However, as with the Boltzmann machine, simulated annealing can be incorporated into the sigmoid belief network learning procedure to reach thermal equilibrium faster, if so desired.

TABLE 11.2   Summary of the Sigmoid Belief Network Learning Procedure

---

Initialization. Initialize the network by setting the weights w_{ji} of the network to random values uniformly distributed in the range $[-a, a]$; a typical value for a is 0.5.

1. Given a set of training cases $\mathcal{T}$, clamp the visible neurons of the network to $\mathbf{x}_\alpha$, where $\mathbf{x}_\alpha \in \mathcal{T}$.

2. For each $\mathbf{x}_\alpha$, perform a separate Gibbs sampling simulation of the network at some operating temperature T, and observe the resulting state vector $\mathbf{x}$ of the whole network. Provided that the simulation is performed long enough, the values of $\mathbf{x}$ for the different cases contained in the training set $\mathcal{T}$ should come from the conditional distribution of the corresponding random vector $\mathbf{X}$, given that particular training set.

3. Compute the ensemble average

$$\rho_{ji} = \sum_{\mathbf{x}_\alpha \in \mathcal{T}} \sum_{\mathbf{x}_\beta} P(\mathbf{X} = \mathbf{x}|\mathbf{X}_\alpha = \mathbf{x}_\alpha) x_j x_i \, \varphi\left(-x_j \sum_{i<j} w_{ji} x_i\right)$$

where the random vector $\mathbf{X}_\alpha$ is a subset of $\mathbf{X}$, and $\mathbf{x} = (\mathbf{x}_\alpha, \mathbf{x}_\beta)$ with $\mathbf{x}_\alpha$ referring to the visible neurons and $\mathbf{x}_\beta$ referring to the hidden neurons; x_j is the jth element of state vector $\mathbf{x}$ (i.e., state of neuron j), and w_{ji} is the synaptic weight from neuron i to neuron j. The sigmoid function $\varphi(\cdot)$ is defined by

$$\varphi(v) = \frac{1}{1 + \exp(-v)}$$

4. Increment each synaptic weight w_{ji} of the network by the amount

$$\Delta w_{ji} = \eta \, \rho_{ji}$$

where η is the learning-rate parameter. This adjustment should move the synaptic weights of the network along the gradient toward a local maximum of the log-likelihood function $L(\mathbf{w})$ in accordance with the maximum likelihood principle.

Unlike the Boltzmann machine, only a single phase is needed for learning in a sigmoid belief network. The reason for this simplification is that normalization of the probability distribution over state vectors is accomplished at the local level of each neuron via the sigmoid function $\varphi(\cdot)$, rather than globally via the difficult to compute partition function Z that involves all possible configurations of states. Once the conditional distribution of the random vector $\mathbf{X}$, given the values of $\mathbf{x}_\alpha$ drawn from the training set $\mathcal{T}$, has been correctly modeled via Gibbs sampling, the role of the negative phase in the Boltzmann machine learning procedure is taken over by the weighting factor $\varphi(-\frac{x_i}{T}\sum_{i<j} w_{ji}x_i)$ involved in computing the ensemble-averaged correlation ρ_{ji} between the states of neurons i and j. When the local minimum of the log-likelihood function $L(\mathbf{w})$ is reached, this weighting factor becomes zero when the network learns a deterministic mapping; otherwise, its average effect comes out to zero.

In Neal (1992), experimental results are presented that show (1) sigmoid belief networks are capable of learning to model nontrivial distributions, (2) these networks can learn at a faster rate than the Boltzmann machine, and (3) this advantage of a sigmoid belief network over the Boltzmann machine is due to the elimination of the negative phase from the learning procedure.

11.9 HELMHOLTZ MACHINE

Sigmoid belief networks provide a powerful multilayer framework for representing and learning higher-order statistical relationships among sensory inputs of interest in an unsupervised manner. The *Helmholtz machine*,[10] first described in Dayan et al. (1995) and Hinton et al. (1995), provides another ingenious multilayer framework for achieving a similar objective without the use of Gibbs sampling.

The Helmholtz machine uses two entirely different sets of synaptic connections, as illustrated in Fig. 11.7 for the case of a network with two layers of two-valued, stochastic neurons. The forward connections, shown as solid lines in Fig. 11.7, constitute

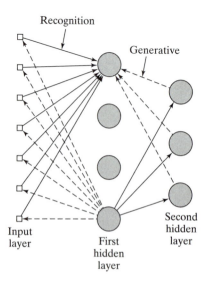

FIGURE 11.7 Architectural graph of Helmholtz machine consisting of interconnected neurons with recognition (solid lines) and generative (dashed lines) connections.

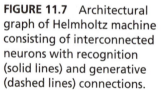

the *recognition model*. The purpose of this model is to *infer* a probability distribution over the underlying causes of the input vector. The backward connections, shown as dashed lines in Fig. 11.7, constitute the *generative model*. The purpose of this second model is to *reconstruct* an approximation to the original input vector from the underlying representations captured by the hidden layers of the network, thereby enabling it to operate in a *self-supervised manner*. Both the recognition and generative models operate in a strictly feedforward fashion, with no feedback; they interact with each other only via the learning procedure.

Hinton et al. (1995) describe a stochastic algorithm, called the *wake-sleep algorithm*, for calculating the recognition and generative weights of the Helmholtz machine. As the name implies, there are two phases to the algorithm: a "wake" phase and a "sleep" phase. In the "wake" phase, the network is driven in the forward direction by the recognition weights. A representation of the input vector is thereby produced in the first hidden layer. This is, in turn, followed by a second representation of that first representation, which is produced in the second hidden layer, and so on for the other hidden layers of the network. The set of representations so produced in the different hidden layers of the network provides a total representation of the input vector by the network. Although the neurons are driven by the recognition weights, it is only the generative weights that are actually learned during the "wake" phase using locally available information. In effect, this phase of the learning process makes each layer of the total representation better at reconstructing the activities formed in the preceding layer.

In the "sleep" phase of the algorithm, the recognition weights are turned off. The network is driven in the backward direction by the generative weights, starting at the outermost hidden layer and working backwards, layer by layer, all the way to the input layer. Because of the fact that the neurons are stochastic, repeating this process would typically give rise to many different "fantasy" vectors on the input layer. These fantasies supply an unbiased sample of the network's generative model of the world. Having produced a fantasy, the simple delta rule (described in Chapter 3) is used to adjust the recognition weights so as to maximize the logarithm of the probability of recovering the hidden activities that actually caused the fantasy. Like the "wake" phase, the "sleep" phase only uses locally available information.

The learning rule for the generative weights (i.e., backward connections) also uses the simple delta rule. However, instead of following the gradient of the log-likelihood function, this rule follows the gradient of a *penalized* log-likelihood function. The penalty term is the Kullback–Leibler divergence between the true *a posteriori* distribution and the actual distribution produced by the recognition model (Hinton et al., 1995); the Kullback–Leibler divergence or relative entropy is discussed in the preceding chapter. In effect, the penalized log-likelihood function acts as a lower bound on the log-likelihood function of the input data, with the lower bound being improved through the learning process. In particular, the learning process tries to adjust the generative weights to bring the true *a posteriori* distribution as close as possible to the distribution actually computed by the recognition model. Unfortunately, learning the recognition model's weights does not precisely correspond to the penalized likelihood function. The wake-sleep learning procedure is not guaranteed to work in all practical situations; it does fail sometimes.

11.10 MEAN-FIELD THEORY

The learning machines considered in the preceding three sections share a common feature: they all use stochastic neurons and may therefore suffer from a slow learning process. In the third and final part of the chapter we study the use of mean-field theory as the mathematical basis for deriving *deterministic approximations* to these stochastic machines to speed up learning. Because the stochastic machines considered herein have different architectures, the theory is applied in correspondingly different ways. In particular, we may identify two specific approaches that have been pursued in literature:

1. Correlations are replaced by their mean-field approximations.
2. An intractable model is replaced by a tractable model via a variational principle.

Approach 2 is highly principled and therefore very appealing. It lends itself for application to the sigmoid belief network (Saul et al., 1996) and the Helmholtz machine (Dayan et al., 1995). However, in the case of the Boltzmann machine, the application of approach 2 is complicated by the need for an upper bound on the partition function Z. For this reason, in Peterson and Anderson (1987), the first approach is used to accelerate the Boltzmann learning rule. In this section we provide a rationale for the first approach. The second approach is considered later in the chapter.

The idea of mean-field approximation is well known in statistical physics (Glauber, 1963). While it cannot be denied that in the context of stochastic machines it is desirable to know the states of all the neurons in the network at all times, we must nevertheless recognize that, in the case of a network with a large number of neurons, the neural states contain vastly more information than we usually require in practice. In fact, to answer the most familiar physical questions about the stochastic behavior of the network, we need only know the average values of neural states or the average products of pairs of neural states.

In a stochastic neuron, the firing mechanism is described by a probabilistic rule. In such a situation, it is rational for us to enquire about the *average* of the state x_j of neuron j. To be precise, we should speak of the average as a "thermal" average, since the synaptic noise is usually modeled in terms of thermal fluctuations. In any event, let $<x_j>$ denote the average (mean) of x_j. The state of neuron j is described by the probabilistic rule:

$$x_j = \begin{cases} +1 & \text{with probability } P(v_j) \\ -1 & \text{with probability } 1 - P(v_j) \end{cases} \tag{11.67}$$

where

$$P(v_j) = \frac{1}{1 + \exp(-v_j/T)} \tag{11.68}$$

where T is the operating temperature. Hence we may express the average $<x_j>$ for some *specified* value of induced local field v_j as follows:

$$<x_j> = (+1)P(v_j) + (-1)[1 - P(v_j)]$$

$$= 2P(v_j) - 1 \tag{11.69}$$

$$= \tanh(v_j/2T)$$

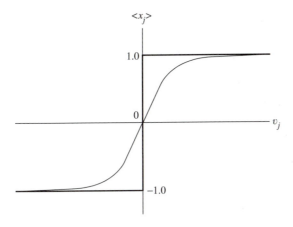

FIGURE 11.8   Graph of the thermal average $<x_j>$ versus the induced local field v_j; heavy solid curve corresponds to normal operation of the McCulloch-Pitts neuron.

where $\tanh(v_j/2T)$ is the hyperbolic tangent of $v_j/2T$. Figure 11.8 shows two plots of the average $<x_j>$ versus the induced local field v_j. The continuous curve is for some temperature T greater than zero, and the plot shown in heavy solid lines is for the limiting case of $T = 0$. In the latter case, Eq. (11.69) takes the limiting form

$$<x_j> \rightarrow \text{sgn}(v_j) \quad \text{as } T \rightarrow 0 \tag{11.70}$$

which corresponds to the activation function of the McCulloch–Pitts neuron.

The discussion so far has focused on the simple case of a single stochastic neuron. In the more general case of a stochastic machine composed of a large assembly of neurons, we have a much more difficult task on our hands. The difficulty arises because of the combination of two factors:

- The probability $P(v_j)$ that neuron j is on is a nonlinear function of the induced local field v_j.
- The induced local field v_j is a random variable, being influenced by the stochastic action of other neurons connected to the inputs of neuron j.

It is generally safe to say that there is no mathematical method that we can use to evaluate the behavior of a stochastic machine in exact terms. But there is an approximation known as the *mean-field approximation* that we can use, which often yields good results. The basic idea of mean-field approximation is to replace the actual fluctuating induced local field v_j for each neuron j in the network by its average $<v_j>$, as shown by

$$v_j \overset{\text{approx}}{=} <v_j> = \left\langle \sum_i w_{ji} x_i \right\rangle = \sum_i w_{ji} <x_i> \tag{11.71}$$

Accordingly, we may compute the average state $<x_j>$ for neuron j embedded in a stochastic machine made up of a total of N neurons, just as we did in Eq. (11.69) for a single stochastic neuron, by writing

$$<x_j> = \tanh\left(\frac{1}{2T}v_j\right) \overset{\text{approx}}{=} \tanh\left(\frac{1}{2T}<v_j>\right) = \tanh\left(\frac{1}{2T}\sum_i w_{ji}<x_j>\right) \quad (11.72)$$

In light of Eq. (11.72), we may formally state the mean-field approximation as:

The average of some function of a random variable is approximated as the function of the average of that random variable.

For $j = 1, 2, ..., N$, Eq. (11.72) represents a set of nonlinear equations with N unknowns $<x_j>$. The solution of this set of nonlinear equations is now a manageable proposition because the unknowns are all *deterministic* rather than stochastic variables, which they are in the original network.

11.11 DETERMINISTIC BOLTZMANN MACHINE

Learning in the Boltzmann machine is *exponential* in the number of neurons because the Boltzmann learning rule requires the computation of correlations between every pair of neurons in the network. Boltzmann learning therefore requires exponential time. Peterson and Anderson (1987) proposed a method for accelerating the Boltzmann learning process. The method involves replacing the correlations in the Boltzmann learning rule of Eq. (11.53) by a mean-field approximation as shown here:

$$<x_j x_i> \overset{\text{approx}}{=} <x_j><x_i>, \quad (i, j) = 1, 2, ..., K \quad (11.73)$$

where the mean quantity $<x_j>$ is itself computed using the mean-field equation (11.72).

The form of Boltzmann learning in which the computation of the correlations is approximated in the manner just described is called the *deterministic Boltzmann learning rule*. Specifically, the standard Boltzmann learning rule of Eq. (11.53) is approximated as:

$$\Delta w_{ji} = \eta(U_j^+ U_i^+ - U_j^- U_i^-) \quad (11.74)$$

where U_j^+ and U_j^- are the average outputs of visible neuron j (on a single pattern) in the clamped and free-running conditions, respectively, and η is the learning-rate parameter. Whereas the Boltzmann machine uses binary stochastic neurons, its deterministic counterpart uses analog deterministic neurons.

The deterministic Boltzmann machine provides a substantial increase in speed over the standard Boltzmann machine by one or two orders of magnitude (Peterson and Anderson, 1987). However, there are some cautionary notes on its practical use:

1. The deterministic Boltzmann learning rule only works in the supervised case, that is, when some of the visible neurons are assigned the role of output neurons. Unsupervised learning does not work at all in the mean-field regime because the mean state is a very impoverished representation of the free-running probability distribution.

2. In supervised learning, the use of deterministic Boltzmann learning is restricted to neural networks with a single hidden layer (Galland, 1993). In theory, there is no reason not to load multiple hidden layers. In practice, however, the use of more than one hidden layer results in the same problem as for the unsupervised case mentioned under point 1.

The deterministic Boltzmann learning rule in Eq. (11.74) has a simple and local form, which makes it a good candidate for implementation in very large scale integration (VLSI) hardware (Alspector et al., 1991; Schneider and Card, 1993). However, in Schneider and Card (1998) it is reported that in the case of continuous learning of capacitive weights, the deterministic Boltzmann machine cannot tolerate weight storage capacitor change decay and offsets in the learning circuitry. The reason is that these internal problems cause the synaptic weight values to drift, resulting in oscillation, which is clearly unacceptable.

11.12 DETERMINISTIC SIGMOID BELIEF NETWORKS

The essence of mean-field approximation described in Section 11.10 is that the average of some function of a random variable can be approximated by the function of the average of that random variable. This viewpoint of mean-field theory works in a limited fashion for the approximation of the Boltzmann machine as discussed in the previous section. In this section we describe another viewpoint of mean-field theory that is well suited to the approximation of a sigmoid belief network. Basically, here we find that an intractable model is approximated by a tractable one via a variational principle (Saul et al., 1996; Jordan et al., 1998). Generally speaking, the tractable model is characterized by a decoupling of the degrees of freedom that make the original model intractable. The decoupling is accomplished by expanding the intractable model to include additional parameters known as *variational parameters* that are designed to fit the problem at hand. The terminology comes from the use of techniques rooted in the calculus of variations (Parisi, 1988).

Lower Bound on the Log-Likelihood Function

The starting point of our discussion is the probabilistic relation of Eq. (11.58), reproduced here in the logarithmic form

$$\log P(\mathbf{X}_\alpha = \mathbf{x}_\alpha) = \log \sum_{\mathbf{x}_\beta} P(\mathbf{X} = \mathbf{x}) \tag{11.75}$$

As in Section 11.8, we partition the random vector $\mathbf{X}$ into $\mathbf{X}_\alpha$ and $\mathbf{X}_\beta$, with $\mathbf{X}_\alpha$ pertaining to the visible neurons and $\mathbf{X}_\beta$ pertaining to the hidden neurons. Realizations of the random vectors $\mathbf{X}$, $\mathbf{X}_\alpha$, and $\mathbf{X}_\beta$ are denoted by the state vectors $\mathbf{x}$, $\mathbf{x}_\alpha$, and $\mathbf{x}_\beta$, respectively. Now, the logarithm of a sum of probabilities required in Eq. (11.75) is difficult to handle. We overcome this difficulty by noting that for any conditional distribution $Q(\mathbf{X}_\beta = \mathbf{x}_\beta | \mathbf{X}_\alpha = \mathbf{x}_\alpha)$, we may rewrite Eq. (11.75) in a different but equivalent form:

$$\log P(\mathbf{X}_\alpha = \mathbf{x}_\alpha) = \log \sum_{\mathbf{x}_\beta} Q(\mathbf{X}_\beta = \mathbf{x}_\beta | \mathbf{X}_\alpha = \mathbf{x}_\alpha) \left[\frac{P(\mathbf{X} = \mathbf{x})}{Q(\mathbf{X}_\beta = \mathbf{x}_\beta | \mathbf{X}_\alpha = \mathbf{x}_\alpha)} \right] \tag{11.76}$$

This equation is formulated in a way to prepare it for the application of Jensen's inequality, which is discussed in the previous chapter. On this application we obtain the lower bound:

$$\log P(\mathbf{X}_\alpha = \mathbf{x}_\alpha) \geq \sum_{\mathbf{x}_\beta} Q(\mathbf{X}_\beta = \mathbf{x}_\beta | \mathbf{X}_\alpha = \mathbf{x}_\alpha) \log \left[\frac{P(\mathbf{X} = \mathbf{x})}{Q(\mathbf{X}_\beta = \mathbf{x}_\beta | \mathbf{X}_\alpha = \mathbf{x}_\alpha)} \right] \quad (11.77)$$

With mean-field theory in mind, henceforth we will refer to the approximation distribution $Q(\mathbf{X}_\beta = \mathbf{x}_\beta | \mathbf{X}_\alpha = \mathbf{x}_\alpha)$ as the *mean-field distribution*.

What we are interested in is a formula for the log-likelihood function. In the case of a sigmoid belief network, the log-likelihood function $L(\mathbf{w})$ is defined where the summation is over all $\mathbf{x}_\alpha$ (determined by the training set $\mathcal{T}$), hence the use of a batch algorithm for the network. We are going to follow a different strategy for the mean-field approximation to the sigmoid belief network. Specifically, we will adopt a sequential mode of operation, where the log-likelihood function is computed on an *example-by-example basis*, as shown by

$$\mathcal{L}(\mathbf{w}) = \log P(\mathbf{X}_\alpha = \mathbf{x}_\alpha) \quad (11.78)$$

where $\mathbf{w}$ is the weight vector of the network. In the case of identically and independently distributed (*iid*) data, the actual log-likelihood function $\mathcal{L}(\mathbf{w})$ is the sum of $\mathcal{L}(\mathbf{w})$ terms, one for each data point. In such a situation, the definitions of $L(\mathbf{w})$ and $\mathcal{L}(\mathbf{w})$ are basically equivalent. In general, the use of $\mathcal{L}(\mathbf{w})$ provides an approximation to $L(\mathbf{w})$.

The sequential or on-line approach to learning has become the standard approach in the design of neural networks, largely because of its simplicity in implementation. Thus, in light of Eq. (11.78), we may write

$$\mathcal{L}(\mathbf{w}) \geq \sum_{\mathbf{x}_\beta} Q(\mathbf{X}_\beta = \mathbf{x}_\beta | \mathbf{X}_\alpha = \mathbf{x}_\alpha) \log \left[\frac{P(\mathbf{X} = \mathbf{x})}{Q(\mathbf{X}_\beta = \mathbf{x}_\beta | \mathbf{X}_\alpha = \mathbf{x}_\alpha)} \right]$$

or equivalently,

$$\mathcal{L}(\mathbf{w}) \geq - \sum_{\mathbf{x}_\beta} Q(\mathbf{X}_\beta = \mathbf{x}_\beta | \mathbf{X}_\alpha = \mathbf{x}_\alpha) \log Q(\mathbf{X}_\beta = \mathbf{x}_\beta | \mathbf{X}_\alpha = \mathbf{x}_\alpha)$$

$$+ \sum_{\mathbf{x}_\beta} Q(\mathbf{X}_\beta = \mathbf{x}_\beta | \mathbf{X}_\alpha = \mathbf{x}_\alpha) \log P(\mathbf{X} = \mathbf{x}) \quad (11.79)$$

The first term on the right-hand side of Eq. (11.79) is the entropy of the mean-field distribution $Q(\mathbf{X}_\beta = \mathbf{x}_\beta | \mathbf{X}_\alpha = \mathbf{x}_\alpha)$; this is *not* to be confused with a condition entropy. The second term is the average of $\log P(\mathbf{X} = \mathbf{x})$ over all possible states of the hidden neurons. At unit temperature, we note from the discussion on the Gibbs distribution presented in Section 11.2 that the energy of a sigmoid belief network is $-\log P(\mathbf{X} = \mathbf{x})$. Since from Eq. (11.61) we have (for $T = 1$)

$$P(\mathbf{X} = \mathbf{x}) = \prod_j \varphi \left(x_j \sum_{i<j} w_{ji} x_i \right)$$

it follows that

$$E = - \log P(\mathbf{X} = \mathbf{x})$$

$$= - \sum_j \log\varphi\left(x_j \sum_{i<j} w_{ji} x_i \right) \qquad (11.80)$$

Using the definition of the sigmoid function

$$\varphi(v) = \frac{1}{1 + \exp(-v)}$$

$$= \frac{\exp(v)}{1 + \exp(v)}$$

we may thus formally express the energy function of a sigmoid belief network as:

$$E = - \sum_i \sum_{\substack{j \\ i<j}} w_{ji} x_i x_j + \sum_j \log\left(1 + x_j \sum_{i<j} w_{ji} x_i \right) \qquad (11.81)$$

Except for a multiplying factor of $1/2$, the first term on the right-hand side of Eq. (11.81) is recognized as the energy function of a Markovian system (e.g., Boltzmann machine). However, the second term is unique to sigmoid belief networks.

The lower bound of Eq. (11.79) is valid for any mean-field distribution $Q(\mathbf{X}_\beta = \mathbf{x}_\beta | \mathbf{X}_\alpha = \mathbf{x}_\alpha)$. To put it to good use, however, we must choose a distribution that enables us to evaluate this bound. We may do so by choosing the *factorial distribution* (Saul et al., 1996)

$$Q(\mathbf{X}_\beta = \mathbf{x}_\beta | \mathbf{X}_\alpha = \mathbf{x}_\alpha) = \prod_{j \in \mathcal{H}} \mu_j^{x_j} (1 - \mu_j)^{1-x_j} \qquad (11.82)$$

where $\mathcal{H}$ denotes the set of all hidden neurons, and the states of the hidden neurons appear as independent *Bernoulli variables* with adjustable means μ_j. (A Bernoulli (θ) is defined as a binary random variable that takes the value 1 with probability θ.) Thus, substituting Eq. (11.82) in (11.79), we obtain (after simplifications)

$$\mathcal{L}(\mathbf{w}) \geq - \sum_{j \in \mathcal{H}} [\mu_j \log\mu_j + (1 - \mu_j)\log(1 - \mu_j)]$$

$$+ \sum_i \sum_{\substack{j \in \mathcal{H} \\ i<j}} w_{ji} \mu_i \mu_j - \sum_{j \in \mathcal{H}} < \log\left[1 + \exp\left(\sum_{i<j} w_{ji} x_i \right) \right] > \qquad (11.83)$$

where the use of $<\cdot>$ signifies an ensemble average over the mean-field distribution and $j \in \mathcal{H}$ signifies that j refers to a hidden neuron. The first term on the right-hand side of Eq. (11.83) is the mean-field entropy, and the second term is the mean-field energy. Both of these terms pertain to the factorial distribution of Eq. (11.82).

Unfortunately, we still have an intractable problem: It is not possible to compute an average of the form $<\log[1 + \exp(z_j)]>$ exactly. This term arises in Eq. (11.83) with

$$z_j = \sum_{i<j} w_{ji} x_i \qquad (11.84)$$

To overcome this difficulty, we again resort to the use of Jensen's inequality for a bound. First, for any variable z_j and any real number ξ_j, we express $<\log[1+\exp(z_j)]>$ in a different but equivalent form as follows:

$$
\begin{aligned}
<\log(1 + e^{z_j})> &= <\log[e^{\xi_j z_j} e^{-\xi_j z_j}(1 + e^{z_j})]> \\
&= \xi_j <z_j> + <\log[e^{-\xi_j z_j} + e^{(1-\xi_j)z_j}]>
\end{aligned}
\tag{11.85}
$$

where $<z_j>$ is the ensemble average of z_j. Next, we apply Jensen's inequality in the opposite direction to what we did before, so as to upper bound the average on the right-hand side of Eq. (11.85), thereby obtaining

$$
1 \quad <\log(1 + e^{z_j})> \leq \xi_j <z_j> + \log <e^{-\xi_j z_j} + e^{(1-\xi_j)z_j}>
\tag{11.86}
$$

Putting $\xi_j = 0$ in Eq. (11.86), we obtain the standard bound:

$$
<\log(1 + e^{z_j})> \leq \log <1 + e^{z_j}>
$$

By permitting the use of nonzero values of ξ_j in Eq. (11.86), we get a tighter bound on the average $<\log(1 + e^{z_j})>$ than would be possible with the standard bound (Seung, 1995), as illustrated in the following example.

Example 11.3 Gaussian-Distributed Variable

To illustrate the utility of the bound described in Eq. (11.86), consider a Gaussian-distributed variable with zero mean and unit variance. For this special case, the exact value of $<\log(1 + e^{z_j})>$ is 0.806. The bound described in Eq. (11.86) yields $[e^{0.5\xi^2} + e^{0.5(1-\xi)^2}]$, which attains its minimum value of 0.818 at $\xi = 0.5$. This bound is a great deal closer to the actual result than the value of 0.974 obtained from the standard bound at $\xi = 0$ (Saul et al., 1996).

∎

Returning to the issue at hand, substituting Eqs. (11.85) and (11.86) in (11.83) yields the lower bound on the instantaneous log-likelihood of the evidence $\mathbf{X}_\alpha = \mathbf{x}_\alpha$ as:

$$
\mathcal{L}(\mathbf{w}) \geq - \sum_{j \in \mathcal{H}} [\mu_j \log \mu_j + (1 - \mu_j)\log(1 - \mu_j)]
$$

$$
+ \sum_{j \in \mathcal{H}} \sum_{i<j} w_{ji}\mu_i(\mu_j - \xi_j) - \sum_{j \in \mathcal{H}} \log <\exp(-\xi_j z_j) + \exp((1 - \xi_j)z_j)>
\tag{11.87}
$$

where z_j is itself defined by Eq. (11.84). This is the desired bound on the log-likelihood function $\mathcal{L}(\mathbf{w})$ computed on an example-by-example basis.

Learning Procedure for the Mean-Field Approximate to a Sigmoid Belief Network

In deriving the bound of Eq. (11.87), we introduced two sets of *variational parameters*: μ_j for $j \in \mathcal{H}$ and ξ_j for all j, without actually specifying them. These are adjustable parameters. Since the objective is to maximize the log-likelihood function $\mathcal{L}(\mathbf{w})$, it is natural for us to seek those values of μ_j and ξ_j that maximize the expression on the right-hand side of Eq. (11.87). To achieve this objective, we use a two-step iterative procedure as described in Saul et al. (1996).

Consider first the situation where the mean values μ_j are fixed and the requirement is to find the parameters ξ_j that yield the tightest bound on the log-likelihood function $\mathscr{L}(\mathbf{w})$. Here we note that the expression on the right-hand side of Eq. (11.87) does *not* couple those terms with ξ_j that belong to different neurons in the network. Hence, minimization of this expression with respect to the ξ_j reduces to N independent minimizations over the interval $[0, 1]$, where N is the total number of neurons in the network.

Consider next the situation where the values of ξ_j are fixed and the requirement is to find the mean values μ_j that yield the tightest bound on the log-likelihood function $\mathscr{L}(\mathbf{w})$. For this purpose we introduce the following definitions:

$$K_{ji} = -\frac{\partial}{\partial \mu_i} \log <\exp(-\xi_j z_j) + \exp((1 - \xi_j)z_j)> \qquad (11.88)$$

where the random variable z_j is itself defined by Eq. (11.84). The partial derivative K_{ji} provides a measure of the parental influence of the state x_i of neuron i on the state x_j of neuron j, given the evidence (example) $\mathbf{x}_\alpha \in \mathcal{T}$. As with the synaptic weight of a sigmoid belief network, K_{ji} is nonzero only when the state x_i is a parent to state x_j. Using the factorial distribution of Eq. (11.82), we may evaluate the ensemble averages of $\exp(-\xi_j z_j)$ and $\exp((1 - \xi_j)z_j)$ and then evaluate the partial derivative K_{ji}, where the formula for computing K_{ji} is given in Table 11.3. With the value of K_{ji} at hand, we may now resume the task of finding the value of parameter μ_j that maximizes the log-likelihood function $\mathscr{L}(\mathbf{w})$ for a fixed ξ_j. In particular, differentiating Eq. (11.87) with respect to μ_j, setting the result equal to zero, and rearranging terms, we obtain

$$\log\left(\frac{\mu_j}{1 - \mu_j}\right) = \sum_{i<j} [w_{ji}\mu_i + w_{ij}(\mu_i - \xi_i) + K_{ij}]$$

Equivalently, we may write

$$\mu_j = \varphi\left(\sum_{i<j} [w_{ji}\mu_i + w_{ij}(\mu_i - \xi_i) + K_{ij}]\right) \quad \text{for } j \in \mathcal{H} \qquad (11.89)$$

where $\varphi(\cdot)$ is the sigmoid function. Equation (11.89) is called the *mean-field equation* for a sigmoid belief network. The argument of the sigmoid function in this equation constitutes the so-called *Markov blanket* of neuron j, which is made up as follows:

- The parents and children of neuron j, represented by the terms $w_{ji}\mu_i$ and $w_{ij}\mu_{ij}$, respectively.
- Other parents of the children of neuron j, inherited through the partial derivative K_{ij}.

The Markov blanket of neuron j is illustrated in Fig. 11.9. The notion of a "Markov blanket" was originated by Pearl (1988); it states that the effective input to neuron j, for example, is composed of terms due to its parents, children and their parents.

While it is granted that the choice of the factorial distribution described in Eq. (11.82) as an approximation to the true *a posteriori* distribution $P(\mathbf{X}_\beta = \mathbf{x}_\beta | \mathbf{X}_\alpha = \mathbf{x}_\alpha)$ is not exact, the mean-field equations (11.89) set the parameters $\{\mu_j\}_{j \in \mathcal{H}}$ to optimum

TABLE 11.3   Learning Procedure for the Mean-Field Approximate to a Sigmoid Belief Network

Initialization. Initialize the network by setting the weights w_{ji} of the network to random values uniformly distributed in the range $[-a, a]$; a typical value for a is 0.5.

Computation. For example $\mathbf{x}_\alpha$ drawn from the training set $\mathcal{T}$, perform the following computations:

1. *Updating of $\{\xi_j\}$ for fixed $\{\mu_j\}$.*

   Fix the mean values $\{\mu_j\}_{j \in \mathcal{H}}$ pertaining to the factorial approximation to the *a posteriori* distribution $P(\mathbf{X}_\beta = \mathbf{x}_\beta \mid \mathbf{X}_\alpha = \mathbf{x}_\alpha)$, and minimize the following bound on the log-likelihood function:

$$B(\mathbf{w}) = -\sum_{j \in \mathcal{H}} [\mu_j \log \mu_j + (1 - \mu_j)\log(1 - \mu_j)] + \sum_i \sum_{\substack{j \in \mathcal{H} \\ i < j}} w_{ji}\mu_i\mu_j$$

$$-\sum_i \sum_{\substack{j \in \mathcal{H} \\ i<j}} w_{ji}\mu_i\xi_j - \sum_{j \in \mathcal{H}} \log < \exp(-\xi_j z_j) + \exp((1 - \xi_j)z_j) >$$

   where

$$z_j = \sum_{i<j} w_{ji}x_i$$

   The minimization of $B(\mathbf{w})$ reduces to N independent minimizations over the interval $[0, 1]$.

2. *Updating of $\{\mu_j\}$ for fixed $\{\xi_j\}$.*

   For fixed values of the parameters $\{\xi_j\}$, iterate the mean-field equations:

$$\mu_j = \varphi\left(\sum_{i<j} [w_{ji}\mu_i + w_{ij}(\mu_i - \xi_i) + K_{ij}] \right)$$

   where

$$K_{ji} = -\frac{\partial}{\partial \mu_i} \log < \exp(-\xi_j z_j) + \exp((1 - \xi_j)z_j) >$$

$$= \frac{(1 - \theta_j)(1 - \exp(-\xi_j w_{ji}))}{1 - \mu_i + \mu_i \exp(-\xi_j w_{ji})} + \frac{\theta_j(1 - \exp((1 - \xi_j)w_{ji}))}{1 - \mu_i + \mu_i \exp((1 - \xi_j)w_{ji})}$$

$$\theta_j = \frac{< \exp((1 - \xi_j)z_j) >}{< \exp(-\xi_j z_j) + \exp((1 - \xi_j)z_j) >}$$

$$z_j = \sum_{i<j} w_{ji}x_i$$

   The function $\varphi(\cdot)$ is the sigmoid function:

$$\varphi(v) = \frac{1}{1 + \exp(-v)}$$

TABLE 11.3   (continued)

3. *Correction to synaptic weights.*

For the updated values of parameters $\{\xi_j\}$ and $\{\mu_j\}$, compute the correction Δw_{ji} to synaptic weight w_{ji}:

$$\Delta w_{ji} = \eta \frac{\partial B(\mathbf{w})}{\partial w_{ji}}$$

where η is the learning-rate parameter and

$$\frac{\partial B(\mathbf{w})}{\partial w_{ji}} = -(\xi_j - \mu_j)\mu_i + \frac{(1 - \theta_j)\xi_j\mu_i\exp(-\xi_j w_{ji})}{1 - \mu_i + \mu_i\exp(-\xi_j w_{ji})} - \frac{\theta_j(1 - \xi_j)\mu_i\exp((1 - \xi_j)w_{ji})}{1 - \mu_i + \mu_i\exp((1 - \xi_j)w_{ji})}$$

where θ_j is already defined. Update the synaptic weights:

$$w_{ji} \leftarrow w_{ji} + \Delta w_{ji}$$

4. *Cycling through the training set $\mathcal{T}$.*

Cycle through all the training examples contained in the training set $\mathcal{T}$, thereby maximizing their likelihood for a fixed number of iterations, or until the onset of overfitting is detected through the use of cross-validation, for example.

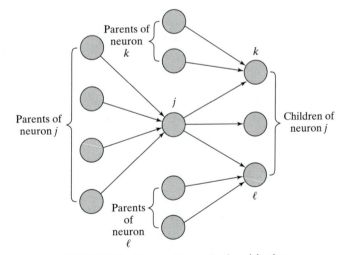

FIGURE 11.9   Illustrating a Markov blanket.

values that make the approximation as accurate as possible. This in turn translates into the tightest mean-field bound on the log-likelihood function $\mathcal{L}(\mathbf{w})$ computed on an example-by-example basis (Saul et al., 1996).

After computing the updated values of parameters $\{\xi_i\}$ and $\{\mu_j\}$, we go on to compute the correction to synaptic weight w_{ji} using the formula:

$$\Delta w_{ji} = \eta \frac{\partial B(\mathbf{w})}{\partial w_{ji}} \tag{11.90}$$

where η is the learning-rate parameter, and $B(\mathbf{w})$ is the lower bound on the log-likelihood function $\mathcal{L}(\mathbf{w})$; that is, $B(\mathbf{w})$ is the expression on the right-hand side of Eq. (11.83). Using this expression, it is a straightforward matter to evaluate the partial derivative $\partial B(\mathbf{w})/\partial w_{ji}$.

A summary of the learning procedure for the mean-field approximate to a sigmoid belief network is presented in Table 11.3. This table includes the formulas for evaluating the partial derivatives K_{ji} and $\partial B(\mathbf{w})/\partial w_{ji}$.

11.13 DETERMINISTIC ANNEALING

We now come to the final topic of the chapter, deterministic annealing. In Section 11.5 we discuss simulated annealing, a stochastic relaxation technique that provides a powerful method for solving nonconvex optimization problems. However, care must be exercised in choosing the annealing schedule. In particular, a global minimum is achieved only if the temperature is decreased at a rate no faster than logarithmically. This requirement makes the use of simulated annealing impractical in many applications. Simulated annealing operates by making random moves on the energy surface (landscape). By contrast, in *deterministic annealing* some form of randomness is incorporated into the energy or cost function itself, which is then deterministically optimized at a sequence of decreasing temperatures (Rose et al., 1990; Rose, 1998); deterministic annealing is *not* to be confused with mean-field annealing (a term that is sometimes used to refer to the deterministic Boltzmann machine).

In what follows we describe the idea of deterministic annealing in the context of an unsupervised learning task: clustering.[11]

Clustering via Deterministic Annealing

Clustering is defined as the partitioning of a given set of data points into subgroups, each of which should be as homogeneous as possible. Clustering is typically a nonconvex optimization problem since virtually all distortion functions used in clustering are nonconvex functions of the input data. Moreover, a plot of the distortion function versus the input is riddled with local minima, making the task of finding the global minimum even more difficult.

In Rose (1991, 1998), a probabilistic framework is described for clustering by *randomization of the partition*, or equivalently, *randomization of the encoding rule*. The main principle used here is that each data point is *associated in probability* with a particular cluster (subgroup). To be specific, let the random vector $\mathbf{X}$ denote a *source (input) vector*, and let the random vector $\mathbf{Y}$ denote the best *reconstruction (output) vector* from a codebook of interest. Individual realizations of these two vectors are denoted by vectors $\mathbf{x}$ and $\mathbf{y}$, respectively.

For clustering we need a *distortion measure*, which is denoted by $d(\mathbf{x}, \mathbf{y})$. It is assumed that $d(\mathbf{x}, \mathbf{y})$ satisfies two desirable properties: (1) it is a convex function of $\mathbf{y}$ for all $\mathbf{x}$, and (2) it is finite whenever its arguments are finite. These mild assumptions are satisfied, for example, by the Euclidean distortion measure

$$d(\mathbf{x}, \mathbf{y}) = \|\mathbf{x} - \mathbf{y}\| \tag{11.91}$$

The *expected distortion* for the randomized pattern is defined by

$$
\begin{aligned}
D &= \sum_{\mathbf{x}} \sum_{\mathbf{y}} P(\mathbf{X} = \mathbf{x}, \mathbf{Y} = \mathbf{y}) d(\mathbf{x}, \mathbf{y}) \\
&= \sum_{\mathbf{x}} P(\mathbf{X} = \mathbf{x}) \sum_{\mathbf{y}} P(\mathbf{Y} = \mathbf{y} | \mathbf{X} = \mathbf{x}) d(\mathbf{x}, \mathbf{y})
\end{aligned}
\tag{11.92}
$$

where $P(\mathbf{X} = \mathbf{x}, \mathbf{Y} = \mathbf{y})$ is the probability of the joint event $\mathbf{X} = \mathbf{x}$ and $\mathbf{Y} = \mathbf{y}$. In the second line of Eq. (11.92) we have used the formula for the probability of a joint event:

$$P(\mathbf{X} = \mathbf{x}, \mathbf{Y} = \mathbf{y}) = P(\mathbf{Y} = \mathbf{y} | \mathbf{X} = \mathbf{x}) P(\mathbf{X} = \mathbf{x}) \tag{11.93}$$

The conditional probability $P(\mathbf{Y} = \mathbf{y} | \mathbf{X} = \mathbf{x})$ is referred to as the *association probability*, that is, the probability of associating the code vector $\mathbf{y}$ with the source vector $\mathbf{x}$.

The expected distortion D is traditionally minimized with respect to the free parameters of the clustering model: the reconstruction vector $\mathbf{y}$ and the association probability $P(\mathbf{Y} = \mathbf{y} | \mathbf{X} = \mathbf{x})$. This form of minimization produces a "hard" clustering solution, hard in the sense that a source vector $\mathbf{x}$ is assigned to the nearest code vector $\mathbf{y}$. In deterministic annealing, on the other hand, the optimization problem is reformulated as that of seeking the probability distribution that minimizes the expected distortion *subject to a specified level of randomness*. For a principled measure of the level of randomness, we use the Shannon entropy defined by (see Section 10.4)

$$H(\mathbf{X}, \mathbf{Y}) = \sum_{\mathbf{x}} \sum_{\mathbf{y}} P(\mathbf{X} = \mathbf{x}, \mathbf{Y} = \mathbf{y}) \log P(\mathbf{X} = \mathbf{x}, \mathbf{Y} = \mathbf{y}) \tag{11.94}$$

The constrained optimization of the expected distortion is then expressed as the minimization of the Lagrangian:

$$F = D - TH \tag{11.95}$$

where T is the Lagrange multiplier. From Eq. (11.95) we observe the following:

- For large values of T the entropy H is maximized.
- For small values of T the expected distortion D is minimized, resulting in a hard (nonrandom) clustering solution.
- For intermediate values of T the minimization of F provides a tradeoff between an increase in the entropy H and a reduction in the expected distortion D.

Most importantly, comparing Eq. (11.95) with (11.11), we may identify the correspondence between the constrained clustering optimization problem and statistical

---

TABLE 11.4 Correspondence between Constrained Clustering and Statistical Physics

Constrained clustering optimization	Statistical physics
Lagrangian, F	Free energy, F
Expected distortion, D	Average energy, $<E>$
Shannon entropy, H	Entropy, H
Lagrange multiplier, T	Temperature, T

---

mechanics listed in Table 11.4. In light of this analogy, henceforth we refer to T as the temperature.

To develop further insight into the Lagrangian F, we note that the joint entropy $H(\mathbf{X}, \mathbf{Y})$ may be decomposed into two terms as follows (see Eq. (10.25)):

$$H(\mathbf{X}, \mathbf{Y}) = H(\mathbf{X}) + H(\mathbf{Y}|\mathbf{X})$$

where $H(\mathbf{X})$ is the source entropy and $H(\mathbf{Y}|\mathbf{X})$ is the conditional entropy of the reconstruction vector $\mathbf{Y}$ given the source vector $\mathbf{X}$. The source entropy $H(\mathbf{X})$ is independent of clustering. Accordingly, we may drop the source entropy $H(\mathbf{X})$ from the definition of the Lagrangian F, and thereby focus on the conditional entropy

$$H(\mathbf{Y}|\mathbf{X}) = - \sum_{\mathbf{x}} P(\mathbf{X} = \mathbf{x}) \sum_{\mathbf{y}} P(\mathbf{Y} = \mathbf{y}|\mathbf{X} = \mathbf{x}) \log P(\mathbf{Y} = \mathbf{y}|\mathbf{X} = \mathbf{x}) \quad (11.96)$$

which highlights the role of the association probability $P(\mathbf{Y} = \mathbf{y}|\mathbf{X} = \mathbf{x})$. Hence, keeping in mind the correspondence between the constrained clustering optimization problem and statistical physics, and invoking the principle of minimal free energy described in Section 11.2, we find that minimizing the Lagrangian F with respect to the association probabilities results in the Gibbs distribution

$$P(\mathbf{Y} = \mathbf{y}|\mathbf{X} = \mathbf{x}) = \frac{1}{Z_{\mathbf{x}}} \exp\left(-\frac{d(\mathbf{x}, \mathbf{y})}{T}\right) \quad (11.97)$$

where $Z_{\mathbf{x}}$ is the partition function for the problem at hand. It is defined by

$$Z_{\mathbf{x}} = \sum_{\mathbf{y}} \exp\left(-\frac{d(\mathbf{x}, \mathbf{y})}{T}\right) \quad (11.98)$$

As the temperature T approaches infinity, we find from Eq. (11.97) that the association probability approaches a uniform distribution. The implication of this statement is that at very high temperatures, each input vector is equally associated with all clusters. Such associations may be viewed as "extremely fuzzy." At the other extreme, as the temperature T approaches zero, the association probability approaches a delta function. Accordingly, at very low temperatures the classification is hard, with each input sample being assigned to the nearest code vector with probability 1.

To find the minimum value of the Lagrangian F, we substitute the Gibbs distribution of Eq. (11.97) in (11.92) and (11.96) and then use the resulting expressions in the formula for the Lagrangian F in Eq. (11.95). The result obtained by so doing is (see Problem 11.22):

$$F^* = \min_{P(\mathbf{Y}=\mathbf{y}|\mathbf{X}=\mathbf{x})} F$$

$$= -T\sum_{\mathbf{x}} P(\mathbf{X}=\mathbf{x})\log Z_{\mathbf{x}} \tag{11.99}$$

To minimize the Lagrangian with respect to the remaining free parameters, namely the code vectors $\mathbf{y}$, we set the gradients of F^* with respect to $\mathbf{y}$ to zero. Hence, we obtain the condition

$$\sum_{\mathbf{x}} P(\mathbf{X}=\mathbf{x}, \mathbf{Y}=\mathbf{y})\frac{\partial}{\partial \mathbf{y}}d(\mathbf{x},\mathbf{y}) = \mathbf{0} \quad \text{for all } \mathbf{y} \in \mathcal{Y} \tag{11.100}$$

where $\mathcal{Y}$ is the set of all code vectors. Using the formula of Eq. (11.93) and normalizing with respect to $P(\mathbf{X}=\mathbf{x})$, we may redefine this minimizing condition as

$$\frac{1}{N}\sum_{\mathbf{x}} P(\mathbf{Y}=\mathbf{y}|\mathbf{X}=\mathbf{x})\frac{\partial}{\partial \mathbf{y}}d(\mathbf{x},\mathbf{y}) = 0 \quad \text{for all } \mathbf{y} \in \mathcal{Y} \tag{11.101}$$

where the association probability $P(\mathbf{Y}=\mathbf{y} \mid \mathbf{X}=\mathbf{x})$ is itself defined by the Gibbs distribution of Eq. (11.97). In Eq. (11.101) we have included the scaling factor $1/N$ merely for completeness, where N is the number of available examples.

We may now describe the deterministic annealing algorithm for clustering (Rose, 1998). Basically, the algorithm consists of minimizing the Lagrangian F^* with respect to the code vectors at a high value of temperature T, and then tracking the minimum while the temperature T is lowered. In other words, deterministic annealing operates with a specific annealing schedule where the temperature is lowered in an orderly fashion. At each value of the temperature T, a two-step iteration central to the algorithm is performed, as described here:

1. The code vectors are fixed, and the Gibbs distribution of Eq. (11.97) for a specific distortion measure $d(\mathbf{x}, \mathbf{y})$ is used to calculate the association probabilities.
2. The associations are fixed, and Eq. (11.101) is used to optimize the distortion measure $d(\mathbf{x}, \mathbf{y})$ with respect to the code vectors $\mathbf{y}$.

This two-step iterative procedure is monotonically nonincreasing in F^*, and is therefore assured of converging to a minimum. At high values of temperature T, the Lagrangian F^* is fairly smooth, and is a convex function of $\mathbf{y}$ under the mild assumptions previously made on the distortion measure $d(\mathbf{x}, \mathbf{y})$. A global minimum of F^* can be found at high temperatures. As the temperature T is lowered, the association probabilities become hard, resulting in a hard clustering solution.

As the temperature T is lowered in the course of going through the annealing schedule, the system undergoes a sequence of phase transitions, which consists of natural cluster splits where the clustering model grows in size (i.e., number of clusters) (Rose et al., 1990; Rose, 1991). This phenomenon is significant for the following reasons:

- It provides a useful tool for *controlling* the size of the clustering model.
- As in ordinary physical annealing, the phase transitions are the *critical points* of the deterministic annealing process where care has to be exercised with the annealing.

- The critical points are *computable*, thereby providing information that can be used to accelerate the algorithm in between phase transitions.
- An *optimum model size* may be identified by coupling a validation procedure with the sequence of solutions produced at various phases, which represent solutions of increasing model size.

Example 11.4

Figures 11.10 and 11.11 illustrate the evolution of the clustering solution via deterministic annealing at various phases as the temperature T is decreased or the reciprocal of temperature, $B = 1/T$, is increased (Rose 1991). The data set used to generate these figures is a mixture of six Gaussian distributions whose centers are marked with X. The centers of the computed clusters are marked with o. Since the clustering solutions at non-zero temperatures are not hard, this random partition is depicted by contours of equal probability—for example, probability 1/3 of belonging to a particular cluster. This process starts with one natural cluster containing the training set (Fig. 11.10a). At the first phase transition, it splits into two clusters (Fig. 11.10b), and then passes through a sequence of phase transitions until it reaches the "natural" set of six clusters. The next phase transition results in an "explosion" when all clusters split. Figure 11.11 shows the phase diagram, displaying the behavior of the average distortion throughout the annealing process and the number of natural clusters at each phase. In this figure the average distortion (normalized with respect to its minimum value) is plotted versus the reciprocal of temperature, namely B (normalized with respect to its minimum value). Both axes are labeled in their relative logarithmic forms.

■

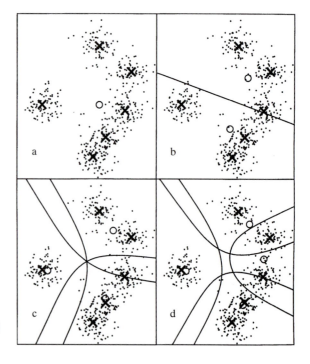

FIGURE 11.10 Clustering at various phases. The lines are equiprobability contours, $p = 1/2$ in (b), and $p = 1/3$ elsewhere:
(a) 1 cluster ($B = 0$),
(b) 2 clusters ($B = 0.0049$),
(c) 3 clusters ($B = 0.0056$),
(d) 4 clusters ($B = 0.0100$),
(e) 5 clusters ($B = 0.0156$),
(f) 6 clusters ($B = 0.0347$), and
(g) 19 clusters ($B = 0.0605$).

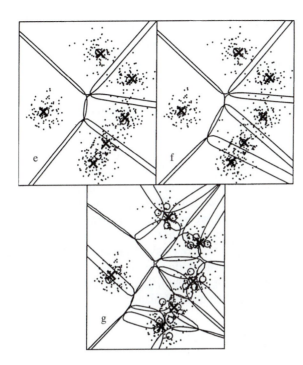

FIGURE 11.10   (continued)

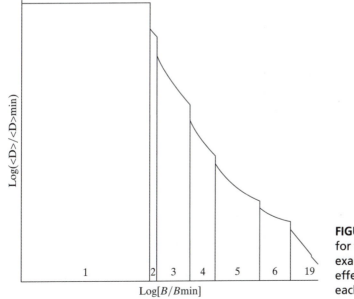

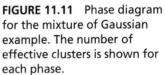

FIGURE 11.11 Phase diagram for the mixture of Gaussian example. The number of effective clusters is shown for each phase.

Analogy with the EM Algorithm

For another important aspect of the deterministic annealing algorithm, suppose we view the association probability $P(\mathbf{Y} = \mathbf{y} \,|\, \mathbf{X} = \mathbf{x})$ as the expected value of a random binary variable $V_{\mathbf{xy}}$ defined as:

$$V_{\mathbf{xy}} = \begin{cases} 1 & \text{if the source vector } \mathbf{x} \text{ is assigned to code vector } \mathbf{y} \\ 0 & \text{otherwise} \end{cases} \qquad (11.102)$$

Then from such a perspective we recognize the two-step iteration of the deterministic annealing algorithm to be a form of the Expectation-Maximization (EM) algorithm described in Chapter 7 for maximum likelihood estimation. In particular, in step 1 that computes the association probabilities, we have the equivalent of the expectation step. In step 2 that minimizes the Lagrangian F^*, we have the equivalent of the maximization step.

In making this analogy, however, note that deterministic annealing is more general than maximum likelihood estimation. We say so because, unlike maximum likelihood estimation, deterministic annealing does *not* make any assumption on the underlying probability distribution of the data. The association probabilities are, in fact, derived from the Lagrangian F^* to be minimized.

11.14 SUMMARY AND DISCUSSION

In this chapter we discuss the use of ideas rooted in statistical mechanics as the mathematical basis for the formulation of optimization techniques and learning machines. The learning machines considered here may be categorized as follows:

- *Stochastic machines*, exemplified by the Boltzmann machine, sigmoid belief networks, and the Helmholtz machine.
- *Deterministic machines*, derived from the Boltzmann machine and sigmoid belief networks by invoking mean-field approximations.

The Boltzmann machine uses hidden and visible neurons that are in the form of stochastic, binary-state units. It cleverly exploits the beautiful properties of the Gibbs distribution, thereby offering some appealing features:

- Through training, the probability distribution exhibited by the neurons is matched to that of the environment.
- The network offers a generalized approach that is applicable to the basic issues of search, representation, and learning (Hinton, 1989).
- The network is guaranteed to find the global minimum of the energy surface with respect to the *states*, provided that the annealing schedule in the learning process is performed slowly enough (Geman and Geman, 1984).

Unfortunately, the annealing schedule is much too slow to be of practical value. However, the learning process can be accelerated for specific classes of Boltzmann machines, for which we do not have to run a sampling algorithm or apply a mean-field approximation. In particular, in Boltzmann machines in which the hidden neurons are in the form of a chain, a tree, or a coupled pair of chains or trees, learning can be performed exactly in polynomial time. This is achieved by using an algorithm from statistical mechanics known as "decimation," which is a simple and elegant procedure that recursively removes links and nodes from the graph, much like solving a resistance inductance capacitance (RLC) circuit (Saul and Jordan, 1995, 1996).

Sigmoid belief networks offer a significant improvement over the Boltzmann machine by eliminating the need for the negative (free-running) phase. They do so by replacing the symmetric connections of the Boltzmann machine with directed acyclic connections. That is, whereas the Boltzmann machine is a recurrent network with an abundance of feedback, sigmoid belief networks have a multilayer architecture with *no* feedback. As the name implies, sigmoid belief networks are closely related to classical belief networks pioneered by Pearl (1988), thereby linking the subject of neural networks with that of *probabilistic reasoning and graphical models* (Jordan, 1998; Jordan et al., 1998).

The Helmholtz machine is different again. Its development is motivated by the idea that vision is inverse graphics (Horn, 1977; Hinton and Ghahramani, 1997). In particular it uses a stochastic, generative model, operating in the backward direction, to convert an abstract representation of a scene into an intensity image. The abstract representation of the scene (i.e., the network's own visual knowledge of the world) is itself learned by a stochastic, recognition model, operating in the forward direction. Through a clever integration of the recognition and generative models (i.e., forward/backward projections), the Helmholtz machine assumes the role of a self-supervising machine, thereby eliminating the need for a teacher.

Turning next to the class of deterministic machines, the deterministic Boltzmann machine is derived from the Boltzmann machine by applying a naive form of mean-field approximation, where the correlation between two random variables is replaced by the product of their mean values. The net result is that the deterministic Boltzmann machine can be considerably faster than the standard stochastic Boltzmann machine. Unfortunately, in practice its use is restricted to a single layer of hidden neurons. In Kappen and Rodriguez (1998), it is argued that, in the correct treatment of mean-field theory for the Boltzmann machine, the correlations need to be computed using the linear response theorem. The essence of this theorem is to replace the clamped and free-running correlations in the Boltzmann learning rule of Eq. (11.53) by their linear response approximations. According to Kappen and Rodriguez, the new learning procedure is applicable to networks with or without hidden neurons.

The deterministic form of sigmoid belief networks is derived by applying another form of mean-field theory, where a rigorous lower bound on the log-likelihood function is derived through the use of Jensen's inequality. Moreover, the theory exploits the virtues of a tractable substructure in a principled manner, making this class of neural networks into an important addition to belief networks.

In this chapter we also discuss two optimization techniques: simulated annealing and deterministic annealing. Simulated annealing distinguishes itself by performing random moves on the energy surface, which can make the annealing schedule very slow, with the result that its use is unrealistic for many applications. By contrast, deterministic annealing incorporates randomness into the cost function, which is then deterministically optimized at each temperature sequentially, starting at a high temperature and then going down. However, note that simulated annealing is guaranteed to reach a global minimum, whereas no such guarantee has yet been found for deterministic annealing.

Although, in this chapter, we emphasize the use of optimization techniques and stochastic machines for solving unsupervised learning tasks, they can all also be used for supervised learning tasks if so desired.

NOTES AND REFERENCES

1. The term "canonical distribution" as a description of Eq. (11.3) was coined by J. Willard Gibbs (1902). On page 33 of Part One (Elementary Principles in Statistical Mechanics) of his collected works, he wrote

   "The distribution represented by...

   $$P = \exp\left(\frac{\psi - \in}{H}\right)$$

   where H and ψ are constants, and H positive, seems to represent the most simple case conceivable, since it has the property that when the system consists of parts with separate energies, the laws of the distribution in phase of the separate parts are of the same nature—a property which enormously simplifies the discussion, and is the foundation of extremely important relations to thermodynamics.....

   When an ensemble of systems is distributed in phase in the manner described, i.e., when the index of probability (P) is a linear function of the energy ($\in$), we shall say that the ensemble is *canonically distributed*, and shall call the divisor of the energy (H) the modulus of distribution."

   In the physics literature, Eq. (11.3) is commonly referred to as the canonical distribution (Reif, 1965) or Gibbs distribution (Landau and Lifschitz, 1980). In the neural network literature it has been referred to as the Gibbs distribution, Boltzmann distribution, and Boltzmann-Gibbs distribution.

2. The idea of introducing temperature and simulating annealing into combinatorial optimization problems is due to Kirkpatrick, Gelatt, and Vacchi (1983) and independently to Černy (1985).

   In a physical context, annealing is a delicate process by nature. In their 1983 paper, Kirkpatrick et al. discuss the notion of "melting" a solid, which involves raising the temperature to a maximum value at which all particles of the solid arrange themselves "randomly" in the liquid phase. Then the temperature is lowered, permitting all particles to arrange themselves in the low-energy ground state of a corresponding lattice. If the cooling is too rapid—that is, the solid is not allowed enough time to reach thermal equilibrium at each temperature value—the resulting crystal will have many defects, or the substance may form a glass with no crystalline order and only metastable locally optimal structures.

   The notion of "melting" may be the right way of thinking about glasses, and perhaps combinatorial optimization problems in a corresponding computational context. However, it is misleading when discussing many other application domains (Beckerman, 1997). For example, in image processing, if we raise the "temperature" so that all particles arrange themselves randomly we have lost the image—it becomes uniformly gray. In a corresponding metallurgical sense, when we anneal either iron or copper we must keep the annealing temperature below the melting point; otherwise we ruin the sample.

There are several important parameters that govern metallurgical annealing:

- *Annealing temperature*, which specifies the temperature to which the metal or alloy is heated.
- *Annealing time*, which specifies the duration of time for which the elevated temperature is maintained.
- *Cooling schedule*, which specifies the rate at which the temperature is lowered.

These parameters have their counterparts in simulated annealing as described in the subsection on annealing schedule.

3. The *Langevin equation* (with time-dependent temperature) provides the basis for another global optimization algorithm that was proposed by Grenander (1983), and subsequently analyzed by Gidas (1985). The Langevin equation is a stochastic differential equation described as (Reif, 1965):

$$\frac{dv(t)}{dt} = -\gamma v(t) + \Gamma(t)$$

where $v(t)$ is the velocity of a particle of mass m immersed in a viscous fluid, γ is a constant equal to the ratio of the coefficient of friction to the mass m, and $\Gamma(t)$ is a fluctuating force per unit mass. The Langevin equation was the first mathematical equation to describe nonequilibrium thermodynamics.

4. For more elaborate and theoretically oriented annealing schedules, see the books by Aarts and Korst (1989, pp. 60–75) and by van Laarhoven and Aarts (1988, pp. 62–71).

5. Gibbs sampling is referred to in statistical physics as a "heat bath" version of the Metropolis algorithm. It is widely used in image processing, neural networks, and statistics, following its formal exposition in the literature by Geman and Geman (1984) and Gelfand and Smith (1990). The latter paper also discusses other approaches to sampling (or Monte Carlo) based approaches to the numerical calculation of estimates of marginal probability distributions. Hastings (1970) presented a generalization of the Metropolis algorithm, of which Gibbs sampling is a special case; its potential for solving numerical problems in statistics was mentioned.

6. The visible neurons of a Boltzmann machine may also be subdivided into input and output neurons. In this second configuration, the Boltzmann machine performs *association* under the supervision of a teacher. The input neurons receive information from the environment and the output neurons report the outcome of the computation to an end user.

7. The formula of Eq. (11.39) applies to a Boltzmann machine whose "on" and "off" states are denoted by $+1$ and -1, respectively. In the case of a machine using 1 and 0 to denote its "on" and "off" states, respectively, we have

$$E(\mathbf{x}) = -\sum_i \sum_{\substack{j \\ i \neq j}} w_{ji} x_i x_j$$

8. Traditionally, the relative entropy or Kullback–Leibler distance has been used as the index of performance for the Boltzmann machine (Ackley et al., 1985; Hinton and Sejnowski, 1986). This criterion provides a measure of the discrepancy between the environment and the network's internal model. It is defined by

$$D_{p^+ \| p^-} = \sum_\alpha p_\alpha^+ \log\left(\frac{p_\alpha^+}{p_\alpha^-}\right)$$

where p_α^+ is the probability that the visible neurons are in state α when the network is in its clamped condition, and p_α^- is the probability that the same neurons are in state α when the network is in its free-running condition. The synaptic weights of the network are adjusted to minimize $D_{p_{\alpha}^{+} \| p_{\alpha}^{-}}$; see Problem 11.10.

The principles of minimum Kullback–Leibler divergence and maximum likelihood are basically equivalent when applied to a training set. To see the equivalence, we note that the Kullback–Leibler divergence between two distributions f and g is given by

$$D_{f \| g} = -H(f) - \sum f \log(g)$$

If the distribution f is specified by a training set, and a model for g is given for optimization, the first term is constant, and the second term is the negative of log-likelihood. Hence, minimum Kullback–Leibler divergence is equivalent to maximum likelihood.

9. Belief networks were originally introduced for the purpose of representing probabilistic knowledge in expert systems (Pearl, 1988). They are also referred to in the literature as *Bayesian networks* or *Bayesian nets*.

10. The Helmholtz machine belongs to a class of neural networks characterized by *forward-backward projections*. The idea of forward-backward projections was originated by Grossberg (1980) in his studies of *adaptive resonance theory*; see also Carpenter and Grossberg (1987). In this model, forward adaptive filtering is combined with backward template matching so that adaptive resonance (i.e., amplification and prolongation of neural activity) takes place. In contrast to Grossberg's adaptive resonance theory, the Helmholtz machine uses a statistical approach to treat self-supervised learning as one of ascertaining a generative model that tries to accurately capture, the underlying structure of the input data.

Another closely related work is that of Luttrell (1994, 1997). In Luttrell (1994), the idea of a *folded Markov chain* (FMC) is developed. Specifically, the forward transitions through a Markov chain are followed by inverse transitions (using Bayes' theorem) in a backward direction through a copy of the same chain. In Luttrell (1997), the relationship between the FMC and Helmholtz machine is discussed.

Other related works include those of Kawato et al., (1993) where forward (recognition) and backward (generative) models are considered in a similar manner to the Helmholtz machine but without a probabilistic perspective, and the proposals by Mumford (1994) for mapping Grenander's generative model onto the brain.

In Dayan and Hinton (1996), a number of different varieties of Helmholtz machine, including a supervised scheme, are suggested.

11. Deterministic annealing has been successfully applied to many learning tasks:

- Vector quantization (Rose et al., 1992; Miller and Rose, 1994)
- Statistical classifier design (Miller et al., 1996)
- Nonlinear regression using mixture of experts (Rao et al., 1997a)
- Hidden Markov models for speech recognition (Rao, et al., 1997b)

A *hidden Markov model* is similar to a Markov chain because in both cases the transition from one state to another is probabilistic. However, they differ from each other in one fundamental respect. In a Markov chain, the production of an output symbol is deterministic. In a hidden Markov model, on the other hand, the output symbols are probabilistic as well, with the result that all symbols are possible at each state. Thus, with each state of a hidden Markov model, we have a probability distribution of all the output symbols. Hidden Markov models are discussed in Rabiner (1989), Rabiner and Juang (1986), and Jelinek (1997).

PROBLEMS

Markov Chains

11.1 The n-step transition probability from state i to state j is denoted by $p_{ij}^{(n)}$. Using the method of induction, show that

$$p_{ij}^{(1+n)} = \sum_k p_{ik} p_{kj}^{(n)}$$

11.2 Figure P11.2 shows the state transition diagram for the *random walk* process, where the transition probability p is greater than zero. Is the infinitely long Markov chain depicted here irreducible? Justify your answer.

11.3 Consider the Markov chain depicted in Fig. P11.3, which is reducible. Identify the classes of states contained in this state transition diagram.

11.4 Calculate the steady-state probabilities of the Markov chain shown in Fig. P11.4.

Simulation techniques

11.5 The Metropolis algorithm and the Gibbs sampler represent two alternative techniques for simulating a large-scale problem of interest. Discuss the basic similarities and differences between them.

11.6 In this problem we consider the use of simulated annealing for solving the *traveling salesman problem* (TSP). You are given the following:

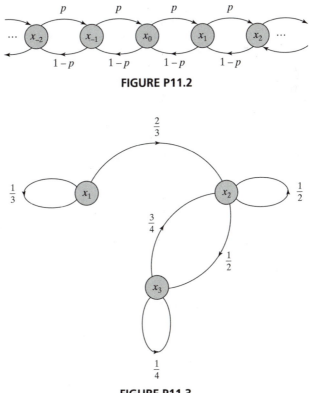

FIGURE P11.2

FIGURE P11.3

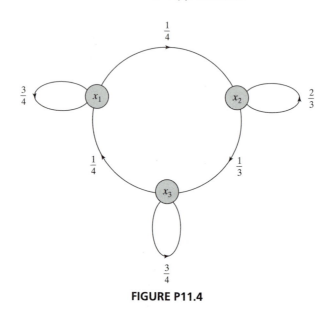

- N cities
- The distance between each pair of cities, d
- A tour represented by a closed path visiting each city once and only once.

The objective is to find a tour (i.e., permutation of the order in which the cities are visited) that is of minimal total length L. In this problem the different possible tours are the configurations, and the total length of a tour is the cost function to be minimized.

(a) Devise an iterative method of generating valid configurations.

(b) The total length of a tour is defined by

$$L_P = \sum_{i=1}^{N} d_{P(i)P(i+1)}$$

where P denotes a permutation with $P(N + 1) = P(1)$. Correspondingly, the partition function is

$$Z = \sum_{P} e^{-L_P/T}$$

where T is a control parameter. Set up a simulated annealing algorithm for the TSP.

Boltzmann machine

11.7 Consider a stochastic, two-state neuron j operating at temperature T. This neuron *flips* from state x_j to state $-x_j$ with probability

$$P(x_j \to -x_j) = \frac{1}{1 + \exp(-\Delta E_j/T)}$$

where ΔE_j is the energy change resulting from such a flip. The total energy of the Boltzmann machine is defined by

$$E = -\frac{1}{2} \sum_{i} \sum_{\substack{j \\ i \neq j}} w_{ji} x_i x_j$$

where w_{ji} is the synaptic weight from neuron i to neuron j, with $w_{ji} = w_{ij}$ and $w_{ii} = 0$.

(a) Show that

$$\Delta E_j = -2x_j v_j$$

where v_j is the induced local field of neuron j.

(b) Hence, show that for an initial state $x_j = -1$, the probability that neuron j is flipped into state $+1$ is $1/(1 + \exp(-2v_j/T))$.

(c) Show that the same formula in part (b) holds for neuron j flipping into state -1 when it is initially in state $+1$.

11.8 Derive the formula given in Eq. (11.49) that defines the derivative of the log-likelihood function $L(\mathbf{w})$ with respect to the synaptic weight w_{ji} for the Boltzmann machine.

11.9 The Gibbs distribution may be derived using a self-contained mathematical approach that does *not* rely on concepts from statistical physics. In particular, a *two-step Markov chain model* of a stochastic machine may be used to formalize the assumptions that yield the unique properties of the Boltzmann machine (Mazaika, 1987). This should not come as a surprise since the simulated annealing, basic to the operation of the Boltzmann machine, is known to have a Markov property of its own (van Laarhoven and Aarts, 1988).

Consider then a transition model between states of a neuron in a stochastic machine that is composed of two random processes:

- The first process decides which state transition should be attempted.
- The second process decides if the transition succeeds.

(a) Expressing the transition probability p_{ji} as the product of two factors, that is,

$$p_{ji} = \tau_{ji} q_{ji} \quad \text{for } j \neq i$$

show that

$$p_{ii} = 1 - \sum_{j \neq i} \tau_{ji} q_{ji}$$

(b) Assume that the attempt rate matrix is symmetric:

$$\tau_{ji} = \tau_{ij}$$

Also assume that the probability of a successful attempt satisfies the property of complementary conditional transition probability:

$$q_{ji} = 1 - q_{ij}$$

By invoking these two assumptions, show that

$$\sum_j \tau_{ji}(q_{ji}\pi_j + q_{ij}\pi_i - \pi_j) = 0$$

(c) Given that $\tau_{ji} \neq 0$, use the result of a part (a) of the problem to show that

$$q_{ji} = \frac{1}{1 + (\pi_i/\pi_j)}$$

(d) Finally, make a change of variables:

$$E_i = -T \log \pi_i + T^*$$

where T and T^* are arbitrary constants. Hence, derive the following results:

(i) $\pi_i = \dfrac{1}{Z} \exp\left(-\dfrac{E_i}{T}\right)$

(ii) $Z = \sum_j \exp\left(-\dfrac{E_j}{T}\right)$

(iii) $q_{ji} = \dfrac{1}{1 + \exp(-\Delta E/T)}$

where $\Delta E = E_j - E_i$.

(e) What conclusions can you draw from these results?

11.10 In Section 11.7 we use maximum likelihood as the criterion for deriving the Boltzmann learning rule, described in Eq. (11.53). In this problem we revisit this learning rule using another criterion. From the discussion presented in Chapter 10, the Kullback–Leibler divergence between two probability distributions p_α^+ and p_α^- is defined by

$$D_{p^+\|p^-} = \sum_\alpha p_\alpha^+ \log\left(\frac{p_\alpha^+}{p_\alpha^-}\right)$$

where the summation is over all possible states α. The probability p_α^+ denotes the probability that the visible neurons are in state α when the network is in its clamped (positive) condition, and the probability p_α^- denotes the probability that the same neurons are in a state α when the network is in its free-running (negative) condition. Using $D_{p^+\|p^-}$, rederive the Boltzmann learning rule.

11.11 Consider a Boltzmann machine whose visible neurons are divided into input neurons and output neurons. The states of these neurons are denoted by α and γ, respectively. The state of the hidden neurons is denoted by β. The Kullback–Leibler divergence for this machine is defined by

$$D_{p^+\|p^-} = \sum_\alpha p_\alpha^+ \sum_\gamma p_{\gamma|\alpha}^+ \log\left(\frac{p_{\gamma|\alpha}^+}{p_{\gamma|\alpha}^-}\right)$$

where p_α^+ is the probability of state α over the input neurons; $p_{\gamma|\alpha}^+$ is the conditional probability that the output neurons are clamped in state α given an input state α; and $p_{\gamma|\alpha}^-$ is the conditional probability that the output neurons are in thermal equilibrium in state γ given that only the input neurons are clamped in state α. As before, the plus and minus superscripts denote the positive (clamped) and negative (free-running) conditions, respectively.

(a) Derive the formula $D_{p^+\|p^-}$ for a Boltmann machine that includes input, hidden, and output neurons.

(b) Show that the Boltzmann learning rule for adjusting the synaptic weight w_{ji} in this network configuration may still be expressed in the same form as that described in Eq. (11.53), with new interpretations for the correlations ρ_{ji}^+ and ρ_{ji}^-.

Sigmoid Belief Networks

11.12 Summarize the similarities and differences between the Boltzmann machine and a sigmoid belief network.

11.13 In Problem 11.9 we demonstrated that the Boltzmann machine is described by a two-step Markov chain model. Does a sigmoid belief network admit a Markov chain model? Justify your answer.

11.14 Let w_{ji} denote the synaptic weight from neuron i to neuron j in a sigmoid belief network that uses $+1$ for the on state of a neuron and -1 for the off state. Let w_{ji}' denote the corresponding synaptic weight of a sigmoid belief network that uses 1 for the on state of a neuron and 0 for the off state. Show that w_{ji} can be converted to w_{ji}' by using the tranformation:

$$w'_{ji} = \frac{w_{ji}}{2} \quad \text{for } 0 < i < j$$

$$w'_{j0} = w_{j0} = \frac{1}{2} \sum_{0<i<j} w_{ji}$$

The last line pertains to bias applied to neuron j.

11.15 In a sigmoid belief network we identify the probability $P(\mathbf{X}_\beta = \mathbf{x}_\beta | \mathbf{X}_\alpha = \mathbf{x}_\alpha)$ as a Gibbs distribution, and the probability $P(\mathbf{X}_\alpha = \mathbf{x}_\alpha)$ as the corresponding partition function. Justify the validity of this twofold identification.

Helmholtz Machine

11.16 The Helmholtz machine is free of feedback in both its recognition and generative models. What would happen to the operation of this network if the use of feedback is permitted in either of these two models?

Deterministic Boltzmann Machine

11.17 The Boltzmann machine performs gradient descent (in weight space) on the probability space, as discussed in Problem 11.10. On what function does the deterministic Boltzmann machine perform its gradient descent? You may refer to Hinton (1989) for a discussion of this issue.

11.18 Consider a recurrent network that is asymmetric in that $w_{ji} \neq w_{ij}$. Show that the deterministic Boltzmann learning algorithm will automatically symmetrize the network provided that, after each weight update, each weight is decayed toward zero by a small amount proportional to its magnitude (Hinton, 1989).

Deterministic Sigmoid Belief Network

11.19 Show that the difference between the expressions on the left- and right-hand sides of Eq. (11.77) is equal to the Kullback–Leibler divergence between the distributions $Q(\mathbf{X}_\beta = \mathbf{x}_\beta | \mathbf{X}_\alpha = \mathbf{x}_\alpha)$ and $P(\mathbf{X}_\beta = \mathbf{x}_\beta | \mathbf{X}_\alpha = \mathbf{x}_\alpha)$.

11.20 The argument of the sigmoid function in Eq. (11.89) defines the induced local field v_j of neuron j in the deterministic sigmoid belief network. In what ways does v_j differ from the corresponding induced local field of a neuron in a multilayer perceptron trained with the back-propagation algorithm?

Deterministic Annealing

11.21 In Section 11.13 we developed the idea of deterministic annealing using an information-theoretic approach. The idea of deterministic annealing may also be developed in a principled manner using the maximum entropy principle that is discussed in Chapter 10. Follow through the rationale of this second approach (Rose, 1998).

11.22 **(a)** Using Eqs. (11.97) and (11.98), derive the result given in Eq. (11.99) that defines the Lagrangian F^* that results from use of the Gibbs distribution for the association probability.

   (b) Using the result from part a of this problem, derive the condition given in Eq. (11.101) for the minimum of F^* with respect to the code vectors $\mathbf{y}$.

   (c) Apply the minimizing condition of Eq. (11.101) to the squared distortion measure of Eq. (11.91) and comment on your result.

11.23 Consider a data set that is a mixture of Gaussian distributions. In what way does the use of deterministic annealing offer an advantage over maximum likelihood estimation in such a situation?

11.24 In this problem we explore the use of deterministic annealing for pattern classification using a neural network (Miller et al., 1996). The output of neuron j in the output layer is denoted by $F_j(\mathbf{x})$ where $\mathbf{x}$ is the input vector. The classification decision is based on the maximum discriminant $F_j(\mathbf{x})$.

(a) For a probabilistic objective function, consider the following:

$$F = \frac{1}{N} \sum_{(\mathbf{x}, \mathscr{C}) \in \mathscr{T}} \sum_j P(\mathbf{x} \in \mathscr{R}_j) F_j(\mathbf{x})$$

where $\mathscr{T}$ is a training set of labeled vectors with x denoting an input vector and $\mathscr{C}$ its class label, and $P(\mathbf{x} \in \mathscr{R}_j)$ is the probability of association between input vector x and class region $\mathscr{R}_j$. Using the maximum entropy principle that is discussed in Chapter 10, formulate the Gibbs distribution for $P(\mathbf{x} \in \mathscr{R}_j)$.

(b) Let $<P_e>$ denote the average misclassification cost. Formulate the Lagrangian for minimization of $<P_e>$ subject to the constraint that the entropy corresponding to the association probabilities $P(\mathbf{x} \in \mathscr{R}_j)$ is equal to some constant value H.

CHAPTER 12

Neurodynamic Programming

12.1 INTRODUCTION

In Chapter 2 we identify two main paradigms of learning: learning with a teacher, and learning without a teacher. The paradigm of learning without a teacher is subdivided into self-organized (unsupervised) learning and reinforcement learning. Different forms of learning with a teacher or supervised learning are covered in Chapters 4 through 7, and different forms of unsupervised learning are discussed in Chapters 8 through 11. In this chapter we discuss *reinforcement learning.*

Supervised learning is a "cognitive" learning problem performed under the tutelage of a teacher: It relies on the availability of an adequate set of input–output examples that are representative of the operating environment. In contrast, reinforcement learning is a "behavioral" learning problem: It is performed through *interaction* between the learning system and its environment, in which the system seeks to achieve a specific goal despite the presence of uncertainties (Barto et al., 1983; Sutton and Barto, 1998). The fact that this interaction is performed without a teacher makes reinforcement learning particularly attractive for dynamic situations where it is costly or difficult (if not impossible) to gather a satisfactory set of input–output examples.

There are two approaches to the study of reinforcement learning,[1] summarized as follows:

1. *Classical approach,* in which learning takes place through a process of punishment and reward with the goal of achieving a *highly skilled behavior.*
2. *Modern approach,* which builds on a mathematical technique known as dynamic programming to decide on a course of action by considering possible future stages without actually experiencing them; the emphasis here is on *planning.*

Our discussion focuses on modern reinforcement learning.

Dynamic programming[2] is a technique that deals with situations where decisions are made in stages, with the outcome of each decision being predictable to some extent before the next decision is made. A key aspect of such situations is that decisions cannot be made in isolation. Rather, the desire for a low cost at the present must be balanced against the undesirability of high costs in the future. This is a *credit assignment problem* because credit or blame must be assigned to each one of a set of interacting decisions. For optimal planning it is necessary to have an efficient tradeoff between immediate and future costs. Such a tradeoff is indeed captured by the formalism of dynamic programming. In particular, dynamic programming addresses the question: How can a system learn to improve long-term performance when this may require sacrificing short-term performance?

Following Bertsekas and Tsitsiklis (1996), we refer to the modern approach to reinforcement learning as *neurodynamic programming*. We do so primarily for two reasons:

- The theoretical foundation is provided by dynamic programming.
- The learning capability is provided by neural networks.

A succinct definition of neurodynamic programming is (Bertsekas and Tsitsiklis, 1996):

> Neurodynamic programming enables a system to learn how to make good decisions by observing its own behavior, and to improve its actions by using a built-in mechanism through reinforcement.

The observation of behavior is attained through the use of Monte Carlo simulation in an off-line manner. The improvement of actions through reinforcement is attained through the use of an iterative optimization scheme.

Organization of the Chapter

Dynamic programming has two main features: an underlying discrete-time dynamic system, and a cost function that is additive over time. These two features are discussed in Section 12.2. This is followed by a formulation of Bellman's optimality equation in Section 12.3, which plays an important role in dynamic programming. In Sections 12.4 and 12.5 we discuss two different methods for computing an optimal policy for dynamic programming, namely policy iteration and value iteration.

In Section 12.6 we present an overview of the issues involved in neurodynamic programming. This overview leads to the discussion of approximate policy iteration and Q-learning, which lend themselves to the use of neural networks for function approximation. These two algorithms are discussed in Sections 12.7 and 12.8, respectively. Section 12.9 presents a computer experiment on the use of Q-learning.

The chapter concludes with some final remarks in Section 12.10.

12.2 MARKOVIAN DECISION PROCESS

Consider a *learning system* or *agent* that interacts with its environment in the manner illustrated in Fig. 12.1. The system operates in accordance with a *finite, discrete-time Markovian decision process* that is characterized as follows:

- The environment evolves probabilistically occupying a finite set of discrete states. Note, however, that the state does *not* contain past statistics, even though these statistics could be useful to the learning system.
- For each environmental state there is a finite set of possible actions that may be taken by the learning system.

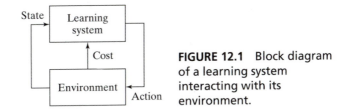

FIGURE 12.1 Block diagram of a learning system interacting with its environment.

- Every time the learning system takes an action, a certain cost is incurred.
- States are observed, actions are taken, and costs are incurred at discrete times.

In the context of our present discussion, the *state* of the environment is defined as *a summary of the entire past experience of the learning system gained from its interaction with the environment, such that the information necessary for the learning system to predict future behavior of the environment is contained in that summary.* The random variable denoting the state at time step n is X_n, and the actual state at time step n is denoted by $\mathbf{x}(n)$. The finite set of states is denoted by $\mathscr{X}$. A surprising aspect of dynamic programming is that its applicability depends very little on the nature of the state. We may therefore proceed without any assumption on the structure of the state space.

For state i, for example, the available set of *actions* (i.e., inputs applied to the environment by the learning system) is denoted by $\mathscr{A}_i = \{a_{ik}\}$, where the second subscript k in action a_{ik} taken by the learning system merely indicates the availability of more than one possible action when the environment is in state i. The transition of the environment from the state i to the new state j, for example, due to action a_{ik} is probabilistic in nature. Most importantly, however, *the transition probability from state i to state j depends entirely on the current state i and the corresponding action a_{ik}.* This is the *Markov property,* which is discussed in Chapter 11. This property is crucial because it means that the current state of the environment provides the necessary information for the learning system to decide what action to take.

The random variable denoting the action taken by the learning system at time step n is denoted by A_n. Let $p_{ij}(a)$ denote the transition probability from state i to state j due to action taken at time step n, where $A_n = a$. By virtue of the Markov property we have

$$p_{ij}(a) = P(X_{n+1} = j \mid X_n = i, A_n = a) \qquad (12.1)$$

The transition probability $p_{ij}(a)$ satisfies two conditions that are imposed on it by probability theory:

1. $p_{ij}(a) \geq 0$ for all i and j (12.2)

2. $\sum_j p_{ij}(a) = 1$ for all i (12.3)

For a given number of states and given transition probabilities the sequence of environmental states resulting from the actions taken by the learning system over time forms a *Markov chain.* Marlov chains are discussed in Chapter 11.

At each transition from one state to another, a *cost* is incurred by the learning system. Thus, at the nth transition from state i to state j under action a_{ik}, the learning system incurs a cost denoted by $\gamma^n g(i, a_{ik}, j)$, where $g(., ., .)$ is a prescribed function, and γ is a scalar with $0 \leq \gamma < 1$ called the *discount factor*. By adjusting γ we are able to control the extent to which the learning system is concerned with long-term versus short-term consequences of its own actions. In the limit, when $\gamma = 0$ the system is myopic in the sense that it is only concerned with immediate consequences of its actions. In what follows, we will ignore this limiting value, that is, confine the discussion to $0 < \gamma < 1$. As γ approaches 1, future costs become more important in determining optimal actions.

Our interest is in the formulation of a *policy*, defined as a *mapping of states into actions*. In other words, a policy is a rule used by the learning system to decide what to do, given knowledge of the current state of the environment. The policy is denoted by

$$\pi = \{\mu_0, \mu_1, \mu_2, ...\} \tag{12.4}$$

where μ_n is a function that maps the state $X_n = i$ into an action $A_n = a$ at time step $n = 0, 1, 2, ...$ This mapping is such that

$$\mu_n(i) \in \mathcal{A}_i \qquad \text{for all states } i \in \mathcal{X}$$

where $\mathcal{A}_i$ denotes the set of all possible actions taken by the learning system in state i. Such policies are said to be *admissible*.

A policy can be nonstationary or stationary. A *nonstationary* policy is time varying, as indicated in Eq. (12.4). When, however, the policy is independent of time, that is,

$$\pi = \{\mu, \mu, \mu, ...\}$$

the policy is said to be *stationary*. In other words, a stationary policy specifies exactly the same action each time a particular state is visited. For a stationary policy, the underlying Markov chain may be stationary or nonstationary; it is possible to use a stationary policy on a nonstationary Markov chain, but this is *not* a wise thing to do. If a stationary policy μ is employed, then the sequence of states $\{X_n, n = 0, 1, 2, ...\}$ forms a Markov chain with transition probabilities $p_{ij}(\mu(i))$, where $\mu(i)$ signifies an action. It is for this reason that the process is referred to as a *Markov decision process*.

Basic Problem

A dynamic programming problem can be of a finite-horizon or infinite-horizon kind. In a *finite-horizon* problem the cost accumulates over a finite number of stages. In an *infinite-horizon problem* the cost accumulates over an infinite number of stages. Infinite-horizon problems provide a reasonable approximation to problems involving a finite but very large number of stages. They are also of particular interest because discounting ensures that the costs for all states are finite for any policy.

The total expected cost in an infinite-horizon problem, starting from an initial state $X_0 = i$ and using a policy $\pi = \{\mu_n\}$, is defined by

$$J^\pi(i) = E\left[\sum_{n=0}^{\infty} \gamma^n g(X_n, \mu_n(X_n), X_{n+1}) \,|\, X_0 = i \right] \tag{12.5}$$

where the expected value is taken with respect to the Markov chain $\{X_1, X_2, ...\}$. The function $J^\pi(i)$ is called the *cost-to-go function* for policy π starting from state i. Its *optimal* value, denoted by $J^*(i)$, is defined by

$$J^*(i) = \min_\pi J^\pi(i) \tag{12.6}$$

When the policy π is stationary, that is, $\pi = \{\mu, \mu, ...\}$, we use the notation $J^\mu(i)$ in place of $J^\pi(i)$, and say that μ is optimal if

$$J^\mu(i) = J^*(i) \qquad \text{for all initial states } i \tag{12.7}$$

We may now sum up the basic problem in dynamic programming as follows:

> *Given a stationary Markovian decision process describing the interaction between a learning system and its environment, find a stationary policy $\pi = \{\mu, \mu, \mu...\}$ that minimizes the cost-to-go function $J^\mu(i)$ for all initial states i.*

Note that during learning the behavior of a learning system may change with time. However, the optimal policy that the learning system seeks will be stationary (Watkins, 1989).

12.3 BELLMAN'S OPTIMALITY CRITERION

The dynamic programming technique rests on a very simple idea known as the *principle of optimality* due to Bellman (1957). Simply stated this principle says (Bellman and Dreyfus, 1962):

> An optimal policy has the property that whatever the initial state and initial decision are, the remaining decisions must constitute an optimal policy with regard to the state resulting from the first decision.

As used here, a "decision" is a choice of control at a particular time, and a "policy" is the entire control sequence or control function.

To formulate the principle of optimality in mathematical terms, consider a finite-horizon problem for which the cost-to-go function is defined by

$$J_0(X_0) = E\left[g_K(X_K) + \sum_{n=0}^{K-1} g_n(X_n, \mu_n(X_n), X_{n+1})\right] \tag{12.8}$$

where K is the *horizon* (i.e., number of stages) and $g_K(X_K)$ is the *terminal cost*. Given X_0, the expectation in Eq. (12.8) is with respect to the remaining states $X_1, ..., X_{K-1}$. With this terminology we may now formally state the principle of optimality as (Bertsekas, 1995b):

> Let $\pi^* = \{\mu_0^*, \mu_1^*, ..., \mu_{K-1}^*\}$ be an optimal policy for the basic finite-horizon problem. Assume that when using the optimal policy π^* a given state X_n occurs with positive probability. Consider the subproblem where the environment is in state X_n at time n, and suppose we wish to minimize the corresponding cost-to-go function

$$J_n(X_n) = E\left[g_K(X_K) + \sum_{k=n}^{K-1} g_k(X_k, \mu_k(X_k), X_{k+1})\right] \tag{12.9}$$

for $n = 0, 1, ..., K-1$. Then the truncated policy $\{\mu_n^*, \mu_{n+1}^*, ..., \mu_{K-1}^*\}$ *is optimal for the sub-problem.*

We may intuitively justify the principle of optimality by the following argument: If the truncated policy $\{\mu_n^*, \mu_{n+1}^*, ..., \mu_{K-1}^*\}$ was not optimal as stated, then once the state X_n is reached at time n we could reduce the cost-to-go function $J_n(X_n)$ simply by switching to a policy that is optimal for the subproblem.

The principle of optimality builds on the engineering notion of "divide and conquer." Basically, an optimal policy for a complex multistage planning or control problem can be constructed by proceeding as follows:

- Construct an optimal policy for the "tail subproblem" involving only the last stage of the system.
- Extend the optimal policy to the "tail subproblem" involving the last two stages of the system.
- Continue the procedure in this fashion until the entire problem has been dealt with.

Dynamic Programming Algorithm

On the basis of the procedure just described, we may now formulate the dynamic programming algorithm, which proceeds backward in time from period $N - 1$ to period 0. Let $\pi = \{\mu_0, \mu_1, ..., \mu_{K-1}\}$ denote an admissible policy. For each $n = 0, 1, ..., K - 1$, let $\pi^n = \{\mu_n, \mu_{n+1}, ..., \mu_{K-1}\}$ and let $J_n^*(X_n)$ be the optimal cost for the $(K - n)$-stage problem that starts at state X_n and time n and ends at time K; that is,

$$J_n^*(X_n) = \min_{\pi^n} \; E_{(X_{n+1},...,X_{K-1})} \left[g_K(X_K) + \sum_{k=n}^{K-1} g_k(X_k, \mu_k(X_k), X_{k+1}) \right] \quad (12.10)$$

which represents the optimal form of Eq. (12.9). Recognizing that $\pi^n = (\mu_n, \pi^{n+1})$ and partially expanding the summation on the right-hand side of Eq. (12.10), we may write

$$J_n^*(X_n) = \min_{(\mu_n, \pi^{n+1})} \; E_{(X_{n+1},...,X_{K-1})} \left[g_n(X_n, \mu_n(X_n), X_{n+1}) \right.$$

$$\left. + g_K(X_K) + \sum_{k=n+1}^{K-1} g_k(X_k, \mu_k(X_k), X_{k+1}) \right]$$

$$= \min_{\mu_n} \; E_{X_{n+1}} \left\{ g_n(X_n), \mu_n(X_n), X_{n+1}) \right. \quad (12.11)$$

$$\left. + \min_{\pi^n} \; E_{(X_{n+2},...,X_{K-1})} \left[g_K(X_K) + \sum_{k=n+1}^{K-1} g_k(X_k, \mu_k(X_k), X_{k+1}) \right] \right\}$$

$$= \min_{\mu_n} \; E_{X_{n+1}} \left[g_n(X_n, \mu_n(X_n), X_{n+1}) + J_{n+1}^*(X_{n+1}) \right]$$

where in the last line we have made use of the defining equation (12.10) with $n + 1$ used in place of n. Now assume that for some n and all X_{n+1}, we have

$$J_{n+1}^*(X_{n+1}) = J_{n+1}(X_{n+1}) \quad (12.12)$$

Then we may rewrite Eq. (12.11) in the form

$$J_n^*(X_n) = \min_{\mu_n} \; \underset{X_{n+1}}{E} \; [g_n(X_n, \mu_n(X_n), X_{n+1}) + J_{n+1}(X_{n+1})] \qquad (12.13)$$

If Eq. (12.12) holds for all X_{n+1}, then clearly the equation

$$J_n^*(X_n) = J_n(X_n)$$

also holds for all X_n. Accordingly, we deduce from Eq. (12.13) that

$$J_n(X_n) = \min_{\mu_n} \; \underset{X_{n+1}}{E} \; [g_n(X_n, \mu_n(X_n), X_{n+1}) + J_{n+1}(X_{n+1})]$$

We may thus formally state the *dynamic programming algorithm* as follows (Bertsekas, 1995b):

For every initial state X_0, the optimal cost $J^*(X_0)$ of the basic finite-horizon problem is equal to $J_0(X_0)$, where the function J_0 is obtained from the last step of the following algorithm:

$$J_n(X_n) = \min_{\mu_n} \; \underset{X_{n+1}}{E} \; [g_n(X_n, \mu_n(X_n), X_{n+1}) + J_{n+1}(X_{n+1})] \qquad (12.14)$$

which runs backward in time, with

$$J_K(X_K) = g_K(X_K)$$

Furthermore, if μ_n^* minimizes the right-hand side of Eq. (12.14) for each X_n and n, then the policy $\pi^* = \{\mu_0^*, \mu_1^*, ..., \mu_{K-1}^*\}$ is optimal.

Bellman's Optimality Equation

In its basic form, the dynamic programming algorithm deals with a finite-horizon problem. We are interested in extending the use of this algorithm to deal with the infinite-horizon discounted problem described by the cost-to-go function of Eq. (12.5) under a stationary policy $\pi = \{\mu, \mu, \mu, ...\}$. With this objective in mind, we do two things:

- Reverse the time index of the algorithm so that it corresponds to the discounted problem.
- Define the cost $g_n(X_n, \mu(X_n), X_{n+1})$ as

$$g_n(X_n, \mu(X_n), X_{n+1}) = \gamma^n g(X_n, \mu(X_n), X_{n+1}) \qquad (12.15)$$

We may now reformulate the dynamic programming algorithm as follows (see Problem 12.4):

$$J_{n+1}(X_0) = \min_{\mu} \; \underset{X_1}{E} \; [g(X_0, \mu(X_0), X_1) + \gamma J_n(X_1)] \qquad (12.16)$$

which starts from the initial conditions

$$J_0(X) = 0 \qquad \text{for all } X$$

The state X_0 is the initial state, X_1 is the new state that results from the action of policy μ, and γ is the discount factor.

Let $J^*(i)$ denote the optimal infinite-horizon cost for the initial state $X_0 = i$. We may then view $J^*(i)$ as the limit of the corresponding K-stage optimal cost $J_K(i)$ as the horizon K approaches infinity; that is,

$$J^*(i) = \lim_{K \to \infty} J_K(i) \qquad \text{for all } i \qquad (12.17)$$

This relation is the connecting link between the finite-horizon and infinite-horizon discounted problems. Putting $n + 1 = K$ and $X_0 = i$ in Eq. (12.16), and then applying Eq. (12.17), we obtain

$$J^*(i) = \min_{\mu} \; \mathop{E}_{X_1} \left[g(i, \mu(i), X_1) + \gamma J^*(X_1) \right] \tag{12.18}$$

To evaluate the optimal infinite-horizon cost $J^*(i)$, we proceed in two stages:

1. We evaluate the expectation of the cost $g(i, \mu(i), X_1)$ with respect to X_1 by writing

$$E[g(i), \mu(i), X_1] = \sum_{j=1}^{N} p_{ij} g(i, \mu(i), j) \tag{12.19}$$

where N is the number of states of the environment, and p_{ij} is the transition probability from the initial state $X_0 = i$ to the new state $X_1 = j$. The quantity defined in Eq. (12.19) is the *immediate expected cost* incurred at state i by following the action recommended by the policy μ. Denoting this cost by $c(i, \mu(i))$, we may write

$$c(i, \mu(i)) = \sum_{j=1}^{N} p_{ij} g(i, \mu(i), j) \tag{12.20}$$

2. We evaluate the expectation of $J^*(X_1)$ with respect to X_1. Here we note that if we know the cost $J^*(X_1)$ for each state X_1 of a finite-state system, we may readily determine the expectation of $J^*(X_1)$ in terms of the transition probabilities of the underlying Markov chain by writing

$$E[J^*(X_1)] = \sum_{j=1}^{N} p_{ij} J^*(j) \tag{12.21}$$

Thus, using Eqs. (12.19) to (12.21) in Eq. (12.16), we obtain the desired result

$$J^*(i) = \min_{\mu} \left(c(i, \mu(i)) + \gamma \sum_{j=1}^{N} p_{ij}(\mu) J^*(j) \right) \qquad \text{for } i = 1, 2, ..., N \tag{12.22}$$

Equation (12.22) is called *Bellman's optimality equation.* It should *not* be viewed as an algorithm. Rather, it represents a system of N equations with one equation per state. The solution of this system of equations defines the optimal cost-to-go functions for the N states of the environment.

There are two basic methods for computing an optimal policy. They are called policy iteration and value iteration. These two methods are described in Sections 12.4 and 12.5, respectively.

12.4 POLICY ITERATION

To set the stage for a description of the policy iteration algorithm, we begin by introducing a concept called the Q-factor due to Watkins (1989). Consider an existing policy μ for which the cost-to-go function $J^\mu(i)$ is known for all states i. The *Q-factor* for each state $i \in \mathcal{X}$ and action $a \in \mathcal{A}_i$ is defined as *the immediate cost plus the sum of the discounted costs of all successor states following policy* μ, as shown by

$$Q^{\mu}(i, a) = c(i, a) + \gamma \sum_{j=1}^{n} p_{ij}(a)J^{\mu}(j) \tag{12.23}$$

where the action $a = \mu(i)$. Note that the Q-factors $Q^{\mu}(i, a)$ contain more information than the cost-to-go function $J^{\mu}(i)$. For example, actions may be ranked on the basis of Q-factors alone, whereas ranking on the basis of cost-to-go function also requires knowledge of the state-transition probabilities and costs.

We may develop insight into the meaning of the Q-factor by visualizing a new system whose states are made up of the original states 1, 2, ..., N and all the possible state-action pairs (i, a), as portrayed in Fig. 12.2. There are two distinct possibilities that can occur:

- The system is in state (i, a), in which case no action is taken. Transition is made automatically to state j, say, with probability $p_{ij}(a)$; and a cost $g(i, a, j)$ is incurred.
- The system is in state i, say, in which case action $a \in \mathcal{A}_i$ is taken. The next state is (i, a), deterministically.

The policy μ is said to be *greedy* with respect to the cost-to-go function $J^{\mu}(i)$ if, for all states, $\mu(i)$ is an action that satisfies the condition

$$Q^{\mu}(i, \mu(i)) = \min_{a \in \mathcal{A}_i} Q^{\mu}(i, a) \qquad \text{for all } i \tag{12.24}$$

The following two observations on Eq. (12.24) are noteworthy:

- It is possible for more than one action to minimize the set of Q-factors for some state, in which case there can be more than one greedy policy with respect to the pertinent cost-to-go function.
- A policy can be greedy with respect to many different cost-to-go functions.

Moreover, the following fact is basic to all dynamic programming methods:

$$Q^{\mu^*}(i, \mu^*(i)) = \min_{a \in \mathcal{A}_i} Q^{\mu^*}(i, a) \tag{12.25}$$

where μ^* is an optimal policy and J^* is the corresponding optimal cost-to-go function.

With the notions of Q-factor and greedy policy at our disposal, we are ready to describe the *policy iteration algorithm*. Specifically, the algorithm operates by alternating between two steps (Bertsekas, 1995b):

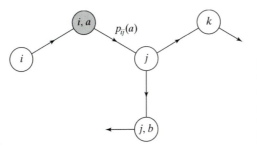

FIGURE 12.2   Illustration of two possible transitions: the transition from state (i, a) to state j is probabilistic, but the transition from state i to (i, a) is deterministic.

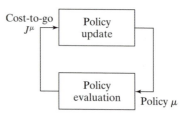

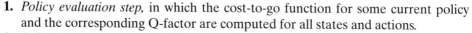

FIGURE 12.3   Block diagram
for policy iteration algorithm.

1. *Policy evaluation step,* in which the cost-to-go function for some current policy
   and the corresponding Q-factor are computed for all states and actions.
2. *Policy improvement step,* in which the current policy is updated in order to be
   greedy with respect to the cost-to-go function computed in step 1.

These two steps are illustrated in Fig. 12.3. To be specific, we start with some initial policy μ_0, and then generate a sequence of new policies $\mu_1, \mu_2, \ldots$. Given the current policy μ_n, we perform the policy evaluation step by computing the cost-to-go function $J^{\mu_n}(i)$ as the solution of the linear system of equations (see Eq. (12.22))

$$J^{\mu_n}(i) = c(i, \mu_n(i)) + \gamma \sum_{j=1}^{N} p_{ij}(\mu_n(i)) J^{\mu_n}(i), \qquad i = 1, 2, \ldots, N \qquad (12.26)$$

in the unknowns $J^{\mu_n}(1), J^{\mu_n}(2), \ldots, J^{\mu_n}(N)$. Using these results, we then compute the Q-factor for state-action pair (i, a) (see Eq. (12.23))

$$Q^{\mu_n}(i, a) = c(i, a) + \gamma \sum_{j=1}^{N} p_{ij}(a) J^{\mu_n}(i), \qquad a \in \mathcal{A}_i \text{ and } i = 1, 2, \ldots, N \quad (12.27)$$

Next we perform the policy improvement step by computing a new policy μ_{n+1} defined by (see Eq. (12.24))

$$\mu_{n+1}(i) = \arg \min_{a \in \mathcal{A}_i} Q^{\mu_n}(i, a), \qquad i = 1, 2, \ldots, N \qquad (12.28)$$

The two-step process just described is repeated with policy μ_{n+1} used in place of μ_n, unless we have

$$J^{\mu_{n+1}}(i) = J^{\mu_n}(i) \qquad \text{for all } i$$

in which case the algorithm is terminated with policy μ_n. With $J^{\mu_{n+1}} \leq J^{\mu_n}$ (see Problem 12.5) we may then say that the policy iteration algorithm will terminate after a finite number of iterations because the underlying Markovian decision process has a finite number of states. Table 12.1 presents a summary of the policy iteration algorithm based on Eqs. (12.26) to (12.28).

12.5   VALUE ITERATION

In the policy iteration algorithm, the cost-to-go function has to be recomputed entirely at each iteration of the algorithm, which is expensive. Even though the cost-to-go function for the new policy may be similar to that for the old policy, unfortunately there is

---

TABLE 12.1   Summary of the Policy Iteration Algorithm

---

1. Start with an arbitrary initial policy μ_0.
2. For $n = 0, 1, 2, ...,$ compute $J^{\mu_n}(i)$ and $Q^{\mu_n}(i, a)$ for all states $i \in \mathcal{X}$ and actions $a \in \mathcal{A}_i$.
3. For each state i, compute

$$\mu_{n+1}(i) = \arg \min_{a \in \mathcal{A}_i} Q^{\mu_n}(i, a)$$

4. Repeat steps 2 and 3 until μ_{n+1} is not an improvement on μ_n, at which point the algorithm terminates with μ_n as the desired policy.

---

no dramatic shortcut for this computation. There is, however, another method for finding the optimal policy that avoids the burdensome task of repeatedly computing the cost-to-go function. This alternative method, based on successive approximations, is known as the value iteration algorithm.

The *value iteration algorithm* involves solving Bellman's optimality equation (12.22) for each of a sequence of finite-horizon problems. In the limit, the cost-to-go function of the finite-horizon problem converges uniformly over all states to the corresponding cost-to-go function of the infinite-horizon problem as the number of iterations of the algorithm approaches infinity (Ross, 1983; Bertsekas, 1995b).

Let $J_n(i)$ denote the cost-to-go function for state i at iteration n of the value iteration algorithm. The algorithm begins with an arbitrary guess $J_0(i)$ for $i = 1, 2, ..., N$. The only restriction on $J_0(i)$ is that it should be bounded; this is automatically true for finite-state problems. If some estimate of the optimal cost-to-go function $J^*(i)$ is available, it should be used as the initial value $J_0(i)$. Once $J_0(i)$ has been chosen, we may compute the sequence of cost-to-go functions $J_1(i), J_2(i), ...,$ using the value iteration algorithm

$$J_{n+1}(i) = \min_{a \in \mathcal{A}_i} \left\{ c(i, a) + \gamma \sum_{j=1}^{N} p_{ij}(a) J_n(j) \right\}, \qquad i = 1, 2, ..., N \qquad (12.29)$$

Application of the update to the cost-to-go function, described in Eq. (12.29) for state i, is referred to as *backing up of i's cost*. This backup is a direct implementation of Bellman's optimality equation (12.22). Note that the values of the cost-to-go functions in Eq. (12.29) for states $i = 1, 2, ..., N$ are backed up simultaneously on each iteration of the algorithm. This method of implementation represents the traditional *synchronous* form of the value iteration algorithm.[3] Thus, starting from arbitrary initial values $J_0(1), J_0(2), ..., J_0(N)$, the algorithm described by Eq. (12.29) converges to the corresponding optimal values $J^*(1), J^*(2), ..., J^*(N)$ as the number of iterations n approaches infinity (Ross, 1983; Bertsekas, 1995b).

Unlike the policy iteration algorithm, an optimal policy is not computed directly in the value iteration algorithm. Rather, the optimal values $J^*(1), J^*(2), ..., J^*(N)$ are first computed using Eq. (12.29). Then a greedy policy with respect to that optimal set is obtained as an optimal policy. That is,

$$\mu^*(i) = \arg \min_{a \in \mathcal{A}_i} Q^*(i, a), \qquad i = 1, 2, ..., N \qquad (12.30)$$

TABLE 12.2 Summary of the Value Iteration Algorithm

---

1. Start with arbitrary initial value $J_0(i)$ for state $i = 1, 2, ..., N$.
2. For $n = 0, 1, 2, ...,$ compute

$$J_{n+1}(i) = \min_{a \in \mathcal{A}_i} \left\{ c(i, a) + \gamma \sum_{j=1}^{N} p_{ij}(a) J_n(j), \right\}, \qquad \begin{array}{c} a \in \mathcal{A}_i \\ i = 1, 2, ..., N \end{array}$$

Continue this computation until

$$|J_{n+1}(i) - J_n(i)| < \epsilon \qquad \text{for each state } i$$

where ϵ is a prescribed tolerance parameter. It is presumed that ϵ is sufficiently small for $J_n(i)$ to be close enough to the optimal cost-to-go function $J^*(i)$. We may thus set

$$J_n(i) = J^*(i) \qquad \text{for all states } i$$

3. Compute the Q-factor

$$Q^*(i, a) = c(i, a) + \gamma \sum_{j=1}^{N} p_{ij}(a) J^*(j) \qquad \begin{array}{c} \text{for } a \in \mathcal{A}_i \text{ and} \\ i = 1, 2, ..., N \end{array}$$

Hence, determine the optimal policy as a greedy policy for $J^*(i)$:

$$\mu^*(i) = \arg \min_{a \in \mathcal{A}_i} Q^*(i, a)$$

---

where

$$Q^*(i, a) = c(i, a) + \gamma \sum_{j=1}^{N} p_{ij}(a) J^*(j), \qquad i = 1, 2, ..., N \qquad (12.31)$$

A summary of the value iteration algorithm, based on Eqs. (12.29) to (12.31), is presented in Table 12.2. This summary includes a stopping criterion for Eq. (12.29).

Example 12.1 Stagecoach Problem

To illustrate the usefulness of the Q-factor in dynamic programming, we consider the *stagecoach problem*. A fortune seeker in Missouri decided to go west to join the gold rush in California in the mid-nineteenth century (Hiller and Lieberman, 1995). The journey required traveling by stagecoach through unsettled country, which posed a serious danger of attack by marauders along the way. The starting point of the journey (Missouri) and the destination (California) were fixed, but there was considerable choice as to which other eight states to travel through en route, as shown in Fig. 12.4. In this figure we have the following:

- A total of 10 states, with each state represented by a letter.
- The direction of travel is from left to right.
- There are four stages (i.e., stagecoach runs) from the point of embarkation in state A (Missouri) to the destination in state J (California).
- In moving from one state to the next, the action taken by the fortune seeker is to move up, straight, or down.
- There are a total of 18 possible routes from state A to state J.

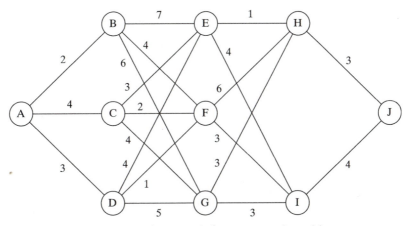

FIGURE 12.4 Flow-graph for stagecoach problem.

Figure 12.4 also includes the cost of a life insurance policy for taking any stagecoach run based on a careful evaluation of the safety of that run. The problem is to find the route from state A to state J with the cheapest insurance policy.

To find the optimum route, we consider a sequence of finite horizon problems, starting from the destination in state J and working backward. This is in accordance with Bellman's principle of optimality described in Section 12.3.

Calculating the Q-factors for the last stage before the destination, we readily find in Fig. 12.5a that the terminal Q-values are as follows:

$$Q(H, \text{down}) \quad = \quad 3$$
$$Q(I, \text{up}) \quad = \quad 4$$

These numbers are indicated on states H and I, respectively, in Fig. 12.5a.

Next, moving back by one more stage and using the Q-values in Fig. 12.5a, we have the following Q values:

$$Q(E, \text{straight}) \quad = \quad 1 + 3 = 4$$
$$Q(E, \text{down}) \quad = \quad 4 + 4 = 8$$
$$Q(F, \text{up}) \quad = \quad 6 + 3 = 9$$
$$Q(F, \text{down}) \quad = \quad 3 + 4 = 7$$
$$Q(G, \text{up}) \quad = \quad 3 + 3 = 6$$
$$Q(G, \text{straight}) \quad = \quad 3 + 4 = 7$$

Since the requirement is to find the route with the smallest insurance policy, the Q-values indicate that only the stage runs $E \rightarrow H$, $F \rightarrow I$, and $G \rightarrow H$ should be retained and the remaining ones pruned, as indicated in Fig. 12.5b.

Moving back one further stage, repeating the calculations of the Q-factors for states B, C, and D in the manner described and retaining only those stage runs from the states B, C, and D that are covered by the lowest insurance costs, we obtain the picture depicted in Fig. 12.5c.

Finally, moving back one last stage and proceeding in the same way as before, we obtain the picture depicted in Fig. 12.5d. From this figure we see that there are indeed three optimal routes, as described here:

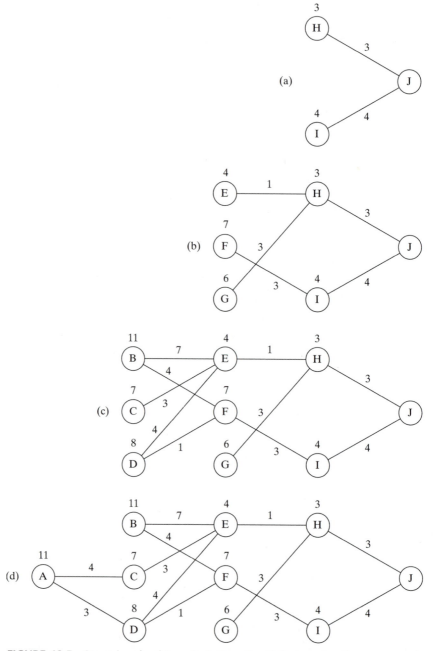

FIGURE 12.5   Steps involved in calculating the Q-factors for the stagecoach problem.

$$A \rightarrow C \rightarrow E \rightarrow H \rightarrow J$$
$$A \rightarrow D \rightarrow E \rightarrow H \rightarrow J$$
$$A \rightarrow D \rightarrow F \rightarrow I \rightarrow J$$

They all yield a total cost of 11.

■

12.6 NEURODYNAMIC PROGRAMMING

The primary objective of dynamic programming is to find an optimal policy, that is, an optimal choice of the action that should be taken by the learning system for each possible state of the environment. In this context there are two practical issues that must be remembered when considering use of the policy iteration or value iteration algorithm to solve a dynamic programming problem:

- *Curse of dimensionality.* For many difficult real-world problems the numbers of possible states and admissible actions are so large that the computational requirements of dynamic programming are overwhelming. For a dynamic programming problem involving a total of N possible states and M admissible actions for each state, each iteration of the value iteration algorithm, for example, requires about $N^2 M$ operations for a stationary policy. This frequently makes it impossible to complete even one iteration of the algorithm when N is very large. For example, backgammon has 10^{20} states, which means that a single iteration of the algorithm would take more than 1000 years using a 1000 MIPS processor (Barto, et al., 1995).

- *Incomplete information.* The policy iteration or value iteration algorithm requires prior knowledge of the underlying Markov decision process. That is, for the computation of an optimal policy to be feasible we require that the state-transition probabilities p_{ij} and the observed costs $g(i, a, j)$ be known. Unfortunately, this prior knowledge is not always available.

In light of one or the other or both of these difficulties, we often have to abandon the quest for an optimal policy and settle for a *suboptimal policy.*

Our interest here is in suboptimal procedures that involve the use of neural networks and/or simulation for the purpose of approximating the optimal cost-to-go function $J^*(i)$ for all $i \in \mathcal{X}$. Specifically, for a specified state i, $J^*(i)$ is replaced with a suitable approximation $\hat{J}(i, \mathbf{w})$, where $\mathbf{w}$ is a parameter vector. The function $\hat{J}(\cdot, \mathbf{w})$ is called the *scoring function* or *approximate cost-to-go function,* and the value $\hat{J}(i, \mathbf{w})$ is called the *score* or *approximate-to-go cost* for state i. Thus, as illustrated in Fig. 12.6, the score $\hat{J}(i, \mathbf{w})$ is the output of the neural network in response to the state i as input. The property that is exploited here is that of *universal approximation,* which, as discussed in previous chapters, is an inherent characteristic of multilayer perceptrons and radial-basis function networks.

Dynamic programming problems of particular interest are those with a large number of states, where the requirement is to find a scoring function $\hat{J}(\cdot, \mathbf{w})$ for which the parameter vector $\mathbf{w}$ has a small dimension. In this form of approximation, referred

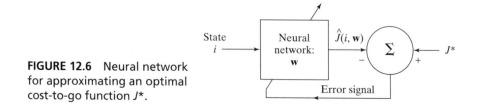

FIGURE 12.6 Neural network for approximating an optimal cost-to-go function J*.

to as a *compact representation,* only the parameter vector **w** and general structure of the scoring function $\hat{J}(\cdot, \mathbf{w})$ are stored. The scores $\hat{J}(i, \mathbf{w})$ for all states $i \in \mathscr{X}$ are generated only when they are needed. The challenge is to find the parameter vector **w** algorithmically, such that for a given neural network structure (e.g., multilayer perceptron), the score $\hat{J}(i, \mathbf{w})$ provides a satisfactory approximation to the optimal value $J^*(i)$ for all $i \in \mathscr{X}$.

From the material presented in Chapters 4 through 7 on learning with a teacher, we know that a neural network, irrespective of its type, requires a set of labeled data that are representative of this task. However, in the context of dynamic programming problems, no such training data (i.e., input–output examples $\{(i, J^*(i))\}$) are available to train the neural network in Fig. 12.6, so as to optimize its design in some statistical sense. The only possibility is to use *Monte Carlo simulation,* where a surrogate model is used for the actual system underlying the Markovian decision process. The result is a novel off-line dynamic programming mode of operation that offers the following potential benefits (Bertsekas and Tsitsiklis, 1996):

1. The use of simulation to evaluate approximately the optimal cost-to-go function is the key idea that distinguishes the methodology of neurodynamic programming from traditional approximation methods in dynamic programming.
2. Simulation permits the use of neurodynamic programming methods to design systems for which no explicit models are available. For such systems, the traditional dynamic programming techniques are inapplicable, as it is cumbersome if not impossible to provide estimates of the state-transition probabilities.
3. Through simulation, it is possible to implicitly identify the most important or most representative states of the system as those states that are most often visited during the simulation. Consequently, the scoring function discovered by the neural network may provide a good approximation to the optimal cost-to-go function for those particular states. The end result may be a good suboptimal policy for a difficult dynamic programming problem.

However, it is important to recognize that once approximations are introduced, convergence of the scoring function $\hat{J}(\cdot, \mathbf{w})$ to the optimal cost-to-go function $J^*(\cdot)$ cannot be expected. That is for the simple reason that $J^*(\cdot)$ may not be within the set of functions represented exactly by the chosen neural network structure.

In the next two sections we discuss two approximate dynamic programming procedures with cost-to-go function approximations. The first procedure, described in Section 12.7, deals with approximate policy iteration, assuming that a Markovian model of the system is available. The second procedure, described in Section 12.8, deals with a procedure called Q-learning, which makes no such assumption.

12.7 APPROXIMATE POLICY ITERATION

Suppose we have a dynamic programming problem for which the numbers of possible states and admissible actions are too large, making the use of a traditional approach impractical. It is assumed that we do have a model of the system; that is, the transition probabilities $p_{ij}(a)$ and the observed costs $g(i, a, j)$ are all known. To deal with this situation, we propose to use an approximation to policy iteration, based on Monte Carlo simulation and the method of least-squares, as described next (Bertsekas and Tsitsiklis, 1996).

Figure 12.7 shows a simplified block diagram of the *approximate policy iteration algorithm*. It is similar to the block diagram of Fig. 12.3 for the traditional policy iteration algorithm, but with an important difference: The policy evaluation step in Fig. 12.3 has been replaced with an *approximate* one. Thus, the approximate policy iteration algorithm proceeds by alternating between an approximate policy evaluation step and a policy improvement step as follows:

1. *Approximate policy evaluation step.* Given the current policy μ, a cost-to-go function $\hat{J}^{\mu}(i, \mathbf{w})$ approximating the actual cost-to-go function $J^{\mu}(i)$ is computed for all states i. The vector $\mathbf{w}$ is the parameter vector of the neural network used to perform the approximation.
2. *Policy improvement step.* Using the approximate cost-to-go function $\hat{J}^{\mu}(i, \mathbf{w})$, an improved policy μ is generated. This new policy is designed to be greedy with respect to $\hat{J}^{\mu}(i, \mathbf{w})$ for all i.

For the approximate policy iteration algorithm to yield satisfactory results, it is important to carefully choose the policy used to initialize the algorithm. This could be done through the use of heuristics. Alternatively, we may start with some weight vector $\mathbf{w}$, and use it to derive a greedy policy, which is in turn used as the initial policy.

Suppose then, in addition to the known transition probabilities and observed costs, we have the following items:

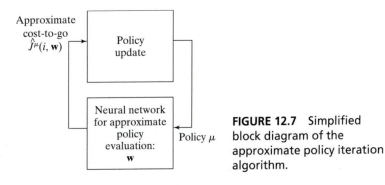

FIGURE 12.7 Simplified block diagram of the approximate policy iteration algorithm.

- A stationary policy μ as the initial policy
- A set of states $\mathcal{X}$ representative of the operating environment
- A set of $M(i)$ samples of the cost-to-go function $J^\mu(i)$ for each state $i \in \mathcal{X}$; one such sample is denoted by $k(i, m)$, where $m = 1, 2, ..., M(i)$

Let $\hat{J}^\mu(i, \mathbf{w})$ denote an approximate representation of the cost-to-go function $J^\mu(i)$. The approximation is performed by a neural network (e.g., multilayer perceptron trained with the back-propagation algorithm). The parameter vector $\mathbf{w}$ of the neural network is determined by using the method of least-squares, that is, minimizing the cost function:

$$\mathcal{E}(\mathbf{w}) = \sum_{i \in \mathcal{X}} \sum_{m=1}^{M(i)} (k(i, m) - \hat{J}^\mu(i, \mathbf{w}))^2 \tag{12.32}$$

Having determined the optimum weight vector $\mathbf{w}$ and therefore the approximate cost-to-go function $\hat{J}^\mu(i, \mathbf{w})$, we next determine the approximate Q-factors using the formula (see Eqs. (12.20) and (12.23))

$$Q(i, a, \mathbf{w}) = \sum_{j \in \mathcal{X}} p_{ij}(a)(g(i, a, j) + \gamma \hat{J}^\mu(j, \mathbf{w})) \tag{12.33}$$

where the $p_{ij}(a)$ is the transition probability from state i to state j under action a (known), $g(i, a, j)$ is the observed cost (also known), and γ is a prescribed discount factor. The iteration is completed by using these approximate Q-factors to determine an improved policy based on the formula (see Eq. (12.28))

$$\mu(i) = \arg \min_{a \in \mathcal{A}_i} Q(i, a, \mathbf{w}) \tag{12.34}$$

It is important to note that Eqs. (12.33) and (12.34) are only used by the simulator to generate actions at the states that are *actually visited* by the simulation, rather than for all states. As such, these two equations need not suffer from the curse of dimensionality.

The block diagram of Fig. 12.8 presents a more detailed picture of the approximate policy iteration algorithm. This diagram consists of four interconnected modules (Bertsekas and Tsitsiklis, 1996):

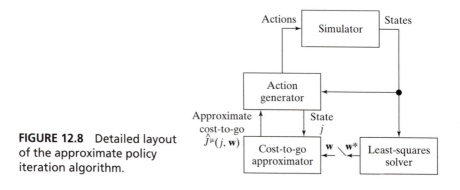

FIGURE 12.8 Detailed layout of the approximate policy iteration algorithm.

1. *Simulator,* which uses the given state-transition probabilities and observed one-step costs to construct a surrogate model of the environment. The simulator generates two things: (a) states in response to actions to mimic the environment, and (b) samples of the cost-to-go function for a given policy.
2. *Action generator,* which generates an improved policy (i.e., sequence of actions) in accordance with Eq. (12.34).
3. *Cost-to-go approximator,* which generates the approximate cost-to-go function $\hat{J}^{\mu}(i, \mathbf{w})$ for state i and parameter vector $\mathbf{w}$, for use in Eqs. (12.33) and (12.34).
4. *Least-squares solver,* which takes samples of the cost-to-go function $J^{\mu}(i)$ supplied by the simulator for policy μ and state i, and computes the optimum parameter vector $\mathbf{w}$ that minimizes the cost function of Eq. (12.32). The link from the least-squares solver to the cost-to-go approximator is switched on only after a policy has been fully evaluated and an optimal parameter vector $\mathbf{w}^*$ has been determined. At that point, the cost-to-go approximation $\hat{J}^{\mu}(i, \mathbf{w})$ is replaced by $\hat{J}^{\mu}(i, \mathbf{w}^*)$.

Table 12.3 presents a summary of the approximate policy iteration algorithm.

Naturally, the operation of this algorithm is subject to errors due to unavoidable imperfections in the design of the simulator and least-squares solver. The neural network used to perform the least-squares approximation of the desired cost-to-go function may lack adequate computing power, hence the first source of error. Optimization

---

TABLE 12.3   Summary of the Approximate Policy Iteration Algorithm

---

Known parameters: transition probabilities $p_{ij}(a)$ and costs $g(i, a, j)$.

Computation:
1. Choose a stationary policy μ as the initial policy.
2. Using a set of samples $\{k(i, m)\}_{m=1}^{M(i)}$ of the cost-to-go function $J^{\mu}(i)$ generated by the simulator, determine the parameter vector $\mathbf{w}$ of the neural network employed as the least-squares solver:

$$\mathbf{w}^* = \min_{\mathbf{w}} \mathscr{E}(\mathbf{w})$$

$$= \min_{\mathbf{w}} \sum_{i \in \mathscr{X}} \sum_{m=1}^{M(i)} (k(i, m) - \hat{J}^{\mu}(i, \mathbf{w}))^2$$

3. For the parameter vector $\mathbf{w}$ determined in step 2, compute the approximate cost-to-go function $\hat{J}^{\mu}(i, \mathbf{w})$ for the states visited. Determine the approximate Q-factors:

$$Q(i, a, \mathbf{w}) = \sum_{j \in \mathscr{X}} p_{ij}(a)(g(i, a, j) + \gamma \hat{J}^{\mu}(j, \mathbf{w}))$$

4. Determine the improved policy

$$\mu(i) = \arg \min_{a \in \mathscr{A}_i} Q(i, a, \mathbf{w})$$

5. Repeat steps 2 through 4.

Note: Steps 3 and 4 apply only to actions at the states that are actually visited, rather than all states.

of the neural network approximator and therefore tuning of the parameter vector $\mathbf{w}$ is based on a desired response provided by the simulator, hence the second source of error. Assuming that all policy evaluations and all policy improvements are performed within certain error tolerances of ϵ and δ, respectively, it is shown in Bertsekas and Tsitsiklis (1996) that the approximate policy iteration algorithm will produce policies whose performances differ from the optimal policies by a factor that decreases to zero as ϵ and δ are reduced. In other words, the approximate policy iteration algorithm is sound with minimal performance guarantees. According to Bertsekas and Tsitsiklis (1996), the approximate policy iteration algorithm tends to initially make rapid and fairly monotonic progress, but a sustained policy oscillation of a random nature may result as a limiting condition. This oscillatory behavior occurs after the approximating cost-to-go function $\hat{J}$ gets within a zone of $O(\delta + 2\gamma\epsilon)/(1 - \gamma)^2)$ of the optimal value J^*, where γ is the discount parameter. Apparently there is a fundamental structure that is common to all variants of approximate policy iteration, which causes an oscillatory behavior.

12.8 Q-LEARNING

The behavioral task of the reinforcement learning system in Fig. 12.1 is how to find an optimal (i.e., minimal cost) policy after trying out various possible sequences of actions, and observing the costs incurred and the state transitions that occur. In this context we may raise the following question: Is there an on-line procedure for learning the optimal policy through experience gained solely on the basis of samples of the form

$$s_n = (i_n, a_n, j_n, g_n) \tag{12.35}$$

where n denotes discrete time, and each sample s_n consists of a four-tuple, described by a trial action a_n on state i_n that results in a transition to state $j_n = i_{n+1}$ at a cost $g_n = g(i_n, a_n, j_n)$? The answer to this fundamental question is an emphatic yes, and it is to be found in a stochastic method called *Q-learning*[4] due to Watkins (1989). Q-learning is an incremental dynamic programming procedure that determines the optimal policy in a step-by-step manner. It is highly suited for solving Markovian decision problems without explicit knowledge of the transition probabilities. However, successful use of Q-learning hinges on the assumption that the state of the environment is *fully observable,* which in turn means that the environment is a fully observable Markov chain.

We recall from Section 12.4 that the Q-factor $Q(i, a)$ for state-action pair (i, a) is defined by Eq. (12.23), and Bellman's optimality equation is defined by Eq. (12.22). By combining these two equations and using the definition of the immediate expected cost $c(i, a)$ given in Eq. (12.20), we obtain

$$Q^*(i, a) = \sum_{j=1}^{N} p_{ij}(a) \left(g(i, a, j) + \gamma \min_{b \in \mathcal{A}_j} Q^*(j, b) \right) \qquad \text{for all } (i, a) \tag{12.36}$$

which can be viewed as a two-step version of Bellman's optimality equation. The solutions to the linear system of equations (12.36) define the optimal Q-factors $Q^*(i, a)$ uniquely for all state-action pairs (i, a).

We may use the value iteration algorithm formulated in terms of the Q-factors to solve this linear system of equations. Thus, for one iteration of the algorithm we have

$$Q(i, a): = \sum_{j=1}^{N} p_{ij}(a) \left(g(i, a, j) + \gamma \min_{b \in \mathcal{A}_j} Q(j, b) \right) \quad \text{for all } (i, a)$$

The small step-size version of this iteration is described by

$$Q(i, a): = (1 - \eta)Q(i, a) + \eta \sum_{j=1}^{N} p_{ij}(a) \left(g(i, a, j) + \gamma \min_{b \in \mathcal{A}_j} Q(j, b) \right) \quad \text{for all } (i, a)$$

$$(12.37)$$

where η is a small *learning-rate parameter* that lies in the range $0 < \eta < 1$.

As it stands, an iteration of the value iteration algorithm described in Eq. (12.37) requires knowledge of the transition probabilities. We may eliminate the need for this prior knowledge by formulating a *stochastic* version of Eq. (12.37). Specifically, the averaging performed in an iteration of Eq. (12.37) over all possible states is replaced by a single sample, thereby resulting in the following update for the Q-factor:

$$Q_{n+1}(i,a) = (1 - \eta_n(i,a))Q_n(i,a) + \eta_n(i,a)[g(i,a, j) + \gamma J_n(j)] \quad \text{for } (i,a) = (i_n, a_n)$$

$$(12.38)$$

where

$$J_n(j) = \min_{b \in \mathcal{A}_j} Q_n(j,b) \quad\quad (12.39)$$

and j is the successor state, and $\eta_n(i, a)$ is the learning-rate parameter at time step n for the state-action pair (i, a). The update equation (12.38) applies to the current state-action pair (i_n, a_n), for which $j = j_n$ in accordance with Eq. (12.35). For all other admissible state-action pairs, the Q-factors remain unchanged as shown by

$$Q_{n+1}(i,a) = Q_n(i,a) \quad\quad \text{for all } (i,a) \neq (i_n, a_n) \quad\quad (12.40)$$

Equations (12.38) to (12.40) constitute one iteration of the *Q-learning algorithm*.

Convergence Theorem[5]

Suppose that the learning-rate parameter $\eta_n(i, a)$ satisfies the conditions

$$\sum_{n=0}^{\infty} \eta_n(i,a) = \infty \quad \text{and} \quad \sum_{n=0}^{\infty} \eta_n^2 (i,a) < \infty \quad\quad \text{for all } (i,a) \quad\quad (12.41)$$

Then, the sequence of Q-factors $\{Q_n(i, a)\}$ generated by the Q-learning algorithm converges with probability 1 to the optimal value $Q^*(i, a)$ for all state-action pairs (i, a) as the number of iterations n approaches infinity, provided that all state-action pairs are visited infinitely often.

An example of a time-varying learning parameter that guarantees convergence of the algorithm is

$$\eta_n = \frac{\alpha}{\beta + n}, \quad\quad n = 1, 2, \ldots \quad\quad (12.42)$$

where α and β are positive numbers.

To sum up, the Q-learning algorithm is a stochastic approximation form of the value iteration policy. It backs up the Q-factor for a *single* state-action pair at each iteration of the algorithm, namely the observed current state and the action actually executed. Most importantly, in the limit, the algorithm converges to the optimal Q-values without forming an explicit model of the underlying Markovian decision processes. Once the optimal Q-values are available, an optimal policy can be determined with relatively little computation using Eq. (12.30).

The convergence of Q-learning to an optimal policy assumes the use of a *look-up table* representation for the Q-factors $Q_n(i, a)$. This method of representation is straightforward and computationally efficient. However, when the input space consisting of state-action pairs is large or the input variables are continuous, the use of a look-up table can be prohibitively expensive due to the requirement for a huge memory. In such a situation, we may resort to the use of a neural network for the purpose of function approximation.

Approximate Q-Learning

Equations (12.38) and (12.39) define the update formulas for the Q-factor for the current state-action pair (i_n, a_n). This pair of equations may be rewritten in the equivalent form

$$Q_{n+1}(i_n, a_n) = Q_n(i_n, a_n)$$
$$+ \eta_n(i_n, a_n)\left[g(i_n, a_n, j_n) + \gamma \min_{b \in \mathcal{A}_{j_n}} Q_n(j_n, b) - Q_n(i_n, a_n)\right] \qquad (12.43)$$

Treating the expression inside the square brackets on the right-hand side of Eq. (12.43) as the error signal involved in updating the current Q-factor $Q_n(i_n, a_n)$, we may identify the target (desired) Q-factor at time step n as:

$$Q_n^{\text{target}}(i_n, a_n) = g(i_n, a_n, j_n) + \gamma \min_{b \in \mathcal{A}_{j_n}} Q_n(j_n, b) \qquad (12.44)$$

where $j_n = i_{n+1}$ is the successor state. Equation (12.44) shows that the successor state j_n plays a critical role in determining the target Q-factor. Using this definition of the target Q-factor, we may reformulate the Q-learning algorithm as:

$$Q_{n+1}(i, a) = Q_n(i, a) + \Delta Q_n(i, a) \qquad (12.45)$$

where the incremental change in the current Q-factor is defined by

$$\Delta Q_n(i, a) = \begin{cases} \eta_n\left(Q_n^{\text{target}}(i, a) - Q_n(i, a)\right) & \text{for } (i, a) = (i_n, a_n) \\ 0, & \text{otherwise} \end{cases} \qquad (12.46)$$

By definition, the "optimal" action a_n at the current state i_n is the particular action at that state for which the Q-factor at time step n is minimum. Hence, given the Q-factors $Q_n(i_n, a)$ for admissible actions $a \in \mathcal{A}_{i_n}$ at state i_n, the optimal action a_n for use in Eq. (12.44) is given by

$$Q_n = \min_{a \in \mathcal{A}_{i_n}} Q_n(i_n, a) \qquad (12.47)$$

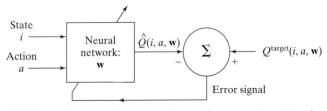

FIGURE 12.9 Neural network layout for approximating the target Q-factor Q^{target} (i, a, **w**).

Let $\hat{Q}_n(i_n, a_n, \mathbf{w})$ denote the approximation to the Q-factor $Q_n(i_n, a_n)$, computed by means of a neural network (e.g., multilayer perceptron trained with the back-propagation algorithm). The current state-action pair (i_n, a_n) is input to the neural network with parameter vector $\mathbf{w}$, producing the output $\hat{Q}_n(i_n, a_n, \mathbf{w})$, as illustrated in Fig. 12.9. At each iteration of the algorithm, the weight vector $\mathbf{w}$ of the neural network is changed slightly in a way that brings the output $\hat{Q}_n(i_n, a_n, \mathbf{w})$ closer to the target value $Q_n^{\text{target}}(i_n, a_n)$.   However, once the weight vector $\mathbf{w}$ is changed, the target value is itself implicitly affected by the change, assuming the modified value $Q_n^{\text{target}}(i_n, a_n, \mathbf{w})$. There is therefore no guarantee that the distance between these two Q-values is reduced on every iteration. This is also the reason why the approximate Q-learning algorithm has the potential of divergence. If the algorithm does not diverge, the weight vector $\mathbf{w}$ provides a means of storing the approximated Q-factor in the trained neural network, because it outputs $\hat{Q}_n(i_n, a_n, \mathbf{w})$ in response to the input (i_n, a_n).

Table 12.4 presents a summary of the approximate Q-learning algorithm.

Exploration

In policy iteration, all potentially important parts of the state space should be explored. In Q-learning we have an additional requirement: all potentially profitable actions should be tried as well. In particular, all admissible state-action pairs should be explored often enough to satisfy the convergence theorem. For a greedy policy denoted by μ, only the state-action pairs $(i, \mu(i))$ are explored. Unfortunately, there is no guarantee that all profitable actions would be tried, even if the entire state-space is explored.

What we need is a strategy that expands on Q-learning by providing a compromise between two conflicting objectives (Thrun, 1992):

- *Exploration,* which ensures that all admissible state-action pairs are explored often enough to satisfy the Q-learning convergence theorem.
- *Exploitation,* which seeks to minimize the cost-to-go function by following a greedy policy.

One way to achieve this compromise is to follow a *mixed nonstationary policy* that switches between an auxiliary Markov process and the original Markov process controlled by a stationary greedy policy determined by Q-learning (Cybenko, 1995). The auxiliary process has the following interpretation: The transition probabilities between

---

TABLE 12.4 Summary of the Approximate Q-Learning Algorithm

---

1. Start with an initial weight vector $\mathbf{w}_0$, resulting in the Q-factor $Q(i_0, a_0, \mathbf{w}_0)$; the weight vector $\mathbf{w}_0$ refers to a neural network used to perfom the approximation.
2. For iteration $n = 1, 2, \ldots$, do the following:
   (a) For the setting $\mathbf{w}$ of the neural network, determine the optimal action

   $$a_n = \min_{a \in \mathcal{A}_{i_n}} Q_n(i_n, a, \mathbf{w})$$

   (b) Determine the target Q-factor

   $$Q_n^{\text{target}}(i_n, a_n, \mathbf{w}) = g(i_n, a_n, j_n) + \gamma \min_{b \in \mathcal{A}_{j_n}} Q_n(j_n, b, \mathbf{w})$$

   (c) Update the Q-factor

   $$Q_{n+1}(i_n, a_n, \mathbf{w}) = Q_n(i_n, a_n, \mathbf{w}) + \Delta Q_n(i_n, a_n, \mathbf{w})$$

   where

   $$\Delta Q_n(i_n, a_n, \mathbf{w}) = \begin{cases} \eta_n(i_n, a_n)(Q_n^{\text{target}}(i_n, a_n, \mathbf{w}) - Q_n(i_n, a_n, \mathbf{w})), & (i, a) = (i_n, a_n) \\ 0, & \text{otherwise} \end{cases}$$

   (d) Apply (i_n, a_n) as input to the neural network producing the output $\hat{Q}_n(i_n, a_n, \mathbf{w})$ as an approximation to the target Q-factor $Q_n^{\text{target}}(i_n, a_n, \mathbf{w})$. Change the weight vector $\mathbf{w}$ slightly in a way that brings $\hat{Q}_n(i_n, a_n, \mathbf{w})$ closer to the target value $Q_n^{\text{target}}(i_n, a_n, \mathbf{w})$.
   (e) Go back to step (a) and repeat the computation.

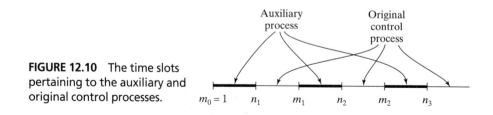

FIGURE 12.10 The time slots pertaining to the auxiliary and original control processes.

possible states are determined by the transition probabilities of the original controlled process with the added ingredient that the corresponding actions are uniformly randomized. The mixture policy starts in any state of the auxiliary process and chooses actions by following it, then switches to the original controlled process, and it goes back and forth in the manner illustrated in Fig. 12.10. The time spent operating on the auxiliary process occupies a fixed number of steps L, say, defined as twice the longest expected time to visit all states of the auxiliary process. The time spent operating on the original controlled process increases progressively with every switch. Let n_k denote the times at which we switch from the auxiliary process to the original controlled process, and m_k denote the times at which we switch back to the auxiliary process, with n_k and m_k defined as follows, respectively:

$$n_k = m_{k-1} + L, \qquad k = 1, 2, \ldots, \text{ and } m_0 = 1$$

and

$$m_k = n_k + kL, \qquad k = 1, 2, \ldots$$

The auxiliary process is constructed in such a way that as $k \to \infty$, there is an infinite number of visits to all states with probability 1, thereby guaranteeing convergence to the optimal Q-factors. Moreover, as $k \to \infty$, the time spent by the mixed policy operating in the auxiliary process becomes an asymptotically small fraction of the time spent operating in the original controlled process, which in turn means that the mixed policy asymptotically converges to a greedy policy. Hence, given the convergence of the Q-factors to their optimal values, the greedy policy must indeed be optimal provided that the policy becomes greedy slowly enough.

12.9 COMPUTER EXPERIMENT

In this computer experiment we revisit the stagecoach problem considered in Example 12.1. This time we use approximate Q-learning to solve the problem. Two approaches were used to implement the algorithm: One approach used a table to represent the Q-values, and the other approach used a neural network.

Figure 12.11 presents the learning histories for the following Q-factors: $Q(A, \text{up})$, $Q(C, \text{straight})$, $Q(E, \text{straight})$, and $Q(I, \text{up})$ using the table method. The dotted lines in Fig. 12.11 represent the desired Q-values. Each trial was a complete route from state i to the destination in state J. The starting state for each trial was chosen at random. The learning-rate parameter $\eta_n(i, a)$ was defined by

$$\eta_n(i, a) = \frac{\alpha v_n(i, a)}{K + v_n(i, a)}$$

where $v_n(i, a)$ is the number of times the state-action pairs (i, a) was visited up to the current time n, $\alpha = 1.6$, and $K = 600$. After a total of 1000 trials were performed, the optimal route was found to be

$$A \to D \to F \to I \to J$$

which is recognized to be one of the optimal routes with an overall cost of 11.

Figure 12.12 presents the corresponding results obtained using a multilayer perceptron with two input nodes, 10 hidden neurons, and one output neuron. One of the input nodes represented the state and the other nodes represented the action taken in moving from one state to the next. The output of the multilayer perceptron represented the Q-value computed by the network. The network was trained using the standard back-propagation algorithm. The target Q-value used at time n was calculated using Eq. (12.44). The learning-rate parameter was set at 0.012, and no momentum was used. The network was trained for 10,000 trials for each state-action pair. Figure 12.12 presents the learning histories for the Q-values: $Q(A, \text{up})$, $Q(C, \text{straight})$, $Q(E, \text{straight})$, and $Q(I, \text{up})$. The optimal route found by the network was

$$A \to D \to E \to H \to J$$

which is also recognized to be one of the optimal routes with a total cost of 11.

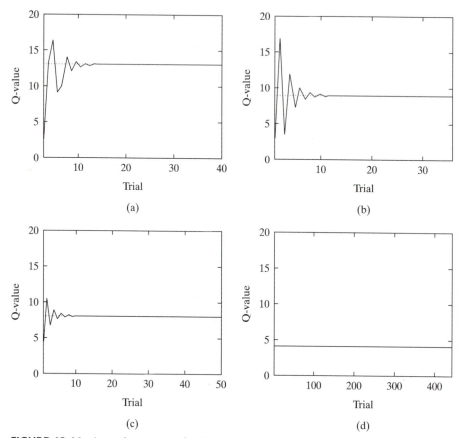

FIGURE 12.11 Learning curves for the stagecoach problem using a look up table (a) Learning curve for Q (A, up). (b) Learning curve for Q (C, straight). (c) Learning curve for Q (E, straight). (d) Learning curve for Q (I, up).

The computational requirements for the two methods of implementation are summarized as follows:

(a) Neural network:
Number of inputs = 2
Number of hidden neurons = 10
Number of output neurons = 1
Total number of synaptic weights and biases = $2 \times 10 + 10 + 10 \times 1 + 1 = 41$

(b) Look-up table:
Number of states = 10
Number of actions = 2 or 3
Size of table = 21

In this experiment the number of possible states is small, with the result that the look-up table requires less storage than the neural network. When, however, the number of states is large as in large-scale problems, the neural network usually gains an advantage over the table method in storage requirement.

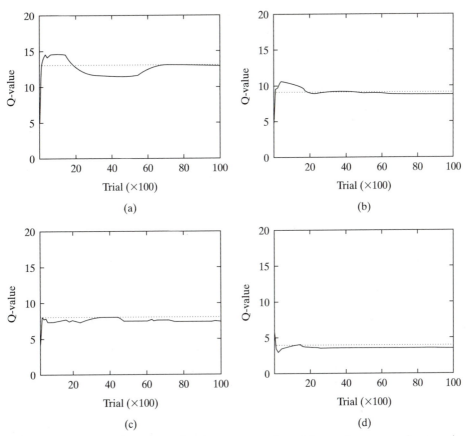

FIGURE 12.12   Learning curves for the stagecoach problem using a neural network. (a) Learning curve for Q (*A*, up). (b) Learning curve for Q (*C*, straight). (c) Learning curve for Q (*E*, straight). (d) Learning curve for Q (*I*, up).

12.10   SUMMARY AND DISCUSSION

Neurodynamic programming, combining the mathematical formalism of classical dynamic programming and the learning capability of neural networks, offers a powerful approach to the solution of behavioral tasks that require planning. In this modern approach to reinforcement learning, a system learns to do two things: make good decisions by observing its own behavior, and improve its actions through a reinforcement mechanism. The underlying decision-making process follows a Markovian model.

In this chapter we have described two neurodynamic programming procedures:

1. *Approximate policy iteration.* Policy iteration alternates between two basic steps:
    - Policy evaluation, in which the cost-to-go function for the current policy is determined.
    - Policy improvement, in which the current policy is updated to be greedy with respect to the current cost-to-go function.

    In approximate policy iteration, simulation and function approximation are combined for the purpose of policy evaluation. To simulate the Markovian model of the system, knowledge of the state-transition probabilities is required. To perform the function approximation, we may use a neural network (e.g., multilayer perceptron, radial-basis function network, or support vector machine), for which it is well suited because of its universal approximation property.

2. *Approximate Q-learning.* In value iteration, the alternative to policy iteration, a Markovian decision problem is solved by using a successive approximation procedure that converges to the optimal policy. Q-learning is an asynchronous form of value iteration formulated to avoid the need for explicit knowledge of the state-transition probabilities. It offers the following attractive properties:

- Q-learning converges to the optimal Q-factors with probability 1, provided that all state-action pairs are visited infinitely often and the learning-rate parameters satisfy the conditions specified in Eq. (12.41).

- Q-learning directly updates estimates of the Q-factors associated with an optimal policy, and thereby avoids the multiple policy evaluation steps involved in policy iteration.

In approximate Q-learning, a neural network is used to approximate estimates of the Q-factors in order to avoid the need for excessive memory requirement when the number of possible states is large. In short, approximate Q-learning is a simulation-based algorithm for solving a Markovian decision problem when a model of the system is unavailable and memory requirement is at a premium. Of course, it can be applied even if a model of the system is available, in which case it provides an alternative to approximate policy iteration.

Neurodynamic programming techniques are particularly effective in solving large-scale problems where planning is of major concern. Traditional approaches to dynamic programming are hardly applicable to problems of this kind because of the enormous size of the state space that would have to be explored. Indeed, neurodynamic programming has been successfully applied to solve difficult real-world problems in many diverse fields, which include backgammon (Tesauro, 1989, 1994), combinatorial optimization (Bertsekas and Tsitsiklis, 1996), elevator dispatching (Crites and Barto, 1996), and dynamic channel allocation (Singh and Bertsekas, 1997; Nie and Haykin, 1996, 1998). In what follows, the application to backgammon is described in some detail.

The development of a neural network-based computer program player to play *backgammon,* first reported in Tesauro (1989) and subsequently improved in Tesauro (1994), is a particularly impressive success story that has been a source of motivation for research in neurodynamic programming. Backgammon is an ancient two-player board game. It is played along an effectively one-dimensional track. The players take turns rolling a pair of dice, and accordingly move their checkers in opposite directions along the track. The legal moves made by each player depend on the outcome of the dice roll and the board configuration. The first player to move all of his or her checkers all the way forward to the ultimate end of the board is the winner. The game can be modeled as a Markov decision process, with a state being defined by a description of the board configuration, the outcome of the dice roll, and the identity of the player making a move. The first version of the neuro-backgammon built by Tesauro (1989) used supervised learning. It was able to learn at a strong intermediate level, given only a "raw" description of the state. Perhaps the most interesting finding reported was a good scaling behavior, in the sense that as the size of the neural network and amount of training experience were increased, substantial improvements in performance were observed. The neural network used in the study was a multilayer perceptron (MLP) trained with the back-propagation algorithm. The best performance was obtained using an MLP with 40 hidden neurons, and the training was performed over a total of 200,000 games. In a subsequent study reported in Tesauro (1994), a form of policy iteration called *optimistic TD*(λ) was used to train the neural network; TD stands for *temporal difference learning* due to Sutton (1988). Optimistic TD(λ) is a simulation-based method for approximating the cost-to-go function J^μ, in which the policy μ is replaced with a new policy μ that is greedy with respect to the approximation to J^μ at every state transition (Bertsekas and Tsitsiklis, 1996). The computer program based on this neurodynamic programming method is commonly referred to as *TD-gammon*. Handcrafted functions of the state (i.e., features) were added by Tesauro to the neural network's input representation, enabling TD-gammon to play at a strong master level, extremely close to the world's best human player. Among the indications contributing to this assessment are numerous tests of TD-gammon playing against several world class human grandmasters (Tesauro, 1995).

NOTES AND REFERENCES

1. The classical approach to reinforcement learning is rooted in psychology, going back to the early work of Thorndike (1911) on animal learning and that of Pavlov (1927) on conditioning. Contributions to classical reinforcement learning also include the work of Widrow et al. (1973); in that paper, the notion of a *critic* was introduced. Classical reinforcement learning is discussed in book form in Hampson (1990).

    Major contributions to modern reinforcement learning include the works of Samuel (1959) on his celebrated checkers playing program, Barto et al. (1983) on adaptive critic systems, Sutton (1988) on temporal difference methods, and Watkins (1989) on Q-learning. The handbook of intelligent control by White and Sofge (1992) presents material on optimal control by White and Jordan, reinforcement learning and adaptive critic methods by Barto, and heuristic dynamic programming by Werbos.

    Bertsekas and Tsitsiklis (1996) present the first treatment of modern reinforcement learning in book form. For a historical account of reinforcement learning, see Sutton and Barto (1998).

2. Dynamic programming was developed by R.E. Bellman in the late 1950s; see Bellman (1957), Bellman and Dreyfus (1962). For a detailed exposition of the subject, see the two-volume book by Bertsekas (1995b).

3. Policy iteration and value iteration are two principal methods of dynamic programming. There are two other dynamic programming methods that deserve to be mentioned: the *Gauss–Seidel method* and *asynchronous dynamic programming* (Barto et al., 1995; Bertsekas, 1995b). In the Gauss–Seidel method, the cost-to-go function is updated at one state at a time in a sequential sweep of all the states, with the competition for each state based on the most recent costs of the other states. Asynchronous dynamic programming differs from the Gauss–Seidel method in that it is not organized in terms of systematic successive sweeps of the set of states.

4. On page 96 of his Ph.D. thesis, Watkins (1989) makes the following remarks on Q-learning:

    "Appendix 1 presents a proof that this learning method does work for finite Markov decision processes. The proof also shows that the learning method will converge rapidly to the optimal action-value function. Although this is a very simple idea, it has not, as far as I know, been suggested previously. However, it must be said that finite Markov decision processes and stochastic dynamic programming have been extensively studied for use in several different fields for over thirty years, and it is unlikely that nobody has considered the Monte-Carlo method before."

In a footnote commentary on these remarks, Barto et al. (1995) point out that although the idea of assigning values to state-action pairs formed the basis of the approach to dynamic programming taken in Denardo (1967), they have not seen algorithms like Q-learning for estimating these values that predate Watkins' 1989 thesis.

5. The outline of a proof of the convergence theorem for Q-learning was presented in Watkins (1989), which was refined later in Watkins and Dayan (1992). More general results on the convergence of Q-learning were presented in Tsitsiklis (1994); see also Bertsekas and Tsitsiklis (1996).

PROBLEMS

Bellman's optimality criterion

12.1 When the discount factor γ approaches 1, computation of the cost-to-go function in Eq. (12.22) becomes longer. Why? Justify your answer.

12.2 In this problem we present another proof of Bellman's optimality equation (12.22) due to Ross (1983).

(a) Let π be any arbitrary policy, and suppose that π chooses action a at time step 0 with probability p_a and $a \in \mathcal{A}_i$. Then,

$$J^\pi(i) = \sum_{a \in \mathcal{A}_i} p_a \left(c(i, a) + \sum_{j=1}^N p_{ij}(a) W^\pi(j) \right)$$

where $W^\pi(j)$ represents the expected cost-to-go function from time step 1 onward, given that policy π is being used and that j is the state at time step 1. Hence, show that

$$J^\pi(i) \leq \max_{a \in \mathcal{A}_i} \left(c(i, a) + \gamma \sum_{j=1}^N p_{ij}(a) J(j) \right)$$

where

$$W^\pi(j) \leq \gamma J(j)$$

(b) Let π be the policy that chooses action a_0 at time step 0 and, if the next state is j, it views the process as originating in state j, following a policy π_j such that

$$J^{\pi_j}(j) \geq J(j) - \epsilon$$

where ϵ is a small positive number. Hence, show that

$$J(i) \geq \max_{a \in \mathcal{A}_i} \left(c(i, a) + \gamma \sum_{j=1}^N p_{ij}(a) J(j) \right) - \gamma \epsilon$$

(c) Using the results derived in parts (a) and (b), prove Eq. (12.22).

12.3 Equation (12.22) represents a linear system of N equations, with one equation per state. Let

$$\mathbf{J}^\pi = \left[J^\mu(1), J^\mu(2), ..., J^\mu(N) \right]^T$$

$$\mathbf{c}(\mu) = \left[c(1, \mu), c(2, \mu), ..., c(N, \mu) \right]^T$$

$$\mathbf{P}(\mu) = \begin{bmatrix} p_{11}(\mu) & p_{12}(\mu) & \cdots & p_{1N}(\mu) \\ p_{21}(\mu) & p_{22}(\mu) & \cdots & p_{2N}(\mu) \\ \vdots & \vdots & & \vdots \\ p_{N1}(\mu) & p_{N2}(\mu) & \cdots & p_{NN}(\mu) \end{bmatrix}$$

Show that Eq. (12.22) may be reformulated in the equivalent matrix form:

$$(\mathbf{I} - \gamma \mathbf{P}(\mu)) \mathbf{J}^\mu = \mathbf{c}(\mu)$$

where $\mathbf{I}$ is the identity matrix. Comment on the uniqueness of the vector $\mathbf{J}^\mu$ representing the cost-to-go functions for the N states.

12.4 In Section 12.3 we derive the dynamic programming algorithm for a finite-horizon problem. In this problem we rederive this algorithm for a discounted problem for which the cost-to-go function is defined by

$$J^\mu(X_0) = \lim_{K \to \infty} \left[\sum_{n=0}^{K-1} \gamma^n g(X_n, \mu(X_n), X_{n+1}) \right]$$

In particular, show that

$$J_K(X_0) = \min_\mu \mathop{E}_{X_1} \left[g(X_0, \mu(X_0), X_1) + \gamma J_{K-1}(X_1) \right]$$

Policy iteration

12.5 In Section 12.4 we say that the cost-to-go function satisfies the statement

$$J^{\mu_{n+1}} \le J^{\mu_n}$$

Justify this statement.

12.6 Discuss the significance of the statement described in Eq. (12.25.)

12.7 Using a two-dimensional picture, illustrate the interaction between the policy update and policy evaluation in the policy iteration algorithm.

Value iteration

12.8 A dynamic programming problem involves a total of N possible states and M admissible actions. Assuming the use of a stationary policy, show that a single iteration of the value iteration algorithm requires on the order of $N^2 M$ operations.

12.9 Table 12.2 presents a summary of the value iteration algorithm formulated in terms of the cost-to-go function $J^\mu(i)$ for states $i \in \mathcal{X}$. Reformulate this algorithm in terms of the Q-factors $Q(i, a)$.

12.10 Policy iteration always terminates finitely, whereas value iteration may require an infinite number of iterations. Discuss other differences between these two methods of dynamic programming.

Q-learning

12.11 Show that

$$J^*(i) = \min_{a \in \mathcal{A}_i} Q(i, a)$$

12.12 The Q-learning algorithm is sometimes referred to as an adaptive form of the value iteration policy. Justify the validity of this description.

12.13 Construct a signal-flow graph for the approximate Q-learning algorithm summarized in Table 12.4.

12.14 The approximate Q-learning algorithm summarized in Table 12.4 assumes lack of knowledge of the state-transition probabilities. Reformulate this algorithm assuming the availability of these probabilities.

CHAPTER 13

Temporal Processing Using Feedforward Networks

13.1 INTRODUCTION

Time constitutes an essential ingredient of the learning process. It can be continuous or discrete. Whatever its form, time is an ordered entity that is basic to many of the cognitive tasks encountered in practice such as vision, speech, signal processing, and motor control. It is through the incorporation of time into the operation of a neural network that it is enabled to follow statistical variations in nonstationary processes such as speech signals, radar signals, signals picked up from the engine of an automobile, and fluctuations in stock market prices, just to mention a few. The question is: How do we build time into the operation of a neural network? The answer to this fundamental question lies in one of two possible ways:

- *Implicit representation.* Time is represented by the effect it has on signal processing in an implicit manner.[1] For example, the input signal is *uniformly sampled*, and the sequence of synaptic weights of each neuron connected to the input layer of the network is *convolved* with a different sequence of input samples. In so doing, the temporal structure of the input signal is embedded in the spatial structure of the network.

- *Explicit representation.* Time is given its own particular representation.[2] For example, the echo-location system of a bat operates by emitting a short frequency modulated (FM) signal, so that the same intensity level is maintained for each frequency channel restricted to a very short period within the FM sweep. Multiple comparisons between several different frequencies encoded by an array of auditory receptors are made for the purpose of extracting accurate distance (range) information about a target (Suga and Kanwal, 1995). When an echo is received from the target with an unknown delay, a neuron (in the auditory system) with a matching delay line responds, thereby providing an estimate of the range to the target.

In this chapter we are concerned with the implicit representation of time, whereby a "static" neural network (e.g., multilayer perceptron) is provided with *dynamic* properties. This, in turn, makes the network responsive to the temporal structure of information-bearing signals.

For a neural network to be dynamic, it must be given *memory*. As pointed out in Chapter 2, memory may be divided into "short-term" and "long-term" memory, depending on the retention time. Long-term memory is built into a neural network through supervised learning, whereby the information content of the training data set is stored (in part or in full) in the synaptic weights of the network. However, if the task at hand has a temporal dimension, we need some form of short-term memory to make the network dynamic. One simple way of building short-term memory into the structure of a neural network is through the use of *time delays*, which can be implemented at the synaptic level inside the network or at the input layer of the network. The use of time delays in neural networks is neurobiologically motivated, since it is well known that signal delays are omnipresent in the brain and play an important role in neurobiological information processing (Braitenberg, 1967, 1977, 1986; Miller, 1987).

Organization of the Chapter

The material in this chapter is organized in three parts. The first part, consisting of Sections 13.2 and 13.3, deals with network structures and models. In Section 13.2 we present a discussion of memory structures, followed by Section 13.3 on a description of two different network architectures for temporal processing of signals.

The second part of the chapter, consisting of Sections 13.4 to 13.6, deals with a class of neural networks known as focused time lagged feedforward networks; the term "focused" refers to the fact that the short-term memory is located entirely at the front end of the network. A computer experiment on this structure is described in Section 13.6.

The third part of the chapter, consisting of Sections 13.7 to 13.9, deals with distributed time lagged feedforward networks, where delay lines are distributed throughout the network. Section 13.7 describes spatio-temporal models of a neuron, followed by a discussion in Section 13.8 on the second class of neural networks just mentioned. In Section 13.9 we describe the "temporal" back-propagation algorithm for the supervised training of distributed time lagged feedforward networks.

The chapter concludes with some final remarks in Section 13.10.

13.2   SHORT-TERM MEMORY STRUCTURES

The primary role of memory is to *transform a static network into a dynamic one*. In particular, by embedding memory into the structure of a static network such as an ordinary multilayer perceptron, the output of the network becomes a function of time. This approach for building a nonlinear dynamical system is straightforward because it provides for a clear separation of responsibilities: The static network accounts for nonlinearity, and the memory accounts for time.

Short-term memory[3] can be implemented in continuous time or in discrete time. Continuous time is denoted by t, and discrete time is denoted by n. The resistance-capacitance circuit in Fig. 13.1 is an example of continuous-time memory, which is characterized by an impulse response (i.e., memory trace) $h(t)$ that decays exponentially with time t. This circuit is responsible for memory, at the synaptic level, in an analog implementation of the additive model of a neuron to be described later in the chapter. In this section we are concerned mainly with discrete-time memory.

A useful tool for dealing with discrete-time systems is the *z-transform*. Let $\{x(n)\}$ denote a discrete-time sequence, which may extend into the infinite past. The z-transform of this sequence, denoted by $X(z)$, is defined by

$$X(z) = \sum_{n=-\infty}^{\infty} x(n)z^{-n} \tag{13.1}$$

where z^{-1} is the *unit delay operator*; that is, z^{-1} operating on $x(n)$ yields its delayed version $x(n-1)$. Suppose $x(n)$ is applied to a discrete-time system of impulse response $h(n)$. The output of the system, $y(n)$, is defined by the *convolution sum*

$$y(n) = \sum_{k=-\infty}^{\infty} h(k)x(n-k) \tag{13.2}$$

For $x(n)$ equal to the unit impulse, $y(n)$ reduces to the *impulse response $h(n)$* of the system. An important property of the z-transform is that *convolution in the time domain is transformed into multiplication in the z-domain* (Oppenheim and Schafer, 1989; Haykin and Van Veen, 1998). If we denote the z-transform of the sequences $\{h(n)\}$ and $\{y(n)\}$ by $H(z)$ and $Y(z)$, respectively, application of the z-transform to Eq. (13.2) yields

$$Y(z) = H(z)X(z) \tag{13.3}$$

or equivalently

$$H(z) = \frac{Y(z)}{X(z)} \tag{13.4}$$

The function $H(z)$ is called the *transfer function* of the system.

Figure 13.2 shows a block diagram of a discrete-time memory consisting of p identical sections connected in cascade; hereafter p is referred to as the *order* of the

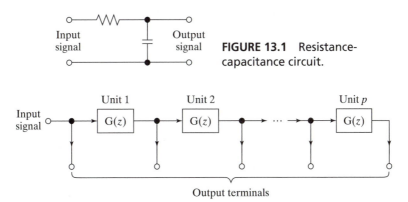

FIGURE 13.1   Resistance-capacitance circuit.

FIGURE 13.2   Generalized tapped delay line memory of order p.

memory. Each delay section, viewed as an operator, is characterized by a transfer function $G(z)$, as indicated in the figure. Equivalently, each section may be described in terms of impulse response $g(n)$, which has the following two properties:

- It is *causal*, that is, $g(n) = 0$ for $n < 0$.
- It is *normalized*, which means that $\sum_{n=0}^{\infty} |g(n)| = 1$.

Henceforth $g(n)$ is referred to as the *generating kernel* of the discrete-time memory.

In light of Fig. 13.2, we may formally define a *discrete-time memory* as a linear time invariant, single input–multiple output (SIMO) system whose generating kernel satisfies these properties. The junction points, to which the output terminals of the memory are connected, are commonly called *taps*. Note that for a memory of order p, there are $p + 1$ taps with one tap belonging to the input.

The attributes of a memory structure are measured in terms of depth and resolution. Let $g_p(n)$ denote the overall impulse response of the memory, defined as p successive convolutions of $g(n)$, or equivalently, the inverse z-transform of $G^p(z)$. *Memory depth*, denoted by D, is defined as the first time moment of $g_p(n)$, as shown by

$$D = \sum_{n=0}^{\infty} n g_p(n) \tag{13.5}$$

A memory of low depth D only holds its information content for a relatively short period of time, whereas a high depth memory holds its information content much further into the past. *Memory resolution,* denoted by R, is defined as the number of taps in the memory structure per unit time. A memory of high resolution R is able to hold information about the input sequence at a fine level, whereas a low resolution memory can only do so at a much coarser level. For a fixed number of taps, the product of memory depth and memory resolution is a constant equal to the memory order p.

Different choices of the generating kernel $g_p(n)$ naturally result in different values for the depth D and resolution R, as illustrated in the following two memory structures.

Tapped delay line memory.   Figure 13.3 shows the block diagram of the simplest and most commonly used form of short-term memory called a *tapped delay line memory*. It consists of p unit delay operators, each of which is characterized by $G(z) = z^{-1}$. That is, the generating kernel is $g(n) = \delta(n - 1)$, where $\delta(n)$ is the unit impulse:

$$\delta(n) = \begin{cases} 1, & n = 0 \\ 0, & n \neq 0 \end{cases} \tag{13.6}$$

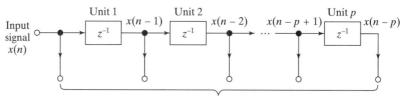

FIGURE 13.3   Ordinary tapped delay line memory of order p.

The overall impulse response of the tapped delay line in Fig. 13.3 is $g_p(n) = \delta(n - p)$. Substituting this $g_p(n)$ in Eq. (13.5) yields the memory depth $D = p$, which is intuitively satisfying. From Fig. 13.3 we see that there is only one tap per unit time; hence, $R = 1$. Thus the memory depth of a tapped delay line increases linearly with its order p, but its memory depth is fixed at unity; the depth-resolution product is constant at p.

   We need an additional degree of freedom in order to exercise control over memory depth. Such a provision is made possible by an alternative to the tapped delay line considered next.

Gamma memory.   Figure 13.4 shows the signal-flow graph of the basic functional block $G(z)$ used in a memory structure called the *gamma memory* (deVries and Principe, 1992). Specifically, each section of this memory structure consists of a feedback loop with unit delay z^{-1} and adjustable parameter μ. The transfer function of each such section is

$$G(z) = \frac{\mu z^{-1}}{1 - (1 - \mu)z^{-1}}$$

$$= \frac{\mu}{z - (1 - \mu)} \tag{13.7}$$

For stability, the only pole of $G(z)$ at $z = 1 - \mu$ must lie inside the unit circle in the z-plane. This in turn requires that

$$0 < \mu < 2 \tag{13.8}$$

The generating kernel of the gamma memory is the inverse z-transform of $G(z)$, that is,

$$g(n) = \mu(1 - \mu)^{n-1}, \qquad n \geq 1 \tag{13.9}$$

The condition of Eq. (13.8) ensures that $g(n)$ decays exponentially to zero as n approaches infinity.

   The overall impulse response of the gamma memory is the inverse z-transform of the overall transfer function

$$G_p(z) = \left(\frac{\mu}{z - (1 - \mu)} \right)^p$$

That is,

$$g_p(n) = \binom{n-1}{p-1} \mu^p (1 - \mu)^{n-p}, \qquad n \geq p \tag{13.10}$$

where $(:)$ is a binomial coefficient defined by $\binom{n}{p} = \frac{n(n-1)\cdots(n-p+1)}{p!}$ for integer values of n and p. The overall impulse response $g_p(n)$ for varying p represents a discrete version of the integrand of the *gamma function* (deVries and Principe, 1992), hence the

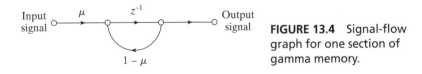

FIGURE 13.4   Signal-flow graph for one section of gamma memory.

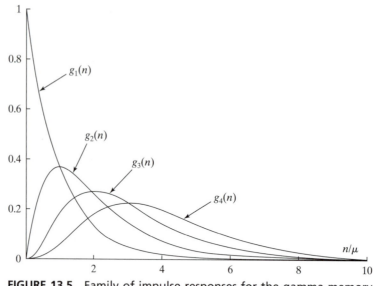

FIGURE 13.5   Family of impulse responses for the gamma memory for order $p = 1, 2, 3, 4$ and $\mu = 0.7$.

name of the memory. Figure 13.5 shows a family of impulse responses $g_p(n)$, normalized with respect to μ, for $p = 1, 2, 3, 4$ and $\mu = 0.7$. Note that the time axis in Fig. 13.5 is scaled by the parameter μ. This scaling has the effect of positioning the peak value of $g_p(n)$ at $n = p$.

The depth of the gamma memory is p/μ and its resolution is μ, for a depth-resolution product of p. Accordingly, by choosing μ to be less than unity, the gamma memory provides improvement in depth (but sacrifices resolution) over the tapped delay line for a specified order p. When $\mu = 1$, these quantities reduce to the respective values assumed by the tapped delay line. Thus the gamma memory includes the tapped delay line as a special case. This observation is also readily ascertained by setting $\mu = 1$ in Eq. (13.9). If μ is greater than 1 but less than 2, then $(1 - \mu)$ in this equation becomes negative but with an absolute value less than 1.

13.3   NETWORK ARCHITECTURES FOR TEMPORAL PROCESSING

Network architectures for temporal processing take more than one form, just as memory structures do. In this section we will describe two feedforward network architectures that have enriched the literature on temporal processing in their own individual ways.

NETtalk

NETtalk, devised by Sejnowski and Rosenberg (1987), was the first demonstration of a massively parallel distributed network that converts English speech to phonemes; a *phoneme* is a basic linguistic unit. Figure 13.6 shows a schematic diagram of the NETtalk system, which is based on a multilayer perceptron with an input layer of 203 sensory nodes, a hidden layer of 80 neurons, and an output layer of 26 neurons. All the

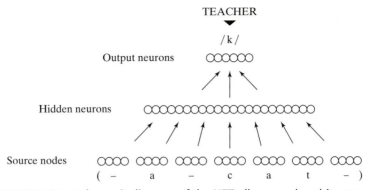

FIGURE 13.6 Schematic diagram of the NETtalk network architecture.

neurons used sigmoid (logistic) activation functions. The synaptic connections in the network were specified by a total of 18,629 weights, including a variable threshold for each neuron; threshold is the negative of bias. The standard back-propagation algorithm was used to train the network.

The network had seven groups of nodes in the input layer, with each group encoding one letter of the input text. Strings of seven letters were thus presented to the input layer at any one time. The desired response for the training process was specified as the correct phoneme associated with the center (i.e., fourth) letter in the seven-letter window. The other six letters (three on either side of the center letter) provided a partial *context* for each decision made by the network. The text was stepped through the window on a letter-by-letter basis. At each step in the process, the network computed a phoneme, and after each word the synaptic weights of the network were adjusted according to how closely the computed pronunciation matched the correct one.

The performance of NETtalk exhibited some similarities with observed human performance, as summarized here (Sejnowski and Rosenberg, 1987).

- The training followed a power law.
- The more words the network learned, the better it was at generalizing and correctly pronouncing new words.
- The performance of the network degraded very slowly as synaptic connections in the network were damaged.
- Relearning after damage to the network was much faster than learning during the original training.

NETtalk was a brilliant illustration in miniature of many aspects of learning, starting out with considerable "innate" knowledge of its input patterns and then gradually acquiring competence at converting English speech to phonemes through practice. However, it did not lead to practical applications.

Time-delay Neural Network

A popular neural network that uses ordinary time delays to perform temporal processing is the so-called *time delay neural network* (TDNN), which was first described in Lang and Hinton (1988) and Waibel et al. (1989). The TDNN is a multilayer feedforward

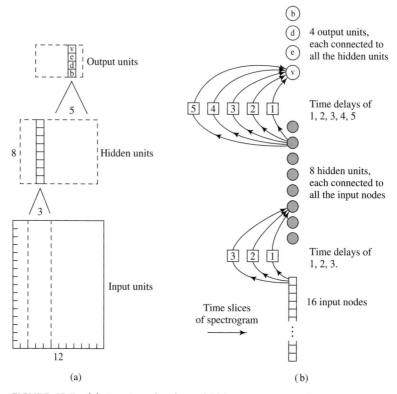

FIGURE 13.7 (a) A network whose hidden neurons and output neurons are replicated across time. (b) Time delay neural network (TDNN) representation. (From K. J. Lang and G. E. Hinton, 1988, with permission)

network whose hidden neurons and output neurons are *replicated across time*. It was devised to capture explicitly the concept of time symmetry as encountered in the recognition of an isolated word (phoneme) using a spectrogram. A *spectrogram* is a two-dimensional image in which the vertical dimension corresponds to frequency and the horizontal dimension corresponds to time; the intensity (darkness) of the image corresponds to signal energy (Rabiner and Schafer, 1978). Figure 13.7a illustrates a single hidden layer version of the TDNN (Lang and Hinton, 1988). The input layer consists of 192 (16 by 12) sensory nodes encoding the spectrogram; the hidden layer contains 10 copies of 8 hidden neurons; and the output layer contains 6 copies of 4 output neurons. The various replicas of a hidden neuron apply the same set of synaptic weights to narrow (three-time-step) windows of the spectrogram; similarly, the various replicas of an output neuron apply the same set of synaptic weights to narrow (five-time-step) windows of the pseudospectrogram computed by the hidden layer. Figure 13.7b presents a *time delay* interpretation of the replicated neural network of Fig. 13.7a—hence the name "time delay neural network". This network has a total of 544 synaptic weights. Lang and Hinton (1988) used the TDNN for the recognition of four isolated words: "bee", "dee", "ee", and "vee", which accounts for the use of four output neurons in Fig. 13.7. A recognition score of 93 percent was obtained on test data different from the training data. In a more elaborate study reported by Waibel et al. (1989), a TDNN with two hidden layers was used for the recognition of three isolated words: "bee", "dee", and "gee". In performance evaluation involving the use of test data from three speakers, the TDNN achieved an average recognition score of 98.5 percent.

The TDNN appears to work best for classifying a temporal pattern that consists of a sequence of fixed dimensional feature vectors such as phonemes. In a practical speech recognizer, however, it is unrealistic to assume that the speech signal can be accurately segmented into its constituent phonemes. Rather, it is essential to adequately model the super-segmented temporal structure of speech patterns. In particular, the speech recognizer has to deal with word and sentence segments that vary significantly in their duration and nonlinear temporal structure. To model these natural characteristics of speech signals, the traditional approach in the speech recognition field has been to use a state transition structure like the hidden Markov model (Rabiner, 1989; Jelinek, 1997). Basically, a *hidden Markov model* (HMM) represents a stochastic process generated by an underlying Markov chain, and a set of observation distributions associated with its hidden states; see note 11 in Chapter 11. Many hybrids of TDNN and HMM have been studied in the literature.[4]

13.4   FOCUSED TIME LAGGED FEEDFORWARD NETWORKS

The prototypical use of a static neural network (e.g., multilayer perceptron, radial-basis function network) is in *structural pattern recognition*. In contrast, *temporal pattern recognition* requires processing of patterns that evolve over time, with the response at a particular instant of time depending not only on the present value of the input but also on its past values. Figure 13.8 shows the block diagram of a *nonlinear filter* built on a static neural network (Mozer, 1994). The network is stimulated through a short-term memory. Specifically, given an input signal consisting of the present value $x(n)$ and the p past values $x(n-1), ..., x(n-p)$ stored in a delay line memory of order p, for example, the free parameters of the neural network are adjusted to minimize the mean-square error between the output of the network, $y(n)$, and the desired response $d(n)$.

The structure of Fig. 13.8 can be implemented at the level of a single neuron or a network of neurons. These two cases are illustrated in Figs. 13.9 and 13.10 respectively.

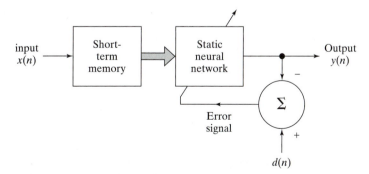

FIGURE 13.8   Nonlinear filter built on a static neural network.

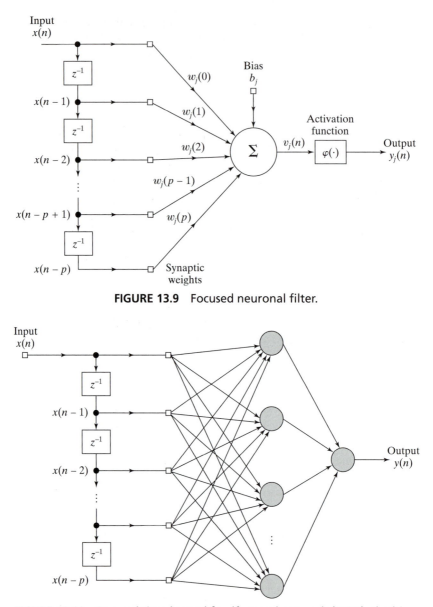

FIGURE 13.9   Focused neuronal filter.

FIGURE 13.10   Focused time lagged feedforward network (TLFN); the bias levels have been omitted for convenience of presentation.

To simplify the presentation, we have used a tapped delay line memory as the short-term memory structure in Figs. 13.9 and 13.10. Clearly, both of these figures could be generalized by using a unit with transfer function $G(z)$ in place of z^{-1}.

The temporal processing unit of Fig. 13.9 is composed of a tapped delay line memory with its taps connected to the synapses of a neuron. The tapped delay line memory captures temporal information contained in the input signal and the neuron embeds that information in its own synaptic weights. The processing unit of Fig. 13.9 is called a *focused neuronal filter*, focused in the sense that the entire memory structure is located at the input end of the unit. The output of the filter, in response to the input $x(n)$ and its past values $x(n - 1), ..., x(n - p)$, is given by

$$y_j(n) = \varphi\left(\sum_{l=0}^{p} w_j(l)x(n - l) + b_j\right) \tag{13.11}$$

where $\varphi(\cdot)$ is the activation function of neuron j, the $w_j(l)$ are its synaptic weights, and b_j is the bias. Note that the input to the activation function consists of a bias plus the *convolution* of sequences of input samples and synaptic weights of the neuron.

Turning next to Fig. 13.10, referred to as a *focused time lagged feedforward network* (TLFN), here we have a more powerful nonlinear filter consisting of a tapped delay line memory of order p and a multilayer perceptron. To train the filter, we may use the standard back-propagation algorithm described in Chapter 4. At time n, the "temporal pattern" applied to the input layer of the network is the signal vector

$$x(n) = [x(n), x(n - 1), ..., x(n - p)]^T$$

which may be viewed as a description of the *state* of the nonlinear filter at time n. An epoch consists of a sequence of states (patterns), the number of which is determined by the memory order p and the size N of the training sample.

The output of the nonlinear filter, assuming that the multilayer perceptron has a single hidden layer as shown in Fig. 13.10, is given by

$$y(n) = \sum_{j=1}^{m_1} w_j y_j(n)$$
$$= \sum_{j=1}^{m_1} w_j \varphi\left(\sum_{l=0}^{p} w_j(l)x(n - l) + b_j\right) + b_o \tag{13.12}$$

where the output neuron in the focused TLFN is assumed to be linear; the synaptic weights of the output neuron are denoted by the set $\{w_j\}_{j=1}^{m_1}$, where m_1 is the size of the hidden layer, and the bias is denoted by b_o.

13.5 COMPUTER EXPERIMENT

In this computer experiment, we investigate the use of the focused TLFN in Fig. 13.10 to simulate a time series representing a difficult frequency modulated signal:

$$x(n) = \sin(n + \sin(n^2)), \qquad n = 0, 1, 2, ...$$

The network was used as a *one-step predictor* with $x(n + 1)$ providing the desired response for an input consisting of the set $\{x(n - l)\}_{l=0}^{p}$. The composition of the network and its parameters were as follows:

Order of tapped delay line memory, p:	20
Hidden layer, m_1:	10 neurons
Activation function of hidden neurons:	logistic
Output layer:	1 neuron
Activation function of output neuron:	linear
Learning-rate parameter (both layers):	0.01
Momentum constant:	none

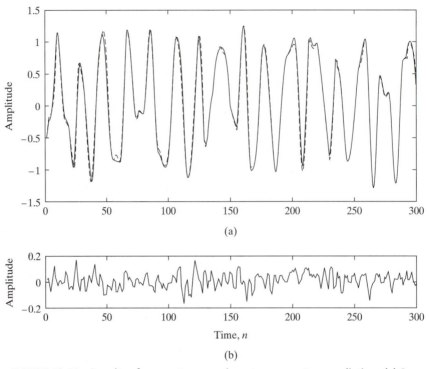

(a)

(b)

FIGURE 13.11 Results of computer experiment on one-step prediction. (a) Superposition of actual (continuous) and predicted (dashed) waveforms. (b) Waveform of prediction error.

The data set used to train the network consisted of 500 randomized patterns, with each pattern consisting of 20 time-ordered samples selected from the time series $\{x(n)\}$.

Figure 13.11a shows a superposition of the one-step prediction performed by the network on test data (not seen before) and the actual waveform. Figure 13.11b displays the waveform of the prediction error defined as the difference between the actual and predicted waveforms. The mean-square value of the prediction error is 1.2×10^{-3}.

13.6 UNIVERSAL MYOPIC MAPPING THEOREM

The nonlinear filter of Fig. 13.9 may be generalized as shown in Fig. 13.12. This generic dynamic structure consists of two functional blocks. The block labeled $\{h_j\}_{j=1}^{L}$ represents *multiple convolutions* in the time domain, that is, a bank of *linear filters* operating in parallel. The h_j are drawn from a large set of real-valued kernels, each one of which represents the impulse response of a linear filter. The block labeled $\mathcal{N}$ represents a static (i.e., memoryless) nonlinear feedforward network such as an ordinary multilayer perceptron. The structure of Fig. 13.12 is a *universal dynamic mapper*. In Sandberg and Xu (1997a) it is shown that any shift-invariant *myopic map* can be uniformly approximated arbitrarily well by a structure of the form depicted in Fig. 13.12 under mild conditions. The requirement that a map be myopic is equivalent to "uniform fading

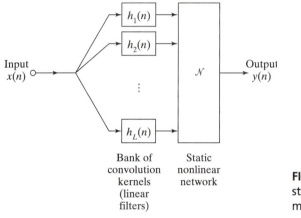

Bank of
convolution
kernels
(linear
filters)

Static
nonlinear
network

FIGURE 13.12 Generic structure for universal myopic mapping theorem.

memory"; it is assumed here that the map is *causal*, which means that an output signal is produced by the map at time $n \geq 0$ only when the input signal is applied at time $n = 0$. By "shift invariant" we mean: If $y(n)$ is the output of the map due to an input $x(n)$, then the output of the map due to the shifted input $x(n - n_0)$ is $y(n - n_0)$ where the time shift n_0 is an integer. In Sandberg and Xu (1997b), it is further shown that for any single-variable, shift-invariant, causal, uniformly-fading-memory map, there is a gamma memory and static neural network, the combination of which approximates the map uniformly and arbitrarily well.

We may now formally state the *universal myopic mapping theorem*[5] as follows (Sandberg and Xu, 1997a, 1997b):

> Any shift-invariant myopic dynamic map can be uniformly approximated arbitrarily well by a structure consisting of two functional blocks: a bank of linear filters feeding a static neural network.

The structure embodied in this theorem may take the form of a focused TLFN. It is also noteworthy that this theorem holds when the input and output signals are functions of a finite number of variables as in image processing, for example.

The universal myopic mapping theorem has profound practical implications. Not only does it provide mathematical justification for NETtalk and its possible gamma memory extension, but it also lays down the framework for the design of more elaborate models of nonlinear dynamical processes. The multiple convolutions at the front end of the structure in Fig. 13.12 may be implemented using linear filters with a finite-duration impulse response (FIR) or infinite-duration impulse response (IIR). As for the static neural network, it can be implemented using a multilayer perceptron, radial-basis function network, or support vector machine trained by means of the algorithms described in Chapters 4, 5, and 6. In other words, we may naturally build on the material presented in those chapters on supervised learning to build nonlinear filters or models of nonlinear dynamic processes. Most importantly, the structure of Fig. 13.12 is *inherently stable*, provided that the linear filters are themselves stable. We thus have a clear-cut separation of roles as to how to take care of short-term memory and memoryless nonlinearity.

13.7   SPATIO-TEMPORAL MODELS OF A NEURON

The focused neuronal filter in Fig. 13.9 has an interesting interpretation as described here. The combination of unit delay elements and associated synaptic weights may be viewed as a *finite-duration impulse response (FIR) filter* of order p, as shown in Fig. 13.13a; the FIR filter is one of the basic building blocks in digital signal processing (Oppenheim and Schafer, 1989; Haykin and Van Veen, 1998). Accordingly, the focused neuronal filter of Fig. 13.9 is, in fact, a nonlinear FIR filter, as shown in Fig. 13.13b. We may build on this representation and thereby extend the processing power of the neuron in a spatial sense through the use of multiple inputs, m_0 in number, as depicted in Fig. 13.14. The spatio-temporal model of Fig. 13.14 is referred to as a *multiple input neuronal filter*.

Yet another way of describing the model in Fig. 13.14 is to think of it as a *distributed neuronal filter*, in the sense that the filtering action is distributed across different points in space. The spatio-temporal characterization of the model is represented as follows:

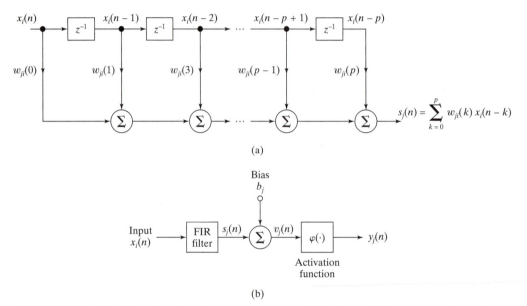

$$s_j(n) = \sum_{k=0}^{p} w_{ji}(k) \, x_i(n-k)$$

(a)

(b)

FIGURE 13.13   (a) Finite-duration impulse response (FIR) filter. (b) Interpretation of neuronal filter as a nonlinear FIR filter.

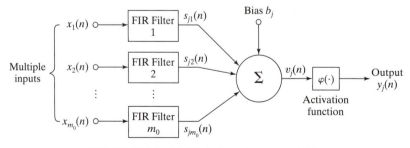

FIGURE 13.14   Multiple input neuronal filter.

- The neuron has m_0 "primary" synapses, each of which consists of a linear discrete-time filter implemented in the form of an FIR filter of order p; the primary synapses account for the spatial dimension of signal processing.
- Each primary synapse has $(p + 1)$ "secondary" synapses that are connected to its respective input and the memory taps of its FIR filter, thereby accounting for the temporal dimension of signal processing.

The synaptic structure of the neuronal filter in Fig. 13.14 is thus tree-like, as depicted in Fig. 13.15. The total number of synaptic weights in the structure is $m_0(p + 1)$.

In mathematical terms, we may describe the spatio-temporal processing performed by the neuronal filter in Fig. 13.14 by expressing its output, $y_j(n)$, as

$$y_j(n) = \varphi\left(\sum_{i=1}^{m_0} \sum_{l=0}^{p} w_{ji}(l)x_i(n - l) + b_j \right) \tag{13.13}$$

where $w_{ji}(l)$ is the weight of the lth secondary synapse belonging to the ith primary synapse, $x_i(n)$ is the input applied to the ith primary synapse at time n, and b_j is the bias applied to the neuron. The induced local field $v_j(n)$ of the neuron, that is, the argument of the activation function φ in Eq. (13.13) may be viewed as the discrete time "approximation" to the continuous-time formula

$$v_j(t) = \sum_{i=1}^{m_0} \int_{-\infty}^{t} h_{ji}(\lambda)x_i(t - \lambda)d\lambda + b_j \tag{13.14}$$

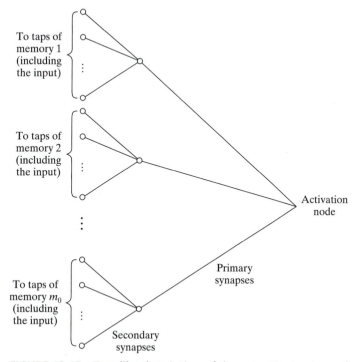

To taps of
memory 1
(including
the input)

To taps of
memory 2
(including
the input)

To taps of
memory m_0
(including
the input)

Activation
node

Primary
synapses

Secondary
synapses

FIGURE 13.15   Tree-like description of the synaptic structure of a multiple input neuronal filter.

The integral in Eq. (13.14) is the *convolution* of the continuous-time input signal $x_i(t)$ and the impulse response $h_{ij}(t)$ characterizing a linear continuous-time filter representing synapse i. Equation (13.14) is the most general way in which the spatio-temporal behavior of the induced local field of a neuron can be described.

Additive Model

Equation (13.14) provides the basis for another commonly used spatio-temporal model of a neuron. Specifically, suppose we simplify the temporal behavior of the neuron by using a scaling parameter to determine the sign and strength of a "typical" synaptic impulse response, in which case we write

$$h_{ji}(t) = w_{ji} \cdot h_j(t) \qquad \text{for all } i \tag{13.15}$$

where $h_j(t)$ models the temporal characteristics of a typical postsynaptic potential, and w_{ji} is a scalar that determines its sign (excitatory or inhibitory) and the overall strength of the connection between neuron j and input i (Shamma, 1989). Thus using Eq. (13.15) in (13.14) and by interchanging the order of integration and summation, we obtain

$$
\begin{aligned}
v_j(t) &= \int_{-\infty}^{t} h_j(\lambda) \left(\sum_{i=1}^{m_0} w_{ji} x_i(t - \lambda) \right) d\lambda + b_j \\
&= h_j(t) * \left(\sum_{i=1}^{m_0} w_{ji} x_i(t) \right) + b_j
\end{aligned}
\tag{13.16}
$$

where the asterisk denotes convolution. The form of the common impulse response $h_j(t)$ depends on the amount of detail required. A popular choice is an exponential function defined by

$$h_j(t) = \frac{1}{\tau_j} \exp\left(-\frac{t}{\tau_j}\right) \tag{13.17}$$

where τ_j is a *time constant* that is a characteristic of neuron j. The time function $h_j(t)$ of Eq. (13.17) is recognized as the impulse response of a simple circuit consisting of resistor R_j and capacitor C_j connected in parallel and fed from a current source; that is,

$$\tau_j = R_j C_j \tag{13.18}$$

Accordingly, we may use Eqs. (13.16) and (13.17) to formulate the model shown in Fig. 13.16. In physical terms, the synaptic weights $w_{j1}, w_{j2}, ..., w_{jm_0}$ are represented by conductances (i.e., reciprocals of resistances), and the respective inputs $x_1(t)$, $x_2(t), ..., x_{m_0}(t)$ are represented by potentials (i.e., voltages). The summing junction is characterized by a low input resistance, unity current gain, and high output resistance; that is, it acts as a summing node for incoming currents. The total current fed into the resistance-capacitance (RC) circuit is therefore

$$\sum_{i=1}^{m_0} w_{ji} x_i(t) + I_j$$

where the first (summation) term is due to the stimuli $x_1(t), x_2(t), ..., x_{m_0}(t)$ acting on the synaptic weights (conductances) $w_{j1}, w_{j2}, ..., w_{jm_0}$, respectively, and the second term is the current source I_j representing the externally applied bias b_j.

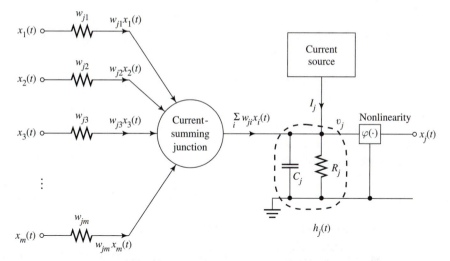

FIGURE 13.16   Additive model of a neuron.

In the neural network literature, the neuronal model shown in Fig. 13.16 is commonly referred to as the *additive model*. This model may be viewed as a lumped circuit approximation of the distributed transmission line model of a biological dendritic neuron (Rall, 1989). The low-pass nature of the RC circuit in Fig. 13.16 may also be justified by the fact that a biological synapse is itself a low-pass filter to an excellent approximation (Scott, 1977).

13.8   DISTRIBUTED TIME LAGGED FEEDFORWARD NETWORKS

The universal myopic mapping algorithm, which provides the mathematical justification for focused TLFNs, is limited to maps that are shift invariant. The implication of this limitation is that the use of focused TLFNs is only suitable for use in stationary (i.e., time-invariant) environments. We may overcome this limitation by using a *distributed time lagged feedforward network (TLFN)*, distributed in the sense that the implicit influence of time is distributed throughout the network. The construction of such a network is based on the multiple input neuronal filter of Fig. 13.14 as the spatio-temporal model of a neuron.

Let $w_{ji}(l)$ denote the weight connected to the lth tap of the FIR filter modeling the synapse that connects the output of neuron i to neuron j. The index l ranges from 0 to p, where p is the order of the FIR filter. According to this model, the signal $s_{ji}(n)$ appearing at the output of the ith synapse of neuron j is given by the *convolution sum*

$$s_{ji}(n) = \sum_{l=0}^{p} w_{ji}(l)x_i(n-l) \qquad (13.19)$$

where n denotes discrete time. We may rewrite Eq. (13.19) in matrix form by introducing the following definitions for the *state vector* and *weight vector* for synapse i, respectively:

$$\mathbf{x}_i(n) = [x_i(n), x_i(n-1), ..., x_i(n-p)]^T \qquad (13.20)$$

$$\mathbf{w}_{ji} = [w_{ji}(0), w_{ji}(1), ..., w_{ji}(p)]^T \qquad (13.21)$$

We may thus express the (scalar) signal $s_{ji}(n)$ as the inner product of the vectors $\mathbf{w}_{ji}(n)$ and $\mathbf{x}_i(n)$; that is,

$$s_{ji}(n) = \mathbf{w}_{ji}^T \mathbf{x}_i(n) \tag{13.22}$$

Equation (13.22) defines the output $s_{ji}(n)$ of the ith synapse of neuron j in the model of Fig. 13.14 in response to the input vector $\mathbf{x}_i(n)$ where $i = 1, 2, ..., m_0$. The vector $\mathbf{x}_i(n)$ is referred to as a "state" in that it represents the condition of the ith synapse at time n. Hence, summing the contributions of the complete set of m_0 synapses depicted in this model (i.e., summing over the index i), we may describe the output $y_j(n)$ of neuron j by the following pair of equations:

$$v_j(n) = \sum_{i=1}^{m_0} s_{ji}(n) + b_j = \sum_{i=1}^{m_0} \mathbf{w}_{ji}^T \mathbf{x}_i(n) + b_j \tag{13.23}$$

$$y_j(n) = \varphi(v_j(n)) \tag{13.24}$$

where $v_j(n)$ denotes the induced local field of neuron j, b_j is the externally applied bias, and $\varphi(\cdot)$ denotes the nonlinear activation function of the neuron. It is assumed that the same form of nonlinearity is used for all the neurons in the network. Note that if the weight vector $\mathbf{w}_{ji}$ and the state vector $\mathbf{x}_i(n)$ are replaced by the scalars w_{ji} and x_i, respectively, and if the operation of inner product is correspondingly replaced by ordinary multiplication, the dynamic model of a neuron described in Eqs. (13.23) and (13.24) reduces to the static model of the ordinary multilayer perceptron described in Chapter 4.

13.9 TEMPORAL BACK-PROPAGATION ALGORITHM

To train a distributed TLFN network, we need a supervised learning algorithm in which the actual response of each neuron in the output layer is compared with a desired (target) response at each time instant. Assume that neuron j lies in the output layer with its actual response denoted by $y_j(n)$ and that the desired response for this neuron is denoted by $d_j(n)$, both of which are measured at time n. We may then define an *instantaneous value* for the sum of squared errors produced by the network as follows:

$$\mathscr{E}(n) = \frac{1}{2} \sum_j e_j^2(n) \tag{13.25}$$

where the index j refers to a neuron in the output layer only, and $e_j(n)$ is the error signal defined by

$$e_j(n) = d_j(n) - y_j(n) \tag{13.26}$$

The goal is to minimize a *cost function* defined as the value of $\mathscr{E}(n)$ computed over all time:

$$\mathscr{E}_{\text{total}} = \sum_n \mathscr{E}(n) \tag{13.27}$$

The algorithm we have in mind for computing an estimate of the optimum weight vector that attains this goal is based on an approximation to the method of steepest descent.

An obvious way of proceeding with this development is to differentiate the cost function of Eq. (13.27) with respect to the weight vector $\mathbf{w}_{ji}$, and so write

$$\frac{\partial \mathcal{E}_{total}}{\partial \mathbf{w}_{ji}} = \sum_n \frac{\partial \mathcal{E}(n)}{\partial \mathbf{w}_{ji}} \tag{13.28}$$

To proceed further with the instantaneous gradient approach, we *unfold the network in time*. The strategy here is first to try to remove all the time delays in the network by expanding it into an equivalent but larger "static" network, and then to apply the standard back-propagation algorithm to compute the instantaneous error gradients. Unfortunately, such an approach is handicapped by the following negative attributes:

- A loss of a sense of symmetry between the forward propagation of states and the backward propagation of terms needed to calculate instantaneous error gradients
- No nice recursive formula for propagating the error terms
- Need for global bookkeeping to keep track of which static weights are actually the same in the equivalent network obtained by unfolding the distributed TLFN

Although using instantaneous gradient estimates is the obvious approach to developing a temporal version of back propagation, from a practical standpoint it is *not* the way to proceed.

To overcome the problems associated with the instantaneous gradient approach, we propose to proceed as follows (Wan, 1990, 1994). First, we recognize that the expansion of the total error gradient into a sum of instantaneous error gradients, as shown in Eq. (13.28), is not unique. In particular, we may consider an alternative way of expressing the partial derivative of the cost function $\mathcal{E}_{total}$ with respect to the weight vector $\mathbf{w}_{ji}(n)$ by writing

$$\frac{\partial \mathcal{E}_{total}}{\partial \mathbf{w}_{ji}} = \sum_n \frac{\partial \mathcal{E}_{total}}{\partial v_j(n)} \frac{\partial v_j(n)}{\partial \mathbf{w}_{ji}} \tag{13.29}$$

where the time index n runs only over $v_j(n)$. We may interpret the partial derivative $\partial \mathcal{E}_{total}/\partial v_j(n)$ as the change in the cost function $\mathcal{E}_{total}$ produced by a change in the induced local field v_j of neuron j at time n. However, it is important to note that

$$\frac{\partial \mathcal{E}_{total}}{\partial v_j(n)} \frac{\partial v_j(n)}{\partial \mathbf{w}_{ji}} \neq \frac{\partial \mathcal{E}(n)}{\partial \mathbf{w}_{ji}}$$

It is only when we take the sum over all n, as in Eqs. (13.28) and (13.29), that the equality holds.

Given the expansion of Eq. (13.29), we may now use the idea of gradient descent in weight space. In particular, we postulate a recursion for updating the tap-weight vector $\mathbf{w}_{ji}(n)$ as shown by

$$\mathbf{w}_{ji}(n + 1) = \mathbf{w}_{ji}(n) - \eta \frac{\partial \mathcal{E}_{total}}{\partial v_j(n)} \frac{\partial v_j(n)}{\partial \mathbf{w}_{ji}(n)} \tag{13.30}$$

where η is the *learning-rate parameter*. From the defining equation (13.23), we find that for any neuron j in the network, the partial derivative of the induced local field $v_j(n)$ with respect to the weight vector $\mathbf{w}_{ji}(n)$ is given by

$$\frac{\partial v_j(n)}{\partial \mathbf{w}_{ji}(n)} = \mathbf{x}_i(n) \tag{13.31}$$

where $\mathbf{x}_i(n)$ is the input vector applied to synapse i of neuron j. Moreover, we may define the *local gradient* for neuron j as

$$\delta_j(n) = -\frac{\partial \mathscr{E}_{\text{total}}}{\partial v_j(n)} \tag{13.32}$$

Accordingly, we may rewrite Eq. (13.30) in the familiar form

$$\mathbf{w}_{ji}(n+1) = \mathbf{w}_{ji}(n) + \eta \delta_j(n) \mathbf{x}_i(n) \tag{13.33}$$

As in the derivation of the standard back-propagation algorithm described in Chapter 4, the explicit form of the local gradient $\delta_j(n)$ depends on whether or not neuron j lies in the output layer or in a hidden layer of the network. These two different cases are considered next.

CASE 1. Neuron j is an output unit

For the output layer, we simply have

$$\begin{aligned}
\delta_j(n) &= \frac{\partial \mathscr{E}_{\text{total}}}{\partial v_j(n)} \\
&= -\frac{\partial \mathscr{E}(n)}{\partial v_j(n)} \\
&= e_j(n) \varphi'(v_j(n))
\end{aligned} \tag{13.34}$$

where $e_j(n)$ is the error signal measured at the output of neuron j, and $\varphi'(\cdot)$ is the derivative of the activation function $\varphi(\cdot)$ with respect to its argument.

CASE 2. Neuron j is a hidden unit

For neuron j located in a hidden layer, we define $\mathcal{A}$ as the set of all neurons whose inputs are fed by neuron j in a forward manner. Let $v_r(n)$ denote the induced local field of neuron r that belongs to the set $\mathcal{A}$. We may then write

$$\begin{aligned}
\delta_j(n) &= -\frac{\partial \mathscr{E}_{\text{total}}}{\partial v_j(n)} \\
&= -\sum_{r \in \mathcal{A}} \sum_k \frac{\partial \mathscr{E}_{\text{total}}}{\partial v_r(k)} \frac{\partial v_r(k)}{\partial v_j(n)}
\end{aligned} \tag{13.35}$$

where we have used the index k in place of n in those positions that are of particular concern. Using the definition of Eq. (13.32) (with index r used in place of j) in Eq. (13.35), we may thus write

$$\begin{aligned}
\delta_j(n) &= \sum_{r \in \mathcal{A}} \sum_n \delta_r(k) \frac{\partial v_r(k)}{\partial v_j(n)} \\
&= \sum_{r \in \mathcal{A}} \sum_n \delta_r(k) \frac{\partial v_r(k)}{\partial y_j(n)} \frac{\partial y_j(n)}{\partial v_j(n)}
\end{aligned} \tag{13.36}$$

where $y_j(n)$ is the output of neuron j. Next we recognize that the partial derivative $\partial y_j(n)/\partial v_j(n)$ is equal to $\varphi'(v_j(n))$, referring to neuron j that lies outside the set $\mathcal{A}$. We may therefore take this term outside the double summation and rewrite Eq. (13.36) as

$$\delta_j(n) = \varphi'(v_j(n)) \sum_{r \in \mathcal{A}} \sum_k \delta_r(k) \frac{\partial v_r(k)}{\partial y_j(n)} \tag{13.37}$$

As defined previously, $v_r(n)$ denotes the induced local field of neuron r fed by the output of neuron j. Hence, adapting the meaning of Eqs. (13.19) and (13.23) to the situation at hand, we may express $v_r(k)$ as

$$v_r(k) = \sum_{j=0}^{m_0} \sum_{l=0}^{p} w_{rj}(l) y_j(n - l) \tag{13.38}$$

In Eq. (13.38) we have included the bias b_r applied to neuron r as the term corresponding to $j = 0$ by defining

$$w_{r0}(l) = b_r \quad \text{and} \quad y_0(n - l) = 1 \qquad \text{for all } l \text{ and } n \tag{13.39}$$

The index p defining the upper limit of the inner summation in Eq. (13.38) is the order of each synaptic filter of neuron r, and every other neuron in the layer in question. The index m_0, defining the upper limit of the outer summation in Eq. (13.38), is the total number of primary synapses belonging to neuron r. Recognizing that the convolution sum with respect to l is commutative, we may rewrite Eq. (13.38) in the equivalent form

$$v_r(k) = \sum_{j=0}^{m_0} \sum_{l=0}^{p} y_j(l) w_{rj}(n - l) \tag{13.40}$$

Differentiating Eq. (13.40) with respect to y_j, we thus obtain

$$\frac{\partial v_r(k)}{\partial y_j(n)} = \begin{cases} w_{rj}(k - l), & n \le k \le n + p \\ 0, & \text{otherwise} \end{cases} \tag{13.41}$$

In light of Eq. (13.41), the partial derivatives $\partial v_r(k)/\partial y_j(n)$ in Eq. (13.37), for which n is outside the range $n \le k \le n + p$, evaluate to zero. For the case of a hidden neuron j, the use of Eq. (13.41) in (13.37) yields

$$\delta_j(n) = \varphi'(v_j(n)) \sum_{r \in \mathcal{A}} \sum_{k=n}^{n+p} \delta_r(k) w_{rj}(k - l)$$

$$= \varphi'(v_j(n)) \sum_{r \in \mathcal{A}} \sum_{l=0}^{p} \delta_r(n+l) w_{rj}(n) \tag{13.42}$$

Define a new $(p + 1)$-by-1 vector

$$\Delta_r(n) = [\delta_r(n), \delta_r(n + 1), ..., \delta_r(n + p)]^T \tag{13.43}$$

Earlier we defined the weight vector $\mathbf{w}_{ji}$ as in Eq. (13.21). By using matrix notation we may therefore rewrite Eq. (13.42) in the compact form

$$\delta_j(n) = \varphi'(v_j(n)) \sum_{r \in \mathcal{A}} \Delta_r^T(n) \mathbf{w}_{rj} \tag{13.44}$$

where $\mathbf{\Delta}_r^T(n)\mathbf{w}_{rj}$ is the inner product of the vectors $\mathbf{\Delta}_r(n)$ and $\mathbf{w}_{rj}$, both of which have dimension $(p + 1)$. Equation (13.44) completes the evaluation of $\delta_j(n)$ for neuron j in the hidden layer.

We are now ready to summarize the weight update equation for *temporal back propagation* as the following pair of relations (Wan, 1990, 1994):

$$\mathbf{w}_{ji}(n + 1) = \mathbf{w}_{ji}(n) + \eta\delta_j(n)\mathbf{x}_i(n) \qquad (13.45)$$

$$\delta_j(n) = \begin{cases} e_j(n)\varphi'(v_j(n)), & \text{neuron } j \text{ in the output layer} \\ \varphi'(v_j(n)) \sum_{r \in \mathcal{A}} \mathbf{\Delta}_r^T(n)\mathbf{w}_{rj}, & \text{neuron } j \text{ in a hidden layer} \end{cases} \qquad (13.46)$$

which may be readily generalized to any number of hidden layers. We immediately observe that these relations represent a *vector generalization* of the standard back-propagation algorithm. If we replace the input vector $\mathbf{x}_i(n)$, the weight vector $\mathbf{w}_{rj}$, and the local gradient vector $\mathbf{\Delta}_r$ by their scalar counterparts, the temporal back-propagation algorithm reduces to the standard form of the back-propagation algorithm derived in Chapter 4.

To compute $\delta_j(n)$ for neuron j located in a hidden layer, we *propagate* the δs from the next layer backward through those synaptic filters whose excitation is derived from neuron j, in accordance with Eq. (13.44). This backward-propagation mechanism is illustrated in Fig. 13.17. The local gradient $\delta_j(n)$ is thus formed not by simply taking a weighted sum but by backward filtering through each primary synapse. In particular, for each new set of input and desired response vectors, the forward filters are incremented one time step and the backward filters are incremented one time step.

We now see the practical benefits gained by using the temporal back-propagation algorithm described herein:

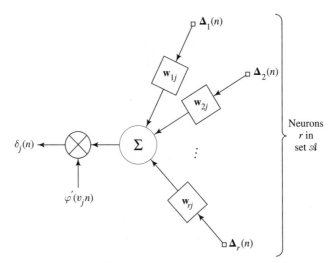

FIGURE 13.17 Back propagation of local gradients through a distributed TLFN.

1. The symmetry between the forward propagation of states and the backward propagation of error terms is preserved, and the sense of parallel distributed processing is thereby maintained.
2. Each unique weight of synaptic filter is used only once in the calculation of the δs; there is no redundant use of terms experienced in the instantaneous gradient method.

In deriving the temporal back-propagation algorithm[6] described in Eqs. (13.45) and (13.46), it is assumed that the synaptic filter weights are fixed for all gradient calculations. This is clearly not a valid assumption during actual adaptation. Accordingly, discrepancies in performance will arise between the temporal back-propagation algorithm and the temporal version obtained using the instantaneous gradient method. However, these discrepancies are usually of a minor nature. For a small learning-rate parameter η, the differences in the learning characteristics of these two algorithms are negligible for all practical purposes.

Causality Constraints

Careful examination of Eq. (13.42) reveals that the computation of $\delta_j(n)$ is *noncausal* because it requires knowledge of future values of the δs and the **w**s. To make this computation causal, we first note that the exact time reference used for adaptation is unimportant. Moreover, the synaptic structures employed in the network are all FIR filters. Accordingly, causality requires the use of additional buffering to store internal states of the network. In what follows we thus require that the adaptation of all weight vectors be based only on the current and past values of error signals. We may therefore immediately set up $\delta_j(n)$ for neuron j in the output layer and so adapt the synaptic filter weights in that layer. For the next layer back (i.e., one hidden layer back from the output layer), causality constraints imply that for neuron j in this layer the computation of the local gradient

$$\delta_j(n - p) = \varphi'(v_j(n - p)) \sum_{r \in \mathcal{A}} \mathbf{\Delta}_r^T(n - p)\mathbf{w}_{rj} \qquad (13.47)$$

is based only on current and past values of the vector $\mathbf{\Delta}_r$; that is,

$$\mathbf{\Delta}_r(n - p) = [\delta_r(n - p), \delta_r(n + 1 - p), ..., \delta_r(n)]^T \qquad (13.48)$$

Equation (13.47) is obtained from the second line of Eq. (13.46) simply by replacing the time index n by $n - p$, where p is the order of each synaptic FIR filter. As pointed out earlier, the states $\mathbf{x}_i(n - p)$ must be stored so that we may compute the product $\delta_j(n - p)\mathbf{x}_i(n - p)$ for the adaptation of the weight vector connecting neuron j in the last hidden layer to neuron i one layer farther back. For a network with multiple hidden layers, we may continue the operation described here for one more layer back (i.e., two layers back from the output layer) simply by making the time shift twice as long. The operation is continued in this fashion until all the computation layers in the network are accounted for.

We may formulate the *causal* form of the temporal back-propagation algorithm as summarized in Table 13.1.

TABLE 13.1 Summary of the Temporal Back-Propagation Algorithm

---

1. Propagate the input signal through the network in the forward direction, layer by layer. Determine the error signal $e_j(n)$ for neuron j in the output layer by subtracting its actual output from the corresponding desired response. Also record the state vector for each synapse in the network.
2. For neuron j in the output layer compute

$$\delta_j(n) = e_j(n)\varphi_j'(n)$$

$$\mathbf{w}_{ji}(n+1) = \mathbf{w}_{ji}(n) + \eta\delta_j(n)\mathbf{x}_i(n)$$

   where $\mathbf{x}_i(n)$ is the state of synapse i of a hidden neuron connected to output neuron j.
3. For neuron j in a hidden layer, compute

$$\delta_j(n - lp) = \varphi'(v_j(n - lp)) \sum_{r \in \mathcal{A}} \boldsymbol{\Delta}_r^T(n - lp)\mathbf{w}_{rj}$$

$$\mathbf{w}_{ji}(n+1) = \mathbf{w}_{ji}(n) + \eta\delta_j(n - lp)\mathbf{x}_i(n - lp)$$

   where p is the order of each synaptic FIR filter, and the index l identifies the hidden layer in question. Specifically, for networks with multiple hidden layers, $l = 1$ corresponds to one layer back from the output layer, $l = 2$ corresponds to two layers back from the output layer, and so on.

Although this algorithm is less aesthetically pleasing than the noncausal form described in Eqs. (13.45) and (13.46), basically the two forms of the algorithm differ from each other only in terms of a simple change of indices.

Summarizing, then, we may state the following:

- The δs are propagated through the layers of the network backward and continuously *without* added delays. This kind of propagation forces the internal values of the δs to be shifted in time.
- To correct for this time shift, the states (i.e., the values of $\mathbf{x}_i(n)$) are stored so as to form the proper product terms needed for adaptation of the weight vectors. In other words, added storage delays are required only for the states, whereas the backward propagation of the deltas is performed without delays.
- The backward propagation of the δs remains symmetric with respect to the forward propagation of the states.
- The order of computations is linear in the number of synaptic weights in the network as in the instantaneous gradient approach.

The distributed TLFN is naturally a more elaborate structure than the focused TLFN described in Section 13.4. Moreover, the temporal back-propagation algorithm needed to train the distributed TLFN is more computationally demanding than the standard back-propagation algorithm that is adequate to train the focused TLFN. In the final analysis, the use of one or the other of these two approaches is determined by whether the temporal processing task that needs to be solved pertains to a stationary environment or a nonstationary one.[7]

13.10 SUMMARY AND DISCUSSION

The need for temporal processing arises in numerous applications that include the following:

- *Prediction* and *modeling* of time series (Box and Jenkins, 1976; Haykin, 1996)
- *Noise cancelation*, where the requirement is to use a primary sensor (supplying a desired signal contaminated with noise) and a reference sensor (supplying a correlated version of the noise) to cancel the effect of noise (Widrow and Stearns, 1985; Haykin, 1996)
- *Adaptive equalization* of an unknown communication channel (Proakis, 1989; Haykin, 1996)
- *Adaptive control* (Narendra and Annaswamy, 1989)
- *System identification* (Ljung, 1987)

We already have well-developed theories for solving these problems when the system under study or the underlying physical mechanism of interest is linear; see the books cited above. However, when the system or physical mechanism is nonlinear, we have a more difficult task on our hands. It is in situations of this kind that neural networks have the potential to provide a viable solution and thereby make a significant difference in their application.

In the context of neural networks, we have two candidate networks for temporal processing:

- *Time lagged feedforward networks*
- *Recurrent networks*

Discussion of recurrent networks is taken up in the next two chapters. In this chapter we describe two classes of time lagged feedforward networks (TLFNs): *focused* and *distributed*. In a focused TLFN, the short-term memory is located entirely at the front end of a static network, which makes it straightforward to design. The training of the focused TLFN is accomplished by using the standard back-propagation algorithm, assuming that a multilayer perceptron is used for implementing the static neural network. In the *universal myopic mapping theorem* due to Sanberg and Xu (1997a, 1997b), we have an existence theorem in the sense that it provides the mathematical justification for the approximation of an arbitrary myopic map (i.e., a causal map with uniform-fading memory) by using a cascade of two functional blocks: a bank of linear filters and a static neural network. Such a structure may be implemented by using the focused TLFN, thereby providing a physical realization of this theorem.

The other class of TLFNs, that is, distributed TLFNs, relies on the use of a spatio-temporal model of a neuron, namely, a multiple-input neuronal filter. This model uses finite-duration impulse response (FIR) filters as synaptic filters. As such, the multiple input neuronal filter provides a powerful functional block in its own right for spatio-temporal signal processing built around a single neuron. To train it, we may use the least-mean-square (LMS) algorithm described in Chapter 3. However, to train a distributed TLFN, we require an elaborate learning algorithm exemplified by the temporal back-propagation algorithm described in Section 13.9. A distinctive feature of distributed TLFNs is the way in which the implicit representation of time is distributed throughout the network, hence the ability to cope with nonstationary (i.e., time-varying) environments. In contrast, in a focused TLFN, by definition, the implicit representation of time is concentrated at the front end of the network, which therefore limits its practical use to stationary (i.e., time-invariant) environments.

NOTES AND REFERENCES

1. For an essay on the role of time in neural processing, see the classic paper entitled "Finding Structure in Time" by Elman (1990).

2. In Hopfield (1995), a method for the explicit representation of time in neural processing is described. In particular, analog information is represented by using the timing of action potentials with respect to an ongoing collective oscillatory pattern of activity, for which neurobiological evidence is cited; action potentials are described in Chapter 1.

3. For a review of short-term memory structures and their role in temporal processing, see Mozer (1994).

4. For a discussion of hybrids of TDNN and HMM for speech recognition, see Bourlard and Morgan (1994), Katagiri and McDermott (1996), and Bengio (1996).

   Some TDNN-HMM hybrids combine the use of a TDNN frame coder (i.e., mapping "acoustic feature detector" into a "phoneme code") and an HMM word/sentence path finder (i.e., mapping "phoneme symbol" into "word/sentence class"), where the coder and path finder are designed separately. In some advanced TDNN-HMM hybrids the squared error loss function for the entire system is used so that a loss related to word/sentence error count can be minimized. An example of this latter scheme is the multi-state TDNN described in Haffner et al. (1991) and Haffner (1994). A simple hybrid of separately designed modules often causes a mismatch between the training and testing performances of the design. The multi-state TDNN performs better in this regard.

   In a fundamental sense, recurrent networks (discussed in Chapter 15) have a larger capability of modeling the temporal structure of speech signals than "replica" networks like the TDNN. However, since the nonstationarity and nonlinearity of speech signals are considerable, even recurrent networks may not be sufficient for accurate speech recognition by themselves.

5. For the origins of the universal myopic mapping theorem, see Sandberg (1991).

6. For an alternative diagrammatic derivation of the temporal back-propagation algorithm, see Wan and Beaufays (1996).

7. In Wan (1994), the temporal back-propagation algorithm was used to perform nonlinear prediction on a nonstationary time series exhibiting chaotic pulsations of an NH_3 laser. This particular time series was part of the Santa Fe Institute Time-Series Competition that was held in the United States in 1992. Wan's solution for this temporal processing task won the competition from a diverse list of submissions that included standard recurrent and feedforward neural networks as well as many traditional linear techniques (Wan, 1994). Chaos is discussed in Chapter 14.

PROBLEMS

Focused time lagged feedforward networks (TLFNs)

13.1 Summarize the main attributes of a focused TLFN used to model a nonlinear dynamical process.

13.2 The focused TLFN depicted in Fig. 13.10 uses a tapped delay line memory for implementing short-term memory. What are the benefits and shortcomings of a focused TLFN that uses a gamma memory for implementing the short-term memory?

13.3 In Chapter 2 we describe qualitatively a dynamic approach for implementing a nonlinear adaptive filter. The method involves the use of a static neural network whose stimulation is derived by feeding the input data through a sliding window. The window is moved on the arrival of each new data sample, with the oldest sample inside the window dropped to make room for the new sample. Discuss how a focused TLFN can be used to implement this form of continuous learning.

Spatio-temporal models of a neuron

13.4 Consider a neuronal filter whose induced local field $v_j(t)$ is defined by Eq. (13.16). Suppose that the time function $h_j(t)$ in this equation is replaced by the shifted unit impulse

$$h_j(t) = \delta(t - \tau_j)$$

where τ_j is a fixed delay. Describe the way in which the neuronal filter is changed by this modification.

13.5 Using the LMS algorithm, formulate a learning algorithm for the multiple-input neuronal filter of Fig. 13.14.

Temporal back-propagation

13.6 Figure P13.6 illustrates the use of a *Gaussian-shaped time window* as a method for temporal processing (Bodenhausen and Waibel, 1991). The time window associated with synapse i of neuron j is denoted by $\theta(n, \tau_{ji}, \sigma_{ji})$, where τ_{ji} and σ_{ji} are measures of *time delay* and *width* of the windows, respectively, as shown by

$$\theta(n, \tau_{ji}, \sigma_{ji}) = \frac{1}{\sqrt{2\pi}\,\sigma_{ji}} \exp\left(-\frac{1}{2\sigma_{ji}^2}(n - \tau_{ji})^2\right)$$

The output of neuron j is thus modeled as

$$y_j(n) = \varphi\left(\sum_{i=0}^{m_0} w_{ji} u_i(n)\right)$$

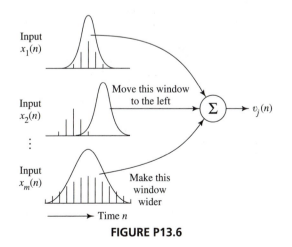

FIGURE P13.6

where $u_i(n)$ is the convolution of the input $x_i(n)$ and the time window $\theta(n, \tau_{ji}, \sigma_{ji})$. The weight w_{ji}, and time delay τ_{ji} of synapse i belonging to neuron j are all to be *learned* in a supervised manner.

This learning may be accomplished by using the standard back-propagation algorithm. Demonstrate this learning process by deriving update equations for w_{ji}, τ_{ji}, and σ_{ji}.

13.7 The material presented in Section 13.9 on the temporal back-propagation algorithm deals with synaptic FIR filters of equal length. How could you handle the case of synaptic FIR filters of different lengths?

13.8 Discuss how the temporal back-propagation algorithm may be used for the training of a distributed TLFN for single-step prediction.

13.9 The discrepancies between the constrained (causal) and unconstrained (noncausal) forms of the temporal back-propagation algorithm are analogous to the standard least-mean-square (LMS) versus delayed LMS algorithm; the LMS algorithm is discussed in Chapter 3. Expand on this analogy.

Computer experiment

13.10 In this problem we use the standard back-propagation algorithm to solve a difficult non-linear prediction problem and compare its performance with that of the LMS algorithm. The time series to be considered is created using a discrete *Volterra model* that has the form

$$x(n) = \sum_i g_i v(n - i) + \sum_i \sum_j g_{ij} v(n - i) v(n - j) + \cdots$$

where g_i, g_{ij}, ... are the Volterra coefficients, the $v(n)$ are samples of a white, independently distributed Gaussian noise sequence, and $x(n)$ is the resulting Volterra model output. The first summation term is the familiar moving average (MA) time series model, and the remaining summation terms are nonlinear components of ever-increasing order. In general, the estimation of the Volterra coefficients is considered to be difficult, primarily because of their nonlinear relationship to the data.

In this problem we consider the simple example

$$x(n) = v(n) + \beta v(n - 1) v(n - 2)$$

The time series has zero mean, is uncorrelated, and therefore has a white spectrum. However, the time series samples are not independent of each other, and therefore a higher-order predictor can be constructed. The variance of the model output is given by

$$\sigma_x^2 = \sigma_v^2 + \beta^2 \sigma_v^4$$

where σ_v^2 is the white noise variance.

(a) Construct a multilayer perceptron with an input layer of 6 nodes, a hidden layer of 16 neurons, and a single output neuron. A tapped delay line memory is used to feed the input layer of the network. The hidden neurons use sigmoid activation functions limited to the interval $[0, 1]$, whereas the output neuron operates as a linear combiner. The network is trained with the standard back-propagation algorithm having the following description:

Learning-rate parameter	$\eta = 0.001$
Momentum constant	$\alpha = 0.6$
Total number of samples processed	100,000
Number of samples per epoch	1,000
Total number of epochs	100

The white noise variance σ_v^2 is set equal to unity. Hence, with $\beta = 0.5$, we find that the output variance of the predictor is $\sigma_x^2 = 1.25$.

Compute the learning curve of the nonlinear predictor, with the variance of the predictor output $x(n)$ plotted as a function of the number of epochs of training samples up to 200 epochs. For the preparation of each epoch used to perform the training, explore the following two modes:

(i)   The time ordering of the training sample is maintained from one epoch to the next in exactly the same form as it is generated.

(ii) The ordering of the training sample is randomized from one pattern (state) to another.

Also, use cross-validation (described in Chapter 4) with a validation set of 1000 samples to monitor the learning behavior of the predictor.

(b) Repeat the experiment using the LMS algorithm designed to perform a linear prediction on an input of six samples. The learning-rate parameter of the algorithm is set equal to $\eta = 10^{-5}$.

(c) Repeat the entire experiment for $\beta = 1, \sigma_x^2 = 2$, and then for $\beta = 2, \sigma_x^2 = 5$.

The results of each experiment should reveal that initially the back-propagation algorithm and the LMS algorithm follow essentially a similar path, and then the back-propagation algorithm continues to improve, finally producing a prediction variance close to the prescribed value of σ_x^2.

CHAPTER 14

Neurodynamics

14.1 INTRODUCTION

In the previous chapter on temporal processing, we studied short-time memory structures and how to operate a static neural network (e.g., multilayer perceptron) as a dynamic mapper by stimulating it via a memory structure. Another important way in which time can be built into the operation of a neural network in an implicit manner is through the use of *feedback*. There are two basic ways of applying feedback to a neural network: *local feedback* at the level of a single neuron inside the network, and *global feedback* encompassing the whole network. Local feedback is a relatively simple matter to deal with, but global feedback has much more profound implications. In the neural network literature, neural networks with one or more feedback loops are referred to as *recurrent networks*. In this chapter and the next one, we focus attention on recurrent networks that use global feedback.

Feedback is like a two-edged sword, in that when it is applied improperly it can produce harmful effects. In particular, the application of feedback can cause a system that is originally stable to become unstable. Our primary interest in this chapter is in the stability of recurrent networks. Other aspects of recurrent networks are considered in the next chapter.

The subject of neural networks viewed as nonlinear dynamical systems, with particular emphasis on the *stability* problem, is referred to as *neurodynamics* (Hirsch, 1989). An important feature of the stability (or instability) of a nonlinear dynamical system is that it is a property of the whole system. As a corollary, *the presence of stability always implies some form of coordination between the individual parts of the system* (Ashby, 1960). It appears that the study of neurodynamics began in 1938 with the work of Nicholas Rashevsky, in whose visionary mind the application of dynamics to biology came into view for the first time.

The stability of a nonlinear dynamical system is a difficult issue to deal with. When we speak of the stability problem, those with an engineering background usually think in terms of the *bounded input–bounded output (BIBO) stability criterion*. According to this criterion, stability means that the output of a system must *not* grow without bound as a result of a bounded input, initial condition, or unwanted disturbance (Brogan, 1985). The BIBO stability criterion is well suited for a linear dynamical system. However, it is useless to apply it to neural networks because all such nonlinear dynamical systems are BIBO stable because of the saturating nonlinearity built into the constitution of a neuron.

When we speak of stability in the context of a nonlinear dynamical system, we usually mean *stability in the sense of Lyapunov*. In a celebrated mémoire dated 1892, Lyapunov (a Russian mathematician and engineer) presented the fundamental concepts of the stability theory known as the *direct method of Lyapunov*.[1] This method is widely used for the stability analysis of linear and nonlinear systems, both time-invariant and time-varying. As such it is directly applicable to the stability analysis of neural networks. Indeed, much of the material presented in this chapter is concerned with the direct method of Lyapunov. However, its application is no easy task.

The study of neurodynamics may follow one of two routes, depending on the application of interest:

- *Deterministic neurodynamics*, in which the neural network model has a deterministic behavior. In mathematical terms, it is described by a set of *nonlinear differential equations* that define the exact evolution of the model as a function of time (Grossberg, 1967; Cohen and Grossberg, 1983; Hopfield, 1984).
- *Statistical neurodynamics*, in which the neural network model is perturbed by the presence of noise. In this case, we have to deal with *stochastic nonlinear differential equations*, thereby expressing the solution in probabilistic terms (Amari et al., 1972; Peretto, 1984; Amari, 1990). The combination of stochasticity and nonlinearity makes the subject more difficult to handle.

In this chapter we restrict ourselves to deterministic neurodynamics.

Organization of the Chapter

The material in this chapter is organized in three parts. In the first part of the chapter, consisting of Sections 14.2 through 14.6, we provide introductory material. Section 14.2 introduces some fundamental concepts in dynamical systems, followed by a discussion of the stability of equilibrium points in Section 14.3. In Section 14.4 we describe various types of attractors that arise in the study of dynamical systems. In Section 14.5 we revisit the additive model of a neuron that was derived in Chapter 13. In Section 14.6 we discuss the manipulation of attractors as a neural network paradigm.

The second part of the chapter, consisting of Sections 14.7 through 14.11, deals with associative memories. Section 14.7 is devoted to a detailed discussion of Hopfield models and the use of discrete Hopfield models as a content-addressable memory. Section 14.8 presents a computer experiment on this application of the Hopfield network. In Section 14.9 we present the Cohen–Grossberg theorem for nonlinear dynamical systems that includes the Hopfield network and other associative memories as special cases. In Section 14.10, we describe another neurodynamical model known as the brain state-in-a-box model that is well suited for clustering. Section 14.11 presents a computer experiment on this second model.

The last part of the chapter, consisting of Sections 14.12 through 14.14, deals with the topic of chaos. Section 14.12 discusses the invariant characteristics of a chaotic process, followed by a discussion of the closely related topic of dynamic reconstruction of a chaotic process in Section 14.13. A computer experiment on dynamic reconstruction is presented in Section 14.14.

The chapter concludes with some final remarks in Section 14.15.

14.2 DYNAMICAL SYSTEMS

In order to proceed with the study of neurodynamics, we need a *mathematical model* for describing the dynamics of a nonlinear system. A model most naturally suited for this purpose is the *state-space model*. According to this model, we think in terms of a set of *state variables* whose values (at any particular instant of time) are supposed to contain sufficient information to predict the future evolution of the system. Let $x_1(t)$, $x_2(t), ..., x_N(t)$ denote the state variables of a nonlinear dynamical system, where continuous time t is the *independent variable* and N is the *order* of the system. For convenience of notation, these state variables are collected into an N-by-1 vector $\mathbf{x}(t)$ called the *state vector* of the system. The dynamics of a large class of nonlinear dynamical systems may then be cast in the form of a system of first-order differential equations written as follows:

$$\frac{d}{dt}x_j(t) = F_j(x_j(t)), \quad j = 1, 2, ..., N \tag{14.1}$$

where the function $F_j(\cdot)$ is, in general, a nonlinear function of its argument. We may put this system of equations in a compact form by using vector notation, as shown by

$$\frac{d}{dt}\mathbf{x}(t) = \mathbf{F}(\mathbf{x}(t)) \tag{14.2}$$

where the nonlinear function $\mathbf{F}$ is vector valued, each element of which operates on a corresponding element of the state vector:

$$\mathbf{x}(t) = [x_1(t), x_2(t), ..., x_N(t)]^T \tag{14.3}$$

A nonlinear dynamical system for which the vector function $\mathbf{F}(\mathbf{x}(t))$ does not depend *explicitly* on time t, as in Eq. (14.2), is said to be *autonomous*; otherwise, it is *nonautonomous*.[2] We will concern ourselves with autonomous systems only.

Regardless of the exact form of the nonlinear function $\mathbf{F}(\cdot)$, the state vector $\mathbf{x}(t)$ must vary with time t; otherwise, $\mathbf{x}(t)$ is constant and the system is no longer dynamic. We may therefore formally define a dynamical system as follows:

A dynamical system is a system whose state varies with time.

Moreover, we may think of $d\mathbf{x}/dt$ as a "velocity" vector, not in a physical but rather in an abstract sense. Then, according to Eq. (14.2), we may refer to the vector function $\mathbf{F}(\mathbf{x})$ as a velocity vector field or simply as a *vector field*.

State Space

It is informative to view the state-space equation (14.2) as describing the *motion* of a point in an N-dimensional *state space*. The state space can be a *Euclidean space* or a subset thereof. It can also be a non-Euclidean space such as a circle, a sphere, a torus, or some other *differentiable manifold*. Our interest, however, is confined to a Euclidean space.

The state space is important because it provides us with a visual/conceptual tool for analyzing the dynamics of a nonlinear system described by Eq. (14.2). It does so by focusing our attention on the *global characteristics* of the motion rather than the detailed aspects of analytic or numeric solutions of the equation.

At a particular instant of time t, the observed state of the system (i.e., the state vector $\mathbf{x}(t)$) is represented by a single point in the N-dimensional state space. Changes in the state of the system with time t are represented as a curve in the state space, with each point on the curve carrying (explicitly or implicitly) a label that records the time of observation. This curve is called a *trajectory* or *orbit* of the system. Figure 14.1 illustrates the trajectory of a two-dimensional system. The instantaneous velocity of the trajectory, (i.e., the velocity vector $d\mathbf{x}(t)/dt$) is represented by the *tangent vector*, shown as a dashed line in Fig. 14.1 for time $t = t_0$. We may thus derive a velocity vector for each point of the trajectory.

The family of trajectories, for different initial conditions, is referred to as the *state portrait* of the system. The state portrait includes *all* those points in the state space where the vector field $\mathbf{F}(\mathbf{x})$ is defined. Note that for an autonomous system there will be only one trajectory passing through an initial state. A useful idea that emerges from the state portrait is the *flow* of a dynamical system, defined as the motion of the space of states within itself. In other words, we may imagine the space of states to flow, just like a fluid, around in itself with each point (state) following a particular trajectory (Abraham and Shaw, 1992). The idea of flow as described here is vividly illustrated in the state portrait of Fig. 14.2.

Given a state portrait of a dynamical system, we may construct a field of velocity (tangent) vectors, one for every point of the state space. The picture so obtained in turn

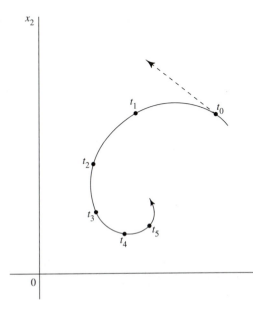

FIGURE 14.1   A two-dimensional trajectory (orbit) of a dynamical system.

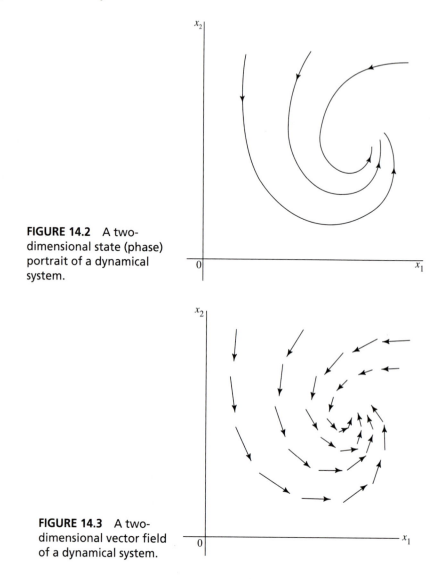

FIGURE 14.2 A two-dimensional state (phase) portrait of a dynamical system.

FIGURE 14.3 A two-dimensional vector field of a dynamical system.

provides a portrayal of the vector field of the system. In Fig. 14.3 we show a number of velocity vectors to develop a feeling for what a full field looks like. The usefulness of a vector field thus lies in the fact that it gives us a visual description of the inherent tendency of a dynamical system to move with a habitual velocity at each specific point of a state space.

Lipschitz Condition

For the state-space equation (14.2) to have a solution and for the solution to be unique, we must impose certain restrictions on the vector function $\mathbf{F}(\mathbf{x})$. For convenience of presentation, we have dropped dependence of the state vector $\mathbf{x}$ on time t, a practice that we follow from time to time. For a solution to exist, it is sufficient that $\mathbf{F}(\mathbf{x})$ be continuous in all of its arguments. However, this restriction by itself does not guarantee uniqueness of the solution. To do so, we must impose a further restriction known as the *Lipschitz condition*. Let $\|\mathbf{x}\|$ denote the *norm* or *Euclidean length* of the vector $\mathbf{x}$. Let $\mathbf{x}$ and $\mathbf{u}$ be a pair of vectors in an open set $\mathcal{M}$ in a normal vector (state) space. Then, according to the Lipschitz condition, there exists a constant K such that (Hirsch and Smale, 1974; E.A. Jackson, 1989)

$$\|\mathbf{F}(\mathbf{x}) - \mathbf{F}(\mathbf{u})\| \leq K\|\mathbf{x} - \mathbf{u}\| \tag{14.4}$$

for all $\mathbf{x}$ and $\mathbf{u}$ in $\mathcal{M}$. A vector function $\mathbf{F}(\mathbf{x})$ that satisfies Eq. (14.4) is said to be *Lipschitz*, and K is called the *Lipschitz constant* for $\mathbf{F}(\mathbf{x})$. Equation (14.4) also implies the continuity of the function $\mathbf{F}(\mathbf{x})$ with respect to $\mathbf{x}$. It follows, therefore, that in the case of autonomous systems, the Lipschitz condition guarantees both the existence and uniqueness of solutions for the state-space equation (14.2). In particular, if all partial derivatives $\partial F_i/\partial x_j$ are finite everywhere, then the function $\mathbf{F}(\mathbf{x})$ satisfies the Lipschitz condition.

Divergence Theorem

Consider a region of volume V and surface S in the state space of an autonomous system and assume a "flow" of points from this region. From our earlier discussion, we recognize that the velocity vector $d\mathbf{x}/dt$ is equal to the vector field $\mathbf{F}(\mathbf{x})$. Provided that the vector field $\mathbf{F}(\mathbf{x})$ within the volume V is "well behaved," we may apply the *divergence theorem* from vector calculus (Jackson, 1975). Let $\mathbf{n}$ denote a unit vector normal to the surface at dS pointing outward from the enclosed volume. Then, according to the divergence theorem, the relation

$$\int_S (\mathbf{F}(\mathbf{x}) \cdot \mathbf{n})dS = \int_V (\nabla \cdot \mathbf{F}(\mathbf{x}))dV \tag{14.5}$$

holds between the volume integral of the divergence of $\mathbf{F}(\mathbf{x})$ and the surface integral of the outwardly directed normal component of $\mathbf{F}(\mathbf{x})$. The quantity on the left-hand side of Eq. (14.5) is recognized as the net *flux* flowing out of the region surrounded by the closed surface S. If this quantity is zero, the system is *conservative*; if it is negative, the system is *dissipative*. In light of Eq. (14.5), we may state equivalently that if the divergence $\nabla \cdot \mathbf{F}(\mathbf{x})$ (which is a scalar) is zero the system is conservative, and if it is negative the system is dissipative.

14.3   STABILITY OF EQUILIBRIUM STATES

Consider an autonomous dynamical system described by the state-space equation (14.2). A constant vector $\bar{\mathbf{x}} \in \mathcal{M}$ is said to be an *equilibrium* (*stationary*) *state* of the system if the following condition is satisfied:

$$\mathbf{F}(\bar{\mathbf{x}}) = \mathbf{0} \tag{14.6}$$

where $\mathbf{0}$ is the null vector. The velocity vector $d\mathbf{x}/dt$ vanishes at the equilibrium state $\bar{\mathbf{x}}$, and therefore the constant function $\mathbf{x}(t) = \bar{\mathbf{x}}$ is a solution of Eq. (14.2). Furthermore, because of the uniqueness property of solutions, no other solution curve can pass through the equilibrium state $\bar{\mathbf{x}}$. The equilibrium state is also referred to as a *singular point*, signifying the fact that in the case of an equilibrium point the trajectory will degenerate into the point itself.

In order to develop a deeper understanding of the equilibrium condition, suppose that the nonlinear function $\mathbf{F}(\mathbf{x})$ is smooth enough for the state-space equation (14.2) to be linearized in the neighborhood of $\bar{\mathbf{x}}$. Specifically, let

$$\mathbf{x}(t) = \bar{\mathbf{x}} + \Delta\mathbf{x}(t) \tag{14.7}$$

where $\Delta\mathbf{x}(t)$ is a small deviation from $\bar{\mathbf{x}}$. Then, retaining the first two terms in the Taylor series expansion of $\mathbf{F}(\mathbf{x})$, we may approximate it as follows

$$\mathbf{F}(\mathbf{x}) \simeq \bar{\mathbf{x}} + \mathbf{A}\,\Delta\mathbf{x}(t) \tag{14.8}$$

The matrix $\mathbf{A}$ is the *Jacobian* of the nonlinear function $\mathbf{F}(\mathbf{x})$, evaluated at the point $\mathbf{x} = \bar{\mathbf{x}}$, as shown by

$$\mathbf{A} = \frac{\partial}{\partial\mathbf{x}}\mathbf{F}(\mathbf{x})\Big|_{\mathbf{x}=\bar{\mathbf{x}}} \tag{14.9}$$

By substituting Eqs. (14.7) and (14.8) in (14.2), and then using the definition of an equilibrium state, we get

$$\frac{d}{dt}\Delta\mathbf{x}(t) \simeq \mathbf{A}\,\Delta\mathbf{x}(t) \tag{14.10}$$

Provided that the Jacobian $\mathbf{A}$ is nonsingular, that is, the inverse matrix $\mathbf{A}^{-1}$ exists, the approximation described in Eq. (14.10) is sufficient to determine the *local* behavior of the trajectories of the system in the neighborhood of the equilibrium state $\bar{\mathbf{x}}$. If $\mathbf{A}$ is nonsingular, the nature of the equilibrium state is essentially determined by its *eigenvalues*, and may therefore be classified in a corresponding fashion. In particular, when the Jacobian matrix $\mathbf{A}$ has m eigenvalues with positive real parts, we say that the equilibrium state $\bar{\mathbf{x}}$ is of *type m*.

For the special case of a *second-order system*, we may classify the equilibrium state as summarized in Table 14.1 and illustrated in Fig. 14.4 (Cook, 1986; Arrowsmith and Place, 1990). Without loss of generality, the equilibrium state is assumed to be at the origin of the state space, that is, $\mathbf{x} = \mathbf{0}$. Note also that in the case of a *saddle point*,

---

TABLE 14.1 Classification of the Equilibrium State of a Second-Order System

Type of Equilibrium State $\bar{\mathbf{x}}$	Eigenvalues of the Jacobian Matrix $\mathbf{A}$
Stable node	Real and negative
Stable focus	Complex conjugate with negative real parts
Unstable node	Real and positive
Unstable focus	Complex conjugate with positive real parts
Saddle point	Real with opposite signs
Center	Conjugate purely imaginary

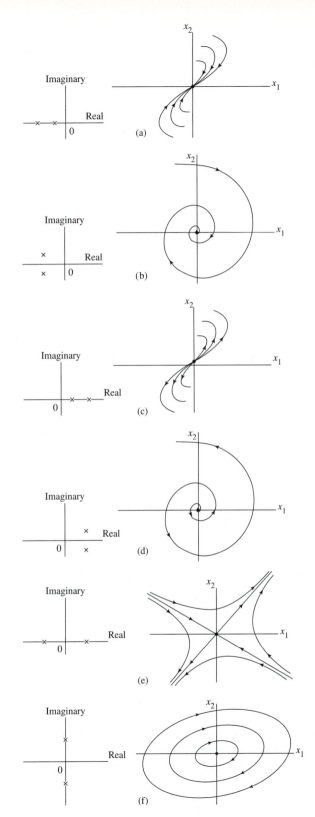

FIGURE 14.4 (a) Stable node. (b) Stable focus. (c) Unstable node. (d) Unstable focus. (e) Saddle point. (f) Center.

shown in Fig. 14.4e, the trajectories going to the saddle point are stable, whereas the trajectories coming from the saddle point are unstable.

Definitions of Stability

Linearization of the state-space equation, as outlined, provides useful information about the *local* stability properties of an equilibrium state. However, for us to be able to investigate the stability of a nonlinear dynamical system in a more detailed fashion, we need precise definitions of the stability and convergence of an equilibrium state.

In the context of an autonomous nonlinear dynamical system with equilibrium state $\bar{\mathbf{x}}$, the definitions of stability and convergence are as follows (Cook, 1986):

DEFINITION 1. The equilibrium state $\bar{\mathbf{x}}$ is said to be uniformly stable if for any given positive ϵ there exists a positive δ such that the condition

$$\|\mathbf{x}(0) - \bar{\mathbf{x}}\| < \delta$$

implies

$$\|\mathbf{x}(t) - \bar{\mathbf{x}}\| < \epsilon$$

for all $t > 0$.

This definition states that a trajectory of the system can be made to stay within a small neighborhood of the equilibrium state $\bar{\mathbf{x}}$ if the initial state $\mathbf{x}(0)$ is close to $\bar{\mathbf{x}}$.

DEFINITION 2. The equilibrium state $\bar{\mathbf{x}}$ is said to be convergent if there exists a positive δ such that the condition

$$\|\mathbf{x}(0) - \bar{\mathbf{x}}\| < \delta$$

implies that

$$\mathbf{x}(t) \to \bar{\mathbf{x}} \quad \text{as } t \to \infty$$

The meaning of this second definition is that if the initial state $\mathbf{x}(0)$ of a trajectory is close enough to the equilibrium state $\bar{\mathbf{x}}$, then the trajectory described by the state vector $\mathbf{x}(t)$ will approach $\bar{\mathbf{x}}$ as time t approaches infinity.

DEFINITION 3. The equilibrium state $\bar{\mathbf{x}}$ is said to be asymptotically stable if it is both stable and convergent.

Here we note that stability and convergence are independent properties. It is only when both properties are satisfied that we have asymptotic stability.

DEFINITION 4. The equilibrium state $\bar{\mathbf{x}}$ is said to be asymptotically stable or globally asymptotically stable if it is stable and all trajectories of the system converge to $\bar{\mathbf{x}}$ as time t approaches infinity.

This definition implies that the system cannot have other equilibrium states, and it requires that every trajectory of the system remain bounded for all time $t > 0$. In other words, global asymptotic stability implies that the system will ultimately settle down to a steady state for any choice of initial conditions.

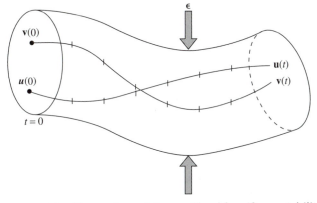

FIGURE 14.5 Illustration of the notion of uniform stability (convergence) of a state vector.

Example 14.1

Let a solution $\mathbf{u}(t)$ of the nonlinear dynamical system described by Eq. (14.2) vary with time t as indicated in Fig. 14.5. For the solution $\mathbf{u}(t)$ to be uniformly stable, we require that $\mathbf{u}(t)$ and any other solution $\mathbf{v}(t)$ remain close to each other for the same values of t (i.e., time "ticks"), as illustrated in Fig. 14.5. This kind of behavior is referred to as an *isochronous correspondence* of the two solutions $\mathbf{v}(t)$ and $\mathbf{u}(t)$ (E.A. Jackson, 1989). The solution $\mathbf{u}(t)$ is convergent provided that, for every other solution $\mathbf{v}(t)$ for which $\|\mathbf{v}(0) - \mathbf{u}(0)\| \leq \delta(\epsilon)$ at time $t = 0$, the solutions $\mathbf{v}(t)$ and $\mathbf{u}(t)$ converge to an equilibrium state as t approaches infinity.

∎

Lyapunov's Theorems

Having defined stability and asymptotic stability of an equilibrium state of a dynamical system, the next issue to be considered is that of determining stability. We may obviously do so by actually finding all possible solutions to the state-space equation of the system; however, such an approach is often difficult if not impossible. A more elegant approach is to be found in *modern stability theory*, founded by Lyapunov. Specifically, we may investigate the stability problem by applying the *direct method of Lyapunov*, which makes use of a continuous scalar function of the state vector, called a Lyapunov function.

Lyapunov's theorems on the stability and asymptotic stability of the state-space equation (14.2) describing an autonomous nonlinear dynamical system with state vector $\mathbf{x}(t)$ and equilibrium state $\bar{\mathbf{x}}$ may be stated as follows:

THEOREM 1. The equilibrium state $\bar{\mathbf{x}}$ is stable if in a small neighborhood of $\bar{\mathbf{x}}$ there exists a positive definite function $V(\mathbf{x})$ such that its derivative with respect to time is negative semidefinite in that region.

THEOREM 2. The equilibrium state $\bar{\mathbf{x}}$ is asymptotically stable if in a small neighborhood of $\bar{\mathbf{x}}$ there exists a positive definite function $V(\mathbf{x})$ such that its derivative with respect to time is negative definite in that region.

A scalar function $V(\mathbf{x})$ that satisfies these requirements is called a *Lyapunov function* for the equilibrium state $\bar{\mathbf{x}}$.

These theorems require the Lyapunov function $V(\mathbf{x})$ to be a positive definite function. Such a function is defined as: The function $V(\mathbf{x})$ is *positive definite* in the state space $\mathscr{L}$ if, for all $\mathbf{x}$ in $\mathscr{L}$, it satisfies the following requirements:

1. The function $V(\mathbf{x})$ has continuous partial derivatives with respect to the elements of the state vector $\mathbf{x}$
2. $V(\bar{\mathbf{x}}) = 0$
3. $V(\mathbf{x}) > 0$ if $\mathbf{x} \neq \bar{\mathbf{x}}$

Given that $V(\mathbf{x})$ is a Lyapunov function, according to Theorem 1 the equilibrium state $\bar{\mathbf{x}}$ is stable if

$$\frac{d}{dt}V(\mathbf{x}) \leq 0 \quad \text{for } \mathbf{x} \in \mathscr{U} - \bar{\mathbf{x}} \tag{14.11}$$

where $\mathscr{U}$ is a small neighborhood around $\bar{\mathbf{x}}$. Furthermore, according to Theorem 2, the equilibrium state $\bar{\mathbf{x}}$ is asymptotically stable if

$$\frac{d}{dt}V(\mathbf{x}) < 0 \quad \text{for } \mathbf{x} \in \mathscr{U} - \bar{\mathbf{x}} \tag{14.12}$$

The important point of this discussion is that Lyapunov's theorems can be applied without having to solve the state-space equation of the system. Unfortunately, the theorems give no indication of how to find a Lyapunov function; it is a matter of ingenuity and trial and error in each case. In many problems of interest, the energy function can serve as a Lyapunov function. The inability to find a suitable Lyapunov function does not, however, prove instability of the system. The existence of a Lyapunov function is sufficient but not necessary for stability.

The Lyapunov function $V(\mathbf{x})$ provides the mathematical basis for the *global* stability analysis of the nonlinear dynamical system described by Eq. (14.2). On the other hand, the use of Eq. (14.10) based on the Jacobian matrix $\mathbf{A}$ provides the basis for the *local* stability analysis of the system. The global stability analysis is much more powerful in its conclusions than local stability analysis; that is, every globally stable system is also locally stable, but not vice versa.

14.4 ATTRACTORS

Dissipative systems are generally characterized by the presence of attracting sets or manifolds of dimensionality lower than that of the state space. By a "manifold" we mean a k-dimensional surface embedded in the N-dimensional state space, which is defined by a set of equations:

$$M_j(x_1, x_2, \ldots, x_N) = 0, \quad \begin{cases} j = 1, 2, \ldots, k \\ k < N \end{cases} \tag{14.13}$$

where $x_1, x_2, \ldots, x_N$ are elements of the N-dimensional state vector of the system, and M_j is some function of these elements. These manifolds are called *attractors*[3] in that they are bounded subsets to which regions of initial conditions of nonzero state space volume converge as time t increases (Ott, 1993).

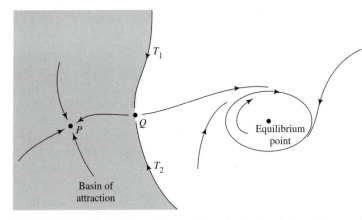

FIGURE 14.6   Illustration of the notion of a basin of attraction, and the idea of a separatrix.

The manifold may consist of a single point in the state space, in which case we speak of a *point attractor*. Alternatively, it may be in the form of a periodic orbit, in which case we speak of a stable *limit cycle*, stable in the sense that nearby trajectories approach it asymptotically. Figure 14.6 illustrates these two types of attractors. Attractors represent the only *equilibrium states* of a dynamical system that may be *observed experimentally*. Note, however, that in the context of attractors an equilibrium state does *not* imply a static equilibrium, nor a steady state. For example, a limit cycle represents a stable state of an attractor, but it varies continuously with time.

In Fig. 14.6 we note that each attractor is encompassed by a distinct region of its own. Such a region is called a *basin (domain) of attraction*. Note also that every initial state of the system is in the basin of some attractor. The boundary separating one basin of attraction from another is called a *separatrix*. In the case of Fig. 14.6, the basin boundary is represented by the union of the trajectory T_1, the saddle point Q, and the trajectory T_2.

A limit cycle constitutes the typical form of an oscillatory behavior that arises when an equilibrium point of a nonlinear system becomes unstable. As such, it can arise in nonlinear systems of any order. Nevertheless, limit cycles are particularly characteristic of second-order systems.

Hyperbolic Attractors

Consider a point attractor whose nonlinear dynamical equations are linearized around the equilibrium state $\bar{x}$ in the manner described in Section 14.2. Let **A** denote the Jacobian matrix of the system evaluated at $x = \bar{x}$. The attractor is said to be a *hyperbolic attractor* if the eigenvalues of the Jacobian matrix **A** all have an absolute value less than 1 (Ott, 1993). For example, the flow of a second-order hyperbolic attractor may have the form shown in Fig. 14.4a or that of Fig. 14.4b; in both cases the eigenvalues of the Jacobian matrix **A** have negative real parts. Hyperbolic attractors are of particular interest in the study of a problem known as the vanishing gradients problem that arises in dynamically driven recurrent networks; this problem is discussed in the next chapter.

14.5 NEURODYNAMICAL MODELS

Having familiarized ourselves with the behavior of nonlinear dynamical systems, we are now ready to discuss some of the important issues involved in neurodynamics, which we do in this and the following sections. We emphasize that there is no universally agreed upon definition of what we mean by neurodynamics. Rather than try to present such a definition, we will instead define the most general properties of the neurodynamical systems considered in this chapter. In particular, the discussion is limited to neurodynamical systems whose state variables are continuous-valued, and whose equations of motion are described by differential equations or difference equations. The systems of interest possess four general characteristics (Peretto and Niez, 1986; Pineda, 1988a):

1. *A large number of degrees of freedom.* The human cortex is a *highly parallel, distributed system* that is estimated to possess about 10 billion neurons, with each neuron modeled by one or more state variables. It is generally believed that both the computational power and the fault-tolerant capability of such a neurodynamical system are the result of the collective dynamics of the system. The system is characterized by a very large number of coupling constants represented by the strengths (efficacies) of the individual synaptic junctions.
2. *Nonlinearity.* A neurodynamical system is nonlinear. In fact, nonlinearity is essential to create a universal computing machine.
3. *Dissipation.* A neurodynamical system is dissipative. It is therefore characterized by the convergence of the state-space volume onto a manifold of lower dimensionality as time goes on.
4. *Noise.* Finally, noise is an intrinsic characteristic of neurodynamical systems. In real-life neurons, membrane noise is generated at synaptic junctions (Katz, 1966).

The presence of noise necessitates the use of a probabilistic treatment of neural activity, adding another level of complexity to the analysis of neurodynamical systems. A detailed treatment of stochastic neurodynamics is beyond the scope of this book. The effect of noise is therefore ignored in the material that follows.

Additive Model

Consider the noiseless, dynamical model of a neuron shown in Fig. 14.7, the mathematical basis of which was discussed in Chapter 13. In physical terms, the synaptic weights $w_{j1}, w_{j2}, ..., w_{jN}$ represent *conductances*, and the respective inputs $x_1(t), x_2(t), ..., x_N(t)$ represent *potentials*; N is the number of inputs. These inputs are applied to a *current-summing junction* characterized as follows:

- Low input resistance
- Unity current gain
- High output resistance

It thus acts as a summing node for the input currents. The total current flowing *toward* the input node of the nonlinear element (activation function) in Fig. 14.7 is therefore

$$\sum_{i=1}^{N} w_{ji} x_i(t) + I_j$$

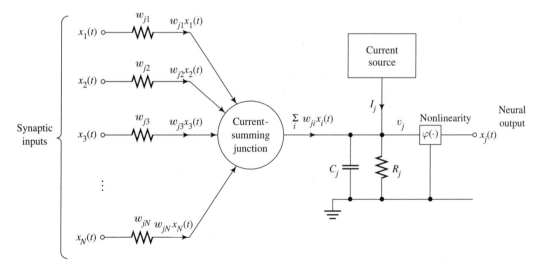

FIGURE 14.7   Additive model of a neuron.

where the first (summation) term is due to the stimuli $x_1(t), x_2(t), ..., x_N(t)$ acting on the synaptic weights (conductances) $w_{j1}, w_{j2}, ..., w_{jN}$, respectively, and the second term is due to the current source I_j representing an externally applied bias. Let $v_j(t)$ denote the induced local field at the input of the nonlinear activation function $\varphi(\cdot)$. We may then express the total current flowing *away* from the input node of the nonlinear element as follows:

$$\frac{v_j(t)}{R_j} + C_j \frac{dv_j(t)}{dt}$$

where the first term is due to leakage resistance R_j and the second term is due to leakage capacitance C_j. From *Kirchoff's current law*, we know that the total current flowing toward any node of an electrical circuit is zero. By applying Kirchoff's current law to the input node of the nonlinearity in Fig. 14.7, we get

$$C_j \frac{dv_j(t)}{dt} + \frac{v_j(t)}{R_j} = \sum_{i=1}^{N} w_{ji} x_i(t) + I_j \tag{14.14}$$

The capacitive term $C_j dv_j(t)/dt$ on the left-hand side of Eq. (14.14) is the simplest way to add dynamics (memory) to the model of a neuron. Given the induced local field $v_j(t)$, we may determine the output of neuron j by using the nonlinear relation

$$x_j(t) = \varphi(v_j(t)) \tag{14.15}$$

The *RC* model described by Eq. (14.14) is commonly referred to as the *additive model*; this terminology is used to discriminate the model from multiplicative (or shunting) models where w_{ji} is dependent on x_i (Grossberg, 1982).

A characteristic feature of the additive model described by Eq. (14.14) is that the signal $x_i(t)$ applied to neuron j by adjoining neuron i is a slowly varying function of time t. The model thus described constitutes the basis of *classical neurodynamics.*[4]

To proceed, consider a *recurrent network* consisting of an interconnection of N neurons, each one of which is assumed to have the same mathematical model described in Eqs. (14.14) and (14.15). Then, ignoring interneuron propagation time delays, we may define the dynamics of the network by the following *system of coupled first-order differential equations*:

$$C_j \frac{dv_j(t)}{dt} = -\frac{v_j(t)}{R_j} + \sum_{i=1}^{N} w_{ji} x_i(t) + I_j, \quad j = 1, 2, ..., N \tag{14.16}$$

which has the same mathematical form as the state equations (14.1), and which follows from a simple rearrangement of terms in Eq. (14.14). It is assumed that the activation function $\varphi(\cdot)$ relating the output $x_j(t)$ of neuron j to its induced local field $v_j(t)$ is a continuous function and therefore differentiable. A commonly used activation function is the logistic function

$$\varphi(v_j) = \frac{1}{1 + \exp(-v_j)}, \quad j = 1, 2, ..., N \tag{14.17}$$

A necessary condition for the learning algorithms described in Sections 14.6 through 14.11 to exist is that the recurrent network described by Eqs. (14.15) and (14.16) possesses fixed points (i.e., point attractors).

Related Model

To simplify the exposition, we assume that the time constant $\tau_j = R_j C_j$ of neuron j in Eq. (14.16) is the same for all j. Then, by normalizing time t with respect to the common value of this time constant, and normalizing the w_{ji} and I_j with respect to the R_j, we may recast the model of Eq. (14.16) as follows:

$$\frac{dv_j(t)}{dt} = -v_j(t) + \sum_i w_{ji} \varphi(v_i(t)) + I_j, \quad j = 1, 2, ..., N \tag{14.18}$$

where we have also incorporated Eq. (14.15). The attractor structure of the system of coupled first-order nonlinear differential equations (14.18) is basically the same as that of a closely related model described by (Pineda, 1987):

$$\frac{dx_j(t)}{dt} = -x_j(t) + \varphi\left(\sum_i w_{ji} x_i(t)\right) + K_j, \quad j = 1, 2, ..., N \tag{14.19}$$

In the additive model described by Eq. (14.18), the induced local fields $v_1(t), v_2(t), ..., v_N(t)$ of the individual neurons constitute the state vector. On the other hand, in the related model of Eq. (14.19), the outputs of the neurons $x_1(t), x_2(t), ..., x_N(t)$ constitute the state vector.

These two neurodynamical models are in fact related to each other by a linear, invertible transformation. Specifically, by multiplying both sides of Eq. (14.19) by w_{kj}, summing with respect to j, and then substituting the transformation

$$v_k(t) = \sum_j w_{kj} x_j(t)$$

we obtain a model of the type described by Eq. (14.18), and so find that the bias terms of the two models are related by

$$I_k = \sum_j w_{kj} K_j$$

The important point to note here is that results concerning the stability of the additive model of Eq. (14.18) are applicable to the related model of Eq. (14.19).

The close relationship between the two neurodynamical models described here is also illustrated in the block diagrams shown in Fig. 14.8. Parts a and b of this figure correspond to the matrix formulations of Eqs. (14.18) and (14.19), respectively; $\mathbf{W}$ is the matrix of synaptic weights, $\mathbf{v}(t)$ is the vector of induced local fields at time t, and $\mathbf{x}(t)$ is

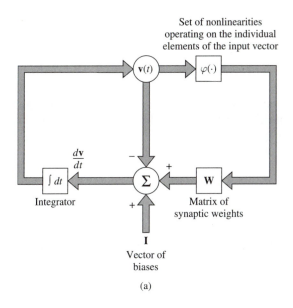

Set of nonlinearities
operating on the individual
elements of the input vector

(a)

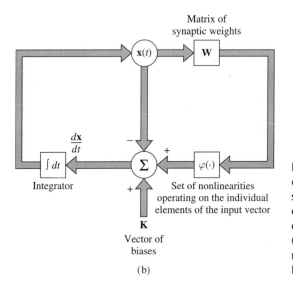

Matrix of
synaptic weights

Set of nonlinearities
operating on the individual
elements of the input vector

(b)

FIGURE 14.8 (a) Block diagram of a neurodynamical system represented by the coupled, first-order differential equations (14.18). (b) Block diagram of related model described by Eqs. (14.19).

the vector of neuronal outputs at time t. The presence of *feedback* in both models is clearly visible in Fig. 14.8.

14.6 MANIPULATION OF ATTRACTORS AS A RECURRENT NETWORK PARADIGM

When the number of neurons, N, is very large, the neurodynamical model described by Eq. (14.16) possesses, except for the effect of noise, the general properties outlined earlier in Section 14.5: very many degrees of freedom, nonlinearity, and dissipation. Accordingly, such a neurodynamical model can have complicated attractor structures and therefore exhibit useful computational capabilities.

The identification of attractors with computational objects (e.g., associative memories, input–output mappers) is one of the foundations of neural network paradigms. In order to implement this idea, we must exercise *control* over the locations of the attractors in the state space of the system. A learning algorithm then takes the form of a nonlinear dynamical equation that manipulates the locations of the attractors for the purpose of encoding information in a desired form, or learning temporal structures of interest. In this way, it is possible to establish an intimate relationship between the physics of the machine and the algorithms of the computation.

One way in which the collective properties of a neural network may be used to implement a computational task is by way of the concept of *energy minimization*. The Hopfield network and the brain-state-in-a-box model, to be considered in Sections 14.7 and 14.10, respectively, are well-known examples of such an approach. Both of these models are energy-minimizing networks; they differ from each other in their areas of application. The Hopfield network is useful as a content addressable memory or an analog computer for solving combinatorial-type optimization problems. The brain-state-in-a-box model, on the other hand, is useful for clustering types of applications. More will be said about these applications in subsequent sections of the chapter.

The Hopfield network and brain-state-in-a-box model are examples of an associative memory with no hidden neurons: An associative memory is an important resource for intelligent behavior. Another neurodynamical model is that of an input–output mapper, the operation of which relies on the availability of hidden neurons. In this latter case, the method of steepest descent is often used to minimize a cost function defined in terms of the network parameters, and thereby to change the attractor locations. This latter application of a neurodynamical model is exemplified by the dynamically driven recurrent networks discussed in the next chapter.

14.7 HOPFIELD MODEL

The *Hopfield network (model)* consists of a set of neurons and a corresponding set of unit delays, forming a *multiple-loop feedback system,* as illustrated in Fig. 14.9. The number of feedback loops is equal to the number of neurons. Basically, the output of each neuron is fed back, via a unit delay element, to each of the other neurons in the network. In other words, there is *no* self-feedback in the network; the reason for avoiding the use of self-feedback is explained later.

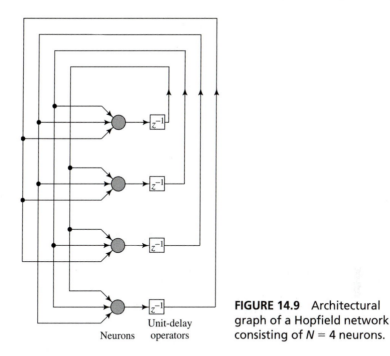

Neurons   Unit-delay
operators

FIGURE 14.9 Architectural graph of a Hopfield network consisting of $N = 4$ neurons.

To study the dynamics of the Hopfield network, we use the neurodynamical model described in Eq. (14.16), which is based on the additive model of a neuron. Recognizing that $x_i(t) = \varphi_i(v_i(t))$, we may rewrite Eq. (14.16) in the form

$$C_j \frac{d}{dt} v_j(t) = -\frac{v_j(t)}{R_j} + \sum_{i=1}^{N} w_{ji} \varphi_i(v_i(t)) + I_j, \quad j = 1, \dots, N \qquad (14.20)$$

To proceed with the discussion, we make the following assumptions:

1. The matrix of synaptic weights is *symmetric*, as shown by

$$w_{ji} = w_{ij} \quad \text{for all } i \text{ and } j \qquad (14.21)$$

2. Each neuron has a *nonlinear* activation of its own—hence the use of $\varphi_i(\cdot)$ in Eq. (14.20).
3. The *inverse* of the nonlinear activation function exists, so we may write

$$v = \varphi_i^{-1}(x) \qquad (14.22)$$

Let the sigmoid function $\varphi_i(v)$ be defined by the hyperbolic tangent function

$$x = \varphi_i(v) = \tanh\left(\frac{a_i v}{2}\right) = \frac{1 - \exp(-a_i v)}{1 + \exp(-a_i v)} \qquad (14.23)$$

which has a slope of $a_i/2$ at the origin as shown by

$$\frac{a_i}{2} = \frac{d\varphi_i}{dv}\Bigg|_{v=0} \qquad (14.24)$$

Henceforth we refer to a_i as the *gain* of neuron i.

The inverse output–input relation of Eq. (14.22) may thus be rewritten in the form

$$v = \varphi_i^{-1}(x) = -\frac{1}{a_i} \log\left(\frac{1-x}{1+x}\right) \tag{14.25}$$

The *standard* form of the inverse output–input relation for a neuron of unity gain is defined by

$$\varphi^{-1}(x) = -\log\left(\frac{1-x}{1+x}\right) \tag{14.26}$$

We may rewrite Eq. (14.25) in terms of this standard relation as

$$\varphi_i^{-1}(x) = \frac{1}{a_i}\varphi^{-1}(x) \tag{14.27}$$

Figure 14.10a shows a plot of the standard sigmoidal nonlinearity $\varphi(v)$, and Fig. 14.10b shows the corresponding plot of the inverse nonlinearity $\varphi^{-1}(x)$.

The energy (Lyapunov) function of the Hopfield network in Fig. 14.9 is defined by (Hopfield, 1984)

$$E = -\frac{1}{2}\sum_{i=1}^{N}\sum_{j=1}^{N} w_{ji}x_ix_j + \sum_{j=1}^{N}\frac{1}{R_j}\int_{0}^{x_j}\varphi_j^{-1}(x)\,dx - \sum_{j=1}^{N} I_jx_j \tag{14.28}$$

The energy function E defined by Eq. (14.28) may have a complicated *landscape* with many minima. The dynamics of the network are described by a mechanism that seeks out those minima.

Hence, differentiating E with respect to time, we get

$$\frac{dE}{dt} = -\sum_{j=1}^{N}\left(\sum_{i=1}^{N} w_{ji}x_i - \frac{v_j}{R_j} + I_j\right)\frac{dx_j}{dt} \tag{14.29}$$

The quantity inside the parentheses on the right-hand side of Eq. (14.29) is recognized as $C_j\,dv_j/dt$ by virtue of the neurodynamical equation (14.20). We may thus simplify Eq. (14.29) to

$$\frac{dE}{dt} = -\sum_{j=1}^{N} C_j\left(\frac{dv_j}{dt}\right)\frac{dx_j}{dt} \tag{14.30}$$

We now recognize the inverse relation that defines v_j in terms of x_j. The use of Eq. (14.22) in (14.30) yields

$$\frac{dE}{dt} = -\sum_{j=1}^{N} C_j\left[\frac{d}{dt}\varphi_j^{-1}(x_j)\right]\frac{dx_j}{dt}$$

$$= -\sum_{j=1}^{N} C_j\left(\frac{dx_j}{dt}\right)^2\left[\frac{d}{dx_j}\varphi_j^{-1}(x_j)\right] \tag{14.31}$$

From Fig. 14.10b we see that the inverse output–input relation $\varphi_j^{-1}(x_j)$ is a monotonically increasing function of the output x_j. It follows therefore that

$$\frac{d}{dx_j}\varphi_j^{-1}(x_j) \geq 0 \quad \text{for all } x_j \tag{14.32}$$

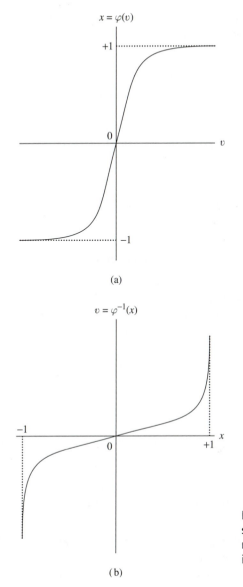

(a)

(b)

FIGURE 14.10   Plots of (a) the standard sigmoidal nonlinearity, and (b) its inverse.

We also note that

$$\left(\frac{dx_j}{dt}\right)^2 \geq 0 \quad \text{for all } x_j \tag{14.33}$$

Hence, all the factors that make up the sum on the right-hand side of Eq. (14.31) are nonnegative. In other words, for the energy function E defined in Eq. (14.28), we have

$$\frac{dE}{dt} \leq 0$$

From the definition of Eq. (14.28), we note that the function E is bounded. Accordingly, we may make the following two statements:

1. The energy function E is a Lyapunov function of the continuous Hopfield model.
2. The model is stable in accordance with Lyapunov's Theorem 1.

In other words, the time evolution of the continuous Hopfield model described by the system of nonlinear first-order differential equations (14.20) represents a trajectory in state space, which seeks out the minima of the energy (Lyapunov) function E and comes to a stop at such fixed points. From Eq. (14.31) we also note that the derivative dE/dt vanishes only if

$$\frac{d}{dt}x_j(t) = 0 \quad \text{for all } j$$

We may thus go one step further and write

$$\frac{dE}{dt} < 0 \quad \text{except at a fixed point} \tag{14.34}$$

Equation (14.34) provides the basis for the following theorem:

The (Lyapunov) energy function E of a Hopfield network is a monotonically decreasing function of time .

Accordingly, the Hopfield network is globally asymptotically stable; the attractor fixed-points are the minima of the energy function, and vice versa.

Relation between the Stable States of the Discrete and Continuous Versions of the Hopfield Model

The Hopfield network may be operated in a continuous mode or discrete mode, depending on the model adopted for describing the neurons. The continuous mode of operation is based on an additive model, as previously described. On the other hand, the discrete mode of operation is based on the McCulloch–Pitts model. We may readily establish the relationship between the stable states of the continuous Hopfield model and those of the corresponding discrete Hopfield model by redefining the input–output relation for a neuron such that we may satisfy two simplifying characteristics:

1. The output of a neuron has the asymptotic values

$$x_j = \begin{cases} +1 & \text{for } v_j = \infty \\ -1 & \text{for } v_j = -\infty \end{cases} \tag{14.35}$$

2. The midpoint of the activation function of a neuron lies at the origin, as shown by

$$\varphi_j(0) = 0 \tag{14.36}$$

Correspondingly, we may set the bias I_j equal to zero for all j.

In formulating the energy function E for a continuous Hopfield model, the neurons are permitted to have self-loops. A discrete Hopfield model, on the other hand, need not have self-loops. We may therefore simplify our discussion by setting $w_{jj} = 0$ for all j in both models.

In light of these observations, we may redefine the energy function of a continuous Hopfield model given in Eq. (14.28) as follows:

$$E = -\frac{1}{2}\sum_{\substack{i=1 \\ i \neq j}}^{N}\sum_{j=1}^{N} w_{ji}x_ix_j + \sum_{j=1}^{N}\frac{1}{R_j}\int_0^{x_j}\varphi_j^{-1}(x)dx \tag{14.37}$$

The inverse function $\varphi_j^{-1}(x)$ is defined by Eq. (14.27). We may thus rewrite the energy function of Eq. (14.37) as follows:

$$E = -\frac{1}{2}\sum_{\substack{i=1 \\ i \neq j}}^{N}\sum_{j=1}^{N} w_{ji}x_ix_j + \sum_{j=1}^{N}\frac{1}{a_jR_j}\int_0^{x_j}\varphi^{-1}(x)dx \tag{14.38}$$

The integral

$$\int_0^{x_j}\varphi^{-1}(x)dx$$

has the standard form plotted in Fig. 14.11. Its value is zero for $x_j = 0$, and positive otherwise. It assumes a very large value as x_j approaches ± 1. If, however, the gain a_j of neuron j becomes infinitely large (i.e., the sigmoidal nonlinearity approaches the idealized hard-limiting form), the second term of Eq. (14.38) becomes negligibly small. In the limiting case when $a_j = \infty$ for all j, the maxima and minima of the continuous Hopfield model become identical with those of the corresponding discrete Hopfield model. In the latter case, the energy (Lyapunov) function is defined simply by

$$E = -\frac{1}{2}\sum_{\substack{i=1 \\ i \neq j}}^{N}\sum_{j=1}^{N} w_{ji}x_ix_j \tag{14.39}$$

where the jth neuron state $x_j = \pm 1$. We conclude, therefore, that the only stable points of the very high-gain, continuous, deterministic Hopfield model correspond to the stable points of the discrete stochastic Hopfield model.

When, however, each neuron j has a large but finite gain a_j, we find that the second term on the right-hand side of Eq. (14.38) makes a noticeable contribution to the

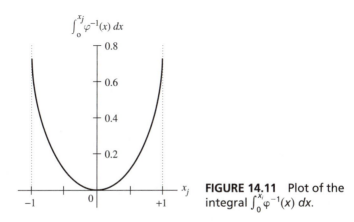

FIGURE 14.11   Plot of the integral $\int_0^{x_j}\varphi^{-1}(x)\,dx$.

energy function of the continuous model. In particular, this contribution is large and positive near all surfaces, edges, and corners of the unit hypercube that defines the state space of the model. On the other hand, the contribution is negligibly small at points that are far removed from the surface. Accordingly, the energy function of such a model has its maxima at corners, but the minima are displaced slightly toward the interior of the hypercube (Hopfield, 1984).

Figure 14.12 depicts the *energy contour map* or *energy landscape* for a continuous Hopfield model using two neurons. The outputs of the two neurons define the two axes of the map. The lower left- and upper right-hand corners of Fig. 14.12 represent stable minima for the limiting case of infinite gain; the minima for the case of finite gain are displaced inward. The flow to the fixed points (i.e., stable minima) may be interpreted as the solution to the minimization of the energy function E defined in Eq. (14.28).

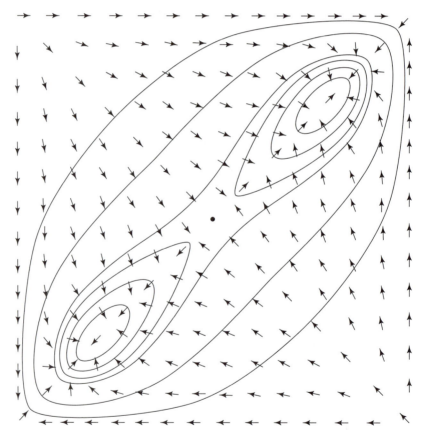

FIGURE 14.12   An energy contour map for a two-neuron, two-stable-state system. The ordinate and abscissa are the outputs of the two neurons. Stable states are located near the lower left and upper right corners, and unstable extrema at the other two corners. The arrows show the motion of the state. This motion is not generally perpendicular to the energy contours. (From J.J. Hopfield, 1984, with permission of the National Academy of Sciences of the U.S.A.)

The Discrete Hopfield Model as a Content-Addressable Memory

The Hopfield network has attracted a great deal of attention in the literature as a *content-addressable memory*. In this application, we know the fixed points of the network a *priori* in that they correspond to the patterns to be stored. However, the synaptic weights of the network that produce the desired fixed points are unknown, and the problem is how to determine them. The primary function of a content-addressable memory is to retrieve a pattern (item) stored in memory in response to the presentation of an incomplete or noisy version of that pattern. To illustrate the meaning of this statement in a succinct way, we can do no better than to quote from Hopfield's 1982 paper:

> Suppose that an item stored in memory is "H.A. Kramers & G.H. Wannier *Physi Rev. 60*, 252 (1941)." A general content-addressable memory would be capable of retrieving this entire memory item on the basis of sufficient partial information. The input "& Wannier (1941)" might suffice. An ideal memory could deal with errors and retrieve this reference even from the input "Wannier, (1941)."

An important property of a content-addressable memory is therefore the ability to retrieve a stored pattern, given a reasonable subset of the information content of that pattern. Moreover, a content-addressable memory is *error-correcting* in the sense that it can override inconsistent information in the cues presented to it.

The essence of a content-addressable memory (CAM) is to map a fundamental memory $\boldsymbol{\xi}_\mu$ onto a fixed (stable) point $\mathbf{x}_\mu$ of a dynamic system, as illustrated in Fig. 14.13. Mathematically, we may express this mapping in the form

$$\boldsymbol{\xi}_\mu \rightleftharpoons \mathbf{x}_\mu$$

The arrow from left to right describes an *encoding* operation, whereas the arrow from right to left describes a *decoding* operation. The attractor fixed points of the state space of the network are the *fundamental memories* or *prototype states* of the network.

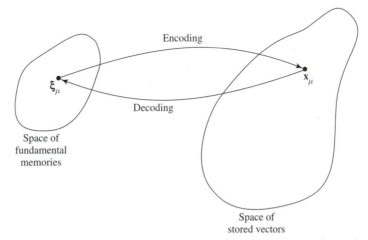

FIGURE 14.13   Illustration of the encoding–decoding performed by a recurrent network.

Suppose now that the network is presented a pattern containing partial but sufficient information about one of the fundamental memories. We may then represent that particular pattern as a starting point in the state space. In principle, provided that the starting point is close to the fixed point representing the memory being retrieved (i.e., it lies inside the basin of attraction belonging to the fixed point), the system should evolve with time and finally converge onto the memory state itself. At that point the entire memory is generated by the network. Consequently, the Hopfield network has an *emergent* property, which helps it retrieve information and cope with errors.

With the Hopfield model using the formal neuron of McCulloch and Pitts (1943) as its basic processing unit, each such neuron has two states determined by the level of the induced local field acting on it. The "on" or "firing" state of neuron i is denoted by the output $x_i = +1$, and the "off" or "quiescent" state is represented by $x_i = -1$. For a network made up of N neurons, the *state* of the network is thus defined by the vector

$$\mathbf{x} = [x_1, x_2, ..., x_N]^T$$

With $x_i = \pm 1$, the state of neuron i represents one *bit* of information, and the N-by-1 state vector $\mathbf{x}$ represents a binary word of N bits of information.

The induced local field v_j of neuron j is defined by

$$v_j = \sum_{i=1}^{N} w_{ji} x_i + b_j \tag{14.40}$$

where b_j is a fixed *bias* applied externally to neuron j. Hence, neuron j modifies its state x_j according to the *deterministic rule*

$$x_j = \begin{cases} +1 & \text{if } v_j > 0 \\ -1 & \text{if } v_j < 0 \end{cases}$$

This relation may be rewritten in the compact form

$$x_j = \text{sgn}[v_j]$$

where sgn is the *signum function*. What if v_j is exactly zero? The action taken here can be quite arbitrary. For example, we may set $x_j = \pm 1$ if $v_j = 0$. However, we will use the following convention: If v_j is zero, neuron j remains in its previous state, regardless of whether it is on or off. The significance of this assumption is that the resulting flow diagram is symmetrical, as will be illustrated later.

There are two phases to the operation of the discrete Hopfield network as a content-addressable memory, namely the storage phase and the retrieval phase, as described here.

1. *Storage Phase.* Suppose that we wish to store a set of N-dimensional vectors (binary words), denoted by $\{\xi_\mu | \mu = 1, 2, ..., M\}$. We call these M vectors *fundamental memories*, representing the patterns to be memorized by the network. Let $\xi_{\mu, i}$ denote the ith element of the fundamental memory ξ_μ, where the class $\mu = 1, 2, ..., M$. According to the *outer product rule* of storage, that is, the generalization of *Hebb's postulate of learning*, the synaptic weight from neuron i to neuron j is defined by

$$w_{ji} = \frac{1}{N} \sum_{\mu=1}^{M} \xi_{\mu, j} \xi_{\mu, i} \tag{14.41}$$

The reason for using $1/N$ as the constant of proportionality is to simplify the mathematical description of information retrieval. Note also that the learning rule of Eq. (14.41) is a "one shot" computation. In the normal operation of the Hopfield network, we set

$$w_{ii} = 0 \quad \text{for all } i \qquad (14.42)$$

which means that the neurons have *no* self-feedback. Let $\mathbf{W}$ denote the *N-by-N synaptic weight matrix* of the network, with w_{ji} as its *ji*th element. We may then combine Eqs. (14.41) and (14.42) into a single equation written in matrix form as follows:

$$\mathbf{W} = \frac{1}{N}\sum_{\mu=1}^{M}\boldsymbol{\xi}_{\mu}\boldsymbol{\xi}_{\mu}^{T} - M\mathbf{I} \qquad (14.43)$$

where $\boldsymbol{\xi}_{\mu}\boldsymbol{\xi}_{\mu}^{T}$ represents the outer product of the vector $\boldsymbol{\xi}_{\mu}$ with itself, and $\mathbf{I}$ denotes the identity matrix. From these defining equations of the synaptic weights/weight matrix, we may reconfirm the following:

- The output of each neuron in the network is fed back to all other neurons.
- There is no self-feedback in the network (i.e., $w_{ii} = 0$).
- The weight matrix of the network is symmetric as shown by (see Eq. (14.21))

$$\mathbf{W}^{T} = \mathbf{W} \qquad (14.44)$$

2. *Retrieval Phase.* During the retrieval phase, an N-dimensional vector $\boldsymbol{\xi}_{\text{probe}}$, called a *probe*, is imposed on the Hopfield network as its state. The probe vector has elements equal to ± 1. It typically represents an incomplete or noisy version of a fundamental memory of the network. Information retrieval then proceeds in accordance with a *dynamical rule* in which each neuron j of the network *randomly* but at some fixed rate examines the induced local field v_j (including any nonzero bias b_j) applied to it. If, at that instant of time, v_j is greater than zero, neuron j will switch its state to $+1$ or remain in that state if it is already there. Similarly, if v_j is less than zero, neuron j will switch its state to -1 or remain in that state if it is already there. If v_j is exactly zero, neuron j is left in its previous state, regardless of whether it is on or off. The state updating from one iteration to the next is therefore deterministic, but the selection of a neuron to perform the updating is done randomly. The *asynchronous* (serial) updating procedure described here is continued until there are no further changes to report. That is, starting with the probe vector $\mathbf{x}$, the network finally produces a time invariant state vector $\mathbf{y}$ whose individual elements satisfy the *condition for stability*:

$$y_i = \text{sgn}\left(\sum_{i=1}^{N} w_{ji}y_i + b_j\right), \quad j = 1, 2, ..., N \qquad (14.45)$$

or, in matrix form,

$$\mathbf{y} = \text{sgn}(\mathbf{W}\mathbf{y} + \mathbf{b}) \qquad (14.46)$$

where $\mathbf{W}$ is the synaptic weight matrix of the network, and $\mathbf{b}$ is the externally applied *bias vector*. The stability condition described here is also referred to as the *alignment condition*. The state vector $\mathbf{y}$ that satisfies it is called a *stable state* or *fixed point* of the state space of the system. We may therefore make the statement that the Hopfield network will always converge to a stable state when the retrieval operation is performed *asynchronously*.[5]

TABLE 14.2   Summary of the Hopfield Model

---

1. *Learning.* Let $\xi_1, \xi_2, ..., \xi_M$ denote a known set of N-dimensional fundamental memories. Use the outer product rule (i.e., Hebb's postulate of learning) to compute the synaptic weights of the network:

$$w_{ji} = \begin{cases} \dfrac{1}{N} \displaystyle\sum_{\mu=1}^{M} \xi_{\mu,j}\, \xi_{\mu,i}, & j \neq i \\[2mm] 0, & j = i \end{cases}$$

   where w_{ji} is the synaptic weight from neuron i to neuron j. The elements of the vector ξ_μ equal ± 1. Once they are computed, the synaptic weights are kept fixed.

2. *Initialization.* Let ξ_{probe} denote an unknown N-dimensional input vector (probe) presented to the network. The algorithm is initialized by setting

$$x_j(0) = \xi_{j,\,\text{probe}}, \quad j = 1, ..., N$$

   where $x_j(0)$ is the state of neuron j at time $n = 0$, and $\xi_{j,\,\text{probe}}$ is the jth element of the probe vector ξ_{probe}.

3. *Iteration Until Convergence.* Update the elements of state vector $\mathbf{x}(n)$ asynchronously (i.e., randomly and one at a time) according to the rule

$$x_j(n+1) = \mathrm{sgn}\!\left[\sum_{i=1}^{N} w_{ji}\, x_i(n) \right], \quad j = 1, 2, ..., N$$

   Repeat the iteration until the state vector $\mathbf{x}$ remains unchanged.

4. *Outputting.* Let $\mathbf{x}_{\text{fixed}}$ denote the fixed point (stable state) computed at the end of step 3. The resulting output vector $\mathbf{y}$ of the network is

$$\mathbf{y} = \mathbf{x}_{\text{fixed}}$$

Step 1 is the storage phase, and steps 2 through 4 constitute the retrieval phase.

Table 14.2 presents a summary of the steps involved in the storage phase and retrieval phase of operating a Hopfield network.

Example 14.2

To illustrate the emergent behavior of the Hopfield model, consider the network of Fig. 14.14a, which consists of three neurons. The weight matrix of the network is

$$\mathbf{W} = \frac{1}{3} \begin{bmatrix} 0 & -2 & +2 \\ -2 & 0 & -2 \\ +2 & -2 & 0 \end{bmatrix}$$

which is legitimate since it satisfies the conditions of Eqs. (14.42) and (14.44). The bias applied to each neuron is assumed to be zero. With three neurons in the network, there are $2^3 = 8$ possible states to consider. Of these eight states, only the two states $(1, -1, 1)$ and $(-1, 1, -1)$ are stable; the remaining six states are all unstable. We say that these two particular states are stable because they both satisfy the alignment condition of Eq. (14.46). For the state vector $(1, -1, 1)$ we have

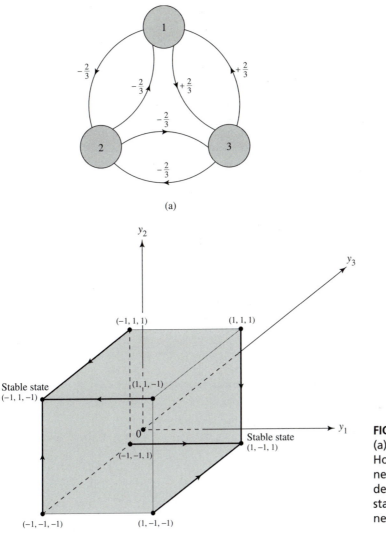

(a)

FIGURE 14.14
(a) Architectural graph of Hopfield network for $N = 3$ neurons. (b) Diagram depicting the two stable states and flow of the network.

$$\mathbf{Wy} = \frac{1}{3}\begin{bmatrix} 0 & -2 & +2 \\ -2 & 0 & -2 \\ +2 & -2 & 0 \end{bmatrix}\begin{bmatrix} +1 \\ -1 \\ +1 \end{bmatrix} = \frac{1}{3}\begin{bmatrix} +4 \\ -4 \\ +4 \end{bmatrix}$$

Hard limiting this result yields

$$\text{sgn}[\mathbf{Wy}] = \begin{bmatrix} +1 \\ -1 \\ +1 \end{bmatrix} = \mathbf{y}$$

Similarly, for the state vector $(-1, 1, -1)$ we have

$$\mathbf{Wy} = \frac{1}{3}\begin{bmatrix} 0 & -2 & +2 \\ -2 & 0 & -2 \\ +2 & -2 & 0 \end{bmatrix}\begin{bmatrix} -1 \\ +1 \\ -1 \end{bmatrix} = \frac{1}{3}\begin{bmatrix} -4 \\ +4 \\ -4 \end{bmatrix}$$

which, after hard limiting, yields

$$\text{sgn}[\mathbf{Wy}] = \begin{bmatrix} -1 \\ +1 \\ -1 \end{bmatrix} = \mathbf{y}$$

Hence, both of these state vectors satisfy the alignment condition.

Moreover, following the asynchronous updating procedure summarized in Table 14.2, we get the flow described in Fig. 14.14b. This flow map exhibits symmetry with respect to the two stable states of the network, which is intuitively satisfying. This symmetry is the result of leaving a neuron in its previous state if the induced local field acting on it is exactly zero.

Figure 14.14b also shows that if the network of Fig. 14.14a is in the initial state $(1, 1, 1)$, $(-1, -1, 1)$, or $(1, -1, -1)$, it will converge onto the stable state $(1, -1, 1)$ after one iteration. If the initial state is $(-1, -1, -1)$, $(-1, 1, 1)$, or $(1, 1, -1)$, it will converge onto the second stable state $(-1, 1, -1)$.

The network therefore has two fundamental memories, $(1, -1, 1)$ and $(-1, 1, -1)$, representing the two stable states. The application of Eq. (14.43) yields the synaptic weight matrix

$$\mathbf{W} = \frac{1}{3}\begin{bmatrix} +1 \\ -1 \\ +1 \end{bmatrix}[+1, -1, +1] + \frac{1}{3}\begin{bmatrix} -1 \\ +1 \\ -1 \end{bmatrix}[-1, +1, -1] - \frac{2}{3}\begin{bmatrix} 1 & 0 & 0 \\ 0 & 1 & 0 \\ 0 & 0 & 1 \end{bmatrix}$$

$$= \frac{1}{3}\begin{bmatrix} 0 & -2 & +2 \\ -2 & 0 & -2 \\ +2 & -2 & 0 \end{bmatrix}$$

which checks with the synaptic weights shown in Fig. 14.14a.

The error correcting capability of the Hopfield network is readily seen by examining the flow map of Fig. 14.14b:

1. If the probe vector $\boldsymbol{\xi}_{\text{probe}}$ applied to the network equals $(-1, -1, 1)$, $(1, 1, 1)$, or $(1, -1, -1)$, the resulting output is the fundamental memory $(1, -1, 1)$. Each of these values of the probe represents a single error, compared to the stored pattern.
2. If the probe vector $\boldsymbol{\xi}_{\text{probe}}$ equals $(1, 1, -1)$, $(-1, -1, -1)$, or $(-1, 1, 1)$, the resulting network output is the fundamental memory $(-1, 1, -1)$. Here again, each of these values of the probe represents a single error, compared to the stored pattern.

■

Spurious States

The weight matrix $\mathbf{W}$ of a discrete Hopfield network is symmetric, as indicated in Eq. (14.44). The eigenvalues of $\mathbf{W}$ are therefore all real. However, for large M, the eigenvalues are ordinarily *degenerate*, which means that there are several eigenvectors with the same eigenvalue. The eigenvectors associated with a degenerate eigenvalue form a subspace. Furthermore, the weight matrix $\mathbf{W}$ has a degenerate eigenvalue with a value of zero, in which case the subspace is called the *null space*. The null space exists by virtue of the fact that the number of fundamental memories, M, is smaller than the number of neurons, N, in the network. The presence of a null subspace is an intrinsic characteristic of the Hopfield network.

An eigenanalysis of the weight matrix $\mathbf{W}$ leads us to take the following viewpoint of the discrete Hopfield network used as a content-addressable memory (Aiyer et al., 1990):

1. The discrete Hopfield network acts as a *vector projector* in the sense that it projects a probe vector onto a subspace $\mathcal{M}$ spanned by the fundamental memory vectors.
2. The underlying dynamics of the network drive the resulting projected vector to one of the corners of a unit hypercube where the energy function is minimized.

The unit hypercube is N-dimensional. The M fundamental memory vectors, spanning the subspace $\mathcal{M}$, constitute a set of fixed points (stable states) represented by certain corners of the unit hypercube. The other corners of the unit hypercube that lie in or near subspace $\mathcal{M}$ are potential locations for *spurious states*, also referred to as *spurious attractors* (Amit, 1989). Spurious states represent stable states of the Hopfield network that are different from the fundamental memories of the network.

In the design of a Hopfield network as a content-addressable memory, we are therefore faced with a tradeoff between two conflicting requirements: (1) the need to preserve the fundamental memory vectors as fixed points in the state space, and (2) the desire to have few spurious states.

Storage Capacity of the Hopfield Network

Unfortunately, the fundamental memories of a Hopfield network are not always stable. Moreover, spurious states representing other stable states that are different from the fundamental memories can arise. These two phenomena tend to decrease the efficiency of the Hopfield network as a content-addressable memory. Here we explore the first of these two phenomena.

Let a probe equal to one of the fundamental memories, ξ_v, be applied to the network. Then, permitting the use of self-feedback for generality and assuming zero bias, we find using Eq. (14.41) that the induced local field of neuron j is

$$
\begin{aligned}
v_j &= \sum_{i=1}^{N} w_{ji} \xi_{v,i} \\
&= \frac{1}{N} \sum_{\mu=1}^{M} \xi_{\mu,j} \sum_{i=1}^{N} \xi_{\mu,i} \xi_{v,i} \\
&= \xi_{v,j} + \frac{1}{N} \sum_{\substack{\mu=1 \\ \mu \neq v}}^{M} \xi_{\mu,j} \sum_{i=1}^{N} \xi_{\mu,i} \xi_{v,i}
\end{aligned}
\tag{14.47}
$$

The first term on the right-hand side of Eq. (14.47) is simply the jth element of the fundamental memory ξ_v; now we can see why the scaling factor $1/N$ was introduced in the definition of the synaptic weight w_{ji} in Eq. (14.41). This term may therefore be viewed as the desired "signal" component of v_j. The second term on the right-hand side of Eq. (14.47) is the result of "crosstalk" between the elements of the fundamental memory ξ_v under test and those of some other fundamental memory ξ_μ. This second term may therefore be viewed as the "noise" component of v_j. We therefore have a situation similar to the classical "signal-in-noise detection problem" in communication theory (Haykin, 1994b).

We assume that the fundamental memories are random, being generated as a sequence of MN Bernoulli trials. The noise term of Eq. (14.47) then consists of a sum of

$N(M - 1)$ independent random variables, taking values ± 1 divided by N. This is a situation where the central limit theorem of probability theory applies. The *central limit theorem* states (Feller, 1968):

> Let $\{X_k\}$ be a sequence of mutually independent random variables with a common distribution. Suppose that X_k has mean μ and variance σ^2, and let $Y = X_1 + X_2 + \cdots + X_n$ Then, as n approaches infinity, the probability distribution of the sum random variable Y approaches a Gaussian distribution.

Hence, by applying the central limit theorem to the noise term in Eq. (14.47), we find that the noise is asymptotically Gaussian distributed. Each of the $N(M - 1)$ random variables constituting the noise term in this equation has a mean of zero and a variance of $1/N^2$. It follows, therefore, that the statistics of the Gaussian distribution are

- Zero mean
- Variance equal to $(M - 1)/N$

The signal component $\xi_{v,j}$ has a value of $+1$ or -1 with equal probability, and therefore a mean of zero and variance of one. The *signal-to-noise ratio* is thus defined by

$$\rho = \frac{\text{variance of signal}}{\text{variance of noise}}$$

$$= \frac{1}{(M - 1)/N} \tag{14.48}$$

$$\simeq \frac{N}{M} \quad \text{for large } M$$

The components of the fundamental memory ξ_v will be *stable* if, and only if, the signal-to-noise ratio ρ is high. Now, the number M of fundamental memories provides a direct measure of the *storage capacity* of the network. Therefore, it follows from Eq. (14.48) that so long as the storage capacity of the network is not overloaded—that is, the number M of fundamental memories is small compared to the number N of neurons in the network—the fundamental memories are stable in a probabilistic sense.

The reciprocal of the signal-to-noise ratio, that is,

$$\alpha = \frac{M}{N} \tag{14.49}$$

is called the *load parameter*. Statistical physics considerations reveal that the quality of memory recall of the Hopfield network deteriorates with increasing load parameter α, and breaks down at the *critical value* $\alpha_c = 0.14$ (Amit, 1989; Müller and Reinhardt, 1990). This critical value is in agreement with the estimate of Hopfield (1982), where it is reported that as a result of computer simulations $0.15N$ states can be recalled simultaneously before errors become severe.

With $\alpha_c = 0.14$, we find from Eq. (14.48) that the critical value of the signal-to-noise ratio is $\rho_c \simeq 7$, or equivalently 8.45 dB. For a signal-to-noise ratio below this critical value, memory recall breaks down.

The critical value

$$M_c = \alpha_c N = 0.14 \, N \qquad (14.50)$$

defines the *storage capacity with errors* on recall. To determine the storage capacity without errors we must use a more stringent criterion defined in terms of probability of error as described next.

Let the jth bit of the probe $\boldsymbol{\xi}_{\text{probe}} = \boldsymbol{\xi}_\nu$ be a symbol 1, that is, $\xi_{\nu,j} = 1$. Then the *conditional probability of bit error on recall* is defined by the shaded area in Fig. 14.15. The rest of the area under this curve is the *conditional probability that bit j of the probe is retrieved correctly*. Using the well-known formula for a Gaussian distribution, this latter conditional probability is given by

$$P(v_j > 0 | \xi_{\nu,j} = +1) = \frac{1}{\sqrt{2\pi}\sigma} \int_0^\infty \exp\left(-\frac{(v_j - \mu)^2}{2\sigma^2}\right) dv_j \qquad (14.51)$$

With $\xi_{\nu,j}$ set to $+1$, and the mean of the noise term in Eq. (14.47) equal to zero, it follows that the mean of the random variable V is $\mu = 1$ and its variance is $\sigma^2 = (M-1)/N$. From the definition of the *error function* commonly used in calculations involving the Gaussian distribution, we have

$$\text{erf}(y) = \frac{2}{\sqrt{\pi}} \int_0^y e^{-z^2} dz \qquad (14.52)$$

where y is a variable defining the upper limit of integration. We may now simplify the expression for the conditional probability of correctly retrieving the jth bit of the fundamental memory $\boldsymbol{\xi}_\nu$ by rewriting Eq. (14.51) in terms of the error function as:

$$P(v_j > 0 | \xi_{\nu,j} = +1) = \frac{1}{2}\left[1 + \text{erf}\left(\sqrt{\frac{\rho}{2}}\right)\right] \qquad (14.53)$$

where ρ is the signal-to-noise ratio defined in Eq. (14.48). Each fundamental memory consists of n bits. Also, the fundamental memories are usually equiprobable. It follows therefore that the *probability of stable patterns* is defined by

$$P_{\text{stab}} = (P(v_j > 0 | \xi_{\nu,j} = +1))^N \qquad (14.54)$$

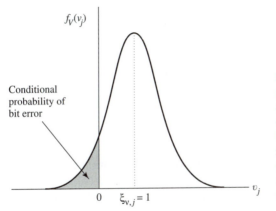

FIGURE 14.15 Conditional probability of bit error, assuming a Gaussian distribution for the induced local field v_j of neuron j; the subscript V in the probability density function $f_V(v_j)$ denotes a random variable with v_j representing a realization of it.

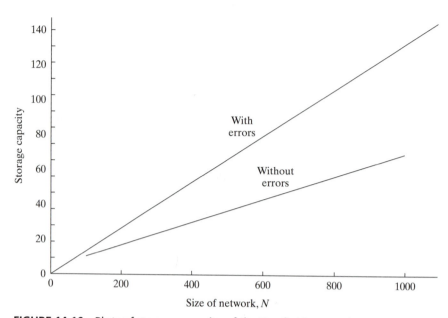

FIGURE 14.16 Plots of storage capacity of the Hopfield network versus network size for two cases: with errors and almost without errors.

We may use this probability to formulate an expression for the capacity of a Hopfield network. Specifically, we define the *storage capacity almost without errors*, M_{max}, as the largest number of fundamental memories that can be stored in the network and yet insist that most of them be recalled correctly. In Problem 14.8 it is shown that this definition of storage capacity yields the formula

$$M_{max} = \frac{N}{2 \log_e N} \tag{14.55}$$

where $\log_e$ denotes the natural logarithm.

Figure 14.16 shows graphs of the storage capacity with errors defined in Eq. (14.50) and the storage capacity almost without errors defined in Eq. (14.55), both plotted versus the network size N. From this figure we note the following points:

- Storage capacity of the Hopfield network scales essentially *linearly* with the size N of the network.
- A major limitation of the Hopfield network is that its storage capacity must be maintained small for the fundamental memories to be recoverable.[6]

14.8 COMPUTER EXPERIMENT I

In this section we use a computer experiment to illustrate the behavior of the discrete Hopfield network as a content-addressable memory. The network used in the experiment consists of $N = 120$ neurons, and therefore $N^2 - N = 12{,}280$ synaptic weights. It was

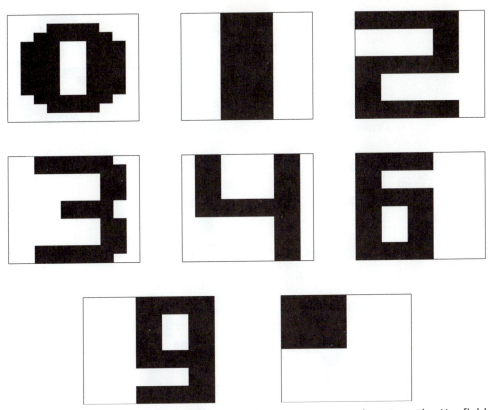

FIGURE 14.17 Set of handcrafted patterns for computer experiment on the Hopfield network.

trained to retrieve the eight digitlike black and white patterns shown in Fig. 14.17, with each pattern containing 120 pixels (picture elements) and designed specially to produce good performance (Lippmann, 1987). The inputs applied to the network assume the value $+1$ for black pixels and -1 for white pixels. The eight patterns of Fig. 14.17 were used as fundamental memories in the storage (learning) phase of the Hopfield network to create the synaptic weight matrix **W**, which was done using Eq. (14.43). The retrieval phase of the network's operation was performed asynchronously, as described in Table 14.2.

During the first stage of the retrieval part of the experiment, the fundamental memories were presented to the network to test its ability to recover them correctly from the information stored in the synaptic weight matrix. In each case, the desired pattern was produced by the network after one iteration.

Next, to demonstrate the error-correcting capability of the Hopfield network, a pattern of interest was distorted by randomly and independently reversing each pixel of the pattern from $+1$ to -1 and vice versa with a probability of 0.25, and then using the corrupted pattern as a probe for the network. The result of this experiment for digit 3 is presented in Fig. 14.18. The pattern in the mid-top part of this figure represents a corrupted version of digit 3, which is applied to the network at zero time. The patterns produced by the network after 5, 10, 15, 20, 25, 30, and 35 iterations are presented in

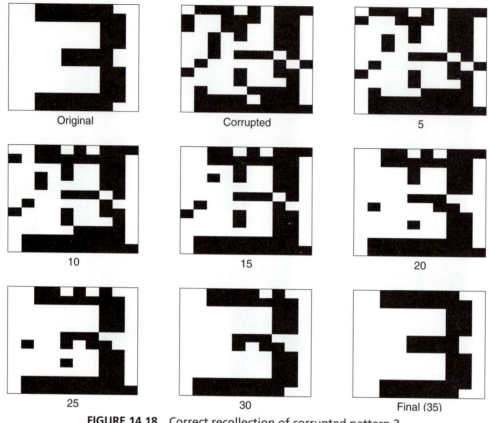

FIGURE 14.18 Correct recollection of corrupted pattern 3.

the rest of the figure. As the number of iterations is increased, we see that the resemblance of the network output to digit 3 is progressively improved. Indeed, after 35 iterations, the network converges onto the exactly correct form of digit 3.

Since, in theory, one quarter of the 120 neurons of the Hopfield network end up changing state for each corrupted pattern, the number of iterations needed for recall, on average, is 30. In our experiment, the number of iterations needed for the recall of the different patterns from their corrupted versions were as follows:

Pattern	Number of patterns needed for recall
0	34
1	32
2	26
3	35
4	25
6	37
"■"	32
9	26

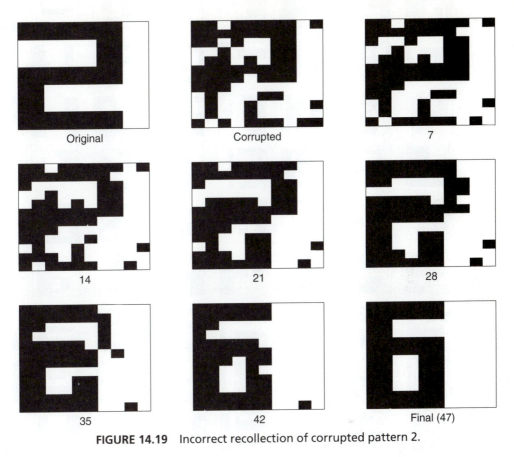

FIGURE 14.19 Incorrect recollection of corrupted pattern 2.

The average number of iterations needed for recall, averaged over the eight patterns, was about 31, which shows that the Hopfield network behaved as expected.

A problem inherent to the Hopfield network arises when the network is presented with a corrupted version of a fundamental memory, and the network then proceeds to converge onto the wrong fundamental memory. This is illustrated in Fig. 14.19, where the network is presented with a corrupted pattern "2," but after 47 iterations it converged to the fundamental memory "6."

As mentioned earlier, there is another problem that arises in the Hopfield network: the presence of spurious states. Figure 14.20 (viewed as a matrix of 14-by-8 network states) presents a listing of 108 spurious attractors found in 43,097 tests of randomly selected digits corrupted with the probability of flipping a bit set at 0.25. The spurious states may be grouped as follows (Amit, 1989):

1. *Reversed fundamental memories.* These spurious states are reversed (i.e., negative) versions of the fundamental memories of the network; see, for example, the state in location 1-by-1 in Fig. 14.20, which represents the negative of digit 6 in Fig. 14.17. To explain this kind of a spurious state, we note that the energy function E is symmetric in the sense that its value remains unchanged if the states of the neurons are reversed (i.e., the state x_i is replaced by $-x_i$ for all i).

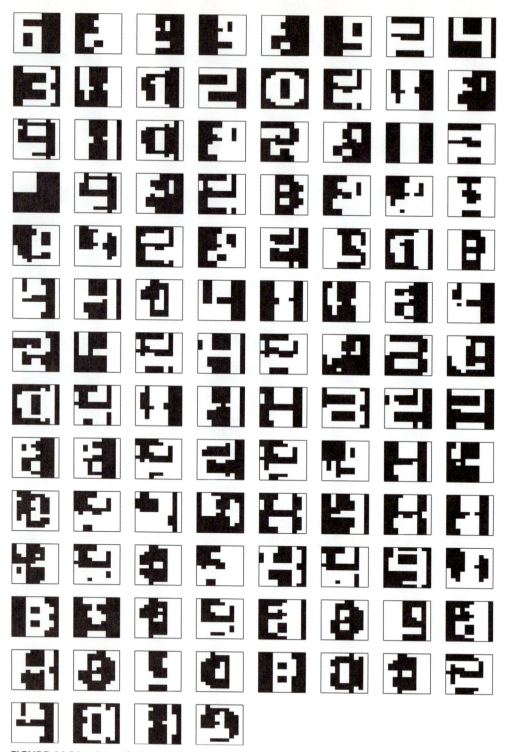

FIGURE 14.20 Compilation of the spurious states produced in the computer experiment on the Hopfield network.

Accordingly, if the fundamental memory ξ_μ corresponds to a particular local minimum of the energy landscape, that same local minimum also corresponds to $-\xi_\mu$. This sign reversal does not pose a problem in the retrieval of information if it is agreed to reverse all the information bits of a retrieved pattern if it is found that a particular bit designated as the "sign" bit is -1 instead of $+1$.

2. *Mixture states.* A mixture spurious state is a linear combination of an *odd* number of stored patterns. For example, consider the state

$$x_i = \text{sgn}(\xi_{1,i} + \xi_{2,i} + \xi_{3,i})$$

which is a three-mixture spurious state. It is a state formed out of three fundamental memories ξ_1, ξ_2, and ξ_3 by a majority rule. The stability condition of Eq. (14.45) is satisfied by such a state for a large network. The state in location row 6, column 4 in Fig. 14.20 represents a three-mixture spurious state formed by a combination of the fundamental memories: ξ_1 = negative of digit 1, ξ_2 = digit 4, and ξ_3 = digit 9.

3. *Spin-glass states.* This kind of a spurious state is so named by analogy with spin-glass models of statistical mechanics. Spin-glass states are defined by local minima of the energy landscape that are *not* correlated with any of the fundamental memories of the network; see, for example, the state in location row 7, column 6 in Fig. 14.20.

14.9   COHEN–GROSSBERG THEOREM

In Cohen and Grossberg (1983), a general principle for assessing the stability of a certain class of neural networks is described by the following system of coupled nonlinear differential equations:

$$\frac{d}{dt}u_j = a_j(u_j)\left[b_j(u_j) - \sum_{i=1}^{N} c_{ji}\varphi_i(u_i) \right], \quad j = 1, ..., N \tag{14.56}$$

According to Cohen and Grossberg, this class of neural networks admits a Lyapunov function defined as

$$E = \frac{1}{2}\sum_{i=1}^{N}\sum_{j=1}^{N} c_{ji}\varphi_i(u_i)\varphi_j(u_j) - \sum_{j=1}^{N}\int_{0}^{u_j} b_j(\lambda)\varphi_j'(\lambda)d\lambda \tag{14.57}$$

where

$$\varphi_j'(\lambda) = \frac{d}{d\lambda}(\varphi_j(\lambda)) \tag{14.58}$$

For the definition of Eq. (14.57) to be valid, however, we require the following conditions to hold:

1. The synaptic weights of the network are "symmetric:"

$$c_{ij} = c_{ji} \tag{14.59}$$

2. The function $a_j(u_j)$ satisfies the condition for "nonnegativity:"

$$a_j(u_j) \geq 0 \tag{14.60}$$

3. The nonlinear input–output function $\varphi_j(u_j)$ satisfies the condition for "monotonicity:"

$$\varphi_j'(u_j) = \frac{d}{du_j} \varphi_j(u_j) \geq 0 \tag{14.61}$$

We may now formally state the *Cohen–Grossberg theorem*:

> Provided that the system of nonlinear differential equations (14.56) satisfies the conditions of symmetry, nonegativity, and monotonicity, the Lyapunov function E of the system defined by Eq. (14.57) satisfies the condition

$$\frac{dE}{dt} \leq 0$$

Once this basic property of the Lyapunov function E is in place, global stability of the system follows from Lyapunov's Theorem 1.

Hopfield Model as a Special Case of the Cohen–Grossberg Theorem

By comparing the general system of Eq. (14.56) with the system of Eq. (14.20) for a continuous Hopfield model, we may make the correspondence between the Hopfield model and the Cohen–Grossberg theorem that are summarized in Table 14.3. The use of this table in Eq. (14.57) yields the following Lyapunov function for the continuous Hopfield model:

$$E = -\frac{1}{2} \sum_{i=1}^{N} \sum_{j=1}^{N} w_{ji} \varphi_i(v_i) \varphi_j(v_j) + \sum_{j=1}^{N} \int_0^{v_j} \left(\frac{v_j}{R_j} - I_j \right) \varphi_j'(v) dv \tag{14.62}$$

where the nonlinear activation function $\varphi_j(\cdot)$ is defined by Eq. (14.23).

We next make the following observations:

1. $\varphi_i(v_i) = x_i$
2. $\int_0^{v_j} \varphi_j'(v) dv = \int_0^{x_j} dx = x_j$
3. $\int_0^{v_j} v \varphi_j'(v) dv = \int_0^{x_j} dx = \int_0^{x_j} \varphi_j^{-1}(x) \, dx$

Basically, relations 2 and 3 result from the use of $x = \varphi_i(v)$. Thus the use of these observations in the Lyapunov function of Eq. (14.62) yields a result identical to that we

TABLE 14.3   Correspondence between the Cohen–Grossberg Theorem and the Hopfield Model

Cohen–Grossberg Theorem	Hopfield Model
u_j	$C_j v_j$
$a_j(u_j)$	1
$b_j(u_j)$	$-(v_j/R_j) + I_j$
c_{ji}	$-w_{ji}$
$\varphi_i(u_i)$	$\varphi_i(v_i)$

defined earlier; see Eq. (14.28). Note, however, that although $\varphi_i(v)$ must be a nondecreasing function of the input v, it does not need to have an inverse in order for the generalized Lyapunov function of Eq. (14.62) to hold.

The Cohen–Grossberg theorem is a general principle of neurodynamics with a wide range of applications (Grossberg, 1990). In the next section we consider another application of this important theorem.

14.10   BRAIN-STATE-IN-A-BOX MODEL

In this section we continue the neurodynamical analysis of an associative memory by studying the *brain-state-in-a-box* (*BSB*) *model*, which was first described by Anderson et al. (1977). The BSB model is basically a *positive feedback system with amplitude limitation*. It consists of a highly interconnected set of neurons that feed back upon themselves. This model operates by using the built-in positive feedback to *amplify* an input pattern until all the neurons in the model are driven into saturation. The BSB model may thus be viewed as a categorization device in that an analog input pattern is given a digital representation defined by a stable state of the model.

Let $\mathbf{W}$ denote a *symmetric weight matrix* whose largest eigenvalues have positive real components. Let $\mathbf{x}(0)$ denote the *initial state vector* of the model, representing an input activation pattern. Assuming that there are N neurons in the model, the state vector of the model has dimension N, and the weight matrix $\mathbf{W}$ is an N-by-N matrix. The BSB algorithm is then completely defined by the following pair of equations:

$$\mathbf{y}(n) = \mathbf{x}(n) + \beta \mathbf{W}\mathbf{x}(n) \tag{14.63}$$

$$\mathbf{x}(n + 1) = \varphi(\mathbf{y}(n)) \tag{14.64}$$

where β is a small positive constant called the *feedback factor* and $\mathbf{x}(n)$ is the state vector of the model at discrete time n. Figure 14.21a shows a block diagram of the combination of Eqs. (14.63) and (14.64); the block labeled $\mathbf{W}$ represents a single-layer linear neural network, as depicted in Fig. 14.21b. The activation function φ is a *piecewise-linear function* that operates on $y_j(n)$, the jth component of the vector $\mathbf{y}(n)$, as follows (see Fig. 14.22):

$$x_j(n + 1) = \varphi(y_j(n))$$

$$= \begin{cases} +1 & \text{if } y_j(n) > +1 \\ y_j(n) & \text{if } -1 \leq y_j(n) \leq +1 \\ -1 & \text{if } y_j(n) < -1 \end{cases} \tag{14.65}$$

Equation (14.65) constrains the state vector of the BSB model to lie within an N-dimensional unit cube centered on the origin.

The algorithm thus proceeds as follows. An activation pattern $\mathbf{x}(0)$ is input into the BSB model as the initial state vector, and Eq. (14.63) is used to compute the vector $\mathbf{y}(0)$. Equation (14.64) is then used to truncate $\mathbf{y}(0)$, obtaining the updated state vector $\mathbf{x}(1)$. Next, $\mathbf{x}(1)$ is cycled through Eqs. (14.63) and (14.64), thereby obtaining $\mathbf{x}(2)$. This procedure is repeated until the BSB model reaches a *stable state* represented by a particular corner of the unit hypercube. Intuitively, positive feedback in the BSB model

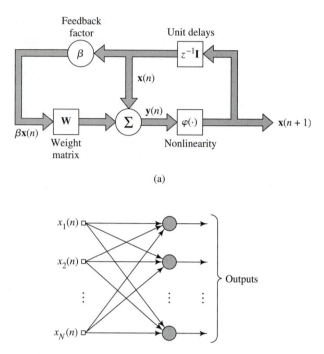

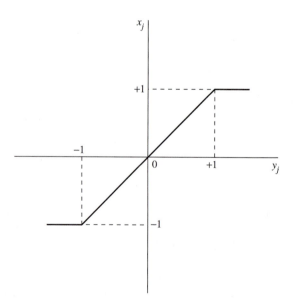

FIGURE 14.21 (a) Block diagram of the brain-state-in-a-box (BSB) model. (b) Signal-flow graph of the linear associator represented by the weight matrix **W**.

FIGURE 14.22 Piecewise-linear activation function used in the BSB model.

causes the initial state vector $\mathbf{x}(0)$ to increase in Euclidean length (norm) with an increasing number of iterations until it hits a wall of the box (unit hypercube), then slides along the wall and eventually ends up in a stable corner of the box, where it keeps on "pushing" but cannot get out of the box (Kawamoto and Anderson, 1985), hence the name of the model.

Lyapunov Function of the BSB Model

The BSB may be redefined as a special case of the neurodynamical model described in Eq. (14.16) as follows (Grossberg, 1990). To see this, we first rewrite the jth component of the BSB algorithm described by Eqs. (14.63) and (14.64) in the form

$$x_j(n + 1) = \varphi\left(\sum_{i=1}^{N} c_{ji}x_i(n)\right), \quad j = 1, 2, ..., N \tag{14.66}$$

The coefficients c_{ji} are defined by

$$c_{ji} = \delta_{ji} + \beta w_{ji} \tag{14.67}$$

where δ_{ji} is the Kronecker delta equal to 1 if $j = i$ and 0 otherwise, and w_{ji} is the ji-th element of the weight matrix $\mathbf{W}$. Equation (14.66) is written in discrete time form. To proceed further, we need to reformulate it in a continuous time form, as shown by

$$\frac{d}{dt}x_j(t) = -x_j(t) + \varphi\left(\sum_{i=1}^{N} c_{ji}x_i(t)\right), \quad j = 1, 2, ..., N \tag{14.68}$$

where the bias I_j is zero for all j. However, for us to apply the Cohen–Grossberg theorem, we have to go one step further and transform Eq. (14.68) into the same form as the additive model. We may do so by introducing a new set of variables,

$$v_j(t) = \sum_{i=1}^{N} c_{ji}x_i(t) \tag{14.69}$$

Then, by virtue of the definition of c_{ji} given in Eq. (14.67), we find that

$$x_j(t) = \sum_{i=1}^{N} c_{ji}v_i(t) \tag{14.70}$$

Correspondingly, we may recast the model of Eq. (14.68) in the equivalent form

$$\frac{d}{dt}v_j(t) = -v_j(t) + \sum_{i=1}^{N} c_{ji}\varphi(v_i(t)), \quad j = 1, 2, ..., N \tag{14.71}$$

We are now ready to apply the Cohen–Grossberg theorem to the BSB model. By comparing Eq. (14.71) with (14.56), we may deduce the correspondence listed in Table 14.4 between the BSB model and the Cohen–Grossberg theorem. Therefore, using the results of Table 14.4 in Eq. (14.57), we find that the Lyapunov function of the BSB model is given by

$$E = -\frac{1}{2}\sum_{j=1}^{N}\sum_{i=1}^{N} c_{ji}\varphi(v_j)\varphi(v_i) + \sum_{j=1}^{N}\int_0^{v_j} v\varphi'(v)dv \tag{14.72}$$

TABLE 14.4   Correspondence between the
Cohen–Grossberg Theorem
and the BSB Model

Cohen–Grossberg Theorem	BSB Model
u_j	v_j
$a_j(u_j)$	1
$b_j(u_j)$	$-v_j$
c_{ji}	$-c_{ji}$
$\varphi_j(u_j)$	$\varphi_j(v_j)$

where $\varphi'(v)$ is the first derivative of the sigmoid function $\varphi(v)$ with respect to its argument. Finally, substituting the definitions of Eqs. (14.65), (14.67), and (14.69) in (14.72), we can define the Lyapunov (energy) function of the BSB model in terms of the original state variables as follows (Grossberg, 1990):

$$
\begin{aligned}
E &= -\frac{\beta}{2}\sum_{i=1}^{N}\sum_{j=1}^{N} w_{ji}x_j x_i \\
&= -\frac{\beta}{2}\mathbf{x}^T\mathbf{W}\mathbf{x}
\end{aligned}
\tag{14.73}
$$

The evaluation of the Lyapunov function for the Hopfield network presented in Section 14.7 assumes the existence of the derivative of the inverse of the model's sigmoidal nonlinearity, which is satisfied by the use of a hyperbolic tangent function. In contrast, this condition is not satisfied in the BSB model when the state variable of the jth neuron in the model is either $+1$ or -1. Despite this difficulty, the Lyapunov function of the BSB model can be evaluated via the Cohen–Grossberg theorem, which clearly illustrates the general applicability of this important theorem.

Dynamics of the BSB Model

In a direct analysis carried out by Golden (1986), it is demonstrated that the BSB model is in fact a gradient descent algorithm that minimizes the energy function E defined by Eq. (14.73). This important property of the BSB model, however, presumes that the weight matrix $\mathbf{W}$ satisfies the following two conditions:

- The weight matrix $\mathbf{W}$ is *symmetric*:

$$
\mathbf{W} = \mathbf{W}^T
$$

- The weight matrix $\mathbf{W}$ is *positive semidefinite*; that is, in terms of the eigenvalues of $\mathbf{W}$, we have

$$
\lambda_{\min} \geq 0
$$

where $\lambda_{\min}$ is the smallest eigenvalue of $\mathbf{W}$.

The energy function E of the BSB model thus decreases with increasing n (number of iterations) whenever the state vector $\mathbf{x}(n + 1)$ at time $n + 1$ is different from the state vector $\mathbf{x}(n)$ at time n. Moreover, the minimum points of the energy function E define the *equilibrium states* of the BSB model that are characterized by

$$\mathbf{x}(n + 1) = \mathbf{x}(n)$$

In other words, like the Hopfield model, the BSB model is an *energy-minimizing network*.

The equilibrium states of the BSB model are defined by certain corners of the unit hypercube and its origin. In the latter case, any fluctuation in the state vector, no matter how small, is amplified by positive feedback in the model, and therefore causes the state of the model to shift away from the origin in the direction of a stable configuration; in other words, the origin is a saddle point. For every corner of the hypercube to serve as a possible equilibrium state of the BSB model, the weight matrix $\mathbf{W}$ has to satisfy a third condition (Greenberg, 1988):

- The weight matrix $\mathbf{W}$ is *diagonal dominant*, which means that

$$w_{jj} \geq \sum_{i \neq j} |w_{ij}| \quad \text{for } j = 1, 2, \ldots, N \tag{14.74}$$

where w_{ij} is the ij-th element of $\mathbf{W}$.

For an equilibrium state $\mathbf{x}$ to be *stable*—that is, for a certain corner of the unit hypercube to be a fixed point *attractor*—there has to be a basin of attraction $\mathcal{N}(\mathbf{x})$ in the unit hypercube such that for all initial state vectors $\mathbf{x}(0)$ in $\mathcal{N}(\mathbf{x})$ the BSB model converges onto $\mathbf{x}$. For every corner of the unit hypercube to be a possible point attractor, the weight matrix $\mathbf{W}$ has to satisfy a fourth condition (Greenberg, 1988):

- The weight matrix $\mathbf{W}$ is *strongly diagonal-dominant*, as shown by

$$w_{jj} \geq \sum_{i \neq j} |w_{ij}| + \alpha \quad \text{for } j = 1, 2, \ldots, N \tag{14.75}$$

where α is a positive constant.

The important point in this discussion is that in the case of a BSB model for which the weight matrix $\mathbf{W}$ is symmetric and positive semidefinite, as is often the case, only some (but not all) of the corners of the unit hypercube act as point attractors. For all the corners of the unit hypercube to act as potential point attractors, the weight matrix $\mathbf{W}$ has to satisfy Eq. (14.75) as well, which of course subsumes the condition of Eq. (14.74).

Clustering

A natural application for the BSB model is *clustering*. This follows from the fact that the stable corners of the unit hypercube act as point attractors with well-behaved basins of attraction, which therefore divide the state space into a corresponding set of well-defined regions. Consequently, the BSB model may be used as an *unsupervised* clustering algorithm, with each stable corner of the unit hypercube representing a "cluster" of related data. The self-amplification provided by positive feedback (in conformity with Principle 1 of self-organization described in Chapter 8) is an important ingredient of this clustering property.

Anderson et al. (1990b) describe the use of the BSB model to cluster and therefore identify radar signals from different emitters. In this application the weight matrix **W**, basic to the operation of the BSB model, is *learned* using the *linear associator (associative memory) with error correction learning* that is described in Chapter 2. To be specific, suppose that information is represented by a set of K training vectors that are associated with themselves as follows:

$$\mathbf{x}_k \rightarrow \mathbf{x}_k, \quad k = 1, 2, ..., K \tag{14.76}$$

Let a training vector $\mathbf{x}_k$ be selected at random. Then the weight matrix **W** is incremented in accordance with the error correction algorithm (see Problem 3.9)

$$\Delta \mathbf{W} = \eta(\mathbf{x}_k - \mathbf{W}\mathbf{x}_k)\mathbf{x}_k \tag{14.77}$$

where η is the learning-rate parameter. The goal of learning the set of stimuli $\mathbf{x}_1, \mathbf{x}_2, ...,$ $\mathbf{x}_K$ is to have the linear associator behave as

$$\mathbf{W}\mathbf{x}_k = \mathbf{x}_k, \quad k = 1, 2, ..., K \tag{14.78}$$

The error correction algorithm described by Eq. (14.77) approximates the ideal condition of Eq. (14.78) in a least-mean-square error sense. The net effect of this learning process is to force the linear associator to develop a particular set of eigenvectors (defined by the training vectors) with eigenvalues equal to unity.

To perform radar clustering, the BSB model uses the linear associator with error correction learning to construct the weight matrix **W** and performs the following computation (Anderson et al., 1990):

$$\mathbf{x}(n + 1) = \varphi(\gamma\mathbf{x}(n) + \beta\mathbf{W}\mathbf{x}(n) + \delta\mathbf{x}(0)) \tag{14.79}$$

which is slightly different from the version of the BSB algorithm described in Eqs. (14.63) and (14.64). The difference is in two respects:

- The decay constant γ in the first term $\gamma\mathbf{x}(n)$ is included to cause the current state to decay slightly; provided that γ is a positive constant less than unity, the errors may eventually decay to zero.
- The third term $\delta\mathbf{x}(0)$ is included to keep the initial state vector $\mathbf{x}(0)$ constantly present; it has the effect of limiting the possible states of the BSB model.

Repeated iteration of the BSB model leads to an activity dominated by the eigenvectors of the weight matrix **W** with the largest positive eigenvalues, and therefore the vectors $\mathbf{x}_1, \mathbf{x}_2, ..., \mathbf{x}_K$ learned by the linear associator. The clustering ability of the BSB model develops largely as a result of signal-related eigenvectors being associated with large eigenvalues, becoming enhanced by positive feedback in the model, and thereby dominating the state of the model after a number of iterations. On the other hand, noise-related eigenvectors are usually associated with small eigenvalues, and therefore have a diminishing influence on the state of the BSB model, provided that the received signal-to-noise ratio is sufficiently high.

In a radar surveillance environment, detailed descriptions of emitters operating in the environment are not known a *priori*. Hundreds of thousands of radar pulses are typically received for processing in fractions of seconds. Hence there is no lack of data; the challenge is how to make sense of the data. The BSB model is able to help by learning the microwave structure of the radar environment through its inherent clustering property. Clusters are formed around the point attractors of the BSB model (i.e., stable corners of the unit hypercube), with each point attractor representing a particular emitter. The BSB model may thus identify received pulses as being produced by a particular emitter.

14.11 COMPUTER EXPERIMENT II

Figure 14.23 presents the results of an experiment performed on a BSB model containing two neurons. The two-by-two weight matrix $\mathbf{W}$ is defined by

$$\mathbf{W} = \begin{bmatrix} 0.035 & -0.005 \\ -0.005 & 0.035 \end{bmatrix}$$

which is symmetric, positive-definite, and satisfies Eq. (14.75).

The four different parts of Fig. 14.23 correspond to four different settings of the initial state $\mathbf{x}(0)$, as follows:

(a) $\mathbf{x}(0) = [\ 0.1, \ \ 0.2]^T$
(b) $\mathbf{x}(0) = [-0.2, \ \ 0.3]^T$
(c) $\mathbf{x}(0) = [-0.8, -0.4]^T$
(d) $\mathbf{x}(0) = [\ 0.6, \ \ 0.1]^T$

The areas shown shaded in this figure are the four basins of attraction characterizing the model. The figure clearly illustrates that when the initial state of the model lies in a particular basin of attraction, the underlying dynamics of the model drive the weight matrix $\mathbf{W}(n)$ with increasing number of iterations n, until the network state $\mathbf{x}(n)$ terminates on the fixed point attractor (i.e., a corner of the two-by-two square) belonging to that basin of attraction. A case of particular interest is the trajectory shown in Fig. 14.23d: The initial condition $\mathbf{x}(0)$ lies in the first quadrant, yet the trajectory terminates on the corner $(+1, -1)$ in the fourth quadrant because that is where the point attractor is for the pertinent basin of attraction.

14.12 STRANGE ATTRACTORS AND CHAOS

Up to this point in our discussion of neurodynamics we have focused attention on the kind of behavior exhibited by nonlinear dynamical systems characterized as fixed point attractors. In this section we consider another class of attractors called strange attractors that characterize certain nonlinear dynamical systems of order greater than 2.

A strange attractor exhibits a chaotic behavior that is highly complex. What makes the study of strange attractors and chaos particularly interesting is the fact that the system in question is *deterministic* in the sense that its operation is governed by *fixed* rules, yet such a system with only a few degrees of freedom can exhibit a behavior so complicated

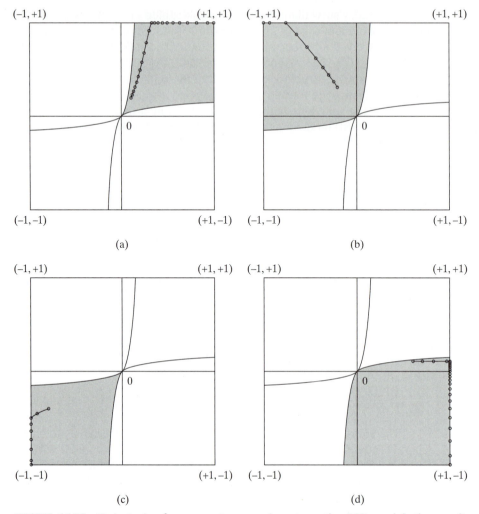

FIGURE 14.23 Trajectories for computer experiment on the BSB model; the results shown in parts (a) through (d) correspond to different initial conditions.

that it looks random. Indeed, the randomness is fundamental in the sense that the second-order statistics of a chaotic time series seem to indicate that it is random. However, unlike a true random phenomenon, the randomness exhibited by a chaotic system does not go away by gathering more information! In principle, the future behavior of a chaotic system is completely determined by the past, but in practice, any uncertainty in the choice of initial conditions, no matter how small, grows exponentially with time. Consequently, even though the dynamic behavior of a chaotic system is predictable in the short term, it is impossible to predict the long-term behavior of the system. A chaotic time series is therefore paradoxical in the sense that its generation is governed by a deterministic dynamical system, and yet it has a randomlike appearance. It is this attribute of a chaotic phenomenon that was originally emphasized by Lorenz with the discovery of an attractor that bears his name (Lorenz, 1963).

In a nonlinear dynamical system, when the orbits in an attractor with neighboring initial conditions tend to move apart with increasing time, the system is said to possess a *strange attractor* and the system itself is said to be *chaotic*. In other words, a fundamental property that makes an attractor "strange" is the *sensitive dependence on initial conditions*. Sensitivity in this context means that if two identical nonlinear systems are started at slightly different initial conditions, namely, $\mathbf{x}$ and $\mathbf{x} + \boldsymbol{\epsilon}$, where $\boldsymbol{\epsilon}$ is a very small quantity, their dynamic states will diverge from each other in state space and their separation will increase exponentially on the average.

Invariant Characteristics of Chaotic Dynamics

Two major features, namely fractal dimensions and Lyapunov exponents, have emerged as the classifiers of a chaotic process. Fractal dimensions characterize the geometric structure of a strange attractor. The term "fractal" was coined by Mandelbrot (1982). Unlike integer dimensions (as in a two-dimensional surface or a three-dimensional object), fractal dimensions are *not* integers. As for Lyapunov exponents, they describe how orbits on the attractor move under the evolution of the dynamics. These two invariant characteristics of chaotic dynamics are discussed in what follows. The term "invariant" signifies the fact that both fractal dimensions and Lyapunov exponents of a chaotic process remain unchanged under smooth nonlinear changes of the coordinate system of the process (Abarbanal, 1996).

Fractal Dimensions

Consider a strange attractor whose dynamics in d-dimensional state space are described by

$$\mathbf{x}(n + 1) = \mathbf{F}(\mathbf{x}(n)), \quad n = 0, 1, 2,... \qquad (14.80)$$

which is the discrete-time version of Eq. (14.2). This is readily seen by setting $t = n\Delta t$, where Δt is the sampling period. Assuming that Δt is sufficiently small, we may correspondingly set

$$\frac{d}{dt}\mathbf{x}(t) = \frac{1}{\Delta t}[\mathbf{x}(n\Delta t + \Delta t) - \mathbf{x}(n\Delta t)]$$

We may thus formulate the discrete-time version of Eq. (14.2) as follows:

$$\frac{1}{\Delta t}[\mathbf{x}(n\Delta t + \Delta t) - \mathbf{x}(n\Delta t)] = \mathbf{F}(\mathbf{x}(n\Delta t)) \text{ for small } \Delta t$$

By putting $\Delta t = 1$ for convenience of presentation and rearranging terms, we get

$$\mathbf{x}(n + 1) = \mathbf{x}(n) + \mathbf{F}(\mathbf{x}(n))$$

which may be cast into the form shown in Eq. (14.80) simply by redefining the vector-valued function $\mathbf{F}(\cdot)$.

Returning to Eq. (14.80), suppose we construct a small sphere of radius r around some location $\mathbf{y}$ on or near an orbit of the attractor. We may then define a *natural distribution* of points for the attractor as follows:

$$\rho(\mathbf{y}) = \lim_{N \to \infty} \frac{1}{N} \sum_{n=1}^{N} \delta(\mathbf{y} - \mathbf{x}(n)) \tag{14.81}$$

where $\delta(\cdot)$ is a d-dimensional delta function, and N is the number of data points. Note the change of notation concerning the use of N. The natural distribution $\rho(\mathbf{y})$ plays a role for a strange attractor that is analogous to that of the probability density function for a random variable. Accordingly, we may define an invariant $\bar{f}$ with respect to a function $f(\mathbf{y})$ under the evolution of the dynamics described as the multifold integral

$$\bar{f} = \int_{-\infty}^{\infty} f(\mathbf{y})\rho(\mathbf{y})d\mathbf{y} \tag{14.82}$$

A function $f(\mathbf{y})$ of interest is one that gives us a measure of how the number of points within the small sphere scales as the radius r of the sphere is reduced to zero. Recognizing that the volume occupied by the d-dimensional sphere is proportional to r^d, we may get a sense of attractor dimension by seeing how the density of points on the attractor behaves at small distances in state space.

The Euclidean distance between the center $\mathbf{y}$ of the sphere and the point $\mathbf{x}(n)$ at time step n is $\|\mathbf{y} - \mathbf{x}(n)\|$. Hence, the point $\mathbf{x}(n)$ lies inside the sphere of radius r provided that

$$\|\mathbf{y} - \mathbf{x}(n)\| < r$$

or equivalently,

$$r - \|\mathbf{y} - \mathbf{x}(n)\| > 0$$

Thus the function $f(\mathbf{x})$ for the situation described here may be written in the general form

$$f(\mathbf{x}) = \left(\frac{1}{N-1} \sum_{\substack{k=1 \\ k \neq n}}^{N} \theta(r - \|\mathbf{y} - \mathbf{x}(k)\|) \right)^{q-1} \tag{14.83}$$

where q is an integer, and $\theta(\cdot)$ is the *Heaviside function* defined by

$$\theta(z) = \begin{cases} 1 & \text{for } z > 0 \\ 0 & \text{for } z < 0 \end{cases}$$

Substituting Eqs. (14.81) and (14.83) in (14.82), we get a new function $C(q, r)$ that depends on q and r, as shown by

$$C(q, r) = \int_{-\infty}^{\infty} \left(\frac{1}{N-1} \sum_{\substack{k=1 \\ k \neq n}}^{N} \theta(r - \|\mathbf{y} - \mathbf{x}(k)\|) \right)^{q-1} \left(\frac{1}{N} \sum_{n=1}^{N} \delta(\mathbf{y} - \mathbf{x}(n)) \right) d\mathbf{y}$$

Hence, using the sifting property of a delta function, namely the relation

$$\int_{-\infty}^{\infty} g(\mathbf{y})\delta(\mathbf{y} - \mathbf{x}(n))d\mathbf{y} = g(\mathbf{x}(n))$$

for some function $g(\cdot)$, and interchanging the order of summation, we may redefine the function $C(q,r)$ as

$$C(q, r) = \frac{1}{N} \sum_{n=1}^{N} \left(\frac{1}{N-1} \sum_{\substack{k=1 \\ k \neq n}}^{N} \theta(r - \|\mathbf{x}(n) - \mathbf{x}(k)\|) \right)^{q-1} \tag{14.84}$$

The function $C(q, r)$ is called the *correlation function*;[7] it is a measure of the probability that two points $\mathbf{x}(n)$ and $\mathbf{x}(k)$ on the attractor are separated by a distance r. The number of data points N in the defining equation (14.84) is assumed to be large.

The correlation function $C(q, r)$ is an invariant of the attractor in its own right. Nevertheless, the customary practice is to focus on the behavior of $C(q, r)$ for small r. This limiting behavior is described by

$$C(q, r) \simeq r^{(q-1)D_q} \tag{14.85}$$

where D_q, called a *fractal dimension* of the attractor, is assumed to exist. Taking the logarithm of both sides of Eq. (14.85), we may formally define D_q as

$$D_q = \lim_{r \to 0} \frac{\log C(q, r)}{(q-1)\log r} \tag{14.86}$$

However, since we usually have a finite number of data points, the radius r must be just small enough to permit enough points to fall inside the sphere. For a prescribed q, we may then determine the fractal dimension D_q as the slope of the part of the function $C(q, r)$ that is *linear* in $\log r$.

For $q = 2$, the definition of the fractal dimension D_q assumes a simple form that lends it to reliable computation. The resulting dimension, D_2, is called the *correlation dimension* of the attractor (Grassberger and Procaccia, 1983). The correlation dimension reflects the complexity of the underlying dynamical system and bounds the degrees of freedom required to describe the system.

Lyapunov Exponents

The Lyapunov exponents are statistical quantities that describe the uncertainty about the future state of an attractor. More specifically, they quantify the exponential rate at which nearby trajectories separate from each other while moving on the attractor. Let $\mathbf{x}(0)$ be an initial condition and $\{\mathbf{x}(n), n = 0, 1, 2, \ldots\}$ be the corresponding orbit. Consider an infinitesimal displacement from the initial condition $\mathbf{x}(0)$ in the direction of a vector $\mathbf{y}(0)$ tangential to the orbit. Then, the evolution of the tangent vector determines the evolution of the infinitesimal displacement of the perturbed orbit $\{\mathbf{y}(n), n = 0, 1, 2, \ldots\}$ from the unperturbed orbit $\{\mathbf{x}(n), n = 0, 1, 2, \ldots\}$. In particular, the ratio $\mathbf{y}(n)/\|\mathbf{y}(n)\|$ defines the infinitesimal displacement of the orbit from $\mathbf{x}(n)$, and the ratio $\|\mathbf{y}(n)\|/\|\mathbf{y}(0)\|$ is the factor by which the infinitesimal displacement *grows* if $\|\mathbf{y}(n)\| > \|\mathbf{y}(0)\|$ or *shrinks* if $\|\mathbf{y}(n)\| < \|\mathbf{y}(0)\|$. For an initial condition $\mathbf{x}(0)$ and initial displacement $\alpha_0 = \mathbf{y}(0)/\|\mathbf{y}(0)\|$, the *Lyapunov exponent* is defined by

$$\lambda(\mathbf{x}(0), \alpha) = \lim_{n \to \infty} \frac{1}{n} \log\left(\frac{\|\mathbf{y}(n)\|}{\|\mathbf{y}(0)\|}\right) \tag{14.87}$$

A d-dimensional chaotic process has a total of d Lyapunov exponents that can be positive, negative, or zero. Positive Lyapunov exponents account for the instability of an orbit throughout the state space. Stated in another way, positive Lyapunov exponents are responsible for the *sensitivity of a chaotic process to initial conditions*. Negative Lyapunov exponents, on the other hand, govern the decay of transients in the orbit. A zero Lyapunov exponent signifies the fact that the underlying dynamics responsible for the generation of chaos are describable by a coupled system of nonlinear differential equations, that is, the chaotic process is a *flow*. A volume in d-dimensional state space behaves as $\exp(L(\lambda_1 + \lambda_2 + \cdots + \lambda_d))$, where L is the number of time steps into the future. It follows therefore that for a *dissipative* process, the sum of all Lyapunov exponents must be negative. This is a necessary condition for a volume in state space to shrink as time progresses, which is a requirement for physical realizability.

Lyapunov Dimension

Given the Lyapunov spectrum λ_1, λ_2, ..., λ_d, Kaplan and Yorke (1979) suggested a *Lyapunov dimension* for a strange attractor as follows:

$$D_L = K + \frac{\sum_{i=1}^{K} \lambda_i}{|\lambda_{K+1}|} \tag{14.88}$$

where K is an integer that satisfies the two conditions:

$$\sum_{i=1}^{K} \lambda_i > 0 \quad \text{and} \quad \sum_{i=1}^{K+1} \lambda_i < 0$$

Ordinarily, the Lyapunov dimension D_L is about the same size as the correlation dimension D_2. This is an important property of a chaotic process. That is, although the Lyapunov and correlation dimensions are defined in entirely different ways, their values for a strange attractor are usually quite close to each other.

Definition of a Chaotic Process

Throughout this section we have spoken of a chaotic process without a formal definition of it. In light of what we now know about Lyapunov exponents, we can offer the following definition:

> *A chaotic process is generated by a nonlinear deterministic system, with at least one positive Lyapunov exponent.*

The positivity of at least one Lyapunov exponent is a necessary condition for sensitivity to initial conditions, which is the hallmark of a strange attractor.

The largest Lyapunov exponent also defines the *horizon of predictability* of a chaotic process. Specifically, the short-term predictability of a chaotic process is approximately equal to the reciprocal of the largest Lyapunov exponent (Abarbanal, 1996).

14.13 DYNAMIC RECONSTRUCTION

Dynamic reconstruction may be defined as the identification of a mapping that provides a model for an unknown dynamical system of dimensionality m. Our interest here is in the dynamic modeling of a time series produced by a physical system that is known to be chaotic. In other words, given a time series $\{y(n)\}_{n=1}^{N}$, we wish to build a model that captures the underlying dynamics responsible for generation of the observable $y(n)$. As we pointed out earlier in the previous section, N denotes the sample size.

The primary motivation for dynamic reconstruction is to make physical sense from such a time series, thereby bypassing the need for a detailed mathematical knowledge of the underlying dynamics. The system of interest is typically much too complex to characterize in mathematical terms. The only information available to us is contained in a time series obtained from measurements on one of the observables of the system.

A fundamental result in dynamic reconstruction theory[8] is a geometric theorem called the delay-embedding theorem due to Takens (1981). Takens considered a noise-free situation, focusing on *delay coordinate maps* or *predictive models* that are constructed from a time series representing an observable from a dynamical system. In particular, Takens showed that if the dynamical system and the observable are generic, then the delay coordinate map from a d-dimensional smooth compact manifold to $\mathbb{R}^{2d+1}$ is a diffeomorphism on that manifold, where d is the dimension of the state space of the dynamical system. (Diffeomorphism is discussed on p. 744.)

For an interpretation of Takens' theorem in signal processing terms, first consider an unknown dynamical system whose evolution in discrete time is described by the nonlinear difference equation

$$\mathbf{x}(n + 1) = \mathbf{F}(\mathbf{x}(n)) \tag{14.89}$$

where $\mathbf{x}(n)$ is the d-dimensional state vector of the system at time n, and $\mathbf{F}(\cdot)$ is a vector-valued function. It is assumed here that the sampling period is normalized to unity. Let the time series $\{y(n)\}$ observable at the output of the system be defined in terms of the state vector $\mathbf{x}(n)$ as follows:

$$y(n) = g(\mathbf{x}(n)) + v(n) \tag{14.90}$$

where $g(\cdot)$ is a scalar-valued function, and $v(n)$ denotes additive noise. The noise $v(n)$ accounts for the combined effects of imperfections and imprecisions in the observable $y(n)$. Equation (14.89) and (14.90) describe the state-space behavior of the dynamical system. According to Takens' theorem, the geometric structure of the multivariable dynamics of the system can be unfolded from the observable $y(n)$ with $v(n) = 0$ in a D-dimensional space constructed from the new vector

$$\mathbf{y}_R(n) = [y(n), y(n - \tau), ..., y(n - (D - 1)\tau)]^T, \tag{14.91}$$

where τ is a positive integer called the *normalized embedding delay*. That is, given the observable $y(n)$ for varying discrete time n, which pertains to a single observable (component) of an unknown dynamical system, dynamic reconstruction is possible using the D-dimensional vector $\mathbf{y}_R(n)$ provided that $D \geq 2d + 1$, where d is the dimension of the state space of the system. Hereafter we refer to this statement as the *delay-embedding theorem*. The condition $D \geq 2d + 1$ is a *sufficient* but not necessary condition for dynamic reconstruction. The procedure for finding a suitable D is called *embedding*, and the minimum integer D that achieves dynamic reconstruction is called the *embedding dimension*; it is denoted by D_E.

The delay-embedding theorem has a powerful implication: Evolution of the points $\mathbf{y}_R(n) \rightarrow \mathbf{y}_R(n + 1)$ in the reconstruction space follows that of the unknown dynamics $\mathbf{x}(n) \rightarrow \mathbf{x}(n + 1)$ in the original state space. That is, many important properties of the unobservable state vector $\mathbf{x}(n)$ are reproduced without ambiguity in the reconstruction space defined by $\mathbf{y}_R(n)$. However, for this important result to be attainable, we need *reliable estimates* of the embedding dimension D_E and the normalized embedding delay τ, as summarized here:

- The sufficient condition $D \geq 2d + 1$ makes it possible to undo the intersections of an orbit of the attractor with itself, which arise from projection of that orbit to lower dimensions. The embedding dimension D_E can be less than $2d + 1$. The recommended procedure is to estimate D_E directly from the observable data. A reliable method for estimating D_E is the *method of false nearest neighbors* described in Abarbanal (1996). In this method we systematically survey the data points and their neighbors in dimension $d = 1$, then $d = 2$, and so on. We thereby establish the condition when apparent neighbors stop being "unprojected" by the addition of more elements to the reconstruction vector $\mathbf{y}_R(n)$, and thus obtain an estimate for the embedding dimension D_E.
- Unfortunately, the delay-embedding theorem has nothing to say on the choice of the normalized embedding delay τ. In fact, it permits the use of any τ so long as the available time series is infinitely long. In practice, however, we always have to work with observable data of finite length N. The proper prescription for choosing τ is to recognize that the normalized embedding delay τ should be large enough for $y(n)$ and $y(n - \tau)$ to be essentially independent of each other so as to serve as coordinates of the reconstruction space, but not so independent as to have no correlation with each other. This requirement is best satisfied by using the particular τ for which the *mutual information* between $y(n)$ and $y(n - \tau)$ attains its first minimum (Fraser, 1989). Mutual information is discussed in Chapter 10.

Recursive Prediction

From the discussion presented, the dynamic reconstruction problem may be interpreted as one representing the signal dynamics properly (the embedding step), as well as the construction of a predictive mapping (the identification step). Thus in practical terms we have the following network topology for dynamic modeling:

- A *short-term memory* (e.g., delay-line memory) structure to perform the embedding, whereby the reconstruction vector $\mathbf{y}_R(n)$ is defined in terms of the observable $y(n)$ and its delayed versions: see Eq. (14.91).
- A multiple input, single output (MISO) adaptive nonlinear system trained as a *one-step predictor* (e.g., neural network) to identify the unknown mapping $f : \mathbb{R}^D \to \mathbb{R}^1$, which is defined by

$$\hat{y}(n + 1) = f(\mathbf{y}_R(n)) \tag{14.92}$$

The predictive mapping described in Eq. (14.92) is the center piece of dynamic modeling: Once it is determined, the evolution $\mathbf{y}_R(n) \to \mathbf{y}_R(n + 1)$ becomes known, which in turn determines the unknown evolution $\mathbf{x}(n) \to \mathbf{x}(n + 1)$.

Presently, we do not have a rigorous theory to help us decide if the nonlinear predictor has successfully identified the unknown mapping f. In linear prediction, minimizing the mean-square value of the prediction error leads to an accurate model. However, a chaotic time series is different. Two trajectories in the same attractor are vastly different on a sample-by-sample basis, so minimizing the mean-square value of the prediction error is a necessary but not a sufficient condition of a successful mapping.

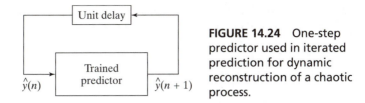

FIGURE 14.24 One-step predictor used in iterated prediction for dynamic reconstruction of a chaotic process.

The dynamic invariants, namely correlation dimension and Lyapunov exponents, measure global properties of the attractor, so they should gauge the success of dynamic modeling. Hence, *a pragmatic approach for testing the dynamic model is to seed it with a point on the strange attractor, and to feed the output back to its input as an autonomous system* as illustrated in Fig. 14.24. Such an operation is called *iterated prediction* or *recursive prediction*. Once the initialization is completed, the output of the autonomous system is a realization of the dynamic reconstruction process. This of course presumes that the predictor has been designed properly in the first place.

We say that dynamic reconstruction performed by means of the autonomous system described in Fig. 14.24 is successful if the following two conditions are satisfied (Haykin and Principe, 1998):

1. *Short-term behavior.* Once the initialization is completed, the reconstructed time series $\{\hat{y}(n)\}$ in Fig. 14.24 closely follows the original time series $\{y(n)\}$ for a period of time, on average equal to the horizon of predictability determined from the Lyapunov spectrum of the process.
2. *Long-term behavior.* The dynamic invariants computed from the reconstructed time series $\{\hat{y}(n)\}$ closely match the corresponding ones computed from the original time series $\{y(n)\}$.

To gauge the long-term behavior of the reconstructed dynamics, we need to estimate (1) the correlation dimension as a measure of attractor complexity, and (2) the Lyapunov spectrum as a framework for assessing sensitivity to initial conditions and for estimating the Lyapunov dimension; see Eq. (14.88). The Lyapunov dimension should have a value close to that of the correlation dimension.

Two Possible Formulations for Recursive Prediction

The reconstruction vector $\mathbf{y}_R(n)$ defined in Eq. (14.91) is of dimension D_E, assuming that the dimension D is set equal to the embedding dimension D_E. The size of the delay line memory required to perform the embedding is τD_E. But the delay line memory is required to provide only D_E outputs (the dimension of the reconstruction space); that is, we use τ equally spaced taps, representing sparse connections.

Alternatively, we may define the reconstruction vector $\mathbf{y}_R(n)$ as a full m-dimensional vector as follows:

$$\mathbf{y}_R(n) = [y(n), y(n-1), ..., y(n-m+1)]^T \tag{14.93}$$

where m is an integer defined by

$$m \geq D_E \tau \tag{14.94}$$

This second formulation of the reconstruction vector $\mathbf{y}_R(n)$ supplies more information to the predictive model than that provided by Eq. (14.91) and may therefore yield a more accurate dynamic reconstruction. However, both formulations share a common feature: their compositions are uniquely defined by knowledge of the embedding dimension D_E. In any event, it is wise to use the minimum permissible value of D, namely D_E, to minimize the effect of additive noise $v(n)$ on the quality of dynamic reconstruction.

Dynamic Reconstruction Is an Ill-Posed Filtering Problem

The dynamic reconstruction problem is, in reality, an *ill-posed inverse problem* for one or more of the following reasons. (The conditions for an inverse problem to be well posed are discussed in Chapter 5.) First, for some unknown reason the existence condition may be violated. Second, there may not be sufficient information in the observable time series to reconstruct the nonlinear dynamics uniquely; hence, the uniqueness criterion is violated. Third, the unavoidable presence of additive noise or some form of imprecision in the observable time series adds uncertainty to the dynamic reconstruction. In particular, if the noise level is too high, it is possible for the continuity criterion to be violated. How then do we make the dynamic reconstruction problem well posed? The answer lies in the inclusion of some form of *prior knowledge* about the input–output mapping as an essential requirement. In other words, some form of constraints (e.g., smoothness of input–output mapping) would have to be imposed on the predictive model designed for solving the dynamic reconstruction problem. One effective way in which this requirement can be satisfied is to invoke Tikhonov's *regularization theory*, which is also discussed in Chapter 5.

Another issue that needs to be considered is the ability of the predictive model to solve the inverse problem with sufficient accuracy. In this context, the use of a neural network to build the predictive model is appropriate. In particular, the universal approximation property of a multilayer perceptron or that of a radial-basis function network means that we can take care of the issue of reconstruction accuracy by using one or the other of these neural networks with an appropriate size. In addition, however, we need the solution to be regularized for the reasons explained. In theory, both multilayer perceptrons and radial-basis function networks lend themselves to the use of regularization; in practice, it is in radial-basis function networks that we find regularization theory included in a mathematically tractable manner as an integral part of their design. Accordingly, in the computer experiment described in the next section, we focus on the regularized radial-basis function (RBF) network (described in Chapter 5) as the basis for solving the dynamic reconstruction problem.

14.14   COMPUTER EXPERIMENT III

To illustrate the idea of dynamic reconstruction, we consider the system of three coupled ordinary differential equations, abstracted by Lorenz (1963) from the Galerkin approximation to the partial differential equations of thermal convection in the lower atmosphere, which stands as a workhorse set of equations for testing ideas in nonlinear dynamics. The equations for the Lorenz attractor are:

$$\frac{dx(t)}{dt} = -\sigma x(t) + \sigma y(t)$$

$$\frac{dy(t)}{dt} = -x(t)z(t) + rx(t) - y(t) \qquad (14.95)$$

$$\frac{dz(t)}{dt} = x(t)y(t) - bz(t)$$

where σ, r, and b are dimensionless parameters. Typical values for these parameters are $\sigma = 10, b = 8/3$, and $r = 28$.

Figure 14.25 shows the results of iterated prediction performed on two RBF networks with 400 centers using a "noisy" time series based on the component $x(t)$ of the Lorenz attractor. The signal-to-noise ratio was +25 dB. In Fig. 14.25a, the design of the network is regularized. In Fig. 14.25b, the design of the network is unregularized. These two parts of Fig. 14.25 clearly demonstrate the practical importance of regularization. Without regularization, the solution to the dynamic reconstruction problem presented in Fig. 14.25b is unacceptable as it fails to approximate the true trajectory of the Lorenz attractor; the unregularized system is just a predictor. On the other hand, the solution to the dynamic reconstruction problem presented in Fig. 14.25a using a regularized form of the RBF network has *learned* the dynamics, in the sense that the output of the network under iterated prediction closely approximates the actual trajectory of the Lorenz attractor in the short term. This is borne out by the results presented in Table 14.5, where we have a summary of Lorenz data for three cases:

(a) Noise-free Lorenz system.
(b) Noisy Lorenz system with signal-to-noise ratio SNR = 25 dB.
(c) Reconstructed data, using the noisy Lorenz time series described under Case b.

The invariants of the reconstructed data using noisy data are close to the corresponding ones pertaining to the noise-free Lorenz data. The deviations in absolute values are due to the residual effect of noise embedded in the reconstructed attractor and to inaccuracies in the estimation procedure. Figure 14.25 clearly shows that there is more to dynamic modeling than just prediction. This figure, and many others not included here, demonstrate the "robustness" of the regularized RBF solution with respect to the point on the attractor that is used to initialize the iterated prediction process.

The following two observations from Fig. 14.25a, pertaining to the use of regularization, are particularly noteworthy:

1. The short-term predictability of the reconstructed time series in Fig. 14.25a is about 60 samples. The theoretical horizon of predictability computed from the Lyapunov spectrum of the noiseless Lorenz attractor is approximately 100 samples. The experimental deviation from the horizon of predictability of the noise-free Lorenz attractor is merely a manifestation of the presence of noise in the actual data used to perform the dynamic reconstruction. The theoretical horizon

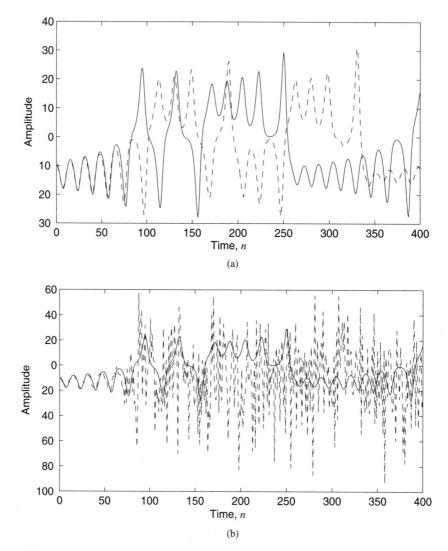

FIGURE 14.25 (a) Regularized iterated prediction ($N = 400$, $m = 20$) on Lorenz data at SNR = +25 dB. (b) Unregularized iterated prediction ($N = 400$, $m = 20$) on Lorenz data at SNR = +25 dB. In both parts (a) and (b), the solid curve is the actual chaotic signal and the dahsed curve is the reconstructed signal.

of predictability computed from the reconstructed data was 61 (Table 14.5), which is quite close to the experimentally observed value of short-term predictability.

2. Once the period of short-term predictability is over, the reconstructed time series in Fig. 14.25a begins to deviate from the noiseless realization of the actual Lorenz attractor. This is basically a manifestation of chaotic dynamics, namely sensitivity to initial conditions. As mentioned previously, sensitivity to initial conditions is a hallmark of chaos.

TABLE 14.5   Summary of Parameters for Experiment on Dynamic Reconstruction Using the Lorenz System

---

(a) *Noise-free Lorenz System*
      Number of samples used: 35,000

   1. Normalized embedding delay, $\tau = 4$
   2. Embedding dimension, $D_E = 3$
   3. Lyapunov exponents:
         $\lambda_1 = 1.5697$
         $\lambda_2 = -0.0314$
         $\lambda_3 = -22.3054$
   4. Horizon of predictability $\simeq 100$ samples

(b) *Noisy Lorenz system: 25 dB SNR*
      Number of samples used: 35,000

   1. Normalized embedding delay, $\tau = 4$
   2. Embedding dimension, $D_E = 5$
   3. Lyapunov exponents:
         $\lambda_1 = 13.2689$
         $\lambda_2 = 5.8562$
         $\lambda_3 = -3.1447$
         $\lambda_4 = -18.0082$
         $\lambda_5 = -47.0572$
   4. Horizon of predictability $\simeq 12$ samples

(c) *Reconstructed system using the noisy Lorenz data under (b)*
      Number of samples generated (recursively): 35,000

   1. Normalized embedding delay, $\tau = 4$
   2. Embedding dimension, $D_E = 3$
   3. Lyapunov exponents:
         $\lambda_1 = 2.5655$
         $\lambda_2 = -0.6275$
         $\lambda_3 = -15.0342$
   4. Horizon of predictability $\simeq 61$ samples

Notes: All of the Lyapunov exponents are expressed in nats per second; a *nat* is a natural unit for measuring information as discussed in Chapter 10. Also, in case b, the effect of noise is to increase the size of the Lyapunov spectrum, and the number and magnitude of positive Lyapunov exponents.

Choice of *m* and λ

The size of the input layer, *m*, is determined in accordance with Eq. (14.94). As explained previously, the recommended method is to use the smallest permissible value of *m* in accordance with the equality sign so as to minimize the effect of noise on dynamic reconstruction.

The estimated value of the normalized embedding delay τ is essentially independent of the presence of noise for moderate to high signal-to-noise ratios. In contrast, the presence of noise has a profound impact on the estimated value of the embedding dimension D_E, which is intuitively satisfying. For example, for the noise-free Lorenz attractor, the correlation dimension is 2.01. We may therefore choose the embedding dimension $D_E = 3$, which is confirmed by the method of false nearest neighbors. The normalized embedding delay is $\tau = 4$. Thus, using Eq. (14.94) with the equality sign yields $m = 12$ for dynamic reconstruction. However, for a noisy Lorenz attractor with SNR $= +25$ dB, the use of the method of false nearest neighbors yields $D_E = 5$, and the use of the method of mutual information yields $\tau = 4$. By substituting these estimated values in Eq. (14.94) with the equality sign, we get $m = 20$ for the noisy dynamic reconstruction reported in Fig. 14.25. Table 14.5 includes the normalized embedding delay τ and embedding dimension D_E for the three cases summarized therein.

As for the regularization parameter λ used in Fig. 14.25a, it was determined from the training data using the generalized cross-validation (GCV); this method is discussed in Chapter 5. The value of λ that was used in Fig. 14.25a, calculated on the basis of GCV, varied between a minimum value of 10^{-14} to a maximum value of 10^{-2} in accordance with the data.

14.15 SUMMARY AND DISCUSSION

Much of the material in this chapter is devoted to the Hopfield model and the brain-state-in-a-box (BSB) model as examples of an associative memory rooted in neurodynamics. These two models share some common features:

- They both employ positive feedback.
- They both have an energy (Lyapunov) function, and the underlying dynamics tend to minimize it in an iterative fashion.
- They both learn in a self-organized manner using Hebb's postulate of learning.
- They are both capable of performing computation using attractor dynamics.

Naturally, they differ in their areas of application.

The BSB model has an inherent clustering capability that may be put to good use for data representation and concept formation. The most interesting application of the BSB model is perhaps as the basic computational unit in a *network of networks* proposed as a plausible model of describing different levels of system organization within the brain (Anderson and Sutton, 1995). In this model, the computational units form local networks that are distributed in a two-dimensional array, hence the term "network of networks." Instead of communicating only average activity from column to column, these local networks are designed to communicate with other local networks by means of *activity patterns (vectors)*. In place of synaptic weights between neurons as in a conventional neural network, we now have a set of interaction matrices that describe the coupling between attractors in two local networks. The local networks form clusters and levels, based on their interconnections, with the result that the *anatomical* connectivity is sparse. That is, the local networks are more densely connected within clusters than across clusters. However, the *functional* connectivity among clusters has a rich dynamic, due in part to temporally correlated activity among local networks.

In contrast, the Hopfield model may be used to solve the following computational problems:

1. *Content-addressable memory*, which involves the recall of a stored pattern by presenting a partial or distorted version of it to the memory. For this application, the usual procedure is to use the "discrete" Hopfield model that is based on the McCulloch–Pitts neuron (i.e., one using a hard limiting activation function). Viewed in a computational context, the construction of a content-addressable memory is a trivial matter. Nevertheless, the Hopfield model of a content-addressable memory is important because it elucidates the *link between dynamics and computation* in a novel way. In particular, the Hopfield model exhibits the following properties that have neurobiological relevance:
   - The dynamics of the model are dominated by a large number of point attractors in a high-dimensional state space.
   - A point attractor (i.e., fundamental memory) of interest may be located by merely initializing the model with an inexact description of that attractor's location and then allowing the dynamics to evolve the state of the model to the nearest point attractor.
   - Learning (i.e., computation of the free parameters of the model) takes place in accordance with Hebb's postulate of learning. Moreover, this learning mechanism permits the insertion of new point attractors into the model if so desired.

2. *Combinatorial optimization problems*, which rank among the most difficult known to mathematicians. This class of optimization problems includes the *traveling salesman problem* (TSP), considered to be a classic. Given the positions of a specified number of cities, assumed to lie in a plane, the problem is to find the shortest tour that starts and finishes at the same city. The TSP is simple to state but hard to solve exactly, in that there is no known method of finding the optimum tour, short of computing the length of every possible tour and then selecting the shortest one. It is said to be *NP-complete* (Hopcroft and Ullman, 1979). In a pioneering paper, Hopfield and Tank (1985) demonstrated how an analog network, based on the system of coupled first-order differential equations (14.20) can be used to represent a solution of the TSP. Specifically, the synaptic weights of the network are determined by distances between the cities visited on the tour, and the optimum solution to the problem is a fixed point of the neurodynamical equations (14.20). Herein lie the difficulties encountered with "mapping" combinatorial optimization problems onto the continuous (analog) Hopfield network. The network acts to minimize a single energy (Lyapunov) function, and yet the typical combinatorial optimization problem requires the minimization of an objective function *subject to* some hard constraints (Gee et al., 1993). If any of these constraints are violated, the solution is considered to be invalid. The early mapping procedures were based on a Lyapunov function constructed in an *ad hoc* manner, usually employing one term for each constraint, as shown by

$$E = E^{\text{opt}} + c_1 E_1^{\text{cns}} + c_2 E_2^{\text{cns}} + \cdots \tag{14.96}$$

The first term, E^{opt}, is the objective function to be minimized (e.g., the length of a TSP tour); it is determined by the problem at hand. The remaining terms $E_1^{\text{cns}}, E_2^{\text{cns}}, \dots$, represent penalty functions whose minimization satisfies the *constraints*. The scalars $c_1, c_2, \dots$, are constant weights assigned to the respective penalty functions $E_1^{\text{cns}}, E_2^{\text{cns}}, \dots$, usually by trial and error. Unfortunately, the many terms in the Lyapunov function of Eq. (14.96) tend to frustrate one another, and the success of the Hopfield network is highly sensitive to the relative values of $c_1, c_2, \ldots$ (Gee et al., 1993). It is therefore not surprising that the network often produces a large number of invalid solutions (Wilson and Pawley, 1988; Ansari and Hou, 1997). In Gee (1993) a number of basic questions concerning the use of the continuous Hopfield network as a tool for solving combinatorial optimization problems is addressed; the main findings reported therein may be summarized as follows:

- Given a combinatorial optimization problem expressed in terms of quadratic 0-1 programming, as in the traveling salesman problem, there is a straightforward method for programming the network for its solution, and the solution found will not violate any of the problem's constraints.

- Building on results from complexity theory and mathematical programming, it is shown that, except when the problem's constraints have special properties producing an integral polytope, it is not possible to force the network to converge to a valid, interpretable solution. In geometric terms, a polytope, that is a bounded polyhedron, is said to be an *integral polytope* if all the vertices of the polytope are 0-1 points. Even when dealing with integral polytopes, if the objective function E^{opt} is quadratic, the problem is NP-complete and there is no guarantee that the network will find the optimum solution; this class of problems includes the traveling salesman problem. However, a valid solution will be found and, given the nature of the descent process to this solution, there is a good chance that the solution will be reliable.

The Hopfield model, considered in this chapter, uses symmetric couplings between its neurons. The dynamics of such a structure are similar to gradient descent dynamics, thereby assuring convergence to a fixed point. However, the dynamics of a brain differ from that of the Hopfield model in two important respects:

- The interneuron connections in the brain are asymmetric.
- Oscillatory and complex nonperiodic behavior are observed in the brain.

Indeed, it is because of these special features of the brain that there has been a long-standing interest in the study of asymmetric networks,[9] predating the Hopfield model.

If we abandon the restriction of symmetry, the next simplest model is the *excitatory-inhibitory network* whose neurons fall into two populations: one with only excitatory outputs, and the other with only inhibitory outputs. The synaptic connections between the two populations are antisymmetric, while the synaptic connections within each population are symmetric. In Seung et al. (1998), the dynamics of such a network are considered. The analysis presented therein exploits the similarity of the underlying dynamics of the inhibitory-excitatory network to *gradient descent–gradient ascent dynamics*, where the equations of motion are gradient descent in some state variables and gradient ascent in the others. Consequently, unlike the gradient descent dynamics characterizing the Hopfield model, the dynamics of the model considered by Seung et. al. can converge to a fixed point or a limit cycle, depending on the choice of network parameters. Thus the antisymmetric model studied in Seung et al. (1998) represents an advance on the symmetric Hopfield model.

NOTES AND REFERENCES

1. The direct method of Lyapunov is also referred in the literature as the second method. For an early account of this pioneering work, see the book by LaSalle and Lefschetz (1961).

The alternative spelling, Liapunov, is frequently used in literature; the difference in spelling arose during transliteration from Russian characters (Brogan, 1985).

2. A *nonautonomous* dynamical system is defined by the state equation

$$\frac{d}{dt}\mathbf{x}(t) = \mathbf{F}(\mathbf{x}(t), t)$$

with the initial condition $\mathbf{x}(t_0) = \mathbf{x}_0$. For a nonautonomous system the vector field $\boldsymbol{F}(\mathbf{x}(t), t)$ depends on time t. Therefore, unlike the case of an autonomous system, we generally cannot set the initial time equal to zero (Parker and Chua, 1989).

3. For a rigorous definition of an attractor we offer the following (Lanford, 1981; Lichtenberg and Lieberman, 1992):

A subset (manifold) M of the state space is called an attractor if:
- M is invariant under the flow
- There is an (open) neighborhood around M that shrinks down to M under the flow
- No part of M is transient
- M cannot be decomposed into two nonoverlapping invariant pieces.

4. Integrate-and-Fire Neuron

The additive model of Eq. (14.14) does not fully capture the essence of what a biological neuron does. In particular, it ignores the timing information encoded into action potentials; action potentials are described briefly in qualitative terms in Chapter 1. Hopfield (1994) describes a dynamical model that accounts for action potentials by considering an integrate-and-fire neuron. The operation of such a neuron is described by the first-order differential equation

$$C\frac{d}{dt}u(t) = -\frac{1}{R}(u(t) - u_0) + i(t) \tag{1}$$

where

$\quad u(t)$ = interior potential of the neuron
$\quad C$ = capacitance of the membrane surrounding the neuron
$\quad R$ = leakage resistance of the membrane
$\quad i(t)$ = electrical current injected into the neuron by another neuron
$\quad u_0$ = potential to which the neuron is reduced when $i(t)$ vanishes.

An action potential is generated each time the interior potential $u(t)$ reaches a threshold value.

The action potentials are treated as Dirac delta (impulse) functions as shown by

$$g_k(t) = \sum_n \delta(t - t_{k,n}) \tag{2}$$

where $t_{k,n}$, $n = 1, 2, 3, \ldots$ denotes the times at which neuron k *fires* action potentials. These times are defined by Eq. (1).

The behavior of total current $i_k(t)$ flowing into neuron k is modeled as

$$\frac{d}{dt}i_k(t) = -\frac{1}{\tau}i_k(t) + \sum_j w_{kj}g_j(t) \tag{3}$$

where w_{kj} is the synaptic weight from neuron j to neuron k, τ is a characteristic time constant of neuron k, and the function $g_j(t)$ is defined in accordance with Eq. (2).

The additive model of Eq. (14.14) may be viewed as a special case of Eq. (3). Specifically, the spiky nature of $g_j(t)$ is ignored by replacing it with the convolution of $g_j(t)$ with a smoothing function. Such a move is justified if, during a reasonable time interval, there are many contributions to the sum on the right-hand side of Eq. (3) due to high connectivity, and all that we are really interested in is the short-term behavior of the firing rate of neuron k.

5. The *Little model* (Little, 1974; Little and Shaw, 1975) uses the same synaptic weights as the Hopfield model. However, they differ from each other in that the Hopfield model uses *asynchronous (serial) dynamics*, whereas the Little model uses *synchronous (parallel) dynamics*. Accordingly, they exhibit different convergence properties (Bruck, 1990; Goles and Martinez, 1990): The Hopfield network will always converge to a stable state, whereas the Little model will always converge to a stable state or a limit cycle of length at most 2. By such a "limit cycle" we mean that the cycles in the state space of the network are of a length less than or equal to 2.

6. **Nonmonotonic Activation Function**
Various proposals have been made in literature for overcoming the limitations of the Hopfield model as a content-addressable memory. Perhaps the most significant improvement suggested to date is that due to Morita (1993), which applies to the continuous (analog) form of the Hopfield model. The modification is confined to the activation function $\varphi(\cdot)$ of a neuron, thereby retaining the simplicity of the network as an associative memory. Specifically, the usual hard-limiting or sigmoid activation function of each neuron in the network is replaced by a *nonmonotonic function*. In mathematical terms this activation function is defined as the product of two factors, as shown by

$$\varphi(v) = \left(\frac{1 - \exp(-av)}{1 + \exp(-av)}\right)\left(\frac{1 + \kappa\exp(b(|v| - c))}{1 + \exp(b(|v| - b))}\right) \tag{1}$$

where v is the induced local field. The first factor on the right-hand side of Eq. (1) is the usual sigmoid function (hyperbolic tangent) used in the continuous version of the Hopfield network; the parameter a is the gain of the neuron. The second factor is responsible for making the activation function $\varphi(v)$ nonmonotonic. Two of the parameters characterizing this second factor, namely b and c, are positive constants; the remaining parameter κ is usually negative. In the experiments performed by Morita (1993), the following parameter values were used:

$$a = 50; \quad b = 15$$
$$c = 0.5; \quad \kappa = -1$$

According to Morita, the exact form of the activation function and the parameters used to describe it are not very critical; the essential factor is the nonmonotonic property of the activation function.

The model of a content-addressable memory described by Morita exhibits two interesting properties (Yoshizawa et al., 1993):

1. For a network made up of N neurons, the storage capacity of the model is about $0.3N$, which (for large N) is much greater than the corresponding value $N/(2\log N)$ of the conventional Hopfield model.

2. The model exhibits *no* spurious states; instead, when it fails to recall a correct memorized pattern, the state of the network is driven into a chaotic behavior. The notion of chaos is discussed in Section 14.13.

7. The idea of a correlation function $C(q, r)$ as defined in Eq. (14.84) was known in statistics from the work of Rényi (1970). However, the use of it to characterize a strange attractor is due to Grassberger and Procaccia (1983). They originally discussed the use of $C(q, r)$ in the context of correlation dimension for $q = 2$.

8. The construction of dynamics using independent coordinates from a time series was first advocated by Packard et al. (1980). However, this paper does not give proof, and uses "derivative" embeddings rather than time-delay embeddings. The idea of time-delay or delay coordinate embeddings is attributed to Ruelle and Takens. Specifically, in 1981, Takens published a mathematically profound paper on time-delay embeddings, which applies to attractors that are surfaces, or like a torus; see also the paper by Mañé (1981) on the same subject published in the same issue. Takens' paper is difficult to read for non-mathematicians, and Mañé's paper is even more difficult to read. The idea of delay coordinate mapping was refined in 1991 by Sauer et al. The approach taken in this latter paper integrates and expands on previous results due to Whitney (1936) and Takens (1981).

9. The treatment of biological neural networks as nonlinear dynamical systems that exhibit oscillatory behavior and traveling waves has a long history (Wilson and Cowan, 1972; Amari, 1977a, 1977b; Amari and Arbib, 1977); see also the discussion in Carpenter et al. (1987).

PROBLEMS

Dynamical Systems

14.1 Restate Lyapunov's theorems for the state vector $\mathbf{x}(0)$ as the equilibrium state of a dynamical system.

14.2 Verify the block diagrams of Figs. 14.8a and 14.8b for the neurodynamical equations (14.18) and (14.19), respectively.

14.3 Consider a general neurodynamical system with an unspecified dependence on internal dynamical parameters, external dynamical stimuli, and state variables. The system is defined by the state equations

$$\frac{dx_j}{dt} = \varphi_j(\mathbf{W}, \mathbf{u}, \mathbf{x}), \quad j = 1, 2, ..., N$$

where the matrix $\mathbf{W}$ represents the internal dynamical parameters of the system, the vector $\mathbf{u}$ represents the external dynamical stimuli, and $\mathbf{x}$ is the state vector whose jth element is denoted by x_j. Assume that trajectories of the system converge onto point attractors for values of $\mathbf{W}$, $\mathbf{u}$, and initial states $\mathbf{x}(0)$ in some operating region of the state space (Pineda, 1988b). Discuss how the system described may be used for the following applications:

(a) Continuous mapper, with $\mathbf{u}$ as input and $\mathbf{x}(\infty)$ as output

(b) Autoassociative memory, with $\mathbf{x}(0)$ as input and $\mathbf{x}(\infty)$ as output

Hopfield Models

14.4 Consider a Hopfield network made up of five neurons, which is required to store the following three fundamental memories:

$$\boldsymbol{\xi}_1 = [+1, +1, +1, +1, +1]^T$$
$$\boldsymbol{\xi}_2 = [+1, -1, -1, +1, -1]^T$$
$$\boldsymbol{\xi}_3 = [-1, +1, -1, +1, +1]^T$$

(a) Evaluate the 5-by-5 synaptic weight matrix of the network.

(b) Use asynchronous updating to demonstrate that all three fundamental memories, ξ_1, ξ_2, and ξ_3, satisfy the alignment condition.

(c) Investigate the retrieval performance of the network when it is presented with a noisy version of ξ_1 in which the second element is reversed in polarity.

14.5 Investigate the use of synchronous updating for the retrieval performance of the Hopfield network described in Problem 14.4.

14.6 (a) Show that

$$\xi_1 = [-1, -1, -1, -1, -1]^T$$

$$\xi_2 = [-1, +1, +1, -1, +1]^T$$

$$\xi_3 = [+1, -1, +1, -1, -1]^T$$

are also fundamental memories of the Hopfield network described in Problem 14.4. How are these fundamental memories related to those of Problem 14.4?

(b) Suppose that the first element of the fundamental memory ξ_3 in Problem 14.4 is masked (i.e., reduced to zero). Determine the resulting pattern produced by the Hopfield network. Compare this result with the original form of ξ_3.

14.7 Consider a simple Hopfield network made up of two neurons. The synaptic weight matrix of the network is

$$\mathbf{W} = \begin{bmatrix} 0 & -1 \\ -1 & 0 \end{bmatrix}$$

The bias applied to each neuron is zero. The four possible states of the network are

$$\mathbf{x}_1 = [+1, +1]^T$$

$$\mathbf{x}_2 = [-1, +1]^T$$

$$\mathbf{x}_3 = [-1, -1]^T$$

$$\mathbf{x}_4 = [+1, -1]^T$$

(a) Demonstrate that states $\mathbf{x}_2$ and $\mathbf{x}_4$ are stable, whereas states $\mathbf{x}_1$ and $\mathbf{x}_3$ exhibit a limit cycle. Do this demonstration using the following tools:

1. The alignment (stability) condition

2. The energy function

(b) What is the length of the limit cycle characterizing states $\mathbf{x}_1$ and $\mathbf{x}_3$?

14.8 In this problem we derive the formula of Eq. (14.55) for the storage capacity almost without errors pertaining to the Hopfield network used as a content-addressable memory.

(a) The asymptotic behavior of the error function is closely described by

$$\text{erf}(y) \simeq 1 - \frac{e^{-y^2}}{\sqrt{\pi}y} \quad \text{for large } y$$

Using this approximation, show that the conditional probability of Eq. (14.53) may be approximated as

$$P(v_j > 0 | \xi_{v,j} = +1) \simeq 1 - \frac{e^{-\rho/2}}{\sqrt{2\pi\rho}}$$

where ρ is the signal-to-noise ratio. Show that the probability of stable patterns is correspondingly approximated as

$$p_{\text{stab}} \simeq 1 - \frac{Ne^{-\rho/2}}{\sqrt{\pi\rho}}$$

(b) The second term in the formula for p_{stab} given in part (a) is the probability that a bit in a fundamental memory is unstable. For the definition of storage capacity almost without errors it is not sufficient to require that this term be small; rather it must be small compared to $1/N$, where N is the size of the Hopfield network. Show that the signal-to-noise ratio must satisfy the condition

$$\rho > 2\log_e N + \frac{1}{2}\log_e(2\pi\rho)$$

(c) Using the result derived in part (b), show that the minimum permissible value of the signal-to-noise ratio for the perfect recall of most of the fundamental memories is

$$\rho_{\min} = 2\log_e N$$

What is the corresponding value of p_{stab}?

(d) Using the results of part (c), show that

$$M_{\max} \simeq \frac{N}{2\log_e N}$$

as described in Eq. (14.55).

(e) The formula derived in part (d) for the storage capacity is based on the premise that *most* of the fundamental memories are stable. For a more stringent definition of storage capacity without errors, we require that *all* of the fundamental memories be retrieved correctly. Using this latter definition, show that the maximum number of fundamental memories that can be stored in the Hopfield network is given by (Amit, 1989)

$$M_{\max} \simeq \frac{N}{4\log_e N}$$

14.9 Show that the energy function of a Hopfield network may be expressed as

$$E = -\frac{N}{2}\sum_{v=1}^{M}m_v^2$$

where m_v denotes overlaps defined by

$$m_v = \frac{1}{N}\sum_{j=1}^{N}x_j\xi_{v,j}, \qquad v = 1, 2, \ldots, M$$

where x_j is the jth element of the state vector $\mathbf{x}$, $\xi_{v,j}$ is the jth element of the fundamental memory $\mathbf{\xi}_v$, and M is the number of fundamental memories.

14.10 A Hopfield network is designed to store the two fundamental memory patterns $(+1, +1, -1, +1, +1)$ and $(+1, -1, +1, -1, +1)$. The synaptic matrix of the network is given by

$$\mathbf{W} = \begin{bmatrix} 0 & 0 & 0 & 0 & 2 \\ 0 & 0 & -2 & 2 & 0 \\ 0 & -2 & 0 & -2 & 0 \\ 0 & 2 & -2 & 0 & 0 \\ 2 & 0 & 0 & 0 & 0 \end{bmatrix}$$

(a) The sum of the eigenvalues of the matrix $\mathbf{W}$ is zero. Why?

(b) The state space of the network is a subspace of $\mathbb{R}^5$. Specify the configuration of this subspace.

(c) Specify the subspace $\mathcal{M}$ spanned by the fundamental memory vectors, and the null subspace $\mathcal{N}$ of the weight matrix $\mathbf{W}$. What are the fixed points (stable states) and spurious states of the network?

(The reader may wish to refer to the paper by deSilva and Attikiouzzel (1992) for a more detailed description of the dynamics of the network described here.)

14.11 Figure P14.11 shows a piecewise-linear form of nonmonotonic activation function. The recalling dynamics of the Hopfield network using this approximation are defined by

$$\frac{d}{dt}\mathbf{v}(t) = -\mathbf{v}(t) + \mathbf{W}\mathbf{x}(t), \quad \mathbf{x}(t) = \text{sgn}(\mathbf{v}(t)) - k\mathbf{v}(t)$$

where $\mathbf{v}(t)$ is the vector of induced local fields, $\mathbf{W}$ is the synaptic weight matrix, $\mathbf{x}(t)$ is the state (output) vector, and $-k$ is a negative constant slope. Let $\bar{\mathbf{v}}$ be an equilibrium state of the network that lies in the quadrant of the fundamental memory $\boldsymbol{\xi}_1$, and let

$$\bar{\mathbf{x}} = \text{sgn}(\bar{\mathbf{v}}) - k\bar{\mathbf{v}}$$

Show that $\bar{\mathbf{x}}$ is characterized by the following three conditions (Yoshizawa et al., 1993):

(a) $\displaystyle\sum_{i=1}^{N} \bar{x}_i \xi_{\mu, i} = 0, \quad \mu = 2, 3, \ldots, M$

(b) $\displaystyle\sum_{i=1}^{N} \bar{x}_i \xi_{1, i} = M$

(c) $\bar{x}_i < 1, \quad i = 1, 2, \ldots, N$

where $\boldsymbol{\xi}_1, \boldsymbol{\xi}_2, \ldots, \boldsymbol{\xi}_M$ are the fundamental memories stored in the network, $\xi_{\mu, i}$ is the ith element of $\boldsymbol{\xi}_\mu$, $\bar{x}_i$ is the ith element of $\bar{\mathbf{x}}$, and N is the number of neurons.

14.12 Consider the simple neurodynamical model described by the system of equations

$$\frac{dv_j}{dt} = -v_j + \sum_i w_{ji}\varphi(v_i) + I_j, \quad j = 1, 2, \ldots, N$$

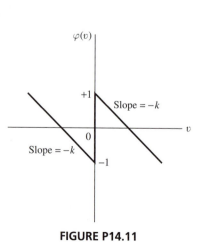

FIGURE P14.11

The system described always converges to a unique point attractor, provided that the synaptic weights w_{ji} satisfy the condition

$$\sum_j \sum_i w_{ji}^2 < \frac{1}{(\max|\varphi'|)^2}$$

where $\varphi' = d\varphi/dv_j$. Explore the validity of this condition. You may refer to the paper (Atiya, 1987) where this condition is derived.

Cohen-Grossberg Theorem

14.13 Consider the Lyapunov function E defined in Eq. (14.57). Show that

$$\frac{dE}{dt} \leq 0$$

provided that the conditions in Eqs. (14.59) to (14.61) are satisfied.

14.14 In Section 14.10 we derived the Lyapunov function of the BSB model by applying the Cohen–Grossberg theorem. In carrying out the derivation, we omitted some of the details leading to Eq. (14.73). Fill in the details.

14.15 Figure P14.15 shows a plot of the nonmonotonic activation function due to Morita (1993) discussed in note 6. This activation function is used in place of the hyperbolic tangent function in the construction of a Hopfield network. Is the Cohen–Grossberg theorem applicable to the associative memory so constructed? Justify your answer.

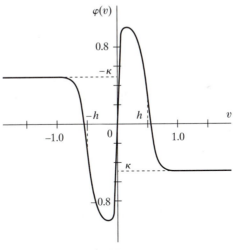

FIGURE P14.15

CHAPTER 15

Dynamically Driven Recurrent Networks

15.1 INTRODUCTION

As mentioned in the previous chapter, *recurrent networks* are neural networks with one or more feedback loops. The feedback can be of a *local* or *global* kind. In this chapter we continue the study of recurrent networks with global feedback.

Given a multilayer perceptron as the basic building block, the application of global feedback can take a variety of forms. We may have feedback from the output neurons of the multilayer perceptron to the input layer. Yet another possible form of global feedback is from the hidden neurons of the network to the input layer. When the multilayer perceptron has two or more hidden layers, the possible forms of global feedback expand even further. The point is that recurrent networks have a rich repertoire of architectural layouts.

Basically, there are two functional uses of recurrent networks:

- *Associative memories*
- *Input–output mapping networks*

The use of recurrent networks as associative memories is considered in detail in Chapter 14. In the present chapter, we will study their use as input–output mapping networks. Whatever the use, an issue of particular concern in the study of recurrent networks is that of *stability;* that issue is also considered in Chapter 14.

By definition, the input space of a mapping network is mapped onto an output space. For this kind of an application, a recurrent network responds *temporally* to an externally applied input signal. We may therefore speak of the recurrent networks considered in this chapter as *dynamically driven recurrent networks.* Moreover, the application of feedback enables recurrent networks to acquire *state* representations, which make them suitable devices for such diverse applications as nonlinear prediction and modeling, adaptive equalization of communication channels, speech processing, plant control, and automobile engine diagnostics. As such, recurrent networks offer an alternative to the dynamically driven feedforward networks described in Chapter 13.

Because of the beneficial effects of global feedback, they may actually fare better in these applications. The use of global feedback has the potential of reducing the memory requirement significantly.

Organization of the Chapter

The chapter is organized in four parts: architectures, theory, learning algorithms, and applications. Part 1, consisting of Section 15.2, deals with recurrent network architectures.

Part 2, consisting of Sections 15.3 to 15.5, deals with theoretical aspects of recurrent networks. Section 15.3 describes the state-space model and the related issues of controllability and observability. Section 15.4 derives an equivalent to the state-space model known as the nonlinear autoregressive with exogenous inputs model. Section 15.5 discusses some theoretical issues pertaining to the computational power of recurrent networks.

Part 3, consisting of Sections 15.6 to 15.12, is devoted to learning algorithms and related issues. It starts with an overview of the subject matter in Section 15.6. Then Section 15.7 discusses back-propagation through time that builds on material presented in Chapter 4. Section 15.8 discusses another popular algorithm: real-time recurrent learning. In Section 15.9 we present a brief review of classical Kalman filter theory, followed by a description of the decoupled extended Kalman filtering algorithm in Section 15.10. A computer experiment on this latter algorithm for recurrent learning is presented in Section 15.11. Gradient-based recurrent learning suffers from the vanishing gradients problem, which is discussed in Section 15.12.

The fourth and last part of the chapter, consisting of Sections 15.13 and 15.14, deals with two important applications of recurrent networks. Section 15.13 discusses system identification. Section 15.14 discusses model-reference adaptive control.

The chapter concludes with some final remarks in Section 15.15.

15.2   RECURRENT NETWORK ARCHITECTURES

As mentioned in the introduction, the architectural layout of a recurrent network takes many different forms. In this section we describe four specific network architectures, each of which highlights a specific form of global feedback.[1] They share the following common features:

- They all incorporate a *static* multilayer perceptron or parts thereof.
- They all exploit the nonlinear mapping capability of the multilayer perceptron.

Input–Output Recurrent Model

Figure 15.1 shows the architecture of a generic recurrent network that follows naturally from a multilayer perceptron. The model has a single input that is applied to a tapped-delay-line memory of q units. It has a single output that is fed back to the input via another tapped-delay-line memory also of q units. The contents of these two tapped-delay-line memories are used to feed the input layer of the multilayer perceptron. The present value of the model input is denoted by $u(n)$, and the corresponding

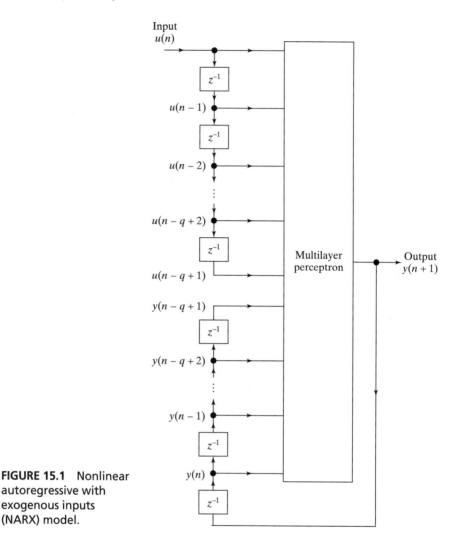

FIGURE 15.1 Nonlinear autoregressive with exogenous inputs (NARX) model.

value of the model output is denoted by $y(n + 1)$; that is, the output is ahead of the input by one time unit. Thus, the signal vector applied to the input layer of the multilayer perceptron consists of a data window made up as follows:

- Present and past values of the input, namely $u(n)$, $u(n - 1)$, ..., $u(n - q + 1)$, which represent *exogenous* inputs originating from outside the network.
- Delayed values of the output, namely, $y(n)$, $y(n - 1)$, ..., $y(n - q + 1)$, on which the model output $y(n + 1)$ is *regressed*.

Thus the recurrent network of Fig. 15.1 is referred to as a *nonlinear autoregressive with exogenous inputs (NARX) model.*[2] The dynamic behavior of the NARX model is described by

$$y(n + 1) = F(y(n), ..., y(n - q + 1), u(n), ..., u(n - q + 1)) \qquad (15.1)$$

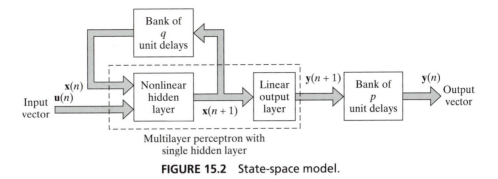

FIGURE 15.2   State-space model.

where F is a nonlinear function of its arguments. Note that in Fig. 15.1 we have assumed that the two delay-line memories in the model are both of size q; they are generally different. The NARX model is explored in greater detail in Section 15.4.

State-Space Model

Figure 15.2 shows the block diagram of another generic recurrent network, called a *state-space model*. The hidden neurons define the *state* of the network. The output of the hidden layer is fed back to the input layer via a bank of unit delays. The input layer consists of a concatenation of feedback nodes and source nodes. The network is connected to the external environment via the source nodes. The number of unit delays used to feed the output of the hidden layer back to the input layer determines the *order* of the model. Let the m-by-1 vector $\mathbf{u}(n)$ denote the input vector, and the q-by-1 vector $\mathbf{x}(n)$ denote the output of the hidden layer at time n. We may then describe the dynamic behavior of the model in Fig. 15.2 by the pair of coupled equations:

$$\mathbf{x}(n + 1) = \mathbf{f}(\mathbf{x}(n), \mathbf{u}(n)) \tag{15.2}$$

$$\mathbf{y}(n) = \mathbf{C}\mathbf{x}(n) \tag{15.3}$$

where $\mathbf{f}(\cdot, \cdot)$ is a nonlinear function characterizing the hidden layer, and $\mathbf{C}$ is the matrix of synaptic weights characterizing the output layer. The hidden layer is nonlinear, but the output layer is linear.

The recurrent network of Fig. 15.2 includes several recurrent architectures as special cases. Consider, for example, the *simple recurrent network* (SRN) described in Elman (1990) and depicted in Fig. 15.3. Elman's network has an architecture similar to that of Fig. 15.2 except for the fact that the output layer may be nonlinear and the bank of unit delays at the output is omitted.

Elman's network contains recurrent connections from the hidden neurons to a layer of *context units* consisting of unit delays. These context units store the outputs of the hidden neurons for one time step, and then feed them back to the input layer. The hidden neurons thus have some record of their prior activations, which enables the network to perform learning tasks that extend over time. The hidden neurons also feed the output neurons that report the response of the network to the externally applied stimulus. Due to the nature of feedback around the hidden neurons, these neurons may

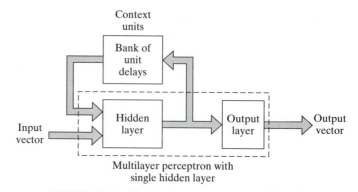

FIGURE 15.3 Simple Recurrent network (SRN).

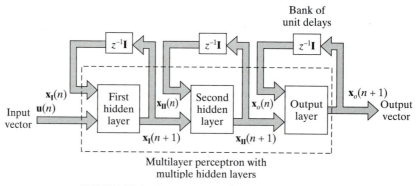

FIGURE 15.4 Recurrent multilayer perceptron.

continue to recycle information through the network over multiple time steps, and thereby discover abstract representations of time. The simple recurrent network is therefore not merely a tape recording of past data.

Elman (1990) discusses the use of the simple recurrent network shown in Fig. 15.3 to discover word boundaries in a continuous stream of phonemes without any built-in representational constraints. The input to the network represents the current phoneme. The output represents the network's best guess as to what the next phoneme is in the sequence. The role of the context units is to provide the network with *dynamic memory* so as to encode the information contained in the sequence of phonemes, which is relevant to the prediction.

Recurrent Multilayer Perceptron

The third recurrent architecture considered here is known as a *recurrent multilayer perceptron* (RMLP) (Puskorius et al., 1996). It has one or more hidden layers, basically for the same reasons that static multilayer perceptrons are often more effective and parsimonious than those using a single hidden layer. Each computation layer of an RMLP has feedback around it, as illustrated in Fig. 15.4 for the case of an RMLP with two hidden layers.[3]

Let the vector $\mathbf{x}_\mathrm{I}(n)$ denote the output of the first hidden layer, $\mathbf{x}_\mathrm{II}(n)$ denote the output of the second hidden layer, and so on. Let the vector $\mathbf{x}_o(n)$ denote the output of the output layer. Then the dynamic behavior of the RMLP, in general, in response to an input vector $\mathbf{u}(n)$ is described by the following system of coupled equations:

$$\mathbf{x}_\mathrm{I}(n + 1) = \boldsymbol{\varphi}_\mathrm{I}(\mathbf{x}_\mathrm{I}(n), \mathbf{u}(n))$$

$$\mathbf{x}_\mathrm{II}(n + 1) = \boldsymbol{\varphi}_\mathrm{II}(\mathbf{x}_\mathrm{II}(n), \mathbf{x}_\mathrm{I}(n + 1))$$

$$\vdots$$

$$\mathbf{x}_o(n + 1) = \boldsymbol{\varphi}_o(\mathbf{x}_o(n), \mathbf{x}_K(n + 1))$$

(15.4)

where $\boldsymbol{\varphi}_\mathrm{I}(\cdot, \cdot), \boldsymbol{\varphi}_\mathrm{II}(\cdot, \cdot), ..., \boldsymbol{\varphi}_o(\cdot, \cdot)$ denote the activation functions characterizing the first hidden layer, second hidden layer, ..., and output layer of the RMLP, respectively; and K denotes the number of hidden layers in the network.

The RMLP described herein subsumes the Elman network of Fig. 15.3 and the state-space model of Fig. 15.2 since the output layer of the RMLP or any of its hidden layers is not constrained to have a particular form of activation function.

Second-Order Network

In describing the state-space model of Fig. 15.2 we used the term "order" to refer to the number of hidden neurons whose outputs are fed back to the input layer via a bank of unit delays.

In yet another context, the term "order" is sometimes used to refer to the way in which the induced local field of a neuron is defined. Consider, for example, a multilayer perceptron where the induced local field v_k of neuron k is defined by

$$v_k = \sum_j w_{a,kj} x_j + \sum_i w_{b,ki} u_i$$

(15.5)

where x_j is the feedback signal derived from hidden neuron j and u_i is the source signal applied to node i in the input layer; the w's represent the pertinent synaptic weights in the network. We refer to a neuron described in Eq. (15.5) as a *first-order neuron*. When, however, the induced local field v_k is combined using multiplications, as shown by

$$v_k = \sum_i \sum_j w_{kij} x_i u_j$$

(15.6)

we refer to the neuron as a *second-order neuron*. The second-order neuron k uses a single weight, w_{kji}, that connects it to the input nodes i and j.

Second-order neurons constitute the basis of *second-order recurrent networks* (Giles et al., 1990), an example of which is shown in Fig. 15.5. The network accepts a time-ordered sequence of inputs and evolves with dynamics defined by the following pair of equations:

$$v_k(n) = b_k + \sum_i \sum_j w_{kij} x_i(n) u_j(n)$$

(15.7)

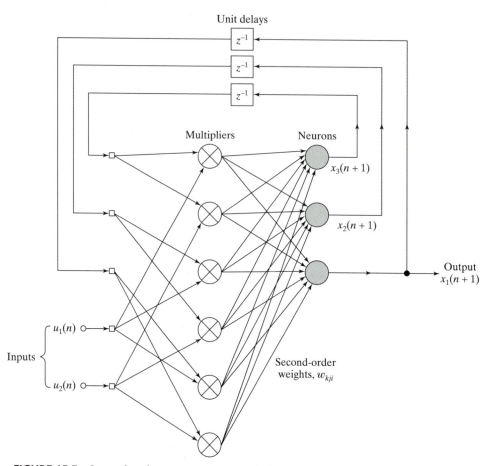

FIGURE 15.5   Second-order recurrent network; bias connections to the neurons are omitted to simplify the presentation. The network has 2 inputs and 3 state neurons, hence the need for 3 x 2 = 6 multipliers.

and

$$x_k(n + 1) = \varphi(v_k(n))$$

$$= \frac{1}{1 + \exp(-v_k(n))} \tag{15.8}$$

where $v_k(n)$ is the induced local field of hidden neuron k, b_k is the associated bias, $x_k(n)$ is the state (output) of neuron k, $u_j(n)$ is the input applied to source node j, and w_{kij} is a weight of second-order neuron k.

A unique feature of the second-order recurrent network in Fig. 15.5 is that the product $x_j(n)u_j(n)$ represents the pair {state, input} and that a positive weight w_{kij} represents the presence of the state transition, {state, input} $\rightarrow$ {next state}, while a negative weight represents the absence of the transition. The state transition is described by

$$\delta(x_i, u_j) = x_k \tag{15.9}$$

In light of this relationship, second-order networks are readily used for representing and learning *deterministic finite-state automata*[4] (DFA); a DFA is an information processing device with a finite number of states. More information on the relationship between neural networks and automata is found in Section 15.5.

The recurrent network architectures discussed in this section emphasize the use of global feedback. As mentioned in the introduction, it is also possible for a recurrent network architecture to have only local feedback. A summary of the properties of this latter class of recurrent networks is presented in Tsoi and Back (1994); see also Problem 15.7.

15.3 STATE-SPACE MODEL

The notion of *state* plays a vital role in the mathematical formulation of a dynamical system. The state of a dynamical system is formally defined as a *set of quantities that summarizes all the information about the past behavior of the system that is needed to uniquely describe its future behavior, except for the purely external effects arising from the applied input (excitation).* Let the q-by-1 vector $\mathbf{x}(n)$ denote the state of a nonlinear discrete-time system. Let the m-by-1 vector $\mathbf{u}(n)$ denote the input applied to the system, and the p-by-1 vector $\mathbf{y}(n)$ denote the corresponding output of the system. In mathematical terms, the dynamic behavior of the system, assumed to be *noise free,* is described by the following pair of nonlinear equations (Sontag, 1996):

$$\mathbf{x}(n+1) = \varphi(\mathbf{W}_a\mathbf{x}(n) + \mathbf{W}_b\mathbf{u}(n)) \tag{15.10}$$

$$\mathbf{y}(n) = \mathbf{C}\mathbf{x}(n) \tag{15.11}$$

where $\mathbf{W}_a$ is a q-by-q matrix, $\mathbf{W}_b$ is a q-by-$(m+1)$ matrix, $\mathbf{C}$ is a p-by-q matrix; and $\varphi : \mathbb{R}^q \to \mathbb{R}^q$ is a diagonal map described by

$$\varphi : \begin{bmatrix} x_1 \\ x_2 \\ \vdots \\ x_q \end{bmatrix} \to \begin{bmatrix} \varphi(x_1) \\ \varphi(x_2) \\ \vdots \\ \varphi(x_q) \end{bmatrix} \tag{15.12}$$

for some memoryless, component-wise nonlinearity $\varphi : \mathbb{R} \to \mathbb{R}$. The spaces $\mathbb{R}^m$, $\mathbb{R}^q$, and $\mathbb{R}^p$ are called the *input space, state space,* and *output space,* respectively. The dimensionality of the state space, namely q, is the *order* of the system. Thus the state-space model of Fig. 15.2 is an *m-input, p-output recurrent model of order q.* Equation (15.10) is the *process equation* of the model and Eq. (15.11) is the *measurement equation.* The process equation (15.10) is a special form of Eq. (15.2).

The recurrent network of Fig. 15.2, based on the use of a static multilayer perceptron and two delay-line memories, provides a method for implementing the nonlinear feedback system described by Eqs. (15.10) to (15.12). Note that in Fig. 15.2 *only those neurons in the multilayer perceptron that feed back their outputs to the input layer via delays are responsible for defining the state of the recurrent network.* This statement therefore excludes the neurons in the output layer from the definition of the state.

For the interpretation of matrices $\mathbf{W}_a$, $\mathbf{W}_b$, and $\mathbf{C}$, and nonlinear function $\varphi(\cdot)$, we may say:

- The matrix $\mathbf{W}_a$ represents the synaptic weights of the q neurons in the hidden layer that are connected to the feedback nodes in the input layer. The matrix $\mathbf{W}_b$ represents the synaptic weights of these hidden neurons that are connected to the source nodes in the input layer. It is assumed that the bias terms for the hidden neurons are absorbed in the weight matrix $\mathbf{W}_b$.
- The matrix $\mathbf{C}$ represents the synaptic weights of the p linear neurons in the output layer that are connected to the hidden neurons. It is assumed that the bias terms for the output neurons are absorbed in the weight matrix $\mathbf{C}$.
- The nonlinear function $\varphi(\cdot)$ represents the sigmoid activation function of a hidden neuron. The activation function typically takes the form of a hyperbolic tangent function:

$$\varphi(x) = \tanh(x) = \frac{1 - e^{-2x}}{1 + e^{-2x}} \tag{15.13}$$

or a logistic function:

$$\varphi(x) = \frac{1}{1 + e^{-x}} \tag{15.14}$$

An important property of a recurrent network described by the state-space model of Eqs. (15.10) and (15.11) is that it can *approximate* a wide class of nonlinear dynamical systems. However, the approximations are only valid on compact subsets of the state space and for finite time intervals, so that interesting dynamical characteristics are not reflected (Sontag, 1992).

Example 15.1

To illustrate the compositions of matrices $\mathbf{W}_a$, $\mathbf{W}_b$ and $\mathbf{C}$, consider the *fully connected recurrent network* shown in Fig. 15.6, where the feedback paths originate from the hidden neurons. In this example we have $m = 2$, $q = 3$, and $p = 1$. The matrices $\mathbf{W}_a$ and $\mathbf{W}_b$ are defined as follows:

$$\mathbf{W}_a = \begin{bmatrix} w_{11} & w_{12} & w_{13} \\ w_{21} & w_{22} & w_{23} \\ w_{31} & w_{32} & w_{33} \end{bmatrix}$$

and

$$\mathbf{W}_b = \begin{bmatrix} b_1 & w_{14} & w_{15} \\ b_2 & w_{24} & w_{25} \\ b_3 & w_{34} & w_{35} \end{bmatrix}$$

where the first column of $\mathbf{W}_b$ consisting of b_1, b_2, and b_3 represents the bias terms applied to neurons 1, 2, and 3, respectively. The matrix C is a row vector defined by

$$\mathbf{C} = [1, \quad 0, \quad 0]$$

■

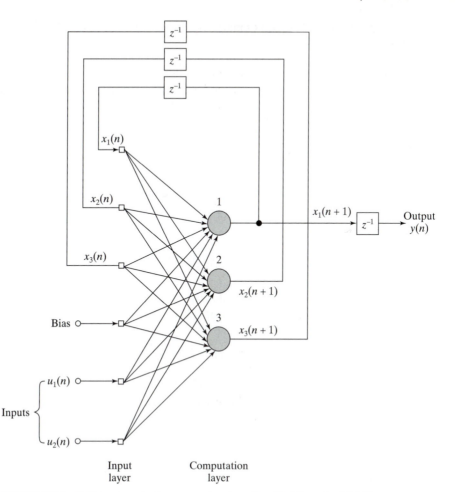

FIGURE 15.6   Fully connected recurrent network with 2 inputs, 2 hidden neurons, and 1 output neuron.

Controllability and Observability

In the study of system theory, stability, controllability, and observability are prominent features, each in its own fundamental way. In this section we discuss controllability and observability since they are usually treated together; stability is discussed in the previous chapter and will therefore not be pursued further.

As mentioned earlier, many recurrent networks can be represented by the state-space model shown in Fig. 15.2, where the state is defined by the output of the hidden layer fed back to the input layer via a set of unit delays. In that context, it is important to know whether or not the recurrent network is controllable and observable. Controllability is concerned with whether or not we can control the dynamic behavior of the recurrent network. Observability is concerned with whether or not we can observe the result of the control applied to the recurrent network. In that sense, observability is the dual of controllability.

A recurrent network is said to be *controllable* if an initial state is steerable to any desired state within a finite number of time steps; the output is irrelevant to this definition. The recurrent network is said to be *observable* if the state of the network can be determined from a finite set of input/output measurements. A rigorous treatment of controllability and observability of recurrent networks is beyond the scope of this book.[5] Here we confine ourselves to *local* forms of controllability and observability, local in the sense that these notions apply in the neighborhood of an equilibrium state of the network (Levin and Narendra, 1993).

A state $\bar{\mathbf{x}}$ is said to be an *equilibrium state* of Eq. (15.10) if, for an input $\mathbf{u}$, it satisfies the condition:

$$\bar{\mathbf{x}} = \varphi(\mathbf{A}\bar{\mathbf{x}} + \mathbf{B}\bar{\mathbf{u}}) \tag{15.15}$$

Without loss of generality, we may set $\bar{\mathbf{x}} = \mathbf{0}$ and $\bar{\mathbf{u}} = \mathbf{0}$. The equilibrium state is then described by

$$\mathbf{0} = \varphi(\mathbf{0})$$

In other words, the equilibrium point is represented by the origin $(\mathbf{0}, \mathbf{0})$.

Also without loss of generality, we may simplify the exposition by limiting ourselves to a single input, single output (SISO) system. We may then rewrite Eqs. (15.10) and (15.11) as follows, respectively:

$$\mathbf{x}(n + 1) = \varphi(\mathbf{W}_a\mathbf{x}(n) + \mathbf{w}_b u(n)) \tag{15.16}$$

$$y(n) = \mathbf{c}^T\mathbf{x}(n) \tag{15.17}$$

where both $\mathbf{w}_b$ and $\mathbf{c}$ are q-by-1 vectors, $u(n)$ is the scalar input, and $y(n)$ is the scalar output. Since φ is continuously differentiable for the sigmoid function of Eq. (15.13) or that of Eq. (15.14), we may *linearize* Eq. (15.16) by expanding it as a Taylor series around the equilibrium point $\bar{\mathbf{x}} = \mathbf{0}$ and $\bar{\mathbf{u}} = \mathbf{0}$ and retaining first-order terms as follows:

$$\delta\mathbf{x}(n + 1) = \varphi'(\mathbf{0})\mathbf{W}_a\delta\mathbf{x}(n) + \varphi'(\mathbf{0})\mathbf{w}_b\delta u(n) \tag{15.18}$$

where $\delta\mathbf{x}(n)$ and $\delta u(n)$ are small displacements applied to the state and input, respectively, and the q-by-q matrix $\varphi'(\mathbf{0})$ is the Jacobian of $\varphi(\mathbf{v})$ with respect to its argument $\mathbf{v}$, evaluated at $\mathbf{v} = \mathbf{0}$. We may thus describe the linearized system by writing

$$\delta\mathbf{x}(n + 1) = \mathbf{A}\delta\mathbf{x}(n) + \mathbf{b}\delta u(n) \tag{15.19}$$

$$\delta y(n) = \mathbf{c}^T\delta\mathbf{x}(n) \tag{15.20}$$

where the q-by-q matrix $\mathbf{A}$ and the q-by-1 vector $\mathbf{b}$ are respectively defined by

$$\mathbf{A} = \varphi'(\mathbf{0})\mathbf{W}_a \tag{15.21}$$

and

$$\mathbf{b} = \varphi'(\mathbf{0})\mathbf{w}_b \tag{15.22}$$

The state equations (15.19) and (15.20) are in the standard linear form. We may therefore make use of well-known results on the controllability and observability of linear dynamical systems that are a standard part of mathematical control theory.

Local Controllability

From the linearized equation (15.19), we readily find that its repeated use yields the following equations:

$$\delta\mathbf{x}(n + 1) = \mathbf{A}\delta\mathbf{x}(n) + \mathbf{b}\delta u(n)$$

$$\delta\mathbf{x}(n + 2) = \mathbf{A}\delta\mathbf{x}(n + 1) + \mathbf{b}\delta u(n + 1)$$

$$\vdots$$

$$\delta\mathbf{x}(n + q) = \mathbf{A}^q\mathbf{b}\delta\mathbf{x}(n) + \mathbf{A}^{q-1}\mathbf{b}\delta u(n + q - 1) + \cdots + \mathbf{A}\mathbf{b}\delta u(n + 1) + \mathbf{b}\delta u(n)$$

where q is the dimensionality of the state space. Accordingly, we may state that (Levin and Narendra, 1993):

> The linearized system represented by Eq. (15.19) is controllable if the matrix
>
> $$\mathbf{M}_c = [\mathbf{A}^{q-1}\mathbf{b}, ..., \mathbf{A}\mathbf{b}, \mathbf{b}] \tag{15.23}$$
>
> is of rank q, that is, full rank, because then the linearized process equation (15.19) would have a unique solution.

The matrix $\mathbf{M}_c$ is called the *controllability matrix* of the linearized system.

Let the recurrent network described by Eqs. (15.16) and (15.17) be driven by a sequence of inputs $\mathbf{u}_q(n)$ defined by

$$\mathbf{u}_q(n) = [u(n), u(n + 1), ..., u(n + q - 1)]^T \tag{15.24}$$

Hence we may consider the mapping

$$\mathbf{G}(\mathbf{x}(n), \mathbf{u}_q(n)) = (\mathbf{x}(n), \mathbf{x}(n + q)) \tag{15.25}$$

where $\mathbf{G} : \mathbb{R}^{2q} \to \mathbb{R}^{2q}$. In Problem 15.4, it is shown that:

- The state $\mathbf{x}(n + q)$ is a nested nonlinear function of its past value $\mathbf{x}(n)$ and the inputs $u(n), u(n + 1), ..., u(n + q - 1)$.
- The Jacobian of $\mathbf{x}(n + q)$ with respect to $\mathbf{u}_q(n)$, evaluated at the origin, is equal to the controllability matrix $\mathbf{M}_c$ of Eq. (15.23).

We may express the Jacobian of the mapping $\mathbf{G}$ with respect to $\mathbf{x}(n)$ and $\mathbf{u}_q(n)$, evaluated at the origin $(\mathbf{0}, \mathbf{0})$, as follows:

$$\mathbf{J}_{(0,0)}^{(c)} = \begin{bmatrix} \left(\dfrac{\partial\mathbf{x}(n)}{\partial\mathbf{x}(n)}\right)_{(0,0)} & \left(\dfrac{\partial\mathbf{x}(n + q)}{\partial\mathbf{x}(n)}\right)_{(0,0)} \\ \left(\dfrac{\partial\mathbf{x}(n)}{\partial\mathbf{u}_q(n)}\right)_{(0,0)} & \left(\dfrac{\partial\mathbf{x}(n + q)}{\partial\mathbf{u}_q(n)}\right)_{(0,0)} \end{bmatrix}$$

$$= \begin{bmatrix} \mathbf{I} & \mathbf{X} \\ \mathbf{0} & \mathbf{M}_c \end{bmatrix} \tag{15.26}$$

where $\mathbf{I}$ is the identity matrix, $\mathbf{0}$ is the null matrix, and the entry $\mathbf{X}$ is of no interest. Because of its special form, the determinant of the Jacobian $\mathbf{J}_{(0,0)}^{(c)}$ is equal to the product of the determinant of the identity matrix $\mathbf{I}$ (which equals 1) and the determinant of the controllability matrix $\mathbf{M}_c$. If $\mathbf{M}_c$ is of full rank, then so is $\mathbf{J}_{(0,0)}^{(c)}$.

To proceed further, we need to invoke the *inverse function theorem*, which may be stated as follows (Vidyasagar, 1993):

Consider the mapping $\mathbf{f} : \mathbb{R}^q \to \mathbb{R}^q$, and suppose that each component of the mapping $\mathbf{f}$ is differentiable with respect to its argument at the equilibrium point $\mathbf{x}_0 \in \mathbb{R}^q$, and let $\mathbf{y}_0 = \mathbf{f}(\mathbf{x}_0)$. Then there exist open sets $\mathcal{U} \subseteq \mathbb{R}^q$ containing $\mathbf{x}_0$ and $\mathcal{V} \subseteq \mathbb{R}^q$ containing $\mathbf{y}_0$ such that $\mathbf{f}$ is a diffeomorphism of $\mathcal{U}$ onto $\mathcal{V}$. If, in addition $\mathbf{f}$ is smooth, then the inverse mapping $\mathbf{f}^{-1} : \mathbb{R}^q \to \mathbb{R}^q$ is also smooth, that is, $\mathbf{f}$ is a smooth diffeomorphism.

The mapping $\mathbf{f} : \mathcal{U} \to \mathcal{V}$ is said to be a *diffeomorphism* of $\mathcal{U}$ onto $\mathcal{V}$ if it satisfies the following three conditions:

1. $\mathbf{f}(\mathcal{U}) = \mathcal{V}$.
2. The mapping $\mathbf{f} : \mathcal{U} \to \mathcal{V}$ is one-to-one (i.e., invertible).
3. Each component of the inverse mapping $\mathbf{f}^{-1} : \mathcal{V} \to \mathcal{U}$ is continuously differentiable with respect to its argument.

Returning to the issue of controllability, we may identify $\mathbf{f}(\mathcal{U}) = \mathcal{V}$ in the inverse function theorem with the mapping defined in Eq. (15.25). By using the inverse function theorem, we may say that if the controllability matrix $\mathbf{M}_c$ is of rank q, then locally there exists an inverse mapping defined by

$$(\mathbf{x}(n), \mathbf{x}(n + q)) = \mathbf{G}^{-1}(\mathbf{x}(n), \mathbf{u}_q(n)) \tag{15.27}$$

Equation (15.27), in effect, states that there exists an input sequence $\{\mathbf{u}_q(n)\}$ that can locally drive the network from state $\mathbf{x}(n)$ to state $\mathbf{x}(n + q)$ in q time steps. Accordingly, we may formally state the *local controllability theorem* as follows (Levin and Narendra, 1993):

Let a recurrent network be defined by Eqs. (15.16) and (15.17), and let its linearized version around the origin (i.e., equilibrium point) be defined by Eqs. (15.19) and (15.20). If the linearized system is controllable, then the recurrent network is locally controllable around the origin.

Local Observability

Using the linearized equations (15.19) and (15.20) repeatedly, we may write

$$\delta y(n) = \mathbf{c}^T \delta \mathbf{x}(n)$$

$$\delta y(n + 1) = \mathbf{c}^T \delta \mathbf{x}(n + 1)$$

$$= \mathbf{c}^T \mathbf{A} \delta \mathbf{x}(n) + \mathbf{c}^T \mathbf{b} \delta u(n)$$

$$\vdots$$

$$\delta y(n + q - 1) = \mathbf{c}^T \mathbf{A}^{q-1} \delta \mathbf{x}(n) + \mathbf{c}^T \mathbf{A}^{q-2} \mathbf{b} \delta u(n) + \cdots + \mathbf{c}^T \mathbf{A} \mathbf{b} \delta u(n + q - 3)$$

$$+ \mathbf{c}^T \mathbf{b} \delta u(n + q - 2)$$

where q is the dimensionality of the state space. Accordingly, we may state that (Levin and Narendra, 1993):

The linearized system described by Eqs. (15.19) and (15.20) is observable if the matrix

$$\mathbf{M}_o = [\mathbf{c}, \mathbf{c}\mathbf{A}^T, ..., \mathbf{c}(\mathbf{A}^T)^{q-1}] \tag{15.28}$$

is of rank q, that is, full rank.

The matrix $\mathbf{M}_o$ is called the *observability matrix* of the linearized system.

Let the recurrent network described by Eqs. (15.16) and (15.17) be driven by a sequence of inputs defined by

$$\mathbf{u}_{q-1}(n) = [u(n), u(n+1), ..., u(n+q-2)]^T \tag{15.29}$$

Correspondingly, let

$$\mathbf{y}_q(n) = [y(n), y(n+1), ..., y(n+q-1)]^T \tag{15.30}$$

denote the vector of outputs produced by the initial state $\mathbf{x}(n)$ and the sequence of inputs $\mathbf{u}_{q-1}(n)$. We may then consider the mapping:

$$\mathbf{H}(\mathbf{u}_{q-1}(n), \mathbf{x}(n)) = (\mathbf{u}_{q-1}(n), \mathbf{y}_q(n)) \tag{15.31}$$

where $\mathbf{H} : \mathbb{R}^{2q-1} \to \mathbb{R}^{2q-1}$. In Problem 15.5 it is shown that the Jacobian of $\mathbf{y}_q(n)$ with respect to $\mathbf{x}(n)$, evaluated at the origin, is equal to the observability matrix $\mathbf{M}_o$ of Eq. (15.28). We may thus express the Jacobian of $\mathbf{H}$ with respect to $\mathbf{u}_{q-1}(n)$ and $\mathbf{x}(n)$, evaluated at the origin $(\mathbf{0}, \mathbf{0})$, as follows:

$$\mathbf{J}_{(0,0)}^{(o)} = \begin{bmatrix} \left(\dfrac{\partial \mathbf{u}_{q-1}(n)}{\partial \mathbf{u}_{q-1}(n)}\right)_{(0,0)} & \left(\dfrac{\partial \mathbf{y}_q(n)}{\partial \mathbf{u}_{q-1}(n)}\right)_{(0,0)} \\ \left(\dfrac{\partial \mathbf{u}_{q-1}(n)}{\partial \mathbf{x}(n)}\right)_{(0,0)} & \left(\dfrac{\partial \mathbf{y}_q(n)}{\partial \mathbf{x}(n)}\right)_{(0,0)} \end{bmatrix} \tag{15.32}$$

$$= \begin{bmatrix} \mathbf{I} & \mathbf{X} \\ \mathbf{0} & \mathbf{M}_o \end{bmatrix}$$

where again the entry $\mathbf{X}$ is of no interest. The determinant of the Jacobian $\mathbf{J}_{(0,0)}^{(o)}$ is equal to the product of the determinant of the identity matrix $\mathbf{I}$ (which equals 1) and the determinant of $\mathbf{M}_o$. If $\mathbf{M}_o$ is of full rank, then so is $\mathbf{J}_{(0,0)}^{(o)}$. Invoking the inverse function theorem, we may therefore say that if the observability matrix $\mathbf{M}_o$ of the linearized system is of full rank, then locally there exists an inverse mapping defined by

$$(\mathbf{u}_{q-1}(n), \mathbf{x}(n)) = \mathbf{H}^{-1}(\mathbf{u}_{q-1}(n), \mathbf{y}_q(n)) \tag{15.33}$$

In effect, this equation states that in the local neighborhood of the origin, $\mathbf{x}(n)$ is some nonlinear function of both $\mathbf{u}_{q-1}(n)$ and $\mathbf{y}_q(n)$, and that nonlinear function is an observer of the recurrent network. We may therefore formally state the *local observability theorem* as follows (Levin and Narendra, 1993):

> Let a recurrent network be defined by Eqs. (15.16) and (15.17), and let its linearized version around the origin (i.e., equilibrium point) be defined by Eqs. (15.19) and (15.20). If the linearized system is observable, then the recurrent network is locally observable around the origin.

Example 15.2

Consider a state-space model with matrix $\mathbf{A} = a\mathbf{I}$, where a is a scalar and $\mathbf{I}$ is the identity matrix. Then the controllability matrix $\mathbf{M}_c$ of Eq. (15.23) reduces to

$$\mathbf{M}_c = a[\mathbf{b}, ..., \mathbf{b}, \mathbf{b}]$$

The rank of this matrix is 1. Hence, the linearized system with this value of matrix $\mathbf{A}$ is not controllable.

Putting $\mathbf{A} = a\mathbf{I}$ in Eq. (15.28), we obtain the observability matrix

$$\mathbf{M}_o = a[\mathbf{c}, \mathbf{c}, ..., \mathbf{c}]$$

whose rank is also 1. The linearized system is also not observable.

■

15.4 NONLINEAR AUTOGRESSIVE WITH EXOGENOUS INPUTS MODEL

Consider a recurrent network with a single input and single output, whose behavior is described by the state equations (15.16) and (15.17). Given this state-space model, we wish to modify it into an input–output model as an equivalent representation of the recurrent network.

Using Eqs. (15.16) and (15.17), we may readily show that the output $y(n + q)$ is expressible in terms of the state $\mathbf{x}(n)$ and the vector of inputs $\mathbf{u}_q(n)$ as follows (see Problem 15.8):

$$y(n + q) = \Phi(\mathbf{x}(n), \mathbf{u}_q(n)) \tag{15.34}$$

where q is the dimensionality of the state space, and $\Phi : \mathbb{R}^{2q} \to \mathbb{R}$. Provided that the recurrent network is observable, we may use the local observability theorem to write

$$\mathbf{x}(n) = \Psi(\mathbf{y}_q(n), \mathbf{u}_{q-1}(n)) \tag{15.35}$$

where $\Psi : \mathbb{R}^{2q-1} \to \mathbb{R}^q$. Hence, substituting Eq. (15.35) in (15.34), we get

$$\begin{aligned}
y(n + q) &= \Phi(\Psi(\mathbf{y}_q(n), \mathbf{u}_{q-1}(n)), \mathbf{u}_q(n)) \\
&= F(\mathbf{y}_q(n), \mathbf{u}_q(n))
\end{aligned} \tag{15.36}$$

where $\mathbf{u}_{q-1}(n)$ is contained in $\mathbf{u}_q(n)$ as its first $(q - 1)$ elements, and the nonlinear mapping $F : \mathbb{R}^{2q} \to \mathbb{R}$ takes care of both Φ and Ψ. Using the definitions of $\mathbf{y}_q(n)$ and $\mathbf{u}_q(n)$ given in Eqs. (15.30) and (15.29), we may rewrite Eq. (15.36) in the expanded form:

$$y(n + q) = F(y(n + q - 1), ..., y(n), u(n + q - 1), ..., u(n))$$

Replacing n with $n - q + 1$, we may equivalently write (Narendra, 1995):

$$y(n + 1) = F(y(n), ..., y(n - q + 1), u(n), ..., u(n - q + 1)) \tag{15.37}$$

Stated in words, some nonlinear mapping $F : \mathbb{R}^{2q} \to \mathbb{R}$ exists whereby the present value of the output $y(n + 1)$ is uniquely defined in terms of its past values $y(n), ..., y(n - q + 1)$ and the present and past values of the input $u(n), ..., u(n - q + 1)$. For this input–output representation to be equivalent to the state-space model of Eqs. (15.16)

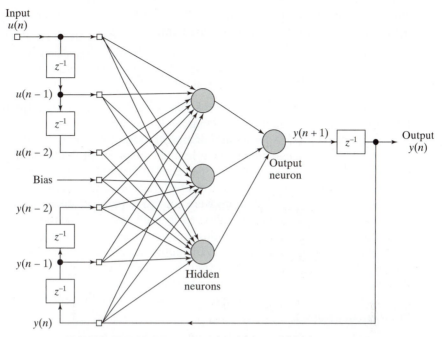

FIGURE 15.7   NARX network with $q = 3$ hidden neurons.

and (15.17), the recurrent network must be observable. The practical implication of this equivalence is that the NARX model of Fig. 15.1, with its global feedback limited to the output neuron, is in fact able to simulate the corresponding fully recurrent state-space model of Fig. 15.2 (assuming that $m = 1$ and $p = 1$) with no difference between their input–output behavior.

Example 15.3

Consider again the fully connected recurrent network of Fig. 15.6. For the purpose of our present discussion, suppose that one of the inputs, $u_2(n)$ say, is reduced to zero, so that we have a single input, single output network. We may then replace this fully connected recurrent network by the NARX model shown in Fig. 15.7, provided that the network is locally observable. This equivalence holds despite the fact that the NARX model has limited feedback that originates only from the output neuron, whereas in the fully connected recurrent network of Fig. 15.6 the feedback around the multilayer perceptron originates from the three hidden/output neurons. ∎

15.5   COMPUTATIONAL POWER OF RECURRENT NETWORKS

Recurrent networks, exemplified by the state-space model of Fig. 15.2 and the NARX model of Fig. 15.1, have an inherent ability to simulate finite-state automata. *Automata* represent abstractions of information processing devices such as computers. Indeed, automata and neural networks share a long history.[6] In his 1967 book (p. 55), Minsky makes the following consequential statement:

> "Every finite-state machine is equivalent to, and can be 'simulated' by, some neural net. That is, given any finite-state machine $\mathcal{M}$, we can build a certain neural net $\mathcal{N}^{\mathcal{M}}$ which, regarded as a black-box machine, will behave precisely like $\mathcal{M}$!"

The early work on recurrent networks used hard threshold logic for the activation function of a neuron rather than soft sigmoid function.

Perhaps the first experimental demonstration of whether or not a recurrent network could learn the contingencies implied by a small finite-state grammar was reported in Cleeremans et al. (1989). Specifically, the simple recurrent network (Fig. 15.3) was presented with strings derived from the grammar and required to predict the next letter at every step. The predictions were context dependent since each letter appeared twice in the grammar and was followed in each case by different successors. It was shown that the network is able to develop internal representations in its hidden neurons that correspond to the states of the automaton (finite-state machine). In Kremer (1995) a formal proof is presented that the simple recurrent network has a computational power as great as that of any finite-state machine.

In a generic sense, the computational power of a recurrent network is embodied in two main theorems:

Theorem I (Siegelmann and Sontag, 1991).

> All Turing machines may be simulated by fully connected recurrent networks built on neurons with sigmoid activation functions.

The *Turing machine* is an abstract computing device invented by Turing (1936). It consists of three functional blocks as depicted in Fig. 15.8: (1) *control unit* that can assume any one of a finite number of possible states; (2) *linear tape* (assumed to be infinite in both directions) that is marked off into discrete squares with each square available to store a single symbol taken from a finite set of symbols; and (3) *read-write head* that moves along the tape and transmits information to and from the control unit (Fischler and Firschein, 1987). For the present discussion it suffices to say that the Turing machine is an abstraction that is functionally as powerful as any computer. This idea is known as the *Church–Turing hypothesis.*

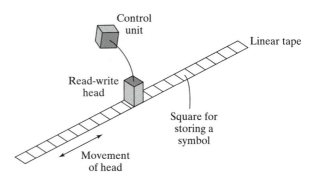

FIGURE 15.8   Turing Machine.

Theorem II (Siegelmann et al., 1997)

> NARX networks with one layer of hidden neurons with bounded, one-sided saturated activation functions and a linear output neuron can simulate fully connected recurrent networks with bounded, one-sided saturated activation functions, except for a linear slow-down.

A "linear slowdown" means that if the fully connected recurrent network with N neurons computes a task of interest in time T, then the total time taken by the equivalent NARX network is $(N + 1)T$. A function $\varphi(\cdot)$ is said to be a *bounded, one-sided saturated (BOSS) function* if it satisfies the following three conditions:

1. The function $\varphi(\cdot)$ has a bounded range; that is, $a \leq \varphi(x) \leq b, a \neq b$, for all $x \in \mathbb{R}$.
2. The function $\varphi(\cdot)$ is saturated on the left side; that is, there exist values s and S such that $\varphi(x) = S$ for all $x \leq s$.
3. The function $\varphi(\cdot)$ is nonconstant; that is, $\varphi(x_1) \neq \varphi(x_2)$ for some x_1 and x_2.

The threshold (Heaviside) and piecewise-linear functions satisfy the BOSS conditions. However, in a strict sense, a sigmoid function is not a BOSS function because it does not satisfy condition 2. Nevertheless, with a minor modification, it can be made into a BOSS function by writing (in the case of a logistic function)

$$\varphi(x) = \begin{cases} \dfrac{1}{1 + \exp(-x)} & \text{for } x > s \\ 0 & \text{for } x \leq s \end{cases}$$

where $s \in \mathbb{R}$. In effect, the logistic function is truncated for $x \leq s$.

As a corollary to Theorems I and II, we may state the following (Giles, 1996):

> NARX networks with one hidden layer of neurons with BOSS activations functions and a linear output neuron are Turing equivalent.

Figure 15.9 presents a portrayal of Theorems I and II and this corollary. It should, however, be noted that when the network architecture is constrained, the computational power of a recurrent network may no longer hold, as described in Sperduti(1997). References to examples of constrained network architectures are presented in note 7.

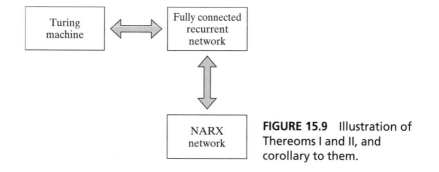

FIGURE 15.9 Illustration of Thereoms I and II, and corollary to them.

15.6 LEARNING ALGORITHMS

We now turn to the issue of training recurrent networks. From Chapter 4 we recall that there are two modes of training an ordinary (static) multilayer perceptron: batch mode and sequential mode. In the batch mode, the sensitivity of the network is computed for the entire training set before adjusting the free parameters of the network. In the sequential mode, on the other hand, parameter adjustments are made after the presentation of each pattern in the training set. Likewise, we have two modes of training a recurrent network, as described here (Williams and Zipser, 1995):

1. *Epochwise training.* For a given epoch, the recurrent network starts running from some initial state until it reaches a new state, at which point the training is stopped and the network is reset to an initial state for the next epoch. The initial state doesn't have to be the same for each epoch of training. Rather, what is important is for the initial state for the new epoch to be different from the state reached by the network at the end of the previous epoch. Consider, for example, the use of a recurrent network to emulate the operation of a finite-state machine, that is, a device whose distinguishable internal configurations (states) are finite in number. In such a situation it is reasonable to use epochwise training since we have a good possibility for a number of distinct initial states and a set of distinct final states in the machine to be emulated by the recurrent network. In epochwise training for recurrent networks the term "epoch" is used in a sense different from that for an ordinary multilayer perceptron. In the current terminology, the epoch for the recurrent network corresponds to one training pattern for the ordinary multilayer perceptron.

2. *Continuous training.* This second method of training is suitable for situations where there are no reset states available and/or on-line learning is required. The distinguishing feature of continuous training is that the network learns while signal processing is being performed by the network. Simply put, the learning process never stops. Consider, for example, the use of a recurrent network to model a nonstationary process such as a speech signal. In this kind of situation, continuous operation of the network offers no convenient times at which to stop the training and begin anew with different values for the free parameters of the network.

Keeping these two modes of training in mind, in the next two sections we will describe different learning algorithms for recurrent networks as summarized here:

- The back-propagation-through-time algorithm, discussed in Section 15.7, operates on the premise that the temporal operation of a recurrent network may be unfolded into a multilayer perceptron. This would then pave the way for application of the standard back-propagation algorithm. Back-propagation through time can be implemented in the epochwise mode, continuous (real-time) mode, or combination thereof.
- The real-time recurrent learning algorithm, discussed in Section 15.8, is derived from the state-space model described by Eqs. (15.10) and (15.11).

These two algorithms share many common features. First, they are both based on the method of gradient descent, whereby the instantaneous value of a cost function (based on a squared-error criterion) is minimized with respect to the synaptic weights of the network. Second, they are both relatively simple to implement, but can be slow to converge. Third, they are related in that the signal-flow graph representation of the back-propagation-through-time algorithm can be obtained from *transposition* of the signal-flow graph representation of a certain form of the real-time recurrent learning algorithm (Lefebvre, 1991; Beaufays and Wan, 1994).

Real-time (continuous) learning, based on gradient descent, uses the *minimum amount of available information,* namely an instantaneous estimate of the gradient of the cost function with respect to the parameter vector to be adjusted. We may accelerate the learning process by exploiting Kalman filter theory, which utilizes information contained in the training data more effectively. In Section 15.10 we describe the decoupled extended Kalman filter, by means of which we are able to tackle dynamic learning tasks that would be very difficult for gradient-descent based methods. A brief review of Kalman filters is presented in Section 15.9. Note that the decoupled extended Kalman filter is applicable to both static feedforward networks as well as recurrent networks.

Some Heuristics

Before proceeding to describe the new learning algorithms mentioned, we will list some heuristics for the improved training of recurrent networks that involve the use of gradient-descent methods (Giles, 1996):

- Lexigraphic order of training samples should be followed, with the shortest strings of symbols being presented to the network first.
- The training should begin with a small training sample, and then its size should be incrementally increased as the training proceeds.
- The synaptic weights of the network should be updated only if the absolute error on the training sample currently being processed by the network is greater than some prescribed criterion.
- The use of weight decay during training is recommended; weight decay, a crude form of complexity regularization, is discussed in Chapter 4.

The first heuristic is of particular interest. If implementable, it provides a procedure for alleviating the vanishing gradients problem that arises in recurrent networks trained by means of gradient-descent methods. This problem is discussed in Section 15.12.

15.7   BACK-PROPAGATION THROUGH TIME

The *back-propagation-through-time (BPTT) algorithm* for training a recurrent network is an extension of the standard back-propagation algorithm.[8] It may be derived by *unfolding* the temporal operation of the network into a layered feedforward network, the topology of which grows by one layer at every time step.

To be specific, let $\mathcal{N}$ denote a recurrent network required to learn a temporal task, starting from time n_0 all the way up to time n. Let $\mathcal{N}^*$ denote the feedforward network that results from unfolding the temporal operation of the recurrent network $\mathcal{N}$. The unfolded network $\mathcal{N}^*$ is related to the original network $\mathcal{N}$ as follows:

1. For each time step in the interval $(n_0, n]$, the network $\mathcal{N}^*$ has a layer containing K neurons, where K is the number of neurons contained in the network $\mathcal{N}$.
2. In every layer of the network $\mathcal{N}^*$ there is a copy of each neuron in the network $\mathcal{N}$.
3. For each time step $l \in [n_0, n]$, the synaptic connection from neuron i in layer l to neuron j in layer $l + 1$ of the network $\mathcal{N}^*$ is a copy of the synaptic connection from neuron i to neuron j in the network $\mathcal{N}$.

These points are illustrated in the following example.

Example 15.4

Consider the two-neuron recurrent network $\mathcal{N}$ shown in Fig. 15.10a. To simplify the presentation we have omitted unit delay operators z^{-1} that should be inserted in each of the synaptic connections (including the self-loops) in Fig. 15.10a. By unfolding the temporal operation of this network in a step-by-step manner, we get the signal-flow graph shown in Fig. 15.10b where the starting time $n_0 = 0$. The graph of Fig. 15.10b represents the layered feedforward $\mathcal{N}^*$, where a new layer is added at each step of the temporal operation.   ∎

Application of the unfolding procedure leads to two basically different implementations of back-propagation through time, depending on whether epochwise training or continuous (real-time) training is used. These two methods of recurrent learning are now described in that order.

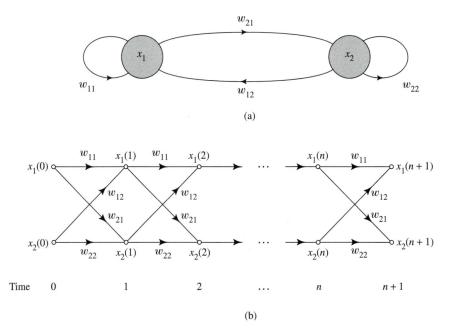

(a)

(b)

FIGURE 15.10   (a) Architectural graph of a two-neuron recurrent network $\mathcal{N}$. (b) Signal-flow graph of the network $\mathcal{N}$ unfolded in time.

Epochwise Back-propagation through Time

Let the data set used to train a recurrent network be partitioned into independent epochs, with each epoch representing a temporal pattern of interest. Let n_0 denote the start time of an epoch and n_1 denote its end time. Given this epoch, we may define the cost function

$$\mathscr{E}_{\text{total}}(n_0, n_1) = \frac{1}{2} \sum_{n=n_0}^{n_1} \sum_{j \in \mathscr{A}} e_j^2(n) \tag{15.38}$$

where $\mathscr{A}$ is the set of indices j pertaining to those neurons in the network for which desired responses are specified, and $e_j(n)$ is the error signal at the output of such a neuron measured with respect to some desired response. We wish to compute sensitivity of the network, that is, the partial derivatives of the cost function $\mathscr{E}_{\text{total}}(n_0, n_1)$ with respect to synaptic weights of the network. To do so, we may use the *epochwise back-propagation-through-time (BPTT) algorithm,* which builds on the batch mode of standard back-propagation learning that is described in Chapter 4. The epochwise BPTT algorithm proceeds as follows (Williams and Peng, 1990):

- First, a single forward pass of the data through the network for the interval (n_0, n_1) is performed. The complete record of input data, network state (i.e., synaptic weights of the network), and desired responses over this interval is *saved.*
- A single backward pass over this past record is performed to compute the values of the local gradients

$$\delta_j(n) = -\frac{\partial \mathscr{E}_{\text{total}}(n_0, n_1)}{\partial v_j(n)} \tag{15.39}$$

for all $j \in \mathscr{A}$ and $n_0 < n \leq n_1$. This computation is performed by using the formula:

$$\delta_j(n) = \begin{cases} \varphi'(v_j(n))e_j(n) & \text{for } n = n_1 \\ \varphi'(v_j(n))\left[e_j(n) + \sum_{k \in \mathscr{A}} w_{jk}\delta_k(n+1) \right] & \text{for } n_0 < n < n_1 \end{cases} \tag{15.40}$$

where $\varphi'(\cdot)$ is the derivative of an activation function with respect to its argument, and $v_j(n)$ is the induced local field of neuron j. It is assumed that all neurons in the network have the same activation function $\varphi(\cdot)$. The use of Eq. (15.40) is repeated, starting from time n_1 and working back, step by step, to time n_0; the number of steps involved here is equal to the number of time steps contained in the epoch.

- Once the computation of back-propagation has been performed back to time $n_0 + 1$, the following adjustment is applied to the synaptic weight w_{ji} of neuron j:

$$\Delta w_{ji} = -\eta \frac{\partial \mathscr{E}_{\text{total}}(n_0, n_1)}{\partial w_{ji}}$$

$$= \eta \sum_{n=n_0+1}^{n_1} \delta_j(n)x_i(n-1) \tag{15.41}$$

where η is the learning-rate parameter and $x_i(n-1)$ is the input applied to the ith synapse of neuron j at time $n-1$.

Comparing the procedure just described for epochwise BPTT with the batch mode of standard back-propagation learning, we see that the basic difference between them is that in the former case the desired responses are specified for neurons in many layers of the network because the actual output layer is replicated many times when the temporal behavior of the network is unfolded.

Truncated Back-propagation through Time

To use back-propagation through time in a real-time fashion, we use the instantaneous value of the sum of squared errors, namely,

$$\mathcal{E}(n) = \frac{1}{2} \sum_{j \in \mathcal{A}} e_j^2(n)$$

as the cost function to be minimized. As with the sequential (stochastic) mode of standard back-propagation learning, we use the negative gradient of the cost function $\mathcal{E}(n)$ to compute the appropriate adjustments to the synaptic weights of the network at each time instant n. The adjustments are made on a continuous basis, while the network is running. However, in order to do this in a computationally feasible manner, we only save the relevant history of input data and network state for a fixed number of time steps, called the *truncation depth*. Henceforth the truncation depth is denoted by h. Any information older than h time steps into the past is considered irrelevant, and may therefore be ignored. If we were not to truncate the computation, thereby permitting it to go back to the starting time, the computation time and storage requirement would grow linearly with time as the network runs, eventually reaching a point where the whole learning process becomes impractical.

This second form of the algorithm is called the *truncated back-propagation-through-time (BPTT(h))* algorithm (Williams and Peng, 1990). The local gradient for neuron j is now defined by

$$\delta_j(l) = -\frac{\partial \mathcal{E}(l)}{\partial v_j(l)} \qquad \begin{array}{l} \text{for all } j \in \mathcal{A} \\ \text{and } n - h < l \le n \end{array} \tag{15.42}$$

which in turn leads to the formula:

$$\delta_j(l) = \begin{cases} \varphi'(v_j(l))e_j(l) & \text{for } l = n \\ \varphi'(v_j(l)) \sum_{k \in \mathcal{A}} w_{kj}(l)\, \delta_k(l+1) & \text{for } n - h < l < n \end{cases} \tag{15.43}$$

Once the computation of back-propagation has been performed back to time $n - h + 1$, the following adjustment is applied to the synaptic weight w_{ji} of neuron j:

$$\Delta w_{ji}(n) = \eta \sum_{l=n-h+1}^{n} \delta_j(l)x_i(l-1) \tag{15.44}$$

where η and $x_i(l-1)$ are as defined previously. Note that the use of $w_{kj}(l)$ in Eq. (15.43) requires that a history of weight values be maintained. The use of w_{kj} in this equation may be justified only if the learning-rate parameter η is small enough to ensure that the weight values do not change significantly from one time step to the next.

In comparing Eq. (15.43) with (15.40), we see that, unlike the epochwise BPTT algorithm, the error signal is only injected into the computation at the current time n. This explains the reason for not keeping a record of past values of the desired responses. In effect, the truncated back-propagation-through-time algorithm treats the computation for all earlier time steps similar to the way in which the stochastic back-propagation algorithm (discussed in Chapter 4) treats the computations for hidden neurons in a multilayer perceptron.

Some Practical Considerations

In real-life applications of BPTT, the use of truncation is not as artificial as it may sound. Unless the recurrent network is unstable, there should be a convergence of the derivatives $\partial\mathcal{E}(l)/\partial v_j(l)$ because computations farther back in time correspond to higher powers of feedback strengths (roughly equal to sigmoid slopes multiplied by weights). In any event, the truncation depth h must be large enough to produce derivatives that closely approximate the actual values. This requirement places a lower bound on the value of h. For example, in the application of dynamically driven recurrent networks to engine idle-speed control, the value $h = 30$ is considered to be a reasonably conservative choice for that learning task to be accomplished (Puskorius et al., 1996).

One other practical matter needs to be discussed. The unfolding procedure described in this section for back-propagation through time provides a useful tool for picturing it in terms of a cascade of similar layers progressing forward in time, thereby helping us to develop an understanding of how the procedure functions. This strong point is unfortunately the cause of its weakness. The procedure works perfectly fine for relatively simple recurrent networks consisting of a few neurons. However, the underlying formulas, particularly Eq. (15.43), become unwieldy when the unfolding procedure is applied to more general architectures that are typical of those encountered in practice. In situations of this kind, the preferred procedure is to use the more general approach described in Werbos (1990,) in which each expression in the forward propagation of a layer gives rise to a corresponding set of back-propagation expressions. An advantage of this approach is its homogeneous treatment of forward and recurrent (feedback) connections.

To describe the mechanics of this particular form of BPTT(h), let F_{-x}^l denote an *ordered derivative* of the network output at node l with respect to x. To derive the back-propagation equations, the forward propagation equations are considered in reverse order. From each equation we derive one or more back-propagation expressions according to the following principle:

$$\text{If } a = \varphi(b, c), \text{ then } F_{-b}^l = \frac{\partial\varphi}{\partial b}F_{-a}^l \quad \text{and} \quad F_{-c}^l = \frac{\partial\varphi}{\partial c}F_{-a}^l \tag{15.45}$$

Example 15.5

To clarify the notion of ordered derivatives, consider a nonlinear system described by the following pair of equations:

$$x_1 = \log u + x_2^3$$

$$y = x_1^2 + 3x_2$$

The variable x_2 influences the output y in two ways: directly via the second equation, and indirectly via the first equation. The ordered derivative of y with respect to x_2 is defined by the total causal impact that includes the direct and indirect effects of x_2 on y, as shown by

$$F_{-x_2} = \frac{\partial y}{\partial x_2} + \frac{\partial y}{\partial x_1} \frac{\partial x_1}{\partial x_2}$$

$$= 3 + (2x_1)(3x_2^2)$$

$$= 3 + 6x_1 x_2^2$$

∎

In programming the ordered derivatives for BPTT(h), the quantity on the right-hand side of each ordered derivative in Eq. (15.45) is added to the previous value of the left-hand side. In this way, the appropriate derivatives are distributed from a given node in the network to all the nodes and synaptic weights that feed it in the forward direction, with due allowance being made for any delays that may be present in each connection. The simplicity of the formulation described herein reduces the need for visualizations such as unfolding in time or signal-flow graphs. In Feldkamp and Puskorius (1998) and Puskorius et al. (1996), this procedure is used to develop a pseudocode for implementing the BPTT(h) algorithm.

15.8 REAL-TIME RECURRENT LEARNING

In this section we describe another learning algorithm referred to as *real-time recurrent learning (RTRL)*.[9] The algorithm derives its name from the fact that adjustments are made to the synaptic weights of a fully connected recurrent network in real time, that is, while the network continues to perform its signal processing function (Williams and Zipser, 1989). Figure 15.11 shows the layout of such a recurrent network. It consists of q neurons with m external inputs. The network has two distinct layers: a *concatenated input-feedback layer* and a *processing layer of computation nodes.* Correspondingly, the synaptic connections of the network are made up of feedforward and feedback connections.

The state-space description of the network is defined by Eqs. (15.10) and (15.11). The process equation (15.10) is reproduced here in the following expanded form:

$$\mathbf{x}(n+1) = \begin{bmatrix} \varphi(\mathbf{w}_1^T \boldsymbol{\xi}(n)) \\ \vdots \\ \varphi(\mathbf{w}_j^T \boldsymbol{\xi}(n)) \\ \vdots \\ \varphi(\mathbf{w}_q^T \boldsymbol{\xi}(n)) \end{bmatrix} \tag{15.46}$$

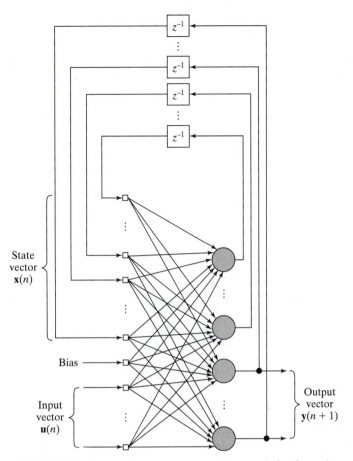

FIGURE 15.11 Fully connected recurrent network for formulation of the RTRL algorithm.

where it is assumed that all the neurons have a common activation function $\varphi(\cdot)$. The $(q + m + 1)$-by-1 vector $\mathbf{w}_j$ is the synaptic weight vector of neuron j in the recurrent network, that is,

$$\mathbf{w}_j = \begin{bmatrix} \mathbf{w}_{a,j} \\ \mathbf{w}_{b,j} \end{bmatrix}, \qquad j = 1, 2, \dots, q \tag{15.47}$$

where $\mathbf{w}_{a,j}$ and $\mathbf{w}_{b,j}$ are the jth columns of the transposed weight matrices $\mathbf{W}_a^T$ and $\mathbf{W}_b^T$, respectively. The $(q + m + 1)$-by-1 vector $\boldsymbol{\xi}(n)$ is defined by

$$\boldsymbol{\xi}(n) = \begin{bmatrix} \mathbf{x}(n) \\ \mathbf{u}(n) \end{bmatrix} \tag{15.48}$$

where $\mathbf{x}(n)$ is the q-by-1 state vector and $\mathbf{u}(n)$ is the $(m + 1)$-by-1 input vector. The first element of $\mathbf{u}(n)$ is +1 and, in a corresponding way, the first element of $\mathbf{w}_{b,j}$ is equal to the bias b_j applied to neuron j.

To simplify the presentation, we introduce three new matrices $\Lambda_j(n)$, $\mathbf{U}_j(n)$, and $\boldsymbol{\Phi}(n)$, described as follows:

1. $\Lambda_j(n)$ is a q-by-$(q + m + 1)$ matrix defined as the partial derivative of the state vector $\mathbf{x}(n)$ with respect to the weight vector $\mathbf{w}_j$:

$$\Lambda_j(n) = \frac{\partial \mathbf{x}(n)}{\partial \mathbf{w}_j}, \qquad j = 1, 2, ..., q \tag{15.49}$$

2. $\mathbf{U}_j(n)$ is a q-by-$(q + m + 1)$ matrix whose rows are all zero, except for the jth row that is equal to the transpose of vector $\boldsymbol{\xi}(n)$:

$$\mathbf{U}_j(n) = \begin{bmatrix} 0 \\ \boldsymbol{\xi}^T(n) \\ 0 \end{bmatrix} \leftarrow j\text{th row}, \qquad j = 1, 2, ..., q \tag{15.50}$$

3. $\boldsymbol{\Phi}(n)$ is a q-by-q diagonal matrix whose kth diagonal element is the partial derivative of the activation function with respect to its argument, evaluated at $\mathbf{w}_j^T \boldsymbol{\xi}(n)$:

$$\boldsymbol{\Phi}(n) = \mathrm{diag}\left(\varphi'(\mathbf{w}_1^T \boldsymbol{\xi}(n)), ..., \varphi'(\mathbf{w}_j^T \boldsymbol{\xi}(n)), ..., \varphi(\mathbf{w}_q^T \boldsymbol{\xi}(n))\right) \tag{15.51}$$

With these definitions, we may now differentiate Eq. (15.46) with respect to $\mathbf{w}_j$. Then, using the chain rule of calculus, we obtain the following recursive equation:

$$\Lambda_j(n + 1) = \boldsymbol{\Phi}(n)[\mathbf{W}_a(n)\Lambda_j(n) + \mathbf{U}_j(n)], \qquad j = 1, 2, ..., q \tag{15.52}$$

This recursive equation describes the *nonlinear state dynamics* (i.e., evolution of the state) of the real-time recurrent learning process.

To complete the description of this learning process, we need to relate the matrix $\Lambda_j(n)$ to the gradient of the error surface with respect to $\mathbf{w}_j$. To do this, we first use the measurement equation (15.11) to define the p-by-1 error vector:

$$\begin{aligned} \mathbf{e}(n) &= \mathbf{d}(n) - \mathbf{y}(n) \\ &= \mathbf{d}(n) - \mathbf{C}\mathbf{x}(n) \end{aligned} \tag{15.53}$$

The instantaneous sum of squared errors at time n is defined in terms of $\mathbf{e}(n)$ by

$$\mathcal{E}(n) = \frac{1}{2}\mathbf{e}^T(n)\mathbf{e}(n) \tag{15.54}$$

The objective of the learning process is to minimize a cost function obtained by summing $\mathcal{E}(n)$ over all time n; that is,

$$\mathcal{E}_{\text{total}} = \sum_n \mathcal{E}(n)$$

To accomplish this objective we may use the method of steepest descent, which requires knowledge of the *gradient matrix,* written as

$$\nabla_{\mathbf{w}} \mathscr{E}_{total} = \frac{\partial \mathscr{E}_{total}}{\partial \mathbf{W}}$$

$$= \sum_n \frac{\partial \mathscr{E}(n)}{\partial \mathbf{W}}$$

$$= \sum_n \nabla_{\mathbf{w}} \mathscr{E}(n)$$

where $\nabla_{\mathbf{w}} \mathscr{E}(n)$ is the gradient of $\mathscr{E}(n)$ with respect to the weight matrix $\mathbf{W} = \{\mathbf{w}_k\}$. We may, if desired, continue with this equation and derive update equations for the synaptic weights of the recurrent network without invoking approximations. However, in order to develop a learning algorithm that can be used to train the recurrent network in *real time,* we must use an instantaneous *estimate* of the gradient, namely $\nabla_{\mathbf{w}} \mathscr{E}(n)$, which results in an *approximation* to the method of steepest descent.

Returning to Eq. (15.54) as the cost function to be minimized, we differentiate it with respect to the weight vector $\mathbf{w}_j$, obtaining

$$\frac{\partial \mathscr{E}(n)}{\partial \mathbf{w}_j} = \left(\frac{\partial \mathbf{e}(n)}{\partial \mathbf{w}_j} \right) \mathbf{e}(n)$$

$$= -\mathbf{C} \left(\frac{\partial \mathbf{x}(n)}{\partial \mathbf{w}_j} \right) \mathbf{e}(n) \qquad (15.55)$$

$$= -\mathbf{C} \mathbf{\Lambda}_j(n) \mathbf{e}(n), \qquad j = 1, 2, ..., q$$

The adjustment applied to synaptic weight vector $\mathbf{w}_j(n)$ of neuron j is therefore determined by

$$\Delta \mathbf{w}_j(n) = -\eta \frac{\partial \mathscr{E}(n)}{\partial \mathbf{w}_j}$$

$$= \eta \, \mathbf{C} \mathbf{\Lambda}_j(n) \mathbf{e}(n), \qquad j = 1, 2, ..., q \qquad (15.56)$$

where η is the learning-rate parameter and $\mathbf{\Lambda}_j(n)$ is itself governed by Eq. (15.52).

The only remaining item is that of specifying the *initial conditions* to start the learning process. For this purpose we set

$$\mathbf{\Lambda}_j(0) = \mathbf{0} \qquad \text{for all } j \qquad (15.57)$$

the implication of which is that initially the recurrent network resides in a constant state.

Table 15.1 presents a summary of the real-time recurrent learning algorithm. The formulation of the algorithm as described here applies to an arbitrary activation function $\varphi(\cdot)$ that is differentiable with respect to its argument. For the special case of a sigmoidal nonlinearity in the form of a hyperbolic tangent function, we have

$$x_j(n + 1) = \varphi(v_j(n))$$

$$= \tanh(v_j(n))$$

TABLE 15.1 Summary of the Real-Time Recurrent Learning Algorithm

---

Parameters:

m = dimensionality of input space
q = dimensionality of state space
p = dimensionality of output space
$\mathbf{w}_j$ = synaptic weight vector of neuron $j, j = 1, 2, ..., q$.

Initialization:

1. Set the synaptic weights of the algorithm to small values selected from a uniform distribution.
2. Set the initial value of the state vector $\mathbf{x}(0) = \mathbf{0}$.
3. Set $\Lambda_j(0) = \mathbf{0}$ for $j = 1, 2, ..., q$.

Computations: Compute for $n = 0, 1, 2, ...,$

$$\Lambda_j(n + 1) = \mathbf{\Phi}(n)[\mathbf{W}_a(n)\Lambda_j(n) + \mathbf{U}_j(n)]$$

$$\mathbf{e}(n) = \mathbf{d}(n) - \mathbf{C}\mathbf{x}(n)$$

$$\Delta\mathbf{w}_j(n) = \eta\mathbf{C}\Lambda_j(n)\mathbf{e}(n)$$

The definitions of $\mathbf{x}(n), \Lambda_j(n), \mathbf{U}_j(n)$, and $\mathbf{\Phi}(n)$ are given in Eqs. (15.46), (15.49), (15.50), and (15.51), respectively.

---

and

$$
\begin{aligned}
\varphi'(v_j(n)) &= \frac{\partial\varphi(v_j(n))}{\partial v_j(n)} \\
&= \operatorname{sech}^2(v_j(n)) \\
&= 1 - x_j^2(n + 1)
\end{aligned}
\tag{15.58}
$$

where $v_j(n)$ is the induced local field of neuron j and $x_j(n + 1)$ is its state at $n + 1$.

The use of the instantaneous gradient $\nabla_\mathbf{w}\mathscr{E}(n)$ means that the real-time recurrent learning algorithm described here deviates from a non-real-time one based on the true gradient $\nabla_\mathbf{w}\mathscr{E}_{\text{total}}$. However, this deviation is exactly analogous to that encountered in the standard back-propagation algorithm used in Chapter 4 to train an ordinary multilayer perceptron, where weight changes are made after each pattern presentation. While the real-time recurrent learning algorithm is not guaranteed to follow the precise negative gradient of the total error function $\mathscr{E}_{\text{total}}(\mathbf{W})$ with respect to the weight matrix $\mathbf{W}$, the practical differences between the real-time and non-real-time versions are often slight; these two versions become nearly identical as the learning-rate parameter η is reduced. The most severe potential consequence of this deviation from the true gradient-following behavior is that the observed trajectory (obtained by plotting $\mathscr{E}(n)$ versus the elements of the weight matrix $\mathbf{W}(n)$) may itself depend on the weight changes produced by the algorithm, which may be viewed as another source of feedback and therefore a cause of instability in the system. We can avoid this effect by using a learning-rate parameter η small enough to make the time scale of the weight changes much smaller than the time scale of the network operation (Williams and Zipser, 1989).

Example 15.6

In this example we formulate the RTRL algorithm for the fully recurrent network shown in Fig. 15.6 with two inputs and single output. The network has three neurons, with the composition of matrices $\mathbf{W}_a$, $\mathbf{W}_b$, and $\mathbf{C}$ as described in Example 15.1.

With $m = 2$ and $q = 3$, we find from Eq. (15.48) that

$$\boldsymbol{\xi}(n) = \begin{bmatrix} x_1(n) \\ x_2(n) \\ x_3(n) \\ 1 \\ u_1(n) \\ u_2(n) \end{bmatrix}$$

Let $\lambda_{j,kl}(n)$ denote the kl-th element of matrix $\boldsymbol{\Lambda}_j(n)$. The use of Eqs. (15.52) and (15.56) then yields, respectively,

$$\lambda_{j,kl}(n + 1) = \varphi'(v_j(n)) \left[\sum_{i=1}^{3} w_{ji}(n) \lambda_{j,kl}(n) + \delta_{kj} \xi_l(n) \right]$$

$$\Delta w_{kl}(n) = \eta(d_1(n)) - x_1(n)) \lambda_{1,kl}(n)$$

where δ_{kj} is the Kronecker delta, which is equal to 1 for $k = j$ and zero otherwise; and $(j, k) = 1, 2, 3$ and $l = 1, 2, ..., 6$. Figure 15.12 presents a *sensitivity graph* determining the evolution of the weight adjustment $\Delta w_{kl}(n)$. Note that $\mathbf{W}_a = \{w_{ji}\}$ for $(j, i) = 1, 2, 3$, and $\mathbf{W}_b = \{w_{jl}\}$ for $j = 1, 2, 3$ and $l = 4, 5, 6$.

∎

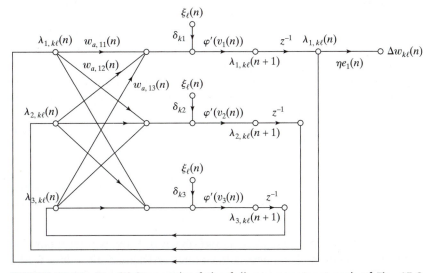

FIGURE 15.12 Sensitivity graph of the fully recurrent network of Fig. 15.6. Note: The three nodes labeled $\xi_l(n)$ are all to be viewed as a single input.

Teacher Forcing

A strategy that is frequently used in the training of recurrent networks is *teacher forcing* (Williams and Zipser, 1989, 1995); in adaptive filtering, teacher forcing is known as the *equation-error method* (Mendel, 1995). Basically, teacher forcing involves replacing the actual output of a neuron, during training of the network, with the corresponding desired response (i.e., target signal) in subsequent computation of the dynamic behavior of the network, whenever that desired response is available. Although teacher forcing is being described under the RTRL algorithm, its use applies to any other learning algorithm. For it to be applicable, however, the neuron in question must feed its output back to the network.

Beneficial effects of teacher forcing include (Williams and Zipser, 1995):

- *Teacher forcing may lead to faster training.* The reason for this improvement is the use of teacher forcing amounts to the assumption that the network has correctly learned all the earlier parts of the task that pertain to the neurons where teacher forcing has been applied.
- *Teacher forcing may serve as a corrective mechanism during training.* For example, the synaptic weights of the network may have the correct values, but somehow the network is currently operating in the wrong region of the state space. Clearly, adjusting the synaptic weights is the wrong strategy in such a situation.

A gradient-based learning algorithm that uses teacher forcing is in actual fact optimizing a cost function different from its unforced counterpart. The teacher forced and unforced versions of the algorithm may therefore yield different solutions, unless the pertinent error signals are zero, in which case learning is unnecessary.

15.9 KALMAN FILTERS

As mentioned previously, continuous learning based on gradient descent, exemplified by the real-time recurrent learning algorithm, is typically slow due to reliance on instantaneous estimates of gradients. We may overcome this serious limitation by viewing the supervised training of a recurrent network as an *optimum filtering problem,* the solution of which *recursively* utilizes information contained in the training data in a manner going back to the first iteration of the learning process. The idea described here is the essence of *Kalman filtering* (Kalman, 1960). Novel features of Kalman filters include:

- The theory is formulated in terms of state-space concepts, providing efficient utilization of the information contained in the input data.
- Estimation of the state is computed recursively; that is, each updated estimate of the state is computed from the previous estimate and the data currently available, so only the previous estimate requires storage.

In this section we present a brief review of Kalman filter theory[10] to pave the way for the derivation of the decoupled extended Kalman filter described in the next section. The development of the theory usually begins with linear dynamical systems. To extend its use to nonlinear dynamical systems, a form of *linearization* is applied to the system; this latter part of the discussion is deferred to the next section.

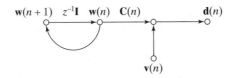

FIGURE 15.13 Signal-flow graph of linear, discrete-time dynamical system for describing the Kalman filter.

Consider then a *linear, discrete-time dynamical system* described by the signal-flow graph shown in Fig. 15.13. The time-domain description of the system presented here is along similar lines to the state-space formalism presented in Section 15.3. In mathematical terms, Fig. 15.13 embodies the following pair of equations:

$$\mathbf{w}(n + 1) = \mathbf{w}(n) \tag{15.59}$$

$$\mathbf{d}(n) = \mathbf{C}(n)\mathbf{w}(n) + \boldsymbol{v}(n) \tag{15.60}$$

The various quantities in the *process equation* (15.59) and the *measurement equation* (15.60) are described as follows:

- $\mathbf{w}(n)$ is the *state vector* of the system.
- $\mathbf{d}(n)$ is the *observation vector.*
- $\mathbf{C}(n)$ is the *measurement matrix.*
- $\boldsymbol{v}(n)$ is the *measurement noise.*

In the process equation (15.59) we have made two simplifying assumptions. First, the process equation is *noiseless*. Second, the transition matrix relating the states of the system at times $n + 1$ and n is equal to the identity matrix. We have also used a new notation for the state in Fig. 15.13 for reasons that will become apparent in the next section.

The Kalman filtering problem may now be stated as follows:

Use the entire observed data, consisting of the set of vectors $\{\mathbf{d}(i)\}_{i=1}^{n}$, to find for each $n \geq 1$ the minimum mean-square error estimate of the state $\mathbf{w}(i)$.

Note that information on the state vector is not available. The problem is called *filtering* if $i = n$, *prediction* if $i > n$, and *smoothing* if $1 \leq i \leq n$. The solution to the problem is derived on the basis of the following assumptions (beyond the assumed linearity of the system):

1. The measurement noise $\boldsymbol{v}(n)$ is a zero mean, white noise process whose covariance matrix is defined by

$$E[\boldsymbol{v}(n)\boldsymbol{v}^{T}(k)] = \begin{cases} \mathbf{R}(n), & n = k \\ \mathbf{0}, & n \neq k \end{cases} \tag{15.61}$$

2. The initial value of the state, $\mathbf{w}(0)$, is uncorrelated with $v(n)$ for all $n \geq 0$.

For an elegant derivation of the Kalman filter, we may use the notion of innovations (Kailath, 1968). Specifically, the *innovations process* associated with the observation vector $\mathbf{d}(n)$ is defined by

$$\boldsymbol{\alpha}(n) = \mathbf{d}(n) - \hat{\mathbf{d}}(n|n - 1) \tag{15.62}$$

where $\hat{\mathbf{d}}(n|n - 1)$ is the *minimum mean-square error estimate* of $\mathbf{d}(n)$, given *all past values* of the observation vector starting at time $n = 1$ and extending up to time $n - 1$.

By the "minimum mean-square error estimate" we mean that particular estimate that minimizes the mean-square error measured with respect to $\mathbf{d}(n)$. The innovations process $\boldsymbol{\alpha}(n)$ may be regarded as a measure of the *new* information contained in $\mathbf{d}(n)$ that is not available in the predictable part $\hat{\mathbf{d}}(n \mid n - 1)$. The innovations process $\boldsymbol{\alpha}(n)$ has some nice properties as summarized here (Kailath, 1968):

1. The innovations process $\boldsymbol{\alpha}(n)$ associated with $\mathbf{d}(n)$ is uncorrelated to all past observations $\mathbf{d}(1), \mathbf{d}(2), ..., \mathbf{d}(n - 1)$, as shown by

$$E[\boldsymbol{\alpha}(n)\mathbf{d}^T(k)] = \mathbf{0} \qquad \text{for } 1 \leq k \leq n - 1$$

2. The innovations process consists of a sequence of random vectors that are uncorrelated with each other, as shown by

$$E[\boldsymbol{\alpha}(n)\boldsymbol{\alpha}^T(k)] = \mathbf{0} \qquad \text{for } 1 \leq k \leq n - 1$$

3. There is a one-to-one correspondence between the sequence of random vectors representing the observed data and the sequence of random vectors representing the innovations process, as shown by

$$\{\mathbf{d}(1), \mathbf{d}(2), ..., \mathbf{d}(n)\} \rightleftharpoons \{\boldsymbol{\alpha}(1), \boldsymbol{\alpha}(2), ..., \boldsymbol{\alpha}(n)\} \tag{15.63}$$

We may now replace the correlated sequence of observed data with the uncorrelated (and therefore simpler) sequence of innovations *without any loss of information*. In so doing, the derivation of the Kalman filter is simplified by expressing the estimate of the state at time i, given the set of innovations $\{\boldsymbol{\alpha}(k)\}_{k=1}^n$. In carrying out the analysis on this basis, we may derive the standard Kalman filter as summarized in Table 15.2. There are three new quantities in this algorithm that need to be defined:

- $\mathbf{K}(n, n - 1)$ is the *error covariance matrix* defined by

$$\mathbf{K}(n, n - 1) = E[\boldsymbol{\epsilon}(n, n - 1)\boldsymbol{\epsilon}^T(n, n - 1)] \tag{15.64}$$

where the *state error* $\boldsymbol{\epsilon}(n, n - 1)$ is itself defined by

$$\boldsymbol{\epsilon}(n, n - 1) = \mathbf{w}(n) - \hat{\mathbf{w}}(n \mid n - 1) \tag{15.65}$$

where $\mathbf{w}(n)$ is the actual state and $\hat{\mathbf{w}}(n \mid n - 1)$ is its one-step prediction based on past values of the observed data up to time $n - 1$.

TABLE 15.2 Summary of the Kalman Filter

Compute for $n = 1, 2, 3, ...$

$$\boldsymbol{\Gamma}(n) = [\mathbf{C}(n)\mathbf{K}(n, n - 1)\mathbf{C}^T(n) + \mathbf{R}(n)]^{-1}$$

$$\mathbf{G}(n) = \mathbf{K}(n, n - 1)\mathbf{C}^T(n)\boldsymbol{\Gamma}(n)$$

$$\boldsymbol{\alpha}(n) = \mathbf{y}(n) - \mathbf{C}(n)\hat{\mathbf{w}}(n \mid n - 1)$$

$$\hat{\mathbf{w}}(n + 1 \mid n) = \hat{\mathbf{w}}(n \mid n - 1) + \mathbf{G}(n)\boldsymbol{\alpha}(n)$$

$$\mathbf{K}(n + 1, n) = \mathbf{K}(n, n - 1) - \mathbf{G}(n)\mathbf{C}(n)\mathbf{K}(n, n - 1)$$

- $\mathbf{\Gamma}(n)$ is the *conversion factor* that relates the *filtered estimation error* $\mathbf{e}(n)$ to the innovations $\boldsymbol{\alpha}(n)$ as shown by

$$\mathbf{e}(n) = \mathbf{R}(n)\mathbf{\Gamma}(n)\boldsymbol{\alpha}(n) \tag{15.66}$$

where

$$\mathbf{e}(n) = \mathbf{d}(n) - \hat{\mathbf{d}}(n|n) \tag{15.67}$$

where $\hat{\mathbf{d}}(n\,|\,n)$ is the estimate of the observation vector $\mathbf{d}(n)$ given all the observed data up to time n.

- $\mathbf{G}(n)$ is the *Kalman gain*, which determines the correction used to update the estimate of the state.

The type of Kalman filter summarized in Table 15.2 is designed to propagate the error covariance matrix $\mathbf{K}(n, n-1)$. This algorithm is therefore called the *covariance Kalman filtering algorithm*.

Square Root Kalman Filter

The covariance Kalman filter is prone to serious numerical difficulties. In particular, the updated matrix $\mathbf{K}(n+1, n)$ is determined by the *Riccati equation*, which is defined by the last line of computation in Table 15.2. The right-hand side of the Riccati equation is the difference between two matrix quantities. Unless the numerical accuracy employed at every iteration of the algorithm is high enough, the updated matrix $\mathbf{K}(n+1, n)$ resulting from this computation may *not* be nonnegative definite. Such a solution is clearly unacceptable because $\mathbf{K}(n+1, n)$ represents a covariance matrix, which by definition is nonnegative definite. The unstable behavior of the Kalman filter, which results from numerical inaccuracies due to the use of finite word length arithmetic, is called the *divergence phenomenon*.

This problem may be overcome by propagating the square root of the error covariance matrix, $\mathbf{K}^{1/2}(n, n-1)$, rather than $\mathbf{K}(n, n-1)$ itself. Specifically, using the *Cholesky factorization*, we may express $\mathbf{K}(n, n-1)$ as follows (Golub and Van Loan, 1996):

$$\mathbf{K}(n, n-1) = \mathbf{K}^{1/2}(n, n-1)\mathbf{K}^{T/2}(n, n-1) \tag{15.68}$$

where $\mathbf{K}^{1/2}(n, n-1)$ is a lower triangular matrix, and $\mathbf{K}^{T/2}(n, n-1)$ is its transpose. In linear algebra, the Cholesky factor $\mathbf{K}^{1/2}(n, n-1)$ is commonly referred to as the *square root* of $\mathbf{K}(n, n-1)$. Thus a Kalman filter based on the Cholesky factorization is called a *square root Kalman filter*.[11] The important point is that the matrix product $\mathbf{K}^{1/2}(n, n-1)\mathbf{K}^{T/2}(n, n-1)$ is much less likely to become indefinite because the product of any square matrix and its transpose is always positive definite.

15.10 DECOUPLED EXTENDED KALMAN FILTER

Our primary interest in the Kalman filter is to exploit its unique properties to perform the supervised training of a recurrent network.[12] Given the architectural complexity of a recurrent network (e.g., recurrent multilayer perceptron), the important issue is how to proceed with this approach in a computationally feasible manner without compromising the application of Kalman filter theory. The answer is found in using a *decoupled* form of the extended Kalman filter, in which the computational complexity is made to suit the requirements of a particular application and of available computational resources (Puskorius and Feldkamp, 1991).

Consider a recurrent network built around a static multilayer perceptron with W synaptic weights and p output nodes. Let the vector $\mathbf{w}(n)$ denote the synaptic weights of the entire network at time n. With adaptive filtering in mind, the state-space equations for the network may be modeled as follows (Singhal and Wu, 1989; Haykin, 1996):

$$\mathbf{w}(n + 1) = \mathbf{w}(n) \tag{15.69}$$

$$\mathbf{d}_o(n) = \mathbf{c}(\mathbf{w}(n), \mathbf{u}(n), \mathbf{v}(n)) + \boldsymbol{v}(n) \tag{15.70}$$

where the weight vector $\mathbf{w}(n)$ plays the role of a state. The second argument $\mathbf{u}(n)$ and third argument $\mathbf{v}(n)$ pertaining to the vector valued function $\mathbf{c}(\cdot,\cdot,\cdot)$ denote the input vector and vector of recurrent node activities, respectively. In effect, Eq. (15.69) states that the model is residing in its "optimum" condition, with the transition matrix that takes the weight vector $\mathbf{w}(n)$ at time n and transforms it into $\mathbf{w}(n + 1)$ at time $n + 1$ being equal to the identity matrix. The optimum condition described here refers to a local or global minimum on the error surface of the recurrent network. The only source of nonlinearity in the model resides in the measurement equation (15.70). The vector $\mathbf{d}_o$ denotes the desired response of the model. With Eq. (15.70) representing the input–output equation of the model, it follows that $\mathbf{c}(\cdot,\cdot,\cdot)$ is the overall nonlinearity from the input layer to the output layer of the multilayer perceptron. The measurement noise vector $\boldsymbol{v}(n)$ in Eq. (15.70) is assumed to be a multivariate white noise process with zero mean and diagonal covariance matrix $\mathbf{R}(n)$.

It is important to note that in applying the extended Kalman filter to a recurrent network, there are two different contexts in which the term "state" is used:

- Evolution of the system via adaptive filtering, which manifests itself in changes to the recurrent network's weights through training; the vector $\mathbf{w}(n)$ takes care of this first notion of state.
- Operation of the recurrent network itself, exemplified by the recurrent node activities on which the function $\mathbf{c}$ depends; the vector $\mathbf{v}(n)$ takes care of this second notion of state.

By comparing the model described in Eqs. (15.69) and (15.70) with the linear dynamical model of Eqs. (15.59) and (15.60), we see that the only difference between these two models is in the nonlinear form of the measurement equation. To prepare the way for the application of Kalman filter theory to the state-space model just described, we must therefore first linearize Eq. (15.70) and recast it in the form

$$\mathbf{d}(n) = \mathbf{C}(n)\mathbf{w}(n) + \boldsymbol{v}(n) \tag{15.71}$$

where $\mathbf{C}(n)$ is the p-by-W measurement matrix of the linearized model, and we have used $\mathbf{d}(n)$ to distinguish it from $\mathbf{d}_o(n)$ in Eq. (15.70). The linearization consists of the partial derivatives of the p outputs of the whole network with respect to the W weights of the model as shown by

$$
\mathbf{C}(n) =
\begin{bmatrix}
\dfrac{\partial c_1}{\partial w_1} & \dfrac{\partial c_1}{\partial w_2} & \cdots & \dfrac{\partial c_1}{\partial w_W} \\[2ex]
\dfrac{\partial c_2}{\partial w_1} & \dfrac{\partial c_2}{\partial w_2} & \cdots & \dfrac{\partial c_2}{\partial w_W} \\[2ex]
\vdots & \vdots & & \vdots \\[2ex]
\dfrac{\partial c_p}{\partial w_1} & \dfrac{\partial c_p}{\partial w_2} & \cdots & \dfrac{\partial c_p}{\partial w_W}
\end{bmatrix}
\tag{15.72}
$$

where $c_i, i = 1, 2, ..., p$ denotes the ith element of the nonlinearity $\mathbf{c}(\mathbf{w}(n), \mathbf{u}(n), \mathbf{v}(n))$. The partial derivatives in Eq. (15.72) are evaluated at $\mathbf{w}(n) = \hat{\mathbf{w}}(n)$, where $\hat{\mathbf{w}}(n)$ is the estimate of the weight vector $\mathbf{w}(n)$ computed by the extended Kalman filter at time n, given the observed data up to time $n - 1$ (Haykin, 1996). In practice, these partial derivatives are computed by using the back-propagation-through-time or real-time recurrent learning algorithm. In effect, the extended Kalman filter builds on one or the other of these two algorithms described in Sections 15.7 and 15.8. This implies that $\mathbf{c}$ must be a function of the recurrent node activities as stated. In fact, for a single-layer recurrent network the matrix $\mathbf{C}(n)$ can be composed from the elements of the matrices $\mathbf{\Lambda}_j(n)$ as computed by the RTRL algorithm in Eq. (15.52). Thus the measurement matrix $\mathbf{C}(n)$ is the dynamic derivative matrix of the network outputs with respect to the network's free parameters. Just as the recurrent node activities of the network at time step $(n + 1)$ are a function of the corresponding values from the previous time step n, in an analogous manner we find that the derivatives of recurrent node activities with respect to the network's free parameters at time step $(n + 1)$ are a function of the corresponding values from the previous time step n as expressed in the RTRL equations.

Suppose now the synaptic weights of the network are partitioned into g groups, with group i containing k_i neurons, for example. The measurement matrix $\mathbf{C}$ defined in Eq. (15.72) is the p-by-W matrix of derivatives of network outputs with respect to all the weights in the network. The dependence of matrix $\mathbf{C}(n)$ on the input vector $\mathbf{u}(n)$ is implicitly defined in Eq. (15.72). The matrix $\mathbf{C}(n)$ thus defined contains all the derivatives that are necessary for any decoupled version of the extended Kalman filter. For example, if the *global extended Kalman filter* (GEKF) is used (i.e., we have no decoupling), $g = 1$ and the whole $\mathbf{C}(n)$ is as defined in Eq. (15.72). On the other hand, if the *decoupled extended Kalman filter* (DEKF) is used, then the "global" measurement matrix $\mathbf{C}(n)$ must be arranged so that the weights corresponding to a given neuron in the network are grouped as a single block within $\mathbf{C}(n)$, where each block is identified by index $i = 1, 2, ..., g$. In the latter case, the matrix $\mathbf{C}(n)$ is merely the *concatenation* of the individual $\mathbf{C}_i$'s, as shown here

$$
\mathbf{C}(n) = [\mathbf{C}_1(n), \mathbf{C}_2(n), ..., \mathbf{C}_g(n)]
$$

In any event, regardless of the level of decoupling employed, the entire matrix $\mathbf{C}(n)$ must be computed as defined in Eq. (15.72).

The stage is now set for the application of the Kalman filtering algorithm summarized in Table 15.2. In particular, for the linearized dynamic model described by Eqs. (15.69) and (15.71), we have (Puskorius and Feldkamp, 1991):

$$\mathbf{\Gamma}(n) = \left[\sum_{i=1}^{g} \mathbf{C}_i(n)\mathbf{K}_i(n, n-1)\mathbf{C}_i^T(n) + \mathbf{R}(n) \right]^{-1} \tag{15.73}$$

$$\mathbf{G}_i(n) = \mathbf{K}_i(n, n-1)\mathbf{C}_i^T(n)\mathbf{\Gamma}(n) \tag{15.74}$$

$$\boldsymbol{\alpha}(n) = \mathbf{d}(n) - \hat{\mathbf{d}}(n|n-1) \tag{15.75}$$

$$\mathbf{w}_i(n+1|n) = \hat{\mathbf{w}}_i(n|n-1) + \mathbf{G}_i(n)\boldsymbol{\alpha}(n) \tag{15.76}$$

$$\mathbf{K}_i(n+1, n) = \mathbf{K}_i(n, n-1) - \mathbf{G}_i(n)\mathbf{C}_i(n)\mathbf{K}_i(n, n-1) \tag{15.77}$$

where $i = 1, 2, ..., g$. The parameter vectors and signal vectors in Eq. (15.73) to (15.77) are described as follows:

$\mathbf{\Gamma}(n) = p$-by-p matrix, denoting the global conversion factor for the entire network

$\mathbf{G}_i(n) = W_i$-by-p matrix, denoting the Kalman gain for group i of neurons

$\boldsymbol{\alpha}(n) = p$-by-1 vector, denoting the innovations defined as the difference between the desired response $\mathbf{d}(n)$ for the linearized system and its estimate $\hat{\mathbf{d}}(n|n-1)$ based on input data available at time $n-1$; the estimate $\hat{\mathbf{d}}(n|n-1)$ is represented by the actual output vector $\mathbf{y}(n)$ of the network residing in state $\{\hat{\mathbf{w}}_i(n|n-1)\}$, which is produced in response to the input $\mathbf{u}(n)$

$\hat{\mathbf{w}}_i(n|n-1) = W$-by-1 vector, denoting the estimate of the weight vector $\mathbf{w}_i(n)$ for group i at time n, given the observed data up to time $n-1$

$\mathbf{K}_i(n, n-1) = k_i$-by-$k_i$ matrix, denoting the error covariance matrix for group i of neurons

The summation included in the definition of the global conversion factor $\mathbf{\Gamma}(n)$ in Eq. (15.73) accounts for the decoupled nature of the extended Kalman filter.

It is important to understand that in the DEKF algorithm the decoupling really determines which particular elements of the global error covariance matrix $\mathbf{K}(n, n-1)$ are to be maintained and updated. In fact, all computational savings are due to ignoring the maintenance and updates associated with those off-diagonal blocks of the global error covariance matrix $\mathbf{K}(n, n-1)$, which would otherwise correspond to coupling of different groups of synaptic weights.

The DEKF algorithm encoded by Eqs. (15.73) to (15.77) minimizes the cost function:

$$\mathscr{E}(n) = \frac{1}{2} \sum_{j=1}^{n} \|\mathbf{e}(j)\|^2 \tag{15.78}$$

where $\mathbf{e}(j)$ is the error vector defined by

$$\mathbf{e}(j) = \mathbf{d}(j) - \mathbf{y}(j), \qquad j = 1, 2, ..., n$$

where $\mathbf{y}(j)$ is the actual output of the network using all available information up to and including time j. Note that, in general, $\mathbf{e}(j) \neq \boldsymbol{\alpha}(j)$.

Artificial Process Noise

The nonlinear dynamical system modeled in Eqs. (15.69) and (15.70) is *unforced*, in that the process equation (15.69) has no external inputs. This deficiency can lead to serious numerical difficulties, and therefore the divergence of the Kalman filter when it operates in a finite precision environment. As explained in Section 15.9, the divergence phenomenon may be overcome through the use of square root filtering.

Another way of circumventing the divergence phenomenon is to use a heuristic mechanism that involves artificially adding *process noise* to the process equation, as shown by

$$\mathbf{w}_i(n + 1) = \mathbf{w}_i(n) + \boldsymbol{\omega}_i(n), \qquad i = 1, 2, ..., g \qquad (15.79)$$

where $\boldsymbol{\omega}_i(n)$ is the process noise. It is assumed that $\boldsymbol{\omega}_i(n)$ is a multivariate white noise of zero mean and diagonal covariance matrix $\mathbf{Q}_i(n)$. The artificially added process noise $\boldsymbol{\omega}_i(n)$ is naturally independent of both the measurement noise $\boldsymbol{v}(n)$ and the initial state of the network. The net effect of adding $\boldsymbol{\omega}_i(n)$ to the process equation (15.79) is to modify the Riccati equation for updating the error covariance matrix as follows (Haykin, 1996):

$$\mathbf{K}_i(n + 1, n) = \mathbf{K}_i(n, n - 1) - \mathbf{G}_i(n)\mathbf{C}_i(n)\mathbf{K}_i(n, n - 1) + \mathbf{Q}_i(n) \qquad (15.80)$$

Provided that $\mathbf{Q}_i(n)$ is large enough for all i, then $\mathbf{K}_i(n + 1, n)$ is assured of remaining nonnegative definite for all n.

In addition to overcoming these numerical difficulties, the artificial insertion of process noise $\boldsymbol{\omega}_i(n)$ into the process equation has the following beneficial effect: There is less likelihood for the algorithm to be trapped at a local minimum during the training process. This in turn results in a significant improvement in training performance in terms of rate of convergence and quality of solution (Puskorius and Feldkamp, 1991).

Summary of the DEKF Algorithm

Table 15.3 presents a summary of the DEKF algorithm based on Eqs. (15.73) to (15.76), and Eq. (15.80). This table also includes details of initialization of the algorithm.

A final comment on the extended Kalman filter is in order. The DEKF algorithm summarized in Table 15.3 refers to an entire family of possible *information-preserving learning procedures,* including the GEKF. As a general rule, we expect the DEKF to produce a performance, in terms of solution quality, that approaches the GEKF but is not expected to surpass it. On the other hand, the DEKF is always computationally less demanding than the GEKF. Notwithstanding this computational advantage, current computer speeds and memory sizes have now made GEKF feasible for some practical problems, especially in off-line training of recurrent networks.

TABLE 15.3   Summary of the DEKF Algorithm

Initialization:

1. Set the synaptic weights of the recurrent network to small values selected from a uniform distribution.
2. Set the diagonal elements of the covariance matrix $\mathbf{Q}(n)$ (characterizing the artificially inserted process noise $\boldsymbol{\omega}(n)$) equal to 10^{-6} to 10^{-2}.
3. Set $\mathbf{K}(1,0) = \delta^{-1}\mathbf{I}, \qquad \delta$ = small positive constant.

Computations:

For $n = 1, 2, ...,$ compute

$$\boldsymbol{\Gamma}(n) = \left[\sum_{i=1}^{g} \mathbf{C}_i(n)\mathbf{K}_i(n, n-1)\,\mathbf{C}_i^T(n) + \mathbf{R}(n) \right]^{-1}$$

$$\mathbf{G}_i(n) = \mathbf{K}_i(n, n-1)\mathbf{C}_i^T(n)\boldsymbol{\Gamma}(n)$$

$$\boldsymbol{\alpha}(n) = \mathbf{d}(n) - \hat{\mathbf{d}}(n|n-1)$$

$$\hat{\mathbf{w}}_i(n+1|n) = \hat{\mathbf{w}}_i(n|n-1) + \mathbf{G}_i(n)\boldsymbol{\alpha}(n)$$

$$\mathbf{K}_i(n+1, n) = \mathbf{K}_i(n, n-1) - \mathbf{G}_i(n)\mathbf{C}_i(n)\mathbf{K}_i(n, n-1) + \mathbf{Q}_i(n)$$

where in the third line, $\hat{\mathbf{d}}(n|n-1)$ is the actual output vector $\mathbf{y}(n)$ of the network produced in response to the input vector $\mathbf{u}(n)$.

Note: For $g = 1$ (i.e., no decoupling), the DEKF algorithm becomes the global extended Kalman filtering (GEKF) algorithm.

Computational Complexity

Table 15.4 presents a comparison of the computational complexity of the three learning algorithms discussed in this chapter: back-propagation through time, real-time recurrent learning, and decoupled extended Kalman filter. The computational complexity of these algorithms increases in the order arranged here.

15.11   COMPUTER EXPERIMENT

In this experiment we revisit the simulation of the nonlinear time series studied in Section 13.5. The time series is defined by the frequency-modulated signal:

$$x(n) = \sin(n + \sin(n^2)) \qquad n = 0, 1, 2, ...$$

We will investigate two different structures for the simulation:

- Recurrent multilayer perceptron (RMLP) consisting of 1 input node, first hidden layer of 10 recurrent neurons, second hidden layer of 10 neurons, and 1 linear output neuron.
- Focused time lagged feedforward network (TLFN), consisting of a tapped-delay-time memory with 20 taps and a multilayer perceptron with 10 hidden neurons and 1 linear output neuron.

The RMLP has slightly more synaptic weights than the focused TLFN, but half the memory (10 recurrent nodes versus 20 taps).

TABLE 15.4   Comparison of the Computational Complexity of Learning
Algorithms for Recurrent Networks

S = number of states
W = number of synaptic weights
L = length of training sequence

1. Back-propagation-through-time (BPTT) algorithm:
   - Time, storage space requirements: $O(WL + SL), O(WL + SL)$
2. Real-time recurrent learning (RTRL) algorithm:
   - Time, storage space requirements: $O(WS^2L), O(WS)$
3. Decoupled extended Kalman filtering (DEKF) algorithm:
   - At the minimum, DEKF incurs the same expense (in both time and space) for computing derivatives via RTRL or BPTT; for BPTT the time and space requirements are scaled by p, the number of network outputs, relative to the standard BPTT for which derivatives of a single scalar error term are computed.
   - In addition, DEKF requires time complexity $O(p^2W + p\sum_{i=1}^g k_i^2)$ and storage space $O(\sum_{i=1}^g k_i^2)$, where g is the number of groups and k_i is the number of neurons in group i. In the limit of a single weight group as in GEKF, these requirements become time and space storage: $O(pW^2)$ and $O(W^2)$, respectively.

The RMLP was trained using the DEKF algorithm. The TLFN was trained using two versions of the extended Kalman filter: (1) the GEKF algorithm (i.e., global version), and (2) the DEKF algorithm (i.e., decoupled version). The details of these two algorithms are:

- *GEKF*:

  δ = parameter used to initialize the error covariance matrix $\mathbf{K}(n, n-1)$

     = 0.01

  $\mathbf{R}(n)$ = covariance matrix of measurement noise $\boldsymbol{v}(n)$: $\mathbf{R}(0) = 100$ at the start of training and then annealed to $\mathbf{R}(n) = 3$ at the end of training

  $\mathbf{Q}(n)$ = covariance matrix of artificial process noise $\boldsymbol{\omega}(n)$: $\mathbf{Q}(0) = 10^{-2}$ at the start of training and then annealed to $\mathbf{Q}(n) = 10^{-6}$ at the end of training

  The annealing of $\mathbf{R}(n)$ and $\mathbf{Q}(n)$ has the effect of accelerating the learning rate as the training progresses.

- *DEKF*:

  g = number of groups

  $= \begin{cases} 21 \text{ for the RMLP} \\ 11 \text{ for the focused TLFN} \end{cases}$

All other parameters are the same as those used for the GEKF.

The training was performed on a sequence of 4000 samples. For the RMLP, subsets of length 100 were used, with the processing of 30,000 subsets over the entire training run. Each data point in the training set of 4000 samples was processed approximately 750 times. For the focused TLFN, each data point in the training set was also processed about 750 times. In both cases the testing was performed on 300 data points.

Figure 15.14 presents the one-step predicted waveform $\hat{y}(n)$ computed by the RMLP trained on the DEKF algorithm. This figure also includes the actual waveform

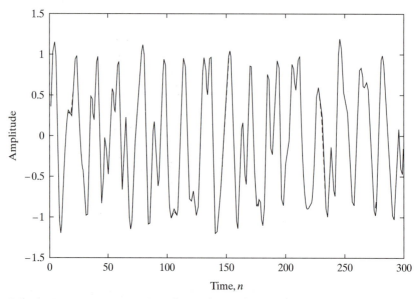

FIGURE 15.14 Superposition of actual waveform (solid) and predicted waveform (dashed) for the computer experiment on modeling; the predicted waveform was computed using the RMLP trained on the DEKF algorithm.

$y(n)$. These two waveforms are hardly distinguishable from each other. Figure 15.15a shows the prediction error

$$e(n) = y(n) - \hat{y}(n)$$

produced by the RMLP. The corresponding prediction errors produced by the focused TLFN trained on the GEKF and DEKF algorithms are shown in Figs. 15.15b and 15.15c, respectively. By comparing the results presented in Fig. 15.15 among themselves and also against the simulation results reported in Section 13.5, we may make the following observations:

1. The most accurate simulation in a mean-square error sense was produced by the RMLP trained on the DEKF algorithm; the variance of the prediction error was 1.1839×10^{-4}, computed over 5980 samples.
2. For the focused TLFN, the most accurate simulation in a mean-square error sense was produced by using GEKF training. For GEKF training the variance of the prediction error was 1.3351×10^{-4}, whereas for DEKF training it was 1.5871×10^{-4}. Both computations were again made using 5980 samples.
3. For the focused TLFN trained on the standard back-propagation algorithm, the variance of the prediction error reported in Section 13.5 was 1.2×10^{-3}, an order of magnitude worse than that obtained with the GEKF and DEKF algorithms.

The superior learning performance of the extended Kalman filter over back-propagation is due to its information-preserving property.

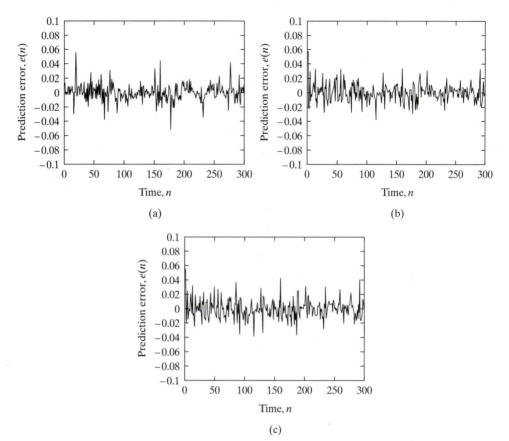

FIGURE 15.15   Prediction error waveforms for three different simulations: (a) RMLP with DEKF training, error variance $= 1.1839 \times 10^{-4}$. (b) Focused TLFN with GEKF training, error variance $= 1.3351 \times 10^{-4}$. (c) Focused TLFN with DEKF training, error variance $= 1.5871 \times 10^{-4}$.

15.12   VANISHING GRADIENTS IN RECURRENT NETWORKS

A problem that may require attention in practical applications of a recurrent network is the *vanishing gradients problem,* which pertains to the training of the network to produce a desired response at the current time that depends on input data in the distant past (Hochreiter, 1991; Bengio et al., 1994). The point is that because of the combined nonlinearities, an infinitesimal change of a temporally distant input may have almost no effect on network training. The problem may arise even if a large change in the temporally distant input has an effect, but the effect is not measurable by the gradient. This vanishing gradients problem makes the learning of long-term dependencies in gradient-based training algorithms difficult if not virtually impossible in certain cases.

In Bengio et al. (1994), it is argued that for many practical applications it is necessary that a recurrent network be able to store state information for an arbitrary duration and to do so in the presence of noise. The long-term storage of definite bits of information in the state variables of the recurrent network is referred to as *information latching*. The information latching must be *robust* so that the stored state information cannot be easily deleted by events that are unrelated to the learning task at hand. In specific terms, we may state the following (Bengio et al., 1994):

> Robust information latching in a recurrent network is accomplished if the states of the network are contained in the reduced attracting set of a hyperbolic attractor.

The notion of a hyperbolic attractor was discussed in Chapter 14. The *reduced attracting set* of a hyperbolic attractor is the set of points in the basin of attraction for which all the eigenvalues of the associated Jacobian have an absolute value less than 1. The implication here is that if a state $\mathbf{x}(n)$ of the recurrent network is in the basin of attraction of a hyperbolic attractor but not in the reduced attracting set, then the size of a ball of uncertainty around $\mathbf{x}(n)$ will grow exponentially with increasing time n, as illustrated in Fig. 15.16a. Therefore, small perturbations (noise) in the input applied to the recurrent network could push the trajectory toward another (possibly wrong) basin of attraction. If, however, the state $\mathbf{x}(n)$ remains in the reduced attracting set of the hyperbolic attractor, a bound on the input can be found that guarantees $\mathbf{x}(n)$ to remain within a certain distance of the attractor, as illustrated in Fig. 15.16b.

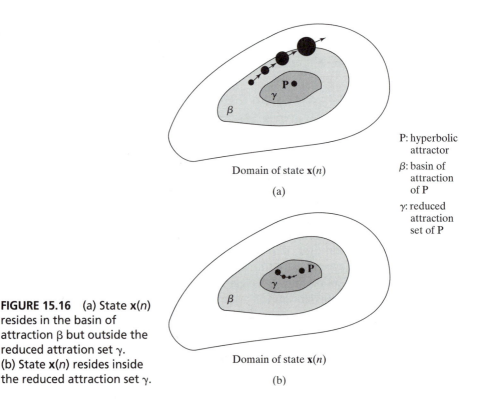

Domain of state $\mathbf{x}(n)$

(a)

P: hyperbolic
attractor

β: basin of
attraction
of P

γ: reduced
attraction
set of P

Domain of state $\mathbf{x}(n)$

(b)

FIGURE 15.16   (a) State $\mathbf{x}(n)$ resides in the basin of attraction β but outside the reduced attration set γ. (b) State $\mathbf{x}(n)$ resides inside the reduced attration set γ.

Long-Term Dependencies

To appreciate the impact of robust information latching on gradient-based learning, we note that the adjustment applied to the weight vector $\mathbf{w}$ of a recurrent network at time n is defined by

$$\Delta \mathbf{w}(n) = -\eta \, \frac{\partial \mathscr{E}_{\text{total}}}{\partial \mathbf{w}}$$

where η is the learning-rate parameter and $\partial \mathscr{E}_{\text{total}}/\partial \mathbf{w}$ is the gradient of the cost function $\mathscr{E}_{\text{total}}$ with respect to $\mathbf{w}$. The cost function $\mathscr{E}_{\text{total}}$ is typically defined by

$$\mathscr{E}_{\text{total}} = \frac{1}{2} \sum_i \| \mathbf{d}_i(n) - \mathbf{y}_i(n) \|^2$$

where $\mathbf{d}_i(n)$ is the desired response and $\mathbf{y}_i(n)$ is the actual response of the network at time n for the ith pattern. Hence, we may write

$$
\begin{aligned}
\Delta \mathbf{w}(n) &= \eta \sum_i \left(\frac{\partial \mathbf{y}_i(n)}{\partial \mathbf{w}} \right) (\mathbf{d}_i(n) - \mathbf{y}_i(n)) \\
&= \eta \sum_i \left(\frac{\partial \mathbf{y}_i(n)}{\partial \mathbf{x}_i(n)} \frac{\partial \mathbf{x}_i(n)}{\partial \mathbf{w}} \right) (\mathbf{d}_i(n) - \mathbf{y}_i(n))
\end{aligned}
\tag{15.81}
$$

where in the second line we have used the chain rule of calculus; the state vector $\mathbf{x}_i(n)$ pertains to the ith pattern (example) in the training sample. In applying algorithms such as back-propagation through time, the partial derivatives of the cost function are computed with respect to independent weights at different time indices. We may expand on the result in Eq. (15.81) by writing

$$\Delta \mathbf{w}(n) = \eta \sum_i \left(\frac{\partial \mathbf{y}_i(n)}{\partial \mathbf{x}_i(n)} \sum_{k=1}^{n} \frac{\partial \mathbf{x}_i(n)}{\partial \mathbf{w}(k)} \right) (\mathbf{d}_i(n) - \mathbf{y}_i(n))$$

Applying the chain rule of calculus a second time yields

$$\Delta \mathbf{w}(n) = \eta \sum_i \left(\frac{\partial \mathbf{y}_i(n)}{\partial \mathbf{x}_i(n)} \sum_{k=1}^{n} \frac{\partial \mathbf{x}_i(n)}{\partial \mathbf{x}_i(k)} \frac{\partial \mathbf{x}_i(k)}{\partial \mathbf{w}(k)} \right) (\mathbf{d}_i(n) - \mathbf{y}_i(n)) \tag{15.82}$$

We now recognize that in light of the state equation (15.2), we have

$$\mathbf{x}_i(n) = \boldsymbol{\varphi}(\mathbf{x}_i(k), \mathbf{u}(n)), \qquad 1 \le k < n$$

Hence, we may interpret $\partial \mathbf{x}_i(n)/\partial \mathbf{x}_i(k)$ as the Jacobian of the nonlinear function $\boldsymbol{\varphi}(\cdot,\cdot)$ expanded over $n - k$ time steps, as shown by

$$
\begin{aligned}
\frac{\partial \mathbf{x}_i(n)}{\partial \mathbf{x}_i(k)} &= \frac{\partial \boldsymbol{\varphi}(\mathbf{x}_i(k), \mathbf{u}(n))}{\partial \mathbf{x}_i(k)} \\
&= \mathbf{J}_{\mathbf{x}}(n, n - k)
\end{aligned}
\tag{15.83}
$$

In Bengio et al. (1994), it is shown that if the input $\mathbf{u}(n)$ is such that the recurrent network remains robustly latched to a hyperbolic attractor after time $n = 0$, then the Jacobian $\mathbf{J}_x(n, k)$ is an exponentially decreasing function of k so that

$$\det(\mathbf{J}_x(n, k)) \to 0 \quad \text{as} \quad k \to \infty \quad \text{for all } n \tag{15.84}$$

The implication of Eq. (15.84) is that a *small* change in the weight vector $\mathbf{w}$ of the network is experienced mostly in the near past (i.e., values of k close to the current time n). There may exist an adjustment $\Delta\mathbf{w}$ to the weight vector $\mathbf{w}$ at time n that would permit the current state $\mathbf{x}(n)$ to move to another possibly better basin of attraction, but the gradient of the cost function $\mathcal{E}_{\text{total}}$ with respect to $\mathbf{w}$ does not carry that information.

To conclude, assuming that hyperbolic attractors are used to store state information in a recurrent network by means of gradient-based learning, we find that either

- the network is *not* robust to the presence of noise in the input signal or
- the network is unable to discover *long-term dependencies* (i.e., relationships between target outputs and inputs that occur in the distant past).

Possible procedures for alleviating the difficulties that arise due to vanishing gradients in recurrent networks include the following:[13]

- Increased temporal span of input–output dependencies by presenting the network, during training, with the shortest strings of symbols first; see the heuristics presented in Section 15.6
- Use of the extended Kalman filter or its decoupled version for a more efficient use of available information than gradient-based learning algorithms; the extended Kalman filter is discussed in Section 15.10
- Use of elaborate optimization methods such as pseudo-Newton and simulated annealing (Bengio et al., 1994); second-order optimization methods and simulated annealing are described in Chapters 4 and 11, respectively

15.13   SYSTEM IDENTIFICATION

System identification is the experimental approach to the modeling of a process or a plant of unknown parameters.[14] It involves the following steps: experimental planning, the selection of a model structure, parameter estimation, and model validation. The procedure of system identification, as pursued in practice, is iterative in nature in that we may have to go back and forth between these steps until a satisfactory model is built.

Suppose then we have an unknown nonlinear dynamical plant, and the requirement is to build a suitably parameterized identification model for it. We have the choice of basing the identification procedure on a state-space model or an input–output model. The decision as to which of these two representations is used hinges on prior information of the inputs and observables of the system. In what follows, both representations are discussed.

System Identification Using the State-Space Model

Suppose that the given plant is described by the state-space model:

$$\mathbf{x}(n + 1) = \mathbf{f}(\mathbf{x}(n), \mathbf{u}(n)) \tag{15.85}$$

$$\mathbf{y}(n) = \mathbf{h}(\mathbf{x}(n)) \tag{15.86}$$

where $\mathbf{f}(\cdot,\cdot)$ and $\mathbf{h}(\cdot)$ are vector-valued nonlinear functions, both of which are assumed to be unknown; Eq. (15.86) is a generalization of Eq. (15.11). We use two neural networks to identify the system, one for dealing with the process equation (15.85) and the other for dealing with the measurement equation (15.86), as depicted in Fig. 15.17.

We recognize that the state $\mathbf{x}(n)$ is the one-step delayed version of $\mathbf{x}(n + 1)$. Let $\hat{\mathbf{x}}(n + 1)$ denote the estimate of $\mathbf{x}(n + 1)$ produced by the first neural network, labeled network I in Fig. 15.17a. This network operates on a concatenated input consisting of the external input $\mathbf{u}(n)$ and the state $\mathbf{x}(n)$ to produce $\hat{\mathbf{x}}(n + 1)$. The estimate $\hat{\mathbf{x}}(n + 1)$ is subtracted from the actual state $\mathbf{x}(n + 1)$ to produce the error vector

$$\mathbf{e}_{\mathrm{I}}(n + 1) = \mathbf{x}(n + 1) - \hat{\mathbf{x}}(n + 1)$$

where $\mathbf{x}(n + 1)$ plays the role of desired response. It is assumed that the actual state $\mathbf{x}(n)$ is physically accessible for it to be used in this way. The error vector $\mathbf{e}_{\mathrm{I}}(n + 1)$ is in

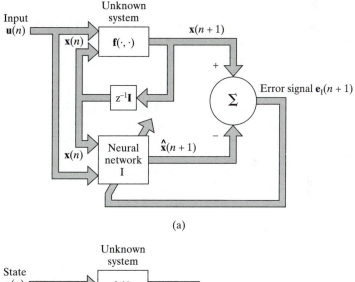

(a)

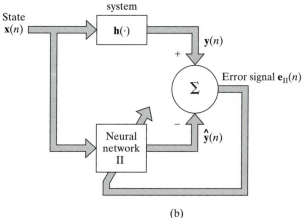

(b)

FIGURE 15.17 State-space solution for the system identification problem.

turn used to adjust the synaptic weights of neural network I, as indicated in Fig. 15.17a, so as to minimize the cost function based on the error vector $\mathbf{e}_I(n + 1)$ in some statistical sense.

The second neural network, labeled network II in Fig. 15.17b, operates on the actual state $\mathbf{x}(n)$ of the unknown plant to produce an estimate $\hat{\mathbf{y}}(n)$ of the actual output $\mathbf{y}(n)$. The estimate $\hat{\mathbf{y}}(n)$ is subtracted from $\mathbf{y}(n)$ to produce the second error vector

$$\mathbf{e}_{II}(n) = \mathbf{y}(n) - \hat{\mathbf{y}}(n)$$

where $\mathbf{y}(n)$ plays the role of desired response. The error vector $\mathbf{e}_{II}(n)$ is then used to adjust the synaptic weights of network II to minimize the Euclidean norm of the error vector $\mathbf{e}_{II}(n)$ in some statistical sense.

The two neural networks shown in Fig. 15.17 operate in a synchronous fashion to provide a state-space solution to the system identification problem (Narendra and Parthasarathy, 1990). Such a model is referred to as a *series-parallel identification model* in recognition of the fact that the actual state of the unknown system (rather than that of the identification model) is fed into the identification model, as depicted in Fig. 15.17a. In light of the discussion presented at the end of Section 15.9, this form of training is an example of teacher forcing.

The series-parallel identification model of Fig. 15.17a should be contrasted against a *parallel identification model* where the $\mathbf{x}(n)$ applied to the neural network I is replaced with $\hat{\mathbf{x}}(n)$; the $\hat{\mathbf{x}}(n)$ is derived from the network's own output $\hat{\mathbf{x}}(n + 1)$ by passing it through a unit delay $z^{-1}\mathbf{I}$. The practical benefit of this alternative model of training is that the neural network model is operated in exactly the same way as the unknown system, that is, the way in which the model will be used after the training is completed. It is therefore likely that the model developed via the parallel training mode may exhibit an autonomous behavior that is superior to the autonomous behavior of the network model developed via the series-parallel training mode. The disadvantage of the parallel training mode, however, is that it may take longer than the series-parallel training mode; see the discussion on teacher forcing in Section 15.9. Specifically, in our present situation, the estimate $\hat{\mathbf{x}}(n)$ of the state used in the parallel training model is ordinarily not as accurate as the actual state $\mathbf{x}(n)$ used in the series-parallel training mode.

Input–Output Model

Suppose next that the unknown plant is only accessible through its output. To simplify the presentation, let the system be of a single input, single output kind. Let $y(n)$ denote the output of the system due to the input $u(n)$ for varying discrete-time n. Then, choosing to work with the NARX model, the identification model takes the form:

$$\hat{y}(n + 1) = \varphi(y(n), ..., y(n - q + 1), u(n), ..., u(n - q + 1))$$

where q is the order of the unknown system. At time $n + 1$, the q past values of the input and the q past values of the output are all available. The model output $\hat{y}(n + 1)$ represents an estimate of the actual output $y(n + 1)$. The estimate $\hat{y}(n + 1)$ is subtracted from $y(n + 1)$ to produce the error signal

$$e(n + 1) = y(n + 1) - \hat{y}(n + 1)$$

where $y(n + 1)$ plays the role of desired response. The error $e(n + 1)$ is used to adjust the synaptic weights of the neural network so as to minimize the error in some statistical sense. The identification model of Fig. 15.18 is of a series-parallel form (i.e., teacher forcing form) because the actual output of the system (rather than that of the identification model) is fed back to the input of the model.

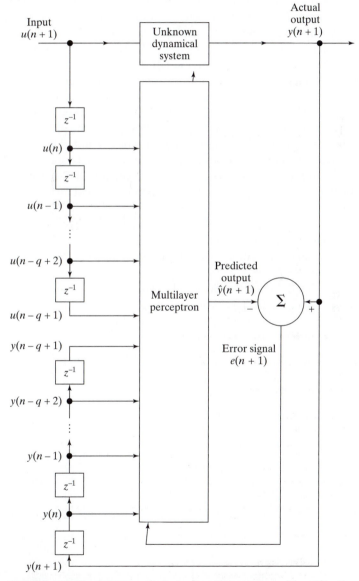

FIGURE 15.18 NARX solution for the system identification problem.

15.14 MODEL REFERENCE ADAPTIVE CONTROL

Another important application of recurrent networks is in the design of *feedback control systems* where the states of a plant are coupled nonlinearly with imposed controls (Puskorius and Feldkamp, 1994; Puskorius et al., 1996). The design of the system is further complicated by other factors such as the presence of unmeasured and random disturbances, the possibility of a nonunique plant inverse, and the presence of plant states that are unobservable.

A control strategy well suited for the use of neural networks is the *model reference adaptive control* (MRAC),[15] where the implicit assumption is that the designer is sufficiently familiar with the system under consideration (Narendra and Annaswany, 1989). Figure 15.19 shows the block diagram of such a system, where adaptivity is used to account for the fact that the dynamics of the plant are unknown. The controller and the plant form a closed loop feedback system, thereby constituting an *externally recurrent* network. The plant receives an input $\mathbf{u}_c(n)$ from the controller along with an external disturbance $\mathbf{u}_d(n)$. Accordingly, the plant evolves in time as a function of the imposed inputs and the plant's own state $\mathbf{x}_p(n)$. The output of the plant, denoted by $\mathbf{y}_p(n+1)$, is a function of $\mathbf{x}_p(n)$. The plant output may also be corrupted by measurement noise.

The controller receives two inputs: an externally specified reference signal $\mathbf{r}(n)$, and $\mathbf{y}_p(n)$ representing a one-step delayed version of the plant output $\mathbf{y}_p(n+1)$. The controller produces a vector of control signals defined by

$$\mathbf{u}_c(n) = \mathbf{f}_1(\mathbf{x}_c(n), \mathbf{y}_p(n), \mathbf{r}(n), \mathbf{w})$$

where $\mathbf{x}_c(n)$ is the controller's own state and $\mathbf{w}$ is a parameter vector that is available for adjustment. The vector-valued function $\mathbf{f}_1(\cdot, \cdot, \cdot, \cdot)$ defines the input–output behavior of the controller.

The desired response $\mathbf{d}(n+1)$ for the plant is supplied by the output of a stable *reference model,* which is produced in response to the reference $\mathbf{r}(n)$. The desired response $\mathbf{d}(n+1)$ is therefore a function of the reference signal $\mathbf{r}(n)$ and the reference model's own state $\mathbf{x}_r(n)$, as shown by

$$\mathbf{d}(n+1) = \mathbf{f}_2(\mathbf{x}_r(n), \mathbf{r}(n))$$

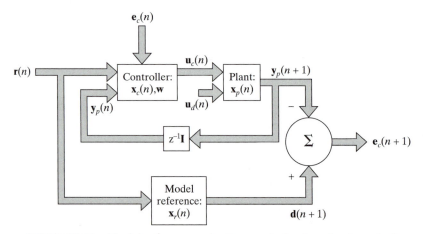

FIGURE 15.19 Model reference adaptive control using direct control.

The vector-valued function $\mathbf{f}_2(\cdot,\cdot)$ defines the input–output behavior of the reference model.

Let the *output error* (i.e., the error between the plant and model reference outputs) be denoted by

$$\mathbf{e}_c(n + 1) = \mathbf{d}(n + 1) - \mathbf{y}_p(n + 1)$$

The design goal is to adjust the parameter vector $\mathbf{w}$ of the controller such that the Euclidean norm of the output error $\mathbf{e}_c(n)$ is minimized over time n.

The method of control used in the MRAC system of Fig. 15.19 is said to be *direct* in the sense that no effort is made to identify the plant parameters, but the parameters of the controller are directly adjusted to improve system performance. Unfortunately, at present, precise methods for adjusting the parameters of the controller based on the output error are not available (Narendra and Parthasarathy, 1990). This is because the unknown plant lies between the controller and the output error. To overcome this difficulty, we may resort to the use of *indirect control,* as shown in Fig. 15.20. In this latter method, a two-step procedure is used to train the controller:

1. A model of the plant P, denoted by $\hat{P}$, is obtained to derive estimates of the differential relationships of the plant output with respect to plant input, prior plant outputs, and prior internal states of the plant. The procedure described in the previous section is used to train a neural network to identify the plant; the model $\hat{P}$ so obtained is called an *identification model.*

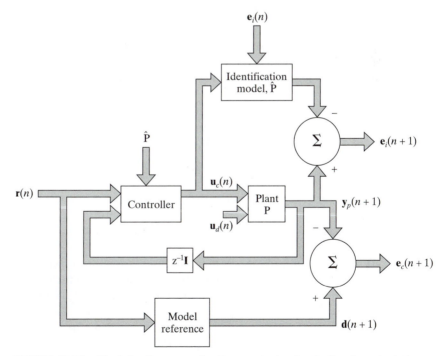

FIGURE 15.20   Model reference adaptive control using indirect control via an identification model.

2. The identification model $\hat{P}$ is used in place of the plant to derive estimates of the dynamic derivatives of the plant output with respect to the adjustable parameter vector of the controller.

In indirect control, the *externally recurrent network* is composed of the controller and an input/output representation of the plant via the identification model $\hat{P}$.

The application of a recurrent network to the controller design in the general structure of Fig. 15.20 has been demonstrated in a series of example control problems ranging from the well-known cart-pole and bioreactor benchmark problems to an automotive subsystem, namely engine idle-speed control (Puskorius and Feldkamp, 1994, Puskorius et al., 1996). The recurrent network used in those studies was a recurrent multilayer perceptron similar to that described in Section 15.2. The training of the network was performed using the DEKF algorithm described in Section 15.11. Note, however, that for the engine idle-speed control a *linear* dynamical system was chosen for the identification model since the imposed controls (over appropriately chosen ranges) appear to influence engine speed monotonically.

15.15 SUMMARY AND DISCUSSION

In this chapter we discuss recurrent networks that involve the use of *global feedback* applied to a static (memoryless) multilayer perceptron. The application of feedback enables neural networks to acquire state representations, making them suitable devices for diverse applications in signal processing and control. Four main network architectures belonging to the class of recurrent networks with global feedback are identified:

- Nonlinear autoregressive with exogenous inputs (NARX) networks using feedback from the output layer to the input layer.
- Fully connected recurrent networks with feedback from the hidden layer to the input layer.
- Recurrent multilayer perceptron with more than one hidden layer, using feedback from the output of each computation layer to its own input.
- Second-order recurrent networks using second-order neurons.

In all of these recurrent networks, the feedback is applied via tapped-delay-line memories.

The first three recurrent networks permit the use of a *state-space framework* for studying their dynamic behavior. This approach, rooted in modern control theory, provides a powerful method for studying the nonlinear dynamics of recurrent networks.

We describe three basic learning algorithms for the training of recurrent networks: back-propagation through time (BPTT), real-time recurrent learning (RTRL), and decoupled extended Kalman filtering (DEKF). The BPTT and RTRL algorithms are gradient based, whereas the DEKF algorithm uses higher-order information more efficiently. It is therefore able to converge much faster than the BPTT and RTRL algorithms, but at the expense of a corresponding increase in computational complexity. Indeed, the DEKF algorithm may be viewed as an enabling technology, which makes it possible to solve difficult signal processing and control problems.

In theory, a recurrent network with global feedback (e.g., recurrent multilayer perceptron trained with the DEKF algorithm) can learn the underlying dynamics of a *nonstationary* environment and do so by storing the knowledge gained from the training sample in a *fixed* set of weights. Most importantly, the network can *track* the statistical variations of the environment provided that two conditions are satisfied.

- The recurrent network does not suffer from underfitting or overfitting.
- The training sample is representative of the nonstationary behavior of the environment.

Throughout this chapter we emphasize the use of recurrent networks for temporal processing. Recurrent networks may also be used to process sequentially ordered data that do not have a straightforward temporal interpretation (e.g., chemical structures represented as trees). In Sperduti and Starita (1997) it is shown that recurrent networks can represent and classify structured patterns that are represented as directed, labeled, acyclic graphs. The main idea behind the approach described therein is the "generalized recursive neuron," which is a structural generalization of a recurrent neuron (i.e., neuron with local feedback). By using such a model, supervised learning algorithms such as back-propagation through time and real-time recurrent learning can be extended to deal with structured patterns.

NOTES AND REFERENCES

1. For other recurrent network architectures, see Jordan (1986), Back and Tsoi (1991), Frasconi et al. (1992), and Robinson and Fallside (1991).

2. The NARX model encompasses an important class of discrete-time nonlinear systems (Leontaritis and Billings, 1985). In the context of neural networks it is discussed in Chen et al. (1990), Narendra and Parthasarathy (1990), Lin et al. (1996), and Sieglemann et al., (1997).

    It has been demonstrated that the NARX model is well suited for modeling nonlinear systems such as heat exchangers (Chen et al., 1990), waste water treatment plants (Su and McAvoy, 1991; Su et al., 1992), catalytic reforming systems in a petroleum refinery (Su et al., 1992), nonlinear oscillations associated with multilegged locomotion in biological systems (Venkataraman, 1994), and grammatical inference (Giles and Horne, 1994).

    The NARX model is also referred to as the nonlinear autoregressive-moving average (NARMA) model, with "moving average" referring to the inputs.

3. The recurrent multilayer perceptron in Fig. 15.4 is a generalization of the recurrent network described in Jordan (1986).

4. Omlin and Giles (1996) show that, using second-order recurrent networks, any known finite-state automata can be mapped into such a network, and the correct classification of temporal sequences of finite length is guaranteed.

5. For a rigorous treatment of controllability and observability, see Zadeh and Desoer (1963), Kailath (1980), Sontag (1990), and Lewis and Syrmos (1995).

6. The first work on neural networks and automata (actually sequential machines-automata implementations), also referenced as the first paper on finite-state automata, artificial intelligence, and recurrent neural networks, was the classic paper by McCulloch and Pitts (1943). The recurrent network (with instantaneous feedback) in the second part of this paper was interpreted as a finite-state automaton in Kleene (1956). Kleene's paper appeared in the book "Automata Studies" edited by Shannon and McCarthy (authors in

this amazing book include Moore, Minsky, von Neumann, Uttley, McCarthy, and Shannon among others). Sometimes Kleene's paper is cited as the first article on finite-state machines (Perrin, 1990). Minsky (1967) discusses automata and neural networks in his book entitled "Computation: Finite and Infinite Machines."

All of the early work on automata and neural networks was concerned with synthesis, that is, how automata are built or designed into neural networks. Because most automata (when implemented as sequential machines) require feedback, the neural networks were necessarily recurrent ones. Note that the early work (with the exception of Minsky) did not make a clear distinction between automata (directed, labeled, acyclic graphs) and sequential machines (logic and feedback delays), and was mostly concerned with finite-state automata. There was little interest (with the exception of Minsky) in moving up the automata hierarchy to push down automata and Turing machines.

After the dark ages of neural networks, research on automata and neural networks started again in the 1980s. This work could be broadly classified into three areas: (1) learning automata, (2) automata synthesis, extraction, and refinement of knowledge, and (3) representation. The first mention of automata and neural networks was in Jordan (1986).

7. A single-layer recurrent network using McCulloch–Pitts neurons cannot simulate any finite-state machine (Goudreau et al., 1994) but Elman's simple recurrent network can (Kremer, 1995). Recurrent networks with only local feedback cannot represent all finite-state machines (Frasconi and Gori, 1996; Giles et al., 1995; Kremer, 1996).

8. The idea behind back-propagation through time is that for every recurrent network it is possible to construct a feedforward network with identical behavior over a particular time interval (Minsky and Papert, 1969). Back-propagation through time was first described in the Ph.D. thesis of Werbos (1974); see also Werbos (1990). The algorithm was rediscovered independently by Rumelhart et al. (1986b). A variant of the back-propagation through time algorithm is described in Williams and Peng (1990). For a review of the algorithm and related issues, see Williams and Zipser (1995).

9. The real-time recurrent learning algorithm was described in the neural network literature for the first time by Williams and Zipser (1989). Its origin may be traced to an earlier paper by McBride and Narendra (1965) on system identification for tuning the parameters of an arbitrary dynamical system.

The derivation given in Williams and Zipser is for a single layer of fully recurrent neurons. It has been extended to more general architectures; see, for example, Kechriotis et al. (1994); Puskorius and Feldkamp (1994).

10. Kalman filter theory owes its origin to the classic paper by Rudolf E. Kalman (1960). It has established itself as an essential part of signal processing and control with numerous applications in highly diverse fields. For a detailed treatment of the standard Kalman filter, its variants, and its extended form dealing with nonlinear dynamical systems, see Grewal and Andrews (1993) and Haykin (1996). The book by Grewal and Andrews is devoted entirely to the theory and practice of Kalman filtering. The book by Haykin discusses Kalman filter theory from the perspective of adaptive filtering. Two other important books on the subject are Jazwinski (1970) and Maybeck (1979, 1982).

11. For a detailed treatment of the square root Kalman filter and efficient methods for its implementation, see Haykin (1996).

12. Singhal and Wu (1989) were perhaps the first to demonstrate the improved mapping performance of a supervised neural network using the extended Kalman filter. Unfortunately, the training algorithm described therein is limited by its computational complexity. To overcome this limitation, Kollias and Anastassiou (1989) and Shah and Palmieri (1990) tried to simplify the application of extended Kalman filtering by parti-

tioning the global problem into a number of subproblems, each of which addresses a single neuron. However, the treatment of each neuron as an identification problem does not rigorously adhere to Kalman filter theory. Moreover, such an approach may lead to unstable behavior during training, and may result in solutions that are inferior to those obtained by other methods (Puskorius and Feldkamp, 1991).

13. Other methods for dealing with the vanishing gradients problem involve bypassing some of the nonlinearities in the recurrent network so as to provide improved learning of long-term dependencies. Examples of this approach include the following:
    - Use of long time delays in the network architecture (El Hihi and Bengio, 1996; Lin et al., 1996; Giles et al., 1997)
    - Hierarchically structuring of the network in multiple levels associated with different time scales (El Hihi and Bengio, 1996)
    - Using gating units to circumvent some of the nonlinearities (Hochreiter and Schmidhuber, 1997)

14. System identification has an extensive literature. For a treatment of the subject in book form, see Ljung (1987), and Ljung and Glad (1994). For an overview of the subject with an emphasis on neural networks, see Sjöberg et al. (1995), and Narendra (1995). The first detailed study of system identification using neural networks appeared in Narendra and Parthasarathy (1990).

15. For detailed treatment of model reference adaptive control, see the book by Landau (1979).

PROBLEMS

State-space model

15.1 Formulate the state-space equations for Elman's simple recurrent network shown in Fig. 15.3.

15.2 Show that the recurrent multilayer perceptron of Fig. 15.4 can be represented by the state-space model:

$$\mathbf{x}(n + 1) = \mathbf{f}(\mathbf{x}(n), \mathbf{u}(n))$$

$$\mathbf{y}(n) = \mathbf{g}(\mathbf{x}(n), \mathbf{u}(n))$$

where $\mathbf{u}(n)$ denotes the input, $\mathbf{y}(n)$ denotes the output, $\mathbf{x}(n)$ denotes the state, and $\mathbf{f}(\cdot,\cdot)$ and $\mathbf{g}(\cdot,\cdot)$ denote vector-valued nonlinear functions.

15.3 Is it possible for a dynamic system to be controllable and unobservable, and vice versa? Justify your answers.

15.4 Referring to the local controllability problem discussed in Section 15.3, show that
    (a) the state $\mathbf{x}(n + q)$ is a nested nonlinear function of its past value $\mathbf{x}(n)$ and the input vector $\mathbf{u}_q(n)$ of Eq. (15.24), and
    (b) the Jacobian of $\mathbf{x}(n + q)$ with respect to $\mathbf{u}_q(n)$, evaluated at the origin, is equal to the controllability matrix $\mathbf{M}_c$ of Eq. (15.23).

15.5 Referring to the local observability problem discussed in Section 15.3, show that the Jacobian of the observation vector $\mathbf{y}_q(n)$ defined in Eq. (15.30) with respect to the state $\mathbf{x}(n)$, evaluated at the origin, is equal to the observability matrix $\mathbf{M}_o$ of Eq. (15.28).

15.6 The process equation of a nonlinear dynamical system is described by

$$\mathbf{x}(n + 1) = \mathbf{f}(\mathbf{x}(n), \mathbf{u}(n))$$

where $\mathbf{u}(n)$ is the input vector at time n and $\mathbf{x}(n)$ is the corresponding state of the system. The input $\mathbf{u}(n)$ appears in the process equation in a nonadditive manner. In this problem

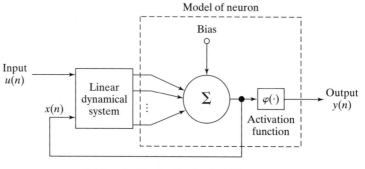

(a) Local activation feedback architecture

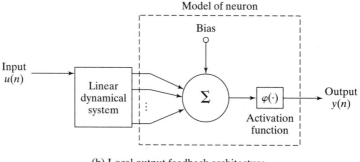

(b) Local output feedback architecture

FIGURE P15.7

we wish to reformulate the process equation so that the input $\mathbf{u}(n)$ appears additively. This is done by writing

$$\mathbf{x}'(n+1) = \mathbf{f}_{\text{new}}(\mathbf{x}'(n)) + \mathbf{u}'(n)$$

Formulate definitions for the vectors $\mathbf{x}'(n)$ and $\mathbf{u}'(n)$, and the function $\mathbf{f}_{\text{new}}(\cdot)$.

15.7 Figure P15.7 presents two examples of recurrent network architectures using local feedback at the neuronal level. The architectures shown in parts a and b of the figure are called *local activation feedback* and *local output feedback,* respectively (Tsoi and Back, 1994). Formulate state-space models for these two recurrent network architectures. Comment on their controllability and observability.

Nonlinear autoregressive with exogenous inputs (NARX) model

15.8 Referring to the NARX model discussed in Section 15.4, show that the use of Eqs. (15.16) and (15.17) leads to the following expression for the output $y(n + q)$ of the NARX model in terms of the state $\mathbf{x}(n)$ and input vector $\mathbf{u}_q(n)$:

$$y(n + q) = \Phi(\mathbf{x}(n), \mathbf{u}_q(n))$$

where $\Phi : \mathbb{R}^{2q} \to \mathbb{R}$, and $\mathbf{u}_q$ is defined in accordance with Eq. (15.29).

15.9 (a) The derivation of the NARX model in Section 15.4 is presented for a single input, single output system. Discuss how the theory described therein can be extended for a multiple input, multiple output system.

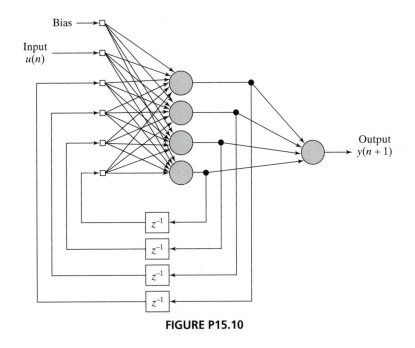

FIGURE P15.10

(b) Construct the NARX equivalent to the two input, single output state-space model in Fig. 15.6.

15.10 Construct the NARX equivalent for the fully recurrent network shown in Fig. P15.10.

15.11 In Section 15.4 we showed that any state-space model can be represented by a NARX model. What about the other way around? Can any NARX model be represented by a state-space model of the form described in Section 15.3? Justify your answer.

Back-propagation through time

15.12 Unfold the temporal behavior of the state-space model shown in Fig. 15.3.

15.13 The truncated BPTT(h) algorithm may be viewed as an approximation to the epochwise BPTT algorithm. The approximation can be improved by incorporating aspects of epochwise BPTT into the truncated BPTT(h) algorithm. Specifically, we may let the network through h' additional steps before performing the next BPTT computation, where $h' < h$. The important feature of the hybrid form of back-propagation through time is that the next backward pass is not performed until time step $n + h'$. In the intervening time, past values of the network input, network state, and desired responses are stored in a buffer, but no processing is performed on them (Williams and Peng, 1990). Formulate the local gradient for neuron j in this hybrid algorithm.

Real-time recurrent learning algorithm

15.14 The dynamics of a *teacher forced recurrent network* during training are as described in Section 15.8, except for this change:

$$\xi_i(n) = \begin{cases} u_i(n) & \text{if } i \in \mathcal{A} \\ d_i(n) & \text{if } i \in \mathcal{C} \\ y_i(n) & \text{if } i \in \mathcal{B} - \mathcal{C} \end{cases}$$

where $\mathcal{A}$ denotes the set of indices i for which ξ_i is an external input, $\mathcal{B}$ denotes the set of indices i for which ξ_i is the output of a neuron, and $\mathcal{C}$ denotes the set of output neurons that are visible.

(a) Show that for this scheme, the partial derivative $\partial y_j(n+1)/\partial w_{kl}(n)$ is given by (Williams and Zipser, 1989)

$$\frac{\partial y_j(n+1)}{\partial w_{kl}(n)} = \varphi'(v_j(n)) \left[\sum_{i \in \mathcal{B} - \mathcal{C}} w_{ji}(n) \frac{\partial y_i(n)}{\partial w_{kl}(n)} + \delta_{kj} \xi_l(n) \right]$$

(b) Derive the training algorithm for a teacher forced recurrent network.

Decoupled Extended Kalman Filtering (DEKF) algorithm

15.15 Describe how the DEKF algorithm can be used to train the simple recurrent network shown in Fig. 15.3. You may also invoke the use of the BPTT algorithm for this training.

15.16 In its usual form, DEKF training is carried out with weight updates, instance by instance. By contrast, in standard back-propagation, simple gradient updates are performed, enabling us to choose to apply the updates immediately or else accumulate the updates for some time and then apply them as a single composite update. Although such an accumulation could be attempted in the DEKF algorithm, doing so would cause inconsistency between the weight vector and the error covariance matrix, which is updated each time a recursion is performed to generate a weight update. Thus the use of DEKF training appears to preclude batch updating. However, it is possible to use *multistream DEKF training,* which allows for multiple training sequences and yet remains consistent with Kalman filter theory, as described in Feldkamp et al. (1997) and Feldkamp and Puskorius (1998).

(a) Consider the training problem with N_{in} inputs, N_{out} outputs, and a fixed training sample of N examples. From the training sample, form $M \le N$ data streams which feed M networks constrained to have identical weights. At each training cycle, one pattern from each stream is presented to its respective network and the N_{out} network outputs for each stream are computed. A single weight update is then computed and applied identically to each stream's network. Derive the multistream form of the DEKF algorithm.

(b) For example, consider the standard XOR problem with four training patterns. Assume that we have a feedforward network that is augmented with a delay-line memory connected to the output layer. We thus effectively have four network outputs: the actual network output fed into the delay-line memory, and three delayed versions of it with each one of them constituting a new network output. Now apply each of the four training patterns in some order to this network structure, but do not perform any weight updates. After the presentation of the fourth training pattern, we have four network outputs that represent the processing of the four training patterns performed through a network with identical weights. If we consider performing a single weight vector update with DEKF on the basis of these four training patterns and four network outputs, we have a four-stream problem. Check this example out.

Second-order recurrent networks

15.17 In this problem we explore the construction of the *parity finite-state automaton* using a second order recurrent network. This automaton recognizes an odd number of 1's in an arbitrary length sequential string of 0's and 1's.

Figure P15.17 shows a two-state automaton. States are represented by circles, and transitions by arrows. S means we start in that state, state A in the case shown here. The thick circle means that whenever we are in that state, shown as state B in the figure, we

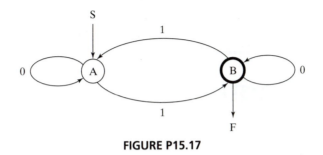

FIGURE P15.17

accept the string. The automaton starts seeing strings in state A and loops back to state A if we see a 0 and loops to state B if we see a 1. Similarly, when in state B, it loops to state B if we see a 0 and loops back to state A if we see a 1. In this way, the automaton is always in state A if we have seen an even (including zero) number of 1's and in state B if we have seen an odd number of 1's.

More formally, we define the states as $Q = \{A, B\}$, $S = A$ as the start state, the input alphabet as $\Sigma = \{0, 1\}$, the accepting state as $F = B$, and the state transition function as:

$$\delta(A, 0) = A$$

$$\delta(A, 1) = B$$

$$\delta(B, 0) = B$$

$$\delta(B, 1) = A$$

These are the equations needed for the application of Eq. (15.9) pertaining to the second-order recurrent network. For more details on finite-state automata, see Hopcroft (1979).

Encode the above transition rules into the second-order recurrent network.

15.18 In Section 15.8 we derive the real-time recurrent learning (RTRL) algorithm for a fully connected recurrent network using first-order neurons. In Section 15.2 we describe a recurrent network using second-order neurons.

Extend the theory described in Section 15.8 by deriving the RTRL algorithm for the training of a second-order recurrent network.

Epilogue

Neural networks represent a *multidisciplinary* subject with roots in the neurosciences, mathematics, statistics, physics, computer science, and engineering, as evidenced by the diversity of topics covered in this book. Their ability to *learn* from data with or without a teacher has endowed them with a powerful property. This learning property has profound theoretical as well as practical implications. In one form or another, the ability of neural networks to learn from examples (representative of their environment) has made them invaluable tools in such diverse applications as modeling, time series analysis, pattern recognition, signal processing, and control. In particular, neural networks have a great deal to offer when the solution of a problem of interest is made difficult by one or more of the following points:

- Lack of physical/statistical understanding of the problem
- Statistical variations in the observable data
- Nonlinear mechanism responsible for the generation of the data

The new wave of neural networks (since the mid 1980s) came into being because learning could be performed at multiple levels. Neural network based learning algorithms have allowed us to eliminate the need for handcrafted feature extraction in handwriting recognizers. Gradient-based learning algorithms inspired by neural networks have allowed us to simultaneously train feature extractors, classifiers, and contextual processors (hidden Markov models and language models) simultaneously. Because of neural networks we have learning all the way down from pixels to symbols.

Learning pervades every level of intelligent machines in an increasing number of applications. It is therefore befitting that this epilogue concludes the book with final remarks on some intelligent machines and the role of neural networks in building them.

INTELLIGENT MACHINES

With no agreed upon scientific definition of intelligence [1] and due to space limitations, we will not venture into a discussion of what intelligence is. Rather, we will confine our brief exposition to intelligent machines in the context of three specific areas of application: pattern classification, control, and signal processing. It is recognized here that there is no "universal" intelligent machine; instead, we have intelligent machines for specific applications.

Much of the research effort on neural networks has focused on pattern classification. Given the practical importance of pattern classification and its rather pervasive nature, and the fact that neural networks are so well suited for the task of pattern classification, this concentration of research effort has been largely the right thing to do. In so doing, we have been able to lay down the foundations of *adaptive pattern classification*. However, we have reached the stage where we think of classification systems in a much broader sense if we are to be successful in solving classification problems of a more complex and sophisticated nature. Figure 1 depicts the layout of a "hypothetical" classification system (Hammerstrom and Rahfuss, 1992). The first level of the system receives sensory data generated by some source of information. The second level extracts a set of features that characterize the sensory data. The third level classifies the features into one or more distinct categories, which are then put into global context by the fourth level. Finally, we may, for example, put the parsed input into some form of a database for an end user. The important features that characterise the system of Fig. 1 are:

- *Recognition*, resulting from the forward flow of information from one level of the system to the next as in a traditional pattern classification system
- *Focusing*, whereby a higher level of the system is able to selectively influence the information processing at a lower level by virtue of knowledge gained from past data

Thus the novelty of the pattern classification system shown in Fig. 1 lies in *knowledge of the target domain* and its exploitation by lower levels of the system to improve overall system performance, given the fundamental constraint of a limited information processing

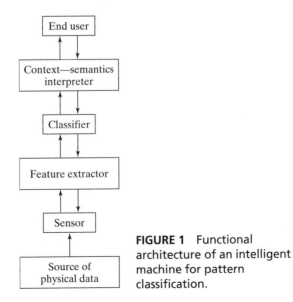

FIGURE 1 Functional architecture of an intelligent machine for pattern classification.

capacity. It is our belief that the evolution of pattern classification using neural networks will be in the direction of creating models that are continually influenced by knowledge of the target domain. We envision a new class of intelligent machines for pattern classification that offer the following attributes:

- Ability to extract *contextual knowledge*, and exploit it through the use of *focusing*
- *Localized* rather than distributed representation of knowledge
- *Sparse* architecture, emphasizing network modularity and hierarchy as principles of neural network design

The realization of such an intelligent machine can only be attained by combining neural networks with other appropriate tools. A useful tool that comes to mind here is the *Viterbi algorithm*, which is a form of dynamic programming designed to deal with *sequential* information processing[2] that is an inherent characteristic of the system described in Fig. 1. (The dynamic programing algorithm is discussed in Chapter 12.)

Control, another area of application naturally suited for neural networks, is also evolving in the direction of *intelligent control*.[3] Autonomy is an important objective of control system designers and intelligent controllers are one way to achieve it. Figure 2 shows a functional architecture for an intelligent autonomous controller with an interface at one end to the process (plant) involving sensing and an interface at the other end to humans and other systems (Antsaklis et al., 1996; Passino, 1996). The system has three functional levels, as summarized here:

1. *Execution level*, which has low-level signal processing and control algorithms for adaptive control and identification.
2. *Coordination level*, which provides the link between the execution and management levels by looking after such matters as tuning, supervision, crisis management, and planning.
3. *Management and organization level*, which provides for the supervision of lower-level functions and management of the interface to the human(s).

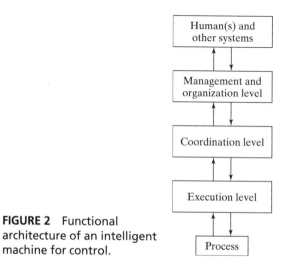

FIGURE 2   Functional architecture of an intelligent machine for control.

Whereas classical control is rooted in the theory of linear differential equations, intelligent control is largely *rule based* because the dependencies involved in its deployment are much too complex to permit an analytical representation. To deal with such dependencies, it is expedient to use the mathematics of fuzzy systems and neural network. The power of *fuzzy systems*[4] lies in their ability (1) to quantify linguistic inputs, and (2) to quickly give a working approximation of complex and often unknown system input–output rules. The power of neural networks is in their ability to *learn* from data. There is a natural synergy between neural networks and fuzzy systems that makes their *hybridization* a powerful tool for intelligent control and other applications.

Turning next to signal processing, we have yet another fertile area for the application of neural networks by virtue of their nonlinear and adaptive characteristics (Haykin, 1996). Many of the physical phenomena responsible for the generation of *information-bearing signals* encountered in practice (e.g., speech signals, radar signals, sonar signals) are governed by *nonlinear dynamics of a nonstationary and complex nature*, defying an exact mathematical description. To exploit the full information content of such signals at all times, we need intelligent machines for signal processing,[5] the design of which addresses the following key issues:

- *Nonlinearity*, which makes it possible to extract the higher-order statistics of the input signals
- *Learning and adaptation*, by means of which the system is enabled to learn the underlying physical mechanism of the environment in which it is embedded and to adapt to slow statistical variations in the environment on a continuing basis
- *Attentional mechanism*, whereby, through interaction with the end user or in a self-organized manner, the system is enabled to focus its computing power around a particular point in an image or a particular location in space for more detailed analysis[6]

Figure 3 shows the functional architecture of an intelligent machine for signal processing that involves three levels of operation:

1. *Low-level processing*, the purpose of which is to preprocess the received signal to prepare it for the second level. The preprocessing involves the use of filtering to reduce the effects of noise and other advanced signal processing operations such as *time-frequency analysis*.[7] The aim of time-frequency analysis is to describe how the spectral content of a signal evolves and to understand what a time varying spectrum is. Specifically, a one-dimensional (temporal) representation of the received signal is transformed into a two-dimensional image with one dimension representing time and the other dimension representing frequency. Time-frequency analysis provides an effective method for highlighting the nonstationary nature of the received signal in a manner far more discernible than in its original temporal form.
2. *Learning and adaptation level*, where memory (of a long-term as well as of a short-term nature) and an attentional mechanism are built into the design of the system. By having the multilayer perceptron, for example, undergo supervised learning with a large enough data set representative of the environment in which the system is embedded, overall statistical information about the environment is

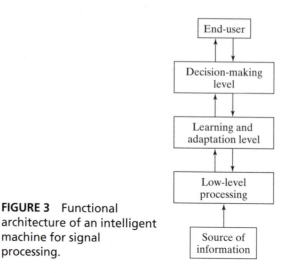

FIGURE 3 Functional architecture of an intelligent machine for signal processing.

stored in the synaptic weights of the network. To account for slow statistical variations in the environment with time, a blind adaptation scheme (i.e., a continuous learning subsystem operating in an unsupervised manner) is appended to the output end of the multilayer perceptron. The learning process also includes provision for an attentional network[8] whereby the system can focus its attention on important features in the received signal by "gating" the flow of information from lower to higher levels as the need arises.

3. *Decision-making level*, where the final decisions are made by the system. The decision could be whether or not a target of interest is present in the received signal as in radar or sonar, or whether the received bit of information corresponds to symbol 1 or symbol 0 as in digital communications; levels of confidence in the decisions made are also provided.

We do not claim that the systems described here are the only ways in which intelligence could be built into pattern classification, control, and signal processing. Rather, they represent systematic ways in which this important objective could be accomplished. Despite their differences in terms of the domain of application, they do share some common features (Valvanis and Saridis, 1992; Passino, 1996):

- There is a bidirectional flow of information, from lower to higher levels and back.
- Higher levels are often concerned with those aspects of the system's behavior that are slower in processing time, broader in scope, and longer in horizon time.
- There is increasing intelligence with decreasing precision as we move from lower to higher levels.
- At the higher levels, there is a decrease in *granularity* (i.e., an increase in model abstraction).

We began the discussion of (artificial) neural networks in Chapter 1 by describing the human brain, the source of motivation for neural networks, as a gigantic information processing machine. It is appropriate to conclude the book with a brief exposition of intelligent machines that are the ultimate in information processing by artificial means. The struggle to build intelligent machines continues.

NOTES AND REFERENCES

1. For a philosophical discussion of intelligence from different perspectives, see Ackerman (1990), Albus (1991), and Kosko (1992).

2. The Viterbi algorithm was originally developed by Viterbi (1967) for solving convolutional decoding problems in communication theory. For a tutorial treatment of the Viterbi algorithm, see Forney (1973).

   For an application in pattern classification that involves the combined use of the convolutional network (described in Chapter 4) and the Viterbi algorithm, see LeCun et al. (1997, 1998).

3. Intelligent control is discussed in the edited books by White and Sofge (1992), Antsaklis and Passino (1993), Gupta and Sinha (1996), and Tzefestas (1997).

4. Fuzzy set theory was originated by Zadeh (1965, 1973) to provide a mathematical tool for dealing with linguistic variables (i.e., concepts described in natural language). For a treatment of fuzzy logic in book form, see Dubois and Prade (1980). In the book by Kosko (1997), a different viewpoint is taken: Fuzzy systems are viewed as function approximators. Therein it is shown that fuzzy systems can model any continuous function or system provided they use enough rules.

5. A special issue of the 1998 Proceedings of the Institute of Electrical and Electronic Engineers (IEEE) is devoted to the subject of intelligent signal processing (Haykin and Kosko, 1998).

6. A self-organizing system for hierarchical focusing or selective attention is described in Fukushima (1988a). The system is a modification of the layered *neocognitron* also pioneered by Fukushima (1975, 1988b). The system is able to focus attention on an individual character in an image composed of multiple characters or on a greatly deformed character that is also contaminated with noise.

   A self-organized attentional mechanism also features in the development of *adaptive resonance theory* (ART) pioneered by Carpenter and Grossberg (1987, 1995). ART for adaptive pattern recognition involves the combination of bottom-up filtering and top-down template matching.

7. For a detailed treatment of the many facets of time-frequency analysis, building on the classical Fourier theory, see the book by Cohen (1995).

   For the theory and applications of the Wigner distribution, an important tool for bilinear/quadratic time-frequency representations, see the book by Mecklenbräuker and Hlawatsch (1997).

   For a different perspective, where we think in terms of scale in place of frequency, see the book by Vetterli and Koračević (1995) on wavelets and the related topic of subband coding.

8. In van de Laar et al. (1997), a neural network model for selective covert visual attention is described. The model is able to learn to focus its attention on important features, depending on the task to be fulfilled, by modulating the flow of information in the preattentive stage.

Bibliography

Aarts, E., and J. Korst, 1989. *Simulated Annealing and Boltzmann Machines: A Stochastic Approach to Combinatorial Optimization and Neural Computing,* New York: Wiley.

Abarbanel, H.D.I., 1996. *Analysis of Observed Chaotic Data,* New York: Springer-Verlag.

Abraham, R.H., and C.D. Shaw, 1992. *Dynamics of the Geometry of Behavior,* Reading, MA: Addison-Wesley.

Abu-Mostafa, Y.S., 1995. "Hints," *Neural computation,* vol. 7, pp. 639–671.

Abu-Mostafa, Y.S., 1990. "Learning from hints in Neural Networks," *Journal of Complexity,* vol. 6, pp. 192–198.

Abu-Mostafa, Y.S., 1989. "The Vapnik-Chervonenkis Dimension: Information Versus Complexity in Learning," *Neural Computation,* vol. 1, pp. 312–317.

Abu-Mostafa, Y.S., and J.M. St. Jacques, 1985. "Information capacity of the Hopfield model," *IEEE Transactions on Information Theory,* vol. IT-31, pp. 461–464.

Ackerman, P.L., 1990. "Intelligence." In S.C. Shapiro, ed., *Encyclopedia of Artificial Intelligence,* pp. 431–440, New York: Wiley (Interscience).

Ackley, D.H., G.E. Hinton, and T.J. Sejnowski, 1985. "A Learning Algorithm for Boltzmann Machines," *Cognitive Science,* vol. 9, pp. 147–169.

Aiyer, S.V.B., N. Niranjan, and F. Fallside, 1990. "A theoretical investigation into the performance of the Hopfield model," *IEEE Transactions on Neural Networks,* vol. 15, pp. 204–215.

Aizerman, M.A., E.M. Braverman, and L.I. Rozonoer, 1964a. "Theoretical foundations of the potential function method in pattern recognition learning," *Automation and Remote Control,* vol. 25, pp. 821–837.

Aizerman, M.A., E.M. Braverman, and L.I. Rozonoer, 1964b. "The probability problem of pattern recognition learning and the method of potential functions," *Automation and Remote Control,* vol. 25, pp. 1175–1193.

Akaike, H., 1974. "A new look at the statistical model identification," *IEEE Transactions on Automatic Control* vol. AC-19, pp. 716–723.

Akaike, H., 1970. "Statistical predictor identification," *Annals of the Institute of Statistical Mathematics,* vol. 22, pp. 202–217.

Albus, J.S., 1991. "Outline for a theory of intelligence," *IEEE Transactions on Systems, Man, and Cybernetics,* vol. 21, pp. 473–509.

Aleksander, I., and H. Morton, 1990. *An Introduction to Neural Computing,* London: Chapman and Hall.

Allport A., 1989. "Visual attention," In *Foundations of Cognitive Science,* M.I. Posner, ed., pp. 631–682, Cambridge, MA: MIT Press.

Al-Mashoug, K.A., and I.S. Reed, 1991. "Including hints in training neural nets," *Neural Computation,* vol. 3, pp. 418–427.

Alspector, J., R.B. Allen, A. Jayakumar, T. Zeppenfeld, and R. Meir, 1991. "Relaxation networks for large supervised learning problems," *Advances in Neural Information Processing Systems,* vol. 3, pp. 1015–1021, San Mateo, CA: Morgan Kaufmann.

Alspector, J., A. Jayakumar, and S. Luna, 1992. "Experimental evaluation of learning in a neural microsystem," *Advances in Neural Information Processing Systems,* vol. 4, pp. 871–878, San Mateo, CA: Morgan Kaufmann.

Alspector, J., R. Meir, B. Yuhas, A. Jayakumar, and D. Lippe, 1993. "A parallel gradient descent method for learning in analog VLSI neural networks," *Advances in Neural Information Processing Systems*, vol. 5, pp. 836–844, San Mateo, CA: Morgan Kaufmann.

Amari, S., 1998. "Natural gradient works efficiently in learning." *Neural Computation*, vol. 10, pp. 251–276.

Amari, S., 1997. Private communication.

Amari, S., 1993. "A universal theorem on learning curves," *Neural Networks*, vol. 6, pp. 161–166.

Amari, S., 1990. "Mathematical foundations of neurocomputing," *Proceedings of the IEEE*, vol. 78, pp. 1443–1463.

Amari, S., 1987. "Differential geometry of a parametric family of invertible systems—Riemanian metric, dual affine connections and divergence," *Mathematical Systems Theory*, vol. 20, pp. 53–82.

Amari, S., 1985. *Differential-Geometrical Methods in Statistics*, New York: Springer-Verlag.

Amari, S., 1983. "Field theory of self-organizing neural nets," *IEEE Transactions on Systems, Man, and Cybernetics* vol. SMC-13, pp. 741–748.

Amari, S., 1980. "Topographic organization of nerve fields," *Bulletin of Mathematical Biology*, vol. 42, pp. 339–364.

Amari, S., 1977a. "Neural theory of association and concept-formation," *Biological Cybernetics*, vol. 26, pp. 175–185.

Amari, S., 1977b. "Dynamics of pattern formation in lateral-inhibition type neural fields," *Biological Cybernetics*, vol. 27, pp. 77–87.

Amari, S., 1972. "Characteristics of random nets of analog neuron-like elements," *IEEE Transactions on Systems, Man, and Cybernetics*, vol. SMC-2, pp. 643–657.

Amari, S., 1967. "A theory of adaptive pattern classifiers," *IEEE Trans. Electronic Computers*, vol. EC-16, pp. 299–307.

Amari, S., and M.A. Arbib, 1977. "Competition and cooperation in neural nets," in J. Metzler, ed., *Systems Neuroscience*, pp. 119–165, New York: Academic Press.

Amari, S., and J.-F. Cardoso, 1997. "Blind source separation—Semiparametric statistical approach," *IEEE Transactions on Signal Processing*, vol. 45, pp. 2692–2700.

Amari, S., T.-P. Chen, and A. Cichoki, 1997. "Stability analysis of learning algorithms for blind source separation," *Neural Networks*, vol. 10, pp. 1345–1351.

Amari, S., A. Cichoki, and H.H. Yang, 1996. "A new learning algorithm for blind signal separation." *Advances in Neural Information Processing Systems*, vol. 8, pp. 757–763, Cambridge, MA: MIT Press.

Amari, S., and K. Maginu, 1988. "Statistical neurodynamics of associative memory," *Neural Networks*, vol. 1, pp. 63–73.

Amari, S., K. Yoshida, and K.-I. Kanatani, 1977. "A mathematical foundation for statistical neurodynamics," *SIAM Journal of Applied Mathematics*, vol. 33, pp. 95–126.

Amari, S., N. Murata, K.-R. Müller, M. Finke, and H. Yang, 1996a. "Statistical theory of overtraining—Is cross-validation asymptotically effective?" *Advances in Neural Information Processing Systems*, vol. 8, pp. 176–182, Cambridge, MA: MIT Press.

Ambros-Ingerson, J., R. Granger, and G. Lynch, 1990. "Simulation of paleo-cortex performs hierarchical clustering," *Science*, vol. 247, pp. 1344–1348.

Amit, D.J., 1989. *Modeling Brain Function: The World of Attractor Neural Networks*, New York: Cambridge University Press.

Anastasio T.J., 1995. "Vestibulo-ocular reflex: Performance and plasticity," In M.A. Arbib, ed., *The Handbook of Brain Theory and Neural Networks*, Cambridge, MA: MIT Press.

Anastasio, T.J., 1993. "Modeling vestibulo-ocular reflex dynamics: From classical analysis to neural networks," in F. Eeckman, ed., *Neural Systems: Analysis and Modeling*, pp. 407–430, Norwell, MA: Kluwer.

Anastasio, T.J., 1991. "A recurrent neural network model of velocity storage in the vestibulo-ocular reflex," *Advances in Neural Information Processing Systems*, vol. 3, pp. 32–38, San Mateo, CA: Morgan Kaufmann.

Anderson, J.A., 1995. *Introduction to Neural Networks*, Cambridge, MA: MIT Press.

Anderson, J.A., 1993. "The BSB model: A simple nonlinear autoassociative neural network," in *Associative Neural Memories* (M. Hassoun, ed.) pp. 77–103, Oxford: Oxford University Press.

Anderson, J.A., 1988. "General introduction," *Neurocomputing: Foundations of Research* (J.A. Anderson and E. Rosenfeld, eds.), pp. xiii–xxi, Cambridge, MA: MIT Press.

Anderson, J.A., 1983. "Cognitive and psychological computation with neural models," *IEEE Transactions on Systems, Man, and Cybernetics*, vol. SMC-13, pp. 799–815.

Anderson, J.A., 1972. "A simple neural network generating an interactive memory," *Mathematical Biosciences,* vol. 14, pp. 197–220.

Anderson, J.A., and G.L. Murphy, 1986. "Concepts in connectionist models," in *Neural Networks for Computing,* J.S. Denker, ed., pp. 17–22, New York: American Institute of Physics.

Anderson, J.A., and E. Rosenfeld, eds., 1988. *Neurocomputing: Foundations of Research,* Cambridge, MA: MIT Press.

Anderson, J.A., A. Pellionisz, and E. Rosenfeld, eds., 1990a. *Neurocomputing 2: Directions for Research,* Cambridge, MA: MIT Press.

Anderson, J.A., J.W. Silverstein, S.A. Ritz, and R.S. Jones, 1977. "Distinctive features, categorical perception, and probability learning: Some applications of a neural model," *Psychological Review,* vol. 84, pp. 413–451.

Anderson, J.A., and J.P. Sutton, 1995. "A network of networks: Computation and neurobiology," *World Congress on Neural Networks,* vol. I, pp. 561–568.

Anderson, J.A., M.T. Gately, P.A. Penz, and D.R. Collins, 1990b. "Radar signal categorization using a neural network," *Proceedings of the IEEE,* vol. 78, pp. 1646–1657.

Anderson, T.W., 1984. *An Introduction to Multivariate Statistical Analysis,* 2nd edition, New York: Wiley.

Andreou, A.G., 1994. "On physical models of neural computation and their analog VLSI implementation," *Proceedings of the 1994 Workshop on Physics and Computation,* IEEE Computer Society Press, pp. 255–264, Los Alamitos, CA.

Andreou, A.G., K.A. Boahen, P.O. Pouliqueen, A. Pasavoic, R.E. Jenkins, and K. Strohbehn, 1991. "Current-mode subthreshold MOS circuits for analog VLSI neural systems," *IEEE Transactions on Neural Networks,* vol. 2, pp. 205–213.

Andreou, A.G., R.C. Meitzler, K. Strohbehn, and K.A. Boahen, 1995. "Analog VLSI neuromorphic image acquisition and pre-processing systems." *Neural Networks,* vol. 8, pp. 1323–1347.

Andrews, R., and J. Diederich, eds., 1996. *Proceedings of the Rule Extraction from Trained Artificial Neural Networks Workshop,* University of Sussex, Brighton, UK.

Ansaklis, P.J., M. Lemmon, and J.A. Stiver, 1996. "Learning to be autonomous," In M.D. Gupta and N.K. Sinha, eds., *Intelligent Control Systems,* pp. 28–62, New York: IEEE Press.

Ansari, N., and E. Hou, 1997. *Computational Intelligence for Optimization,* Norwell, MA: Kluwer.

Anthony, M., and N. Biggs, 1992. *Computational Learning Theory,* Cambridge: Cambridge University Press.

Antsaklis, P.J., and K.M. Passino, eds., 1993. *An Introduction to Intelligent and Automatic Control,* Norwell, MA: Kluwer.

Arbib, M.A., 1989. *The Metaphorical Brain,* 2nd edition, New York: Wiley.

Arbib, M.A., 1987. *Brains, Machines, and Mathematics,* 2nd edition, New York: Springer-Verlag.

Arbib, M.A., ed. 1995. *The Handbook of Brain Theory and Neural Networks,* Cambridge, MA: MIT Press.

Arrowsmith, D.K., and C.M. Place, 1990. *An Introduction to Dynamical Systems,* Cambridge: Cambridge University Press.

Artola, A., and W. Singer, 1987. "Long-term potentiation and NMDA receptors in rat visual cortex," *Nature,* vol. 330, pp. 649–652.

Ash, R.E., 1965. *Information Theory,* New York: Wiley.

Ashby, W.R., 1960. *Design for a Brain,* 2nd edition, New York: Wiley.

Ashby, W.R., 1952. *Design for a Brain,* New York: Wiley.

Aspray, W., and A. Burks, 1986. *Papers of John von Neumann on Computing and Computer Theory,* Charles Babbage Institute Reprint Series for the History of Computing, vol. 12. Cambridge, MA: MIT Press.

Åström, K.J., and T.J. McAvoy, 1992. "Intelligent control: An overview and evaluation," In *Handbook of Intelligent Control,* D.A. White and D.A. Sofge, eds., New York: Van Nostrand Reinhold.

Atherton, D.P., 1981. *Stability of Nonlinear Systems,* Chichester, UK: Research Studies Press.

Atick, J.J., 1992. "Could information theory provide an ecological theory of sensory processing?" *Network: Computation in Neural Systems,* vol. 3, pp. 213–251.

Atick, J.J., and A.N. Redlich, 1992. "What does the retina know about natural scenes," *Neural Computation,* vol. 4, pp. 196–210.

Atick, J.J., and A.N. Redlich, 1990. "Towards a theory of early visual processing," *Neural Computation,* vol. 2, pp. 308–320.

Atick, J.J., P.A. Griffin, and A.N. Redlich, 1996. "Statistical approach to shape from shading: Reconstruction of three-dimensional face surfaces from single two-dimensional images," *Neural Computation,* vol. 8, pp. 1321–1340.

Atiya, A.F., 1987, "Learning on a general network," In *Neural Information Processing Systems,* D.Z. Anderson, ed., pp. 22–30, New York: American Institute of Physics.

Atiya, A.F., and Y.S. Abu-Mostafa, 1993, "An analog feedback associative memory," *IEEE Transactions on Neural Networks,* vol. 4, pp. 117–126.

Attneave, F., 1954. "Some informational aspects of visual perception," *Psychological Review,* vol. 61, pp. 183–193.

Back, A.D., and A.S. Weigend, 1998. "A first application of independent component analysis to extracting structure from stock returns," *International Journal of Neural Systems,* vol. 9, Special Issue on Data Mining in Finance, to appear.

Back, A.D., and A.C. Tsoi, 1991. "FIR and IIR synapses, a new neural network architecture for time series modeling," *Neural Computation,* vol. 3, pp. 375–385.

Back, A.D., and A.C. Tsoi, 1998. "A low-sensitivity recurrent neural network," *Neural Computation,* vol. 10, pp. 165–188.

Baldi, P., and K. Hornik, 1989. "Neural networks and principal component analysis: Learning from examples without local minimum," *Neural Networks,* vol. 1, pp. 53–58.

Bantine, W.L., and A.S. Weigend, 1994. "Computing second derivatives in feed-forward networks: A review," *IEEE Transactions on Neural Networks,* vol. 5, pp. 480–488.

Baras, J.S., and A. LaVigna, 1990. "Convergence of Kohonen's learning vector quantization," *International Joint Conference on Neural Networks,* vol. III, pp. 17–20, San Diego, CA.

Barlow, H.B., 1989. "Unsupervised learning," *Neural Computation,* vol. 1, pp. 295–311.

Barlow, H.B., 1985. "Cognitronics: methods for acquiring and holding cognitive knowledge," Unpublished manuscript.

Barlow, H.B., 1959. "Sensory mechanisms, the reduction of redundancy, and intelligence," in *The Mechanisation of Thought Processes, National Physical Laboratory Symposium No. 10,* Her Majesty's Stationary Office, London.

Barlow, H., and P. Földiák, 1989. "Adaptation and decorrelation in the cortex," in *The Computing Neuron,* R. Durbin, C. Miall, and G. Mitchison, eds., pp. 54–72. Reading, MA: Addison-Wesley.

Barnard, E., and D. Casasent, 1991. "Invariance and neural nets," *IEEE Transactions on Neural Networks,* vol. 2, pp. 498–508.

Barron, A.R., 1992. "Neural net approximation," in *Proceedings of the Seventh Yale Workshop on Adaptive and Learning Systems,* pp. 69–72, New Haven, CT.: Yale University.

Barron, A.R., 1993. "Universal approximation bounds for superpositions of a sigmoidal function," *IEEE Transactions on Information Theory,* vol. 39, pp. 930–945.

Bartlett, P.L., 1997. "For valid generalization, the size of the weights is more important than the size of the network," *Advances in Neural Information Processing Systems,* vol. 9, pp. 134–140, Cambridge, MA: MIT Press.

Barto, A.G., 1992. "Reinforcement learning and adaptive critic methods," in *Handbook of Intelligent Control,* D.A. White and D.A. Sofge, eds., pp. 469–491, New York: Van Nostrand Reinhold.

Barto, A.G., S.J. Bradtke, and S. Singh, 1995. "Learning to act using real-time dynamic programming," *Artificial Intelligence,* vol. 72, pp. 81–138.

Barto, A.G., R.S. Sutton, and C.W. Anderson, 1983. "Neuronlike adaptive elements that can solve difficult learning control problems," *IEEE Transactions on Systems, Man, and Cybernetics,* vol. SMC-13, pp. 834–846.

Basar, E., ed., 1990. *Chaos in Brain Function,* New York: Springer-Verlag.

Bashkirov, O.A., E.M. Braverman, and I.B. Muchnik, 1964. "Potential function algorithms for pattern recognition learning machines," *Automation and Remote Control,* vol. 25, pp. 629–631.

Battiti, R., 1992. "First- and second-order methods for learning: Between steepest descent and Newton's method," *Neural Computation,* vol. 4, pp. 141–166.

Bauer, H.-U., and K.R. Pawelzik, 1992. "Quantifying the neighborhood preservation of self-organizing feature maps," *IEEE Transactions on Neural Networks,* vol. 3, pp. 570–579.

Bauer, H.-U., R. Der, and M. Hermman, 1996. "Controlling the magnification factor of self-organizing feature maps," *Neural Computation,* vol. 8, pp. 757–771.

Baum, E.B., 1991. "Neural net algorithms that learn in polynomial time from examples and queries," *IEEE Transactions on Neural Networks,* vol. 2, pp. 5–19.

Baum, E.B., and D. Haussler, 1989. "What size net gives valid generalization?" *Neural Computation,* vol. 1, pp. 151–160.

Baum, E.B., and F. Wilczek, 1988. "Supervised learning of probability distributions by neural networks," in D.Z. Anderson, ed., pp. 52–61, New York: American Institute of Physics.

Beaufays, F., and E.A. Wan, 1994. "Relating real-time backpropagation and backpropagation-through-time: An application of flow graph interreciprocity," *Neural Computation,* vol. 6, pp. 296–306.

Becker, S., 1996. "Mutual information maximization: models of cortical self-organization," *Network: Computation in Neural Systems,* vol. 7, pp. 7–31.

Becker, S., 1993. "Learning to categorize objects using temporal coherence," *Advances in Neural Information Processing Systems,* vol. 5, pp. 361–368, San Mateo, CA: Morgan Kaufmann.

Becker, S., 1991. "Unsupervised learning procedures for neural networks," *International Journal of Neural Systems,* vol. 2, pp. 17–33.

Becker, S., and G.E. Hinton, 1993. "Learning mixture models of spatial coherence," *Neural Computation,* vol. 5, pp. 267–277.

Becker, S., and G.E. Hinton, 1992. "A self-organizing neural network that discovers surfaces in random-dot stereograms," *Nature (London),* vol. 355, pp. 161–163.

Beckerman, M., 1997. *Adaptive Cooperative Systems,* New York: Wiley (Interscience).

Bell, A.J., and T.J. Sejnowski, 1995. "An information-maximization approach to blind separation and blind deconvolution," *Neural Computation,* vol. 6, pp. 1129–1159.

Bellman, R., 1961. *Adaptive Control Processes: A Guided Tour,* Princeton, NJ: Princeton University Press.

Bellman, R., 1957. *Dynamic Programming,* Princeton, NJ: Princeton University Press.

Bellman, R., and S.E. Dreyfus, 1962. *Applied Dynamic Programming,* Princeton, NJ: Princeton University Press.

Bengio, Y., 1996. *Neural Networks for Speech and Sequence Recognition.* London: International Thomson Computer Press.

Bengio, Y., P. Simard, and P. Frasconi, 1994. "Learning long-term dependencies with gradient descent is difficult," *IEEE Transactions on Neural Networks,* vol. 5, pp. 157–166.

Benveniste, A., M. Métivier, and P. Priouret, 1987. *Adaptive Algorithms and Stochastic Approximation,* New York: Springer-Verlag.

Bertero, M., T.A. Poggio, and V. Torre, 1988. "Ill-posed problems in early vision," *Proceedings of the IEEE,* vol. 76, pp. 869–889.

Bertsekas, D.P., 1995. *Dynamic Programming and Optimal Control,* vol. I and vol. II, Belmont, MA: Athenas Scientific.

Bertsekas, D.P., 1995. *Nonlinear Programming,* Belmont, MA: Athenas Scientific.

Bertsekas, D.P., and J.N. Tsitsiklis, 1996. *Neuro-Dynamic Programming,* Belmont, MA: Athenas Scientific.

Beymer, D., and T. Poggio, 1996. "Image representations for visual learning," *Science,* vol. 272, pp. 1905–1909.

Bienenstock, E.L., L.N. Cooper, and P.W. Munro, 1982. "Theory for the development of neuron selectivity: Orientation specificity and binocular interaction in visual cortex," *Journal of Neuroscience,* vol. 2, pp. 32–48.

Bishop, C.M., 1992. "Exact calculation of the Hessian matrix for the multi-layer perceptron," *Neural Computation,* vol. 4, pp. 494–501.

Bishop, C.M., 1995. *Neural Networks for Pattern Recognition.* Oxford: Clarendon Press.

Black, I.B., 1991. *Information in the Brain: A Molecular Perspective,* Cambridge, MA: MIT Press.

Blake, A., 1983. "The least-disturbance principle and week constraints," *Pattern Recognition Letters,* vol. 1, pp. 393–399.

Bliss, T.V.P., and T. Lomo, 1973. "Long-lasting potentiation of synaptic transmission in the dentate area of the anaesthetized rabbit following stimulation of the perforatant path," *J. Physiol,* vol. 232, pp. 331–356.

Blumer, A., A. Ehrenfeucht, D. Haussler, and M.K. Warmuth, 1989. "Learnability and the Vapnik–Chervonenkis Dimension," *Journal of the Association for Computing Machinery,* vol. 36, pp. 929–965.

Blumer, A., A. Ehrenfeucht, D. Haussler, and M.K. Warmuth, 1987. "Occam's razor," *Information Processing Letters,* vol. 24, pp. 377–380.

Boahen, K.A., 1996. "A retinomorphic vision system," *IEEE Micro,* vol. 16, no. 5, pp. 30–39.

Boahen, K.A., and A.G. Andreou, 1992. "A contrast sensitive silicon retina with reciprocal synapses," *Advances in Neural Information Processing Systems,* vol. 4, pp. 764–772. San Mateo, CA: Morgan Kaufmann.

Boahen, K.A., P.O. Pouliqueen, A.G. Andreou, and R.E. Jenkins, 1989. "A heterassociative memory using current-mode analog VLSI circuits," *IEEE Transactions on Circuits and Systems,* vol. CAS-36, pp. 747–755.

Bodenhausen, U., and A. Waibel, 1991. "The tempo 2 algorithm: Adjusting time-delays by supervised learning," *Advances in Neural Information Processing Systems,* vol. 3, pp. 155–161, San Mateo, CA: Morgan Kaufmann.

Boltzmann, L., 1872. "Weitere studien über das Wärmegleichgewicht unter gasmolekülen," *Sitzungsberichte der Mathematisch-Naturwissenschaftlichen Classe der Kaiserlichen Akademie der Wissenschaften,* vol. 66, pp. 275–370.

Boser, B., I. Guyon, and V.N. Vapnik, 1992. "A training algorithm for optimal margin classifiers," *Fifth Annual Workshop on Computational Learning Theory,* pp. 144–152. San Mateo, CA: Morgan Kaufmann.

Boser, B.E., E. Säckinger, J. Bromley, Y. LeCun, and L.D. Jackel, 1992. "Hardware requirements for neural network pattern classifiers," *IEEE Micro,* vol. 12, pp. 32–40.

Bourlard, H.A., and N. Morgan, 1994. *Connectionist Speech Recognition: A Hybrid Approach,* Boston: Kluwer.

Bourlard, H.A., and C.J. Wellekens, 1990. "Links between Markov models sand multilayer perceptrons," *IEEE Transactions on Pattern Analysis and Machine Intelligence,* vol. PAMI-12, pp. 1167–1178.

Box, G.E.P., and G.M. Jenkins, 1976. *Time Series Analysis: Forecasting and Control,* San Francisco: Holden Day.

Braitenberg, V., 1967. "Is the cerebella cortex a biological clock in the millisecond range?" in *The Cerebellum. Progress in Brain Research,* C.A. Fox and R.S. Snider, eds., vol. 25 pp. 334–346, Amsterdam: Elsevier.

Braitenberg, V., 1990. "Reading the structure of brains," *Network: Computation in Neural Systems,* vol. 1, pp. 1–12.

Braitenberg, V., 1986. "Two views of the cerebral cortex," in *Brain Theory,* G. Palm and A. Aertsen, eds., pp. 81–96. New York: Springer-Verlag.

Braitenberg, V., 1984. *Vehicles: Experiments in Synthetic Psychology,* Cambridge, MA: MIT Press.

Braitenberg, V., 1977. *On the Texture of Brains,* New York: Springer-Verlag.

Bregman, A.S., 1990. *Auditory Scene Analysis: The Perceptual Organization of Sound,* Cambridge, MA: MIT Press.

Breiman, L., 1996a. "Bagging predictors." *Machine Learning,* vol. 24, pp. 123–140.

Breiman, L., 1996b. "Bias, variance, and arcing classifiers," *Technical Report 460,* Statistics Department, University of California, Berkeley, Calif.

Breiman, L., J. Friedman, R. Olshen, and C. Stone, 1984. *Classification and Regression Trees,* New York: Chapman and Hall.

Bridle, J.S., 1990a. "Probabilistic interpretation of feedforward classification network outputs, with relationships to statistical pattern recognition," in *Neuro-computing: Algorithms, Architectures and Applications,* F. Fougelman-Soulie and J. Hérault, eds., New York: Springer-Verlag.

Bridle, J.S., 1990b. "Training stochastic model recognition algorithms as networks can lead to maximum mutual information estimation of parameters," *Advances in Neural Information Processing Systems,* vol. 2, pp. 211–217, San Mateo, CA: Morgan Kaufmann.

Brodal, A., 1981. *Neurological Anatomy in Relation to Clinical Medicine,* 3rd edition, New York: Oxford University Press.

Brodmann, K., 1909. *Vergleichende Lokalisationslehre der Grosshirnrinde,* Leipzig: J.A. Barth.

Brogan, W.L., 1985. *Modern Control Theory,* 2nd edition, Englewood Cliffs, NJ: Prentice-Hall.

Broomhead, D.S., and D. Lowe, 1988. "Multivariable functional interpolation and adaptive networks," *Complex Systems,* vol. 2, pp. 321–355.

Brown, T.H., E.W. Kairiss, and C.L. Keenan, 1990. "Hebbian synapses: Biophysical mechanisms and algorithms," *Annual Review of Neuroscience,* vol. 13, pp. 475–511.

Bruck, J., 1990. "On the convergence properties of the Hopfield model," *Proceedings of the IEEE,* vol. 78, pp. 1579–1585.

Bryson, A.E., Jr., and Y.C. Ho, 1969. *Applied Optimal Control,* Blaisdell. (Revised printing, 1975, Hemisphere Publishing, Washington, DC).

Burg, J.P., 1975. *Modern Spectral Estimation,* Ph.D. Thesis, Stanford University, Stanford, Calif.

Burges, C.J.C., 1998. "A tutorial on support vector machines for pattern recognition," *Data Mining and Knowledge Discovery,* to appear.

Cacoullos, T., 1966. "Estimation of a multivariate density," *Annals of the Institute of Statistical Mathematics (Tokyo),* vol. 18, pp. 179–189.

Caianiello, E.R., 1961. "Outline of a theory of thought-processes and thinking machines," *Journal of Theoretical Biology,* vol. 1, pp. 204–235.

Cameron, S.H., 1960. Tech. Report 60–600, *Proceedings of the Bionics Symposium,* pp. 197–212, Wright Air Development Division, Dayton, Ohio.

Cardoso, J.-F., 1998a. "Blind signal separation: A review," *Proceedings of the IEEE,* vol. 86, to appear.

Cardoso, J.-F., 1998b. "Multidimensional independent component analysis," *Proceedings IEEE ICASSP,* Seattle, WA, May.

Cardoso, J.-F., 1997. "Infomax and maximum likelihood for blind source separation," *IEEE Signal Processing Letters,* vol. 4, pp. 112–114.

Cardoso, J.-F., 1996. "Entropic contrasts for source separation," Presented at the NIPS 96 *Workshop on Blind Signal Processing* organized by A. Cichoki at Snomass, Colo. To appear as a chapter in the book *Unsupervised Adaptive Filtering*, S. Haykin, ed., New York: Wiley.

Cardoso, J.-F., and B. Laheld, 1996. "Equivariant adaptive source separation," *IEEE Transactions on Signal Processing*, vol. 44, pp. 3017–3030.

Cardoso, J.-F., and A. Souloumia, 1993. "Blind beamforming for non-Gaussian signals," *IEE Proceedings (London), Part F,* vol. 140, pp. 362–370.

Carpenter, G.A., and S. Grossberg, 1987. "A massively parallel architecture for a self-organizing neural pattern recognition machine," *Computer Vision, Graphics, and Image Processing*, vol. 37, pp. 54–115.

Carpenter, G.A., M.A. Cohen, and S. Grossberg, 1987. Technical comments on "Computing with neural networks," *Science*, vol. 235, pp. 1226–1227.

Carpenter, G.A., and S. Grossberg, 1995. "Adaptive resonance theory (ART)," in M.A. Arbib, ed., *The Handbook of Brain Theory and Neural Networks*, pp. 79–82, Cambridge, MA: MIT Press.

Casdagli, M., 1989. "Nonlinear prediction of chaotic time-series," *Physica*, vol. 35D, pp. 335–356.

Černy, V., 1985. "Thermodynamic approach to the travelling salesman problem," *Journal of Optimization Theory and Applications*, vol. 45, pp. 41–51.

Changeux, J.P., and A. Danchin, 1976. "Selective stabilization of developing synapses as a mechanism for the specification of neural networks," *Nature*, vol. 264, pp. 705–712.

Chatterjee, C., V.P. Roychowdhhury, and E.K.P. Chong, 1998. "On relative convergence properties of principal component algorithms," *IEEE Transactions on Neural Networks*, vol. 9, pp. 319–329.

Chen, H., and R.-W Liu, 1992. "Adaptive distributed orthogonalization processing for principal components analysis," *International Conference on Acoustics, Speech, and Signal Processing*, vol. 2, pp. 293–296, San Francisco.

Chen, S., 1995. "Nonlinear time series modelling and prediction using Gaussian RBF networks with enhanced clustering and RLS learning," *Electronics Letters*, vol. 31, No. 2, pp. 117–118.

Chen, S., S. Billings, and P. Grant, 1990. "Non-linear system identification using neural networks," *International Journal of Control*, vol. 51, pp. 1191–1214.

Chen, S., B. Mulgrew, and S. McLaughlin, 1992. "Adaptive Bayesian feedback equalizer based on a radial basis function network," *IEEE International Conference on Communications*, vol. 3, pp. 1267–1271, Chicago.

Cherkassky, V., and F. Mulier, 1995. "Self-organization as an iterative kernel smoothing process," *Neural Computation*, vol. 7, pp. 1165–1177.

Cherkassky, V., and F. Mulier, 1998. *Learning from Data: Concepts, Theory and Methods,* New York: Wiley.

Cherry, E.C., 1953. "Some experiments on the recognition of speech, with one and with two ears," *Journal of the Acoustical Society of America*, vol. 25, pp. 975–979.

Cherry, E.C., and W.K. Taylor, 1954. "Some further experiments upon the recognition of speech, with one and with two ears," *Journal of Acoustical Society of America*, vol. 26, pp. 554–559.

Chester, D.L., 1990. "Why two hidden layers are better than one," *International Joint Conference on Neural Networks*, vol. I, pp. 265–268, Washington, D.C.

Chinrungrueng, C., and C.H. Séquin (1994). "Optimal adaptive k-means algorithm with dynamic adjustment of learning rate," *IEEE Transactions on Neural Networks*, vol. 6, pp. 157–169.

Choey, M., and A.S. Weigend, 1996. "Nonlinear trading models through Sharp ratio maximization," in A. Weigend, Y.S. Abu-Mostafa, and A.-P.N. Refenes, eds., *Decision Technologies for Financial Engineering*, pp. 3–22, Singapore: World Scientific.

Churchland, P.S., 1986. *Neurophilosophy: Toward a Unified Science of the Mind/Brain,* Cambridge, MA: MIT Press.

Churchland, P.S., and T.J. Sejnowski, 1992. *The Computational Brain,* Cambridge, MA: MIT Press.

Cichocki, A., and R. Unbehauen, 1996. "Robust neural networks with on-line learning for blind identification and blind separation of sources," *IEEE Transactions on Circuits and Systems-1: Fundamental Theory and Applications*, vol. 43, pp. 894–906.

Cichocki, A., R. Ubenhauen, and E. Rummert, 1994. "Robust learning algorithms for blind separation of signals," *Electronics Letters*, vol. 30, pp. 1386–1387.

Cichocki, A., and L. Moszcsynski, L., 1992. "New learning algorithm for blind separation of sources," *Electronics Letters*, vol. 28, pp. 1986–1987.

Cleeremans, A., D. Servan-Schreiber, and J.L. McClelland, 1989. "Finite state automata and simple recurrent networks," *Neural Computation*, vol. 1, pp. 372–381.

Cohen, L., 1995. *Time-Frequency Analysis,* Englewood Cliffs, NJ: Prentice-Hall.

Cohen, M.A., 1992a. "The synthesis of arbitrary stable dynamics in non-linear neural networks II: Feedback and universality," *International Joint Conference on Neural Networks,* vol. I, pp. 141–146, Baltimore.

Cohen, M.A., 1992b. "The construction of arbitrary stable dynamics in nonlinear neural networks," *Neural Networks,* vol. 5, pp. 83–103.

Cohen, M.A., and S. Grossberg, 1983. "Absolute stability of global pattern formation and parallel memory storage by competitive neural networks," *IEEE Transactions on Systems, Man, and Cybernetics,* vol. SMC-13, pp. 815–826.

Cohn, D., and G. Tesauro, 1992. "How tight are the Vapnik–Chervonenkis bounds?" *Neural Computation,* vol. 4, pp. 249–269.

Comon, P., 1995. "Supervised classification, a probabilistic approach," *European Symposium on Artificial Neural Networks,* pp. 111–128, Brussels, Belgium.

Comon, P., 1994. "Independent component analysis: A new concept?" *Signal Processing,* vol. 36, pp. 287–314.

Comon, P., 1991. "Independent component analysis," *Proceedings of International Signal Processing Workshop on Higher-order Statistics,* pp. 111–120, Chamrousse, France.

Constantine-Paton, M., H.T. Cline, and E. Debski, 1990. "Patterned activity, synaptic convergence, and the NMDA receptor in developing visual pathways," *Annual Review of Neuroscience,* vol. 13, pp. 129–154.

Cook, A.S., 1971. "The complexity of theorem-proving procedures," *Proceedings of the 3rd Annual ACM Symposium on Theory of Computing,* pp. 151–158, New York.

Cook, P.A., 1986. *Nonlinear Dynamical Systems,* London: Prentice-Hall International.

Cooper, L.N., 1973. "A possible organization of animal memory and learning," *Proceedings of the Nobel Symposium on Collective Properties of Physical Systems,* B. Lundquist and S. Lundquist, eds., pp. 252–264, New York: Academic Press.

Cormen, T.H., C.E. Leiserson, and R.R. Rivest, 1990. *Introduction to Algorithms.* Cambridge, MA: MIT Press.

Cortes, C., and V. Vapnik, 1995. "Support vector networks," *Machine Learning,* vol. 20, pp. 273–297.

Cottrell, M., and J.C. Fort, 1986. "A stochastic model of retinotopy: A self organizing process," *Biological Cybernetics,* vol. 53, pp. 405–411.

Cottrell, M., J.C. Fort, and G. Pagés, 1997. "Theoretical aspects of the SOM algorithm," *Proceedings of the Workshop on Self-Organizing Maps,* Espoo, Finland.

Cottrell, G.W., and J. Metcalfe, 1991. "EMPATH: Face, emotion, and gender recognition using holons," *Advances in Neural Information Processing Systems,* vol. 3, pp. 564–571, San Mateo, CA: Morgan Kaufmann.

Cottrell, G.W., P. Munro, and D. Zipser, 1987. "Learning internal representations from grey-scale images: An example of extensional programming," *Proceedings of the 9th Annual Conference of the Cognitive science Society,* pp. 461–473.

Courant, R., and D. Hilbert, 1970. *Methods of Mathematical Physics,* vol. I and II, New York: Wiley Interscience.

Cover, T.M., 1968. "Capacity problems for linear machines," In L. Kanal, ed., *Pattern Recognition,* pp. 283–289, Washington, DC: Thompson Book Co.

Cover, T.M., 1965. "Geometrical and statistical properties of systems of linear inequalities with applications in pattern recognition," *IEEE Transactions on Electronic Computers,* vol. EC-14, pp. 326–334.

Cover, T.M., and P.E. Hart, 1967. "Nearest neighbor pattern classification," *IEEE Transactions on Information Theory,* vol. IT-13, pp. 21–27.

Cover, T.M., and J.A. Thomas, 1991. *Elements of Information Theory,* New York: Wiley.

Cowan, J.D., 1990. "Neural networks: The early days," *Advances in Neural Information Processing Systems,* vol. 2, pp. 828–842, San Mateo, CA: Morgan Kaufmann.

Cowan, J.D., 1968. "Statistical mechanics of nervous nets," in *Neural Networks,* E.R. Caianiello, ed., pp. 181–188, Berlin: Springer-Verlag.

Cowan, J.D. 1967. "A Mathematical Theory of Central Nervous Activity," Ph.D. Thesis, University of London.

Cowan, J.D. 1965. "The problem of organismic reliability," *Progress in Brain Research,* vol. 17, pp. 9–63.

Cowan, J.D., and M.H. Cohen, 1969. "The role of statistical mechanics in neurobiology," *Journal of the Physical Society of Japan,* vol. 26, pp. 51–53.

Cowan, J.D., and D.H. Sharp, 1988. "Neural nets," *Quarterly Reviews of Biophysics,* vol. 21, pp. 365–427.

Cragg, B.G., and H.N.V. Tamperley, 1955. "Memory: The analogy with ferromagnetic hysteresis," *Brain,* vol. 78, part II, pp. 304–316.

Cragg, B.G., and H.N.V. Tamperley, 1954. "The organization of neurons: A cooperative analogy," *EEG Clinical Neurophysiology,* vol. 6, pp. 85–92.

Craik, K.J.W., 1943. *The Nature of Explanation,* Cambridge: Cambridge University Press.

Craven, P., and G. Wahba, 1979. "Smoothing noisy data with spline functions: Estimating the correct degree of smoothing by the method of generalized cross-validation," *Numerische Mathematik.*, vol. 31, pp. 377–403.

Crick, F., 1989. "The recent excitement about neural networks," *Nature,* vol. 337, pp. 129–132.

Crites, R.H., and A.G. Barto, 1996. "Improving elevator performance using reinforcement learning," *Advances in Neural Information Processing Systems,* vol. 8, pp. 1017–1023, Cambridge, MA: MIT Press.

Crutchfield, J.P., J.D. Farmer, N.H. Packard, and R.S. Shaw, 1986. "Chaos," *Scientific American,* vol. 255(6), pp. 38–49.

Cybenko, G., 1995. "Q-learning: A tutorial and extensions." Presented at *Mathematics of Artificial Neural Networks,* Oxford University, England, July 1995.

Cybenko, G., 1989. "Approximation by superpositions of a sigmoidal function," *Mathematics of Control, Signals, and Systems,* vol. 2, pp. 303–314.

Cybenko, G., 1988. "Approximation by superpositions of a sigmoidal function," Urbana, IL.: University of Illinois.

Darken, C., and J. Moody, 1992. "Towards faster stochastic gradient search," *Advances in Neural Information Processing Systems,* vol. 4, pp. 1009–1016, San Mateo, CA: Morgan Kaufmann.

Darmois, G., 1953. "Analyse generale des liaisons stochastiques," *Rev. Inst. Internat. Stat.,* vol. 21, pp. 2–8.

Dasarathy, B.V., ed., 1991. *Nearest Neighbor (NN) Norms: NN Pattern Classification Techniques,* Los Alamitos, CA: IEEE Computer Society Press.

Daubechies, I., 1990. "The wavelet transform, time-frequency," *IEEE Transactions on Information Theory,* vol. IT-36, pp. 961–1005.

Daubechies, I., 1992. *Ten Lectures on Wavelets,* SIAM.

Davis, P.J., 1963. *Interpolation and Approximation,* New York: Blaisdell.

Dayan, P., and G.E. Hinton, 1996. "Varieties of Helmholtz machine," *Neural Networks,* vol. 9, pp. 1385–1403.

Dayan, P., G.E. Hinton, R.M. Neal, and R.S. Zemel, 1995. "The Helmholtz machine," *Neural Computation,* vol. 7, pp. 889–904.

Debnath, L., and P. Mikusiński, 1990. *Introduction to Hilbert Spaces with Applications,* New York: Academic Press.

Deco, G., W. Finnoff, and H.G. Zimmermann, 1995. "Unsupervised mutual information criteria for elimination of overtraining in supervised multilayer networks," *Neural Computation,* vol. 7, pp. 86–107.

Deco, G., and D. Obradovic, 1996. *An Information-Theoretic Approach to Neural Computing,* New York: Springer.

de Figueiredo, R.J.P., 1980. "Implications and applications of Kolmogorov's superposition theorem," *IEEE Transactions on Automatic Control,* vol. AC-25, pp. 1227–1230.

de Figueiredo, R.J.P., and G. Chen, 1993. *Nonlinear Feedback Control Systems,* New York: Academic Press.

DeMers, D., and G. Cottrell, 1993. "Non-linear dimensionality reduction," *Advances in Neural Information Processing Systems,* vol. 5, pp. 580–587. San Mateo, CA: Morgan Kaufmann.

Dempster, A.P., N.M. Laird, and D.B. Rubin, 1977. "Maximum likelihood from incomplete data via the EM algorithm," (with discussion), *Journal of the Royal Statistical Society.*, B, vol. 39, pp. 1–38.

Denardo, E.V., 1967. "Contraction mappings in the theory underlying dynamic programming," *SIAM,* Review, vol. 9, pp. 165–177.

DeSieno, D., 1988. "Adding a conscience to competitive learning," *IEEE International Conference on Neural Networks,* vol. I, pp. 117–124, San Diego, CA.

deSilva, C.J.S., and Y. Attikiouzel, 1992. "Hopfield networks as discrete dynamical systems," *International Joint Conference on Neural Networks,* vol. III, pp. 115–120, Baltimore.

deVries, B., and J.C. Principe, 1992. "The gamma model—A new neural model for temporal processing," *Neural Networks,* vol. 5, pp. 565–576.

Devroye, L., 1991. "Exponential inequalities in nonparametric estimation," in *Nonparametric Functional Estimation and Related Topics,* G. Roussas, ed., pp. 31–44. Boston: Kluwer.

Diamantaras, K.I., and S.Y. Kung, 1996. *Principal Component Neural Networks: Theory and Applications,* New York: Wiley.

Dohrmann, C.R., H.R. Busby, and D.M. Trujillo, 1988. "Smoothing noisy data using dynamic programming and generalized cross-validation," *Journal of Biomechanical Engineering,* vol. 110, pp. 37–41.

Domany, E., J.L. van Hemmen, and K. Schulten, eds., 1991. *Models of Neural Networks,* New York: Springer-Verlag.

Dony, R.D., and S. Haykin, 1997. "Image segmentation using a mixture of principal components representation," *IEE Proceedings (London), Image and Signal Processing,* vol. 144, pp. 73–80.

Dony, R.D., and S. Haykin, 1995. "Optimally adaptive transform coding," *IEEE Transactions on Image Processing,* vol. 4, pp. 1358–1370.

Dorny, C.N., 1975. *A Vector Space Approach to Models and Optimization,* New York: Wiley (Interscience).

Douglas, S.C., and S. Haykin, 1997. "On the relationship between blind deconvolution and blind source separation," *Thirty-First Asilomar Conference on Signals, Systems, and Computers,* Pacific Grove, California, November.

Doyle, J.C., K. Glover, P. Khargonekar, and B. Francis, 1989. "State-space solutions to standard H_2 and H_∞ control problems," *IEEE Transactions on Automatic Control,* vol. AC-34, pp. 831–847.

Drucker, H., C. Cortes, L.D. Jackel, and Y. LeCun, 1994. "Boosting and other ensemble methods." *Neural Computation,* vol. 6, pp. 1289–1301.

Drucker, H., R.E. Schapire, and P. Simard, 1993. "Improving performance in neural networks using a boosting algorithm," *Advances in Neural Information Processing Systems,* vol. 5, pp. 42–49, Cambridge, MA: MIT Press.

Dubois, D., and H. Prade, 1980. *Fuzzy Sets and Systems: Theory and Applications,* New York: Academic Press.

Duda, R.O., and P.E. Hart, 1973. *Pattern Classification and Scene Analysis,* New York: Wiley.

Dunford, N., and J.T. Schwartz, 1966. *Linear Operators, Part 1,* New York: Wiley.

Durbin, R., C. Miall, and G. Mitchison, eds., 1989. *The Computing Neuron,* Reading, MA: Addison-Wesley.

Durbin, R., and D.E. Rumelhart, 1989. "Product units: A computationally powerful and biologically plausible extension to backpropagation networks," *Neural Computation,* vol. 1, pp. 133–142.

Durbin, R., and D. Willshaw, 1987. "An analogue approach to the travelling salesman problem using an elastic net method," *Nature,* vol. 326, pp. 689–691.

Dyn, N., 1987. "Interpolation of scattered data by radial functions," in *Topics in Multivariate Approximation,* C.K. Chui, L.L. Schumaker, and F.I. Uteras, eds., pp. 47–61, Orlando, FL: Academic Press.

Edelman, G.M., 1987. *Neural Darwinism,* New York: Basil Books.

Edelman, G.M., 1973. "Antibody structure and molecular immunology," *Science,* vol. 180, pp. 830–840.

Eeckman, F.H., 1988. "The sigmoid nonlinearity in prepyriform cortex," *Neural Information Processing Systems,* pp. 242–248, New York: American Institute of Physics.

Eeckman, F.H., and W.J. Freeman, 1986. "The sigmoid nonlinearity in neural computation: An experimental approach," *Neural Networks for Computing,* J.S. Denker, ed., pp. 135–145, New York: American Institute of Physics.

Eggermont, J.J., 1990. *The Correlative Brain: Theory and Experiment in Neural Interaction,* New York: Springer-Verlag.

El Hihi, S., and Y. Bengio, 1996. "Hierarchical recurrent neural networks for long-term dependencies," *Advances in Neural Information Processing Systems,* vol. 8, pp. 493–499, MIT Press.

Elman, J.L., 1990. "Finding structure in time," *Cognitive Science,* vol. 14, pp. 179–211.

Elman, J.L., E.A. Bates, M.H. Johnson, A. Karmiloff-Smith, D. Parisi, and K. Plunkett, 1996. *Rethinking Innateness: A Connectionist Perspective on Development,* Cambridge, MA: MIT Press.

Erwin, E., K. Obermayer, and K. Schulten, 1995. "Models of orientation and ocular dominance columns in the visual cortex: A critical comparison," *Neural Computation,* vol. 7, pp. 425–468.

Erwin, E., K. Obermayer, and K. Schulten, 1992a. "I: Self-organizing maps: Stationary states, metastability and convergence rate," *Biological Cybernetics,* vol. 67, pp. 35–45.

Erwin, E., K. Obermayer, and K. Schulten, 1992b. "II: Self-organizing maps: Ordering, convergence properties and energy functions," *Biological Cybernetics,* vol. 67, pp. 47–55.

Faggin, F., 1991. "VLSI implementation of neural networks," Tutorial Notes, *International Joint Conference on Neural Networks,* Seattle.

Faggin, F., and C. Mead, 1990. "VLSI Implementation of Neural Networks," *An Introduction to Neural and Electronic Networks,* S.F. Zornetzer, J.L. Davis, and C. Lau, eds., pp. 275–292, New York: Academic Press.

Fahlman, S.E., and C. Lebiere, 1990. "The cascade-correlation learning architecture," *Advances in Neural Information Processing Systems,* vol. 2, pp. 524–532, San Mateo, CA: Morgan Kaufmann.

Farmer, J.D., and J. Sidorowich, 1987. "Predicting chaotic time series," *Physical Review Letters,* vol. 59, pp. 845–848.

Feldkamp, L.A., and G.V. Puskorius, 1997. "Adaptive behavior from fixed weight networks," *Information Sciences,* vol. 98, pp. 217–235.

Feldkamp, L.A., and G.V. Puskorius, 1998. "A signal processing framework based on dynamic neural networks with application to problems in adaptation, filtering and classification," *Proceedings of the IEEE,* vol. 86, to appear.

Feldkamp, L.A., G.V. Puskorius, and P.C. Moore, 1997. "Adaptation from fixed weight networks," *Information Sciences,* vol. 98, pp. 217–235.

Feldman, J.A., 1992. "Natural computation and artificial intelligence," Plenary Lecture presented at the *International Joint Conference on Neural Networks,* Baltimore.

Feller, W., 1968. *An Introduction to Probability Theory and its Applications,* vol. 1, 3rd edition, New York: John Wiley; 1st edition, 1950.

Fischler, M.A., and O. Firschein, 1987. *Intelligence: The Eye, The Brain, and The Computer,* Reading, MA: Addison-Wesley.

Fisher, R.A., 1925. "Theory of statistical estimation," *Proceedings of the Cambridge Philosophical Society,* vol. 22, pp. 700–725.

Fix, E., and J.L. Hodges, 1951. "Discriminatory analysis: Nonparametric discrimination: Consistency properties," *USAF School of Aviation Medicine,* Project 21-49-004, Report no. 4, pp. 261–279, Randolph Field, Texas.

Fletcher, R., 1987. *Practical Methods of Optimization,* 2nd edition, New York: Wiley.

Fodor, J.A., 1983. *Modularity of Mind,* Cambridge, MA: MIT Press.

Fodor, J.A., and Z.W. Pylyshyn, 1988. "Connectionism and cognitive architecture: a critical analysis," *Cognition,* vol. 28, pp. 3–72.

Földiak, P., 1989. "Adaptive network for optimal linear feature extractions," *IEEE International Joint Conference on Neural Networks,* vol. I, pp. 401–405, Washington, DC.

Forcada, M.L., and R.C. Carrasco, 1995. "Learning the initial state of a second-order recurrent neural network during regular-language inference," *Neural Computation,* vol. 7, pp. 923–930.

Forney, G.D., Jr., 1973. "The Viterbi algorithm," *Proceedings of the IEEE,* vol. 61, pp. 268–278.

Forte, J.C., and G. Pagés, 1995. "On the a.s. convergence of the Kohonen algorithm with a general neighborhood function," *Annals of Applied Probability,* vol. 5, pp. 1177–1216.

Forte, J.C., and G. Pagés, 1996. "Convergence of stochastic algorithm: From the Kushner and Clark theorem to the Lyapunov functional," *Advances in Applied Probability,* vol. 28, pp. 1072–1094.

Frasconi, P., M. Gori, and G. Soda, 1992. "Local feedback multilayered networks," *Neural Computation,* vol. 4, pp. 120–130.

Frasconi, P., and M. Gori, 1996. "Computational capabilities of local-feedback recurrent networks acting as finite-state machines," *IEEE Transactions on Neural Networks,* vol. 7, pp. 1521–1524.

Fraser, A.M., 1989. "Information and entropy in strange attractors," *IEEE Transactions on Information Theory,* vol. 35, pp. 245–262.

Freeman, J.A., and D.M. Sakpura, 1991. *Neural Networks: Algorithms, Applications, and Programming Techniques,* Reading, MA: Addison-Wesley.

Freeman, W.J., 1995. *Societies of Brains.* Hillsdale, NJ: Lawrence Erlbaum.

Freeman, W.J., 1992. "Tutorial on neurobiology: From single neurons to brain chaos," *International Journal of Bifurcation and Chaos in Applied Sciences and Engineering,* vol. 2, pp. 451–482.

Freeman, W.J., 1991. "The physiology of perception," *Scientific American,* vol. 264 (2), pp. 78–85.

Freeman, W.J., 1988. "Why neural networks don't yet fly: Inquiry into the neurodynamics of biological intelligence," *IEEE International Conference on Neural Networks,* vol. II, pp. 1–7, San Diego, CA.

Freeman, W.J., 1987. "Simulation of chaotic EEG patterns with a dynamic model of the olfactory system," *Biological Cybernetics,* vol. 56, pp. 139–150.

Freeman, W.J., 1975. *Mass Action in the Nervous System,* New York: Academic Press.

Frégnac, Y., and D. Schulz, 1994. "Models of synaptic plasticity and cellular analogs of learning in the developing and adult vertebrate visual cortex," *Advances in Neural and Behavioral Development,* vol. 4, pp. 149–235, Norwood, NJ: Neural Ablex.

Freund, Y., 1995. "Boosting a week learning algorithm by majority," *Information Computation,* vol. 121, pp. 256–285.

Freund, Y., and R.E. Schapire, 1997. "A decision-theoretic generalization of on-line learning and an application to boosting." *Journal of Computer and System Sciences,* vol. 55, pp. 119–139.

Freund, Y., and R.E. Schapire, 1996a. "Experiments with a new boosting algorithm," *Machine Learning: Proceedings of the Thirteenth International Conference,* pp. 148–156, Bari, Italy.

Freund, Y., and R.E. Schapire, 1996b. "Game theory, On-line prediction and boosting," *Proceedings of the Ninth Annual Conference on Computational Learning Theory,* pp. 325–332, Desenzano del Garda, Italy.

Friedman, J.H., 1995. "An overview of prediction learning and function approximation," In V. Cherkassky, J.H. Friedman, and H. Wechsler, eds., *From Statistics to Neural Networks: Theory and Pattern Recognition Applications,* New York: Springer-Verlag.

Fukunaga, K., 1990. *Statistical Pattern Recognition,* 2nd edition, New York: Academic Press.

Fukushima, K., 1995. "Neocognitron: A model for visual pattern recognition," in M.A. Arbib, ed., *The Handbook of Brain Theory and Neural Networks,* Cambridge, MA: MIT Press.

Fukushima, K., 1988a. "A hierarchical neural network model for selective attention," in *Neural Computers,* R. Eckmiller and C. von der Malsburg, eds., pp. 81–90, NATO ASI Series, New York: Springer-Verlag.

Fukushima, K., 1988b. "Neocognitron: A hierarchical neural network capable of visual pattern recognition." *Neural Networks,* vol. 1, pp. 119–130.

Fukushima, K., 1980. "Neocognitron: A self-organizing neural network model for a mechanism of pattern recognition unaffected by shift in position," *Biological Cybernetics,* vol. 36, 193–202.

Fukushima, K., 1975. "Cognitron: A self-organizing multi-layered neural network," *Biological Cybernetics,* vol. 20, pp. 121–136.

Fukushima, K., S. Miyake, and T. Ito, 1983. "Neocognitron: A neural network model for a mechanism of visual pattern recognition," *IEEE Transactions on Systems, Man, and Cybernetics,* vol. SMC-13, pp. 826–834.

Funahashi, K., 1989. "On the approximate realization of continuous mappings by neural networks," *Neural Networks,* vol. 2, pp. 183–192.

Gabor, D., 1954. "Communication theory and cybernetics," *IRE Transactions on Circuit Theory,* vol. CT-1, pp. 19–31.

Gabor, D., W.P.L. Wilby, and R. Woodcock, 1960. "A universal non-linear filter, predictor, and simulator which optimizes itself by a learning process," *Proceedings of the Institution of Electrical Engineers,* London, vol. 108, pp. 422–435.

Galland, C.C., 1993. "The limitations of deterministic Boltzmann machine learning," *Network,* vol. 4, pp. 355–379.

Gallant, A.R., and H. White, 1988. "There exists a neural network that does not make avoidable mistakes," *IEEE International Conference on Neural Networks,* vol. I, pp. 657–664, San Diego, CA.

Gallant, A.R., and H. White, 1992. "On learning the derivatives of an unknown mapping with multilayer feedforward networks," *Neural Networks,* vol. 5, pp. 129–138.

Gallant, S.I., 1993. *Neural Network Learning and Expert Systems,* Cambridge, MA: MIT Press.

Gallistel, C.R., 1990. *The Organization of Learning,* Cambridge, MA: MIT Press.

Gardner, E., 1987. "Maximum storage capacity in neural networks," *Electrophysics Letters,* vol. 4, pp. 481–485.

Garey, M.R., and D.S. Johnson, 1979. *Computers and Intractability,* New York: W.H. Freeman.

Gee, A.H., 1993. "Problem solving with optimization networks," Ph.D. dissertation, University of Cambridge.

Gee, A.H., S.V.B. Aiyer, and R. Prager, 1993. "An analytical framework for optimizing neural networks." *Neural Networks,* vol. 6, pp. 79–97.

Geisser, S., 1975. "The predictive sample reuse method with applications," *Journal of the American Statistical Association,* vol. 70, pp. 320–328.

Gelfand, A.E., and A.F.M. Smith, 1990. "Sampling-based approaches to calculating marginal densities," *Journal of the American Statistical Association,* vol. 85, pp. 398–409.

Geman, S., and D. Geman, 1984. "Stochastic relaxation, Gibbs distributions, and the Bayesian restoration of images," *IEEE Transactions on Pattern Analysis and Machine Intelligence,* vol. PAMI-6, pp. 721–741.

Geman, S., E. Bienenstock, and R. Doursat, 1992. "Neural networks and the bias/variance dilemma," *Neural Computation,* vol. 4, pp. 1–58.

Gersho, A., 1982. "On the structure of vector quantizers," *IEEE Transactions on Information Theory,* vol. IT-28, pp. 157–166.

Gersho, A., and R.M. Gray, 1992. *Vector Quantization and Signal Compression,* Norwell, MA: Kluwer.

Gerstein, G.L., P. Bedenbaugh, and A.M.H.J. Aersten, 1989. "Neural assemblies," *IEEE Transactions on Biomedical Engineering,* vol. 36, pp. 4–14.

Gibbs, J.W., 1902. "Elementary principles in statistical mechanics," reproduced in vol. 2 of *Collected Works of J. Willard Gibbs in Two Volumes,* New York: Longmans, Green and Co., 1928.

Gibson, G.J., and C.F.N. Cowan, 1990. "On the decision regions of multilayer perceptrons," *Proceedings of the IEEE,* vol. 78, pp. 1590–1599.

Gidas, B., 1985. "Global optimization via the Langevin equation," *Proceedings of 24th Conference on Decision and Control,* pp. 774–778, Ft. Lauderdale, FL.

Giles, C.L., 1996. "Dynamically driven recurrent neural networks: Models, learning algorithms, and applications," Tutorial #4, *International Conference on Neural Networks,* Washington, DC.

Giles, C.L., D. Chen, G.Z. Sun, H.H. Chen, Y.C. Lee, and M.W. Goudreau, 1995. "Constructive learning of recurrent neural networks: Limitations of recurrent cascade correlation with a simple solution," *IEEE Transactions on Neural Networks,* vol. 6, pp. 829–836.

Giles, C.L., T. Lin, and B.G. Horne, 1997. "Remembering the past: The role of embedded memory in recurrent neural network architectures," *Neural Networks for Signal Processing,* VII, Proceedings of the 1997 IEEE Workshop, IEEE Press, p. 34.

Giles, C.L., and T. Maxwell, 1987. "Learning, invariance, and generalization in higher-order neural networks," *Applied Optics,* vol. 26, pp. 4972–4978.

Giles, C.L., and B.G. Horne, 1994. "Representation of learning in recurrent neural network architectures," *Proceedings of the Eighth Yale Workshop on Adaptive and Learning Systems,* pp. 128–134, Yale University, New Haven, Ct.

Giles, C.L., C.B. Miller, D. Chen, H.H. Chen, G.Z. Sun, and Y.C. Lee, 1992. "Learning and extracting finite state automata with second-order recurrent neural networks," *Neural Computation,* vol. 4, pp. 393–405.

Giles, C.L., G.Z. Sun, H.H. Chen, Y.C. Lee, and D. Chen, 1990. "Higher order recurrent networks and grammatical inference," *Advances in Neural Information Processing Systems,* vol. 2, pp. 380–387, San Mateo, CA: Morgan Kaufmann.

Gill, P., S. Hammarling, W. Murray, M. Saunders, and M. Wright, 1986. "User's guide for LSSOL," Technical Report 86–1, Systems Optimization Laboratory, Stanford University, Stanford, CA.

Gill, P., and W. Murray, 1991. "Inertia-controlling methods for general quadratic programming," *SIAM Review,* vol. 33, pp. 1–36.

Girosi, F., and G. Anzellotti, 1992. "Rates of convergence of approximation by translates," *A.I. Memo 1288,* Artificial Intelligence Laboratory, MIT Cambridge, MA:

Girosi, F., M. Jones, and T. Poggio, 1995. "Regularization theory and neural networks architectures," *Neural Computation,* vol. 7, pp. 219–269.

Girosi, F., and T. Poggio, 1990. "Networks and the best approximation property," *Biological Cybernetics,* vol. 63, pp. 169–176.

Glauber, R.J., 1963. "Time-dependent statistics of the Ising model," *Journal of Mathematical Physics,* vol. 4, pp. 294–307.

Goggin, S.D.D., K.M. Johnson, and K. Gustafson, 1989. "Primary and recency effects due to momentum in back-propagation learning," *OCS Technical Report 89–25,* Boulder, CO.: University of Colorado.

Golden, R.M., 1996. *Mathematical Methods for Neural Network Analysis and Design,* Cambridge, MA: MIT Press.

Golden, R.M., 1986. "The 'Brain-State-in-a-Box' neural model is a gradient descent algorithm," *Journal of Mathematical Psychology,* vol. 30, pp. 73–80.

Goles, E., and S. Martinez, 1990. *Neural and Automata Networks,* Dondrecht, The Netherlands: Kluwer.

Golub, G.H., and C.G. Van Loan, 1996. *Matrix Computations,* 3rd edition, Baltimore: Johns Hopkins University Press.

Goodman, R.M., C.M. Higgins, J.W. Miller, and P. Smyth, 1992. "Rule-based neural networks for classification and probability estimation," *Neural Computation,* vol. 4, pp. 781–804.

Gori, M., and A. Tesi, 1992. "On the problem of local minima in backpropagation," *IEEE Transactions on Pattern Analysis and Machine Intelligence,* vol. 14, pp. 76–86.

Gorin, A., 1992. "Network structure, generalization and adaptive language acquisition," *Proceedings of the Seventh Yale Workshop on Adaptive and Learning Systems,* pp. 155–160, Yale University, New Haven, CT.

Goudreau, M.W., and C.L. Giles, 1995. "Using recurrent neural networks to learn the structure of interconnection networks," *Neural Networks,* vol. 8, pp. 793–804.

Goudreau, M.W., C.L. Giles, S.T. Chakradhar, and D. Chen, 1994. "First-order vs. second-order single-layer recurrent neural networks," *IEEE Transactions on Neural Networks,* vol. 5, pp. 511–513.

Granger, R., J. Whitson, J. Larson, and G. Lynch, 1994. "Non-Hebbian properties of LTP enable high-capacity encoding of temporal sequences," *Proceedings of the National Academy of Sciences of the U.S.A.,* to appear.

Grassberger, I., and I. Procaccia, 1983. "Measuring the strangeness of strange attractors," *Physica D,* vol. 9, pp. 189–208.

Graubard, S.R., ed., 1988. *The Artificial Intelligence Debate: False Starts, Real Foundations,* Cambridge, MA: MIT Press.

Gray, R.M., 1990. *Entropy and Information Theory,* New York: Springer-Verlag.

Gray, R.M., 1988. *Probability, Random Processes, and Ergodic Properties,* New York: Springer-Verlag.

Gray, R.M., 1984. "Vector quantization," *IEEE ASSP Magazine,* vol. 1, pp. 4–29.

Gray, R.M., and L.D. Davisson, 1986. *Random Processes: A Mathematical Approach for Engineers,* Englewood Cliffs, NJ: Prentice-Hall.

Green, M., and D.J.N. Limebeer, 1995. *Linear Robust Control,* Englewood Cliffs, NJ: Prentice-Hall.

Greenberg, H.J., 1988. "Equilibria of the brain-state-in-a-box (BSB) neural model," *Neural Networks,* vol. 1, pp. 323–324.

Gregory, R.L., 1970. *The Intelligent Eye,* Wiedefeld and Nicholson, London.

Grenander, U., 1983. *Tutorial in Pattern Theory,* Brown University, Providence, R.I.

Grewal, M.S., and A.P. Andrews, 1993. *Kalman Filtering: Theory and Practice,* Englewood Cliffs, NJ: Prentice-Hall.

Griffiths, L.J., and C.W. Jim, 1982. "An alternative approach to linearly constrained optimum beamforming," *IEEE Transactions on Antennas and Propagation,* vol. AP-30, pp. 27–34.

Grossberg, S., 1990. "Content-addressable memory storage by neural networks: A general model and global Liapunov method," In *Computational Neuroscience,* E.L. Schwartz, ed., pp. 56–65, Cambridge, MA: MIT Press.

Grossberg, S., 1988a. "Competitive learning: From interactive activation to adaptive resonance," in *Neural Networks and Natural Intelligence,* S. Grossberg, ed., Cambridge, MA: MIT Press.

Grossberg, S., 1988b. *Neural Networks and Natural Intelligence,* Cambridge, MA: MIT Press.

Grossberg, S., 1988c. "Nonlinear neural networks: Principles, mechanisms, and architectures," *Neural Networks,* vol. 1, pp. 17–61.

Grossberg, S., 1982. *Studies of Mind and Brain,* Boston: Reidel.

Grossberg, S., 1980. "How does a brain build a cognitive code?" *Psychological Review,* vol. 87, pp. 1–51.

Grossberg, S., 1978a. "Decision, patterns, and oscillations in the dynamics of competitive systems with application to Volterra-Lotka systems," *J. Theoretical Biology,* vol. 73, pp. 101–130.

Grossberg, S., 1978b. "Competition, decision, and consensus," *J. Mathematical Analysis and Applications,* vol. 66, pp. 470–493.

Grossberg, S., 1977. "Pattern formation by the global limits of a nonlinear competitive interaction in n dimensions," *J. Mathematical Biology,* vol. 4, pp. 237–256.

Grossberg, S., 1976a. "Adaptive pattern classification and universal recoding: I. Parallel development and coding of neural detectors," *Biological Cybernetics,* vol. 23, pp. 121–134;

Grossberg, S., 1976b. "Adaptive pattern classification and universal recoding: II. Feedback, expectation, olfaction, illusions," *Biological Cybernetics,* vol. 23, pp. 187–202.

Grossberg, S., 1972. "Neural expectation: Cerebellar and retinal analogs of cells fired by learnable or unlearned pattern classes," *Kybernetik,* vol. 10, pp. 49–57.

Grossberg, S., 1969a. "A prediction theory for some nonlinear functional-difference equations," *Journal of Mathematical Analysis and Applications,* vol. 22, pp. 490–522.

Grossberg, S., 1969b. "On learning and energy-entropy dependence in recurrent and nonrecurrent signed networks," *Journal of Statistical Physics,* vol. 1, pp. 319–350.

Grossberg, S., 1968. "A prediction theory for some nonlinear functional-difference equations," *Journal of Mathematical Analysis and Applications,* vol. 21, pp. 643–694, vol. 22, pp. 490–522.

Grossberg, S., 1967. "Nonlinear difference—differential equations in prediction and learning theory," *Proceedings of the National Academy of Sciences,* USA, vol. 58, pp. 1329–1334.

Gupta, M.M., and N.K. Sinha, eds. 1996. *Intelligent Control Systems: Theory and Applications,* New York: IEEE Press.

Guyon, I., 1990. *Neural Networks and Applications,* Computer Physics Reports, Amsterdam: Elsevier.

Haffner, P., 1994. "A new probabilistic framework for connectionist time alignment," *Proceedings of ICSLP 94,* pp. 1559–1562, Yokohama, Japan.

Haffner, P., M. Franzini, and A. Waibel, 1991. "Integrating time alignment and neural networks for high performance continuous speech recognition," *Proceedings of IEEE ICASSP 91,* pp. 105–108.

Haft, M., and J.L. van Hemmen, 1998. "Theory and implementations of infomax filters for the retina," *Network: Computations in Neural Systems,* vol. 9, pp. 39–71.

Hagiwara, M., 1992. "Theoretical derivation of momentum term in back-propagation," *International Joint Conference on Neural Networks,* vol. I, pp. 682–686, Baltimore.

Hajek, B., 1985. "A tutorial survey of theory and applications of simulated annealing," *Proceedings of the 24th Conference on Decision and Control,* IEEE Press, pp. 755–760, Ft. Lauderdale, Fla.

Hajek, B., 1988. "Cooling schedules for optimal annealing," *Mathematics of Operations Research,* vol. 13, pp. 311–329.

Hammerstrom, D., 1993a. "Neural networks at work," *IEEE Spectrum,* vol. 30, no. 6, pp. 26–32.

Hammerstrom, D., 1993b. "Working with neural networks," *IEEE Spectrum,* vol. 30, no. 7, pp. 46–53.

Hammerstrom, D., and S. Rahfuss, 1992. "Neurocomputing hardware: Present and future," *Swedish National Conference on Connectionism,* Skovade, Sweden, September.

Hampshire, J.B., and B. Pearlmutter, 1990. "Equivalence proofs for multilayer perceptron classifiers and Bayesian discriminant function," *Proceedings of the 1990 Connectionist Models Summer School,* pp. 159–172, San Mateo, CA: Morgan Kaufmann.

Hampson, S.E., 1990. *Connectionistic Problem Solving: Computational Aspects of Biological Learning,* Berlin: Birkhäuser.

Hancock, P.J.B., R.J. Baddeley, and L.S. Smith, 1992. "The principal components of natural images," *Network,* vol. 3, pp. 61–70.

Hanson, L.K., and P. Solamon, 1990. "Neural network ensembles," *IEEE Transactions on Pattern Analysis and Machine Intelligence,* vol. PAMI-12, pp. 993–1002.

Härdle, W., 1990. *Applied Nonparametric Regression,* Cambridge: Cambridge University Press.

Hardy, R.L., 1971. "Multiquadric equations of topography and other irregular surfaces," *Journal of Geophysics Research,* vol. 76, pp. 1905–1915.

Harel, D., 1987. "*Algorithmics: The Spirit of Computing,*" Reading, MA: Addison-Wesley.

Hartline, H.K., 1940. "The receptive fields of optic nerve fibers," *American Journal of Physiology,* vol. 130, pp. 690–699.

Hartman, E., 1991. "A high storage capacity neural network content-addressable memory," *Network,* vol. 2, pp. 315–334.

Hartman, E.J., J.D. Keeler, and J.M. Kowalski, 1990. "Layered neural networks with Gaussian hidden units as universal approximators," *Neural Computation,* vol. 2, pp. 210–215.

Hashem, S., 1997. "Optimal linear combinations of neural networks." *Neural Networks,* vol. 10, pp. 599–614.

Hassibi, B., A.H. Sayed, and T. Kailath, 1998. *Indefinite Quadratic Estimation and Control: A Unified Approach to H_2 and H_∞ Theories,* SIAM.

Hassibi, B., A.H. Sayed, and T. Kailath, 1996. "The H^∞ optimality of the LMS algorithm," *IEEE Transactions on Signal Processing,* vol. 44, pp. 267–280.

Hassibi, B., A.H. Sayed, and T. Kailath, 1993. "LMS is H^∞ optimal," *Proceedings of the IEEE Conference on Decision and Control,* pp. 74–79, San Antonio, Texas.

Hassibi, B., D.G. Stork, and G.J. Wolff, 1992. "Optimal brain surgeon and general network pruning," *IEEE International Conference on Neural Networks,* vol. 1, pp. 293–299, San Francisco.

Hassibi, B., and T. Kailath, 1995. "H^∞ optimal training algorithms and their relation to back propagation," *Advances in Neural Information Proccessing Systems,* vol. 7, pp. 191–198.

Hastie, T., and W. Stuetzle, 1989. "Principal curves," *Journal of the American Statistical Association,* vol. 84, pp. 502–516.

Hastings, W.K., 1970. "Monte Carlo sampling methods using Markov chains and their applications," *Biometrika,* vol. 87, pp. 97–109.

Haussler, D., 1988. "Quantifying inductive bias: AI learning algorithms and Valiant's learning framework," *Artificial Intelligence,* vol. 36, pp. 177–221.

Hawkins, R.D., and G.H. Bower, eds., 1989. *Computational Models of Learning in Simple Neural Systems,* San Diego, CA: Academic Press.

Haykin, S., 1996a. *Adaptive Filter Theory,* 3rd edition, Englewood Cliffs, NJ: Prentice-Hall.

Haykin, S., 1996b. "Neural networks expand SP's horizons," *IEEE Signal Processing Magazine,* vol. 13, no. 2, pp. 24–29.

Haykin, S., ed. 1994a. *Blind Deconvolution,* Englewood Cliffs, NJ: Prentice-Hall.

Haykin, S., 1994b. *Communication Systems,* 3rd edition, New York: John Wiley.

Haykin, S., 1992. "Blind equalization formulated as a self-organized learning process," *Proceedings of the Twenty-Sixth Asilomar Conference on Signals, Systems, and Computers,* pp. 346–350, Pacific Grove, CA.

Haykin, S., and C. Deng, 1991. "Classification of radar clutter using neural networks," *IEEE Transactions on Neural Networks,* vol. 2, pp. 589–600.

Haykin, S., and B. Kosko, eds. 1998. Special Issue of *Proceedings of the IEEE on Intelligent Signal Processing,* vol. 88, to appear.

Haykin, S., and J. Principe, 1998. "Making sense of a complex world: Using neural networks to dynamically model chaotic events such as sea clutter," *IEEE Signal Processing Magazine,* vol. 15, to appear.

Haykin, S., W. Stehwien, P. Weber, C. Deng, and R. Mann, 1991. "Classification of radar clutter in air traffic control environment," *Proceedings of the IEEE,* vol. 79, pp. 741–772.

Haykin, S., P. Yee, and E. Derbez, 1997. "Optimum nonlinear filtering," *IEEE Transactions on Signal Processing,* vol. 45, pp. 2774–2786.

Haykin, S., and B. Van Veen, 1998. *Signals and Systems,* New York: Wiley.

Hebb, D.O., 1949. *The Organization of Behavior: A Neuropsychological Theory,* New York: Wiley.

Hecht-Nielsen, R., 1995. "Replicator neural networks for universal optimal source coding," *Science,* vol. 269, pp. 1860–1863.

Hecht-Nielson, R., 1990. *Neurocomputing,* Reading, MA: Addison-Wesley.

Hecht-Nielson, R., 1987. "Kolmogorov's mapping neural network existence theorem," *First IEEE International Conference on Neural Networks,* vol. III, pp. 11–14, San Diego, CA.

Helstrom, C.W., 1968. *Statistical Theory of Signal Detection,* 2nd edition, Pergamon Press.

Herault, J., and C. Jutten, 1994. *Reseaux Neuronaux et Traitement du Signal,* Paris: Hermes Publishers.

Herault, J., and C. Jutten, 1986. "Space or time adaptive signal processing by neural network models," in J.S. Denker, ed., *Neural Networks for Computing.* Proceedings of the AIP Conference, American Institute of Physics, New York, pp. 206–211.

Herault, J., C. Jutten, and B. Ans, 1985. "Detection de grandeurs primitives dans un message composite par une architecture de calcul neuromimetique un apprentissage non supervise." *Procedures of GRETSI,* Nice, France.

Hertz, J., A. Krogh, and R.G. Palmer, 1991. *Introduction to the Theory of Neural Computation,* Reading, MA: Addison-Wesley.

Hestenes, M.R., and E. Stiefel, 1952. "Methods of conjugate gradients for solving linear systems," *Journal of Research of the National Bureau of Standards,* vol. 49, pp. 409–436.

Hetherington, P.A., and M.L. Shapiro, 1993. "Simulating Hebb cell assemblies: The necessity for partitioned dendritic trees and a post-not-pre LTD rule," *Network,* vol. 4, pp. 135–153.

Hiller, F.S., and G.J. Lieberman, 1995. *Introduction to Operations Research,* 6th edition, New York: McGraw-Hill.

Hinton, G.E., 1989. "Connectionist learning procedures," *Artificial Intelligence,* vol. 40, pp. 185–234.

Hinton, G.E., 1989. "Deterministic Boltzmann machine learning performs steepest descent in weight-space," *Neural Computation,* vol. 1, pp. 143–150.

Hinton, G.E., 1981. "Shape representation in parallel systems," *Proceedings of the 7th International Joint Conference on Artificial Intelligence,* Vancouver, British Columbia.

Hinton, G.E., P. Dayan, B.J. Frey, and R.M. Neal, 1995. "The 'wake-sleep' algorithm for unsupervised neural networks," *Science,* vol. 268, pp. 1158–1161.

Hinton, G.E., and Z. Ghahramani, 1997. "Generative models for discovering sparse distributed representations," *Philosophical Transactions of the Royal Society, Series B,* vol. 352, pp. 1177–1190.

Hinton, G.E., and S.J. Nowlan, 1990. "The bootstrap Widrow-Hoff rule as a cluster-formation algorithm," *Neural Computation,* vol. 2, pp. 355–362.

Hinton, G.E., and S.J. Nowlan, 1987. "How learning can guide evolution," *Complex Systems,* vol. 1, pp. 495–502.

Hinton, G.E., and T.J. Sejnowski, 1986. "Learning and relearning in Boltzmann machines," in *Parallel Distributed Processing: Explorations in Microstructure of Cognition,* D.E. Rumelhart and J.L. McClelland, eds., Cambridge, MA: MIT Press.

Hinton, G.E., and T.J. Sejnowski, 1983. "Optimal perceptual inference," *Proceedings of IEEE Computer Society Conference on Computer Vision and Pattern Recognition,* pp. 448–453, Washington, DC.

Hirsch, M.W., 1989. "Convergent activation dynamics in continuous time networks," *Neural Networks,* vol. 2, pp. 331–349.

Hirsch, M.W., 1987. "Convergence in neural nets," *First IEEE International Conference on Neural Networks,* vol. II, pp. 115–125, San Diego, CA.

Hirsch, M.W., and S. Smale, 1974. *Differential Equations, Dynamical Systems, and Linear Algebra,* New York: Academic Press.

Hochreiter, S., 1991. *Untersuchungen zu dynamischen neuronalen Netzen,* Diploma Thesis, Technische Universität München, Germany.

Hochreiter, S., and J. Schmidhuber, 1997. "LSTM can solve hard long time lag problems," *Advances in Neural Information Processing Systems,* vol. 9, pp. 473–479. Cambridge, MA: MIT Press.

Hodgkin, A.L., and A.F. Huxley, 1952. "A quantitative description of membrane current and its application to conduction and excitation in nerve," *Journal of Physiology,* vol. 117, pp. 500–544.

Holden, S.B., and M. Niranjan, 1995. "On the practical applicability of Vapnik-Chervonenkis dimension bounds," *Neural Computation,* vol. 7, pp. 1265–1288.

Holland, J.H., 1992. *Adaptation in Natural and Artificial Systems,* Cambridge, MA: MIT Press.

Hopcroft, J., and U. Ullman, 1979. *Introduction to Automata Theory, Languages and Computation,* Reading, MA: Addison-Wesley.

Hopfield, J.J., 1995. "Pattern recognition computation using action potential timing for stimulus representation," *Nature,* vol. 376, pp. 33–36.

Hopfield, J.J., 1994. "Neurons, dynamics and computation," *Physics Today,* vol. 47, pp. 40–46, February.

Hopfield, J.J., 1987a. "Networks, Computations, Logic, and Noise," *IEEE International Conference on Neural Networks,* vol. I, pp. 107–141, San Diego, CA.

Hopfield, J.J., 1987b. "Learning algorithms and probability distributions in feed-forward and feed-back networks," *Proceedings of the National Academy of Sciences, USA,* vol. 84, pp. 8429–8433.

Hopfield, J.J., 1984. "Neurons with graded response have collective computational properties like those of two-state neurons," *Proceedings of the National Academy of Sciences, USA,* vol. 81, pp. 3088–3092.

Hopfield, J.J., 1982. "Neural networks and physical systems with emergent collective computational abilities," *Proceedings of the National Academy of Sciences, USA,* vol. 79, pp. 2554–2558.

Hopfield, J.J., and D.W. Tank, 1986. "Computing with neural circuits: A model," *Science,* vol. 233, pp. 625–633.

Hopfield, J.J., and T.W. Tank, 1985. "'Neural' computation of decisions in optimization problems," *Biological Cybernetics,* vol. 52, pp. 141–152.

Hopfield, J.J., D.I. Feinstein, and R.G. Palmer, 1983. "'Unlearning' has a stabilizing effect in collective memories," *Nature,* vol. 304, pp. 158–159.

Horn, B.K.P., 1977. "Understanding image intensities," *Artificial Intelligence,* vol. 8, pp. 201–237.

Hornik, K., M. Stinchcombe, and H. White, 1990. "Universal approximation of an unknown mapping and its derivatives using multilayer feedforward networks," *Neural Networks,* vol. 3, pp. 551–560.

Hornik, K., M. Stinchcombe, and H. White, 1989. "Multilayer feedforward networks are universal approximators," *Neural Networks,* vol. 2, pp. 359–366.

Hotteling, H., 1933. "Analysis of a complex of statistical variables into principal components," *Journal of Educational Psychology,* vol. 24, pp. 417–441, 498–520.

Hubel, D.H., 1988. *Eye, Brain, and Vision,* New York: Scientific American Library.

Hubel, D.H., and T.N. Wiesel, 1977. "Functional architecture of macaque visual cortex," *Proceedings of the Royal Society,* B, vol. 198, pp. 1–59, London.

Hubel, D.H., and T.N. Wiesel, 1962. "Receptive fields, binocular interaction and functional architecture in the cat's visual cortex," *Journal of Physiology,* vol. 160, pp. 106–154, London.

Huber, P.J. 1985. "Projection pursuit," *Annals of Statistics,* vol. 13, pp. 435–475.

Huber, P.J., 1981. *Robust Statistics,* New York: Wiley.

Huber, P.J., 1964. "Robust estimation of a location parameter," *Annals of Mathematical Statistics,* vol. 35, pp. 73–101.

Hush, D.R, 1997. "Learning from examples: From theory to practice," Tutorial #4, *1997 International Conference on Neural Networks,* Houston, June.

Hush, D.R., and B.G. Horne, 1993. "Progress in supervised neural networks: What's new since Lippmann?" *IEEE Signal Processing Magazine,* vol. 10, pp. 8–39.

Hush, D.R., and J.M. Salas, 1988. "Improving the learning rate of back-propagation with the gradient reuse algorithm," *IEEE International Conference on Neural Networks,* vol. I, pp. 441–447, San Diego, CA.

Illingsworth, V., E.L. Glaser, and I.C. Pyle, 1989. *Dictionary of Computing,* New York: Oxford University Press.

Intrator, N., 1992. "Feature extraction using an unsupervised neural network," *Neural Computation,* vol. 4, pp. 98–107.

Jaakkola, T., and M.I. Jordan, 1996. "Computing upper and lower bounds on likelihoods in intractable networks," in E. Horwitz, ed., *Workshop on Uncertainty in Artificial Intelligence,* Portland, Or.

Jackson, E.A., 1989. *Perspectives of Nonlinear Dynamics,* vol. 1, Cambridge: Cambridge University Press.

Jackson, E.A., 1990. *Perspectives of Nonlinear Dynamics,* vol. 2, Cambridge: Cambridge University Press.

Jackson, I.R.H., 1989, "An order of convergence for some radial basis functions," *IMA Journal of Numerical Analysis,* vol. 9, pp. 567–587.

Jackson, J.D., 1975. *Classical Electrodynamics,* 2nd edition, New York: Wiley.

Jacobs, R.A., 1990. "Task Decomposition Through Computation in a Modular Connectionist Architecture," Ph.D. Thesis, University of Massachusetts.

Jacobs, R.A., 1988. "Increased rates of convergence through learning rate adaptation," *Neural Networks,* vol. 1, pp. 295–307.

Jacobs, R.A., and M.I. Jordan, 1993. "Learning piecewise control strategies in a modular neural network architecture," *IEEE Transactions on Systems, Man, and Cybernetics,* vol. 23, pp. 337–345.

Jacobs, R.A., and M.I. Jordan, 1991. "A competitive modular connectionist architecture," *Advances in Neural Information Processing Systems,* vol. 3, pp. 767–773, San Mateo, CA: Morgan Kaufmann.

Jacobs, R.A., M.I. Jordan, S.J. Nowlan, and G.E. Hinton, 1991a. "Adaptive mixtures of local experts," *Neural Computation,* vol. 3, pp. 79–87.

Jacobs, R.A., M.I. Jordan, and A.G. Barto, 1991b. "Task decomposition through competition in a modular connectionist architecture: The what and where vision tasks," *Cognitive Science,* vol. 15, pp. 219–250.

Jayant, N.S., and P. Noll, 1984. *Digital Coding of Waveforms,* Englewood Cliffs, NJ: Prentice-Hall.

Jaynes, E.T., 1982. "On the rationale of maximum-entropy methods," *Proceedings of the IEEE,* vol. 70, pp. 939–952.

Jaynes, E.T., 1957. "Information theory and statistical mechanics," *Physical Review,* vol. 106, pp. 620–630; "Information theory and statistical mechanic II," *Physical Review,* vol. 108, pp. 171–190.

Jazwinski, A.H., 1970. *Stochastic Processes and Filtering Theory,* New York: Academic Press.

Jelinek, F., 1997. *Statistical Methods for Speech Recognition,* Cambridge, MA: MIT Press.

Johansson, E.M., F.U. Dowla, and D.M. Goodman, 1990. "Back-propagation learning for multi-layer feed-forward neural networks using the conjugate gradient method," Report UCRL-JC-104850, Lawrence Livermore National Laboratory, CA.

Johnson, D.S., C.R. Aragon, L.A. McGeoch, and C. Schevon, 1989. "Optimization by simulated annealing: An experimental evaluation," *Operations Research,* vol. 37, pp. 865–892.

Jolliffe, I.T., 1986. *Principal Component Analysis,* New York: Springer-Verlag.

Jones, J.P., and L.A. Palmer, 1987a. "The two-dimensional spatial structure of simple receptive fields in cat striate cortex," *Journal of Neurophysiology,* vol. 58, pp. 1187–1211.

Jones, J.P., and L.A. Palmer, 1987b. "An evaluation of the two-dimensional Gabor filter model of simple receptive fields in cat striate cortex," *Journal of Neurophysiology,* vol. 58, pp. 1233–1258.

Jones, J.P., A. Steponski, and L.A. Palmer, 1987. "The two-dimensional spectral structure of simple receptive fields in cat striate cortex," *Journal of Neurophysiology,* vol. 58, pp. 1212–1232.

Jordan, M.I., 1994. "A statistical approach to decision tree modeling," *Proceedings of the Seventh Annual ACM Conference on Computational Learning Theory,* New York: ACM Press.

Jordan, M.I., 1986. "Attractor dynamics and parallelism in a connectionist sequential machine," *The Eighth Annual Conference of the Cognitive Science Society,* pp. 531–546, Amherst, MA.

Jordan, M.I., ed., 1998. *Learning in Graphical Models,* Boston: Kluwer.

Jordan, M.I., Z. Ghahramani, T.S. Jakkolla, and L.K. Saul, 1998. "An introduction to variational methods for graphical models," In M.I. Jordan, ed., *Learning in Graphical Models,* Boston: Kluwer.

Jordan, M.I., and R.A. Jacobs, 1995. "Modular and Hierarchical Learning Systems," in M.A. Arbib, ed., *The Handbook of Brain Theory and Neural Networks,* pp. 579–583, Cambridge, MA: MIT Press.

Jordan, M.I., and R.A. Jacobs, 1994. "Hierarchical mixtures of experts and the EM algorithm," *Neural Computation,* vol. 6, pp. 181–214.

Jordan, M.I., and R.A. Jacobs, 1992. "Hierarchies of adaptive experts," *Advances in Neural Information Processing Systems,* vol. 4, pp. 985–992, San Mateo, CA: Morgan Kaufmann.

Joseph, R.D., 1960. "The number of orthants in n-space intersected by an s-dimensional subspace," Technical Memo 8, Project PARA, Cornell Aeronautical Lab., Buffalo, N.Y.

Jutten, C., and J. Herault, 1991. "Blind separation of sources, Part I: An adaptive algorithm based on neuromimetic architecture," *Signal Processing,* vol. 24, pp. 1–10.

Kaas, J.H., M.M. Merzenich, and H.P. Killackey, 1983. "The reorganization of somatosensory cortex following peripheral nerve damage in adult and developing mammals," *Annual Review of Neurosciences,* vol. 6, pp. 325–356.

Kailath, T., 1980. *Linear Systems,* Englewood Cliffs, NJ: Prentice-Hall.

Kailath, T., 1974. "A view of three decades of linear filtering theory," *IEEE Transactions of Information Theory,* vol. IT-20, pp. 146–181.

Kailath, T., 1971. "RKHS approach to detection and estimation problems—Part I: Deterministic signals in Gaussian noise," *IEEE Transactions of Information Theory,* vol. IT-17, pp. 530–549.

Kailath, T., 1968. "An innovations approach to least-squares estimation: Part 1. Linear filtering in additive white noise," *IEEE Transactions of Automatic Control,* vol. AC-13, pp. 646–655.

Kalman, R.E., 1960. "A new approach to linear filtering and prediction problems," *Transactions of the ASME, Journal of Basic Engineering,* vol. 82, pp. 35–45.

Kandel, E.R., and J.H. Schwartz, 1991. *Principles of Neural Science,* 3rd ed., New York: Elsevier.

Kangas, J., T. Kohonen, and J. Laaksonen, 1990. "Variants of self-organizing maps," *IEEE Transactions on Neural Networks* 1, 93–99.

Kanter, I., and H. Sompolinsky, 1987. "Associative recall of memory without errors," *Physical Review A,* vol. 35, pp. 380–392.

Kaplen, J.L., and J.A. Yorke, 1979. "Chaotic behavior of multidimensional difference equations," in H.-O Peitgen and H.-O Walker, eds., *Functional Differential Equations and Approximations of Fixed Points,* pp. 204–227, Berlin: Springer.

Kappen, H.J., and F.B. Rodriguez, 1998. "Efficient learning in Boltzmann machines using linear response theory," *Neural Computation,* vol. 10, to appear.

Karhunen, K., 1947. "Uber lineare methoden in der Wahrscheinlichkeitsrechnung," *Annales Academiae Scientiarum Fennicae, Series AI: Mathematica-Physica,* vol. 37, pp. 3–79, (Transl.: RAND Corp., Santa Monica, CA, Rep. T-131, Aug. 1960).

Karhunen, J., and J. Joutsensalo, 1995. "Generalizations of principal component analysis, optimization problems, and neural networks," *Neural Networks,* vol. 8, pp. 549–562.

Karpinski, M., and A. Macintyre, 1997. "Polynomial bounds for VC dimension of sigmoidal and general Pfaffian neuronal networks," *Journal of Computer and System Sciences,* vol. 54, pp. 169–176.

Katagiri, S., and E. McDermott, 1996. "Discriminative training—Recent progress in speech recognition," in C.H. Chen, L.F. Pau, and P.S.P. Wang, eds., *Handbook of Pattern Recognition and Computer Vision,* 2nd edition, Singapore: World Scientific Publishing.

Katz, B., 1966. *Nerve, Muscle and Synapse,* New York: McGraw-Hill.

Kawamoto, A.H., and J.A. Anderson, 1985. "A neural network model of multistable perception," *Acta Psychologica,* vol. 59, pp. 35–65.

Kawato, M., H. Hayakama, and T. Inui, 1993. "A forward-inverse optics model of reciprocal connections between visual cortical areas," *Network,* vol. 4, pp. 415–422.

Kay, J., 1992. "Feature discovery under contextual supervision using mutual information," *International Joint Conference on Neural Networks,* vol. IV, pp. 79–84, Baltimore.

Kearns, M., 1996. "A bound on the error of cross validation using the approximation and estimation rates, with consequences for the training-test split," *Advances in Neural Information Processing Systems,* vol. 8, pp. 183–189, Cambridge, MA: MIT Press.

Kearns, M., and L.G. Valiant, 1989. "Cryptographic limitations on learning Boolean formulae and finite automata," *Proceedings of the Twenty-First Annual ACM Symposium on Theory of Computing,* pp. 433–444, New York.

Kearns, M.J., and U.V. Vazirani, 1994. *An Introduction to Computational Learning Theory,* Cambridge, MA: MIT Press.

Kechriotis, G., E. Zervas, and E.S. Manolakos, 1994. "Using recurrent neural networks for adaptive communication channel equalization," *IEEE Transactions on Neural Networks,* vol. 5, pp. 267–278.

Keeler, J.D., 1986. "Basins of attraction of neural network models," in *Neural Networks for Computing,* J.S. Denker, ed., pp. 259–264, New York: American Institute of Physics.

Keller, J.B., 1976. "Inverse problems," *American Mathematical Monthly,* vol. 83, pp. 107–118.

Kelso, S.R., A.H. Ganong, and T.H. Brown, 1986. "Hebbian synapses in hippocampus," *Proceedings of the National Academy of Sciences, USA,* vol. 83, pp. 5326–5330.

Kennel, M.B., R. Brown, and H.D.I. Abarbanel, 1992. "Determining minimum embedding dimension using a geometrical construction," *Physical Review A,* vol. 45, pp. 3403–3411.

Kerlirzin, P., and F. Vallet, 1993. "Robustness in multilayer perceptrons," *Neural Computation,* vol. 5, pp. 473–482.

Kirkpatrick, S., 1984. "Optimization by simulated annealing: Quantitative Studies," *Journal of Statistical Physics,* vol. 34, pp. 975–986.

Kirkpatrick, S., and D. Sherrington, 1978. "Infinite-ranged models of spin-glasses," *Physical Review,* Series B, vol. 17, pp. 4384–4403.

Kirkpatrick, S., C.D. Gelatt, Jr., and M.P. Vecchi, 1983. "Optimization by simulated annealing," *Science,* vol. 220, pp. 671–680.

Kirsch, A., 1996. *An Introduction to the Mathematical Theory of Inverse Problems,* New York: Springer-Verlag.

Kleene, S.C., 1956. "Representation of events in nerve nets and finite automata," in C.E. Shannon and J. McCarthy, eds., *Automata Studies,* Princeton, NJ: Princeton University Press.

Kmenta, J., 1971. *Elements of Econometrics,* New York: Macmillan.

Knudsen, E.I., S. duLac, and S.D. Esterly, 1987. "Computational maps in the brain," *Annual Review of Neuroscience,* vol. 10, pp. 41–65.

Koch, C., and I. Segev, eds., 1989. *Methods in Neuronal Modeling: From Synapses to Networks,* Cambridge, MA: MIT Press.

Koch, C., T. Poggio, and V. Torre, 1983. "Nonlinear interactions in a dendritic tree: Localization, timing, and role in information processing," *Proceedings of the National Academy of Sciences, USA,* vol. 80, pp. 2799–2802.

Koch, C., and B. Mathur, 1996. "Neuromorphic vision chips," *IEEE Spectrum,* vol. 33, no. 5, pp. 38–46.

Kohonen, T., 1997a. "Exploration of very large databases by self-organizing maps," *1997 International Conference on Neural Networks,* vol. I, pp. PL1–PL6, Houston.

Kohonen, T., 1997b. *Self-Organizing Maps,* 2nd edition, Berlin: Springer-Verlag.

Kohonen, T., 1996. "Emergence of invariant–feature detectors in the adaptive-subspace self-organizing maps," *Biological Cybernetics,* vol. 75, pp. 281–291.

Kohonen, T., 1993. "Physiological interpretation of the self-organizing map algorithm." *Neural Networks,* vol. 6, pp. 895–905.

Kohonen, T., 1993. "Things you haven't heard about the self-organizing map," *Proceedings of the IEEE International Conference on neural networks,* pp. 1147–1156, San Francisco.

Kohonen, T., 1990a. "The self-organizing map," *Proceedings of the Institute of Electrical and Electronics Engineers,* vol. 78, pp. 1464–1480.

Kohonen, T., 1990b. "Improved versions of learning vector quantization," *IEEE International Joint Conference on Neural Networks,* vol. I, pp. 545–550, San Diego, CA.

Kohonen, T., 1988a. "An introduction to neural computing," *Neural Networks,* vol. 1, pp. 3–16.

Kohonen, T., 1988b. *Self-Organization and Associative Memory,* 3rd edition, New York: Springer-Verlag.

Kohonen, T., 1988c. "The 'neural' phonetic typewriter." *Computer,* vol. 21, pp. 11–22.

Kohonen, T., 1986. "Learning vector quantization for pattern recognition," *Technical Report TKK-F-A601,* Helsinki University of Technology, Finland.

Kohonen, T., 1982. "Self-organized formation of topologically correct feature maps," *Biological Cybernetics,* vol. 43, pp. 59–69.

Kohonen, T., 1972. "Correlation matrix memories," *IEEE Transactions on Computers,* vol. C-21, pp. 353–359.

Kohonen, T., and E. Oja, 1976. "Fast adaptive formation of orthogonalizing filters and associative memory in recurrent networks for neuron-like elements," *Biological Cybernetics,* vol. 21, pp. 85–95.

Kohonen, T., E. Oja, O. Simula, A. Visa, and J. Kangas, 1996. "Engineering applications of the self-organizing map," *Proceedings of the IEEE,* vol. 84, pp. 1358–1384.

Kohonen, T., E. Reuhkala, K. Mäkisara, and L. Vainio, 1976. "Associative recall of images," *Biological Cybernetics,* vol. 22, pp. 159–168.

Kohonen, T., G. Barna, and R. Chrisley, 1988. "Statistical pattern recognition with neural networks: Benchmarking studies," *IEEE International Conference on Neural Networks,* vol. I, pp. 61–68, San Diego, CA.

Kohonen, T., J. Kangas, J. Laaksonen, and K. Torkkola, 1992. "LVQ-PAK: The learning vector quantization Program Package," Helsinki University of Technology, Finland.

Koiran, P., and E.D. Sontag, 1996. "Neural networks with quadratic VC dimension," *Advances in Neural Information Processing Systems,* vol. 8, pp. 197–203, Cambridge, MA: MIT Press.

Kolen, J.F., and J.B. Pollack, 1990. "Backpropagation is sensitive to initial conditions," *Complex Systems,* vol. 4, pp. 269–280.

Kollias, S., and D. Anastassiou, 1989. "An adaptive least squares algorithm for the efficient training of artificial neural networks," *IEEE Transactions on Circuits and Systems,* vol. 36, pp. 1092–1101.

Kollias, S., and D. Anastassiou, 1988. "Adaptive training of multilayer neural networks using a least squares estimation technique," *IEEE International Conference on Neural Networks,* vol. 1, pp. 383–390, San Diego.

Kolmogorov, A.N., 1942. "Interpolation and extrapolation of stationary random sequences," translated by the Rand Corporation, Santa Monica, CA., April 1962.

Kosko, B., 1997. *Fuzzy Engineering,* Upper Saddle River, NJ: Prentice-Hall.

Kosko, B., 1992. *Neural Networks and Fuzzy Systems,* Englewood Cliffs, NJ: Prentice-Hall.

Kosko, B., 1988. "Bidirectional associative memories," *IEEE Transactions on Systems, Man, and Cybernetics,* vol. 18, pp. 49–60.

Kotilainen, P., 1993. "Simulations and implementations of neural networks for principal component analysis," *Electronics Lab Report 1–93,* Tampre University of Technology, Finland.

Kraaijveld, M.A., and R.P.W. Duin, 1991. "Generalization capabilities of minimal kernel-based networks," *International Joint Conference on Neural Networks,* vol. I, pp. 843–848, Seattle.

Kramer, A.H., and A. Sangiovanni-Vincentelli, 1989. "Efficient parallel learning algorithms for neural networks," *Advances in neural Information Processing Systems,* vol. 1, pp. 40–48, San Mateo, CA: Morgan Kaufmann.

Kremer, S.C., 1996. "Comments on constructive learning of recurrent neural networks: Limitations of recurrent cascade correlation and a simple solution," *IEEE Transactions on Neural Networks,* vol. 7, pp. 1047–1049.

Kremer, S.C., 1995. "On the computational power of Elman-style recurrent networks," *IEEE Transactions on Neural Networks,* vol. 6, pp. 1000–1004.

Kreyszig, E., 1988. *Advanced Engineering Mathematics,* 6th ed., New York: Wiley.

Krishnamurthy, A.K., S.C. Ahalt, D.E. Melton, and P. Chen, 1990. "Neural networks for vector quantization of speech and images," *IEEE Journal of Selected Areas in Communications,* vol. 8, pp. 1449–1457.

Krzyżak, A., T. Linder, and G. Lugosi, 1996. "Nonparametric estimation and classification using radial basis functions," *IEEE Transactions on Neural Networks,* vol. 7, pp. 475–487.

Kuan, C.-M., and K. Hornik, 1991. "Convergence of learning algorithms with constant learning rates," *IEEE Transactions on Neural Networks,* vol. 2, pp. 484–489.

Kuan, C.-M., K. Hornik, and H. White, 1994. "A convergence result for learning in recurrent neural networks," *Neural Computation,* vol. 6, pp. 420–440.

Kuffler, S.W., J.G. Nicholls, and A.R. Martin, 1984. *From Neuron to Brain: A Cellular Approach to the Function of the Nervous System,* 2nd edition, Sunderland, MA: Sinauer Associates.

Kullback, S., 1968. *Information Theory and Statistics,* Gloucester, MA: Peter Smith.

Kung, S.Y., and K.I. Diamantaras, 1990. "A neural network learning algorithm for adaptive principal component extraction (APEX)." *IEEE International Conference on Acoustics, Speech, and Signal Processing,* vol. 2, pp. 861–864, Albuquerque.

Kushner, H.J., and D.S. Clark, 1978. *Stochastic Approximation Methods for Constrained and Unconstrained Systems,* New York: Springer-Verlag.

Lacoume, J.L., P.O. Amblard, and P. Comon, 1997. *Statistiques d'ordre Superieur pour le Traitement du Signal,* Masson Publishers.

Lancoz, C., 1964. *Linear Differential Operators,* London: Van Nostrand.

Landau, Y.D., 1979. *Adaptive Control: The Model Reference Approach,* New York: Marcel Dekker.

Landau, L.D., and E.M. Lifshitz, 1980. *Statistical Physics: Part 1,* 3rd edition, London: Pergamon Press.

Lanford, O.E., 1981. "Strange attractors and turbulence," in H.L. Swinney and J.P. Gollub, eds., *Hydrodynamic Instabilities and the Transition to Turbulence,* New York: Springer-Verlag.

Lang, K.J., and G.E. Hinton, 1988. "The development of the time-delay neural network architecture for speech recognition," Technical Report CMU-CS-88-152, Carnegie-Mellon University, Pittsburgh, PA.

Lapedes, A., and R. Farber, 1986. "Programming a massively parallel, computation universal system: Static Behavior," In *Neural Networks for Computing,* J.S. Denker, ed., pp. 283–298, New York: American Institute of Physics.

Larson, J., and G. Lynch, 1989. "Theta pattern stimulation and the induction of LTP: The sequence in which synapses are stimulated determines the degree to which they potentiate," *Brain Research,* vol. 489, pp. 49–58.

LaSalle, J., and S. Lefschetz, 1961. *Stability by Liapunov's Direct Method with Applications,* New York: Academic Press.

LeCun, Y., 1993. *Efficient Learning and Second-order Methods, A Tutorial at NIPS 93,* Denver.

LeCun, Y., 1989. "Generalization and network design strategies," Technical Report CRG-TR-89-4, Department of Computer Science, University of Toronto, Canada.

LeCun, Y., 1985. "Une procedure d'apprentissage pour reseau a seuil assymetrique." *Cognitiva,* vol. 85, pp. 599–604.

LeCun, Y., and Y. Bengio, 1995. "Convolutional networks for images, speech, and time series," in M.A. Arbib, ed., *The Handbook of Brain Theory and Neural Networks,* Cambridge, MA: MIT Press.

LeCun, Y., B. Boser, J.S. Denker, D. Henderson, R.E. Howard, W. Hubbard, and L.D. Jackel, 1990a. "Handwritten digit recognition with a back-propagation network," *Advances in Neural Information Processing,* vol. 2, pp. 396–404, San Mateo, CA: Morgan Kaufmann.

LeCun, Y., L. Bottou, and Y. Bengio, 1997. "Reading checks with multilayer graph transformeer networks," *IEEE International Conference on Acoustics, Speech and Signal Processing,* pp. 151–154, Munich, Germany.

LeCun, Y., L. Bottou, Y. Bengio, and P. Haffner, 1998. "Gradient-based learning applied to document recognition," *Proceedings of the IEEE,* vol. 86, to appear.

LeCun, Y., J.S. Denker, and S.A. Solla, 1990. "Optimal brain damage," *Advances in Neural Information Processing Systems,* vol. 2, pp. 598–605, San Mateo, CA: Morgan Kaufmann.

LeCun, Y., I. Kanter, and S.A. Solla, 1991. "Second order properties of error surfaces: Learning time and generalization," *Advances in Neural Information Processing Systems,* vol. 3, pp. 918–924, Cambridge, MA: MIT Press.

Lee, D.D., and H.S. Seung, 1997. "Unsupervised learning by convex and conic coding," *Advances in Neural Information Processing Systems,* vol. 9, pp. 515–521, Cambridge, MA: MIT Press.

Lee, T., 1997. *Independent Component Analysis: Theory and Applications,* Ph.D. Thesis, Technische Universität, Berlin, Germany.

Lee, T.-C., A.M. Peterson and J.J-C. Tsai, 1990. "A multilayer feed-forward neural network with dynamically adjustable structures," *IEEE International Conference on Systems, Man, and Cybernetics,* pp. 367–369, Los Angeles.

Lee, Y., and R.P. Lippmann, 1990. "Practical characteristics of neural networks and conventional pattern classifiers on artificial and speech problems," *Advances in Neural Information Processing Systems,* vol. 2, pp. 168–177, San Mateo, CA: Morgan Kaufmann.

Lee, Y., S. Oh, and M. Kim, 1991. "The effect of initial weights on premature saturation in back-propagation learning," *International Joint Conference on Neural Networks,* vol. I, pp. 765–770, Seattle.

Lee, Y.C., G. Doolen, H.H. Chan, G.Z. Sen, T. Maxwell, H.Y Lee, and C.L. Giles, 1986. "Machine learning using a higher order correlation network," *Physica,* D22, pp. 276–289.

Lefebvre, W.C., 1991. *An Object Oriented Approach for the Analysis of Neural Networks,* Master's Thesis, University of Florida, Gainsville, Fl.

Leon-Garcia, A., 1994. *Probability and Random Processes for Electrical Engineering,* 2nd edition, Reading, MA: Addison-Wesley.

Leontaritis, I., and S. Billings, 1985. "Input–output parametric models for nonlinear systems: Part I: Deterministic nonlinear systems," *International Journal of Control,* vol. 41, pp. 303–328.

Levin, A.V., and K.S. Narendra, 1996. "Control of nonlinear dynamical systems using neural networks—Part II: Observability, identification, and control," *IEEE Transactions on Neural Networks,* vol. 7, pp. 30–42.

Levin, A.V., and K.S. Narendra, 1993. "Control of nonlinear dynamical systems using neural networks—Controllability and stabilization," *IEEE Transactions on Neural Networks,* vol. 4, pp. 192–206.

Levine, M., 1985. *Man and Machine Vision.* New York: McGraw-Hill.

Lewis, F.L., and V.L. Syrmas, 1995. *Optimal Control,* 2nd edition, New York: Wiley (Interscience).

Lewis, F.L., A. Yesildirek, and K. Liu, 1996. "Multilayer neural-net robot controller with guaranteed tracking performance," *IEEE Transactions on Neural Networks,* vol. 7, pp. 1–12.

Lichtenberg, A.J., and M.A. Lieberman, 1992. *Regular and Chaotic Dynamics,* 2nd edition, New York: Springer-Verlag.

Light, W.A., 1992a. "Some aspects of radial basis function approximation," in *Approximation Theory, Spline Functions and Applications,* S.P. Singh, ed., NATO ASI vol. 256, pp. 163–190, Boston: Kluwer Academic Publishers.

Light, W., 1992b. "Ridge functions, sigmoidal functions and neural networks," in E.W. Cheney, C.K. Chui, and L.L. Schumaker, eds., *Approximation Theory VII,* pp. 163–206, Boston: Academic Press.

Lin, J.K., D.G. Grier, and J.D. Cowan, 1997. "Faithful representation of separable distributions," *Neural Computation,* vol. 9, pp. 1305–1320.

Lin, S., 1965. "Computer solutions of the traveling salesman problem," *Bell System Technical Journal,* vol. 44, pp. 2245–2269.

Lin, T., B.G. Horne, P. Tino, and C.L. Giles, 1996. "Learning long-term dependencies in NARX recurrent neural networks," *IEEE Transactions on Neural Networks,* vol. 7, pp. 1329–1338.

Linde, Y., A. Buzo, and R. M. Gray, 1980. "An algorithm for vector quantizer design," *IEEE Transactions on Communications,* vol. COM-28, pp. 84–95.

Linsker, R., 1993. "Deriving receptive fields using an optimal encoding criterion," *Advances in Neural Information Processing Systems,* vol. 5, pp. 953–960, San Mateo, CA: Morgan Kaufmann.

Linsker, R., 1990a. "Designing a sensory processing system: What can be learned from principal components analysis?" *Proceedings of the International Joint Conference on Neural Networks,* vol. 2, pp. 291–297, Washington, DC.

Linsker, R., 1990b. "Self-organization in a perceptual system: How network models and information theory may shed light on neural organization," Chapter 10 in *Connectionist Modeling and Brain Function: The Developing Interface,* S.J. Hanson and C.R. Olson, eds., pp. 351–392, Cambridge, MA: MIT Press.

Linsker, R., 1990c. "Perceptual neural organization: Some approaches based on network models and information theory," *Annual Review of Neuroscience,* vol. 13, pp. 257–281.

Linsker, R., 1989a. "An application of the principle of maximum information preservation to linear systems," *Advances in Neural Information Processing Systems,* vol. 1, pp. 186–194, San Mateo, CA: Morgan Kaufmann.

Linsker, R., 1989b. "How to generate ordered maps by maximizing the mutual information between input and output signals," *Neural computation,* vol. 1, pp. 402–411.

Linsker, R., 1988a. "Self-organization in a perceptual network," *Computer,* vol. 21, pp. 105–117.

Linsker, R., 1988b. "Towards an organizing principle for a layered perceptual network," in *Neural Information Processing Systems,* D.Z. Anderson, ed., pp. 485–494, New York: American Institute of Physics.

Linsker, R., 1987. "Towards an organizing principle for perception: Hebbian synapses and the principle of optimal neural encoding," *IBM Research Report RC12820,* IBM Research, Yorktown Heights, NY.

Linsker, R., 1986. "From basic network principles to neural architecture" (series), *Proceedings of the National Academy of Sciences, USA,* vol. 83, pp. 7508–7512, 8390–8394, 8779–8783.

Lippmann, R.P., 1987. "An introduction to computing with neural nets," *IEEE ASSP Magazine,* vol. 4, pp. 4–22.

Lippmann, R.P., 1989a. "Review of neural networks for speech recognition," *Neural Computation,* vol. 1, pp. 1–38.

Lippmann, R.P., 1989b. "Pattern classification using neural networks," *IEEE Communications Magazine,* vol. 27, pp. 47–64.

Little, W.A., 1974. "The existence of persistent states in the brain," *Mathematical Biosciences,* vol. 19, pp. 101–120.

Little, W.A., and G.L. Shaw, 1978. "Analytic study of the memory storage capacity of a neural network," *Mathematical Biosciences,* vol. 39, pp. 281–290.

Little, W.A., and G.L. Shaw, 1975. "A statistical theory of short and long term memory," *Behavioral Biology,* vol. 14, pp. 115–133.

Livesey, M., 1991. "Clamping in Boltzmann machines," *IEEE Transactions on Neural Networks,* vol. 2, pp. 143–148.

Ljung, L., 1987. *System Identification: Theory for the User.* Englewood Cliffs, NJ: Prentice-Hall.

Ljung, L., 1977. "Analysis of recursive stochastic algorithms," *IEEE Transactions on Automatic Control,* vol. AC-22, pp. 551–575.

Ljung, L., and T. Glad, 1994. *Modeling of Dynamic Systems,* Englewood Cliffs, NJ: Prentice-Hall.

Lloyd, S.P., 1957. "Least squares quantization in PCM," unpublished Bell Laboratories technical note. Published later under the same title in *IEEE Transactions on Information Theory,* vol. IT-28, pp. 127–135, 1982.

Lo, Z.-P., M. Fujita, and B. Bavarian, 1991. "Analysis of neighborhood interaction in Kohonen neural networks," *6th International Parallel Processing Symposium Proceedings,* pp. 247–249, Los Alamitos, CA.

Lo, Z.-P., Y. Yu and B. Bavarian, 1993. "Analysis of the convergence properties of topology preserving neural networks," *IEEE Transactions on Neural Networks,* vol. 4, pp. 207–220.

Lockery, S.R., Y. Fang, and T.J. Sejnowski, 1990. "A dynamical neural network model of sensorimotor transformations in the leech," *International Joint Conference on Neural Networks,* vol. I, pp. 183–188, San Diego, CA.

Loéve, M., 1963. *Probability Theory,* 3rd edition, New York: Van Nostrand.

Lorentz, G.G., 1976. "The 13th problem of Hilbert," *Proceedings of Symposia in Pure Mathematics,* vol. 28, pp. 419–430.

Lorentz, G.G., 1966. *Approximation of Functions,* Orlando, FL: Holt, Rinehart & Winston.

Lorenz, E.N., 1963. "Deterministic non-periodic flows," *Journal of Atmospheric Sciences,* vol. 20, pp. 130–141.

Lowe, D., 1989. "Adaptive radial basis function nonlinearities, and the problem of generalisation," *First IEE International Conference on Artificial Neural Networks,* pp. 171–175, London.

Lowe, D., 1991a. "What have neural networks to offer statistical pattern processing?" *Proceedings of the SPIE Conference on Adaptive Signal Processing,* pp. 460–471, San Diego, CA.

Lowe, D., 1991b. "On the iterative inversion of RBF networks: A statistical interpretation," *Second IEE International Conference on Artificial Neural Networks,* pp. 29–33, Bournemouth, England.

Lowe, D., and A.R. Webb, 1991a. "Time series prediction by adaptive networks: A dynamical systems perspective," *IEE Proceedings (London), Part F,* vol. 138, pp. 17–24.

Lowe, D., and A.R. Webb, 1991b. "Optimized feature extraction and the Bayes decision in feed-forward classifier networks," *IEEE Transactions on Pattern Analysis and Machine Intelligence,* PAMI-13, 355–364.

Lowe, D., and A.R. Webb, 1990. "Exploiting prior knowledge in network optimization: an illustration from medical prognosis," *Network,* vol. 1, pp. 299–323.

Lowe, D., and M.E. Tipping, 1996. "Neuroscale: Novel topographic feature extraction using RBF networks," *Neural Information Processing Systems,* vol. 9, pp. 543–549, Cambridge, MA: MIT Press.

Luenberger, D.G., 1984. *Linear and Nonlinear Programming,* 2nd edition, Reading, MA: Addison-Wesley.

Lui, H.C., 1990. "Analysis of decision contour of neural network with sigmoidal nonlinearity," *International Joint Conference on Neural Networks,* vol. I, pp. 655–659, Washington, DC.

Luo, Z., 1991. "On the convergence of the LMS algorithm with adaptive learning rate for linear feedforward networks," *Neural Computation,* vol. 3, pp. 226–245.

Luo, F., and R. Unbehauen, 1997. *Applied Neural Networks for Signal Processing,* New York: Cambridge University Press.

Luttrell, S.P., 1997. "A unified theory of density models and auto-encoders," *Technical Report 97303,* Defence Research Agency, Great Malvern, UK.

Luttrell, S.P., 1994. "A Bayesian analysis of self-organizing maps," *Neural Computation,* vol. 6, pp. 767–794.

Luttrell, S.P., 1991a. "Code vector density in topographic mappings: Scalar case," *IEEE Transactions on Neural Networks,* vol. 2, pp. 427–436.

Luttrell, S.P., 1991b. "Self-supervised training of hierarchical vector quantizers," *2nd International Conference on Artificial Neural Networks,* pp. 5–9, Bournemouth, England.

Luttrell, S.P., 1989a. "Hierarchical vector quantization," *IEE Proceedings (London),* vol. 136 (Part I), pp. 405–413.

Luttrell, S.P., 1989b. "Self-organization: A derivation from first principle of a class of learning algorithms," *IEEE Conference on Neural Networks,* pp. 495–498, Washington, DC.

Maass, W., 1993. "Bounds for the computational power and learning complexity of analog neural nets," *Proceedings of the 25th Annual ACM Symposium on the Theory of Computing,* pp. 335–344, New York: ACM Press.

Maass. W., 1993. "Vapnik–Chervonenkis dimension of neural networks," in M.A. Arbib, ed., *The Handbook of Brain Theory and Neural Networks,* Cambridge, MA: MIT Press.

Mach, E., 1865. "Über die Wirkung der räumlichen Vertheilung des Lichtreizes auf die Netzhaut, I. Sitzungsberichte der Mathematisch-Naturwissenschaftslichen Klasse der Kaiserlichen Akademie der Wissenschaften," vol. 52, pp. 303–322.

MacKay, D., 1992a. "Bayesian interpolation," *Neural Computation,* vol. 4, pp. 415–447.

MacKay, D., 1992b. "A practical Bayesian framework for back-propagation networks," *Neural Computation,* vol. 4, pp. 448–472.

MacKay, D.J.C., and K.D. Miller, 1990. "Analysis of Linsker's simulations of Hebbian rules," *Neural Computation,* vol. 2, pp. 173–187.

Macintyre, A.J., and E.D. Sontag, 1993. "Fitness results for sigmoidal 'neuronal' networks," *Proceedings of the 25th Annual ACM Symposium on the Theory of Computing,* pp. 325–334, New York: ACM Press.

MacQueen, J., 1967. "Some methods for classification and analysis of multivariate observation," in *Proceedings of the 5th Berkeley Symposium on Mathematical Statistics and Probability,* L.M. LeCun and J. Neyman, eds., vol. 1, pp. 281–297, Berkeley: University of California Press.

Madhuranath, H., and S. Haykin, 1998. "Improved Activation Functions for Blind Separation: Details of Algebraic Derivations," *CRL Internal Report No. 358,* Communications Research Laboratory, McMaster University, Hamilton, Ontario.

Mahowald, M.A., and C. Mead, 1989. "Silicon retina," in *Analog VLSI and Neural Systems* (C. Mead), Chapter 15. Reading, MA: Addison-Wesley.

Mandelbrot, B.B., 1982. *The Fractal Geometry of Nature,* San Francisco: Freeman.

Mañé, R., 1981. "On the dimension of the compact invariant sets of certain non-linear maps," in D. Rand and L.S. Young, eds., *Dynamical Systems and Turbulence,* Lecture Notes in Mathematics, vol. 898, pp. 230–242, Berlin: Springer-Verlag.

Marr, D., 1982. *Vision,* New York: W.H. Freeman and Company.

Martinetz, T.M., H.J. Ritter, and K.J. Schulten, 1990. "Three-dimensional neural net for learning visuomotor coordination of a robot arm," *IEEE Transactions on Neural Networks,* vol. 1, pp. 131–136.

Mason, S.J., 1953. "Feedback theory—Some properties of signal-flow graphs," *Proceedings of the Institute of Radio Engineers,* vol. 41, pp. 1144–1156.

Mason, S.J., 1956. "Feedback theory—Further properties of signal-flow graphs," *Proceedings of the Institute of Radio Engineers,* vol. 44, pp. 920–926.

Maybeck, P.S., 1982. *Stochastic Models, Estimation, and Control,* vol. 2, New York: Academic Press.

Maybeck, P.S., 1979. *Stochastic Models, Estimation, and Control,* vol. 1, New York: Academic Press.

Mazaika, P.K., 1987. "A mathematical model of the Boltzmann machine," *IEEE First International Conference on Neural Networks,* vol. III, pp. 157–163, San Diego, CA.

McBride, L.E., Jr., and K.S. Narendra, 1965. "Optimization of time-varying systems," *IEEE Transactions on Automatic Control,* vol. AC-10, pp. 289–294.

McCullagh, P., and J.A. Nelder, 1989. *Generalized Linear Models,* 2nd edition, London: Chapman and Hall.

McCulloch, W.S., 1988. *Embodiments of Mind,* Cambridge, MA: MIT Press.

McCulloch, W.S., and W. Pitts, 1943. "A logical calculus of the ideas immanent in nervous activity," *Bulletin of Mathematical Biophysics,* vol. 5, pp. 115–133.

McEliece, R.J., E.C. Posner, E.R. Rodemich, and S.S. Venkatesh, 1987. "The capacity of the Hopfield associative memory," *IEEE Transactions on Information Theory,* vol. IT-33, pp. 461–482.

McLachlan, G.J., and K.E. Basford, 1988. *Mixture Models: Inference and Applications to Clustering,* New York: Marcel Dekker.

McLachlan, G.J., and T. Krishnan, 1997. *The EM Algorithm and Extensions,* New York: Wiley (Interscience).

McQueen, J., 1967. "Some methods for classification and analysis of multivariate observations," *Proceedings of the 5th Berkeley Symposium on Mathematical Statistics and Probability,* vol. 1, pp. 281–297, Berkeley, CA: University of California Press.

Mead, C.A., 1990. "Neuromorphic electronic systems," *Proceedings of the Institute of Electrical and Electronics Engineers,* vol. 78, pp. 1629–1636.

Mead, C.A., 1989. *Analog VLSI and Neural Systems,* Reading, MA: Addison-Wesley.

Mead, C.A., and M.A. Mahowald, 1988. "A silicon model of early visual processing," *Neural Networks,* vol. 1, pp. 91–97.

Mead, C.A., X. Arreguit, and J. Lazzaro, 1991. "Analog VLSI model of binaural hearing," *IEEE Transactions on Neural Networks,* vol. 2, pp. 232–236.

Mecklenbräuker, W., and F. Hlawatsch, eds., 1997. *The Wigner Distribution,* New York: Elsevier.

Memmi, D., 1989. "Connectionism and artificial intelligence," *Neuro-Nimes'89 International Workshop on Neural Networks and their Applications,* pp. 17–34, Nimes, France.

Mendel, J.M., 1995. *Lessons in Estimation Theory for Signal Processing, Communications, and Control.* Englewood Cliffs, NJ: Prentice-Hall.

Mendel, J.M., and R.W. McLaren, 1970. "Reinforcement-learning control and pattern recognition systems," in *Adaptive, Learning, and Pattern Recognition Systems: Theory and Applications,* vol. 66, J.M. Mendel and K.S. Fu, eds., pp. 287–318, New York: Academic Press.

Mennon, A., K. Mehrotra, C.K. Mohan, and S. Ranka, 1996. "Characterization of a class of sigmoid functions with applications to neural networks," *Neural Networks,* vol. 9, pp. 819–835.

Mercer, J., 1909. "Functions of positive and negative type, and their connection with the theory of integral equations," *Transactions of the London Philosophical Society (A),* vol. 209, pp. 415–446.

Mesulam, M.M., 1985. "Attention, confusional states, and neglect," in *Principles of Behavioral Neurology,* M.M. Mesulam, ed., Philadelphia: F.A. Davis.

Metropolis, N., A. Rosenbluth, M. Rosenbluth, A. Teller, and E. Teller, 1953. Equations of state calculations by fast computing machines, *Journal of Chemical Physics,* vol. 21, pp. 1087–1092.

Mhaskar, H.N., 1996. "Neural networks for optimal approximation of smooth and analytic functions," *Neural Computation,* vol. 8, pp. 1731–1742.

Mhaskar, H.N., and C.A. Micchelli, 1992. "Approximation by superposition of sigmoidal and radial basis functions," *Advances in Applied Mathematics,* vol. 13, pp. 350–373.

Micchelli, C.A., 1986. "Interpolation of scattered data: Distance matrices and conditionally positive definite functions," *Constructive Approximation,* vol. 2, pp. 11–22.

Miller, D., A.V. Rao, K. Rose, and A. Gersho, 1996. "A global optimization technique for statistical classifier design," *IEEE Transactions on Signal Processing,* vol. 44, pp. 3108–3122.

Miller, K.D., J.B. Keller, and M.P. Stryker, 1989. "Ocular dominance column development: Analysis and simulation," *Science,* vol. 245, pp. 605–615.

Miller, D., and K. Rose, 1996. "Hierarchical, unsupervised learning with growing via phase transitions," *Neural Computation,* vol. 8, pp. 425–450.

Miller, D., and K. Rose, 1994. "Combined source-channel vector quantization using deterministic annealing," *IEEE Transactions on Communications,* vol. 42, pp. 347–356.

Miller, R., 1987. "Representation of brief temporal patterns, Hebbian synapses, and the left-hemisphere dominance for phoneme recognition," *Psychobiology,* vol. 15, pp. 241–247.

Minai, A.A., and R.J. Williams, 1990. "Back-propagation heuristics: A study of the extended delta-bar-delta algorithm," *IEEE International Joint Conference on Neural Networks,* vol. I, pp. 595–600, San Diego, CA.

Minsky, M.L., 1986. Society of Mind, New York: Simon and Schuster.

Minsky, M.L., 1967. *Computation: Finite and Infinite Machines.* Englewood Cliffs, NJ: Prentice-Hall.

Minsky, M.L., 1961. "Steps towards artificial intelligence," *Proceedings of the Institute of Radio Engineers,* vol. 49, pp. 8–30 (Reprinted in: Feigenbaum, E.A., and J. Feldman, eds., *Computers and Thought,* pp. 406–450, New York: McGraw-Hill.)

Minsky, M.L., 1954. "Theory of neural-analog reinforcement systems and its application to the brain-model problem," Ph.D. thesis, Princeton University, Princeton, NJ.

Minsky, M.L., and S.A. Papert, 1988. *Perceptrons,* expanded edition, Cambridge, MA: MIT Press.

Minsky, M.L., and S.A. Papert, 1969. *Perceptrons,* Cambridge, MA: MIT Press.

Minsky, M.L., and O.G. Selfridge, 1961. "Learning in random nets," *Information Theory, Fourth London Symposium,* London: Butterworths.

Mitchell, T.M., 1997. *Machine Learning.* New York: McGraw-Hill.

Mitchison, G., 1989. "Learning algorithms and networks of neurons," in *The Computing Neuron* (R. Durbin, C. Miall, and G. Michison, eds), pp. 35–53, Reading, MA: Addison-Wesley.

Møller, M.F., 1993. "A scaled conjugate gradient algorithm for fast supervised learning," *Neural Networks,* vol. 6, pp. 525–534.

Moody, J., and C.J. Darken, 1989. "Fast learning in networks of locally-tuned processing units," *Neural Computation,* vol. 1, pp. 281–294.

Moody, J., and L. Wu, 1996. "Optimization of trading systems and portfolios," in A. Weigend, Y. Abu-Mostafa, and A.-P.N. Refenes, eds., *Decision Technologies for Financial Engineering,* pp. 23–35, Singapore: World Scientific.

Moody, J.E., and T. Rögnvaldsson, 1997. "Smoothing regularizers for projective basis function networks," *Advances in Neural Information Processing Systems,* vol. 9, pp. 585–591.

Moray, N., 1959. "Attention in dichotic listening: Affective cues and the influence of instructions," *Quarterly Journal of Experimental Psychology,* vol. 27, pp. 56–60.

Morgan, N., and H. Bourlard, 1990. "Continuous speech recognition using multilayer perceptrons with hidden Markov models," *IEEE International Conference on Acoustics, Speech, and Signal Processing,* vol. 1, pp. 413–416, Albuquerque.

Morita, M., 1993. "Associative memory with nonmonotonic dynamics," *Neural Networks,* vol. 6, pp. 115–126.

Morozov, V.A., 1993. *Regularization Methods for Ill-Posed Problems,* Boca Raton, FL: CRC Press.

Morse, P.M., and H. Feshbach, 1953. *Methods of Theoretical Physics,* Part 1, New York: McGraw-Hill.

Mozer, M.C., 1994. "Neural net architectures for temporal sequence processing," in A.S. Weigend and N.A. Gershenfeld, eds., *Time Series Prediction: Forecasting the Future and Understanding the Past,* pp. 243–264, Reading, MA: Addison-Wesley.

Mpitsos, G.J., 1990. "Chaos in brain function and the problem of nonstationarity: A commentary," in *Chaos in Brain Function,* E. Basar, ed., pp. 162–176. New York: Springer-Verlag.

Müller, B., and J. Reinhardt, 1990. *Neural Networks: An Introduction,* New York: Springer-Verlag.

Muller, D., and G. Lynch, 1988. "Long-term potentiation differentially affects two components of synaptic responses in hippocampus," *Proceedings of the National Academy of Sciences,* USA, vol. 85, pp. 9346–9350.

Mumford, D., 1994. "Neural architectures for pattern-theoretic problems," in C. Koch and J. Davis, eds., *Large-Scale Theories of the Cortex,* pp. 125–152, Cambridge, MA: MIT Press.

Murray, M.K., and J.W. Rice, 1993. *Differential Geometry and Statistics,* New York: Chapman and Hall.

Murtagh, B., and M. Saunders, 1978. "Large-scale linearly constrained optimization," *Mathematical Programming,* vol. 14, pp. 41–72.

Muselli, M., 1997. "On convergence properties of pocket algorithm," *IEEE Transactions on Neural Networks,* vol. 8, pp. 623–629.

Nadal, J.-P., and N. Parga, 1997. "Redundancy reduction and independent component analysis: Conditions on cumulants and adaptive approaches," *Neural Computation,* vol. 9, pp. 1421–1456.

Nadal, J.-P., and N. Parga, 1994. "Nonlinear neurons in the low-noise limit: A factorial code maximizes information transfer," *Network,* vol. 5, pp. 565–581.

Nadaraya, É.A., 1965. "On nonparametric estimation of density functions and regression curves," *Theory of Probability and its Applications,* vol. 10, pp. 186–190.

Nadaraya, É.A., 1964. "On estimating regression," *Theory of Probability and its Applications*, vol. 9, pp. 141–142.

Naftaly, U., N. Intrator, and D. Horn, 1997. "Optimal ensemble averaging of neural networks," *Network*, vol. 8, pp. 283–296.

Nakano, K., 1972. "Association—a model of associative memory," *IEEE Transactions on Systems, Man, and Cybernetics*, vol. SMC-2, pp. 380–388.

Narendra, K.S., 1995. *Neural Networks for Identification and Control*, NIPS 95, Tutorial Program, pp. 1–46, Denver.

Narendra, K.S., and A.M. Annaswamy, 1989. *Stable Adaptive Systems*, Englewood Cliffs, NJ: Prentice-Hall.

Narendra, K.S., and K. Parthasarathy, 1990. "Identification and control of dynamical systems using neural networks," *IEEE Transactions on Neural Networks*, vol. 1, pp. 4–27.

Natarajan, B.K., 1991. *Machine Learning: A Theoretical Approach*, San Mateo, CA: Morgan Kaufmann.

Neal, R.M., 1995. *Bayesian Learning for Neural Networks*, Ph.D. Thesis, University of Toronto, Canada.

Neal, R.M., 1993. "Bayesian learning via stochastic dynamics," *Advances in Neural Information Processing Systems*, vol. 5, pp. 475–482, San Mateo, CA: Morgan Kaufmann.

Neal, R.M., 1992. "Connectionist learning of belief networks," *Artificial Intelligence*, vol. 56, pp. 71–113.

Newcomb, S., 1886. "A generalized theory of the combination of observations so as to obtain the best result," *American Journal of Mathematics*, vol. 8, pp. 343–366.

Newell, A., and H.A. Simon, 1972. *Human Problem Solving*, Englewood Cliffs, NJ: Prentice-Hall.

Ng, K., and R.P. Lippmann, 1991. "Practical characteristics of neural network and conventional pattern classifiers," *Advances in Neural Information Processing Systems*, vol. 3, pp. 970–976, San Mateo, CA: Morgan Kaufmann.

Nguyen, D., and B. Widrow, 1989. "The truck backer-upper: An example of self-learning in neural networks," *International Joint Conference on Neural Networks*, vol. II, pp. 357–363, Washington, DC.

Nie, J., and S. Haykin, 1998. "A Q-learning-based dynamic channel assignment technique for mobile communication systems," *IEEE Transactions on Vehicular Technology*, to appear.

Nie, J., and S. Haykin, 1996. "A dynamic channel assignment policy through Q-learning," *CRL Report No. 334*, Communications Research Laboratory, McMaster University, Hamilton, Ontario.

Nilsson, N.J., 1980. *Principles of Artificial Intelligence*, New York: Springer-Verlag.

Nilsson, N.J., 1965. *Learning Machines: Foundations of Trainable Pattern-Classifying Systems*, New York: McGraw-Hill.

Niyogi, P., and F. Girosi, 1996. "On the relationship between generalization error, hypothesis complexity, and sample complexity for radial basis functions," *Neural Computation*, vol. 8, pp. 819–842.

Novikoff, A.B.J., 1962. "On convergence proofs for perceptrons," in *Proceedings of the Symposium on the Mathematical Theory of Automata*, pp. 615–622, Brooklyn, NY: Polytechnic Institute of Brooklyn.

Nowlan, S.J., 1990. "Maximum likelihood competitive learning," *Advances in Neural Information Processing Systems*, vol. 2, pp. 574–582, San Mateo, CA: Morgan Kaufmann.

Nowlan, S.J., and G.E. Hinton, 1992. "Adaptive soft weight tying using Gaussian mixtures," *Advances in Neural Information Processing Systems*, vol. 4, pp. 993–1000, San Mateo, CA: Morgan Kaufmann.

Nowlan, S.J., and G.E. Hinton, 1991. "Evaluation of adaptive mixtures of competing experts," *Advances in Neural Information Processing Systems*, vol. 3, pp. 774–780, San Mateo, CA: Morgan Kaufmann.

Obermayer, K., H. Ritter, and K. Schulten, 1991. "Development and spatial structure of cortical feature maps: A model study," *Advances in Neural Information Processing Systems*, vol. 3, pp. 11–17, San Mateo, CA: Morgan Kaufmann.

Oja, E., 1992a. "Principal components, minor components, and linear neural networks," *Neural Networks*, vol. 5, 927–936.

Oja, E., 1992b. "Self-organizing maps and computer vision," in *Neural Networks for Perception*, vol. 1, H. Wechsler, ed., vol. 1, pp. 368–385, San Diego, CA: Academic Press.

Oja, E., 1991. "Data compression, feature extraction, and autoassociation in feedforward neural networks," *Artificial Neural Networks*, vol. 1, pp. 737–746, Amsterdam: North-Holland.

Oja, E., 1989. "Neural networks, principal components, and subspaces," *International Journal of Neural Systems*, vol. 1, 61–68.

Oja, E., 1983. *Subspace Methods of Pattern Recognition*, Letchworth, England: Research Studies Press.

Oja, E., 1982. "A simplified neuron model as a principal component analyzer," *Journal of Mathematical Biology*, vol. 15, pp. 267–273.

Oja, E., and J. Karhunen, 1985. "A stochastic approximation of the eigenvectors and eigenvalues of the expectation of a random matrix," *Journal of Mathematical Analysis and Applications*, vol. 106, pp. 69–84.

Oja, E., and T. Kohonen, 1988. "The subspace learning algorithm as formalism for pattern recognition and neural networks," *IEEE International Conference on Neural Networks,* vol. I, pp. 277–284, San Diego, CA.

Omlin, C.W., and C.L. Giles, 1996. "Constructing deterministic finite-state automata in recurrent neural networks," *Journal of the Association for Computing Machinery,* vol. 43, pp. 937–972.

Oppenheim, A.V., and R.W. Schafer, 1989. *Discrete-Time Signal Processing,* Englewood Cliffs, NJ: Prentice-Hall.

Orlando, J., R. Mann, and S. Haykin, 1990. "Classification of sea-ice using a dual-polarized radar," *IEEE Journal of Oceanic Engineering,* vol. 15, pp. 228–237.

Osherson, D.N., S. Weinstein, and M. Stoli, 1990. "Modular learning," *Computational Neuroscience,* E.L. Schwartz, ed., pp. 369–377, Cambridge, MA: MIT Press.

Osuna, E., 1998. "Support Vector Machines: Training and Applications," Ph.D. Thesis, Operations Research Center, MIT.

Osuna, E., and F. Girosi, 1998. "Reducing the run-time complexity of support vector machines," ICPR 98, Brisbane, Australia.

Osuna, E., R. Freund, and F. Girosi, 1997. "An improved training algorithm for support vector machines," *Neural Networks for Signal Processing* VII, Proceedings of the 1997 IEEE Workshop, pp. 276–285, Amelia Island, FL.

Ott, E., 1993. *Chaos in Dynamical Systems,* Cambridge, MA: Cambridge University Press.

Packard, N.H., J.P. Crutchfield, J.D. Farmer, and R.S. Shaw, 1980. "Geometry from a time series," *Physical Review Letters,* vol. 45, pp. 712–716.

Palm, G., 1982. *Neural Assemblies: An Alternative Approach,* New York: Springer-Verlag.

Palmieri, F., and S.A. Shah, 1990. "Fast training of multilayer perceptrons using multi-linear parameterization," *International Joint Conference on Neural Networks,* vol. I, pp. 696–699, Washington, DC.

Palmieri, F., J. Zhu, and C. Chang, 1993. "Anti-Hebbian learning in topologically constrained linear networks: A tutorial," *IEEE Transactions on Neural Networks,* vol. 5, pp. 748–761.

Papoulis, A., 1984. *Probability, Random Variables, and Stochastic Processes,* 2nd edition, New York: McGraw-Hill.

Parisi, G., 1988. *Statistical Field Theory,* Reading, MA: Addison-Wesley.

Park, J., and I.W. Sandberg, 1991. "Universal approximation using radial-basis-function networks," *Neural Computation,* vol. 3, pp. 246–257.

Parker, D.B., 1987. "Optimal algorithms for adaptive networks: Second order back propagation, second order direct propagation, and second order Hebbian learning," *IEEE 1st International Conference on Neural Networks,* vol. 2, pp. 593–600, San Diego, CA.

Parker, D.B., 1985. "Learning-logic: Casting the cortex of the human brain in silicon," *Technical Report TR-47,* Center for Computational Research in Economics and Management Science, Cambridge, MA: MIT Press.

Parker, T.S., and L.O., Chua, 1989. *Practical Numerical Algorithms for Chaotic Systems,* New York: Springer.

Parzen, E., 1962. "On estimation of a probability density function and mode," *Annals of Mathematical Statistics,* vol. 33, pp. 1065–1076.

Passino, K.N., 1996. "Toward bridging the perceived gap between conventional and intelligent control," in M.D. Gupta and N.K. Sinha, eds., *Intelligent Control Systems,* pp. 3–27, New York: IEEE Press.

Pavlov, I.P., 1927. *Conditional Reflexes: An Investigation of the Physiological Activity of the Cerebral Cortex,* (Translation from the Russian by G.V. Anrep), New York: Oxford University Press.

Pearl, J., 1988. *Probabilistic Reasoning in Intelligent Systems,* San Mateo, CA: Morgan Kaufmann. (Revised 2nd printing, 1991).

Pearlmutter, B.A., 1989. "Learning state-space trajectories in recurrent neural networks," *Neural Computation,* vol. 1, pp. 263–269.

Pearson, K., 1901. "On lines and planes of closest fit to systems of points in space," *Philosophical Magazine,* vol. 2, pp. 559–572.

Peretto, P., 1984. "Collective properties of neural networks: A statistical physics approach," *Biological Cybernetics,* vol. 50, pp. 51–62.

Peretto, P., and J.-J Niez, 1986. "Stochastic dynamics of neural networks," *IEEE Transactions on Systems, Man, and Cybernetics,* vol. SMC-16, pp. 73–83.

Perrin, D., 1990. "Finite automata," in J. van Leeuwen, ed., *Handbook of Theoretical Computer Science, Volume B: Formal Models and Semantics,* Chapter 1, pp. 3–57, Cambridge, MA: MIT Press.

Perrone, M.P., 1993. "Improving regression estimation: Averaging methods for variance reduction with extensions, to general convex measure optimization," Ph.D. Thesis, Brown University, Rhode Island.

Personnaz, L., I. Guyon, and G. Dreyfus, 1985. "Information storage and retrieval in spin-glass like neural networks," *Journal of Physique, Letters,* Orsay, France, vol. 46, L-359–L-365.

Peterson, C., 1991. "Mean field theory neural networks for feature recognition, content addressable memory and optimization," *Connection Science,* vol. 3, pp. 3–33.

Peterson, C., and J.R. Anderson, 1987. "A mean field theory learning algorithm for neural networks," *Complex Systems,* vol. 1, pp. 995–1019.

Peterson, C., and E. Hartman, 1989. "Explorations of the mean field theory learning algorithm," *Neural Networks,* vol. 2, pp. 475–494.

Peterson, C., and B. Söderberg, 1989. "A new method of mapping optimization problems onto neural networks," *International Journal of Neural Systems,* vol. 1, pp. 3–22.

Pham, D.T., and P. Garrat, 1997. "Blind separation of mixture of independent sources through a quasi-maximum likelihood approach," *IEEE Transactions on Signal Processing,* vol. 45, pp. 1712–1725.

Pham, D.T., P. Garrat, and C. Jutten, 1992. "Separation of a mixture of independent sources through a maximum likelihood approach," *Proceedings of EUSIPCO,* pp. 771–774.

Phillips, D., 1962. "A technique for the numerical solution of certain integral equations of the first kind," *Journal of Association for Computing Machinery,* vol. 9, pp. 84–97.

Pineda, F.J., 1989, "Recurrent backpropagation and the dynamical approach to adaptive neural computation," *Neural Computation,* vol. 1, pp. 161–172.

Pineda, F.J., 1988a. "Generalization of backpropagation to recurrent and higher order neural networks," in *Neural Information Processing Systems,* D.Z. Anderson, ed., pp. 602–611, New York: American Institute of Physics.

Pineda, F.J., 1988b. "Dynamics and architecture in neural computation," *Journal of Complexity,* vol. 4, pp. 216–245.

Pineda, F.J., 1987. "Generalization of back-propagation to recurrent neural networks," *Physical Review Letters,* vol. 59, pp. 2229–2232.

Pitts, W., and W.S. McCulloch, 1947, "How we know universals: The perception of auditory and visual forms," *Bulletin of Mathematical Biophysics,* vol. 9, pp. 127–147.

Plumbley, M.D., and F. Fallside, 1989. "Sensory adaptation: An information-theoretic viewpoint," *International Joint Conference on Neural Networks,* vol. 2, p. 598, Washington, DC.

Plumbley, M.D., and F. Fallside, 1988. "An information-theoretic approach to unsupervised connectionist models," in *Proceedings of the 1988 Connectionist Models Summer School,* D. Touretzky, G. Hinton, and T. Sejnowski, eds., pp. 239–245. San Mateo, CA: Morgan Kaufmann.

Poggio, T., 1990. "A theory of how the brain might work," *Cold Spring Harbor Symposium on Quantitative Biology,* vol. 5, pp. 899–910.

Poggio, T., and D. Beymer, 1996. "Learning to see," *IEEE Spectrum,* vol. 33, no. 5, pp. 60–69.

Poggio, T., and S. Edelman, 1990. "A network that learns to recognize three-dimensional objects," *Nature,* vol. 343, pp. 263–266.

Poggio, T., and F. Girosi, 1990a. "Networks for approximation and learning," *Proceedings of the IEEE,* vol. 78, pp. 1481–1497.

Poggio, T., and F. Girosi, 1990b. "Regularization algorithms for learning that are equivalent to multilayer networks," *Science,* vol. 247, pp. 978–982.

Poggio, T., and C. Koch, 1985. "Ill-posed problems in early vision: From computational theory to analogue networks," *Proceedings of the Royal Society of London,* Series B, vol. 226, pp. 303–323.

Poggio, T., V. Torre, and C. Koch, 1985. "Computational vision and regularization theory," *Nature,* vol. 317, pp. 314–319.

Polak, E., and G. Ribiére, 1969. "Note sur la convergence de methods de directions conjuguees," *Revue Française Information Recherche Operationnelle* vol. 16, pp. 35–43.

Pöppel G., and U. Krey, 1987. "Dynamical learning process for recognition of correlated patterns in symmetric spin glass models," *Europhysics Letters,* vol. 4, pp. 979–985.

Powell, M.J.D., 1992. "The theory of radial basis function approximation in 1990," in W. Light, ed., *Advances in Numerical Analysis Vol. II: Wavelets, Subdivision Algorithms, and Radial Basis Functions,* pp. 105–210, Oxford: Oxford Science Publications.

Powell, M.J.D., 1988. "Radial basis function approximations to polynomials," *Numerical Analysis 1987 Proceedings,* pp. 223–241, Dundee, UK.

Powell, M.J.D., 1985. "Radial basis functions for multivariable interpolation: A review," *IMA Conference on Algorithms for the Approximation of Functions and Data,* pp. 143–167, RMCS, Shrivenham, England.

Powell, M.J.D., 1977. "Restart procedures for the conjugate gradient method," *Mathematical Programming*, vol. 12, pp. 241–254.

Preisendorfer, R.W., 1988. *Principal Component Analysis in Meteorology and Oceanography*, New York: Elsevier.

Press, W.H., B.P. Flannery, S.A. Teukolsky, and W.T. Vetterling, 1988. *Numerical Recipes in C: The Art of Scientific Computing*, Cambridge: Cambridge University Press.

Proakis, J.G., 1989. *Digital Communications*, 2nd edition, New York: McGraw-Hill.

Prokhorov, D.V., and D.C. Wunsch, II, 1997. "Adaptive critic designs," *IEEE Transactions on Neural Networks*, vol. 8, pp. 997–1007.

Puskorius, G.V., and L.A. Feldkamp, 1994. "Neurocontrol of nonlinear dynamical systems with Kalman filter-trained recurrent networks," *IEEE Transactions on Neural Networks*, vol. 5, pp. 279–297.

Puskorius, G.V., and L.A. Feldkamp, 1992. "Model reference adaptive control with recurrent networks trained by the dynamic DEKF algorithm," *International Joint Conference on Neural Networks*, vol. II, pp. 106–113, Baltimore.

Puskorius, G.V., L.A. Feldkamp, and L.I. Davis, Jr., 1996. "Dynamic neural network methods applied to on-vehicle idle speed control," *Proceedings of the IEEE*, vol. 84, pp. 1407–1420.

Puskorius, G.V., and L.A. Feldkamp, 1991. "Decoupled extended Kalman filter training of feedforward layered networks," *International Joint Conference on Neural Networks*, vol. 1, pp. 771–777, Seattle.

Rabiner, L.R., 1989. "A tutorial on hidden Markov models," *Proceedings of the IEEE*, vol. 73, pp. 1349–1387.

Rabiner, L.R., and B.H. Juang, 1986. "An introduction to hidden Markkov models," *IEEE ASSP Magazine*, vol. 3, pp. 4–16.

Rall, W., 1989. "Cable theory for dendritic neurons," in *Methods in Neuronal Modeling*, C. Koch and I. Segev, eds., pp. 9–62, Cambridge, MA: MIT Press.

Rall, W., 1990. "Some historical notes," in *Computational Neuroscience*, E.L. Schwartz, Ed., pp. 3–8, Cambridge: MIT Press.

Ramón y Cajál, S., 1911, *Histologie du Systéms Nerveux de l'homme et des vertébrés*, Paris: Maloine.

Rao, A., D. Miller, K. Rose, and A. Gersho, 1997a. "Mixture of experts regression modeling by deterministic annealing." *IEEE Transactions on Signal Processing*, vol. 45, pp. 2811–2820.

Rao, A., K. Rose, and A. Gersho, 1997b. "A deterministic annealing approach to discriminative hidden Markov model design," *Neural Networks for Signal Processing VII, Proceedings of the 1997 IEEE Workshop*, pp. 266–275, Amelia Island, FL.

Rao, C.R., 1973. *Linear Statistical Inference and Its Applications*, New York: Wiley.

Rashevsky, N., 1938. *Mathematical Biophysics*, Chicago: University of Chicago Press.

Raviv, Y., and N. Intrator, 1996. "Bootstrapping with noise: An effective regularization technique," *Connection Science*, vol. 8, pp. 355–372.

Reed, R., 1993. "Pruning algorithms—A survey." *IEEE Transactions on Neural Networks*, vol. 4, pp. 740–747.

Reeke, G.N. Jr., L.H. Finkel, and G.M. Edelman, 1990. "Selective recognition automata," in *An Introduction to Neural and Electronic Networks*, S.F. Zornetzer, J.L. Davis, and C. Lau, eds,, pp. 203–226, New York: Academic Press.

Reif, 1965. *Fundamentals of Statistical and Thermal Physics*, New York: McGraw-Hill.

Renals, S., 1989. "Radial basis function network for speech pattern classification," *Electronics Letters*, vol. 25, pp. 437–439.

Rényi, A. 1960. "On measures of entropy and information," *Proceedings of the 4th Berkeley Symposium on Mathematics, Statistics, and Probability*, pp. 547–561.

Rényi, A., 1970. *Probability Theory*, North-Holland, Amsterdam.

Richard, M.D., and R.P. Lippmann, 1991. "Neural network classifiers estimate Bayesian a posteriori probabilities," *Neural Computation*, vol. 3, pp. 461–483.

Riesz, F., and B. Sz-Nagy, 1955. *Functional Analysis*, 2nd edition, New York: Frederick Ungar.

Ripley, B.D., 1996. *Pattern Recognition and Neural Networks*, Cambridge: Cambridge University Press.

Rissanen, J., 1978. "Modeling by shortest data description," *Automatica*, vol. 14, pp. 465–471.

Rissanen, J., 1989. *Stochastic Complexity in Statistical Inquiry*, Singapore: World Scientific.

Ritter, H., 1991. "Asymptotic level density for a class of vector quantization processes," *IEEE Transactions on Neural Networks*, vol. 2, pp. 173–175.

Ritter, H., 1995. "Self-organizing feature maps: Kohonen maps," in M.A. Arbib, ed., *The Handbook of Brain Theory and Neural Networks*, pp. 846–851, Cambridge, MA: MIT Press.

Ritter, H., and T. Kohonen, 1989. "Self-organizing semantic maps," *Biological Cybernetics*, vol. 61, pp. 241–254.

Ritter, H., and K. Schulten, 1988. "Convergence properties of Kohonen's topology conserving maps: Fluctuations, stability, and dimension selection," *Biological Cybernetics,* vol. 60, pp. 59–71.

Ritter, H., T.M. Martinetz, and K.J. Schulten, 1989. "Topology-conserving maps for learning visuo-motor-coordination," *Neural Networks,* vol. 2, pp. 159–168.

Ritter, H., T. Martinetz, and K. Schulten, 1992. *Neural Computation and Self-Organizing Maps: An Introduction,* Reading, MA: Addison-Wesley.

Robbins, H., and S. Monro, 1951. "A stochastic approximation method," *Annals of Mathematical Statistics,* vol. 22, pp. 400–407.

Robinson, D.A., 1992. "Signal processing by neural networks in the control of eye movements," *Computational Neuroscience Symposium,* pp. 73–78, Indiana University-Purdue University at Indianapolis.

Rochester, N., J.H. Holland, L.H. Haibt, and W.L. Duda, 1956. "Tests on a cell assembly theory of the action of the brain, using a large digital computer," *IRE Transactions on Information Theory,* vol. IT-2, pp. 80–93.

Rose, K., 1998. "Deterministic annealing for clustering, compression, classification, regression, and related optimization problems," *Proceedings of the IEEE,* vol. 86, to appear.

Rose, K., 1991. *Deterministic Annealing, Clustering, and Optimization,* Ph.D. Thesis, California Institute of Technology, Pasadena, CA.

Rose, K., E. Gurewitz, and G.C. Fox, 1992. "Vector quantization by deterministic annealing," *IEEE Transactions on Information Theory,* vol. 38, pp. 1249–1257.

Rose, K., E. Gurewitz, and G.C. Fox, 1990. "Statistical mechanics and phase transitions in clustering," *Physical Review Letters,* vol. 65, pp. 945–948.

Rosenblatt, F., 1962. *Principles of Neurodynamics,* Washington, DC: Spartan Books.

Rosenblatt, F., 1960a. "Perceptron simulation experiments," *Proceedings of the Institute of Radio Engineers,* vol. 48, pp. 301–309.

Rosenblatt, F., 1960b. "On the convergence of reinforcement procedures in simple perceptrons," Cornell Aeronautical Laboratory Report, VG-1196–G-4, Buffalo, NY.

Rosenblatt, F., 1958. "The Perceptron: A probabilistic model for information storage and organization in the brain," *Psychological Review,* vol. 65, pp. 386–408.

Rosenblatt, M., 1970. "Density estimates and Markov sequences," in M. Puri, ed., *Nonparametric Techniques in Statistical Inference,* pp. 199–213, London: Cambridge University Press.

Rosenblatt, M., 1956. "Remarks on some nonparametric estimates of a density function," *Annals of Mathematical Statistics.,* vol. 27, pp. 832–837.

Ross, S.M., 1983. *Introduction to Stochastic Dynamic Programming,* New York: Academic Press.

Roth, Z., and Y. Baram, 1996. "Multi-dimensional density shaping by sigmoids," *IEEE Transactions on Neural Networks,* vol. 7, pp. 1291–1298.

Roussas, G., ed., 1991. *Nonparametric Functional Estimation and Related Topics,* The Netherlands: Kluwer.

Roy, S., and J.J. Shynk, 1990. "Analysis of the momentum LMS algorithm," *IEEE Transactions on Acoustics, Speech, and Signal Processing,* vol. ASSP-38, pp. 2088–2098.

Rubner, J., and K. Schulten, 1990. "Development of feature detectors by self-organization," *Biological Cybernetics,* vol. 62, pp. 193–199.

Rubner, J., and P. Tavan, 1989. "A self-organizing network for principal component analysis," *Europhysics Letters,* vol. 10, pp. 693–698.

Rueckl, J.G., K.R. Cave, and S.M. Kosslyn, 1989. "Why are 'what' and 'where' processed by separate cortical visual systems? A computational investigation," *J. Cognitive Neuroscience,* vol. 1, pp. 171–186.

Rumelhart, D.E., and J.L. McClelland, eds., 1986. *Parallel Distributed Processing: Explorations in the Microstructure of Cognition,* vol. 1, Cambridge, MA: MIT Press.

Rumelhart, D.E., and D. Zipser, 1985. "Feature discovery by competitive learning," *Cognitive Science,* vol. 9, pp. 75–112.

Rumelhart, D.E., G.E. Hinton, and R.J. Williams, 1986a. "Learning representations of back-propagation errors," *Nature (London),* vol. 323, pp. 533–536.

Rumelhart, D.E., G. E. Hinton, and R.J. Williams, 1986b. "Learning internal representations by error propagation," in D.E. Rumelhart and J.L. McCleland, eds., vol 1, Chapter 8, Cambridge, MA: MIT Press.

Russell, S.J., and P. Novig, 1995. *Artificial Intelligence: A Modern Approach,* Upper Saddle River, NJ: Prentice-Hall.

Russo, A.P., 1991. *Neural Networks for Sonar Signal Processing,* Tutorial No. 8, *IEEE Conference on Neural Networks for Ocean Engineering,* Washington, DC.

Ruyck, D.W., S.K. Rogers, M. Kabrisky, M.E. Oxley, and B.W. Suter, 1990. "The multilayer perceptron as an approximation to a Bayes optimal discriminant function," *IEEE Transactions of Neural Networks,* vol. 1, pp. 296–298.

Saarinen, S., R.B. Bramley, and G. Cybenko, 1992. "Neural networks, backpropagation, and automatic differentiation," in *Automatic Differentiation of Algorithms: Theory, Implementation, and Application,* A. Griewank and G.F. Corliss, eds., pp. 31–42, Philadelphia: SIAM.

Saarinen, S., R. Bramley, and G. Cybenko, 1991. "The numerical solution of neural network training problems," *CRSD Report No. 1089,* Center for Supercomputing Research and Development, University of Illinois, Urbana, IL.

Säckinger, E., B.E. Boser, J. Bromley, Y. LeCun, and L.D. Jackel, 1992a. "Application of the ANNA neural network chip to high-speed character recognition," *IEEE Transactions on Neural Networks,* vol. 3, pp. 498–505.

Säckinger, E., B.E. Boser, and L.D. Jackel, 1992b. "A neurocomputer board based on the ANNA neural network chip," *Advances in Neural Information Processing Systems,* vol. 4, pp. 773–780, San Mateo, CA: Morgan Kaufmann.

Saerens, M., and A. Soquet, 1991. "Neural controller based on back-propagation algorithm," *IEE Proceedings (London), Part F,* vol. 138, pp. 55–62.

Sage, A.P., ed., 1990. *Concise Encyclopedia of Information Processing in Systems and Organizations,* New York: Pergamon.

Salomon, R., and J.L. van Hemmen, 1996. "Accelerating backpropagation through dynamic self-adaptation," *Neural Networks,* vol. 9, pp. 589–601.

Samuel, A.L., 1959. "Some studies in machine learning using the game of checkers," *IBM Journal of Research and Development,* vol. 3, pp. 211–229.

Sandberg, I.W., 1991. "Structure theorems for nonlinear systems," *Multidimensional Systems and Signal Processing,* vol. 2, pp. 267–286.

Sandberg, I.W., L. Xu, 1997a. "Uniform approximation of multidimensional myopic maps," *IEEE Transactions on Circuits and Systems,* vol. 44, pp. 477–485.

Sandberg, I.W., and L. Xu, 1997b. "Uniform approximation and gamma networks," *Neural Networks,* vol. 10, pp. 781–784.

Sanger, T.D., 1990. "Analysis of the two-dimensional receptive fields learned by the Hebbian algorithm in response to random input," *Biological Cybernetics,* vol. 63, pp. 221–228.

Sanger, T.D., 1989a. "An optimality principle for unsupervised learning," *Advances in Neural Information Processing Systems,* vol. 1, pp. 11–19, San Mateo, CA: Morgan Kaufmann.

Sanger, T.D., 1989b. "Optimal unsupervised learning in a single-layer linear feedforward neural network," *Neural Networks,* vol. 12, pp. 459–473.

Sanner, R.M., and J.,-J.E. Slotine, 1992. "Gaussian networks for direct adaptive control," *IEEE Transactions on Neural Networks,* vol. 3, pp. 837–863.

Sauer, N., 1972. "On the densities of families of sets," *Journal of Combinatorial Theory, Series A,* vol. 13, pp. 145–172.

Sauer, T., J.A. Yorke, and M. Casdagli, 1991. "Embedology," *Journal of Statistical Physics,* vol. 65, pp. 579–617.

Saul, L.K., T. Jakkolla, and M.I. Jordan, 1996. "Mean field theory for sigmoid belief networks," *Journal of Artificial Intelligence Research,* vol. 4, pp. 61–76.

Saul, L.K., and M.I. Jordan, 1996. "Exploiting tractable substructures in intractable networks," *Advances in Neural Information Processing Systems,* vol. 8, pp. 486–492, Cambridge, MA: MIT Press.

Saul, L.K., and M.I. Jordan, 1995. "Boltzmann chains and hidden Markov models," *Advances in Neural Information Processing Systems,* vol. 7, pp. 435–442.

Schapire, R.E., 1997. "Using output codes to boost multiclass learning problems," *Machine Learning: Proceedings of the Fourteenth International Conference,* Nashville, TN.

Schapire, R.E., 1990. "The strength of weak learnability," *Machine Learning,* vol. 5, pp. 197–227.

Schapire R.E., Y. Freund, and P. Bartlett, 1997. "Boosting the margin: A new explanation for the effectiveness of voting methods," *Machine Learning: Proceedings of the Fourteenth International Conference,* Nashville, TN.

Schiffman, W.H., and H.W. Geffers, 1993. "Adaptive control of dynamic systems by back propagation networks," *Neural Networks,* vol. 6, pp. 517–524.

Schneider, C.R., and H.C. Card, 1998. "Analog hardware implementation issues in deterministic Boltzmann machines," *IEEE Transactions on Circuits and Systems II,* vol. 45, to appear.

Schneider, C.R., and H.C. Card, 1993. "Analog CMOS deterministic Boltzmann circuits," *IEEE Journal Solid-State Circuits,* vol. 28, pp. 907–914.

Schölkopf, B., 1997. *Support Vector Learning,* Munich, Germany: R. Oldenbourg Verlag.

Schölkopf, B., P. Simard, V. Vapnik, and A.J. Smola, 1997. "Improving the accuracy and speed of support vector machines," *Advances in Neural Information Processing Systems,* vol. 9, pp. 375–381.

Schölkopf, B., A. Smola, and K.-R. Müller, 1998. "Nonlinear component analysis as a kernel eigenvalue problem," *Neural Computation,* vol. 10, to appear.

Schölkopf, B., K.-K Sung, C.J.C. Burges, F. Girosi, P. Niyogi, T. Poggio, and V. Vapnik, 1997. "Comparing support vector machines with Gaussian kernels to radial basis function classifiers," *IEEE Transactions on Signal Processing,* vol. 45, pp. 2758–2765.

Schraudolph, N.N., and T.J. Sejnowski, 1996. "Tempering back propagation networks: Not all weights are created equal," *Advances in Neural Information Processing Systems,* vol. 8, pp. 563–569, Cambridge, MA: MIT Press.

Schumaker, L.L., 1981, *Spline Functions: Basic Theory,* New York: Wiley.

Schurmars, D., 1997. "Alternative metrics for maximum margin classification," *NIPS Workshop on Support Vector Machines,* Beckenbridge, CO.

Schuster, H.G., 1988. *Deterministic Chaos: An Introduction,* Weinheim, Germany: VCH.

Scofield, C.L., and L.N. Cooper, 1985. "Development and properties of neural networks," *Contemporary Physics,* vol. 26, pp. 125–145.

Scott, A.C., 1977. *Neurophysics,* New York: Wiley.

Segee, B.E., and M.J. Carter, 1991. "Fault tolerance of pruned multilayer networks," *International Joint Conference on Neural Networks,* vol. II, pp. 447–452, Seattle.

Sejnowski, T.J., 1977a. "Strong covariance with nonlinearly interacting neurons," *Journal of Mathematical Biology,* vol. 4, pp. 303–321.

Sejnowski, T.J., 1977b. "Statistical constraints on synaptic plasticity," *Journal of Theoretical Biology,* vol. 69, pp. 385–389.

Sejnowski, T.J., 1976. "On global properties of neuronal interaction," *Biological Cybernetics,* vol. 22, pp. 85–95.

Sejnowski, T.J., and P.S. Churchland, 1989. "Brain and cognition," in *Foundations of Cognitive Science,* M.I. Posner, ed., pp. 301–356, Cambridge, MA: MIT Press.

Sejnowski, T.J., P.K. Kienker, and G.E. Hinton, 1986. "Learning symmetry groups with hidden units: Beyond the perceptron," *Physica,* vol. 22D, pp. 260–275.

Sejnowski, T.J., C. Koch, and P.S. Churchland, 1988. "Computational neuroscience," *Science,* vol. 241, pp. 1299–1306.

Sejnowski, T.J., and C.R. Rosenberg, 1987. "Parallel networks that learn to pronounce English text," *Complex Systems,* vol. 1, pp. 145–168.

Sejnowski, T.J., B.P. Yuhas, M.H. Goldstein, Jr., and R.E. Jenkins, 1990. "Combining visual and acoustic speech signals with a neural network improves intelligibility." *Advances in Neural Information Processing Systems,* vol. 2, pp. 232–239, San Mateo, CA: Morgan Kaufmann.

Selfridge, O.G., R.S. Sutton, and C.W. Anderson, 1988. "Selected bibliography on connectionism," *Evolution, Learning, and Cognition,* Y.C. Lee, Ed., pp. 391–403, River Edge, NJ: World Scientific Publishing, Inc.

Seung, H., 1995. "Annealed theories of learning," in J.-H Oh, C. Kwon, and S. Cho, eds., *Neural Networks: The Statistical Mechanics Perspective,* Singapore: World Scientific.

Seung, H.S., T.J. Richardson, J.C. Lagarias, and J.J. Hopfield, 1998. "Saddle point and Hamiltonian structure in excitatory-inhibitory networks," *Advances in Neural Information Processing Systems,* vol. 10, to appear.

Shah, S., and F. Palmieri, 1990. "MEKA—A fast, local algorithm for training feedforward neural networks," *International Joint Conference on Neural Networks,* vol. 3, pp. 41–46, San Diego, CA.

Shamma, S., 1989. "Spatial and temporal processing in central auditory networks," in *Methods in Neural Modeling,* C. Koch and I. Segev, Eds., Cambridge, MA: MIT Press.

Shanno, D.F., 1978. "Conjugate gradient methods with inexact line searches," *Mathematics of Operations Research,* vol. 3, pp. 244–256.

Shannon, C.E., 1948. "A mathematical theory of communication," *Bell System Technical Journal,* vol. 27, pp. 379–423, 623–656.

Shannon, C.E., and W. Weaver, 1949. *The Mathematical Theory of Communication,* Urbana, IL.: The University of Illinois Press.

Shannon, C.E., and J. McCarthy, eds., 1956. *Automata Studies,* Princeton, NJ: Princeton University Press.

Shepherd, G.M., 1988. *Neurobiology,* 2nd edition, New York: Oxford University Press.

Shepherd, G.M., 1978. "Microcircuits in the nervous system," *Scientific American,* vol. 238, pp. 92–103.

Shepherd, G.M., ed., 1990a. *The Synaptic Organization of the Brain,* 3rd edition, New York: Oxford University Press.

Shepherd, G.M., 1990b. "The significance of real neuron architectures for neural network simulations," in *Computational Neuroscience,* E.L. Schwartz, ed., pp. 82–96, Cambridge: MIT Press.

Shepherd, G.M., and C. Koch, 1990. "Introduction to synaptic circuits," in *The Synaptic Organization of the Brain,* G.M. Shepherd, ed., pp. 3–31. New York: Oxford University Press.

Sherrington, C.S., 1906. *The Integrative Action of the Nervous System,* New York: Oxford University Press.

Sherrington, C.S., 1933. *The Brain and Its Mechanism,* London: Cambridge University Press.

Sherrington, D., and S. Kirkpatrick, 1975. "Spin-glasses," *Physical Review Letters,* vol. 35, p. 1972.

Shewchuk, J.R., 1994. *An Introduction to the Conjugate Gradient Method Without the Agonizing Pain,* School of Computer Science, Carnegie Mellon University, Pittsburgh, PA, August 4, 1994.

Shore, J.E., and R.W. Johnson, 1980. "Axiomatic derivation of the principle of maximum entropy and the principle of minimum cross-entropy," *IEEE Transactions on Information Theory,* vol. IT-26, pp. 26–37.

Shynk, J.J., 1990. "Performance surfaces of a single-layer perceptron," *IEEE Transactions or Neural Networks,* 1, 268–274.

Shynk, J.J. and N.J. Bershad, 1991. Steady-state analysis of a single-layer perceptron based on a system identification model with bia terms," *IEEE Transactions on Circuits and Systems,* vol. CAS-38, pp. 1030–1042.

Shustorovich, A., 1994. "A subspace projection approach to feature extraction: The two-dimensional Gabor transform for character recognition," *Neural Networks,* vol. 7, pp. 1295–1301.

Shustorovich, A., and C. Thrasher, 1996. "Neural network positioning and classification of handwritten characters," *Neural Networks,* vol. 9, pp. 685–693.

Shynk, J.J., and N.J. Bershad, 1992. "Stationary points and performance surfaces of a perceptron learning algorithm for a nonstationary data model," *International Joint Conference on Neural Networks,* vol. 2, pp. 133–139, Baltimore.

Shynk, J.J., and N.J. Bershad, 1991. "Steady-state analysis of a single-layer perceptron based on a system identification model with bias terms," *IEEE Transactions on Circuits and Systems,* vol. CAS-38, pp. 1030–1042.

Siegelmann, H.T., B.G. Horne, and C.L. Giles, 1997 "Computational capabilities of recurrent NARX neural networks," *Systems, Man, and Cybernetics, Part B: Cybernetics,* vol. 27, pp. 208–215.

Siegelmann, H.T., and E.D. Sontag, 1991. "Turing computability with neural nets," *Applied Mathematics Letters,* vol. 4, pp. 77–80.

Simard, P., Y. LeCun, and J. Denker, 1993. "Efficient pattern recognition using a new transformation distance," *Advances in Neural Information Processing Systems,* vol. 5, pp. 50–58, San Mateo, CA: Morgan Kaufmann.

Simard, P., B. Victorri, Y. LeCun, and J. Denker, 1992. "Tangent prop — A formalism for specifying selected invariances in an adaptive network," *Advances in Neural Information Systems,* vol. 4, pp. 895–903, San Mateo, CA: Morgan Kaufmann.

Simmons, J.A. 1989. "A view of the world through the bat's ear: The formation of acoustic images in echolocation," *Cognition,* vol. 33, pp. 155–199.

Simmons, J.A., P.A. Saillant, and S.P. Dear, 1992. "Through a bat's ear," *IEEE Spectrum,* vol. 29(3), pp. 46–48.

Singh, S.P., ed., 1992. *Approximation Theory, Spline Functions and Applications,* Dordrecht, The Netherlands: Kluwer.

Singh, S., and D. Bertsekas, 1997. "Reinforcement learning for dynamic channel allocation in cellular telephone systems," *Advances in Neural Information Processing Systems,* vol. 9, pp. 974–980, Cambridge, MA: MIT Press.

Singhal, S., and L. Wu, 1989. "Training feed-forward networks with the extended Kalman filter," *IEEE International Conference on Acoustics, Speech, and Signal Processing,* pp. 1187–1190, Glasgow, Scotland.

Singleton, R.C., 1962. "A test for linear separability as applied to self-organizing machines," in M.C. Yovitz, G.T. Jacobi, and G.D. Goldstein, eds., *Self Organizing Systems,* pp. 503–524, Washington DC: Spartan Books.

Sjöberg, J., Q. Zhang, L. Ljung, A. Benveniste, B. Delyon, P.-Y. Glorennec, H. Hjalmarsson, and A. Juditsky, 1995. "Nonlinear black-box modeling in system identification: A unified overview," *Automatica,* vol. 31, pp. 1691–1724.

Slepian, D., 1973. *Key papers in the development of information theory,* New York: IEEE Press.

Sloane, N.J.A., and A.D. Wyner, 1993. *Claude Shannon: Collected Papers,* New York: IEEE Press.

Smith, M., 1993. *Neural Networks for Statistical Modeling,* New York: Van Nostrand Reinhold.

Smola, A.J., and B. Schölkopf, 1998. "From regularization operators to support vector kernels," *Advances in Neural Information Processing Systems,* vol. 10, to appear.

Smolensky, P., 1988. "On the proper treatment of connectionism," *Behavioral and Brain Sciences*, vol. 11, pp. 1–74.

Sontag, E.D., 1996. "Recurrent neural networks: Some learning and systems-theoretic aspects," Department of Mathematics, Rutgers University, New Brunswick, NJ.

Sontag, E.D., 1992. "Feedback stabilization using two-hidden-layer nets," *IEEE Transactions on Neural Networks*, vol. 3, pp. 981–990.

Sontag, E.D., 1990. *Mathematical Control Theory: Deterministic Finite Dimensional Systems*, New York: Springer-Verlag.

Sontag, E.D., 1989. "Sigmoids distinguish more efficiently than Heavisides," *Neural Computation*, vol. 1, pp. 470–472.

Southwell, R.V., 1946. *Relaxation Methods in Theoretical Physics*, New York: Oxford University Press.

Specht, D.F., 1991. "A general regression neural network," *IEEE Transactions on Neural Networks*, vol. 2, pp. 568–576.

Sperduti, A., 1997. "On the computational power of recurrent neural networks for structures," *Neural Networks*, vol. 10, pp. 395–400.

Sperduti, A., and A. Starita, 1997. "Supervised neural networks for the classification of structures," *IEEE Transactions on Neural Networks*, vol. 8, pp. 714–735.

Sprecher, D.A., 1965. "On the structure of continuous functions of several variables," *Transactions of the American Mathematical Society*, vol. 115, pp. 340–355.

Steinbuch, K., 1961. "Die Lernmatrix." *Kybernetik*, vol. 1, pp.36–45.

Stent, G.S., 1973. "A physiological mechanism for Hebb's postulate of learning," *Proceedings of the National Academy of Sciences, USA*, vol. 70, pp. 997–1001.

Sterling, P., 1990. "Retina," in *The Synaptic Organization of the Brain*, G.M. Shepherd, ed., 3rd edition, pp. 170–213, New York: Oxford University Press.

Stevenson, M., R. Winter, and B. Widrow, 1990. "Sensitivity of layered neural networks to errors in the weights," *International Joint Conference on Neural Networks*, vol. 1, pp. 337–340, Washington, DC.

Stone, M., 1978. "Cross-validation: A review," *Mathematische Operationsforschung Statistischen, Serie Statistics*, vol. 9, pp. 127–139.

Stone, M., 1974. "Cross-validatory choice and assessment of statistical predictions," *Journal of the Royal Statistical Society*, vol. B36, pp. 111–133.

Stork, D., 1989. "Is backpropagation biologically plausible?" *International Joint Conference on Neural Networks*, vol. 2, pp. 241–246, Washington, DC.

Strang, G., 1980. *Linear Algebra and its Applications*, New York: Academic Press.

Stuart, A., and K. Ord, 1994. *Kendall's Advanced Theory of Statistics*, vol. I, 6th edition, New York: Halsted Press.

Su, H.-T., and T. McAvoy, 1991. "Identification of chemical processes using recurrent networks," *Proceedings of the 10th American Controls Conference*, vol. 3, pp. 2314–2319, Boston.

Su, H.-T., T. McAvoy, and P. Werbos, 1992. "Long-term predictions of chemical processes using recurrent neural networks: A parallel training approach," *Industrial Engineering and Chemical Research*, vol. 31, pp. 1338–1352.

Suga, N., 1990a. "Cortical computational maps for auditory imaging," *Neural Networks*, vol. 3, pp. 3–21.

Suga, N., 1990b. "Computations of velocity and range in the bat auditory system for echo location," in *Computational Neuroscience*, E.L. Schwartz, ed., pp. 213–231, Cambridge, MA: MIT Press.

Suga, N., 1990c. "Biosonar and neural computation in bats," *Scientific American*, vol. 262, pp. 60–68.

Suga, N., 1985. "The extent to which bisonar information is represented in the bat auditory cortex," in *Dynamic Aspects of Neocortical Function*, G.M. Edelman, W.E. Gall, and W.M. Cowan, eds. pp. 653–695, New York: Wiley (Interscience).

Suga, N., and J.S. Kanwal, 1995. "Echolocation: Creating computational maps," in M.A. Arbib, ed., *The Handbook of Brain Theory and Neural Networks*, Cambridge, MA: MIT Press.

Sutton, J.P., and J.A. Anderson, 1995. "Computational and neurobiological features of a network of networks," in J.M. Bower, ed., *The Neurobiology of Computation*, pp. 317–322, Boston: Kluwer.

Sutton, R.S., 1988. "Learning to predict by the methods of temporal differences," *Machine Learning*, vol. 3, pp. 9–44.

Sutton, R.S., 1986. "Two problems with back-propagation and other steepest-descent learning procedures for networks," *Proceedings of the Eighth Annual Conference of the Cognitive Science Society*, pp. 823–831. Hillsdale, NJ: Lawrence Erlbaum.

Sutton, R.S., ed., 1992. Special Issue on Reinforcement Learning, *Machine Learning,* vol. 8, pp. 1–395.

Sutton, R.S., 1984. "Temporal credit assignment in reinforcement learning," Ph.D. Dissertation, University of Massachusetts, Amherst, MA.

Sutton, R.S., and A.G. Barto, 1998. *Reinforcement Learning: An Introduction,* Cambridge, MA: MIT Press.

Suykens, J.A.K., J.P.L. Vandewalle, and B.L.R. DeMoor, 1996. *Artificial Neural Networks for Modeling and Control of Non-Linear Systems,* Dordrecht, The Netherlands: Kluwer.

Swindlehurst, A.L., M.J. Goris, and B. Ottersten, 1997. "Some experiments with array data collected in actual urban and suburban environments," *IEEE Workshop on Signal Processing Advances in Wireless Communications,* pp. 301–304, Paris, France.

Takahashi, Y., 1993. "Generalization and approximation capabilities of multilayer networks," *Neural Computation,* vol. 5, pp. 132–139.

Takens, F., 1981. "On the numerical determination of the dimension of an attractor," in D. Rand and L.S. Young, eds., *Dynamical Systems and Turbulence,* Annual Notes in Mathematics, vol. 898, pp. 366–381, Berlin: Springer-Verlag.

Tapia, R.A., and J.R. Thompson, 1978. *Nonparametric Probability Density Estimation,* Baltimore: The Johns Hopkins University Press.

Taylor, J.G., 1997. "Neural computation: The historical background," in E. Fiesler and R. Beale, eds., *Handbook of Neural Computation,* New York: Oxford University Press.

Taylor, W.K., 1964. "Cortico-thalamic organization and memory," *Proceedings of the Royal Society, London, Series B,* vol. 159, pp. 466–478.

Taylor, W.K. 1956. "Electrical simulation of some nervous system functional activities," *Information Theory,* vol. 3, E.C. Cherry, ed., pp. 314–328, London: Butterworths.

Tesauro, G., 1995. "Temporal difference learning and TD-gamma," *Communications of the Association for Computing Machinery,* vol. 38, pp. 58–68.

Tesauro, G., 1994. "TD-Gammon, A self-teaching Backgammon program, achieves master-level play," *Neural Computation,* vol. 6, pp. 215–219.

Tesauro, G., 1992. "Practical issues in temporal difference learning," *Machine Learning,* vol. 8, pp. 257–277.

Tesauro, G., 1989. "Neurogammon wins computer olympiad," *Neural Computation,* vol. 1, pp. 321–323.

Tesauro, G., and T.J. Sejnowski, 1989. "A parallel network that learns to play backgamma," *Artificial Intelligence,* vol. 39, pp. 357–390.

Tesauro, G., and R. Janssens, 1988. "Scaling relationships in back-propagation learning," *Complex Systems,* vol. 2, pp. 39–44.

Teyler, T.J., 1986. "Memory: Electrophysiological analogs," in *Learning and Memory: A Biological View,* J.L. Martinez, Jr. and R.S. Kesner, eds., pp. 237–265, New York: Academic Press.

Thorndike, E.L., 1911. *Animal Intelligence,* Darien, CT: Hafner.

Thrun, S.B., 1992. "The role of exploration in learning control," in *Handbook of Intelligent Control,* D.A. White and D.A. Sofge, eds., pp. 527–559, New York: Van Nostrand Reinhold.

Tikhonov, A.N., 1973. "On regularization of ill-posed problems," *Doklady Akademii Nauk USSR,* vol. 153, pp. 49–52.

Tikhonov, A.N., 1963. "On solving incorrectly posed problems and method of regularization," *Doklady Akademii Nauk USSR,* vol. 151, pp. 501–504.

Tikhonov, A.N., and V.Y. Arsenin, 1977. *Solutions of Ill-posed Problems,* Washington, DC: W.H. Winston.

Titterington, D.M., A.F.M. Smith, and V.E. Makov, 1985. *Statistical Analysis of Finite Mixture Distributions,* New York: Wiley.

Touretzky, D.S., and D.A. Pomerleau, 1989. "What is hidden in the hidden layers?" *Byte,* vol. 14, pp. 227–233.

Tsitsiklis, J.N., 1994. "Asynchronous stochastic approximation and Q-learning," *Machine Learning,* vol. 16, pp. 185–202.

Tsoi, A.C., and A.D. Back, 1994. "Locally recurrent globally feedforward networks: A critical review," *IEEE Transactions on Neural Networks,* vol. 5, pp. 229–239.

Turing, A.M., 1952. "The chemical basis of morphogenesis," *Philosophical Transactions of the Royal Society, B,* vol. 237, pp. 5–72.

Turing, A.M., 1950. "Computing machinery and intelligence," *Mind,* vol. 59, pp. 433–460.

Turing, A.M., 1936. "On computable numbers with an application to the Entscheidungs problem," *Proceedings of the London Mathematical Society,* Series 2, vol. 42, pp. 230–265. Correction published in vol. 43, pp. 544–546.

Tsoi, A.C., and A. Back, 1994. "Locally recurrent globally feedforward networks: A critical review," *IEEE Transactions on Neural Networks,* vol. 5, pp. 229–239.

Tzefestas, S.G., ed., 1997. *Methods and Applications of Intelligent Control,* Boston: Kluwer.

Udin, S.B., and J.W. Fawcett, 1988. "Formation of topographic maps," *Annual Review of Neuroscience,* vol. 2, pp. 289–327.

Ukrainec, A.M., and S. Haykin, 1996. "A modular neural network for enhancement of cross-polar radar targets," *Neural Networks,* vol. 9, pp. 143–168.

Ukrainec, A., and S. Haykin, 1992. "Enhancement of radar images using mutual information based unsupervised neural networks," *Canadian Conference on Electrical and Computer Engineering,* pp. MA6.9.1–MA6.9.4, Toronto, Canada.

Uttley, A.M., 1979. *Information Transmission in the Nervous System,* London: Academic Press.

Uttley, A.M., 1970. "The informon: A network for adaptive pattern recognition," *Journal of Theoretical Biology,* vol. 27, pp. 31–67.

Uttley, A.M., 1966. "The transmission of information and the effect of local feedback in theoretical and neural networks," *Brain Research,* vol. 102, pp. 23–35.

Uttley, A.M., 1956. "A theory of the mechanism of learning based on the computation of conditional probabilities," *Proceedings of the First International Conference on Cybernetics,* Namur, Gauthier-Villars, Paris.

Vaillant, R., C. Monrocq, and Y. LeCun, 1994. "Original approach for the localization of objects in images," *IEE Proceedings (London) on Vision, Image and Signal Processing,* vol. 141, pp. 245–250.

Valavanis, K.P., and G.N. Saridis, 1992. *Intelligent Robotic Systems: Theory, Design, and Applications,* Norwell, MA: Kluwer.

Valiant, L.G., 1984. "A theory of the learnable," *Communications of the Association for Computing Machinery,* vol. 27, pp. 1134–1142.

Vanderbei, R., 1994. "Interior point methods: Algorithms and formulations," *ORSA Journal on Computing,* vol. 6, pp. 32–34.

Van Essen, D.C., C.H. Anderson, and D.J. Felleman, 1992. "Information processing in the primate visual system: An integrated systems perspective," *Science,* vol. 255, pp. 419–423.

van de Laar, P., T. Heskes, and S. Gielen, 1997. "Task-dependent learning of attention," *Neural Networks,* vol. 10, pp. 981–992.

van Laarhoven, P.J.M., and E.H.L. Aarts, 1988. *Simulated Annealing: Theory and Applications,* Boston: Kluwer Academic Publishers.

Van Trees, H.L., 1968. *Detection, Estimation, and Modulation Theory,* Part I, New York: Wiley.

Van Hulle, M.M., 1997. "Nonparametric density estimation and regression achieved with topographic maps maximizing the information-theoretic entropy of their outputs," *Biological Cybernetics,* vol. 77, pp. 49–61.

Van Hulle, M.M., 1996. "Topographic map formation by maximizing unconditional entropy: A plausible strategy for "on-line" unsupervised competitive learning and nonparametric density estimation," *IEEE Transactions on Neural Networks,* vol. 7, pp. 1299–1305.

Van Veen, B., 1992. "Minimum variance beamforming," in S. Haykin and A. Steinhardt, eds., *Adaptive Radar Detection and Estimation,* New York: Wiley (Interscience).

Vapnik, V.N., 1998. *Statistical Learning Theory,* New York: Wiley.

Vapnik, V.N., 1995. *The Nature of Statistical Learning Theory,* New York: Springer-Verlag.

Vapnik, V.N., 1992. "Principles of risk minimization for learning theory," *Advances in Neural Information Processing Systems,* vol. 4, pp. 831–838, San Mateo, CA: Morgan Kaufmann.

Vapnik, V.N., 1982. *Estimation of Dependences Based on Empirical Data,* New York: Springer-Verlag.

Vapnik, V.N., and A.Ya. Chervonenkis, 1971. "On the uniform convergence of relative frequencies of events to their probabilities," *Theoretical Probability and Its Applications,* vol. 17, pp. 264–280.

Vapnik, V.N., and A. Ya. Chervonenkis, 1964. "A note on a class of perceptrons," *Automation and Remote Control,* vol. 25, pp. 103–109.

Velmans, M., 1995. "Consciousness, Theories of." In M.A. Arbib, ed., *The Handbook of Brain Theory and Neural Networks,* pp. 247–250, Cambridge, MA: MIT Press.

Venkataraman, S., 1994. "On encoding nonlinear oscillations in neural networks for locomotion," *Proceedings of the 8th Yale Workshop on Adaptive and Learning Systems,* pp. 14–20, New Haven, CT.

Venkatesh, S.J., G. Panche, D. Psaltis, and G. Sirat, 1990. "Shaping attraction basins in neural networks," *Neural Networks,* vol. 3, pp. 613–623.

Vetterli, M., and J. Kovačević, 1995. *Wavelets and Subband Coding,* Englewood Cliffs, NJ: Prentice-Hall.

Vidyasagar, M., 1997. *A Theory of Learning and Generalization,* London: Springer-Verlag.

Vidyasagar, M., 1993. *Nonlinear Systems Analysis,* 2nd edition, Englewood Cliffs, NJ: Prentice-Hall.

Viterbi, A.J., 1967. "Error bounds for convolutional codes and an asymptotically optimum decoding algorithm," *IEEE Transactions on Information Theory,* vol. IT-13, pp. 260–269.

von der Malsburg, C., 1990a. "Network self-organization," in *An Introduction to Neural and Electronic Networks,* S.F. Zornetzer, J.L. Davis, and C. Lau, eds., pp. 421–432, San Diego, CA: Academic Press.

von der Malsburg, C., 1990b. "Considerations for a visual architecture," in *Advanced Neural Computers,* R. Eckmiller, ed., pp. 303–312, Amsterdam: North-Holland.

von der Malsburg, C., 1981. "The correlation theory of brain function," *Internal Report 81–2,* Department of Neurobiology, Max-Plak-Institute for Biophysical Chemistry, Göttingen, Germany.

von der Malsburg, C., 1973. "Self-organization of orientation sensitive cells in the striate cortex," *Kybernetik,* vol. 14, pp. 85–100.

von der Malsburg, C., and W. Schneider, 1986. "A neural cocktail party processor," *Biological Cybernetics,* vol. 54, pp. 29–40.

von Neumann, J., 1986. *Papers of John von Neumann on Computing and Computer Theory,* W. Aspray and A. Burks, eds., Cambridge, MA: MIT Press.

von Neumann, J., 1958. *The Computer and the Brain,* New Haven, CT: Yale University Press.

von Neumann, J., 1956. "Probabilistic logics and the synthesis of reliable organisms from unreliable components," in *Automata Studies,* C.E. Shannon and J. McCarthy, eds., pp. 43–98, Princeton, NJ: Princeton University Press.

Wahba, G., 1990. *Spline Models for Observational Data,* SIAM.

Wahba, G., D.R. Johnson, F. Gao, and J. Gong, 1995. "Adaptive tuning of numerical weather prediction models: Randomized GCV in three and four dimensional data assimilation," *Monthly Weather Review,* vol. 123, pp. 3358–3369.

Waibel, A., T. Hanazawa, G. Hinton, K. Shikano, and K.J. Lang, 1989. "Phoneme recognition using time-delay neural networks," *IEEE Transactions on Acoustics, Speech, and Signal Processing,* vol. ASSP-37, pp. 328–339.

Waltz, D., 1997. "Neural nets and AI: Time for a synthesis," plenary talk, *International Conference on Neural Networks,* vol. 1, p. xiii, Houston.

Waltz, M.D., and K.S. Fu, 1965. "A heuristic approach to reinforcement learning control systems," *IEEE Transactions on Automatic Control,* vol. AC-10, pp. 390–398.

Wan, E.A., 1994. "Time series prediction by using a connectionist network with internal delay lines," in *Time Series Prediction: Forecasting the Future and Understanding the Past,* A.S. Weigend and N.A. Gershenfield, eds., pp. 195–217. Reading, MA: Addison-Wesley.

Wan, E.A., 1990. "Temporal backpropagation for FIR neural networks," *IEEE International Joint Conference on Neural Networks,* vol. I, pp. 575–580, San Diego, CA.

Wan, E.A., and F. Beaufays, 1996. "Diagrammatic derivation of gradient algorithms for neural networks," *Neural Computation,* vol. 8, pp. 182–201.

Watanabe, H., Yamaguchi, and S. Katagiri, 1997. "Discriminative metric design for robust pattern recognition," *IEEE Transactions on Signal Processing,* vol. 45, pp. 2655–2662.

Waterhouse, S., D. MacKay, and A. Robinson, 1996. "Bayesian methods for mixtures of experts," *Advances in Neural Information Processing Systems,* vol. 8, pp. 351–357, Cambridge, MA: MIT Press.

Watkins, C.J.C.H., 1989. *Learning from Delayed Rewards,* Ph.D. Thesis, University of Cambridge, England.

Watkins, C.J.C.H., and P. Dayan, 1992. "Q-learning," *Machine Learning,* vol. 8, pp. 279–292.

Watrous, R.L. 1987. "Learning algorithms for connectionist networks: Applied gradient methods of nonlinear optimization," *First IEEE International Conference on Neural Networks,* vol. 2, pp. 619–627, San Diego, CA.

Watson, G.S., 1964. "Smooth regression analysis," *Sankhyā: The Indian Journal of Statistics, Series A,* vol. 26, pp. 359–372.

Webb, A.R., 1994. "Functional approximation by feed-forward networks: A least-squares approach to generalisation," *IEEE Transactions on Neural Networks,* vol. 5, pp. 480–488.

Webb, A.R., and D. Lowe, 1990. "The optimal internal representation of multilayer classifier networks performs nonlinear discriminant analysis," *Neural Networks,* vol. 3, pp. 367–375.

Weigend, A.S., B. Huberman, and D. Rumelhart, 1990. "Predicting the future: A connectionist approach," *International Journal of Neural Systems,* vol. 3, pp. 193–209.

Weigend, A.S., D.E. Rumelhart, and B.A. Huberman, 1991. "Generalization by weight-elimination with application to forecasting, *Advances in Neural Information Processing Systems,* vol. 3, pp. 875–882, San Mateo, CA: Morgan Kaufmann.

Weigend, A.S., and N.A. Gershenfield, eds., 1994. *Time Series Prediction: Forecasting the Future and Understanding the Past,* vol. 15, Santa Fe Institute Studies in the Sciences of Complexity, Reading, MA: Addison-Wesley.

Weierstrass, K., 1885. "Uber die analytische Darstellbarkeit sogenannter willkurlicher Funktionen einer reellen veranderlichen," *Sitzungsberichte der Akademie der Wissenschaften, Berlin,* pp. 633–639, 789–905.

Werbos, P.J., 1992. "Neural networks and the human mind: New mathematics fits humanistic insight," *IEEE International Conference on Systems, Man, and Cybernetics,* vol. 1, pp. 78–83, Chicago.

Werbos, P.J., 1990. "Backpropagation through time: What it does and how to do it," *Proceedings of the IEEE,* vol. 78, pp. 1550–1560.

Werbos, P.J., 1989. "Backpropagation and neurocontrol: A review and prospectus," *International Joint Conference on Neural Networks,* vol. I, pp. 209–216, Washington, DC.

Werbos, P.J., 1974. "Beyond regression: New tools for prediction and analysis in the behavioral sciences," Ph.D. Thesis, Harvard University, Cambridge, MA.

Wettschereck, D., and T. Dietterich, 1992. "Improving the performance of radial basis function networks by learning center locations," *Advances in Neural Information Processing Systems,* vol. 4, pp. 1133–1140, San Mateo, CA: Morgan Kaufmann.

White, D.A., and D.A. Sofge, eds,, 1992. *Handbook of Intelligent Control: Neural, Fuzzy, and Adaptive Approaches,* New York: Van Nostrand Reinhold.

White, H., 1992. *Artificial Neural Networks: Approximation and Learning Theory,* Cambridge, MA: Blackwell.

White, H., 1990. "Connectionist nonparametric regression: Multilayer feedforward networks can learn arbitrary mappings," *Neural Networks,* vol. 3, pp. 535–549.

White, H., 1989a. "Learning in artificial neural networks: A statistical perspective," *Neural Computation,* vol. 1, pp. 425–464.

White, H., 1989b. "Some asymptotic results for learning in single hidden-layer feedforward network models," *Journal of the American Statistical Society,* vol. 84, pp. 1003–1013.

Whitney, H., 1936. "Differentiable manifolds," *Annals of Mathematics,* vol. 37, pp. 645–680.

Whittaker, E.T., 1923. "On a new method of graduation," *Proceedings of the Edinburgh Mathematical Society,* vol. 41, pp. 63–75.

Widrow, B., 1962. "Generalization and information storage in networks of adeline 'neurons'," in M.C. Yovitz, G.T. Jacobi, and G.D., Goldstein, eds., *Self-Organizing Systems,* pp. 435–461, Washington, DC: Spartan Books.

Widrow, B., J.M. McCool, M.G. Larimore, and C.R. Johnson, Jr., 1976. "Stationary and nonstationary learning characteristics of the LMS adaptive filter," *Proceedings of the IEEE,* vol. 64, pp.1151–1162.

Widrow, B., J.R. Glover, Jr., J.M. McCool, J. Kaunitz, C.S. Williams, R.H. Hearn, J.R. Zeidler, J. Dong, Jr., and R.C.Goodlin, 1975. "Adaptive noise cancelling: Principles and applications," *Proceedings of the IEEE,* vol 63, pp. 1692–1716.

Widrow, B., N.K. Gupta, and S. Maitra, 1973. "Punish/reward: Learning with a critic in adaptive threshold systems," *IEEE Transactions of Systems, Man, and Cybernetics,* vol. SMC-3, pp. 455–465.

Widrow, B., and M.E. Hoff, Jr., 1960. "Adaptive switching circuits," *IRE WESCON Convention Record,* pp. 96–104.

Widrow, B., and M.A. Lehr, 1990. "30 years of adaptive neural networks: Perceptron, madaline, and backpropagation," *Proceedings of the Institute of Electrical and Electronics Engineers,* vol. 78, pp. 1415–1442.

Widrow, B., P.E. Mantey, L.J. Griffiths, and B.B. Goode, 1967. "Adaptive antenna systems," *Proceedings of the IEEE,* vol. 55, pp. 2143–2159.

Widrow, B., and S.D. Stearns, 1985. *Adaptive Signal Processing,* Englewood Cliffs, NJ: Prentice-Hall.

Widrow, B., and E. Walach, 1996. *Adaptive Inverse Control,* Upper Saddle River, NJ: Prentice-Hall.

Wieland, A., and R. Leighton, 1987. "Geometric analysis of neural network capabilities," first *IEEE International Conference on Neural Networks,* vol. III, pp. 385–392, San Diego, CA.

Wiener, N., 1961. *Cybernetics,* 2nd edition New York: Wiley.

Wiener, N., 1958. *Nonlinear Problems in Random Theory,* New York: Wiley.

Wiener, N., 1949. *Extrapolation, Interpolation, and Smoothing of Stationary Time Series with Engineering Applications,* Cambridge, MA: MIT Press. (This was originally issued as a classified National Defense Research Report, February 1942).

Wiener, N., 1948. *Cybernetics: Or Control and Communication in the Animal and the Machine,* New York: Wiley.

Wilks, S.S., 1962. *Mathematical Statistics,* New York: Wiley.

Williams, R.J., 1992. "Simple statistical gradient-following algorithms for connectionist reinforcement learning," *Machine Learning,* vol. 8, pp. 229–256.

Williams, R.J., 1988. "Toward a theory of reinforcement-learning connectionist systems," *Technical Report NU-CCS-88-3,* College of Computer Science, Northeastern University, Boston.

Williams, R.J., 1985. "Feature discovery through error-correction learning," *Technical Report ICS-8501.* University of California, San Diego, CA.

Williams, R.J., and J. Peng, 1990. "An efficient gradient-based algorithm for on-line training of recurrent network trajectories," *Neural Computation,* vol. 2, pp. 490–501.

Williams, R.J., and D. Zipser, 1995. "Gradient-based learning algorithms for recurrent networks and their computational complexity," in Y. Chauvin and D.E. Rumelhart, eds., *Backpropagation: Theory, Architectures, and Applications,* pp. 433–486, Hillsdale, NJ: Lawrence Erlbaum.

Williams, R.J., and D. Zipser, 1989. "A learning algorithm for continually running fully recurrent neural networks," *Neural Computation,* vol. 1, pp. 270–280.

Willshaw, D.J., O.P. Buneman, and H.C. Longuet-Higgins, 1969. "Non-holographic associative memory," *Nature (London),* vol. 222, pp. 960–962.

Willshaw, D.J., and C. von der Malsburg, 1976. "How patterned neural connections can be set up by self-organization," *Proceedings of the Royal Society of London Series B,* vol. 194, pp. 431–445.

Wilson, G.V., and G.S. Pawley, 1988. "On the stability of the travelling salesman problem algorithm of Hopfield and Tank," *Biological Cybernetics,* vol. 58, pp. 63–70.

Wilson, H.R., and J.D. Gowan, 1972. "Excitatory and inhibitory interactions in localized populations of model neurons," *Journal of Biophysics,* vol. 12, pp. 1–24.

Winder, R.O., 1961. "Single stage threshold logic," *Switching Circuit Theory and Logical Design,* AIEE Special Publications, vol. S-134, pp. 321–332.

Winograd, S., and J.D. Cowan, 1963. *Reliable Computation in the Presence of Noise,* Cambridge, MA: MIT Press.

Wolpert, D.H., 1992. "Stacked generalization," *Neural Networks,* vol. 5, pp. 241–259.

Wood, N.L., and N. Cowan, 1995. "The cocktail party phenomenon revisited: Attention and memory in the classic selective listening procedure of Cherry (1953)," *Journal of Experimental Psychology: General,* vol. 124, pp. 243–262.

Woods, W.A., 1986. "Important issues in knowledge representation," *Proceedings of the Institute of Electrical and Electronics Engineers,* vol. 74, pp. 1322–1334.

Wu, C.F.J., 1983. "On the convergence properties of the EM algorithm," *Annals of Statistics,* vol. 11, pp. 95–103.

Wylie, C.R., and L.C. Barrett, 1982. *Advanced Engineering Mathematics,* 5th edition, New York: McGraw-Hill.

Xu, L., A. Krzyżak, and A. Yuille, 1994. "On radial basis function nets and kernel regression: Statistical consistency, convergence rates, and receptive field size," *Neural Networks,* vol. 7, pp. 609–628.

Xu, L., E. Oja, and C.Y. Suen, 1992. "Modified Hebbian learning for curve and surface fitting," *Neural Networks,* vol. 5, pp. 441–457.

Yang, H., and S. Amari, 1997. "Adaptive online learning algorithms for blind separation: Maximum entropy and minimum mutual information," *Neural Computation,* vol. 9, pp. 1457–1482.

Yee, P.V., 1998. *Regularized Radial Basis Function Networks: Theory and Applications to Probability Estimation, Classification, and Time Series Prediction,* Ph.D. Thesis, McMaster University, Hamilton, Ontario.

Yockey, H.P., 1992. *Information Theory and Molecular Biology,* Cambridge: Cambridge University Press.

Yoshizawa, S., M. Morita, and S. Amari, 1993. "Capacity of associative memory using a nonmonotonic neuron model," *Neural Networks,* vol. 6, pp. 167–176.

Zadeh, L.A., 1973. "Outline of a new approach to the analysis of complex systems and decision processes," *IEEE Transactions on Systems, Man, and Cybernetics,* vol. SMC-3, pp. 28–44.

Zadeh, L.A., 1965. "Fuzzy sets," *Information and Control,* vol. 8, pp. 338–353.

Zadeh, L.A., 1953. "A contribution to the theory of nonlinear systems," *J. Franklin Institute,* vol. 255, pp. 387–401.

Zadeh, L.A., and C.A. Desoer, 1963. *Linear System Theory: The State Space Approach,* New York: McGraw-Hill.

Zames, G., 1981. "Feedback and optimal sensitivity: Model reference transformations, multiplicative semi-norms, and approximate inverses," *IEEE Transactions on Automatic Control,* vol. AC-26, pp. 301–320.

Zames, G., and B.A. Francis, 1983. "Feedback, minimax, sensitivity, and optimal robustness," *IEEE Transactions on Automatic Control,* vol. AC-28, pp. 585–601.

Zeevi, A.J., R. Meir, and V. Majorov, 1998. "Error bounds for functional approximation and estimation using mixtures of experts," *IEEE Transactions on Information Theory,* vol. 44, pp. 1010–1025.

Zeki, S., 1993. *A Vision of the Brain,* Oxford: Blackwell Scientific Publications.

Zipser, D., and D.E. Rumelhart, 1990. "The neurobiological significance of the new learning models," in *Computational Neuroscience,* E.L. Schwartz, ed., pp. 192–200, Cambridge, MA: MIT Press.

Index